AF782801

SI Prefixes

Multiple	*Exponential Form*	*Prefix*	*SI Symbol*
1 000 000 000	10^9	giga	G
1 000 000	10^6	mega	M
1 000	10^3	kilo	k
Submultiple			
0.001	10^{-3}	milli	m
0.000 001	10^{-6}	micro	μ
0.000 000 001	10^{-9}	nano	n

Conversion Factors (FPS) to (SI)

Quantity	*Unit of Measurement (FPS)*	*Equals*	*Unit of Measurement (SI)*
Force	lb		4.4482 N
Mass	slug		14.5938 kg
Length	ft		0.3048 m

Conversion Factors (FPS)

1 ft = 12 in. (inches)
1 mi. (mile) = 5280 ft
1 kip (kilopound) = 1000 lb
1 ton = 2000 lb

Dynamics

Engineering Mechanics
Dynamics

Eighth Edition

R. C. Hibbeler

Prentice-Hall, Upper Saddle River, New Jersey 07458

Library of Congress Cataloging-in-Publication Data

Hibbeler, R. C.
Engineering mechanics. Dynamics / R. C. Hibbeler.—8th ed.
p. cm.
Includes bibliographical references and index.
ISBN 0–13–578261–9
1. Dynamics. I. Title.
TA352.H5 1997
620.1′054—dc21 97–46929
CIP

Editor in Chief: *Marcia Horton*
Associate Editor: *Alice Dworkin*
Production Editor: *Rose Kernan*
Marketing Manager: *Mary Mullally*
Managing Editor: *Bayani Mendoza de Leon*
Director of Production and Manufacturing: *David W. Riccardi*
Text Designer: *Elmsford Publishing*
Cover Designer: *Joseph Sengotta*
Cover Art (Photo): *Tony Stone Images*
Photo Researcher: *Julie Tesser*
Photo Editor: *Melinda Reo*
Copyeditor: *Rose Kernan*
Manufacturing Buyer: *Julia Meehan*
Art Manager: *Gus Vibal*
Art Director: *Amy Rosen*
Creative Director: *Paula Maylahn*
Text Composition: *York Graphic Services*
Art Studio: *Precision Graphics, Susan Sibille*
Electronic Imaging Specialist: *Robert Handago*

Published by Prentice-Hall, Inc.
Simon & Schuster / A Viacom Company
Upper Saddle River, New Jersey 07458

PRINTED IN THE UNITED STATES OF AMERICA

10 9 8 7 6 5 4 3

ISBN 0-13-578261-9

Prentice-Hall International (UK) Limited, London
Prentice-Hall of Australia Pty. Limited, Sydney
Prentice-Hall Canada Inc., Toronto
Prentice-Hall Hispanoamericana, S.A., Mexico
Prentice-Hall of India Private Limited, New Delhi
Prentice-Hall of Japan, Inc., Tokyo
Simon & Schuster Asia Pte, Ltd., Singapore
Editora Prentice-Hall do Brasil, Ltda., Rio de Janeiro
Prentice-Hall, Upper Saddle River, New Jersey

To the Student

With the hope that this work
will stimulate an interest in Engineering Mechanics
and provide an acceptable guide to its understanding.

Preface

The main purpose of this book is to provide the student with a clear and thorough presentation of the theory and applications of engineering mechanics. To achieve this objective, the author has by no means worked alone; to a large extent, this book, through its eight editions, has been shaped by the comments and suggestions of hundreds of reviewers in the teaching profession as well as many of the author's students.

New Features

Significant improvements have been made to this the eighth edition. The following is a list of some of the more important ones:

- **Photographs.** Many photographs are used throughout the book to show how the principles of engineering mechanics apply to real-world situations. In some sections, photographs have been used to enhance the example problems, and they provide a practical aid in understanding the theory.
- **Artwork.** Throughout the book, the artwork has been further enhanced in a multicolor presentation in order to provide the reader with a more realistic and understandable sense of the material. Motion of both particles and rigid bodies is depicted, along with time-lapsed positions of mechanisms, so that students have a full understanding of their kinematic behavior. Particular attention has been given to rendering each body such that its view, its dimensions, and the vectors applied to it can be easily understood.
- **Improved Pedagogy**. The "procedure for analysis" sections, along with a new feature, "important points," are presented using a bulleted list format in order to aid in problem solving and review. Also, clarity throughout the text has been improved, new examples have been provided, and many new problems have been added.
- **Problems.** The problem sets have been revised so that instructors can select both design and analysis problems having a wide range of difficulty. Apart from the author, three other professionals have checked all the problems for clarity and accuracy of the solutions. At the end of some chapters, design projects have now been included.

- **Review Material.** A new Appendix D has been added that provides practice for solving problems for the Fundamentals in Engineering Examination. Partial solutions and answers are given to all these problems, providing students with further applications of the theory.

In addition to the many improvements, the hallmarks of the book remain the same: Where necessary, a strong emphasis is placed on drawing a free-body diagram, and the importance of selecting an appropriate coordinate system and associated sign convention for vector components is stressed when the equations of mechanics are applied.

Contents

The book is divided into eleven chapters, in which the principles introduced are first applied to simple, then to more complicated situations. The kinematics of a particle is discussed in Chapter 12, followed by a discussion of particle kinetics in Chapter 13 (equation of motion), Chapter 14 (work and energy), and Chapter 15 (impulse and momentum). The concepts of particle dynamics contained in these four chapters are then summarized in a "review" section, and the student is given the chance to identify and solve a variety of problems. A similar sequence of presentation is given for the planar motion of a rigid body: Chapter 16 (planar kinematics), Chapter 17 (equations of motion), Chapter 18 (work and energy), and Chapter 19 (impulse and momentum), followed by a summary and review set of problems for these chapters.

If time permits, some of the material involving three-dimensional rigid-body motion may be included in the course. The kinematics and kinetics of this motion are discussed in Chapters 20 and 21, respectively. Chapter 22 (vibrations) may be included if the student has the necessary mathematical background. Sections of the book which are considered to be beyond the scope of the basic dynamics course are indicated by a star (★) and may be omitted. Note that this material also provides a suitable reference for basic principles when it is discussed in more advanced courses.

Alternative Coverage. At the discretion of the instructor, it is possible to cover Chapters 12 through 19 in the following order with no loss in continuity: Chapters 12 and 16 (kinematics), Chapters 13 and 17 (equations of motion), Chapters 14 and 18 (work and energy), and Chapters 15 and 19 (impulse and momentum).

Special Features

Organization and Approach. The contents of each chapter are organized into well-defined sections which contain an explanation of specific topics, illustrative example problems, and a set of homework problems. The topics within each section are placed into subgroups defined by boldface titles. The purpose of this is to present a structured method for introducing each new definition or concept and to make the book convenient for later reference and review.

Chapter Contents. Each chapter begins with a photo to illustrate a broad-range application of the material within the chapter. A bulleted list of the chapter contents is provided to give a general overview of the material that will be covered.

Procedures for Analysis. Found after many of the sections of the book, this unique feature provides the student with a logical and orderly method to follow when applying the theory. The example problems are solved using this outlined method in order to clarify its numerical application. It is to be understood, however, that once the relevant principles have been mastered and enough confidence and judgment have been gained, the student can then develop his or her own procedures for solving problems.

Important Points. This feature provides a review or summary of the most important concepts in a section and highlights the most significant points that should be realized when applying the theory to solve problems.

Conceptual Understanding. Through the use of photographs placed throughout the book, the theory is applied in a simplified way in order to illustrate some of its more important conceptual features and instill the physical meaning of many of the terms used in the equations. These simplified applications increase interest in the subject matter and better prepare the student to understand the examples and solve problems.

Example Problems. All the example problems are presented in a concise manner and in a style that is easy to understand. New examples have been added throughout the text, and some now include photographs to enhance the reality of the problem.

Homework Problems

- ***General Analysis and Design Problems.*** The majority of problems in the book depict realistic situations encountered in engineering practice. Some of these problems involve actual products used in industry and are stated as such. It is hoped that this realism will both stimulate the student's interest in engineering mechanics and provide a means for developing the skill to reduce any such problem from its physical description to a model or symbolic representation to which the principles of mechanics may be applied.

Throughout the book, there is an approximate balance of problems using either SI or FPS units. Furthermore, in any set, an attempt has been made to arrange the problems in order of increasing difficulty. (Review problems at the end of Chapters 15 and 19 are presented in random order.) The answers to all but every fourth problem are listed in the back of the book. To alert the user to a problem without a reported answer, an asterisk (*) is placed before the problem number.

- ***Computer Problems.*** An effort has been made to include some problems that may be solved using a numerical procedure executed on either a desktop computer or a programmable pocket calculator. Suitable numerical techniques along with associated computer programs are given in Appendix B. The intent here is to broaden the student's capacity for using other forms of mathematical analysis without sacrificing the time needed to focus on the application of the principles of mechanics. Problems of this type, which either can or must be solved using numerical procedures, are identified by a "square" symbol (■) preceding the problem number.
- ***Design Projects.*** At the end of some of the chapters, design projects have been included. It is felt that this type of assignment should be given only after the student has developed a basic understanding of the subject matter. These projects focus on solving a problem by specifying the geometry of a structure or mechanical object needed for a specific purpose. A kinematic and/or kinetic analysis is required, and in many cases safety and cost issues must be addressed.

Appendices. The appendices provide a source of mathematical formula and numerical analysis needed to solve the problems in the book. Appendix D provides a set of problems typically found on the Fundamentals of Engineering Examination. By including a partial solution to all the problems, the student is given a chance to further practice his or her skills.

Ancillaries For the Student

Workbook and Study Guide. This useful book has been improved from past editions. It now provides students with over 150 additional examples. Each example is only partially completed, so students gain practice in developing their problem-solving skills. All solutions are given in the back of the book.

The New York Times/Prentice Hall *Themes of the Times* Supplement. This newspaper-format supplement brings together a collection of recent engineering articles from the pages of *The New York Times*. This free supplement, available in quantity through your local representative, encourages students to make connections between what is taught in the classroom and the world around them.

Web Site. The contents of the web site reflect the input of more than 75 professors. Visit the Hibbeler Web Site at www.prenhall.com/hibbeler.

Acknowledgments

The author has endeavored to write this book so that it will appeal to both the student and instructor. Through the years, many people have helped in its development, and I will always be grateful for their valued suggestions and comments. Specifically, I wish to personally thank the following individuals who have provided their comments related to this edition:

C. Ammerman, Colorado School of Mines
G. Batson, Clarkson University
E. L. Bernstein, Alabama Agricultural and Mechanical University
P. C. Chan, New Jersey Institute of Technology
E. Collins, Florida State University
A. Dollar, Illinois Institute of Technology
R. Downer, University of Vermont
W. L. Elban, Loyola College
J. R. Filler, University of Idaho
L. L. Friel, Montana Tech
A. Hassan, Purdue University
F. Holly, University of Iowa
H. Huntley, University of Michigan—Dearborn
D. Junge, University of Alaska
M. E. Kamara, University of Alabama at Birmingham
J. Ligon, Michigan Technological University
A. May, University of New Mexico
J. A. Nathanson, Union County Community College
M. Ness, Seattle Central Community College
J. Panerali, University of Nebraska—Lincoln
J. Ramos, United States Military Academy, West Point
P. Reinhall, University of Washington
L. Spainhour, FAMU/FSU

J. Steffen, Valparaiso University
D. Thorpe, Utah State University
C. Robert Ullrich, University of Louisville
D. Vanden Brink, Western Michigan University
R. Weber, Marquette University
T. Wicker, University of Wisconsin—Madison

A particular note of thanks is given to Kurt Norlin and Tom Smith, and to Will Liddell, Jr. of Auburn University at Montgomery, for specific help and support, to graduate students Jason Huffman and Eric Gaudet III, and to Susan Sibille for help with the artwork. I should also like to acknowledge the encouragement and support from the editorial staff at Prentice Hall and the proofreading assistance of my wife, Conny (Cornelie), during the time it has taken to prepare this manuscript for publication.

Lastly, many thanks are extended to all my students and to members of the teaching profession who have freely taken the time to send me their suggestions and comments. Since this list is too long to mention, it is hoped that those who have given help in this manner will accept this anonymous recognition.

I would greatly appreciate hearing from you if at any time you have any comments or suggestions regarding the contents of this edition.

Russell Charles Hibbeler
hibbeler@bellsouth.net

Contents

Dynamics

▶ 12

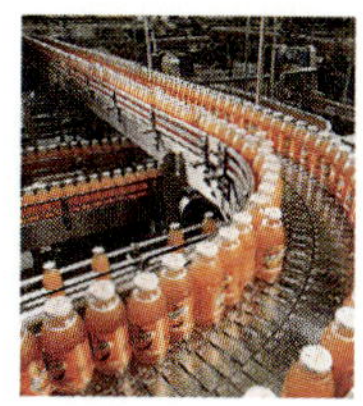

▶ 13

14

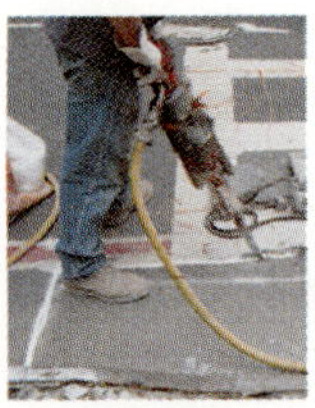

15

Review 1

16

17

18

Planar Kinetics of a Rigid Body: Work and Energy 431

19

Planar Kinetics of a Rigid Body: Impulse and Momentum 465

Review 2

Planar Kinematics and Kinetics of a Rigid Body 500

20

Three-Dimensional Kinematics of a Rigid Body 515

21

Three-Dimensional Kinetics of a Rigid Body 545

Dynamics

Although each of these planes is rather large, from a distance their motion can be modeled as if each plane were a particle

CHAPTER 12

Kinematics of a Particle

Chapter Objectives

- To introduce the concepts of position, displacement, velocity, and acceleration.
- To study particle motion along a straight line and represent this motion graphically.
- To investigate particle motion along a curved path using different coordinate systems.
- To present an analysis of dependent motion of two particles.
- To examine the principles of relative motion of two particles using translating axes.

12.1 Introduction

Mechanics is a branch of the physical sciences that is concerned with the state of rest or motion of bodies subjected to the action of forces. The mechanics of rigid bodies is divided into two areas: statics and dynamics. *Statics* is concerned with the equilibrium of a body that is either at rest or moves with constant velocity. The foregoing treatment is concerned with *dynamics* which deals with the accelerated motion of a body. Here the subject of dynamics will be presented in two parts: *kinematics,* which treats only the geometric aspects of the motion, and *kinetics,* which is the analysis of the forces causing the motion. To develop these principles, the dynamics of a particle will be discussed first, followed by topics in rigid-body dynamics in two and then three dimensions.

Historically, the principles of dynamics developed when it was possible to make an accurate measurement of time. Galileo Galilei (1564–1642) was one of the first major contributors to this field. His work consisted of experiments using pendulums and falling bodies. The most significant contributions in dynamics, however, were made by Isaac Newton (1642–1727), who is noted for his formulation of the three fundamental laws of motion and the law of universal gravitational attraction. Shortly after these laws were postulated, important techniques for their application were developed by Euler, D'Alembert, Lagrange, and others.

There are many problems in engineering whose solutions require application of the principles of dynamics. Typically the structural design of any vehicle, such as an automobile or airplane, requires consideration of the motion to which it is subjected. This is also true for many mechanical devices, such as motors, pumps, movable tools, industrial manipulators, and machinery. Furthermore, predictions of the motions of artificial satellites, projectiles, and spacecraft are based on the theory of dynamics. With further advances in technology, there will be an even greater need for knowing how to apply the principles of this subject.

Problem Solving. Dynamics is considered to be more involved than statics since both the forces applied to a body and its motion must be taken into account. Also, many applications require using calculus, rather than just algebra and trigonometry. In any case, the most effective way of learning the principles of dynamics is *to solve problems.* To be successful at this, it is necessary to present the work in a logical and orderly manner as suggested by the following sequence of steps:

1. Read the problem carefully and try to correlate the actual physical situation with the theory studied.
2. Draw any necessary diagrams and tabulate the problem data.
3. Establish a coordinate system and apply the relevant principles, generally in mathematical form.
4. Solve the necessary equations algebraically as far as practical; then, use a consistent set of units and complete the solution numerically. Report the answer with no more significant figures than the accuracy of the given data.
5. Study the answer using technical judgment and common sense to determine whether or not it seems reasonable.
6. Once the solution has been completed, review the problem. Try to think of other ways of obtaining the same solution.

In applying this general procedure, do the work as neatly as possible. Being neat generally stimulates clear and orderly thinking, and vice versa.

12.2 Rectilinear Kinematics: Continuous Motion

We will begin our study of dynamics by discussing the kinematics of a particle that moves along a rectilinear or straight line path. Recall that a *particle* has a mass but negligible size and shape. Therefore we must limit application to those objects that have dimensions that are of no consequence in the analysis of the motion. In most problems, one is interested in bodies of finite size, such as rockets, projectiles, or vehicles. Such objects may be considered as particles, provided motion of the body is characterized by motion of its mass center and any rotation of the body is neglected.

Rectilinear Kinematics. The kinematics of a particle is characterized by specifying, at any given instant, the particle's position, velocity, and acceleration.

Position. The straight-line path of a particle will be defined using a single coordinate axis s, Fig. 12–1a. The origin O on the path is a fixed point, and from this point the *position vector* $\mathbf{r}$ is used to specify the location of the particle P at any given instant. Notice that $\mathbf{r}$ is *always* along the s axis, and so its direction never changes. What will change is its magnitude and its sense or arrowhead direction. For analytical work it is therefore convenient to represent $\mathbf{r}$ by an *algebraic scalar s,* representing the *position coordinate* of the particle, Fig. 12–1a. The magnitude of s (and $\mathbf{r}$) is the distance from O to P, usually measured in meters (m) or feet (ft), and the sense (or arrowhead direction of $\mathbf{r}$) is defined by the algebraic sign on s. Although the choice is arbitrary, in this case s is positive since the coordinate axis is positive to the right of the origin. Likewise, it is negative if the particle is located to the left of O.

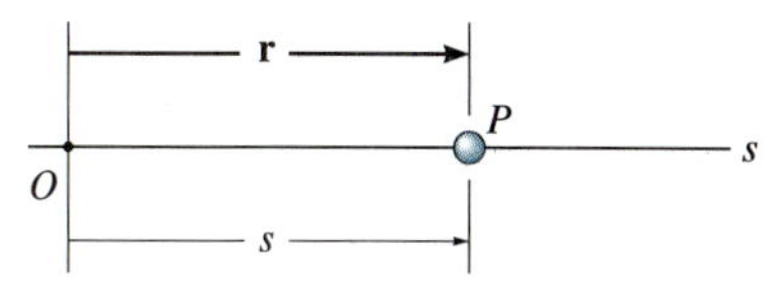

Position

(a)

Displacement. The *displacement* of the particle is defined as the *change* in its *position.* For example, if the particle moves from P to P', Fig. 12–1b, the displacement is $\Delta\mathbf{r} = \mathbf{r}' - \mathbf{r}$. Using algebraic scalars to represent $\Delta\mathbf{r}$, we also have

$$\Delta s = s' - s$$

Here Δs is *positive* since the particle's final position is to the *right* of its initial position, i.e., $s' > s$. Likewise, if the final position is to the *left* of its initial position, Δs is *negative.*

Since the displacement of a particle is a *vector quantity,* it should be distinguished from the distance the particle travels. Specifically, the *distance traveled* is a *positive scalar* which represents the total length of path over which the particle travels.

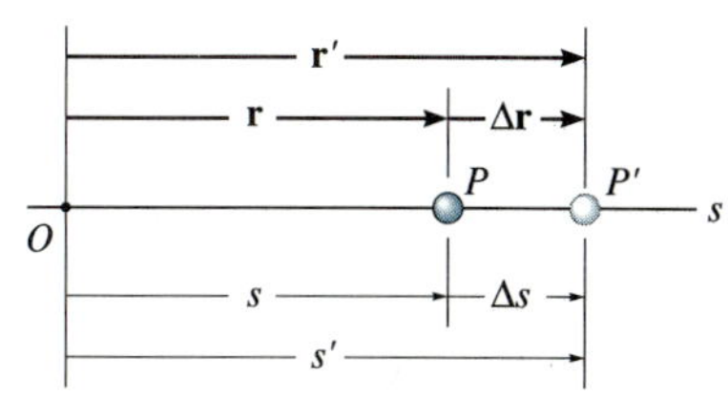

Displacement

(b)

Fig. 12–1

Velocity. If the particle moves through a displacement $\Delta\mathbf{r}$ from P to P' during the time interval Δt, Fig. 12–1*b*, the *average velocity* of the particle during this time interval is

$$\mathbf{v}_{avg} = \frac{\Delta\mathbf{r}}{\Delta t}$$

If we take smaller and smaller values of Δt, the magnitude of $\Delta\mathbf{r}$ becomes smaller and smaller. Consequently, the *instantaneous velocity* is defined as $\mathbf{v} = \lim_{\Delta t\to 0} (\Delta\mathbf{r}/\Delta t)$, or

$$\mathbf{v} = \frac{d\mathbf{r}}{dt}$$

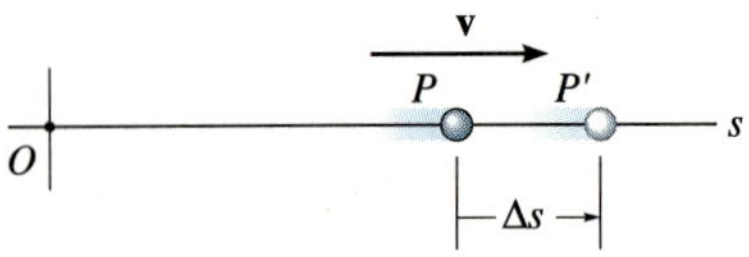

Velocity

(c)

Representing $\mathbf{v}$ as an algebraic scalar, Fig. 12–1*c*, we can also write

$$(\overset{+}{\rightarrow}) \qquad v = \frac{ds}{dt} \qquad (12\text{–}1)$$

Since Δt or dt is always positive, the sign used to define the *sense* of the velocity is the same as that of Δs or ds. For example, if the particle is moving to the *right,* Fig. 12–1*c*, the velocity is *positive;* whereas if it is moving to the *left,* the velocity is *negative.* (This is emphasized here by the arrow written at the left of Eq. 12–1.) The *magnitude* of the velocity is known as the *speed,* and it is generally expressed in units of m/s or ft/s.

Occasionally, the term "average speed" is used. The *average speed* is always a positive scalar and is defined as the total distance traveled by a particle, s_T, divided by the elapsed time Δt; i.e.,

$$(v_{sp})_{avg} = \frac{s_T}{\Delta t}$$

For example the particle in Fig. 12–1*d* travels along the path of length s_T in time Δt, so its average speed is $(v_{sp})_{avg} = s_T/\Delta t$, but its average velocity is $v_{avg} = -\Delta s/\Delta t$.

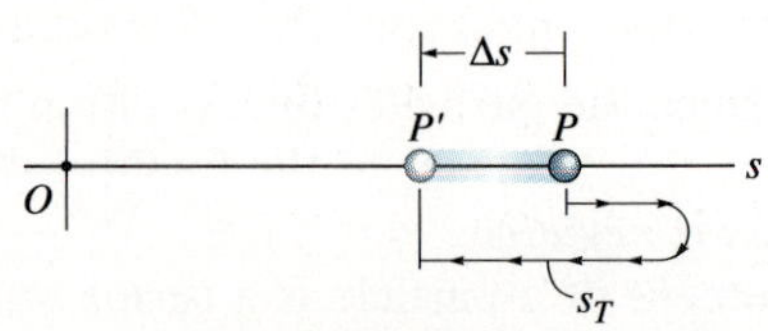

Average velocity and
Average speed
(d)

Fig. 12–1

Acceleration. Provided the velocity of the particle is known at the two points P and P', the *average acceleration* of the particle during the time interval Δt is defined as

$$\mathbf{a}_{\text{avg}} = \frac{\Delta \mathbf{v}}{\Delta t}$$

Here $\Delta \mathbf{v}$ represents the difference in the velocity during the time interval Δt, i.e., $\Delta \mathbf{v} = \mathbf{v}' - \mathbf{v}$, Fig. 12–1*e*.

The *instantaneous acceleration* at time t is found by taking smaller and smaller values of Δt and corresponding smaller and smaller values of $\Delta \mathbf{v}$, so that $\mathbf{a} = \lim_{\Delta t \to 0} (\Delta \mathbf{v}/\Delta t)$ or, using algebraic scalars,

$$(\overset{+}{\rightarrow}) \qquad a = \frac{dv}{dt} \qquad (12\text{–}2)$$

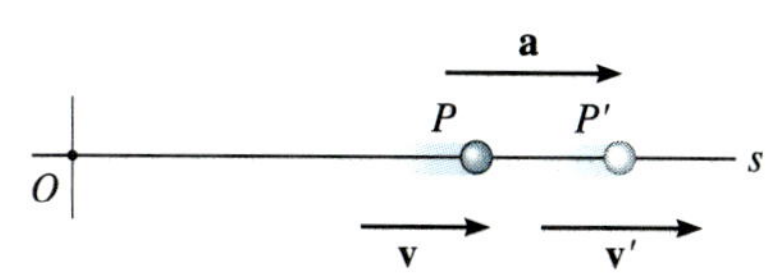

Acceleration

(e)

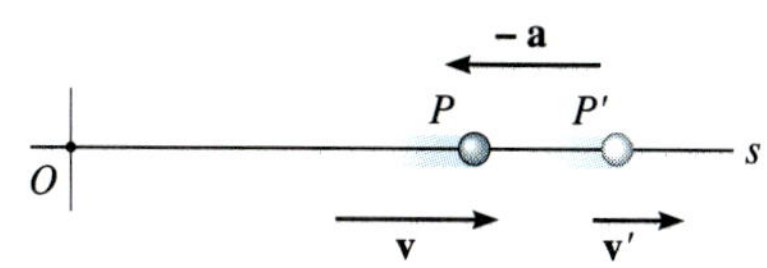

Deceleration

(f)

Substituting Eq. 12–1 into this result, we can also write

$$(\overset{+}{\rightarrow}) \qquad a = \frac{d^2s}{dt^2}$$

Both the average and instantaneous acceleration can be either positive or negative. In particular, when the particle is *slowing down*, or its speed is decreasing, it is said to be *decelerating*. In this case, v' in Fig. 12–1*f* is *less* than v, and so $\Delta v = v' - v$ will be negative. Consequently, a will also be negative, and therefore it will act to the *left*, in the opposite *sense* to v. Also, note that when the *velocity* is *constant*, the *acceleration is zero* since $\Delta v = v - v = 0$. Units commonly used to express the magnitude of acceleration are m/s^2 or ft/s^2.

A differential relation involving the displacement, velocity, and acceleration along the path may be obtained by eliminating the time differential dt between Eqs. 12–1 and 12–2. Realize that although we can then establish another equation, by doing so it will *not* be independent of Eqs. 12–1 and 12–2. Show that

$$(\overset{+}{\rightarrow}) \qquad a\,ds = v\,dv \qquad (12\text{–}3)$$

Constant Acceleration, $a = a_c$. When the acceleration is constant, each of the three kinematic equations $a_c = dv/dt$, $v = ds/dt$, and $a_c\,ds = v\,dv$ may be integrated to obtain formulas that relate a_c, v, s, and t.

Velocity as a Function of Time. Integrate $a_c = dv/dt$, assuming that initially $v = v_0$ when $t = 0$.

$$\int_{v_0}^{v} dv = \int_0^t a_c dt$$

$$(\overset{+}{\rightarrow}) \qquad \boxed{\begin{array}{c} v = v_0 + a_c t \\ \text{Constant Acceleration} \end{array}} \qquad (12\text{–}4)$$

Position as a Function of Time. Integrate $v = ds/dt = v_0 + a_c t$, assuming that initially $s = s_0$ when $t = 0$.

$$\int_{s_0}^{s} ds = \int_0^t (v_0 + a_c t)\, dt$$

$$(\overset{+}{\rightarrow}) \qquad \boxed{\begin{array}{c} s = s_0 + v_0 t + \frac{1}{2} a_c t^2 \\ \text{Constant Acceleration} \end{array}} \qquad (12\text{–}5)$$

Velocity as a Function of Position. Either solve for t in Eq. 12–4 and substitute into Eq. 12–5, or integrate $v\,dv = a_c\,ds$, assuming that initially $v = v_0$ at $s = s_0$.

$$\int_{v_0}^{v} v\, dv = \int_{s_0}^{s} a_c\, ds$$

$$(\overset{+}{\rightarrow}) \qquad \boxed{\begin{array}{c} v^2 = v_0^2 + 2a_c(s - s_0) \\ \text{Constant Acceleration} \end{array}} \qquad (12\text{–}6)$$

This equation is not independent of Eqs. 12–4 and 12–5 since it can be obtained by eliminating t between these equations.

The magnitudes and signs of s_0, v_0, and a_c, used in the above three equations are determined from the chosen origin and positive direction of the s axis as indicated by the arrow written at the left of each equation. Also, it is important to remember that these equations are useful *only when the acceleration is constant and when* $t = 0$, $s = s_0$, $v = v_0$. A common example of constant accelerated motion occurs when a body falls freely toward the earth. If air resistance is neglected and the distance of fall is short, then the *downward* acceleration of the body when it is close to the earth is constant and approximately 9.81 m/s^2 or 32.2 ft/s^2. The proof of this is given in Example 13–2.

Important Points

- Dynamics is concerned with bodies that have accelerated motion.
- Kinematics is a study of the geometry of the motion.
- Kinetics is a study of the forces that cause the motion.
- Rectilinear kinematics refers to straight-line motion.
- Speed refers to the magnitude of velocity.
- Average speed is the total distance traveled divided by the total time. This is different from the average velocity which is the displacement divided by the time.
- The acceleration, $a = dv/dt$, is negative when the particle is slowing down or decelerating.
- A particle can have an acceleration and yet have zero velocity.
- The relationship $a\, ds = v\, dv$ is derived from $a = dv/dt$ and $v = ds/dt$, by eliminating dt.

Procedure for Analysis

The equations of rectilinear kinematics should be applied using the following procedure.

Coordinate System

- Establish a position coordinate s along the path and specify its *fixed origin* and positive direction.
- Since motion is along a straight line, the particle's position, velocity, and acceleration can be represented as algebraic scalars. For analytical work the sense of s, v, and a is then determined from their *algebraic signs*.
- The positive sense for each scalar can be indicated by an arrow shown alongside each kinematic equation as it is applied.

Kinematic Equations

- If a relationship is known between any *two* of the four variables a, v, s and t, then a third variable can be obtained by using one of the kinematic equations, $a = dv/dt$, $v = ds/dt$ or $a\, ds = v\, dv$, which relates all three variables.*
- Whenever integration is performed, it is important that the position and velocity be known at a given instant in order to evaluate either the constant of integration if an indefinite integral is used, or the limits of integration if a definite integral is used.
- Remember that Eqs. 12–4 through 12–6 have only a limited use. Never apply these equations unless it is absolutely certain that the *acceleration is constant*.

*Some standard differentiation and integration formulas are given in Appendix A.

During the time this rocket undergoes rectilinear motion, its altitude as a function of time can be measured and expressed as $s = s(t)$. Its velocity can then be found using $v = ds/dt$, and its acceleration can be determined from $a = dv/dt$.

EXAMPLE 12–1

The car in Fig. 12–2 moves in a straight line such that for a short time its velocity is defined by $v = (3t^2 + 2t)$ ft/s, where t is in seconds. Determine its position and acceleration when $t = 3$ s. When $t = 0, s = 0$.

Fig. 12–2

Solution

Coordinate System. The position coordinate extends from the fixed origin O to the car, positive to the right.

Position. Since $v = f(t)$, the car's position can be determined from $v = ds/dt$, since this equation relates v, s, and t. Noting that $s = 0$ when $t = 0$, we have*

$$(\overset{+}{\rightarrow}) \qquad v = \frac{ds}{dt} = (3t^2 + 2t)$$

$$\int_0^s ds = \int_0^t (3t^2 + 2t)\,dt$$

$$s\Big|_0^s = t^3 + t^2\Big|_0^t$$

$$s = t^3 + t^2$$

When $t = 3$ s,

$$s = (3)^3 + (3)^2 = 36 \text{ ft} \qquad \textit{Ans.}$$

Acceleration. Knowing $v = f(t)$, the acceleration is determined from $a = dv/dt$, since this equation relates a, v, and t.

$$(\overset{+}{\rightarrow}) \qquad a = \frac{dv}{dt} = \frac{d}{dt}(3t^2 + 2t)$$

$$= 6t + 2$$

When $t = 3$ s,

$$a = 6(3) + 2 = 20 \text{ ft/s}^2 \rightarrow \qquad \textit{Ans.}$$

The formulas for constant acceleration *cannot* be used to solve this problem. Why?

*The *same result* can be obtained by evaluating a constant of integration C rather than using definite limits on the integral. For example, integrating $ds = (3t^2 + 2t)\,dt$ yields $s = t^3 + t^2 + C$. Using the condition that at $t = 0, s = 0$, then $C = 0$.

EXAMPLE 12–2

A small projectile is fired vertically *downward* into a fluid medium with an initial velocity of 60 m/s. Due to the resistance of the fluid the projectile experiences a deceleration equal to $a = (-0.4v^3)\ \text{m/s}^2$, where v is in m/s. Determine the projectile's velocity and position 4 s after it is fired.

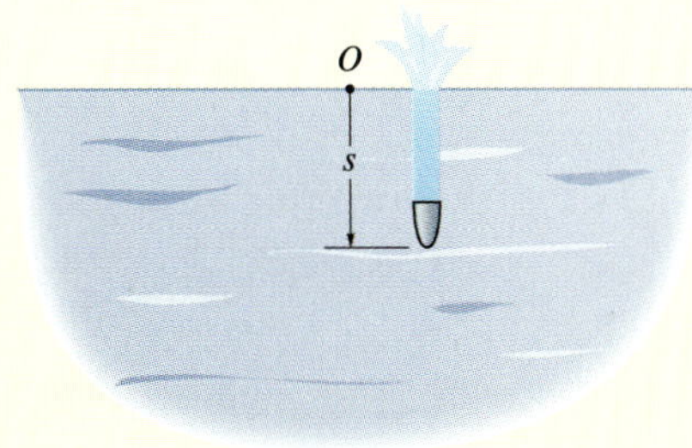

Fig. 12–3

Solution

Coordinate System. Since the motion is downward, the position coordinate is positive downward, with origin located at O, Fig. 12–3.

Velocity. Here $a = f(v)$ and so we must determine the velocity as a function of time using $a = dv/dt$, since this equation relates v, a, and t. (Why not use $v = v_0 + a_c t$?) Separating the variables and integrating, with $v_0 = 60$ m/s when $t = 0$, yields

$$(+\downarrow) \qquad a = \frac{dv}{dt} = -0.4v^3$$

$$\int_{60}^{v} \frac{dv}{-0.4v^3} = \int_0^t dt$$

$$\frac{1}{-0.4}\left(\frac{1}{-2}\right)\frac{1}{v^2}\bigg|_{60}^{v} = t - 0$$

$$\frac{1}{0.8}\left[\frac{1}{v^2} - \frac{1}{(60)^2}\right] = t$$

$$v = \left\{\left[\frac{1}{(60)^2} + 0.8t\right]^{-1/2}\right\}\ \text{m/s}$$

Here the positive root is taken, since the projectile is moving downward. When $t = 4$ s,

$$v = 0.559\ \text{m/s}\downarrow \qquad \textit{Ans.}$$

Position. Knowing $v = f(t)$, we can obtain the projectile's position from $v = ds/dt$, since this equation relates s, v, and t. Using the initial condition $s = 0$, when $t = 0$, we have

$$(+\downarrow) \qquad v = \frac{ds}{dt} = \left[\frac{1}{(60)^2} + 0.8t\right]^{-1/2}$$

$$\int_0^s ds = \int_0^t \left[\frac{1}{(60)^2} + 0.8t\right]^{-1/2} dt$$

$$s = \frac{2}{0.8}\left[\frac{1}{(60)^2} + 0.8t\right]^{1/2}\bigg|_0^t$$

$$s = \frac{1}{0.4}\left\{\left[\frac{1}{(60)^2} + 0.8t\right]^{1/2} - \frac{1}{60}\right\}\ \text{m}$$

When $t = 4$ s,

$$s = 4.43\ \text{m} \qquad \textit{Ans.}$$

EXAMPLE 12-3

During a test a rocket is traveling upward at 75 m/s, and when it is 40 m from the ground its engine fails. Determine the maximum height s_B reached by the rocket and its speed just before it hits the ground. While in motion the rocket is subjected to a constant downward acceleration of 9.81 m/s^2 due to gravity. Neglect the effect of air resistance.

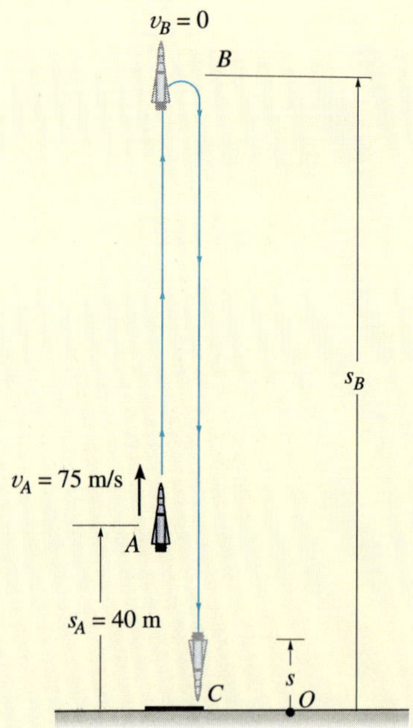

Fig. 12–4

Solution

Coordinate System. The origin O for the position coordinate s is taken at ground level with positive upward, Fig. 12–4.

Maximum Height. Since the rocket is traveling *upward*, $v_A = +75$ m/s when $t = 0$. At the maximum height $s = s_B$ the velocity $v_B = 0$. For the entire motion, the acceleration is $a_c = -9.81$ m/s^2 (negative since it acts in the *opposite* sense to positive velocity or positive displacement). Since a_c is *constant* the rocket's position may be related to its velocity at the two points A and B on the path by using Eq. 12–6, namely,

$$(+\uparrow) \qquad v_B^2 = v_A^2 + 2a_c(s_B - s_A)$$
$$0 = (75 \text{ m/s})^2 + 2(-9.81 \text{ m/s}^2)(s_B - 40 \text{ m})$$
$$s_B = 327 \text{ m} \qquad \textit{Ans.}$$

Velocity. To obtain the velocity of the rocket just before it hits the ground, we can apply Eq. 12–6 between points B and C, Fig. 12–4.

$$(+\uparrow) \qquad v_C^2 = v_B^2 + 2a_c(s_C - s_B)$$
$$= 0 + 2(-9.81 \text{ m/s}^2)(0 - 327 \text{ m})$$
$$v_C = -80.1 \text{ m/s} = 80.1 \text{ m/s} \downarrow \qquad \textit{Ans.}$$

The negative root was chosen since the rocket is moving *downward.*

Similarly, Eq. 12–6 may also be applied between points A and C, i.e.,

$$(+\uparrow) \qquad v_C^2 = v_A^2 + 2a_c(s_C - s_A)$$
$$= (75 \text{ m/s})^2 + 2(-9.81 \text{ m/s}^2)(0 - 40 \text{ m})$$
$$v_C = -80.1 \text{ m/s} = 80.1 \text{ m/s} \downarrow$$

Note: It should be realized that the rocket is subjected to a *deceleration* from A to B of 9.81 m/s^2, and then from B to C it is *accelerated* at this rate. Furthermore, even though the rocket momentarily comes to *rest* at B ($v_B = 0$) the acceleration at B is 9.81 m/s^2 downward!

EXAMPLE 12–4

A metallic particle is subjected to the influence of a magnetic field as it travels downward through a fluid that extends from plate A to plate B, Fig. 12–5. If the particle is released from rest at the midpoint C, $s = 100$ mm, and the acceleration is $a = (4s)\ \text{m/s}^2$, where s is in meters, determine the velocity of the particle when it reaches plate B, $s = 200$ mm, and the time it needs to travel from C to B.

Solution

Coordinate System. As shown in Fig. 12–5, s is taken positive downward, measured from plate A.

Velocity. Since $a = f(s)$, the velocity as a function of position can be obtained by using $v\,dv = a\,ds$. Why not use the formulas for constant acceleration? Realizing that $v = 0$ at $s = 100$ mm $= 0.1$ m, we have

$(+\downarrow)$
$$v\,dv = a\,ds$$
$$\int_0^v v\,dv = \int_{0.1}^s 4s\,ds$$
$$\frac{1}{2}v^2\Big|_0^v = \frac{4}{2}s^2\Big|_{0.1}^s$$
$$v = 2(s^2 - 0.01)^{1/2} \qquad (1)$$

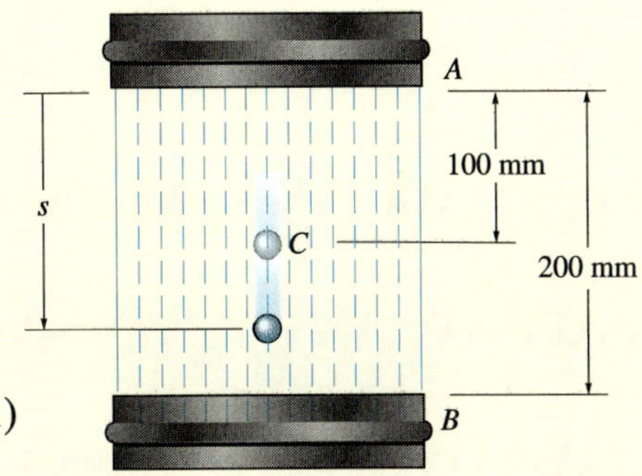

Fig. 12–5

At $s = 200$ mm $= 0.2$ m,

$$v_B = 0.346\ \text{m/s} = 346\ \text{mm/s}\downarrow \qquad \textit{Ans.}$$

The positive root is chosen since the particle is traveling downward, i.e., in the $+s$ direction.

Time. The time for the particle to travel from C to B can be obtained using $v = ds/dt$ and Eq. 1, where $s = 0.1$ m when $t = 0$. From Appendix A,

$(+\downarrow)$
$$ds = v\,dt$$
$$= 2(s^2 - 0.01)^{1/2}\,dt$$
$$\int_{0.1}^s \frac{ds}{(s^2 - 0.01)^{1/2}} = \int_0^t 2\,dt$$
$$\ln\left(\sqrt{s^2 - 0.01} + s\right)\Big|_{0.1}^s = 2t\Big|_0^t$$
$$\ln\left(\sqrt{s^2 - 0.01} + s\right) + 2.30 = 2t$$

At $s = 200$ mm $= 0.2$ m,

$$t = \frac{\ln\left(\sqrt{(0.2)^2 - 0.01} + 0.2\right) + 2.30}{2} = 0.658\ \text{s} \qquad \textit{Ans.}$$

EXAMPLE 12–5

A particle moves along a horizontal path with a velocity of $v = (3t^2 - 6t)$ m/s, where t is the time in seconds. If it is initially located at the origin O, determine the distance traveled in 3.5 s, and the particle's average velocity and average speed during the time interval.

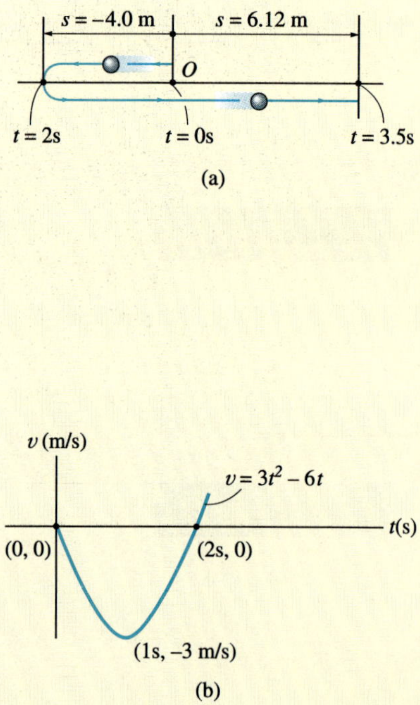

Fig. 12–6

Solution

Coordinate System. Here we will assume positive motion to the right, measured from the origin O, Fig. 12-6a.

Distance Traveled. Since $v = f(t)$, the position as a function of time may be found by integrating $v = ds/dt$ with $t = 0, s = 0$.

$$(\overset{+}{\rightarrow}) \qquad ds = v\,dt$$
$$= (3t^2 - 6t)\,dt$$
$$\int_0^s ds = 3\int_0^t t^2\,dt - 6\int_0^t t\,dt$$
$$s = (t^3 - 3t^2)\text{ m} \qquad (1)$$

In order to determine the distance traveled in 3.5 s, it is necessary to investigate the path of motion. The graph of the velocity function, Fig. 12–6*b*, reveals that for $0 \le t < 2$ s the velocity is *negative*, which means the particle is traveling to the *left*, and for $t > 2$ s the velocity is *positive*, and hence the particle is traveling to the *right*. Also, $v = 0$ at $t = 2$ s. The particle's position when $t = 0$, $t = 2$ s, and $t = 3.5$ s can be determined from Eq. 1. This yields

$$s|_{t=0} = 0 \qquad s|_{t=2\text{s}} = -4.0\text{ m} \qquad s|_{t=3.5\text{s}} = 6.125\text{ m}$$

The path is shown in Fig. 12–6*a*. Hence, the distance traveled in 3.5 s is

$$s_T = 4.0 + 4.0 + 6.125 = 14.125\text{ m} = 14.1\text{ m} \qquad \textit{Ans.}$$

Velocity. The *displacement* from $t = 0$ to $t = 3.5$ s is

$$\Delta s = s|_{t=3.5\text{s}} - s|_{t=0} = 6.12 - 0 = 6.12\text{ m}$$

and so the average velocity is

$$v_{\text{avg}} = \frac{\Delta s}{\Delta t} = \frac{6.12}{3.5 - 0} = 1.75\text{ m/s} \rightarrow \qquad \textit{Ans.}$$

The average speed is defined in terms of the *distance traveled* s_T. This positive scalar is

$$(v_{\text{sp}})_{\text{avg}} = \frac{s_T}{\Delta t} = \frac{14.125}{3.5 - 0} = 4.04\text{ m/s} \qquad \textit{Ans.}$$

Problems

12-1. A ball is thrown downward from a 50-ft tower with an initial speed of 18 ft/s. Determine the speed at which it hits the ground and the time of travel.

12-2. A car has an initial speed of 25 m/s and a constant deceleration of 3 m/s^2. Determine the velocity of the car when $t = 4$ s. What is the displacement of the car during the 4-s time interval? How much time is needed to stop the car?

12-3. If a particle has an initial velocity of $v_0 = 12$ ft/s to the right, at $s_0 = 0$, determine its position when $t = 10$ s, if $a = 2\ \text{ft/s}^2$ to the left.

***12-4.** A particle travels along a straight line with a velocity $v = (12 - 3t^2)$ m/s, where t is in seconds. When $t = 1$ s, the particle is located 10 m to the left of the origin. Determine the acceleration when $t = 4$ s, the displacement from $t = 0$ to $t = 10$ s, and the distance the particle travels during this time period.

12-5. Starting from rest, a particle moving in a straight line has an acceleration of $a = (2t - 6)\ \text{m/s}^2$, where t is in seconds. What is the particle's velocity when $t = 6$ s, and what is its position when $t = 11$ s?

12-6. A particle moves along a straight line such that its position is defined by $s = (t^3 - 3t^2 + 2)$ m. Determine the average velocity, the average speed, and the acceleration of the particle when $t = 4$ s.

12-7. A particle moves along a straight line such that its position is defined by $s = (2t^3 + 3t^2 - 12t - 10)$ m. Determine the velocity, average velocity, and the average speed of the particle when $t = 3$ s.

***12-8.** The velocity of a particle traveling in a straight line is given by $v = (6t - 3t^2)$ m/s, where t is in seconds. If $s = 0$ when $t = 0$, determine the particle's deceleration and position when $t = 3$ s. How far has the particle traveled during the 3-s time interval, and what is its average speed?

12-9. A particle moves along a straight line such that its position is defined by $s = (t^2 - 6t + 5)$ m. Determine the average velocity, the average speed, and the acceleration of the particle when $t = 6$ s.

12-10. A particle travels along a straight line such that in 2 s it moves from an initial position $s_A = +0.5$ m to a position $s_B = -1.5$ m. Then in another 4 s it moves from s_B to $s_C = +2.5$ m. Determine the particle's average velocity and average speed during the 6-s time interval.

12-11. A particle is moving along a straight line such that its position is defined by $s = (10t^2 + 20)$ mm, where t is in seconds. Determine (a) the displacement of the particle during the time interval from $t = 1$ s to $t = 5$ s, (b) the average velocity of the particle during this time interval, and (c) the acceleration when $t = 1$ s.

***12-12.** The acceleration of a particle as it moves along a straight line is given by $a = (2t - 1)\ \text{m/s}^2$, where t is in seconds. If $s = 1$ m and $v = 2$ m/s when $t = 0$, determine the particle's velocity and position when $t = 6$ s. Also, determine the total distance the particle travels during this time period.

12-13. A car is to be hoisted by elevator to the fourth floor of a parking garage, which is 48 ft above the ground. If the elevator can accelerate at 0.6 ft/s^2, decelerate at 0.3 ft/s^2, and reach a maximum speed of 8 ft/s, determine the shortest time to make the lift, starting from rest and ending at rest.

12-14. The position of a particle on a straight line is given by $s = (t^3 - 9t^2 + 15t)$ ft, where t is in seconds. Determine the position of the particle when $t = 6$ s and the total distance it travels during the 6-s time interval. *Hint:* Plot the path to determine the total distance traveled.

12-15. A particle has an initial speed of 27 m/s. If it experiences a deceleration of $a = (-6t)\ \text{m/s}^2$, where t is in seconds, determine the distance traveled before it stops.

■***12-16.** A particle has an initial speed of 27 m/s. If it experiences a deceleration of $a = (-6t)\ \text{m/s}^2$, where t is in seconds, determine its velocity when it travels 10 m. How much time does this take?

12-17. A car starts from rest and moves along a straight line with an acceleration of $a = (3s^{-1/3})\ \text{m/s}^2$, where s is in meters. Determine the car's velocity and position when $t = 6$ s.

12-18. A car starts from rest and moves along a straight line with an acceleration of $a = (3s^{-1/3})\ \text{m/s}^2$, where s is in meters. Determine the car's acceleration when $t = 4$ s.

12-19. Car B is traveling a distance d ahead of car A. Both cars are traveling at 60 ft/s when the driver of B suddenly applies the brakes, causing his car to decelerate at 12 ft/s^2. It takes the driver of car A 0.75 s to react (this is the normal reaction time for drivers). When he applies his brakes, he decelerates at 15 ft/s^2. Determine the minimum distance d between the cars so as to avoid a collision.

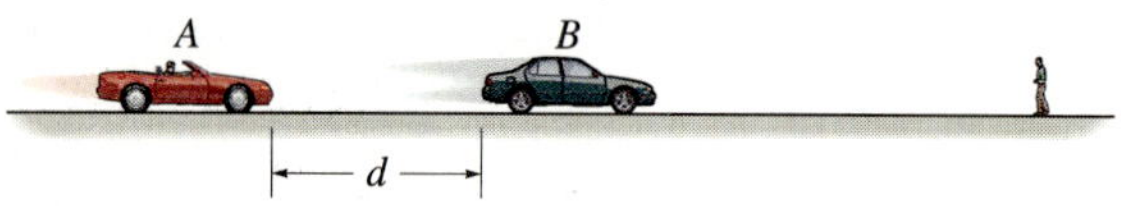

Prob. 12–19

***12-20.** A particle is moving along a straight line such that its speed is defined as $v = (-4s^2)$ m/s, where s is in meters. If $s = 2$ m when $t = 0$, determine the velocity and acceleration as functions of time.

12-21. A particle is moving along a straight line such that its acceleration is defined as $a = (4s^2)$ m/s^2, where s is in meters. If $v = -100$ m/s when $s = 10$ m and $t = 0$, determine the particle's velocity as a function of position.

12-22. A particle is moving along a straight line such that its acceleration is defined as $a = (-2v)$ m/s^2, where v is in meters per second. If $v = 20$ m/s when $s = 0$ and $t = 0$, determine the particle's position, velocity, and acceleration as functions of time.

12-23. A particle is moving along a straight line such that its acceleration is defined as $a = (-2v)$ m/s^2, where v is in meters per second. If $v = 20$ m/s when $s = 0$ and $t = 0$, determine the particle's velocity as a function of position and the distance the particle moves before it stops.

***12-24.** A particle, initially at the origin, moves along a straight line through a fluid medium such that its velocity is defined as $v = 1.8(1 - e^{-0.3t})$ m/s, where t is in seconds. Determine the displacement of the particle during the first 3 s.

■**12-25.** A particle moves along a straight line with an acceleration of $a = 5/(3s^{1/3} + s^{5/2})$ m/s^2, where s is in meters. Determine the particle's velocity when $s = 2$ m, if it starts from rest when $s = 1$ m. Use Simpson's rule to evaluate the integral.

12-26. A sandbag is dropped from a balloon which is ascending vertically at a constant speed of 6 m/s. If the bag is released with the same upward velocity of 6 m/s when $t = 0$ and hits the ground when $t = 8$ s, determine the speed of the bag as it hits the ground and the altitude of the balloon at this instant.

12-27. A particle is moving along a straight line such that when it is at the origin it has a velocity of 4 m/s. If it begins to decelerate at the rate of $a = (-1.5v^{1/2})$ m/s^2, where v is in m/s, determine the distance it travels before it stops.

***12-28.** A particle is moving along a straight line such that when it is at the origin it has a velocity of 4 m/s. If it begins to decelerate at the rate of $a = (-1.5v^{1/2})$ m/s^2, where v is in m/s, determine the particle's position and velocity when $t = 2$ s.

12-29. A particle is moving with a velocity of v_0 when $s = 0$ and $t = 0$. If it is subjected to a deceleration of $a = -kv^3$, where k is a constant, determine its velocity and position as functions of time.

12-30. A car can have an acceleration and a deceleration of 5 m/s^2. If it starts from rest, and can have a maximum speed of 60 m/s, determine the shortest time it can travel a distance of 1200 m when it stops.

12-31. A ball A is thrown vertically upward from the top of a 30-m-high building with an initial velocity of 5 m/s. At the same instant another ball B is thrown upward from the ground with an initial velocity of 20 m/s. Determine the height from the ground and the time at which they pass.

***12-32.** A motorcycle starts from rest at $t = 0$ and travels along a straight road with a constant acceleration of 6 ft/s^2 until it reaches a speed of 50 ft/s. Afterwards it maintains this speed. Also, when $t = 0$, a car located 6000 ft down the road is traveling toward the motorcycle at a constant speed of 30 ft/s. Determine the time and the distance traveled by the motorcycle when they pass each other.

12-33. If the effects of atmospheric resistance are accounted for, a freely falling body has an acceleration defined by the equation $a = 9.81[1 - v^2(10^{-4})]$ m/s^2, where v is in m/s and the positive direction is downward. If the body is released from rest at a *very high altitude*, determine (a) the velocity when $t = 5$ s, and (b) the body's terminal or maximum attainable velocity (as $t \rightarrow \infty$).

12-34. As a body is projected to a high altitude above the earth's *surface,* the variation of the acceleration of gravity with respect to altitude y must be taken into account. Neglecting air resistance, this acceleration is determined from the formula $a = -g_0[R^2/(R + y)^2]$, where g_0 is the constant gravitational acceleration at sea level, R is the radius of the earth, and the positive direction is measured upward. If $g_0 = 9.81\ \text{m/s}^2$ and $R =$ 6356 km, determine the minimum initial velocity (escape velocity) at which a projectile should be shot vertically from the earth's surface so that it does not fall back to the earth. *Hint:* This requires that $v = 0$ as $y \rightarrow \infty$.

12-35. Accounting for the variation of gravitational acceleration a with respect to altitude y (see Prob. 12-34), derive an equation that relates the velocity of a freely falling particle to its altitude. Assume that the particle is released from rest at an altitude y_0 from the earth's surface. With what velocity does the particle strike the earth if it is released from rest at an altitude $y_0 = 500$ km? Use the numerical data in Prob. 12-34.

***12-36.** When a particle falls through the air, its initial acceleration $a = g$ diminishes until it is zero, and thereafter it falls at a constant or terminal velocity v_f. If this variation of the acceleration can be expressed as $a = (g/v_f^2)(v_f^2 - v^2)$, determine the time needed for the velocity to become $v < v_f$. Initially the particle falls from rest.

12.3 Rectangular Kinematics: Erratic Motion

When a particle's motion during a time period is erratic, it may be difficult to obtain a continuous mathematical function to describe its position, velocity, or acceleration. Instead, the motion may best be described graphically using a series of curves that can be generated experimentally from computer output. If the resulting graph describes the relationship between any two of the variables, a, v, s, t, a graph describing the relationship between the other variables can be established by using the kinematic equations $a = dv/dt, v = ds/dt, a\,ds = v\,dv$. Several situations occur frequently.

Commuter trains are subjected to erratic motion as they move from one station to the next. By specifying their velocity–time graph, it is possible to establish their position–time graph, which is necessary for scheduling. Also, the acceleration–time graph can be determined. This graph provides a means for measuring passenger discomfort.

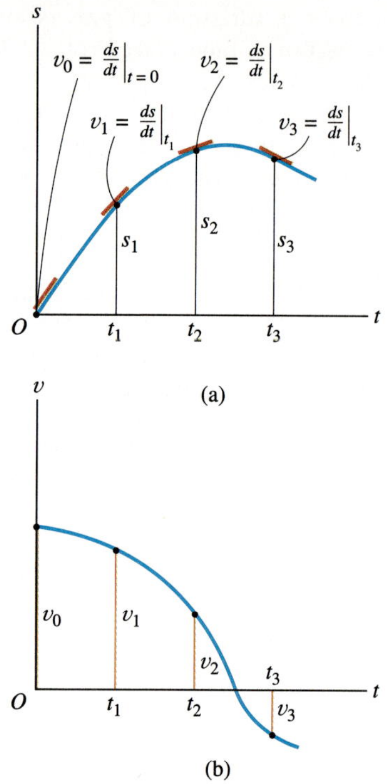

Fig. 12–7

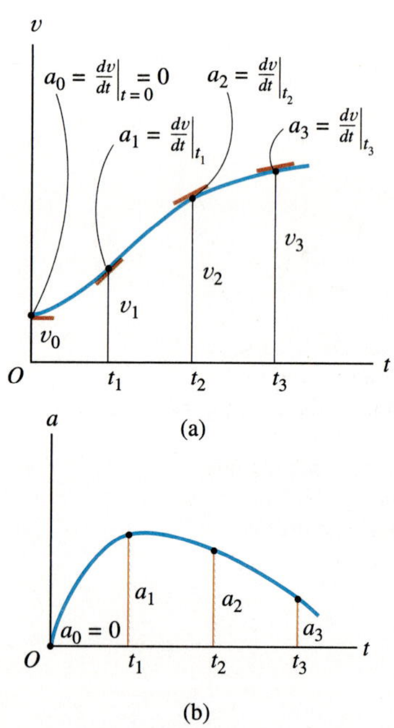

Fig. 12–8

Given the s–t Graph, Construct the v–t Graph. If the position of a particle can be *determined experimentally* during a time period t, the s–t graph for the particle can be plotted, Fig. 12–7*a*. To determine the particle's velocity as a function of time, i.e., the v–t graph, we must use $v = ds/dt$ since this equation relates v, s, and t. Therefore, the velocity at any instant is determined by measuring the *slope* of the s–t graph, i.e.,

$$\frac{ds}{dt} = v$$

slope of
s–t graph = velocity

For example, measurement of the slopes v_0, v_1, v_2, v_3 at the intermediate points $(0, 0)$, (t_1, s_1), (t_2, s_2), (t_3, s_3) on the s–t graph, Fig. 12–7*a*, gives the corresponding points on the v–t graph shown in Fig. 12–7*b*.

It may also be possible to establish the v–t graph *mathematically*, provided the segments of the s–t graph can be expressed in the form of equations $s = f(t)$. Corresponding equations describing the segments of the v–t graph are then determined by time *differentiation*, since $v = ds/dt$.

Given the v–t Graph, Construct the a–t Graph. When the particle's v–t graph is known, as in Fig. 12–8*a*, the acceleration as a function of time, i.e., the a–t graph, can be determined using $a = dv/dt$. (Why?) Hence, the acceleration at any instant is determined by measuring the slope of the v–t graph, i.e.,

$$\frac{dv}{dt} = a$$

slope of
v–t graph = acceleration

For example, measurement of the slopes a_0, a_1, a_2, a_3 at the intermediate points $(0, 0)$, (t_1, v_1), (t_2, v_2), (t_3, v_3) on the v–t graph, Fig. 12–8*a*, yields the corresponding points on the a–t graph shown in Fig. 12–8*b*.

Any segments of the a–t graph can also be determined *mathematically*, provided the equations of the corresponding segments of the v–t graph are known, $v = g(t)$. This is done by simply taking the time *derivative* of $v = g(t)$, since $a = dv/dt$.

Since differentiation reduces a polynomial of degree n to that of degree $n - 1$, then if the s–t graph is parabolic (a second-degree curve), the v–t graph will be a sloping line (a first-degree curve), and the a–t graph will be a constant or a horizontal line (a zero-degree curve).

EXAMPLE 12-6

A bicycle moves along a straight road such that its position is described by the graph shown in Fig. 12–9*a*. Construct the v–t and a–t graphs for $0 \le t \le 30$ s.

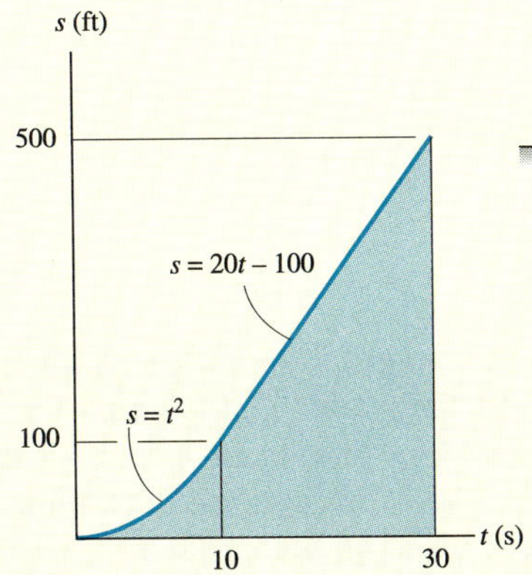

(a)

Fig. 12–9

Solution

v–t Graph. Since $v = ds/dt$, the v–t graph can be determined by differentiating the equations defining the s–t graph, Fig. 12–9*a*. We have

$$0 \le t < 10 \text{ s}; \qquad s = t^2 \qquad v = \frac{ds}{dt} = 2t$$

$$10 \text{ s} < t \le 30 \text{ s}; \qquad s = 20t - 100 \qquad v = \frac{ds}{dt} = 20$$

The results are plotted in Fig. 12–9*b*. We can also obtain specific values of v by measuring the *slope* of the s–t graph at a given instant. For example, at $t = 20$ s, the slope of the s–t graph is determined from the straight line from 10 s to 30 s, i.e.,

$$t = 20 \text{ s}; \qquad v = \frac{\Delta s}{\Delta t} = \frac{500 - 100}{30 - 10} = 20 \text{ ft/s}$$

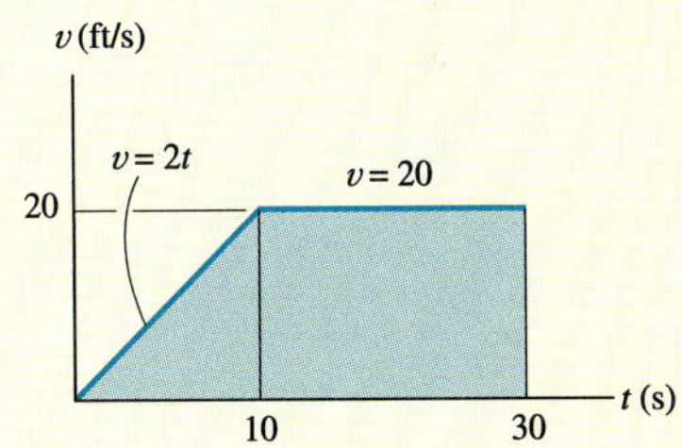

(b)

a–t Graph. Since $a = dv/dt$, the a–t graph can be determined by differentiating the equations defining the lines of the v–t graph. This yields

$$0 \le t < 10 \text{ s}; \qquad v = 2t \qquad a = \frac{dv}{dt} = 2$$

$$10 < t \le 30 \text{ s}; \qquad v = 20 \qquad a = \frac{dv}{dt} = 0$$

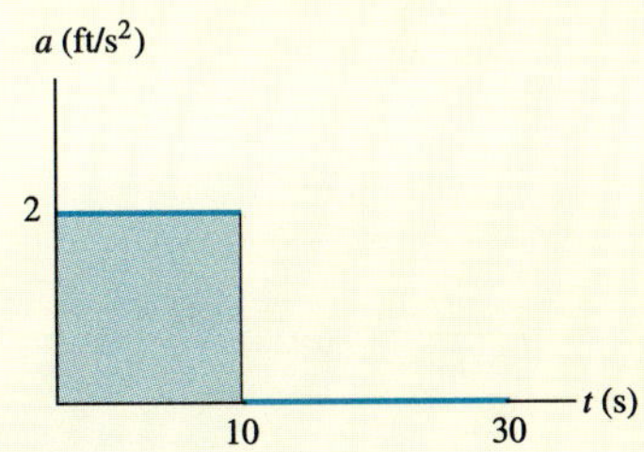

(c)

The results are plotted in Fig. 12–9*c*. Show that $a = 2 \text{ ft/s}^2$ when $t = 5$ s by measuring the slope of the v–t graph.

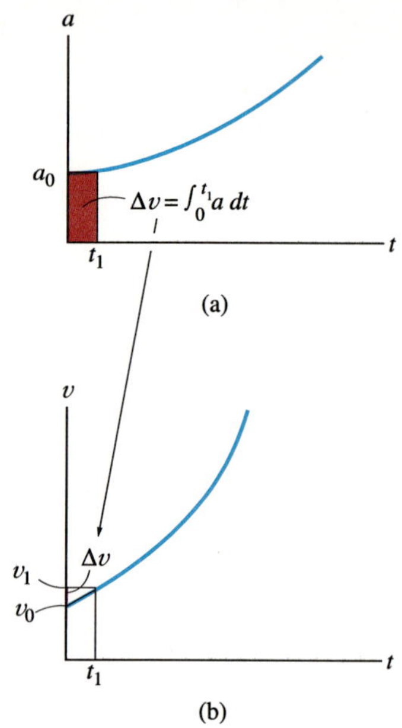

Fig. 12–10

Given the a–t Graph, Construct the v–t Graph. If the a–t graph is given, Fig. 12–10a, the v–t graph may be constructed using $a = dv/dt$, written in integrated form as

$$\Delta v = \int a\,dt$$

$$\text{change in velocity} = \text{area under } a\text{–}t \text{ graph}$$

Hence, to construct the v–t graph, we begin by first knowing the particle's initial velocity v_0 and then add to this small increments of area (Δv) determined from the a–t graph. In this manner, successive points, $v_1 = v_0 + \Delta v$, etc., for the v–t graph are determined, Fig. 12–10b. Notice that an algebraic addition of the area increments is necessary, since areas lying above the t axis correspond to an increase in v ("positive" area), whereas those lying below the axis indicate a decrease in v ("negative" area).

If segments of the a–t graph can be described by a series of equations, then each of these equations may be *integrated* to yield equations describing the corresponding segments of the v–t graph. Hence, if the a–t graph is linear (a first-degree curve), integration will yield a v–t graph that is parabolic (a second-degree curve), etc.

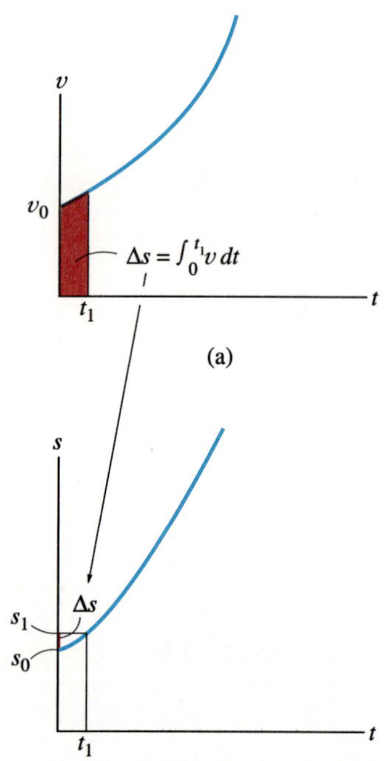

Fig. 12–11

Given the v–t Graph, Construct the s–t Graph. When the v–t graph is given, Fig. 12–11a, it is possible to determine the s–t graph using $v = ds/dt$, written in integrated form

$$\Delta s = \int v\,dt$$

$$\text{displacement} = \text{area under } v\text{–}t \text{ graph}$$

In the same manner as stated above, we begin by knowing the particle's initial position s_0 and add (algebraically) to this small area increments Δs determined from the v–t graph, Fig. 12–11b.

If it is possible to describe segments of the v–t graph by a series of equations, then each of these equations may be *integrated* to yield equations that describe corresponding segments of the s–t graph.

EXAMPLE 12-7

The test car in Fig. 12–12a starts from rest and travels along a straight track such that it accelerates at a constant rate for 10 s and then decelerates at a constant rate. Draw the v–t and s–t graphs and determine the time t' needed to stop the car. How far has the car traveled?

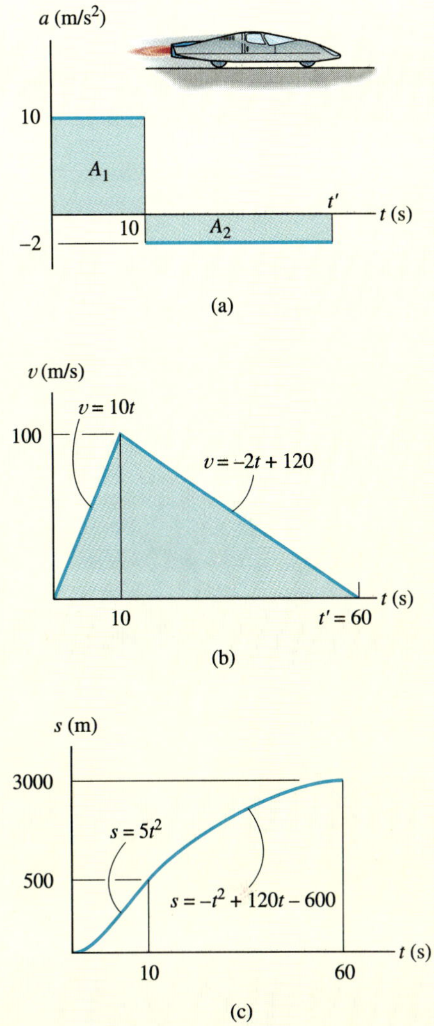

Fig. 12–12

Solution

v–t Graph. Since $dv = a\,dt$, the v–t graph is determined by integrating the straight-line segments of the a–t graph. Using the *initial condition* $v = 0$ when $t = 0$, we have

$$0 \le t < 10 \text{ s}; \qquad a = 10; \qquad \int_0^v dv = \int_0^t 10\,dt, \qquad v = 10t$$

When $t = 10$ s, $v = 10(10) = 100$ m/s. Using this as the *initial condition* for the next time period, we have

$$10 \text{ s} < t \le t'; \qquad a = -2; \qquad \int_{100}^v dv = \int_{10}^t -2\,dt, \qquad v = -2t + 120$$

When $t = t'$ we require $v = 0$. This yields, Fig. 12–12b,

$$t' = 60 \text{ s} \qquad \textit{Ans.}$$

A more direct solution for t' is possible by realizing that the area under the a–t graph is equal to the change in the car's velocity. We require $\Delta v = 0 = A_1 + A_2$, Fig. 12–12a. Thus

$$0 = 10 \text{ m/s}^2(10 \text{ s}) + (-2 \text{ m/s}^2)(t' - 10 \text{ s}) = 0$$

$$t' = 60 \text{ s} \qquad \textit{Ans.}$$

s–t Graph. Since $ds = v\,dt$, integrating the equations of the v–t graph yields the corresponding equations of the s–t graph. Using the *initial condition* $s = 0$ when $t = 0$, we have

$$0 \le t \le 10 \text{ s}; \qquad v = 10t; \qquad \int_0^s ds = \int_0^t 10t\,dt, \qquad s = 5t^2$$

When $t = 10$ s, $s = 5(10)^2 = 500$ m. Using this *initial condition,*

$$10 \text{ s} \le t \le 60 \text{ s}; \qquad v = -2t + 120; \qquad \int_{500}^s ds = \int_{10}^t (-2t + 120)\,dt$$

$$s - 500 = -t^2 + 120t - [-(10)^2 + 120(10)]$$

$$s = -t^2 + 120t - 600$$

When $t' = 60$ s, the position is

$$s = -(60)^2 + 120(60) - 600 = 3000 \text{ m} \qquad \textit{Ans.}$$

The s–t graph is shown in Fig. 12–12c. Note that a direct solution for s is possible when $t' = 60$ s, since the *triangular area* under the v–t graph would yield the displacement $\Delta s = s - 0$ from $t = 0$ to $t' = 60$ s. Hence,

$$\Delta s = \tfrac{1}{2}(60)(100) = 3000 \text{ m} \qquad \textit{Ans.}$$

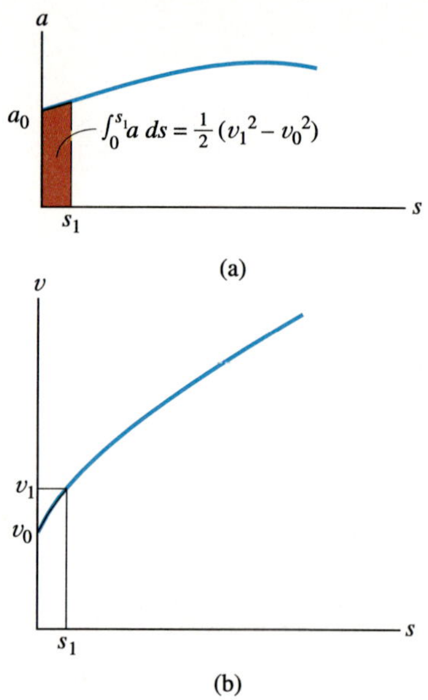

Fig. 12–13

Given the *a–s* Graph, Construct the *v–s* Graph.

In some cases an a–s graph for the particle can be constructed, so that points on the v–s graph can be determined by using $v\,dv = a\,ds$. Integrating this equation between the limits $v = v_0$ at $s = s_0$ and $v = v_1$ at $s = s_1$, we have,

$$\frac{1}{2}(v_1^2 - v_0^2) = \int_{s_0}^{s_1} a\,ds$$

area under
a–s graph

Thus, the initial small segment of area under the a–s graph, $\int_{s_0}^{s_1} a\,ds$, shown colored in Fig. 12–13a, equals one-half the difference in the squares of the speed, $\frac{1}{2}(v_1^2 - v_0^2)$. Therefore, if the area is determined and the initial value of v_0 at $s_0 = 0$ is known, then $v_1 = (2\int_{s_0}^{s_1} a\,ds + v_0^2)^{1/2}$, Fig. 12–13$b$. Successive points on the v–s graph can be constructed in this manner starting from the initial velocity v_0.

Another way to construct the v–s graph is to first determine the equations which define the segments of the a–s graph. Then the corresponding equations defining the segments of the v–s graph can be obtained directly from integration, using $v\,dv = a\,ds$.

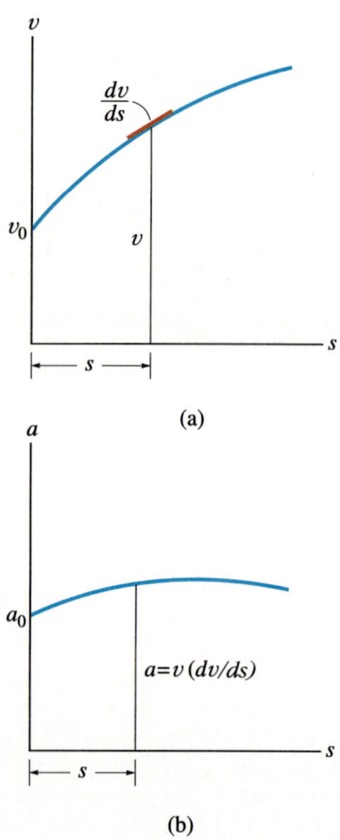

Fig. 12–14

Given the *v–s* Graph, Construct the *a–s* Graph.

If the v–s graph is known, the acceleration a at any position s can be determined using $a\,ds = v\,dv$, written as

$$a = v\left(\frac{dv}{ds}\right)$$

acceleration = velocity times
slope of
v–s graph

Thus, at any point (s, v) in Fig. 12–14a, the slope dv/ds of the v–s graph is measured. Then since v and dv/ds are known, the value of a can be calculated, Fig. 12–14b.

We can also determine the segments describing the a–s graph analytically, provided the equations of the corresponding segments of the v–s graph are known. As above, this requires integration using $a\,ds = v\,dv$.

EXAMPLE 12-8

The v–s graph describing the motion of a motorcycle is shown in Fig. 12–15*a*. Construct the a–s graph of the motion and determine the time needed for the motorcycle to reach the position $s = 400$ ft.

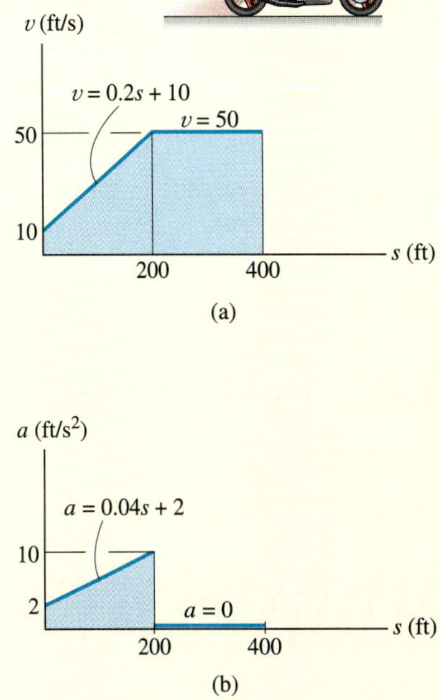

Fig. 12–15

Solution

a–s Graph. Since the equations for segments of the v–s graph are given, the a–s graph can be determined using $a\,ds = v\,dv$.

$0 \leq s < 200$ ft; $\quad v = 0.2s + 10$

$$a = v\frac{dv}{ds} = (0.2s + 10)\frac{d}{ds}(0.2s + 10) = 0.04s + 2$$

200 ft $< s \leq 400$ ft; $\quad v = 50;$

$$a = v\frac{dv}{ds} = (50)\frac{d}{ds}(50) = 0$$

The results are plotted in Fig. 12–15*b*.

Time. The time can be obtained using the v–s graph and $v = ds/dt$, because this equation relates v, s, and t. For the first segment of motion, $s = 0$ at $t = 0$, so

$$0 \leq s < 200\text{ ft}; \quad v = 0.2s + 10; \quad dt = \frac{ds}{v} = \frac{ds}{0.2s + 10}$$

$$\int_0^t dt = \int_0^s \frac{ds}{0.2s + 10}$$

$$t = 5\ln(0.2s + 10) - 5\ln 10$$

At $s = 200$ ft, $t = 5\ln[0.2(200) + 10] - 5\ln 10 = 8.05$ s. Therefore, for the second segment of motion,

$$200\text{ ft} < s \leq 400\text{ ft}; \quad v = 50; \quad dt = \frac{ds}{v} = \frac{ds}{50}$$

$$\int_{8.05}^t dt = \int_{200}^s \frac{ds}{50}$$

$$t - 8.05 = \frac{s}{50} - 4$$

$$t = \frac{s}{50} + 4.05$$

Therefore, at $s = 400$ ft,

$$t = \frac{400}{50} + 4.05 = 12.0\text{ s} \qquad \textit{Ans.}$$

Problems

12-37. The speed of a train during the first minute has been recorded as follows:

t (s)	0	20	40	60
v (m/s)	0	16	21	24

Plot the v–t graph, approximating the curve as straight-line segments between the given points. Determine the total distance traveled.

12-38. A two-stage missile is fired vertically from rest with the acceleration shown. In 15 s the first stage A burns out and the second stage B ignites. Plot the v–t and s–t graphs which describe the motion of the second stage for $0 \le t \le 20$ s.

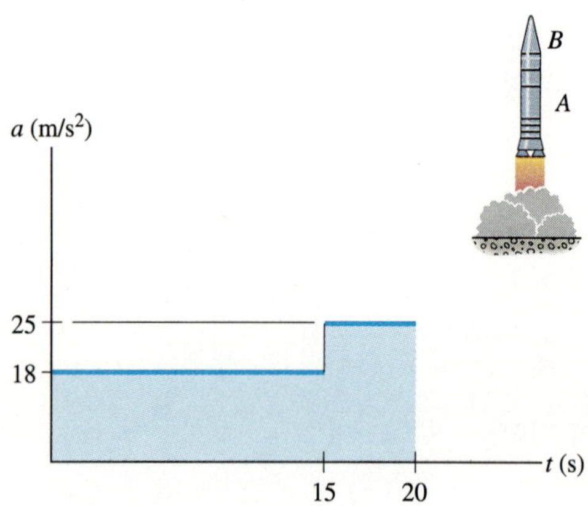

Prob. 12–38

12-39. If the position of a particle is defined as $s = (5t - 3t^2)$ ft, where t is in seconds, construct the s–t, v–t, and a–t graphs for $0 \le t \le 10$ s.

***12-40.** If the position of a particle is defined by $s = [2 \sin (\pi/5)t + 4]$ m, where t is in seconds, construct the s–t, v–t, and a–t graphs for $0 \le t \le 10$ s.

12-41. A particle travels along a curve defined by the equation $s = (t^3 - 3t^2 + 2t)$ m, where t is in seconds. Draw the s–t, v–t, and a–t graphs for the particle for $0 \le t \le 3$ s.

12-42. An airplane lands on the runway, originally traveling at 110 ft/s when $s = 0$. If it is subjected to the deceleration shown, determine the time t' needed to stop the plane and construct the s–t graph for the motion.

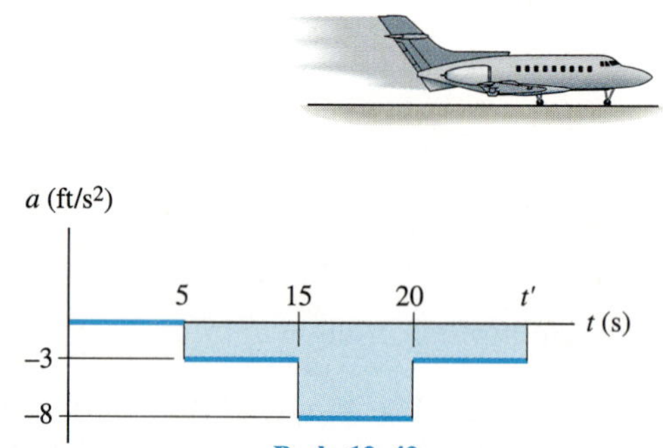

Prob. 12–42

12-43. A car starting from rest moves along a straight track with an acceleration as shown. Determine the time t for the car to reach a speed of 50 m/s and construct the v–t graph that describes the motion until the time t.

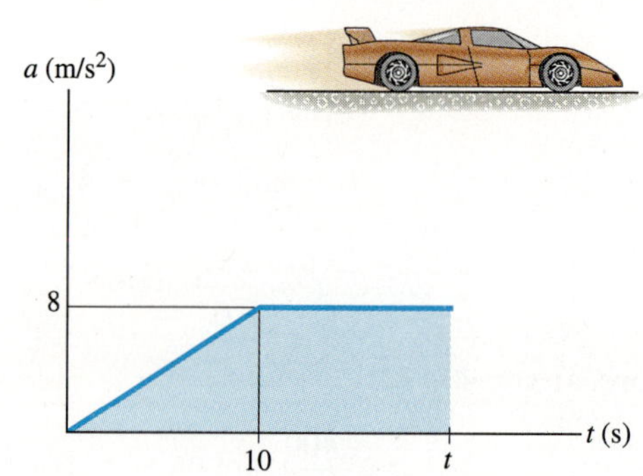

Prob. 12–43

*12-44. A car travels up a hill with the speed shown. Determine the total distance the car moves until it stops ($t = 60$ s). Plot the a–t graph.

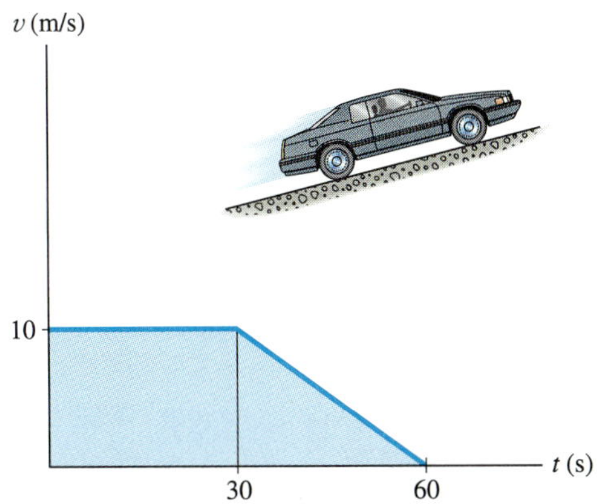

Prob. 12–44

12-45. From experimental data, the motion of a jet plane while traveling along a runway is defined by the v–t graph shown. Construct the s–t and a–t graphs for the motion.

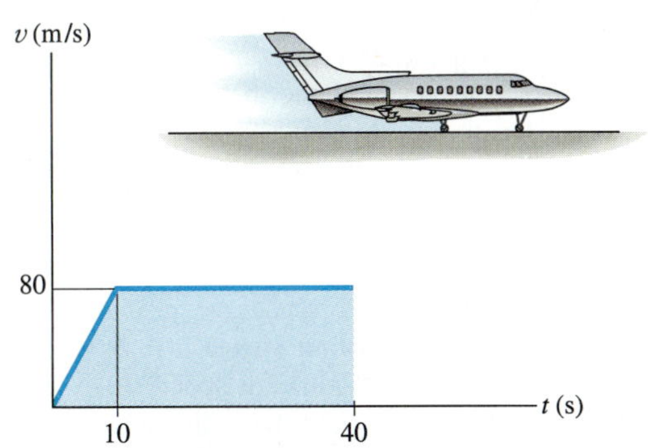

Prob. 12–45

12-46. A car travels along a straight road with the speed shown by the v–t graph. Determine the total distance the car travels until it stops when $t = 48$ s. Also plot the s–t and a–t graphs.

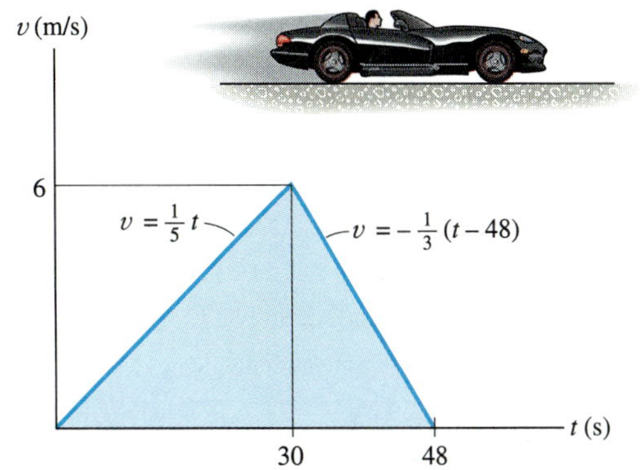

Prob. 12–46

12-47. The v–t graph for the motion of a train as it moves from station A to station B is shown. Draw the a–t graph and determine the average speed and the distance between the stations.

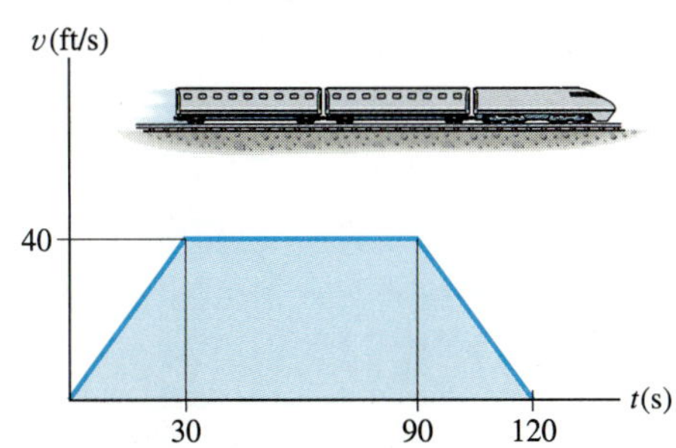

Prob. 12–47

***12-48.** The s–t graph for a train has been experimentally determined. From the data, construct the v–t and a–t graphs for the motion; $0 \le t \le 40$ s. For $0 \le t \le 30$ s, the curve is $s = (0.4t^2)$ m, and then it becomes straight for $t \ge 30$ s.

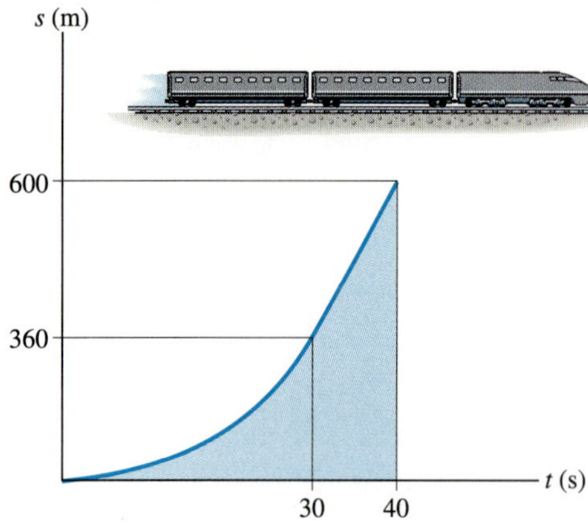

Prob. 12–48

12-49. The v–t graph for the motion of a car as it moves along a straight road is shown. Draw the a–t graph and determine the maximum acceleration during the 30-s time interval. The car starts from rest at $s = 0$.

12-50. The v–t graph for the motion of a car as it moves along a straight road is shown. Draw the s–t graph and determine the average speed and the distance traveled for the 30-s time interval. The car starts from rest at $s = 0$.

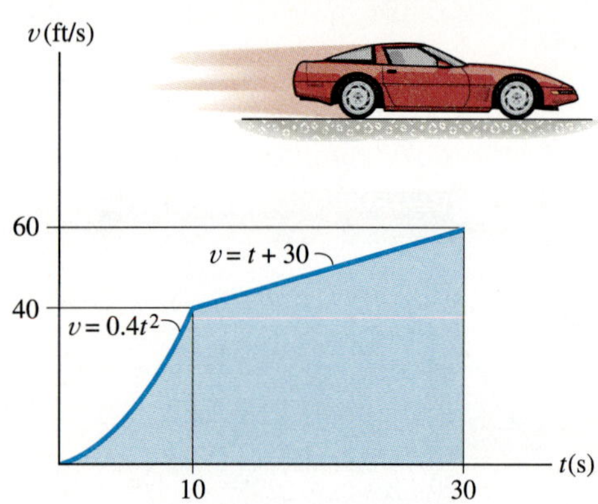

Probs. 12–49/50

12-51. A missile starting from rest travels along a straight track and for 10 s has an acceleration as shown. Draw the v–t graph that describes the motion and find the distance traveled in 10 s.

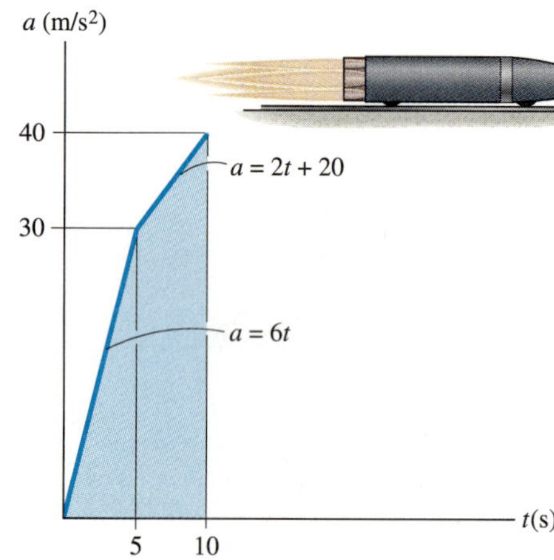

Prob. 12–51

***12-52.** A man riding upward in a freight elevator accidentally drops a package off the elevator when it is 100 ft from the ground. If the elevator maintains a constant upward speed of 4 ft/s, determine how high the elevator is from the ground the instant the package hits the ground. Draw the v–t curve for the package during the time it is in motion. Assume that the package was released with the same upward speed as the elevator.

12-53. Two cars start from rest side by side and travel along a straight road. Car A accelerates at 4 m/s^2 for 10 s and then maintains a constant speed. Car B accelerates at 5 m/s^2 until reaching a constant speed of 25 m/s and then maintains this speed. Construct the a–t, v–t, and s–t graphs for each car until $t = 15$ s. What is the distance between the two cars when $t = 15$ s?

12-54. A two-stage rocket is fired vertically from rest at $s = 0$ with an acceleration as shown. After 30 s the first stage A burns out, and the second stage B ignites. Plot the v–t and s–t graphs which describe the motion of the second stage for $0 \le t \le 60$ s.

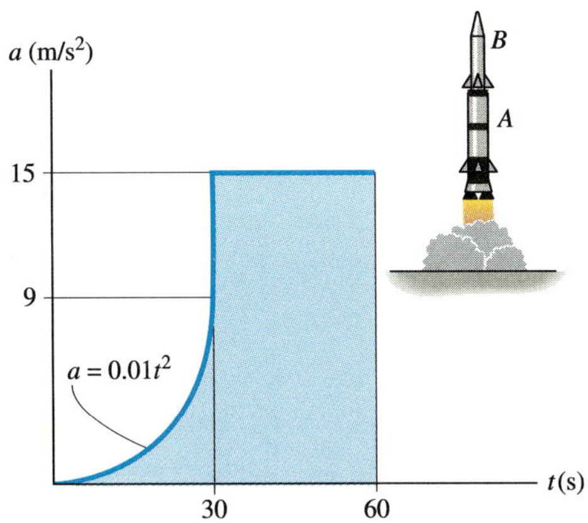

Prob. 12–54

***12-56.** A bicyclist starting from rest travels along a straight road and for 10 s has an acceleration as shown. Draw the v–t graph that describes the motion and find the distance traveled in 10 s.

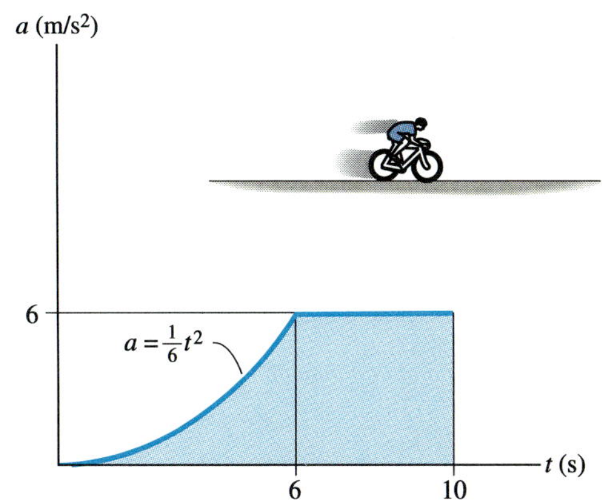

Prob. 12–56

12-55. The a–t graph for a motorcycle traveling along a straight road has been estimated as shown. Determine the time needed for the motorcycle to reach a maximum speed of 100 ft/s and the distance traveled in this time. Draw the v–t and s–t graphs. The motorcycle starts from rest at $s = 0$.

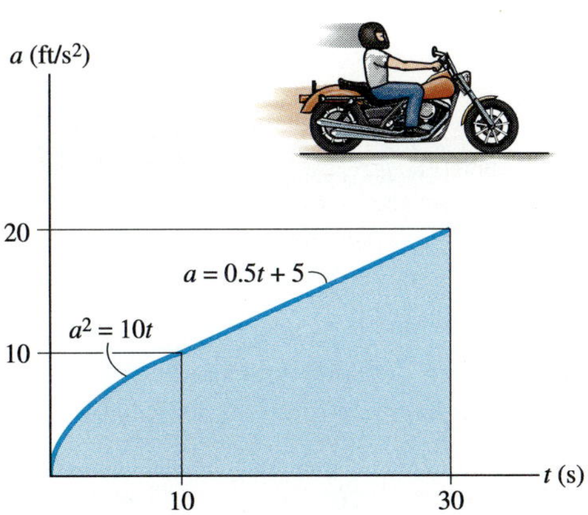

Prob. 12–55

12-57. The v–t graph of a car while traveling along a road is shown. Draw the s–t and a–t graphs for the motion.

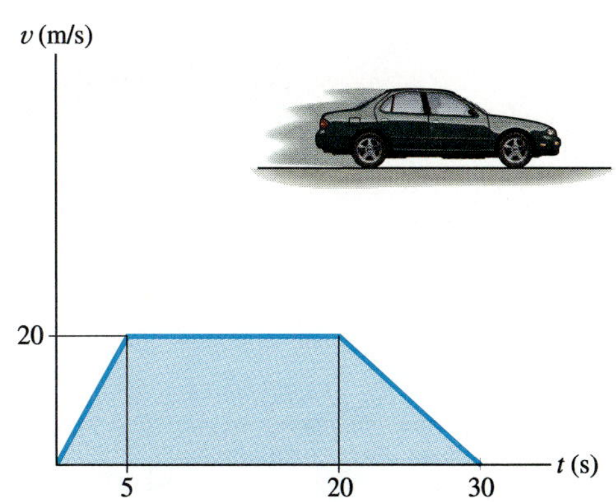

Prob. 12–57

12-58. A motorcyclist at A is traveling at 60 ft/s when he wishes to pass the truck T which is traveling at a constant speed of 60 ft/s. To do so the motorcyclist accelerates at 6 ft/s^2 until reaching a maximum speed of 85 ft/s. If he then maintains this speed, determine the time needed for him to reach a point located 100 ft in front of the truck. Draw the v–t and s–t graphs for the motorcycle during this time.

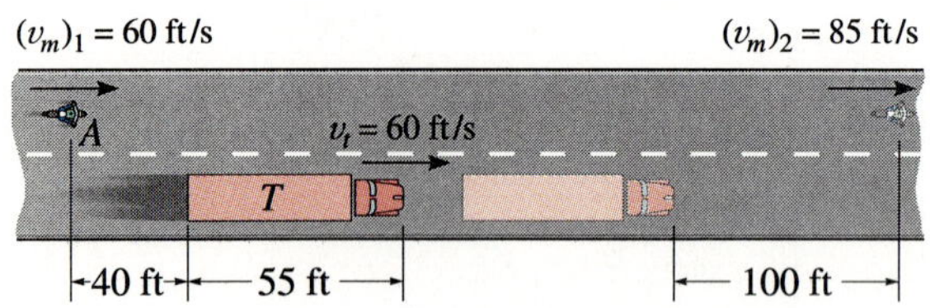

Prob. 12–58

12-59. The jet car is originally traveling at a speed of 20 m/s when it is subjected to the acceleration shown in the graph. Determine the car's maximum speed and the time t when it stops.

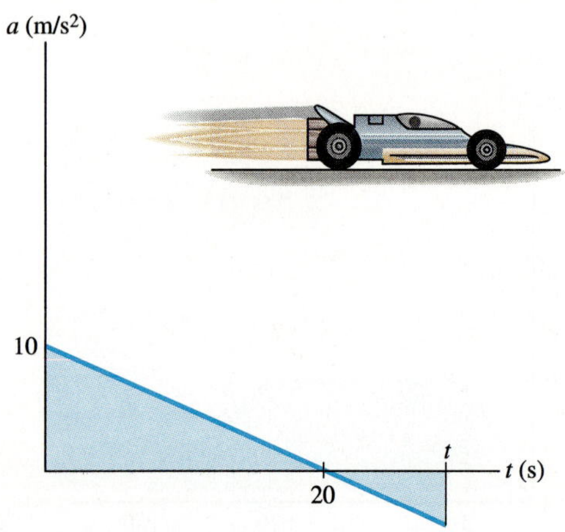

Prob. 12–59

***12-60.** The a–t graph for a car is shown. Construct the v–t and s–t graphs if the car starts from rest at $t = 0$. At what time t' does the car stop?

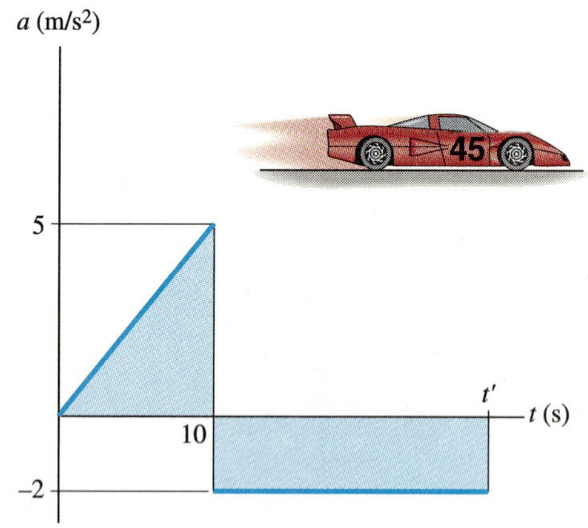

Prob. 12–60

12-61. The a–s graph for a train traveling along a straight track is given for the first 400 m of its motion. Plot the v–s graph. $v = 0$ at $s = 0$.

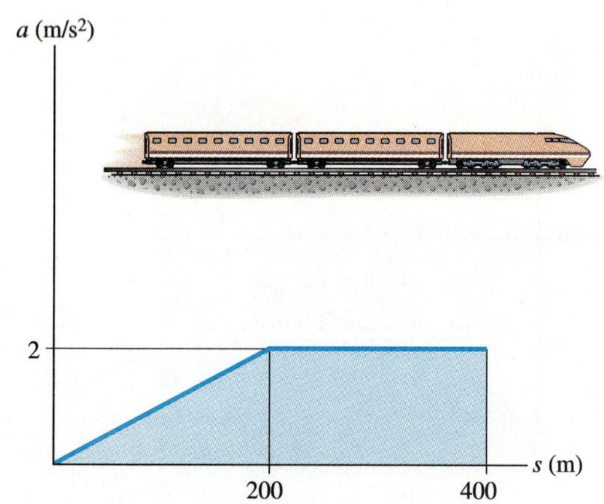

Prob. 12–61

12-62. The v–s graph for a test vehicle is shown. Determine its acceleration when $s = 100$ m and when $s = 175$ m.

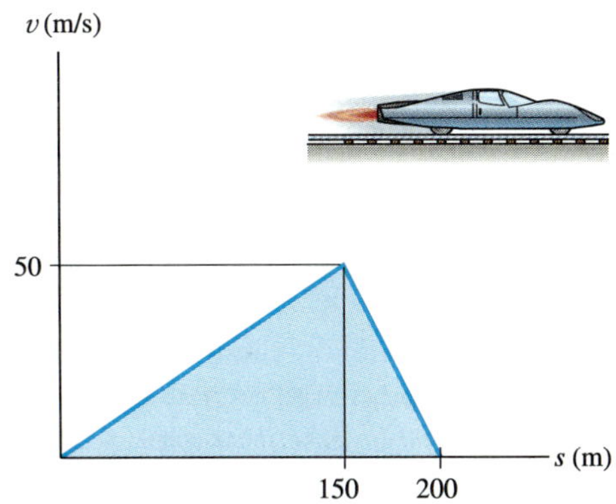

Prob. 12–62

12-63. The rocket sled starts from rest at $s = 0$ and is subjected to an acceleration as shown by the a–s graph. Draw the v–s graph and determine the time needed to travel 500 ft.

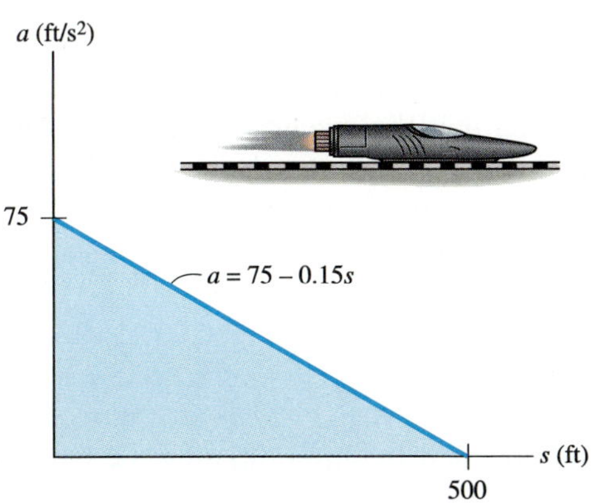

Prob. 12–63

***12-64.** The test car starts from rest and is subjected to a constant acceleration of $a_c = 15\ \text{ft/s}^2$ for $0 \le t < 10$ s. The brakes are then applied, which causes a deceleration at the rate shown until the car stops. Determine the car's maximum speed and the time t when it stops.

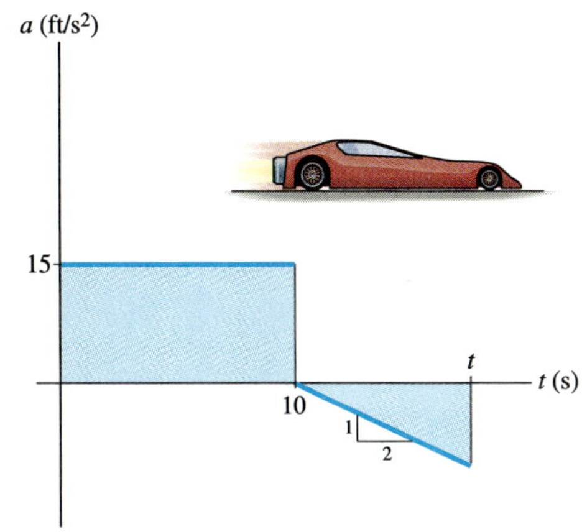

Prob. 12–64

12-65. The v–s graph was determined experimentally to describe the straight-line motion of a rocket sled. Determine the acceleration of the sled when $s = 100$ m, and when $s = 200$ m.

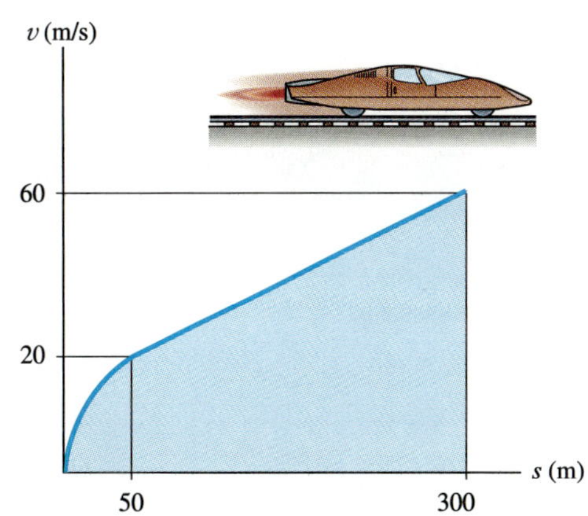

Prob. 12–65

12.4 General Curvilinear Motion

Curvilinear motion occurs when the particle moves along a curved path. Since this path is often described in three dimensions, vector analysis will be used to formulate the particle's position, velocity, and acceleration.* In this section the general aspects of curvilinear motion are discussed, and in subsequent sections three types of coordinate systems often used to analyze this motion will be introduced.

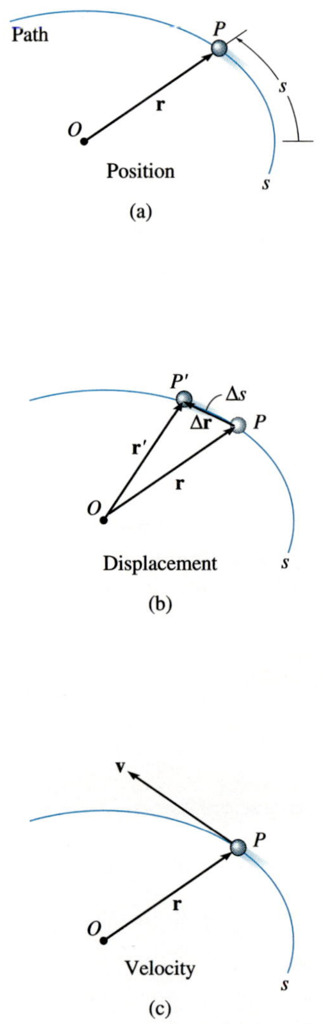

Fig. 12–16

Position. Consider a particle located at point P on a space curve defined by the path function s, Fig. 12–16*a*. The position of the particle, measured from a fixed point O, will be designated by the *position vector* $\mathbf{r} = \mathbf{r}(t)$. This vector is a function of time since, in general, both its magnitude and direction change as the particle moves along the curve.

Displacement. Suppose that during a small time interval Δt the particle moves a distance Δs along the curve to a new position P', defined by $\mathbf{r}' = \mathbf{r} + \Delta\mathbf{r}$, Fig. 12–16*b*. The *displacement* $\Delta\mathbf{r}$ represents the change in the particle's position and is determined by vector subtraction; i.e., $\Delta\mathbf{r} = \mathbf{r}' - \mathbf{r}$.

Velocity. During the time Δt, the *average velocity* of the particle is defined as

$$\mathbf{v}_{avg} = \frac{\Delta\mathbf{r}}{\Delta t}$$

The *instantaneous velocity* is determined from this equation by letting $\Delta t \to 0$, and consequently the direction of $\Delta\mathbf{r}$ *approaches* the *tangent* to the curve at point P. Hence, $\mathbf{v} = \lim_{\Delta t \to 0} (\Delta\mathbf{r}/\Delta t)$ or

$$\mathbf{v} = \frac{d\mathbf{r}}{dt} \qquad (12\text{–}7)$$

Since $d\mathbf{r}$ will be tangent to the curve at P, the *direction* of $\mathbf{v}$ is also *tangent to the curve*, Fig. 12–16*c*. The *magnitude* of $\mathbf{v}$, which is called the *speed*, may be obtained by noting that the magnitude of the displacement $\Delta\mathbf{r}$ is the length of the straight line segment from P to P', Fig. 12–16*b*. Realizing that this length, Δr, approaches the arc length Δs as $\Delta t \to 0$, we have $v = \lim_{\Delta t \to 0} (\Delta r/\Delta t) = \lim_{\Delta t \to 0} (\Delta s/\Delta t)$, or

$$v = \frac{ds}{dt} \qquad (12\text{–}8)$$

Thus, the *speed* can be obtained by differentiating the path function s with respect to time.

*A summary of some of the important concepts of vector analysis is given in Appendix C.

Acceleration. If the particle has a velocity $\mathbf{v}$ at time t and a velocity $\mathbf{v}' = \mathbf{v} + \Delta\mathbf{v}$ at $t + \Delta t$, Fig. 12–16*d*, then the *average acceleration* of the particle during the time interval Δt is

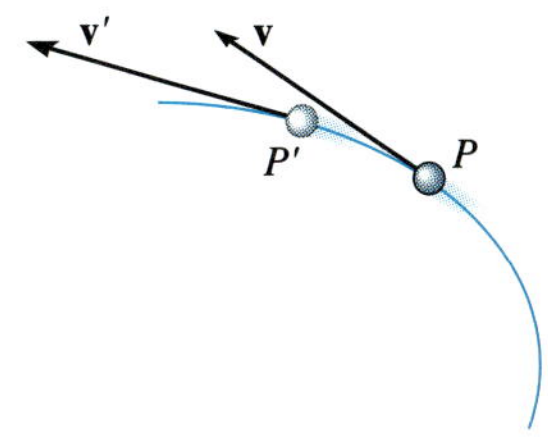

(d)

$$\mathbf{a}_{\text{avg}} = \frac{\Delta\mathbf{v}}{\Delta t}$$

where $\Delta\mathbf{v} = \mathbf{v}' - \mathbf{v}$. To study this time rate of change, the two velocity vectors in Fig. 12–16*d* are plotted in Fig. 12–16*e* such that their tails are located at the fixed point O' and their arrowheads touch points on the curve. This curve is called a *hodograph,* and when constructed, it describes the locus of points for the arrowhead of the velocity vector in the same manner as the *path s* describes the locus of points for the arrowhead of the position vector, Fig. 12–16*a*.

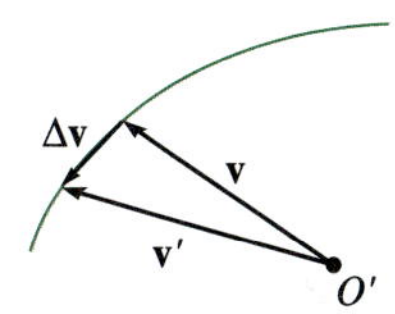

(e)

To obtain the *instantaneous acceleration,* let $\Delta t \to 0$ in the above equation. In the limit $\Delta\mathbf{v}$ will approach the *tangent to the hodograph,* and so $\mathbf{a} = \lim_{\Delta t \to 0} (\Delta\mathbf{v}/\Delta t)$, or

$$\mathbf{a} = \frac{d\mathbf{v}}{dt} \tag{12–9}$$

Substituting Eq. 12–7 into this result, we can also write

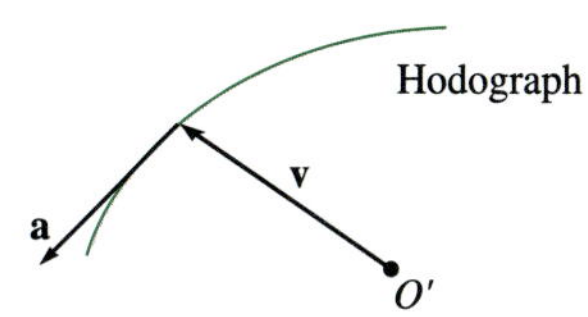

(f)

$$\mathbf{a} = \frac{d^2\mathbf{r}}{dt^2}$$

By definition of the derivative, $\mathbf{a}$ acts *tangent to the hodograph,* Fig. 12–16*f*, and therefore, *in general,* $\mathbf{a}$ *is not tangent to the path of motion,* Fig. 12–16*g*. To clarify this point, realize that $\Delta\mathbf{v}$ and consequently $\mathbf{a}$ must account for the change made in *both* the magnitude *and* direction of the velocity $\mathbf{v}$ as the particle moves from P to P', Fig. 12–16*d*. Just a magnitude change increases (or decreases) the "length" of $\mathbf{v}$, and this in itself would allow $\mathbf{a}$ to remain tangent to the path. However, in order for the particle to follow the path, the directional change always "swings" the velocity vector toward the "inside" or "concave side" of the path, and therefore $\mathbf{a}$ *cannot* remain tangent to the path. In summary, $\mathbf{v}$ is always tangent to the *path* and $\mathbf{a}$ is always tangent to the *hodograph.*

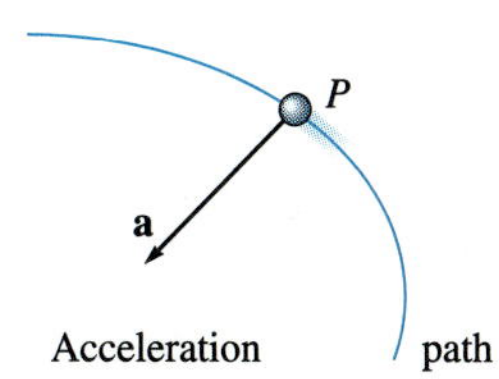

(g)

Fig. 12–16

12.5 Curvilinear Motion: Rectangular Components

Occasionally the motion of a particle can best be described along a path that is represented using a fixed x, y, z frame of reference.

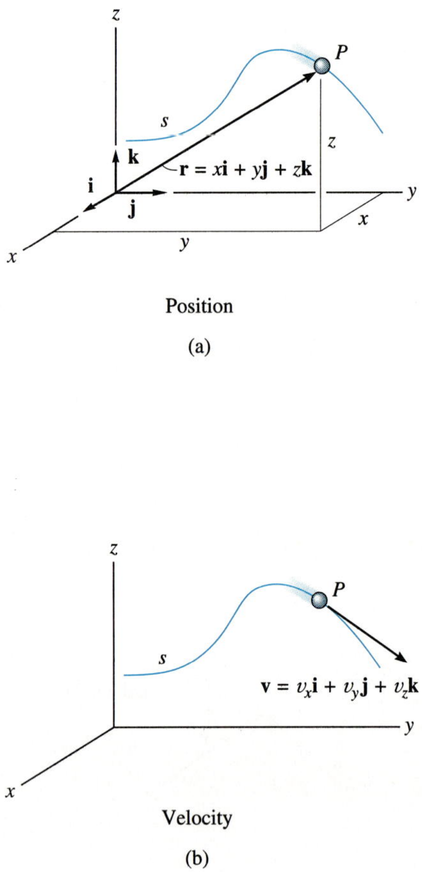

Fig. 12–17

Position. If at a given instant the particle P is at point (x, y, z) on the curved path s, Fig. 12–17a, its location is then defined by the *position vector*

$$\mathbf{r} = x\mathbf{i} + y\mathbf{j} + z\mathbf{k} \qquad (12\text{–}10)$$

Because of the particle motion and the shape of the path, the x, y, z components of $\mathbf{r}$ are generally all functions of time; i.e., $x = x(t)$, $y = y(t)$, $z = z(t)$, so that $\mathbf{r} = \mathbf{r}(t)$.

In accordance with the discussion in Appendix C, the *magnitude* of $\mathbf{r}$ is *always positive* and defined from Eq. C–3 as

$$r = \sqrt{x^2 + y^2 + z^2}$$

The *direction* of $\mathbf{r}$ is specified by the components of the unit vector $\mathbf{u}_r = \mathbf{r}/r$.

Velocity. The first time derivative of $\mathbf{r}$ yields the velocity $\mathbf{v}$ of the particle. Hence,

$$\mathbf{v} = \frac{d\mathbf{r}}{dt} = \frac{d}{dt}(x\mathbf{i}) + \frac{d}{dt}(y\mathbf{j}) + \frac{d}{dt}(z\mathbf{k})$$

When taking this derivative, it is necessary to account for changes in *both* the magnitude and direction of each of the vector's components. The derivative of the $\mathbf{i}$ component of $\mathbf{v}$ is therefore

$$\frac{d}{dt}(x\mathbf{i}) = \frac{dx}{dt}\mathbf{i} + x\frac{d\mathbf{i}}{dt}$$

The second term on the right side is zero, since the x, y, z reference frame is *fixed,* and therefore the *direction* (and the *magnitude*) of $\mathbf{i}$ does not change with time. Differentiation of the $\mathbf{j}$ and $\mathbf{k}$ components may be carried out in a similar manner, which yields the final result,

$$\mathbf{v} = \frac{d\mathbf{r}}{dt} = v_x\mathbf{i} + v_y\mathbf{j} + v_z\mathbf{k} \qquad (12\text{–}11)$$

where

$$v_x = \dot{x} \quad v_y = \dot{y} \quad v_z = \dot{z} \qquad (12\text{–}12)$$

The "dot" notation $\dot{x}, \dot{y}, \dot{z}$ represents the first time derivatives of the parametric equations $x = x(t)$, $y = y(t)$, $z = z(t)$, respectively.

The velocity has a *magnitude* defined as the positive value of

$$v = \sqrt{v_x^2 + v_y^2 + v_z^2}$$

and a *direction* that is specified by the components of the unit vector $\mathbf{u}_v = \mathbf{v}/v$. This direction is *always tangent to the path,* as shown in Fig. 12–17*b*.

Acceleration. The acceleration of the particle is obtained by taking the first time derivative of Eq. 12–11 (or the second time derivative of Eq. 12–10). Using dots to represent the derivatives of the components, we have

$$\mathbf{a} = \frac{d\mathbf{v}}{dt} = a_x\mathbf{i} + a_y\mathbf{j} + a_z\mathbf{k} \qquad (12\text{–}13)$$

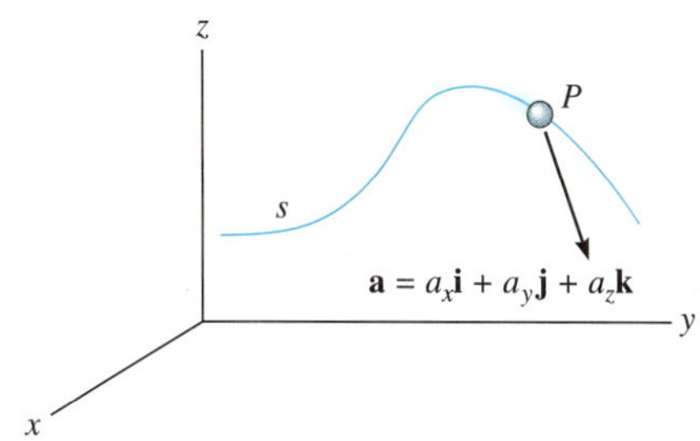

Acceleration

(c)

Fig. 12–17

where

$$\begin{aligned} a_x &= \dot{v}_x = \ddot{x} \\ a_y &= \dot{v}_y = \ddot{y} \\ a_z &= \dot{v}_z = \ddot{z} \end{aligned} \qquad (12\text{–}14)$$

Here a_x, a_y, a_z represent, respectively, the first time derivatives of the functions $v_x = v_x(t)$, $v_y = v_y(t)$, $v_z = v_z(t)$, or the second time derivatives of the functions $x = x(t)$, $y = y(t)$, $z = z(t)$.

The acceleration has a *magnitude* defined by the positive value of

$$a = \sqrt{a_x^2 + a_y^2 + a_z^2}$$

and a *direction* specified by the components of the unit vector $\mathbf{u}_a = \mathbf{a}/a$. Since $\mathbf{a}$ represents the time rate of *change* in velocity, in general $\mathbf{a}$ will *not* be tangent to the path, Fig. 12–17*c*.

Important Points

- Curvilinear motion can cause changes in *both* the magnitude and direction of the position, velocity, and acceleration vectors.
- The velocity vector is always directed *tangent* to the path.
- In general, the acceleration vector is *not* tangent to the path, but rather, it is tangent to the hodograph.
- If the motion is described using rectangular coordinates, then the components along each of the axes do not change direction, only their magnitude and sense (algebraic sign) will change.

Procedure for Analysis

Coordinate System

- A rectangular coordinate system can be used to solve problems for which the motion can conveniently be expressed in terms of its x, y, z components.

Kinematic Quantities

- Since *rectilinear motion* occurs along *each* coordinate axis, the motion of each component is found using $v = ds/dt$ and $a = dv/dt$; or in cases where the motion is not expressed as a function of time, the equation $a\,ds = v\,dv$ can be used.
- Once the x, y, z components of $\mathbf{v}$ and $\mathbf{a}$ have been determined, the magnitudes of these vectors are found from the Pythagorean theorem, Eq. C–3, and their directions from the components of their unit vectors, Eqs. C–4 and C–5.

As the airplane takes off, its path of motion can be established by knowing its horizontal position $x = x(t)$, and its vertical position or altitude $y = y(t)$, both of which can be found from navigation equipment. By plotting the results from these equations the path can be shown, and by taking the time derivatives, the velocity and acceleration of the plane at any instant can be determined.

EXAMPLE 12–9

At any instant the horizontal position of the weather balloon in Fig. 12–18*a* is defined by $x = (8t)$ ft, where t is in seconds. If the equation of the path is $y = x^2/10$, determine (*a*) the distance of the balloon from the station at A when $t = 2$ s, (*b*) the magnitude and direction of the velocity when $t = 2$ s, and (*c*) the magnitude and direction of the acceleration when $t = 2$ s.

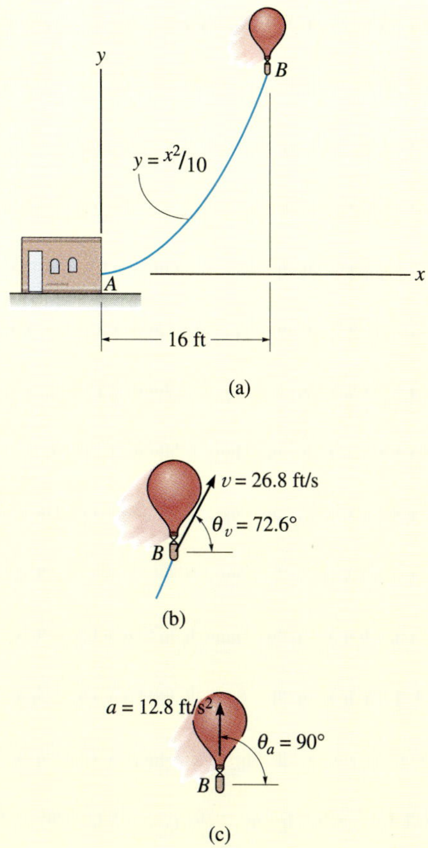

Fig. 12–18

Solution

Position. When $t = 2$ s, $x = 8(2)$ ft $= 16$ ft, and so

$$y = (16)^2/10 = 25.6 \text{ ft}$$

The straight-line distance from A to B is therefore

$$r = \sqrt{(16)^2 + (25.6)^2} = 30.2 \text{ ft} \qquad \textit{Ans.}$$

Velocity. Using Eqs. 12–12 and application of the chain rule of calculus the components of velocity when $t = 2$ s are

$$v_x = \dot{x} = \frac{d}{dt}(8t) = 8 \text{ ft/s} \rightarrow$$

$$v_y = \dot{y} = \frac{d}{dt}(x^2/10) = 2x\dot{x}/10 = 2(16)(8)/10 = 25.6 \text{ ft/s} \uparrow$$

When $t = 2$ s, the magnitude of velocity is therefore

$$v = \sqrt{(8)^2 + (25.6)^2} = 26.8 \text{ ft/s} \qquad \textit{Ans.}$$

The direction is tangent to the path, Fig. 12–18*b*, where

$$\theta_v = \tan^{-1}\frac{v_y}{v_x} = \tan^{-1}\frac{25.6}{8} = 72.6° \qquad \textit{Ans.}$$

Acceleration. The components of acceleration are determined from Eqs. 12–14 and application of the chain rule, noting that $\ddot{x} = d^2(8t)/dt^2 = 0$. We have

$$a_x = \dot{v}_x = 0$$

$$a_y = \dot{v}_y = \frac{d}{dt}(2x\dot{x}/10) = 2(\dot{x})\dot{x}/10 + 2x(\ddot{x})/10$$

$$= 2(8)^2/10 + 2(16)(0)/10 = 12.8 \text{ ft/s}^2 \uparrow$$

Thus

$$a = \sqrt{(0)^2 + (12.8)^2} = 12.8 \text{ ft/s}^2 \qquad \textit{Ans.}$$

The direction of **a**, as shown in Fig. 12–18*c*, is

$$\theta_a = \tan^{-1}\frac{12.8}{0} = 90° \qquad \textit{Ans.}$$

Note: It is also possible to obtain v_y and a_y by first expressing $y = f(t) = (8t)^2/10 = 6.4t^2$ and then taking successive time derivatives.

EXAMPLE 12–10

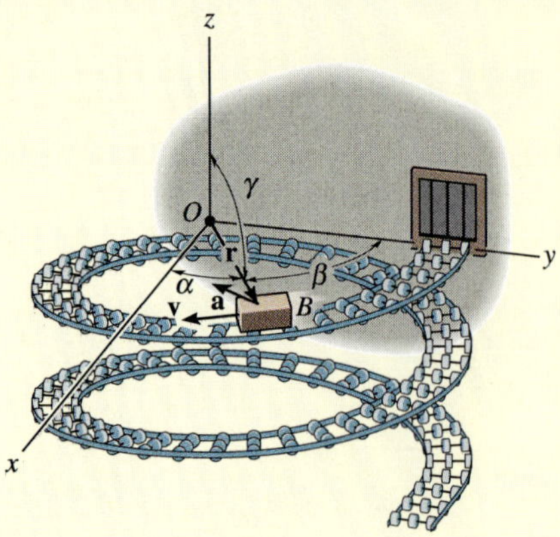

Fig. 12–19

The motion of a box B moving along the spiral conveyor shown in Fig. 12–19 is defined by the position vector $\mathbf{r} = \{0.5 \sin (2t)\mathbf{i} + 0.5 \cos (2t)\mathbf{j} - 0.2t\mathbf{k}\}$ m, where t is in seconds and the arguments for sine and cosine are in radians (π rad $= 180°$). Determine the location of the box when $t = 0.75$ s and the magnitudes of its velocity and acceleration at this instant.

Solution

Position. Evaluating $\mathbf{r}$ when $t = 0.75$ s yields

$$\mathbf{r}\big|_{t=0.75\text{ s}} = \{0.5 \sin (1.5 \text{ rad})\mathbf{i} + 0.5 \cos (1.5 \text{ rad})\mathbf{j} - 0.2(0.75)\mathbf{k}\} \text{ m}$$
$$= \{0.499\mathbf{i} + 0.0354\mathbf{j} - 0.150\mathbf{k}\} \text{ m} \qquad \textit{Ans.}$$

The distance of the box from the origin O is

$$r = \sqrt{(0.499)^2 + (0.0354)^2 + (-0.150)^2} = 0.522 \text{ m} \qquad \textit{Ans.}$$

The direction of $\mathbf{r}$ is obtained from the components of the unit vector,

$$\mathbf{u}_r = \frac{\mathbf{r}}{r} = \frac{0.499}{0.522}\mathbf{i} + \frac{0.0354}{0.522}\mathbf{j} - \frac{0.150}{0.522}\mathbf{k}$$
$$= 0.955\mathbf{i} + 0.0678\mathbf{j} - 0.287\mathbf{k}$$

Hence, the coordinate direction angles α, β, γ, Fig. 12–19, are

$$\alpha = \cos^{-1}(0.955) = 17.2° \qquad \textit{Ans.}$$
$$\beta = \cos^{-1}(0.0678) = 86.1° \qquad \textit{Ans.}$$
$$\gamma = \cos^{-1}(-0.287) = 107° \qquad \textit{Ans.}$$

Velocity. The velocity is defined by

$$\mathbf{v} = \frac{d\mathbf{r}}{dt} = \frac{d}{dt}[0.5 \sin (2t)\mathbf{i} + 0.5 \cos (2t)\mathbf{j} - 0.2t\mathbf{k}]$$
$$= \{1 \cos (2t)\mathbf{i} - 1 \sin (2t)\mathbf{j} - 0.2\mathbf{k}\} \text{ m/s}$$

Hence, when $t = 0.75$ s the magnitude of velocity, or the speed, is

$$v = \sqrt{v_x^2 + v_y^2 + v_z^2}$$
$$= \sqrt{[1 \cos (1.5 \text{ rad})]^2 + [-1 \sin (1.5 \text{ rad})]^2 + (-0.2)^2}$$
$$= 1.02 \text{ m/s} \qquad \textit{Ans.}$$

The velocity is tangent to the path as shown in Fig. 12–19. Its coordinate direction angles can be determined from $\mathbf{u}_v = \mathbf{v}/v$.

Acceleration. The acceleration $\mathbf{a}$ of the box, which is shown in Fig. 12–19, is *not* tangent to the path. Show that

$$\mathbf{a} = \frac{d\mathbf{v}}{dt} = \{-2 \sin (2t)\mathbf{i} - 2 \cos (2t)\mathbf{j}\} \text{ m/s}^2$$

At $t = 0.75s$, $a = 2 \text{ m/s}^2$ *Ans.*

12.6 Motion of a Projectile

The free-flight motion of a projectile is often studied in terms of its rectangular components, since the projectile's acceleration *always* acts in the vertical direction. To illustrate the kinematic analysis, consider a projectile launched at point (x_0, y_0), as shown in Fig. 12–20. The path is defined in the x–y plane such that the initial velocity is $\mathbf{v}_0$, having components $(\mathbf{v}_0)_x$ and $(\mathbf{v}_0)_y$. When air resistance is neglected, the only force acting on the projectile is its weight, which causes the projectile to have a *constant downward acceleration* of approximately $a_c = g = 9.81\ \text{m/s}^2$ or $g = 32.2\ \text{ft/s}^2$.*

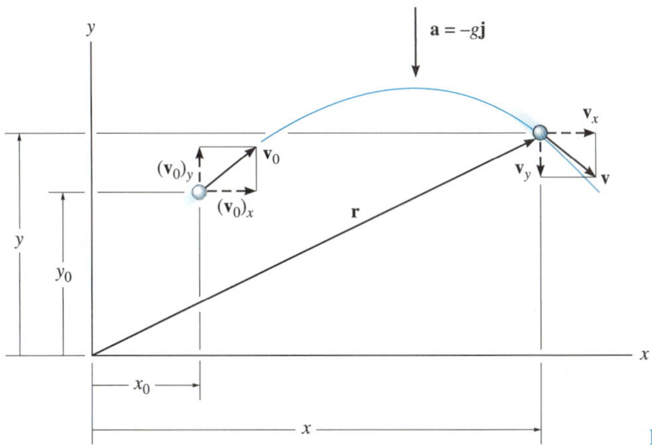

Fig. 12–20

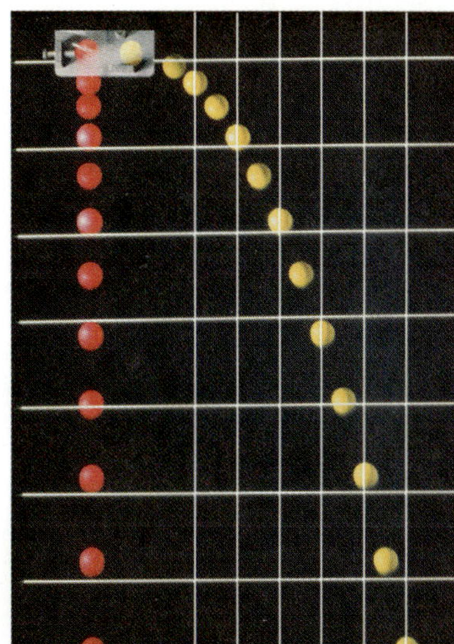

Each picture in this sequence is taken after the same time interval. The red ball falls from rest, whereas the yellow ball is given a horizontal velocity when released. Notice both balls are subjected to the same downward acceleration since they remain at the same elevation at any instant. This acceleration causes the difference in elevation to increase between successive photos. Also, note the horizontal distance between successive photos of the yellow ball is constant since the velocity in the horizontal direction remains constant.

Horizontal Motion. Since $a_x = 0$, application of the constant acceleration equations, 12–4 to 12–6, yields

$$(\overset{+}{\rightarrow})\quad v = v_0 + a_c t; \qquad v_x = (v_0)_x$$
$$(\overset{+}{\rightarrow})\quad x = x_0 + v_0 t + \tfrac{1}{2} a_c t^2; \qquad x = x_0 + (v_0)_x t$$
$$(\overset{+}{\rightarrow})\quad v^2 = v_0^2 + 2a_c(s - s_0); \qquad v_x = (v_0)_x$$

The first and last equations indicate that *the horizontal component of velocity always remains constant during the motion.*

Vertical Motion. Since the positive y axis is directed upward, then $a_y = -g$. Applying Eqs. 12–4 to 12–6, we get

$$(+\uparrow)\quad v = v_0 + a_c t; \qquad v_y = (v_0)_y - gt$$
$$(+\uparrow)\quad y = y_0 + v_0 t + \tfrac{1}{2} a_c t^2; \qquad y = y_0 + (v_0)_y t - \tfrac{1}{2} g t^2$$
$$(+\uparrow)\quad v^2 = v_0^2 + 2a_c(y - y_0); \qquad v_y^2 = (v_0)_y^2 - 2g(y - y_0)$$

Recall that the last equation can be formulated on the basis of eliminating the time t between the first two equations, and therefore *only two of the above three equations are independent of one another.*

*This assumes that the earth's gravitational field does not vary with altitude.

To summarize, problems involving the motion of a projectile can have at most three unknowns since only three independent equations can be written; that is, *one* equation in the *horizontal direction* and *two* in the *vertical direction*. Once $\mathbf{v}_x$ and $\mathbf{v}_y$ are obtained, the resultant velocity $\mathbf{v}$, which is *always tangent* to the path, is defined by the *vector sum* as shown in Fig. 12–20.

Procedure for Analysis

Free-flight projectile motion problems can be solved using the following procedure.

Coordinate System

- Establish the fixed x, y coordinate axes and sketch the trajectory of the particle. Between any *two points* on the path specify the given problem data and the *three unknowns*. In all cases the acceleration of gravity acts downward. The particle's initial and final velocities should be represented in terms of their x and y components.
- Remember that positive and negative position, velocity, and acceleration components always act in accordance with their associated coordinate directions.

Kinematic Equations

- Depending upon the known data and what is to be determined, a choice should be made as to which three of the following four equations should be applied between the two points on the path to obtain the most direct solution to the problem.

Horizontal Motion

- The *velocity* in the horizontal or x direction is *constant*, i.e., $(v_x) = (v_0)_x$, and

$$x = x_0 + (v_0)_x t$$

Vertical Motion

- In the vertical or y direction *only two* of the following three equations can be used for solution.

$$v_y = (v_0)_y - a_c t$$
$$y = y_0 + (v_0)_y t + \tfrac{1}{2} a_c t^2$$
$$v_y^2 = (v_0)_y^2 + 2a_c(y - y_0)$$

- For example, if the particle's final velocity v_y is not needed, then the first and third of these equations (for y) will not be useful.

Gravel falling off the end of this conveyor belt follows a path that can be predicted using the equations of constant acceleration. In this way the location of the accumulated pile can be determined. Rectilinear coordinates are used for the analysis since the acceleration is only in the vertical direction.

EXAMPLE 12-11

A sack slides off the ramp, shown in Fig. 12–21, with a horizontal velocity of 12 m/s. If the height of the ramp is 6 m from the floor, determine the time needed for the sack to strike the floor and the range R where sacks begin to pile up.

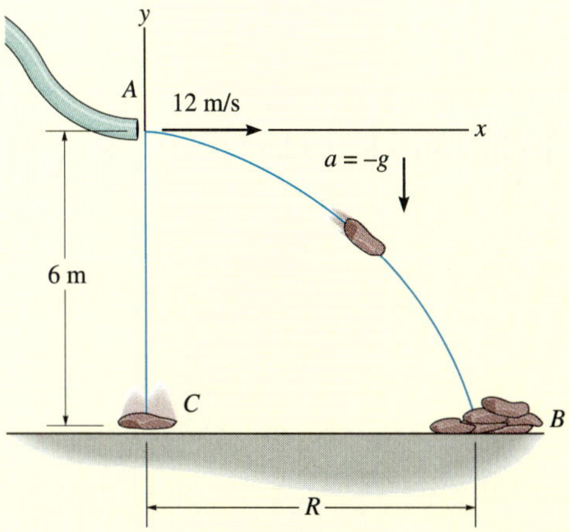

Fig. 12–21

Solution

Coordinate System. The origin of coordinates is established at the beginning of the path, point A, Fig. 12–21. The initial velocity of a sack has components $(v_A)_x = 12$ m/s and $(v_A)_y = 0$. Also, between points A and B the acceleration is $a_y = -9.81 \text{ m/s}^2$. Since $(v_B)_x = (v_A)_x = 12$ m/s, the three unknowns are $(v_B)_y$, R, and the time of flight t_{AB}. Here we do not need to determine $(v_B)_y$.

Vertical Motion. The vertical distance from A to B is known, and therefore we can obtain a direct solution for t_{AB} by using the equation

$(+\uparrow)$
$$y = y_0 + (v_0)_y t_{AB} + \tfrac{1}{2} a_c t_{AB}^2$$
$$-6 \text{ m} = 0 + 0 + \tfrac{1}{2}(-9.81 \text{ m/s}^2) t_{AB}^2$$
$$t_{AB} = 1.11 \text{ s} \qquad \textit{Ans.}$$

This calculation also indicates that if a sack were released *from rest* at A, it would take the same amount of time to strike the floor at C, Fig. 12–21.

Horizontal Motion. Since t has been calculated, R is determined as follows:

$(\overset{+}{\rightarrow})$
$$x = x_0 + (v_0)_x t_{AB}$$
$$R = 0 + 12 \text{ m/s } (1.11 \text{ s})$$
$$R = 13.3 \text{ m} \qquad \textit{Ans.}$$

EXAMPLE 12–12

The chipping machine is designed to eject wood chips at $v_O = 25$ ft/s as shown in Fig. 12–22. If the tube is oriented at 30° from the horizontal, determine how high, h, the chips strike the pile if they land on the pile 20 ft from the tube.

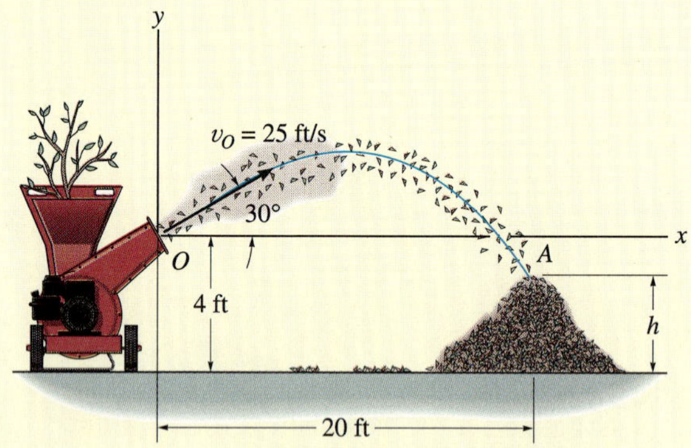

Fig. 12–22

Solution

Coordinate System. When the motion is analyzed between points O and A, the three unknowns are represented as the height h, time of flight t_{OA}, and vertical component of velocity $(v_A)_y$. (Note that $(v_A)_x = (v_O)_x$.) With the origin of coordinates at O, Fig. 12–22, the initial velocity of a chip has components of

$$(v_O)_x = (25 \cos 30°) \text{ ft/s} = 21.65 \text{ ft/s} \rightarrow$$
$$(v_O)_y = (25 \sin 30°) \text{ ft/s} = 12.5 \text{ ft/s} \uparrow$$

Also, $(v_A)_x = (v_O)_x = 21.65$ ft/s and $a_y = -32.2 \text{ ft/s}^2$. Since we do not need to determine $(v_A)_y$, we have

Horizontal Motion

$$(\overset{+}{\rightarrow}) \qquad x_A = x_O + (v_O)_x t_{OA}$$
$$20 \text{ ft} = 0 + (21.65 \text{ ft/s}) t_{OA}$$
$$t_{OA} = 0.9238 \text{ s}$$

Vertical Motion. Relating t_{OA} to the initial and final elevations of a chip, we have

$$(+\uparrow) \quad y_A = y_O + (v_O)_y t_{OA} + \tfrac{1}{2} a_c t_{OA}^2$$
$$(h - 4 \text{ ft}) = 0 + (12.5 \text{ ft/s})(0.9238 \text{ s}) + \tfrac{1}{2}(-32.2 \text{ ft/s}^2)(0.9238 \text{ s})^2$$
$$h = 1.81 \text{ ft}$$

Ans.

EXAMPLE 12–13

The track for this racing event was designed so that riders jump off the slope at 30°, from a height of 1 m. During a race it was observed that the rider shown in Fig. 12–23*a* remained in mid air for 1.5 s. Determine the speed at which he was traveling off the slope, the horizontal distance he travels before striking the ground, and the maximum height he attains. Neglect the size of the bike and rider.

(a)

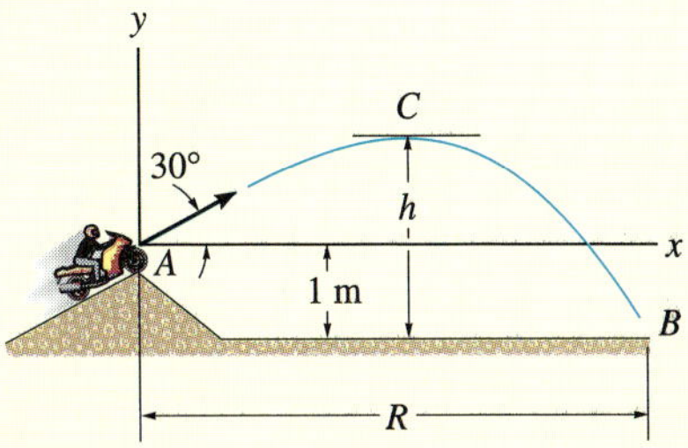

(b)

Fig. 12–23

Solution

Coordinate System. As shown in Fig. 12–23*b*, the origin of the coordinates is established at *A*. Between the end points of the path *AB* the three unknowns are the initial speed v_A, range *R*, and the vertical component of velocity v_B.

Vertical Motion. Since the time of flight and the vertical distance between the ends of the path are known, we can determine v_A.

$(+\uparrow)$
$$(s_B)_y = (s_A)_y + (v_A)_y t_{AB} + \tfrac{1}{2}a_c t^2_{AB}$$
$$-1 = 0 + v_A \sin 30^\circ\,(1.5) + \tfrac{1}{2}(-9.81)(1.5)^2$$
$$v_A = 13.38 \text{ m/s} = 13.4 \text{ m/s} \qquad \textit{Ans.}$$

Horizontal Motion. The range *R* can now be determined.

$(\overset{+}{\rightarrow})$
$$(s_B)_x = (s_A)_x + (v_A)_x t_{AB}$$
$$R = 0 + 13.38 \cos 30^\circ(1.5)$$
$$= 17.4 \text{ m} \qquad \textit{Ans.}$$

In order to find the maximum height *h* we will consider the path *AC*, Fig. 12–23*b*. Here the three unknowns become the time of flight t_{AC}, the horizontal distance from *A* to *C*, and the height *h*. At the maximum height $(v_C)_y = 0$, and since v_A is known, we can determine *h directly* without considering t_{AC} using the following equation.

$$(v_C)_y^2 = (v_A)_y^2 + 2a_c\,[(s_C)_y - (s_A)_y]$$
$$(0)^2 = (13.38 \sin 30^\circ)^2 + 2\,(-9.81)\,[(h-1) - 0]$$
$$h = 3.28 \text{ m} \qquad \textit{Ans.}$$

Show that the bike will strike the ground at *B* with a velocity having components of

$$(v_B)_x = 11.6 \text{ m/s} \rightarrow,\ (v_B)_y = 8.02 \text{ m/s} \downarrow$$

Problems

12-66. If the velocity of a particle is defined as $\mathbf{v}(t) = \{0.8t^2\mathbf{i} + 12t^{1/2}\mathbf{j} + 5\mathbf{k}\}$ m/s, determine the magnitude and coordinate direction angles α, β, γ of the particle's acceleration when $t = 2$ s.

12-67. A particle travels along a path such that its position is $\mathbf{r} = \{(2 \sin t)\mathbf{i} + 2(1 - \cos t)\mathbf{j}\}$ ft, where t is in seconds and the arguments for the sine and cosine are in radians. Find the equation which describes the path in terms of x and y and show that the magnitudes of the particle's velocity and acceleration are constant. What is the magnitude of the particle's displacement from $t = 0$ to $t = 2$ s?

■***12-68.** A particle is traveling with a velocity of $\mathbf{v} = \{3\sqrt{t}e^{-0.2t}\mathbf{i} + 4e^{-0.8t^2}\mathbf{j}\}$ m/s, where t is in seconds. Determine the magnitude of the particle's displacement from $t = 0$ to $t = 3$ s. Use Simpson's rule with $n = 100$ to evaluate the integrals. What is the magnitude of the particle's acceleration when $t = 2$ s?

12-69. The position of particles A and B is described by $\mathbf{r}_A = \{2t\mathbf{i} + (t^2 - 1)\mathbf{j}\}$ ft and $\mathbf{r}_B = \{(t + 2)\mathbf{i} + (2t^2 - 5)\mathbf{j}\}$ ft, respectively, where t is in seconds. Determine the point where the particles collide and their speeds just before the collision.

12-70. The car travels from A to B, and then from B to C, as shown in the figure. Determine the magnitude of the displacement of the car and the distance traveled.

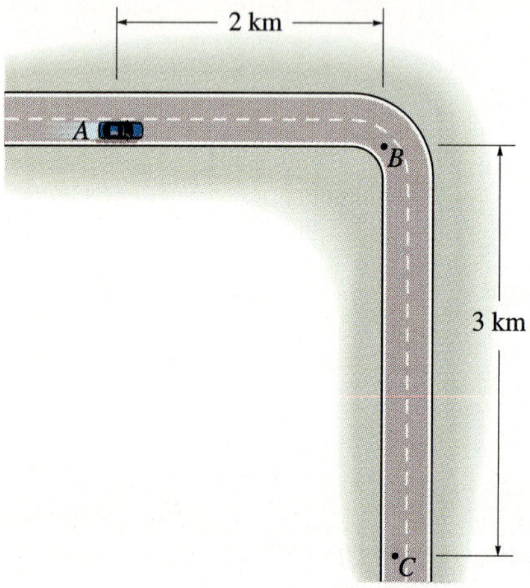

Prob. 12–70

12-71. A particle travels along the curve from A to B in 1 s. If it takes 3 s for it to go from A to C, determine its *average velocity* when it goes from B to C.

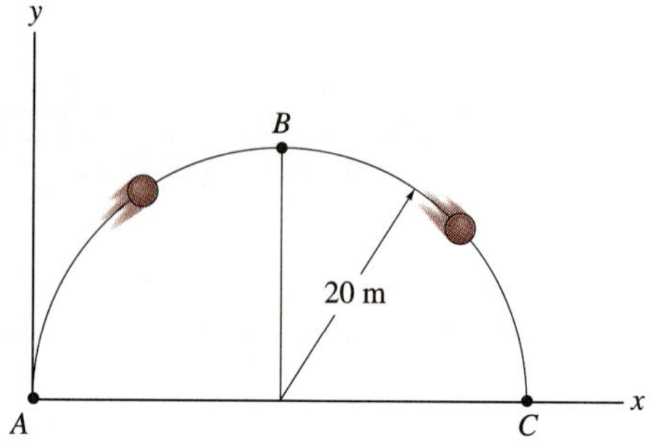

Prob. 12–71

***12-72.** A particle moves with curvilinear motion in the x–y plane such that the y component of motion is described by $y = (7t^3)$ m, where t is in seconds. If the particle starts from rest at the origin when $t = 0$, and maintains a *constant* acceleration in the x direction of 12 m/s^2, determine the particle's speed when $t = 2$ s.

12-73. The roller coaster car travels down the helical path at constant speed such that the parametric equations that define its position are $x = c \sin kt$, $y = c \cos kt$, $z = h - bt$, when c, h, and b are constants. Determine the magnitudes of its velocity and acceleration.

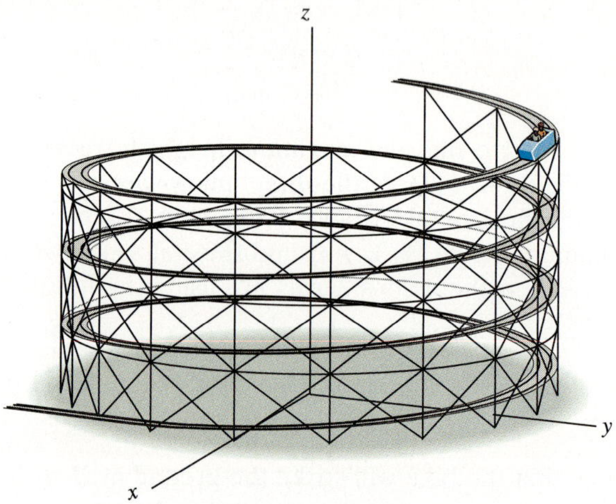

Prob. 12–73

12-74. The path of a particle is defined by $x + y = 1$. If the component of velocity along the x axis is $v_x = \cos kt$, where k is a constant, determine the x and y components of acceleration.

12-75. The path of a particle is defined by $y^2 = 4kx$, and the component of velocity along the y axis is $v_y = ct$, where both k and c are constants. Determine the x and y components of acceleration.

***12-76.** A particle travels along the circular path $x^2 + y^2 = r^2$, where r is the radius. If the component of velocity along the x axis is $v_x = t^2$, determine the x and y components of velocity and acceleration.

12-77. The motorcycle travels with constant speed v_0 along the path that, for a short distance, takes the form of a sine curve. Determine the x and y components of its velocity at any instant on the curve.

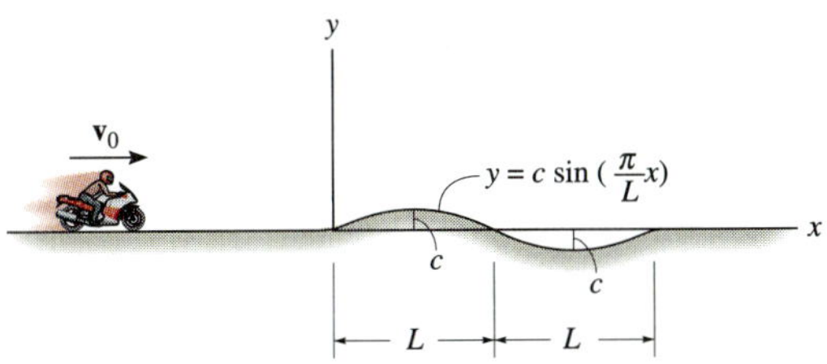

Prob. 12–77

12-78. The flight path of the helicopter as it takes off from A is defined by the parametric equations $x = (2t^2)$ m and $y = (0.04t^3)$ m, where t is the time in seconds. Determine the distance the helicopter is from point A and the magnitudes of its velocity and acceleration when $t = 10$ s.

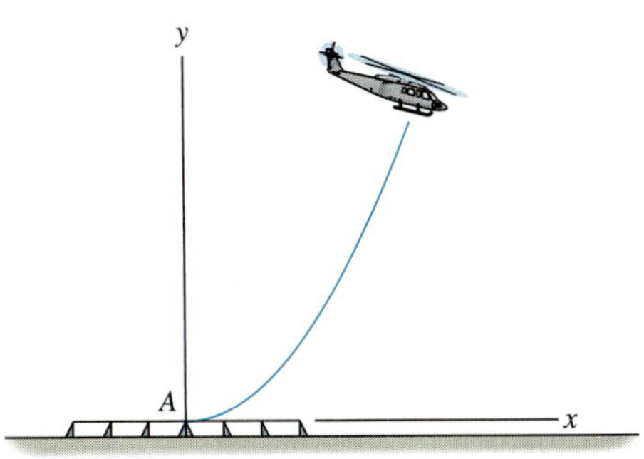

Prob. 12–78

12-79. At the instant shown particle A is traveling to the right at 10 ft/s and has an acceleration of 2 ft/s^2. Determine the initial speed v_0 of particle B so that when it is fired at the same instant from the angle shown it strikes A. Also, at what speed does it strike A?

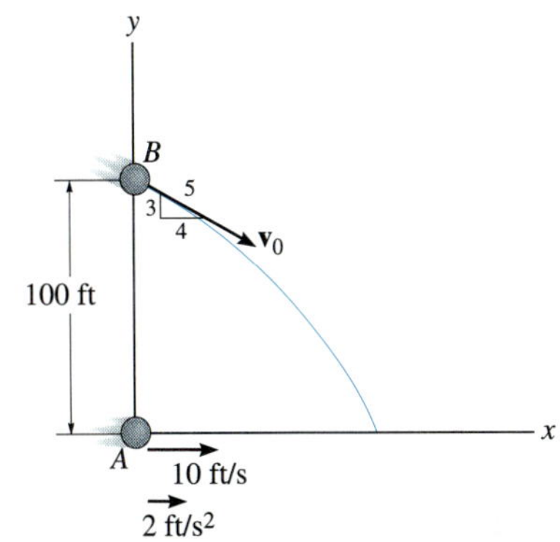

Prob. 12–79

***12-80.** The pitcher throws the baseball horizontally with a speed of 140 ft/s from a height of 5 ft. If the batter is 60 ft away, determine the time needed for the ball to arrive at the batter and the height h at which it passes the batter.

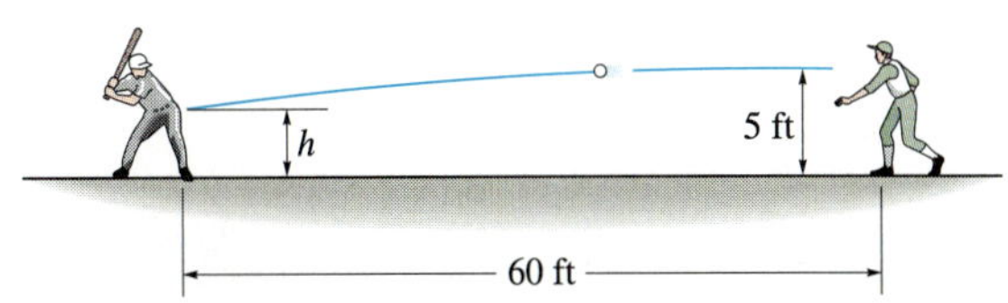

Prob. 12–80

12-81. The nozzle of a garden hose discharges water at the rate of 15 m/s. If the nozzle is held at ground level and directed $\theta = 30°$ from the ground, determine the maximum height reached by the water and the horizontal distance from the nozzle to where the water strikes the ground.

12-82. The projectile is launched with a velocity $\mathbf{v}_0$. Determine the range R, the maximum height h attained, and the time of flight. Express the results in terms of the angle θ and v_0. The acceleration due to gravity is g.

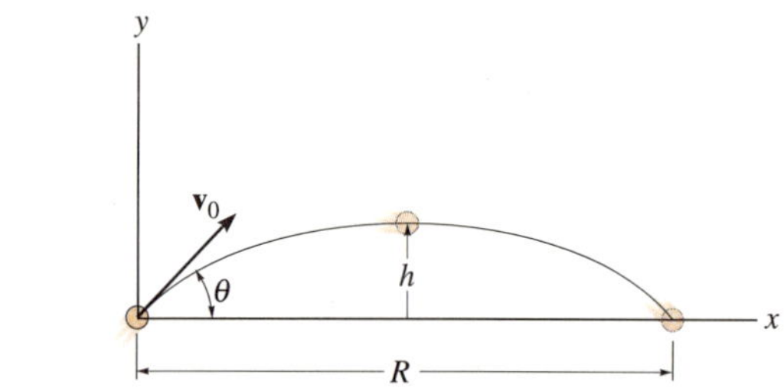

Prob. 12–82

12-83. Show that if a projectile is fired at an angle θ from the horizontal with an initial velocity $\mathbf{v}_0$, the *maximum* range the projectile can travel is given by $R_{max} = v_0^2/g$, where g is the acceleration of gravity. What is the angle θ for this condition?

***12-84.** The catapult is used to launch a ball such that it strikes the wall of the building at the maximum height of its trajectory. If it takes 1.5 s to travel from A to B, determine the velocity $\mathbf{v}_A$ at which it was launched, the angle of release θ, and the height h.

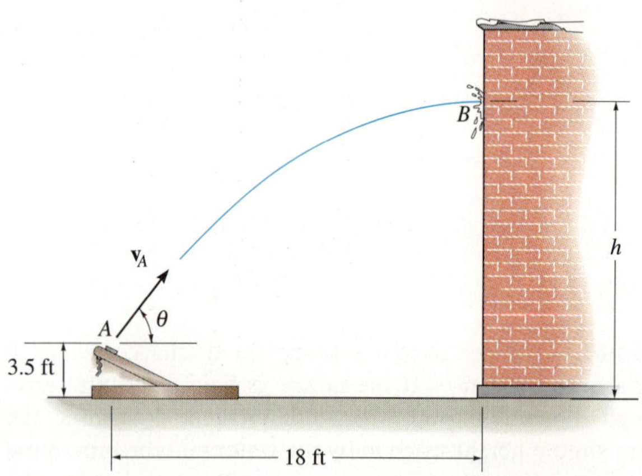

Prob. 12–84

12-85. From a videotape, it was observed that a pro football player kicked a football 126 ft during a measured time of 3.6 seconds. Determine the initial speed of the ball and the angle θ at which it was kicked.

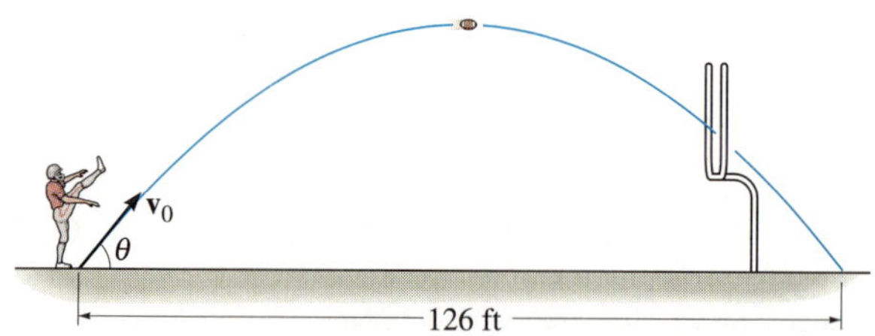

Prob. 12–85

■**12-86.** The fireman standing on the ladder wishes to direct the flow of water from his hose to the fire at B. Determine two possible angles θ_1 and θ_2 at which this can be done. Water flows from the hose at $v_A = 300$ ft/s.

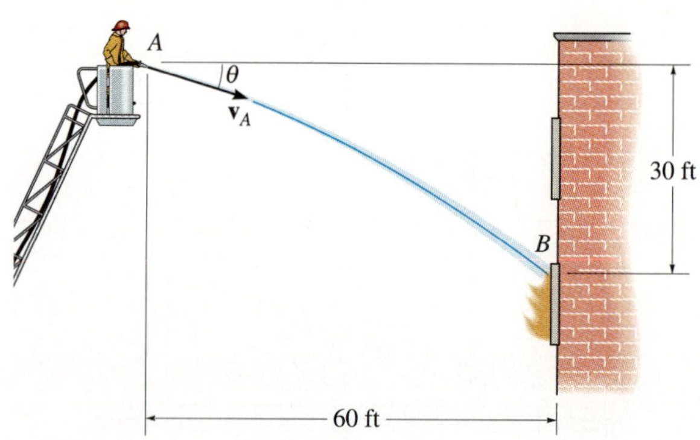

Prob. 12–86

12-87. The fireman standing on the ladder directs the flow of water from his hose to the fire at B. Determine the velocity of the water at A if it is observed that the hose is held at $\theta = 20°$.

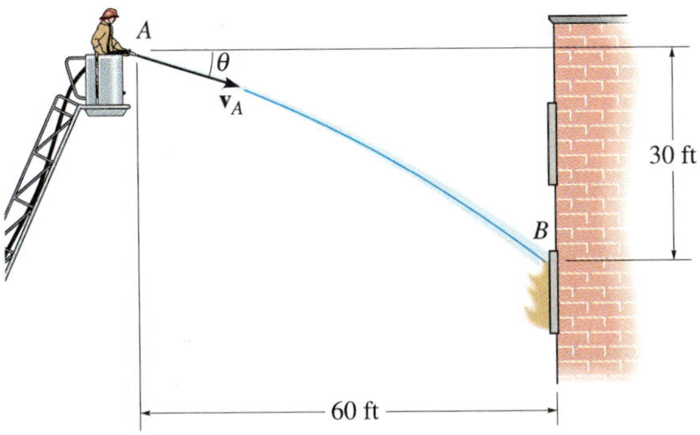

Prob. 12–87

*12-88. A projectile is given a velocity $\mathbf{v}_0$ at an angle ϕ above the horizontal. Determine the distance d to where it strikes the sloped ground. The acceleration due to gravity is g.

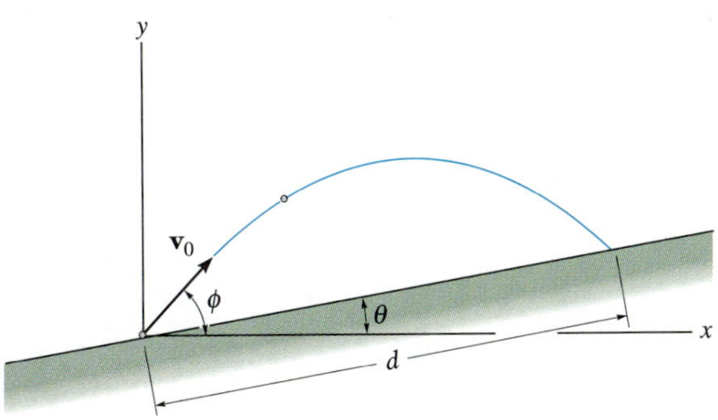

Prob. 12–88

12-89. A projectile is given a velocity $\mathbf{v}_0$. Determine the angle ϕ at which it should be launched so that d is a maximum. The acceleration due to gravity is g.

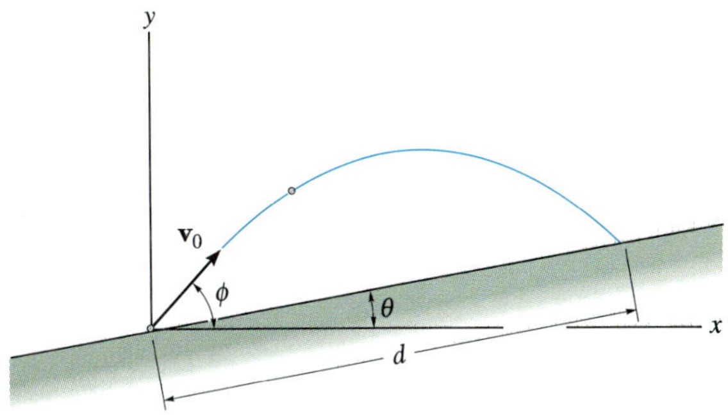

Prob. 12–89

12-90. A ball bounces on the 30° inclined plane such that it rebounds perpendicular to the incline with a velocity of $v_A = 40$ ft/s. Determine the distance R to where it strikes the plane at B.

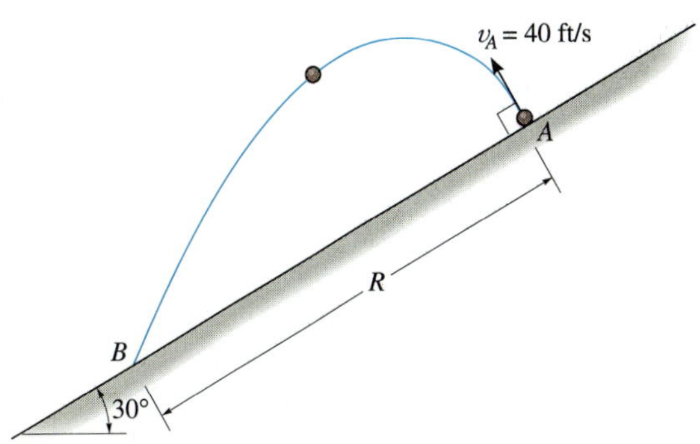

Prob. 12–90

12-91. It is observed that the skier leaves the ramp A at an angle $\theta_A = 25°$ with the horizontal. If he strikes the ground at B, determine his initial speed v_A and the time of flight t_{AB}.

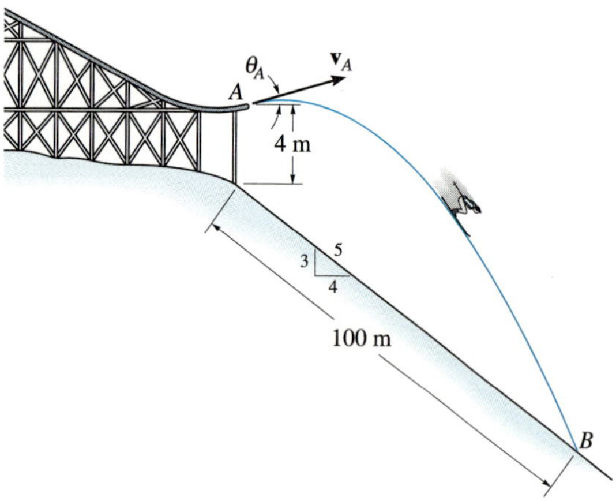

Prob. 12–91

***12-92.** The drinking fountain is designed such that the nozzle is located from the edge of the basin as shown. Determine the maximum and minimum speed at which water can be ejected from the nozzle so that it does not splash over the sides of the basin at B and C.

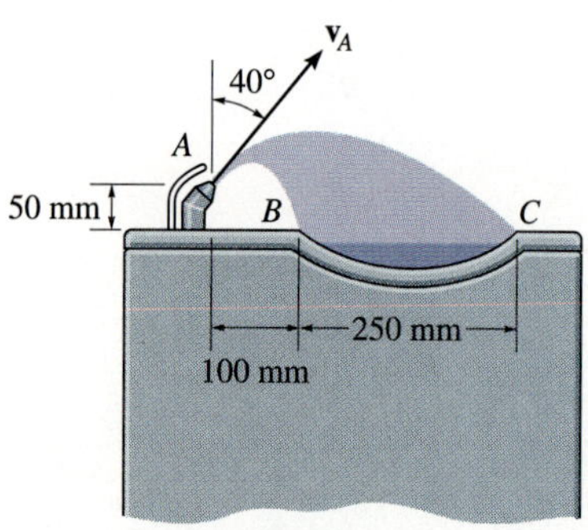

Prob. 12–92

12-93. The stones are thrown off the conveyor with a horizontal velocity of 10 ft/s as shown. Determine the distance d down the slope to where the stones hit the ground at B.

12-94. The stones are thrown off the conveyor with a horizontal velocity of 10 ft/s as shown. Determine the speed at which the stones hit the ground at B.

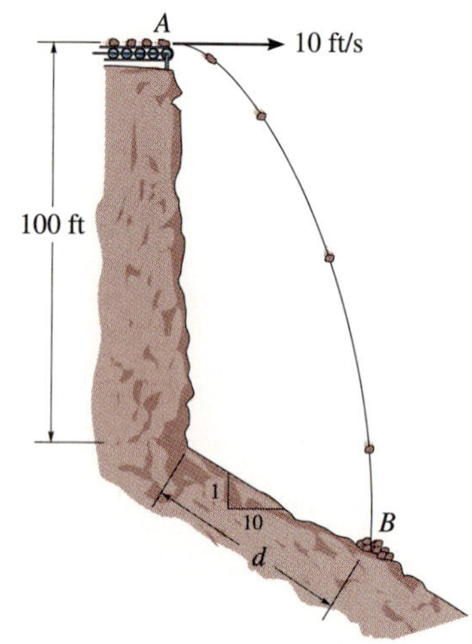

Probs. 12–93/94

12-95. The buckets on the conveyor travel with a speed of 15 ft/s. Each bucket contains a block which falls out of the bucket when $\theta = 120°$. Determine the distance s to where the block strikes the conveyor. Neglect the size of the block.

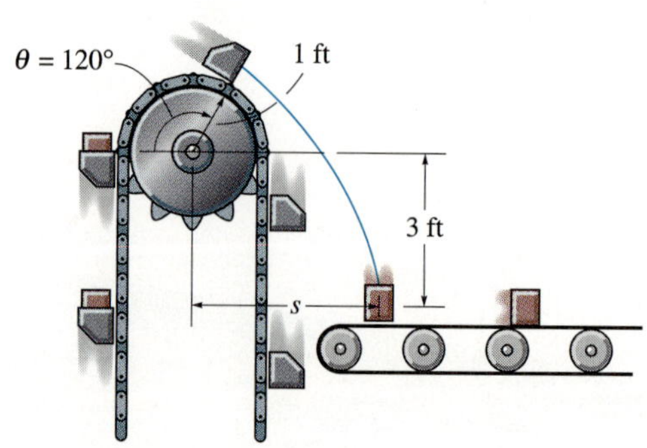

Prob. 12–95

***12-96.** The missile at A takes off from rest and rises vertically to B, where its fuel runs out in 8 s. If the acceleration varies with time as shown, determine the missile's height h_B and speed v_B. If by internal controls the missile is then suddenly pointed 45° as shown, and allowed to travel in free flight, determine the maximum height attained, h_C, and the range R to where it crashes at D.

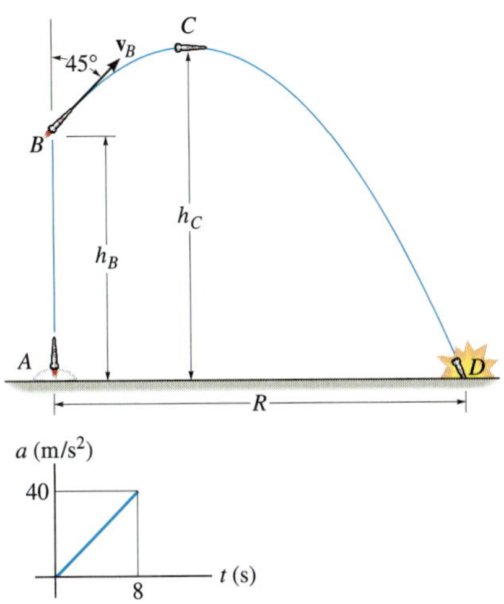

Prob. 12–96

12-97. The water sprinkler, positioned at the base of a hill, releases a stream of water with a velocity of 15 ft/s as shown. Determine the point $B(x, y)$ where the water strikes the ground on the hill. Assume that the hill is defined by the equation $y = (0.05x^2)$ ft and neglect the size of the sprinkler.

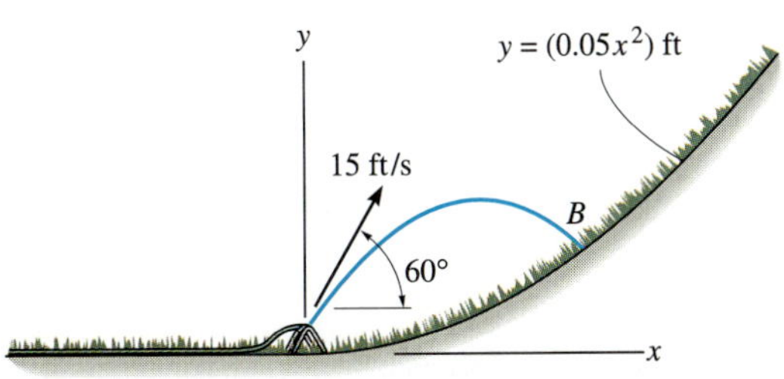

Prob. 12–97

12-98. The ball is thrown from the tower with a velocity of 20 ft/s as shown. Determine the x and y coordinates to where the ball strikes the slope. Also, determine the speed at which the ball hits the ground.

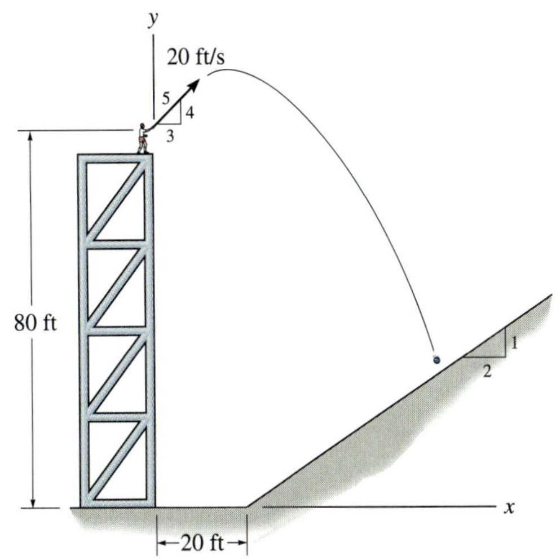

Prob. 12–98

12-99. The projectile is launched from a height h with a velocity $\mathbf{v}_0$. Determine the range R.

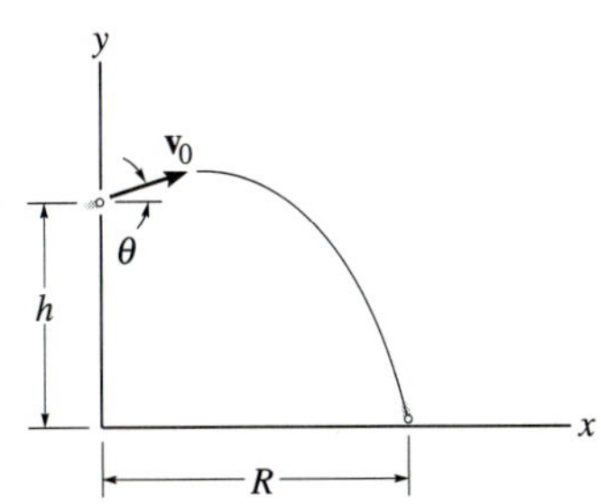

Prob. 12–99

12.7 Curvilinear Motion: Normal and Tangential Components

When the path along which a particle is moving is *known,* it is often convenient to describe the motion using n and t coordinates which act normal and tangent to the path, respectively, and at the instant considered have their *origin located at the particle.*

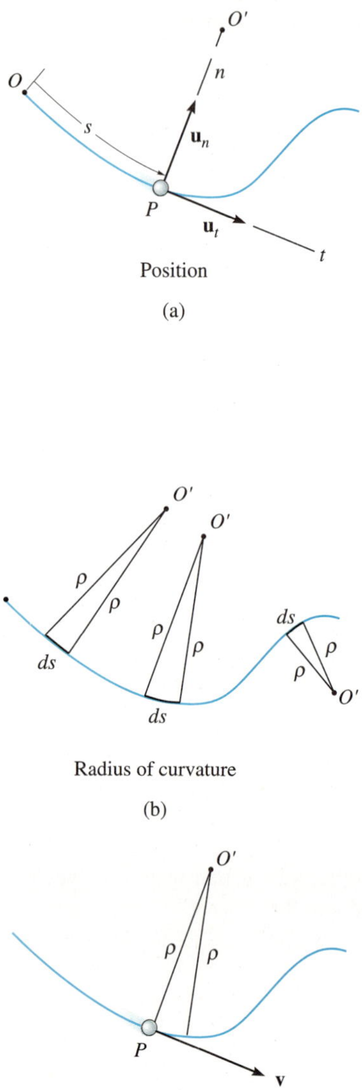

Position

(a)

Radius of curvature

(b)

Velocity

(c)

Fig. 12–24

Planar Motion. Consider the particle P shown in Fig. 12–24*a*, which is moving in a plane along a fixed curve, such that at a given instant it is at position s, measured from point O. We will now consider a coordinate system that has its origin at a *fixed point* on the curve, and at the instant considered this origin happens to *coincide* with the location of the particle. The t axis is *tangent* to the curve at P and is positive in the direction of *increasing* s. We will designate this positive direction with the unit vector $\mathbf{u}_t$. A unique choice for the *normal axis* can be made by noting that geometrically the curve is constructed from a series of differential arc segments ds, Fig. 12–24*b*. Each segment ds is formed from the arc of an associated circle having a *radius of curvature* ρ (rho) and *center of curvature* O'. The normal axis n is perpendicular to the t axis and is directed from P *toward* the center of curvature O', Fig. 12–24*a*. This positive direction, which is *always* on the concave side of the curve, will be designated by the unit vector $\mathbf{u}_n$. The plane which contains the n and t axes is referred to as the *osculating plane,* and in this case it is fixed in the plane of motion.*

Velocity. Since the particle is moving, s is a function of time. As indicated in Sec. 12.4, the particle's velocity $\mathbf{v}$ has a *direction* that is *always tangent to the path,* Fig. 12–24*c*, and a *magnitude* that is determined by taking the time derivative of the path function $s = s(t)$, i.e., $v = ds/dt$ (Eq. 12–8). Hence

$$\mathbf{v} = v\mathbf{u}_t \qquad (12\text{–}15)$$

where

$$v = \dot{s} \qquad (12\text{–}16)$$

*The osculating plane may also be defined as that plane which has the greatest contact with the curve at a point. It is the limiting position of a plane contacting both the point and the arc segment ds. As noted above, the osculating plane is always coincident with a plane curve; however, each point on a three-dimensional curve has a unique osculating plane.

Acceleration. The acceleration of the particle is the time rate of change of the velocity. Thus,

$$\mathbf{a} = \dot{\mathbf{v}} = \dot{v}\mathbf{u}_t + v\dot{\mathbf{u}}_t \qquad (12\text{–}17)$$

In order to determine the time derivative $\dot{\mathbf{u}}_t$, note that as the particle moves along the arc ds in time dt, $\mathbf{u}_t$ preserves its magnitude of unity; however, its *direction* changes, and becomes $\mathbf{u}'_t$, Fig. 12–24*d*. As shown in Fig. 12–24*e*, we require $\mathbf{u}'_t = \mathbf{u}_t + d\mathbf{u}_t$. Here $d\mathbf{u}_t$ stretches between the arrowheads of $\mathbf{u}_t$ and $\mathbf{u}'_t$, which lie on an infinitesimal arc of radius $u_t = 1$. Hence, $d\mathbf{u}_t$ has a *magnitude* of $du_t = (1)\,d\theta$, and its *direction* is defined by $\mathbf{u}_n$. Consequently, $d\mathbf{u}_t = d\theta\mathbf{u}_n$, and therefore the time derivative becomes $\dot{\mathbf{u}}_t = \dot{\theta}\mathbf{u}_n$. Since $ds = \rho\, d\theta$, Fig. 12–24*d*, then $\dot{\theta} = \dot{s}/\rho$, and therefore

$$\dot{\mathbf{u}}_t = \dot{\theta}\mathbf{u}_n = \frac{\dot{s}}{\rho}\mathbf{u}_n = \frac{v}{\rho}\mathbf{u}_n$$

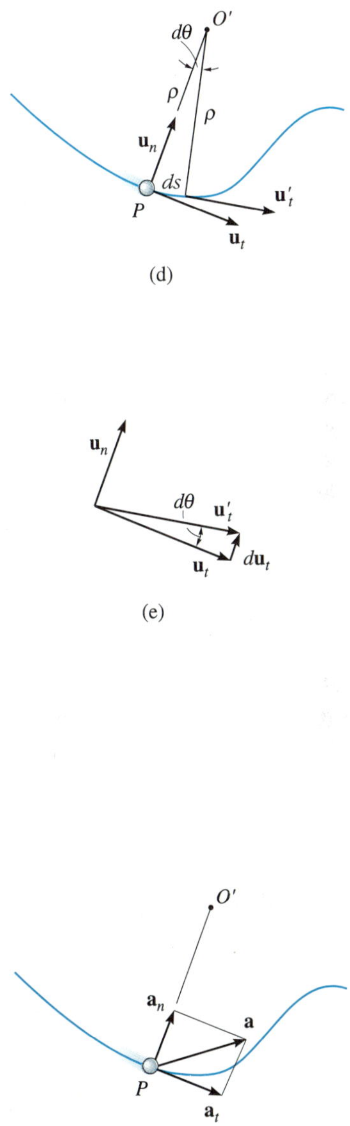

(d)

(e)

Acceleration

(f)

Fig. 12–24

Substituting into Eq. 12–17, **a** can be written as the sum of its two components,

$$\boxed{\mathbf{a} = a_t\mathbf{u}_t + a_n\mathbf{u}_n} \qquad (12\text{–}18)$$

where

$$\boxed{a_t = \dot{v}} \quad \text{or} \quad \boxed{a_t\,ds = v\,dv} \qquad (12\text{–}19)$$

and

$$\boxed{a_n = \frac{v^2}{\rho}} \qquad (12\text{–}20)$$

These two mutually perpendicular components are shown in Fig. 12–24*f*, in which case the *magnitude* of acceleration is the positive value of

$$a = \sqrt{a_t^2 + a_n^2} \qquad (12\text{–}21)$$

To summarize these concepts, consider the following two special cases of motion.

1. If the particle moves along a straight line, then $\rho \to \infty$ and from Eq. 12–20, $a_n = 0$. Thus $a = a_t = \dot{v}$, and we can conclude that the *tangential component of acceleration represents the time rate of change in the magnitude of the velocity.*
2. If the particle moves along a curve with a constant speed, then $a_t = \dot{v} = 0$ and $a = a_n = v^2/\rho$. Therefore, the *normal component of acceleration represents the time rate of change in the direction of the velocity.* Since $\mathbf{a}_n$ *always* acts towards the center of curvature, this component is sometimes referred to as the *centripetal acceleration.*

As a result of these interpretations, a particle moving along the curved path in Fig. 12–25 will have accelerations directed as shown.

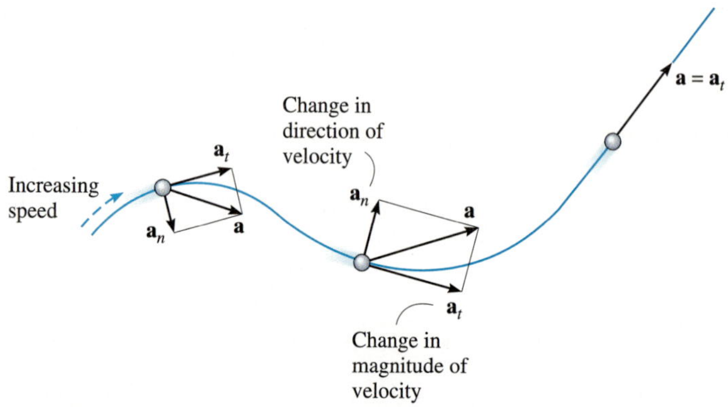

Fig. 12–25

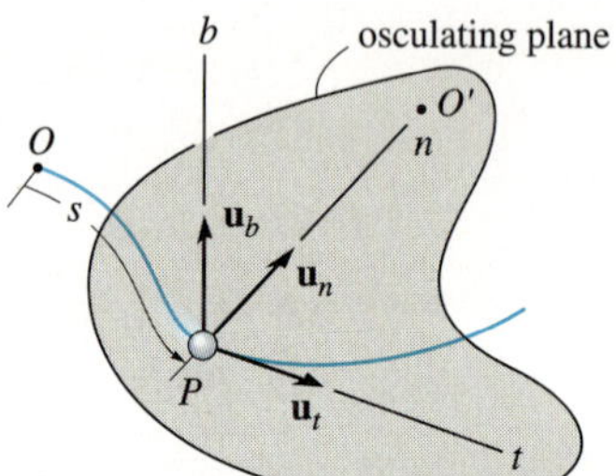

Fig. 12–26

Three-Dimensional Motion. If the particle is moving along a space curve, Fig. 12–26, then at a given instant the t axis is uniquely specified; however, an infinite number of straight lines can be constructed normal to the tangent axis at P. As in the case of planar motion, we will choose the positive n axis directed from P toward the path's center of curvature O'. This axis is referred to as the *principal normal* to the curve at P. With the n and t axes so defined, Eqs. 12–15 to 12–21 can be used to determine $\mathbf{v}$ and $\mathbf{a}$. Since $\mathbf{u}_t$ and $\mathbf{u}_n$ are always perpendicular to one another and lie in the osculating plane, for spatial motion a third unit vector, $\mathbf{u}_b$, defines a *binomial axis b* which is perpendicular to $\mathbf{u}_t$ and $\mathbf{u}_n$, Fig. 12–26.

Since the three unit vectors are related to one another by the vector cross product, e.g., $\mathbf{u}_b = \mathbf{u}_t \times \mathbf{u}_n$, Fig. 12–26, it may be possible to use this relation to establish the direction of one of the axes, if the directions of the other two are known. For example, no motion occurs in the $\mathbf{u}_b$ direction, and so if this direction and $\mathbf{u}_t$ are known, then $\mathbf{u}_n$ can be determined, where in this case $\mathbf{u}_n = \mathbf{u}_b \times \mathbf{u}_t$, Fig. 12–26. Remember, though, that $\mathbf{u}_n$ is always on the concave side of the curve.

Procedure for Analysis

Coordinate System

- Provided the *path* of the particle is *known,* we can establish a set of n and t coordinates having a *fixed origin* which is coincident with the particle at the instant considered.
- The positive tangent axis acts in the direction of motion and the positive normal axis is directed toward the path's center of curvature.
- The n and t axes are particularly advantageous for studying the velocity and acceleration of the particle, because the t and n components of **a** are expressed by Eqs. 12–19 and 12–20, respectively.

Velocity

- The particle's *velocity* is always tangent to the path.

$$v = \dot{s}$$

- The magnitude of velocity is found from the time derivative of the path function.

Tangential Acceleration

- The tangential component of acceleration is the result of the time rate of change in the magnitude of velocity. This component acts in the positive s direction if the particle's speed is increasing or in the opposite direction if the speed is decreasing.
- The relations between a_t, v, t and s are the same as for rectilinear motion, namely,

$$a_t = \dot{v} \qquad a_t\, ds = v\, dv$$

- If a_t is *constant,* $a_t = (a_t)_c$, the above equations, when integrated, yield

$$s = s_0 + v_0 t + \tfrac{1}{2}(a_t)_c t^2$$
$$v = v_0 + (a_t)_c t$$
$$v^2 = v_0^2 + 2(a_t)_c(s - s_0)$$

Normal Acceleration

- The normal component of acceleration is the result of the time rate of change in the direction of the particle's velocity. This component is *always* directed toward the center of curvature of the path, i.e., along the positive n axis.
- The magnitude of this component is determined from

$$a_n = \frac{v^2}{\rho}$$

- If the path is expressed as $y = f(x)$, the radius of curvature ρ at any point on the path is determined from the equation

$$\rho = \frac{[1 + (dy/dx)^2]^{3/2}}{|d^2y/dx^2|}$$

The derivation of this result is given in any standard calculus text.

Motorists traveling along this clover-leaf interchange experience a normal acceleration due to the change in direction of their velocity. A tangential component of acceleration occurs when the cars' speed is increased or decreased.

EXAMPLE 12–14

When the skier reaches point A along the parabolic path in Fig. 12–27*a*, he has a speed of 6 m/s which is increasing at 2 m/s^2. Determine the direction of his velocity and the magnitude of his acceleration at this instant. Neglect the size of the skier in the calculation.

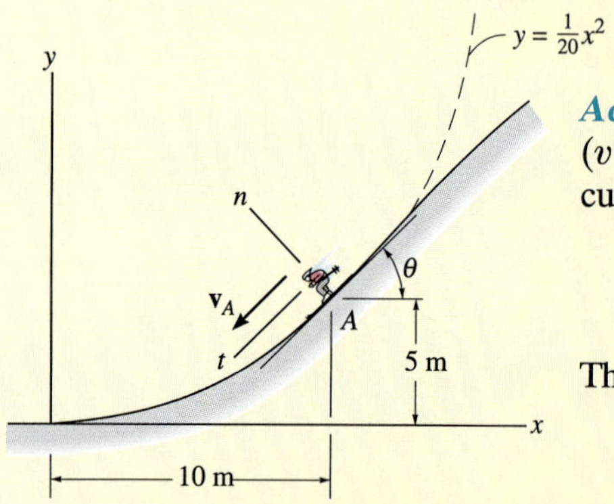

(a)

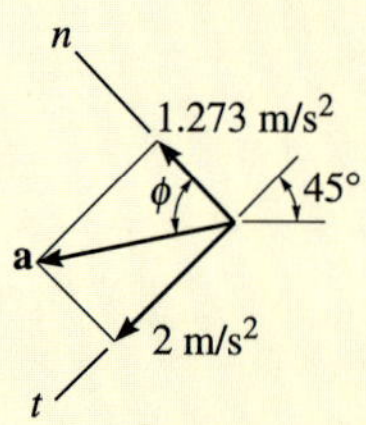

(b)

Fig. 12–27

Solution

Coordinate System. Although the path has been expressed in terms of its x and y coordinates, we can still establish the origin of the n, t axes at the fixed point A on the path and determine the components of $\mathbf{v}$ and $\mathbf{a}$ along these axes, Fig. 12–27*a*.

Velocity. By definition, the velocity is always directed tangent to the path. Since $y = \frac{1}{20}x^2$, $dy/dx = \frac{1}{10}x$, then $dy/dx|_{x=10} = 1$. Hence, at A, $\mathbf{v}$ makes an angle of $\theta = \tan^{-1} 1 = 45°$ with the x axis, Fig. 12–27. Therefore,

$$v_A = 6 \text{ m/s} \qquad 45° \measuredangle \mathbf{v}_A \qquad \textit{Ans.}$$

Acceleration. The acceleration is determined from $\mathbf{a} = \dot{v}\mathbf{u}_t + (v^2/\rho)\mathbf{u}_n$. However, it is first necessary to determine the radius of curvature of the path at A (10 m, 5 m). Since $d^2y/dx^2 = \frac{1}{10}$, then

$$\rho = \frac{[1 + (dy/dx)^2]^{3/2}}{|d^2y/dx^2|} = \frac{[1 + (\frac{1}{10}x)^2]^{3/2}}{|\frac{1}{10}|}\bigg|_{x=10\text{ m}} = 28.28 \text{ m}$$

The acceleration becomes

$$\mathbf{a}_A = \dot{v}\mathbf{u}_t + \frac{v^2}{\rho}\mathbf{u}_n$$

$$= 2\mathbf{u}_t + \frac{(6 \text{ m/s})^2}{28.28 \text{ m}}\mathbf{u}_n$$

$$= \{2\mathbf{u}_t + 1.273\mathbf{u}_n\} \text{ m/s}^2$$

As shown in Fig. 12–27*b*,

$$a = \sqrt{(2)^2 + (1.273)^2} = 2.37 \text{ m/s}^2$$

$$\phi = \tan^{-1}\frac{2}{1.273} = 57.5°$$

Thus, $45° + 90° + 57.5° = 192°$ so that,

$$a = 2.37 \text{ m/s}^2 \qquad a_A \measuredangle 192° \qquad \textit{Ans.}$$

Note: By using n, t coordinates, we were able to readily solve this problem since the n and t components account for the *separate* changes in the magnitude and direction of $\mathbf{v}$.

EXAMPLE 12–15

A race car C travels around the horizontal circular track that has a radius of 300 ft, Fig. 12–28. If the car increases its speed at a constant rate of 7 ft/s^2, starting from rest, determine the time needed for it to reach an acceleration of 8 ft/s^2. What is its speed at this instant?

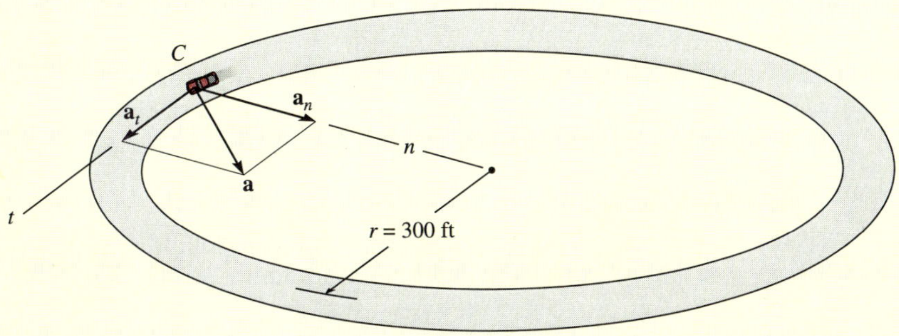

Fig. 12–28

Solution

Coordinate System. The origin of the n and t axes is coincident with the car at the instant considered. The t axis is in the direction of motion, and the positive n axis is directed toward the center of the circle. This coordinate system is selected since the path is known.

Acceleration. The magnitude of acceleration can be related to its components using $a = \sqrt{a_t^2 + a_n^2}$. Here $a_t = 7$ ft/s^2. Since $a_n = v^2/\rho$, the velocity as a function of time is

$$v = v_0 + (a_t)_c t$$
$$v = 0 + 7t$$

Thus

$$a_n = \frac{v^2}{\rho} = \frac{(7t)^2}{300} = 0.163t^2 \text{ ft/s}^2$$

The time needed for the acceleration to reach 8 ft/s^2 is therefore

$$a = \sqrt{a_t^2 + a_n^2}$$
$$8 = \sqrt{(7)^2 + (0.163t^2)^2}$$

Solving for the positive value of t yields

$$0.163t^2 = \sqrt{(8)^2 - (7)^2}$$
$$t = 4.87 \text{ s} \qquad \textit{Ans.}$$

Velocity. The speed at time $t = 4.87$ s is

$$v = 7t = 7(4.87) = 34.1 \text{ ft/s} \qquad \textit{Ans.}$$

EXAMPLE 12–16

(a)

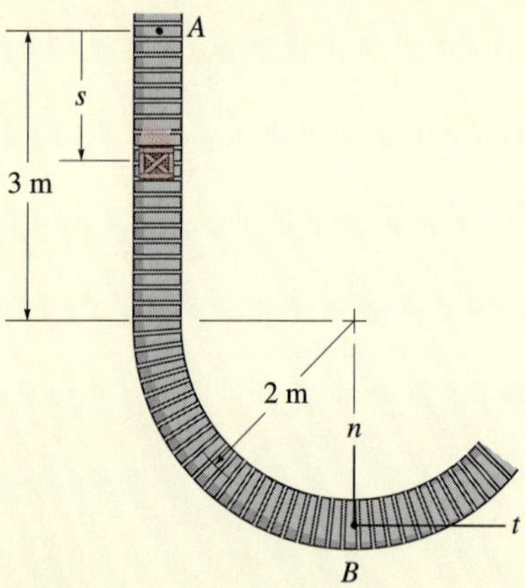

(b)

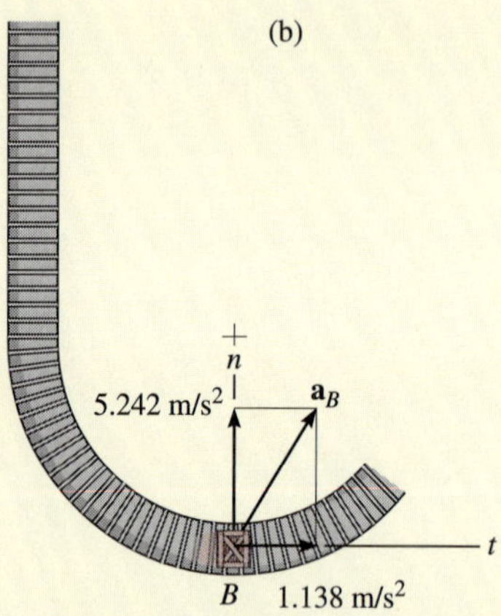

(c)

Fig. 12–29

The boxes in Fig. 12–29*a* travel along the industrial conveyor. If a box as in Fig. 12–29*b* starts from rest at A and increases its speed such that $a_t = (0.2t)\ \text{m/s}^2$, where t is in seconds, determine the magnitude of its acceleration when it arrives at point B.

Solution

Coordinate System. The position of the box at any instant is defined from the fixed point A using the position or path coordinate s, Fig. 12–29*b*. The acceleration is to be determined at B, so the origin of the n, t axes is at this point.

Acceleration. To determine the acceleration components $a_t = \dot{v}$ and $a_n = v^2/\rho$, it is first necessary to formulate v and $\dot{v}$ so that they may be evaluated at B. Since $v_A = 0$ when $t = 0$, then

$$a_t = \dot{v} = 0.2t \qquad (1)$$

$$\int_0^v dv = \int_0^t 0.2t\, dt$$

$$v = 0.1t^2 \qquad (2)$$

The time needed for the box to reach point B can be determined by realizing that the position of B is $s_B = 3 + 2\pi(2)/4 = 6.142$ m, Fig. 12–29*b*, and since $s_A = 0$ when $t = 0$ we have

$$v = \frac{ds}{dt} = 0.1t^2$$

$$\int_0^{6.142} ds = \int_0^{t_B} 0.1t^2\, dt$$

$$6.142 = 0.0333t_B^3$$

$$t_B = 5.690 \text{ s}$$

Substituting into Eqs. 1 and 2 yields

$$(a_B)_t = \dot{v}_B = 0.2(5.690) = 1.138\ \text{m/s}^2$$

$$v_B = 0.1(5.69)^2 = 3.238\ \text{m/s}$$

At B, $\rho_B = 2$ m, so that

$$(a_B)_n = \frac{v_B^2}{\rho_B} = \frac{(3.238\ \text{m/s})^2}{2\ \text{m}} = 5.242\ \text{m/s}^2$$

The magnitude of $\mathbf{a}_B$, Fig. 12–29*c*, is therefore

$$a_B = \sqrt{(1.138)^2 + (5.242)^2} = 5.36\ \text{m/s}^2 \qquad \textit{Ans.}$$

Problems

*12-100. A particle is moving along a curved path at a constant speed of 60 ft/s. The radii of curvature of the path at points P and P' are 20 and 50 ft, respectively. If it takes the particle 20 s to go from P to P', determine the acceleration of the particle at P and P'.

12-101. A car travels along a horizontal curved road that has a radius of 600 m. If the speed is uniformly increased at a rate of 2000 km/h^2, determine the magnitude of the acceleration at the instant the speed of the car is 60 km/h.

12-102. The car travels along the curve having a radius of 300 m. If its speed is uniformly increased from 15 m/s to 27 m/s in 3 s, determine the magnitude of its acceleration at the instant its speed is 20 m/s.

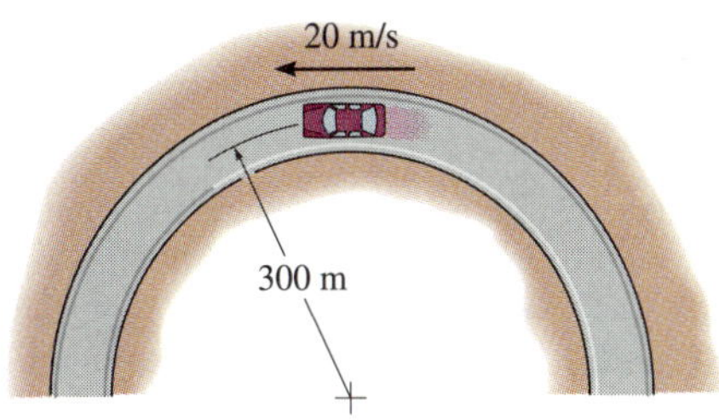

Prob. 12–102

12-103. A boat is traveling along a circular curve having a radius of 100 ft. If its speed at $t = 0$ is 15 ft/s and is increasing at $\dot{v} = (0.8t)$ ft/s^2, determine the magnitude of its acceleration at the instant $t = 5$ s.

***12-104.** A boat is traveling along a circular path having a radius of 20 m. Determine the magnitude of the boat's acceleration when the speed is $v = 5$ m/s and the rate of increase in the speed is $\dot{v} = 2$ m/s^2.

■**12-105.** Starting from rest, a bicyclist travels around a horizontal circular path, $\rho = 10$ m, at a speed of $v = (0.09t^2 + 0.1t)$ m/s, where t is in seconds. Determine the magnitudes of his velocity and acceleration when he has traveled $s = 3$ m.

12-106. The truck travels in a circular path having a radius of 50 m at a speed of 4 m/s. For a short distance from $s = 0$, its speed is increased by $\dot{v} = (0.05s)$ m/s^2, where s is in meters. Determine its speed and the magnitude of its acceleration when it has moved $s = 10$ m.

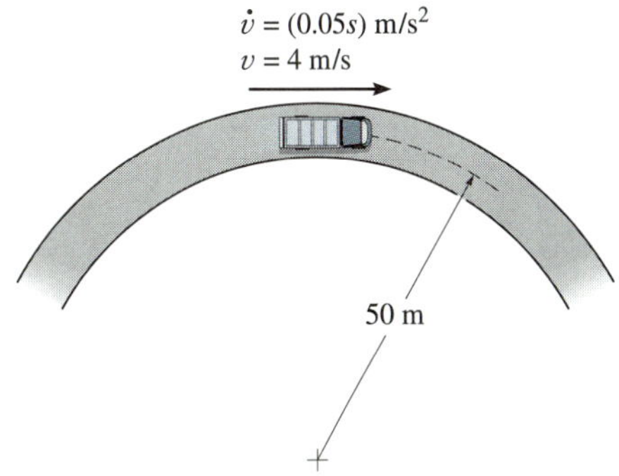

Prob. 12–106

12-107. The satellite S travels around the earth in a circular path with a constant speed of 20 Mm/h. If the acceleration is 2.5 m/s^2, determine the altitude h. Assume the earth's diameter to be 12 713 km.

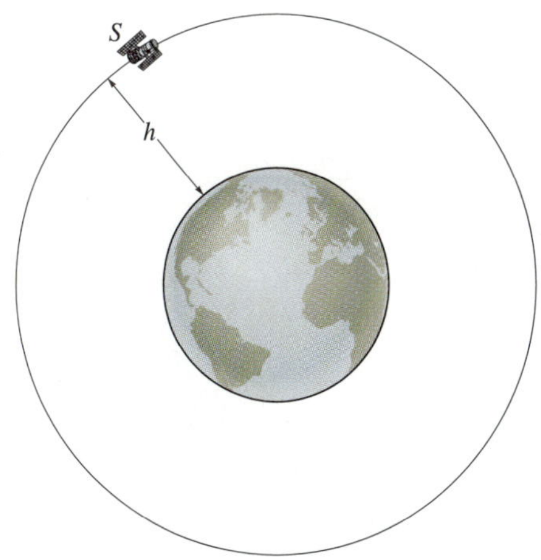

Prob. 12–107

***12-108.** A particle P moves along the curve $y = (x^2 - 4)$ m with a constant speed of 5 m/s. Determine the point on the curve where the maximum magnitude of acceleration occurs and compute its value.

12-109. A car moves along a circular track of radius 250 ft, and its speed for a short period of time $0 \le t \le 2$ s is $v = 3(t + t^2)$ ft/s, where t is in seconds. Determine the magnitude of its acceleration when $t = 2$ s. How far has it traveled in $t = 2$ s?

■**12-110.** The car travels along the curved path such that its speed is increased by $\dot{v} = (0.5e^t)$ m/s^2, where t is in seconds. Determine the magnitudes of its velocity and acceleration after the car has traveled $s = 18$ m starting from rest. Neglect the size of the car.

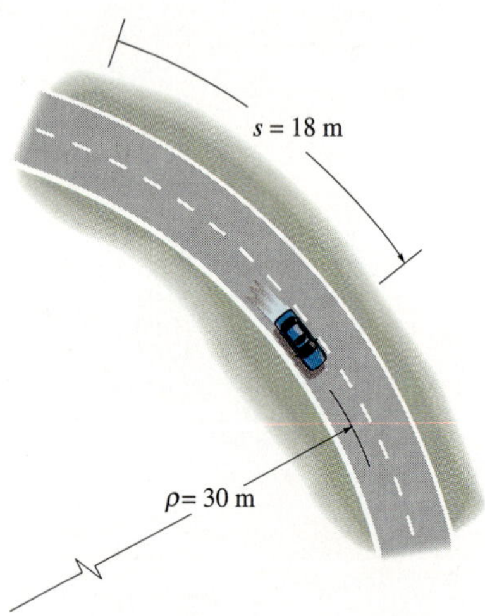

Prob. 12–110

12-111. The Ferris wheel turns such that the speed of the passengers is increased by $\dot{v} = (4t)$ ft/s^2, where t is in seconds. If the wheel starts from rest when $\theta = 0°$, determine the magnitudes of the velocity and acceleration of the passengers when the wheel turns $\theta = 30°$.

Prob. 12–111

***12-112.** A package is dropped from the plane which is flying with a constant horizontal velocity of $v_A = 150$ ft/s. Determine the normal and tangential components of acceleration and the radius of curvature of the path of motion (a) at the moment the package is released at A, where it has a horizontal velocity of $v_A = 150$ ft/s, and (b) *just before* it strikes the ground at B.

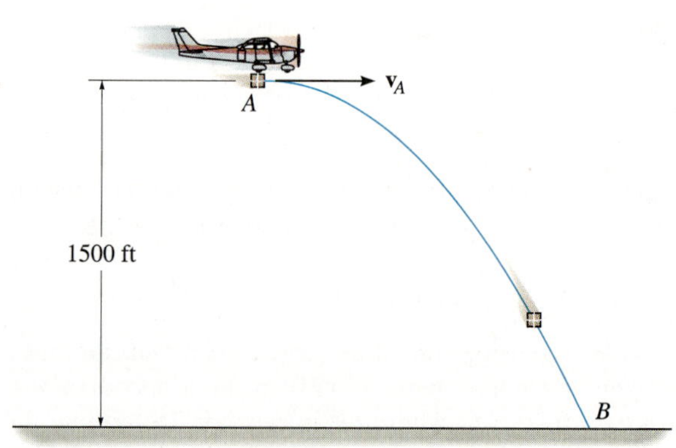

Prob. 12–112

12-113. A particle moves along the curve $y = \sin x$ with a constant speed $v = 2$ m/s. Determine the normal and tangential components of its velocity and acceleration at any instant.

12-114. The particle travels with a constant speed of 300 mm/s along the curve. Determine the particle's acceleration when it is located at point (200 mm, 100 mm) and sketch this vector on the curve.

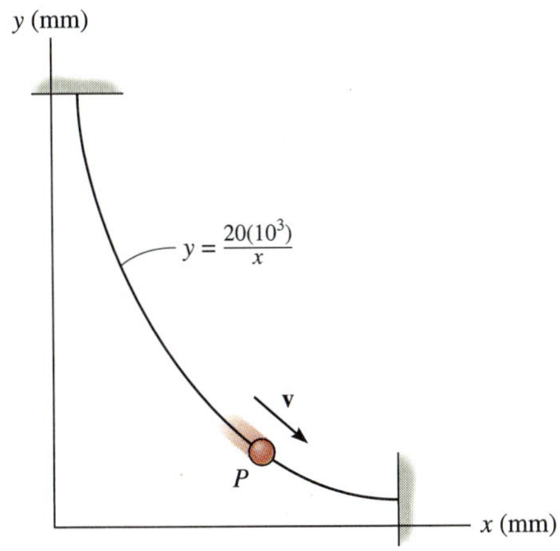

Prob. 12–114

■**12-115.** The jet plane is traveling with a speed of 120 m/s which is decreasing at 40 m/s^2 when it reaches point A. Determine the magnitude of its acceleration when it is at this point. Also, specify the direction of flight, measured from the x axis.

***12-116.** The jet plane is traveling with a constant speed of 110 m/s along the curved path. Determine the magnitude of the acceleration of the plane at the instant it reaches point A ($y = 0$).

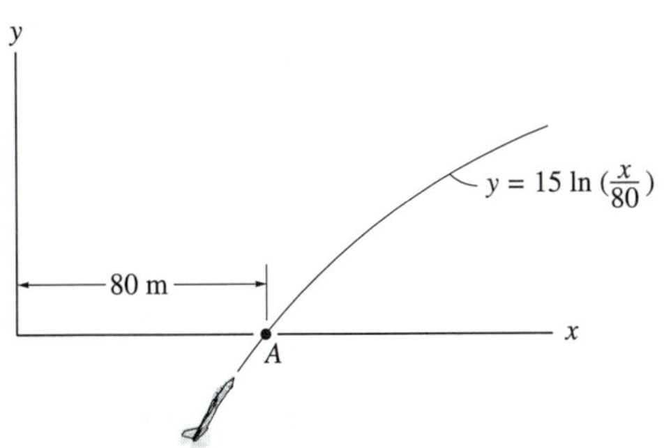

Probs. 12–115/116

12-117. A train is traveling with a constant speed of 14 m/s along the curved path. Determine the magnitude of the acceleration of the front of the train, B, at the instant it reaches point A ($y = 0$).

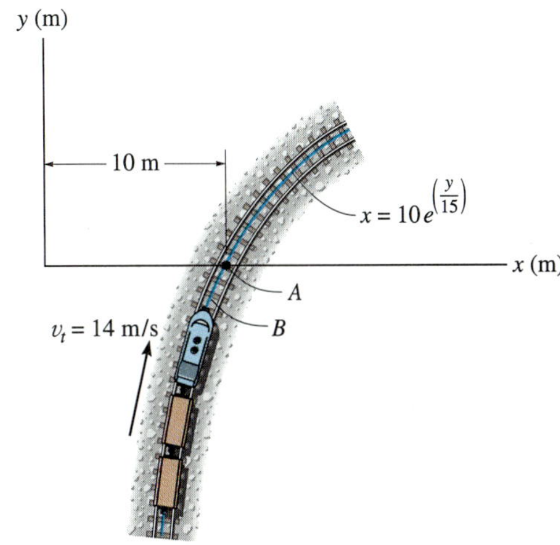

Prob. 12–117

12-118. When the motorcyclist is at A, he increases his speed along the vertical circular path at the rate of $\dot{v} = (0.3t)$ ft/s^2, where t is in seconds. If he starts from rest at A, determine the magnitudes of his velocity and acceleration when he reaches B.

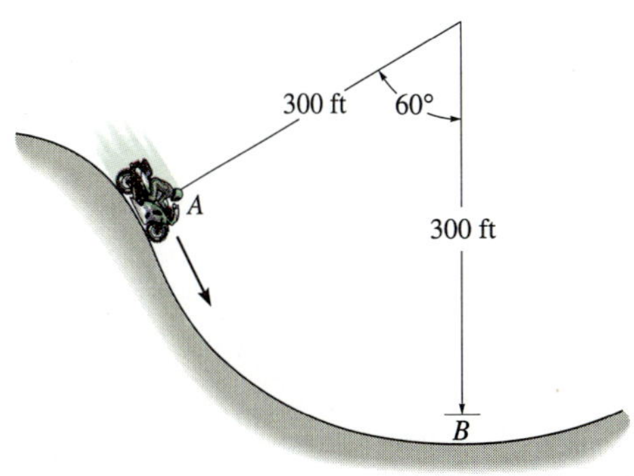

Prob. 12–118

12-119. When the car reaches point A it has a speed of 4 m/s, which is increasing at a constant rate of 2 m/s^2. Determine the time required to reach point B and the magnitudes of its velocity and acceleration.

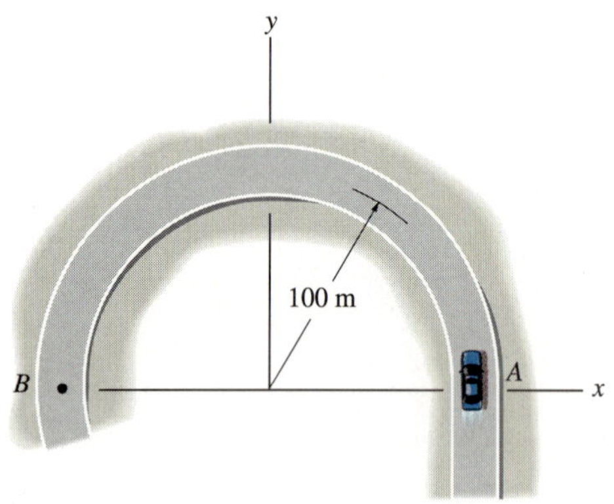

Prob. 12–119

***12-120.** The motorcyclist travels along the curve at a constant speed of 30 ft/s. Determine his acceleration when he is located at point A. Neglect the size of the motorcycle and rider for the calculation.

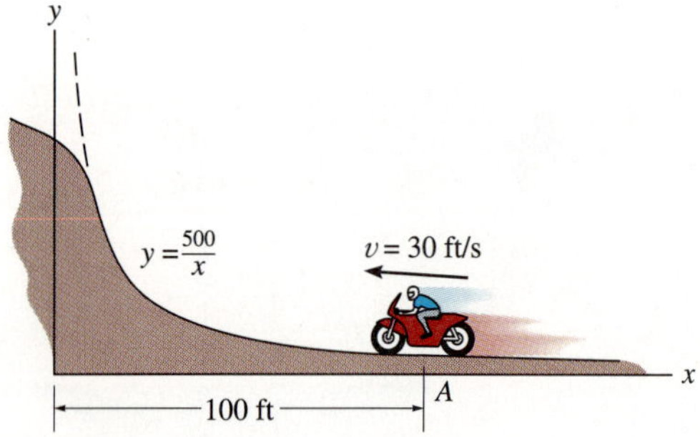

Prob. 12–120

12-121. The box of negligible size is sliding down along a curved path defined by the parabola $y = 0.4x^2$. When it is at A ($x_A = 2$ m, $y_A = 1.6$ m), the speed is $v_B = 8$ m/s and the increase in speed is $dv_B/dt = 4$ m/s^2. Determine the magnitude of the acceleration of the box at this instant.

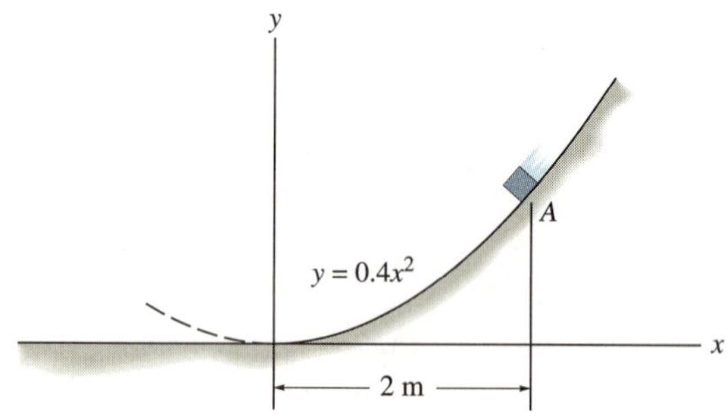

Prob. 12–121

12-122. The ball is ejected horizontally from the tube with a speed of 8 m/s. Find the equation of the path, $y = f(x)$, and then find the ball's velocity and the normal and tangential components of acceleration when $t = 0.25$ s.

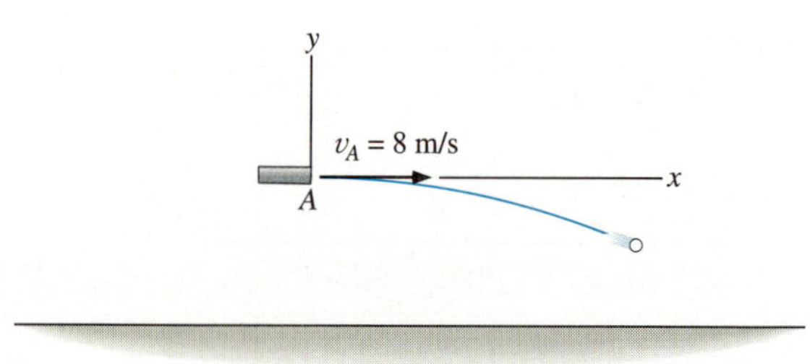

Prob. 12–122

12-123. Cars move around the "traffic circle" which is in the shape of an ellipse. If the speed limit is posted at 60 km/h, determine the maximum acceleration experienced by the passengers.

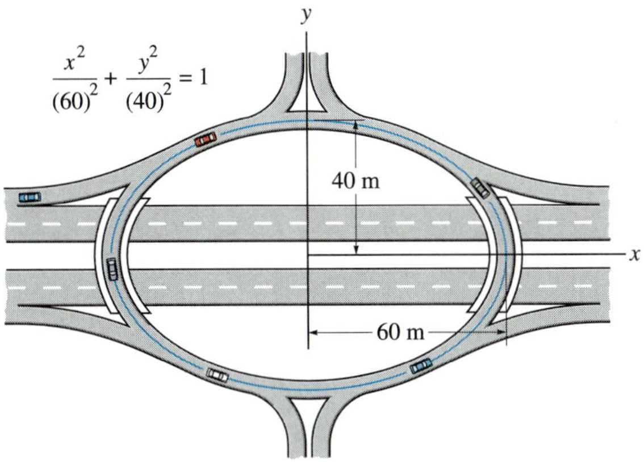

Prob. 12–123

***12-124.** Cars move around the "traffic circle" which is in the shape of an ellipse. If the speed limit is posted at 60 km/h, determine the minimum acceleration experienced by the passengers.

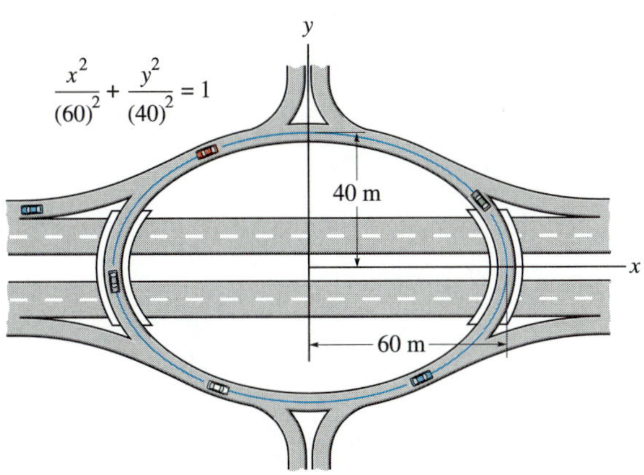

Prob. 12–124

12-125. The race car travels around the circular track with a speed of 16 m/s. When it reaches point A it increases its speed at $\dot{v} = (\frac{4}{3}v^{1/4})$ m/s^2, where v is in m/s. Determine the velocity and acceleration of the car when it reaches point B. Also, how much time is required for it to travel from A to B?

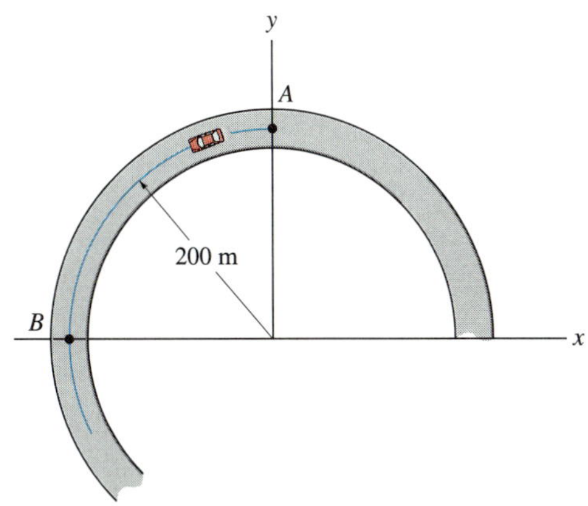

Prob. 12–125

12-126. A race car has an initial speed $v_A = 15$ m/s when it is at A. If it increases its speed along the circular track, at the rate $a_t = (0.4s)$ m/s^2, where s is in meters, determine the car's normal and tangential components of acceleration at $s = 10$ m. Take $\rho = 150$ m.

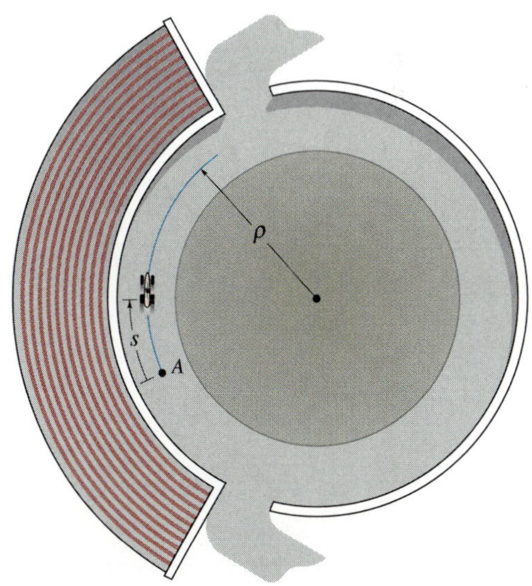

Prob. 12–126

12-127. The race car has an initial speed $v_A = 15$ m/s at A. If it increases its speed along the circular track at the rate $a_t = (0.4s)$ m/s^2, where s is in meters, determine the time needed for the car to travel 20 m. Take $\rho = 150$ m.

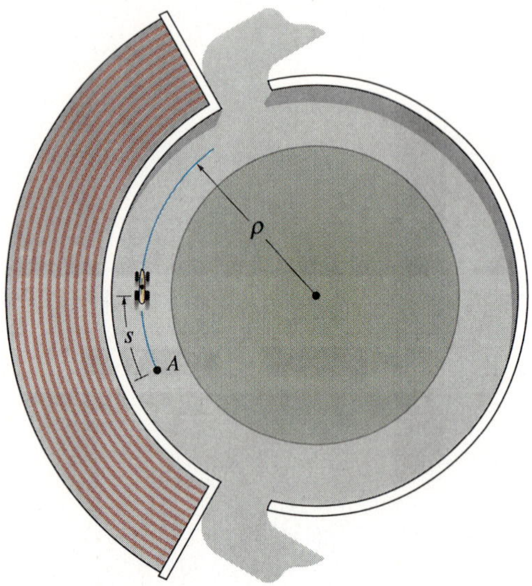

Prob. 12–127

***12-128.** A spiral transition curve is used on railroads to connect a straight portion of the track with a curved portion. If the spiral is defined by the equation $y = (10^{-6})x^3$, where x and y are in feet, determine the magnitude of the acceleration of a train engine moving with a constant speed of 40 ft/s when it is at point $x = 600$ ft.

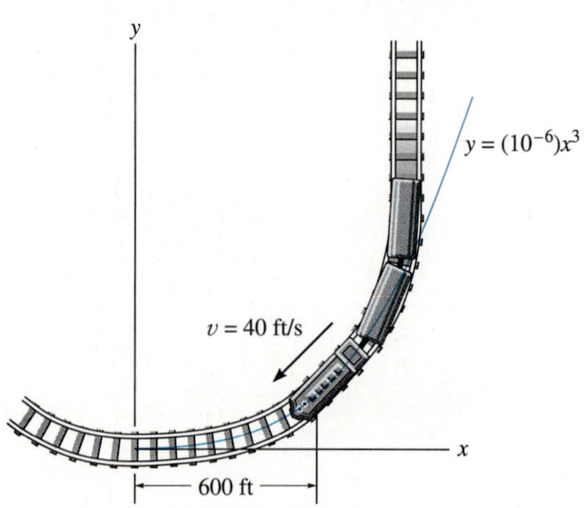

Prob. 12–128

12-129. A particle travels along the path $y = a + bx + cx^2$, where a, b, c are constants. If the speed of the particle is constant, $v = v_0$, determine the x and y components of velocity and the normal component of acceleration when $x = 0$.

■**12-130.** The motorcycle is traveling at 1 m/s when it is at A. If the speed is then increased at $\dot{v} = 0.1$ m/s^2, determine its speed and acceleration at the instant $t = 5$ s.

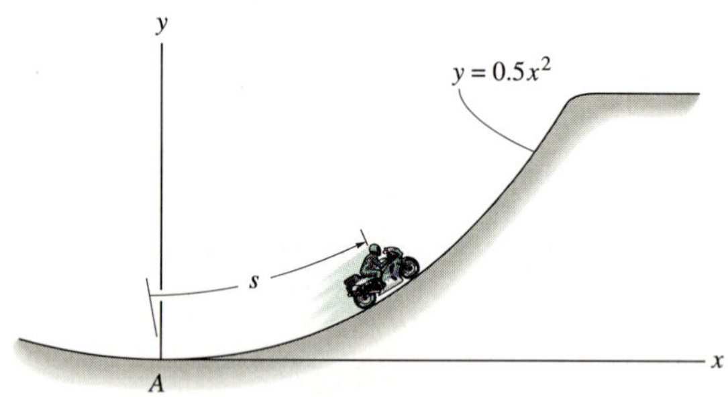

Prob. 12–130

■**12-131.** The car travels around the circular track having a radius of $r = 300$ m such that when it is at point A it has a velocity of 5 m/s, which is increasing at the rate of $\dot{v} = (0.06t)$ m/s^2, where t is in seconds. Determine the magnitudes of its velocity and acceleration when it has traveled one-third the way around the track.

***12-132.** The car travels around the portion of a circular track having a radius of $r = 500$ ft such that when it is at point A it has a velocity of 2 ft/s, which is increasing at the rate of $\dot{v} = (0.002s)$ ft/s^2, where t is in seconds. Determine the magnitudes of its velocity and acceleration when it has traveled three-fourths the way around the track.

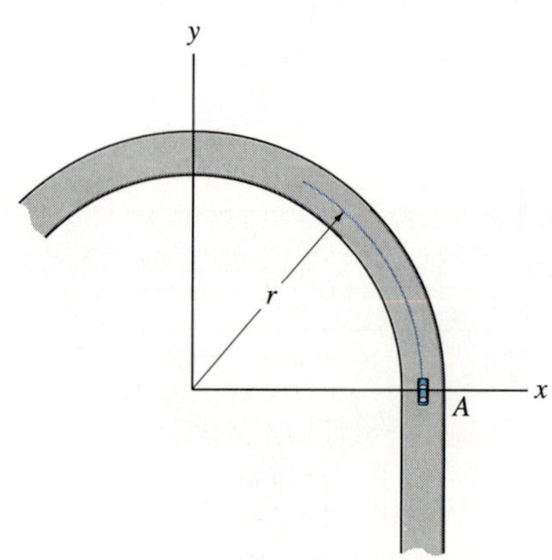

Probs. 12–131/132

12-133. The car travels around the portion of the circular track having a radius of $r = 500$ ft such that when it is at point A it has a velocity of 2 ft/s, which is increasing at the rate of $\dot{v} = (0.002s)$ ft/s^2, where s is in feet. Determine the time required for the car to travel three-fourths the way around the track.

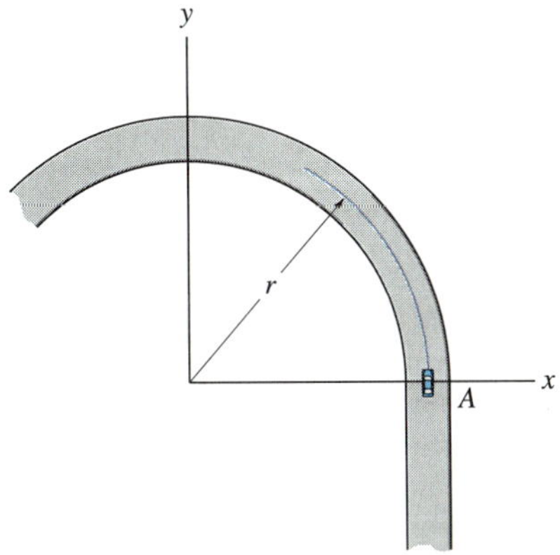

Prob. 12–133

12-134. The motion of a particle along a fixed path is defined by the parametric equations $r = 8$ ft, $\theta = (4t)$ rad, and $z = (6t^2)$ ft, where t is in seconds. Determine the unit vector that specifies the direction of the binormal axis to the osculating plane with respect to a set of fixed x, y, z coordinate axes when $t = 2$ s. *Hint:* Formulate the particle's velocity $\mathbf{v}_P$ and acceleration $\mathbf{a}_P$ in terms of their $\mathbf{i}, \mathbf{j}, \mathbf{k}$ components. Note that $x = r \cos \theta$ and $y = r \sin \theta$. The binormal is parallel to $\mathbf{v}_P \times \mathbf{a}_P$. Why?

12-135. A particle P travels along an elliptical spiral path such that its position vector $\mathbf{r}$ is defined by $\mathbf{r} = \{2 \cos (0.1t)\mathbf{i} + 1.5 \sin (0.1t)\mathbf{j} + (2t)\mathbf{k}\}$ m, where t is in seconds and the arguments for the sine and cosine are given in radians. When $t = 8$ s, determine the coordinate direction angles α, β, and γ, which the binormal axis to the osculating plane makes with the x, y, and z axes. *Hint:* Solve for the velocity $\mathbf{v}_P$ and acceleration $\mathbf{a}_P$ of the particle in terms of their $\mathbf{i}, \mathbf{j}, \mathbf{k}$ components. The binormal is parallel to $\mathbf{v}_P \times \mathbf{a}_P$. Why?

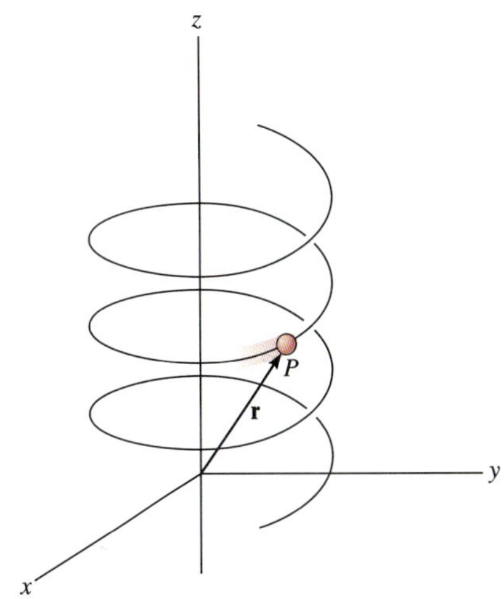

Prob. 12–135

12.8 Curvilinear Motion: Cylindrical Components

In some engineering problems it is often convenient to express the path of motion in terms of cylindrical coordinates, r, θ, z. If motion is restricted to the plane, the polar coordinates r and θ are used.

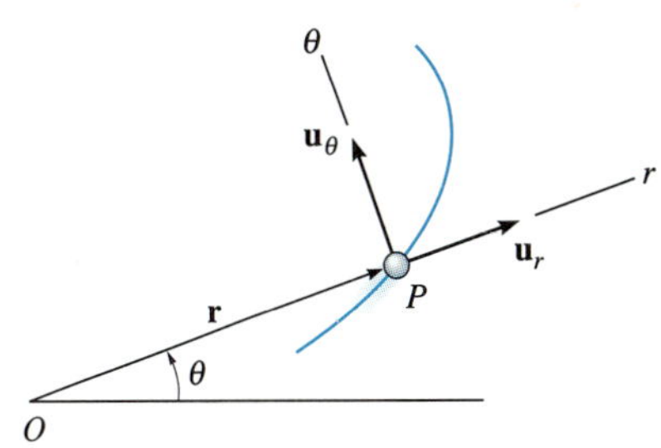

Position

(a)

Fig. 12–30

Polar Coordinates. We can specify the location of particle P shown in Fig. 12–30a using both the *radial coordinate r*, which extends outward from the fixed origin O to the particle, and a *transverse coordinate* θ, which is the counterclockwise angle between a fixed reference line and the r axis. The angle is generally measured in degrees or radians, where 1 rad $= 180°/\pi$. The positive directions of the r and θ coordinates are defined by the unit vectors $\mathbf{u}_r$ and $\mathbf{u}_\theta$, respectively. Here $\mathbf{u}_r$ or the radial direction $+r$ extends from P along increasing r, when θ is held fixed, and $\mathbf{u}_\theta$ or $+\theta$ extends from P in a direction that occurs when r is held fixed and θ is increased. Note that these directions are perpendicular to one another.

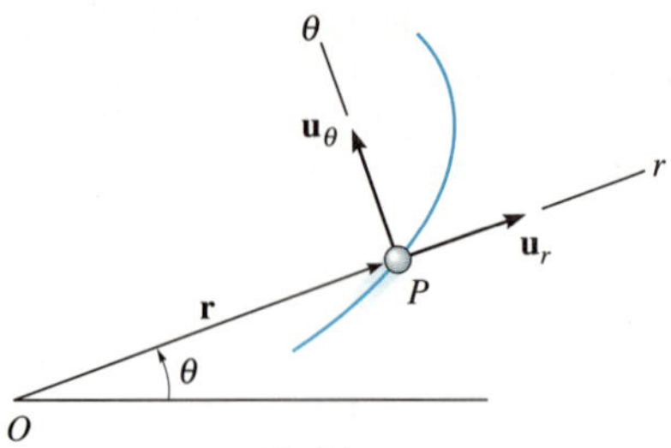

Position

(a)

Position. At any instant the position of the particle, Fig. 12–30*a*, is defined by the position vector

$$\mathbf{r} = r\mathbf{u}_r \tag{12–22}$$

Velocity. The instantaneous velocity $\mathbf{v}$ is obtained by taking the time derivative of $\mathbf{r}$. Using a dot to represent time differentiation, we have

$$\mathbf{v} = \dot{\mathbf{r}} = \dot{r}\mathbf{u}_r + r\dot{\mathbf{u}}_r$$

To evaluate $\dot{\mathbf{u}}_r$, notice that $\mathbf{u}_r$ changes only its direction with respect to time, since by definition the magnitude of this vector is always unity. Hence, during the time Δt, a change Δr will not cause a change in the direction of $\mathbf{u}_r$; however, a change $\Delta\theta$ will cause $\mathbf{u}_r$ to become $\mathbf{u}'_r$, where $\mathbf{u}'_r = \mathbf{u}_r + \Delta\mathbf{u}_r$, Fig. 12–30*b*. The time change in $\mathbf{u}_r$ is then $\Delta\mathbf{u}_r$. For small angles $\Delta\theta$ this vector has a magnitude $\Delta u_r \approx 1(\Delta\theta)$ and acts in the $\mathbf{u}_\theta$ direction. Therefore, $\Delta\mathbf{u}_r = \Delta\theta\mathbf{u}_\theta$, and so

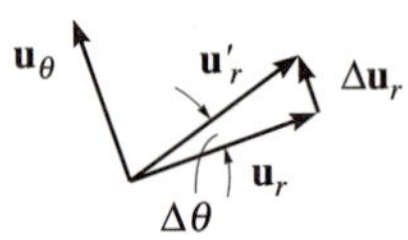

(b)

$$\dot{\mathbf{u}}_r = \lim_{\Delta t \to 0} \frac{\Delta\mathbf{u}_r}{\Delta t} = \left(\lim_{\Delta t \to 0} \frac{\Delta\theta}{\Delta t}\right)\mathbf{u}_\theta$$

$$\dot{\mathbf{u}}_r = \dot{\theta}\mathbf{u}_\theta \tag{12–23}$$

Substituting into the above equation for $\mathbf{v}$, the velocity can be written in component form as

$$\mathbf{v} = v_r\mathbf{u}_r + v_\theta\mathbf{u}_\theta \tag{12–24}$$

where

$$\begin{aligned} v_r &= \dot{r} \\ v_\theta &= r\dot{\theta} \end{aligned} \tag{12–25}$$

These components are shown graphically in Fig. 12–30*c*. The *radial component* $\mathbf{v}_r$ is a measure of the rate of increase or decrease in the length of the radial coordinate, i.e., $\dot{r}$; whereas the *transverse component* $\mathbf{v}_\theta$ can be interpreted as the rate of motion along the circumference of a circle having a radius r. In particular, the term $\dot{\theta} = d\theta/dt$ is called the *angular velocity*, since it indicates the time rate of change of the angle θ. Common units used for this measurement are rad/s.

Since $\mathbf{v}_r$ and $\mathbf{v}_\theta$ are mutually perpendicular, the *magnitude* of velocity or speed is simply the positive value of

$$v = \sqrt{(\dot{r})^2 + (r\dot{\theta})^2} \tag{12–26}$$

and the *direction* of $\mathbf{v}$ is, of course, tangent to the path at P, Fig. 12–30*c*.

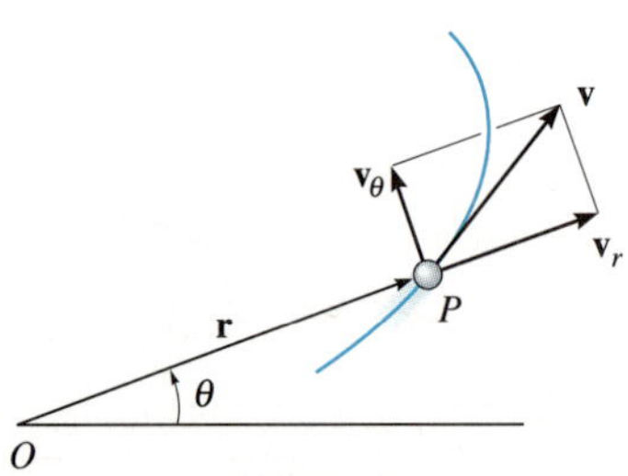

Velocity

(c)

Fig. 12–30

Acceleration. Taking the time derivatives of Eq. 12–24, using Eqs. 12–25, we obtain the particle's instantaneous acceleration,

$$\mathbf{a} = \dot{\mathbf{v}} = \ddot{r}\mathbf{u}_r + \dot{r}\dot{\mathbf{u}}_r + \dot{r}\dot{\theta}\mathbf{u}_\theta + r\ddot{\theta}\mathbf{u}_\theta + r\dot{\theta}\dot{\mathbf{u}}_\theta$$

To evaluate the term involving $\dot{\mathbf{u}}_\theta$, it is necessary only to find the change made in the direction of $\mathbf{u}_\theta$ since its magnitude is always unity. During the time Δt, a change Δr will not change the direction of $\mathbf{u}_\theta$, although a change $\Delta\theta$ will cause $\mathbf{u}_\theta$ to become $\mathbf{u}'_\theta$, where $\mathbf{u}'_\theta = \mathbf{u}_\theta + \Delta\mathbf{u}_\theta$, Fig. 12–30*d*. The time change in $\mathbf{u}_\theta$ is thus $\Delta\mathbf{u}_\theta$. For small angles this vector has a magnitude $\Delta u_\theta \approx 1(\Delta\theta)$ and acts in the $-\mathbf{u}_r$ direction; i.e., $\Delta\mathbf{u}_\theta = -\Delta\theta\mathbf{u}_r$. Thus,

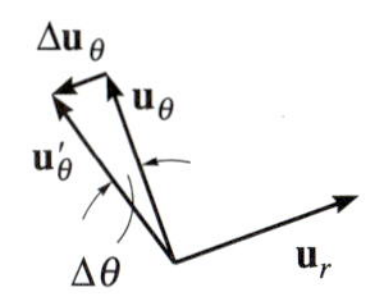

(d)

$$\dot{\mathbf{u}}_\theta = \lim_{\Delta t \to 0} \frac{\Delta\mathbf{u}_\theta}{\Delta t} = -\left(\lim_{\Delta t \to 0} \frac{\Delta\theta}{\Delta t}\right)\mathbf{u}_r$$

$$\dot{\mathbf{u}}_\theta = -\dot{\theta}\mathbf{u}_r \qquad (12\text{–}27)$$

Substituting this result and Eq. 12–23 into the above equation for **a**, we can write the acceleration in component form as

$$\boxed{\mathbf{a} = a_r\mathbf{u}_r + a_\theta\mathbf{u}_\theta} \qquad (12\text{–}28)$$

where

$$\boxed{\begin{aligned} a_r &= \ddot{r} - r\dot{\theta}^2 \\ a_\theta &= r\ddot{\theta} + 2\dot{r}\dot{\theta} \end{aligned}} \qquad (12\text{–}29)$$

The term $\ddot{\theta} = d^2\theta/dt^2 = d/dt(d\theta/dt)$ is called the *angular acceleration* since it measures the change made in the angular velocity during an instant of time. Common units for this measurement are rad/s^2.

Since $\mathbf{a}_r$ and $\mathbf{a}_\theta$ are always perpendicular, the *magnitude* of acceleration is simply the positive value of

$$a = \sqrt{(\ddot{r} - r\dot{\theta}^2)^2 + (r\ddot{\theta} + 2\dot{r}\dot{\theta})^2} \qquad (12\text{–}30)$$

The *direction* is determined from the vector addition of its two components. In general, **a** will *not* be tangent to the path, Fig. 12–30*e*.

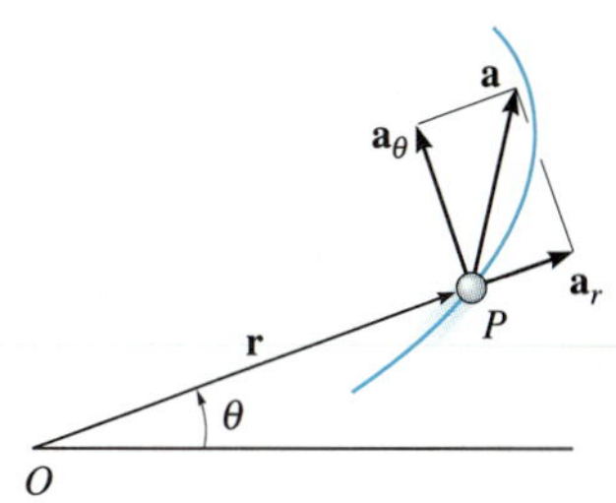

Acceleration

(e)

Fig. 12–30

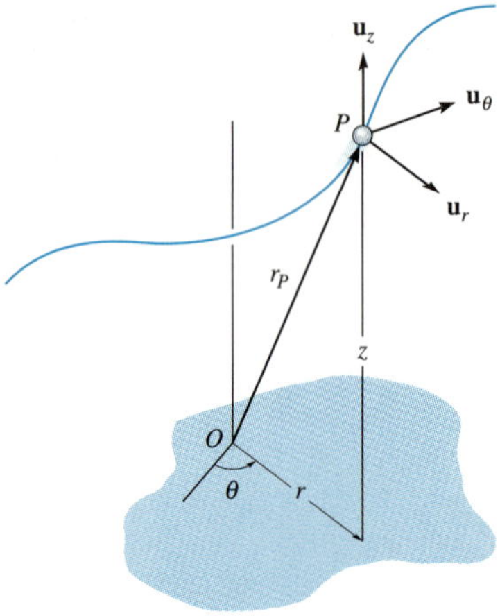

Fig. 12–31

Cylindrical Coordinates. If the particle P moves along a space curve as shown in Fig. 12–31, then its location may be specified by the three *cylindrical coordinates, r, θ, z*. The z coordinate is identical to that used for rectangular coordinates. Since the unit vector defining its direction, $\mathbf{u}_z$, is constant, the time derivatives of this vector are zero, and therefore the position, velocity, and acceleration of the particle can be written in terms of its cylindrical coordinates as follows:

$$\mathbf{r}_P = r\mathbf{u}_r + z\mathbf{u}_z$$

$$\mathbf{v} = \dot{r}\mathbf{u}_r + r\dot{\theta}\mathbf{u}_\theta + \dot{z}\mathbf{u}_z \qquad (12\text{–}31)$$

$$\mathbf{a} = (\ddot{r} - r\dot{\theta}^2)\mathbf{u}_r + (r\ddot{\theta} + 2\dot{r}\dot{\theta})\mathbf{u}_\theta + \ddot{z}\mathbf{u}_z \qquad (12\text{–}32)$$

Time Derivatives. The equations of kinematics require that we obtain the time derivatives $\dot{r}, \ddot{r}, \dot{\theta}$, and $\ddot{\theta}$ in order to evaluate the r and θ components of $\mathbf{v}$ and $\mathbf{a}$. Two types of problems generally occur:

The spiral motion of this boy can be followed by using cylindrical components. Here the radial coordinate r is constant, the transverse coordinate θ will increase with time as the boy rotates about the vertical, and his altitude z will decrease with time.

1. If the coordinates are specified as time parametric equations, $r = r(t)$ and $\theta = \theta(t)$, then the time derivatives can be found directly. For example, consider

$$r = 4t^2 \qquad \theta = (8t^3 + 6)$$
$$\dot{r} = 8t \qquad \dot{\theta} = 24t^2$$
$$\ddot{r} = 8 \qquad \ddot{\theta} = 48t$$

2. If the time-parametric equations are not given, then it will be necessary to specify the path $r = f(\theta)$ and find the *relationship* between the time derivatives using the chain rule of calculus. Consider the following examples.

$$r = 5\theta^2$$
$$\dot{r} = 10\theta\dot{\theta}$$
$$\ddot{r} = 10[(\dot{\theta})\dot{\theta} + \theta(\ddot{\theta})]$$
$$= 10\dot{\theta}^2 + 10\theta\ddot{\theta}$$

or

$$r^2 = 6\theta^3$$
$$2r\dot{r} = 18\theta^2\dot{\theta}$$
$$2[(\dot{r})\dot{r} + r(\ddot{r})] = 18[(2\theta\dot{\theta})\dot{\theta} + \theta^2(\ddot{\theta})]$$
$$\dot{r}^2 + r\ddot{r} = 9(2\theta\dot{\theta}^2 + \theta^2\ddot{\theta})$$

If two of the *four* time derivatives $\dot{r}$, $\ddot{r}$, $\dot{\theta}$, and $\ddot{\theta}$ are *known,* then the other two can be obtained from the equations for first and second time derivatives of $r = f(\theta)$. See Example 12–19. In some problems, however, two of these time derivatives may *not* be known; instead the magnitude of the particle's velocity or acceleration may be specified. If this is the case, Eqs. 12–26 and 12–30 [$v^2 = \dot{r}^2 + (r\dot{\theta})^2$ and $a^2 = (\ddot{r} - r\dot{\theta}^2)^2 + (r\ddot{\theta} + 2\dot{r}\dot{\theta})^2$] may be used to obtain the necessary relationships involving $\dot{r}$, $\ddot{r}$, $\dot{\theta}$, and $\ddot{\theta}$. See Example 12–20.

Procedure for Analysis

Coordinate System

- Polar coordinates are a suitable choice for solving problems for which data regarding the angular motion of the radial coordinate r is given to describe the particle's motion. Also, some paths of motion can conveniently be described in terms of these coordinates.
- To use polar coordinates, the origin is established at a fixed point, and the radial line r is directed to the particle.
- The transverse coordinate θ is measured counterclockwise from a fixed reference line to the radial line.

Velocity and Acceleration

- Once r and the four time derivatives $\dot{r}$, $\ddot{r}$, $\dot{\theta}$, and $\ddot{\theta}$ have been evaluated at the instant considered, their values can be substituted into Eqs. 12–25 and 12–29 to obtain the radial and transverse components of **v** and **a.**
- If it is necessary to take the time derivatives of $r = f(\theta)$, it is very important to use the chain rule of calculus.
- Motion in three dimensions requires a simple extension of the above procedure to include $\dot{z}$ and $\ddot{z}$.

Besides the examples which follow, further examples involving the calculation of a_r and a_θ can be found in the "kinematics" sections of Examples 13–10 through 13–12.

EXAMPLE 12–17

The amusement park ride shown in Fig. 12–32*a* consists of a chair that is rotating in a horizontal circular path of radius r such that the arm OB has an angular velocity $\dot{\theta}$ and angular acceleration $\ddot{\theta}$. Determine the radial and transverse components of velocity and acceleration of the passenger. Neglect his size in the calculation.

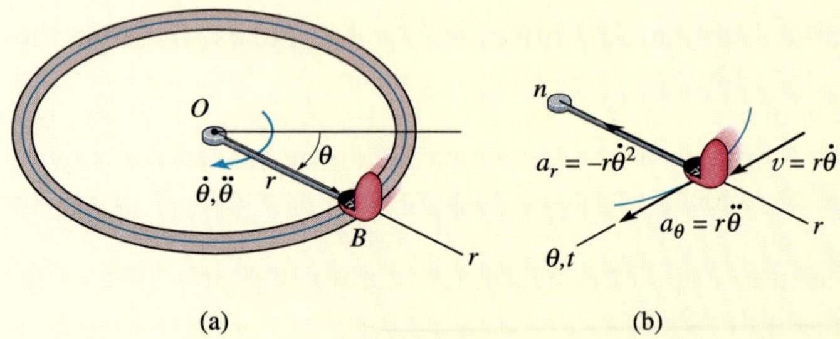

Fig. 12–32

Solution

Coordinate System. Since the angular motion of the arm is reported, polar coordinates are chosen for the solution, Fig. 12–32*a*. Here θ is not related to r, since the radius is constant for all θ.

Velocity and Acceleration. Equations 12–25 and 12–29 will be used for the solution, and so it is first necessary to specify the first and second time derivatives of r and θ. Since r is *constant,* we have

$$r = r \qquad \dot{r} = 0 \qquad \ddot{r} = 0$$

Thus

$$v_r = \dot{r} = 0 \qquad \textit{Ans.}$$

$$v_\theta = r\dot{\theta} \qquad \textit{Ans.}$$

$$a_r = \ddot{r} - r\dot{\theta}^2 = -r\dot{\theta}^2 \qquad \textit{Ans.}$$

$$a_\theta = r\ddot{\theta} + 2\dot{r}\dot{\theta} = r\ddot{\theta} \qquad \textit{Ans.}$$

These results are shown in Fig. 12–32*b*. Also shown are the n, t axes, which in this special case of circular motion happen to be *colinear* with the r and θ axes, respectively. In particular note that $v = v_\theta = v_t = r\dot{\theta}$. Also,

$$-a_r = a_n = \frac{v^2}{\rho} = \frac{(r\dot{\theta})^2}{r} = r\dot{\theta}^2$$

$$a_\theta = a_t = \frac{dv}{dt} = \frac{d}{dt}(r\dot{\theta}) = \frac{dr}{dt}\dot{\theta} + r\frac{d\dot{\theta}}{dt} = 0 + r\ddot{\theta}$$

EXAMPLE 12–18

The rod OA in Fig. 12–33a is rotating in the horizontal plane such that $\theta = (t^3)$ rad. At the same time, the collar B is sliding outward along OA so that $r = (100t^2)$ mm. If in both cases t is in seconds, determine the velocity and acceleration of the collar when $t = 1$ s.

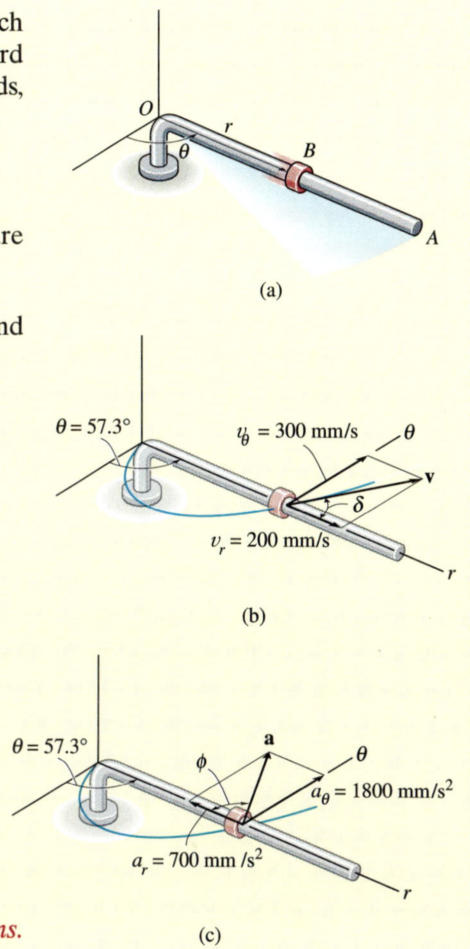

Fig. 12–33

Solution

Coordinate System. Since time-parametric equations of the path are given, it is not necessary to relate r to θ.

Velocity and Acceleration. Determining the time derivatives and evaluating when $t = 1$ s, we have

$$r = 100t^2\Big|_{t=1\,\text{s}} = 100 \text{ mm} \qquad \theta = t^3\Big|_{t=1\,\text{s}} = 1 \text{ rad} = 57.3^\circ$$

$$\dot{r} = 200t\Big|_{t=1\,\text{s}} = 200 \text{ mm/s} \qquad \dot{\theta} = 3t^2\Big|_{t=1\,\text{s}} = 3 \text{ rad/s}$$

$$\ddot{r} = 200\Big|_{t=1\,\text{s}} = 200 \text{ mm/s}^2 \qquad \ddot{\theta} = 6t\Big|_{t=1\,\text{s}} = 6 \text{ rad/s}^2$$

As shown in Fig. 12–33b,

$$\begin{aligned}\mathbf{v} &= \dot{r}\mathbf{u}_r + r\dot{\theta}\mathbf{u}_\theta \\ &= 200\mathbf{u}_r + 100(3)\mathbf{u}_\theta \\ &= \{200\mathbf{u}_r + 300\mathbf{u}_\theta\} \text{ mm/s}\end{aligned}$$

The magnitude of **v** is

$$v = \sqrt{(200)^2 + (300)^2} = 361 \text{ mm/s} \qquad \textit{Ans.}$$

$$\delta = \tan^{-1}\left(\frac{300}{200}\right) = 56.3^\circ \qquad \delta + 57.3^\circ = 114^\circ \qquad \textit{Ans.}$$

As shown in Fig. 12–33c,

$$\begin{aligned}\mathbf{a} &= (\ddot{r} - r\dot{\theta}^2)\mathbf{u}_r + (r\ddot{\theta} + 2\dot{r}\dot{\theta})\mathbf{u}_\theta \\ &= [200 - 100(3)^2]\mathbf{u}_r + [100(6) + 2(200)3]\mathbf{u}_\theta \\ &= \{-700\mathbf{u}_r + 1800\mathbf{u}_\theta\} \text{ mm/s}^2\end{aligned}$$

The magnitude of **a** is

$$a = \sqrt{(700)^2 + (1800)^2} = 1930 \text{ mm/s}^2 \qquad \textit{Ans.}$$

$$\phi = \tan^{-1}\left(\frac{1800}{700}\right) = 68.7^\circ \qquad (180^\circ - \phi) + 57.3^\circ = 169^\circ \qquad \textit{Ans.}$$

EXAMPLE 12–19

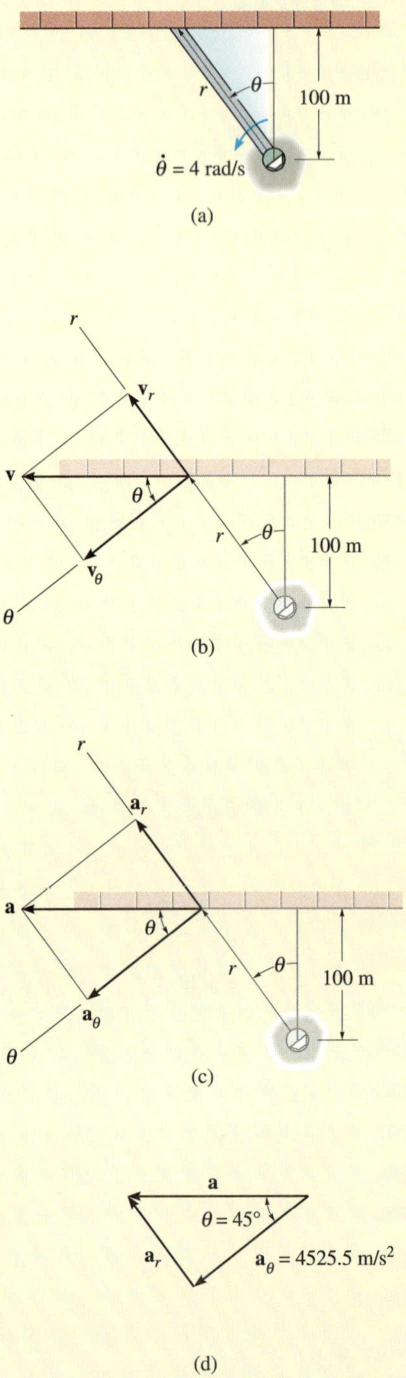

Fig. 12–34

The searchlight in Fig. 12–34*a* casts a spot of light along the face of a wall that is located 100 m from the searchlight. Determine the magnitudes of the velocity and acceleration at which the spot appears to travel across the wall at the instant $\theta = 45°$. The searchlight is rotating at a constant rate of $\dot{\theta} = 4$ rad/s.

Solution

Coordinate System. Polar coordinates will be used to solve this problem since the angular rate of the searchlight is given. To find the necessary time derivatives it is first necessary to relate r to θ. From Fig. 12–34*a*, this relation is

$$r = 100/\cos\theta = 100\sec\theta$$

Velocity and Acceleration. Using the chain rule of calculus, noting that $d(\sec\theta) = \sec\theta\tan\theta\, d\theta$, and $d(\tan\theta) = \sec^2\theta\, d\theta$, we have

$$\dot{r} = 100(\sec\theta\tan\theta)\dot{\theta}$$

$$\ddot{r} = 100(\sec\theta\tan\theta)\dot{\theta}(\tan\theta)\dot{\theta} + 100\sec\theta(\sec^2\theta)\dot{\theta}(\dot{\theta}) + 100\sec\theta\tan\theta(\ddot{\theta})$$

$$= 100\sec\theta\tan^2\theta(\dot{\theta})^2 + 100\sec^3\theta(\dot{\theta})^2 + 100(\sec\theta\tan\theta)\ddot{\theta}$$

Since $\dot{\theta} = 4$ rad/s = constant, then $\ddot{\theta} = 0$, and the above equations, when $\theta = 45°$, become

$$r = 100\sec 45° = 141.4$$

$$\dot{r} = 400\sec 45°\tan 45° = 565.7$$

$$\ddot{r} = 1600(\sec 45°\tan^2 45° + \sec^3 45°) = 6788.2$$

As shown in Fig. 12–34*b*,

$$\mathbf{v} = \dot{r}\mathbf{u}_r + r\dot{\theta}\mathbf{u}_\theta$$

$$= 565.7\mathbf{u}_r + 141.4(4)\mathbf{u}_\theta$$

$$= \{565.7\mathbf{u}_r + 565.7\mathbf{u}_\theta\}\text{ m/s}$$

$$v = \sqrt{v_r^2 + v_\theta^2} = \sqrt{(565.7)^2 + (565.7)^2}$$

$$= 800\text{ m/s} \qquad \textit{Ans.}$$

As shown in Fig. 12–34*c*,

$$\mathbf{a} = (\ddot{r} - r\dot{\theta}^2)\mathbf{u}_r + (r\ddot{\theta} + 2\dot{r}\dot{\theta})\mathbf{u}_\theta$$

$$= [6788.2 - 141.4(4)^2]\mathbf{u}_r + [141.4(0) + 2(565.7)4]\mathbf{u}_\theta$$

$$= \{4525.5\mathbf{u}_r + 4525.5\mathbf{u}_\theta\}\text{ m/s}^2$$

$$a = \sqrt{a_r^2 + a_\theta^2} = \sqrt{(4525.5)^2 + (4525.5)^2}$$

$$= 6400\text{ m/s}^2 \qquad \textit{Ans.}$$

Note: It is also possible to find a without having to calculate $\ddot{r}$ (or a_r). As shown in Fig. 12–34*d*, since $a_\theta = 4525.5$ m/s², then by vector resolution, $a = 4525.5/\cos 45° = 6400$ m/s².

EXAMPLE 12-20

Due to the rotation of the forked rod, the cylindrical peg A in Fig. 12–35a travels around the slotted path, a portion of which is in the shape of a cardioid, $r = 0.5(1 - \cos\theta)$ ft, where θ is in radians. If the peg's velocity is $v = 4$ ft/s and its acceleration is $a = 30$ ft/s^2 at the instant $\theta = 180°$, determine the angular velocity $\dot{\theta}$ and angular acceleration $\ddot{\theta}$ of the fork.

Fig. 12–35

Solution

Coordinate System. This path is most unusual, and mathematically it is best expressed using polar coordinates, as done here, rather than rectangular coordinates. Also, $\dot{\theta}$ and $\ddot{\theta}$ must be determined so r, θ coordinates are an obvious choice.

Velocity and Acceleration. Determining the time derivatives of r using the chain rule of calculus yields

$$r = 0.5(1 - \cos\theta)$$
$$\dot{r} = 0.5(\sin\theta)\dot{\theta}$$
$$\ddot{r} = 0.5(\cos\theta)\dot{\theta}(\dot{\theta}) + 0.5(\sin\theta)\ddot{\theta}$$

Evaluating these results at $\theta = 180°$, we have

$$r = 1 \text{ ft} \qquad \dot{r} = 0 \qquad \ddot{r} = -0.5\dot{\theta}^2$$

Since $v = 4$ ft/s, using Eq. 12–26 to determine $\dot{\theta}$ yields

$$v = \sqrt{(\dot{r})^2 + (r\dot{\theta})^2}$$
$$4 = \sqrt{(0)^2 + (1\dot{\theta})^2}$$
$$\dot{\theta} = 4 \text{ rad/s} \qquad \textit{Ans.}$$

In a similar manner, $\ddot{\theta}$ can be found using Eq. 12–30.

$$a = \sqrt{(\ddot{r} - r\dot{\theta}^2)^2 + (r\ddot{\theta} + 2\dot{r}\dot{\theta})^2}$$
$$30 = \sqrt{[-0.5(4)^2 - 1(4)^2]^2 + [1\ddot{\theta} + 2(0)(4)]^2}$$
$$(30)^2 = (-24)^2 + \ddot{\theta}^2$$
$$\ddot{\theta} = 18 \text{ rad/s}^2 \qquad \textit{Ans.}$$

Vectors **a** and **v** are shown in Fig. 12–35b.

Problems

*12-136. The time rate of change of acceleration is referred to as the *jerk*, which is often used as a means of measuring passenger discomfort. Calculate this vector, $\dot{\mathbf{a}}$, in terms of its cylindrical components, using Eq. 12–32.

12-137. An airplane is flying in a straight line with a velocity of 200 mi/h and an acceleration of 3 mi/h^2. If the propeller has a diameter of 6 ft and is rotating at an angular rate of 120 rad/s, determine the magnitudes of velocity and acceleration of a particle located on the tip of the propeller.

12-138. A particle is moving along a circular path having a radius of 4 in. such that its position as a function of time is given by $\theta = \cos 2t$, where θ is in radians and t is in seconds. Determine the magnitude of the acceleration of the particle when $\theta = 30°$.

12-139. A train is traveling along the circular curve of radius $r = 600$ ft. At the instant shown, its angular rate of rotation is $\dot{\theta} = 0.02$ rad/s, which is decreasing at $\ddot{\theta} = -0.001 \text{ rad/s}^2$. Determine the magnitudes of the train's velocity and acceleration at this instant.

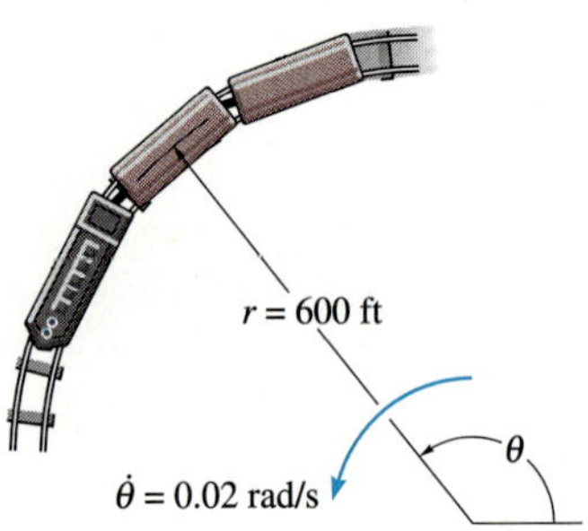

Prob. 12–139

*12-140. If a particle moves along a path such that $r = (2 \cos t)$ ft and $\theta = (t/2)$ rad, where t is in seconds, plot the path $r = f(\theta)$ and determine the particle's radial and transverse components of velocity and acceleration.

12-141. If a particle moves along a path such that $r = (2 \sin t^2)$ m and $\theta = t^2$ rad, where t is in seconds, plot the path $r = f(\theta)$ and determine the particle's radial and transverse components of velocity and acceleration as functions of time.

12-142. A particle is moving along a circular path having a 400-mm radius. Its position as a function of time is given by $\theta = (2t^2)$ rad, where t is in seconds. Determine the magnitude of the particle's acceleration when $\theta = 30°$. The particle starts from rest when $\theta = 0°$.

12-143. If a particle moves along a path such that $r = (e^{at})$ m and $\theta = t$, where t is in seconds, plot the path $r = f(\theta)$, and determine the particle's radial and transverse components of velocity and acceleration.

*12-144. A car is traveling along the circular curve having a radius $r = 400$ ft. At the instant shown, its angular rate of rotation is $\dot{\theta} = 0.025$ rad/s, which is decreasing at the rate $\ddot{\theta} = -0.008 \text{ rad/s}^2$. Determine the radial and transverse components of the car's velocity and acceleration at this instant and sketch these components on the curve.

12-145. A car is traveling along the circular curve of radius $r = 400$ ft with a constant speed of $v = 30$ ft/s. Determine the angular rate of rotation $\dot{\theta}$ of the radial line r and the magnitude of the car's acceleration.

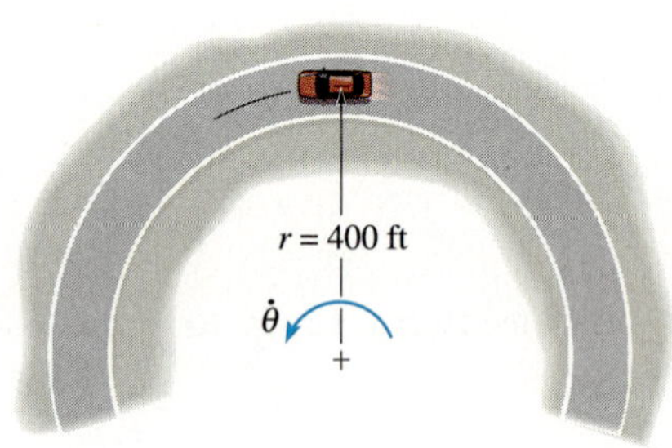

Probs. 12–144/145

12-146. A particle is moving along a circular path having a radius of 6 in. such that its position as a function of time is given by $\theta = \sin 3t$, where θ is in radians, the argument for the sine is in degrees, and t is in seconds. Determine the acceleration of the particle at $\theta = 30°$. The particle starts from rest at $\theta = 0°$.

12-147. A particle travels around a lituus, defined by the equation $r^2\theta = a^2$, where a is a constant. Determine the particle's radial and transverse components of velocity and acceleration as a function of θ and its time derivatives.

*12-148. A particle travels around a limaçon, defined by the equation $r = b - a\cos\theta$, where a and b are constants. Determine the particle's radial and transverse components of velocity and acceleration as a function of θ and its time derivatives.

12-149. A particle travels along a portion of the "four-leaf rose" defined by the equation $r = (5\cos 2\theta)$ m. If the angular velocity of the radial coordinate line is $\dot{\theta} = (3t^2)$ rad/s, where t is in seconds, determine the radial and transverse components of the particle's velocity and acceleration at the instant $\theta = 30°$. When $t = 0$, $\theta = 0°$.

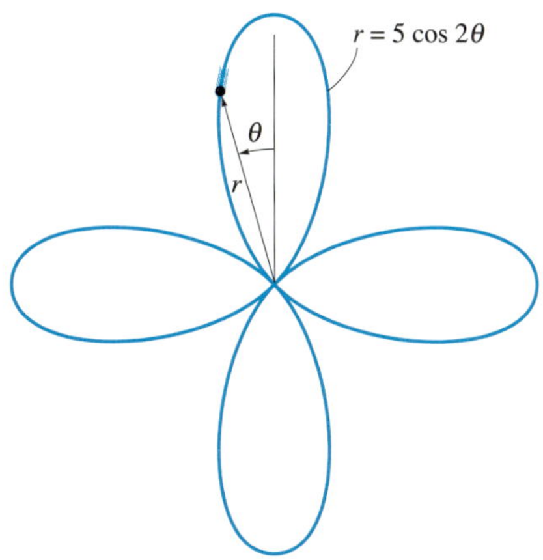

Prob. 12–149

12-150. A block moves outward along the slot in the platform with a speed of $\dot{r} = (4t)$ m/s, where t is in seconds. The platform rotates at a constant rate of 6 rad/s. If the block starts from rest at the center, determine the magnitudes of its velocity and acceleration when $t = 1$ s.

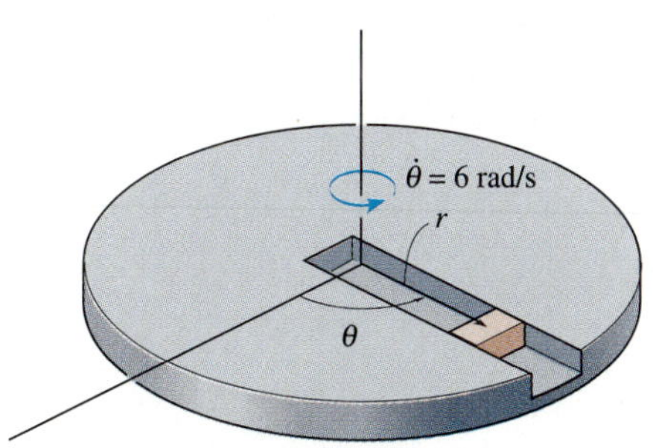

Prob. 12–150

12-151. The small washer is sliding down the cord OA. When it is at the midpoint, its speed is 200 mm/s and its acceleration is 10 mm/s^2. Express the velocity and acceleration of the washer at this point in terms of its cylindrical components.

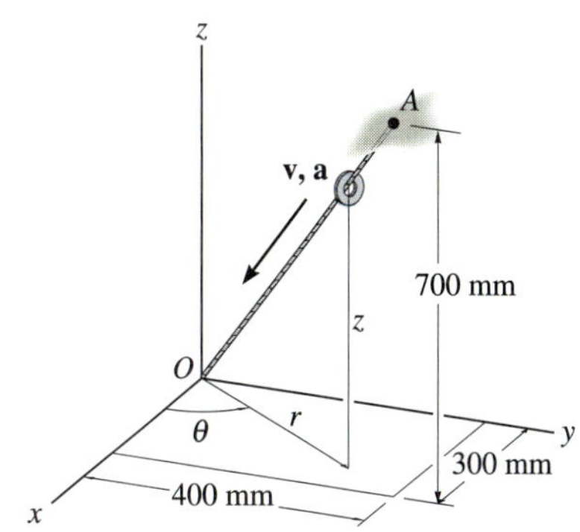

Prob. 12–151

*12-152. The airplane on the amusement park ride moves along a path defined by the equations $r = 4$ m, $\theta = (0.2t)$ rad, and $z = (0.5\cos\theta)$ m, where t is in seconds. Determine the cylindrical components of the velocity and acceleration of the airplane when $t = 6$ s.

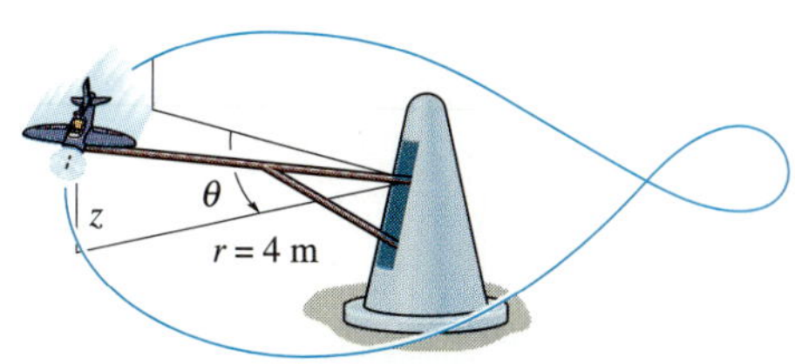

Prob. 12–152

12-153. The automobile is traveling from a parking deck down along a cylindrical spiral ramp at a constant speed of $v = 1.5$ m/s. If the ramp descends a distance of 12 m for every full revolution, $\theta = 2\pi$ rad, determine the magnitude of the car's acceleration as it moves along the ramp, $r = 10$ m. *Hint:* For part of the solution, note that the tangent to the ramp at any point is at an angle of $\phi = \tan^{-1}(12/[2\pi(10)]) = 10.81°$ from the horizontal. Use this to determine the velocity components v_θ and v_z, which in turn are used to determine $\dot{\theta}$ and $\dot{z}$.

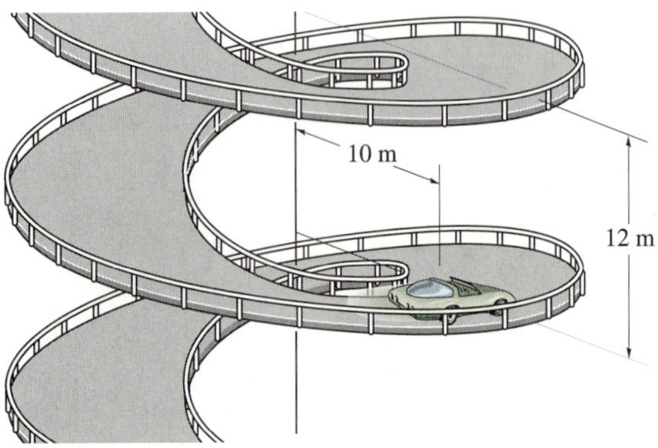

Prob. 12–153

12-154. Because of telescopic action, the end of the industrial robotic arm extends along the path of the limaçon $r = (1 + 0.5 \cos\theta)$ m. At the instant $\theta = \pi/4$, the arm has an angular rotation $\dot{\theta} = 0.6$ rad/s, which is increasing at $\ddot{\theta} = 0.25$ rad/s^2. Determine the radial and transverse components of the velocity and acceleration of the object A held in its grip at this instant.

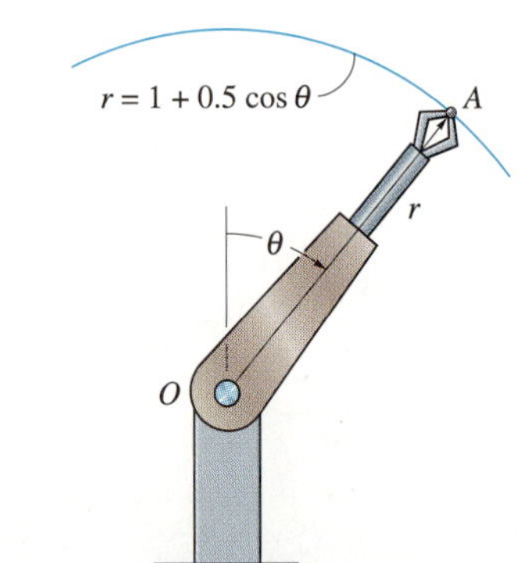

Prob. 12–154

12-155. The boy slides down the slide at a constant speed of 2 m/s. If the slide is in the form of a helix, defined by the equations $r = 1.5$ m and $z = -\theta/\pi$, determine the boy's angular velocity about the z axis, $\dot{\theta}$, and the magnitude of his acceleration.

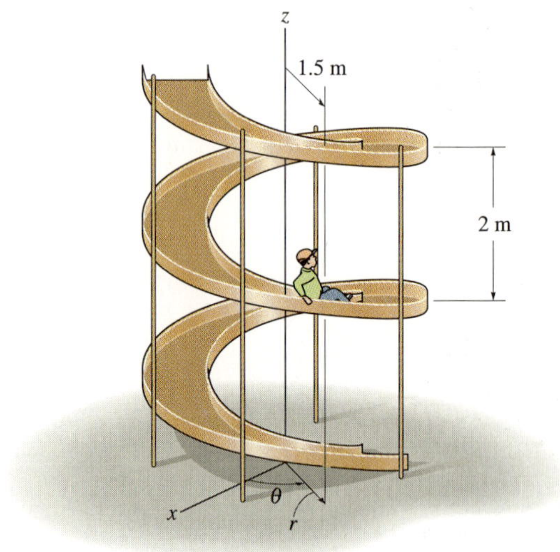

Prob. 12–155

***12-156.** The motion of particle B is controlled by the rotation of the grooved link OA. If the link is rotating at a constant angular rate of $\dot{\theta} = 6$ rad/s, determine the magnitudes of the velocity and acceleration of B at the instant $\theta = \pi/2$ rad. The spiral path is defined by the equation $r = (40\theta)$ mm, where θ is in radians.

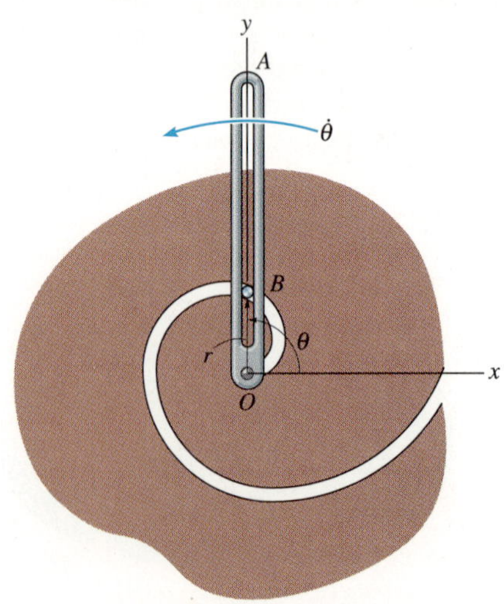

Prob. 12–156

12-157. The slotted fork is rotating about O at a constant rate of $\dot{\theta} = 3$ rad/s. Determine the radial and transverse components of the velocity and acceleration of the pin A at the instant $\theta = 360°$. The path is defined by the spiral groove $r = (5 + \theta/\pi)$ in., where θ is in radians.

12-158. The slotted fork is rotating about O at $\dot{\theta} = 3$ rad/s, which is increasing at $\ddot{\theta} = 2$ rad/s^2 when $\theta = 360°$. Determine the radial and transverse components of the velocity and acceleration of the pin A at this instant. The path is defined by the spiral groove $r = (5 + \theta/\pi)$ in., where θ is in radians.

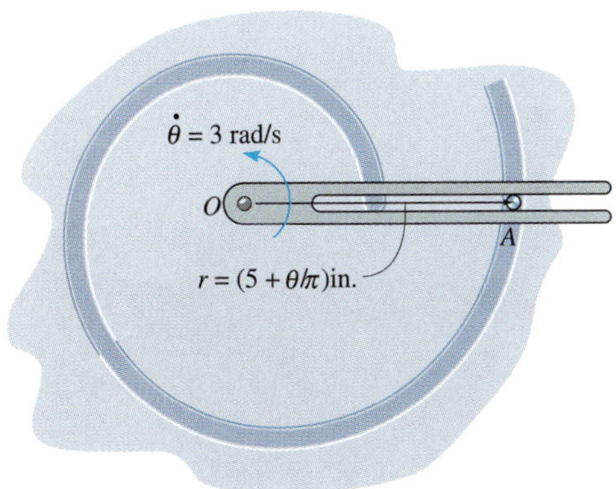

Probs. 12–157/158

12-159. The partial surface of the cam is that of a logarithmic spiral $r = (40e^{0.05\theta})$ mm, where θ is in radians. If the cam is rotating at a constant angular rate of $\dot{\theta} = 4$ rad/s, determine the magnitudes of the velocity and acceleration of the follower rod at the instant $\theta = 30°$.

***12-160.** Solve Prob. 12-159, if the cam has an angular acceleration of $\ddot{\theta} = 2$ rad/s^2 when its angular velocity is $\dot{\theta} = 4$ rad/s at $\theta = 30°$.

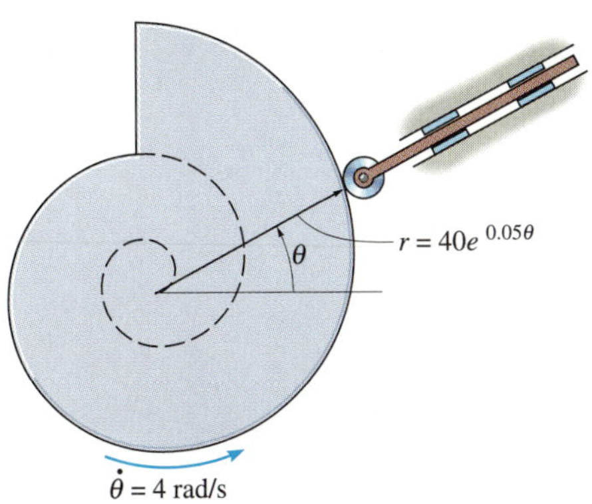

Probs. 12–159/160

12-161. The rod OA rotates counterclockwise with a constant angular velocity of $\dot{\theta} = 5$ rad/s. Two pin-connected slider blocks, located at B, move freely on OA and the curved rod whose shape is a limaçon described by the equation $r = 100(2 - \cos\theta)$ mm. Determine the speed of the slider blocks at the instant $\theta = 120°$.

12-162. Determine the magnitude of the acceleration of the slider blocks in Prob. 12-161 when $\theta = 120°$.

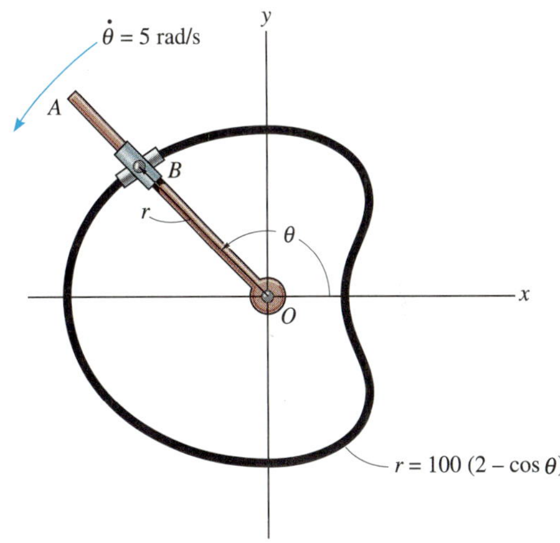

Probs. 12–161/162

12-163. A particle P moves along the spiral path $r = (10/\theta)$ ft, where θ is in radians. If it maintains a constant speed of $v = 20$ ft/s, determine the magnitudes v_r and v_θ as functions of θ and evaluate each at $\theta = 1$ rad.

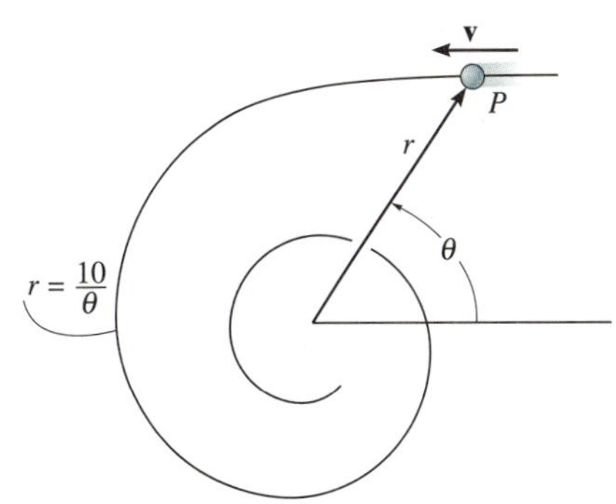

Prob. 12–163

***12-164.** A cameraman standing at A is following the movement of a race car, B, which is traveling along a straight track at a constant speed of 80 ft/s. Determine the angular rate at which he must turn in order to keep the camera directed on the car at the instant $\theta = 60°$.

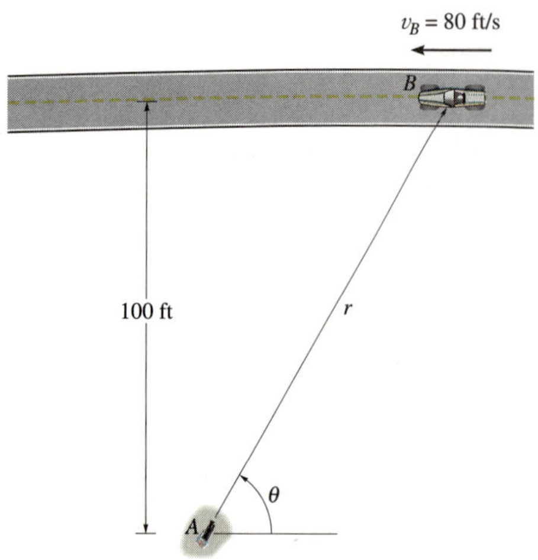

Prob. 12–164

■**12-165.** The double collar C is pin connected such that one collar slides over a fixed rod and the other slides over a rotating rod AB. If the angular velocity of AB is given as $\dot{\theta} = (e^{0.5t^2})$ rad/s, where t is in seconds, and the path defined by the fixed rod is $r = |0.4 \sin \theta + 0.2|$ m, determine the radial and transverse components of the collar's velocity and acceleration when $t = 1$ s. When $t = 0$, $\theta = 0°$. Use Simpson's rule to determine θ at $t = 1$ s.

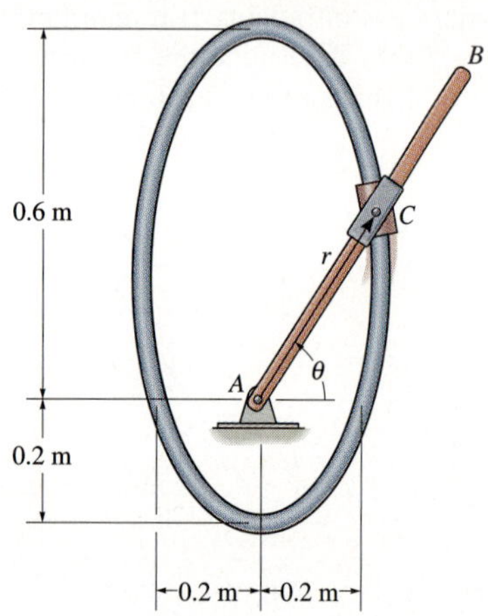

Prob. 12–165

12-166. For a short time the bucket of the backhoe traces the path of the cardioid $r = 25(1 - \cos \theta)$ ft. Determine the magnitudes of the velocity and acceleration of the bucket when $\theta = 120°$ if the boom is rotating with an angular velocity of $\dot{\theta} = 2$ rad/s and an angular acceleration of $\ddot{\theta} = 0.2$ rad/s^2 at the instant shown.

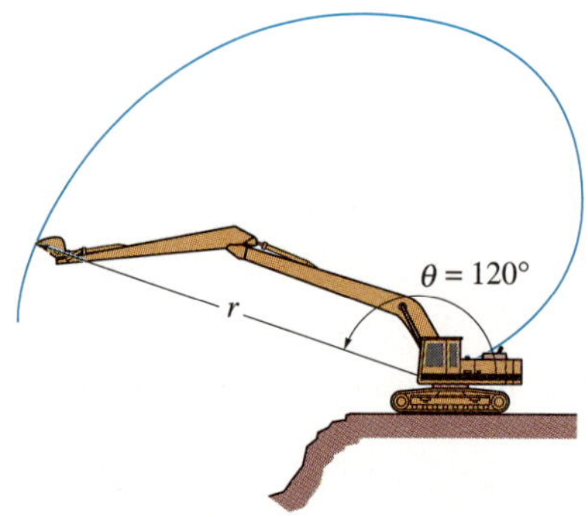

Prob. 12–166

12-167. The car travels along a road which for a short distance is defined by $r = (200/\theta)$ ft, where θ is in radians. If it maintains a constant speed of $v = 35$ ft/s, determine the radial and transverse components of its velocity when $\theta = \pi/3$ rad.

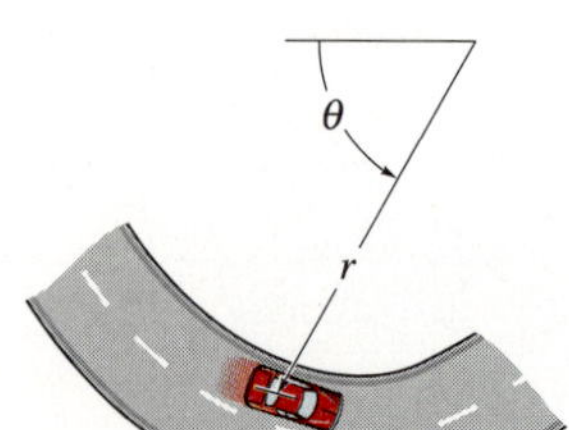

Prob. 12–167

***12-168.** The pin follows the path described by the equation $r = (0.2 + 0.15 \cos \theta)$ m. At the instant $\theta = 30°$, $\dot{\theta} = 0.7$ rad/s and $\ddot{\theta} = 0.5$ rad/s^2. Determine the magnitudes of the pin's velocity and acceleration at this instant. Neglect the size of the pin.

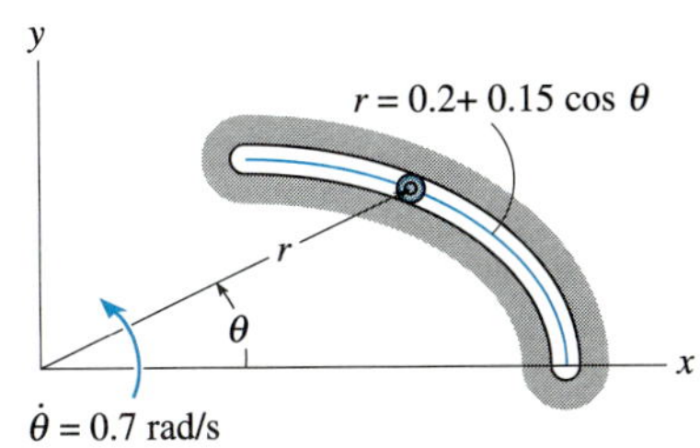

Prob. 12–168

12-169. The mechanism of a machine is constructed so that for a short time the roller at A follows the surface of the cam described by the equation $r = (0.3 + 0.2 \cos \theta)$ m. If $\dot{\theta} = 0.5$ rad/s and $\ddot{\theta} = 0$, determine the magnitudes of the roller's velocity and acceleration at the instant $\theta = 30°$. Neglect the size of the roller. Also determine the velocity components $(\mathbf{v}_A)_x$ and $(\mathbf{v}_A)_y$ of the roller at this instant. The rod to which the roller is attached remains vertical and can slide up or down along the guides while the guides move horizontally to the left.

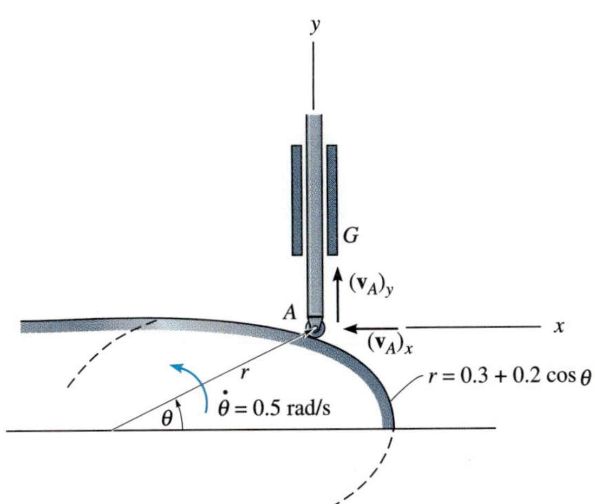

Prob. 12–169

12-170. The crate slides down the section of the spiral ramp such that $r = (0.5z)$ ft and $z = (100 - 0.1t^2)$ ft, where t is in seconds. If the rate of rotation about the z axis is $\dot{\theta} = 0.04\pi t$ rad/s, determine the magnitudes of the velocity and acceleration of the crate at the instant $z = 10$ ft.

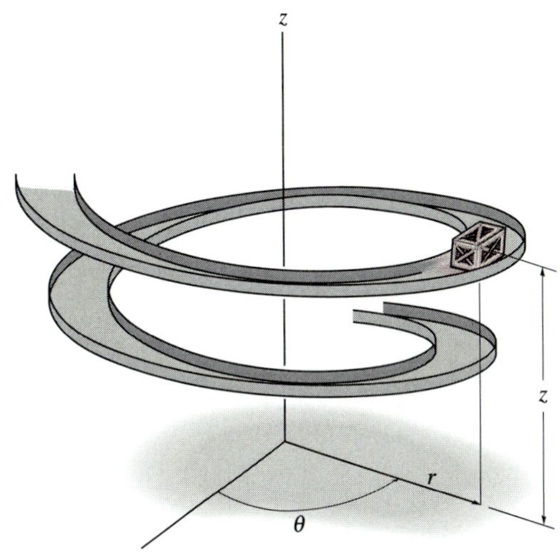

Prob. 12–170

12-171. A double collar C is pin connected together such that one collar slides over a fixed rod and the other slides over a rotating rod. If the geometry of the fixed rod for a short distance can be defined by a lemniscate, $r^2 = (4 \cos 2\theta)$ ft^2, determine the collar's radial and transverse components of velocity and acceleration at the instant $\theta = 0°$ as shown. Rod OA is rotating at a constant rate of $\dot{\theta} = 6$ rad/s.

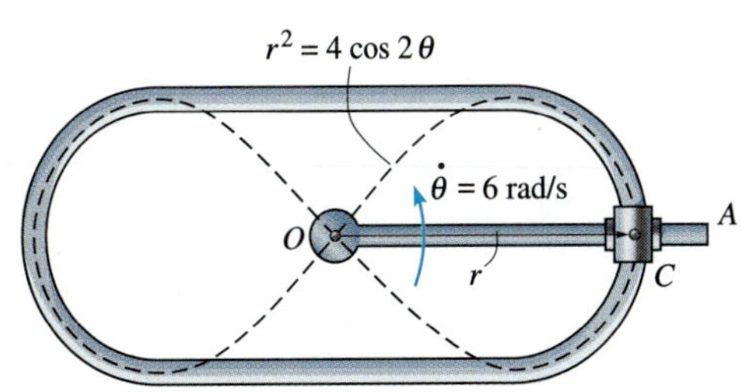

Prob. 12–171

12.9 Absolute Dependent Motion Analysis of Two Particles

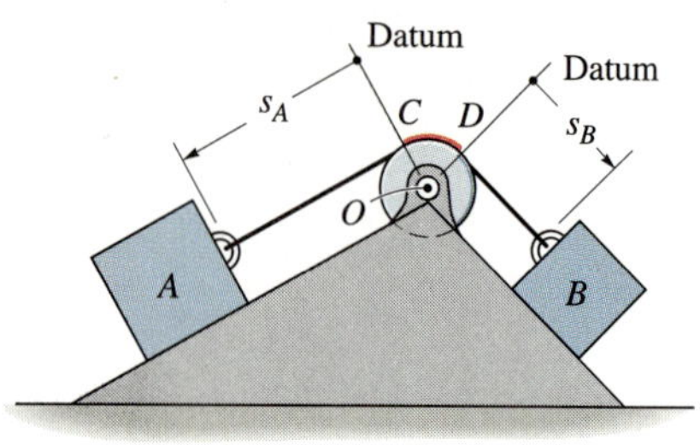

Fig. 12–36

In some types of problems the motion of one particle will *depend* on the corresponding motion of another particle. This dependency commonly occurs if the particles are interconnected by inextensible cords which are wrapped around pulleys. For example, the movement of block A downward along the inclined plane in Fig. 12–36 will cause a corresponding movement of block B up the other incline. We can show this mathematically by first specifying the location of the blocks using *position coordinates* s_A and s_B. Note that each of the coordinate axes is (1) referenced from a *fixed* point (O) or *fixed* datum line, (2) measured along each inclined plane in the direction of motion of block A and block B, and (3) has a positive sense from C to A and D to B. If the total cord length is l_T, the position coordinates are related by the equation

$$s_A + l_{CD} + s_B = l_T$$

Here l_{CD} is the length of the cord passing over arc CD. Taking the time derivative of this expression, realizing that l_{CD} and l_T remain constant, while s_A and s_B measure the lengths of the changing segments of the cord, we have

$$\frac{ds_A}{dt} + \frac{ds_B}{dt} = 0 \qquad \text{or} \qquad v_B = -v_A$$

The negative sign indicates that when block A has a velocity downward, i.e., in the direction of positive s_A, it causes a corresponding upward velocity of block B; i.e., B moves in the negative s_B direction.

In a similar manner, time differentiation of the velocities yields the relation between the accelerations, i.e.,

$$a_B = -a_A$$

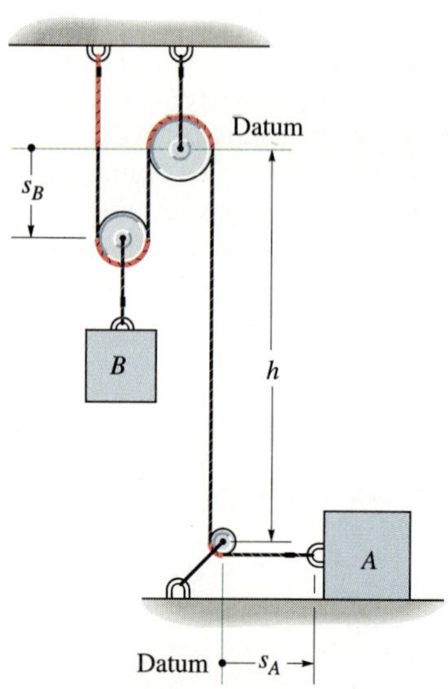

(a)

Fig. 12–37

A more complicated example involving dependent motion of two blocks is shown in Fig. 12–37*a*. In this case, the position of block A is specified by s_A, and the position of the *end* of the cord from which block B is suspended is defined by s_B. Here we have chosen coordinate axes which are (1) referenced from fixed points or datums, (2) measured in the direction of motion of each block, and (3) positive to the right (s_A) and positive downward (s_B). During the motion, the red colored segments of the cord in Fig. 12–37*a* remain constant. If l represents the total length of cord minus these segments, then the position coordinates can be related by the equation

$$2s_B + h + s_A = l$$

Since l and h are constant during the motion, the two time derivatives yield

$$2v_B = -v_A \qquad 2a_B = -a_A$$

Hence, when B moves downward ($+s_B$), A moves to the left ($-s_A$) with two times the motion.

This example can also be worked by defining the position of block B from the center of the bottom pulley (a fixed point), Fig. 12–37*b*. In this case

$$2(h - s_B) + h + s_A = l$$

Time differentiation yields

$$2v_B = v_A \qquad 2a_B = a_A$$

Here the signs are the same. Why?

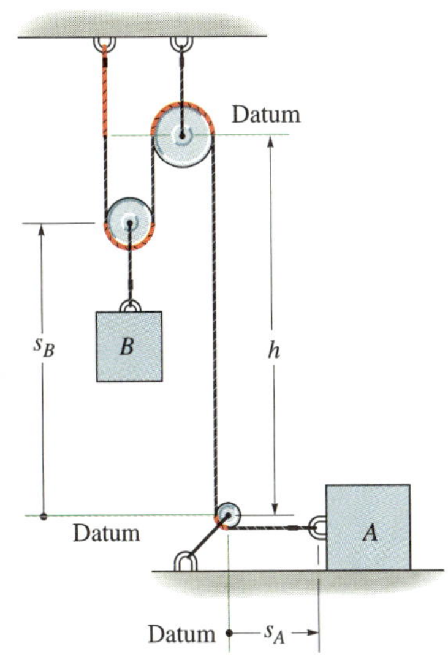

(b)

Fig. 12–37

Procedure for Analysis

The above method of relating the dependent motion of one particle to that of another can be performed using algebraic scalars or position coordinates provided each particle moves along a rectilinear path. When this is the case, only the magnitudes of the velocity and acceleration of the particles will change, not their line of direction. The following procedure is required.

Position-Coordinate Equation

- Establish position coordinates which have their origin located at a *fixed* point or datum.
- The coordinates are directed along the path of motion and extend to a point having the same motion as each of the particles.
- It is *not necessary* that the *origin* be the *same* for each of the coordinates; however, it is *important* that each coordinate axis selected be directed along the *path of motion* of the particle.
- Using geometry or trigonometry, relate the coordinates to the total length of the cord, l_T, or to that portion of cord, l, which *excludes* the segments that do not change length as the particles move—such as arc segments wrapped over pulleys.
- If a problem involves a *system* of two or more cords wrapped around pulleys, then the position of a point on one cord must be related to the position of a point on another cord using the above procedure. Separate equations are written for a fixed length of each cord of the system and the positions of the two particles are then related by these equations (see Examples 12–22 and 12–23).

Time Derivatives

- Two successive time derivatives of the position-coordinate equations yield the required velocity and acceleration equations which relate the motions of the particles.
- The signs of the terms in these equations will be consistent with those that specify the positive and negative sense of the position coordinates.

The motion of the traveling block on this oil rig depends upon the motion of the cable connected to the winch which operates it. It is important to be able to relate these motions in order to determine the power requirements of the winch and the force in the cable caused by accelerated motion.

EXAMPLE 12-21

Determine the speed of block A in Fig. 12–38 if block B has an upward speed of 6 ft/s.

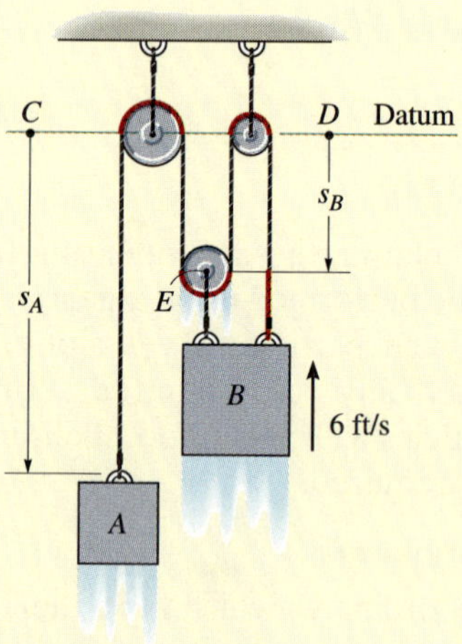

Fig. 12–38

Solution

Position-Coordinate Equation. There is *one cord* in this system having segments which are changing length. Position coordinates s_A and s_B will be used since each is measured from a fixed point (C or D) and extends along each block's *path of motion.* In particular, s_B is directed to point E since motion of B and E is the *same.*

The red colored segments of the cord in Fig. 12–38 remain at a constant length and do not have to be considered as the blocks move. The remaining length of cord, l, is also constant and is related to the changing position coordinates s_A and s_B by the equation

$$s_A + 3s_B = l$$

Time Derivative. Taking the time derivative yields

$$v_A + 3v_B = 0$$

so that when $v_B = -6$ ft/s (upward),

$$v_A = 18 \text{ ft/s} \downarrow \qquad \textit{Ans.}$$

EXAMPLE 12–22

Determine the speed of block A in Fig. 12–39 if block B has an upward speed of 6 ft/s.

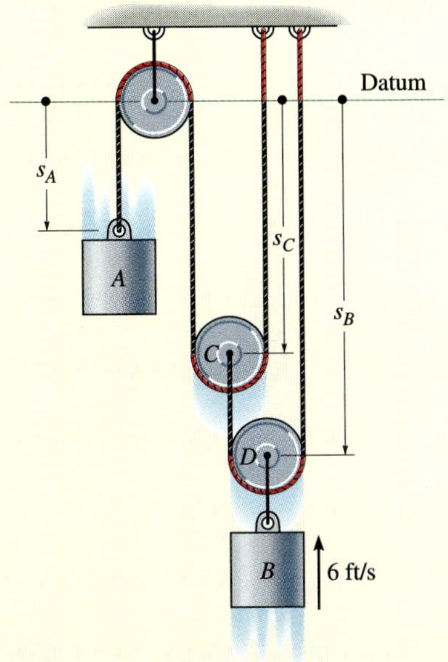

Fig. 12–39

Solution

Position-Coordinate Equation. As shown, the positions of blocks A and B are defined using coordinates s_A and s_B. Since the system has *two* cords which change length, it will be necessary to use a third coordinate, s_C, in order to relate s_A to s_B. In other words, the length of one of the cords can be expressed in terms of s_A and s_C, and the length of the other cord can be expressed in terms of s_B and s_C.

The red colored segments of the cords in Fig. 12–39 do not have to be considered in the analysis. Why? For the remaining cord lengths, say l_1 and l_2, we have

$$s_A + 2s_C = l_1 \qquad s_B + (s_B - s_C) = l_2$$

Eliminating s_C yields an equation defining the positions of both blocks, i.e.,

$$s_A + 4s_B = 2l_2 + l_1$$

Time Derivative. The time derivative gives

$$v_A + 4v_B = 0$$

so that when $v_B = -6$ ft/s (upward),

$$v_A = +24 \text{ ft/s} = 24 \text{ ft/s} \downarrow \qquad \textit{Ans.}$$

EXAMPLE 12-23

Determine the speed with which block B rises in Fig. 12–40 if the end of the cord at A is pulled down with a speed of 2 m/s.

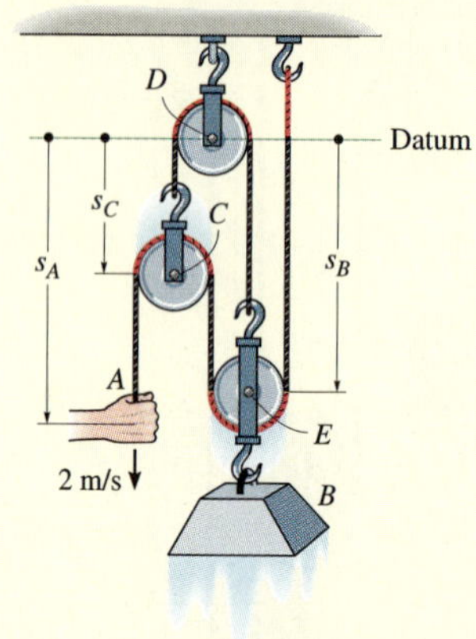

Fig. 12–40

Solution

Position-Coordinate Equation. The position of point A is defined by s_A, and the position of block B is specified by s_B since point E on the pulley will have the same motion as the block. Both coordinates are measured from a horizontal datum passing through the fixed pin at pulley D. Since the system consists of *two* cords, the coordinates s_A and s_B cannot be related directly. Instead, by establishing a third position coordinate, s_C, we can now express the length of one of the cords in terms of s_A and s_C, and the length of the other cord in terms of s_A, s_B, and s_C.

Excluding the red colored segments of the cords in Fig. 12–40, the remaining constant cord lengths l_1 and l_2 (along with the hook and link dimensions) can be expressed as

$$s_C + s_B = l_1$$

$$(s_A - s_C) + (s_B - s_C) + s_B = l_2$$

Eliminating s_C yields

$$s_A + 4s_B = l_2 + 2l_1$$

As required, this equation relates the position s_B of block B to the position s_A of point A.

Time Derivative. The time derivative gives

$$v_A + 4v_B = 0$$

so that when $v_A = 2$ m/s (downward),

$$v_B = -0.5 \text{ m/s} = 0.5 \text{ m/s} \uparrow \qquad \textit{Ans.}$$

EXAMPLE 12-24

A man at A is hoisting a safe S as shown in Fig. 12–41 by walking to the right with a constant velocity $v_A = 0.5$ m/s. Determine the velocity and acceleration of the safe when it reaches the window elevation at E. The rope is 30 m long and passes over a small pulley at D.

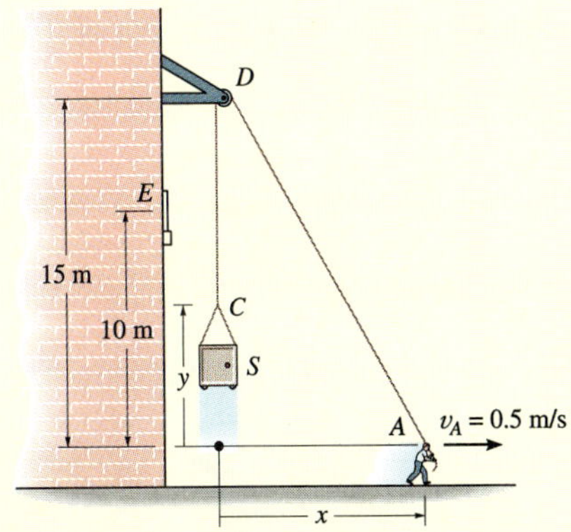

Fig. 12–41

Solution

Position-Coordinate Equation. This problem is unlike the previous examples since rope segment DA changes both direction and magnitude. However, the ends of the rope, which define the positions of S and A, are specified by means of the x and y coordinates measured from a fixed point and *directed along the paths of motion* of the ends of the rope.

The x and y coordinates may be related since the rope has a fixed length $l = 30$ m, which at all times is equal to the length of segment DA plus CD. Using the Pythagorean theorem to determine l_{DA}, we have $l_{DA} = \sqrt{(15)^2 + x^2}$; also, $l_{CD} = 15 - y$. Hence,

$$l = l_{DA} + l_{CD}$$

$$30 = \sqrt{(15)^2 + x^2} + (15 - y)$$

$$y = \sqrt{225 + x^2} - 15 \qquad (1)$$

Time Derivatives. Taking the time derivative, using the chain rule, where $v_S = dy/dt$ and $v_A = dx/dt$, yields

$$v_S = \frac{dy}{dt} = \left[\frac{1}{2}\frac{2x}{\sqrt{225 + x^2}}\right]\frac{dx}{dt}$$

$$= \frac{x}{\sqrt{225 + x^2}}v_A \qquad (2)$$

At $y = 10$ m, x is determined from Eq. 1, i.e., $x = 20$ m. Hence, from Eq. 2 with $v_A = 0.5$ m/s,

$$v_S = \frac{20}{\sqrt{225 + (20)^2}}(0.5) = 0.4 \text{ m/s} = 400 \text{ mm/s}\uparrow \qquad \textit{Ans.}$$

The acceleration is determined by taking the time derivative of Eq. 2. Since v_A is constant, then $a_A = dv_A/dt = 0$, and we have

$$a_S = \frac{d^2y}{dt^2} = \left[\frac{-x(dx/dt)}{(225 + x^2)^{3/2}}\right]xv_A + \left[\frac{1}{\sqrt{225 + x^2}}\right]\left(\frac{dx}{dt}\right)v_A + \left[\frac{1}{\sqrt{225 + x^2}}\right]x\frac{dv_A}{dt} = \frac{225v_A^2}{(225 + x^2)^{3/2}}$$

At $x = 20$ m, with $v_A = 0.5$ m/s, the acceleration becomes

$$a_S = \frac{225(0.5 \text{ m/s})^2}{[225 + (20 \text{ m})^2]^{3/2}} = 0.00360 \text{ m/s}^2 = 3.60 \text{ mm/s}^2\uparrow \qquad \textit{Ans.}$$

Note that the constant velocity at A causes the other end C of the rope to have an acceleration since $\mathbf{v}_A$ causes segment DA to change its direction as well as its length.

12.10 Relative-Motion Analysis of Two Particles Using Translating Axes

Throughout this chapter the absolute motion of a particle has been determined using a single fixed reference frame for measurement. There are many cases, however, where the path of motion for a particle is complicated, so that it may be feasible to analyze the motion in parts by using two or more frames of reference. For example, the motion of a particle located at the tip of an airplane propeller, while the plane is in flight, is more easily described if one observes first the motion of the airplane from a fixed reference and then superimposes (vectorially) the circular motion of the particle measured from a reference attached to the airplane. Any type of coordinates—rectangular, cylindrical, etc.—may be chosen to describe these two different motions.

In this section only *translating frames of reference* will be considered for the analysis. Relative-motion analysis of particles using rotating frames of reference will be treated in Secs. 16.8 and 20.4, since such an analysis depends on prior knowledge of the kinematics of line segments.

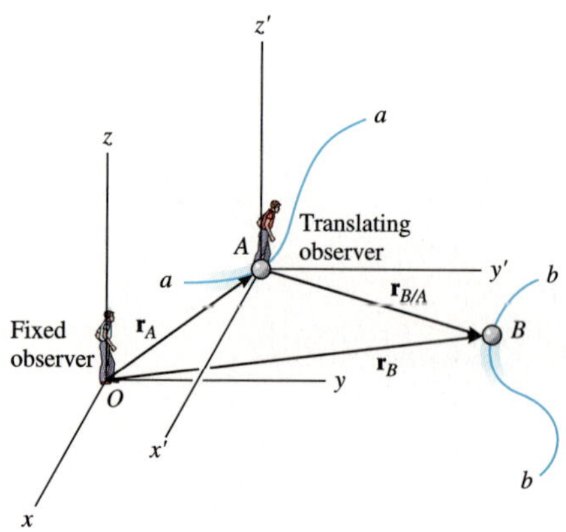

(a)

Fig. 12–42

Position. Consider particles A and B, which move along the arbitrary paths aa and bb, respectively, as shown in Fig. 12–42*a*. The *absolute position* of each particle, $\mathbf{r}_A$ and $\mathbf{r}_B$, is measured from the common origin O of the *fixed* x, y, z reference frame. The origin of a second frame of reference x', y', z' is attached to and moves with particle A. The axes of this frame are *only permitted to translate* relative to the fixed frame. The *relative position* of "B with respect to A" is designated by a *relative-position vector* $\mathbf{r}_{B/A}$. Using vector addition, the three vectors shown in Fig. 12–42*a* can be related by the equation*

$$\mathbf{r}_B = \mathbf{r}_A + \mathbf{r}_{B/A} \qquad (12\text{–}33)$$

Velocity. An equation that relates the velocities of the particles can be determined by taking the time derivative of Eq. 12–33, i.e.,

$$\mathbf{v}_B = \mathbf{v}_A + \mathbf{v}_{B/A} \qquad (12\text{–}34)$$

Here $\mathbf{v}_B = d\mathbf{r}_B/dt$ and $\mathbf{v}_A = d\mathbf{r}_A/dt$ refer to *absolute velocities,* since they are observed from the fixed frame; whereas the *relative velocity* $\mathbf{v}_{B/A} = d\mathbf{r}_{B/A}/dt$ is observed from the translating frame. It is important to note

*An easy way to remember the setup of this equation, and others like it, is to note the "cancellation" of the subscript A between the two terms, i.e., $\mathbf{r}_B = \mathbf{r}_{A} + \mathbf{r}_{B/A}$.

that since the x', y', z' axes translate, the *components* of $\mathbf{r}_{B/A}$ will *not* change direction and therefore the time derivative of this vector's components will only have to account for the change in the vector's magnitude. Equation 12–34 therefore states that the velocity of B is equal to the velocity of A plus (vectorially) the relative velocity of "B with respect to A," as measured by the *translating observer* fixed in the x', y', z' reference, Fig. 12–42b.

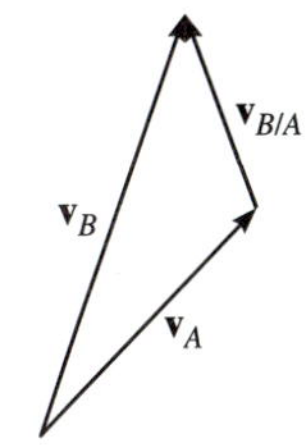

(b)

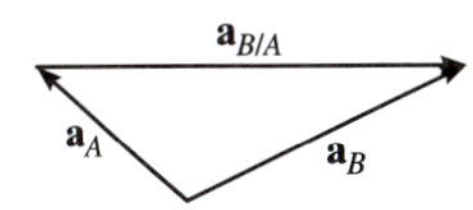

(c)

Fig. 12–42

Acceleration. The time derivative of Eq. 12–34 yields a similar vector relationship between the *absolute* and *relative accelerations* of particles A and B.

$$\mathbf{a}_B = \mathbf{a}_A + \mathbf{a}_{B/A} \tag{12–35}$$

Here $\mathbf{a}_{B/A}$ is the acceleration of B as seen by the observer located at A and translating with the x', y', z' reference frame. The vector addition is shown in Fig. 12–42c.

Procedure for Analysis

- When applying the relative-position equation, $\mathbf{r}_B = \mathbf{r}_A + \mathbf{r}_{B/A}$, it is first necessary to specify the locations of the fixed x, y, z and translating x', y', z' axes.
- Usually, the origin A of the translating axes is located at a point having a *known position*, $\mathbf{r}_A$, Fig. 12–42a.
- A graphical representation of the vector addition $\mathbf{r}_B = \mathbf{r}_A + \mathbf{r}_{B/A}$ can be shown, and both the known and unknown quantities labeled on this sketch.
- Since vector addition forms a triangle, there can be at most *two unknowns*, represented by the magnitudes and/or directions of the vector quantities.
- These unknowns can be solved for either graphically, using trigonometry (law of sines, law of cosines), or by resolving each of the three vectors $\mathbf{r}_B$, $\mathbf{r}_A$, and $\mathbf{r}_{B/A}$ into rectangular or Cartesian components, thereby generating a set of scalar equations.
- The relative-motion equations $\mathbf{v}_B = \mathbf{v}_A + \mathbf{v}_{B/A}$ and $\mathbf{a}_B = \mathbf{a}_A + \mathbf{a}_{B/A}$ are applied in the same manner as explained above, except in this case the origin O of the fixed x, y, z axes does not have to be specified, Figs. 12–42b and 12–42c.

The pilots of these jet planes flying close to one another must be aware of their relative positions and velocities at all times in order to avoid a collision.

EXAMPLE 12–25

A train, traveling at a constant speed of 60 mi/h, crosses over a road as shown in Fig. 12–43*a*. If automobile A is traveling at 45 mi/h along the road, determine the magnitude and direction of the relative velocity of the train with respect to the automobile.

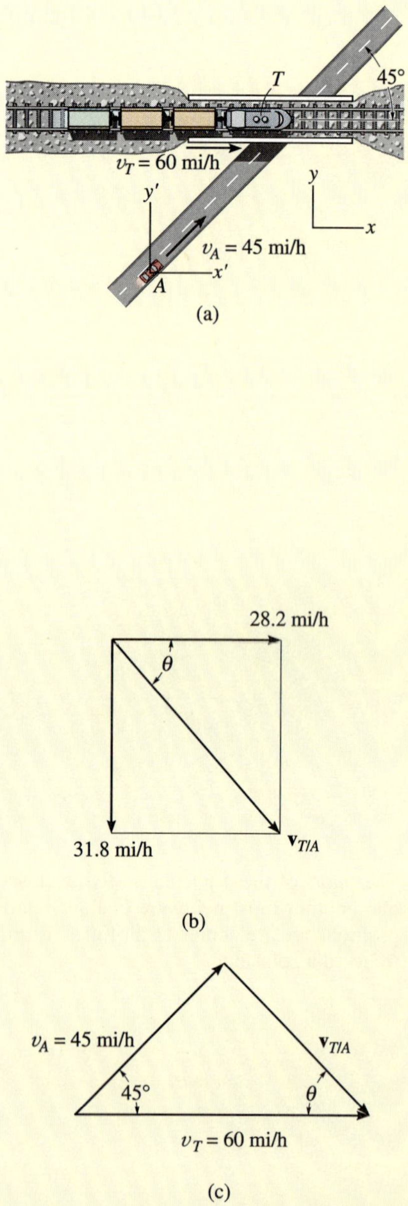

Fig. 12–43

Solution I

Vector Analysis. The relative velocity $\mathbf{v}_{T/A}$ is measured from the translating x', y' axes attached to the automobile, Fig. 12–43*a*. It is determined from $\mathbf{v}_T = \mathbf{v}_A + \mathbf{v}_{T/A}$. Since $\mathbf{v}_T$ and $\mathbf{v}_A$ are known in *both* magnitude and direction, the unknowns become the x and y components of $\mathbf{v}_{T/A}$. Using the x, y axes in Fig. 12–43*a* and a Cartesian vector analysis, we have

$$\mathbf{v}_T = \mathbf{v}_A + \mathbf{v}_{T/A}$$

$$60\mathbf{i} = (45 \cos 45^\circ\,\mathbf{i} + 45 \sin 45^\circ\,\mathbf{j}) + \mathbf{v}_{T/A}$$

$$\mathbf{v}_{T/A} = \{28.2\mathbf{i} - 31.8\mathbf{j}\}\ \text{mi/h} \qquad \textit{Ans.}$$

The magnitude of $\mathbf{v}_{T/A}$ is thus

$$v_{T/A} = \sqrt{(28.2)^2 + (-31.8)^2} = 42.5\ \text{mi/h} \qquad \textit{Ans.}$$

From the direction of each component, Fig. 12–43*b*, the direction of $\mathbf{v}_{T/A}$ defined from the x axis is

$$\tan\theta = \frac{(v_{T/A})_y}{(v_{T/A})_x} = \frac{31.8}{28.2}$$

$$\theta = 48.5^\circ \qquad \textit{Ans.}$$

Note that the vector addition shown in Fig. 12–43*b* indicates the correct sense for $\mathbf{v}_{T/A}$. This figure anticipates the answer and can be used to check it.

Solution II

Scalar Analysis. The unknown components of $\mathbf{v}_{T/A}$ can also be determined by applying a scalar analysis. We will assume these components act in the *positive* x and y directions. Thus,

$$\mathbf{v}_T = \mathbf{v}_A + \mathbf{v}_{T/A}$$

$$\begin{bmatrix} 60\ \text{mi/h} \\ \rightarrow \end{bmatrix} = \begin{bmatrix} 45\ \text{mi/h} \\ \measuredangle^{45^\circ} \end{bmatrix} + \begin{bmatrix} (v_{T/A})_x \\ \rightarrow \end{bmatrix} + \begin{bmatrix} (v_{T/A})_y \\ \uparrow \end{bmatrix}$$

Resolving each vector into its x and y components yields

$$(\overset{+}{\rightarrow}) \qquad 60 = 45 \cos 45^\circ + (v_{T/A})_x + 0$$

$$(+\uparrow) \qquad 0 = 45 \sin 45^\circ + 0 + (v_{T/A})_y$$

Solving, we obtain the previous results,

$$(v_{T/A})_x = 28.2\ \text{mi/h} = 28.2\ \text{mi/h} \rightarrow$$

$$(v_{T/A})_y = -31.8\ \text{mi/h} = 31.8\ \text{mi/h} \downarrow \qquad \textit{Ans.}$$

EXAMPLE 12–26

Plane A in Fig. 12–44*a* is flying along a straight-line path, whereas plane B is flying along a circular path having a radius of curvature of $\rho_B = 400$ km. Determine the velocity and acceleration of B as measured by the pilot of A.

Solution

Velocity. The x, y axes are located at an arbitrary fixed point. Since the motion relative to plane A is to be determined the *translating frame of reference* x', y' is attached to it, Fig. 12–44*a*. Applying the relative-velocity equation in scalar form since the velocity vectors of both planes are parallel at the instant shown, we have

$$(+\uparrow) \qquad v_B = v_A + v_{B/A}$$

$$600 = 700 + v_{B/A}$$

$$v_{B/A} = -100 \text{ km/h} = 100 \text{ km/h} \downarrow \qquad \textit{Ans.}$$

The vector addition is shown in Fig. 12–44*b*.

Acceleration. Plane B has both tangential and normal components of acceleration, since it is flying along a *curved path.* From Eq. 12–20, the magnitude of the normal component is

$$(a_B)_n = \frac{v_B^2}{\rho} = \frac{(600 \text{ km/h})^2}{400 \text{ km}} = 900 \text{ km/h}^2$$

Applying the relative-acceleration equation, we have

$$\mathbf{a}_B = \mathbf{a}_A + \mathbf{a}_{B/A}$$

$$900\mathbf{i} - 100\mathbf{j} = 50\mathbf{j} + \mathbf{a}_{B/A}$$

Thus,

$$\mathbf{a}_{B/A} = \{900\mathbf{i} - 150\mathbf{j}\} \text{ km/h}^2$$

From Fig. 12–44*c*, the magnitude and direction of $\mathbf{a}_{B/A}$ are therefore

$$a_{B/A} = 912 \text{ km/h}^2 \qquad \theta = \tan^{-1}\frac{150}{900} = 9.46° \; ⦪\theta \qquad \textit{Ans.}$$

Notice that the solution to this problem is possible using a translating frame of reference, since the pilot in plane A is "translating." Observation of plane A with respect to the pilot of plane B, however, must be obtained using a *rotating* set of axes attached to plane B. (This assumes, of course, that the pilot of B is fixed in the rotating frame, so he does not turn his eyes to follow the motion of A.) The analysis for this case is given in Example 16–21.

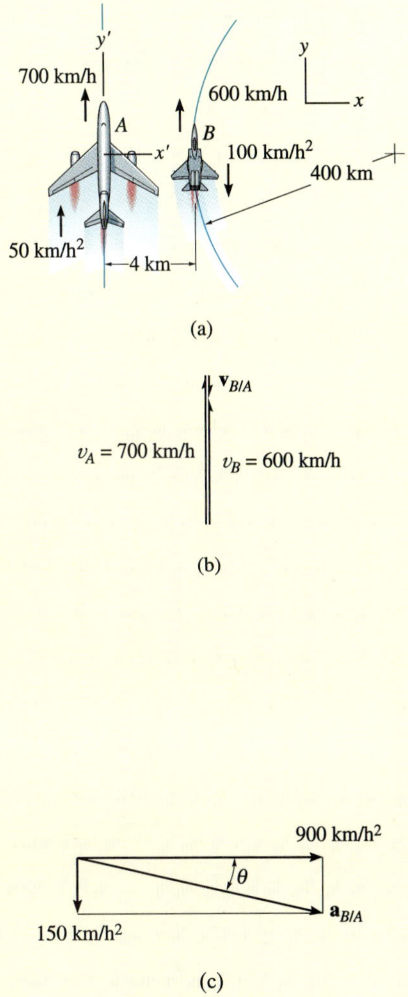

Fig. 12–44

EXAMPLE 12-27

At the instant shown in Fig. 12–45 cars A and B are traveling with speeds of 18 m/s and 12 m/s, respectively. Also at this instant, A has a decrease in speed of 2 m/s^2, and B has an increase in speed of 3 m/s^2. Determine the velocity and acceleration of B with respect to A.

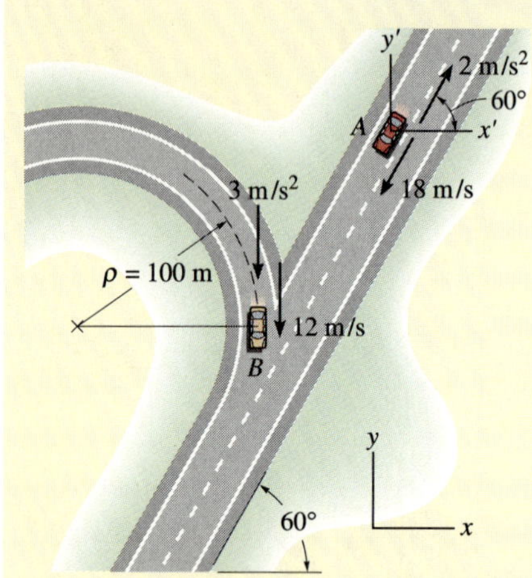

(a)

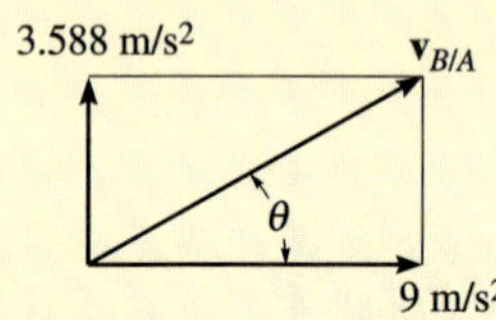

(b)

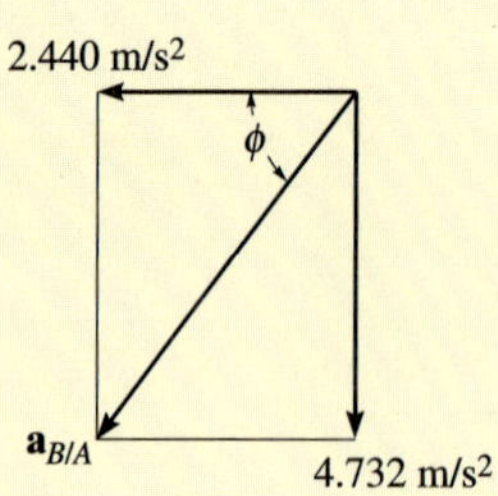

(c)

Fig. 12–45

Solution

Velocity. The fixed x, y axes are established at a point on the ground and the translating x', y' axes are attached to car A, Fig. 12–45*a*. Why? The relative velocity is determined from $\mathbf{v}_B = \mathbf{v}_A + \mathbf{v}_{B/A}$. What are the two unknowns? Using a Cartesian vector analysis, we have

$$\mathbf{v}_B = \mathbf{v}_A + \mathbf{v}_{B/A}$$
$$-12\mathbf{j} = (-18 \cos 60°\mathbf{i} - 18 \sin 60°\mathbf{j}) + \mathbf{v}_{B/A}$$
$$\mathbf{v}_{B/A} = \{9\mathbf{i} + 3.588\mathbf{j}\} \text{ m/s}$$

Thus,

$$v_{B/A} = \sqrt{(9)^2 + (3.588)^2} = 9.69 \text{ m/s} \qquad \textit{Ans.}$$

Noting that $\mathbf{v}_{B/A}$ has $+\mathbf{i}$ and $+\mathbf{j}$ components, Fig. 12–45*b*, its direction is

$$\tan \theta = \frac{(v_{B/A})_y}{(v_{B/A})_x} = \frac{3.588}{9}$$
$$\theta = 21.7° \measuredangle \qquad \textit{Ans.}$$

Acceleration. Car B has both tangential and normal components of acceleration. Why? The magnitude of the normal component is

$$(a_B)_n = \frac{v_B^2}{\rho} = \frac{(12 \text{ m/s})^2}{100 \text{ m}} = 1.440 \text{ m/s}^2$$

Applying the equation for relative acceleration yields

$$\mathbf{a}_B = \mathbf{a}_A + \mathbf{a}_{B/A}$$
$$(-1.440\mathbf{i} - 3\mathbf{j}) = (2 \cos 60°\mathbf{i} + 2 \sin 60°\mathbf{j}) + \mathbf{a}_{B/A}$$
$$\mathbf{a}_{B/A} = \{-2.440\mathbf{i} - 4.732\mathbf{j}\} \text{ m/s}^2$$

Here $\mathbf{a}_{B/A}$ has $-\mathbf{i}$ and $-\mathbf{j}$ components. Thus, from Fig. 12–45*c*,

$$a_{B/A} = \sqrt{(2.440)^2 + (4.732)^2} = 5.32 \text{ m/s}^2 \qquad \textit{Ans.}$$
$$\tan \phi = \frac{(a_{B/A})_y}{(a_{B/A})_x} = \frac{4.732}{2.440}$$
$$\phi = 62.7° \qquad \textit{Ans.}$$

Is it possible to obtain the relative acceleration of $\mathbf{a}_{A/B}$ using this method? Refer to the comment made at the end of Example 12–26.

Problems

*12-172. If the end of the cable at A is pulled down with a speed of 2 m/s, determine the speed at which block B rises.

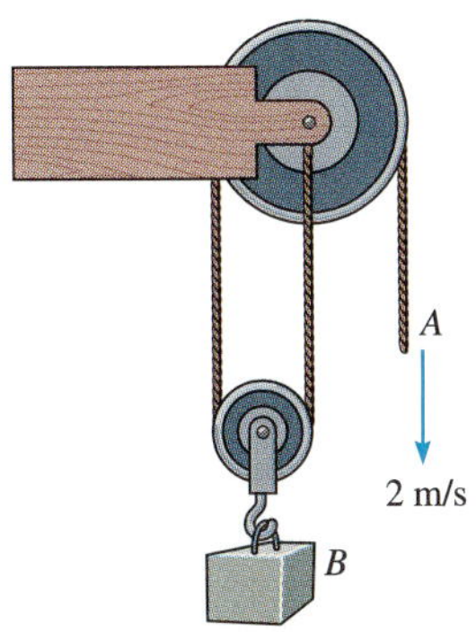

Prob. 12–172

12-173. If the end of the cable at A is pulled down with a speed of 2 m/s, determine the speed at which block B rises.

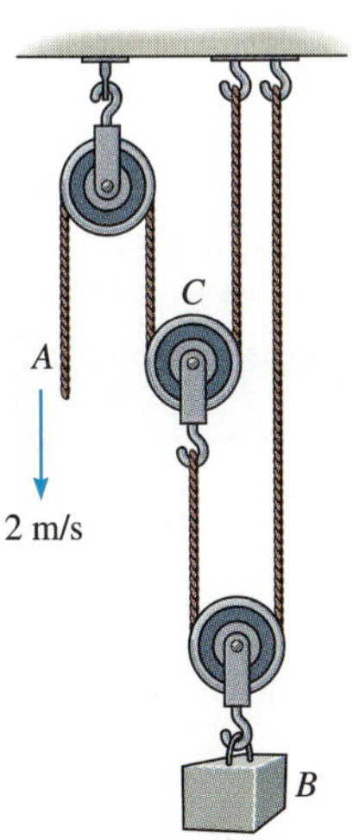

Prob. 12–173

12-174. If the end of the cable at A is pulled down with a speed of 2 m/s, determine the speed at which block B rises.

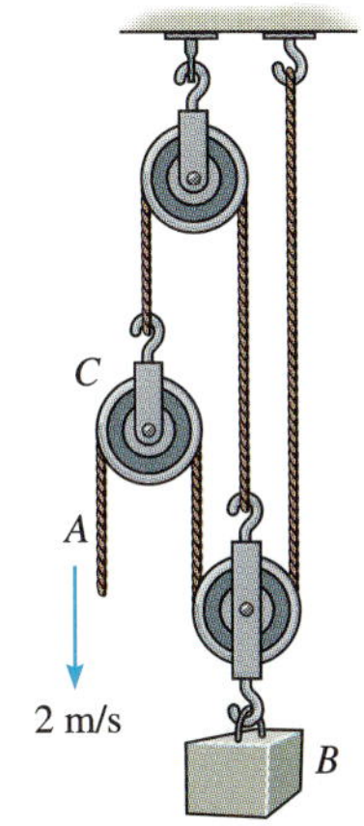

Prob. 12–174

12-175. The mine car C is being pulled up the incline using the motor M and the rope-and-pulley arrangement shown. Determine the speed v_P at which a point P on the cable must be traveling toward the motor to move the car up the plane with a constant speed of $v = 2$ m/s.

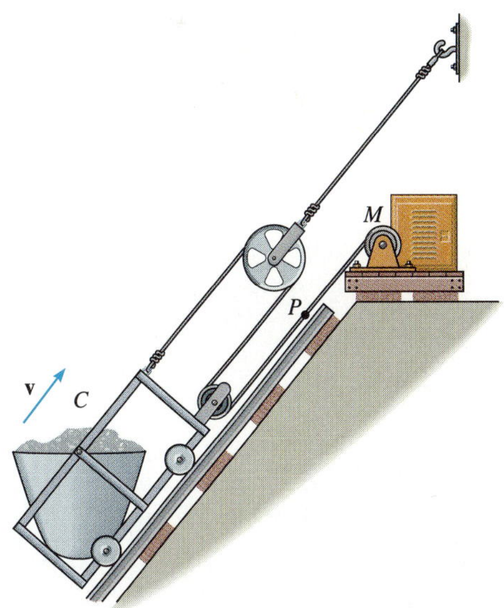

Prob. 12–175

***12-176.** Determine the displacement of the log if the truck at C pulls the cable 4 ft to the right.

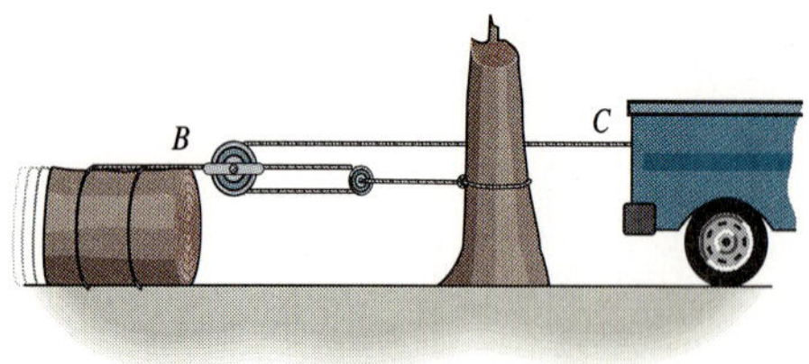

Prob. 12–176

12-177. If the hydraulic cylinder at H draws rod BC in by 8 in., determine how far the slider at A moves.

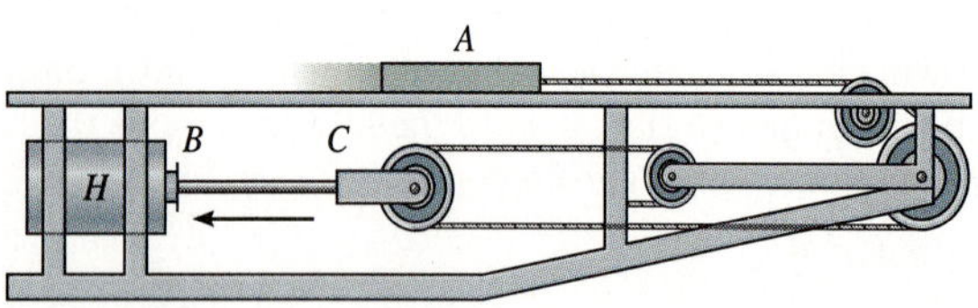

Prob. 12–177

12-178. If the hydraulic cylinder at H draws in rod BC at 2 ft/s, determine the speed of the slider at A.

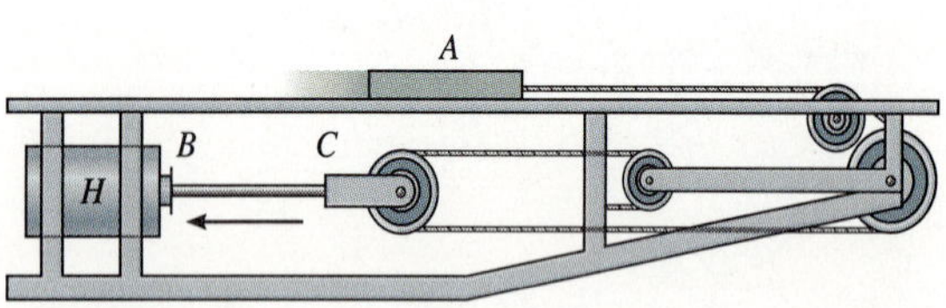

Prob. 12–178

12-179. The cable at B is pulled downwards at 4 ft/s, and is slowing at 2 ft/s^2. Determine the velocity and acceleration of block A at this instant.

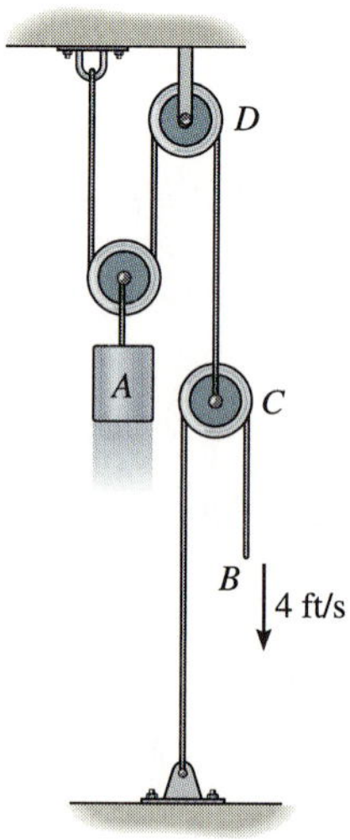

Prob. 12–179

***12-180.** The hoist is used to lift the load at D. If the end A of the chain is traveling downward at $v_A = 5$ ft/s and the end B is traveling upward at $v_B = 2$ ft/s, determine the velocity of the load at D.

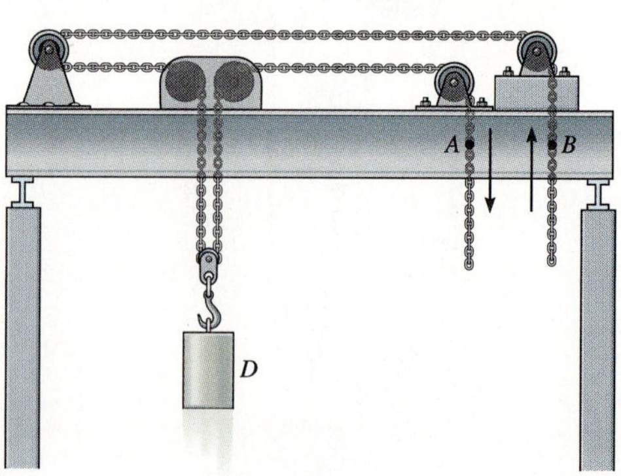

Prob. 12–180

12-181. If block A is moving downward with a speed of 4 ft/s while C is moving up at 2 ft/s, determine the speed of block B.

12-182. If block A is moving downward at 6 ft/s while block C is moving down at 18 ft/s, determine the relative velocity of block B with respect to C.

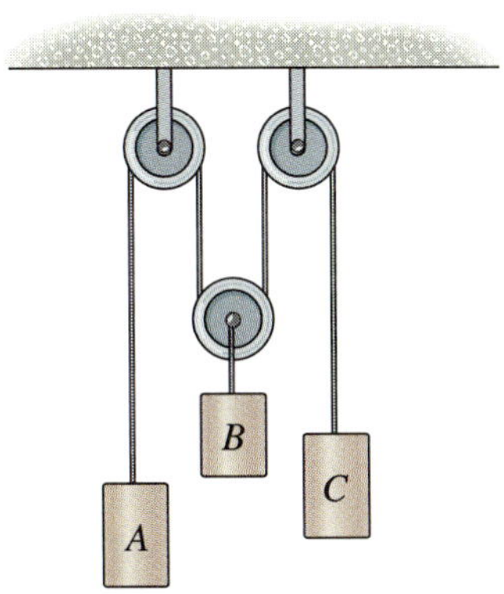

Probs. 12–181/182

12-183. The motor draws in the cable at C with a constant velocity of $v_C = 4$ m/s. The motor draws in the cable at D with a constant acceleration of $a_D = 8$ m/s^2. If $v_D = 0$ when $t = 0$, determine (a) the time needed for block A to rise 3 m, and (b) the relative velocity of block A with respect to block B when this occurs.

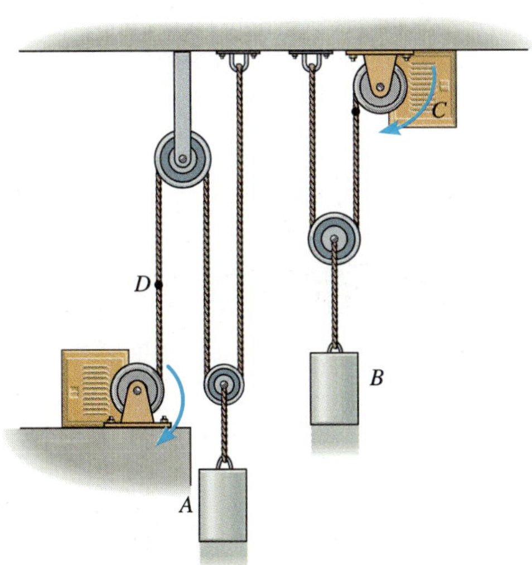

Prob. 12–183

***12-184.** If motors at A and B draw in their attached cables with an acceleration of $a = (0.2t)$ m/s^2, where t is in seconds, determine the speed of the block when it reaches a height of $h = 4$ m, starting from rest. Also, how much time does it take to reach this height?

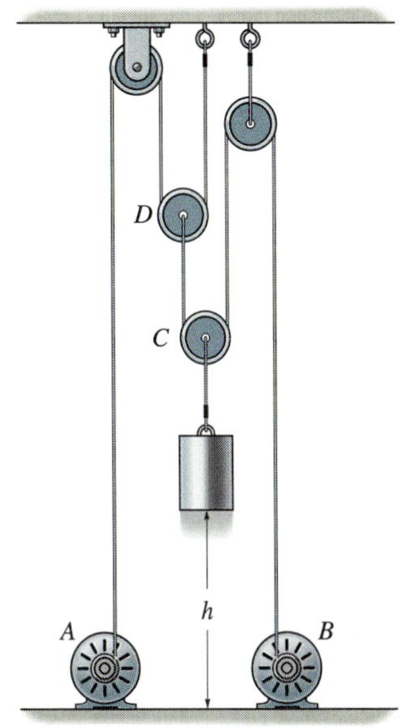

Prob. 12–184

12-185. If the end A of the cable is moving upwards at $v_A = 14$ m/s, determine the speed of block B.

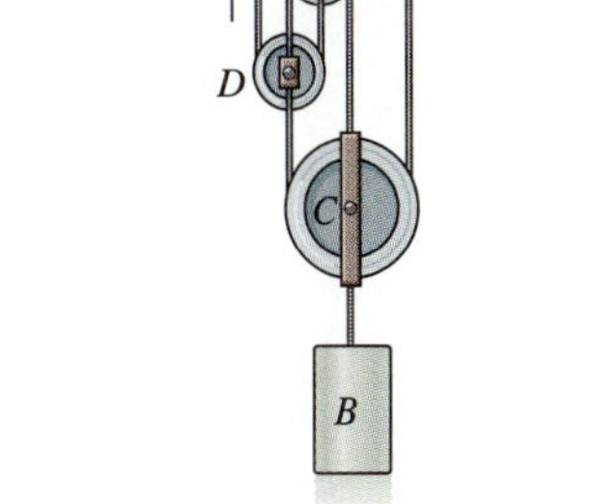

Prob. 12–185

■**12-186.** If the truck is traveling at a constant speed of $v_T = 6$ ft/s, determine the speed of the crate for any angle θ of the rope. The rope has a length of 100 ft and passes over a pulley of negligible size at A. *Hint:* Relate the coordinates x_T and x_C to the length of the rope and take the time derivative. Then substitute the trigonometric relation between x_C and θ.

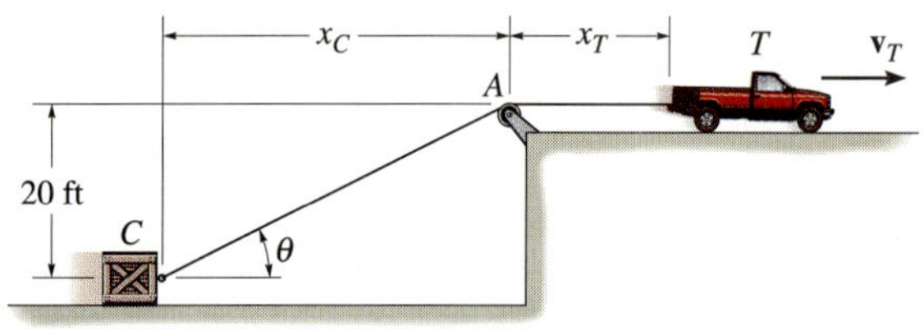

Prob. 12–186

12-187. The motion of the collar at A is controlled by a motor at B such that when the collar is at $s_A = 3$ ft it is moving upwards at 2 ft/s and slowing down at 1 ft/s^2. Determine the velocity and acceleration of the cable as it is drawn into the motor B at this instant.

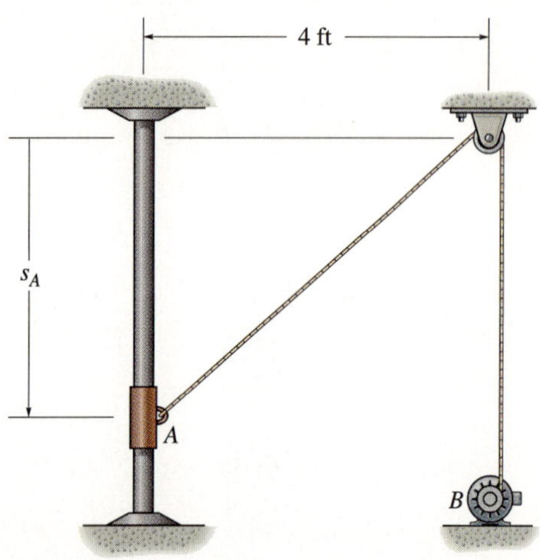

Prob. 12–187

***12-188.** The roller at A is moving upward with a velocity of $v_A = 3$ ft/s and has an acceleration of $a_A = 4$ ft/s^2 when $s_A = 4$ ft. Determine the velocity and acceleration of block B at this instant.

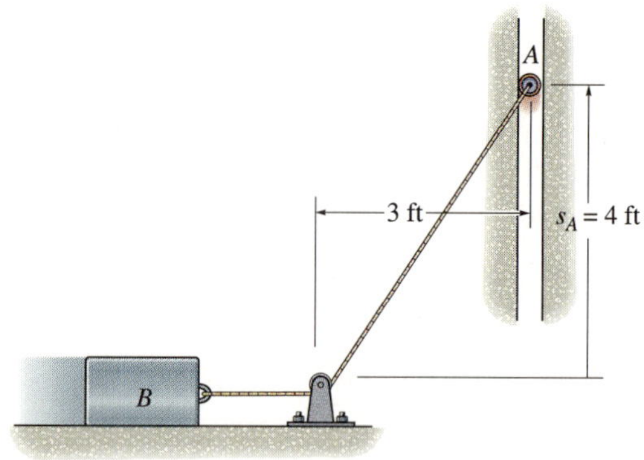

Prob. 12–188

12-189. If the collar at A moves with a constant velocity of 2 ft/s, determine the speed of the collar at B when $s_A = 12$ ft. The 33-ft-long cord wraps over the small pulley at A and is fixed at C.

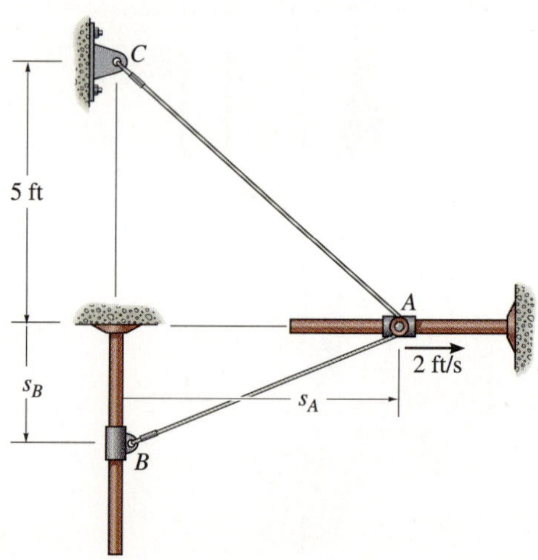

Prob. 12–189

12-190. The cord is attached to the pin at C and passes over the two pulleys at A and D. The pulley at A is attached to the smooth collar that travels along the vertical rod. Determine the velocity and acceleration of the end of the cord at B if at the instant $s_A = 4$ ft the collar is moving upwards at 5 ft/s, which is decreasing at 2 ft/s^2.

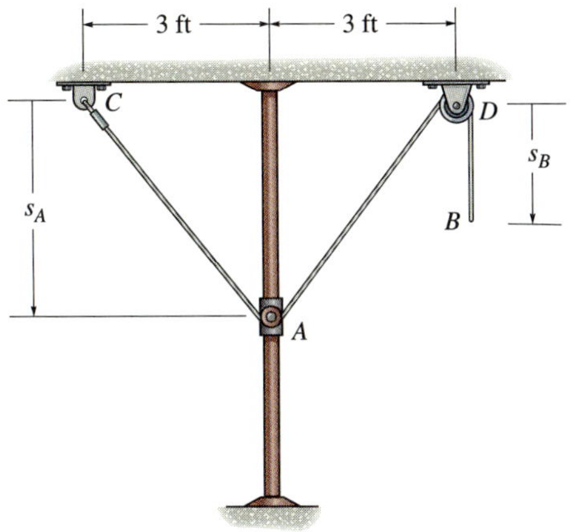

Prob. 12–190

12-191. The 16-ft-long cord is attached to the pin at C and passes over the two pulleys at A and D. The pulley at A is attached to the smooth collar that travels along the vertical rod. When $s_B = 6$ ft, the end of the cord at B is pulled downwards with a velocity of 4 ft/s and is given an acceleration of 3 ft/s^2. Determine the velocity and acceleration of the collar at this instant.

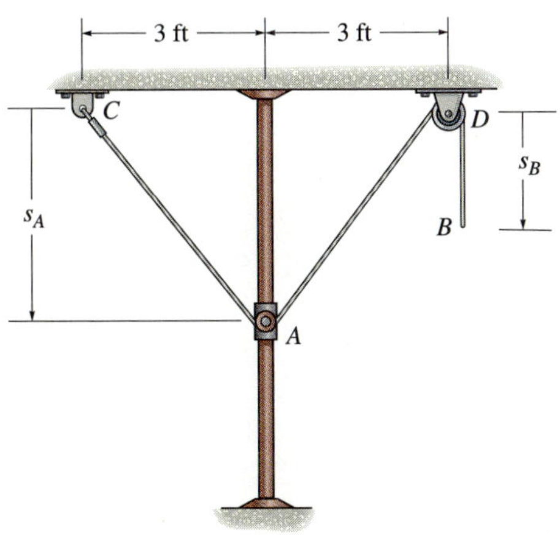

Prob. 12–191

***12-192.** Collars A and B are connected to the cord that passes over the small pulley at C. When A is located at D, B is 24 ft to the left of D. If A moves at a constant speed of 2 ft/s to the right, determine the speed of B when A is 4 ft to the right of D.

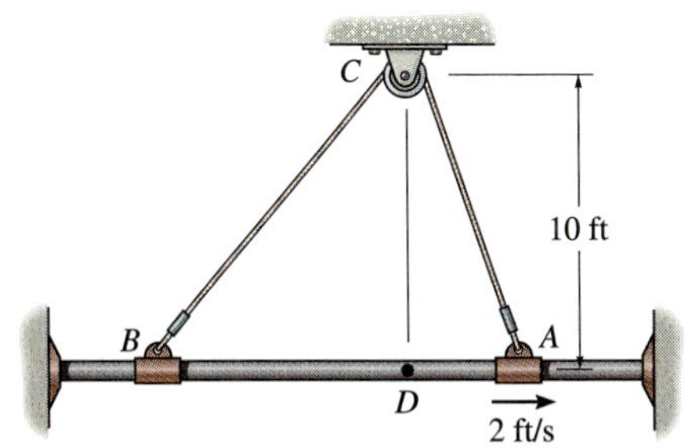

Prob. 12–192

12-193. The motorboat can travel with a speed of $v_B = 12$ ft/s in still water. If it heads for C at the opposite side of the river, and the river flows with a speed of $v_R = 5$ ft/s, determine the resultant velocity of the boat. How far downstream, d, is the boat carried when it reaches the other side at D?

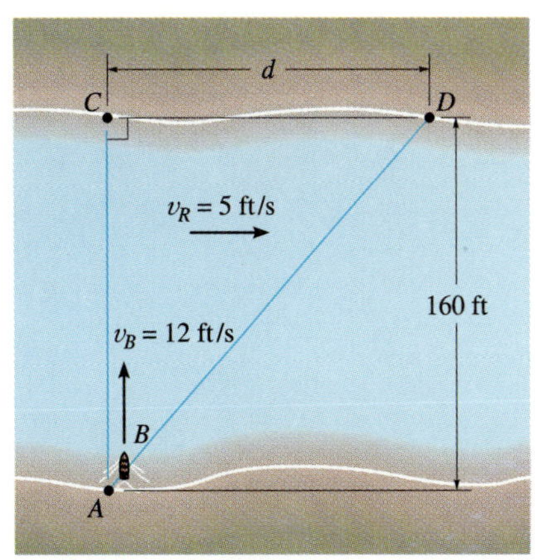

Prob. 12–193

12-194. The motor at C pulls in the cable with an acceleration $a_C = (3t^2)\ \text{m/s}^2$, where t is in seconds. The motor at D draws in its cable at $a_D = 5\ \text{m/s}^2$. If both motors start at the same instant from rest when $d = 3$ m, determine (a) the time needed for $d = 0$, and (b) the relative velocity of block A with respect to block B when this occurs.

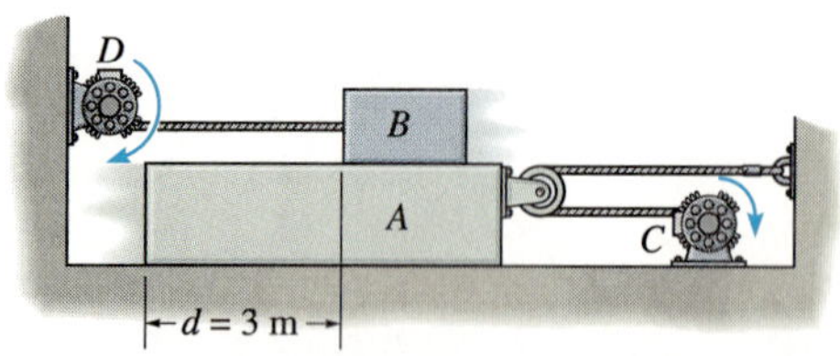

Prob. 12–194

12-195. Sand falls from rest 0.5 m vertically onto a chute. If the sand then slides with a velocity of $v_C = 2$ m/s down the chute, determine the relative velocity of the sand just falling on the chute at A with respect to the sand sliding down the chute. The chute is inclined at an angle of 40° with the horizontal.

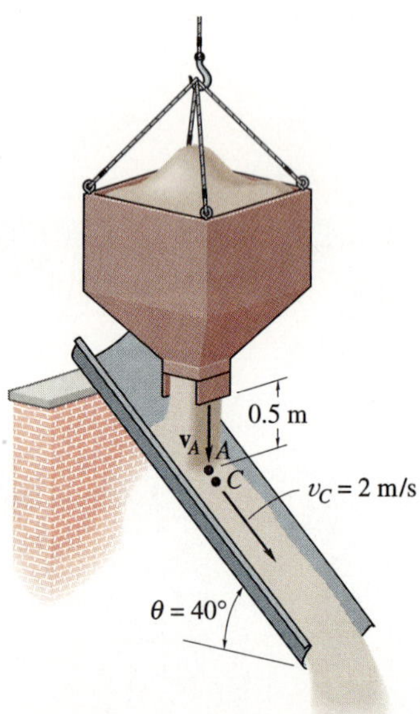

Prob. 12–195

***12-196.** Two planes, A and B, are flying at the same altitude. If their velocities are $v_A = 600$ km/h and $v_B = 700$ km/h, such that the angle between their straight-line paths is $\theta = 35°$, determine the velocity of plane B with respect to plane A.

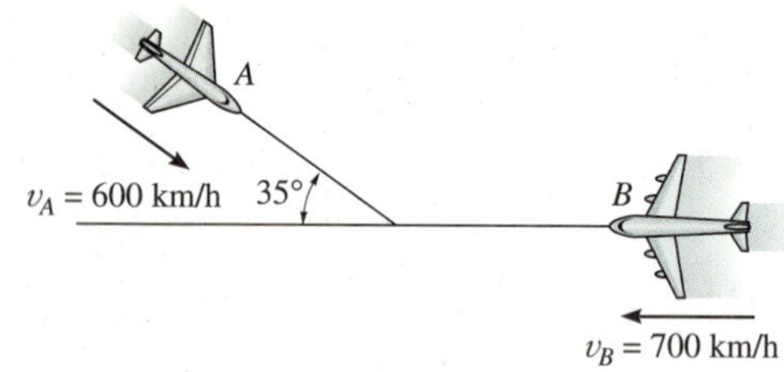

Prob. 12–196

12-197. At the instant shown, cars A and B are traveling at speeds of 55 mi/h and 40 mi/h, respectively. If B is increasing its speed by 1200 mi/h^2, while A maintains a constant speed, determine the velocity and acceleration of B with respect to A. Car B moves along a curve having a radius of curvature of 0.5 mi.

12-198. At the instant shown, cars A and B are traveling at speeds of 55 mi/h and 40 mi/h, respectively. If B is decreasing its speed at 1500 mi/h^2 while A is increasing its speed at 800 mi/h^2, determine the acceleration of B with respect to A. Car B moves along a curve having a radius of curvature of 0.75 mi.

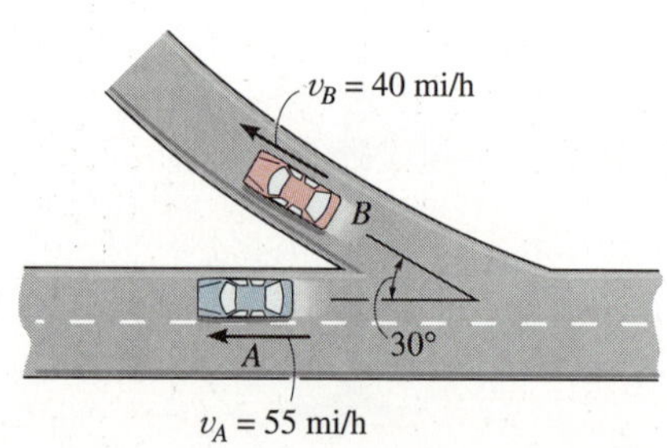

Probs. 12–197/198

12-199. At the instant shown, the car at A is traveling at 10 m/s around the curve while increasing its speed at 5 m/s^2. The car at B is traveling at 18.5 m/s along the straightaway and increasing its speed at 2 m/s^2. Determine the relative velocity and relative acceleration of A with respect to B at this instant.

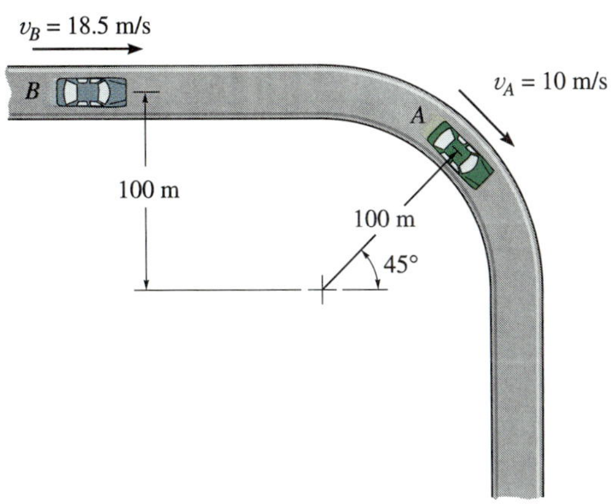

Prob. 12–199

***12-200.** An aircraft carrier is traveling forward with a velocity of 50 km/h. At the instant shown, the plane at A has just taken off and has attained a forward horizontal air speed of 200 km/h, measured from still water. If the plane at B is traveling along the runway of the carrier at 175 km/h in the direction shown, determine the velocity of A with respect to B.

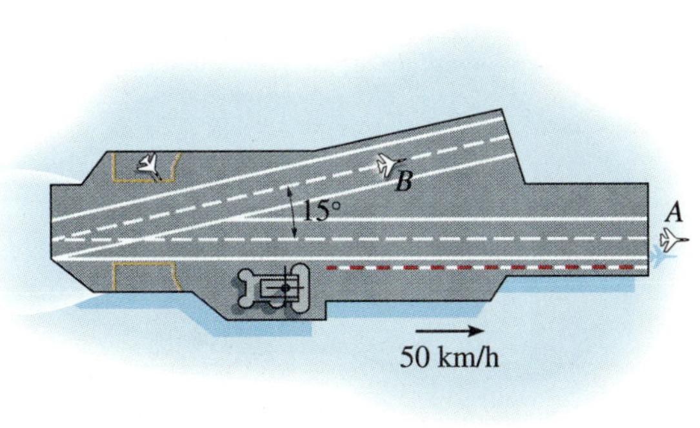

Prob. 12–200

12-201. At the instant shown, cars A and B are traveling at speeds of 30 mi/h and 20 mi/h, respectively. If B is increasing its speed by 1200 mi/h^2, while A maintains a constant speed, determine the velocity and acceleration of B with respect to A.

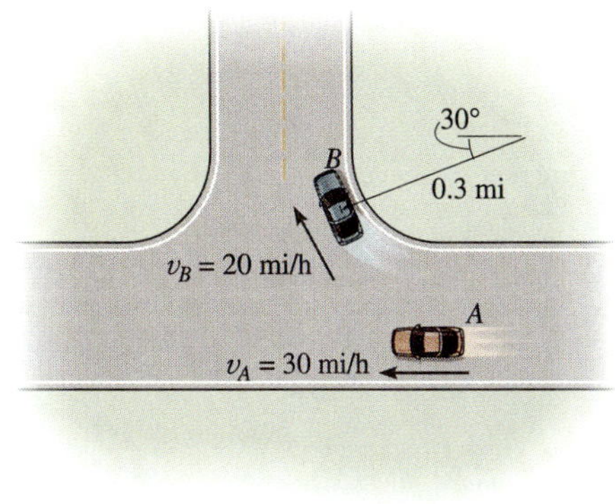

Prob. 12–201

12-202. At the instant shown, car A has a speed of 20 km/h, which is being increased at the rate of 300 km/h^2 as the car enters an expressway. At the same instant, car B is decelerating at 250 km/h^2 while traveling forward at 100 km/h. Determine the velocity and acceleration of A with respect to B.

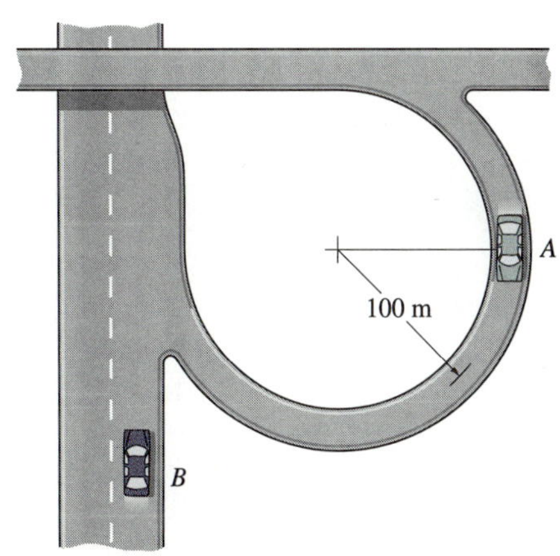

Prob. 12–202

12-203. At the instant shown car A is traveling with a velocity of 30 m/s and has an acceleration of 2 m/s^2 along the highway. At the same instant B is traveling on the trumpet interchange curve with a speed of 15 m/s, which is decreasing at 0.8 m/s^2. Determine the relative velocity and relative acceleration of B with respect to A at this instant.

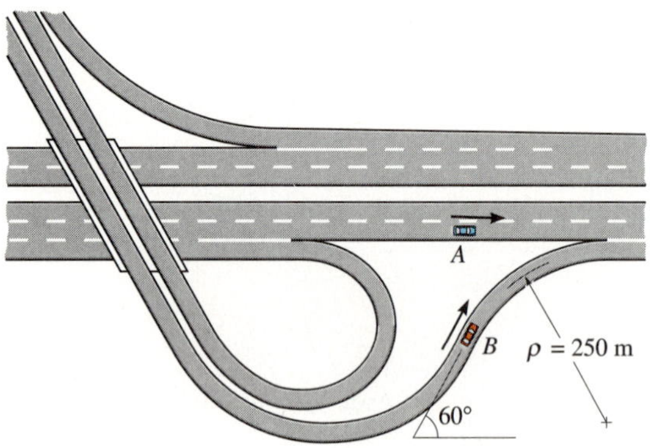

Prob. 12–203

***12-204.** The two cyclists A and B travel at the same constant speed v. Determine the speed of A with respect to B if A travels along the circular track, while B travels along the diameter of the circle.

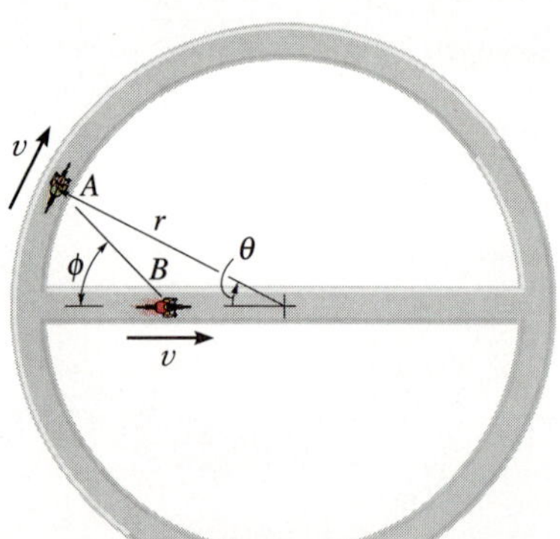

Prob. 12–204

12-205. The airplane has a speed relative to the wind of 100 mi/h. If the wind relative to the ground is 10 mi/h, determine the angle θ at which the plane must be directed in order to travel in the direction of the runway. Also, what is its speed relative to the runway?

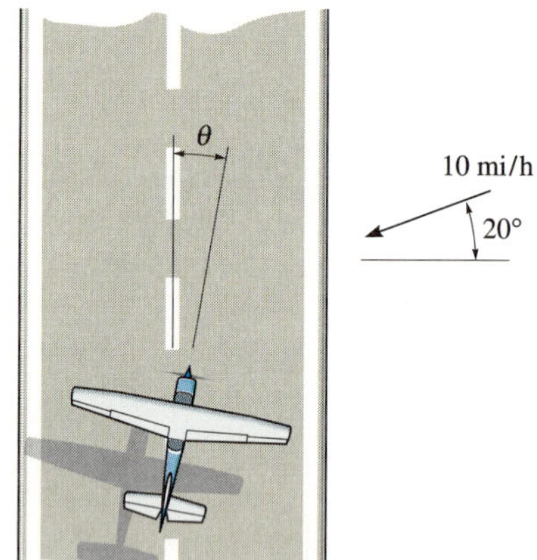

Prob. 12–205

12-206. The boy A is moving in a straight line away from the building at a constant speed of 4 ft/s. The boy C throws the ball B horizontally when A is at $d = 10$ ft. At what speed must C throw the ball so that A can catch it? Also determine the relative speed of the ball with respect to boy A at the instant the catch is made.

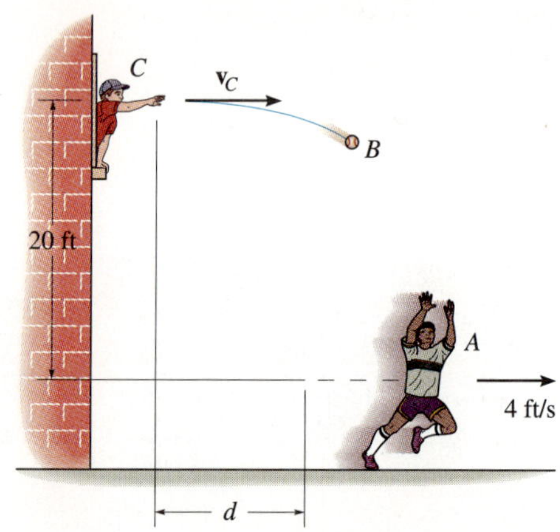

Prob. 12–206

12-207. The boy A is moving in a straight line away from the building at a constant speed of 4 ft/s. At what horizontal distance d must he be from C in order to make the catch if the ball is thrown with a horizontal velocity of $v_C = 10$ ft/s? Also determine the relative speed of the ball with respect to the boy A at the instant the catch is made.

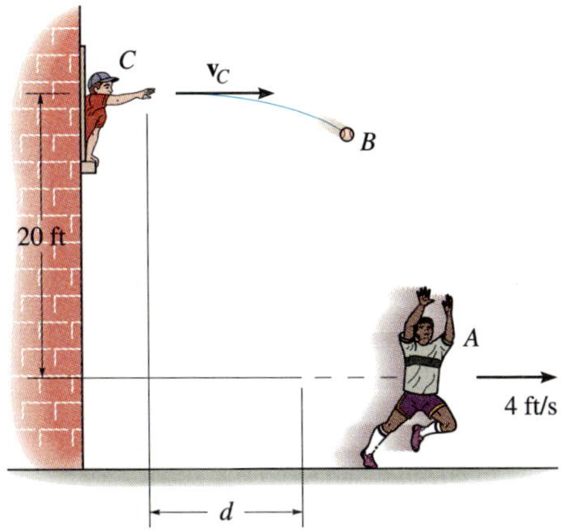

Prob. 12–207

***12-208.** At a given instant, two particles A and B are moving with a speed of 8 m/s along the paths shown. If B is decelerating at 6 m/s^2 and the speed of A is increasing at 5 m/s^2, determine the acceleration of A with respect to B at this instant.

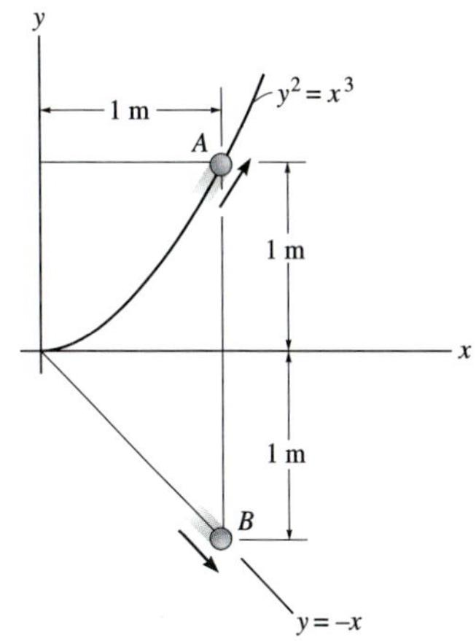

Prob. 12–208

Design Project

12–1D. DESIGN OF A MARBLE-SORTING DEVICE

Marbles in a manufacturing plant are produced with two diameters, namely, 0.5 in. and 0.75 in. If they both roll off the production chute at 0.5 ft/s, design a device that can be used to sort them out and allow them to fall into the separate hoppers. Submit a drawing of the device, and show the path the marbles take and the placement of the hoppers relative to the end of the chute.

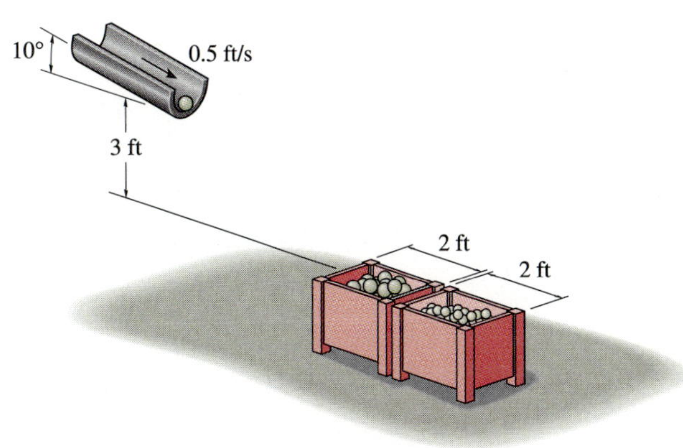

Prob. 12–1D

The design of conveyors for a bottling plant requires knowledge of the forces that act on them and the ability to predict the motion of the bottles they transport.

CHAPTER
13

Kinetics of a Particle: Force and Acceleration

Chapter Objectives

- To state Newton's Laws of Motion and Gravitational Attraction and to define mass and weight.
- To analyze the accelerated motion of a particle using the equation of motion with different coordinate systems.
- To investigate central-force motion and apply it to problems in space mechanics.

13.1 Newton's Laws of Motion

Many of the earlier notions about dynamics were dispelled after 1590 when Galileo performed experiments to study the motions of pendulums and falling bodies. The conclusions drawn from these experiments gave some insight as to the effects of forces acting on bodies in motion. The general laws of motion of a body subjected to forces were not known, however, until 1687, when Isaac Newton first presented three basic laws governing the motion of a particle. In a slightly reworded form, Newton's three laws of motion can be stated as follows:

First Law: *A particle originally at rest, or moving in a straight line with a constant velocity, will remain in this state provided the particle is not subjected to an unbalanced force.*

Second Law: *A particle acted upon by an unbalanced force* **F** *experiences an acceleration* **a** *that has the same direction as the force and a magnitude that is directly proportional to the force.**

Third Law: *The mutual forces of action and reaction between two particles are equal, opposite, and collinear.*

*Stated another way, the unbalanced force acting on the particle is proportional to the time rate of change of the particle's linear momentum. See footnote † on next page.

The first and third laws were used extensively in developing the concepts of statics. Although these laws are also considered in dynamics, Newton's second law of motion forms the basis for most of this study, since this law relates the accelerated motion of a particle to the forces that act on it.

Measurements of force and acceleration can be recorded in a laboratory so that in accordance with the second law, if a known unbalanced force **F** is applied to a particle, the acceleration **a** of the particle may be measured. Since the force and acceleration are directly proportional, the constant of proportionality, m, may be determined from the ratio $m = F/a$.* The positive scalar m is called the *mass* of the particle. Being constant during any acceleration, m provides a quantitative measure of the resistance of the particle to a change in its velocity.

If the mass of the particle is m, Newton's second law of motion may be written in mathematical form as

$$\mathbf{F} = m\mathbf{a}$$

This equation, which is referred to as the *equation of motion,* is one of the most important formulations in mechanics.† As previously stated, its validity is based solely on *experimental evidence.* In 1905, however, Albert Einstein developed the theory of relativity and placed limitations on the use of Newton's second law for describing general particle motion. Through experiments it was proven that *time* is not an absolute quantity as assumed by Newton; as a result, the equation of motion fails to predict the exact behavior of a particle, especially when the particle's speed approaches the speed of light (0.3 Gm/s). Developments of the theory of quantum mechanics by Erwin Schrödinger and others indicate further that conclusions drawn from using this equation are also invalid when particles are the size of an atom and move close to one another. For the most part, however, these requirements regarding particle speed and size are not encountered in engineering problems, so their effects will not be considered in this book.

*Recall that the units of force in the SI system and mass in the FPS system are derived from this equation, where $\mathrm{N} = \mathrm{kg\ m/s^2}$ and $\mathrm{slug} = \mathrm{lb\ s^2/ft}$ (see Sec. 1.3 of *Statics*). If, however, the units of force, mass, length, and time were *all* selected arbitrarily, then it is necessary to write $F = kma$, where k (a dimensionless constant) would have to be determined experimentally in order to preserve the equality.

†Since m is constant, we can also write $\mathbf{F} = d(m\mathbf{v})/dt$, where $m\mathbf{v}$ is the particle's linear momentum.

Newton's Law of Gravitational Attraction. Shortly after formulating his three laws of motion, Newton postulated a law governing the mutual attraction between any two particles. In mathematical form this law can be expressed as

$$F = G\frac{m_1 m_2}{r^2} \tag{13–1}$$

where

F = force of attraction between the two particles

G = universal constant of gravitation; according to experimental evidence $G = 66.73(10^{-12})\ \text{m}^3/(\text{kg}\cdot\text{s}^2)$

m_1, m_2 = mass of each of the two particles

r = distance between the centers of the two particles

Any two particles or bodies have a mutually attractive gravitational force acting between them. In the case of a particle located at or near the surface of the earth, however, the only gravitational force having any sizable magnitude is that between the earth and the particle. This force is termed the "weight" and, for our purpose, it will be the only gravitational force considered.

Mass and Weight. *Mass* is a property of matter by which we can compare the response of one body with that of another. As indicated above, this property manifests itself as a gravitational attraction between two bodies and provides a quantitative measure of the resistance of matter to a change in velocity. It is an *absolute* quantity since the measurement of mass can be made at any location. The weight of a body, however, is *not absolute* since it is measured in a gravitational field, and hence its magnitude depends on where the measurement is made. From Eq. 13–1, we can develop a general expression for finding the weight W of a particle having a mass $m_1 = m$. Let m_2 be the mass of the earth and r the distance between the earth's center and the particle. Then, if $g = Gm_2/r^2$, we have

$$W = mg$$

By comparison with $F = ma$, we term g the acceleration due to gravity. For most engineering calculations g is measured at a point on the surface of the earth at sea level, and at a latitude of 45°, considered the "standard location."

The mass and weight of a body are measured differently in the SI and FPS systems of units, and the method of defining these units should be thoroughly understood.

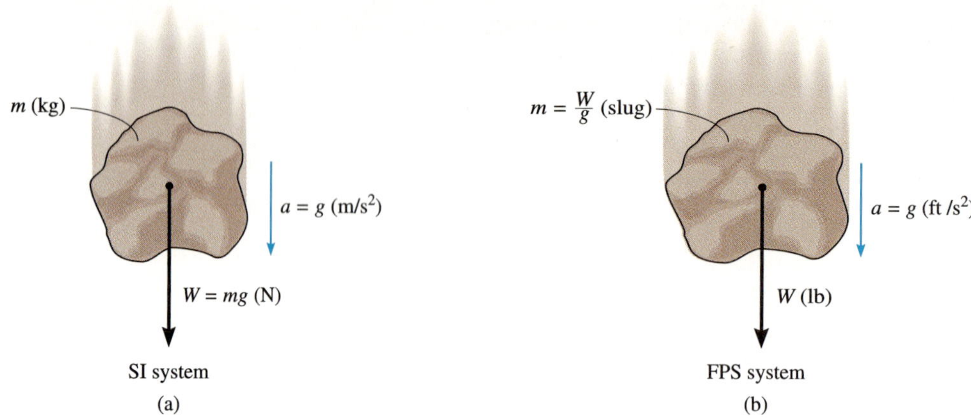

Fig. 13–1

SI System of Units. In the SI system the mass of the body is specified in kilograms, and the weight must be calculated using the equation of motion, $F = ma$. Hence, if a body has a mass m (kg) and is located at a point where the acceleration due to gravity is g (m/s^2), then the weight is expressed in *newtons* as $W = mg$ (N), Fig. 13–1*a*. In particular, if the body is located at the "standard location," the acceleration due to gravity is $g = 9.806\,65\ \text{m/s}^2$. For calculations, the value $g = 9.81\ \text{m/s}^2$ will be used, so that

$$W = mg\ (\text{N}) \qquad (g = 9.81\ \text{m/s}^2) \tag{13–2}$$

Therefore, a body of mass 1 kg has a weight of 9.81 N; a 2-kg body weighs 19.62 N; and so on.

FPS System of Units. In the FPS system the weight of the body is specified in pounds, and the mass must be calculated from $F = ma$. Hence, if a body has a weight W (lb) and is located at a point where the acceleration due to gravity is g (ft/s^2), then the mass is expressed in *slugs* as $m = W/g$ (slug), Fig. 13–1*b*. Since the acceleration of gravity at the standard location is approximately 32.2 ft/s^2(= 9.81 m/s^2), the mass of the body measured in slugs is

$$m = \frac{W}{g}\ (\text{slug}) \qquad (g = 32.2\ \text{ft/s}^2) \tag{13–3}$$

Therefore, a body weighing 32.2 lb has a mass of 1 slug; a 64.4-lb body has a mass of 2 slugs; and so on.

13.2 The Equation of Motion

When more than one force acts on a particle, the resultant force is determined by a vector summation of all the forces; i.e., $\mathbf{F}_R = \Sigma \mathbf{F}$. For this more general case, the equation of motion may be written as

$$\Sigma \mathbf{F} = m\mathbf{a} \tag{13–4}$$

To illustrate application of this equation, consider the particle P shown in Fig. 13–2*a*, which has a mass m and is subjected to the action of two forces, $\mathbf{F}_1$ and $\mathbf{F}_2$. We can graphically account for the magnitude and direction of each force acting on the particle by drawing the particle's *free-body diagram,* Fig. 13–2*b*. Since the *resultant* of these forces *produces* the vector $m\mathbf{a}$, its magnitude and direction can be represented graphically on the *kinetic diagram,* shown in Fig. 13–2*c*.* The equal sign written between the diagrams symbolizes the *graphical* equivalency between the free-body diagram and the kinetic diagram; i.e., $\Sigma \mathbf{F} = m\mathbf{a}$.† In particular, note that if $\mathbf{F}_R = \Sigma \mathbf{F} = \mathbf{0}$, then the acceleration is also zero, so that the particle will either remain at *rest* or move along a straight-line path with *constant velocity.* Such are the conditions of *static equilibrium,* Newton's first law of motion.

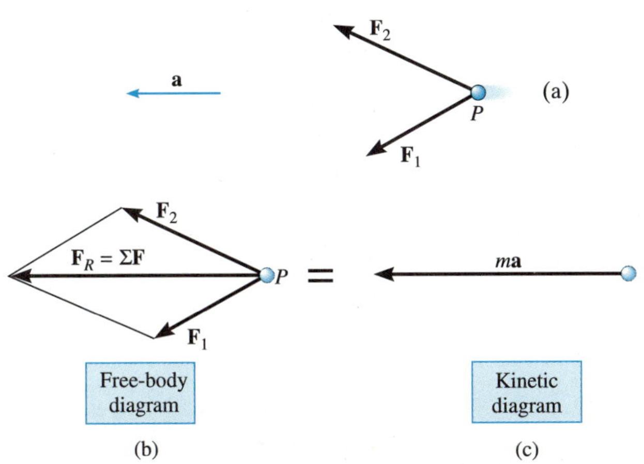

Fig. 13–2

*Recall the free-body diagram considers the particle to be free of its surroundings and shows all the forces acting on the particle. The kinetic diagram pertains to the particle's motion as caused by the forces.

†The equation of motion can also be rewritten in the form $\Sigma \mathbf{F} - m\mathbf{a} = \mathbf{0}$. The vector $-m\mathbf{a}$ is referred to as the *inertia force vector.* If it is treated in the same way as a "force vector," then the state of "equilibrium" created is referred to as *dynamic equilibrium.* This method for application is often referred to as the *D'Alembert principle,* named after the French mathematician Jean le Rond d'Alembert.

Inertial Frame of Reference. Whenever the equation of motion is applied, it is required that measurements of the acceleration be made from a *Newtonian* or *inertial frame of reference. Such a coordinate system does not rotate and is either fixed or translates in a given direction with a constant velocity (zero acceleration).* This definition ensures that the particle's *acceleration* measured by observers in two different inertial frames of reference will always be the *same.* For example, consider the particle P moving with an absolute acceleration $\mathbf{a}_P$ along a straight path as shown in Fig. 13–3. If the observer is *fixed* in the inertial x, y frame of reference, this acceleration, $\mathbf{a}_P$, will be measured by the observer regardless of the direction and magnitude of the velocity $\mathbf{v}_O$ of the frame of reference. On the other hand, if the observer is *fixed* in the noninertial x', y' frame of reference, Fig. 13–3, the observer will not measure the particle's acceleration as $\mathbf{a}_P$. Instead, if the frame is *accelerating* at $\mathbf{a}_{O'}$ the particle will appear to have an acceleration of $\mathbf{a}_{P/O'} = \mathbf{a}_P - \mathbf{a}_{O'}$. Also, if the frame is *rotating,* as indicated by the curl, then the particle will appear to move along a *curved path,* in which case it will appear to have other components of acceleration (see Sec. 16.8). In any case, the measured acceleration from this observer cannot be used in Newton's law of motion to determine the forces acting on the particle.

When studying the motions of rockets and satellites, it is justifiable to consider the inertial reference frame as fixed to the stars, whereas dynamics problems concerned with motions on or near the surface of the earth may be solved by using an inertial frame which is assumed fixed to the earth. Even though the earth both rotates about its own axis and revolves about the sun, the accelerations created by these rotations are relatively small and can be neglected in most computations.

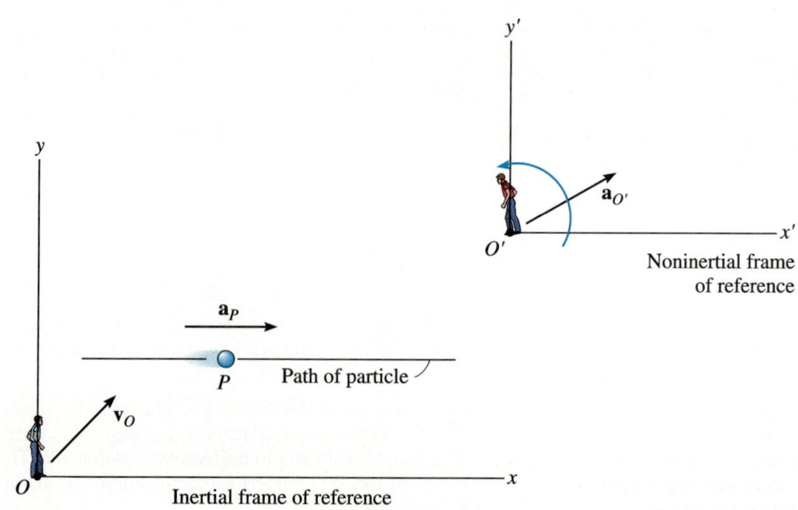

Fig. 13–3

Fig. 1

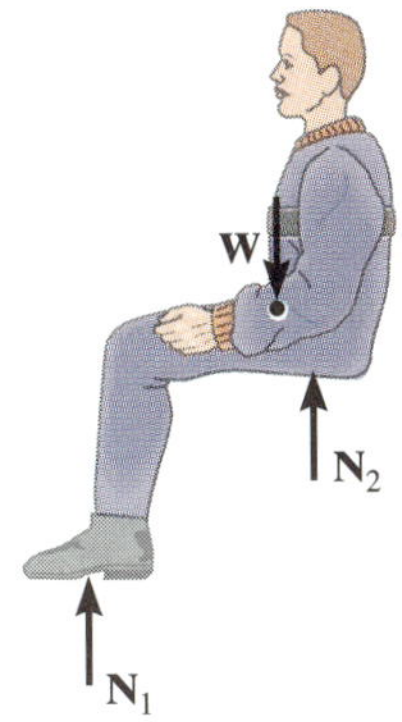

At rest or constant velocity

Fig. 1

We are all familiar with the sensation one feels when sitting in a car that is subjected to a forward acceleration. Often people think this is caused by a "force" which acts on them and tends to push them back in their seats; however, this is not the case. Instead, this sensation occurs due to their inertia or the resistance of their mass to a change in velocity.

Consider the passenger in Fig. 1 who is strapped to the seat of a rocket sled. Provided the sled is at rest or is moving with constant velocity, then no force is exerted on his back as shown on his free-body diagram.

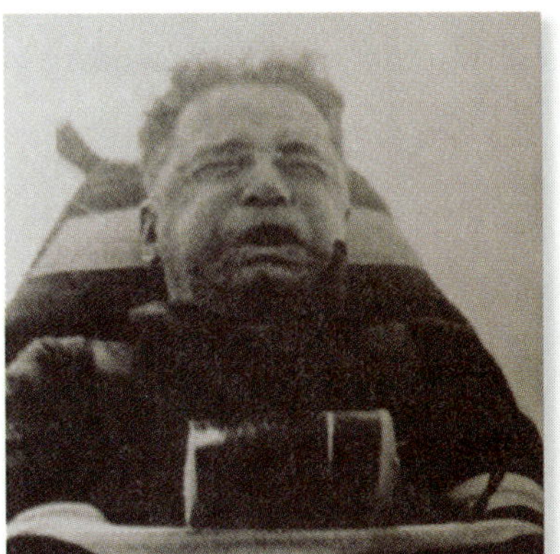

Fig. 2

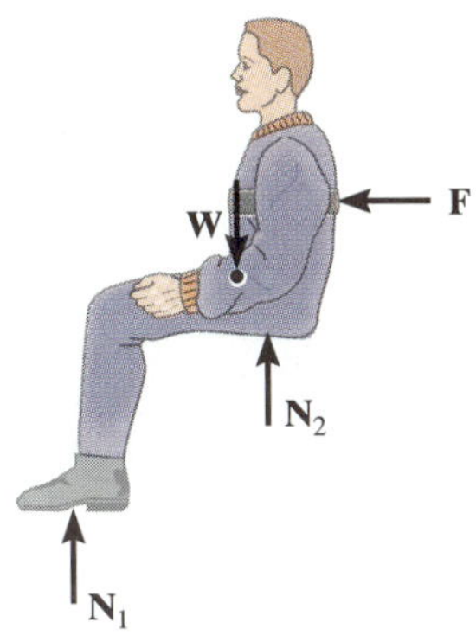

Acceleration

Fig. 2

When the thrust of the rocket engine causes the sled to accelerate, then the seat upon which he is sitting exerts a force **F** on him which pushes him forward, Fig. 2. In the photo, notice that the inertia of his head resists this change in motion (acceleration), and so his head moves back against the seat and his face, which is nonrigid, tends to distort.

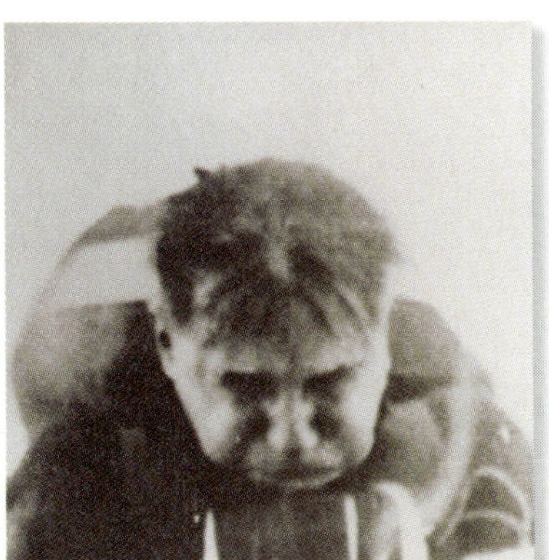

Fig. 3

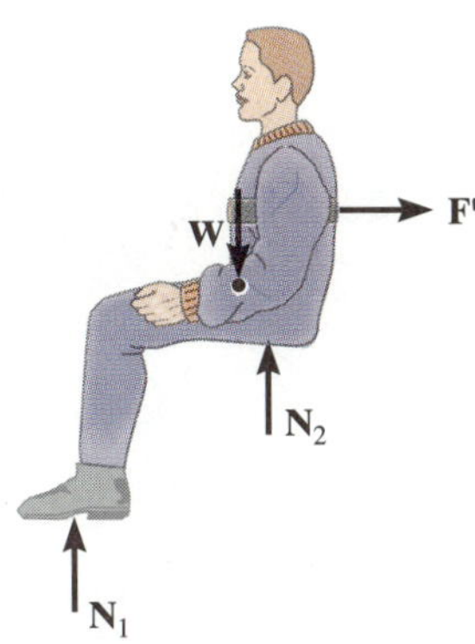

Deceleration

Fig. 3

Upon deceleration, Fig. 3, the force of the seat **F**′ tends to pull him to a stop, and so his head leaves contact with the back of the seat and his face distorts forward, again due to his inertia or tendency to continue to move forward. No force is pulling him forward, although this is the sensation he receives.

13.3 Equation of Motion for a System of Particles

The equation of motion will now be extended to include a system of n particles isolated within an enclosed region in space, as shown in Fig. 13–4*a*. In particular, there is no restriction in the way the particles are connected, and as a result the following analysis will apply equally well to the motion of a solid, liquid, or gas system. At the instant considered, the arbitrary ith particle, having a mass m_i, is subjected to a system of internal forces and a resultant external force. The *resultant internal force*, represented symbolically as $\mathbf{f}_i$, is determined from the forces which the other particles exert on the ith particle. Usually these forces are developed by direct contact, although the summation extends over all n particles within the dashed boundary. The *resultant external force* $\mathbf{F}_i$ represents, for example, the effect of gravitational, electrical, magnetic, or contact forces between the ith particle and adjacent bodies or particles *not* included within the system.

The free-body and kinetic diagrams for the ith particle are shown in Fig. 13–4*b*. Applying the equation of motion yields

$$\Sigma \mathbf{F} = m\mathbf{a}; \qquad \mathbf{F}_i + \mathbf{f}_i = m_i\mathbf{a}_i$$

When the equation of motion is applied to each of the other particles of the system, similar equations will result. If all these equations are added together *vectorially*, we obtain

$$\Sigma \mathbf{F}_i + \Sigma \mathbf{f}_i = \Sigma m_i\mathbf{a}_i$$

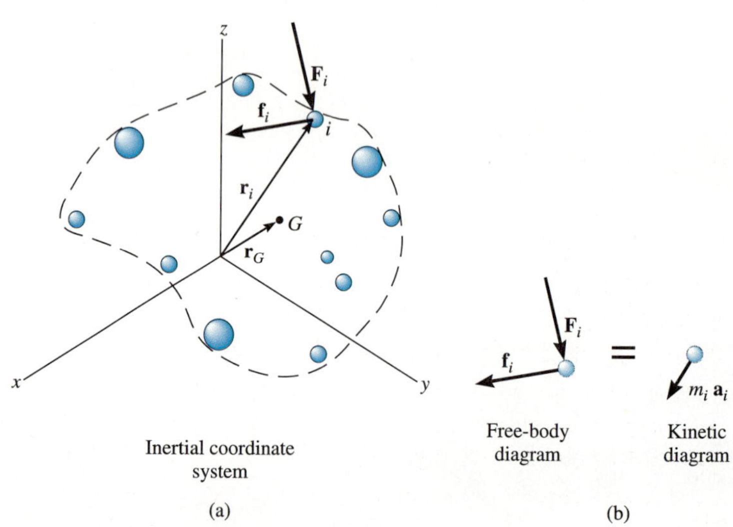

Fig. 13–4

The summation of the internal forces, if carried out, will equal zero, since internal forces between particles all occur in equal but opposite collinear pairs. Consequently, only the sum of the external forces will remain, and therefore the equation of motion, written for the system of particles, becomes

$$\Sigma \mathbf{F}_i = \Sigma m_i \mathbf{a}_i \qquad (13\text{–}5)$$

If $\mathbf{r}_G$ is a position vector which locates the *center of mass* G of the particles, Fig. 13–4*a*, then by definition of the center of mass, $m\mathbf{r}_G = \Sigma m_i \mathbf{r}_i$, where $m = \Sigma m_i$ is the total mass of all the particles. Differentiating this equation twice with respect to time, assuming that no mass is entering or leaving the system, yields

$$m\mathbf{a}_G = \Sigma m_i \mathbf{a}_i$$

Substituting this result into Eq. 13–5, we obtain

$$\boxed{\Sigma \mathbf{F} = m\mathbf{a}_G} \qquad (13\text{–}6)$$

Hence, the sum of the external forces acting on the system of particles is equal to the total mass of the particles times the acceleration of its center of mass G. Since in reality all particles must have a finite size to possess mass, Eq. 13–6 justifies application of the equation of motion to a *body* that is represented as a single particle.

Important Points

- The equation of motion is based on experimental evidence and is valid only when applied from an inertial frame of reference.
- The equation of motion states that the unbalanced force on a particle causes it to accelerate.
- An inertial frame of reference has axes that either translate with constant velocity or are at rest.
- Mass is a property of matter that provides a quantitative measure of its resistance to a change in velocity. It is an absolute quantity.
- Weight is a force that is caused by the earth's gravitation. It is not absolute; rather it depends on the altitude of the mass from the earth's surface.

13.4 Equations of Motion: Rectangular Coordinates

When a particle is moving relative to an inertial x, y, z frame of reference, the forces acting on the particle, as well as its acceleration, may be expressed in terms of their **i, j, k** components, Fig. 13–5. Applying the equation of motion, we have

$$\Sigma \mathbf{F} = m\mathbf{a}$$

$$\Sigma F_x \mathbf{i} + \Sigma F_y \mathbf{j} + \Sigma F_z \mathbf{k} = m(a_x \mathbf{i} + a_y \mathbf{j} + a_z \mathbf{k})$$

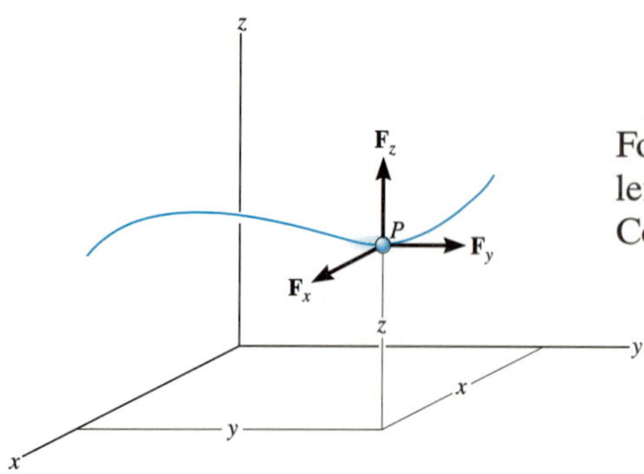

Fig. 13–5

For this equation to be satisfied, the respective **i, j, k** components on the left side must equal the corresponding components on the right side. Consequently, we may write the following three scalar equations:

$$\begin{aligned} \Sigma F_x &= ma_x \\ \Sigma F_y &= ma_y \\ \Sigma F_z &= ma_z \end{aligned} \qquad (13\text{–}7)$$

In particular, if the particle is constrained to move only in the x–y plane, then the first two of these equations are used to specify the motion.

Procedure for Analysis

The equations of motion are used to solve problems which require a relationship between the forces acting on a particle and the accelerated motion they cause.

Free-Body Diagram

- Select the inertial coordinate system. Most often, rectangular or x, y, z coordinates are chosen to analyze problems for which the particle has *rectilinear motion.*
- Once the coordinates are established, draw the particle's free-body diagram. Drawing this diagram is *very important* since it provides a graphical representation that accounts for *all the forces* ($\Sigma\mathbf{F}$) which act on the particle, and thereby makes it possible to resolve these forces into their x, y, z components.
- The direction and sense of the particle's acceleration **a** should also be established. If the senses of its components are unknown, for mathematical convenience assume that they are in the *same direction* as the *positive* inertial coordinate axes.
- The acceleration may be represented as the $m\mathbf{a}$ vector on the kinetic diagram.*
- Identify the unknowns in the problem.

*It is a convention in this text always to use the kinetic diagram as a graphical aid when developing the proofs and theory. The particle's acceleration or its components will be shown as blue colored vectors near the free-body diagram in the examples.

Equations of Motion

- If the forces can be resolved directly from the free-body diagram, apply the equations of motion in their scalar component form.
- If the geometry of the problem appears complicated, which often occurs in three dimensions, Cartesian vector analysis can be used for the solution.
- *Friction.* If the particle contacts a rough surface, it may be necessary to use the *frictional equation,* which relates the coefficient of kinetic friction μ_k to the magnitudes of the frictional and normal forces $\mathbf{F}_f$ and $\mathbf{N}$ acting at the surfaces of contact, i.e., $F_f = \mu_k N$. Remember that $\mathbf{F}_f$ always acts on the free-body diagram such that it opposes the motion of the particle relative to the surface it contacts.
- *Spring.* If the particle is connected to an *elastic spring* having negligible mass, the spring force F_s can be related to the deformation of the spring by the equation $F_s = ks$. Here k is the spring's stiffness measured as a force per unit length, and s is the stretch or compression defined as the difference between the deformed length l and the undeformed length l_0, i.e., $s = l - l_0$.

Kinematics

- If the velocity or position of the particle is to be found, it will be necessary to apply the proper kinematic equations once the particle's acceleration is determined from $\Sigma\mathbf{F} = m\mathbf{a}$.
- If *acceleration is a function of time,* use $a = dv/dt$ and $v = ds/dt$ which, when integrated, yield the particle's velocity and position.
- If *acceleration is a function of displacement,* integrate $a\,ds = v\,dv$ to obtain the velocity as a function of position.
- If *acceleration is constant,* use $v = v_0 + a_c t$, $s = s_0 + v_0 t + \frac{1}{2}a_c t^2$, $v^2 = v_0^2 + 2a_c(s - s_0)$ to determine the velocity or position of the particle.
- If the problem involves the dependent motion of several particles, use the method outlined in Sec. 12.9 to relate their accelerations.
- In all cases, make sure the positive inertial coordinate directions used for writing the kinematic equations are the same as those used for writing the equations of motion; otherwise, simultaneous solution of the equations will result in errors.
- If the solution for an unknown vector component yields a negative scalar, it indicates that the component acts in the direction opposite to that which was assumed.

EXAMPLE 13–1

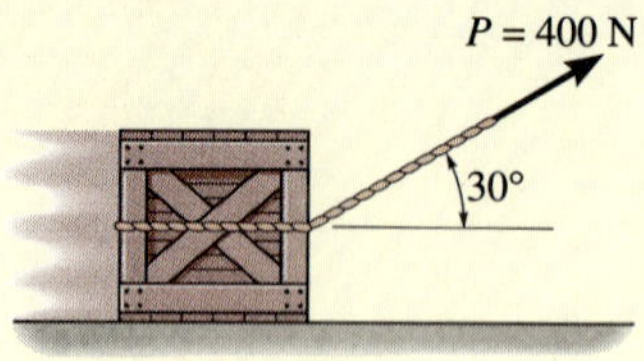

(a)

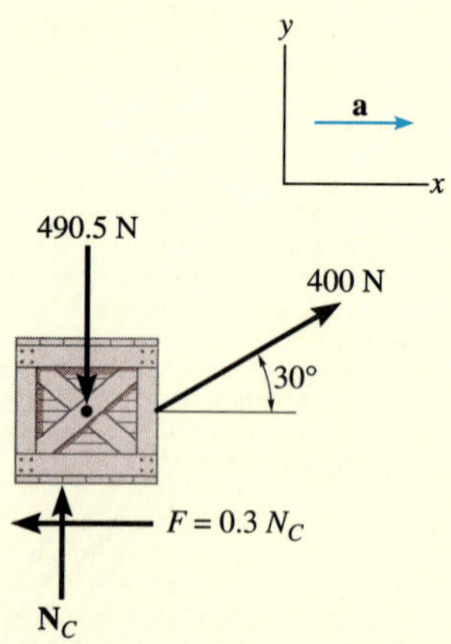

(b)

Fig. 13–6

The 50-kg crate shown in Fig. 13–6*a* rests on a horizontal plane for which the coefficient of kinetic friction is $\mu_k = 0.3$. If the crate is subjected to a 400-N towing force as shown, determine the velocity of the crate in 3 s starting from rest.

Solution

Using the equations of motion, we can relate the crate's acceleration to the force causing the motion. The crate's velocity can then be determined using kinematics.

Free-Body Diagram. The weight of the crate is $W = mg = 50\text{ kg}\,(9.81\text{ m/s}^2) = 490.5\text{ N}$. As shown in Fig. 13–6*b*, the frictional force has a magnitude $F = \mu_k N_C$ and acts to the left, since it opposes the motion of the crate. The acceleration **a** is assumed to act horizontally, in the positive *x* direction. There are two unknowns, namely N_C and *a*. (We can also use the alternative procedure of drawing the crate's free-body *and* kinetic diagrams, Fig. 13–6*c*, prior to applying the equations of motion.)

Equations of Motion. Using the data shown on the free-body diagram, we have

$$\overset{+}{\rightarrow}\Sigma F_x = ma_x; \qquad 400 \cos 30^\circ - 0.3\, N_C = 50a \qquad (1)$$

$$+\uparrow \Sigma F_y = ma_y; \qquad N_C - 490.5 + 400 \sin 30^\circ = 0 \qquad (2)$$

Solving Eq. 2 for N_C, substituting the result into Eq. 1, and solving for *a* yields

$$N_C = 290.5\text{ N}$$

$$a = 5.19\text{ m/s}^2$$

Kinematics. Note that the acceleration is *constant,* since the applied force **P** is constant. Since the initial velocity is zero, the velocity of the crate in 3 s is

$$(\overset{+}{\rightarrow}) \qquad v = v_0 + a_c t$$

$$= 0 + 5.19(3)$$

$$= 15.6\text{ m/s} \rightarrow \qquad \textit{Ans.}$$

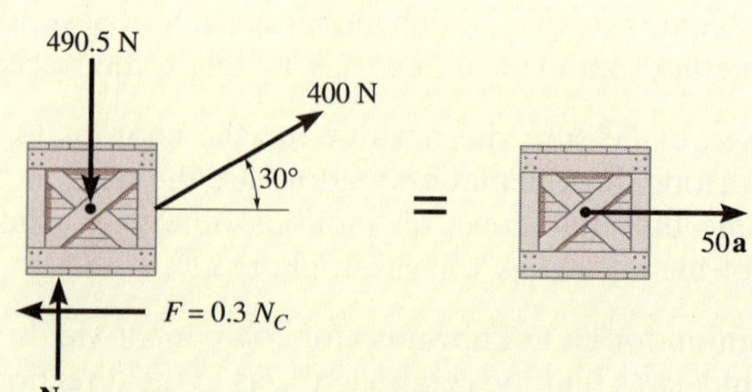

(c)

EXAMPLE 13–2

A 10-kg projectile is fired vertically upward from the ground, with an initial velocity of 50 m/s, Fig. 13–7*a*. Determine the maximum height to which it will travel if (*a*) atmospheric resistance is neglected; and (*b*) atmospheric resistance is measured as $F_D = (0.01v^2)$ N, where v is the speed at any instant, measured in m/s.

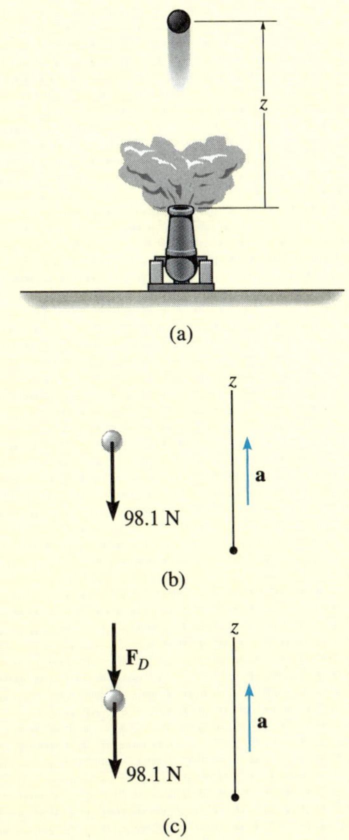

Fig. 13–7

Solution

In both cases the known force on the projectile can be related to its acceleration using the equation of motion. Kinematics can then be used to relate the projectile's acceleration to its position.

Part (a) Free-Body Diagram. As shown in Fig. 13–7*b*, the projectile's weight is $W = mg = 10(9.81) = 98.1$ N. We will assume the unknown acceleration **a** acts upward in the *positive* z direction.

Equation of Motion

$+\uparrow \Sigma F_z = ma_z;$ $\qquad -98.1 = 10a, \qquad a = -9.81 \text{ m/s}^2$

The result indicates that the projectile, like every object having free-flight motion near the earth's surface, is subjected to a *constant* downward acceleration of 9.81 m/s².

Kinematics. Initially, $z_0 = 0$ and $v_0 = 50$ m/s, and at the maximum height $z = h$, $v = 0$. Since the acceleration is *constant*, then

$(+\uparrow)$
$$v^2 = v_0^2 + 2a_c(z - z_0)$$
$$0 = (50)^2 + 2(-9.81)(h - 0)$$
$$h = 127 \text{ m} \qquad \textit{Ans.}$$

Part (b) Free-Body Diagram. Since the force $F_D = (0.01v^2)$ N tends to retard the upward motion of the projectile, it acts downward as shown on the free-body diagram, Fig. 13–7*c*.

Equation of Motion*

$+\uparrow \Sigma F_z = ma_z;$ $\qquad -0.01v^2 - 98.1 = 10a, \qquad a = -0.001v^2 - 9.81$

Kinematics. Here the acceleration is *not constant* since F_D depends on the velocity. Since $a = f(v)$, we can relate a to position using

$(+\uparrow)\ a\,dz = v\,dv;\quad (-0.001v^2 - 9.81)dz = v\,dv$

Separating the variables and integrating, realizing that initially $z_0 = 0$, $v_0 = 50$ m/s (positive upward), and at $z = h$, $v = 0$, we have

$$\int_0^h dz = -\int_{50}^{0} \frac{v\,dv}{0.001v^2 + 9.81} = -500 \ln (v^2 + 9810)\Big|_{50}^{0}$$

$$h = 114 \text{ m} \qquad \textit{Ans.}$$

The answer indicates a lower elevation than that obtained in part (*a*) due to atmospheric resistance.

*Note that if the projectile were fired downward, with z positive downward, the equation of motion would then be $-0.01v^2 + 98.1 = 10a$.

EXAMPLE 13-3

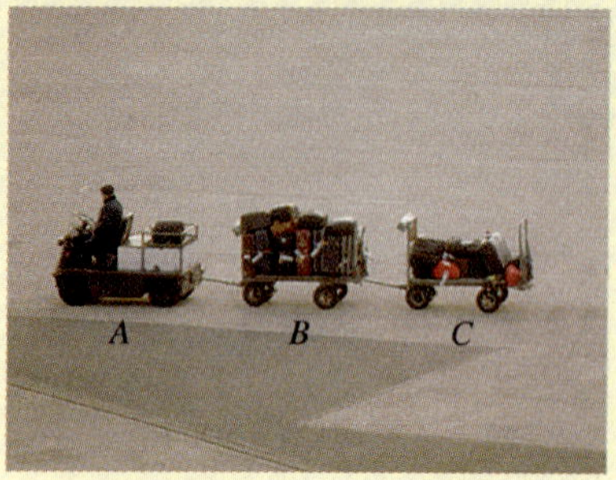

The baggage truck A shown in the photo has a weight of 900 lb and tows a 550-lb cart B and a 325-lb cart C. For a short time the driving frictional force developed at the wheels of the truck is $F_A = (40t)$ lb, where t is in seconds. If the truck starts from rest, determine its speed in 2 seconds. Also, what is the horizontal force acting on the coupling between the truck and cart B at this instant? Neglect the size of the truck and carts.

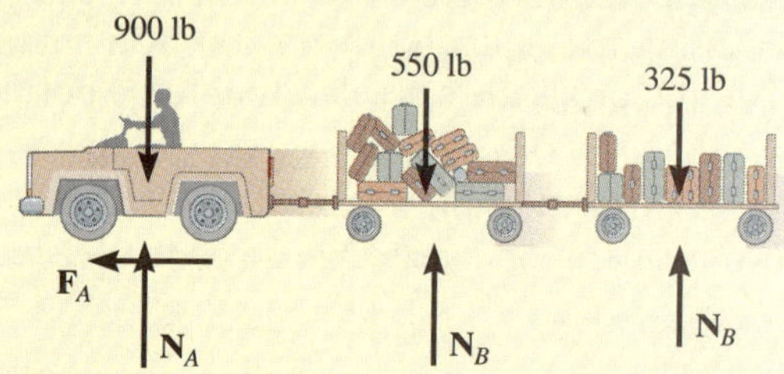

(a)

Solution

Free-Body Diagram. As shown in Fig. 13–8*a*, it is the frictional driving force that gives both the truck and carts an acceleration. Here we have considered all three vehicles.

Equation of Motion. Only motion in the horizontal direction has to be considered.

$$\overset{+}{\leftarrow} \Sigma F_x = ma_x; \qquad 40t = \left(\frac{900 + 550 + 325}{32.2}\right)a$$

$$a = 0.7256t$$

Kinematics. Since the acceleration is a function of time, the velocity of the truck is obtained using $a = dv/dt$ with the initial condition that $v_0 = 0$ at $t = 0$. We have

$$\int_0^v dv = \int_0^2 0.7256t\,dt; \quad v = 0.3628t^2 \Big|_0^2 = 1.45 \text{ ft/s} \qquad \textit{Ans.}$$

Free-Body Diagram. In order to determine the force between the truck and cart B, we can consider a free-body diagram of the truck so that we can "expose" the coupling force $\mathbf{T}$ as external to the free-body diagram, Fig. 13–8*b*.

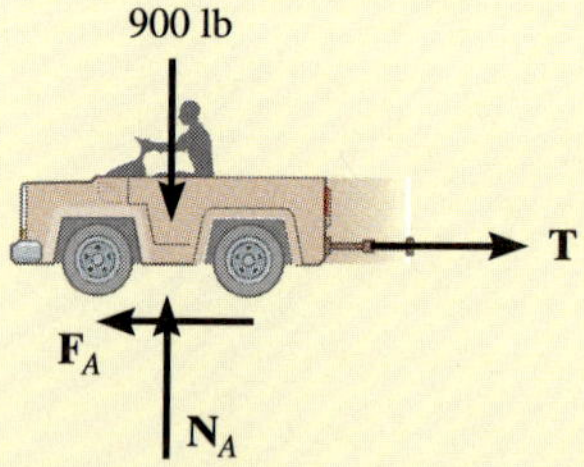

(b)

Fig. 13–8

Equation of Motion. When $t = 2$ s, then

$$\overset{+}{\leftarrow} \Sigma F_x = ma_x; \qquad 40(2) - T = \left(\frac{900}{32.2}\right)[0.7256(2)]$$

$$T = 39.4 \text{ lb} \qquad \textit{Ans.}$$

Try and obtain this same result by considering a free-body diagram of carts B and C.

EXAMPLE 13-4

A smooth 2-kg collar C, shown in Fig. 13–9a, is attached to a spring having a stiffness $k = 3$ N/m and an unstretched length of 0.75 m. If the collar is released from rest at A, determine its acceleration and the normal force of the rod on the collar at the instant $y = 1$ m.

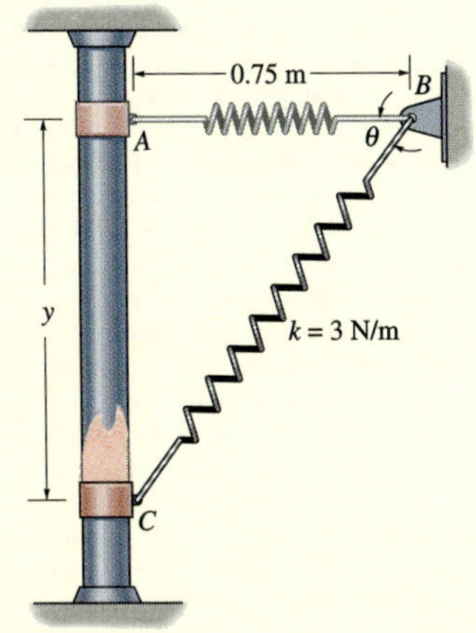

(a)

Solution

Free-Body Diagram. The free-body diagram of the collar when it is located at the arbitrary position y is shown in Fig. 13–9b. Note that the weight is $W = 2(9.81) = 19.62$ N. Furthermore, the collar is *assumed* to be accelerating so that "**a**" acts downward in the *positive* y direction. There are four unknowns, namely, N_C, F_s, a, and θ.

Equations of Motion

$$\overset{+}{\rightarrow}\Sigma F_x = ma_x; \qquad -N_C + F_s \cos\theta = 0 \qquad (1)$$

$$+\downarrow \Sigma F_y = ma_y; \qquad 19.62 - F_s \sin\theta = 2a \qquad (2)$$

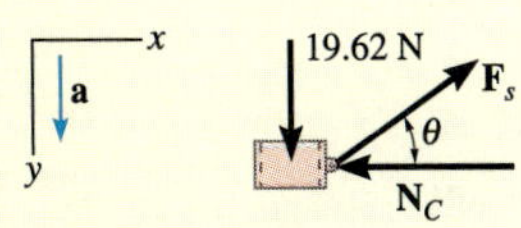

(b)

Fig. 13–9

From Eq. 2 it is seen that the acceleration depends on the magnitude and direction of the spring force. Solution for N_C and a is possible once F_s and θ are known.

The magnitude of the spring force is a function of the stretch s of the spring; i.e., $F_s = ks$. Here the unstretched length is $AB = 0.75$ m, Fig. 13–9a; therefore, $s = CB - AB = \sqrt{y^2 + (0.75)^2} - 0.75$. Since $k = 3$ N/m, then

$$F_s = ks = 3\left(\sqrt{y^2 + (0.75)^2} - 0.75\right) \qquad (3)$$

From Fig. 13–9a, the angle θ is related to y by trigonometry.

$$\tan\theta = \frac{y}{0.75} \qquad (4)$$

Substituting $y = 1$ m into Eqs. 3 and 4 yields $F_s = 1.50$ N and $\theta = 53.1°$. Substituting these results into Eqs. 1 and 2, we obtain

$$N_C = 0.900 \text{ N} \qquad \textit{Ans.}$$

$$a = 9.21 \text{ m/s}^2 \downarrow \qquad \textit{Ans.}$$

EXAMPLE 13–5

The 100-kg block A shown in Fig. 13–10a is released from rest. If the masses of the pulleys and the cord are neglected, determine the speed of the 20-kg block B in 2 s.

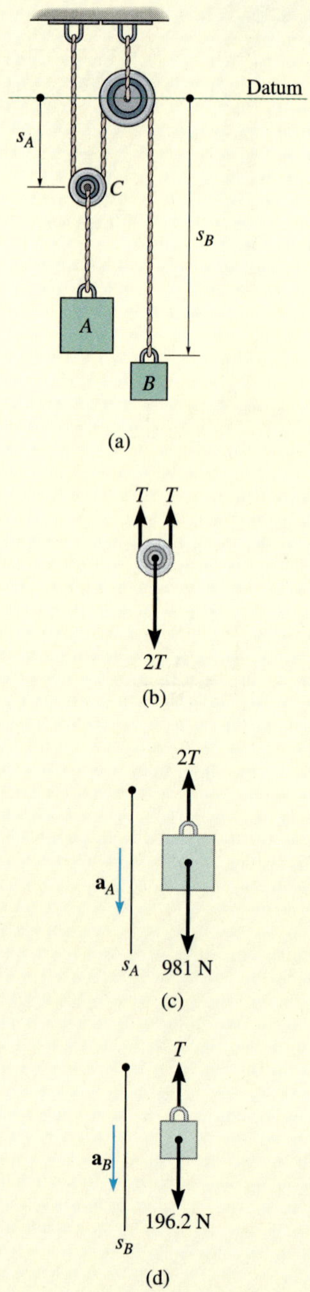

Fig. 13–10

Solution

Free-Body Diagrams. Since the mass of the pulleys is *neglected,* then for pulley C, $ma = 0$ and we can apply $\Sigma F_y = 0$ as shown in Fig. 13–10b. The free-body diagrams for blocks A and B are shown in Fig. 13–10c and d, respectively. One can see that for A to remain static requires $T = 490.5$ N, whereas for B to remain static requires $T = 196.2$ N. Hence A will move down while B moves up. Here we will *assume* both blocks accelerate downward, in the direction of $+s_A$ and $+s_B$. The three unknowns are T, a_A, and a_B.

Equations of Motion

Block A (Fig. 13–10c):

$$+\downarrow \Sigma F_y = ma_y; \qquad 981 - 2T = 100a_A \qquad (1)$$

Block B (Fig. 13–10d):

$$+\downarrow \Sigma F_y = ma_y; \qquad 196.2 - T = 20a_B \qquad (2)$$

Kinematics. The necessary third equation is obtained by relating a_A to a_B using a dependent motion analysis. Using the technique developed in Sec. 12.9, the coordinates s_A and s_B measure the positions of A and B from the fixed datum, Fig. 13–10a. It is seen that

$$2s_A + s_B = l$$

where l is constant and represents the total vertical length of cord. Differentiating this expression twice with respect to time yields

$$2a_A = -a_B \qquad (3)$$

Notice that in writing Eqs. 1 to 3, the *positive direction was always assumed downward.* It is very important to be *consistent* in this assumption since we are seeking a simultaneous solution of equations. The solution yields

$$T = 327.0 \text{ N}$$
$$a_A = 3.27 \text{ m/s}^2$$
$$a_B = -6.54 \text{ m/s}^2$$

Hence when block A accelerates *downward,* block B accelerates *upward.* Since a_B is constant, the velocity of block B in 2 s is thus

$$(+\downarrow) \qquad v = v_0 + a_B t$$
$$= 0 + (-6.54)(2)$$
$$= -13.1 \text{ m/s} \qquad \textit{Ans.}$$

The negative sign indicates that block B is moving upward.

Problems

13-1. Determine the gravitational attraction between two spheres which are just touching each other. Each sphere has a mass of 10 kg and a radius of 200 mm.

13-2. The 10-lb block has an initial velocity of 10 ft/s on the smooth plane. If a force $F = (2.5t)$ lb, where t is in seconds, acts on the block for 3 s, determine the final velocity of the block and the distance the block travels during this time.

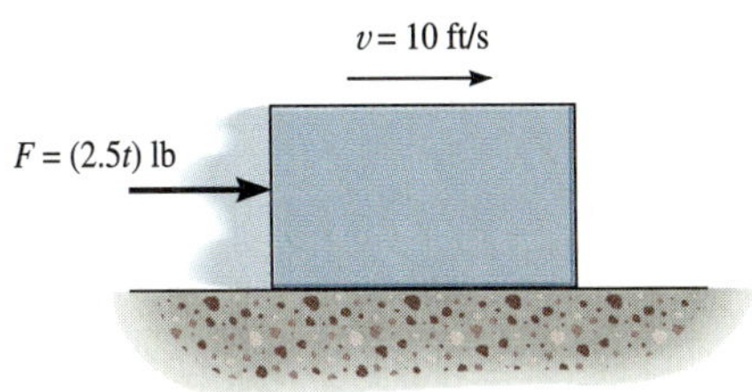

Prob. 13–2

13-3. The 300-kg bar B, originally at rest, is being towed over a series of small rollers. Determine the force in the cable when $t = 5$ s, if the motor M is drawing in the cable for a short time at a rate of $v = (0.4t^2)$ m/s, where t is in seconds $(0 \le t \le 6 \text{ s})$. How far does the bar move in 5 s? Neglect the mass of the cable, pulley, and the rollers.

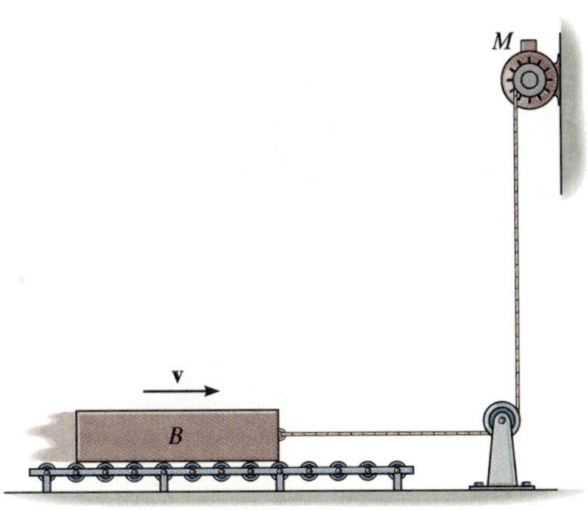

Prob. 13–3

***13-4.** The van is traveling at 20 km/h when the coupling of the trailer at A fails. If the trailer has a mass of 250 kg and coasts 45 m before coming to rest, determine the constant horizontal force F created by rolling friction which causes the trailer to stop.

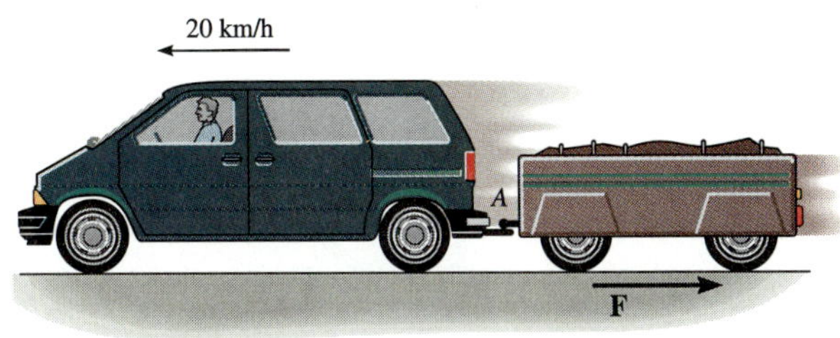

Prob. 13–4

13-5. A crate having a mass of 60 kg falls horizontally off the back of a truck which is traveling at 80 km/h. Determine the coefficient of kinetic friction between the road and the crate if the crate slides 45 m on the ground with no tumbling along the road before coming to rest. Assume that the initial speed of the crate along the road is 80 km/h.

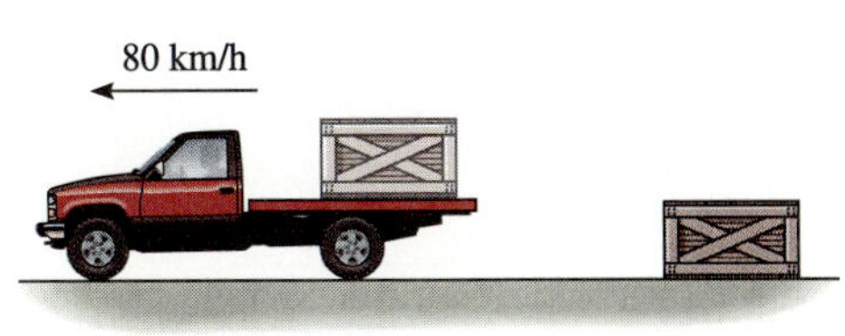

Prob. 13–5

13-6. The crane lifts the 700-kg bin with an initial acceleration of 3 m/s^2. Determine the force in each of the supporting cables due to this motion.

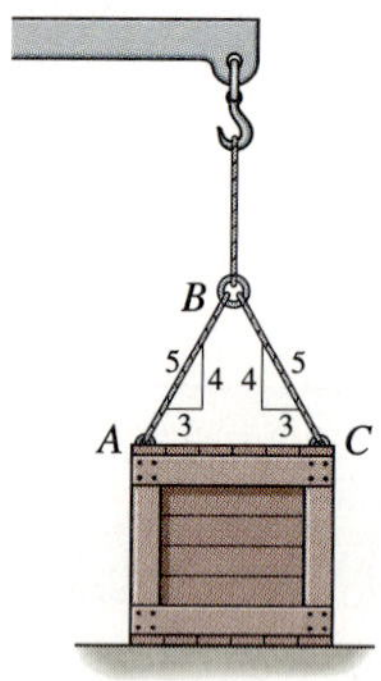

Prob. 13–6

13-7. The block has a mass m and is given a velocity $\mathbf{v}_0$ up the plane. If the angle of inclination is steep enough so that the block is able to slide down, determine the velocity of the block when it returns to the point at which it was launched. The coefficient of kinetic friction between the block and the plane is μ_k.

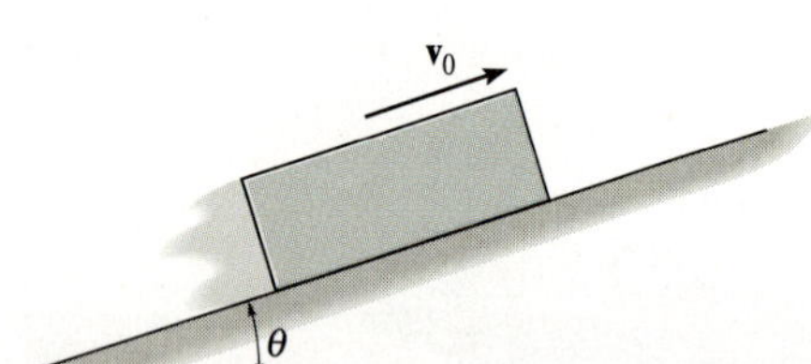

Prob. 13–7

***13-8.** The 200-kg crate is suspended from the cable of a crane. Determine the force in the cable when $t = 2$ s if the crate is moving upward with (a) a constant velocity of 2 m/s, and (b) a speed of $v = (0.2t^2 + 2)$ m/s, where t is in seconds.

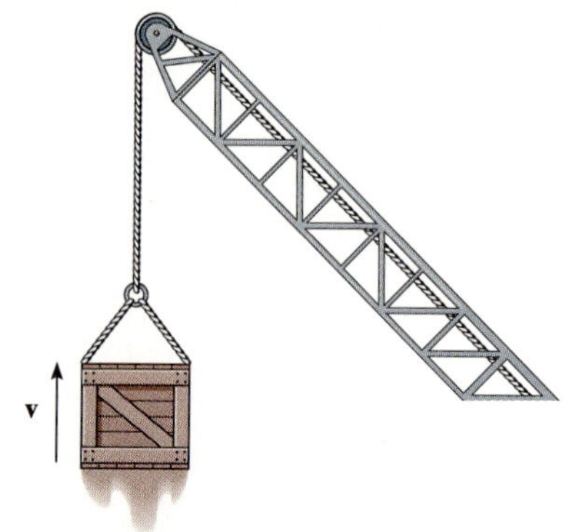

Prob. 13–8

13-9. The elevator E has a mass of 500 kg, and the counterweight at A has a mass of 150 kg. If the motor supplies a constant force of 5 kN on the cable at B, determine the speed of the elevator when $t = 3$ s, starting from rest. Neglect the mass of the pulleys and cable.

13-10. The elevator E has a mass of 500 kg and the counterweight at A has a mass of 150 kg. If the elevator attains a speed of 10 m/s after it rises 40 m, determine the constant force developed in the cable at B. Neglect the mass of the pulleys and cable.

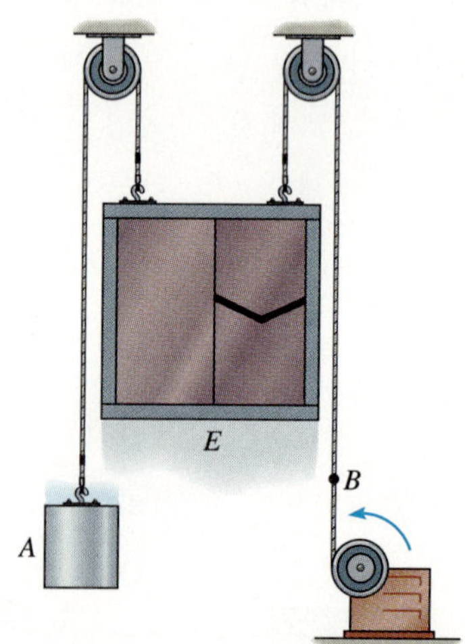

Probs. 13–9/10

13-11. The two boxcars A and B have a weight of 20 000 lb and 30 000 lb, respectively. If they are freely coasting down the incline when the brakes are applied to all the wheels of car A, determine the force in the coupling C between the two cars. The coefficient of kinetic friction between the wheels of A and the tracks is $\mu_k = 0.5$. The wheels of car B are free to roll. Neglect their mass in the calculation. *Suggestion:* Solve the problem by representing single resultant normal forces acting on A and B, respectively.

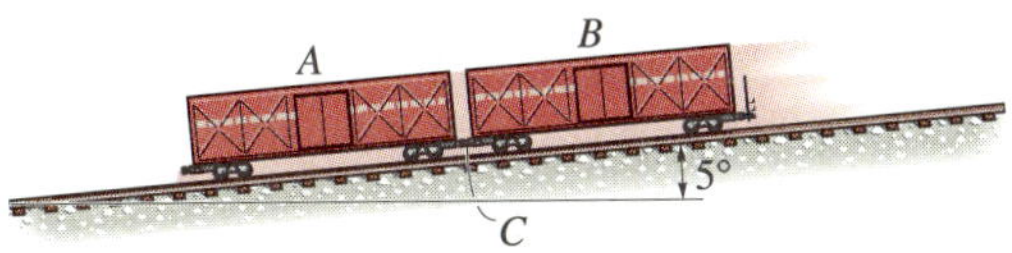

Prob. 13–11

***13-12.** The 6-lb particle is subjected to the action of its weight and forces $\mathbf{F}_1 = \{2\mathbf{i} + 6\mathbf{j} - 2t\mathbf{k}\}$ lb, $\mathbf{F}_2 = \{t^2\mathbf{i} - 4t\mathbf{j} - 1\mathbf{k}\}$ lb, and $\mathbf{F}_3 = \{-2t\mathbf{i}\}$ lb, where t is in seconds. Determine the distance the ball is from the origin 2 s after being released from rest.

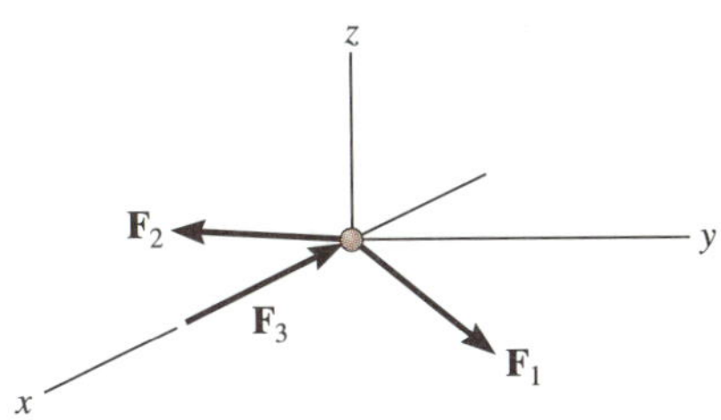

Prob. 13–12

13-13. The 2-lb particle A is acted upon by its weight and the force system $\mathbf{F}_1 = \{2\mathbf{i} + 6\mathbf{j} - 2\mathbf{k}\}$ lb, $\mathbf{F}_2 = \{3\mathbf{i} - 1\mathbf{k}\}$ lb, and $\mathbf{F}_3 = \{1\mathbf{i} - t^2\mathbf{j} - 2\mathbf{k}\}$ lb, where t is in seconds. Determine the distance the particle is from the origin 3 s after it has been released from rest.

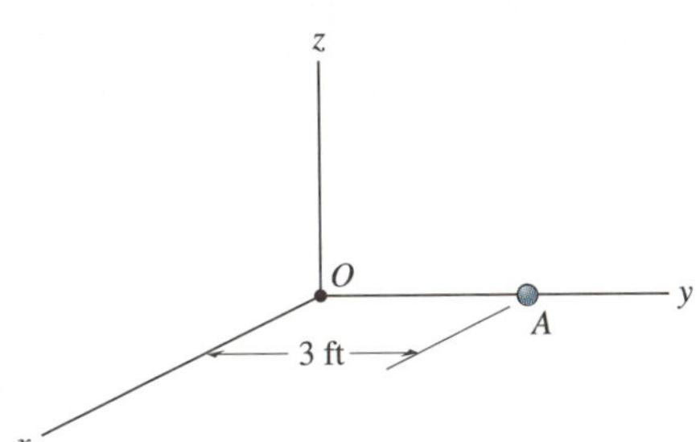

Prob. 13–13

13-14. The 3.5-Mg engine is suspended from a spreader beam AB having a negligible mass and is hoisted by a crane which gives it an acceleration of 4 m/s^2 when it has a velocity of 2 m/s. Determine the force in chains CA and CB during the lift.

13-15. The 3.5-Mg engine is suspended from a spreader beam having a negligible mass and is hoisted by a crane which exerts a force of 40 kN on the hoisting cable. Determine the distance the engine is hoisted in 4 s, starting from rest.

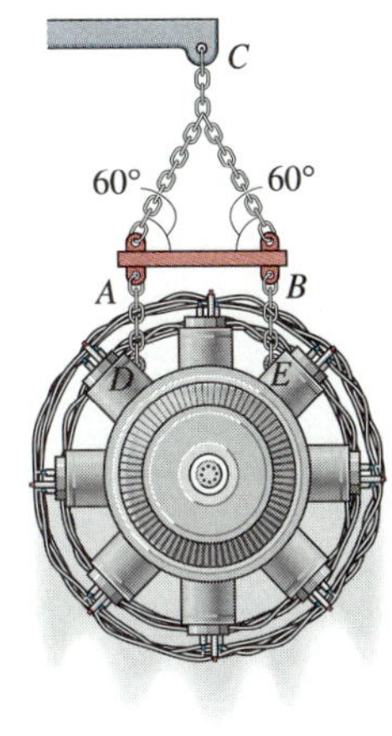

Probs. 13–14/15

***13-16.** The 3.5-Mg engine is suspended from a 500-kg spreader beam and hoisted by a crane which gives it an acceleration of 4 m/s^2 when it has a velocity of 2 m/s. Determine the force in chains AC and AD during the lift.

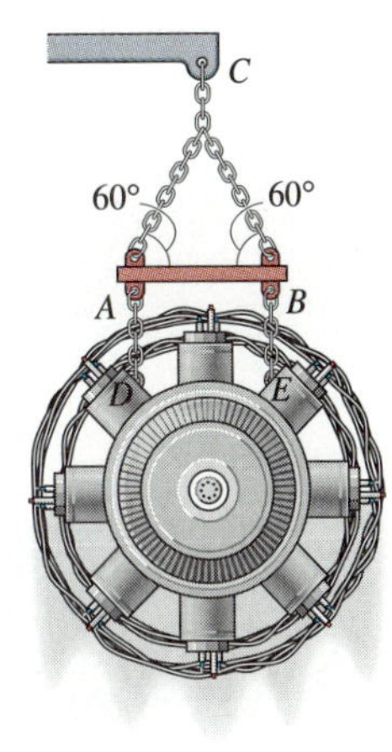

Prob. 13–16

13-17. The bullet of mass m is given a velocity due to gas pressure caused by the burning of powder within the chamber of the gun. Assuming this pressure creates a force of $F = F_0 \sin(\pi t/t_0)$ on the bullet, determine the velocity of the bullet at any instant it is in the barrel. What is the bullet's maximum velocity? Also, determine the position of the bullet in the barrel as a function of time.

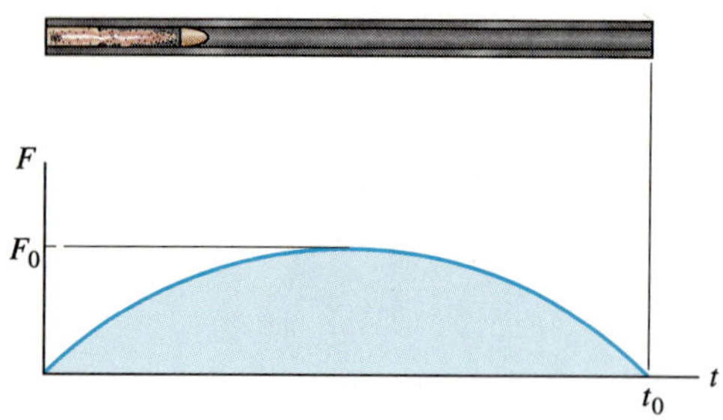

Prob. 13–17

13-18. The man pushes on the 60-lb crate with a force **F**. The force is always directed down at 30° from the horizontal as shown, and its magnitude is increased until the crate begins to slide. Determine the crate's initial acceleration if the static coefficient of friction is $\mu_s = 0.6$ and the kinetic coefficient of friction is $\mu_k = 0.3$.

Prob. 13–18

13-19. A force of $F = 15$ lb is applied to the cord. Determine how high the 30-lb block A rises in 2 s starting from rest. Neglect the weight of the pulleys and cord.

***13-20.** Determine the constant force F which must be applied to the cord in order to cause the 30-lb block A to have a speed of 12 ft/s when it has been displaced 3 ft upward starting from rest. Neglect the weight of the pulleys and cord.

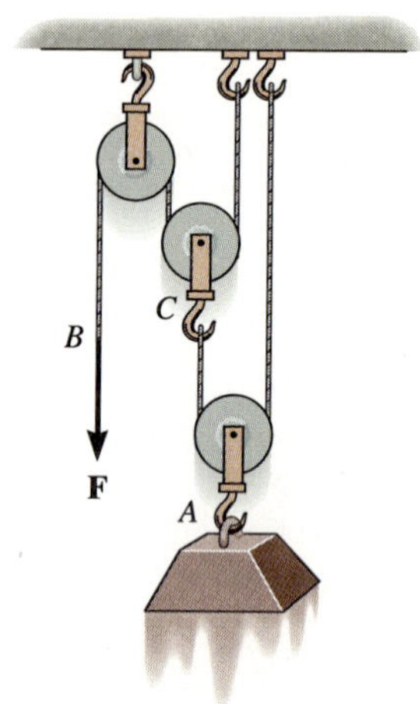

Probs. 13–19/20

13-21. The 400-lb cylinder at A is hoisted using the motor and the pulley system shown. If the speed of point B on the cable is increased at a constant rate from zero to $v_B = 10$ ft/s in $t = 5$ s, determine the tension in the cable at B to cause the motion.

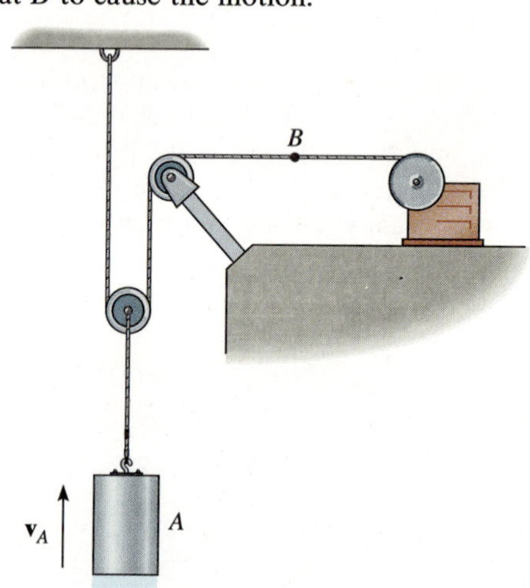

Prob. 13–21

13-22. The 10-lb block A is traveling to the right at $v_A = 2$ ft/s at the instant shown. If the coefficient of kinetic friction is $\mu_k = 0.2$ between the surface and A, determine the velocity of A when it has moved 4 ft. Block B has a weight of 20 lb.

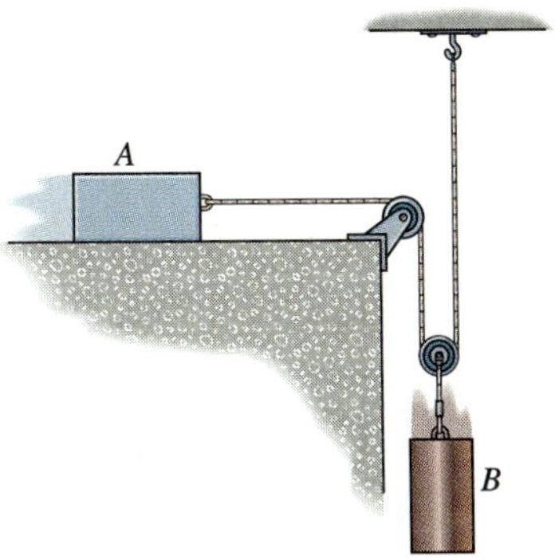

Prob. 13–22

13-23. The motor M pulls in its attached rope causing an acceleration of 6 m/s^2. Determine this towing force. The coefficient of kinetic friction between the 50-kg crate and the plane is $\mu_k = 0.3$. Neglect the mass of the pulleys and rope.

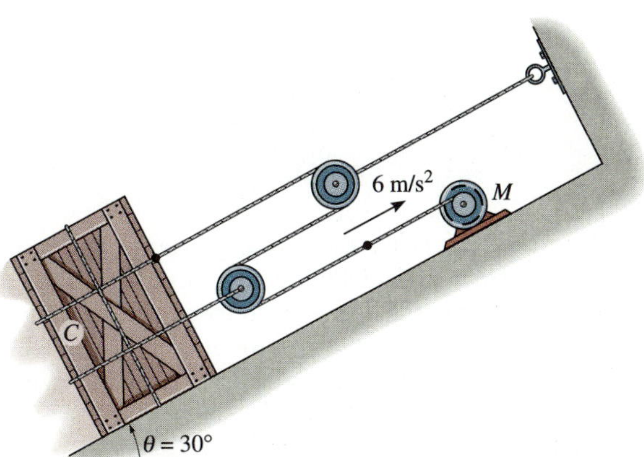

Prob. 13–23

***13-24.** At a given instant the 5-lb weight A is moving downward with a speed of 4 ft/s. Determine its speed 2 s later. Block B has a weight of 6 lb, and the coefficient of kinetic friction between it and the horizontal plane is $\mu_k = 0.3$. Neglect the mass of the pulleys and cord.

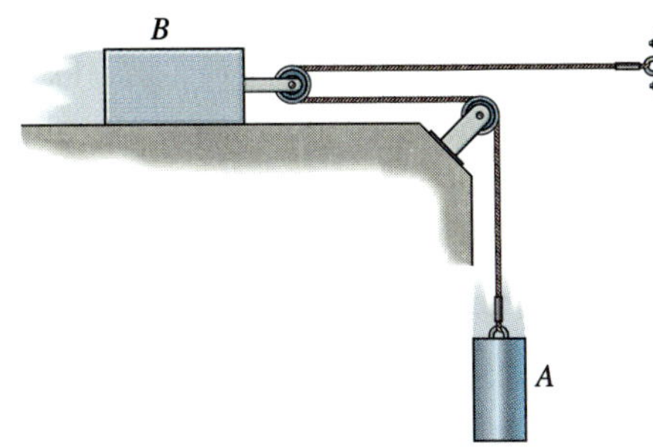

Prob. 13–24

13-25. A freight elevator, including its load, has a mass of 500 kg. It is prevented from rotating due to the track and wheels mounted along its sides. If the motor M develops a constant tension $T = 1.50$ kN in its attached cable, determine the velocity of the elevator when it has moved upward 3 m starting from rest. Neglect the mass of the pulleys and cables.

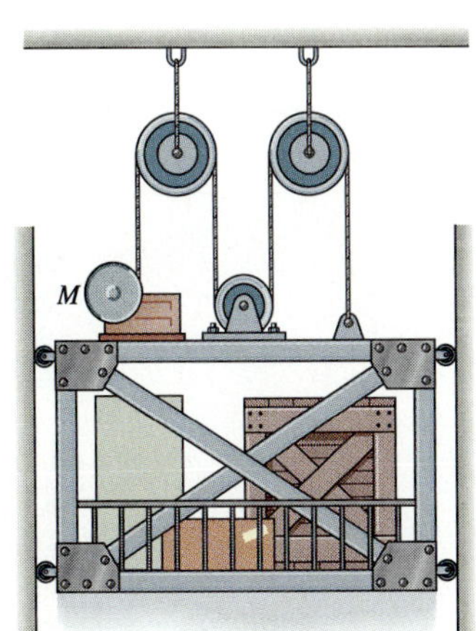

Prob. 13–25

13-26. A freight elevator, including its load, has a mass of 500 kg. It is prevented from rotating by using the track and wheels mounted along its sides. Starting from rest, in $t = 2$ s, the motor M draws in the cable with a speed of 6 m/s, *measured relative to the elevator.* Determine the constant acceleration of the elevator and the tension in the cable. Neglect the mass of the pulleys and cables.

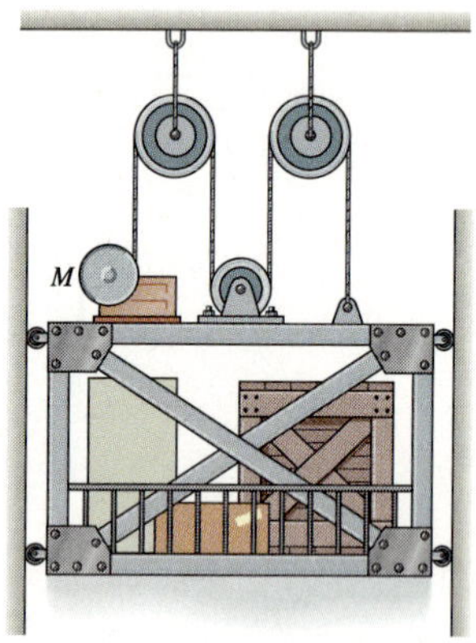

Prob. 13–26

13-27. At the instant shown the 100-lb block A is moving down the plane at 5 ft/s while being attached to the 50-lb block B. If the coefficient of kinetic friction is $\mu_k = 0.2$, determine the acceleration of A and the distance A slides before it stops. Neglect the mass of the pulleys and cables.

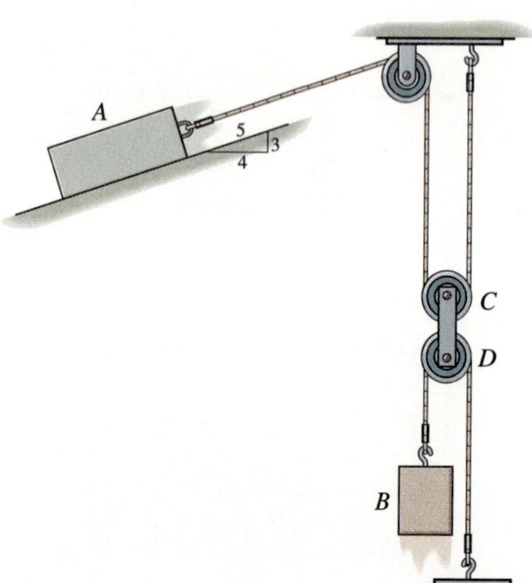

Prob. 13–27

***13-28.** Block B has a mass m and is released from rest when it is on top of cart A, which has a mass of $3m$. Determine the tension in cord CD needed to hold the cart from moving while B is sliding down A. Neglect friction.

13-29. Block B has a mass m and is released from rest when it is on top of cart A, which has a mass of $3m$. Determine the tension in cord CD needed to hold the cart from moving while B is sliding down A. The coefficient of kinetic friction between A and B is μ_k.

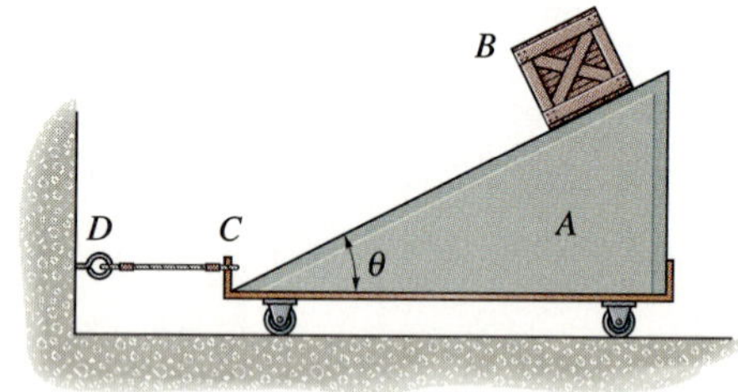

Probs. 13–28/29

■**13-30.** The tanker has a weight of $800(10^6)$ lb and is traveling forward at $v_0 = 3$ ft/s in still water when the engines are shut off. If the drag resistance of the water is proportional to the speed of the tanker at any instant and can be approximated by $F_D = (400(10^3)v)$ lb, where v is in ft/s, determine the time needed for the tanker's speed to become 1.5 ft/s. Given the initial velocity of $v_0 = 3$ ft/s, through what distance must the tanker travel before it stops?

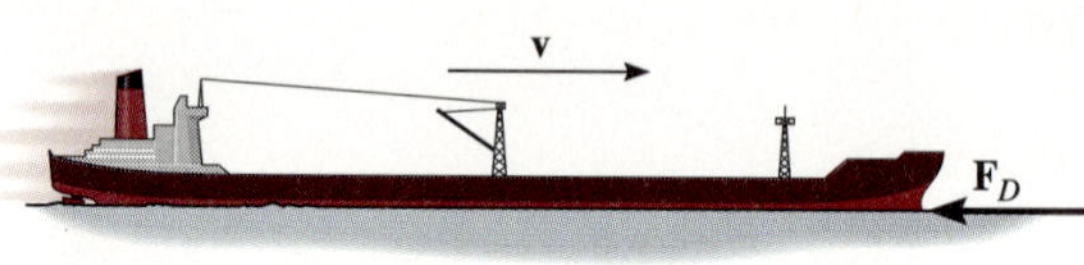

Prob. 13–30

13-31. The 2-kg shaft CA passes through a smooth journal bearing at B. Initially, the springs, which are coiled loosely around the shaft, are unstretched when no force is applied to the shaft. In this position $s = s' = 250$ mm and the shaft is originally at rest. If a horizontal force of $F = 5$ kN is applied, determine the speed of the shaft at the instant $s = 50$ mm, $s' = 450$ mm. The ends of the springs are attached to the bearing at B and the caps at C and A.

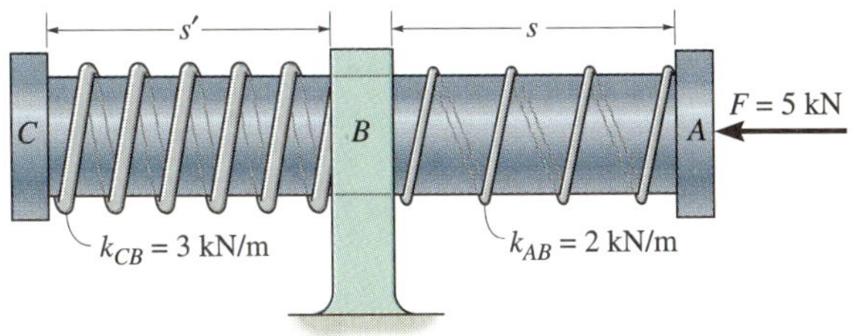

Prob. 13–31

***13-32.** The spring mechanism is used as a shock absorber for railroad cars. Determine the maximum compression of spring HI if the fixed bumper R of a 5-Mg railroad car, rolling freely at 2 m/s, strikes the plate P. Bar AB slides along the guide paths CE and DF. The ends of all springs are attached to their respective members and are originally unstretched.

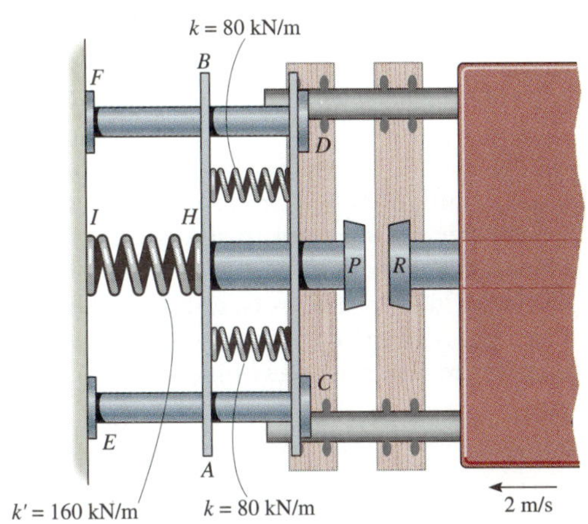

Prob. 13–32

13-33. The 10-lb block A and the 20-lb block B are initially at rest. If a force of $P = 20$ lb is applied to B as shown, determine the acceleration of each block. The coefficient of kinetic friction between any two surfaces is $\mu_k = 0.2$, and the coefficient of static friction is $\mu_s = 0.3$.

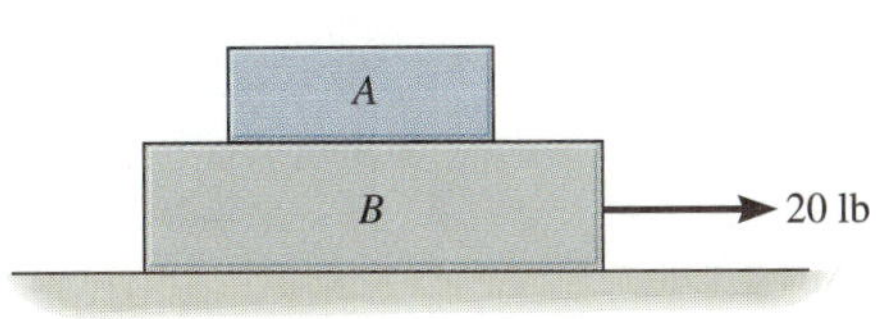

Prob. 13–33

13-34. The 10-kg block A rests on the 50-kg plate B in the position shown. Neglecting the mass of the rope and pulley, and using the coefficients of kinetic friction indicated, determine the time needed for block A to slide 0.5 m *on the plate* when the system is released from rest.

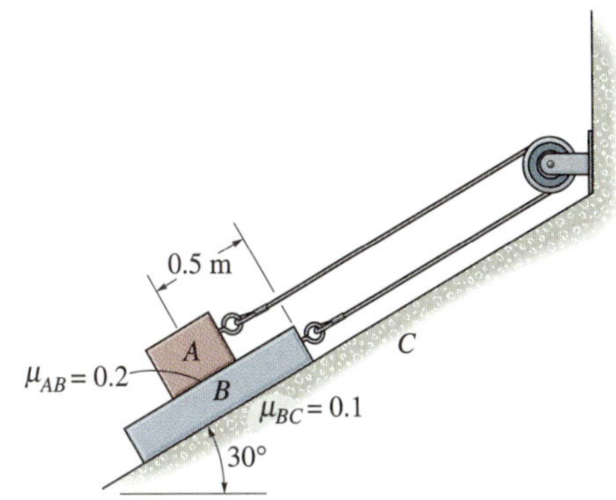

Prob. 13–34

13-35. The 30-lb crate is being hoisted upward with a constant acceleration of 6 ft/s^2. If the uniform beam AB has a weight of 200 lb, determine the components of reaction at A. Neglect the size and mass of the pulley at B. *Hint:* First find the tension in the cable, then analyze the forces in the beam using statics.

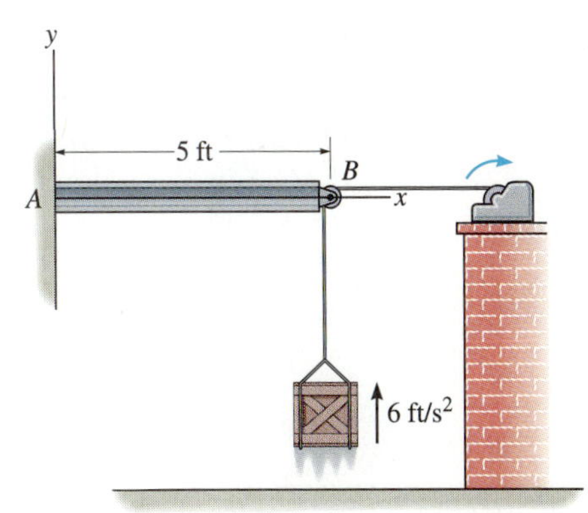

Prob. 13–35

***13-36.** If cylinders B and C have a mass of 15 kg and 10 kg, respectively, determine the required mass of A so that it does not move when all the cylinders are released. Neglect the mass of the pulleys and the cords.

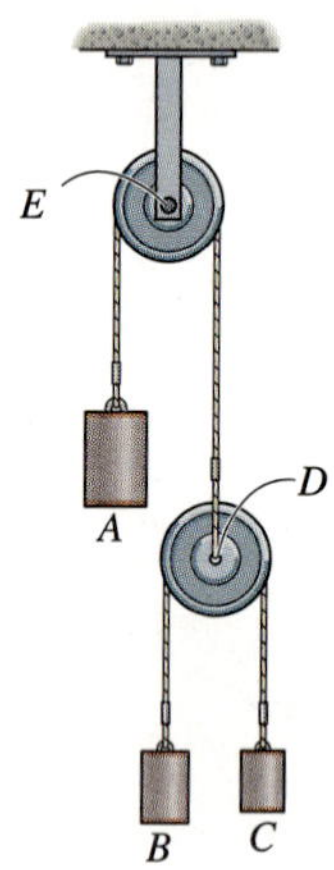

Prob. 13–36

13-37. Determine the acceleration of block A when the system is released. The coefficient of kinetic friction and the weight of each block are indicated. Neglect the mass of the pulleys and cord.

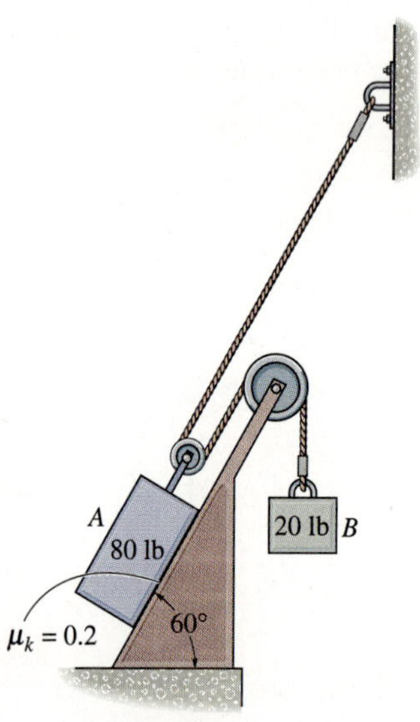

Prob. 13–37

13-38. The 2-lb collar C fits loosely on the smooth shaft. If the spring is unstretched when $s = 0$ and the collar is given a velocity of 15 ft/s, determine the velocity of the collar when $s = 1$ ft.

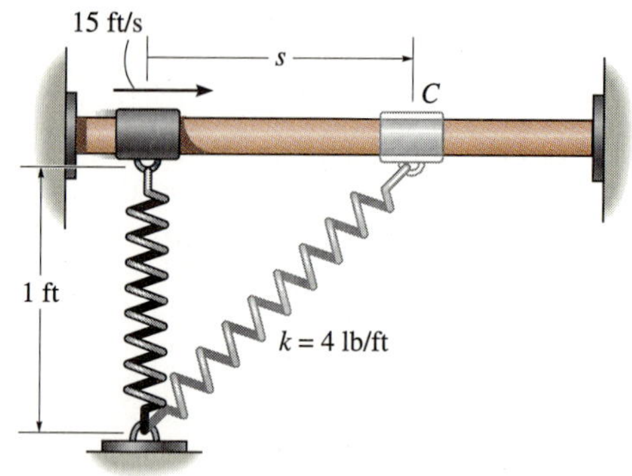

Prob. 13–38

13-39. An electron of mass m is discharged with an initial horizontal velocity of $\mathbf{v}_0$. If it is subjected to two fields of force for which $F_x = F_0$ and $F_y = 0.3F_0$, where F_0 is constant, determine the equation of the path, and the speed of the electron at any time t.

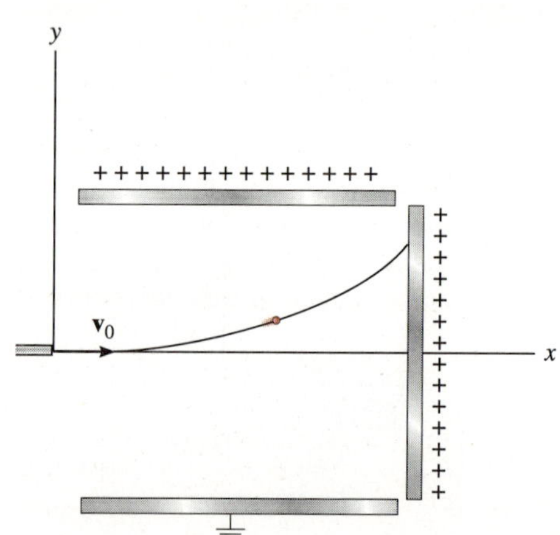

Prob. 13–39

*13-40. In the cathode-ray tube, electrons having a mass m are emitted from a source point S and begin to travel horizontally with an initial velocity $\mathbf{v}_0$. While passing between the grid plates a distance l, they are subjected to a vertical force having a magnitude eV/w, where e is the charge of an electron, V the applied voltage acting across the plates, and w the distance between the plates. After passing clear of the plates, the electrons then travel in straight lines and strike the screen at A. Determine the deflection d of the electrons in terms of the dimensions of the voltage plate and tube. Neglect gravity and the slight vertical deflection which occurs between the plates.

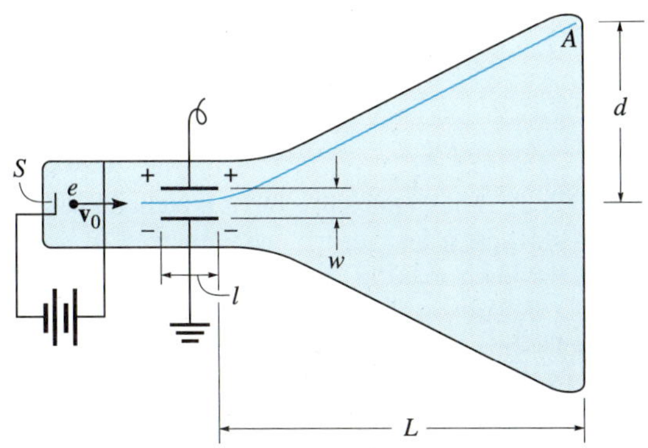

Prob. 13–40

13-41. If a horizontal force $P = 12$ lb is applied to block A determine the acceleration of block B. Neglect friction.

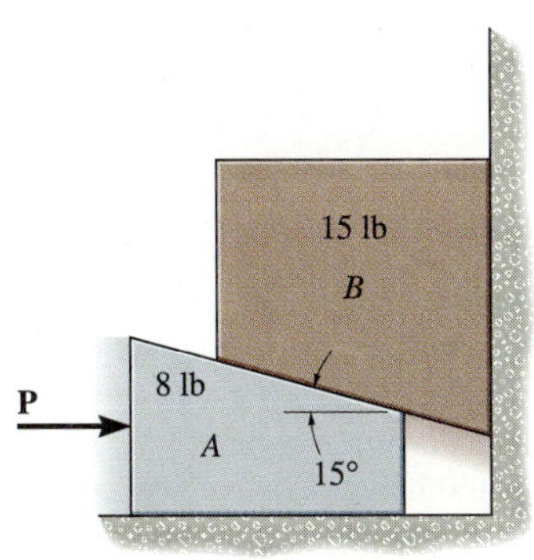

Prob. 13–41

13-42. Blocks A and B each have a mass m. Determine the largest horizontal force P which can be applied to B so that A will not move relative to B. All surfaces are smooth.

13-43. Blocks A and B each have a mass m. Determine the largest horizontal force P which can be applied to B so that A will not slip up B. The coefficient of static friction between A and B is μ_s. Neglect any friction between B and C.

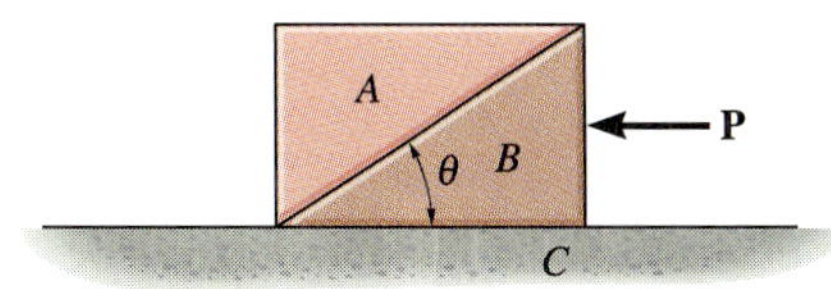

Probs. 13–42/43

■*13-44. Determine the force P that is needed to pull down the cord with an acceleration of 4 ft/s^2 at the instant $\theta = 30°$. The block has a weight of 5 lb and starts from rest. Neglect the mass of the pulleys and the cord.

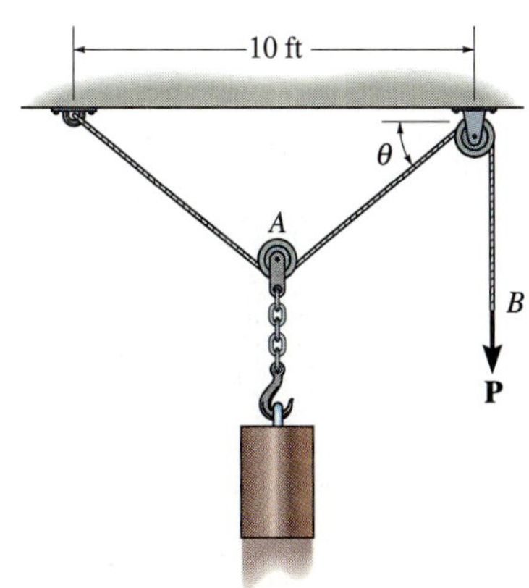

Prob. 13–44

13-45. Block A has a mass of 10 kg and is hoisted using the rope and pulley system shown. If the collar is moving to the right at 10 m/s and has a deceleration of 5 m/s^2 at the instant shown, determine the tension in the cable connected to the collar at this instant.

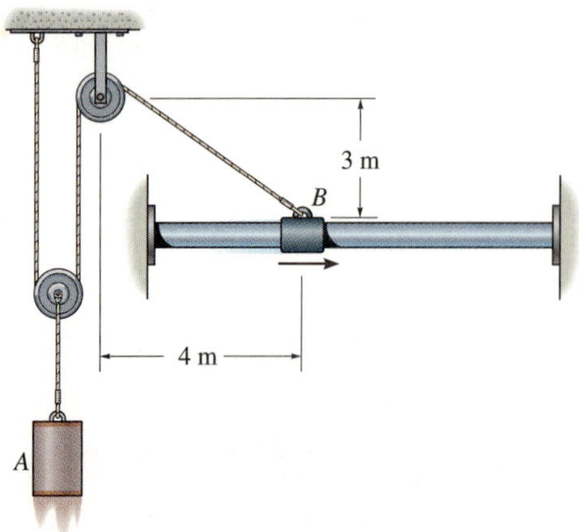

Prob. 13–45

■**13-46.** The tractor is used to lift the 150-kg load B with the 24-m-long rope, boom, and pulley system. If the tractor is traveling to the right at a constant speed of 4 m/s, determine the tension in the rope when $s_A = 5$ m. When $s_A = 0$, $s_B = 0$.

13-47. The tractor is used to lift the 150-kg load B with the 24-m-long rope, boom, and pulley system. If the tractor is traveling to the right with an acceleration of 3 m/s^2 and has a velocity of 4 m/s at the instant $s_A = 5$ m, determine the tension in the rope at this instant. When $s_A = 0$, $s_B = 0$.

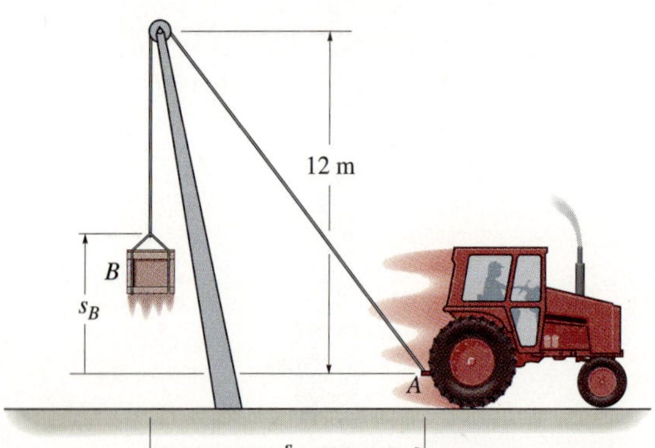

Probs. 13–46/47

***13-48.** Cylinder B has a mass m and is hoisted using the cord and pulley system shown. Determine the magnitude of force **F** as a function of the cylinder's vertical position y so that when **F** is applied the cylinder rises with a constant acceleration $\mathbf{a}_B$. Neglect the mass of the cord, pulleys, hook and chain.

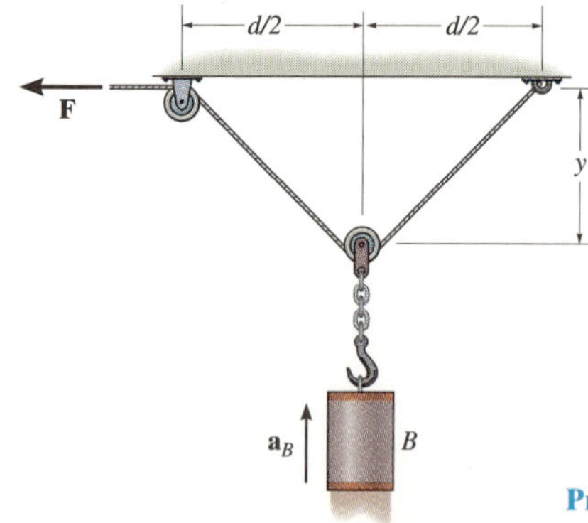

Prob. 13–48

13-49. Block A has a mass m_A and is attached to a spring having a stiffness k and unstretched length l_0. If another block B, having a mass m_B, is pressed against A so that the spring deforms a distance d, determine the distance both blocks slide on the smooth surface before they begin to separate. What is their velocity at this instant?

13-50. Block A has a mass m_A and is attached to a spring having a stiffness k and unstretched length l_0. If another block B, having a mass m_B, is pressed against A so that the spring deforms a distance d, show that for separation to occur it is necessary that $d > 2\mu_k g(m_A + m_B)/k$, where μ_k is the coefficient of kinetic friction between the blocks and the ground. Also, what is the distance the blocks slide on the surface before they separate?

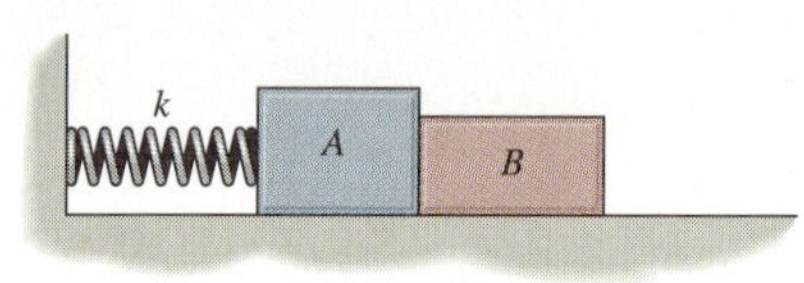

Probs. 13–49/50

13-51. The weight of a particle varies with altitude such that $W = m(gr_0^2)/r^2$, where r_0 is the radius of the earth and r is the distance from the earth's center. If the particle is fired vertically with a velocity v_0 from the earth's surface, determine its velocity as a function of position r. What is the smallest velocity v_0 required to escape the earth's gravitational field, what is r_{max}, and what is the time required to reach this altitude?

13.5 Equations of Motion: Normal and Tangential Coordinates

When a particle moves over a curved path which is known, the equation of motion for the particle may be written in the tangential, normal, and binormal directions. We have

$$\Sigma \mathbf{F} = m\mathbf{a}$$

$$\Sigma F_t \mathbf{u}_t + \Sigma F_n \mathbf{u}_n + \Sigma F_b \mathbf{u}_b = m\mathbf{a}_t + m\mathbf{a}_n$$

Here $\Sigma F_t, \Sigma F_n, \Sigma F_b$ represent the sums of all the force components acting on the particle in the tangential, normal, and binormal directions, respectively, Fig. 13–11. Note that there is no motion of the particle in the binormal direction, since the particle is constrained to move along the path. The above equation is satisfied provided

$$\begin{aligned} \Sigma F_t &= ma_t \\ \Sigma F_n &= ma_n \\ \Sigma F_b &= 0 \end{aligned} \qquad (13\text{–}8)$$

Recall that $a_t (= dv/dt)$ represents the time rate of change in the magnitude of velocity. Consequently, if $\Sigma \mathbf{F}_t$ acts in the direction of motion, the particle's speed will increase, whereas if it acts in the opposite direction, the particle will slow down. Likewise, $a_n \; (= v^2/\rho)$ represents the time rate of change in the velocity's direction. Since this vector *always* acts in the positive n direction, i.e., toward the path's center of curvature, then $\Sigma \mathbf{F}_n$, which causes $\mathbf{a}_n$, also acts in this direction. For example, when the particle is constrained to travel in a circular path with a constant speed, there is a normal force exerted on the particle by the constraint in order to change the direction of the particle's velocity ($\mathbf{a}_n$). Since this force is always directed toward the center of the path, it is often referred to as the *centripetal force*.

The centrifuge is used to subject a passenger to very large normal accelerations caused by high rotations. Realize that these accelerations are caused by the unbalanced normal force exerted on the passenger by the seat of the centrifuge.

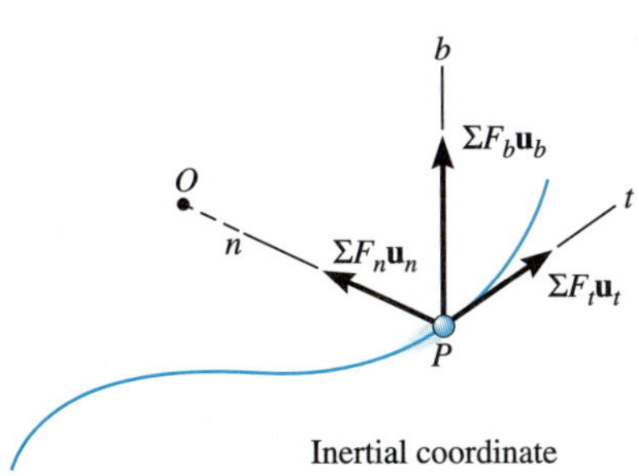

Fig. 13–11

Procedure for Analysis

When a problem involves the motion of a particle along a *known curved path,* normal and tangential coordinates should be considered for the analysis since the acceleration components can be readily formulated. The method for applying the equations of motion, which relate the forces to the acceleration, has been outlined in the procedure given in Sec. 13.4. Specifically, for t, n, b coordinates it may be stated as follows:

Free-Body Diagram

- Establish the inertial t, n, b coordinate system at the particle and draw the particle's free-body diagram.
- The particle's normal acceleration $\mathbf{a}_n$ *always* acts in the positive n direction.
- If the tangential acceleration $\mathbf{a}_t$ is unknown, assume it acts in the positive t direction.
- Identify the unknowns in the problem.

Equations of Motion

- Apply the equations of motion, Eqs. 13–8.

Kinematics

- Formulate the tangential and normal components of acceleration; i.e., $a_t = dv/dt$ or $a_t = v\,dv/ds$ and $a_n = v^2/\rho$.
- If the path is defined as $y = f(x)$, the radius of curvature at the point where the particle is located can be obtained from $\rho = [1 + (dy/dx)^2]^{3/2}/|d^2y/dx^2|$.

EXAMPLE 13–6

Determine the banking angle θ for the race track so that the wheels of the racing cars shown in Fig. 13–12a will not have to depend upon friction to prevent any car from sliding up or down the track. Assume the cars have negligible size a mass m, and travel around the curve of radius ρ with a speed v.

(a)

Solution

Before looking at the following solution, give some thought as to why it should be solved using t, n, b coordinates.

Free-Body Diagram. As shown in Fig. 13–12b, and as stated in the problem, no frictional force acts on the car. Here $\mathbf{N}_C$ represents the *resultant* of the ground on all four wheels. Since a_n can be calculated, the unknowns are N_C and θ.

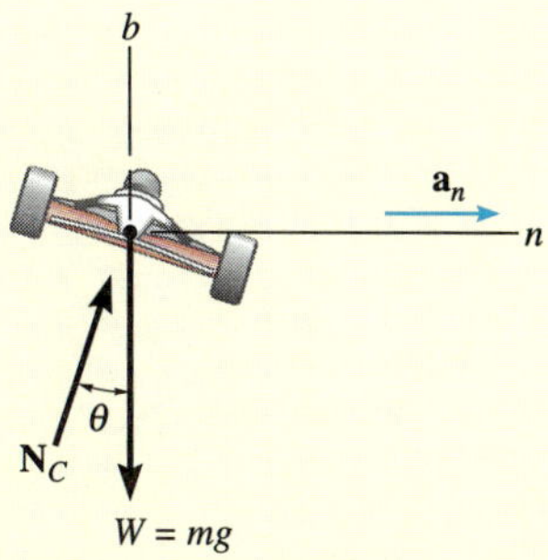

(b)

Fig. 13–12

Equations of Motion. Using the n, b axes shown,

$$+\uparrow \Sigma F_b = 0; \qquad N_C \cos\theta - mg = 0 \qquad (1)$$

$$\overset{+}{\rightarrow}\Sigma F_n = ma_n; \qquad N_C \sin\theta = m\frac{v^2}{\rho} \qquad (2)$$

Eliminating N_C and m from these equations by dividing Eq. 2 by Eq. 1, we obtain

$$\tan\theta = \frac{v^2}{g\rho}$$

$$\theta = \tan^{-1}\left(\frac{v^2}{g\rho}\right) \qquad \textit{Ans.}$$

Notice that the result is independent of the mass of the car. Also, a force summation in the tangential direction is of no consequence to the solution. If it were considered, then $a_t = dv/dt = 0$, since the car moves with *constant speed.* A further analysis of this problem is discussed in Prob. 21–48.

EXAMPLE 13–7

The 3-kg disk D is attached to the end of a cord as shown in Fig. 13–13a. The other end of the cord is attached to a ball-and-socket joint located at the center of a platform. If the platform is rotating rapidly, and the disk is placed on it and released from rest as shown, determine the time it takes for the disk to reach a speed great enough to break the cord. The maximum tension the cord can sustain is 100 N, and the coefficient of kinetic friction between the disk and the platform is $\mu_k = 0.1$.

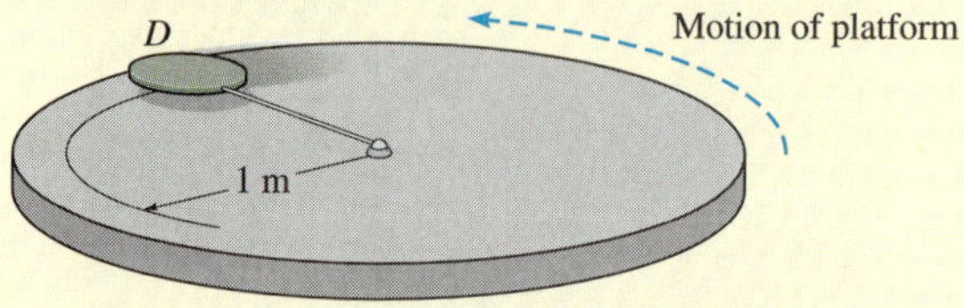

(a)

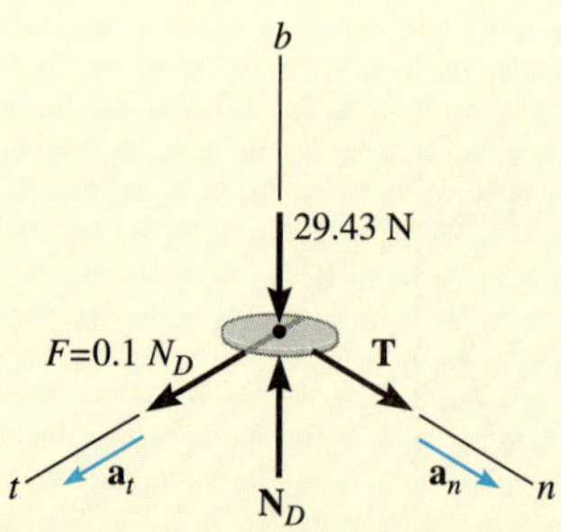

(b)

Fig. 13–13

Solution

Free-Body Diagram. The frictional force has a magnitude $F = \mu_k N_D = 0.1 N_D$ and a sense of direction that opposes the *relative motion* of the disk with respect to the platform. It is this force that gives the disk a tangential component of acceleration causing v to increase, thereby causing T to increase until it reaches 100 N. The weight of the disk is $W = 3(9.81) = 29.43$ N. Since a_n can be related to v, the unknowns are N_D, a_t, and v.

Equations of Motion

$\Sigma F_b = 0;$ $\qquad N_D - 29.43 = 0 \qquad (1)$

$\Sigma F_t = ma_t;$ $\qquad 0.1 N_D = 3a_t \qquad (2)$

$\Sigma F_n = ma_n;$ $\qquad T = 3\left(\frac{v^2}{1}\right) \qquad (3)$

Setting $T = 100$ N, Eq. 3 can be solved for the critical speed v_{cr} of the disk needed to break the cord. Solving all the equations, we obtain

$$N_D = 29.43 \text{ N}$$
$$a_t = 0.981 \text{ m/s}^2$$
$$v_{cr} = 5.77 \text{ m/s}$$

Kinematics. Since a_t is *constant,* the time needed to break the cord is

$$v_{cr} = v_0 + a_t t$$
$$5.77 = 0 + (0.981)t$$
$$t = 5.89 \text{ s} \qquad \textit{Ans.}$$

EXAMPLE 13–8

Design of the ski jump shown in the photo requires knowing the type of forces that will be exerted on the skier and his approximate trajectory. If in this case the jump can be approximated by the parabola shown in Fig. 13–14*a*, determine the normal force on the 150-lb skier the instant he arrives at the end of the jump, point *A*, where his velocity is 65 ft/s. Also, what is his acceleration at this point?

Solution

Why consider using *n*, *t* coordinates to solve this problem?

Free-Body Diagram. The free-body diagram for the skier when he is at *A* is shown in Fig. 13–14*b*. Since the path is *curved,* there are two components of acceleration, $\mathbf{a}_n$ and $\mathbf{a}_t$. Since a_n can be calculated, the unknowns are a_t and N_A.

Equations of Motion

$$+\uparrow \Sigma F_n = ma_n; \qquad N_A - 150 = \frac{150}{32.2}\left(\frac{(65)^2}{\rho}\right) \qquad (1)$$

$$\overset{+}{\leftarrow} \Sigma F_t = ma_t; \qquad 0 = \frac{150}{32.2}a_t \qquad (2)$$

The radius of curvature ρ for the path must be determined at point $A(0, -200 \text{ ft})$. Here $y = \frac{1}{200}x^2 - 200$, $dy/dx = \frac{1}{100}x$, $d^2y/dx^2 = \frac{1}{100}$, so that at $x = 0$,

$$\rho = \left.\frac{[1 + (dy/dx)^2]^{3/2}}{|d^2y/dx^2|}\right|_{x=0} = \frac{[1 + (0)^2]^{3/2}}{\left|\frac{1}{100}\right|} = 100 \text{ ft}$$

Substituting into Eq. 1 and solving for N_A, we have

$$N_A = 347 \text{ lb} \qquad \textit{Ans.}$$

Kinematics. From Eq. 2,

$$a_t = 0$$

Thus,

$$a_n = \frac{v^2}{\rho} = \frac{(65)^2}{100} = 42.2 \text{ ft/s}^2$$

$$a_A = a_n = 42.2 \text{ ft/s}^2 \uparrow \qquad \textit{Ans.}$$

Show that when the skier is in mid-air his acceleration is 32.2 ft/s^2.

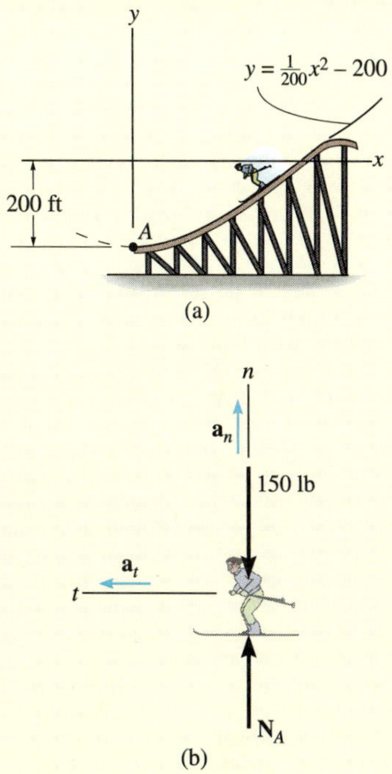

Fig. 13–14

EXAMPLE 13–9

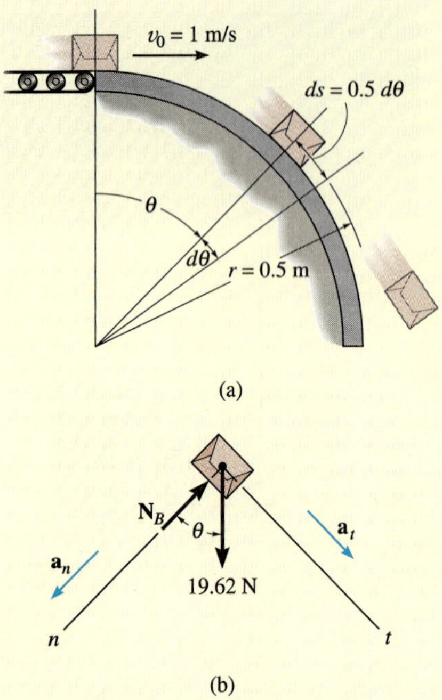

Fig. 13–15

Packages, each having a mass of 2 kg, are delivered from a conveyor to a smooth circular ramp with a velocity of $v_0 = 1$ m/s as shown in Fig. 13–15*a*. If the radius of the ramp is 0.5 m, determine the angle $\theta = \theta_{max}$ at which each package begins to leave the surface.

Solution

Free-Body Diagram. The free-body diagram for a package, when it is located at the *general position* θ, is shown in Fig. 13–15*b*. The package must have a tangential acceleration $\mathbf{a}_t$, since its *speed* is always *increasing* as it slides downward. The weight is $W = 2(9.81) = 19.62$ N. Specify the three unknowns.

Equations of Motion

$$+\swarrow \Sigma F_n = ma_n; \qquad -N_B + 19.62 \cos\theta = 2\frac{v^2}{0.5} \qquad (1)$$

$$+\searrow \Sigma F_t = ma_t; \qquad 19.62 \sin\theta = 2a_t \qquad (2)$$

At the instant $\theta = \theta_{max}$, the package leaves the surface of the ramp so that $N_B = 0$. Therefore, there are three unknowns, v, a_t, and θ.

Kinematics. The third equation for the solution is obtained by noting that the magnitude of tangential acceleration a_t may be related to the speed of the package v and the angle θ. Since $a_t\, ds = v\, dv$ and $ds = r\, d\theta = 0.5\, d\theta$, Fig. 13–15*a*, we have

$$a_t = \frac{v\, dv}{0.5\, d\theta} \qquad (3)$$

To solve, substitute Eq. 3 into Eq. 2 and separate the variables. This gives

$$v\, dv = 4.905 \sin\theta\, d\theta$$

Integrate both sides, realizing that when $\theta = 0°$, $v_0 = 1$ m/s.

$$\int_1^v v\, dv = 4.905 \int_{0°}^{\theta} \sin\theta\, d\theta$$

$$\left.\frac{v^2}{2}\right|_1^v = -4.905 \cos\theta \Big|_{0°}^{\theta}$$

$$v^2 = 9.81(1 - \cos\theta) + 1$$

Substituting into Eq. 1 with $N_B = 0$ and solving for $\cos\theta_{max}$ yields

$$19.62 \cos\theta_{max} = \frac{2}{0.5}[9.81(1 - \cos\theta_{max}) + 1]$$

$$\cos\theta_{max} = \frac{43.24}{58.86}$$

$$\theta_{max} = 42.7° \qquad \textit{Ans.}$$

Problems

*13-52. Determine the mass of the sun, knowing that the distance from the earth to the sun is $149.6(10^6)$ km. *Hint:* Use Eq. 13–1 to represent the force of gravity acting on the earth.

13-53. The sports car, having a mass of 1700 kg, is traveling horizontally along a 20° banked track which is circular and has a radius of curvature of $\rho = 100$ m. If the coefficient of static friction between the tires and the road is $\mu_s = 0.2$, determine the *maximum constant speed* at which the car can travel without sliding up the slope. Neglect the size of the car.

13-54. Using the data in Prob. 13-53, determine the *minimum speed* at which the car can travel around the track without sliding down the slope.

Probs. 13–53/54

13-55. The 1.40-Mg helicopter is traveling at a constant speed of 40 m/s along the horizontal curved path while banking at $\theta = 30°$. Determine the force acting normal to the blade, i.e., in the y' direction, and the radius of curvature of the path.

***13-56.** The 1.40-Mg helicopter is traveling at a constant speed of 33 m/s along the horizontal curved path having a radius of curvature of $\rho = 300$ m. Determine the force the blade exerts on the frame and the bank angle θ.

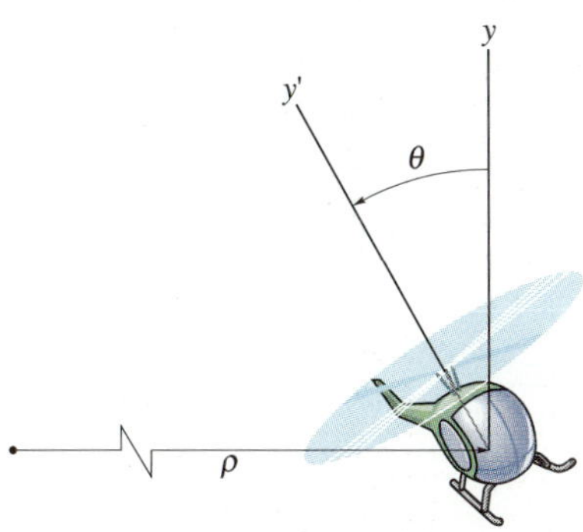

Probs. 13–55/56

13-57. Determine the constant speed of the satellite S so that it circles the earth with an orbit of radius $r = 20$ Mm. The mass of the earth is $5.976(10^{24})$ kg. *Hint:* Use Eq. 13–1.

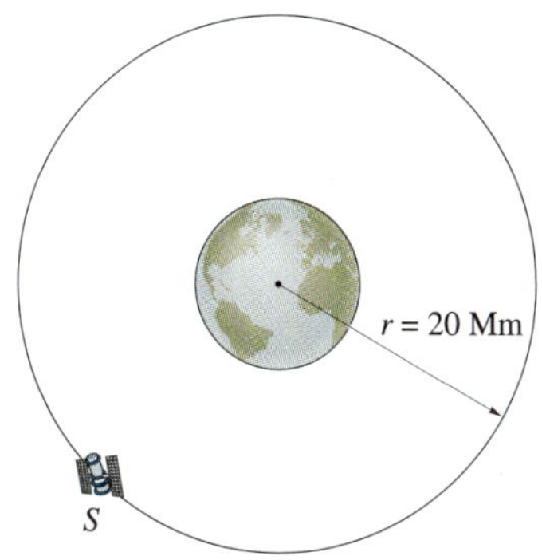

Prob. 13–57

13-58. Prove that if the block is released from rest at point B of a smooth path of *arbitrary shape,* the speed it attains when it reaches point A is equal to the speed it attains when it falls freely through a distance h; i.e., $v = \sqrt{2gh}$.

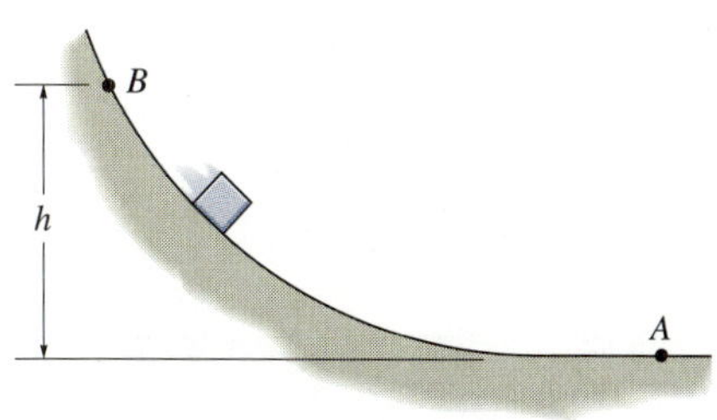

Prob. 13–58

13-59. The sled and rider have a total mass of 80 kg and start from rest at $A(10 \text{ m}, 0)$. If the sled descends the smooth slope, which may be approximated by a parabola, determine the normal force that the ground exerts on the sled at the instant it arrives at point B. Neglect the size of the sled and rider. *Hint:* Use the result of Prob. 13-58.

***13-60.** The sled and rider have a total mass of 80 kg and start from rest at $A(10 \text{ m}, 0)$. If the sled descends the smooth slope which may be approximated by a parabola, determine the normal force that the ground exerts on the sled at the instant it arrives at point C. Neglect the size of the sled and rider. *Hint:* Use the result of Prob. 13-58.

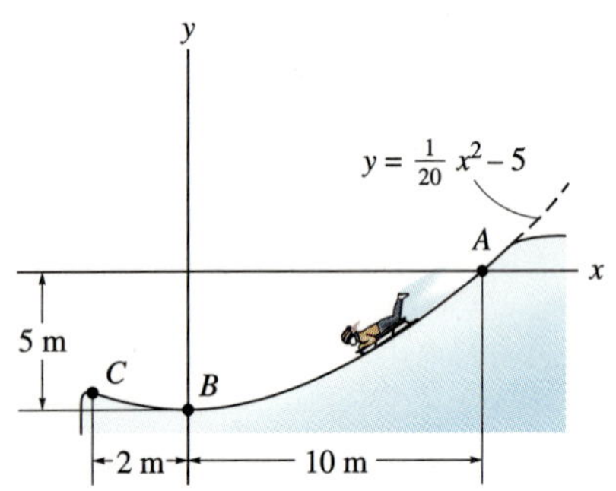

Probs. 13–59/60

13-61. The plane is traveling at a constant speed of 800 ft/s along the curve $y = 20(10^{-6})x^2 + 5000$, where x and y are in feet. If the pilot has a weight of 180 lb, determine the normal and tangential components of the force the seat exerts on the pilot when the plane is at its lowest point.

13-62. The jet plane is traveling at a constant speed of 1000 ft/s along the curve $y = 20(10^{-6})x^2 + 5000$, where x and y are in feet. If the pilot has a weight of 180 lb, determine the normal and tangential components of the force the seat exerts on the pilot when $y = 10\,000$ ft.

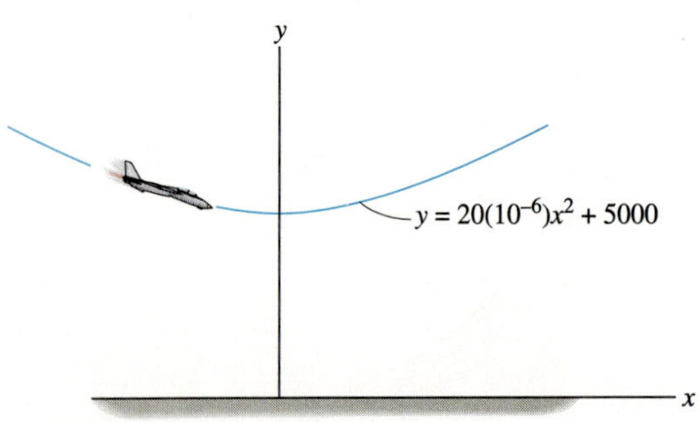

Probs. 13–61/62

13-63. The snowmobile with passenger has a 200-kg mass and is traveling up the hill at a constant speed of 6 m/s. Determine the resultant normal force and the resultant frictional traction force exerted on the snowmobile at the instant it reaches point A. Neglect the size of the snowmobile.

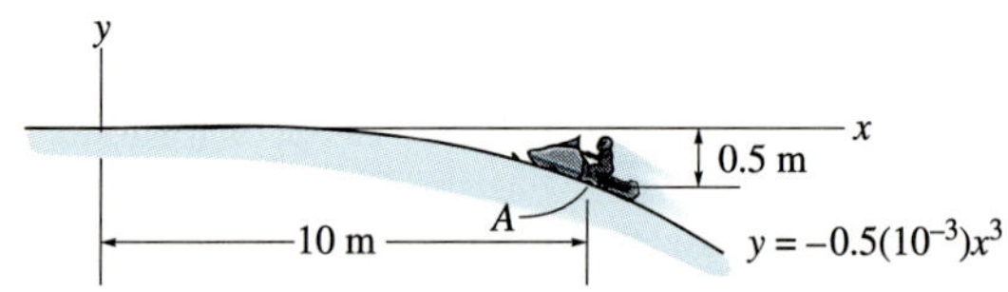

Prob. 13–63

***13-64.** The 8-kg sack slides down the smooth ramp. If it has a speed of 1.5 m/s when $y = 0.2$ m, determine the normal reaction the ramp exerts on the sack and the rate of increase in the speed of the sack at this instant.

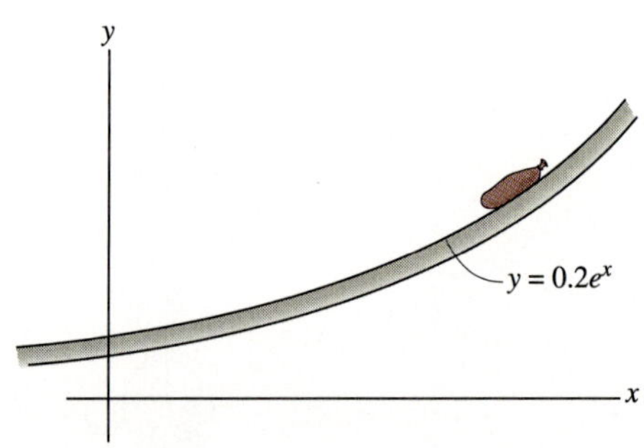

Prob. 13–64

13-65. The 200-kg snowmobile with passenger is traveling down the hill at a constant speed of 6 m/s. Determine the resultant normal force and the resultant frictional force exerted on the tracks at the instant it reaches point A. Neglect the size of the snowmobile.

13-66. The 200-kg snowmobile with passenger is traveling down the hill such that when it is at point A, it is traveling at 4 m/s and increasing its speed at 2 m/s^2. Determine the resultant normal force and the resultant frictional force exerted on the tracks at this instant. Neglect the size of the snowmobile.

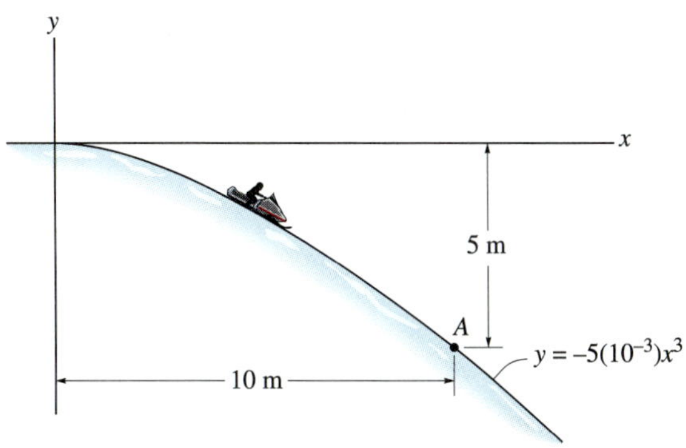

Probs. 13–65/66

13-67. The roller coaster car and passenger have a total weight of 600 lb and starting from rest at A travel down a track that has the shape shown. Determine the normal force of the tracks on the car when the car is at point B, where it has a velocity of 15 ft/s. Neglect friction and the size of the car and passenger.

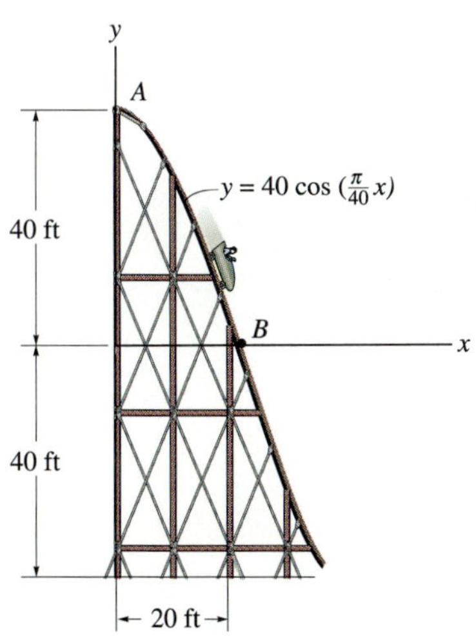

Prob. 13–67

***13-68.** If the ball has a mass of 30 kg and a speed $v = 4$ m/s at the instant it is at its lowest point, $\theta = 0°$, determine the tension in the cord at this instant. Also, determine the angle θ to which the ball swings at the instant it momentarily stops. Neglect the size of the ball.

13-69. The ball has a mass of 30 kg and a speed $v = 4$ m/s at the instant it is at its lowest point, $\theta = 0°$. Determine the tension in the cord and the rate at which the ball's speed is decreasing at the instant $\theta = 20°$. Neglect the size of the ball.

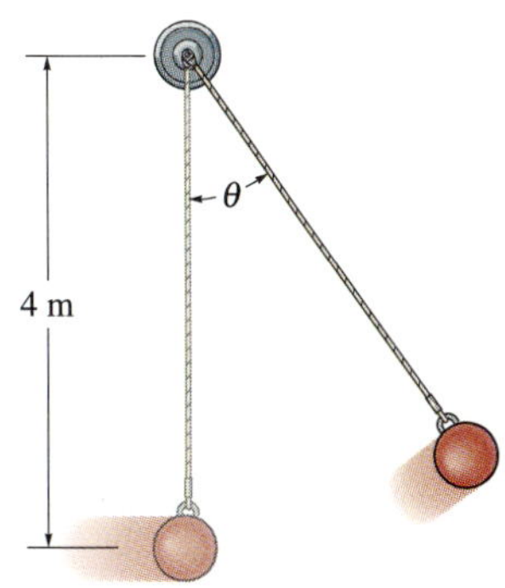

Probs. 13–68/69

13-70. The 5-kg pendulum bob B is released from rest when $\theta = 0°$. Determine the initial tension in the cord and also at the instant the bob reaches point D, $\theta = 45°$. Neglect the size of the bob.

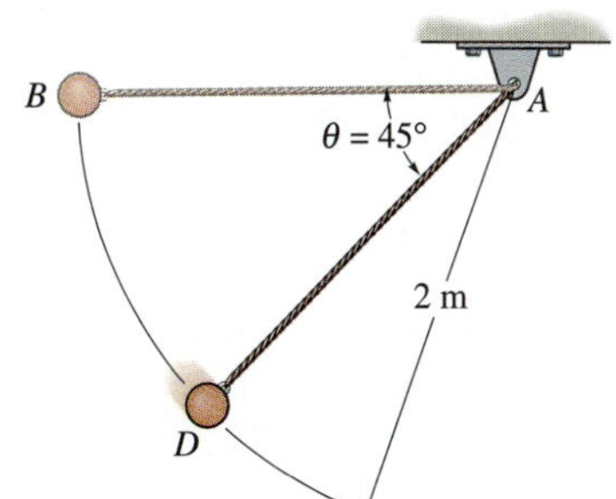

Prob. 13–70

13-71. The ball has a mass m and is attached to the cord of length l. The cord is tied at the top to a swivel and the ball is given a velocity $\mathbf{v}_0$. Show that the angle θ which the cord makes with the vertical as the ball travels around the circular path must satisfy the equation $\tan\theta\sin\theta = v_0^2/gl$. Neglect air resistance and the size of the ball.

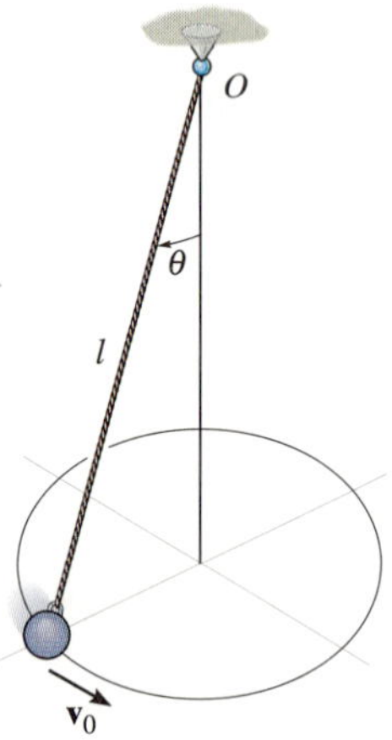

Prob. 13–71

***13-72.** The smooth block B, having a mass of 0.2 kg, is attached to the vertex A of the right circular cone using a light cord. The cone is rotating at a constant angular rate about the z axis such that the block attains a speed of 0.5 m/s. At this speed, determine the tension in the cord and the reaction which the cone exerts on the block. Neglect the size of the block.

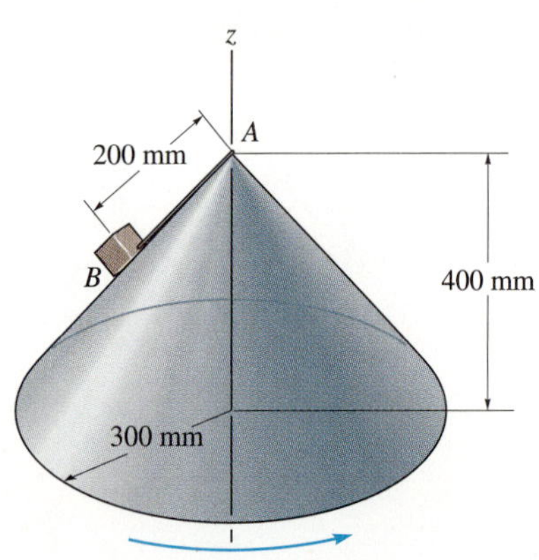

Prob. 13–72

13-73. The rotational speed of the disk is controlled by a 30-g smooth contact arm AB which is spring-mounted on the disk. When the disk is *at rest*, the center of mass G of the arm is located 150 mm from the center O, and the preset compression in the spring is 20 mm. If the initial gap between B and the contact at C is 10 mm, determine the (controlling) speed v_G of the arm's mass center, G, which will close the gap. The disk rotates in the horizontal plane. The spring has a stiffness of $k = 50$ N/m, and its ends are attached to the contact arm at D and to the disk at E.

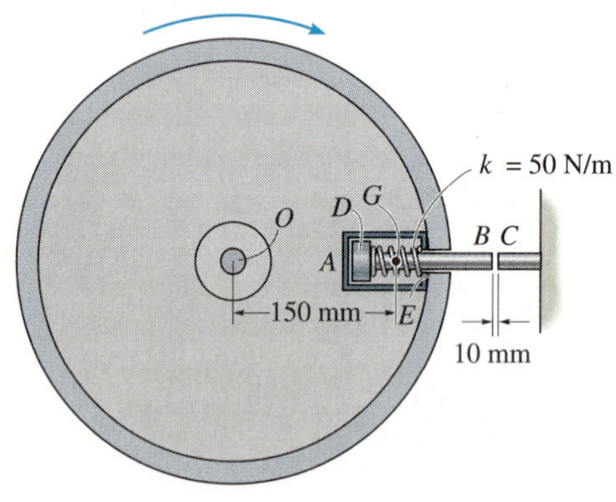

Prob. 13–73

13-74. The collar A, having a mass of 0.75 kg, is attached to a spring having a stiffness of $k = 200$ N/m. When rod BC rotates about the vertical axis, the collar slides outward along the smooth rod DE. If the spring is unstretched when $s = 0$, determine the constant speed of the collar in order that $s = 100$ mm. Also, what is the normal force of the rod on the collar? Neglect the size of the collar.

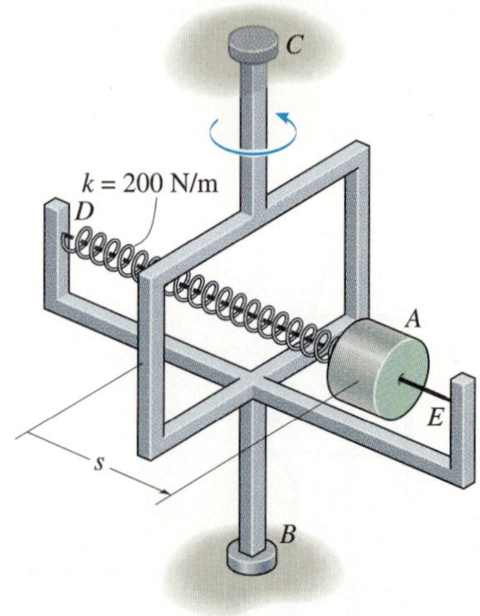

Prob. 13–74

13-75. The 10-lb suitcase slides down the curved ramp for which the coefficient of kinetic friction is $\mu_k = 0.2$. If at the instant it reaches point A it has a speed of 5 ft/s, determine the normal force on the suitcase and the rate of increase of its speed.

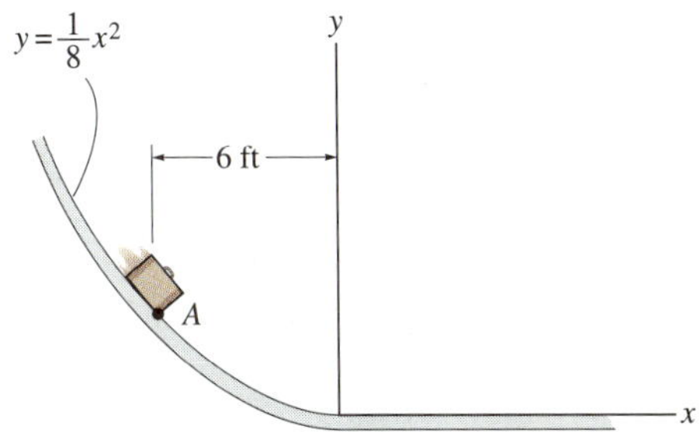

Prob. 13–75

***13-76.** The 6-kg block is confined to move along the smooth parabolic path. The attached spring restricts the motion and, due to the roller guide, always remains horizontal as the block descends. If the spring has a stiffness of $k = 10$ N/m, and an unstretched length of 0.5 m, determine the normal force of the path on the block at the instant $x = 1$ m and the block has a speed of 4 m/s. Also, what is the rate of increase in speed of the block at this point? Neglect the mass of the roller guide and the spring.

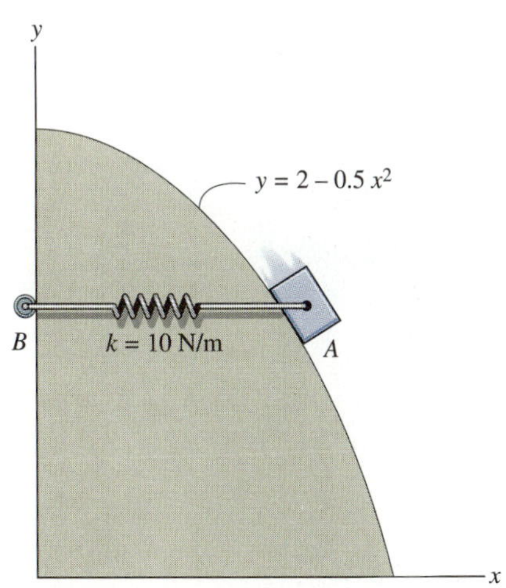

Prob. 13–76

13-77. The box has a mass m and slides down the smooth chute having the shape of a parabola. If it has an initial velocity of v_0 at the origin, determine its velocity as a function of x. Also, what is the normal force on the box, and the tangential acceleration as a function of x?

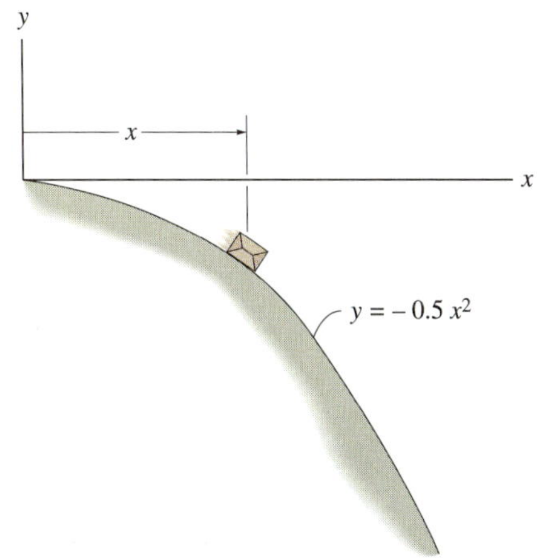

Prob. 13–77

13-78. The 35-kg box has a speed of 2 m/s when it is at A on the smooth ramp. If the surface is in the shape of a parabola, determine the normal force on the box at the instant $x = 3$ m. Also, what is the rate of increase in its speed at this instant?

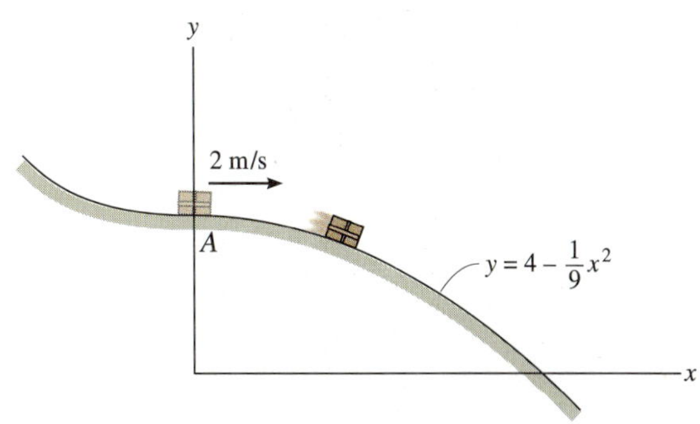

Prob. 13–78

13-79. A collar having a mass of 0.75 kg and negligible size slides over the surface of a horizontal circular rod for which the coefficient of kinetic friction is $\mu_k = 0.3$. If the collar is given a speed of 4 m/s and then released at $\theta = 0°$, determine how far, s, it slides on the rod before coming to rest.

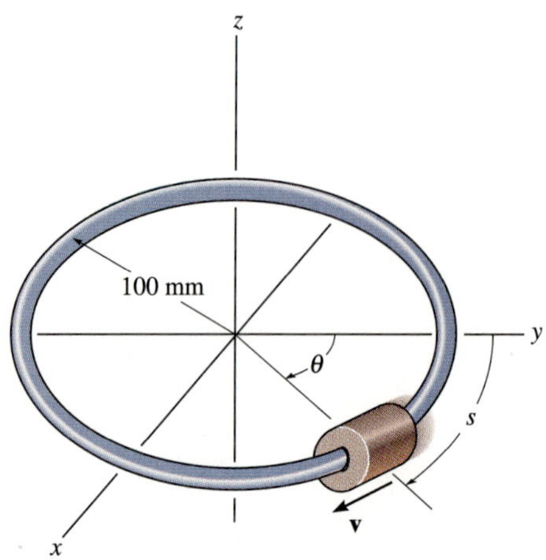

Prob. 13–79

***13-80.** The 5-lb collar slides on the smooth rod, so that when it is at A it has a speed of 10 ft/s. If the spring to which it is attached has an unstretched length of 3 ft and a stiffness of $k = 10$ lb/ft, determine the normal force on the collar and the acceleration of the collar at this instant.

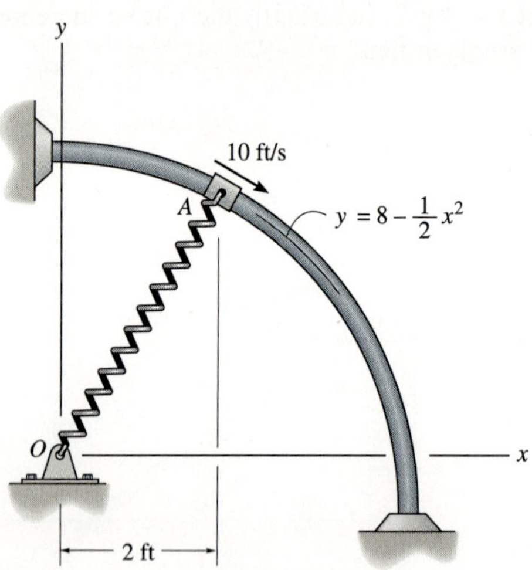

Prob. 13–80

13-81. The 5-lb packages ride on the surface of the conveyor belt. If the belt starts from rest and increases to a constant speed of 2 ft/s in 2 s, determine the maximum angle θ so that none of the packages slip on the inclined surface AB of the belt. The coefficient of static friction between the belt and a package is $\mu_s = 0.3$. At what angle ϕ do the packages first begin to slip off the surface of the belt after the belt is moving at its constant speed of 2 ft/s? Neglect the size of the packages.

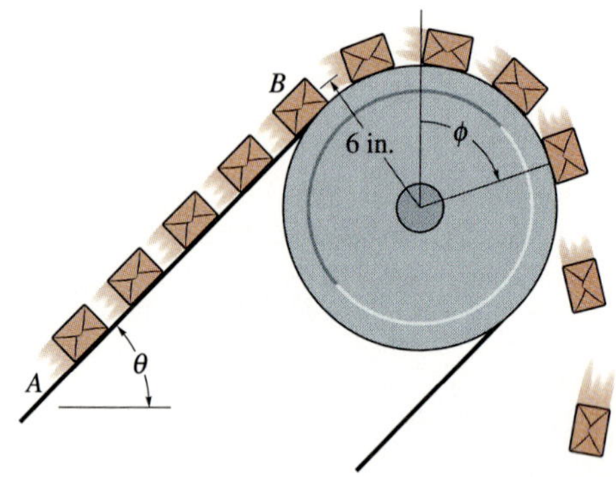

Prob. 13–81

13-82. A ball having a mass of 2 kg and negligible size moves within a smooth vertical circular slot. If it is released from rest when $\theta = 10°$, determine the force of the slot on the ball when the ball arrives at points A and B.

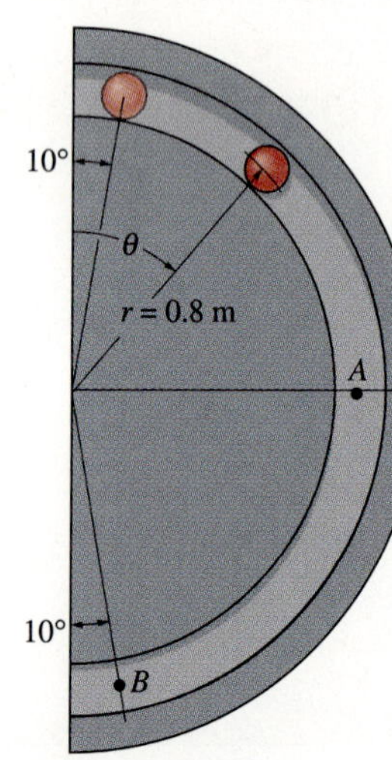

Prob. 13–82

13.6 Equations of Motion: Cylindrical Coordinates

When all the forces acting on a particle are resolved into cylindrical components, i.e., along the unit-vector directions $\mathbf{u}_r$, $\mathbf{u}_\theta$, $\mathbf{u}_z$, Fig. 13–16, the equation of motion may be expressed as

$$\Sigma\mathbf{F} = m\mathbf{a}$$

$$\Sigma F_r\mathbf{u}_r + \Sigma F_\theta\mathbf{u}_\theta + \Sigma F_z\mathbf{u}_z = ma_r\mathbf{u}_r + ma_\theta\mathbf{u}_\theta + ma_z\mathbf{u}_z$$

To satisfy this equation, the respective $\mathbf{u}_r$, $\mathbf{u}_\theta$, $\mathbf{u}_z$ components on the left side must equal the corresponding components on the right side. Consequently, we may write the following three scalar equations of motion:

$$\begin{aligned}\Sigma F_r &= ma_r\\ \Sigma F_\theta &= ma_\theta\\ \Sigma F_z &= ma_z\end{aligned} \qquad (13\text{–}9)$$

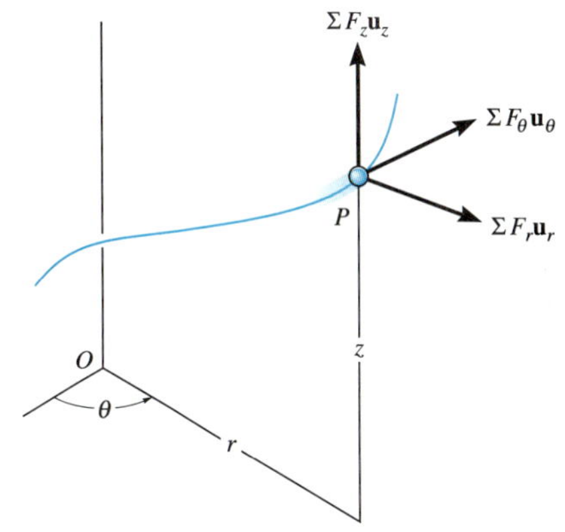

Fig. 13–16

If the particle is constrained to move only in the r–θ plane, then only the first two of Eqs. 13–9 are used to specify the motion.

Tangential and Normal Forces. The most straightforward type of problem involving cylindrical coordinates requires the determination of the resultant force components ΣF_r, ΣF_θ, ΣF_z causing a particle to move with a *known* acceleration. If, however, the particle's accelerated motion is not completely specified at the given instant, then some information regarding the directions or magnitudes of the forces acting on the particle must be known or computed in order to solve Eqs. 13–9. For example, the force **P** causes the particle in Fig. 13–17*a* to move along a path $r = f(\theta)$. The *normal force* **N** which the path exerts on the particle is always *perpendicular to the tangent of the path,* whereas the frictional force **F** always acts along the tangent in the opposite direction of motion. The *directions* of **N** and **F** can be specified relative to the radial coordinate by using the angle ψ (psi), Fig. 13–17*b*, which is defined between the *extended* radial line and the tangent to the curve.

As the car of weight W descends the spiral track, the resultant normal force which the track exerts on the car can be represented by its three cylindrical components. $-\mathbf{N}_r$ creates a radial acceleration, $-\mathbf{a}_r$, $\mathbf{N}_\theta$ creates a transverse acceleration $\mathbf{a}_\theta$, and the difference, $\mathbf{W} - \mathbf{N}_z$ creates an azimuthal acceleration $-\mathbf{a}_z$.

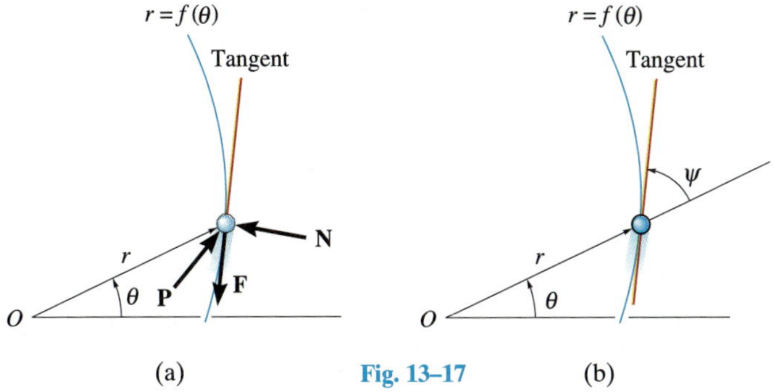

Fig. 13–17

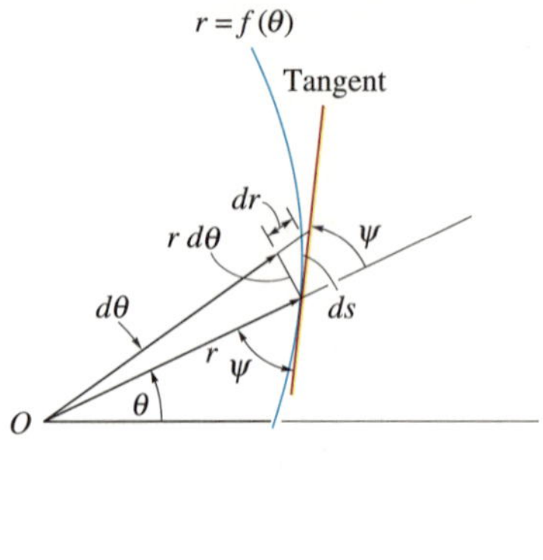

(c)

Fig. 13–17

This angle can be obtained by noting that when the particle is displaced a distance ds along the path, Fig. 13–17c, the component of displacement in the radial direction is dr and the component of displacement in the transverse direction is $r\,d\theta$. Since these two components are mutually perpendicular, the angle ψ can be determined from $\tan\psi = r\,d\theta/dr$, or

$$\tan\psi = \frac{r}{dr/d\theta} \tag{13–10}$$

If ψ is calculated as a positive quantity, it is measured from the *extended radial line* to the tangent in a counterclockwise sense or in the positive direction of θ. If it is negative, it is measured in the opposite direction to positive θ. For example, consider $r = a(1 + \cos\theta)$, the cardioid shown in Fig. 13–18. Because $dr/d\theta = -a\sin\theta$, then when $\theta = 30°$, $\tan\psi = a(1 + \cos 30°)/(-a\sin 30°) = -3.732$, or $\psi = -75°$, measured clockwise, as shown in the figure.

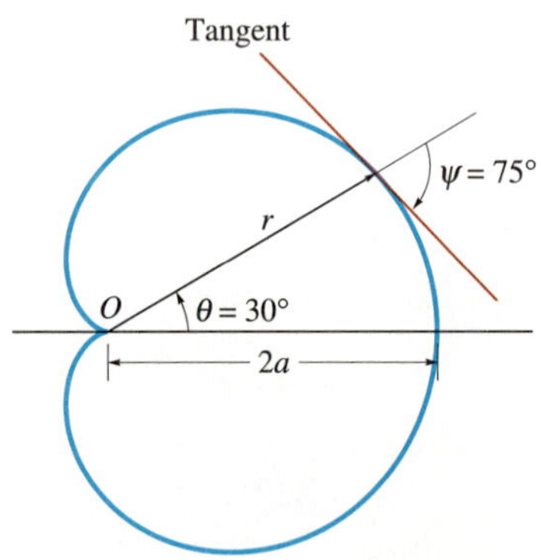

Fig. 13–18

Procedure for Analysis

Cylindrical or polar coordinates are a suitable choice for the analysis of a problem for which data regarding the angular motion of the radial line r are given, or in cases where the path can be conveniently expressed in terms of these coordinates. Once these coordinates have been established, the equations of motion can be applied in order to relate the forces acting on the particle to its acceleration components. The method for doing this has been outlined in the procedure for analysis given in Sec. 13.4. The following is a summary of this procedure.

Free-Body Diagram

- Establish the r, θ, z inertial coordinate system and draw the particle's free-body diagram.
- Assume that $\mathbf{a}_r$, $\mathbf{a}_\theta$, $\mathbf{a}_z$ act in the *positive directions* of r, θ, z if they are unknown.
- Identify all the unknowns in the problem.

Equations of Motion

- Apply the equations of motion, Eqs. 13–9.

Kinematics

- Use the methods of Sec. 12.8 to determine r and the time derivatives $\dot{r}$, $\ddot{r}$, $\dot{\theta}$, $\ddot{\theta}$, $\ddot{z}$, and evaluate the acceleration components $a_r = \ddot{r} - r\dot{\theta}^2$, $a_\theta = r\ddot{\theta} + 2\dot{r}\dot{\theta}$, $a_z = \ddot{z}$.
- If any of the acceleration components is computed as a negative quantity, it indicates that it acts in its negative coordinate direction.
- When taking the time derivatives of $r = f(\theta)$, it is very important to use the chain rule of calculus.

EXAMPLE 13-10

The 2-lb block in Fig. 13–19*a* moves on a smooth horizontal track, such that its path is specified in polar coordinates by the parametric equations $r = (10t^2)$ ft and $\theta = (0.5t)$ rad, where t is in seconds. Determine the magnitude of the tangential force **F** causing the motion at the instant $t = 1$ s.

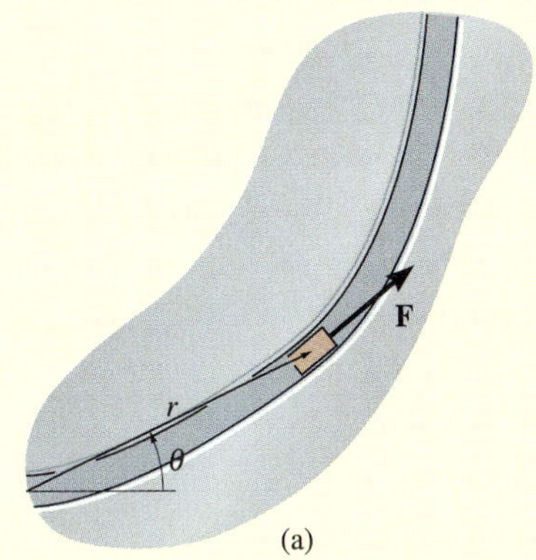

(a)

Solution

Free-Body Diagram. As shown on the block's free-body diagram, Fig. 13–19*b*, the normal force of the track on the block, **N**, and the tangential force **F** are located at an angle ψ from the r and θ axes. This angle can be obtained from Eq. 13–10. To do so, we must first express the path as $r = f(\theta)$ by eliminating the parameter t between r and θ. This yields $r = 40\theta^2$. Also, when $t = 1$ s, $\theta = 0.5(1\text{ s}) = 0.5$ rad. Thus,

$$\tan\psi = \frac{r}{dr/d\theta} = \left.\frac{40\theta^2}{40(2\theta)}\right|_{\theta=0.5\text{ rad}} = 0.25$$
$$\psi = 14.04^\circ$$

Because ψ is a positive quantity, it is measured counterclockwise from the r axis to the tangent (the same direction as θ) as shown in Fig. 13–19*b*. There are presently four unknowns: F, N, a_r and a_θ.

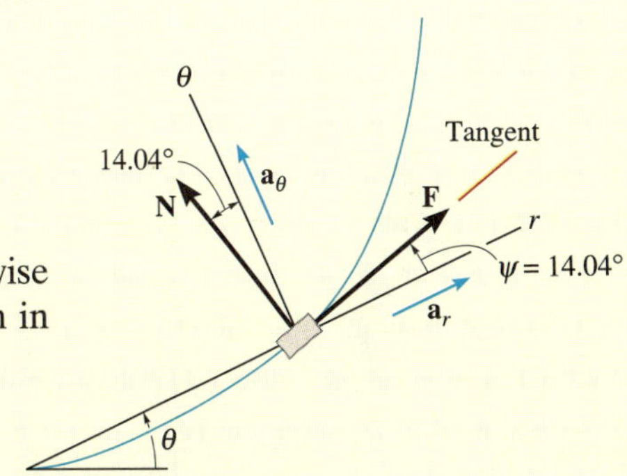

(b)

Fig. 13–19

Equations of Motion

$$+\nearrow\Sigma F_r = ma_r; \qquad F\cos 14.04^\circ - N\sin 14.04^\circ = \frac{2}{32.2}a_r \qquad (1)$$

$$\nwarrow+\Sigma F_\theta = ma_\theta; \qquad F\sin 14.04^\circ + N\cos 14.04^\circ = \frac{2}{32.2}a_\theta \qquad (2)$$

Kinematics. Since the motion is specified, the coordinates and the required time derivatives can be calculated and evaluated at $t = 1$ s.

$$r = 10t^2\Big|_{t=1\text{ s}} = 10\text{ ft} \qquad \theta = 0.5t\Big|_{t=1\text{ s}} = 0.5\text{ rad}$$
$$\dot{r} = 20t\Big|_{t=1\text{ s}} = 20\text{ ft/s} \qquad \dot{\theta} = 0.5\text{ rad/s}$$
$$\ddot{r} = 20\text{ ft/s}^2 \qquad \ddot{\theta} = 0$$
$$a_r = \ddot{r} - r\dot{\theta}^2 = 20 - 10(0.5)^2 = 17.5\text{ ft/s}^2$$
$$a_\theta = r\ddot{\theta} + 2\dot{r}\dot{\theta} = 10(0) + 2(20)(0.5) = 20\text{ ft/s}^2$$

Substituting into Eqs. 1 and 2 and solving, we get

$$F = 1.36\text{ lb} \qquad \textit{Ans.}$$
$$N = 0.942\text{ lb}$$

EXAMPLE 13-11

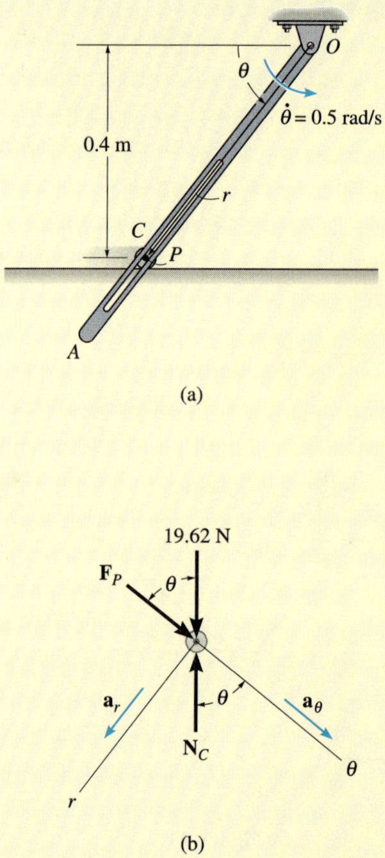

Fig. 13–20

The smooth 2-kg cylinder C in Fig. 13–20a has a peg P through its center which passes through the slot in arm OA. If the arm rotates in the *vertical plane* at a constant rate $\dot{\theta} = 0.5$ rad/s, determine the force that the arm exerts on the peg at the instant $\theta = 60°$.

Solution

Why is it a good idea to use polar coordinates to solve this problem?

Free-Body Diagram. The free-body diagram for the cylinder is shown in Fig. 13–20b. The force on the peg, $\mathbf{F}_P$, acts perpendicular to the slot in the arm. As usual, $\mathbf{a}_r$ and $\mathbf{a}_\theta$ are assumed to act in the directions of *positive* r and θ, respectively. Identify the four unknowns.

Equations of Motion. Using the data in Fig. 13–20b, we have

$$+\swarrow \Sigma F_r = ma_r; \qquad 19.62 \sin\theta - N_C \sin\theta = 2a_r \qquad (1)$$

$$+\searrow \Sigma F_\theta = ma_\theta; \qquad 19.62 \cos\theta + F_P - N_C \cos\theta = 2a_\theta \qquad (2)$$

Kinematics. From Fig. 13–20a, r can be related to θ by the equation

$$r = \frac{0.4}{\sin\theta} = 0.4 \csc\theta$$

Since $d(\csc\theta) = -(\csc\theta \cot\theta)\, d\theta$ and $d(\cot\theta) = -(\csc^2\theta)\, d\theta$, then r and the necessary time derivatives become

$$\dot{\theta} = 0.5 \qquad r = 0.4 \csc\theta$$

$$\ddot{\theta} = 0 \qquad \dot{r} = -0.4(\csc\theta \cot\theta)\dot{\theta}$$

$$= -0.2 \csc\theta \cot\theta$$

$$\ddot{r} = -0.2(-\csc\theta \cot\theta)(\dot{\theta}) \cot\theta - 0.2 \csc\theta(-\csc^2\theta)\dot{\theta}$$

$$= 0.1 \csc\theta(\cot^2\theta + \csc^2\theta)$$

Evaluating these formulas at $\theta = 60°$, we get

$$\dot{\theta} = 0.5 \qquad r = 0.462$$

$$\ddot{\theta} = 0 \qquad \dot{r} = -0.133$$

$$\ddot{r} = 0.192$$

$$a_r = \ddot{r} - r\dot{\theta}^2 = 0.192 - 0.462(0.5)^2 = 0.0770$$

$$a_\theta = r\ddot{\theta} + 2\dot{r}\dot{\theta} = 0 + 2(-0.133)(0.5) = -0.133$$

Substituting these results into Eqs. 1 and 2 with $\theta = 60°$ and solving yields

$$N_C = 19.4 \text{ N} \qquad F_P = -0.356 \text{ N} \qquad \textit{Ans.}$$

The negative sign indicates that $\mathbf{F}_P$ acts opposite to that shown in Fig. 13–20b.

EXAMPLE 13-12

A can C, having a mass of 0.5 kg, moves along a grooved horizontal slot shown in Fig. 13–21a. The slot is in the form of a spiral, which is defined by the equation $r = (0.1\theta)$ m, where θ is in radians. If the arm OA is rotating at a constant rate $\dot{\theta} = 4$ rad/s in the horizontal plane, determine the force it exerts on the can at the instant $\theta = \pi$ rad. Neglect friction and the size of the can.

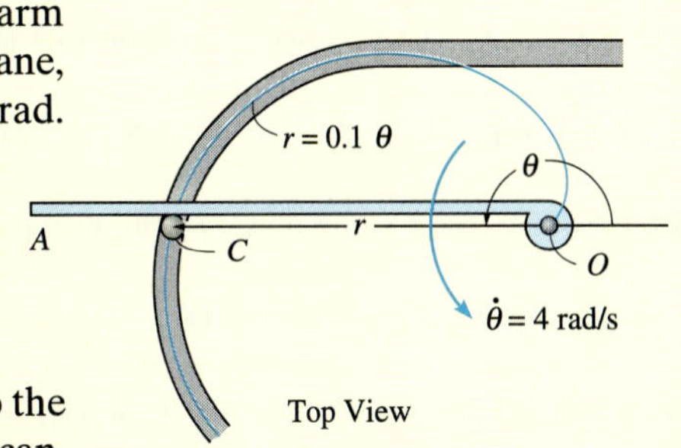

(a)

Solution

Free-Body Diagram. The driving force $\mathbf{F}_C$ acts perpendicular to the arm OA, whereas the normal force of the wall of the slot on the can, $\mathbf{N}_C$, acts perpendicular to the tangent to the curve at $\theta = \pi$ rad, Fig. 13–21b. As usual, $\mathbf{a}_r$ and $\mathbf{a}_\theta$ are assumed to act in the *positive directions* of r and θ, respectively. Since the path is specified, the angle ψ which the extended radial line r makes with the tangent, Fig. 13–21c, can be determined from Eq. 13–10. We have $r = 0.1\theta$, so that $dr/d\theta = 0.1$, and therefore

$$\tan\psi = \frac{r}{dr/d\theta} = \frac{0.1\theta}{0.1} = \theta$$

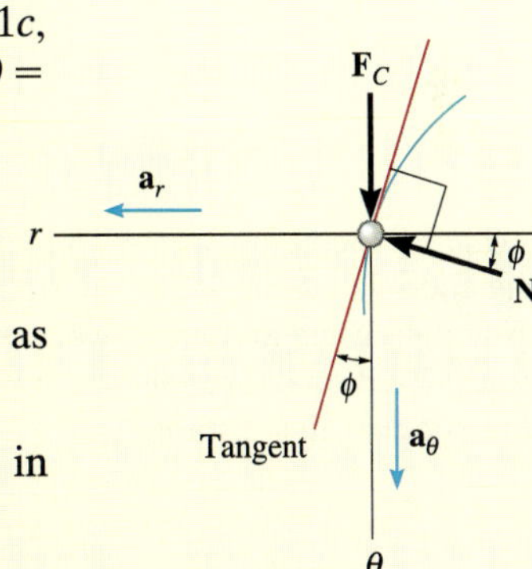

(b)

When $\theta = \pi$, $\psi = \tan^{-1}\pi = 72.3°$, so that $\phi = 90° - \psi = 17.7°$, as shown in Fig. 13–21$c$. Identify the four unknowns in Fig. 13–21b.

Equations of Motion. Using $\phi = 17.7°$ and the data shown in Fig. 13–21b, we have

$$\overset{+}{\leftarrow} \Sigma F_r = ma_r; \qquad N_C \cos 17.7° = 0.5a_r \qquad (1)$$

$$+\downarrow \Sigma F_\theta = ma_\theta; \qquad F_C - N_C \sin 17.7° = 0.5a_\theta \qquad (2)$$

Kinematics. The time derivatives of r and θ are

$$\dot{\theta} = 4 \text{ rad/s} \qquad r = 0.1\theta$$

$$\ddot{\theta} = 0 \qquad \dot{r} = 0.1\dot{\theta} = 0.1(4) = 0.4 \text{ m/s}$$

$$\ddot{r} = 0.1\ddot{\theta} = 0$$

At the instant $\theta = \pi$ rad,

$$a_r = \ddot{r} - r\dot{\theta}^2 = 0 - 0.1(\pi)(4)^2 = -5.03 \text{ m/s}^2$$

$$a_\theta = r\ddot{\theta} + 2\dot{r}\dot{\theta} = 0 + 2(0.4)(4) = 3.20 \text{ m/s}^2$$

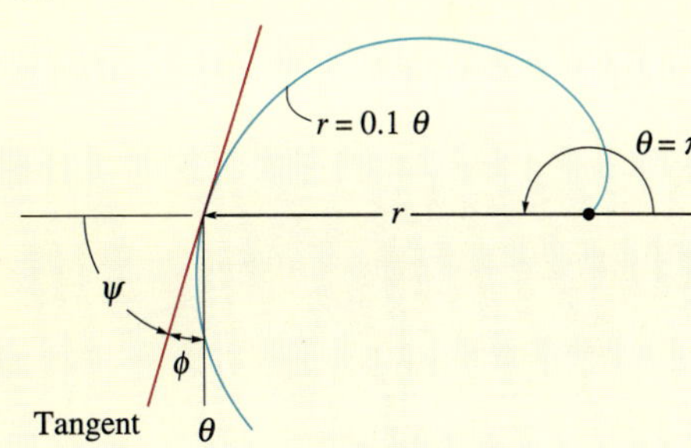

(c)

Fig. 13–21

Substituting these results into Eqs. 1 and 2 and solving yields

$$N_C = -2.64 \text{ N}$$

$$F_C = 0.800 \text{ N} \qquad \textit{Ans.}$$

What does the negative sign for N_C indicate?

Problems

13-83. A particle, having a mass of 1.5 kg, moves along a path defined by the equations $r = (4 + 3t)$ m, $\theta = (t^2 + 2)$ rad, and $z = (6 - t^3)$ m, where t is in seconds. Determine the r, θ, and z components of force which the path exerts on the particle when $t = 2$ s.

***13-84.** The path of motion of a 5-lb particle in the horizontal plane is described in terms of polar coordinates as $r = (2t + 1)$ ft and $\theta = (0.5t^2 - t)$ rad, where t is in seconds. Determine the magnitude of the unbalanced force acting on the particle when $t = 2$ s.

13-85. Determine the magnitude of the unbalanced force acting on a 5-kg particle at the instant $t = 2$ s, if the particle is moving along a horizontal path defined by the equations $r = (2t + 10)$ m and $\theta = (1.5t^2 - 6t)$ rad, where t is in seconds.

13-86. The 4-kg spool slides along the rotating rod. At the instant shown, the angular rate of rotation of the rod is $\dot{\theta} = 6$ rad/s, which is increasing at $\ddot{\theta} = 2$ rad/s^2. At this same instant, the spool is moving outward along the rod at 3 m/s, which is increasing at 1 m/s^2 when $r = 0.5$ m. Determine the radial frictional force and the normal force of the rod on the spool at this instant.

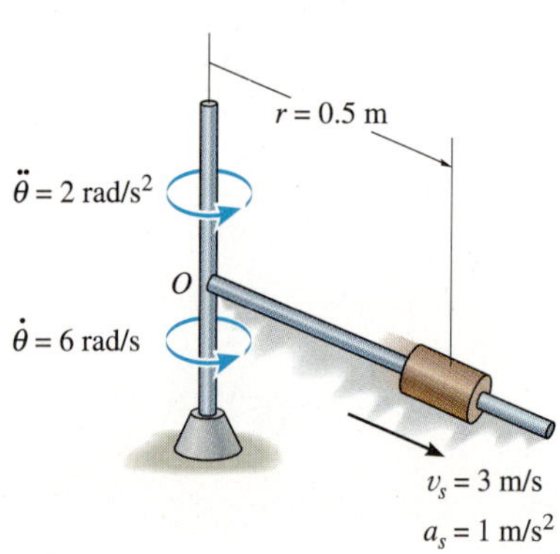

Prob. 13–86

13-87. The girl has a mass of 50 kg. She is seated on the horse of the merry-go-round which undergoes constant rotational motion $\dot{\theta} = 1.5$ rad/s. If the path of the horse is defined by $r = 4$ m, $z = (0.5 \sin \theta)$ m, determine the maximum and minimum force F_z the horse exerts on her during the motion.

Prob. 13–87

***13-88.** A car of a roller coaster travels along a track which for a short distance is defined by a conical spiral, $r = \frac{3}{4}z$, $\theta = -1.5z$, where r and z are in meters and θ in radians. If the angular motion $\dot{\theta} = 1$ rad/s is always maintained, determine the r, θ, z components of reaction exerted on the car by the track at the instant $z = 6$ m. The car and passengers have a total mass of 200 kg.

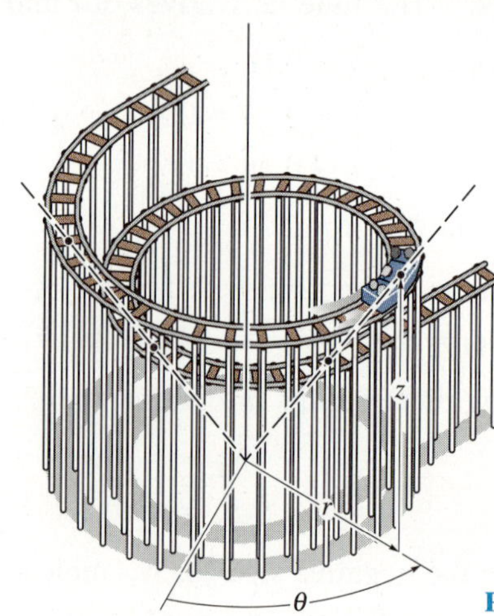

Prob. 13–88

13-89. Using a forked rod, a smooth cylinder C having a mass of 0.5 kg is forced to move along the *horizontal slotted* path $r = (0.5\theta)$ m, where θ is in radians. If the angular position of the arm is $\theta = (0.5t^2)$ rad, where t is in seconds, determine the force of the rod on the cylinder and the normal force of the slot on the cylinder at the instant $t = 2$ s. The cylinder is in contact with only *one edge* of the rod and slot at any instant.

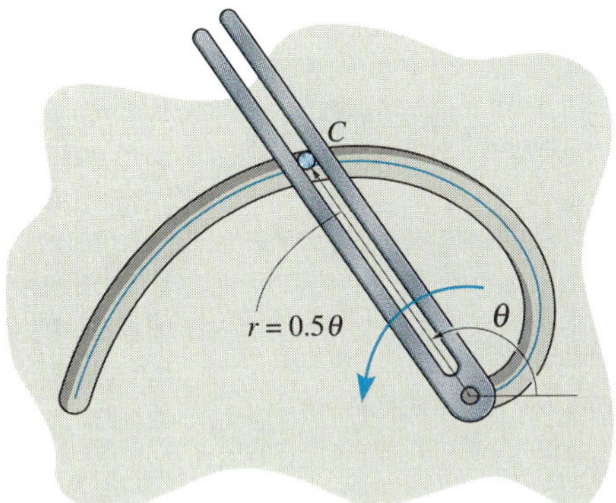

Prob. 13–89

13-90. Using a forked rod, a smooth cylinder C having a mass of 0.5 kg is forced to move along the *vertical slotted* path $r = (0.5\theta)$ m, where θ is in radians. If the angular position of the arm is $\theta = (0.5t^2)$ rad, where t is in seconds, determine the force of the rod on the cylinder and the normal force of the slot on the cylinder at the instant $t = 2$ s. The cylinder is in contact with only *one edge* of the rod and slot at any instant.

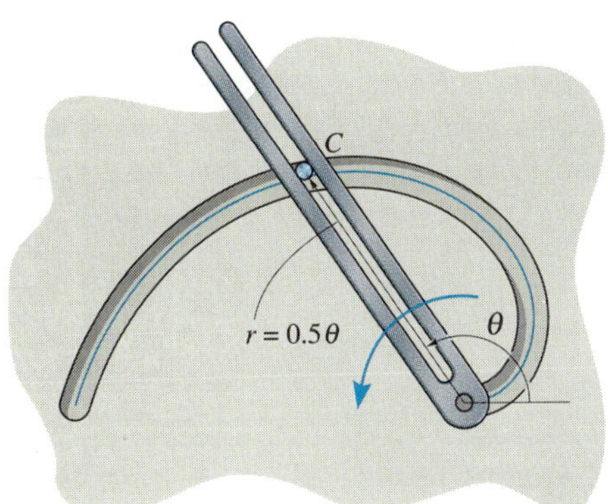

Prob. 13–90

13-91. The particle has a mass of 0.5 kg and is confined to move along the smooth vertical slot due to the rotation of the arm OA. Determine the force of the rod on the particle and the normal force of the slot on the particle when $\theta = 30°$. The rod is rotating with a constant angular velocity of $\dot{\theta} = 2$ rad/s. Assume the particle contacts only one side of the slot at any instant.

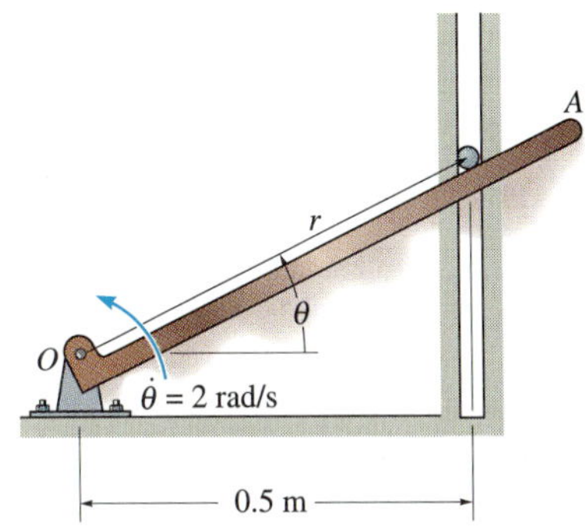

Prob. 13–91

***13-92.** A smooth can C, having a mass of 3 kg, is lifted from a feed at A to a ramp at B by a rotating rod. If the rod maintains a constant angular velocity of $\dot{\theta} = 0.5$ rad/s, determine the force which the rod exerts on the can at the instant $\theta = 30°$. Neglect the effects of friction in the calculation and the size of the can so that $r = (1.2 \cos \theta)$ m. The ramp from A to B is circular, having a radius of 600 mm.

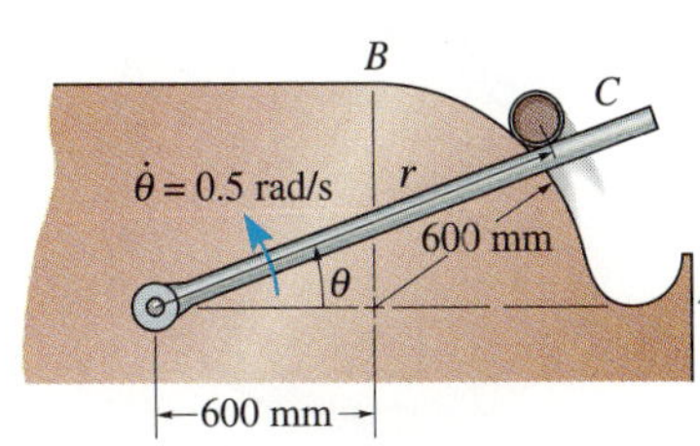

Prob. 13–92

13-93. The spool, which has a mass of 2 kg, slides along the smooth *horizontal* spiral rod, $r = (0.4\theta)$ m, where θ is in radians. If its angular rate of rotation is constant and equals $\dot{\theta} = 6$ rad/s, determine the horizontal tangential force P needed to cause the motion and the horizontal normal force component that the spool exerts on the rod at the instant $\theta = 45°$.

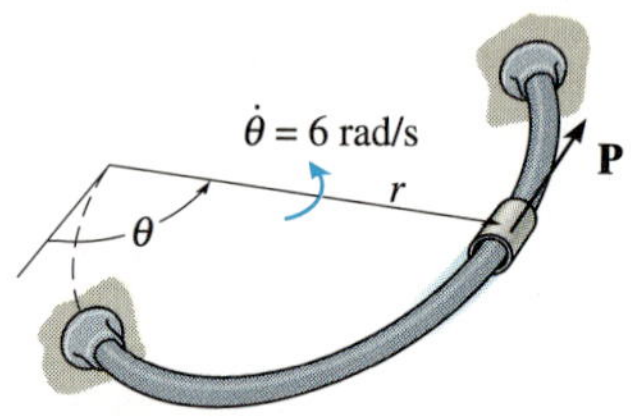

Prob. 13–93

13-94. The forked rod is used to move the smooth 2-lb particle around the horizontal path in the shape of a limaçon, $r = (2 + \cos\theta)$ ft. If at all times $\dot{\theta} = 0.5$ rad/s, determine the force which the rod exerts on the particle at the instant $\theta = 90°$. The fork and path contact the particle on only one side.

13-95. Solve Prob. 13-94 at the instant $\theta = 60°$.

***13-96.** The forked rod is used to move the smooth 2-lb particle around the horizontal path in the shape of a limaçon, $r = (2 + \cos\theta)$ ft. If $\theta = (0.5t^2)$ rad, where t is in seconds, determine the force which the rod exerts on the particle at the instant $t = 1$ s. The fork and path contact the particle on only one side.

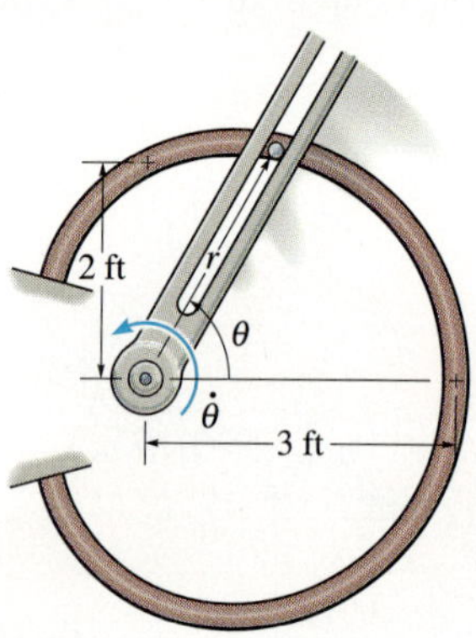

Probs. 13–94/95/96

13-97. The 2-lb collar slides along the smooth *horizontal* spiral rod, $r = (2\theta)$ ft, where θ is in radians. If its angular rate of rotation is constant and equals $\dot{\theta} = 4$ rad/s, determine the tangential force P needed to cause the motion and the normal force that the spool exerts on the rod at the instant $\theta = 90°$.

13-98. Solve Prob. 13-97 if the spiral rod is *vertical.*

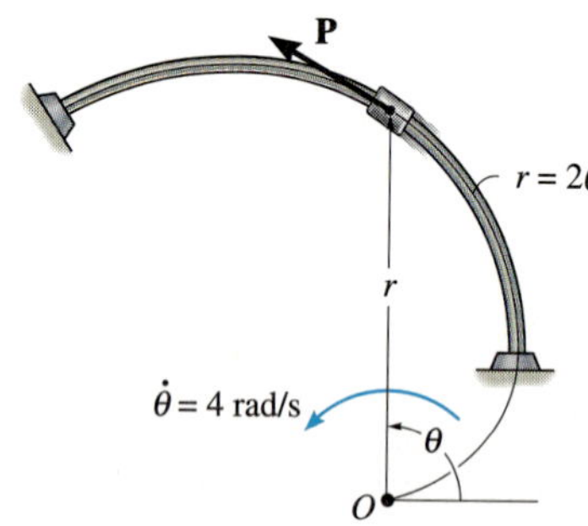

Probs. 13–97/98

13-99. Determine the normal and frictional driving forces that the partial spiral track exerts on the 200-kg motorcycle at the instant $\theta = \frac{5}{3}\pi$ rad, $\dot{\theta} = 0.4$ rad/s, and $\ddot{\theta} = 0.8$ rad/s^2. Neglect the size of the motorcycle.

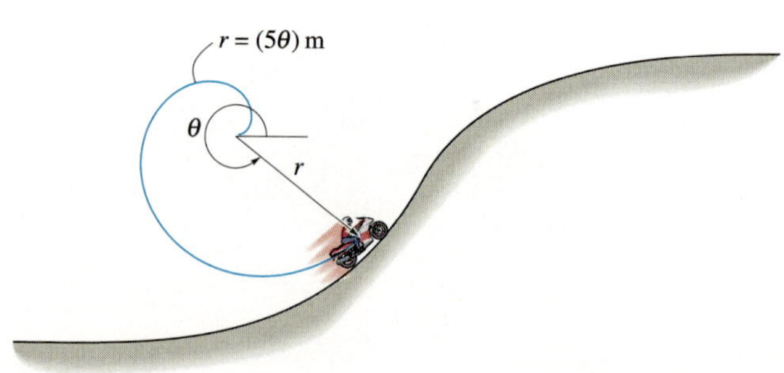

Prob. 13–99

***13-100.** Using air pressure, the 0.5-kg ball is forced to move through the tube lying in the horizontal plane and having the shape of a logarithmic spiral. If the tangential force exerted on the ball due to the air is 6 N, determine the rate of increase in the ball's speed at the instant $\theta = \pi/2$. Also, what is the angle ψ from the extended radial coordinate r to the line of action of the 6-N force?

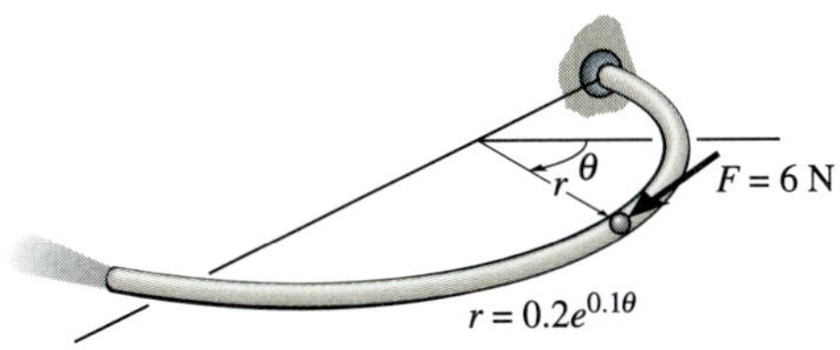

Prob. 13–100

13-101. The ball has a mass of 2 kg and a negligible size. It is originally traveling around the horizontal circular path of radius $r_0 = 0.5$ m such that the angular rate of rotation is $\dot{\theta}_0 = 1$ rad/s. If the attached cord ABC is drawn down through the hole at a constant speed of 0.2 m/s, determine the tension the cord exerts on the ball at the instant $r = 0.25$ m. Also, compute the angular velocity of the ball at this instant. Neglect the effects of friction between the ball and horizontal plane. *Hint:* First show that the equation of motion in the θ direction yields $a_\theta = r\ddot{\theta} + 2\dot{r}\dot{\theta} = (1/r)(d(r^2\dot{\theta})/dt) = 0$. When integrated, $r^2\dot{\theta} = c$, where the constant c is determined from the problem data.

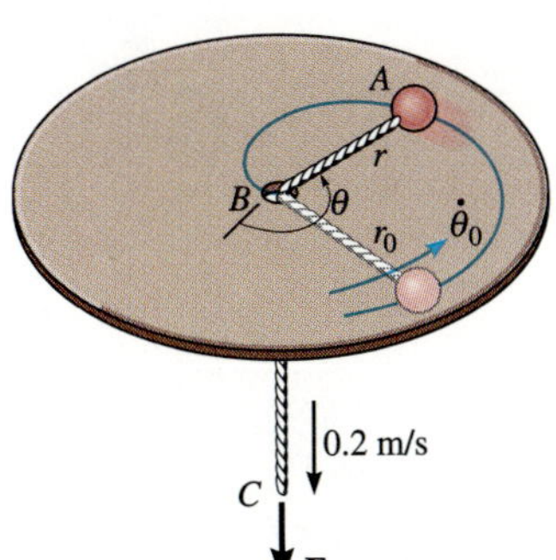

Prob. 13–101

13-102. The 0.5-lb ball is guided along the vertical circular path $r = 2r_c \cos\theta$ using the arm OA. If the arm has an angular velocity $\dot{\theta} = 0.4$ rad/s and an angular acceleration $\ddot{\theta} = 0.8$ rad/s^2 at the instant $\theta = 30°$, determine the force of the arm on the ball. Neglect friction and the size of the ball. Set $r_c = 0.4$ ft.

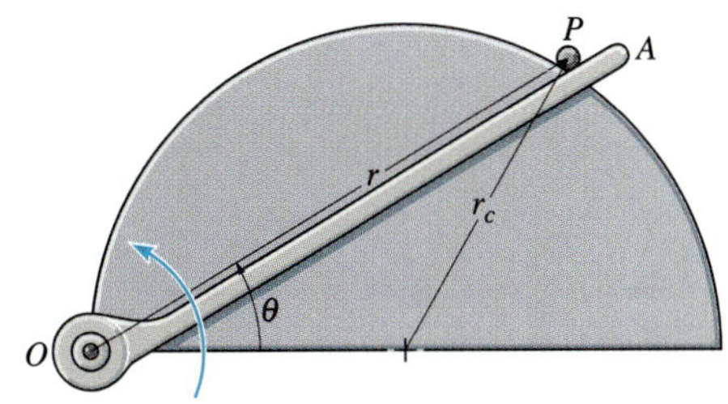

Prob. 13–102

13-103. The ball of mass m is guided along the vertical circular path $r = 2r_c \cos\theta$ using the arm OA. If the arm has a constant angular velocity $\dot{\theta}_0$, determine the angle $\theta \le 45°$ at which the ball starts to leave the surface of the semicylinder. Neglect friction and the size of the ball.

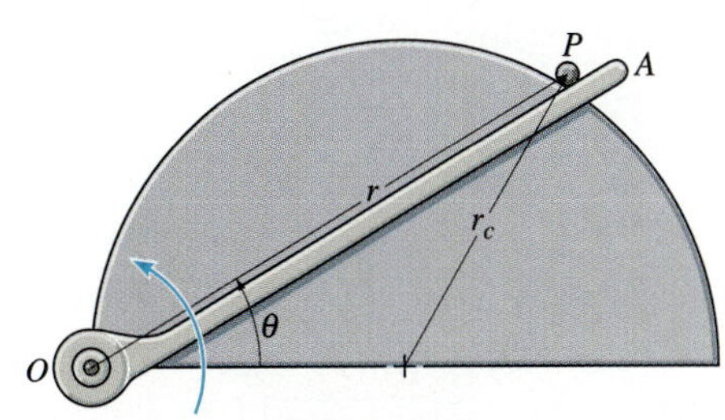

Prob. 13–103

***13-104.** Using a forked rod, a smooth cylinder P, having a mass of 0.4 kg, is forced to move along the *vertical slotted* path $r = (0.6\theta)$ m, where θ is in radians. If the cylinder has a constant speed of $v_C = 2$ m/s, determine the force of the rod and the normal force of the slot on the cylinder at the instant $\theta = \pi$ rad. Assume the cylinder is in contact with only *one edge* of the rod and slot at any instant. *Hint:* To obtain the time derivatives necessary to compute the cylinder's acceleration components a_r and a_θ, take the first and second time derivatives of $r = 0.6\theta$. Then, for further information, use Eq. 12–26 to determine $\dot{\theta}$. Also, take the time derivative of Eq. 12–26, noting that $\dot{v}_C = 0$, to determine $\ddot{\theta}$.

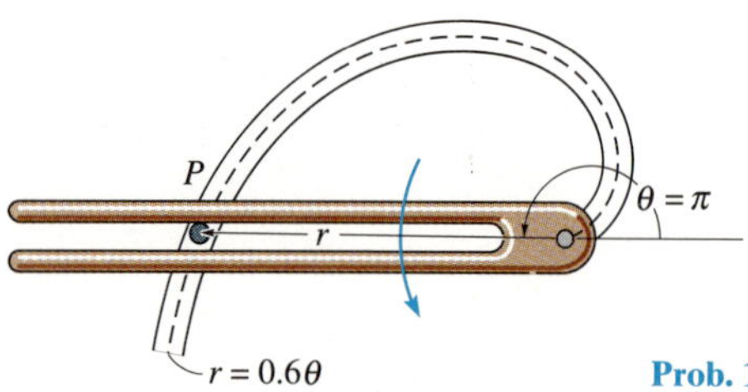

Prob. 13–104

13-105. A ride in an amusement park consists of a cart which is supported by small wheels. Initially the cart is traveling in a circular path of radius $r_0 = 16$ ft such that the angular rate of rotation is $\dot{\theta}_0 = 0.2$ rad/s. If the attached cable OC is drawn inward at a constant speed of $\dot{r} = -0.5$ ft/s, determine the tension it exerts on the cart at the instant $r = 4$ ft. The cart and its passengers have a total weight of 400 lb. Neglect the effects of friction. *Hint:* First show that the equation of motion in the θ direction yields $a_\theta = r\ddot{\theta} + 2\dot{r}\dot{\theta} = (1/r)d(r^2\dot{\theta})/dt = 0$. When integrated, $r^2\dot{\theta} = c$, where the constant c is determined from the problem data.

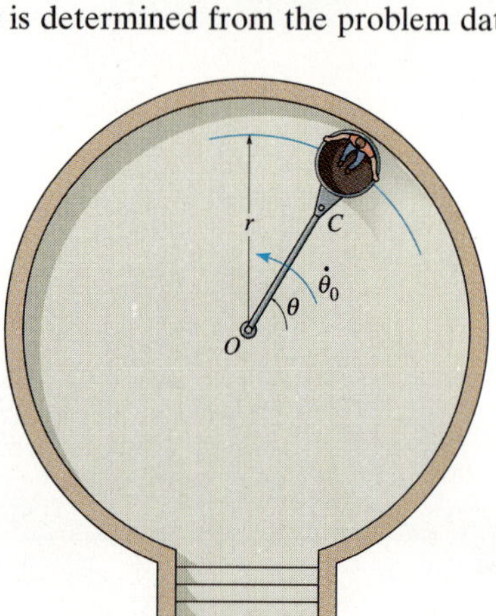

Prob. 13–105

13-106. The smooth surface of the vertical cam is defined in part by the curve $r = (0.2\cos\theta + 0.3)$ m. If the forked rod is rotating with a constant angular velocity of $\dot{\theta} = 4$ rad/s, determine the force the cam and the rod exert on the 2-kg roller when $\theta = 30°$. The attached spring has a stiffness $k = 30$ N/m and an unstretched length of 0.1 m.

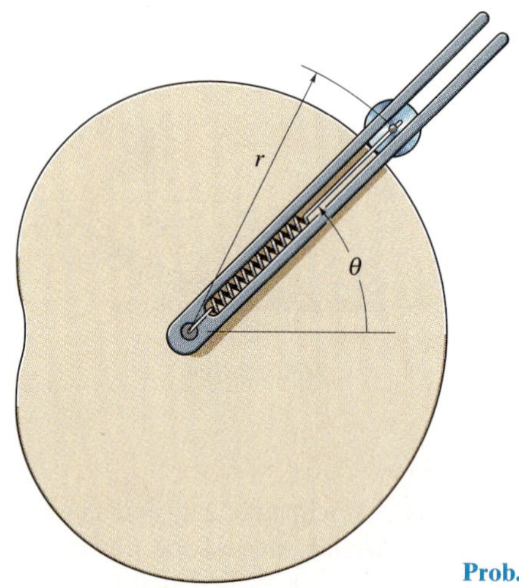

Prob. 14–106

13-107. The smooth surface of the vertical cam is defined in part by the curve $r = (0.2\cos\theta + 0.3)$ m. The forked rod is rotating with an angular acceleration of $\ddot{\theta} = 2$ rad/s^2, and when $\theta = 45°$ the angular velocity is $\dot{\theta} = 6$ rad/s. Determine the force the cam and the rod exert on the 2-kg roller at this instant. The attached spring has a stiffness $k = 100$ N/m and an unstretched length of 0.1 m.

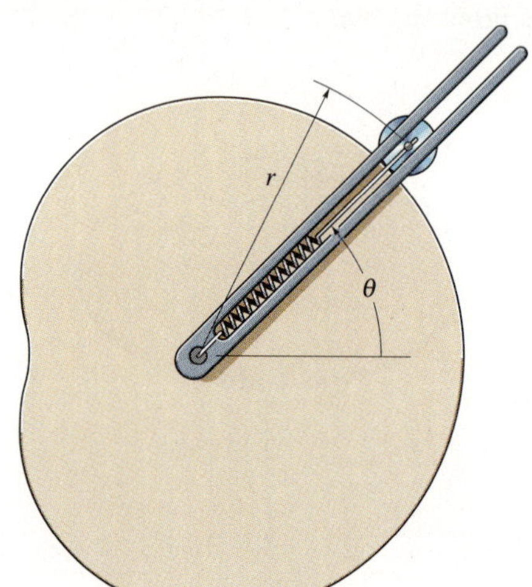

Prob. 13–107

***13-108.** The 4.20-Mg car C is traveling along a portion of a road defined by the cardioid $r = 400(1 + \cos\theta)$ m. If the car maintains a constant speed $v_C = 20$ m/s, determine the radial and transverse components of the frictional force which must be exerted by the road on all the wheels in order to maintain the motion when $\theta = 0°$. *Hint:* To determine the time derivatives necessary to compute the car's acceleration components a_r and a_θ, take the first and second time derivatives of $r = 400(1 + \cos\theta)$. Then, for further information, use Eq. 12–26 to determine $\dot{\theta}$. Also, take the time derivative of Eq. 12–26, noting that $\dot{v}_C = 0$, to determine $\ddot{\theta}$.

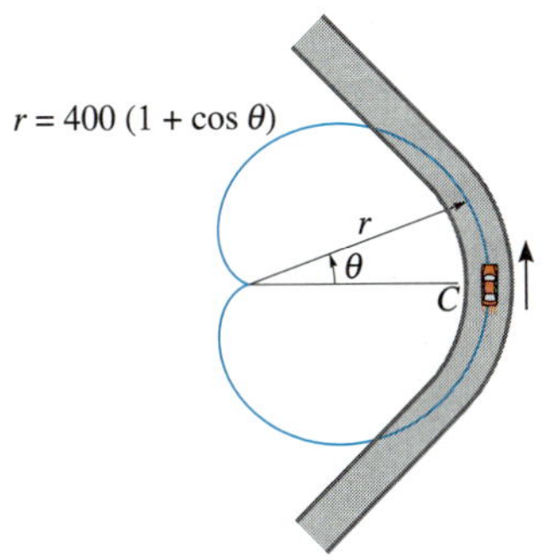

Prob. 13–108

13-109. The pilot of an airplane executes a vertical loop which in part follows the path of a "four-leaved rose," $r = (-600\cos 2\theta)$ ft, where θ is in radians. If his speed at A is a constant $v_P = 80$ ft/s, determine the vertical reaction the seat of the plane exerts on the pilot when the plane is at A. He weighs 130 lb. See hint related to Prob. 13–108.

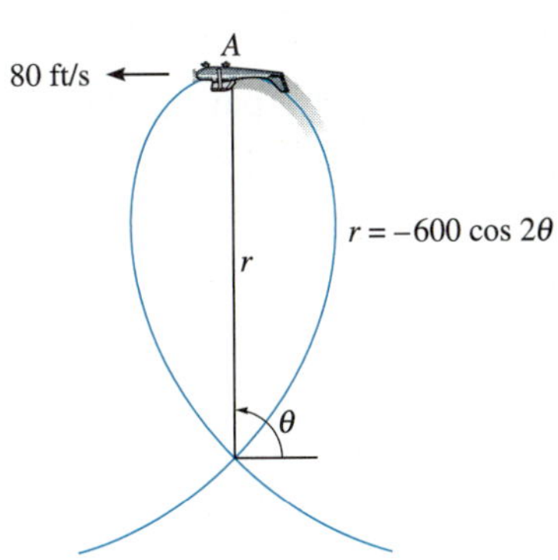

Prob. 13–109

13-110. The pilot of an airplane executes a vertical loop which in part follows the path of a cardioid, $r = 600(1 + \cos\theta)$ ft, where θ is in radians. If his speed at A ($\theta = 0°$) is a constant $v_P = 80$ ft/s, determine the vertical force the belt of his seat must exert on him to hold him to his seat when the plane is upside down at A. He weighs 150 lb. See hint related to Prob. 13–108.

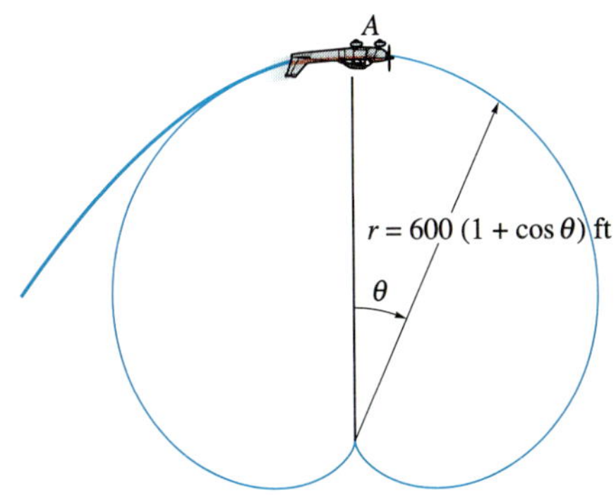

Prob. 13–110

13-111. The tube rotates in the horizontal plane at a constant rate of $\dot{\theta} = 4$ rad/s. If a 0.2-kg ball B starts at the origin O with an initial radial velocity of $\dot{r} = 1.5$ m/s and moves outward through the tube, determine the radial and transverse components of the ball's velocity at the instant it leaves the outer end at C, $r = 0.5$ m. *Hint:* Show that the equation of motion in the r direction is $\ddot{r} - 16r = 0$. The solution is of the form $r = Ae^{-4t} + Be^{4t}$. Evaluate the integration constants A and B, and determine the time t when $r = 0.5$ m. Proceed to obtain v_r and v_θ.

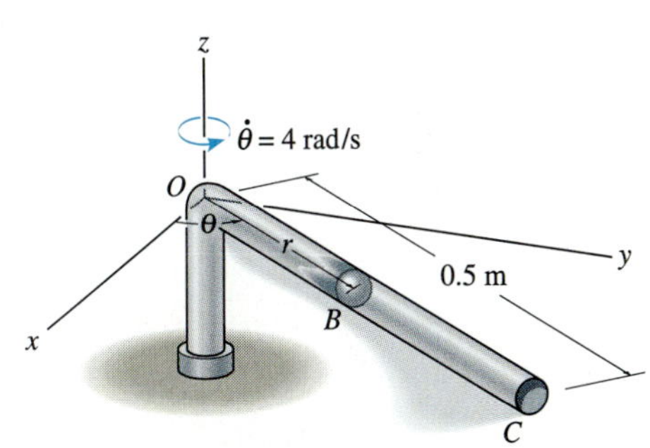

Prob. 13–111

★13.7 Central-Force Motion and Space Mechanics

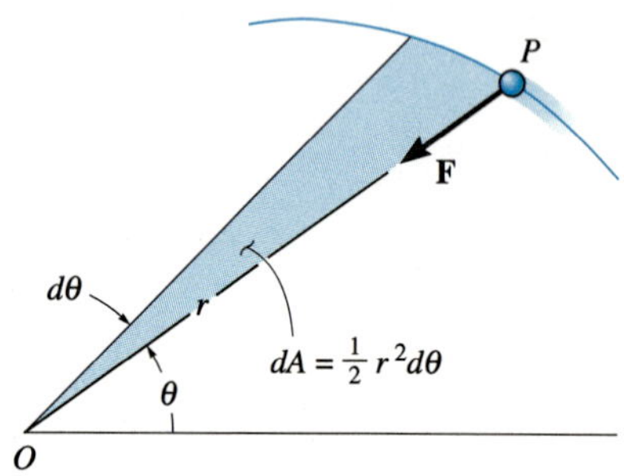

(a)

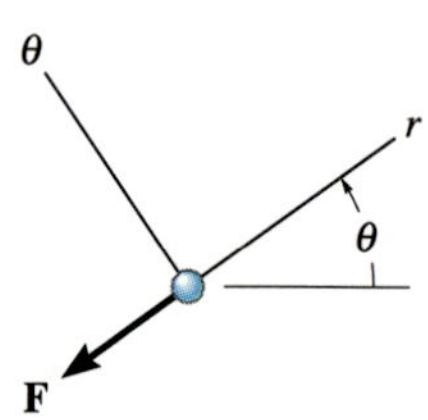

(b)

Fig. 13–22

If a particle is moving only under the influence of a force having a line of action which is always directed toward a fixed point, the motion is called *central-force motion.* This type of motion is commonly caused by electrostatic and gravitational forces.

In order to determine the motion, we will consider the particle P shown in Fig. 13–22a, which has a mass m and is acted upon only by the central force **F**. The free-body diagram for the particle is shown in Fig. 13–22b. Using polar coordinates (r, θ), the equations of motion, Eqs. 13–9, become

$$-F = m\left[\frac{d^2r}{dt^2} - r\left(\frac{d\theta}{dt}\right)^2\right]$$

$$0 = m\left(r\frac{d^2\theta}{dt^2} + 2\frac{dr}{dt}\frac{d\theta}{dt}\right) \qquad (13\text{–}11)$$

The second of these equations may be written in the form

$$\frac{1}{r}\left[\frac{d}{dt}\left(r^2\frac{d\theta}{dt}\right)\right] = 0$$

so that integrating yields

$$r^2\frac{d\theta}{dt} = h \qquad (13\text{–}12)$$

Here h is a constant of integration. From Fig. 13–22a notice that the shaded area described by the radius r, as r moves through an angle $d\theta$, is $dA = \frac{1}{2}r^2 d\theta$. If the *areal velocity* is defined as

$$\frac{dA}{dt} = \frac{1}{2}r^2\frac{d\theta}{dt} = \frac{h}{2} \qquad (13\text{–}13)$$

then, it is seen that the areal velocity for a particle subjected to central-force motion is *constant.* In other words, the particle will sweep out equal segments of area per unit of time as it travels along the path. To obtain the *path of motion,* $r = f(\theta)$, the independent variable t must be eliminated from Eqs. 13–11. Using the chain rule of calculus and Eq. 13–12, the time derivatives of Eqs. 13–11 may be replaced by

$$\frac{dr}{dt} = \frac{dr}{d\theta}\frac{d\theta}{dt} = \frac{h}{r^2}\frac{dr}{d\theta}$$

$$\frac{d^2r}{dt^2} = \frac{d}{dt}\left(\frac{h}{r^2}\frac{dr}{d\theta}\right) = \frac{d}{d\theta}\left(\frac{h}{r^2}\frac{dr}{d\theta}\right)\frac{d\theta}{dt} = \left[\frac{d}{d\theta}\left(\frac{h}{r^2}\frac{dr}{d\theta}\right)\right]\frac{h}{r^2}$$

Substituting a new dependent variable (xi) $\xi = 1/r$ into the second equation, we have

$$\frac{d^2r}{dt^2} = -h^2\xi^2\frac{d^2\xi}{d\theta^2}$$

Also, the square of Eq. 13–12 becomes

$$\left(\frac{d\theta}{dt}\right)^2 = h^2\xi^4$$

Substituting these last two equations into the first of Eqs. 13–11 yields

$$-h^2\xi^2\frac{d^2\xi}{d\theta^2} - h^2\xi^3 = -\frac{F}{m}$$

or

$$\frac{d^2\xi}{d\theta^2} + \xi = \frac{F}{mh^2\xi^2} \qquad (13\text{–}14)$$

This differential equation defines the path over which the particle travels when it is subjected to the central force* **F**.

For application, the force of gravitational attraction will be considered. Some common examples of central-force systems which depend on gravitation include the motion of the moon and artificial satellites about the earth, and the motion of the planets about the sun. As a typical problem in space mechanics, consider the trajectory of a space satellite or space vehicle launched into free-flight orbit with an initial velocity $\mathbf{v}_0$, Fig. 13–23. It will be assumed that this velocity is initially *parallel* to the tangent at the surface of the earth, as shown in the figure.† Just after the satellite is released into free flight, the only force acting on it is the gravitational force of the earth. (Gravitational attractions involving other bodies such as the moon or sun will be neglected, since for orbits close to the earth their effect is small in comparison with the earth's gravitation.) According to Newton's law of gravitation, force **F** will always act between the mass centers of the earth and the satellite, Fig. 13–23. From Eq. 13–1, this force of attraction has a magnitude of

$$F = G\frac{M_e m}{r^2}$$

where M_e and m represent the mass of the earth and the satellite, respectively, G is the gravitational constant, and r is the distance between

This satellite is subjected to a central force and as such its orbital motion can be closely predicted using the equations developed in this section.

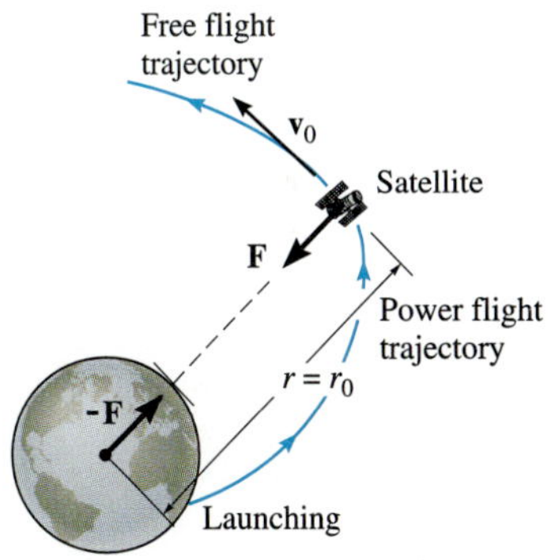

Fig. 13–23

*In the derivation, **F** is considered positive when it is directed toward point O. If **F** is oppositely directed, the right side of Eq. 13–14 should be negative.

†The case where $\mathbf{v}_0$ acts at some initial angle θ to the tangent is best described using the conservation of angular momentum (see Prob. 15–116).

the mass centers. Setting $\xi = 1/r$ in the foregoing equation and substituting the result into Eq. 13–14, we obtain

$$\frac{d^2\xi}{d\theta^2} + \xi = \frac{GM_e}{h^2} \quad (13\text{–}15)$$

This second-order ordinary differential equation has constant coefficients and is nonhomogeneous. The solution is represented as the sum of the complementary and particular solutions. The complementary solution is obtained when the term on the right is equal to zero. It is

$$\xi_c = C\cos(\theta - \phi)$$

where C and ϕ are constants of integration. The particular solution is

$$\xi_p = \frac{GM_e}{h^2}$$

Thus, the complete solution to Eq. 13–15 is

$$\xi = \xi_c + \xi_p$$

$$= \frac{1}{r} = C\cos(\theta - \phi) + \frac{GM_e}{h^2} \quad (13\text{–}16)$$

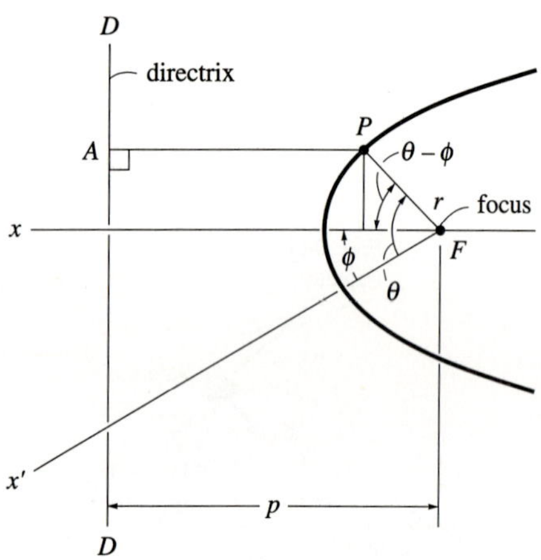

Fig. 13–24

The validity of this result may be checked by substitution into Eq. 13–15.

Equation 13–16 represents the *free-flight trajectory* of the satellite. It is the equation of a conic section expressed in terms of polar coordinates. As shown in Fig. 13–24, a *conic section* is defined as the locus of point P, which moves in a plane in such a way that the ratio of its distance from a fixed point F to its distance from a fixed line is constant. The fixed point is called the *focus*, and the fixed line DD is called the *directrix*. The constant ratio is called the *eccentricity* of the conic section and is denoted by e. Thus,

$$e = \frac{FP}{PA}$$

which may be written in the form

$$FP = r = e(PA) = e[p - r\cos(\theta - \phi)]$$

or

$$\frac{1}{r} = \frac{1}{p}\cos(\theta - \phi) + \frac{1}{ep}$$

Comparing this equation with Eq. 13–16, it is seen that the eccentricity of the conic section for the trajectory is

$$e = \frac{Ch^2}{GM_e} \quad (13\text{–}17)$$

and the fixed distance from the focus to the directrix is

$$p = \frac{1}{C} \tag{13–18}$$

Provided the polar angle θ is measured from the x axis (an axis of symmetry since it is perpendicular to the directrix), the angle ϕ is zero, Fig. 13–24, and therefore Eq. 13–16 reduces to

$$\frac{1}{r} = C \cos \theta + \frac{GM_e}{h^2} \tag{13–19}$$

The constants h and C are determined from the data obtained for the position and velocity of the satellite at the end of the *power-flight trajectory*. For example, if the initial height or distance to the space vehicle is r_0 (measured from the center of the earth) and its initial speed is v_0 at the beginning of its free flight, Fig. 13–25, then the constant h may be obtained from Eq. 13–12. When $\theta = \phi = 0°$, the velocity $\mathbf{v}_0$ has no radial component; therefore, from Eq. 12–25, $v_0 = r_0(d\theta/dt)$, so that

$$h = r_0^2 \frac{d\theta}{dt}$$

or

$$\boxed{h = r_0 v_0} \tag{13–20}$$

To determine C, use Eq. 13–19 with $\theta = 0°$, $r = r_0$, and substitute Eq. 13–20 for h:

$$\boxed{C = \frac{1}{r_0}\left(1 - \frac{GM_e}{r_0 v_0^2}\right)} \tag{13–21}$$

The equation for the free-flight trajectory therefore becomes

$$\boxed{\frac{1}{r} = \frac{1}{r_0}\left(1 - \frac{GM_e}{r_0 v_0^2}\right) \cos \theta + \frac{GM_e}{r_0^2 v_0^2}} \tag{13–22}$$

The type of path taken by the satellite is determined from the value of the eccentricity of the conic section as given by Eq. 13–17. If

$e = 0$	free-flight trajectory is a circle	(13–23)
$e = 1$	free-flight trajectory is a parabola	
$e < 1$	free-flight trajectory is an ellipse	
$e > 1$	free-flight trajectory is a hyperbola	

Each of these trajectories is shown in Fig. 13–25. From the curves it is seen that when the satellite follows a parabolic path, it is "on the border" of never returning to its initial starting point. The initial launch velocity, $\mathbf{v}_0$, required for the satellite to follow a parabolic path is called the *escape velocity*. The speed, v_e, can be determined by using the second of Eqs. 13–23 with Eqs. 13–17, 13–20, and 13–21. It is left as an exercise to show that

$$v_e = \sqrt{\frac{2GM_e}{r_0}} \tag{13–24}$$

The speed v_c required to launch a satellite into a *circular orbit* can be found using the first of Eqs. 13–23. Since e is related to h and C, Eq. 13–17, C must be zero to satisfy this equation (from Eq. 13–20, h cannot be zero); and therefore, using Eq. 13–21, we have

$$v_c = \sqrt{\frac{GM_e}{r_0}} \tag{13–25}$$

Provided r_0 represents a minimum height for launching, in which frictional resistance from the atmosphere is neglected, speeds at launch which are less than v_c will cause the satellite to reenter the earth's atmosphere and either burn up or crash, Fig. 13–25.

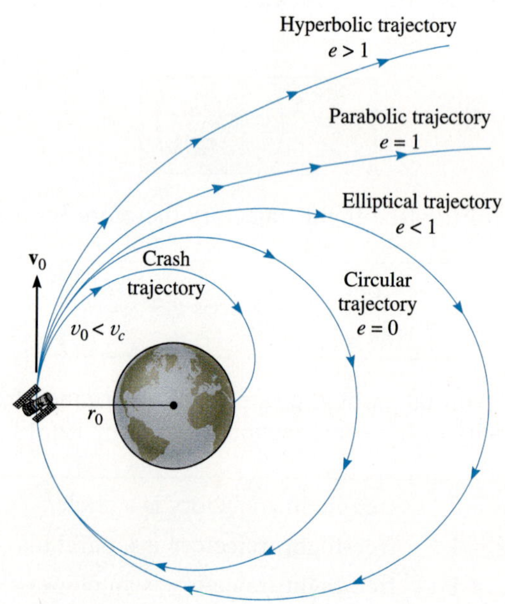

Fig. 13–25

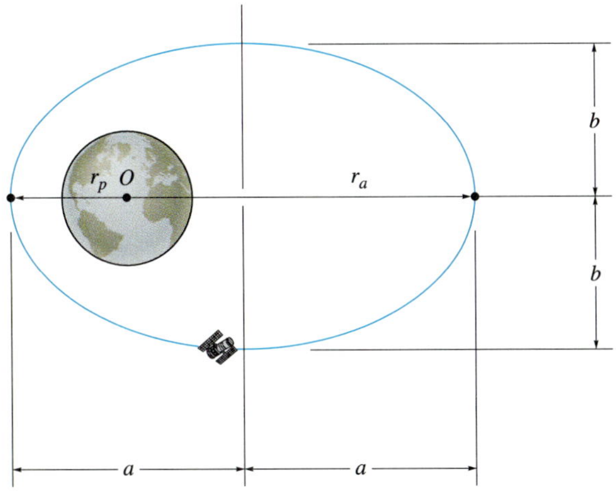

Fig. 13–26

All the trajectories attained by planets and most satellites are elliptical, Fig. 13–26. For a satellite's orbit about the earth, the *minimum distance* from the orbit to the center of the earth O (which is located at one of the foci of the ellipse) is r_p and can be found using Eq. 13–22 with $\theta = 0°$. Therefore;

$$\boxed{r_p = r_0} \tag{13–26}$$

This minimum distance is called the *perigee* of the orbit. The *apogee* or maximum distance r_a can be found using Eq. 13–22 with $\theta = 180°$.* Thus,

$$\boxed{r_a = \frac{r_0}{(2GM_e/r_0v_0^2) - 1}} \tag{13–27}$$

With reference to Fig. 13–26, the semimajor axis a of the ellipse is

$$a = \frac{r_p + r_a}{2} \tag{13–28}$$

Using analytical geometry, it can be shown that the minor axis b is determined from the equation

$$b = \sqrt{r_p r_a} \tag{13–29}$$

*Actually, the terminology perigee and apogee pertains only to orbits about the *earth*. If any other heavenly body is located at the focus of an elliptical orbit, the minimum and maximum distances are referred to respectively as the *periapsis* and *apoapsis* of the orbit.

Furthermore, by direct integration, the area of an ellipse is

$$A = \pi ab = \frac{\pi}{2}(r_p + r_a)\sqrt{r_p r_a} \tag{13–30}$$

The areal velocity has been defined by Eq. 13–13, $dA/dt = h/2$. Integrating yields $A = hT/2$, where T is the *period* of time required to make one orbital revolution. From Eq. 13–30, the period is

$$T = \frac{\pi}{h}(r_p + r_a)\sqrt{r_p r_a} \tag{13–31}$$

In addition to predicting the orbital trajectory of earth satellites, the theory developed in this section is valid, to a surprisingly close approximation, at predicting the actual motion of the planets traveling around the sun. In this case the mass of the sun, M_s, should be substituted for M_e when the appropriate formulas are used.

The fact that the planets do indeed follow elliptic orbits about the sun was discovered by the German astronomer Johannes Kepler in the early seventeenth century. His discovery was made *before* Newton had developed the laws of motion and the law of gravitation, and so at the time it provided important proof as to the validity of these laws. Kepler's laws, developed after 20 years of planetary observation, are summarized as follows:

1. Every planet moves in its orbit such that the line joining it to the sun sweeps over equal areas in equal intervals of time, whatever the line's length.

2. The orbit of every planet is an ellipse with the sun placed at one of its foci.

3. The square of the period of any planet is directly proportional to the cube of the minor axis of its orbit.

A mathematical statement of the first and second laws is given by Eqs. 13–13 and 13–22. The third law can be shown from Eq. 13–31 using Eqs. 13–19, 13–28, and 13–29.

EXAMPLE 13–13

A satellite is launched 600 km from the surface of the earth, with an initial velocity of 30 Mm/h acting parallel to the tangent at the surface of the earth, Fig. 13–27. Assuming that the radius of the earth is 6378 km and that its mass is $5.976(10^{24})$ kg, determine (*a*) the eccentricity of the orbital path, and (*b*) the velocity of the satellite at apogee.

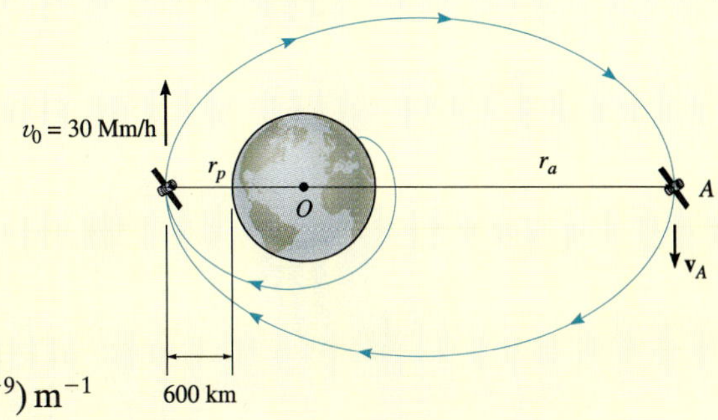

Fig. 13–27

Solution

Part (a). The eccentricity of the orbit is obtained using Eq. 13–17. The constants h and C are first determined from Eqs. 13–20 and 13–21. Since

$$r_p = r_0 = 6378 \text{ km} + 600 \text{ km} = 6.978(10^6) \text{ m}$$
$$v_0 = 30 \text{ Mm/h} = 8333.3 \text{ m/s}$$

then

$$h = r_p v_0 = 6.978(10^6)(8333.3) = 58.15(10^9) \text{ m}^2/\text{s}$$

$$C = \frac{1}{r_p}\left(1 - \frac{GM_e}{r_p v_0^2}\right)$$

$$= \frac{1}{6.978(10^6)}\left\{1 - \frac{66.73(10^{-12})[5.976(10^{24})]}{6.978(10^6)(8333.3)^2}\right\} = 25.4(10^{-9}) \text{ m}^{-1}$$

Hence,

$$e = \frac{Ch^2}{GM_e} = \frac{2.54(10^{-8})[58.15(10^9)]^2}{66.73(10^{-12})[5.976(10^{24})]} = 0.215 < 1 \qquad \textit{Ans.}$$

From Eq. 13–23, observe that the orbit is an *ellipse.*

Part (b). If the satellite were launched at the apogee A shown in Fig. 13–27, with a velocity $\mathbf{v}_A$, the same orbit would be maintained provided that

$$h = r_p v_0 = r_a v_A = 58.15(10^9) \text{ m}^2/\text{s}$$

Using Eq. 13–27, we have

$$r_a = \frac{r_p}{\dfrac{2GM_e}{r_p v_0^2} - 1} = \frac{6.978(10^6)}{\left\{\dfrac{2[66.73(10^{-12})][5.976(10^{24})]}{6.978(10^6)(8333.3)^2} - 1\right\}} = 10.804(10^6)$$

Thus,

$$v_A = \frac{58.15(10^9)}{10.804(10^6)} = 5382.2 \text{ m/s} = 19.4 \text{ Mm/h} \qquad \textit{Ans.}$$

Problems

In the following problems, except where otherwise indicated, assume that the radius of the earth is 6378 km, the earth's mass is $5.976(10^{24})$ kg, the mass of the sun is $1.99(10^{30})$ kg, and the gravitational constant is $G = 66.73(10^{-12})\ m^3/(kg \cdot s^2)$.

***13-112.** The rocket is in circular orbit about the earth at an altitude of $h = 4$ Mm. Determine the minimum increment in speed it must have in order to escape the earth's gravitational field.

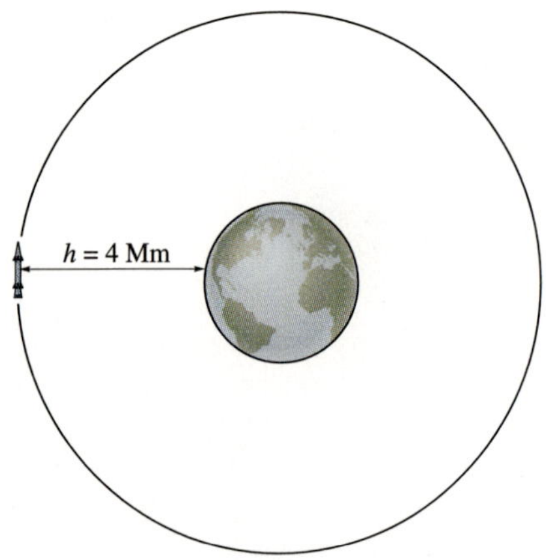

Prob. 13–112

13-113. Prove Kepler's third law of motion. *Hint:* Use Eqs. 13–19, 13–28, 13–29, and 13–31.

13-114. The satellite is moving in an elliptical orbit with an eccentricity $e = 0.25$. Determine its speed when it is at its maximum distance A and minimum distance B from the earth.

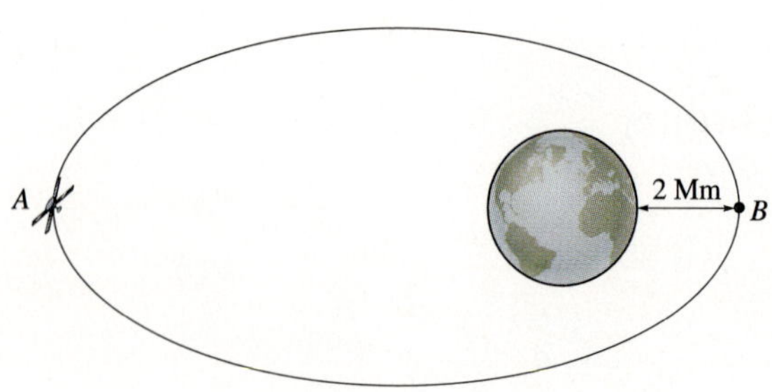

Prob. 13–114

13-115. A satellite is to be placed into an elliptical orbit about the earth such that at the perigee of its orbit it has an *altitude* of 800 km, and at apogee its *altitude* is 2400 km. Determine its required launch velocity tangent to the earth's surface at perigee and the period of its orbit.

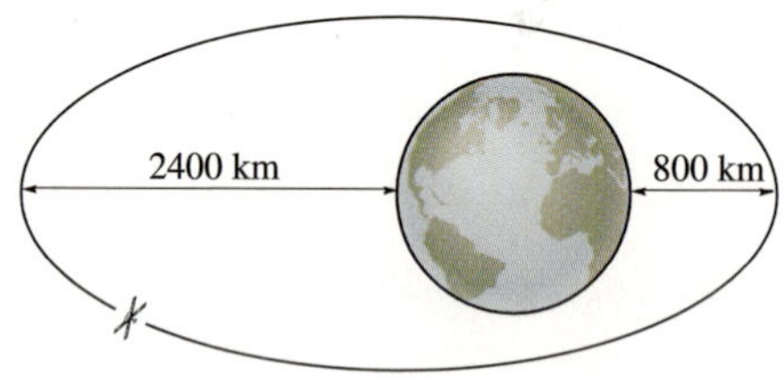

Prob. 13–115

***13-116.** Determine the constant speed of the satellite S so that it circles the earth with an orbit of radius $r = 15$ Mm. *Hint:* Use Eq. 13–1.

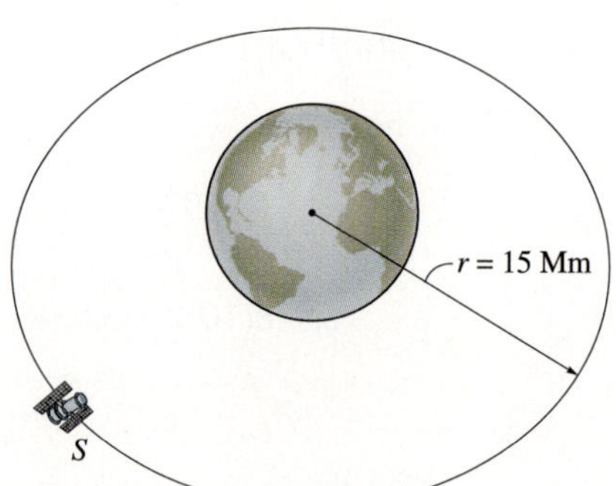

Prob. 13–116

13-117. A satellite is launched with an initial velocity $v_0 = 2500$ mi/h parallel to the surface of the earth. Determine the required altitude (or range of altitudes) above the earth's surface for launching if the free-flight trajectory is to be (a) circular, (b) parabolic, (c) elliptical, and (d) hyperbolic. Take $G = 34.4(10^{-9})\ (\text{lb}\cdot\text{ft}^2)/\text{slug}^2$, $M_e = 409(10^{21})$ slug, the earth's radius $r_e = 3960$ mi, and 1 mi = 5280 ft.

13-118. The elliptical path of a satellite has an eccentricity $e = 0.130$. If the satellite has a speed of 15 Mm/h when it is at perigee, P, determine its speed when it arrives at apogee, A. Also, how far is it from the earth's surface when it is at A?

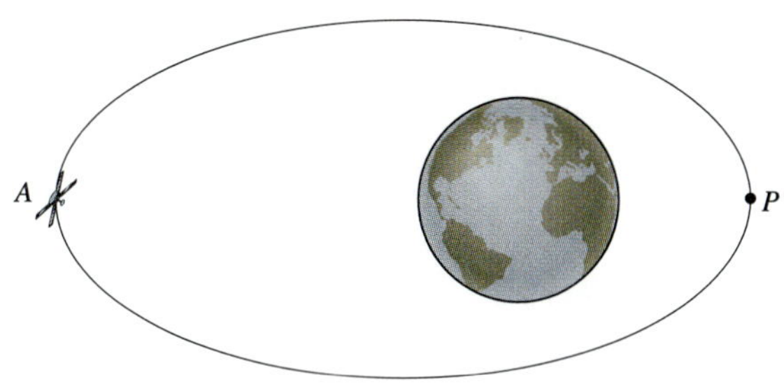

Prob. 13–118

13-119. The earth has an eccentricity $e = 0.0821$ in its orbit around the sun. Knowing that the earth's minimum distance from the sun is $151.3(10^6)$ km, find the speed at which a rocket is traveling when it is at this distance. Determine the equation in polar coordinates which describes the rocket's orbit about the sun.

***13-120.** The rocket is in a free-flight elliptical orbit about the earth such that the eccentricity of its orbit is e and its perigee is r_0. Determine the minimum increment of speed it should have in order to escape the earth's gravitational field when it is at this point along its orbit.

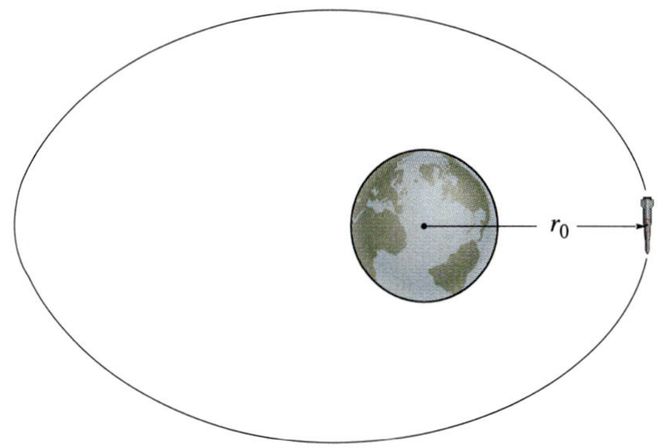

Prob. 13–120

13-121. A communications satellite is to be placed into an equatorial circular orbit around the earth so that it always remains directly over a point on the earth's surface. If this requires the period to be 24 hours (approximately), determine the radius of the orbit and the satellite's velocity.

13-122. The rocket is traveling in free flight along an elliptical trajectory $A'A$. The planet has no atmosphere, and its mass is 0.60 times that of the earth's. If the rocket has the apogee and perigee shown, determine the rocket's velocity when it is at point A. Take $G = 34.4(10^{-9})\ (\text{lb}\cdot\text{ft}^2)/\text{slug}^2$, $M_e = 409(10^{21})$ slug, 1 mi = 5280 ft.

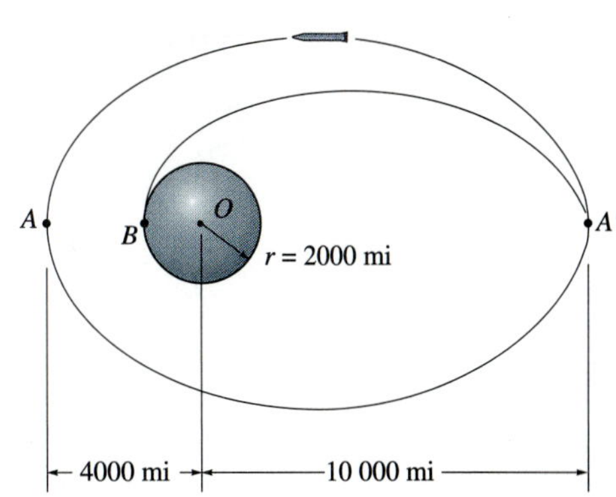

Prob. 13–122

13-123. If the rocket in Prob. 13-122 is to land on the surface of the planet, determine the required free-flight speed it must have at A' so that the landing occurs at B. How long does it take for the rocket to land, in going from A' to B? The planet has no atmosphere, and its mass is 0.6 times that of the earth's. Take $G = 34.4(10^{-9})$ $(\text{lb}\cdot\text{ft}^2)/\text{slug}^2$, $M_e = 409(10^{21})$ slug, 1 mi = 5280 ft.

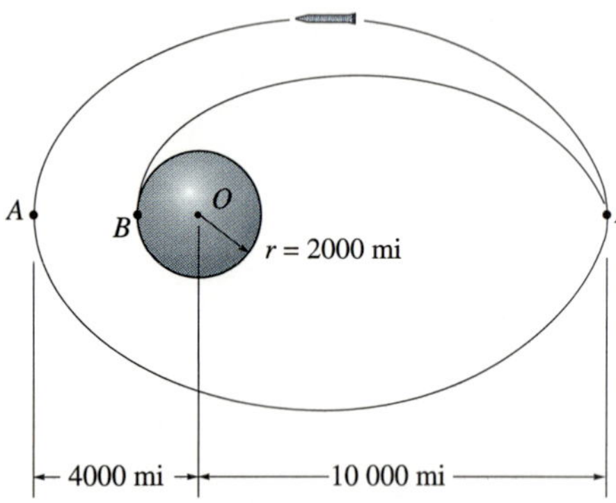

Prob. 13–123

***13-124.** A satellite is launched with an initial velocity $v_0 = 4000$ km/h parallel to the surface of the earth. Determine the required altitude (or range of altitudes) above the earth's surface for launching if the free-flight trajectory is to be (a) circular, (b) parabolic, (c) elliptical, and (d) hyperbolic.

13-125. The rocket is traveling in a free-flight elliptical orbit about the earth such that $e = 0.76$ as shown. Determine its speed when it is at point A. Also determine the sudden change in speed the rocket must experience at B in order to travel in free flight along the orbit indicated by the dashed path.

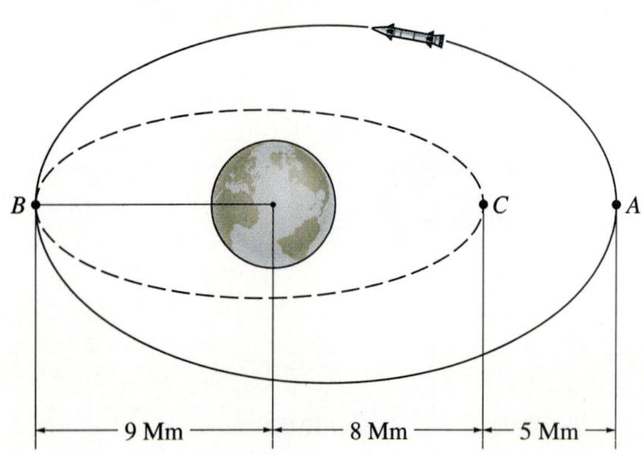

Prob. 13–125

13-126. A satellite is in an elliptical orbit around the earth such that $e = 0.156$. If its perigee is 5 Mm, determine its velocity at this point and also the distance OB when it is at point B, located 135° away as shown.

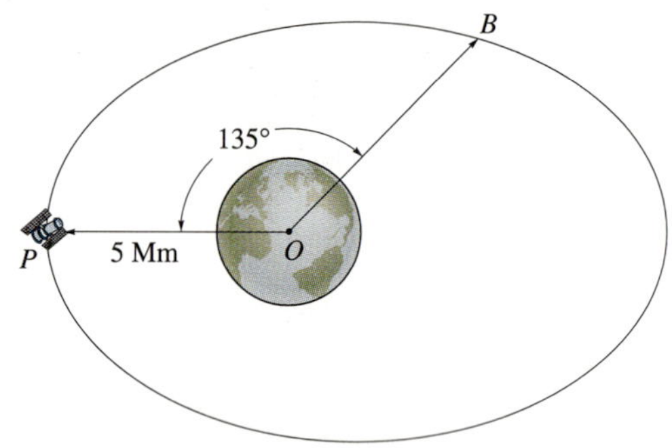

Prob. 13–126

13-127. A rocket is in free-flight elliptical orbit around the planet Venus. Knowing that the periapsis and apoapsis of the orbit are 8 Mm and 26 Mm, respectively, determine (a) the speed of the rocket at point A', (b) the required speed it must attain at A just after braking so that it undergoes an 8-Mm free-flight circular orbit around Venus, and (c) the periods of both the circular and elliptical orbits. The mass of Venus is 0.816 times the mass of the earth.

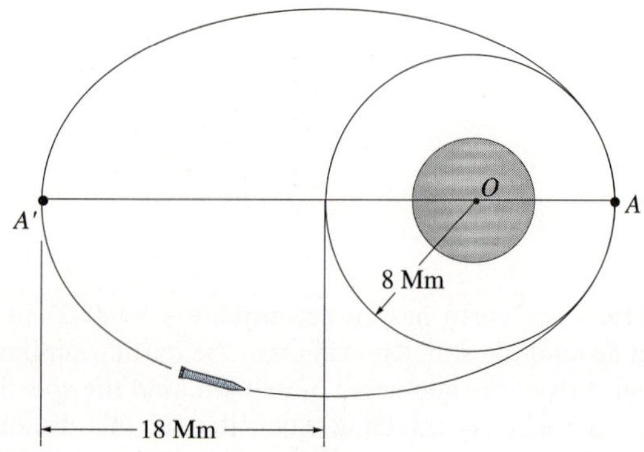

Prob. 13–127

Design Projects

13–1D. DESIGN OF A RAMP CATAPULT

The block B has a mass of 20 kg and is to be catapulted from the table. Design a catapulting mechanism that can be attached to the table and to the container of the block, using cables and pulleys. Neglect the mass of the container, assume the operator can exert a constant tension of 120 N on a single cable during operation, and that the maximum movement of his arm is 0.5 m. The coefficient of kinetic friction between the table and container is $\mu_k = 0.2$. Submit a drawing of your design, and calculate the maximum range R to where the block will strike the ground. Compare your value with that of others in the class.

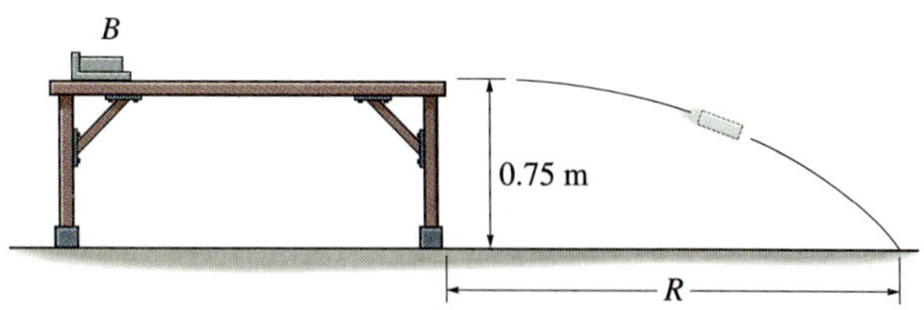

Prob. 13–1D

13–2D. DESIGN OF A WATER-BALLOON LAUNCHER

Design a method for launching a 0.25-lb water balloon. Hold a contest with other students to see who can launch the balloon the farthest or hit a target. Materials should consist of a single rubber band of specified length and stiffness, and if necessary no more than three pieces of wood of specified size. Submit a report to show your calculations of where the balloon is predicted to strike the ground from the point at which it was launched. Compare this with the actual value R and discuss why the two distances may be different.

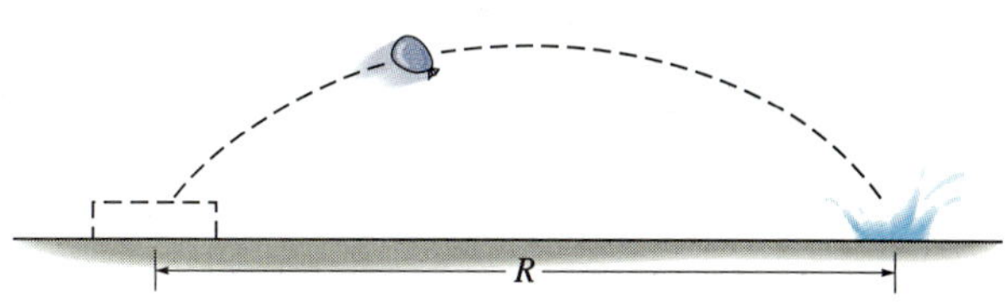

Prob. 13–2D

In order to properly design the loop of this roller coaster it is necessary to ensure that the cars have enough energy to be able to make the loop without leaving the tracks.

CHAPTER
14

Kinetics of a Particle: Work and Energy

Chapter Objectives

- To develop the principle of work and energy and apply it to solve problems that involve force, velocity, and displacement.
- To study problems that involve power and efficiency.
- To introduce the concept of a conservative force and apply the theorem of conservation of energy to solve kinetic problems.

14.1 The Work of a Force

In mechanics a force **F** does *work* on a particle only when the particle undergoes a *displacement in the direction of the force.* For example, consider the force **F** acting on the particle in Fig. 14–1. If the particle moves along the path s from position **r** to a new position **r′**, the displacement is then $d\mathbf{r} = \mathbf{r}' - \mathbf{r}$. The magnitude of $d\mathbf{r}$ is represented by ds, the differential segment along the path. If the angle between the tails of $d\mathbf{r}$ and **F** is θ, Fig. 14–1, then the work dU which is done by **F** is a

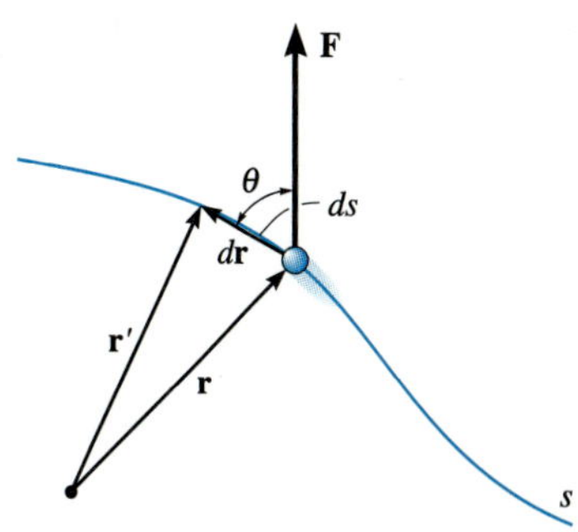

Fig. 14–1

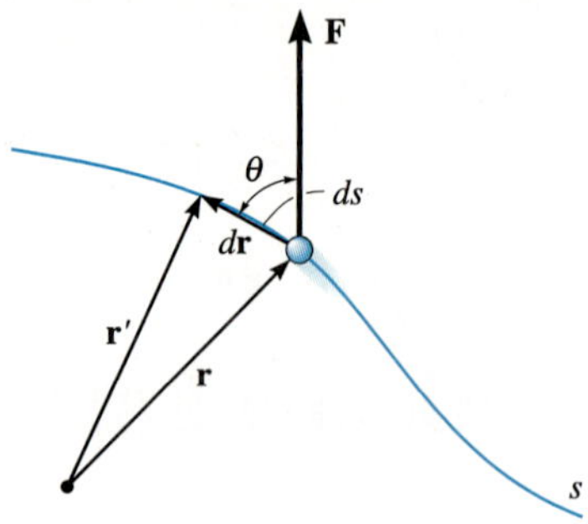

Fig. 14–1

scalar quantity, defined by

$$dU = F\,ds\cos\theta$$

By definition of the dot product (see Eq. C–14) this equation may also be written as

$$dU = \mathbf{F}\cdot d\mathbf{r}$$

This result may be interpreted in one of two ways: either as the product of F and the component of displacement in the direction of the force, i.e., $ds\cos\theta$, or as the product of ds and the component of force in the direction of displacement, i.e., $F\cos\theta$. Note that if $0° \le \theta < 90°$, then the force component and the displacement have the *same sense* so that the work is *positive;* whereas if $90° < \theta \le 180°$, these vectors have an *opposite sense,* and therefore the work is *negative*. Also, $dU = 0$ if the force is *perpendicular* to displacement, since $\cos 90° = 0$, or if the force is applied at a *fixed point,* in which case the displacement is zero.

The basic unit for work in the SI system is called a joule (J). This unit combines the units of force and displacement. Specifically, 1 *joule* of work is done when a force of 1 newton moves 1 meter along its line of action (1 J = 1 N·m). The moment of a force has this same combination of units (N·m); however, the concepts of moment and work are in no way related. A moment is a vector quantity, whereas work is a scalar. In the FPS system work is generally defined by writing the units as ft·lb, which is distinguished from the units for a moment, written as lb·ft.

Work of a Variable Force. If the particle undergoes a finite displacement along its path from $\mathbf{r}_1$ to $\mathbf{r}_2$ or s_1 to s_2, Fig. 14–2*a*, the work is determined by integration. If $\mathbf{F}$ is expressed as a function of position, $F = F(s)$, we have

$$U_{1-2} = \int_{\mathbf{r}_1}^{\mathbf{r}_2} \mathbf{F}\cdot d\mathbf{r} = \int_{s_1}^{s_2} F\cos\theta\,ds \qquad (14\text{–}1)$$

If the working component of the force, $F\cos\theta$, is plotted versus s, Fig. 14–2*b*, the integral in this equation can be interpreted as the *area under the curve* from position s_1 to position s_2.

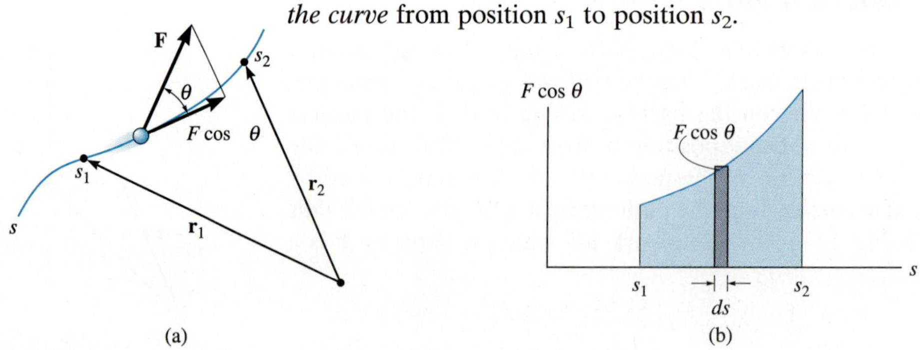

Fig. 14–2

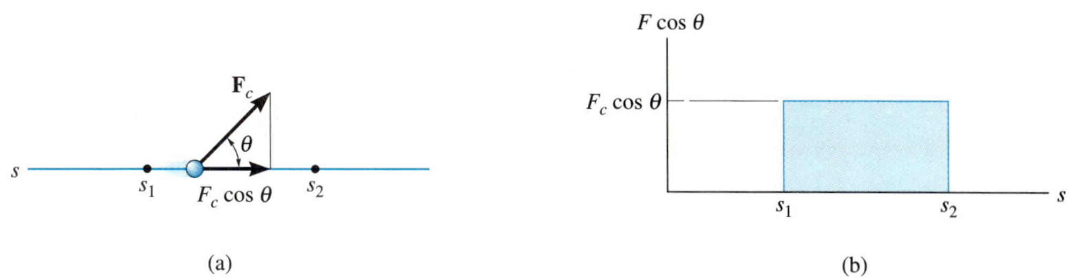

Fig. 14–3

Work of a Constant Force Moving Along a Straight Line. If the force $\mathbf{F}_c$ has a constant magnitude and acts at a constant angle θ from its straight-line path, Fig. 14–3*a*, then the component of $\mathbf{F}_c$ in the direction of displacement is $F_c \cos\theta$. The work done by $\mathbf{F}_c$ when the particle is displaced from s_1 to s_2 is determined from Eq. 14–1, in which case

$$U_{1-2} = F_c \cos\theta \int_{s_1}^{s_2} ds$$

or

$$U_{1-2} = F_c \cos\theta\,(s_2 - s_1) \qquad (14\text{–}2)$$

Here the work of $\mathbf{F}_c$ represents the *area of the rectangle* in Fig. 14–3*b*.

Work of a Weight. Consider a particle which moves up along the path s shown in Fig. 14–4 from position s_1 to position s_2. At an intermediate point, the displacement $d\mathbf{r} = dx\,\mathbf{i} + dy\,\mathbf{j} + dz\,\mathbf{k}$. Since $\mathbf{W} = -W\mathbf{j}$, applying Eq. 14–1 yields

$$U_{1-2} = \int \mathbf{F}\cdot d\mathbf{r} = \int_{\mathbf{r}_1}^{\mathbf{r}_2} (-W\mathbf{j})\cdot(dx\,\mathbf{i} + dy\,\mathbf{j} + dz\,\mathbf{k})$$

$$= \int_{y_1}^{y_2} -W\,dy = -W(y_2 - y_1)$$

or

$$U_{1-2} = -W\,\Delta y \qquad (14\text{–}3)$$

Thus, the work done is equal to the magnitude of the particle's weight times its vertical displacement. In the case shown in Fig. 14–4 the work is *negative,* since W is downward and Δy is upward. Note, however, that if the particle is displaced *downward* $(-\Delta y)$, the work of the weight is *positive*. Why?

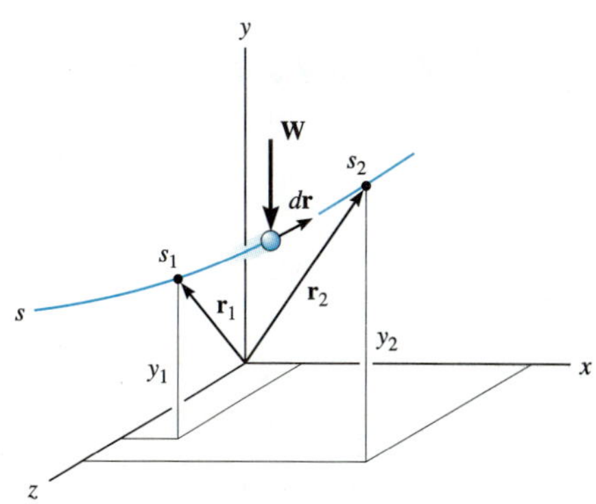

Fig. 14–4

Work of a Spring Force. The magnitude of force developed in a linear elastic spring when the spring is displaced a distance s from its unstretched position is $F_s = ks$, where k is the spring stiffness. If the spring is elongated or compressed from a position s_1 to a further position s_2, Fig. 14–5*a*, the work done *on the spring* by $\mathbf{F}_s$ is *positive,* since in each case the force and displacement are in the *same direction.* We require

$$U_{1-2} = \int_{s_1}^{s_2} F_s\, ds = \int_{s_1}^{s_2} ks\, ds$$
$$= \tfrac{1}{2}ks_2^2 - \tfrac{1}{2}ks_1^2$$

This equation represents the trapezoidal area under the line $F_s = ks$, Fig. 14–5*b*.

If a particle (or body) is attached to a spring, then the force $\mathbf{F}_s$ exerted on the particle is *opposite* to that exerted on the spring, Fig. 14–5*c*. Consequently, the force will do *negative work* on the particle when the particle is moving so as to further elongate (or compress) the spring. Hence, the above equation becomes

$$U_{1-2} = -\left(\tfrac{1}{2}ks_2^2 - \tfrac{1}{2}ks_1^2\right) \tag{14–4}$$

When this equation is used, a mistake in sign can be eliminated if one simply notes the direction of the spring force acting on the particle and compares it with the direction of displacement of the particle—if both are in the *same direction, positive work* results; if they are *opposite* to one another, the *work is negative.*

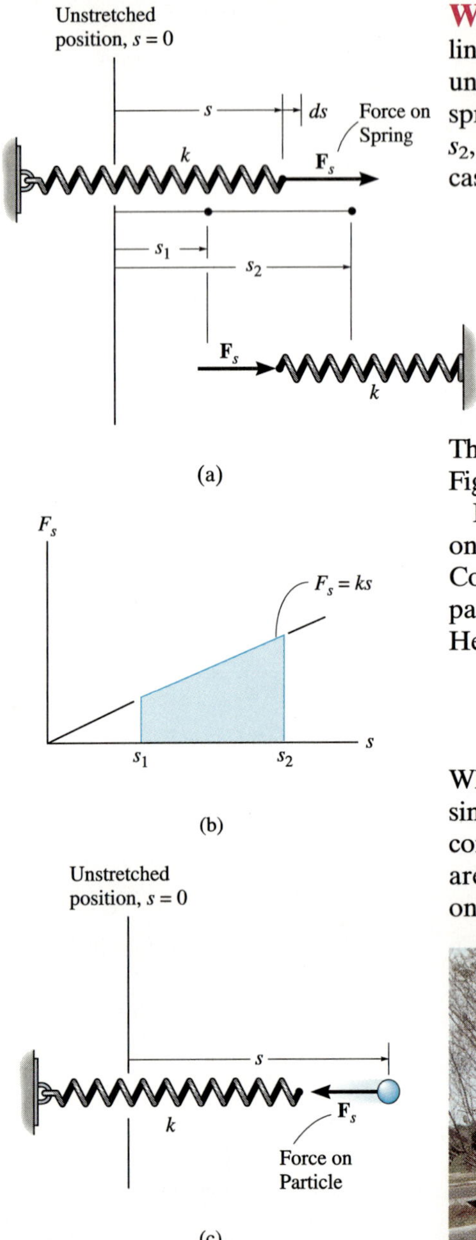

Fig. 14–5

The forces acting on the cart as it is pulled a distance s up the incline, are shown on its free-body diagram. The constant towing force $\mathbf{T}$ does positive work of $U_T = (T\cos\phi)s$, the weight does negative work of $U_W = W\,(s\sin\theta)$, and the normal force $\mathbf{N}$ does no work since there is no displacement of this force along its line of action.

EXAMPLE 14-1

The 10-kg block shown in Fig. 14–6*a* rests on the smooth incline. If the spring is originally stretched 0.5 m, determine the total work done by all the forces acting on the block when a horizontal force $P = 400$ N pushes the block up the plane $s = 2$ m.

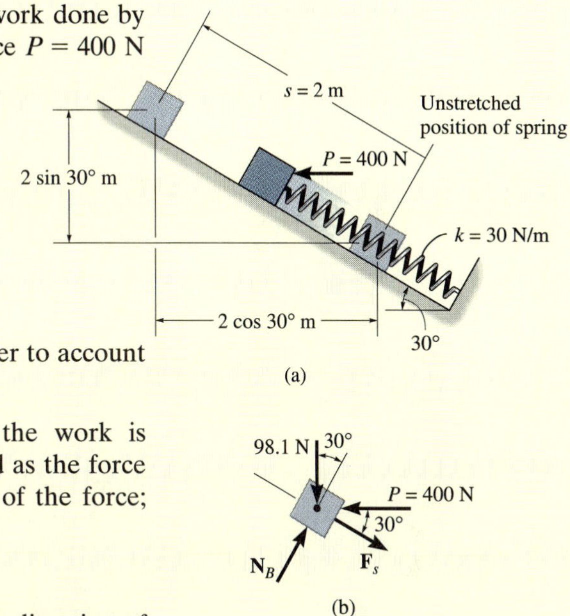

Fig. 14–6

Solution

First the free-body diagram of the block is drawn in order to account for all the forces that act on the block, Fig. 14–6*b*.

Horizontal Force P. Since this force is *constant,* the work is determined using Eq. 14–2. The result can be calculated as the force times the component of displacement in the direction of the force; i.e.,

$$U_P = 400\text{ N }(2\text{ m}\cos 30^\circ) = 692.8\text{ J}$$

or the displacement times the component of force in the direction of displacement, i.e.,

$$U_P = 400\text{ N}\cos 30^\circ(2\text{ m}) = 692.8\text{ J}$$

Spring Force $\mathbf{F}_s$. In the initial position the spring is stretched $s_1 = 0.5$ m and in the final position it is stretched $s_2 = 0.5 + 2 = 2.5$ m. We require the work to be negative since the force and displacement are in opposite directions. The work of $\mathbf{F}_s$ is thus

$$U_s = -\left[\tfrac{1}{2}(30\text{ N/m})(2.5\text{ m})^2 - \tfrac{1}{2}(30\text{ N/m})(0.5\text{ m})^2\right] = -90\text{ J}$$

Weight W. Since the weight acts in the opposite direction to its vertical displacement, the work is negative; i.e.,

$$U_W = -98.1\text{ N }(2\text{ m}\sin 30^\circ) = -98.1\text{ J}$$

Note that it is also possible to consider the component of weight in the direction of displacement; i.e.,

$$U_W = -(98.1\sin 30^\circ\text{ N})2\text{ m} = -98.1\text{ J}$$

Normal Force $\mathbf{N}_B$. This force does *no work* since it is *always* perpendicular to the displacement.

Total Work. The work of all the forces when the block is displaced 2 m is thus

$$U_T = 692.8 - 90 - 98.1 = 505\text{ J} \qquad \textit{Ans.}$$

14.2 Principle of Work and Energy

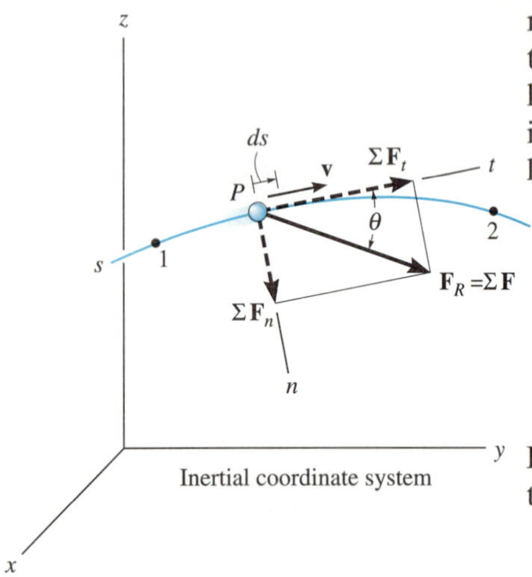

Fig. 14–7

Consider a particle P in Fig. 14–7, which at the instant considered is located on the path as measured from an inertial coordinate system. If the particle has a mass m and is subjected to a system of external forces represented by the resultant $\mathbf{F}_R = \Sigma \mathbf{F}$, then the equation of motion for the particle in the tangential direction is $\Sigma F_t = ma_t$. Applying the kinematic equation $a_t = v\,dv/ds$ and integrating both sides, assuming initially that the particle has a position $s = s_1$ and a speed $v = v_1$, and later at $s = s_2$, $v = v_2$, yields

$$\Sigma \int_{s_1}^{s_2} F_t ds = \int_{v_1}^{v_2} mv\,dv$$

$$\Sigma \int_{s_1}^{s_2} F_t ds = \tfrac{1}{2}mv_2^2 - \tfrac{1}{2}mv_1^2 \qquad (14\text{–}5)$$

From Fig. 14–7, $\Sigma F_t = \Sigma F \cos\theta$ and since work is defined from Eq. 14–1, the final result may be written as

$$\Sigma U_{1\text{-}2} = \tfrac{1}{2}mv_2^2 - \tfrac{1}{2}mv_1^2 \qquad (14\text{–}6)$$

This equation represents the *principle of work and energy* for the particle. The term on the left is the sum of the work done by *all* the forces acting on the particle as the particle moves from point 1 to point 2. The two terms on the right side, which are of the form $T = \frac{1}{2}mv^2$, define the particle's final and initial *kinetic energy,* respectively. These terms are always *positive* scalars. Furthermore, Eq. 14–6 must be dimensionally homogeneous so that the kinetic energy has the same units as work, e.g., joules (J) or ft·lb.

When Eq. 14–6 is applied, it is often symbolized in the form

$$T_1 + \Sigma U_{1\text{-}2} = T_2 \qquad (14\text{–}7)$$

which states that the particle's initial kinetic energy plus the work done by all the forces acting on the particle as it moves from its initial to its final position is equal to the particle's final kinetic energy.

As noted from the derivation, the principle of work and energy represents an integrated form of $\Sigma F_t = ma_t$, obtained by using the kinematic equation $a_t = v\,dv/ds$. As a result, this principle will provide a convenient *substitution* for $\Sigma F_t = ma_t$ when solving those types of kinetic problems which involve force, velocity, and displacement, since these variables are involved in the terms of Eq. 14–7. For example, if a particle's initial speed is known and the work of all the forces acting on the particle can be determined, then Eq. 14–7 provides a *direct means* of obtaining the final speed v_2 of the particle after it undergoes a

If an oncoming car strikes these crash barrels, the car's kinetic energy will be transformed into work, which causes the barrels, and to some extent the car, to be deformed. By knowing the amount of energy absorbed by each barrel it is possible to design a crash cushion such as this.

specified displacement. If instead v_2 is determined by means of the equation of motion, a two-step process is necessary; i.e., apply $\Sigma F_t = ma_t$ to obtain a_t, then integrate $a_t = v\, dv/ds$ to obtain v_2. Note that the principle of work and energy cannot be used, for example, to determine forces directed *normal* to the path of motion, since these forces do no work on the particle. Instead $\Sigma F_n = ma_n$ must be applied. For curved paths, however, the magnitude of the normal force is a function of speed. Hence, it may be easier to obtain this speed using the principle of work and energy, and then substitute this quantity into the equation of motion $\Sigma F_n = mv^2/\rho$ to obtain the normal force.

Procedure for Analysis

The principle of work and energy is used to solve kinetic problems that involve *velocity, force,* and *displacement,* since these terms are involved in the equation. For application it is suggested that the following procedure be used.

Work (Free-Body Diagram)

- Establish the inertial coordinate system and draw a free-body diagram of the particle in order to account for all the forces that do work on the particle as it moves along its path.

Principle of Work and Energy

- Apply the principle of work and energy, $T_1 + \Sigma U_{1-2} = T_2$.
- The kinetic energy at the initial and final points is always positive, since it involves the speed squared ($T = \frac{1}{2}mv^2$).
- A force does work when it moves through a displacement in the direction of the force.
- Work is *positive* when the force component is in the *same direction* as its displacement, otherwise it is negative.
- Forces that are functions of displacement must be integrated to obtain the work. Graphically, the work is equal to the area under the force-displacement curve.
- The work of a weight is the product of the weight magnitude and the vertical displacement, $U_W = \pm Wy$. It is positive when the weight moves downwards.
- The work of a spring is of the form $U_s = \frac{1}{2}\, ks^2$, where k is the spring stiffness and s is the stretch or compression of the spring.

Numerical application of this procedure is illustrated in the examples following Sec. 14.3.

14.3 Principle of Work and Energy for a System of Particles

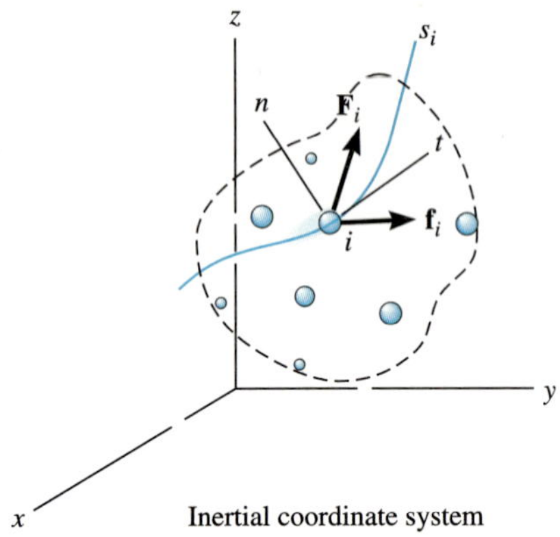

Inertial coordinate system

Fig. 14–8

The principle of work and energy can be extended to include a system of n particles isolated within an enclosed region of space as shown in Fig. 14–8. Here the arbitrary ith particle, having a mass m_i, is subjected to a resultant external force $\mathbf{F}_i$ and a resultant internal force $\mathbf{f}_i$ which each of the other particles exerts on the ith particle. Using Eq. 14–5, which applies in the tangential direction, the principle of work and energy written for the ith particle is thus

$$\tfrac{1}{2}m_i v_{i1}^2 + \int_{s_{i1}}^{s_{i2}} (F_i)_t\,ds + \int_{s_{i1}}^{s_{i2}} (f_i)_t\,ds = \tfrac{1}{2}m_i v_{i2}^2$$

Similar equations result if the principle of work and energy is applied to each of the other particles of the system. Since both work and kinetic energy are scalars, the results may be added together algebraically, so that

$$\Sigma\tfrac{1}{2}m_i v_{i1}^2 + \Sigma\int_{s_{i1}}^{s_{i2}} (F_i)_t\,ds + \Sigma\int_{s_{i1}}^{s_{i2}} (f_i)_t\,ds = \Sigma\tfrac{1}{2}m_i v_{i2}^2$$

We can write this equation symbolically as

$$\Sigma T_1 + \Sigma U_{1-2} = \Sigma T_2 \qquad (14\text{–}8)$$

This equation states that the system's initial kinetic energy (ΣT_1) plus the work done by all the external and internal forces acting on the particles of the system (ΣU_{1-2}) is equal to the system's final kinetic energy (ΣT_2). To maintain this balance of energy, strict accountability of the work done by all the forces must be made. In this regard, note that although the internal forces on adjacent particles occur in equal but opposite collinear pairs, the total work done by each of these forces will, in general, *not cancel* out since the paths over which corresponding particles travel will be *different.* There are, however, two important exceptions to this rule which often occur in practice. If the particles are contained within the boundary of a *translating rigid body,* the internal forces all undergo the same displacement, and therefore the internal work will be zero. Also, particles connected by inextensible cables make up a system that has internal forces which are displaced by an equal amount. In this case, adjacent particles exert equal but opposite internal forces that have components which undergo the same displacement, and therefore the work of these forces cancels. On the other hand, note that if the body is assumed to be *nonrigid,* the particles of the body are displaced along *different paths,* and some of the energy due to force interactions would be given off and lost as heat or stored in the body if permanent deformations occur. We will discuss these effects briefly at the end of this section and in Sec. 15–4. Throughout this text, however, the principle of work and energy will be applied to the solution of problems only when direct accountability of these energy losses does not have to be made.

The procedure for analysis outlined in Sec. 14.2 provides a method for applying Eq. 14–8; however, only one equation applies for the entire system. If the particles are connected by cords, other equations can generally be obtained by using the kinematic principles outlined in Sec. 12.9 in order to relate the particle's speeds. See Example 14–6.

Work of Friction Caused by Sliding. A special class of problems will now be investigated which requires a careful application of Eq. 14–8. These problems all involve cases where a body is sliding over the surface of another body in the presence of friction. Consider, for example, a block which is translating a distance s over a rough surface as shown in Fig. 14–9*a*. If the applied force $\mathbf{P}$ just balances the *resultant* frictional force $\mu_k N$, Fig. 14–9*b*, then due to equilibrium a constant velocity $\mathbf{v}$ is maintained, and one would expect Eq. 14–8 to be applied as follows:

$$\tfrac{1}{2}mv^2 + Ps - \mu_k Ns = \tfrac{1}{2}mv^2$$

Indeed this equation is satisfied if $P = \mu_k N$; however, as one realizes from experience, the sliding motion will *generate heat,* a form of energy which seems not to be accounted for in the work–energy equation. In order to explain this paradox and thereby more closely represent the nature of friction, we should actually model the block so that both surfaces of contact are *deformable* (nonrigid.)* Recall that the rough portions at the bottom of the block act as "teeth," and when the block slides these teeth *deform slightly* and either break off or vibrate due to interlocking effects and pulling away from "teeth" at the contacting surface, Fig. 14–9*c*. As a result, frictional forces that act on the block at these points are displaced slightly, due to the localized deformations, and then they are replaced by other frictional forces as other points of contact are made. At any instant, the *resultant* $\mathbf{F}$ of all these frictional forces remains essentially constant, i.e., $\mu_k N$; however, due to the localized deformations, the actual displacement s' of $\mu_k N$ is *not* the same displacement s as the applied force $\mathbf{P}$. Instead, s' will be *less* than s $(s' < s)$, and therefore the *external work* done by the resultant frictional force will be $\mu_k Ns'$ and not $\mu_k Ns$. The remaining amount of work, $\mu_k N(s - s')$, manifests itself as an increase in *internal energy,* which in fact causes the block's temperature to rise.

In summary then, Eq. 14–8 can be applied to problems involving sliding friction; however, it should be fully realized that the work of the resultant frictional force is not represented by $\mu_k Ns$; instead, this term represents *both* the external work of friction ($\mu_k Ns'$) and internal work $[\mu_k N(s - s')]$ which is converted into various forms of internal energy, such as heat.†

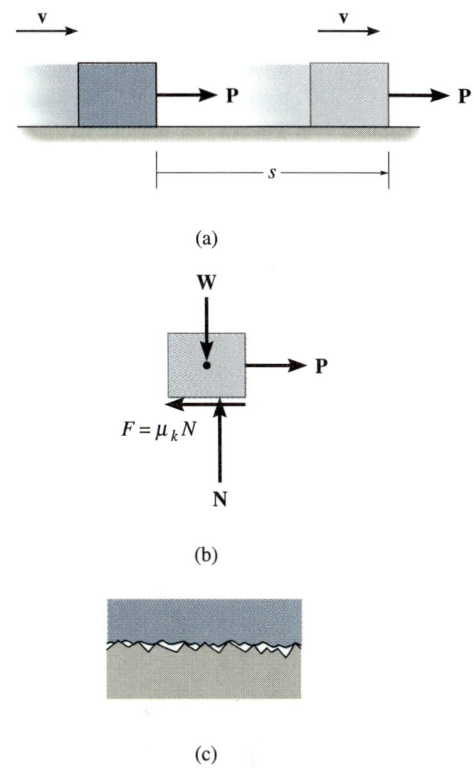

Fig. 14–9

*See Chapter 8 of *Engineering Mechanics: Statics.*

†See B. A. Sherwood and W. H. Bernard, "Work and Heat Transfer in the Presence of Sliding Friction," *Am. J. Phys.* 52, 1001 (1984).

EXAMPLE 14-2

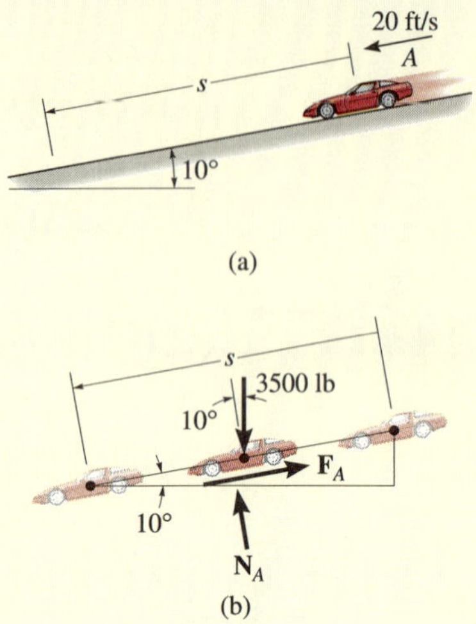

Fig. 14–10

The 3500-lb automobile shown in Fig. 14–10*a* is traveling down the 10° inclined road at a speed of 20 ft/s. If the driver jams on the brakes, causing his wheels to lock, determine how far *s* his tires skid on the road. The coefficient of kinetic friction between the wheels and the road is $\mu_k = 0.5$.

Solution I

This problem can be solved using the principle of work and energy, since it involves force, velocity, and displacement.

Work (Free-Body Diagram). As shown in Fig. 14–10*b*, the normal force $\mathbf{N}_A$ does no work since it never undergoes displacement along its line of action. The weight, 3500 lb, is displaced $s \sin 10°$ and does positive work. Why? The frictional force $\mathbf{F}_A$ does both external and internal work when it is *thought* to undergo a displacement *s*. This work is negative since it is in the opposite direction to displacement. Applying the equation of equilibrium normal to the road, we have

$$+\nwarrow \Sigma F_n = 0; \qquad N_A - 3500 \cos 10° \text{ lb} = 0 \qquad N_A = 3446.8 \text{ lb}$$

Thus,

$$F_A = 0.5N_A = 1723.4 \text{ lb}$$

Principle of Work and Energy

$$T_1 + \Sigma U_{1-2} = T_2$$

$$\frac{1}{2}\left(\frac{3500 \text{ lb}}{32.2 \text{ ft/s}^2}\right)(20 \text{ ft/s})^2 + \{3500 \text{ lb}(s \sin 10°) - (1723.4 \text{ lb})\, s\} = 0$$

Solving for *s* yields

$$s = 19.5 \text{ ft} \qquad \textit{Ans.}$$

Solution II

If this problem is solved by using the equation of motion, *two steps* are involved. First, from the free-body diagram, Fig. 14–10*b*, the equation of motion is applied along the incline. This yields

$$+\swarrow \Sigma F_s = ma_s; \quad 3500 \sin 10° \text{ lb} - 1723.4 \text{ lb} = \frac{3500 \text{ lb}}{32.2 \text{ ft/s}^2}\, a$$

$$a = -10.3 \text{ ft/s}^2$$

Then, using the integrated form of $a\, ds = v\, dv$ (kinematics), since *a* is constant, we have

$$(+\swarrow) \quad v^2 = v_0^2 + 2a_c(s - s_0);$$

$$(0)^2 = (20 \text{ ft/s})^2 + 2(-10.3 \text{ ft/s}^2)(s - 0)$$

$$s = 19.5 \text{ ft} \qquad \textit{Ans.}$$

EXAMPLE 14-3

For a short time the crane in Fig. 14–11*a* lifts the 2.50-Mg beam with a force of $F = (28 + 3s^2)$ kN. Determine the speed of the beam when it has risen $s = 3$ m. Also, how much time does it take to attain this height?

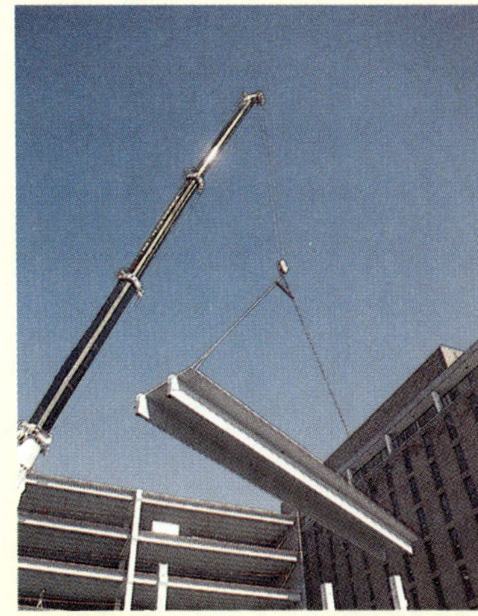

(a)

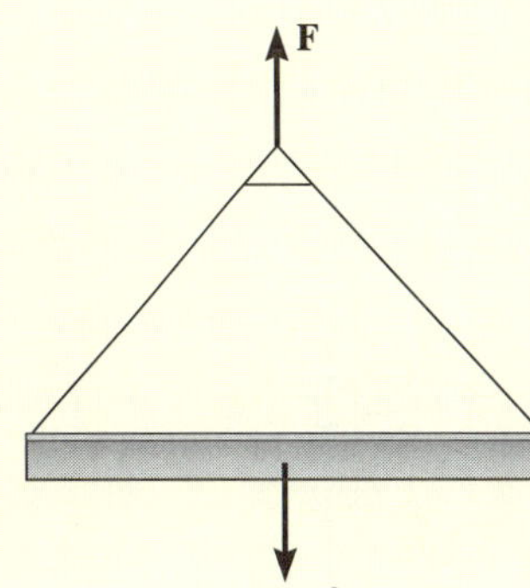

(b)

Fig. 14–11

Solution

We can solve part of this problem using the principle of work and energy since it involves force, velocity, and displacement. Kinematics must be used to determine the time.

Work (Free-Body Diagrams). As shown on the free-body diagram, Fig. 14–11*b*, the towing force **F** does positive work, which must be determined by integration since this force is a variable. Also, the weight is constant and will do negative work since the displacement is upwards.

Principles of Work and Energy

$$T_1 + \Sigma U_{1\text{-}2} = T_2$$

$$0 + \int_0^s (28 + 3s^2)(10^3)\,ds - (2.50)(10^3)(9.81)s = \tfrac{1}{2}(2.50)(10^3)v^2$$

$$28(10^3)s + (10^3)s^3 - 24.525(10^3)s = 1.25(10^3)v^2$$

$$v = (2.78s + 0.8s^3)^{\frac{1}{2}}$$

When $s = 3$ m,

$$v = 5.47 \text{ m/s} \qquad \textit{Ans.}$$

Kinematics. Since we were able to express the velocity as a function of displacement, the time can be determined using $v = ds/dt$. In this case,

$$(2.78s + 0.8s^3)^{\frac{1}{2}} = \frac{ds}{dt}$$

$$t = \int_0^3 \frac{ds}{(2.78s + 0.8s^3)^{\frac{1}{2}}}$$

The integration can be performed numerically using a pocket calculator. The result is

$$t = 1.78\text{s} \qquad \textit{Ans.}$$

EXAMPLE 14-4

The platform P, shown in Fig. 14–12*a*, has negligible mass and is tied down so that the 0.4-m-long cords keep a 1-m long spring compressed 0.6 m when *nothing* is on the platform. If a 2-kg block is placed on the platform and released from rest after the platform is pushed down 0.1 m, Fig. 14–12*b*, determine the maximum height h the block rises in the air, measured from the ground.

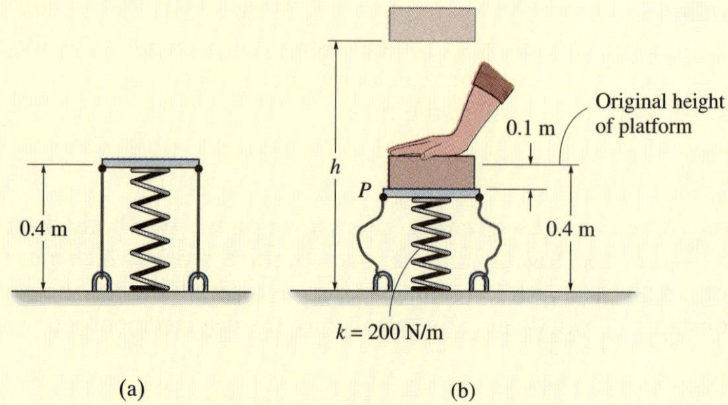

Fig. 14–12

Solution

Work (Free-Body Diagram). Since the block is released from rest and later reaches its maximum height, the initial and final velocities are zero. The free-body diagram of the block when it is still in contact with the platform is shown in Fig. 14–12*c*. Note that the weight does negative work and the spring force does positive work. Why? In particular, the *initial compression* in the spring is $s_1 = 0.6\text{ m} + 0.1\text{ m} = 0.7\text{ m}$. Due to the cords, the spring's *final compression* is $s_2 = 0.6\text{ m}$ (after the block leaves the platform). The bottom of the block rises from a height of $(0.4\text{ m} - 0.1\text{ m}) = 0.3\text{ m}$ to a final height h.

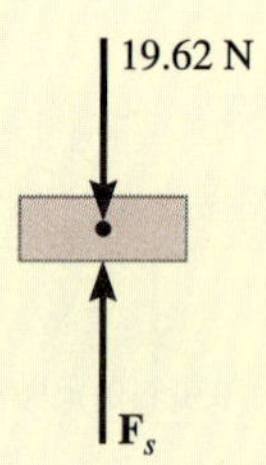

(c)

Principle of Work and Energy

$$T_1 + \Sigma U_{1-2} = T_2$$

$$\tfrac{1}{2}mv_1^2 + \{-(\tfrac{1}{2}ks_2^2 - \tfrac{1}{2}ks_1^2) - W\,\Delta y\} = \tfrac{1}{2}mv_2^2$$

Note that here $s_1 = 0.7\text{ m} > s_2 = 0.6\text{ m}$ and so the work of the spring as determined from Eq. 14–4 will indeed be positive once the calculation is made. Thus,

$$0 + \{-[\tfrac{1}{2}(200\text{ N/m})(0.6\text{ m})^2 - \tfrac{1}{2}(200\text{ N/m})(0.7\text{ m})^2] - (19.62\text{ N})[h - (0.3\text{ m})]\} = 0$$

Solving yields

$$h = 0.963\text{ m} \qquad \textit{Ans.}$$

EXAMPLE 14-5

Packages having a mass of 2 kg are delivered from a conveyor to a smooth circular ramp with a velocity of $v_0 = 1$ m/s as shown in Fig. 14–13*a*. If the radius of the ramp is 0.5 m, determine the angle $\theta = \theta_{max}$ at which each package begins to leave the surface.

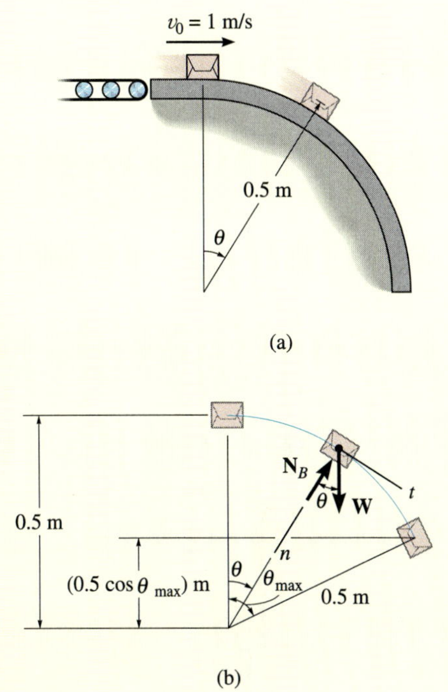

Fig. 14–13

Solution

Work (Free-Body Diagram). The free-body diagram of the block is shown at the intermediate location θ. The weight $W = 2(9.81) = 19.62$ N does positive work during the displacement. If a package is assumed to leave the surface when $\theta = \theta_{max}$ then the weight moves through a vertical displacement of $[0.5 - 0.5 \cos \theta_{max}]$ m, as shown in the figure.

Principle of Work and Energy

$$T_1 + \Sigma U_{1-2} = T_2$$

$$\tfrac{1}{2}(2\text{ kg})(1\text{ m/s})^2 + \{19.62\text{ N}(0.5 - 0.5\cos\theta_{max})\text{m}\} = \tfrac{1}{2}(2\text{ kg})v_2^2$$

$$v_2^2 = 9.81(1 - \cos\theta_{max}) + 1 \qquad (1)$$

Equation of Motion. There are two unknowns in Eq. 1, θ_{max} and v_2. A second equation relating these two variables may be obtained by applying the equation of motion in the *normal direction* to the forces on the free-body diagram. (The principle of work and energy has replaced application of $\Sigma F_t = ma_t$ as noted in the derivation.) Thus,

$$+\swarrow \Sigma F_n = ma_n; \quad -N_B + 19.62\text{ N}\cos\theta = (2\text{ kg})\left(\frac{v^2}{0.5\text{ m}}\right)$$

When the package leaves the ramp at $\theta = \theta_{max}$, $N_B = 0$ and $v = v_2$; hence, this equation becomes

$$\cos\theta_{max} = \frac{v_2^2}{4.905} \qquad (2)$$

Eliminating the unknown v_2^2 between Eqs. 1 and 2 gives

$$4.905\cos\theta_{max} = 9.81(1 - \cos\theta_{max}) + 1$$

Solving, we have

$$\cos\theta_{max} = 0.735$$

$$\theta_{max} = 42.7^\circ \qquad \textit{Ans.}$$

This problem has also been solved in Example 13–9. If the two methods of solution are compared, it will be apparent that a work–energy approach yields a more direct solution.

EXAMPLE 14–6

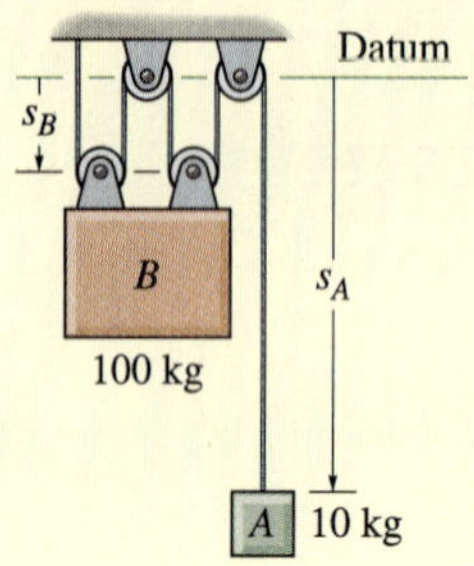

(a)

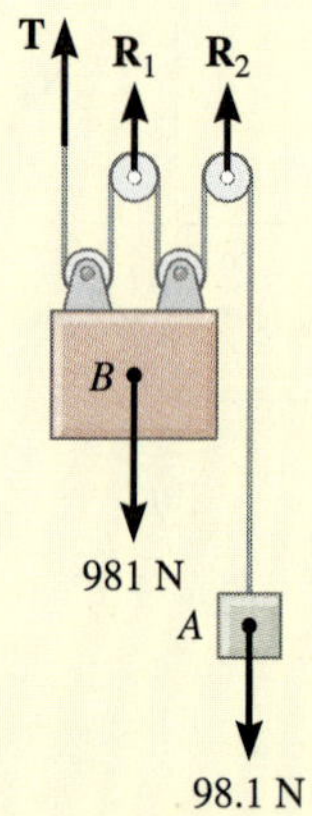

(b)

Fig. 14–14

The blocks A and B shown in Fig. 14–14a have a mass of 10 kg and 100 kg, respectively. Determine the distance B travels from the point where it is released from rest to the point where its speed becomes 2 m/s.

Solution

This problem may be solved by considering the blocks separately and applying the principle of work and energy to each block. However, the work of the (unknown) cable tension can be eliminated from the analysis by considering blocks A and B together as a *system*. The solution will require simultaneous solution of the equations of work and energy *and* kinematics. To be consistent with our sign convention, we will assume both blocks move in the positive *downward* direction.

Work (Free-Body Diagram). As shown on the free-body diagram of the system, Fig. 14–14b, the cable force **T** and reactions $\mathbf{R}_1$ and $\mathbf{R}_2$ do *no work*, since these forces represent the reactions at the supports and consequently do not move while the blocks are being displaced. The weights both do positive work since, as stated above, they are both assumed to move downward.

Principle of Work and Energy. Realizing the blocks are released from rest, we have

$$\Sigma T_1 + \Sigma U_{1-2} = \Sigma T_2$$

$$\{\tfrac{1}{2}m_A(v_A)_1^2 + \tfrac{1}{2}m_B(v_B)_1^2\} + \{W_A\,\Delta s_A + W_B\,\Delta s_B\} = \{\tfrac{1}{2}m_A(v_A)_2^2 + \tfrac{1}{2}m_B(v_B)_2^2\}$$

$$\{0 + 0\} + \{98.1\text{ N}(\Delta s_A) + 981\text{ N}(\Delta s_B)\} = \{\tfrac{1}{2}(10\text{ kg})(v_A)_2^2 + \tfrac{1}{2}(100\text{ kg})(2\text{ m/s})^2\} \quad (1)$$

Kinematics. Using the methods of kinematics discussed in Sec. 12.9, it may be seen from Fig. 14–14a that at any given instant the total length l of all the vertical segments of cable may be expressed in terms of the position coordinates s_A and s_B as

$$s_A + 4s_B = l$$

Hence, a change in position yields the displacement equation

$$\Delta s_A + 4\,\Delta s_B = 0$$

$$\Delta s_A = -4\,\Delta s_B \quad (2)$$

As required, both of these displacements are positive downward. Taking the time derivative yields

$$v_A = -4v_B = -4(2\text{ m/s}) = -8\text{ m/s}$$

Retaining the negative sign in Eq. 2 and substituting into Eq. 1 yields

$$\Delta s_B = 0.883\text{ m}\downarrow \qquad \textit{Ans.}$$

Problems

14-1. A woman having a mass of 70 kg stands in an elevator which has a downward acceleration of 4 m/s^2 starting from rest. Determine the work done by her weight and the work of the normal force which the floor exerts on her when the elevator descends 6 m. Explain why the work of these forces is different.

14-2. A 1500-lb crate is pulled along the ground at a constant speed for a distance of 25 ft, using a cable that makes an angle of 15° with the horizontal. Determine the tension in the cable and the work done by this force. The coefficient of kinetic friction between the ground and the crate is $\mu_k = 0.55$.

14-3. The 20-lb crate has a velocity of $v_A = 12$ ft/s when it is at A. Determine its velocity after it slides $s = 6$ ft down the plane. The coefficient of kinetic friction between the crate and the plane is $\mu_k = 0.2$.

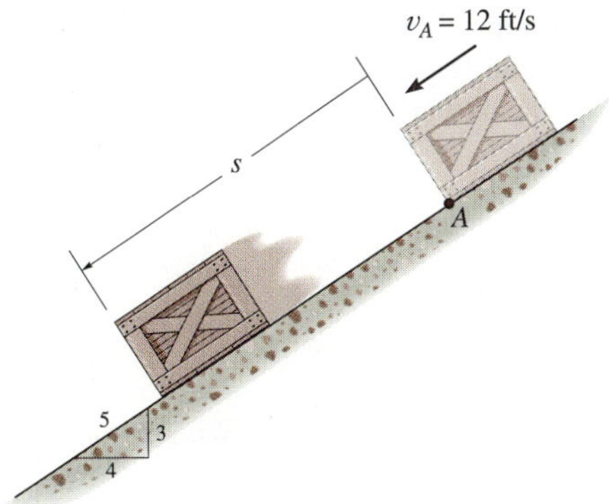

Prob. 14–3

***14-4.** A car is equipped with a bumper B designed to absorb collisions. The bumper is mounted to the car using pieces of flexible tubing T. Upon collision with a rigid barrier at A, a constant horizontal force **F** is developed which causes a car deceleration of $3g = 29.43$ m/s^2 (the highest safe deceleration for a passenger without a seatbelt). If the car and passenger have a total mass of 1.5 Mg and the car is initially coasting with a speed of 1.5 m/s, determine the magnitude of **F** needed to stop the car and the deformation x of the bumper tubing.

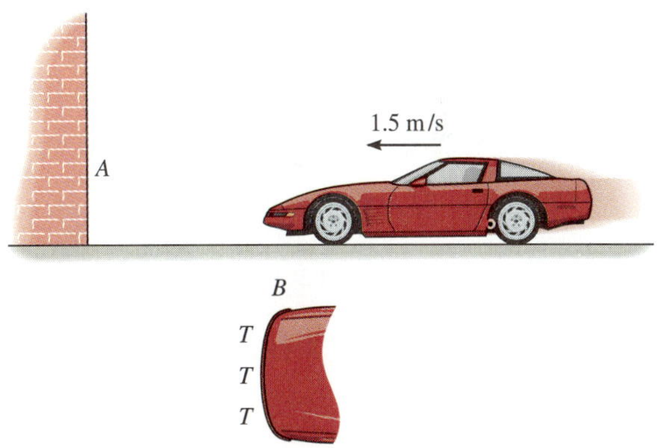

Prob. 14–4

14-5. The smooth plug has a weight of 20 lb and is pushed against a series of Belleville spring washers so that the compression in the spring is $s = 0.05$ ft. If the force of the spring on the plug is $F = (3s^{1/3})$ lb, where s is given in feet, determine the speed of the plug after it moves away from the spring. Neglect friction.

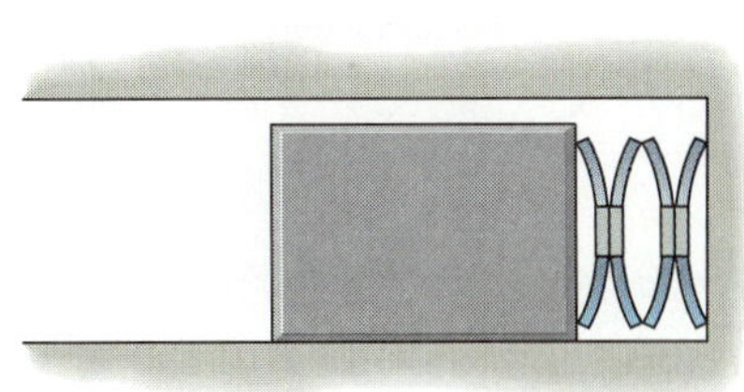

Prob. 14–5

14-6. When a 7-kg projectile is fired from a cannon barrel that has a length of 2 m, the explosive force exerted on the projectile, while it is in the barrel, varies in the manner shown. Determine the approximate muzzle velocity of the projectile at the instant it leaves the barrel. Neglect the effects of friction inside the barrel and assume the barrel is horizontal.

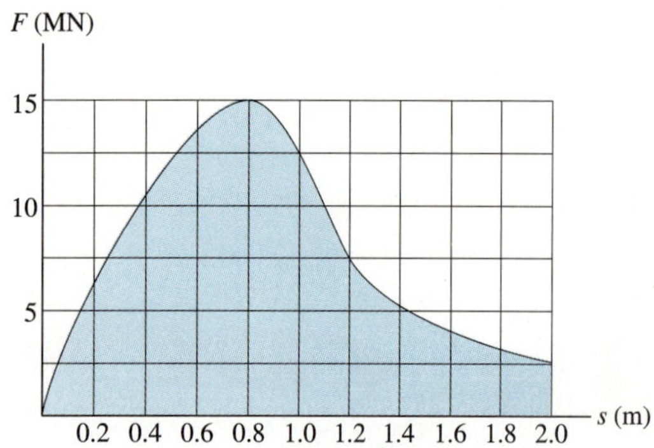

Prob. 14–6

14-7. The 100-kg crate is subjected to forces $F_1 = 800$ N and $F_2 = 1.5$ kN, as shown. If it is originally at rest, determine the distance it slides in order to attain a speed of $v = 6$ m/s. The coefficient of kinetic friction between the crate and the surface is $\mu_k = 0.2$.

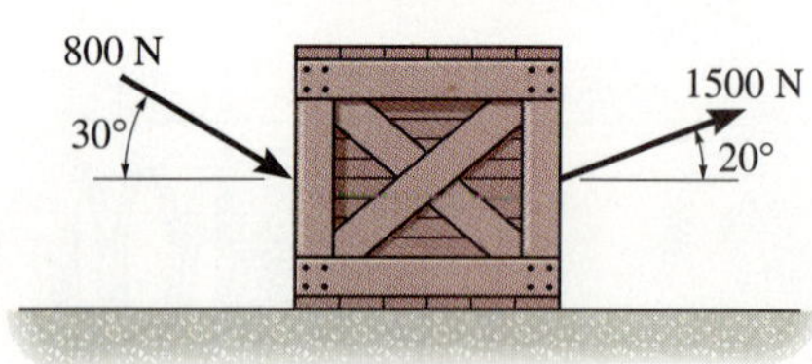

Prob. 14–7

***14-8.** Determine the required height h of the roller coaster so that when it is essentially at rest at the crest of the hill it will reach a speed of 100 km/h when it comes to the bottom. Also, what should be the minimum radius of curvature ρ for the track at B so that the passengers do not experience a normal force greater than $4\,mg = (39.24\,m)$ N? Neglect the size of the car and passenger.

Prob. 14–8

14-9. When the driver applies the brakes of a light truck traveling 40 km/h, it skids 3 m before stopping. How far will the truck skid if it is traveling 80 km/h when the brakes are applied?

Prob. 14–9

14-10. The 2-kg block is subjected to a force having a constant direction and a magnitude $F = [300/(1 + s)]$ N, where s is in meters. When $s = 4$ m, the block is moving to the left with a speed of 8 m/s. Determine its speed when $s = 12$ m. The coefficient of kinetic friction between the block and the ground is $\mu_k = 0.25$.

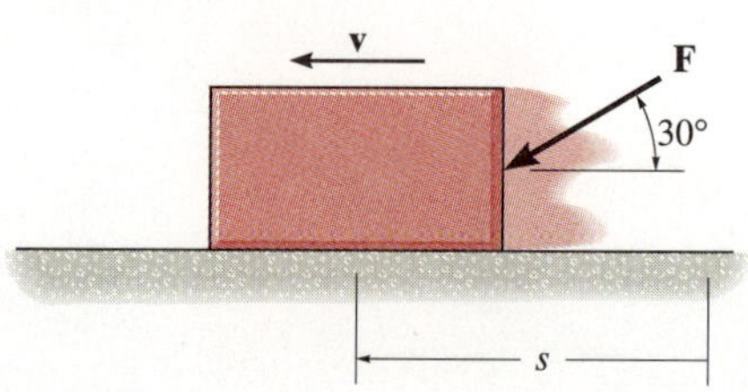

Prob. 14–10

14-11. The 6-lb block is released from rest at A and slides down the smooth parabolic surface shown. Determine how far it compresses the spring.

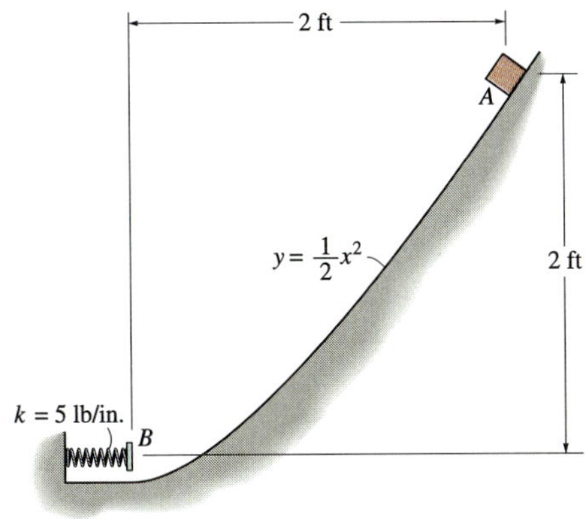

Prob. 14–11

***14-12.** As indicated by the derivation, the principle of work and energy is valid for observers in *any* inertial reference frame. Show that this is so, by considering the 10-kg block which rests on the smooth surface and is subjected to a horizontal force of 6 N. If observer A is in a *fixed* frame x, determine the final speed of the block if it has an initial speed of 5 m/s and travels 10 m, both directed to the right and measured from the fixed frame. Compare the result with that obtained by an observer B, attached to the x' axis and moving at a constant velocity of 2 m/s relative to A. *Hint:* The distance the block travels will first have to be computed for observer B before applying the principle of work and energy.

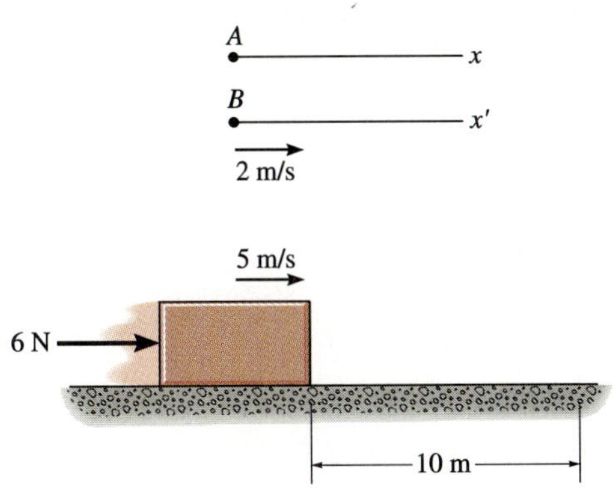

Prob. 14–12

14-13. Determine the velocity of the 60-lb block A if the two blocks are released from rest and the 40-lb block B moves 2 ft up the incline. The coefficient of kinetic friction between both blocks and the inclined planes is $\mu_k = 0.10$.

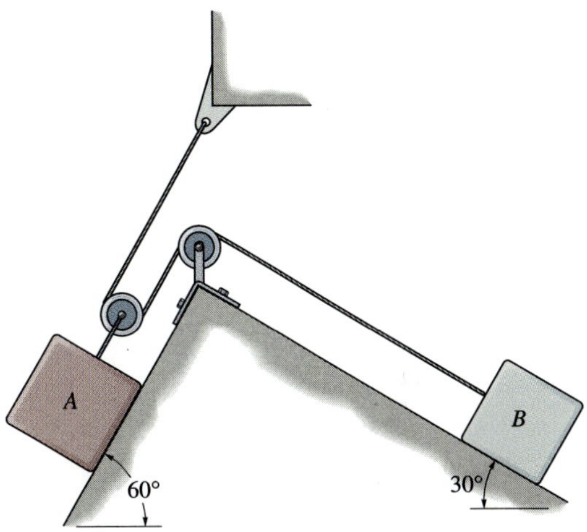

Prob. 14–13

14-14. Determine the velocity of the 20-kg block A after it is released from rest and moves 2 m down the plane. Block B has a mass of 10 kg and the coefficient of kinetic friction between the plane and block A is $\mu_k = 0.2$. Also, what is the tension in the cord?

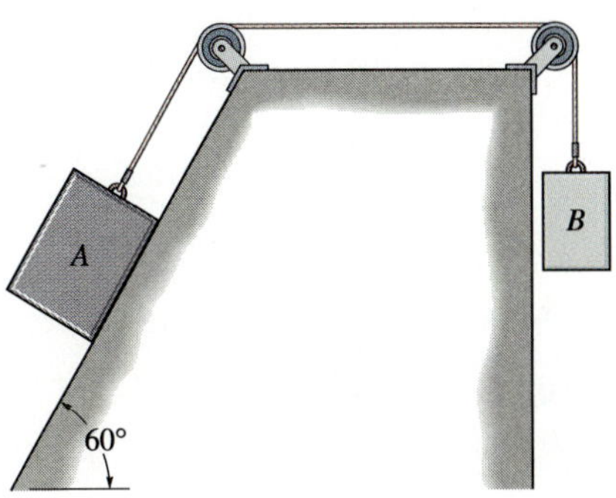

Prob. 14–14

14-15. The 3-lb block A rests on a surface for which the coefficient of kinetic friction is $\mu_k = 0.3$. Determine the distance the 8-lb cylinder B must descend so that A has a speed of $v_A = 5$ ft/s starting from rest.

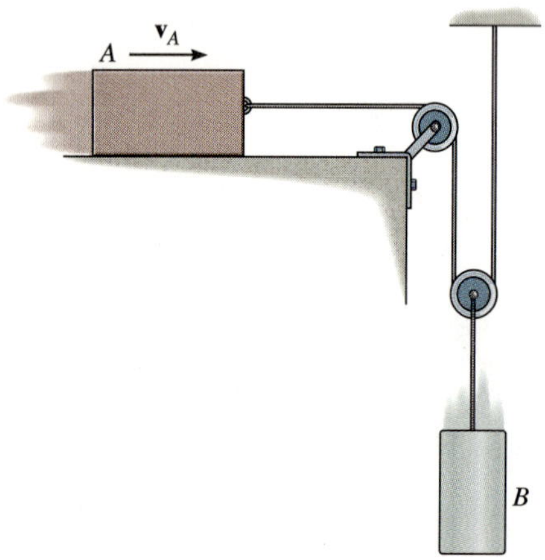

Prob. 14–15

*14-16.** The 100-lb block slides down the inclined plane for which the coefficient of kinetic friction is $\mu_k = 0.25$. If it is moving at 10 ft/s when it reaches point A, determine the maximum deformation of the spring needed to momentarily arrest the motion.

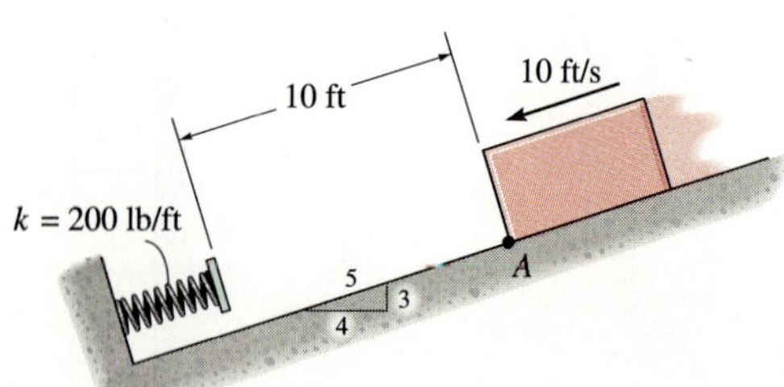

Prob. 14–16

14-17. The collar has a mass of 20 kg and rests on the smooth rod. Two springs are attached to it and the ends of the rod as shown. Each spring has an uncompressed length of 1 m. If the collar is displaced $s = 0.5$ m and released from rest, determine its velocity at the instant it returns to the point $s = 0$.

14-18. The collar has a mass of 20 kg and rests on the smooth rod. Two springs are attached to it and the ends of the rod as shown. Each spring has an uncompressed length of 1.5 m. If the collar is held in the center position of the rod, $s = 0$, and then displaced $s = 0.5$ m and released from rest, determine its velocity at the instant it returns to the point $s = 0$.

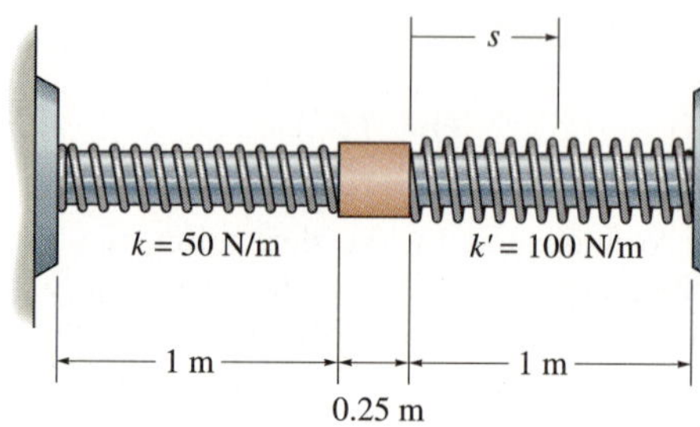

Probs. 14–17/18

14-19. The collar has a mass of 30 kg and is supported on the rod having a coefficient of kinetic friction $\mu_k = 0.4$. The attached spring has an unstretched length of 0.2 m and a stiffness $k = 50$ N/m. Determine the speed of the collar after the applied force $F = 200$ N causes it to be displaced $s = 1.5$ m from point A. When $s = 0$ the collar is held at rest.

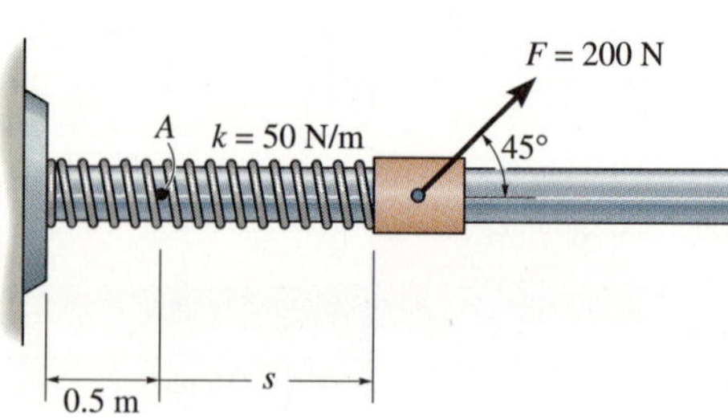

Prob. 14–19

***14-20.** The 5-lb block is released from rest at A and slides down the smooth circular surface AB. It then continues to slide along the horizontal rough surface until it strikes the spring. Determine how far it compresses the spring before stopping.

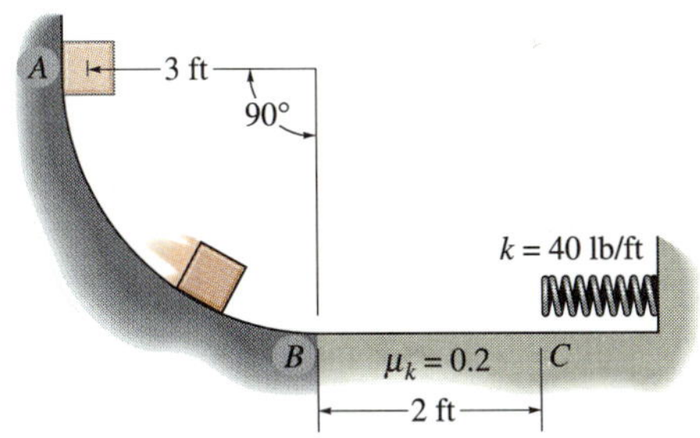

Prob. 14–20

14-21. The 2-lb box slides on the smooth curved ramp. If the box has a velocity of 30 ft/s at A, determine the velocity of the box and normal force acting on the ramp when the box is located at B and C. Assume the radius of curvature of the path at C is still 5 ft.

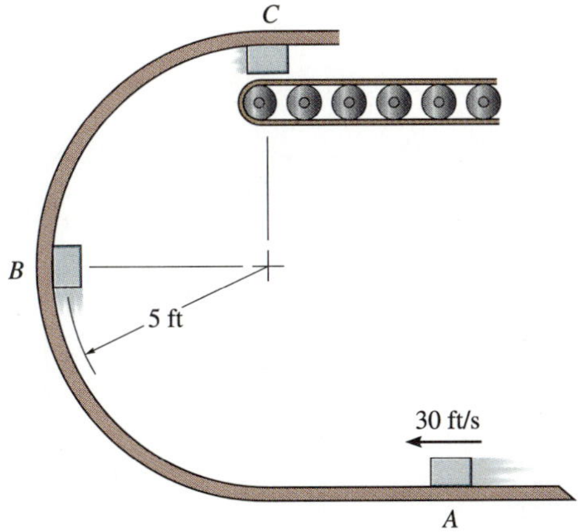

Prob. 14–21

14-22. The conveyor belt delivers each 12-kg crate to the ramp at A such that the crate's velocity is $v_A = 2.5$ m/s, directed down *along* the ramp. If the coefficient of kinetic friction between each crate and the ramp is $\mu_k = 0.3$, determine the speed at which each crate slides off the ramp at B. Assume that no tipping occurs.

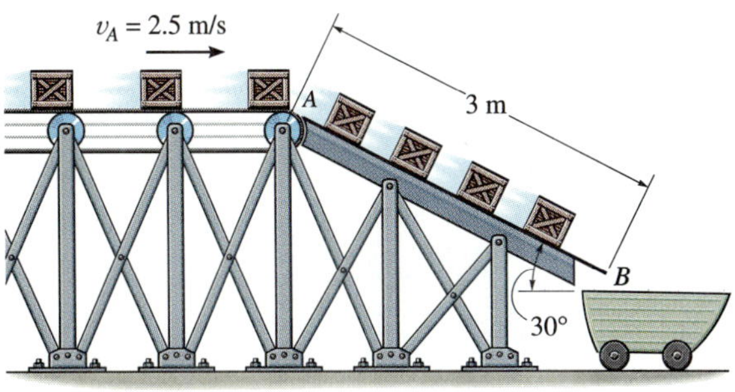

Prob. 14–22

14-23. Packages having a weight of 50 lb are delivered to the chute at $v_A = 3$ ft/s using a conveyor belt. Determine their speeds when they reach points B, C, and D. Also calculate the normal force of the chute on the packages at B and C. Neglect friction and the size of the packages.

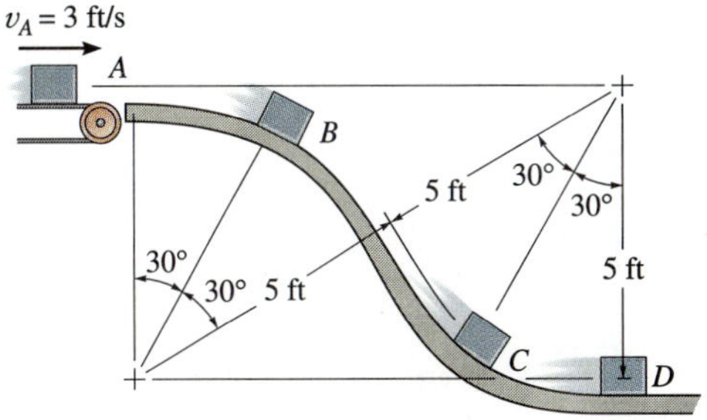

Prob. 14–23

***14-24.** The 2-lb block slides down the smooth parabolic surface, such that when it is at A it has a speed of 10 ft/s. Determine the magnitude of the block's velocity and acceleration when it reaches point B, and the maximum height y_{max} reached by the block.

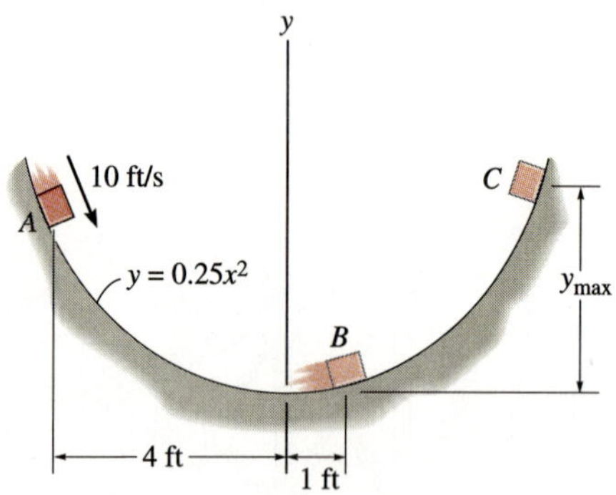

Prob. 14–24

14-25. When the 150-lb skier is at point A he has a speed of 5 ft/s. Determine his speed when he reaches point B on the smooth slope. For this distance the slope follows the cosine curve shown. Also, what is the normal force on his skis at B and his rate of increase in speed? Neglect friction and air resistance.

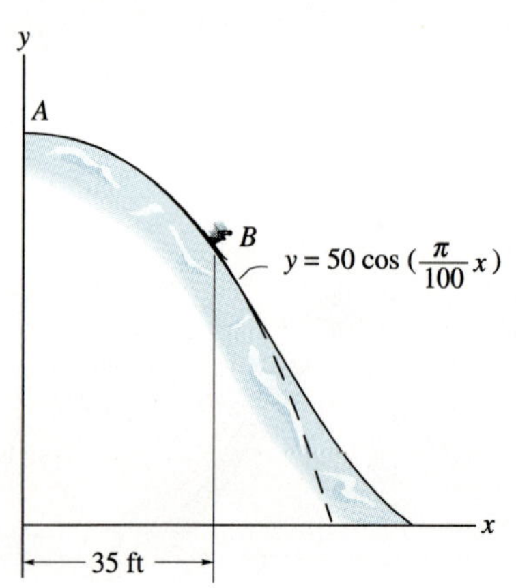

Prob. 14–25

14-26. The skier starts from rest at A and travels down the ramp. If friction and air resistance can be neglected, determine his speed v_B when he reaches B. Also, find the distance s to where he strikes the ground at C, if he makes the jump traveling horizontally at B. Neglect the skier's size. He has a mass of 70 kg.

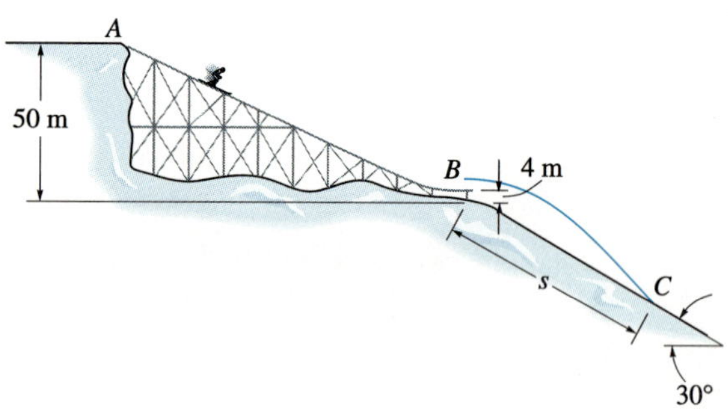

Prob. 14–26

14-27. When the 12-lb block A is released from rest it lifts the two 15-lb weights B and C. Determine the maximum distance A will fall before its motion is momentarily stopped. Neglect the weight of the cord and the size of the pulleys.

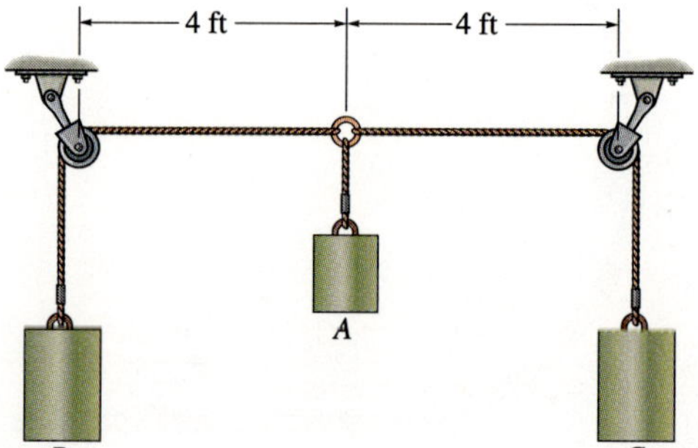

Prob. 14–27

*14-28. The 2-lb brick slides down a smooth roof, such that when it is at A it has a velocity of 5 ft/s. Determine the speed of the block just before it leaves the surface at B, the distance d from the wall to where it strikes the ground, and the speed at which it hits the ground.

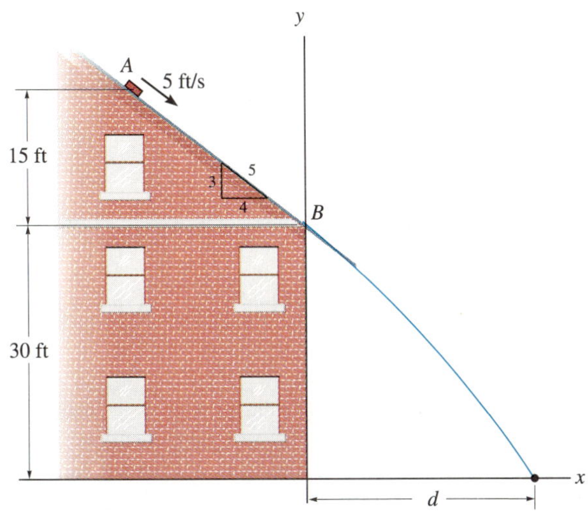

Prob. 14–28

14-29. The 10-lb block is pressed against the spring so as to compress it 2 ft when it is at A. If the plane is smooth, determine the distance d, measured from the wall, to where the block strikes the ground. Neglect the size of the block.

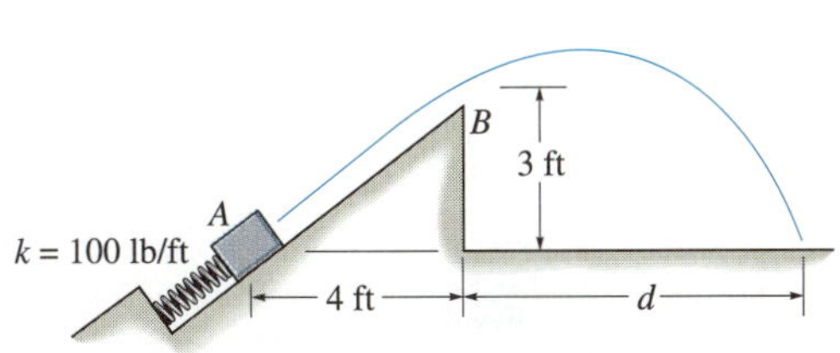

Prob. 14–29

14-30. The 120-lb man acts as a human cannonball by being "fired" from the spring-loaded cannon shown. If the greatest acceleration he can experience is $a = 10g = 322\ \text{ft/s}^2$, determine the required stiffness of the spring which is compressed 2 ft at the moment of firing. With what velocity will he exit the cannon barrel when the cannon is fired? When the spring is compressed $s = 2$ ft then $d = 8$ ft. Neglect friction and assume the man holds himself in a rigid position throughout the motion.

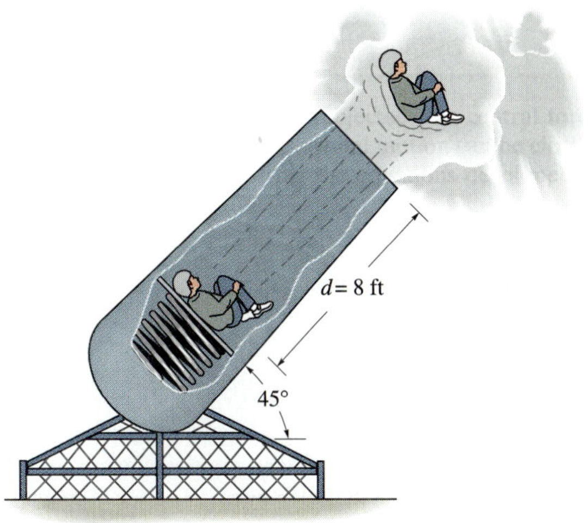

Prob. 14–30

14-31. Marbles having a mass of 5 g fall from rest at A through the glass tube and accumulate in the can at C. Determine the placement R of the can from the end of the tube and the speed at which the marbles fall into the can. Neglect the size of the can.

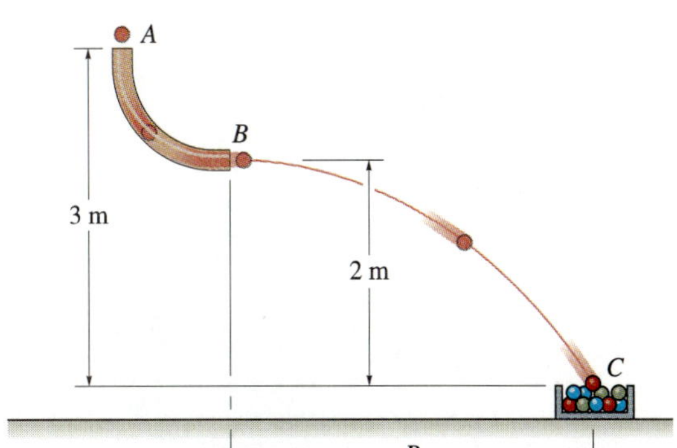

Prob. 14–31

***14-32.** The 100-kg stone is being dragged across the smooth surface by means of a truck T. If the towing cable passes over a small pulley at A, determine the amount of work that must be done by the truck in order to increase the cable angle θ from $\theta_1 = 30°$ to $\theta_2 = 45°$. The truck exerts a constant force $F = 500$ N on the cable at B. Neglect the mass of the pulley and cable.

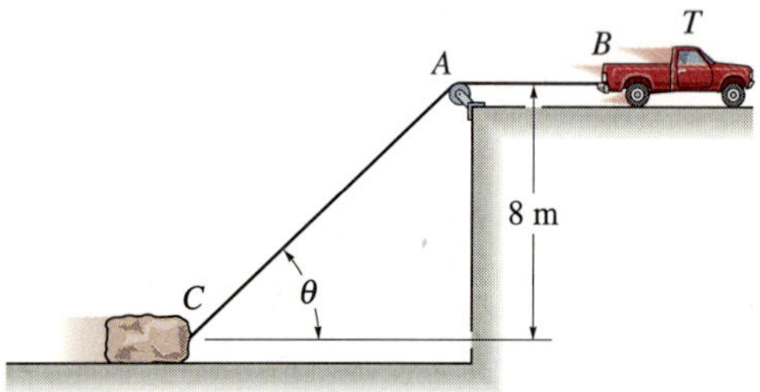

Prob. 14–32

14-33. A rocket of mass m is fired vertically from the surface of the earth, i.e., at $r = r_1$. Assuming no mass is lost as it travels upward, determine the work it must do against gravity to reach a distance r_2. The force of gravity is $F = GM_em/r^2$ (Eq. 13–1), where M_e is the mass of the earth and r the distance between the rocket and the center of the earth.

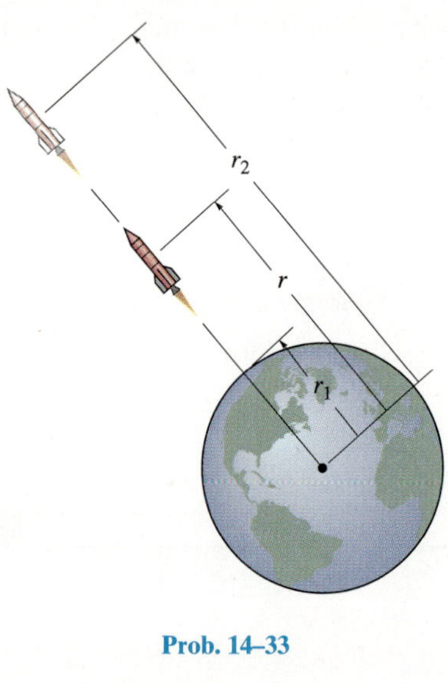

Prob. 14–33

14-34. The spring has a stiffness $k = 50$ lb/ft and an *unstretched length* of 2 ft. As shown, it is confined by the plate and wall using cables so that its length is 1.5 ft. A 4-lb block is given a speed v_A when it is at A, and it slides down the incline having a coefficient of kinetic friction $\mu_k = 0.2$. If it strikes the plate and pushes it forward 0.25 ft before stopping, determine its speed at A. Neglect the mass of the plate and spring.

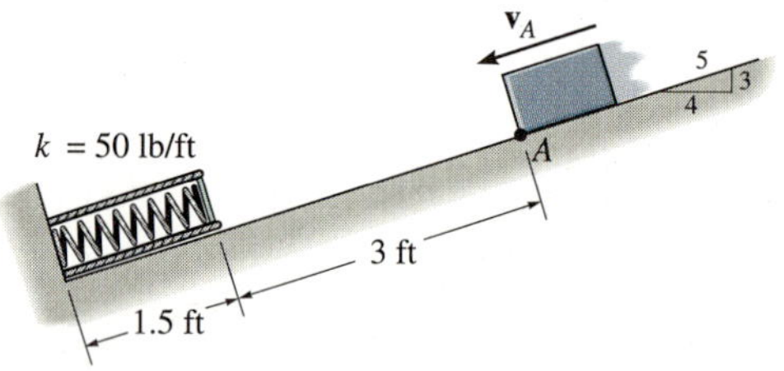

Prob. 14–34

14-35. The block has a mass of 0.8 kg and moves within the smooth vertical slot. If it starts from rest when the *attached* spring is in the unstretched position at A, determine the *constant* vertical force F which must be applied to the cord so that the block attains a speed $v_B = 2.5$ m/s when it reaches B; $s_B = 0.15$ m. Neglect the size and mass of the pulley. *Hint:* The work of $\mathbf{F}$ can be determined by finding the difference Δl in cord lengths AC and BC and using $U_F = F\,\Delta l$.

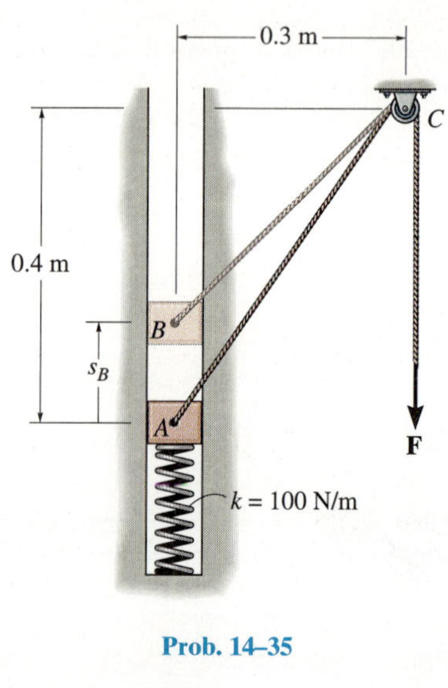

Prob. 14–35

*14-36. A 2-lb block rests on the smooth semicylindrical surface. An elastic cord having a stiffness $k = 2$ lb/ft is attached to the block at B and to the base of the semicylinder at point C. If the block is released from rest at $A(\theta = 0°)$, determine the unstretched length of the cord so the block begins to leave the semicylinder at the instant $\theta = 45°$. Neglect the size of the block.

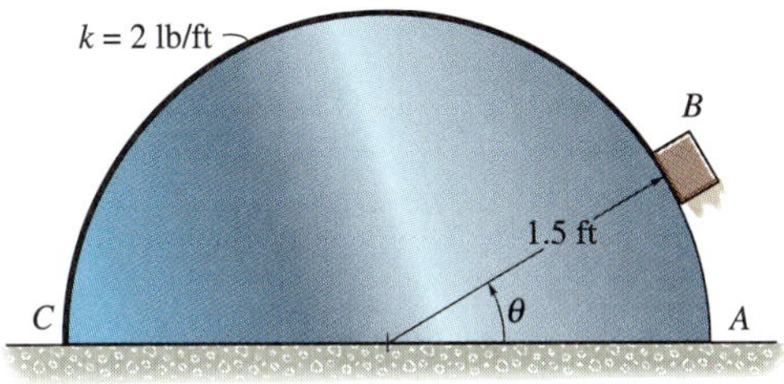

Prob. 14–36

14-37. The collar has a mass of 5 kg and is moving at 8 m/s when $x = 0$ and a force of $F = 60$ N is applied to it. The direction θ of this force varies such that $\theta = 10x$, where x is in meters and θ is clockwise, measured in degrees. Determine the speed of the collar when $x = 3$ m. The coefficient of kinetic friction between the collar and the rod is $\mu_k = 0.3$.

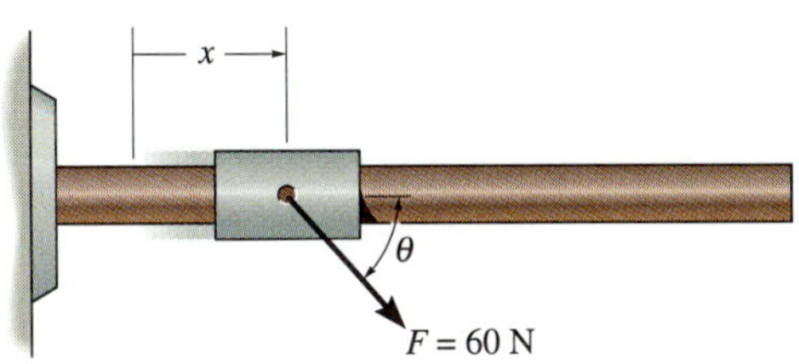

Prob. 14–37

14-38. The 0.5-kg ball of negligible size is fired up the vertical circular track using the spring plunger. The plunger keeps the spring compressed 0.08 m when $s = 0$. Determine how far s the plunger was pulled back and released if the ball begins to leave the track when $\theta = 135°$.

14-39. The 0.5-kg ball of negligible size is fired up the vertical circular track using the spring plunger. The plunger keeps the spring compressed 0.08 m when $s = 0$. Determine how far s it must be pulled back and released so that the ball will just make it around the loop and land on the platform at B. What is the ball's speed when it reaches the platform?

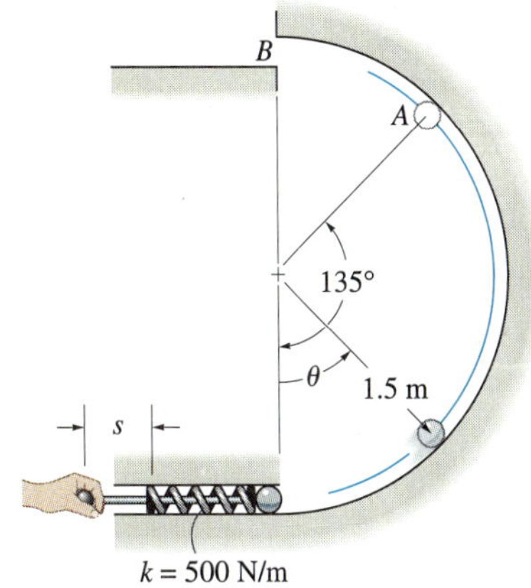

Probs. 14–38/39

*14-40. The "flying car" is a ride at an amusement park which consists of a car having wheels that roll along a track mounted inside a rotating drum. By design the car cannot fall off the track, however motion of the car is developed by applying the car's brake, thereby gripping the car to the track and allowing it to move with a constant speed of the track, $v_t = 3$ m/s. If the rider applies the brake when going from B to A and then releases it at the top of the drum, A, so that the car coasts freely down along the track to B ($\theta = \pi$ rad), determine the speed of the car at B and the normal reaction which the drum exerts on the car at B. Neglect friction during the motion from A to B. The rider and car have a total mass of 250 kg and the center of mass of the car and rider moves along a circular path having a radius of 8 m.

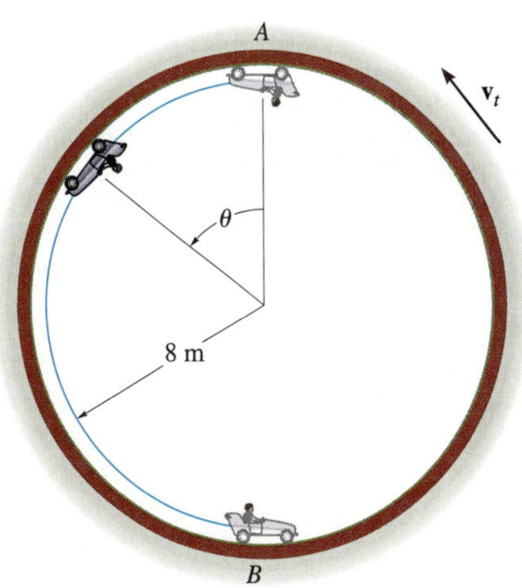

Prob. 14–40

14.4 Power and Efficiency

Power. *Power* is defined as the amount of work performed per unit of time. Hence, the *power* generated by a machine or engine that performs an amount of work dU within the time interval dt is

$$P = \frac{dU}{dt} \tag{14–9}$$

Provided the work dU is expressed by $dU = \mathbf{F} \cdot d\mathbf{r}$, then it is also possible to write

$$P = \frac{dU}{dt} = \frac{\mathbf{F} \cdot d\mathbf{r}}{dt} = \mathbf{F} \cdot \frac{d\mathbf{r}}{dt}$$

or

$$P = \mathbf{F} \cdot \mathbf{v} \tag{14–10}$$

The power output of this locomotive comes from the driving frictional force **F** developed at its wheels. It is this force that overcomes the frictional resistance of the cars in tow and is able to lift the weight of the train up the grade.

Hence, power is a *scalar,* where in the formulation $\mathbf{v}$ represents the velocity of the point which is acted upon by the force **F**.

The basic units of power used in the SI and FPS systems are the watt (W) and horsepower (hp), respectively. These units are defined as

$$1 \text{ W} = 1 \text{ J/s} = 1 \text{ N} \cdot \text{m/s}$$
$$1 \text{ hp} = 550 \text{ ft} \cdot \text{lb/s}$$

For conversion between the two systems of units, 1 hp = 746 W.

The term "power" provides a useful basis for determining the type of motor or machine which is required to do a certain amount of work in a given time. For example, two pumps may each be able to empty a reservoir if given enough time; however, the pump having the larger power will complete the job sooner.

Efficiency. The *mechanical efficiency* of a machine is defined as the ratio of the output of useful power produced by the machine to the input of power supplied to the machine. Hence,

$$\epsilon = \frac{\text{power output}}{\text{power input}} \tag{14–11}$$

If energy applied to the machine occurs during the *same time interval* at which it is removed, then the efficiency may also be expressed in terms of the ratio of output energy to input energy; i.e.

$$\epsilon = \frac{\text{energy output}}{\text{energy input}} \qquad (14\text{–}12)$$

Since machines consist of a series of moving parts, frictional forces will always be developed within the machine, and as a result, extra energy or power is needed to overcome these forces. Consequently, *the efficiency of a machine is always less than 1.*

The power requirements of this elevator depend upon the vertical force **F** that acts on the elevator and causes it to move upwards. If the velocity of the elevator is **v**, then the power output is $P = \mathbf{F}\cdot\mathbf{v}$.

Procedure for Analysis

The power supplied to a body can be computed using the following procedure.

- First determine the external force **F** acting on the body which causes the motion. This force is usually developed by a machine or engine placed either within or external to the body.
- If the body is accelerating, it may be necessary to draw its free-body diagram and apply the equation of motion ($\Sigma\mathbf{F} = m\mathbf{a}$) to determine **F**.
- Once **F** and the velocity **v** of the point where **F** is applied have been found, the power is determined by multiplying the force magnitude by the component of velocity acting in the direction of **F**, i.e., $P = \mathbf{F}\cdot\mathbf{v} = Fv\cos\theta$.
- In some problems the power may be found by calculating the work done by **F** per unit of time ($P_{avg} = \Delta U/\Delta t$, or $P = dU/dt$).

EXAMPLE 14-7

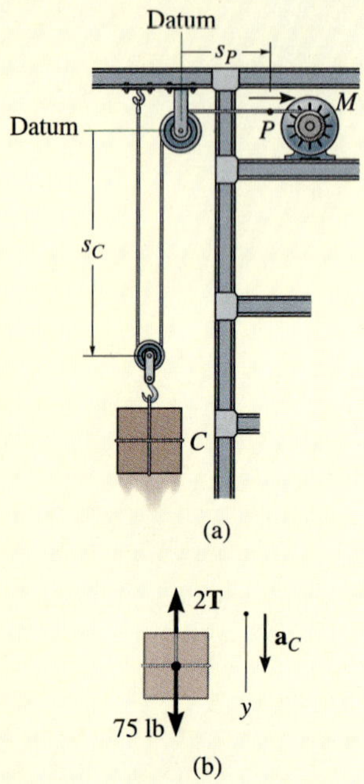

Fig. 14–15

The motor M of the hoist shown in Fig. 14–15a operates with an efficiency of 0.85. Determine the power that must be supplied to the motor to lift the 75-lb crate C at the instant point P on the cable has an acceleration of 4 ft/s^2 and a velocity of 2 ft/s. Neglect the mass of the pulley and cable.

Solution

In order to compute the power output of the motor, it is first necessary to determine the tension in the cable since this force is developed by the motor.

From the free-body diagram, Fig. 14–15b, we have

$$+\downarrow \Sigma F_y = ma_y; \qquad -2T + 75\text{ lb} = \frac{75\text{ lb}}{32.2\text{ ft/s}^2}a_C \tag{1}$$

The acceleration of the crate can be obtained by using kinematics to relate it to the known acceleration of point P, Fig. 14–15a. Using the methods of Sec. 12.9, the coordinates s_C and s_P in Fig. 14–15a can be related to a constant portion of cable length l which is changing in the vertical and horizontal directions. We have $2s_C + s_P = l$. Taking the second time derivative of this equation yields

$$2a_C = -a_P \tag{2}$$

Since $a_P = +4\text{ ft/s}^2$, then $a_C = (-4\text{ ft/s}^2)/2 = -2\text{ ft/s}^2$. What does the negative sign indicate? Substituting this result into Eq. 1 and *retaining* the negative sign since the acceleration in *both* Eqs. 1 and 2 is considered positive downward, we have

$$-2T + 75\text{ lb} = \frac{75\text{ lb}}{32.2\text{ ft/s}^2}(-2\text{ ft/s}^2)$$
$$T = 39.8\text{ lb}$$

The power output, measured in units of horsepower, required to draw the cable in at a rate of 2 ft/s is therefore

$$P = \mathbf{T}\cdot\mathbf{v} = (39.8\text{ lb})(2\text{ ft/s})[1\text{ hp}/(550\text{ ft}\cdot\text{lb/s})]$$
$$= 0.145\text{ hp}$$

This *power output* requires that the motor provide a *power input* of

$$\text{power input} = \frac{1}{\epsilon}(\text{power output})$$
$$= \frac{1}{0.85}(0.145\text{ hp}) = 0.170\text{ hp} \qquad \textit{Ans.}$$

Since the velocity of the crate is constantly changing, notice that this power requirement is *instantaneous*.

EXAMPLE 14-8

The sports car shown in Fig. 14–16*a* has a mass of 2 Mg and an engine running efficiency of $\epsilon = 0.63$. As it moves forward, the wind creates a drag resistance on the car of $F_D = 1.2v^2$ N, where v is the velocity in m/s. If the car is traveling at a constant speed of 50 m/s, determine the maximum power supplied by the engine.

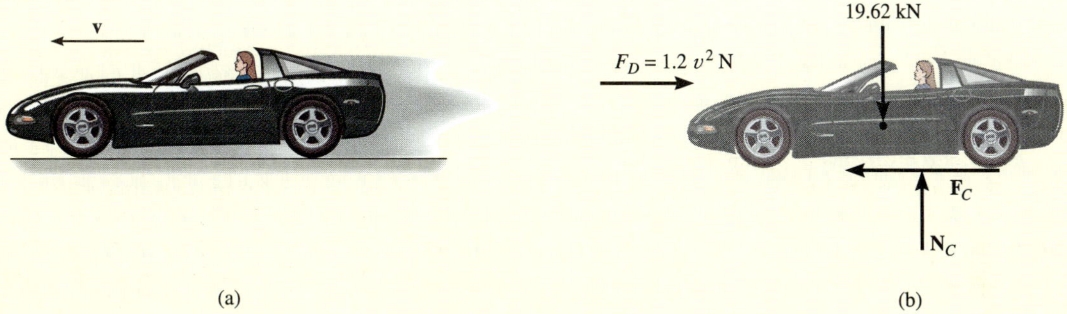

Fig. 14–16

Solution

As shown on the free-body diagram, Fig. 14–16*b*, the normal force $\mathbf{N}_C$ and frictional force $\mathbf{F}_C$ represent the *resultant forces* of all four wheels. In particular, the unbalanced frictional force drives or pushes the car *forward.* This effect is, of course, created by the rotating motion of the rear wheels on the pavement and is developed by the power of the engine.

Applying the equation of motion in the x direction, we have

$$\overset{+}{\leftarrow} \Sigma F_x = ma_x; \qquad F_C - 1.2v^2\text{ N} = (2000\text{ kg})\frac{dv}{dt}$$

Since the car is traveling with *constant velocity,* $dv/dt = 0$. Hence, with $v = 50$ m/s,

$$F_C = 1.2(50\text{ m/s})^2 = 3000\text{ N}$$

The power output of the car is manifested by the driving (frictional) force $\mathbf{F}_C$. Thus

$$P = \mathbf{F}_C \cdot \mathbf{v} = (3000\text{ N})(50\text{ m/s}) = 150\text{ kW}$$

The power supplied by the engine (power input) is therefore

$$\text{power input} = \frac{1}{\epsilon}(\text{power output}) = \frac{1}{0.63}(150\text{ kW}) = 238\text{ kW} \qquad \textit{Ans.}$$

Problems

14-41. The diesel engine of a 400-Mg train increases the train's speed uniformly from rest to 10 m/s in 100 s along a horizontal track. Determine the average power developed.

14-42. A spring having a stiffness of 5 kN/m is compressed 400 mm. The stored energy in the spring is used to drive a machine which requires 90 W of power. Determine how long the spring can supply energy at the required rate.

14-43. An electric streetcar has a weight of 15 000 lb and accelerates along a horizontal straight road from rest such that the power is always 100 hp. Determine how far it must travel to reach a speed of 40 ft/s.

***14-44.** The Milkin Aircraft Co. manufactures a turbojet engine that is placed in a plane having a weight of 13 000 lb. If the engine develops a constant thrust of 5200 lb, determine the power outputs of the plane when it is just ready to take off and when it is flying at 600 mi/h.

14-45. A truck has a weight of 25 000 lb and an engine which transmits a power of 350 hp to *all* the wheels. Assuming that the wheels do not slip on the ground, determine the angle θ of the largest incline the truck can climb at a constant speed of $v = 50$ ft/s.

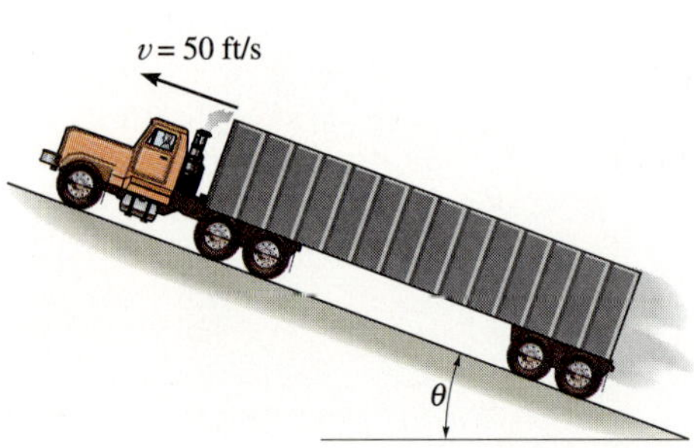

Prob. 14–45

14-46. A loaded truck weighs $16(10^3)$ lb and accelerates uniformly on a level road from 15 ft/s to 30 ft/s during 4 s. If the frictional resistance to motion is 325 lb, determine the maximum power that must be delivered to the wheels.

14-47. An automobile having a weight of 3500 lb travels up a 7° slope at a constant speed of $v = 40$ ft/s. If friction and wind resistance are neglected, determine the power developed by the engine if the automobile has a mechanical efficiency of $\epsilon = 0.65$.

***14-48.** The escalator steps move with a constant speed of 0.6 m/s. If the steps are 125 mm high and 250 mm in length, determine the power of a motor needed to lift an average mass of 150 kg per step. There are 32 steps.

14-49. If the escalator in Prob. 14-48 is *not moving,* determine the constant speed at which a man having a mass of 80 kg must walk up the steps to generate 100 W of power—the same amount that is needed to power a standard light bulb.

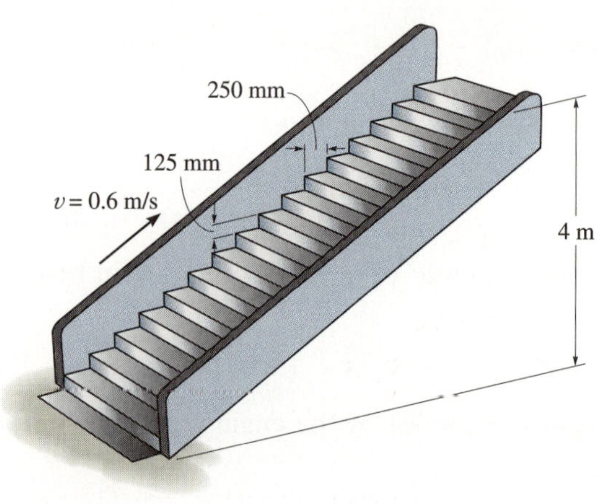

Probs. 14–48/49

14-50. An electrically powered train car draws 30 kW of power. If the car weighs 40 000 lb and starts from rest, determine the maximum speed it attains in 30 s. The mechanical efficiency is $\epsilon = 0.8$.

14-51. Determine the final velocity of the train in Prob. 14-50 if the wind resistance is $F_w = (0.6v)$ lb, where v is in ft/s.

***14-52.** Determine the power output of the draw-works motor M necessary to lift the 600-lb drill pipe upward with a constant speed of 4 ft/s. The cable is tied to the top of the oil rig, wraps around the lower pulley, then around the top pulley, and then to the motor.

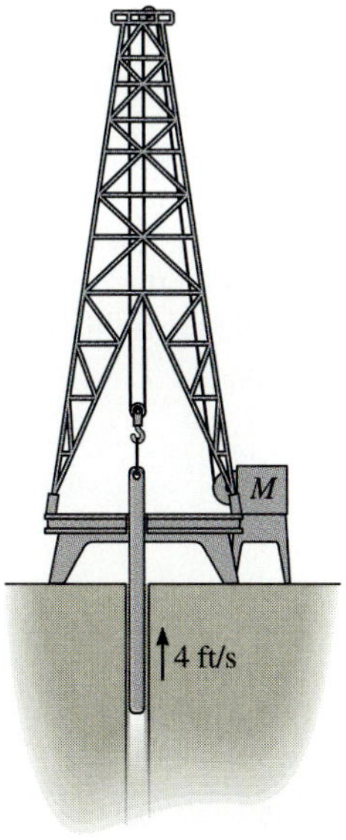

Prob. 14–52

14-53. The 50-lb crate is hoisted by the motor M. If the crate starts from rest and by constant acceleration attains a speed of 12 ft/s after rising $s = 10$ ft, determine the power that must be supplied to the motor at the instant $s = 10$ ft. The motor has an efficiency $\epsilon = 0.65$. Neglect the mass of the pulley and cable.

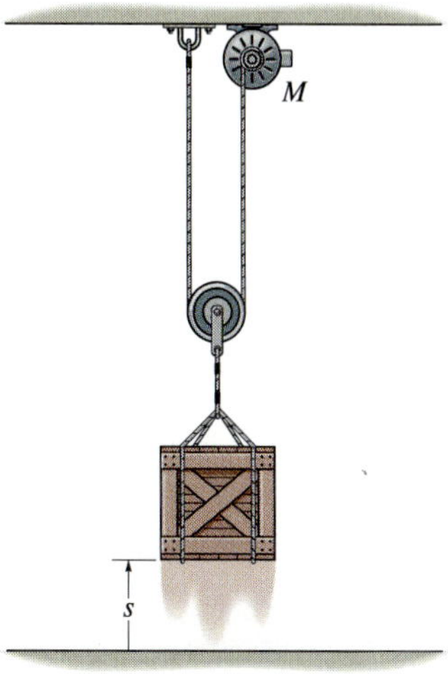

Prob. 14–53

14-54. The 50-lb crate is given a speed of 10 ft/s in $t = 4$ s starting from rest. If the acceleration is constant, determine the power that must be supplied to the motor when $t = 2$ s. The motor has an efficiency $\epsilon = 0.65$. Neglect the mass of the pulley and cable.

Prob. 14–54

14-55. The elevator E and its freight have a total mass of 400 kg. Hoisting is provided by the motor M and the 60-kg block C. If the motor has an efficiency of $\epsilon = 0.6$, determine the power that must be supplied to the motor when the elevator is hoisted upward at a constant speed of $v_E = 4$ m/s.

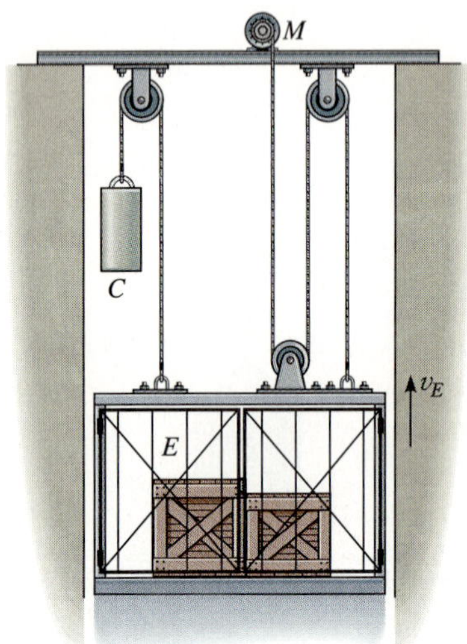

Prob. 14–55

***14-56.** The 50-kg crate is hoisted up the 30° incline by the pulley system and motor M. If the crate starts from rest and by constant acceleration attains a speed of 4 m/s after traveling 8 m along the plane, determine the power that must be supplied to the motor at this instant. Neglect friction along the plane. The motor has an efficiency of $\epsilon = 0.74$.

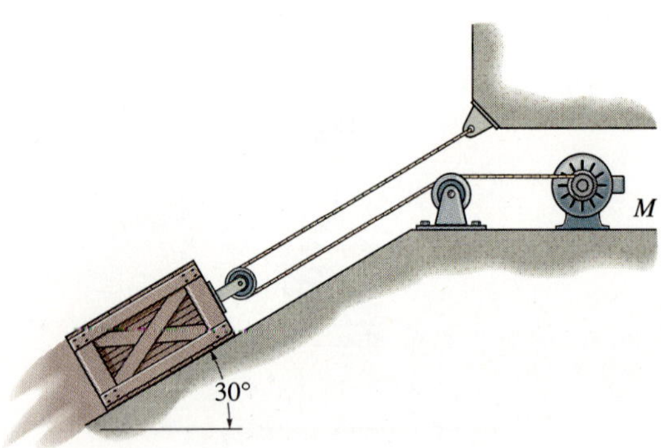

Prob. 14–56

14-57. The 50-lb load is hoisted by the pulley system and motor M. If the crate starts from rest and by constant acceleration attains a speed of 15 ft/s after rising $s = 6$ ft, determine the power that must be supplied to the motor at this instant. The motor has an efficiency of $\epsilon = 0.76$. Neglect the mass of the pulleys and cable.

14-58. The 50-lb load is hoisted by the pulley system and motor M. If the motor exerts a constant force of 30 lb on the cable, determine the power that must be supplied to the motor if the load has been hoisted $s = 10$ ft starting from rest. The motor has an efficiency of $\epsilon = 0.76$.

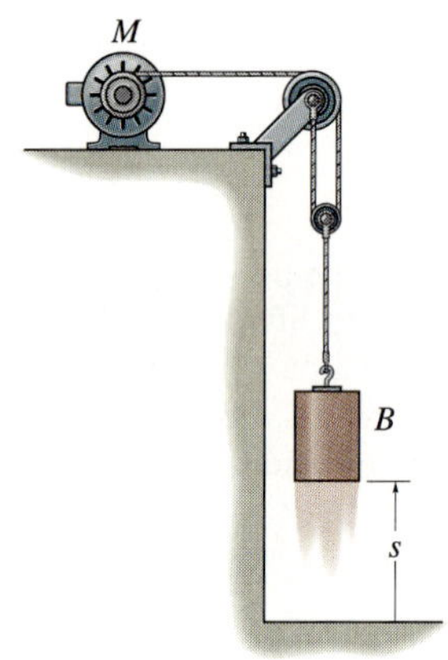

Probs. 14–57/58

14-59. It has been found that the best strategy for maximum bicycle pedaling power is to initially provide a large acceleration, then increase the power gradually over the next 30 s in order to reach maximum cruising speed. Using the biomechanical power curve shown, determine the maximum speed attained by the rider and his bicycle, which have a total mass of 92 kg, as the rider ascends the 20° slope starting from rest.

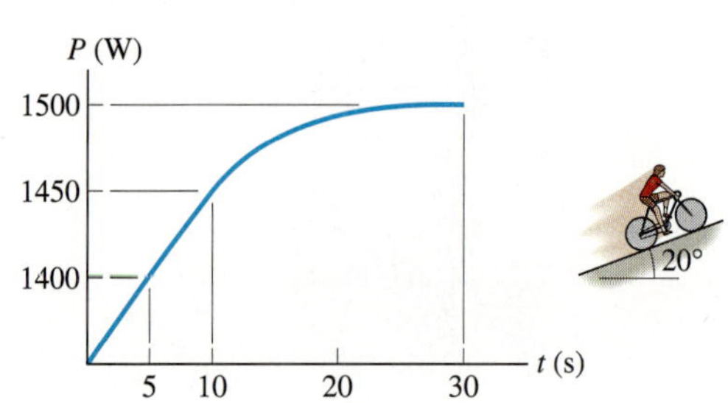

Prob. 14–59

*14-60. A rocket having a total mass of 8 Mg is fired vertically from rest. If the engines provide a constant thrust of $T = 300$ kN, determine the power output of the engines as a function of time. Neglect the effect of drag resistance and the loss of fuel mass and weight.

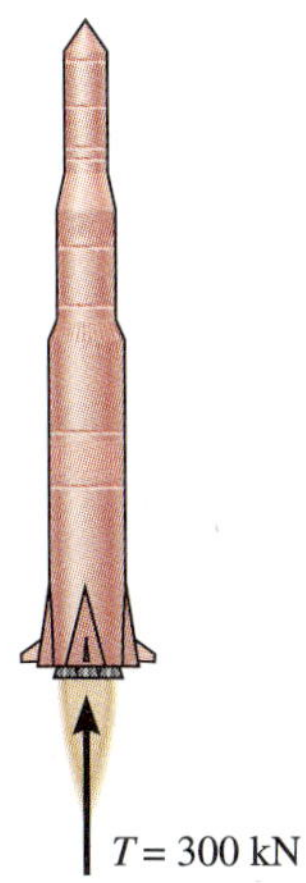

Prob. 14–60

14-61. The block has a weight of 80 lb and rests on the floor for which $\mu_k = 0.4$. If the motor draws in the cable at a constant rate of 6 ft/s, determine the output of the motor at the instant $\theta = 30°$. Neglect the mass of the cable and pulleys.

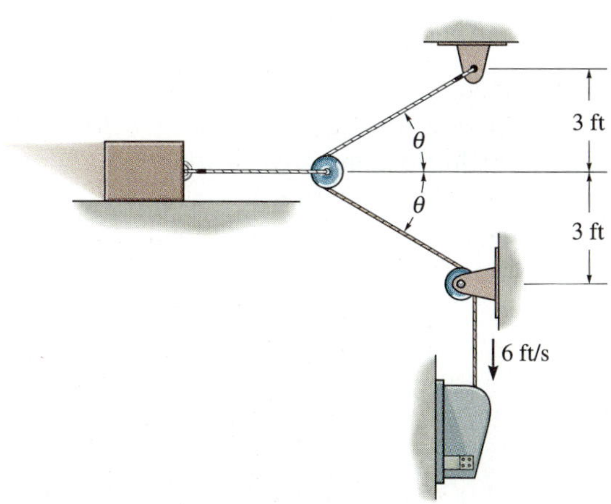

Prob. 14–61

14-62. The block has a mass of 150 kg and rests on a surface for which the coefficients of static and kinetic friction are $\mu_s = 0.5$ and $\mu_k = 0.4$, respectively. If a force $F = (60t^2)$ N, where t is in seconds, is applied to the cable, determine the power developed by the force when $t = 5$ s. *Hint:* First determine the time needed for the force to cause motion.

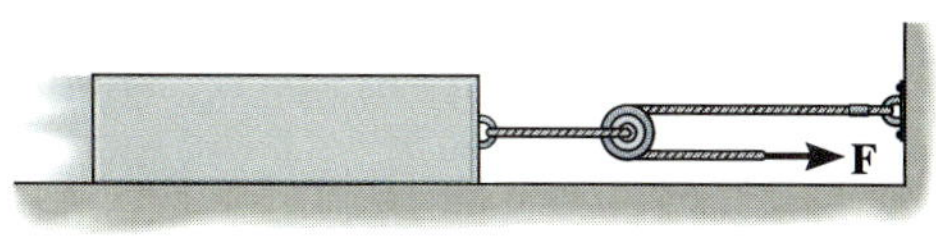

Prob. 14–62

14-63. The 50-lb block rests on the rough surface for which the coefficient of kinetic friction is $\mu_k = 0.2$. A force $F = (40 + s^2)$ lb, where s is in ft, acts on the block in the direction shown. If the spring is originally unstretched ($s = 0$) and the block is at rest, determine the power developed by the force the instant the block has moved $s = 1.5$ ft.

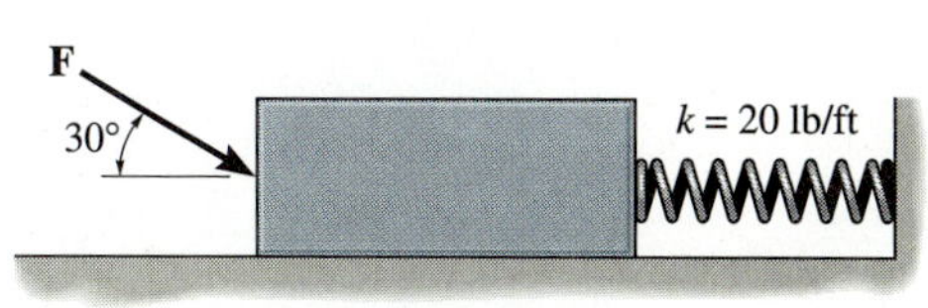

Prob. 14–63

14.5 Conservative Forces and Potential Energy

Conservative Force. When the work done by a force in moving a particle from one point to another is *independent of the path* followed by the particle, then this force is called a *conservative force.* The weight of a particle and the force of an elastic spring are two examples of conservative forces often encountered in mechanics. The work done by the weight of a particle is *independent of the path* since it depends only on the particle's *vertical displacement.* The work done by a spring force *acting on a particle* is *independent of the path* of the particle, rather it depends only on the extension or compression s of the spring.

In contrast to a conservative force, consider the force of friction exerted *on a moving object* by a fixed surface. The work done by the frictional force *depends on the path*—the longer the path, the greater the work. Consequently, *frictional forces are nonconservative.* The work is dissipated from the body in the form of heat.

Potential Energy. Energy may be defined as the capacity for doing work. When energy comes from the *motion* of the particle, it is referred to as *kinetic energy.* When it comes from the *position* of the particle, measured from a fixed datum or reference plane, it is called potential energy. Thus, *potential energy* is a measure of the amount of work a conservative force will do when it moves from a given position to the datum. In mechanics, the potential energy due to gravity (weight) or an elastic spring is important.

Gravitational Potential Energy. If a particle is located a distance y *above* an arbitrarily selected datum, as shown in Fig. 14–17, the particle's weight $\mathbf{W}$ has positive *gravitational potential energy,* V_g, since $\mathbf{W}$ has the capacity of doing positive work when the particle is moved back down to the datum. Likewise, if the particle is located a distance y *below* the datum, V_g is negative since the weight does negative work when the particle is moved back up to the datum. At the datum $V_g = 0$.

In general, if y is *positive upward,* the gravitational potential energy of the particle of weight W is thus*

$$V_g = Wy \tag{14–13}$$

*Here the weight is assumed to be *constant.* This assumption is suitable for small differences in elevation Δy. If the elevation change is significant, however, a variation of weight with elevation must be taken into account (see Prob. 14–97).

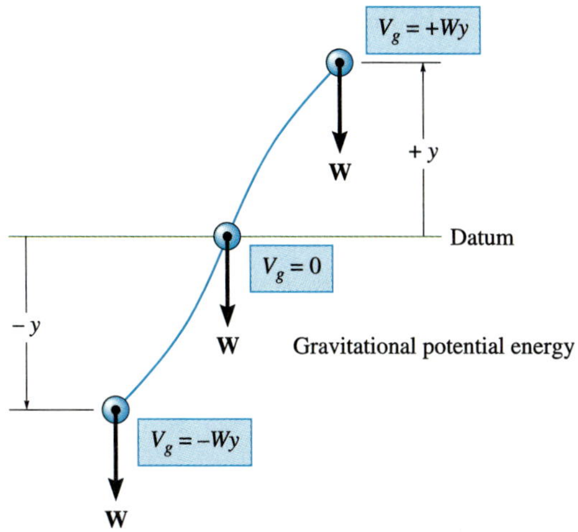

Fig. 14–17

Elastic Potential Energy. When an elastic spring is elongated or compressed a distance s from its unstretched position, the elastic potential energy V_e due to the spring's configuration can be expressed as

$$V_e = +\tfrac{1}{2}ks^2 \tag{14–14}$$

Here V_e is *always positive* since, in the deformed position, the force of the spring has the *capacity* for always doing positive work on the particle when the spring is returned to its unstretched position, Fig. 14–18.

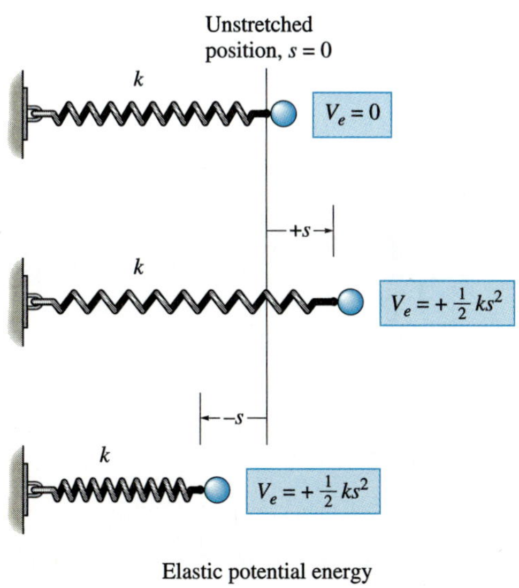

Fig. 14–18

Potential Function. In the general case, if a particle is subjected to both gravitational and elastic forces, the particle's potential energy can be expressed as a *potential function,* which is the algebraic sum

$$V = V_g + V_e \tag{14–15}$$

Measurement of V depends on the location of the particle with respect to a selected datum in accordance with Eqs. 14–13 and 14–14.

If the particle is located at an arbitrary point (x, y, z) in space, this potential function is then $V = V(x, y, z)$. The work done by a conservative force in moving the particle from point (x_1, y_1, z_1) to point (x_2, y_2, z_2) is measured by the *difference* of this function, i.e.,

$$U_{1-2} = V_1 - V_2 \tag{14–16}$$

For example, the potential function for a particle of weight W suspended from a spring can be expressed in terms of its position, s, measured from a datum located at the unstretched length of the spring, Fig. 14–19. We have

$$\begin{aligned} V &= V_g + V_e \\ &= -Ws + \tfrac{1}{2}ks^2 \end{aligned}$$

If the particle moves from s_1 to a lower position s_2, then applying Eq. 14–16 it can be seen that the work of $\mathbf{W}$ and $\mathbf{F}_s$ is

$$\begin{aligned} U_{1-2} = V_1 - V_2 &= \left(-Ws_1 + \tfrac{1}{2}ks_1^2\right) - \left(-Ws_2 + \tfrac{1}{2}ks_2^2\right) \\ &= W(s_2 - s_1) - \left(\tfrac{1}{2}ks_2^2 - \tfrac{1}{2}ks_1^2\right) \end{aligned}$$

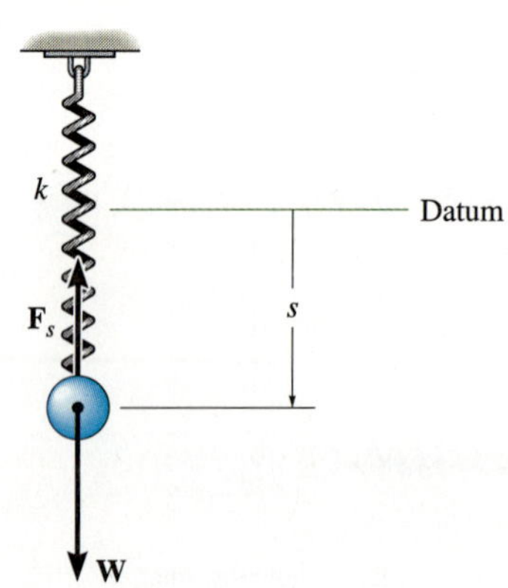

Fig. 14–19

When the displacement along the path is infinitesimal, i.e., from point (x, y, z) to $(x + dx, y + dy, z + dz)$, Eq. 14–16 becomes

$$\begin{aligned} dU &= V(x, y, z) - V(x + dx, y + dy, z + dz) \\ &= -dV(x, y, z) \end{aligned} \tag{14–17}$$

Provided both the force and displacement are defined using rectangular coordinates, then the work can also be expressed as

$$\begin{aligned} dU &= \mathbf{F}\cdot d\mathbf{r} = (F_x\mathbf{i} + F_y\mathbf{j} + F_z\mathbf{k})\cdot(dx\mathbf{i} + dy\mathbf{j} + dz\mathbf{k}) \\ &= F_x dx + F_y dy + F_z dz \end{aligned}$$

Substituting this result into Eq. 14–17 and expressing the differential $dV(x, y, z)$ in terms of its partial derivatives yields

$$F_x dx + F_y dy + F_z dz = -\left(\frac{\partial V}{\partial x}dx + \frac{\partial V}{\partial y}dy + \frac{\partial V}{\partial z}dz\right)$$

Since changes in x, y, and z are all independent of one another, this equation is satisfied provided

$$F_x = -\frac{\partial V}{\partial x}, \qquad F_y = -\frac{\partial V}{\partial y}, \qquad F_z = -\frac{\partial V}{\partial z} \tag{14–18}$$

Thus,

$$\begin{aligned} \mathbf{F} &= -\frac{\partial V}{\partial x}\mathbf{i} - \frac{\partial V}{\partial y}\mathbf{j} - \frac{\partial V}{\partial z}\mathbf{k} \\ &= -\left(\frac{\partial}{\partial x}\mathbf{i} + \frac{\partial}{\partial y}\mathbf{j} + \frac{\partial}{\partial z}\mathbf{k}\right)V \end{aligned}$$

or

$$\mathbf{F} = -\nabla V \tag{14–19}$$

where ∇ (del) represents the vector operator $\nabla = (\partial/\partial x)\mathbf{i} + (\partial/\partial y)\mathbf{j} + (\partial/\partial z)\mathbf{k}$.

Equation 14–19 relates a force **F** to its potential function V and thereby provides a mathematical criterion for proving that **F** is conservative. For example, the gravitational potential function for a weight located a distance y above a datum is $V_g = Wy$. To prove that **W** is conservative, it is necessary to show that it satisfies Eq. 14–19 (or Eq. 14–18), in which case

$$F_y = -\frac{\partial V}{\partial y}; \qquad F = -\frac{\partial}{\partial y}(Wy) = -W$$

The negative sign indicates that **W** acts downward, opposite to positive y, which is upward.

14.6 Conservation of Energy

The weight of the sacks resting on this platform causes potential energy to be stored in the supporting springs. As each sack is removed, the platform will *rise* slightly since some of the potential energy within the springs will be transferred into an increase in gravitational potential energy of the remaining sacks. Such a device is useful for removing the sacks without having to bend over to pick them up as they are unloaded.

When a particle is acted upon by a system of *both* conservative and nonconservative forces, the portion of the work done by the *conservative forces* can be written in terms of the difference in their potential energies using Eq. 14–16, i.e., $(\Sigma U_{1-2})_{\text{cons.}} = V_1 - V_2$. As a result, the principle of work and energy can be written as

$$T_1 + V_1 + (\Sigma U_{1-2})_{\text{noncons.}} = T_2 + V_2 \qquad (14\text{–}20)$$

Here $(\Sigma U_{1-2})_{\text{noncons.}}$ represents the work of the nonconservative forces acting on the particle. If *only conservative forces* are applied to the body, this term is zero and then we have

$$T_1 + V_1 = T_2 + V_2 \qquad (14\text{–}21)$$

This equation is referred to as the *conservation of mechanical energy* or simply the *conservation of energy*. It states that during the motion the sum of the particle's kinetic and potential energies remains *constant*. For this to occur, kinetic energy must be transformed into potential energy, and vice versa. For example, if a ball of weight $\mathbf{W}$ is dropped from a height h above the ground (datum), Fig. 14–20, the potential energy of the ball is maximum before it is dropped, at which time its kinetic energy is zero. The total mechanical energy of the ball in its initial position is thus

$$E = T_1 + V_1 = 0 + Wh = Wh$$

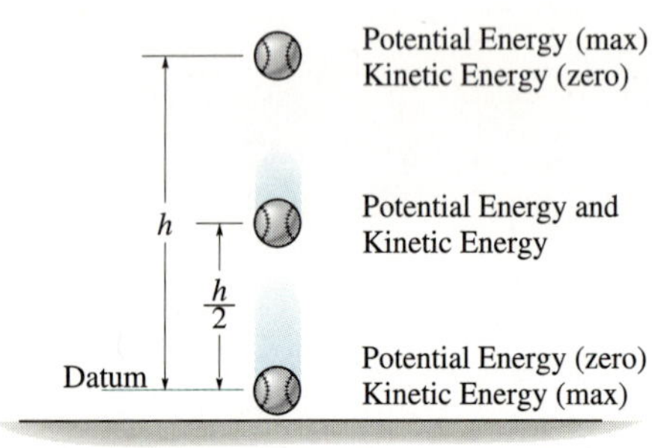

Fig. 14–20

When the ball has fallen a distance $h/2$, its speed can be determined by using $v^2 = v_0^2 + 2a_c(y - y_0)$, which yields $v = \sqrt{2g(h/2)} = \sqrt{gh}$. The energy of the ball at the mid-height position is therefore

$$E = V_2 + T_2 = W\frac{h}{2} + \frac{1}{2}\frac{W}{g}(\sqrt{gh})^2 = Wh$$

Just before the ball strikes the ground, its potential energy is zero and its speed is $v = \sqrt{2gh}$. Here, again, the total energy of the ball is

$$E = V_3 + T_3 = 0 + \frac{1}{2}\frac{W}{g}(\sqrt{2gh})^2 = Wh$$

Note that when the ball comes in contact with the ground, it deforms somewhat, and provided the ground is hard enough, the ball will rebound off the surface, reaching a new height h', which will be less than the height h from which it was first released. Neglecting air friction, the difference in height accounts for an energy loss, $E_l = W(h - h')$, which occurs during the collision. Portions of this loss produce noise, localized deformation of the ball and ground, and heat.

System of Particles. If a system of particles is *subjected only to conservative forces,* then an equation similar to Eq. 14–21 can be written for the particles. Applying the ideas of the preceding discussion, Eq. 14–8 ($\Sigma T_1 + \Sigma U_{1-2} = \Sigma T_2$) becomes

$$\Sigma T_1 + \Sigma V_1 = \Sigma T_2 + \Sigma V_2 \tag{14–22}$$

Here, the sum of the system's initial kinetic and potential energies is equal to the sum of the system's final kinetic and potential energies. In other words, $\Sigma T + \Sigma V = \text{const.}$

It is important to remember that only problems involving conservative force systems (weights and springs) may be solved by using the conservation of energy theorem. As stated previously, friction or other drag-resistant forces, which depend upon velocity or acceleration, are nonconservative. A portion of the work done by such forces is transformed into thermal energy, and consequently this energy dissipates into the surroundings and may not be recovered.

Procedure for Analysis

The conservation of energy equation is used to solve problems involving *velocity, displacement,* and *conservative force systems.* It is generally *easier to apply* than the principle of work and energy because the energy equation just requires specifying the particle's kinetic and potential energies at only *two points* along the path, rather than determining the work when the particle moves through a *displacement.* For application it is suggested that the following procedure be used.

Potential Energy

- Draw two diagrams showing the particle located at its initial and final points along the path.
- If the particle is subjected to a vertical displacement, establish the fixed horizontal datum from which to measure the particle's gravitational potential energy V_g.
- Data pertaining to the elevation y of the particle from the datum and the extension or compression s of any connecting springs can be determined from the geometry associated with the two diagrams.
- Recall $V_g = Wy$, where y is positive upward from the datum and negative downward from the datum; also $V_e = \frac{1}{2}ks^2$, which is *always positive.*

Conservation of Energy

- Apply the equation $T_1 + V_1 = T_2 + V_2$.
- When determining the kinetic energy, $T = \frac{1}{2}mv^2$, the particle's speed v must be measured from an inertial reference frame.

EXAMPLE 14–9

The gantry structure in the photo is used to test the response of an airplane during a crash. As shown in Fig. 14–21*a*, the plane, having a mass of 8 Mg, is hoisted back until $\theta = 60°$, and then the pull-back cable *AC* is released when the plane is at rest. Determine the speed of the plane just before crashing into the ground, $\theta = 15°$. Also, what is the maximum tension developed in the supporting cable during the motion? Neglect the effect of lift caused by the wings during the motion and the size of the airplane.

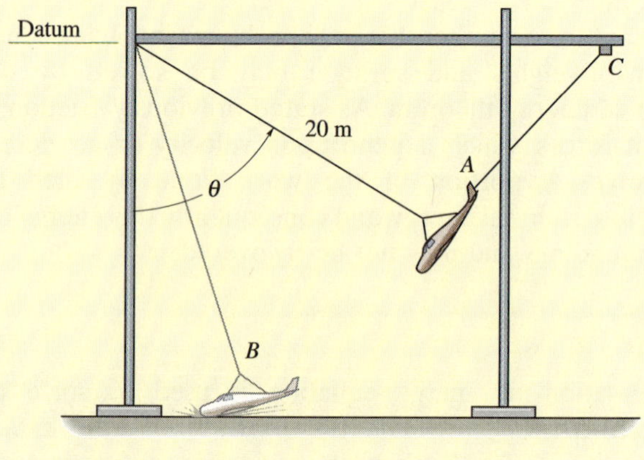

(a)

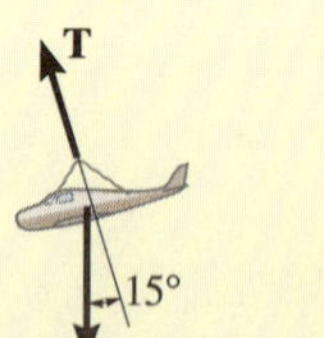

(b)

Fig. 14–21

Solution

Since the force of the cable does *no work* on the plane, it must be obtained using the equation of motion. First, however, we must determine the plane's speed at *B*.

Potential Energy. For convenience, the datum has been established at the top of the gantry.

Conservation of Energy

$$T_A + V_A = T_B + V_B$$

$$0 - 8000\text{ kg }(9.81\text{ m/s}^2)(20\cos 60°\text{ m}) = \frac{1}{2}(8000\text{ kg})v_B^2 - 8000\text{ kg }(9.81\text{ m/s}^2)(20\cos 15°\text{ m})$$

$$v_B = 13.5\text{ m/s} \qquad \textit{Ans.}$$

Equation of Motion. Using the data tabulated on the free-body diagram when the plane is at *B*, Fig. 14–21*b*, we have

$$+\nwarrow \Sigma F_n = ma_n;$$

$$T - 8000\,(9.81)\text{N}\cos 15° = (8000\text{ kg})\frac{(13.5\text{ m/s})^2}{20\text{ m}}$$

$$T = 149\text{ kN} \qquad \textit{Ans.}$$

EXAMPLE 14-10

The ram R shown in Fig. 14–22a has a mass of 100 kg and is released from rest 0.75 m from the top of a spring, A, that has a stiffness $k_A =$ 12 kN/m. If a second spring B, having a stiffness $k_B = 15$ kN/m, is "nested" in A, determine the maximum displacement of A needed to stop the downward motion of the ram. The unstretched length of each spring is indicated in the figure. Neglect the mass of the springs.

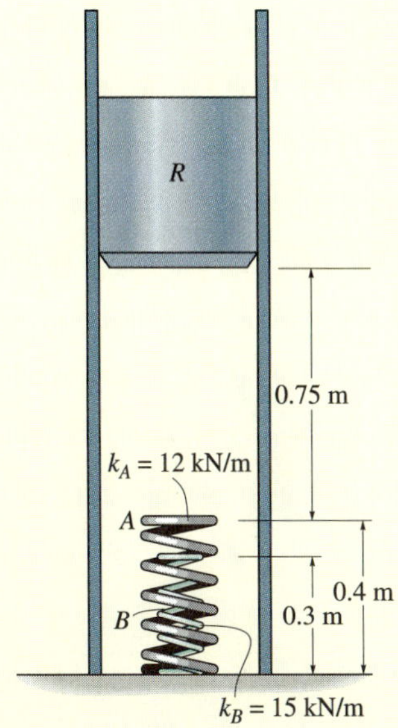

(a)

Solution

Potential Energy. We will *assume* that the ram compresses *both* springs at the instant it comes to rest. The datum is located through the center of gravity of the ram at its initial position, Fig. 14–22b. When the kinetic energy is reduced to zero ($v_2 = 0$), A is compressed a distance s_A and B compresses $s_B = s_A - 0.1$ m.

Conservation of Energy

$$T_1 + V_1 = T_2 + V_2$$

$$0 + 0 = 0 + \{\tfrac{1}{2}k_A s_A^2 + \tfrac{1}{2}k_B(s_A - 0.1)^2 - Wh\}$$

$$0 + 0 = 0 + \{\tfrac{1}{2}(12\,000 \text{ N/m})s_A^2 + \tfrac{1}{2}(15\,000 \text{ N/m})(s_A - 0.1\text{m})^2 - 981 \text{ N}(0.75 \text{ m} + s_A)\}$$

Rearranging the terms,

$$13\,500 s_A^2 - 2481 s_A - 660.75 = 0$$

Using the quadratic formula and solving for the positive root,* we have

$$s_A = 0.331 \text{ m} \qquad \textit{Ans.}$$

Since $s_B = 0.331$ m $- 0.1$ m $= 0.231$ m, which is positive, the assumption that *both* springs are compressed by the ram is correct.

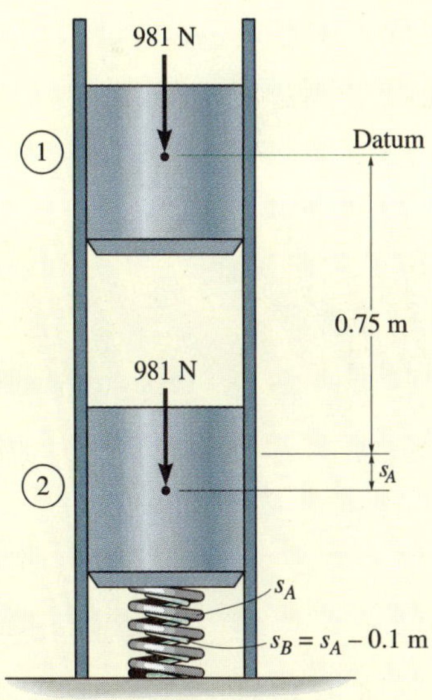

(b)

Fig. 14–22

*The second root, $s_A = -0.148$ m, does not represent the physical situation. Since positive s is measured downward, the negative sign indicates that spring A would have to be "extended" by an amount of 0.148 m to stop the ram.

EXAMPLE 14–11

A smooth 2-kg collar C, shown in Fig. 14–23a, fits loosely on the vertical shaft. If the spring is unstretched when the collar is in the position A, determine the speed at which the collar is moving when $y = 1$ m, if (a) it is released from rest at A, and (b) it is released at A with an *upward* velocity $v_A = 2$ m/s.

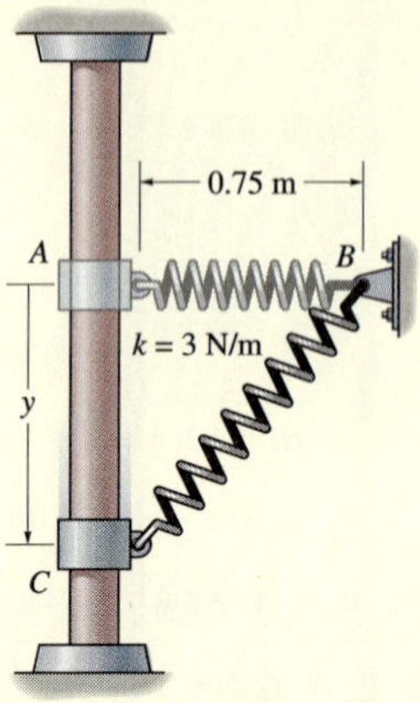

(a)

Solution

Part (a)

Potential Energy. For convenience, the datum is established through AB, Fig. 14–23b. When the collar is at C, the gravitational potential energy is $-(mg)y$, since the collar is *below* the datum, and the elastic potential energy is $\frac{1}{2}ks_{CB}^2$. Here $s_{CB} = 0.5$ m, which represents the *stretch* in the spring as shown in the figure.

Conservation of Energy

$$T_A + V_A = T_C + V_C$$

$$0 + 0 = \tfrac{1}{2}mv_C^2 + \{\tfrac{1}{2}ks_{CB}^2 - mgy\}$$

$$0 + 0 = \{\tfrac{1}{2}(2\text{ kg})v_C^2\} + \{\tfrac{1}{2}(3\text{ N/m})(0.5\text{ m})^2 - 2(9.81)\text{ N}(1\text{ m})\}$$

$$v_C = 4.39\text{ m/s}\downarrow \qquad \textit{Ans.}$$

This problem can also be solved by using the equation of motion or the principle of work and energy. Note that in *both* of these methods the variation of the magnitude and direction of the spring force must be taken into account (see Example 13–4). Here, however, the above method of solution is clearly advantageous since the calculations depend *only* on data calculated at the initial and final points of the path.

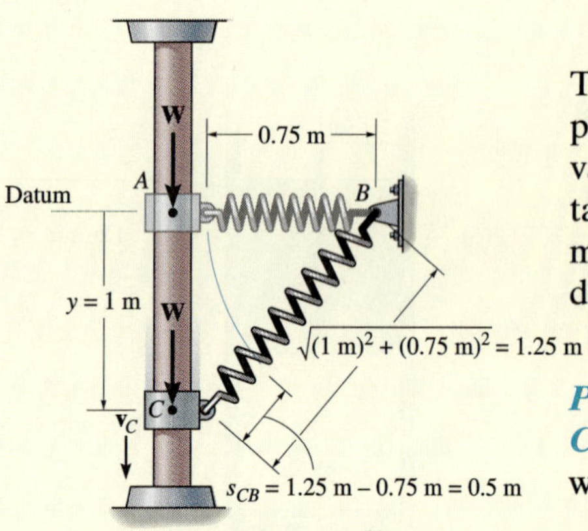

(b)

Fig. 14–23

Part (b)

Conservation of Energy. If $v_A = 2$ m/s, using the data in Fig. 14–23b, we have

$$T_A + V_A = T_C + V_C$$

$$\tfrac{1}{2}mv_A^2 + 0 = \tfrac{1}{2}mv_C^2 + \{\tfrac{1}{2}ks_{CB}^2 - mgy\}$$

$$\tfrac{1}{2}(2\text{ kg})(2\text{ m/s})^2 + 0 = \tfrac{1}{2}(2\text{ kg})v_C^2 + \{\tfrac{1}{2}(3\text{ N/m})(0.5\text{ m})^2 - 2(9.81)\text{ N}(1\text{ m})\}$$

$$v_C = 4.82\text{ m/s}\downarrow \qquad \textit{Ans.}$$

Note that the kinetic energy of the collar depends only on the *magnitude* of velocity, and therefore it is immaterial if the collar is moving up or down at 2 m/s when released at A.

Problems

*14-64. Solve Prob. 14-8 using the conservation of energy equation.

14-65. Solve Prob. 14-11 using the conservation of energy equation.

14-66. Solve Prob. 14-17 using the conservation of energy equation.

14-67. Solve Prob. 14-21 using the conservation of energy equation.

***14-68.** Solve Prob. 14-36 using the conservation of energy equation.

14-69. Solve Prob. 14-23 using the conservation of energy equation.

14-70. The bob of the pendulum has a mass of 0.2 kg and is released from rest when it is in the horizontal position shown. Determine its speed and the tension in the cord at the instant the bob passes through its lowest position.

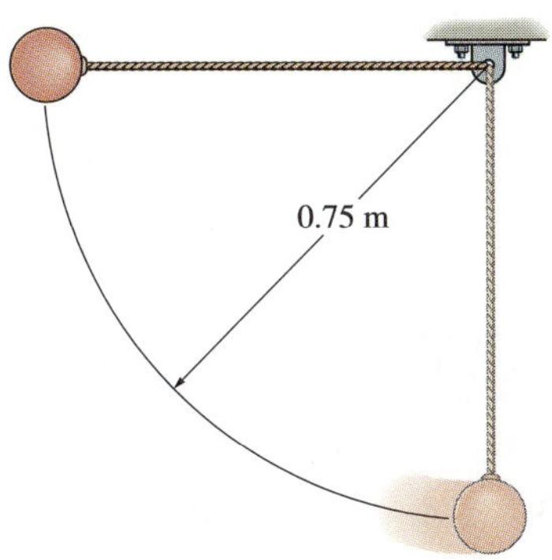

Prob. 14–70

14-71. The block has a weight of 1.5 lb and slides along the smooth chute AB. It is released from rest at A, which has coordinates of $A(5\text{ ft}, 0, 10\text{ ft})$. Determine the speed at which it slides off at B, which has coordinates of $B(0, 8\text{ ft}, 0)$.

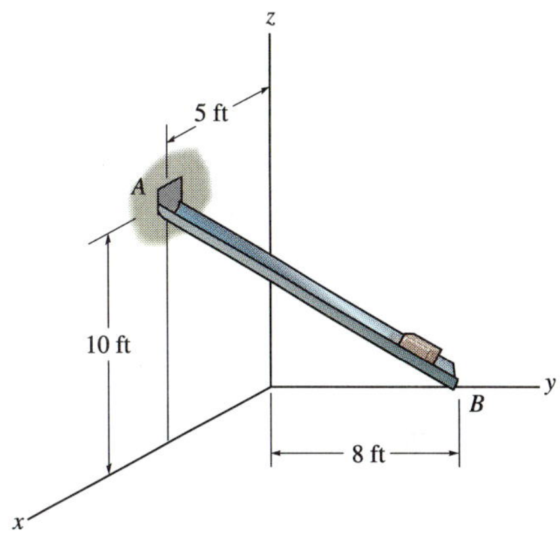

Prob. 14–71

***14-72.** The blocks A and B weigh 10 lb and 30 lb, respectively. They are connected together by a light cord and ride in the frictionless grooves. Determine the speed of both blocks after they are released from rest and block A moves 6 ft along the plane.

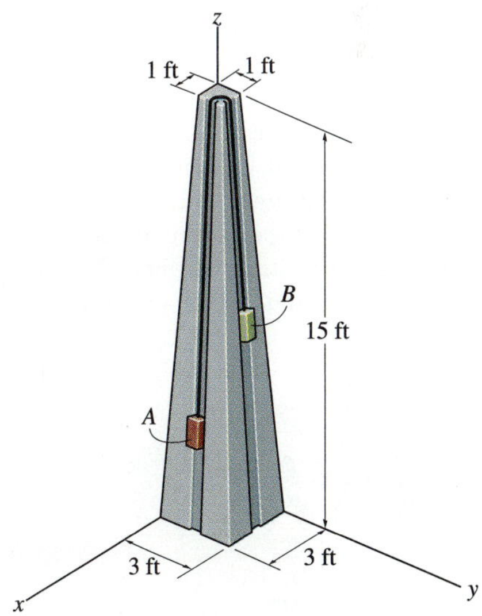

Prob. 14–72

14-73. Block A has a weight of 1.5 lb and slides in the smooth horizontal slot. If the block is drawn back to $s = 1.5$ ft and released from rest, determine its speed at the instant $s = 0$. Each of the two springs has a stiffness of $k = 150$ lb/ft and an unstretched length of 0.5 ft.

14-74. The 2-lb block A slides in the smooth horizontal slot. When $s = 0$ the block is given an initial velocity of 60 ft/s to the right. Determine the maximum horizontal displacement s of the block. Each of the two springs has a stiffness of $k = 150$ lb/ft and an unstretched length of 0.5 ft.

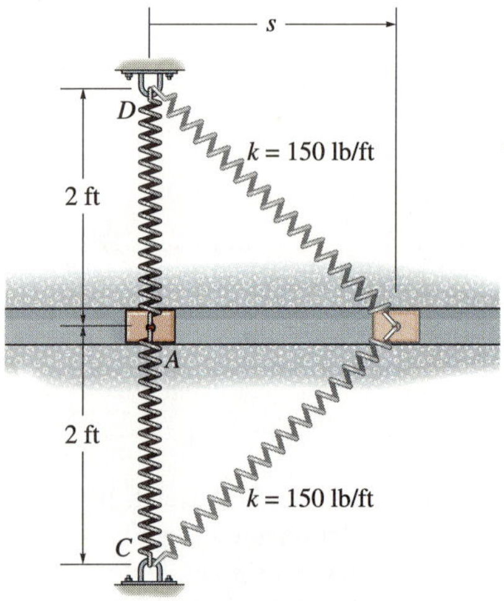

Probs. 14–73/74

14-75. The firing mechanism of a pinball machine consists of a plunger P having a mass of 0.25 kg and a spring of stiffness $k = 300$ N/m. When $s = 0$, the spring is compressed 50 mm. If the arm is pulled back such that $s = 100$ mm and released, determine the speed of the 0.3-kg pinball B *just before* the plunger strikes the stop, i.e., $s = 0$. Assume all surfaces of contact to be smooth. The ball moves in the horizontal plane. Neglect friction, the mass of the spring, and the rolling motion of the ball.

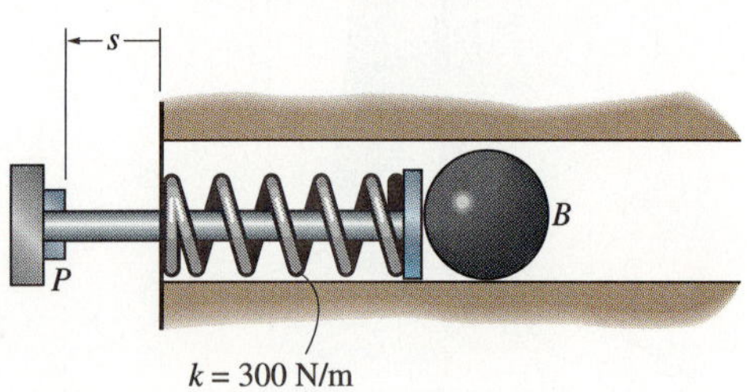

Prob. 14–75

***14-76.** Determine the smallest amount the spring at B must be compressed against the 0.5-lb block so that when it is released from B it slides along the smooth surface and reaches point A.

14-77. If the spring is compressed 3 in. against the 0.5-lb block and it is released from rest, determine the normal force of the smooth surface on the block when it reaches point $x = 0.5$ ft.

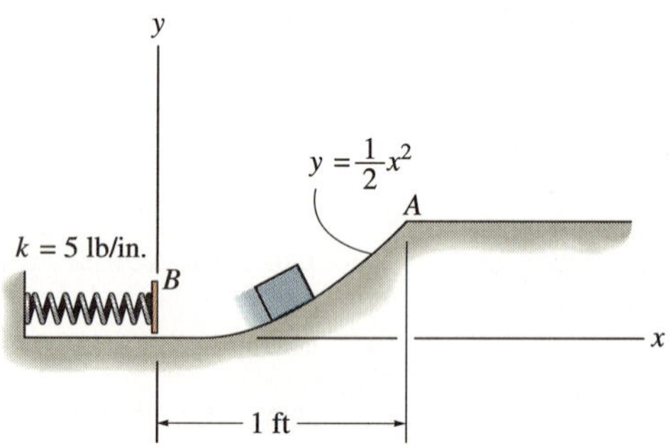

Probs. 14–76/77

14-78. The steel ingot has a mass of 1800 kg. It travels along the conveyor at a speed $v = 0.5$ m/s when it collides with the "nested" spring assembly. If the stiffness of the outer spring is $k_A = 5$ kN/m, determine the required stiffness k_B of the inner spring so that the motion of the ingot is stopped at the moment the front, C, of the ingot is 0.3 m from the wall.

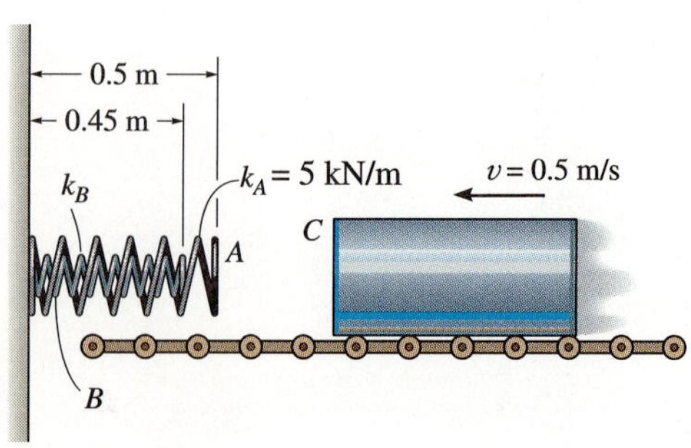

Prob. 14–78

14-79. The 0.75-kg bob of a pendulum is fired from rest at position A by a spring which has a stiffness of $k = 6$ kN/m and is compressed 125 mm. Determine the speed of the bob and the tension in the cord when the bob is at positions B and C. Point B is located on the path where the radius of curvature is still 0.6 m, i.e., just before the cord becomes horizontal.

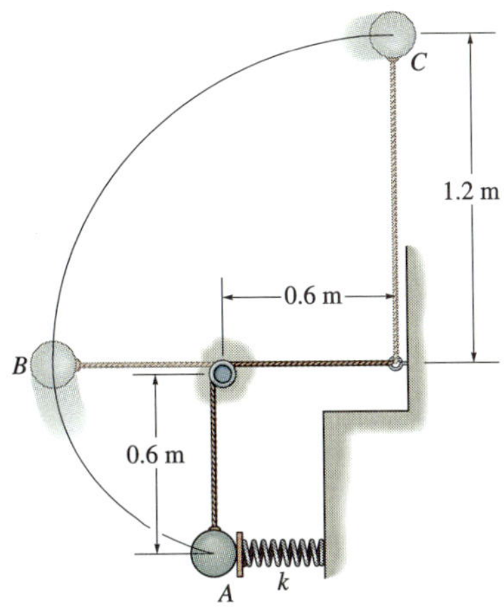

Prob. 14–79

***14-80.** The 0.75-kg bob of a pendulum is fired from rest at position A. If the spring is compressed 50 mm and released, determine (a) its stiffness k so that the speed of the bob is zero when it reaches point B, where the radius of curvature is still 0.6 m, and (b) the stiffness k so that when the bob reaches point C the tension in the cord is zero.

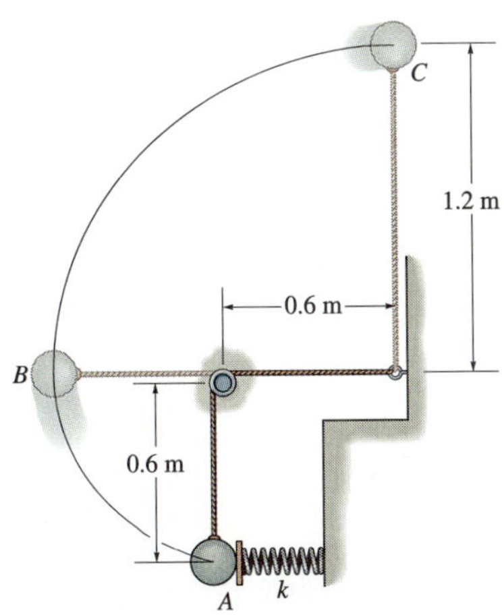

Prob. 14–80

14-81. Tarzan has a mass of 100 kg and from rest swings from the cliff by rigidly holding on to the tree vine, which is 10 m measured from the supporting limb A to his center of mass. Determine his speed just after the vine strikes the lower limb at B. Also, with what force must he hold on to the vine just before and just after the vine contacts the limb at B?

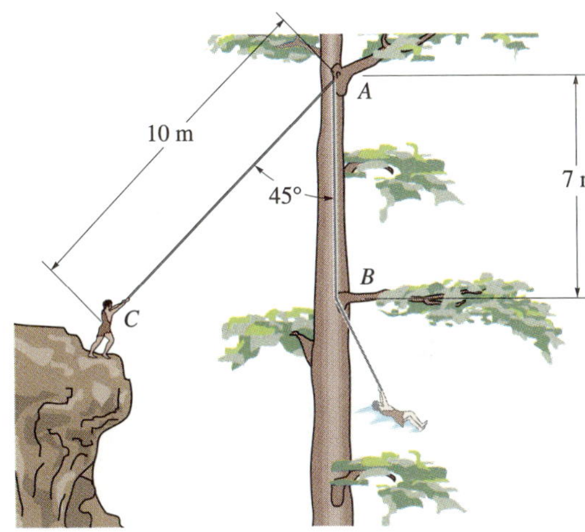

Prob. 14–81

14-82. The car C and its contents have a weight of 600 lb, whereas block B has a weight of 200 lb. If the car is released from rest, determine its speed when it travels 30 ft down the 20° incline. *Suggestion:* To measure the gravitational potential energy, establish separate datums at the initial elevations of B and C.

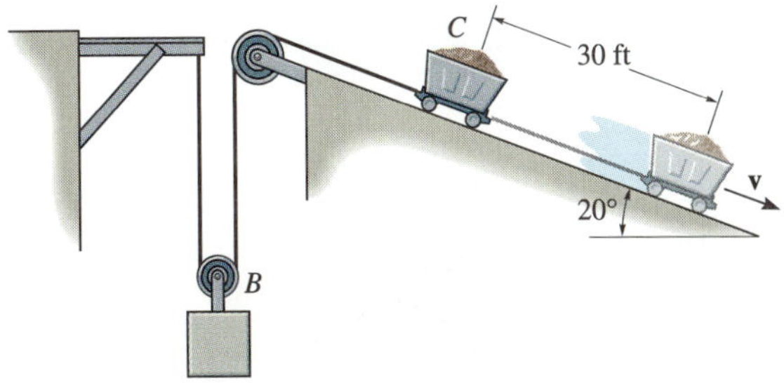

Prob. 14–82

14-83. The roller-coaster car has a speed of 15 ft/s when it is at the crest of a vertical parabolic track. Determine the car's velocity and the normal force it exerts on the track when it reaches point B. Neglect friction and the mass of the wheels. The total weight of the car and the passengers is 350 lb.

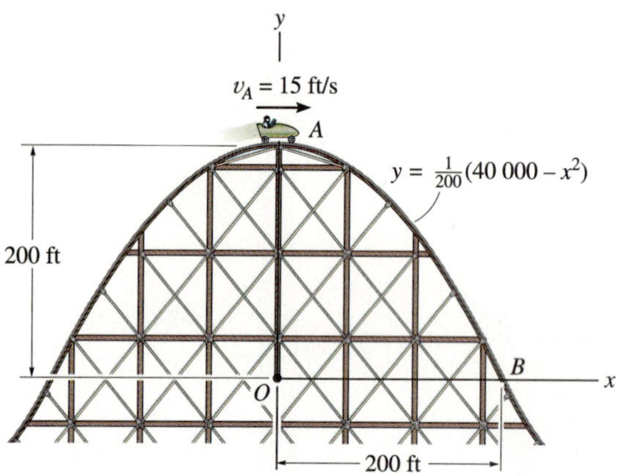

Prob. 14–83

14-85. A block having a mass of 20 kg is attached to four springs. If each spring has a stiffness of $k = 2$ kN/m and an unstretched length of 150 mm, determine the *maximum* downward vertical displacement s_{max} of the block if it is released from rest when $s = 0$.

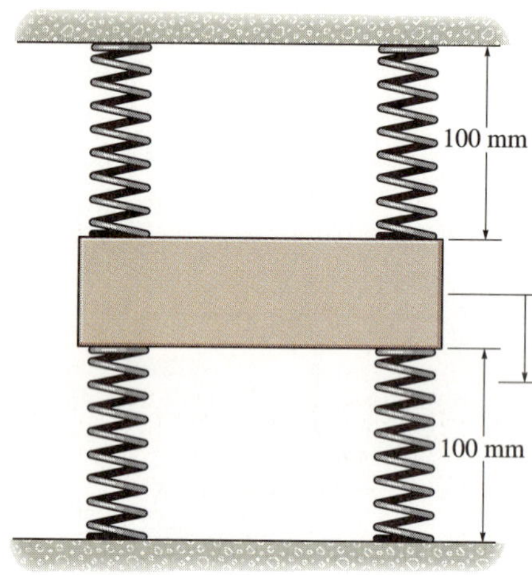

Prob. 14–85

***14-84.** Two equal-length springs having a stiffness $k_A = 300$ N/m and $k_B = 200$ N/m are "nested" together in order to form a shock absorber. If a 2-kg block is dropped from an at-rest position 0.6 m above the top of the springs, determine their deformation when the block momentarily stops.

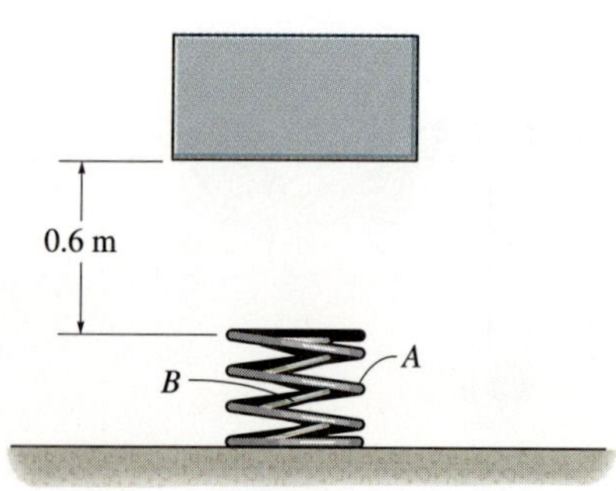

Prob. 14–84

14-86. When the 6-kg box reaches point A it has a speed of $v_A = 2$ m/s. Determine the angle θ at which it leaves the smooth circular ramp and the distance s to where it falls into the cart. Neglect friction.

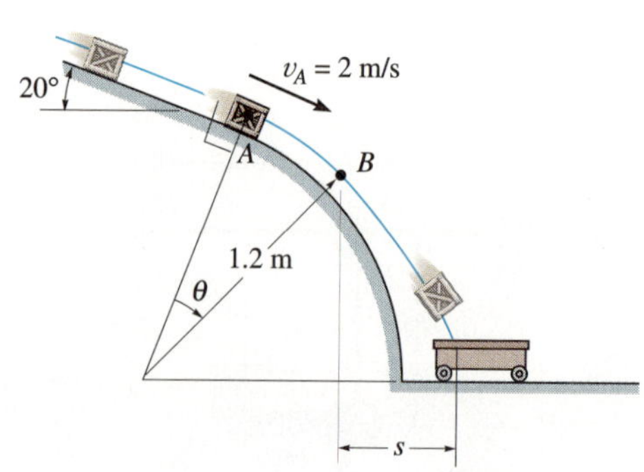

Prob. 14–86

14-87. The 2-lb box has a velocity of 5 ft/s when it begins to slide down the smooth inclined surface at A. Determine the point C (x, y) where it strikes the lower incline.

***14-88.** The 2-lb box has a velocity of 5 ft/s when it begins to slide down the smooth inclined surface at A. Determine its speed just before hitting the surface at C and the time to travel from A to C. The coordinates of point C are $x = 17.66$ ft, and $y = 8.832$ ft.

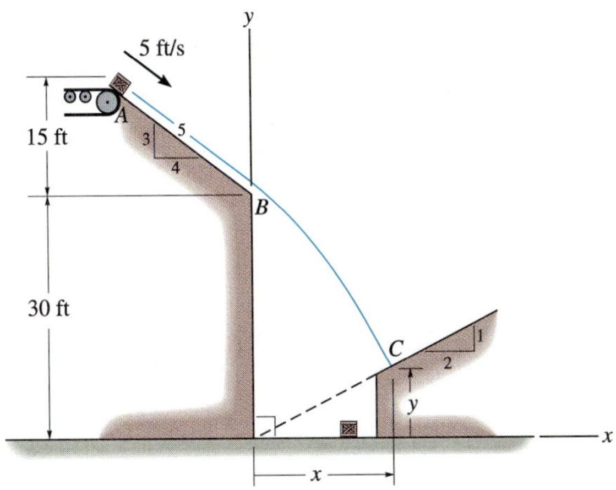

Probs. 14–87/88

14-89. The 2-kg ball of negligible size is fired from point A with an initial velocity of 10 m/s up the smooth inclined plane. Determine the distance from point C to where it hits the horizontal surface at D. Also, what is its velocity when it strikes the surface?

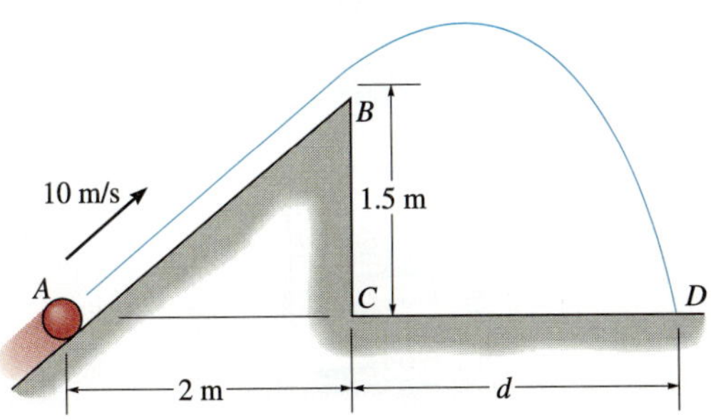

Prob. 14–89

14-90. The ride at an amusement park consists of a gondola which is lifted to a height of 120 ft at A. If it is released from rest and falls along the parabolic track, determine the speed at the instant $y = 20$ ft. Also determine the normal reaction of the tracks on the gondola at this instant. The gondola and passenger have a total weight of 500 lb. Neglect the effects of friction.

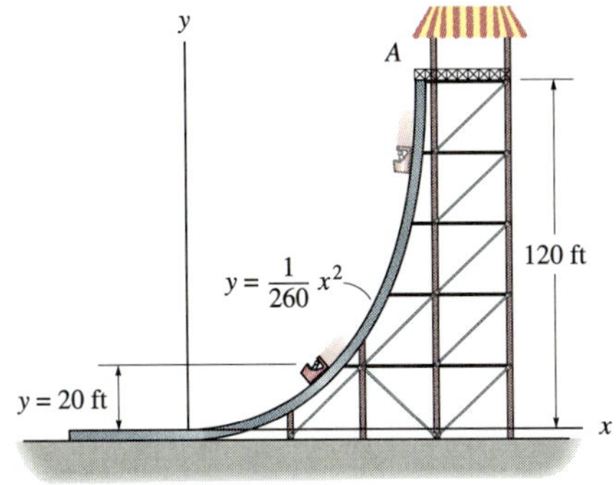

Prob. 14–90

14-91. The Raptor is an outside loop roller coaster in which riders are belted into seats resembling ski-lift chairs. Determine the minimum speed v_0 at which the cars should coast down from the top of the hill, so that passengers can just make the loop without leaving contact with their seats. Neglect friction, the size of the car and passenger, and assume each passenger and car has a mass m.

***14-92.** The Raptor is an outside loop roller coaster in which riders are belted into seats resembling ski-lift chairs. If the cars travel at $v_0 = 4$ m/s when they are at the top of the hill, determine their speed when they are at the top of the loop and the reaction of the 70-kg passenger on his seat at this instant. The car has a mass of 50 kg. Take $h = 12$ m, $\rho = 5$ m. Neglect friction and the size of the car and passenger.

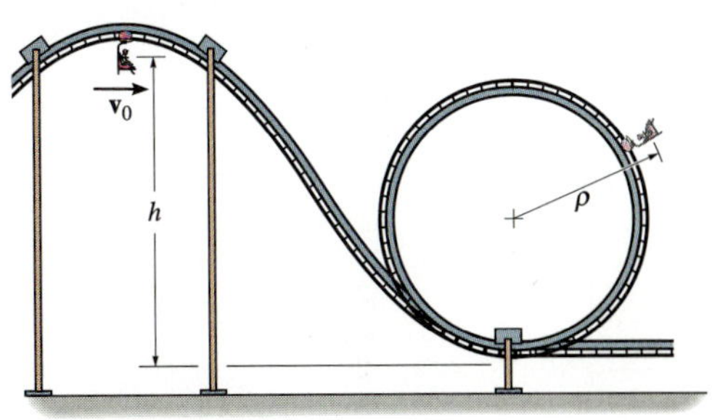

Probs. 14–91/92

14-93. The 2-lb collar has a speed of 5 ft/s at A. The attached spring has an unstretched length of 2 ft and a stiffness of $k = 10$ lb/ft. If the collar moves over the smooth rod, determine its speed when it reaches point B, the normal force of the rod on the collar, and the rate of decrease in its speed.

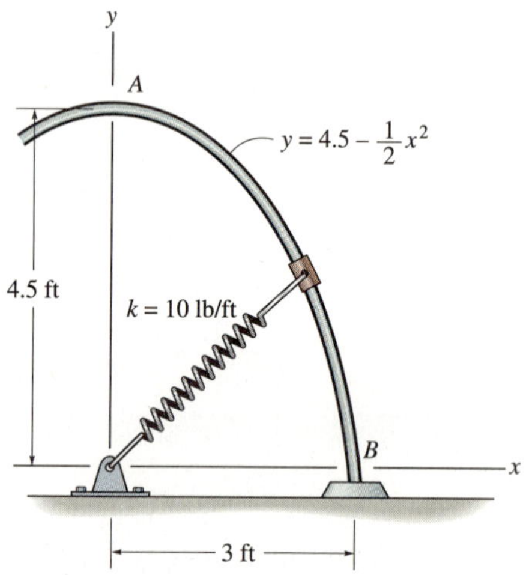

Prob. 14–93

14-94. The 5-lb collar rides on the smooth rod and is attached to the spring that has an unstretched length of 4 ft and a stiffness of $k = 10$ lb/ft. If the collar has a velocity of 5 ft/s when it is at A, determine the speed of the collar when it reaches point B. Also, what is the normal force on the collar when it is at point B?

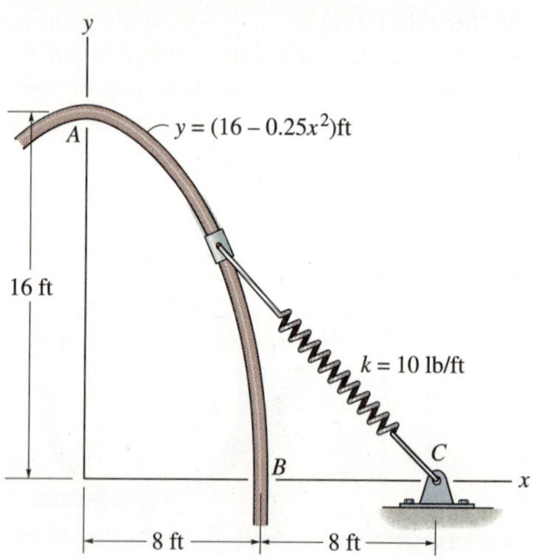

Prob. 14–94

14-95. A tank car is stopped by two spring bumpers A and B, having a stiffness of $k_A = 15(10^3)$ lb/ft and $k_B = 20(10^3)$ lb/ft, respectively. Bumper A is attached to the car, whereas bumper B is attached to the wall. If the car has a weight of $25(10^3)$ lb and is freely coasting at 3 ft/s, determine the maximum deflection of each spring at the instant the bumpers stop the car.

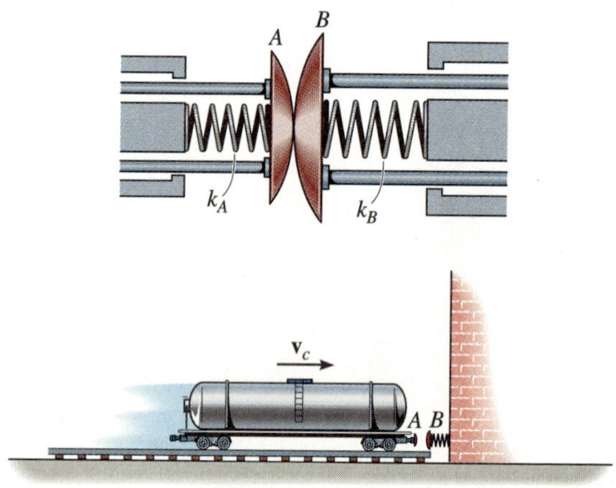

Prob. 14–95

***14-96.** The block has a mass of 20 kg and is released from rest when $s = 0.5$ m. If the mass of the bumpers A and B can be neglected, determine the maximum deformation of each spring due to the collision.

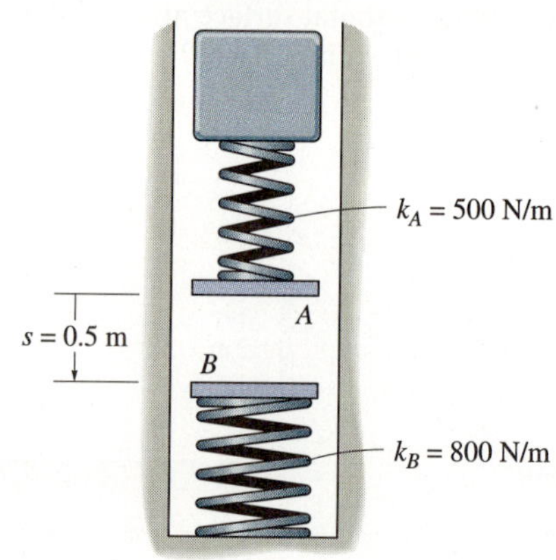

Prob. 14–96

14-97. If the mass of the earth is M_e, show that the gravitational potential energy of a body of mass m located a distance r from the center of the earth is $V_g = -GM_e m/r$. Recall that the gravitational force acting between the earth and the body is $F = G(M_e m/r^2)$, Eq. 13–1. For the calculation, locate the datum an "infinite" distance from the earth. Also, prove that **F** is a conservative force.

14-98. A 60-kg satellite is traveling in free flight along an elliptical orbit such that at A, where $r_A = 20$ Mm, it has a speed $v_A = 40$ Mm/h. What is the speed of the satellite when it reaches point B, where $r_B = 80$ Mm? *Hint:* See Prob. 14-97, where $M_e = 5.976(10^{24})$ kg and $G = 66.73(10^{-12})\ \text{m}^3/(\text{kg}\cdot\text{s}^2)$.

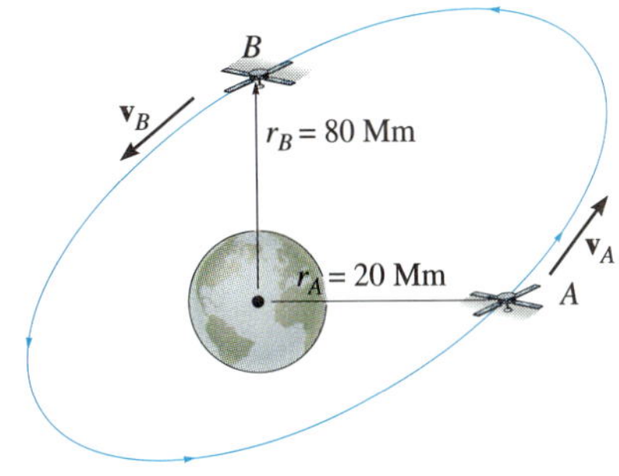

Prob. 14–98

Design Projects

14–1D. DESIGN OF A CAR BUMPER

The body of an automobile is to be protected by a spring-loaded bumper, which is attached to the automobile's frame. Design the bumper so that it will stop a 3500-lb car traveling freely at 5 mi/h and not deform the springs more than 3 in. Submit a sketch of your design showing the placement of the springs and their stiffness. Plot the load-deflection diagram for the bumper during a direct collision with a rigid wall, and also plot the deceleration of the car as a function of the springs' displacement.

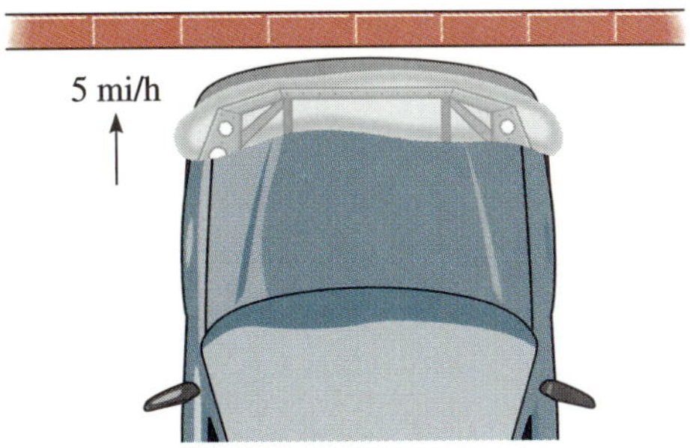

Prob. 14–1D

14–2D. DESIGN OF AN ELEVATOR HOIST

It is required that an elevator and its contents, having a maximum weight of 500 lb, be lifted $y = 20$ ft, starting from rest and then stopping after 6 seconds. A single motor and cable-winding drum can be mounted anywhere and used for the operation. During any lift or descent the acceleration should not exceed 10 ft/s^2. Design a cable-and-pulley system for the elevator, and estimate the material cost if the cable is \$1.30/ft and pulleys are \$3.50 each. Submit a drawing of your design, and include plots of the power output required of the motor and the elevator's speed versus the height y traveled.

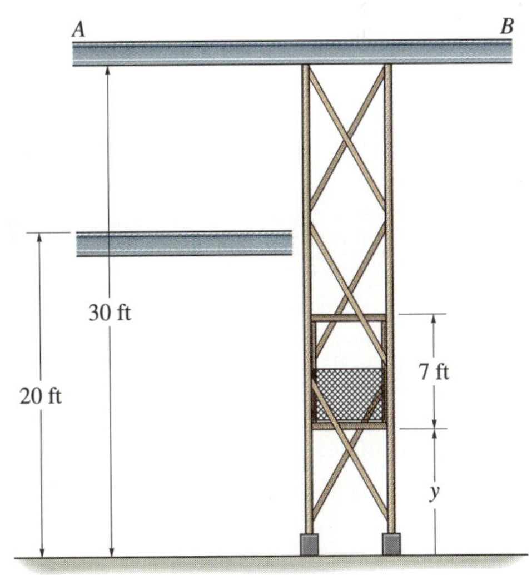

Prob. 14–2D

The velocities of the vehicles involved in this accident can be estimated using the principles of impulse and momentum.

CHAPTER 15

Kinetics of a Particle: Impulse and Momentum

Chapter Objectives

- To develop the principle of linear impulse and momentum for a particle.
- To study the conservation of linear momentum for particles.
- To analyze the mechanics of impact.
- To introduce the concept of angular impulse and momentum.
- To solve problems involving steady fluid streams and propulsion with variable mass.

15.1 Principle of Linear Impulse and Momentum

In this section we will integrate the equation of motion with respect to time and thereby obtain the principle of impulse and momentum. It will then be shown that the resulting equation will be useful for solving problems involving force, velocity, and time.

The equation of motion for a particle of mass m can be written as

$$\Sigma \mathbf{F} = m\mathbf{a} = m\frac{d\mathbf{v}}{dt} \qquad (15\text{–}1)$$

where $\mathbf{a}$ and $\mathbf{v}$ are both measured from an inertial frame of reference. Rearranging the terms and integrating between the limits $\mathbf{v} = \mathbf{v}_1$ at $t = t_1$ and $\mathbf{v} = \mathbf{v}_2$ at $t = t_2$, we have

$$\Sigma \int_{t_1}^{t_2} \mathbf{F}\, dt = m\int_{\mathbf{v}_1}^{\mathbf{v}_2} d\mathbf{v}$$

or

$$\Sigma \int_{t_1}^{t_2} \mathbf{F}\, dt = m\mathbf{v}_2 - m\mathbf{v}_1 \qquad (15\text{–}2)$$

The impulse tool is used to remove the dent in the automobile fender. To do so its end is first screwed into a hole drilled in the fender, then the weight is gripped and jerked upwards, striking the stop ring. The impulse developed is transferred along the shaft of the tool and pulls suddenly on the dent.

This equation is referred to as the *principle of linear impulse and momentum*. From the derivation it can be seen that it is simply a time integration of the equation of motion. It provides a *direct means* of obtaining the particle's final velocity $\mathbf{v}_2$ after a specified time period when the particle's initial velocity is known and the forces acting on the particle are either constant or can be expressed as functions of time. By comparison, if $\mathbf{v}_2$ was determined using the equation of motion, a two-step process would be necessary; i.e., apply $\Sigma\mathbf{F} = m\mathbf{a}$ to obtain **a**, then integrate $\mathbf{a} = d\mathbf{v}/dt$ to obtain $\mathbf{v}_2$.

Linear Momentum. Each of the two vectors of the form $\mathbf{L} = m\mathbf{v}$ in Eq. 15–2 is referred to as the particle's *linear momentum*. Since m is a positive scalar, the linear-momentum vector has the same direction as **v**, and its magnitude mv has units of mass–velocity, e.g., kg·m/s, or slug·ft/s.

Linear Impulse. The integral $\mathbf{I} = \int \mathbf{F}\, dt$ in Eq. 15–2 is referred to as the *linear impulse*. This term is a vector quantity which measures the effect of a force during the time the force acts. Since time is a positive scalar, the impulse acts in the same direction as the force, and its magnitude has units of force–time, e.g., N·s or lb·s.* If the force is expressed as a function of time, the impulse may be determined by direct evaluation of the integral. In particular, the magnitude of the impulse $\mathbf{I} = \int_{t_1}^{t_2} \mathbf{F}\, dt$ can be represented experimentally by the shaded area under the curve of force versus time, Fig. 15–1. When the force is constant in both magnitude and direction, the resulting impulse becomes $\mathbf{I} = \int_{t_1}^{t_2} \mathbf{F}_c\, dt = \mathbf{F}_c(t_2 - t_1)$, which represents the shaded rectangular area shown in Fig. 15–2.

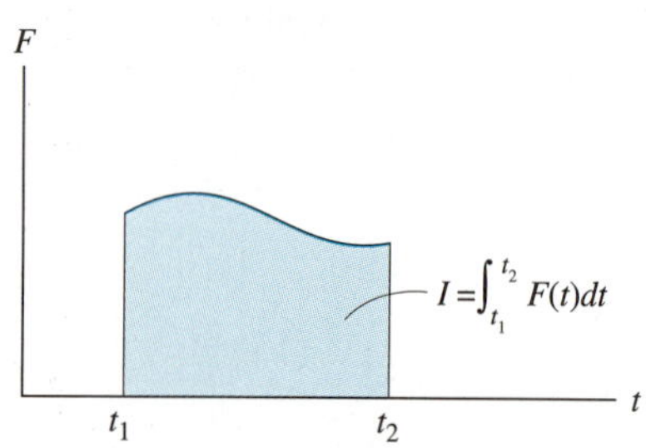

Variable Force

Fig. 15–1

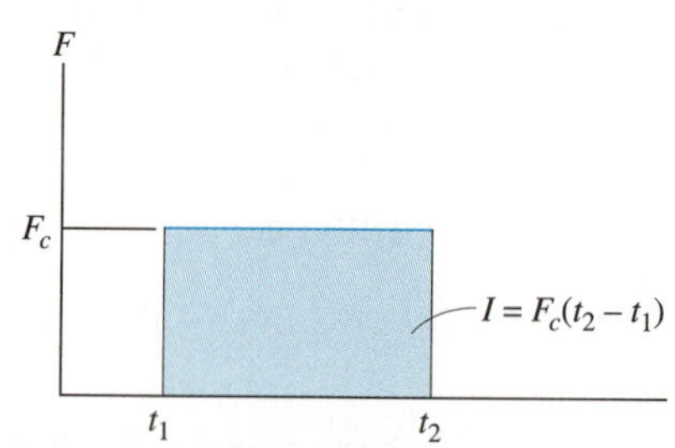

Constant Force

Fig. 15–2

*Although the units for impulse and momentum are defined differently, one can show that Eq. 15–2 is dimensionally homogeneous.

Principle of Linear Impulse and Momentum. For problem solving, Eq. 15–2 will be rewritten in the form

$$m\mathbf{v}_1 + \Sigma\int_{t_1}^{t_2} \mathbf{F}\,dt = m\mathbf{v}_2 \qquad (15\text{–}3)$$

which states that the initial momentum of the particle at t_1 plus the sum of all the impulses applied to the particle from t_1 to t_2 is equivalent to the final momentum of the particle at t_2. These three terms are illustrated graphically on the *impulse and momentum diagrams* shown in Fig. 15–3. The two *momentum diagrams* are simply outlined shapes of the particle which indicate the direction and magnitude of the particle's initial and final momenta, $m\mathbf{v}_1$ and $m\mathbf{v}_2$, respectively, Fig. 15–3. Similar to the free-body diagram, the *impulse diagram* is an outlined shape of the particle showing all the impulses that act on the particle when it is located at some intermediate point along its path. In general, whenever the magnitude or direction of a force *varies* with time, the impulse is represented on the impulse diagram as $\int_{t_1}^{t_2} \mathbf{F}\,dt$. If the force is *constant,* the impulse applied to the particle is $\mathbf{F}_c(t_2 - t_1)$, and it acts in the same direction as $\mathbf{F}_c$.

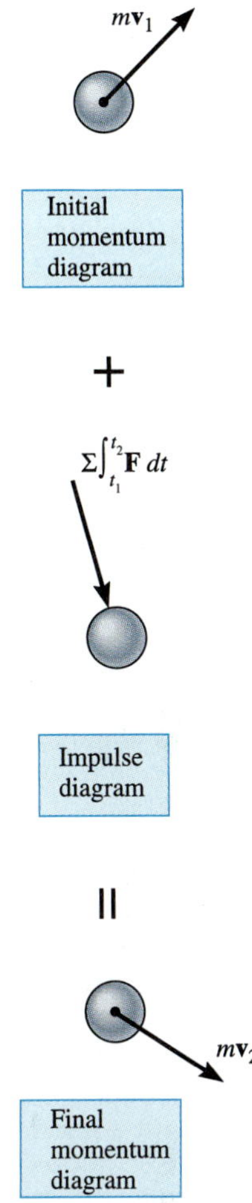

Fig. 15–3

Scalar Equations. If each of the vectors in Eq. 15–3 is resolved into its x, y, z components, we can write symbolically the following three scalar equations:

$$\begin{aligned} m(v_x)_1 + \Sigma\int_{t_1}^{t_2} F_x\,dt &= m(v_x)_2 \\ m(v_y)_1 + \Sigma\int_{t_1}^{t_2} F_y\,dt &= m(v_y)_2 \\ m(v_z)_1 + \Sigma\int_{t_1}^{t_2} F_z\,dt &= m(v_z)_2 \end{aligned} \qquad (15\text{–}4)$$

These equations represent the principle of linear impulse and momentum for the particle in the x, y, z directions, respectively.

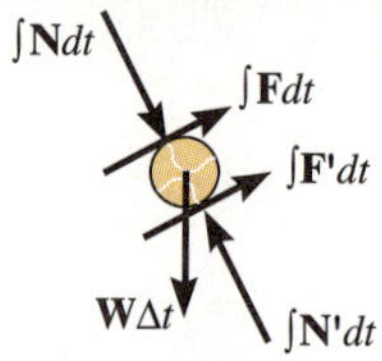

As the wheels of the pitching machine rotate, they apply frictional impulses to the ball, thereby giving it a linear momentum. These impulses are shown on the impulse diagram. Here both the frictional and normal impulses vary with time. By comparison, the weight impulse is constant and is very small since the time Δt the ball is in contact with the wheels is very small.

Procedure for Analysis

The principle of linear impulse and momentum is used to solve problems involving *force, time,* and *velocity,* since these terms are involved in the formulation. For application it is suggested that the following procedure be used.

Free-Body Diagram

- Establish the x, y, z inertial frame of reference and draw the particle's free-body diagram in order to account for all the forces that produce impulses on the particle.
- The direction and sense of the particle's initial and final velocities should be established.
- If a vector is unknown, assume that the sense of its components is in the direction of the positive inertial coordinate(s).
- As an alternative procedure, draw the impulse and momentum diagrams for the particle as discussed in reference to Fig. 15–3.*

Principle of Impulse and Momentum

- In accordance with the established coordinate system apply the principle of linear impulse and momentum, $m\mathbf{v}_1 + \Sigma \int_{t_1}^{t_2} \mathbf{F}\, dt = m\mathbf{v}_2$. If motion occurs in the x–y plane, the two scalar component equations can be formulated by either resolving the vector components of **F** from the free-body diagram, or by using the data on the impulse and momentum diagrams.
- Realize that all the forces acting on the particle's free-body diagram will create an impulse, even though some of these forces will do no work.
- Forces that are functions of time must be integrated to obtain the impulse. Graphically, the impulse is equal to the area under the force–time curve.
- If the problem involves the dependent motion of several particles, use the method outlined in Sec. 12.9 to relate their velocities. Make sure the positive coordinate directions used for writing these kinematic equations are the *same* as those used for writing the equations of impulse and momentum.

*This procedure will be followed when developing the proofs and theory in the text.

EXAMPLE 15-1

The 100-kg stone shown in Fig. 15–4*a* is originally at rest on the smooth horizontal surface. If a towing force of 200 N, acting at an angle of 45°, is applied to the stone for 10 s, determine the final velocity and the normal force which the surface exerts on the stone during the time interval.

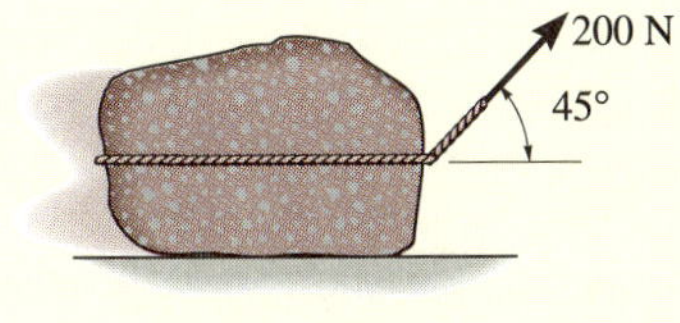

(a)

Solution

This problem can be solved using the principle of impulse and momentum since it involves force, velocity, and time.

Free-Body Diagram. See Fig. 15–4*b*. Since all the forces acting are *constant,* the impulses are simply the product of the force magnitude and 10 s $[\mathbf{I} = \mathbf{F}_c(t_2 - t_1)]$. Note the alternative procedure of drawing the stone's impulse and momentum diagrams, Fig. 15–4*c*.

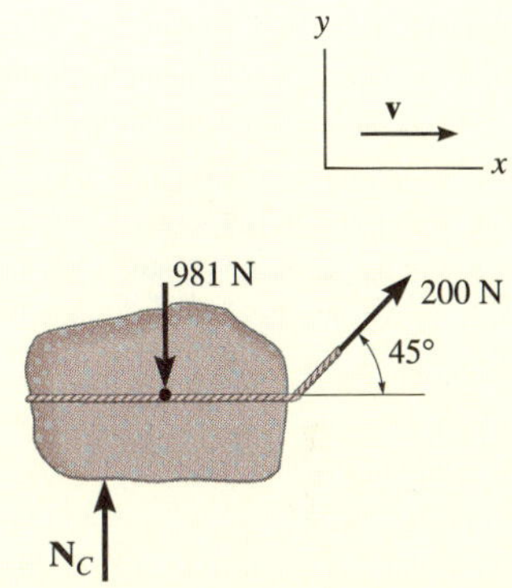

(b)

Principle of Impulse and Momentum. Resolving the vectors in Fig. 15–4*b* along the *x*, *y* axes and applying Eqs. 15–4 yields

$$(\overset{+}{\rightarrow}) \qquad m(v_x)_1 + \Sigma\int_{t_1}^{t_2} F_x\,dt = m(v_x)_2$$

$$0 + 200\text{ N}(10\text{ s})\cos 45^\circ = (100\text{ kg})v_2$$

$$v_2 = 14.1\text{ m/s} \qquad \textit{Ans.}$$

$$(+\uparrow) \qquad m(v_y)_1 + \Sigma\int_{t_1}^{t_2} F_y\,dt = m(v_y)_2$$

$$0 + N_C(10\text{ s}) - 981\text{ N}(10\text{ s}) + 200\text{ N}(10\text{ s})\sin 45^\circ = 0$$

$$N_C = 840\text{ N} \qquad \textit{Ans.}$$

Since no motion occurs in the *y* direction, direct application of the equilibrium equation $\Sigma F_y = 0$ gives the same result for N_C.

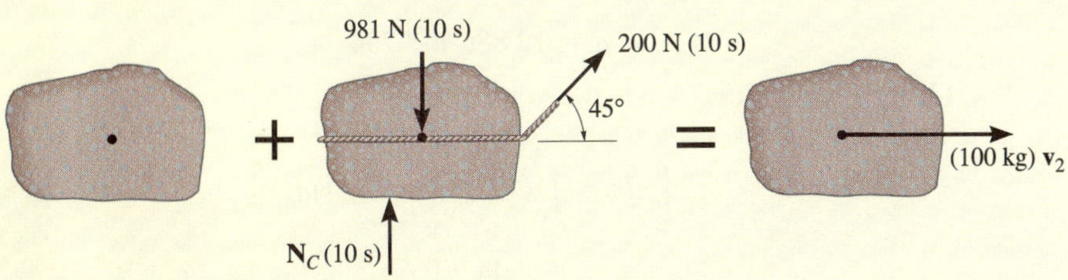

(c)

Fig. 15–4

EXAMPLE 15-2

The 50-lb crate shown in Fig. 15–5*a* is acted upon by a force having a variable magnitude $P = (20t)$ lb, where t is in seconds. Determine the crate's velocity 2 s after **P** has been applied. The initial velocity is $v_1 = 3$ ft/s down the plane, and the coefficient of kinetic friction between the crate and the plane is $\mu_k = 0.3$.

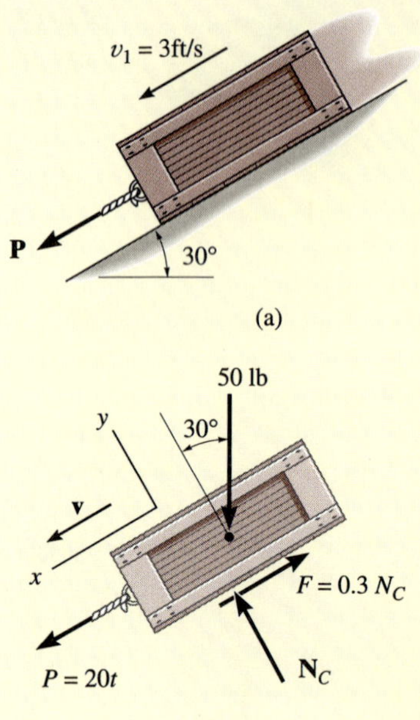

Fig. 15–5

Solution

Free-Body Diagram. See Fig. 15–5*b*. Since the magnitude of force $P = 20t$ varies with time, the impulse it creates must be determined by integrating over the 2-s time interval. The weight, normal force, and frictional force (which acts opposite to the direction of motion) are all *constant*, so that the impulse created by each of these forces is simply the magnitude of the force times 2 s.

Principle of Impulse and Momentum. Applying Eqs. 15–4 in the x direction, we have

$$(+\swarrow) \qquad m(v_x)_1 + \Sigma\int_{t_1}^{t_2} F_x\,dt = m(v_x)_2$$

$$\frac{50\text{ lb}}{32.2\text{ ft/s}^2}(3\text{ ft/s}) + \int_0^2 20t\,dt - 0.3N_C(2\text{ s}) + (50\text{ lb})(2\text{ s})\sin 30^\circ = \frac{50\text{ lb}}{32.2\text{ ft/s}^2}v_2$$

$$4.66 + 40 - 0.6N_C + 50 = 1.55v_2$$

The equation of equilibrium can be applied in the y direction. Why?

$$+\nwarrow\Sigma F_y = 0; \qquad N_C - 50\cos 30^\circ\text{ lb} = 0$$

Solving,

$$N_C = 43.3\text{ lb}$$

$$v_2 = 44.2\text{ ft/s} \swarrow \qquad \textit{Ans.}$$

Note: We can also solve this problem using the equation of motion. From Fig. 15–5*b*,

$$+\swarrow\Sigma F_x = ma_x; \quad 20t - 0.3(43.3) + 50\sin 30^\circ = \frac{50}{32.2}a$$

$$a = 12.88t + 7.734$$

Using kinematics

$$+\swarrow dv = a\,dt; \qquad \int_3^v dv = \int_0^2 (12.88t + 7.734)dt$$

$$v = 44.2\text{ ft/s} \qquad \textit{Ans.}$$

By comparison, application of the principle of impulse and momentum eliminates the need for using kinematics ($a = dv/dt$) and thereby yields an easier method for solution.

EXAMPLE 15-3

Blocks A and B shown in Fig. 15–6a have a mass of 3 kg and 5 kg, respectively. If the system is released from rest, determine the velocity of block B in 6 s. Neglect the mass of the pulleys and cord.

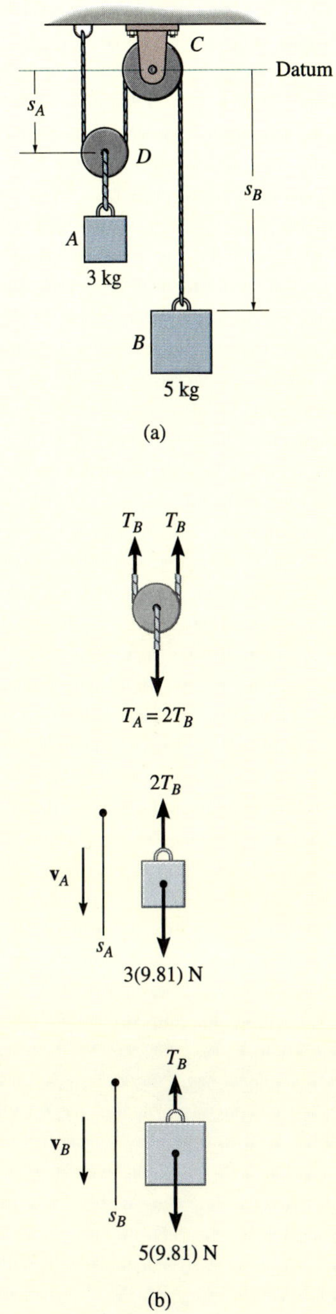

Fig. 15–6

Solution

Free-Body Diagram. See Fig. 15–6b. Since the weight of each block is constant, the cord tensions will also be constant. Furthermore, since the mass of pulley D is neglected, the cord tension $T_A = 2T_B$. Note that the blocks are both assumed to be traveling downward in the positive coordinate directions, s_A and s_B.

Principle of Impulse and Momentum

Block A:

$$(+\downarrow) \qquad m(v_A)_1 + \Sigma\int_{t_1}^{t_2} F_y\,dt = m(v_A)_2$$

$$0 - 2T_B(6\text{ s}) + 3(9.81)\text{ N}(6\text{ s}) = (3\text{ kg})(v_A)_2 \qquad (1)$$

Block B:

$$(+\downarrow) \qquad m(v_B)_1 + \Sigma\int_{t_1}^{t_2} F_y\,dt = m(v_B)_2$$

$$0 + 5(9.81)\text{ N}(6\text{ s}) - T_B(6\text{ s}) = (5\text{ kg})(v_B)_2 \qquad (2)$$

Kinematics. Since the blocks are subjected to dependent motion, the velocity of A may be related to that of B by using the kinematic analysis discussed in Sec. 12.9. A horizontal datum is established through the fixed point at C, Fig. 15–6a, and the position coordinates, s_A and s_B, are related to the constant total length l of the vertical segments of the cord by the equation

$$2s_A + s_B = l$$

Taking the time derivative yields

$$2v_A = -v_B \qquad (3)$$

As indicated by the negative sign, when B moves downward A moves upward.* Substituting this result into Eq. 1 and solving Eqs. 1 and 2 yields

$$(v_B)_2 = 35.8\text{ m/s}\downarrow \qquad \textit{Ans.}$$

$$T_B = 19.2\text{ N}$$

*Realize that the *positive* (downward) direction for $\mathbf{v}_A$ and $\mathbf{v}_B$ is *consistent* in Figs. 15–6a and 15–6b and in Eqs. 1 to 3. Why is this important?

15.2 Principle of Linear Impulse and Momentum for a System of Particles

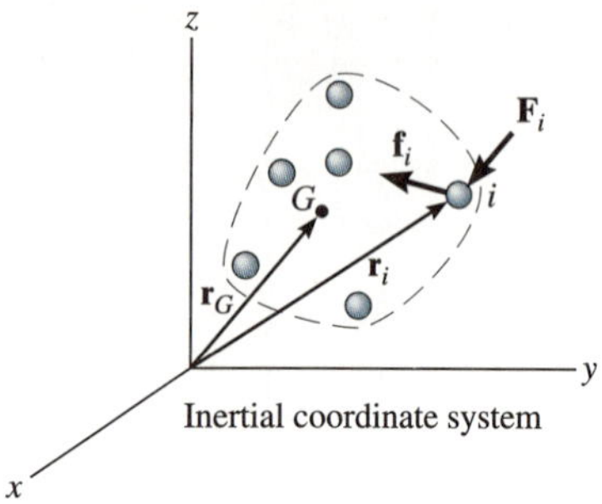

Fig. 15–7

The principle of linear impulse and momentum for a system of particles moving relative to an inertial reference, Fig. 15–7, is obtained from the equation of motion applied to all the particles in the system, i.e.,

$$\Sigma \mathbf{F}_i = \Sigma m_i \frac{d\mathbf{v}_i}{dt} \tag{15–5}$$

The term on the left side represents only the sum of the *external forces* acting on the system of particles. Recall that the internal forces $\mathbf{f}_i$ acting between particles do not appear with this summation, since by Newton's third law they occur in equal but opposite collinear pairs and therefore cancel out. Multiplying both sides of Eq. 15–5 by dt and integrating between the limits $t = t_1$, $\mathbf{v}_i = (\mathbf{v}_i)_1$ and $t = t_2$, $\mathbf{v}_i = (\mathbf{v}_i)_2$ yields

$$\Sigma m_i(\mathbf{v}_i)_1 + \Sigma \int_{t_1}^{t_2} \mathbf{F}_i \, dt = \Sigma m_i(\mathbf{v}_i)_2 \tag{15–6}$$

This equation states that the initial linear momenta of the system plus the impulses of all the *external forces* acting on the system from t_1 to t_2 are equal to the system's final linear momenta.

Since the location of the mass center G of the system is determined from $m\mathbf{r}_G = \Sigma m_i \mathbf{r}_i$, where $m = \Sigma m_i$ is the total mass of all the particles, Fig. 15–7, then taking the time derivatives, we have

$$m\mathbf{v}_G = \Sigma m_i \mathbf{v}_i$$

which states that the total linear momentum of the system of particles is equivalent to the linear momentum of a "fictitious" aggregate particle of mass $m = \Sigma m_i$ moving with the velocity of the mass center of the system. Substituting into Eq. 15–6 yields

$$m(\mathbf{v}_G)_1 + \Sigma \int_{t_1}^{t_2} \mathbf{F} \, dt = m(\mathbf{v}_G)_2 \tag{15–7}$$

Here the initial linear momentum of the aggregate particle plus the external impulses acting on the system of particles from t_1 to t_2 is equal to the aggregate particle's final linear momentum. Since in reality all particles must have finite size to possess mass, the above equation justifies application of the principle of linear impulse and momentum to a rigid body represented as a single particle.

Problems

15-1. A 20-lb block slides down a 30° inclined plane with an initial velocity of 2 ft/s. Determine the velocity of the block in 3 s if the coefficient of kinetic friction between the block and the plane is $\mu_k = 0.25$.

15-2. The baseball has a horizontal speed of 35 m/s when it is struck by the bat B. If it then travels away at an angle of 60° from the horizontal and reaches a maximum height of 50 m, measured from the height of the bat, determine the magnitude of the net impulse of the bat on the ball. The ball has a mass of 400 g. Neglect the weight of the ball during the time the bat strikes the ball.

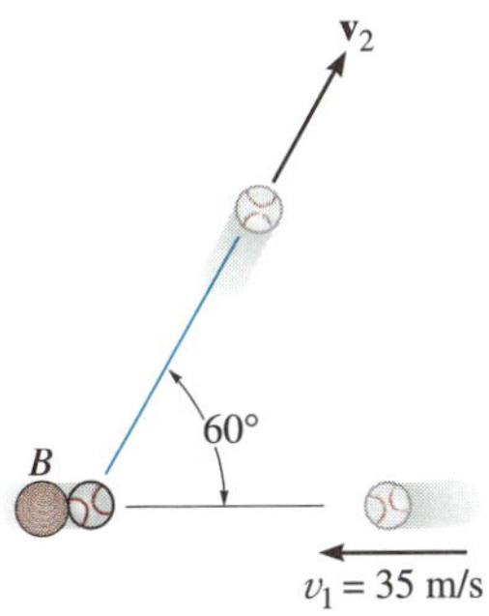

Prob. 15–2

15-3. A 5-lb block is given an initial velocity of 10 ft/s up a 45° smooth slope. Determine the time it will take to travel up the slope before it stops.

***15-4.** The 12-Mg "jump jet" is capable of taking off vertically from the deck of a ship. If its jets exert a constant vertical force of 150 kN on the plane, determine its velocity and how high it goes in $t = 6$ s, starting from rest. Neglect the loss of fuel during the lift.

Prob. 15–4

15-5. The choice of a seating material for moving vehicles depends upon its ability to resist shock and vibration. From the data shown in the graphs, determine the impulses created by a falling weight onto a sample of urethane foam and CONFOR foam.

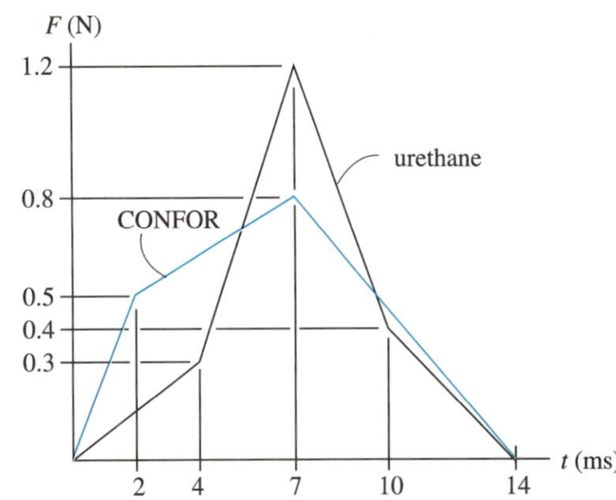

Prob. 15–5

15-6. The 28-Mg bulldozer is originally at rest. Determine its speed when $t = 4$ s if the horizontal traction $\mathbf{F}$ varies with time as shown in the graph.

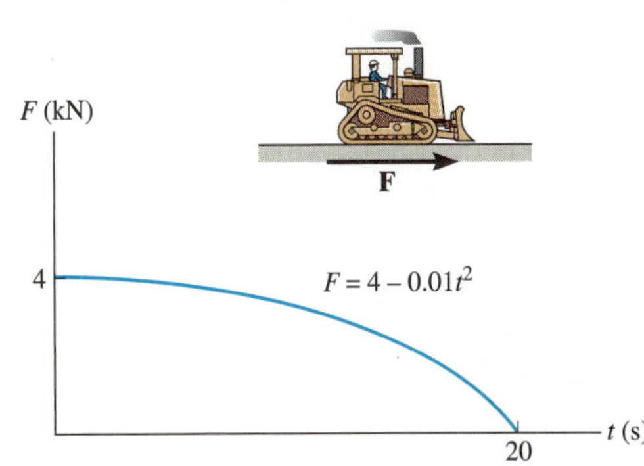

Prob. 15–6

15-7. A hammer head H having a weight of 0.25 lb is moving vertically downward at 40 ft/s when it strikes the head of a nail of negligible mass and drives it into a block of wood. Find the impulse on the nail if it is assumed that the grip at A is loose, the handle has a negligible mass, and the hammer stays in contact with the nail while it comes to rest. Neglect the impulse caused by the weight of the hammer head during contact with the nail.

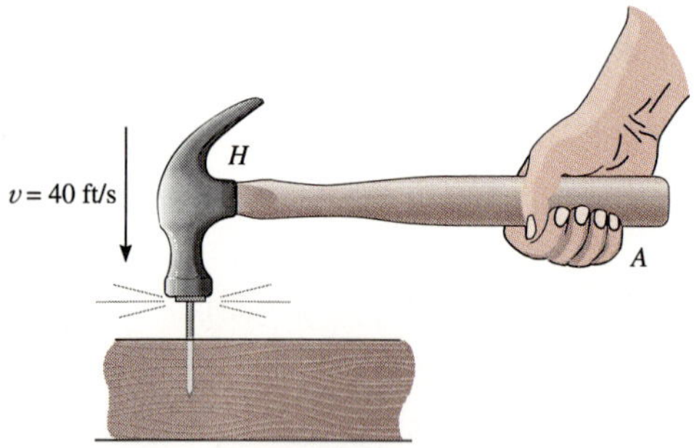

Prob. 15–7

***15-8.** A solid-fueled rocket can be made using a fuel grain with either a hole (a), or starred cavity (b), in the cross section. From experiment the engine thrust-time curves (T vs. t) for the same amount of propellant using these geometries are shown. Determine the total impulse in both cases.

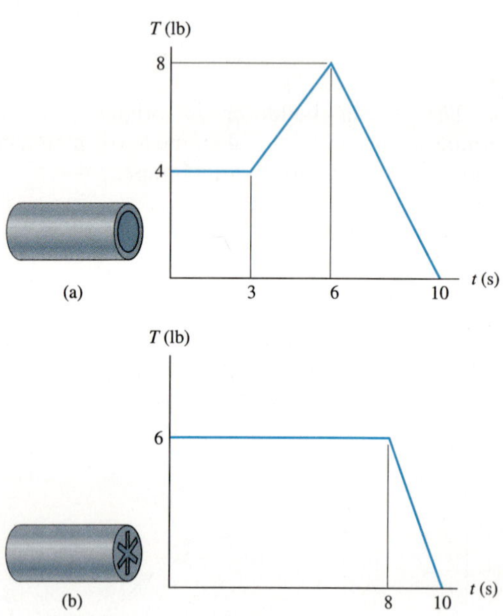

Prob. 15–8

15-9. The jet plane has a mass of 250 Mg and a horizontal velocity of 100 m/s when $t = 0$. If *both* engines provide a horizontal thrust which varies as shown in the graph, determine the plane's velocity in $t = 15$ s. Neglect air resistance and the loss of fuel during the motion.

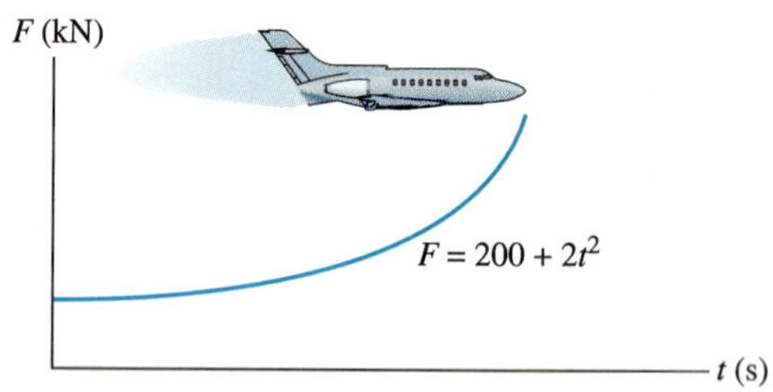

Prob. 15–9

15-10. The particle P is acted upon by its weight of 3 lb and forces $\mathbf{F}_1$ and $\mathbf{F}_2$, where t is in seconds. If the particle originally has a velocity of $\mathbf{v}_1 = \{3\mathbf{i} + 1\mathbf{j} + 6\mathbf{k}\}$ ft/s, determine its speed after 2 s.

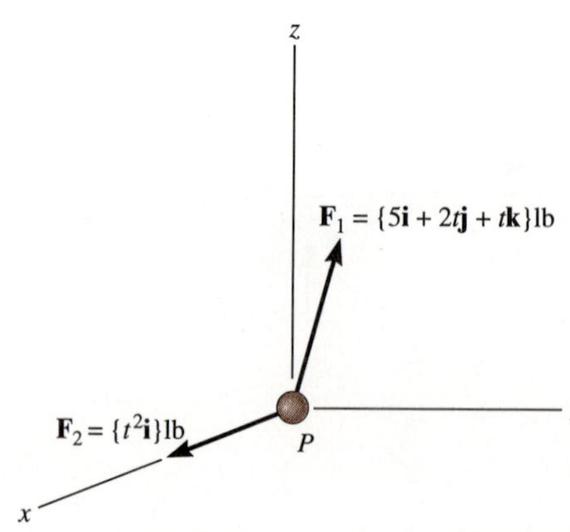

Prob. 15–10

15-11. The 20-lb cabinet is subjected to the force $F = (3 + 2t)$ lb, where t is in seconds. If the cabinet is initially moving down the plane with a speed of 6 ft/s, determine how long it will take the force to bring the cabinet to rest. **F** always acts parallel to the plane.

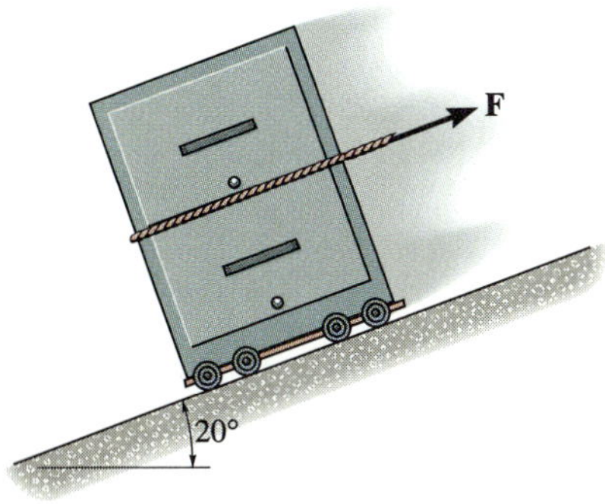

Prob. 15–11

***15-12.** The fuel-element assembly of a nuclear reactor has a weight of 600 lb. Suspended in a vertical position and initially at rest, it is given an upward speed of 5 ft/s in 0.3 s using a crane hook H. Determine the average tension in cables AB and AC during this time interval.

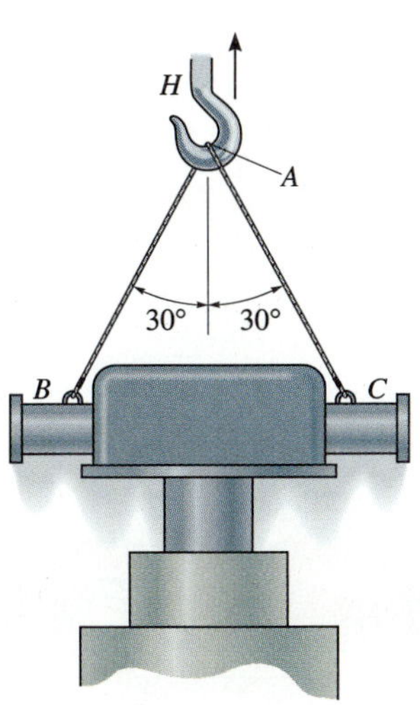

Prob. 15–12

15-13. From experiments, the time variation of the vertical force on a runner's foot as he strikes and pushes off the ground is shown in the graph. These results are reported for a 1-lb *static* load, i.e., in terms of unit weight. If a runner weighs 175 lb, determine the approximate vertical impulse he exerts on the ground if the impulse occurs in 210 ms.

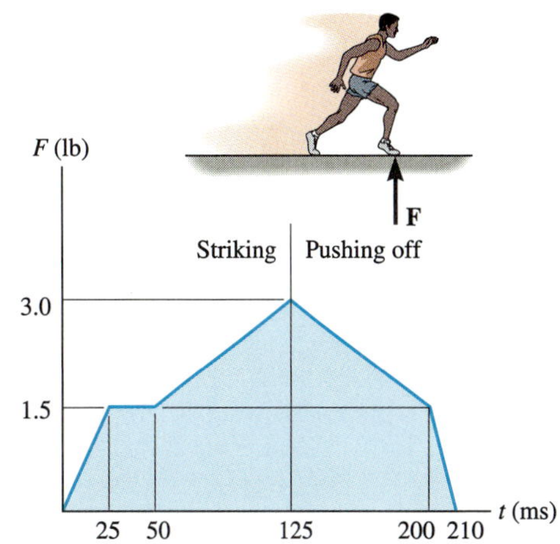

Prob. 15–13

15-14. When the 0.4-lb ball is fired, it leaves the ground at an angle of 40° from the horizontal and strikes the ground at the same elevation a distance of 130 ft away. Determine the impulse given to the ball.

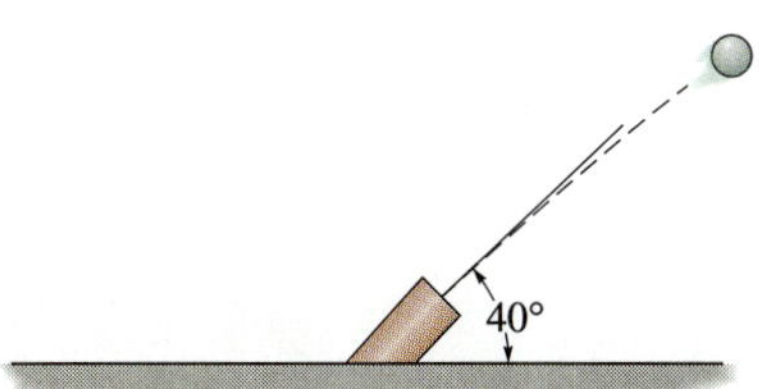

Prob. 15–14

15-15. A 50-kg crate rests against a stop block s, which prevents the crate from moving down the plane. If the coefficients of static and kinetic friction between the plane and the crate are $\mu_s = 0.3$ and $\mu_k = 0.2$, respectively, determine the time needed for the force **F** to give the crate a speed of 2 m/s up the plane. The force always acts parallel to the plane and has a magnitude of $F = (300t)$ N, where t is in seconds. *Hint:* First determine the time needed to overcome static friction and start the crate moving.

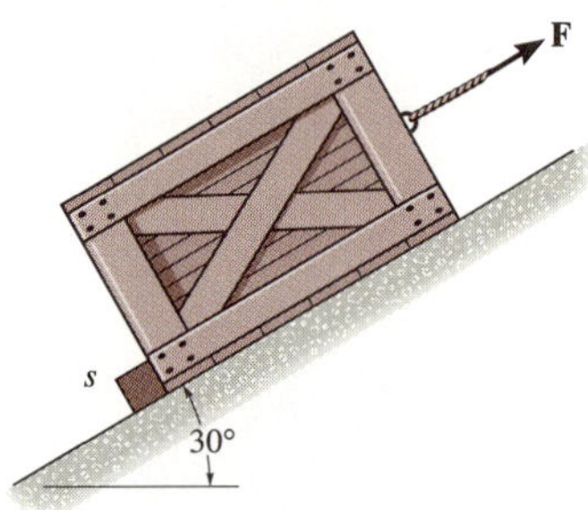

Prob. 15–15

*15-16. A train consists of a 50-Mg engine and three cars, each having a mass of 30 Mg. If it takes 80 s for the train to increase its speed uniformly to 40 km/h, starting from rest, determine the force T developed at the coupling between the engine E and the first car A. The wheels of the engine provide a resultant frictional tractive force **F** which gives the train forward motion, whereas the car wheels roll freely. Also, determine F acting on the engine wheels.

Prob. 15–16

15-17. The 5.5-Mg humpback whale is stuck on the shore due to changes in the tide. In an effort to rescue the whale, a 12-Mg tugboat is used to pull it free using an inextensible rope tied to its tail. To overcome the frictional force of the sand on the whale, the tug backs up so that the rope becomes slack and then the tug proceeds forward at 3 m/s. If the tug then turns the engines off, determine the average frictional force **F** on the whale if sliding occurs for 1.5 s before the tug stops after the rope becomes taut. Also, what is the average force on the rope during the tow?

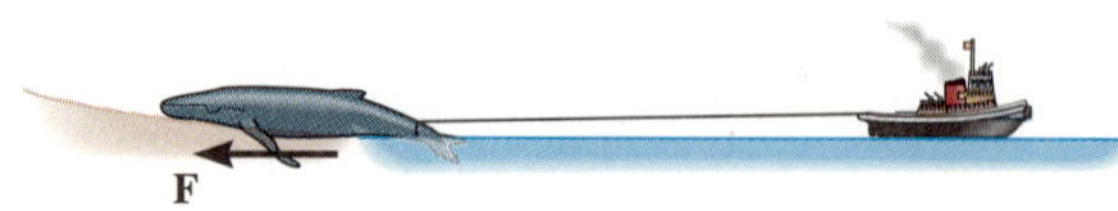

Prob. 15–17

15-18. The automobile has a weight of 2700 lb and is traveling forward at 4 ft/s when it crashes into the wall. If the impact occurs in 0.06 s, determine the average impulsive force acting on the car. Assume the brakes are *not applied*. If the coefficient of kinetic friction between the wheels and the pavement is $\mu_k = 0.3$, calculate the impulsive force on the wall if the brakes *were applied* during the crash. The brakes are applied to all four wheels so that all the wheels slip.

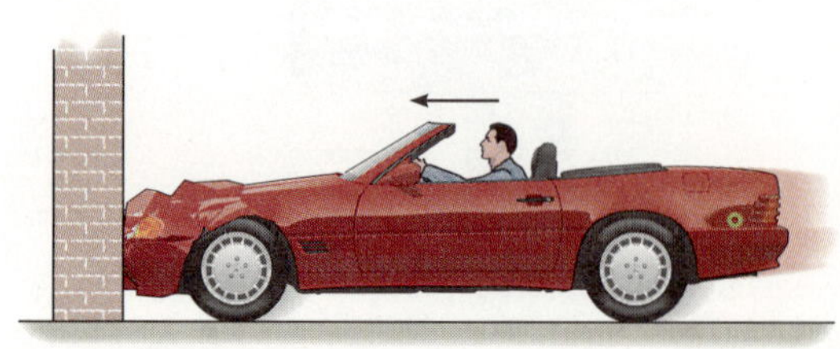

Prob. 15–18

15-19. In case of emergency, the gas actuator is used to move a 75-kg block B by exploding a charge C near a pressurized cylinder of negligible mass. As a result of the explosion, the cylinder fractures and the released gas forces the front part of the cylinder, A, to move B forward, giving it a speed of 200 mm/s in 0.4 s. If the coefficient of kinetic friction between B and the floor is $\mu_k = 0.5$, determine the impulse that the actuator imparts to B.

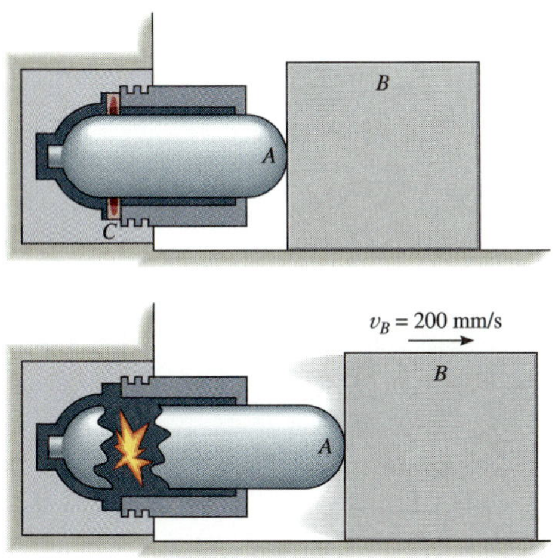

Prob. 15–19

***15-20.** The force acting on a projectile having a mass m as it passes horizontally through the barrel of the cannon is $F = C \sin(\pi t/t')$. Determine the projectile's velocity when $t = t'$. If the projectile reaches the end of the barrel at this instant, determine the length s.

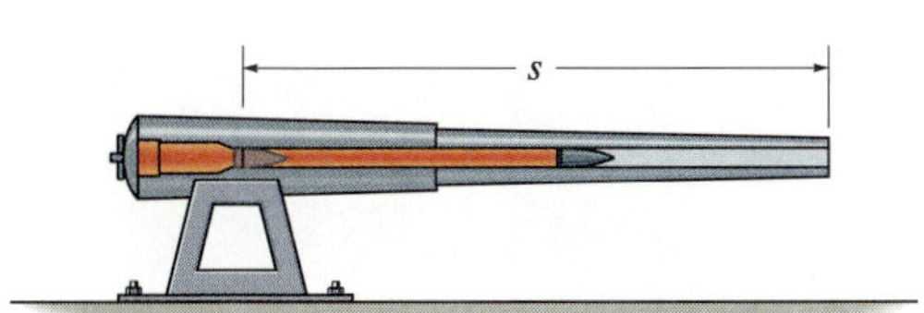

Prob. 15–20

15-21. A 30-lb block is initially moving along a smooth horizontal surface with a speed of $v_1 = 6$ ft/s to the left. If it is acted upon by a force **F**, which varies in the manner shown, determine the velocity of the block in 15 s.

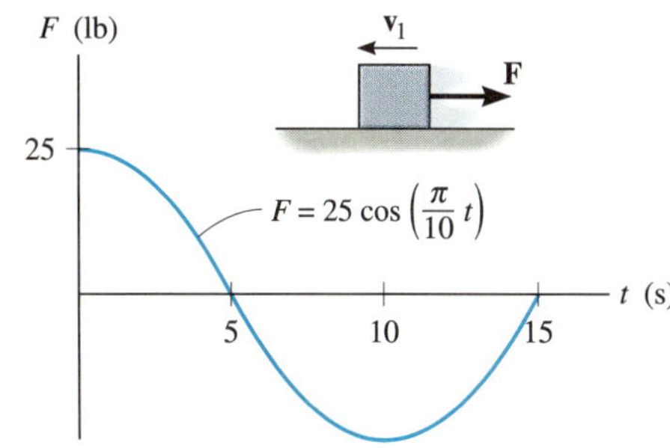

Prob. 15–21

15-22. The 2-lb block has an initial velocity of $v_1 = 10$ ft/s in the direction shown. If a force of $\mathbf{F} = \{0.5\mathbf{i} + 0.2\mathbf{j}\}$ lb acts on the block for $t = 5$ s, determine the final speed of the block. Neglect friction.

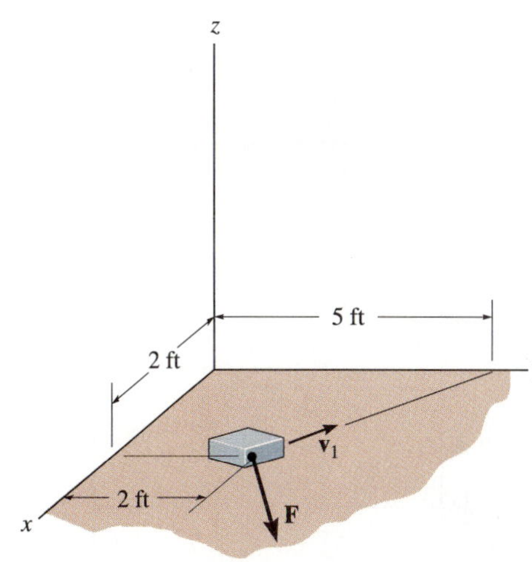

Prob. 15–22

15-23. Determine the velocities of blocks A and B 2 s after they are released from rest. Neglect the mass of the pulleys and cables.

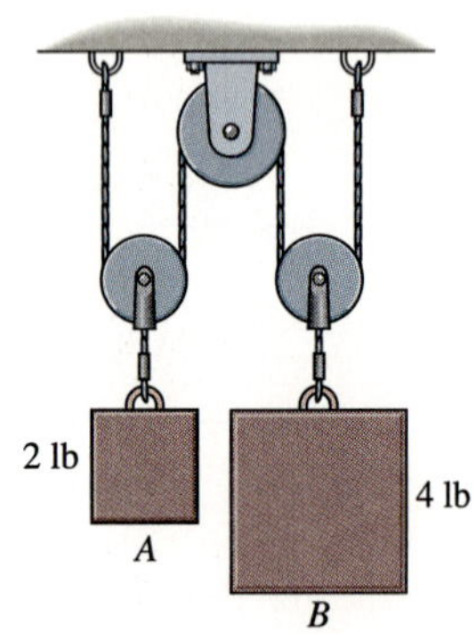

Prob. 15–23

***15-24.** The 40-kg slider block is moving to the right with a speed of 1.5 m/s when it is acted upon by the forces $\mathbf{F}_1$ and $\mathbf{F}_2$. If these loadings vary in the manner shown on the graph, determine the speed of the block at $t = 6$ s. Neglect friction and the mass of the pulleys and cords.

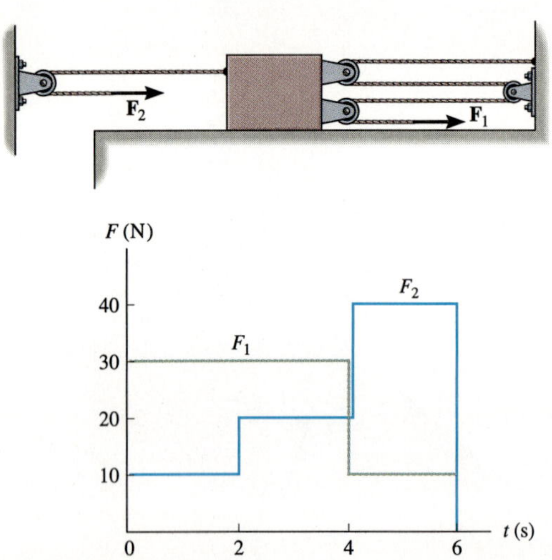

Prob. 15–24

15-25. Block A has a mass of 3 kg and B has a mass of 5 kg. If the system is released from rest, determine the velocity of each block in $t = 4$ s. Neglect the mass of the pulleys.

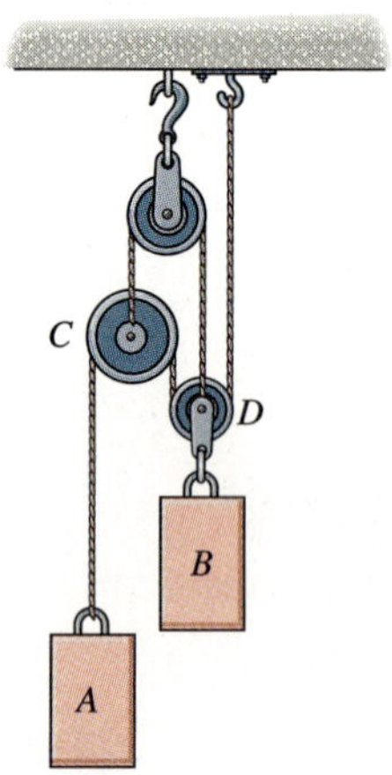

Prob. 15–25

15-26. A jet plane having a mass of 7 Mg takes off from an aircraft carrier such that the engine thrust varies as shown by the graph. If the carrier is traveling forward with a speed of 40 km/h, determine the plane's airspeed after 5 s.

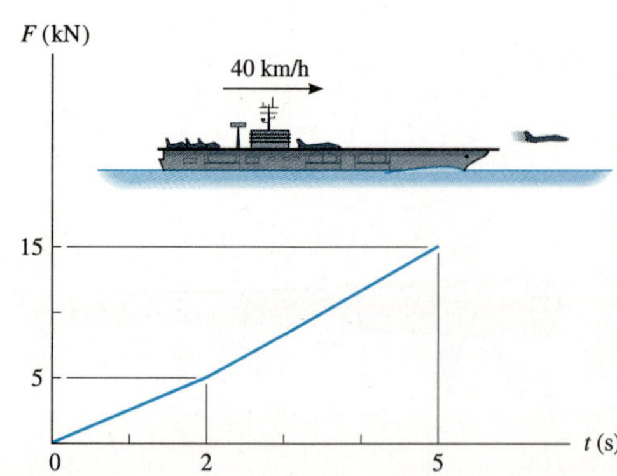

Prob. 15–26

15-27. The 50-kg block is hoisted up the incline using the cable and motor arrangement shown. The coefficient of kinetic friction between the block and the surface is $\mu_k = 0.4$. If the block is initially moving up the plane at $v_0 = 2$ m/s, and at this instant ($t = 0$) the motor develops a tension in the cord of $T = (300 + 120\sqrt{t})$ N, where t is in seconds, determine the velocity of the block when $t = 2$ s.

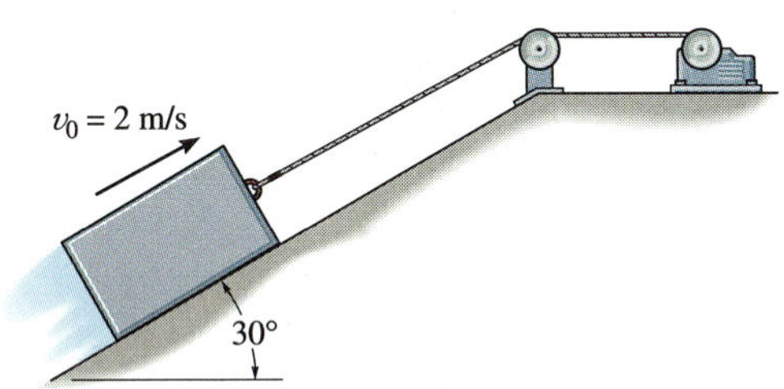

Prob. 15–27

***15-28.** The log has a mass of 500 kg and rests on the ground for which the coefficients of static and kinetic friction are $\mu_s = 0.5$ and $\mu_k = 0.4$, respectively. The winch delivers a towing force **T** to its cable at A which varies as shown in the graph. Determine the speed of the log when $t = 5$ s. Originally the cable tension is zero. *Hint:* First determine the force needed to begin moving the log.

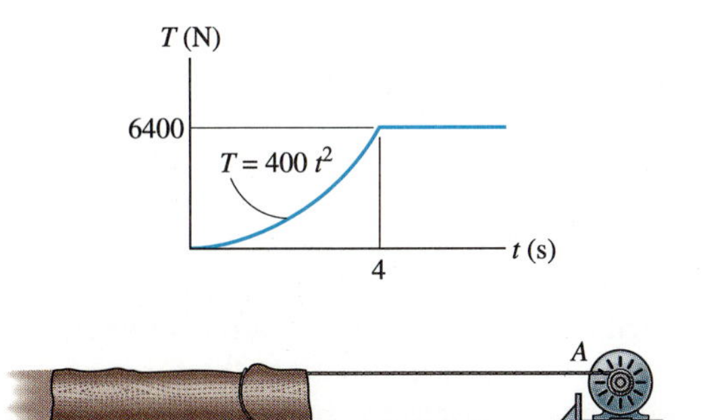

Prob. 15–28

15-29. The log has a mass of 500 kg and rests on the ground for which the coefficients of static and kinetic friction are $\mu_s = 0.5$ and $\mu_k = 0.4$, respectively. The winch delivers a towing force **T** to its cable at A which varies as shown in the graph. How much time is required to give the log a speed of 15 m/s? Originally the cable tension is zero. *Hint:* First determine the force needed to begin moving the log.

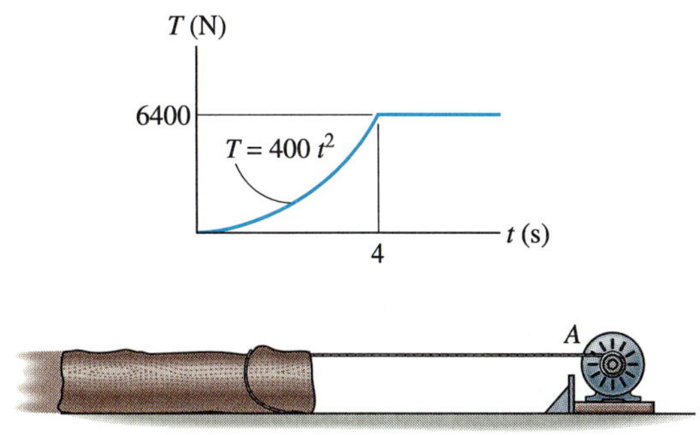

Prob. 15–29

15-30. The motor pulls on the cable at A with a force $F = (30 + t^2)$ lb, where t is in seconds. If the 17-lb crate is originally at rest at $t = 0$, determine its speed in $t = 4$ s. Neglect the mass of the cable and pulleys. *Hint:* First find the time needed to begin lifting the crate.

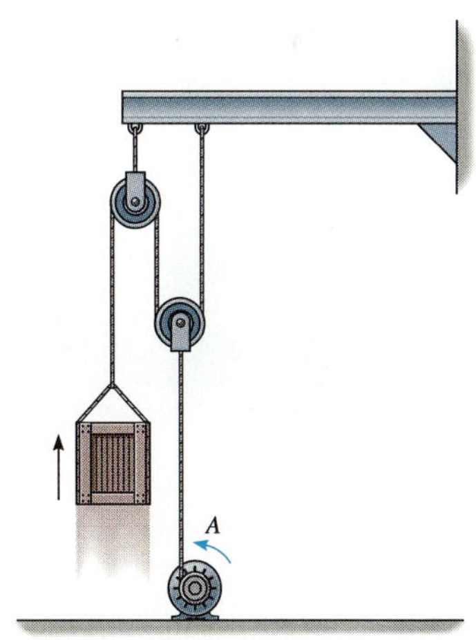

Prob. 15–30

15-31. As a 4-lb sphere falls vertically from rest through a liquid, the drag force exerted on it is $F_D = (0.06v)$ lb, where v is the speed measured in ft/s. Using the differential form of the impulse and momentum principle ($\Sigma F\ dt = m\ dv$), determine the time required for the sphere to attain one fourth of its terminal velocity (the terminal velocity occurs when $t \rightarrow \infty$).

15.3 Conservation of Linear Momentum for a System of Particles

When the sum of the *external impulses* acting on a system of particles is *zero*, Eq. 15–6 reduces to a simplified form, namely,

$$\Sigma m_i(\mathbf{v}_i)_1 = \Sigma m_i(\mathbf{v}_i)_2 \tag{15–8}$$

This equation is referred to as the *conservation of linear momentum*. It states that the linear momenta for a system of particles remain constant during the time period t_1 to t_2. Substituting $m\mathbf{v}_G = \Sigma m_i\mathbf{v}_i$ into Eq. 15–8, we can also write

$$(\mathbf{v}_G)_1 = (\mathbf{v}_G)_2 \tag{15–9}$$

which indicates that the velocity $\mathbf{v}_G$ of the mass center for the system of particles does not change when no external impulses are applied to the system.

The conservation of linear momentum is often applied when particles collide or interact. For application, a careful study of the free-body diagram for the *entire* system of particles should be made in order to identify the forces which create either external or internal impulses and thereby determine in what direction(s) linear momentum is conserved. As stated earlier, the *internal impulses* for the system will always cancel out, since they occur in equal but opposite collinear pairs. If the time period over which the motion is studied is *very short,* some of the external impulses may also be neglected or considered approximately equal to zero. The forces causing these negligible impulses are called *nonimpulsive forces.* By comparison, forces which are very large and act for a very short period of time produce a significant change in momentum and are called *impulsive forces.* They, of course, cannot be neglected in the impulse–momentum analysis.

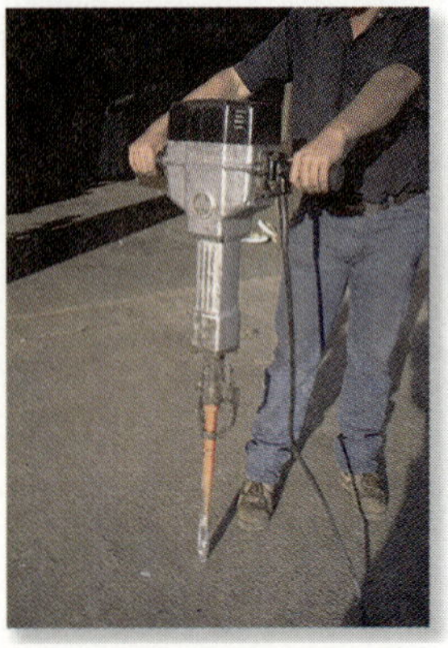

The hammer in the top photo applies an impulsive force to the stake. During this time the weight of the stake can be considered nonimpulsive, and provided the stake is driven into soft ground, the impulse of the ground acting on the stake can also be considered nonimpulsive. By contrast, if the stake is used in a concrete chipper to break concrete, then two impulsive forces act on the stake. One at its top due to the chipper and the other on its bottom due to the rigidity of the concrete.

Impulsive forces normally occur due to an explosion or the striking of one body against another, whereas nonimpulsive forces may include the weight of a body, the force imparted by a slightly deformed spring having a relatively small stiffness, or for that matter, any force that is very small compared to other larger (impulsive) forces. When making this distinction between impulsive and nonimpulsive forces, it is important to realize that this only applies during the time t_1 to t_2. To illustrate, consider the effect of striking a tennis ball with a racket as shown in the photo. During the *very short* time of interaction, the force of the racket on the ball is impulsive since it changes the ball's momentum drastically. By comparison, the ball's weight will have a negligible effect on the change in momentum, and therefore it is nonimpulsive. Consequently, it can be neglected from an impulse–momentum analysis during this time. If an impulse–momentum analysis is considered during the much longer time of flight after the racket–ball interaction, then the impulse of the ball's weight is important since it, along with air resistance, causes the change in the momentum of the ball.

Procedure for Analysis

Generally, the principle of linear impulse and momentum or the conservation of linear momentum is applied to a *system of particles* in order to determine the final velocities of the particles *just after* the time period considered. By applying these equations to the entire system, the internal impulses acting within the system, which may be unknown, are *eliminated* from the analysis. For application it is suggested that the following procedure be used.

Free-Body Diagram

- Establish the x, y, z inertial frame of reference and draw the free-body diagram for each particle of the system in order to identify the internal and external forces.
- The conservation of linear momentum applies to the system in a given direction when no external forces or if nonimpulsive forces act on the system in that direction.
- Establish the direction and sense of the particles' initial and final velocities. If the sense is unknown, assume it is along a positive inertial coordinate axis.
- As an alternative procedure, draw the impulse and momentum diagrams for each particle of the system.

Momentum Equations

- Apply the principle of linear impulse and momentum or the conservation of linear momentum in the appropriate directions.
- If it is necessary to determine the *internal impulse* $\int F dt$ acting on only one particle of a system, then the particle must be *isolated* (free-body diagram), and the principle of linear impulse and momentum must be applied *to the particle.*
- After the impulse is calculated, and provided the time Δt for which the impulse acts is known, then the *average impulsive force* F_{avg} can be determined from $F_{avg} = \int F\, dt/\Delta t$.

EXAMPLE 15-4

The 15-Mg boxcar A is coasting at 1.5 m/s on the horizontal track when it encounters a 12-Mg tank car B coasting at 0.75 m/s toward it as shown in Fig. 15–8*a*. If the cars meet and couple together, determine (*a*) the speed of both cars just after the coupling, and (*b*) the average force between them if the coupling takes place in 0.8 s.

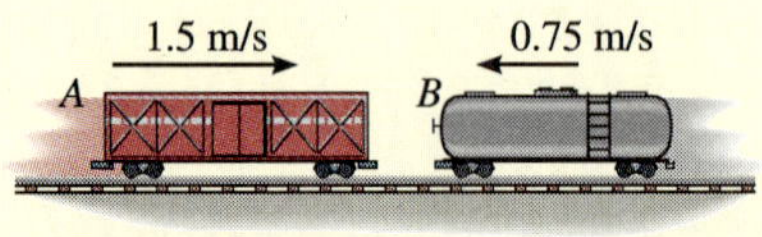

(a)

Solution

Part (a)

Free-Body Diagram.*** Here we have considered *both* cars as a single system, Fig. 15–8*b*. By inspection, momentum is conserved in the x direction since the coupling force $\mathbf{F}$ is *internal* to the system and will therefore cancel out. It is assumed both cars, when coupled, move at $\mathbf{v}_2$ in the positive x direction.

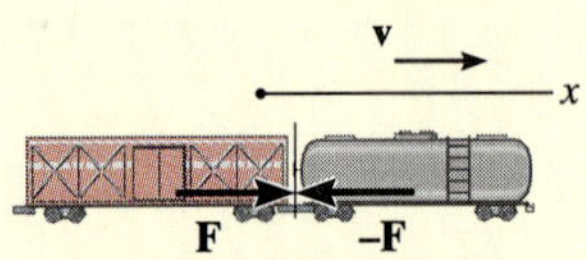

(b)

Conservation of Linear Momentum

$$(\overset{+}{\rightarrow}) \qquad m_A(v_A)_1 + m_B(v_B)_1 = (m_A + m_B)v_2$$

$$(15\,000 \text{ kg})(1.5 \text{ m/s}) - 12\,000 \text{ kg}(0.75 \text{ m/s}) = (27\,000 \text{ kg})v_2$$

$$v_2 = 0.5 \text{ m/s} \rightarrow \qquad \textit{Ans.}$$

Part (b). The average (impulsive) coupling force, $\mathbf{F}_{avg}$, can be determined by applying the principle of linear momentum to *either one* of the cars.

Free-Body Diagram. As shown in Fig. 15–8*c*, by isolating the boxcar the coupling force is *external* to the car.

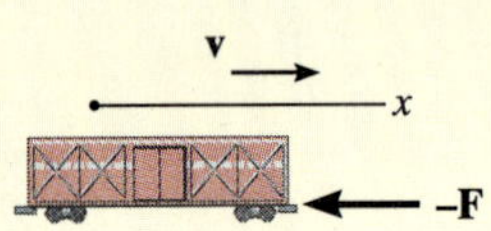

(c)

Fig. 15–8

Principle of Impulse and Momentum. Since $\int F\,dt = F_{avg}\,\Delta t = F_{avg}(0.8)$, we have

$$(\overset{+}{\rightarrow}) \qquad m_A(v_A)_1 + \Sigma \int F\,dt = m_A v_2$$

$$(15\,000 \text{ kg})(1.5 \text{ m/s}) - F_{avg}(0.8 \text{ s}) = (15\,000 \text{ kg})(0.5 \text{ m/s})$$

$$F_{avg} = 18.8 \text{ kN} \qquad \textit{Ans.}$$

Solution was possible here since the boxcar's final velocity was obtained in Part (*a*). Try solving for F_{avg} by applying the principle of impulse and momentum to the tank car.

*Only horizontal forces are shown on the free-body diagram.

EXAMPLE 15-5

The 1200-lb cannon shown in Fig. 15–9*a* fires an 8-lb projectile with a muzzle velocity of 1500 ft/s relative to the ground. If firing takes place in 0.03 s, determine (*a*) the recoil velocity of the cannon just after firing, and (*b*) the average impulsive force acting on the projectile. The cannon support is fixed to the ground, and the horizontal recoil of the cannon is absorbed by two springs.

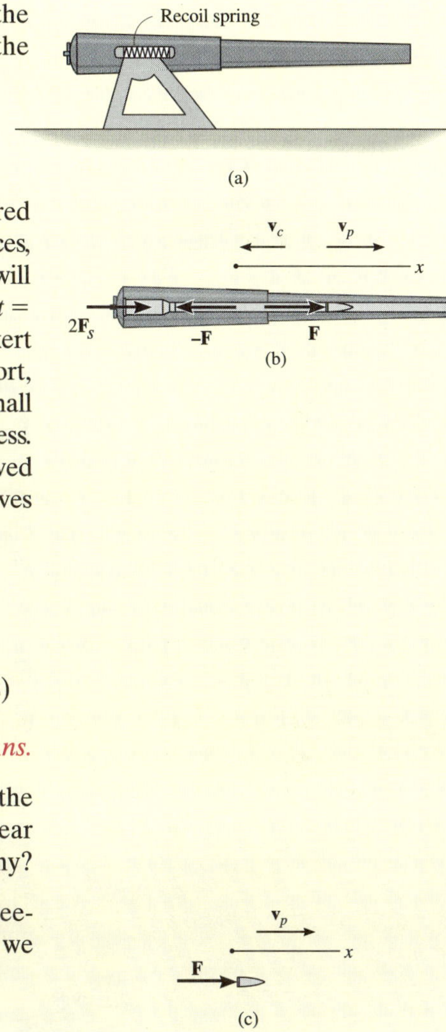

Fig. 15–9

Solution

Part (a)

Free-Body Diagram.* As shown in Fig. 15–9*b*, we have considered the projectile and cannon as a single system, since the impulsive forces, **F**, between the cannon and projectile are *internal* to the system and will therefore cancel from the analysis. Furthermore, during the time $\Delta t = 0.03$ s, the two recoil springs which are attached to the support each exert a *nonimpulsive force* $\mathbf{F}_s$ on the cannon. This is because Δt is very short, so that during this time the cannon only moves through a very small distance† s. Consequently, $F_s = ks \approx 0$, where k is the spring's stiffness. Hence it may be concluded that momentum for the system is conserved in the *horizontal direction*. Here we will assume that the cannon moves to the left, while the projectile moves to the right after firing.

Conservation of Linear Momentum

$$(\overset{+}{\rightarrow}) \quad m_c(v_c)_1 + m_p(v_p)_1 = -m_c(v_c)_2 + m_p(v_p)_2$$

$$0 + 0 = -\frac{1200\text{ lb}}{32.2\text{ ft/s}^2}(v_c)_2 + \frac{8\text{ lb}}{32.2\text{ ft/s}^2}(1500\text{ ft/s})$$

$$(v_c)_2 = 10\text{ ft/s} \leftarrow \qquad \textit{Ans.}$$

Part (b). The average impulsive force exerted by the cannon on the projectile can be determined by applying the principle of linear impulse and momentum to the projectile (or to the cannon). Why?

Principle of Impulse and Momentum. Using the data on the free-body diagram, Fig. 15–9*c*, noting that $\int F\,dt = F_{\text{avg}}\,\Delta t = F_{\text{avg}}(0.03)$, we have

$$(\overset{+}{\rightarrow}) \quad m(v_p)_1 + \Sigma \int F\,dt = m(v_p)_2$$

$$0 + F_{\text{avg}}(0.03\text{ s}) = \frac{8\text{ lb}}{32.2\text{ ft/s}^2}(1500\text{ ft/s})$$

$$F_{\text{avg}} = 12.4(10^3)\text{ lb} = 12.4\text{ kip} \qquad \textit{Ans.}$$

*Only horizontal forces are shown on the free-body diagram.

†If the cannon is firmly fixed to its support (no springs), the reactive force of the support on the cannon must be considered as an external impulse to the system, since the support would allow no movement of the cannon.

EXAMPLE 15-6

The 350-Mg tugboat T shown in Fig. 15–10a is used to pull the 50-Mg barge B with a rope R. If the barge is initially at rest and the tugboat is coasting freely with a velocity of $(v_T)_1 = 3$ m/s while the rope is *slack*, determine the velocity of the tugboat *directly after* the rope becomes taut. Assume the rope does not stretch. Neglect the frictional effects of the water.

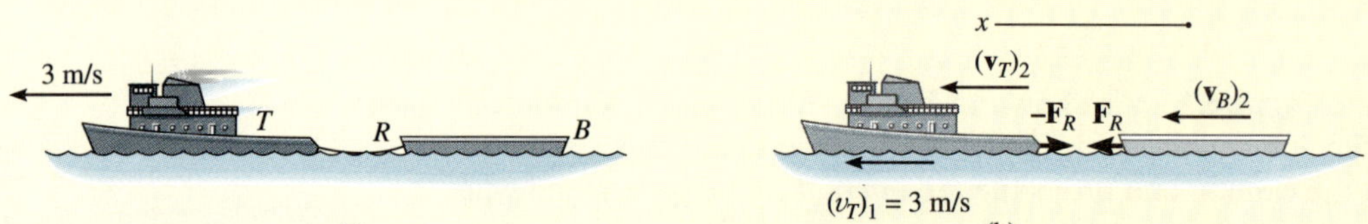

(a) (b)

Solution

Free-Body Diagram.* As shown in Fig. 15–10b, we have considered the entire system (tugboat and barge). Hence, the impulsive force created between the tugboat and the barge is *internal* to the system, and therefore momentum of the system is conserved during the instant of towing.

The alternative procedure of drawing the system's impulse and momentum diagrams is shown in Fig. 15–10c.

Conservation of Momentum. Noting that $(v_B)_2 = (v_T)_2$, we have

$$(\overset{+}{\leftarrow}) \qquad m_T(v_T)_1 + m_B(v_B)_1 = m_T(v_T)_2 + m_B(v_B)_2$$

$$350(10^3)\,\text{kg}(3\text{ m/s}) + 0 = 350(10^3)\,\text{kg}(v_T)_2 + 50(10^3)\,\text{kg}(v_T)_2$$

Solving,

$$(v_T)_2 = 2.62\text{ m/s} \leftarrow \qquad \textit{Ans.}$$

This value represents the tugboat's velocity *just after* the towing impulse. Use this result and show that the towing *impulse* is 131 kN·s.

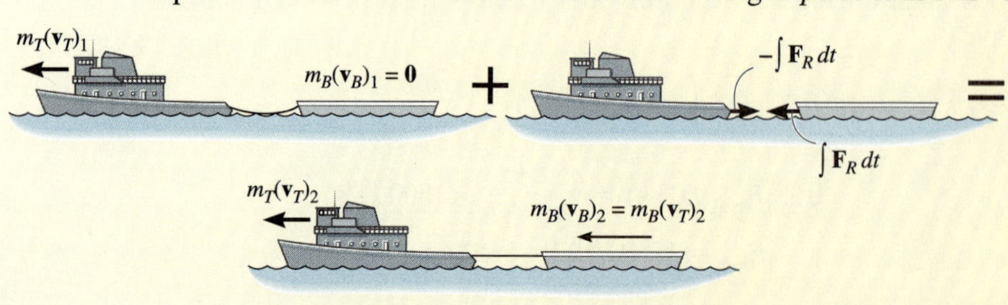

(c)

Fig. 15–10

*Only horizontal forces are shown on the free-body diagram.

EXAMPLE 15–7

An 800-kg rigid pile P shown in Fig. 15–11a is driven into the ground using a 300-kg hammer H. The hammer falls from rest at a height $y_0 =$ 0.5 m and strikes the top of the pile. Determine the impulse which the hammer imparts on the pile if the pile is surrounded entirely by loose sand so that after striking, the hammer does *not* rebound off the pile.

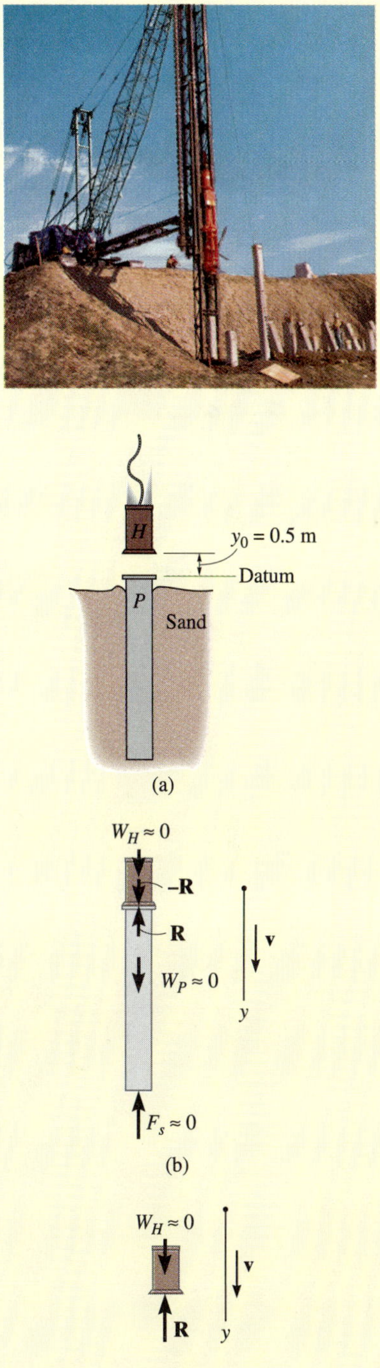

Fig. 15–11

Solution

Conservation of Energy. The velocity at which the hammer strikes the pile can be determined using the conservation of energy equation applied to the hammer. With the datum at the top of the pile, Fig. 15–11a, we have

$$T_0 + V_0 = T_1 + V_1$$

$$\tfrac{1}{2}m_H(v_H)_0^2 + W_H y_0 = \tfrac{1}{2}m_H(v_H)_1^2 + W_H y_1$$

$$0 + 300(9.81)\text{ N}(0.5\text{ m}) = \frac{1}{2}(300\text{ kg})(v_H)_1^2 + 0$$

$$(v_H)_1 = 3.13\text{ m/s}$$

Free-Body Diagram. From the physical aspects of the problem, the free-body diagram of the hammer and pile, Fig. 15–11b, indicates that during the *short time* occurring *just before* to *just after* the *collision,* the weights of the hammer and pile and the resistance force $\mathbf{F}_s$ of the sand are all *nonimpulsive*. The impulsive force **R** is internal to the system and therefore cancels. Consequently, momentum is conserved in the vertical direction during this short time.

Conservation of Momentum. Since the hammer does not rebound off the pile just after collision, then $(v_H)_2 = (v_P)_2 = v_2$.

$$(+\downarrow) \qquad m_H(v_H)_1 + m_P(v_P)_1 = m_H v_2 + m_P v_2$$

$$(300\text{ kg})(3.13\text{ m/s}) + 0 = (300\text{ kg})v_2 + (800\text{ kg})v_2$$

$$v_2 = 0.854\text{ m/s}$$

Principle of Impulse and Momentum. The impulse which the pile imparts to the hammer can now be determined since $\mathbf{v}_2$ is known. From the free-body diagram for the hammer, Fig. 15–11c, we have

$$(+\downarrow) \qquad m_H(v_H)_1 + \Sigma\int_{t_1}^{t_2} F_y\,dt = m_H v_2$$

$$(300\text{ kg})(3.13\text{ m/s}) - \int R\,dt = (300\text{ kg})(0.854\text{ m/s})$$

$$\int R\,dt = 683\text{ N}\cdot\text{s} \qquad \textit{Ans.}$$

The equal but opposite impulse acts on the pile. Try finding this impulse by applying the principle of impulse and momentum to the pile.

EXAMPLE 15-8

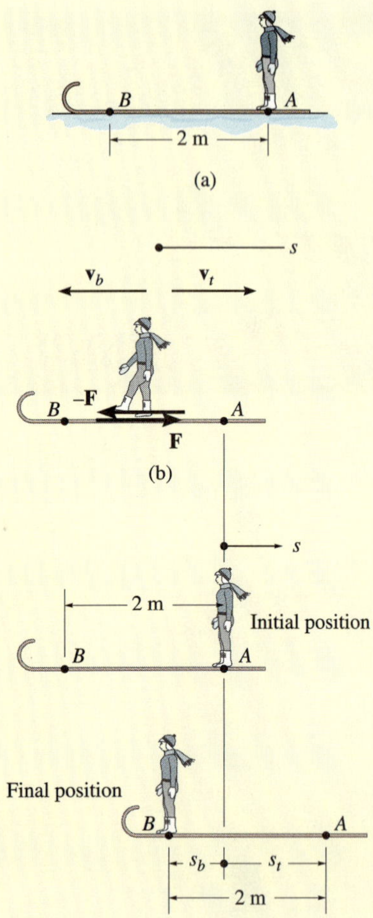

Fig. 15–12

A boy having a mass of 40 kg stands on the back of a 15-kg toboggan which is originally at rest, Fig. 15–12*a*. If he walks to the front *B* and stops, determine the distance the toboggan moves. Neglect friction between the bottom of the toboggan and the ground (ice).

Solution I

Free-Body Diagram. The unknown frictional force of the boy's shoes on the bottom of the toboggan can be *excluded* from the analysis if the toboggan and boy on it are considered as a single system. In this way the frictional force **F** becomes *internal* and the conservation of momentum applies, Fig. 15–12*b*.

Conservation of Momentum. Since both the initial and final momenta of the system are zero (because the initial and final velocities are zero), the system's momentum must also be zero when the boy is at some intermediate point between *A* and *B*. Thus

$$(\overset{+}{\rightarrow}) \qquad -m_b v_b + m_t v_t = 0 \qquad (1)$$

Here the two unknowns v_b and v_t represent the velocities of the boy moving to the left and the toboggan moving to the right. Both are measured from a *fixed inertial reference* on the ground.

At any instant the *position* of point *A* on the toboggan and the *position* of the boy must be determined by integration. Since $v = ds/dt$, then $-m_b ds_b + m_t ds_t = 0$. Assuming the initial position of point *A* to be at the origin, Fig. 15–12*c*, then at the final position we have $-m_b s_b + m_t s_t = 0$. Since $s_b + s_t = 2$ m, or $s_b = (2 - s_t)$ then

$$-m_b(2 - s_t) + m_t s_t = 0 \qquad (2)$$

$$s_t = \frac{2m_b}{m_b + m_t} = \frac{2(40)}{40 + 15} = 1.45 \text{ m} \qquad \textit{Ans.}$$

Solution II

The problem may also be solved by considering the relative motion of the boy with respect to the toboggan, $\mathbf{v}_{b/t}$. This velocity is related to the velocities of the boy and toboggan by the equation $\mathbf{v}_b = \mathbf{v}_t + \mathbf{v}_{b/t}$, Eq. 12–34. Since positive motion is assumed to be to the right in Eq. 1, $\mathbf{v}_b$ and $\mathbf{v}_{b/t}$ are negative, because the boy's motion is to the left. Hence, in scalar form, $-v_b = v_t - v_{b/t}$, and Eq. 1 then becomes $m_b(v_t - v_{b/t}) + m_t v_t = 0$. Integrating gives

$$m_b(s_t - s_{b/t}) + m_t s_t = 0$$

Realizing $s_{b/t} = 2$ m, we obtain Eq. 2.

Problems

*15-32. A railroad car having a mass of 15 Mg is coasting at 1.5 m/s on a horizontal track. At the same time another car having a mass of 12 Mg is coasting at 0.75 m/s in the opposite direction. If the cars meet and couple together, determine the speed of both cars just after the coupling. Find the difference between the total kinetic energy before and after coupling has occurred, and explain qualitatively what happened to this energy.

15-33. A ballistic pendulum consists of a 4-kg wooden block originally at rest, $\theta = 0°$. When a 2-g bullet strikes and becomes embedded in it, it is observed that the block swings upward to a maximum angle of $\theta = 6°$. Estimate the speed of the bullet.

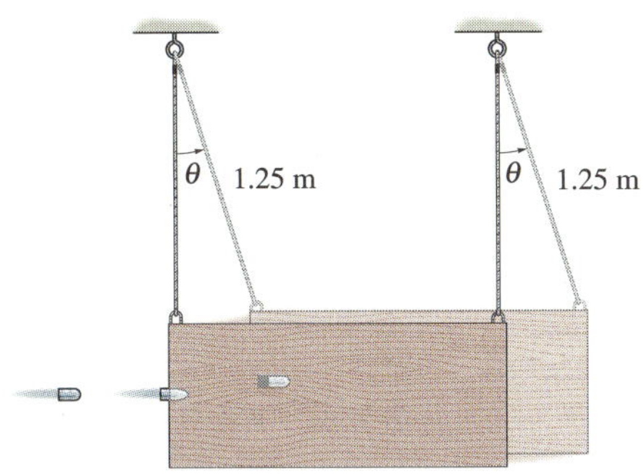

Prob. 15–33

15-34. The cart has a mass of 3 kg and rolls freely down the slope. When it reaches the bottom, a spring loaded gun fires a 0.5-kg ball out the back with a horizontal velocity of $v_{b/c} = 0.6$ m/s, measured relative to the cart. Determine the final velocity of the cart.

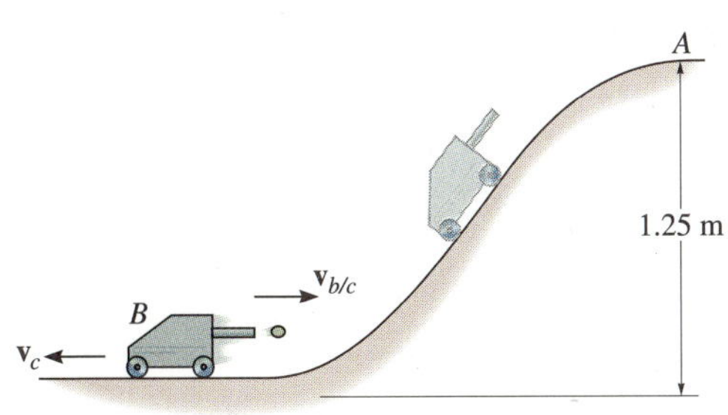

Prob. 15–34

15-35. The two blocks A and B each have a mass of 5 kg and are suspended from parallel cords. A spring, having a stiffness of $k = 60$ N/m, is attached to B and is compressed 0.3 m against A and B as shown. Determine the maximum angles θ and ϕ of the cords when the blocks are released from rest and the spring becomes unstretched.

*15-36. Block A has a mass of 4 kg and B has a mass of 6 kg. A spring, having a stiffness of $k = 40$ N/m, is attached to B and is compressed 0.3 m against A and B as shown. Determine the maximum angles θ and ϕ of the cords after the blocks are released from rest and the spring becomes unstretched.

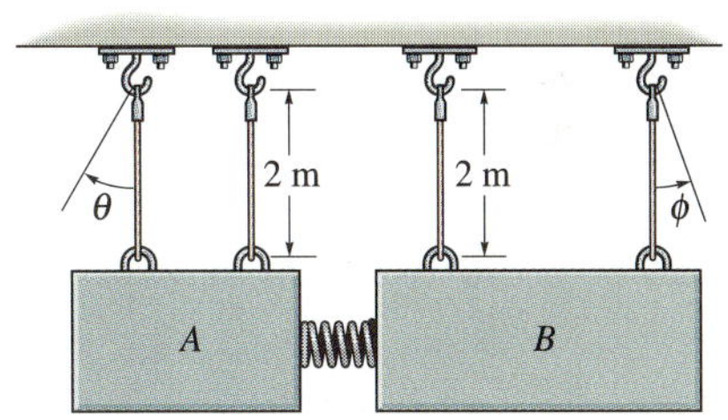

Probs. 15–35/36

15-37. A man wearing ice skates throws an 8-kg block with an initial velocity of 2 m/s, measured relative to himself, in the direction shown. If he is originally at rest and completes the throw in 1.5 s while keeping his legs rigid, determine the horizontal velocity of the man just after releasing the block. What is the vertical reaction of both his skates on the ice during the throw? The man has a mass of 70 kg. Neglect friction and the motion of his arms.

Prob. 15–37

15-38. The barge weighs 45 000 lb and supports two automobiles A and B, which weigh 4000 lb and 3000 lb, respectively. If the automobiles start from rest and drive towards each other, accelerating at $a_A = 4\ \text{ft/s}^2$ and $a_B = 8\ \text{ft/s}^2$ until they reach a constant speed of 6 ft/s relative to the barge, determine the speed of the barge just before the automobiles collide. How much time does this take? Originally the barge is at rest. Neglect water resistance.

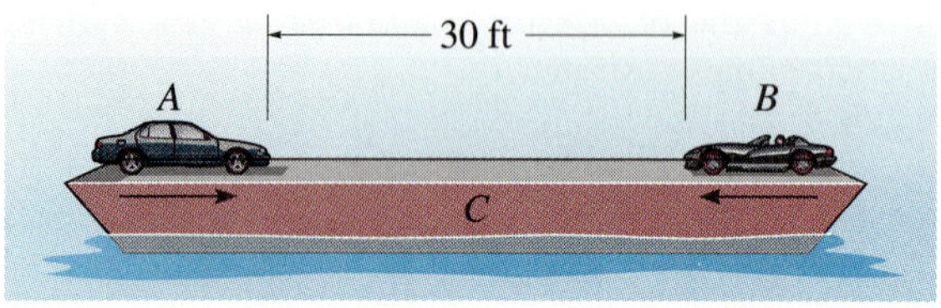

Prob. 15–38

15-39. A 40-lb box slides from rest down the smooth ramp onto the surface of a 20-lb cart. Determine the speed of the box at the instant it stops sliding on the cart. If someone ties the cart to the ramp at B, determine the horizontal impulse the box will exert at C in order to stop its motion. Neglect friction and the size of the box.

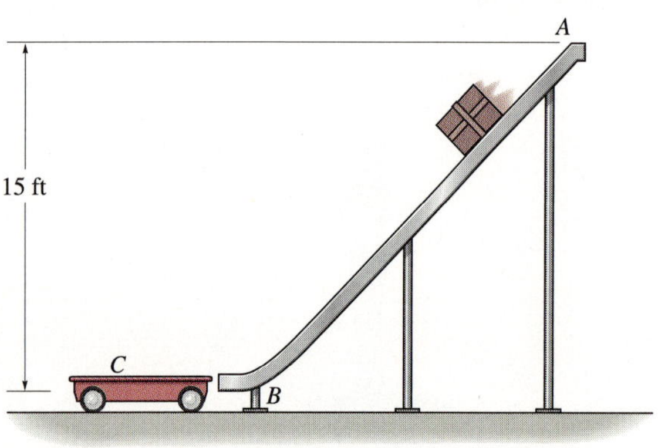

Prob. 15–39

***15-40.** A 0.03-lb bullet traveling at 1300 ft/s strikes the 10-lb wooden block and exits the other side at 50 ft/s as shown. Determine the speed of the block just after the bullet exits the block, and also determine how far the block slides before it stops. The coefficient of kinetic friction between the block and the surface is $\mu_k = 0.5$.

15-41. A 0.03-lb bullet traveling at 1300 ft/s strikes the 10-lb wooden block and exits the other side at 50 ft/s as shown. Determine the speed of the block just after the bullet exits the block. Also, determine the average normal force on the block if the bullet passes through it in 1 ms, and the time the block slides before it stops. The coefficient of kinetic friction between the block and the surface is $\mu_k = 0.5$.

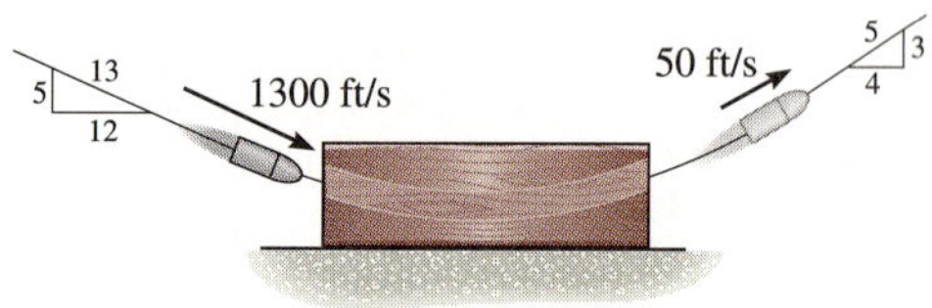

Probs. 15–40/41

15-42. The 5-kg spring-loaded gun rests on the smooth surface. It fires a ball having a mass of 1 kg with a velocity of $v' = 6$ m/s relative to the gun in the direction shown. If the gun is originally at rest, determine the horizontal distance d the ball is from the gun at the instant the ball strikes the ground at D. Neglect the size of the gun.

15-43. The 5-kg spring-loaded gun rests on the smooth surface. It fires a ball having a mass of 1 kg with a velocity of $v' = 6$ m/s relative to the gun in the direction shown. If the gun is originally at rest, determine the distance the ball is from the gun at the instant the ball reaches its highest elevation C. Neglect the size of the gun.

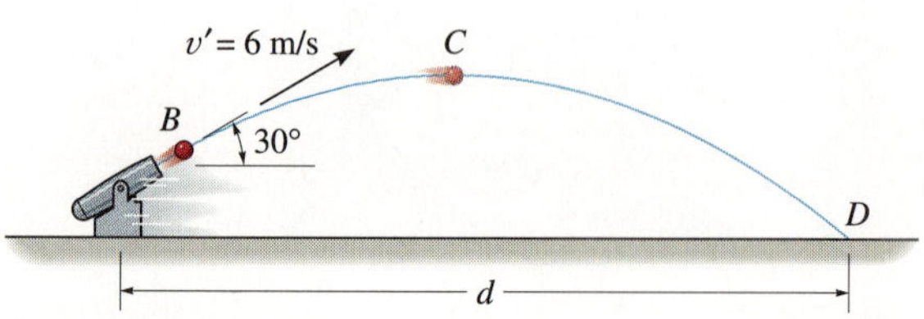

Probs. 15–42/43

***15-44.** A 100-lb boy walks forward over the surface of the 60-lb cart with a constant speed of 3 ft/s relative to the cart. Determine the cart's speed and its displacement at the moment he is about to step off. Neglect the mass of the wheels and assume the cart and boy are originally at rest.

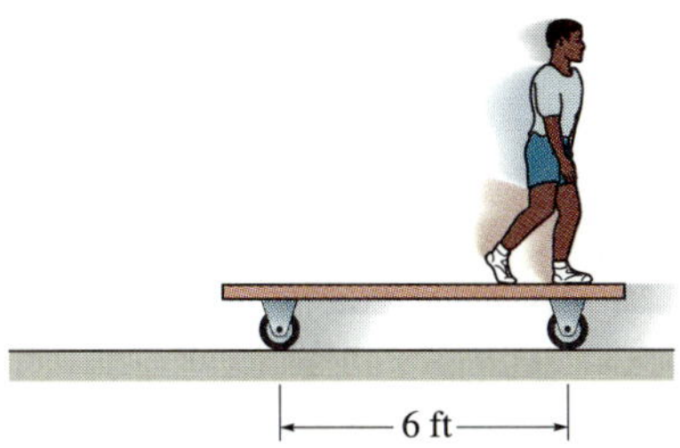

Prob. 15–44

15-45. The block of mass m is traveling at v_1 in the direction θ_1 shown at the top of the smooth slope. Determine its speed v_2 and its direction θ_2 when it reaches the bottom.

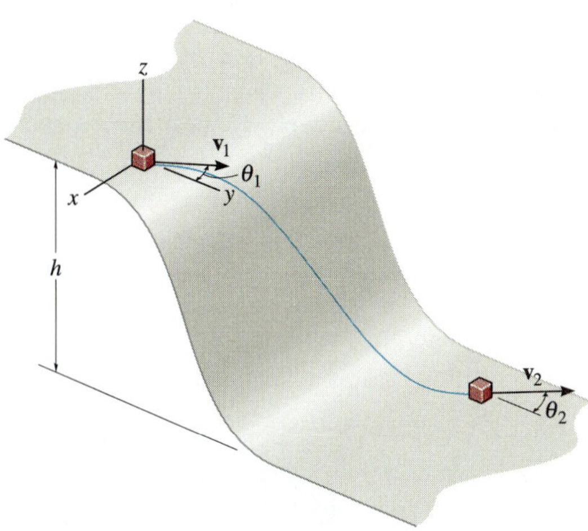

Prob. 15–45

15-46. Two boxes A and B, each having a weight of 160 lb, sit on the 500-lb conveyor which is free to roll on the ground. If the belt starts from rest and begins to run with a speed of 3 ft/s, determine the final speed of the conveyor if (a) the boxes are not stacked and A falls off then B falls off, and (b) A is stacked on top of B and both fall off together.

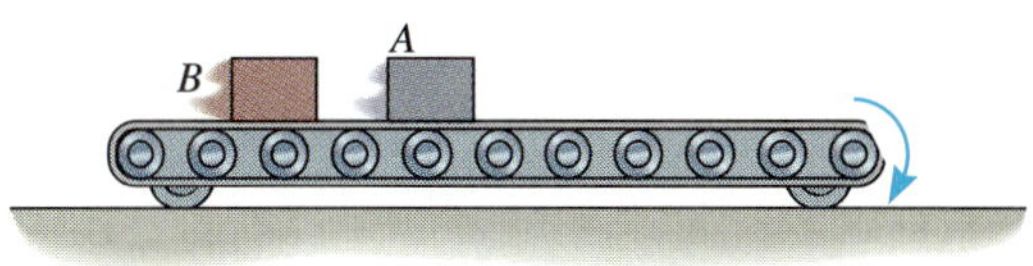

Prob. 15–46

15-47. The winch on the back of the jeep A is turned on and pulls in the tow rope at $v_{rel} = 2$ m/s. If both the 1.25-Mg car B and the 2.5-Mg jeep A are free to roll, determine their velocities at the instant they meet. If the rope is 5 m long, how long will this take?

Prob. 15–47

***15-48.** The block B has a weight of 75 lb and rests at the end of the 50-lb cart. If the cart is free to roll, and the rope is pulled in at 4 ft/s relative to the cart, determine how far d the cart has moved when the block has moved 8 ft on the cart. The coefficient of kinetic friction between the cart and the block is $\mu_k = 0.4$.

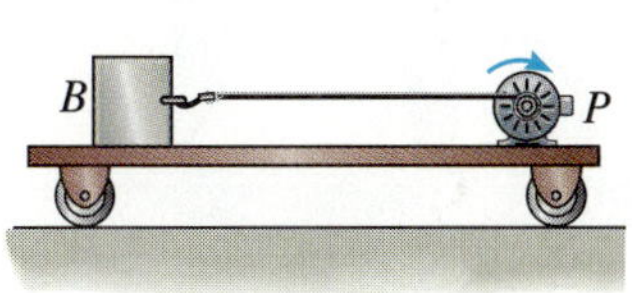

Prob. 15–48

15-49. The 20-lb cart B is supported on rollers of negligible size. If a 10-lb suitcase A is thrown horizontally on it at 10 ft/s, determine the length of time that A slides relative to B, and the final velocity of A and B. The coefficient of kinetic friction between A and B is $\mu_k = 0.4$.

15-50. The 20-lb cart B is supported on rollers of negligible size. If a 10-lb suitcase A is thrown horizontally on it at 10 ft/s, determine the time t and the distance B moves before A stops relative to B. The coefficient of kinetic friction between A and B is $\mu_k = 0.4$.

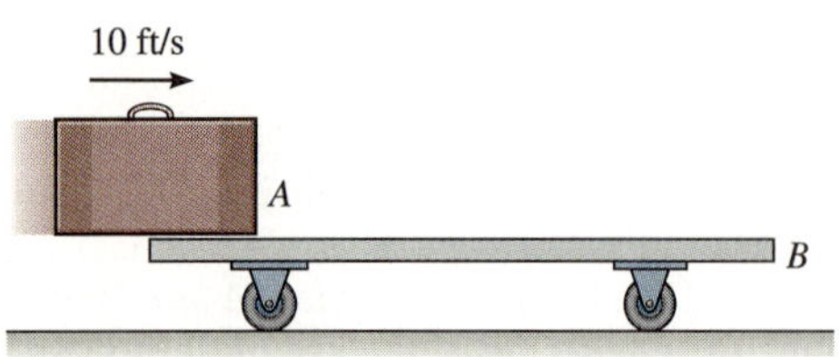

Probs. 15–49/50

***15-52.** The free-rolling ramp has a weight of 120 lb. The crate whose weight is 80 lb slides from rest at A, 15 ft down the ramp to B. Determine the ramp's speed when the crate reaches B. Assume that the ramp is smooth, and neglect the mass of the wheels.

15-53. The free-rolling ramp has a weight of 120 lb. If the 80-lb crate is released from rest at A, determine the distance the ramp moves when the crate slides 15 ft down the ramp and reaches the bottom B.

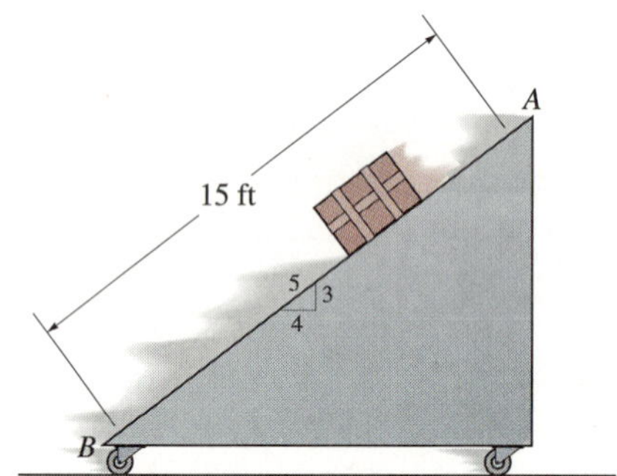

Probs. 15–52/53

15-51. A boy A having a weight of 80 lb and a girl B having a weight of 65 lb stand motionless at the ends of the toboggan, which has a weight of 20 lb. If they exchange positions, A going to B and then B going to A's original position, determine the final position of the toboggan just after the motion. Neglect friction.

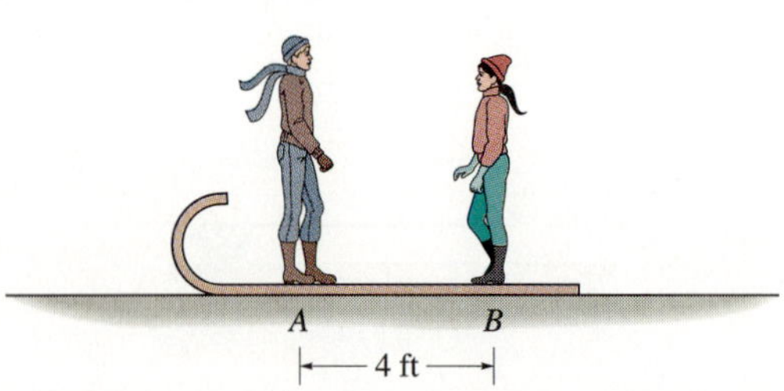

Prob. 15–51

15-54. A tugboat T having a mass of 19 Mg is tied to a barge B having a mass of 75 Mg. If the rope is "elastic" such that it has a stiffness $k = 600$ kN/m, determine the maximum stretch in the rope during the initial towing. Originally both the tugboat and barge are moving in the same direction with speeds $(v_T)_1 = 15$ km/h and $(v_B)_1 = 10$ km/h, respectively. Neglect the resistance of the water.

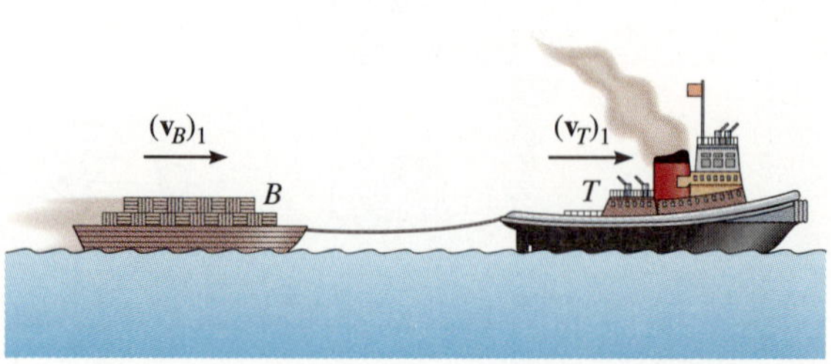

Prob. 15–54

15.4 Impact

Impact occurs when two bodies collide with each other during a very *short* period of time, causing relatively large (impulsive) forces to be exerted between the bodies. The striking of a hammer on a nail, or a golf club on a ball, are common examples of impact loadings.

In general, there are two types of impact. *Central impact* occurs when the direction of motion of the mass centers of the two colliding particles is along a line passing through the mass centers of the particles. This line is called the *line of impact,* Fig. 15–13*a*. When the motion of one or both of the particles is at an angle with the line of impact, Fig. 15–13*b*, the impact is said to be *oblique impact.*

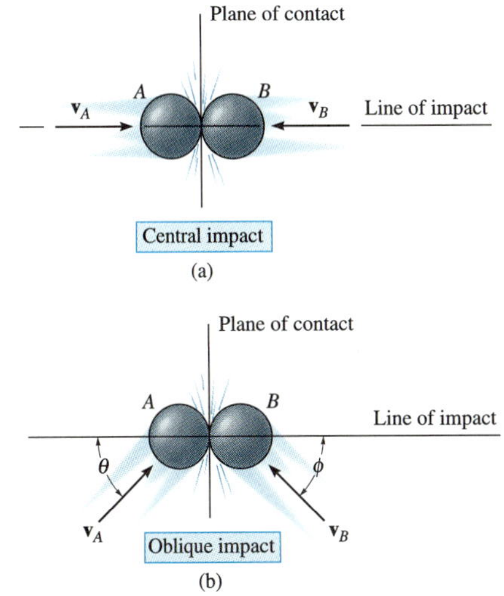

Fig. 15–13

Central Impact. To illustrate the method for analyzing the mechanics of impact, consider the case involving the central impact of two *smooth* particles A and B shown in Fig. 15–14.

- The particles have the initial momenta shown in Fig. 15–14*a*. Provided $(v_A)_1 > (v_B)_1$, collision will eventually occur.
- During the collision the particles must be thought of as *deformable* or nonrigid. The particles will undergo a *period of deformation* such that they exert an equal but opposite deformation impulse $\int \mathbf{P}\,dt$ on each other, Fig. 15–14*b*.
- Only at the instant of *maximum deformation* will both particles move with a common velocity **v**, since their relative motion is zero, Fig. 15–14*c*.
- Afterward a *period of restitution* occurs, in which case the particles will either return to their original shape or remain permanently deformed. The equal but opposite *restitution impulse* $\int \mathbf{R}\,dt$ pushes the particles apart from one another, Fig. 15–14*d*. In reality, the physical properties of any two bodies are such that the deformation impulse is *always greater* than that of restitution, i.e., $\int P\,dt > \int R\,dt$.
- Just after separation the particles will have the final momenta shown in Fig. 15–14*e*, where $(v_B)_2 > (v_A)_2$.

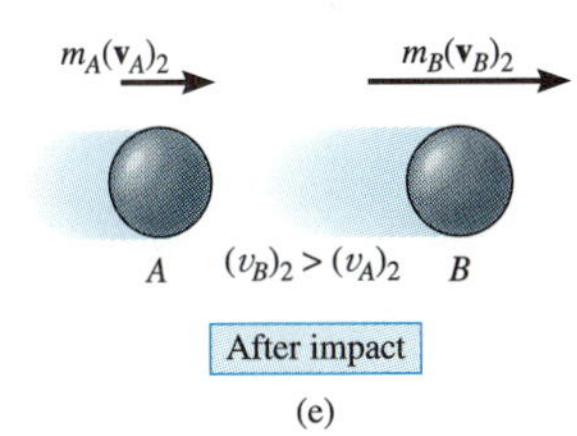

Fig. 15–14

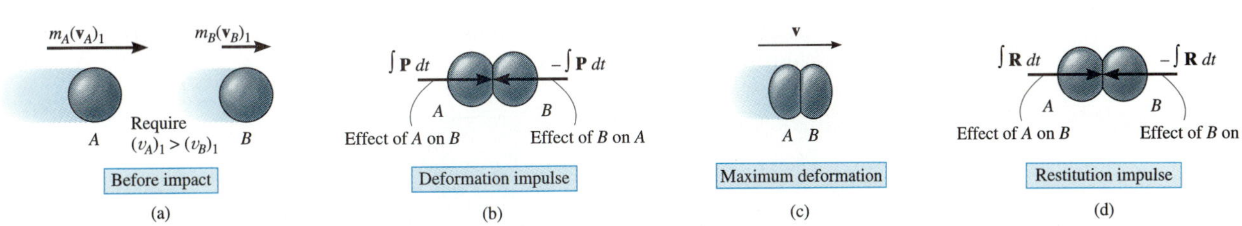

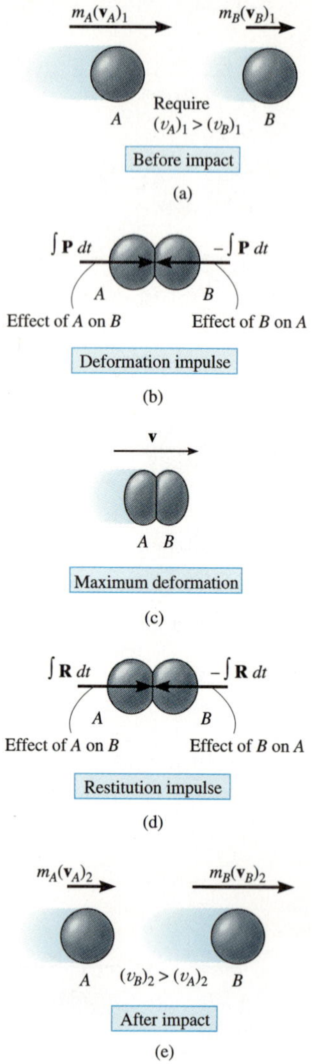

Fig. 15–14

In most problems the initial velocities of the particles will be *known*, and it will be necessary to determine their final velocities $(v_A)_2$ and $(v_B)_2$. In this regard, *momentum* for the *system of particles* is *conserved* since during collision the internal impulses of deformation and restitution *cancel*. Hence, referring to Fig. 15–14*a* and Fig. 15–14*e* we require

$$(\overset{+}{\rightarrow}) \qquad m_A(v_A)_1 + m_B(v_B)_1 = m_A(v_A)_2 + m_B(v_B)_2 \qquad (15\text{–}10)$$

In order to obtain a second equation necessary to solve for $(v_A)_2$ and $(v_B)_2$, we must apply the principle of impulse and momentum to *each particle*. For example, during the deformation phase for particle *A*, Figs. 15–14*a*, 15–14*b*, and 15–14*c*, we have

$$(\overset{+}{\rightarrow}) \qquad m_A(v_A)_1 - \int P\,dt = m_A v$$

For the restitution phase, Figs. 15–14*c*, 15–14*d*, and 15–14*e*,

$$(\overset{+}{\rightarrow}) \qquad m_A v - \int R\,dt = m_A(v_A)_2$$

The ratio of the restitution impulse to the deformation impulse is called the *coefficient of restitution, e*. From the above equations, this value for particle *A* is

$$e = \frac{\int R\,dt}{\int P\,dt} = \frac{v - (v_A)_2}{(v_A)_1 - v}$$

In a similar manner, we can establish *e* by considering particle *B*, Fig. 15–14. This yields

$$e = \frac{\int R\,dt}{\int P\,dt} = \frac{(v_B)_2 - v}{v - (v_B)_1}$$

If the unknown v is eliminated from the above two equations, the coefficient of restitution can be expressed in terms of the particles' initial and final velocities as

$$(\overset{+}{\rightarrow}) \qquad e = \frac{(v_B)_2 - (v_A)_2}{(v_A)_1 - (v_B)_1} \qquad (15\text{–}11)$$

Provided a value for e is specified, Eqs. 15–10 and 15–11 may be solved simultaneously to obtain $(v_A)_2$ and $(v_B)_2$. In doing so, however, it is important to carefully establish a sign convention for defining the positive direction for both $\mathbf{v}_A$ and $\mathbf{v}_B$ and then use it *consistently* when writing *both* equations. As noted from the application shown, and indicated symbolically by the arrow in parentheses, we have defined the positive direction to the right when referring to the motions of both A and B. Consequently, if a negative value results from the solution of either $(v_A)_2$ or $(v_B)_2$, it indicates motion is to the left.

Coefficient of Restitution. With reference to Figs. 15–14*a* and 15–14*e*, it is seen that Eq. 15–11 states that e is equal to the ratio of the relative velocity of the particles' separation *just after impact,* $(v_B)_2 - (v_A)_2$, to the relative velocity of the particles' approach *just before impact,* $(v_A)_1 - (v_B)_1$. By measuring these relative velocities experimentally, it has been found that e varies appreciably with impact velocity as well as with the size and shape of the colliding bodies. For these reasons the coefficient of restitution is reliable only when used with data which closely approximate the conditions which were known to exist when measurements of it were made. In general e has a value between zero and one, and one should be aware of the physical meaning of these two limits.

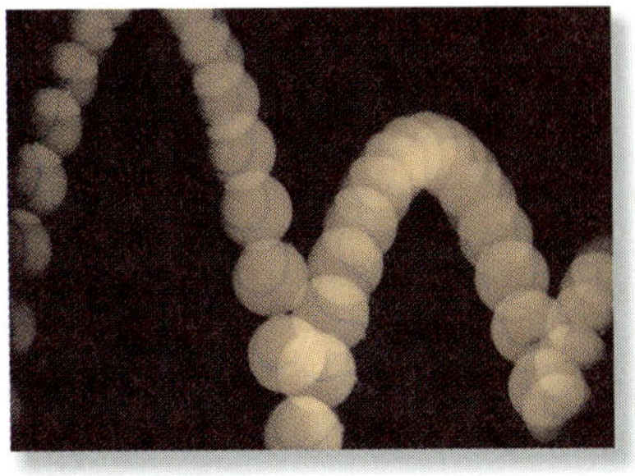

The quality of a manufactured tennis ball is measured by the height of its bounce, which can be related to its coefficient of restitution. Using the mechanics of oblique impact, engineers can design a separation device to maintain quality control of tennis balls after production.

Elastic Impact ($e = 1$): If the collision between the two particles is *perfectly elastic,* the deformation impulse ($\int \mathbf{P}\,dt$) is equal and opposite to the restitution impulse ($\int \mathbf{R}\,dt$). Although in reality this can never be achieved, $e = 1$ for an elastic collision.

Plastic Impact ($e = 0$): The impact is said to be *inelastic or plastic* when $e = 0$. In this case there is no restitution impulse given to the particles ($\int \mathbf{R}\,dt = \mathbf{0}$), so that after collision both particles couple or stick *together* and move with a common velocity.

From the above derivation it should be evident that the principle of work and energy cannot be used for the analysis of impact problems since it is not possible to know how the *internal forces* of deformation and restitution vary or displace during the collision. By knowing the particle's velocities before and after collision, however, the energy loss during collision can be calculated on the basis of the difference in the particle's kinetic energy. This energy loss, $\Sigma U_{1-2} = \Sigma T_2 - \Sigma T_1$, occurs because some of the initial kinetic energy of the particle is transformed into thermal energy as well as creating sound and localized deformation of the material when the collision occurs. In particular, if the impact is *perfectly elastic,* no energy is lost in the collision; whereas if the collision is *plastic,* the energy lost during collision is a maximum.

Procedure for Analysis (Central Impact)

In most cases the *final velocities* of two smooth particles are to be determined *just after* they are subjected to direct central impact. Provided the coefficient of restitution, the mass of each particle, and each particle's initial velocity *just before* impact are known, the solution to the problem can be obtained using the following two equations:

- The conservation of momentum applies to the system of particles, $\Sigma m v_1 = \Sigma m v_2$.
- The coefficient of restitution, $e = [(v_B)_2 - (v_A)_2]/[(v_A)_1 - (v_B)_1]$, relates the relative velocities of the particles along the line of impact, just before and just after collision.

When applying these two equations, the sense of an unknown velocity can be assumed. If the solution yields a negative magnitude, the velocity acts in the opposite sense.

Oblique Impact. When oblique impact occurs between two smooth particles, the particles move away from each other with velocities having unknown directions as well as unknown magnitudes. Provided the initial velocities are known, four unknowns are present in the problem. As shown in Fig. 15–15*a*, these unknowns may be represented either as $(v_A)_2$, $(v_B)_2$, θ_2, and ϕ_2, or as the x and y components of the final velocities.

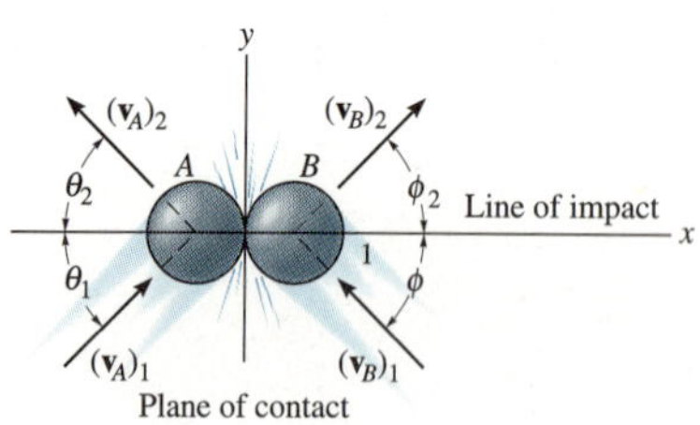

(a)

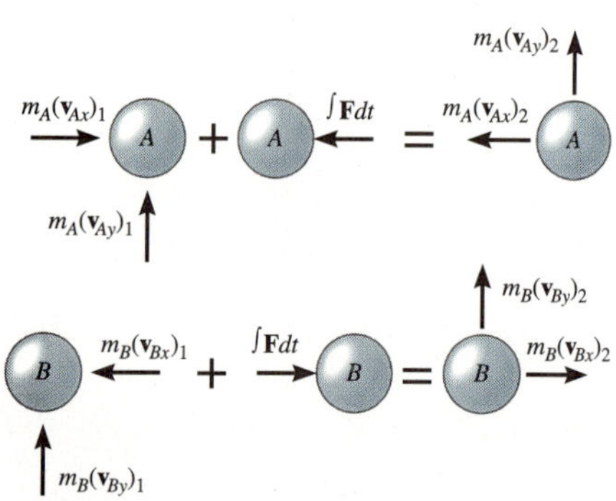

(b)

Fig. 15–15

Procedure for Analysis (Oblique Impact)

If the y axis is established within the plane of contact and the x axis along the line of impact, the impulsive forces of deformation and restitution act *only in the x direction,* Fig. 15–15*b*. Resolving the velocity or momentum vectors into components along the x and y axes, Fig. 15–15*b*, it is possible to write four independent scalar equations in order to determine $(v_{Ax})_2$, $(v_{Ay})_2$, $(v_{Bx})_2$, and $(v_{By})_2$.

- Momentum of the system is conserved *along the line of impact, x* axis, so that $\Sigma m(v_x)_1 = \Sigma m(v_x)_2$.
- The coefficient of restitution, $e = [(v_{Bx})_2 - (v_{Ax})_2]/[(v_{Ax})_1 - (v_{Bx})_1]$, relates the relative-velocity *components* of the particles *along the line of impact* (x axis).
- Momentum of particle A is conserved along the y axis, perpendicular to the line of impact, since no impulse acts on particle A in this direction.
- Momentum of particle B is conserved along the y axis, perpendicular to the line of impact, since no impulse acts on particle B in this direction.

Application of these four equations is illustrated numerically in Example 15–11.

EXAMPLE 15–9

The bag A, having a weight of 6 lb, is released from rest at the position $\theta = 0°$, as shown in Fig. 15–16a. After falling to $\theta = 90°$, it strikes an 18-lb box B. If the coefficient of restitution between the bag and box is $e = 0.5$, determine the velocities of the bag and box just after impact and the loss of energy during collision.

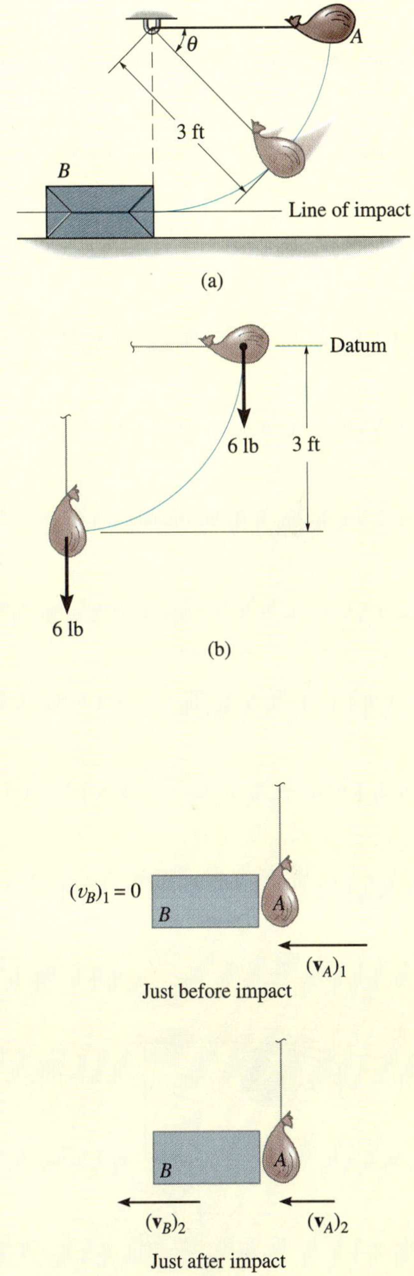

Fig. 15–16

Solution

This problem involves central impact. Why? Before analyzing the mechanics of the impact, however, it is first necessary to obtain the velocity of the bag *just before* it strikes the box.

Conservation of Energy. With the datum at $\theta = 0°$, Fig. 15–16b, we have

$$T_0 + V_0 = T_1 + V_1$$

$$0 + 0 = \frac{1}{2}\left(\frac{6\text{ lb}}{32.2\text{ ft/s}^2}\right)(v_A)_1^2 - 6\text{ lb}(3\text{ ft});\quad (v_A)_1 = 13.9\text{ ft/s}$$

Conservation of Momentum. After impact we will assume A and B travel to the left. Applying the conservation of momentum to the system, we have

$$(\overset{+}{\leftarrow})\qquad m_B(v_B)_1 + m_A(v_A)_1 = m_B(v_B)_2 + m_A(v_A)_2$$

$$0 + \frac{6\text{ lb}}{32.2\text{ ft/s}^2}(13.9\text{ ft/s}) = \frac{18\text{ lb}}{32.2\text{ ft/s}^2}(v_B)_2 + \frac{6\text{ lb}}{32.2\text{ ft/s}^2}(v_A)_2$$

$$(v_A)_2 = 13.9 - 3(v_B)_2 \qquad (1)$$

Coefficient of Restitution. Realizing that for separation to occur after collision $(v_B)_2 > (v_A)_2$, Fig. 15–16c, we have

$$(\overset{+}{\leftarrow})\quad e = \frac{(v_B)_2 - (v_A)_2}{(v_A)_1 - (v_B)_1};\qquad 0.5 = \frac{(v_B)_2 - (v_A)_2}{13.9\text{ ft/s} - 0}$$

$$(v_A)_2 = (v_B)_2 - 6.95 \qquad (2)$$

Solving Eqs. 1 and 2 simultaneously yields

$$(v_A)_2 = -1.74\text{ ft/s} = 1.74\text{ ft/s}\rightarrow \quad\text{and}\quad (v_B)_2 = 5.21\text{ ft/s}\leftarrow \qquad \textit{Ans.}$$

Loss of Energy. Applying the principle of work and energy to the bag and box just before and just after collision, we have

$$\Sigma U_{1\text{-}2} = T_2 - T_1;$$

$$\Sigma U_{1\text{-}2} = \left[\frac{1}{2}\left(\frac{18\text{ lb}}{32.2\text{ ft/s}^2}\right)(5.21\text{ ft/s})^2 + \frac{1}{2}\left(\frac{6\text{ lb}}{32.2\text{ ft/s}^2}\right)(1.74\text{ ft/s})^2\right] - \left[\frac{1}{2}\left(\frac{6\text{ lb}}{32.2\text{ ft/s}^2}\right)(13.9\text{ ft/s})^2\right]$$

$$\Sigma U_{1\text{-}2} = -10.1\text{ ft}\cdot\text{lb} \qquad \textit{Ans.}$$

Why is there an energy loss?

EXAMPLE 15–10

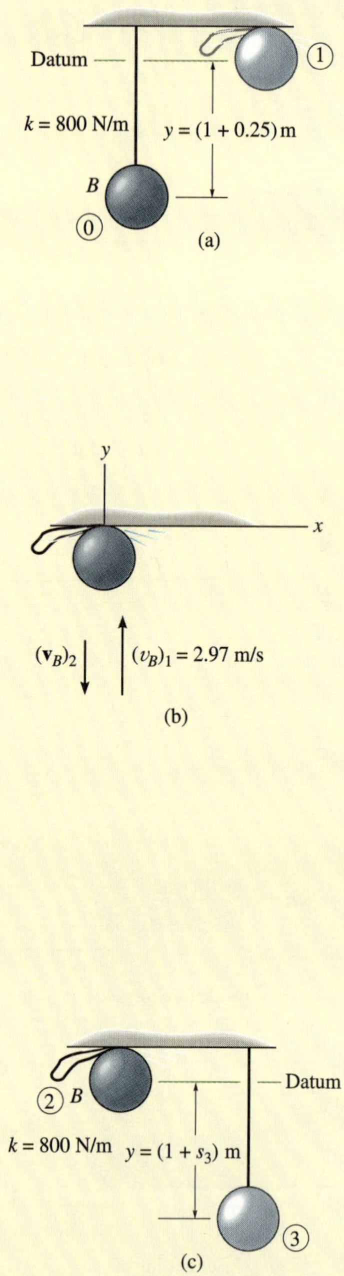

Fig. 15–17

The ball B shown in Fig. 15–17a has a mass of 1.5 kg and is suspended from the ceiling by a 1-m-long elastic cord. If the cord is *stretched* downward 0.25 m and the ball is released from rest, determine how far the cord stretches after the ball rebounds from the ceiling. The stiffness of the cord is $k = 800$ N/m, and the coefficient of restitution between the ball and ceiling is $e = 0.8$. The ball makes a central impact with the ceiling.

Solution

First we must obtain the velocity of the ball *just before* it strikes the ceiling using energy methods, then consider the impulse and momentum between the ball and ceiling, and finally again use energy methods to determine the stretch in the cord.

Conservation of Energy. With the datum located as shown in Fig. 15–17a, realizing that initially $y = y_0 = (1 + 0.25)\text{ m} = 1.25$ m, we have

$$T_0 + V_0 = T_1 + V_1$$

$$\tfrac{1}{2}m(v_B)_0^2 - W_B y_0 + \tfrac{1}{2}ks^2 = \tfrac{1}{2}m(v_B)_1^2 + 0$$

$$0 - 1.5(9.81)\text{ N}(1.25\text{ m}) + \tfrac{1}{2}(800\text{ N/m})(0.25\text{ m})^2 = \tfrac{1}{2}(1.5\text{ kg})(v_B)_1^2$$

$$(v_B)_1 = 2.97\text{ m/s}\uparrow$$

The interaction of the ball with the ceiling will now be considered using the principles of impact.* Since an unknown portion of the mass of the ceiling is involved in the impact, the conservation of momentum for the ball–ceiling system will not be written. The "velocity" of this portion of ceiling is zero since it remains at rest *both* before and after impact.

Coefficient of Restitution. Fig. 15–17b.

$$(+\uparrow)\quad e = \frac{(v_B)_2 - (v_A)_2}{(v_A)_1 - (v_B)_1};\quad 0.8 = \frac{(v_B)_2 - 0}{0 - 2.97\text{ m/s}}$$

$$(v_B)_2 = -2.37\text{ m/s} = 2.37\text{ m/s}\downarrow$$

Conservation of Energy. The maximum stretch s_3 in the cord may be determined by again applying the conservation of energy equation to the ball just after collision. Assuming that $y = y_3 = (1 + s_3)$ m, Fig. 15–17c, then

$$T_2 + V_2 = T_3 + V_3$$

$$\tfrac{1}{2}m(v_B)_2^2 + 0 = \tfrac{1}{2}m(v_B)_3^2 - W_B y_3 + \tfrac{1}{2}ks_3^2$$

$$\tfrac{1}{2}(1.5\text{ kg})(2.37\text{ m/s})^2 = 0 - 9.81(1.5)\text{ N}(1\text{ m} + s_3) + \tfrac{1}{2}(800\text{ N/m})s_3^2$$

$$400s_3^2 - 14.72s_3 - 18.94 = 0$$

Solving this quadratic equation for the positive root yields

$$s_3 = 0.237\text{ m} = 237\text{ mm}\qquad \textit{Ans.}$$

*The weight of the ball is considered a nonimpulsive force.

EXAMPLE 15-11

Two smooth disks A and B, having a mass of 1 kg and 2 kg, respectively, collide with the velocities shown in Fig. 15–18*a*. If the coefficient of restitution for the disks is $e = 0.75$, determine the x and y components of the final velocity of each disk just after collision.

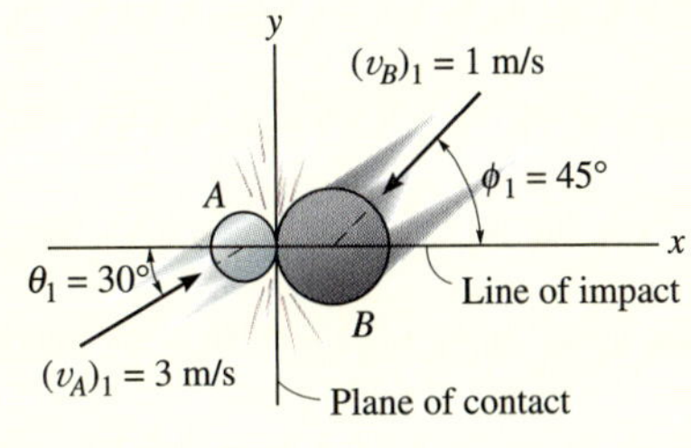

(a)

Solution

The problem involves *oblique impact*. Why? In order to seek a solution, we have established the x and y axes along the line of impact and the plane of contact, respectively, Fig. 15–18*a*.

Resolving each of the initial velocities into x and y components, we have

$$(v_{Ax})_1 = 3\cos 30° = 2.60 \text{ m/s} \qquad (v_{Ay})_1 = 3\sin 30° = 1.50 \text{ m/s}$$
$$(v_{Bx})_1 = -1\cos 45° = -0.707 \text{ m/s} \quad (v_{By})_1 = -1\sin 45° = -0.707 \text{ m/s}$$

The four unknown velocity components after collision are assumed to act in the positive directions, Fig. 15–18*b*. Since the impact occurs only in the x direction (line of impact), the conservation of momentum for *both* disks can be applied in this direction. Why?

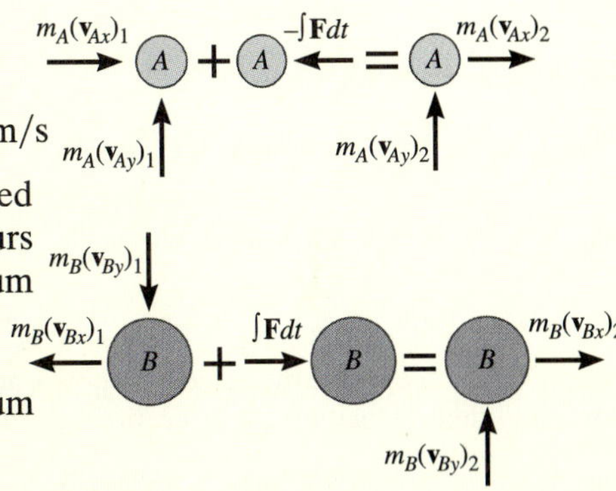

(b)

Conservation of "x" Momentum. In reference to the momentum diagrams, we have

$$(\overset{+}{\rightarrow}) \qquad m_A(v_{Ax})_1 + m_B(v_{Bx})_1 = m_A(v_{Ax})_2 + m_B(v_{Bx})_2$$
$$1\text{ kg}(2.60 \text{ m/s}) + 2\text{ kg}(-0.707 \text{ m/s}) = 1\text{ kg}(v_{Ax})_2 + 2\text{ kg}(v_{Bx})_2$$
$$(v_{Ax})_2 + 2(v_{Bx})_2 = 1.18 \qquad (1)$$

Coefficient of Restitution (x). Both disks are *assumed* to have components of velocity in the $+x$ direction after collision, Fig. 15–18*b*.

$$(\overset{+}{\rightarrow})\ e = \frac{(v_{Bx})_2 - (v_{Ax})_2}{(v_{Ax})_1 - (v_{Bx})_1}; \qquad 0.75 = \frac{(v_{Bx})_2 - (v_{Ax})_2}{2.60 \text{ m/s} - (-0.707 \text{ m/s})}$$
$$(v_{Bx})_2 - (v_{Ax})_2 = 2.48 \qquad (2)$$

Solving Eqs. 1 and 2 for $(v_{Ax})_2$ and $(v_{Bx})_2$ yields

$$(v_{Ax})_2 = -1.26 \text{ m/s} = 1.26 \text{ m/s} \leftarrow \qquad (v_{Bx})_2 = 1.22 \text{ m/s} \rightarrow \qquad \textit{Ans.}$$

Conservation of "y" Momentum. The momentum of *each disk* is *conserved* in the y direction (plane of contact), since the disks are smooth and therefore *no* external impulse acts in this direction. From Fig. 15–18*b*,

$$(+\uparrow)\ m_A(v_{Ay})_1 = m_A(v_{Ay})_2; \qquad (v_{Ay})_2 = 1.50 \text{ m/s} \uparrow \qquad \textit{Ans.}$$
$$(+\uparrow)\ m_B(v_{By})_1 = m_B(v_{By})_2; \ (v_{By})_2 = -0.707 \text{ m/s} = 0.707 \text{ m/s} \downarrow \ \textit{Ans.}$$

Show that when the velocity components are summed, one obtains the results shown in Fig. 15–18*c*.

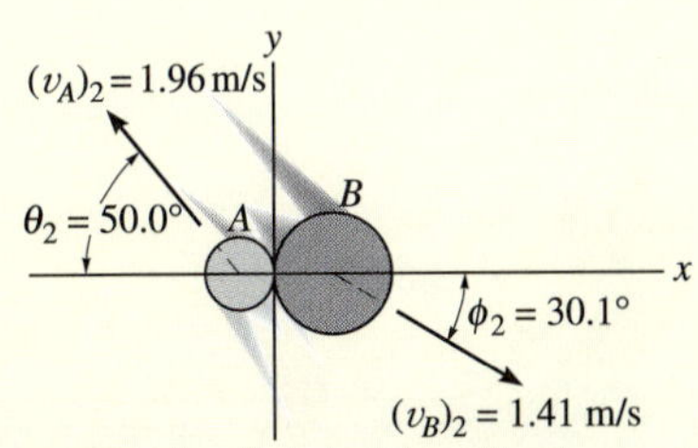

(c)

Fig. 15–18

Problems

15-55. An ivory ball having a mass of 200 g is released from rest at a height of 400 mm above a very large fixed metal surface. If the ball rebounds to a height of 325 mm above the surface, determine the coefficient of restitution between the ball and the surface.

***15-56.** Block B has a mass of 0.75 kg and is sliding forward on the *smooth* surface with a velocity $(v_B)_1 = 4\text{ m/s}$ when it strikes the 2-kg block A, which is originally at rest. If the coefficient of restitution between the blocks is $e = 0.5$, determine the velocities of A and B just after collision.

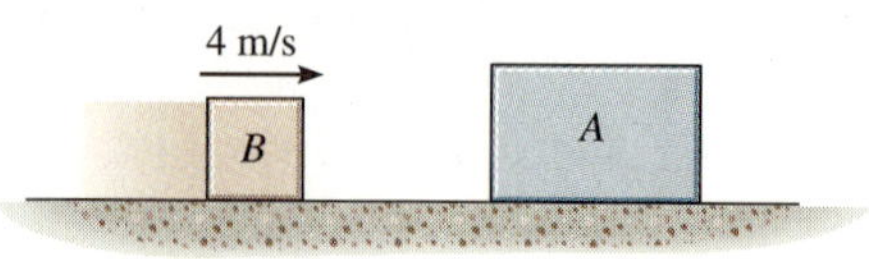

Prob. 15–56

15-57. Disks A and B have a mass of 2 kg and 4 kg, respectively. If they have the velocities shown, and $e = 0.4$, determine their velocities just after direct central impact.

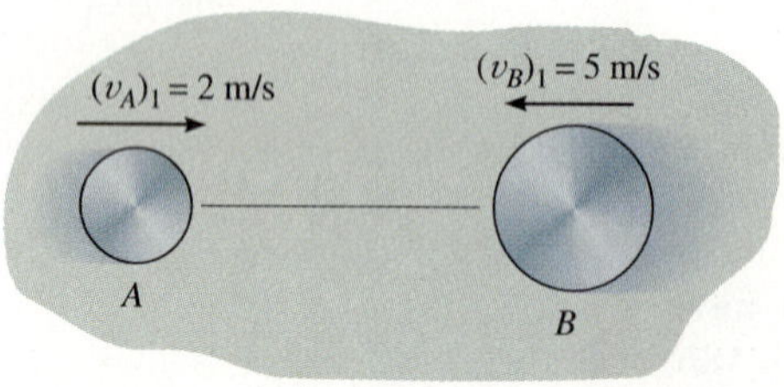

Prob. 15–57

15-58. The three balls each weigh 0.5 lb and have a coefficient of restitution of $e = 0.85$. If ball A is released from rest and strikes ball B and then ball B strikes ball C, determine the velocity of each ball after the second collision has occurred. Neglect the size of each ball.

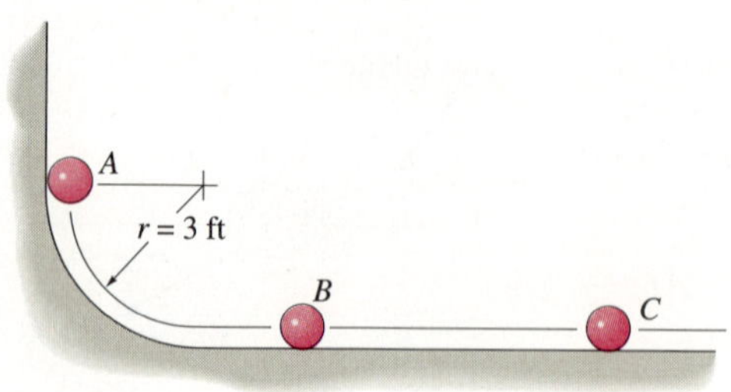

Prob. 15–58

15-59. If two disks A and B have the same mass and are subjected to direct central impact such that the collision is perfectly elastic ($e = 1$), prove that the kinetic energy before collision equals the kinetic energy after collision. The surface upon which they slide is smooth.

***15-60.** Each ball has a mass m and the coefficient of restitution between the balls is e. If they are moving towards one another with a velocity v, determine their speeds after collision. Also, determine their common velocity when they reach the state of maximum deformation. Neglect the size of each ball.

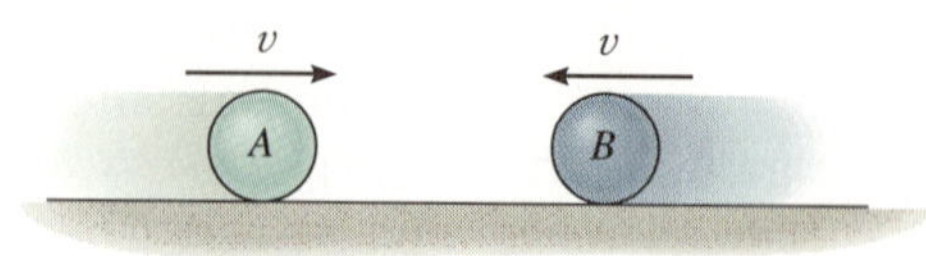

Prob. 15–60

15-61. The 1-lb ball A is thrown so that when it strikes the 10-lb block B it is traveling horizontally at 20 ft/s. If the coefficient of restitution between A and B is $e = 0.6$, and the coefficient of kinetic friction between the plane and the block is $\mu_k = 0.4$, determine the time before block B stops sliding.

15-62. The 1-lb ball A is thrown so that when it strikes the 10-lb block B it is traveling horizontally at 20 ft/s. If the coefficient of restitution between A and B is $e = 0.6$, and the coefficient of kinetic friction between the plane and the block is $\mu_k = 0.4$, determine the distance block B slides on the plane before stopping.

15-63. The 1-lb ball A is thrown so that when it strikes the 10-lb block B it is traveling horizontally at 20 ft/s. Determine the average normal force exerted between A and B if the impact occurs in 0.02 s. The coefficient of restitution between A and B is $e = 0.6$.

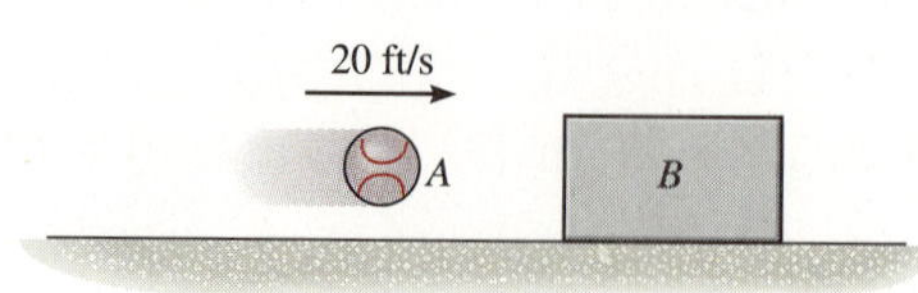

Probs. 15–61/62/63

*15-64. The 0.2-lb ball bearing travels over the edge A with a velocity of $v_A = 3$ ft/s. Determine the speed at which it rebounds from the smooth inclined plane at B. Take $e = 0.8$.

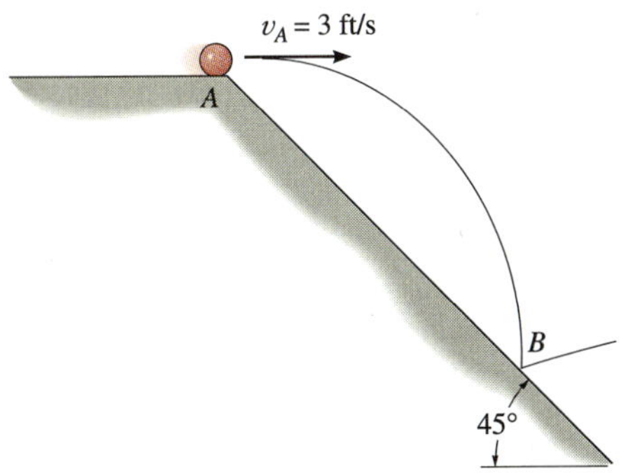

Prob. 15–64

15-65. A ball has a mass m and is dropped onto a surface from a height h. If the coefficient of restitution is e between the ball and the surface, determine the time needed for the ball to stop bouncing.

15-66. To test the manufactured properties of 2-lb steel balls, each ball is released from rest as shown and strikes the 45° smooth inclined surface. If the coefficient of restitution is to be $e = 0.8$, determine the distance s to where the ball must strike the horizontal plane at A. At what speed does the ball strike point A?

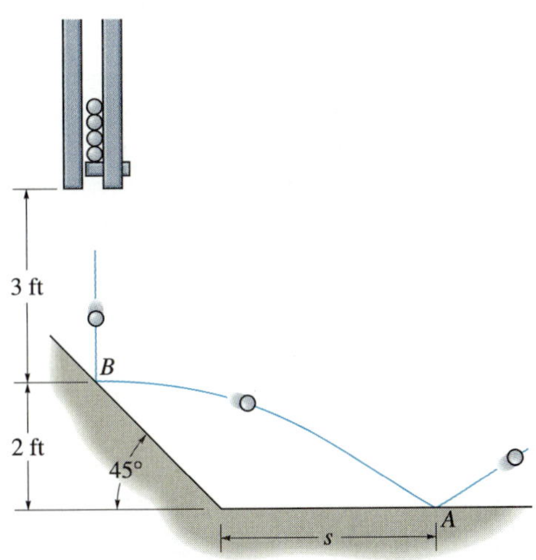

Prob. 15–66

15-67. The 2-kg ball is thrown at the suspended 20-kg block with a velocity of 4 m/s. If the coefficient of restitution between the ball and the block is $e = 0.8$, determine the maximum height h to which the block will swing before it momentarily stops.

*15-68. The 2-kg ball is thrown at the suspended 20-kg block with a velocity of 4 m/s. If the time of impact between the ball and the block is 0.005 s, determine the average normal force exerted on the block during this time. Take $e = 0.8$.

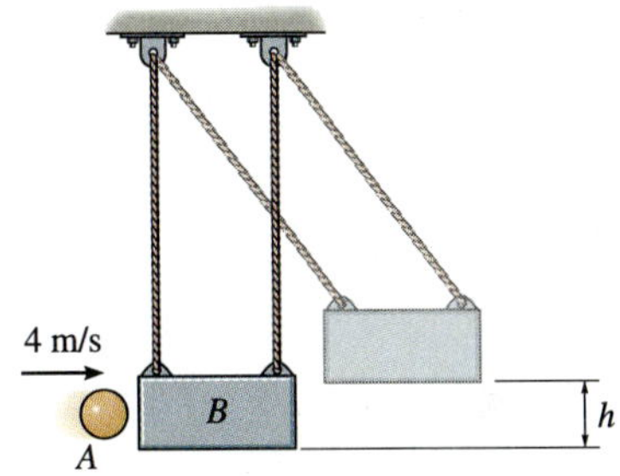

Probs. 15–67/68

15-69. The slider block B is confined to move within the smooth slot. It is connected to two springs, each of which has a stiffness of $k = 30$ N/m. They are originally stretched 0.5 m when $s = 0$ as shown. Determine the maximum distance, s_{max}, block B moves after it is hit by block A which is originally traveling at $(v_A)_1 = 8$ m/s. Take $e = 0.4$ and the mass of each block to be 1.5 kg.

15-70. In Prob. 15-69 determine the average net force exerted between the two blocks A and B during impact if the impact occurs in 0.005 s.

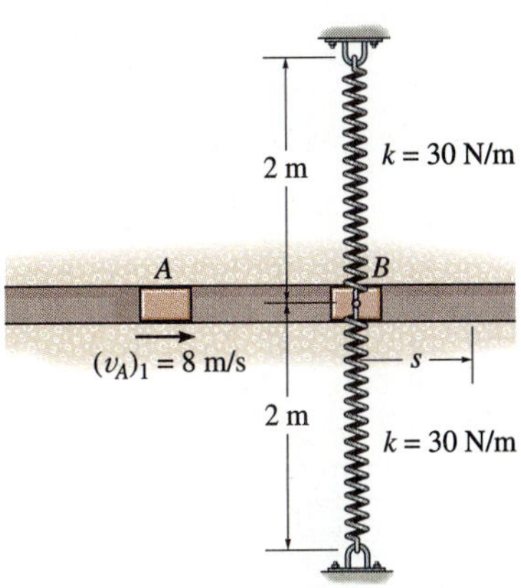

Probs. 15–69/70

15-71. The drop hammer H has a weight of 900 lb and falls from rest $h = 3$ ft onto a forged anvil plate P that has a weight of 500 lb. The plate is mounted on a set of springs which have a combined stiffness of $k_T = 500$ lb/ft. Determine (a) the velocity of P and H just after collision and (b) the maximum compression in the springs caused by the impact. The coefficient of restitution between the hammer and the plate is $e = 0.6$. Neglect friction along the vertical guideposts A and B.

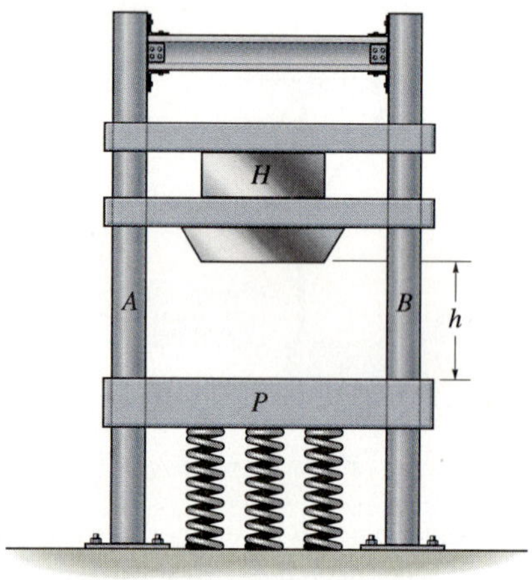

Prob. 15–71

***15-72.** The 8-lb ball is released from rest 10 ft from the surface of a flat plate P which weighs 6 lb. Determine the maximum compression in the spring if the impact is perfectly elastic.

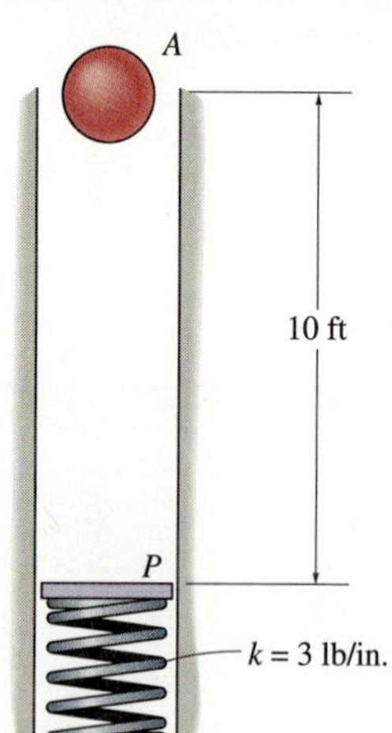

Prob. 15–72

15-73. The pile P has a mass of 800 kg and is being driven into *loose sand* using the 300-kg counterweight C which is dropped a distance of 0.5 m from the top of the pile. Determine the initial speed of the pile just after it is struck by the counterweight. The coefficient of restitution between the counterweight and the pile is $e = 0.1$. Neglect the impulses due to the weights of the pile and counterweight and the impulse due to the sand during the impact.

15-74. The pile P has a mass of 800 kg and is being driven into *loose sand* using the 300-kg counterweight which is dropped a distance of 0.5 m from the top of the pile. Determine the distance the pile is driven into the sand after one blow if the sand offers a frictional resistance against the pile of 18 kN. The coefficient of restitution between the counterweight and the pile is $e = 0.1$. Neglect the impulses due to the weights of the pile and counterweight and the impulse due to the sand during the impact.

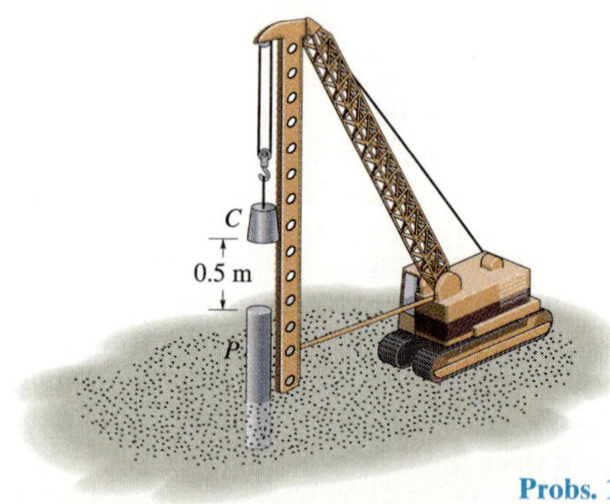

Probs. 15–73/74

15-75. The 10-lb collar B is at rest, and when it is in the position shown the spring is unstretched. If another 1-lb collar A strikes it so that B slides 4 ft on the smooth rod before momentarily stopping, determine the velocity of A just after impact, and the average force exerted between A and B during the impact if the impact occurs in 0.002 s. The coefficient of restitution between A and B is $e = 0.5$.

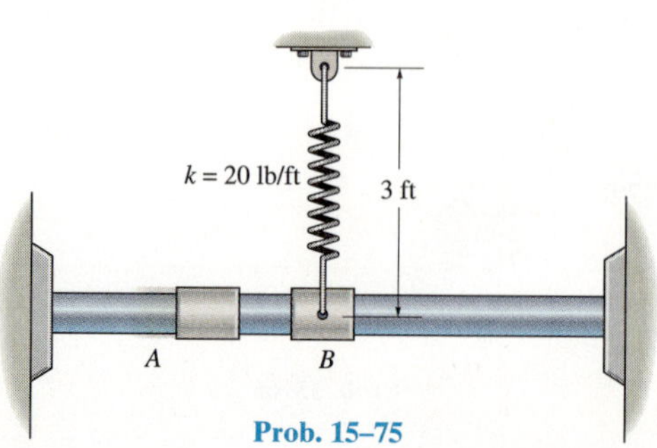

Prob. 15–75

*15-76. The ball is ejected from the tube with a horizontal velocity of $v_1 = 8$ ft/s as shown. If the coefficient of restitution between the ball and the ground is $e = 0.8$, determine (a) the velocity of the ball just after it rebounds from the ground and (b) the maximum height to which the ball rises after the first bounce.

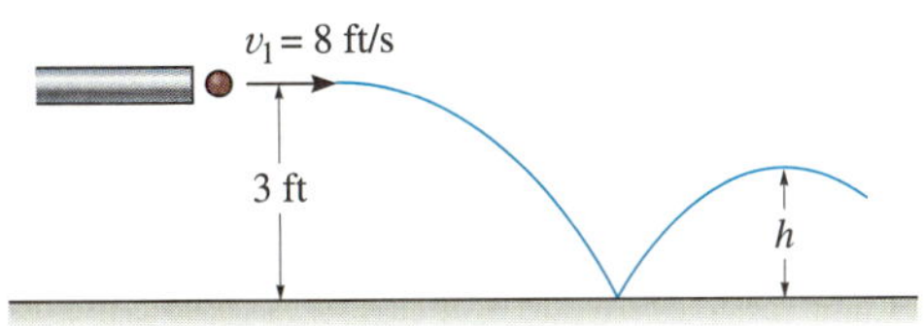

Prob. 15–76

15-77. A pitching machine throws the 0.5-kg ball towards the wall with an initial velocity $v_A = 10$ m/s as shown. Determine (a) the velocity at which it strikes the wall at B, (b) the velocity at which it rebounds from the wall if $e = 0.5$, and (c) the distance s from the wall to where it strikes the ground at C.

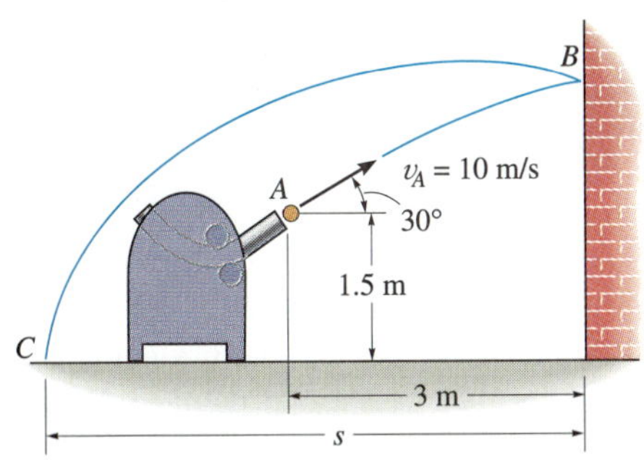

Prob. 15–77

15-78. The 5-lb box B is dropped from rest 5 ft from the top of the 10-lb plate P, which is supported by the spring having a stiffness of $k = 30$ lb/ft. If $e = 0.6$ between the box and plate, determine the maximum compression imparted to the spring. Neglect the mass of the spring.

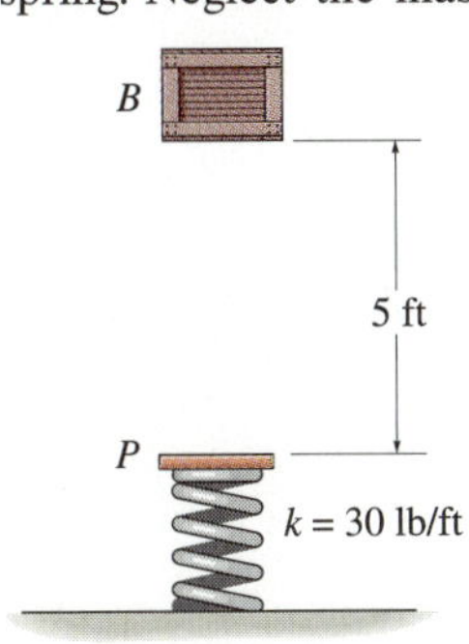

Prob. 15–78

15-79. The 200-g billiard ball is moving with a speed of 2.5 m/s when it strikes the side of the pool table at A. If the coefficient of restitution between the ball and the side of the table is $e = 0.6$, determine the speed of the ball just after striking the table twice, i.e., at A, then at B. Neglect the size of the ball.

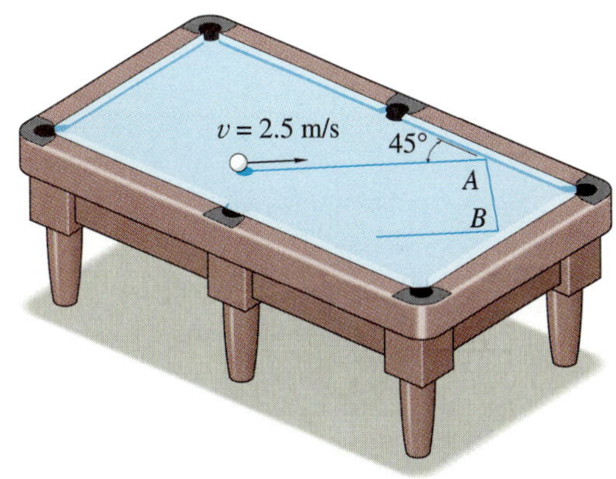

Prob. 15–79

*15-80. Two smooth billiard balls A and B each have a mass of 200 g. If A strikes B with a velocity $(v_A)_1 = 1.5$ m/s as shown, determine their final velocities just after collision. Ball B is originally at rest and the coefficient of restitution is $e = 0.85$. Neglect the size of each ball.

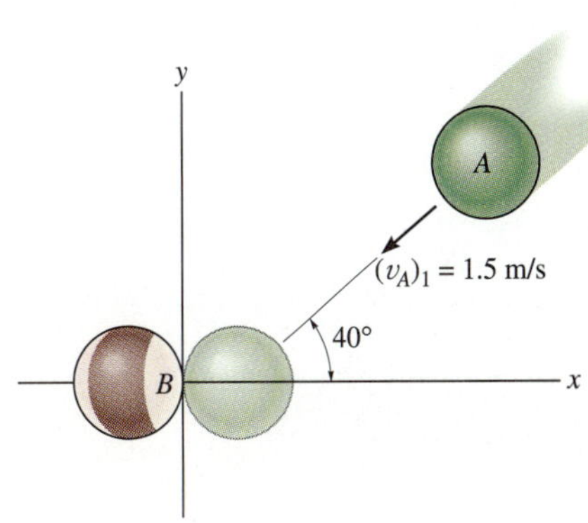

Prob. 15–80

15-81. The two hockey pucks A and B each have a mass of 250 g. If they collide at O and are deflected along the colored paths, determine their speeds just after impact. Assume that the icy surface over which they slide is smooth. *Hint:* Since the y' axis is *not* along the line of impact, apply the conservation of momentum along the x' and y' axes.

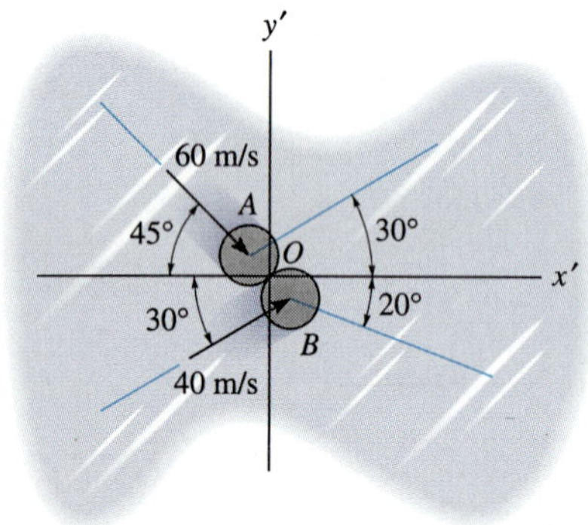

Prob. 15–81

15-83. Two cars A and B, each having a weight of 4000 lb, collide on the icy pavement of an intersection. The direction of motion of each car after collision is measured from snow tracks as shown. If the driver in car A states that he was going 44 ft/s (30 mi/h) just before collision and that after collision he applied the brakes so that his car skidded 10 ft before stopping, determine the approximate speed of car B just before the collision. Assume that the coefficient of kinetic friction between the car wheels and the pavement is $\mu_k = 0.15$. *Note:* The line of impact has not been defined; however, this information is not needed for the solution.

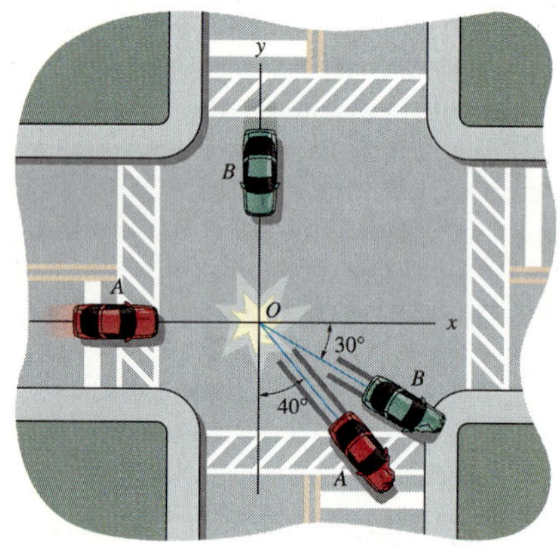

Prob. 15–83

15-82. The two disks A and B have a mass of 3 kg and 5 kg, respectively. If they collide with the initial velocities shown, determine their velocities just after impact. The coefficient of restitution is $e = 0.65$.

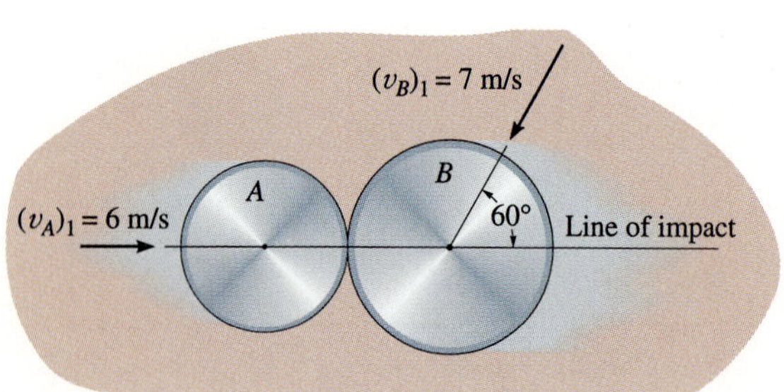

Prob. 15–82

***15-84.** Two coins A and B have the initial velocities shown just before they collide at point O. If they have weights of $W_A = 13.2(10^{-3})$ lb and $W_B = 6.60(10^{-3})$ lb and the surface upon which they slide is smooth, determine their speeds just after impact. The coefficient of restitution is $e = 0.65$.

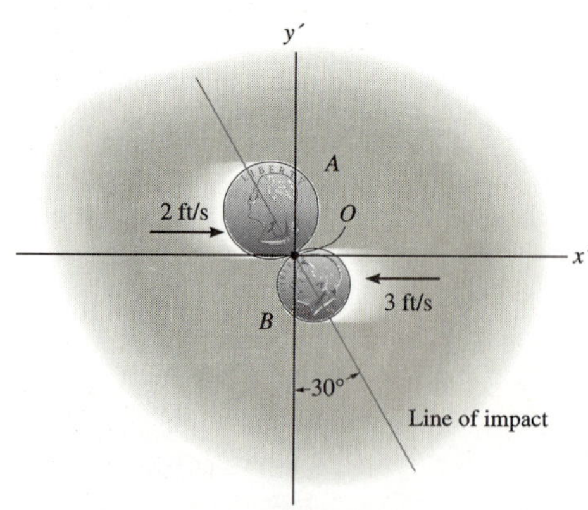

Prob. 15–84

15-85. The 2-kg ball is thrown so that it is traveling horizontally at 10 m/s when it strikes the 6-kg block as it is traveling down the inclined plane at 1 m/s. If the coefficient of restitution between the ball and the block is $e = 0.6$, determine the speeds of the ball and the block just after the impact. Also, what distance does B slide up the plane before it stops? The coefficient of kinetic friction between the block and the plane is $\mu_k = 0.4$.

15-86. The 2-kg ball is thrown so that it is traveling horizontally at 10 m/s when it strikes the 6-kg block as it is traveling down the smooth inclined plane at 1 m/s. If the coefficient of restitution between the ball and the block is $e = 0.6$, and the impact occurs in 0.006 s, determine the average impulsive force between the ball and block.

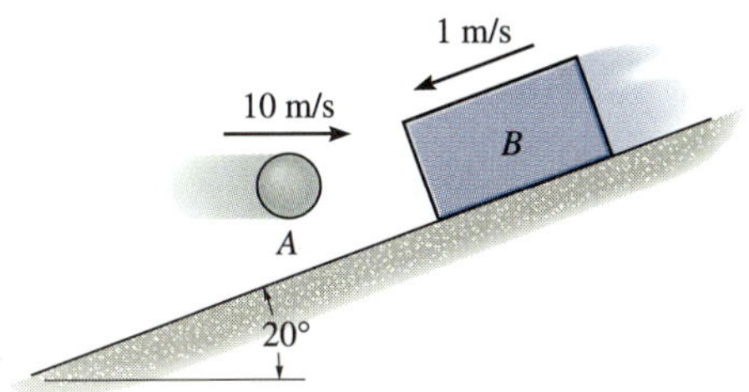

Probs. 15–85/86

15-87. Two disks A and B each have a weight of 2 lb and the initial velocities shown just before they collide. If the coefficient of restitution is $e = 0.5$, determine their speeds just after impact.

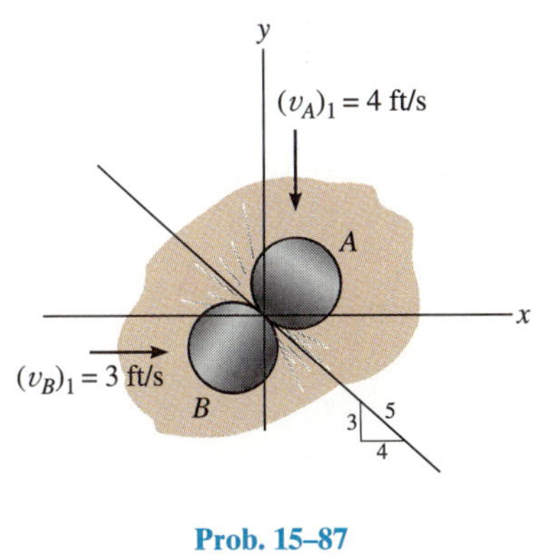

Prob. 15–87

***15-88.** The two billiard balls A and B are originally in contact with one another when a third ball C strikes each of them at the same time as shown. If ball C remains at rest after the collision, determine the coefficient of restitution. All the balls have the same mass. Neglect the size of each ball.

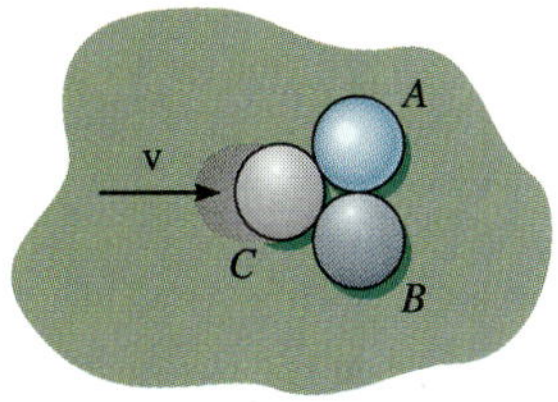

Prob. 15–88

15-89. Ball A strikes ball B with an initial velocity of $(\mathbf{v}_A)_1$ as shown. If both balls have the same mass and the collision is perfectly elastic, determine the angle θ after collision. Ball B is originally at rest. Neglect the size of each ball.

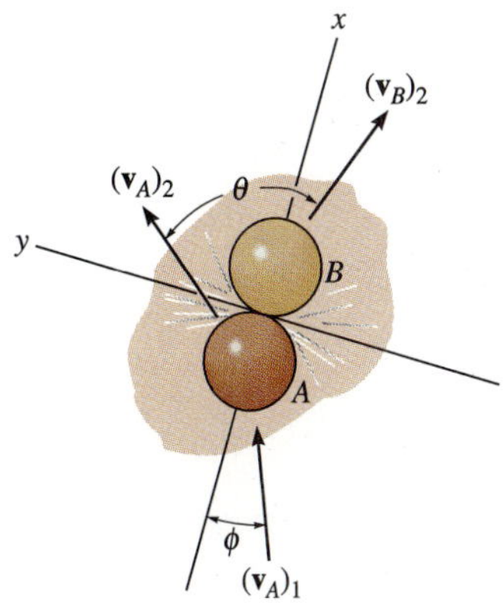

Prob. 15–89

15.5 Angular Momentum

The *angular momentum* of a particle about point O is defined as the "moment" of the particle's linear momentum about O. Since this concept is analogous to finding the moment of a force about a point, the angular momentum, $\mathbf{H}_O$, is sometimes referred to as the *moment of momentum.*

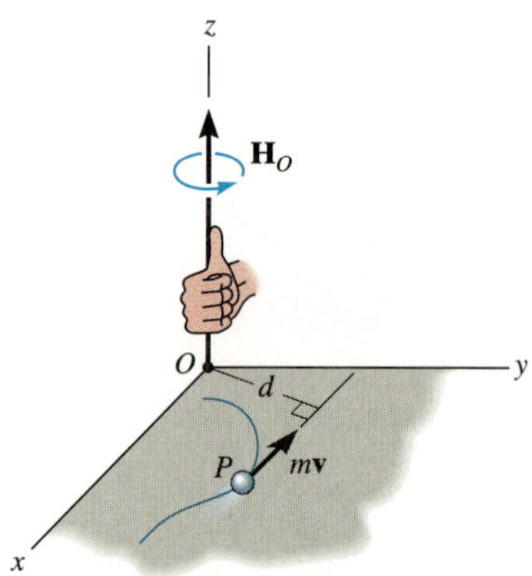

Fig. 15–19

Scalar Formulation. If a particle is moving along a curve lying in the x–y plane, Fig. 15–19, the angular momentum at any instant can be determined about point O (actually the z axis) by using a scalar formulation. The *magnitude* of $\mathbf{H}_O$ is

$$(H_O)_z = (d)(mv) \qquad (15\text{–}12)$$

Here d is the moment arm or perpendicular distance from O to the line of action of $m\mathbf{v}$. Common units for $(H_O)_z$ are $\text{kg}\cdot\text{m}^2/\text{s}$ or $\text{slug}\cdot\text{ft}^2/\text{s}$. The *direction* of $\mathbf{H}_O$ is defined by the right-hand rule. As shown, the curl of the fingers of the right hand indicates the sense of rotation of $m\mathbf{v}$ about O, so that in this case the thumb (or $\mathbf{H}_O$) is directed perpendicular to the x–y plane along the $+z$ axis.

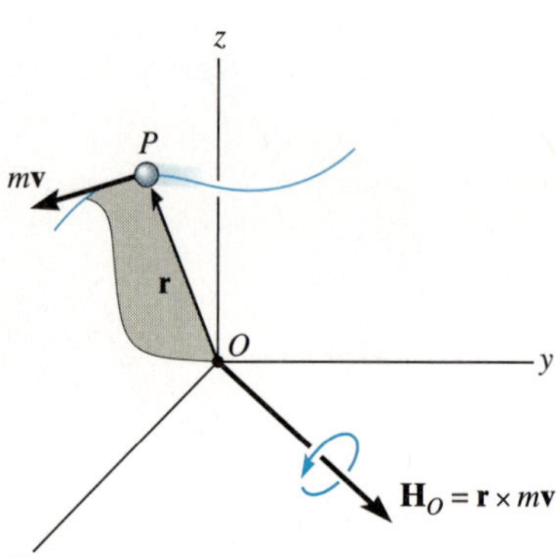

Fig. 15–20

Vector Formulation. If the particle is moving along a space curve, Fig. 15–20, the vector cross product can be used to determine the *angular momentum* about O. In this case

$$\mathbf{H}_O = \mathbf{r} \times m\mathbf{v} \qquad (15\text{–}13)$$

Here $\mathbf{r}$ denotes a position vector drawn from point O to the particle P. As shown in the figure, $\mathbf{H}_O$ is *perpendicular* to the shaded plane containing $\mathbf{r}$ and $m\mathbf{v}$.

In order to evaluate the cross product, $\mathbf{r}$ and $m\mathbf{v}$ should be expressed in terms of their Cartesian components, so that the angular momentum is determined by evaluating the determinant:

$$\mathbf{H}_O = \begin{vmatrix} \mathbf{i} & \mathbf{j} & \mathbf{k} \\ r_x & r_y & r_z \\ mv_x & mv_y & mv_z \end{vmatrix} \qquad (15\text{–}14)$$

15.6 Relation Between Moment of a Force and Angular Momentum

The moments about point O of all the forces acting on the particle in Fig. 15–21*a* may be related to the particle's angular momentum by using the equation of motion. If the mass of the particle is constant, we may write

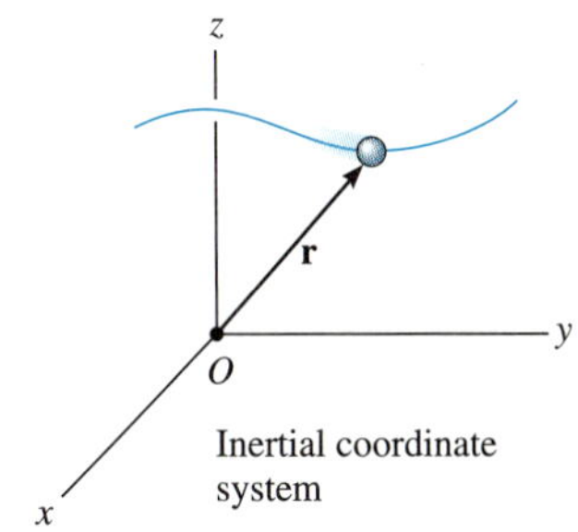

(a)

Fig. 15–21

$$\Sigma \mathbf{F} = m\dot{\mathbf{v}}$$

The moments of the forces about point O can be obtained by performing a cross-product multiplication of each side of this equation by the position vector **r**, which is measured in the x, y, z inertial frame of reference. We have

$$\Sigma \mathbf{M}_O = \mathbf{r} \times \Sigma \mathbf{F} = \mathbf{r} \times m\dot{\mathbf{v}}$$

From Appendix C, the derivative of $\mathbf{r} \times m\mathbf{v}$ can be written as

$$\dot{\mathbf{H}}_O = \frac{d}{dt}(\mathbf{r} \times m\mathbf{v}) = \dot{\mathbf{r}} \times m\mathbf{v} + \mathbf{r} \times m\dot{\mathbf{v}}$$

The first term on the right side, $\dot{\mathbf{r}} \times m\mathbf{v} = m(\dot{\mathbf{r}} \times \dot{\mathbf{r}}) = \mathbf{0}$, since the cross product of a vector with itself is zero. Hence, the above equation becomes

$$\Sigma \mathbf{M}_O = \dot{\mathbf{H}}_O \qquad (15\text{–}15)$$

This equation states that *the resultant moment about point O of all the forces acting on the particle is equal to the time rate of change of the particle's angular momentum about point O.* This result is similar to Eq. 15–1, i.e.,

$$\Sigma \mathbf{F} = \dot{\mathbf{L}} \qquad (15\text{–}16)$$

Here $\mathbf{L} = m\mathbf{v}$, so *the resultant force acting on the particle is equal to the time rate of change of the particle's linear momentum.*

From the derivations, it is seen that Eqs. 15–15 and 15–16 are actually another way of stating Newton's second law of motion. In other sections of this book it will be shown that these equations have many practical applications when extended and applied to the solution of problems involving either a system of particles or a rigid body.

System of Particles. An equation having the same form as Eq. 15–15 may be derived for the system of particles shown in Fig. 15–21*b*. The forces acting on the arbitrary *i*th particle of the system consist of a resultant *external force* $\mathbf{F}_i$ and a resultant *internal force* $\mathbf{f}_i$. Expressing the moments of these forces about point O, using the form of Eq. 15–15, we have

$$(\mathbf{r}_i \times \mathbf{F}_i) + (\mathbf{r}_i \times \mathbf{f}_i) = (\dot{\mathbf{H}}_i)_O$$

Here $\mathbf{r}_i$ is the position vector drawn from the origin O of an inertial frame of reference to the *i*th particle, and $(\dot{\mathbf{H}}_i)_O$ is the time rate of change in the angular momentum of the *i*th particle about O. Similar equations can be written for each of the other particles of the system. When the results are summed vectorially, the result is

$$\Sigma(\mathbf{r}_i \times \mathbf{F}_i) + \Sigma(\mathbf{r}_i \times \mathbf{f}_i) = \Sigma(\dot{\mathbf{H}}_i)_O$$

The second term is zero since the internal forces occur in equal but opposite collinear pairs, and hence the moment of each pair about point O is zero. Dropping the index notation, the above equation can be written in a simplified form as

$$\Sigma \mathbf{M}_O = \dot{\mathbf{H}}_O \qquad (15\text{–}17)$$

which states that *the sum of the moments about point O of all the external forces acting on a system of particles is equal to the time rate of change of the total angular momentum of the system about point O.* Although O has been chosen here as the origin of coordinates, it actually can represent any *fixed point* in the inertial frame of reference.

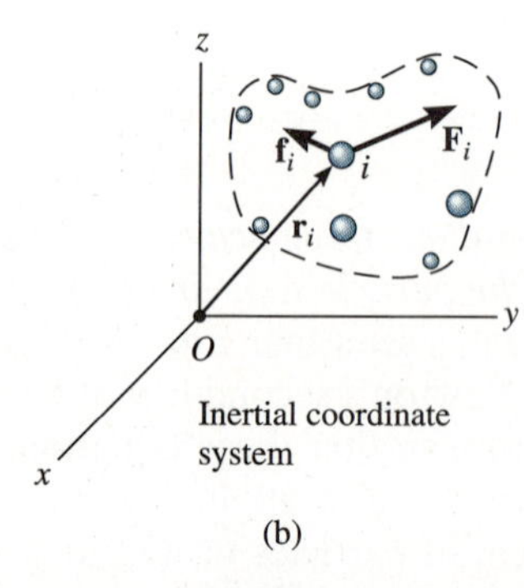

(b)

Fig. 15–21

EXAMPLE 15–12

The box shown in Fig. 15–22*a* has a mass m and is traveling down the smooth circular ramp such that when it is at the angle θ it has a speed v. Determine its angular momentum about point O at this instant and the rate of increase in its speed, i.e., a_t.

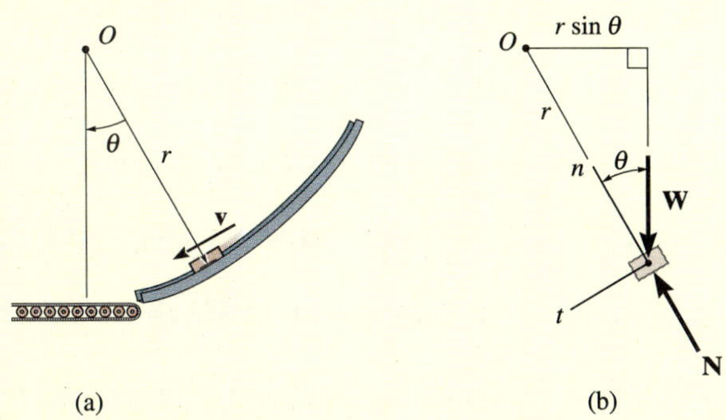

Fig. 15–22

Solution

Since **v** is tangent to the path, applying Eq. 15–12 the angular momentum is

$$H_O = rmv \curvearrowright \qquad \textit{Ans.}$$

The rate of increase in its speed (dv/dt) can be found by applying Eq. 15–15. From the free-body diagram of the block, Fig. 15–22*b*, it is seen that only the weight $W = mg$ contributes a moment about point O. We have

$$\curvearrowleft +\Sigma M_O = \dot{H}_O; \qquad mg(r \sin\theta) = \frac{d}{dt}(rmv)$$

Since r and m are constant,

$$mgr \sin\theta = rm\frac{dv}{dt}$$

$$\frac{dv}{dt} = g\sin\theta \qquad \textit{Ans.}$$

This same result can, of course, be obtained from the equation of motion applied in the tangential direction, Fig. 15–22*b*, i.e.,

$$+\swarrow \Sigma F_t = ma_t; \qquad mg\sin\theta = m\left(\frac{dv}{dt}\right)$$

$$\frac{dv}{dt} = g\sin\theta \qquad \textit{Ans.}$$

15.7 Angular Impulse and Momentum Principles

Principle of Angular Impulse and Momentum. If Eq. 15–15 is rewritten in the form $\Sigma \mathbf{M}_O\, dt = d\mathbf{H}_O$ and integrated, we have, assuming that at time $t = t_1$, $\mathbf{H}_O = (\mathbf{H}_O)_1$ and at time $t = t_2$, $\mathbf{H}_O = (\mathbf{H}_O)_2$,

$$\Sigma \int_{t_1}^{t_2} \mathbf{M}_O\, dt = (\mathbf{H}_O)_2 - (\mathbf{H}_O)_1$$

or

$$(\mathbf{H}_O)_1 + \Sigma \int_{t_1}^{t_2} \mathbf{M}_O\, dt = (\mathbf{H}_O)_2 \qquad (15\text{–}18)$$

This equation is referred to as the *principle of angular impulse and momentum.* The initial and final angular momenta $(\mathbf{H}_O)_1$ and $(\mathbf{H}_O)_2$ are defined as the moment of the linear momentum of the particle ($\mathbf{H}_O = \mathbf{r} \times m\mathbf{v}$) at the instants t_1 and t_2, respectively. The second term on the left side, $\Sigma \int \mathbf{M}_O\, dt$, is called the *angular impulse.* It is determined by integrating, with respect to time, the moments of all the forces acting on the particle over the time period t_1 to t_2. Since the moment of a force about point O is $\mathbf{M}_O = \mathbf{r} \times \mathbf{F}$, the angular impulse may be expressed in vector form as

$$\text{angular impulse} = \int_{t_1}^{t_2} \mathbf{M}_O\, dt = \int_{t_1}^{t_2} (\mathbf{r} \times \mathbf{F})\, dt \qquad (15\text{–}19)$$

Here $\mathbf{r}$ is a position vector which extends from point O to any point on the line of action of $\mathbf{F}$.

In a similar manner, using Eq. 15–18, the principle of angular impulse and momentum for a system of particles may be written as

$$\Sigma(\mathbf{H}_O)_1 + \Sigma \int_{t_1}^{t_2} \mathbf{M}_O\, dt = \Sigma(\mathbf{H}_O)_2 \qquad (15\text{–}20)$$

Here the first and third terms represent the angular momenta of the system of particles $[\Sigma \mathbf{H}_O = \Sigma(\mathbf{r}_i \times m\mathbf{v}_i)]$ at the instants t_1 and t_2. The second term is the sum of the angular impulses given to all the particles from t_1 to t_2. Recall that these impulses are created only by the moments of the external forces acting on the system where, for the ith particle, $\mathbf{M}_O = \mathbf{r}_i \times \mathbf{F}_i$.

Vector Formulation. Using impulse and momentum principles, it is therefore possible to write two equations which define the particle's motion, namely, Eqs. 15–3 and 15–18, restated as

$$m\mathbf{v}_1 + \Sigma \int_{t_1}^{t_2} \mathbf{F}\, dt = m\mathbf{v}_2$$
$$(\mathbf{H}_O)_1 + \Sigma \int_{t_1}^{t_2} \mathbf{M}_O\, dt = (\mathbf{H}_O)_2 \qquad (15\text{–}21)$$

Scalar Formulation. In general, the above equations may be expressed in x, y, z component form, yielding a total of six independent scalar equations. If the particle is confined to move in the x–y plane, three independent scalar equations may be written to express the motion, namely,

$$m(v_x)_1 + \Sigma \int_{t_1}^{t_2} F_x\, dt = m(v_x)_2$$
$$m(v_y)_1 + \Sigma \int_{t_1}^{t_2} F_y\, dt = m(v_y)_2 \qquad (15\text{–}22)$$
$$(H_O)_1 + \Sigma \int_{t_1}^{t_2} M_O\, dt = (H_O)_2$$

The first two of these equations represent the principle of linear impulse and momentum in the x and y directions, which has been discussed in Sec. 15.1, and the third equation represents the principle of angular impulse and momentum about the z axis.

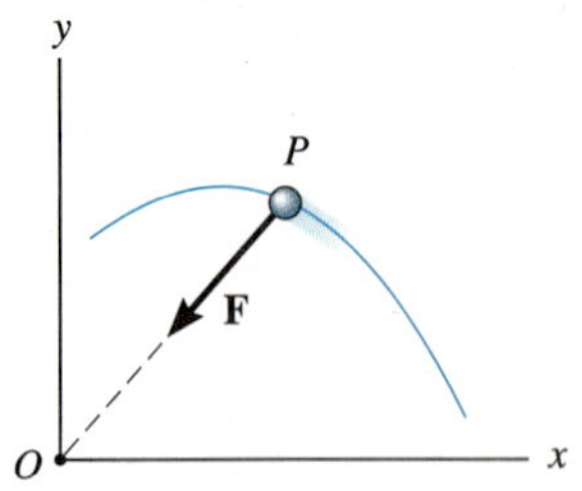

Fig. 15–23

Conservation of Angular Momentum. When the angular impulses acting on a particle are all zero during the time t_1 to t_2, Eq. 15–18 reduces to the following simplified form:

$$(\mathbf{H}_O)_1 = (\mathbf{H}_O)_2 \tag{15–23}$$

This equation is known as the *conservation of angular momentum.* It states that from t_1 to t_2 the particle's angular momentum remains constant. Obviously, if no external impulse is applied to the particle, both linear and angular momentum will be conserved. In some cases, however, the particle's angular momentum will be conserved and linear momentum may not. An example of this occurs when the particle is subjected *only* to a *central force* (see Sec. 13.7). As shown in Fig. 15–23, the impulsive central force **F** is always directed toward point O as the particle moves along the path. Hence, the angular impulse (moment) created by **F** about the z axis passing through point O is always zero, and therefore angular momentum of the particle is conserved about this axis.

From Eq. 15–20, we can also write the conservation of angular momentum for a system of particles, namely,

$$\Sigma(\mathbf{H}_O)_1 = \Sigma(\mathbf{H}_O)_2 \tag{15–24}$$

In this case the summation must include the angular momenta of all particles in the system.

Provided air resistance is neglected, the passengers on this amusement-park ride are subjected to a conservation of angular momentum about the axis of rotation. As shown on the free-body diagram, the line of action of the normal force **N** of the seat on the passenger passes through the axis, and the passenger's weight **W** is parallel to it. No angular impulse acts around the z axis.

Procedure for Analysis

When applying the principles of angular impulse and momentum, or the conservation of angular momentum, it is suggested that the following procedure be used.

Free-Body Diagram

- Draw the particle's free-body diagram in order to determine any axis about which angular momentum may be conserved. For this to occur, the moments of the forces (or impulses) must be parallel or pass through the axis so as to create zero moment throughout the time period t_1 to t_2.
- The direction and sense of the particle's initial and final velocities should also be established.
- An alternative procedure would be to draw the impulse and momentum diagrams for the particle.

Momentum Equations

- Apply the principle of angular impulse and momentum, $(\mathbf{H}_O)_1 + \Sigma \int_{t_1}^{t_2} \mathbf{M}_O \, dt = (\mathbf{H}_O)_2$, or if appropriate, the conservation of angular momentum, $(\mathbf{H}_O)_1 = (\mathbf{H}_O)_2$.

EXAMPLE 15-13

The 5-kg block of negligible size rests on the smooth horizontal plane, Fig. 15–24*a*. It is attached at A to a slender rod of negligible mass. The rod is attached to a ball-and-socket joint at B. If a moment $M = (3t)$ N·m, where t is in seconds, is applied to the rod and a horizontal force $P = 10$ N is applied to the block, determine the speed of the block in 4 s starting from rest.

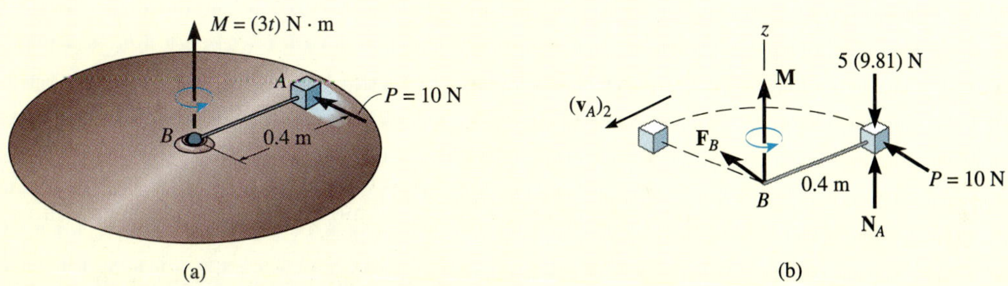

Fig. 15–24

Solution

Free-Body Diagram. If we consider the system of both the rod and block, Fig. 15–24*b*, then the resultant force reaction $\mathbf{F}_B$ at the ball-and-socket can be eliminated from the analysis by applying the principle of angular impulse and momentum about the z axis. If this is done, the angular impulses created by the weight and normal reaction $\mathbf{N}_A$ are also eliminated, since they act parallel to the z axis and therefore create zero moment about this axis.

Principle of Angular Impulse and Momentum

$$(H_z)_1 + \Sigma \int_{t_1}^{t_2} M_z\, dt = (H_z)_2$$

$$(H_z)_1 + \int_{t_1}^{t_2} M\, dt + r_{BA}P(\Delta t) = (H_z)_2$$

$$0 + \int_0^4 3t\, dt + (0.4 \text{ m})(10 \text{ N})(4 \text{ s}) = 5 \text{ kg}(v_A)_2(0.4 \text{ m})$$

$$24 + 16 = 2(v_A)_2$$

$$(v_A)_2 = 20 \text{ m/s} \qquad \textit{Ans.}$$

EXAMPLE 15-14

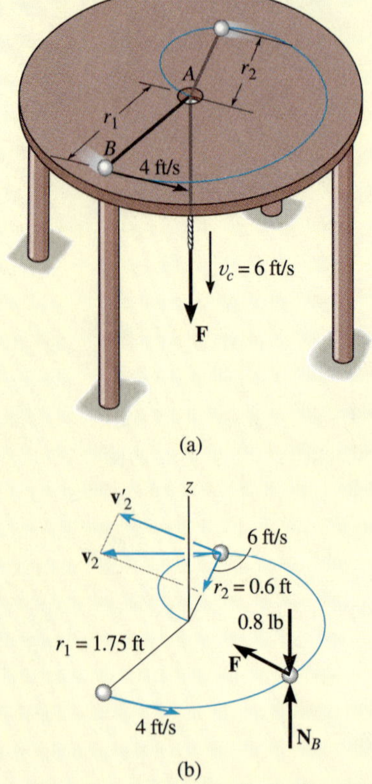

Fig. 15–25

The 0.8-lb ball B, shown in Fig. 15–25a, is attached to a cord which passes through a hole at A in a smooth table. When the ball is $r_1 =$ 1.75 ft from the hole, it is rotating around in a circle such that its speed is $v_1 = 4$ ft/s. By applying a force $\mathbf{F}$ the cord is pulled downward through the hole with a constant speed $v_c = 6$ ft/s. Determine (a) the speed of the ball at the instant it is $r_2 = 0.6$ ft from the hole, and (b) the amount of work done by $\mathbf{F}$ in shortening the radial distance from r_1 to r_2. Neglect the size of the ball.

Solution

Part (a)

Free-Body Diagram. As the ball moves from r_1 to r_2, Fig. 15–25b, the cord force $\mathbf{F}$ on the ball always passes through the z axis, and the weight and $\mathbf{N}_B$ are parallel to it. Hence the moments, or angular impulses created by these forces, are all *zero* about this axis. Therefore, the conservation of angular momentum applies about the z axis.

Conservation of Angular Momentum. The ball's velocity $\mathbf{v}_2$ is resolved into two components. The radial component, 6 ft/s, is known; however, it produces zero angular momentum about the z axis. Thus,

$$\mathbf{H}_1 = \mathbf{H}_2$$

$$r_1 m_B v_1 = r_2 m_B v_2'$$

$$1.75 \text{ ft}\left(\frac{0.8 \text{ lb}}{32.2 \text{ ft/s}^2}\right)4 = 0.6 \text{ ft}\left(\frac{0.8 \text{ lb}}{32.2 \text{ ft/s}^2}\right)v_2'$$

$$v_2' = 11.67 \text{ ft/s}$$

The speed of the ball is thus

$$v_2 = \sqrt{(11.67)^2 + (6)^2}$$
$$= 13.1 \text{ ft/s} \qquad \textit{Ans.}$$

Part (b). The only force that does work on the ball is **F**. (The normal force and weight do not move vertically.) The initial and final kinetic energies of the ball can be determined so that from the principle of work and energy we have

$$T_1 + \Sigma U_{1-2} = T_2$$

$$\frac{1}{2}\left(\frac{0.8 \text{ lb}}{32.2 \text{ ft/s}^2}\right)(4 \text{ ft/s})^2 + U_F = \frac{1}{2}\left(\frac{0.8 \text{ lb}}{32.2 \text{ ft/s}^2}\right)(13.1 \text{ ft/s})^2$$

$$U_F = 1.94 \text{ ft}\cdot\text{lb} \qquad \textit{Ans.}$$

EXAMPLE 15–15

The 2-kg disk shown in Fig. 15–26*a* rests on a smooth horizontal surface and is attached to an elastic cord that has a stiffness $k_c = 20$ N/m and is initially unstretched. If the disk is given a velocity $(v_D)_1 = 1.5$ m/s, perpendicular to the cord, determine the rate at which the cord is being stretched and the speed of the disk at the instant the cord is stretched 0.2 m.

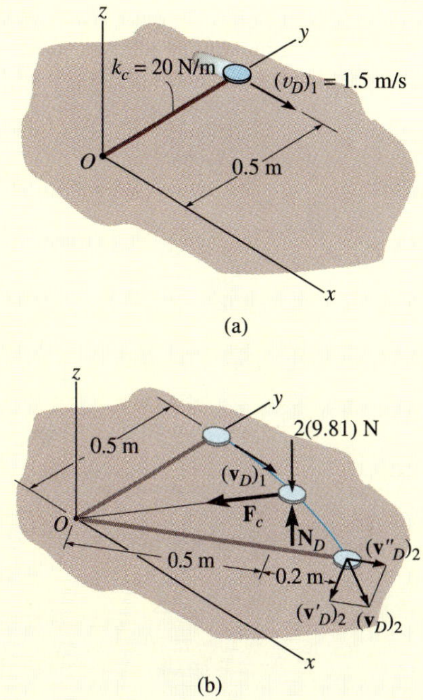

Fig. 15–26

Solution

Free-Body Diagram. After the disk has been launched, it slides along the path shown in Fig. 15–26*b*. By inspection, angular momentum about point *O* (or the *z* axis) is *conserved*, since none of the forces produce an angular impulse about this axis. Also, when the distance is 0.7 m, only the component $(\mathbf{v}'_D)_2$ produces angular momentum of the disk about *O*.

Conservation of Angular Momentum. The component $(\mathbf{v}'_D)_2$ can be obtained by applying the conservation of angular momentum about *O* (the *z* axis), i.e.,

$$(\mathbf{H}_O)_1 = (\mathbf{H}_O)_2$$

$$(\circlearrowleft+) \qquad r_1 m_D (v_D)_1 = r_2 m_D (v'_D)_2$$

$$0.5 \text{ m}(2 \text{ kg})(1.5 \text{ m/s}) = 0.7 \text{ m}(2 \text{ kg})(v'_D)_2$$

$$(v'_D)_2 = 1.07 \text{ m/s}$$

Conservation of Energy. The speed of the disk may be obtained by applying the conservation of energy equation at the point where the disk was launched and at the point where the cord is stretched 0.2 m.

$$T_1 + V_1 = T_2 + V_2$$

$$\tfrac{1}{2}(2 \text{ kg})(1.5 \text{ m/s})^2 + 0 = \tfrac{1}{2}(2 \text{ kg})(v_D)_2^2 + \tfrac{1}{2}(20 \text{ N/m})(0.2 \text{ m})^2$$

Thus,

$$(v_D)_2 = 1.36 \text{ m/s} \qquad \textit{Ans.}$$

Having determined $(v_D)_2$ and its component $(v'_D)_2$, the rate of stretch of the cord $(v''_D)_2$ is determined from the Pythagorean theorem,

$$(v''_D)_2 = \sqrt{(v_D)_2^2 - (v'_D)_2^2}$$

$$= \sqrt{(1.36)^2 - (1.07)^2}$$

$$= 0.838 \text{ m/s} \qquad \textit{Ans.}$$

Problems

15-90. Determine the angular momentum $\mathbf{H}_O$ of each of the particles about point O.

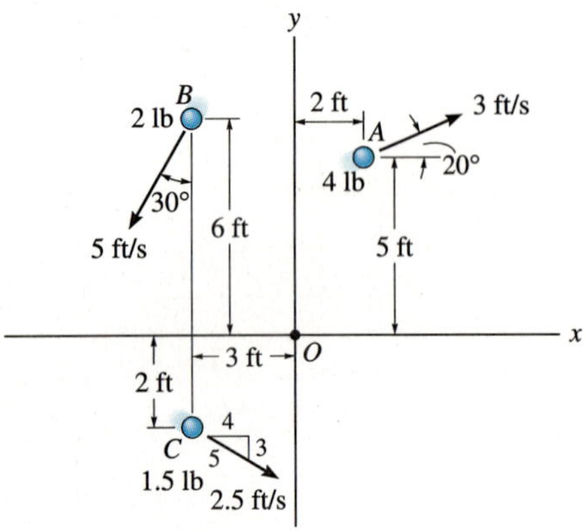

Prob. 15–90

15-91. Determine the angular momentum $\mathbf{H}_O$ of the particle about point O.

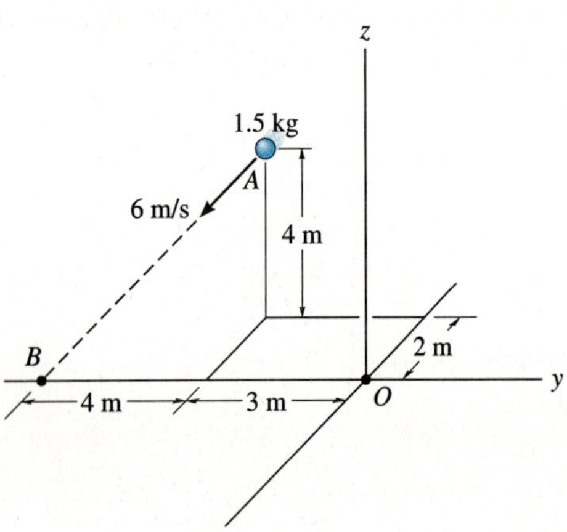

Prob. 15–91

***15-92.** Determine the angular momentum $\mathbf{H}_O$ of each of the particles about point O.

15-93. Determine the angular momentum $\mathbf{H}_P$ of each of the particles about point P.

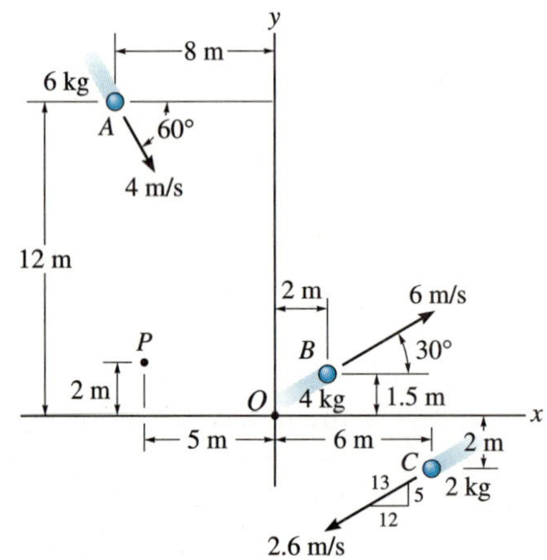

Probs. 15–92/93

15-94. Determine the angular momentum $\mathbf{H}_O$ of the particle about point O.

15-95. Determine the angular momentum $\mathbf{H}_P$ of the particle about point P.

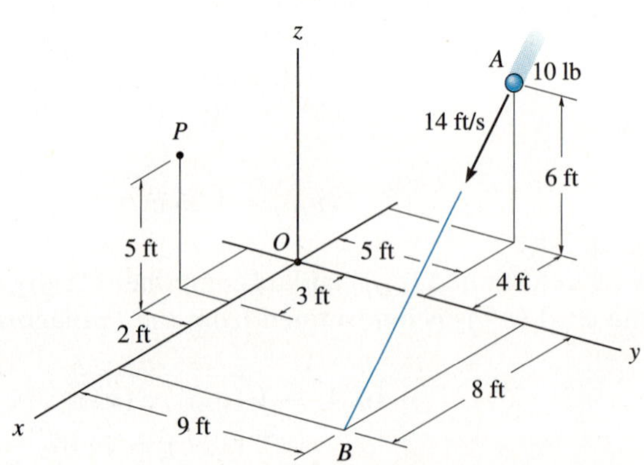

Probs. 15–94/95

***15-96.** Determine the total angular momentum $\mathbf{H}_O$ for the system of three particles about point O. All the particles are moving in the x–y plane.

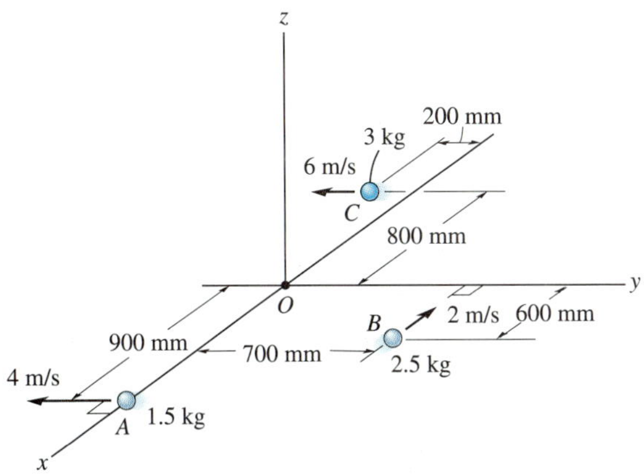

Prob. 15–96

15-97. Determine the angular momentum $\mathbf{H}_O$ of each of the two particles about point O. Use a scalar solution.

15-98. Determine the angular momentum $\mathbf{H}_P$ of each of the two particles about point P. Use a scalar solution.

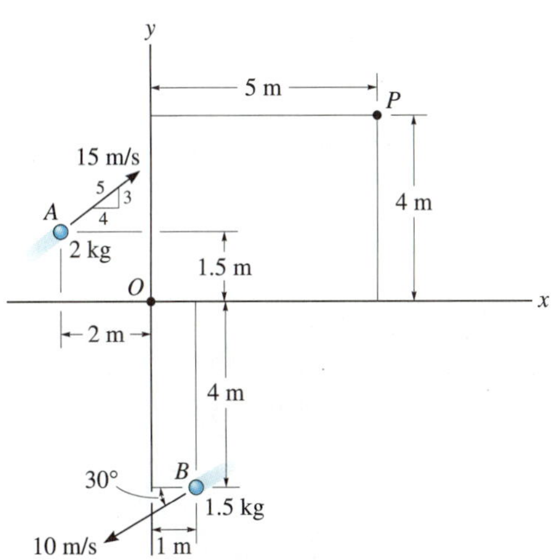

Probs. 15–97/98

15-99. The projectile having a mass of 3 kg is fired from a cannon with a muzzle velocity of $v_0 = 500$ m/s. Determine the projectile's angular momentum about point O at the instant it is at the maximum height of its trajectory.

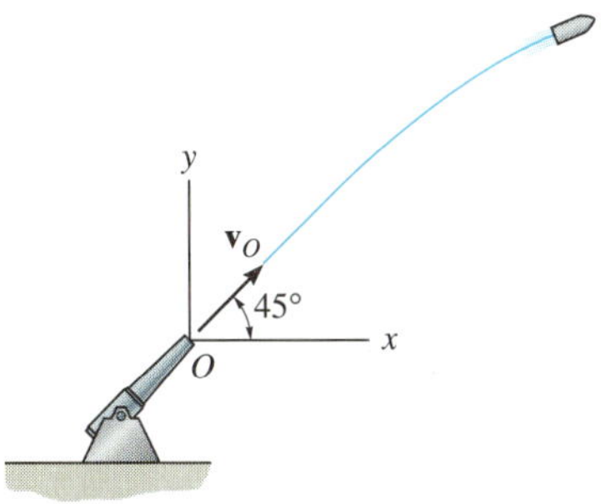

Prob. 15–99

***15-100.** The 3-lb ball located at A is released from rest and travels down the curved path. If the ball exerts a normal force of 5 lb on the path when it reaches point B, determine the angular momentum of the ball about the center of curvature, point O. *Hint:* Neglect the size of the ball. The radius of curvature at point B must be determined.

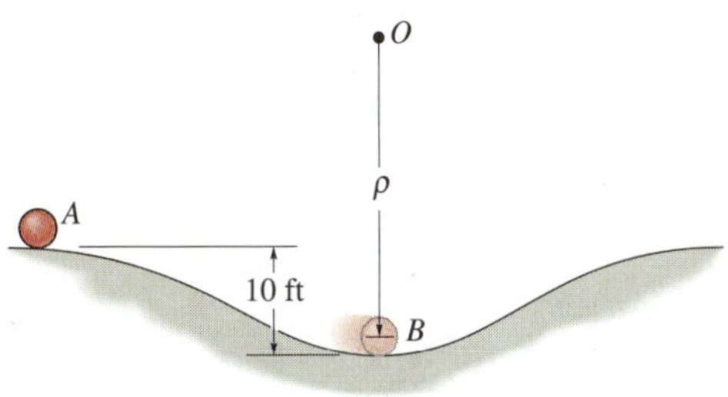

Prob. 15–100

15-101. The two blocks A and B each have a mass of 400 g. The blocks are fixed to the horizontal rods, and their initial velocity is 2 m/s in the direction shown. If a couple moment of $M = 0.6$ N·m is applied about CD of the frame, determine the speed of the blocks when $t = 3$ s. The mass of the frame is negligible, and it is free to rotate about CD. Neglect the size of the blocks.

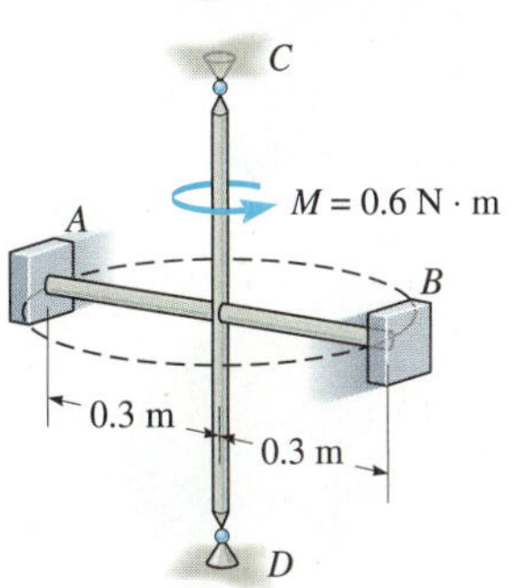

Prob. 15–101

15-102. The four 5-lb spheres are rigidly attached to the crossbar frame having a negligible weight. If a couple moment $M = (0.5t + 0.8)$ lb·ft, where t is in seconds, is applied as shown, determine the speed of each of the spheres in 4 seconds starting from rest. Neglect the size of the spheres.

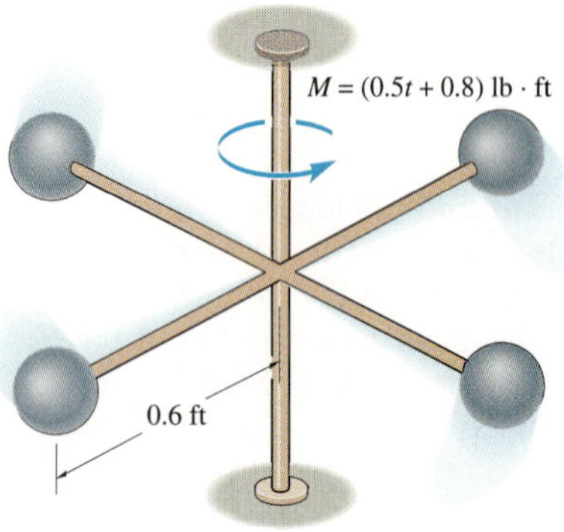

Prob. 15–102

15-103. An earth satellite of mass 700 kg is launched into a free-flight trajectory about the earth with an initial speed of $v_A = 10$ km/s when the distance from the center of the earth is $r_A = 15$ Mm. If the launch angle at this position is $\phi_A = 70°$, determine the speed v_B of the satellite and its closest distance r_B from the center of the earth. The earth has a mass $M_e = 5.976(10^{24})$ kg. *Hint:* Under these conditions, the satellite is subjected only to the earth's gravitational force, $F = GM_e m_s/r^2$, Eq. 13–1. For part of the solution, use the conservation of energy (see Prob. 14-97).

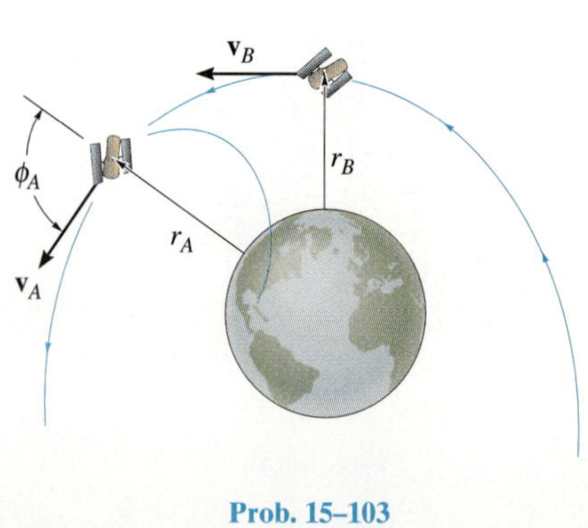

Prob. 15–103

***15-104.** A box having a weight of 8 lb is moving around in a circle of radius $r_A = 2$ ft with a speed of $(v_A)_1 = 5$ ft/s while connected to the end of a rope. If the rope is pulled inward with a constant speed of $v_r = 4$ ft/s, determine the speed of the box at the instant $r_B = 1$ ft. How much work is done after pulling in the rope from A to B? Neglect friction and the size of the box.

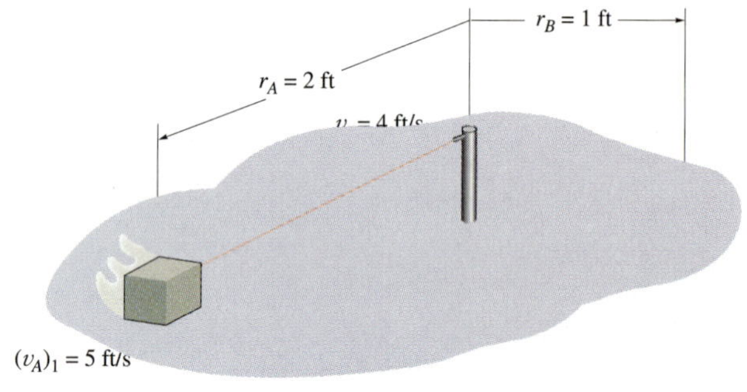

Prob. 15–104

15-105. The 800-lb roller-coaster car starts from rest on the track having the shape of a cylindrical helix. If the helix descends 8 ft for every one revolution, determine the speed of the car when $t = 4$ s. Also, how far has the car descended in this time? Neglect friction and the size of the car.

15-106. The 800-lb roller-coaster car starts from rest on the track having the shape of a cylindrical helix. If the helix descends 8 ft for every one revolution, determine the time required for the car to attain a speed of 60 ft/s. Neglect friction and the size of the car.

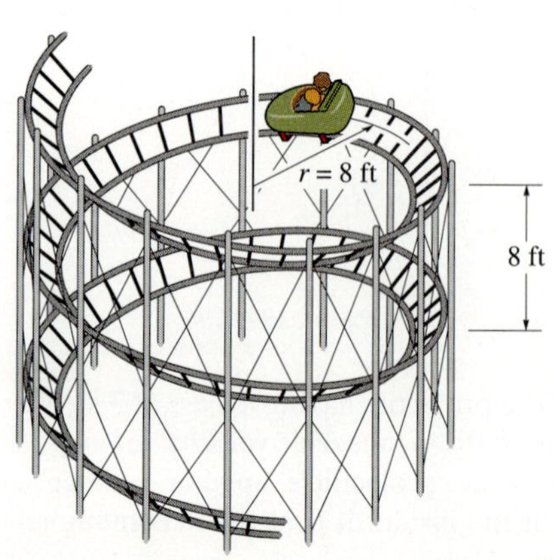

Probs. 15–105/106

■**15-107.** The 5-lb ball B is rotating in a circle as shown when the distance AB is 3 ft. If 1.5 ft of cord is pulled through the hole at A, determine the speed of the ball when it moves in a circular path at C.

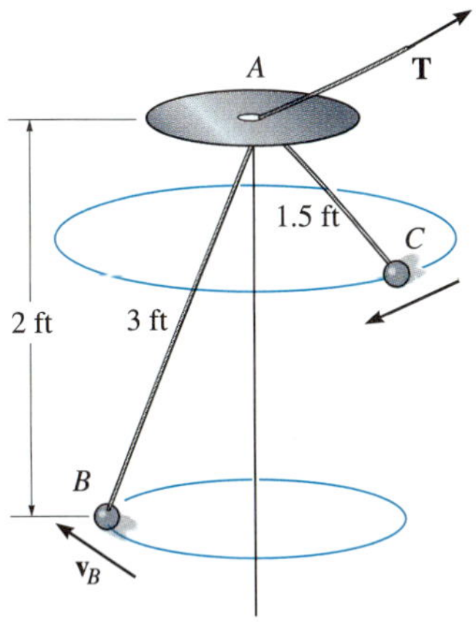

Prob. 15–107

***15-108.** A gymnast having a mass of 80 kg holds the two rings with his arms down in the position shown as he swings downward. His center of mass is located at point G_1. When he is at the bottom position of his swing, his velocity is $(v_G)_1 = 5$ m/s. At this lower position he *suddenly* lets his arms come up, shifting his center of mass to position G_2. Determine his new velocity in the upswing and the angle θ to which he swings before coming to rest. Treat his body as a particle.

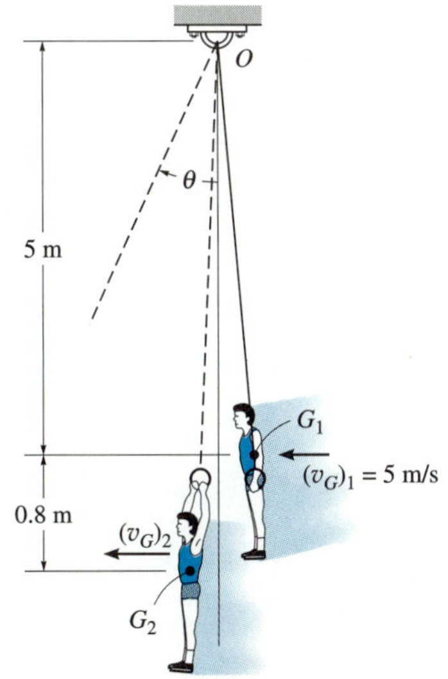

Prob. 15–108

15-109. A small particle having a mass m is placed inside the semicircular tube. The particle is placed at the position shown and released. Apply the principle of angular momentum about point O ($\Sigma M_O = \dot{H}_O$), and show that the motion of the particle is governed by the differential equation $\ddot{\theta} + (g/R)\sin\theta = 0$.

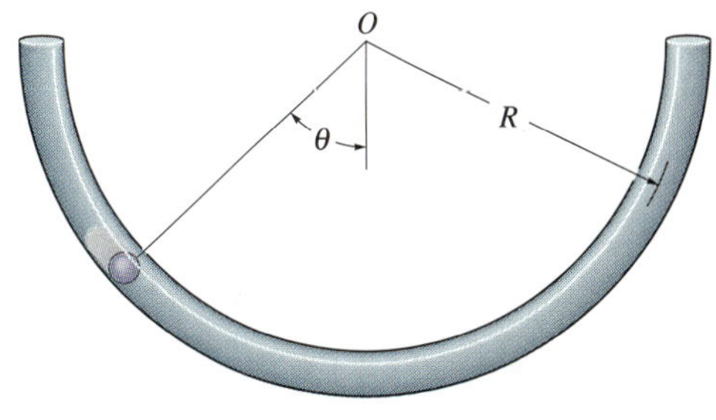

Prob. 15–109

15-110. A toboggan and rider, having a total mass of 150 kg, enter horizontally tangent to a 90° circular curve with a velocity of $v_A = 70$ km/h. If the track is flat and banked at an angle of 60°, determine the speed v_B and the angle θ of "descent," measured from the horizontal in a vertical x–z plane, at which the toboggan exits at B. Neglect friction in the calculation.

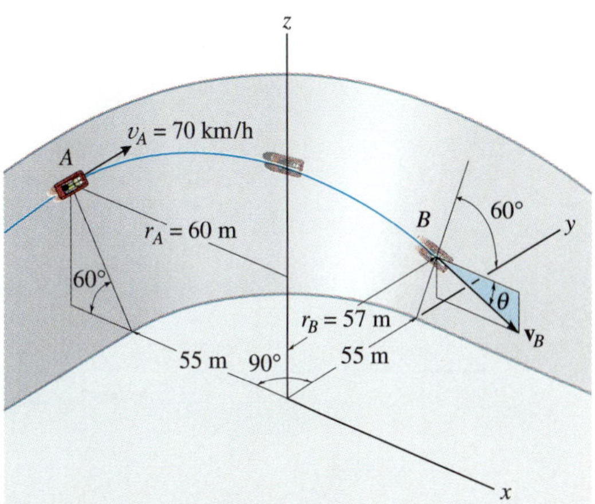

Prob. 15–110

*15.8 Steady Fluid Streams

Knowledge of the forces developed by steadily moving fluid streams is of importance in the design and analysis of turbines, pumps, blades, and fans. To illustrate how the principle of impulse and momentum may be used to determine these forces, consider the diversion of a steady stream of fluid (liquid or gas) by a fixed pipe, Fig. 15–27*a*. The fluid enters the pipe with a velocity $\mathbf{v}_A$ and exits with a velocity $\mathbf{v}_B$. The impulse and momentum diagrams for the fluid stream are shown in Fig. 15–27*b*. The force $\Sigma\mathbf{F}$, shown on the impulse diagram, represents the resultant of all the external forces acting on the fluid stream. It is this loading which gives the fluid stream an impulse whereby the original momentum of the fluid is changed in both its magnitude and direction. Since the flow is steady, $\Sigma\mathbf{F}$ will be *constant* during the time interval *dt*. During this time the fluid stream is in motion, and as a result a small amount of fluid, having a mass *dm*, is about to enter the pipe with a velocity $\mathbf{v}_A$ at time *t*. If this element of mass and the mass of fluid in the pipe are considered as a "closed system," then at time $t + dt$ a corresponding element of mass *dm* must leave the pipe with a velocity $\mathbf{v}_B$. Also, the fluid stream *within* the pipe section has a mass *m* and an *average velocity* $\mathbf{v}$ which is constant during the time interval *dt*. Applying the principle of linear impulse and momentum to the fluid stream, we have

$$dm\,\mathbf{v}_A + m\mathbf{v} + \Sigma\mathbf{F}\,dt = dm\,\mathbf{v}_B + m\mathbf{v}$$

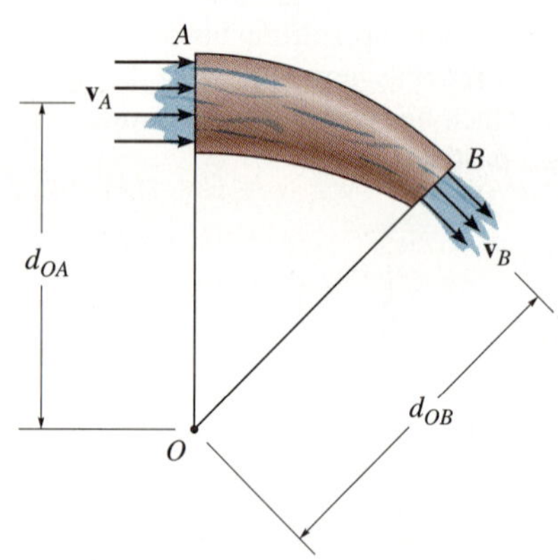

(a)

Fig. 15–27

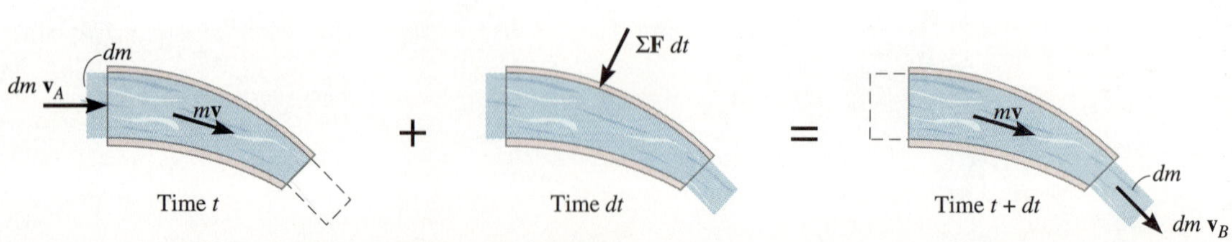

(b)

Force Resultant. Solving for the resultant force yields

$$\Sigma \mathbf{F} = \frac{dm}{dt}(\mathbf{v}_B - \mathbf{v}_A) \tag{15–25}$$

Provided the motion of the fluid can be represented in the x–y plane, it is usually convenient to express this vector equation in the form of two scalar component equations, i.e.,

$$\Sigma F_x = \frac{dm}{dt}(v_{Bx} - v_{Ax})$$
$$\Sigma F_y = \frac{dm}{dt}(v_{By} - v_{Ay}) \tag{15–26}$$

The term dm/dt is called the *mass flow* and indicates the constant amount of fluid which flows either into or out of the pipe per unit of time. If the cross-sectional areas and densities of the fluid at the entrance A and exit B are A_A, ρ_A and A_B, ρ_B, respectively, Fig. 15–27c, then *continuity of mass* requires that $dm = \rho\, dV = \rho_A(ds_A\, A_A) = \rho_B(ds_B\, A_B)$. Hence, during the time dt, since $v_A = ds_A/dt$ and $v_B = ds_B/dt$, we have

$$\frac{dm}{dt} = \rho_A v_A A_A = \rho_B v_B A_B = \rho_A Q_A = \rho_B Q_B \tag{15–27}$$

Here $Q = vA$ is the volumetric *flow rate,* which measures the volume of fluid flowing per unit of time.

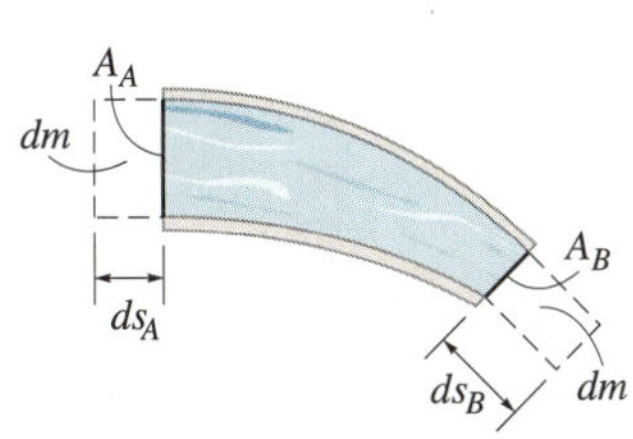

(c)

The conveyor belt must supply frictional forces to the gravel that falls upon it in order to change the momentum of the gravel stream, so that it begins to travel along the belt.

The air on one side of this fan is essentially at rest, and as it passes through the blades its momentum is increased. To change the momentum of the air flow in this manner, the blades must exert a horizontal thrust on the air stream. As the blades turn faster, the equal but opposite thrust of the air on the blades could overcome the rolling resistance of the wheels on the ground and begin to move the frame of the fan.

Moment Resultant. In some cases it is necessary to obtain the support reactions on the fluid-carrying device. If Eq. 15–25 does not provide enough information to do this, the principle of angular impulse and momentum must be used. The formulation of this principle applied to fluid streams can be obtained from Eq. 15–17, $\Sigma \mathbf{M}_O = \dot{\mathbf{H}}_O$, which states that the moment of all the external forces acting on the system about point O is equal to the time rate of change of angular momentum about O. In the case of the pipe shown in Fig. 15–27*a*, the flow is steady in the x–y plane; hence we have

$$(\curvearrowleft +) \qquad \Sigma M_O = \frac{dm}{dt}(d_{OB}v_B - d_{OA}v_A) \qquad (15\text{–}28)$$

where the moment arms d_{OB} and d_{OA} are directed from O to the *geometric center* or *centroid* of the openings at A and B.

Procedure for Analysis

Problems involving steady flow can be solved using the following procedure.

Kinematic Diagram.

- If the device is *moving,* a *kinematic diagram* may be helpful for determining the entrance and exit velocities of the fluid flowing onto the device, since a *relative-motion analysis* of velocity will be involved.
- The measurement of velocities v_A and v_B must be made by an observer fixed in an inertial frame of reference.
- Once the velocity of the fluid flowing onto the device is determined, the mass flow is calculated using Eq. 15–27.

Free-Body Diagram.

- Draw a free-body diagram of the device which is directing the fluid in order to establish the forces $\Sigma \mathbf{F}$ that act on it. These external forces will include the support reactions, the weight of the device and the fluid contained within it, and the static pressure forces of the fluid at the entrance and exit sections of the device.*

Equations of Steady Flow.

- Apply the equations of steady flow, Eqs. 15–26 and 15–28, using the appropriate components of velocity and force shown on the kinematic and free-body diagrams.

*In the SI system pressure is measured using the *pascal* (Pa), where 1 Pa = 1 N/m^2.

EXAMPLE 15–16

Determine the components of reaction which the fixed pipe joint at A exerts on the elbow in Fig. 15–28*a*, if water flowing through the pipe is subjected to a static gauge pressure of 100 kPa at A. The discharge at B is $Q_B = 0.2\ \text{m}^3/\text{s}$. Water has a density $\rho_w = 1000\ \text{kg/m}^3$, and the water-filled elbow has a mass of 20 kg and center of mass at G.

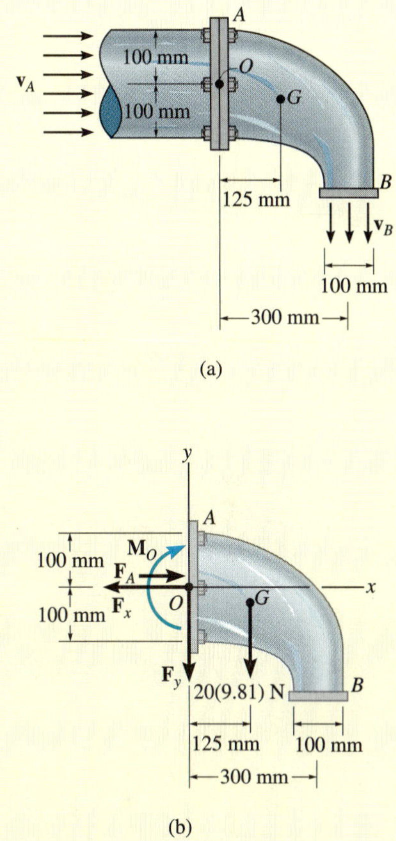

Fig. 15–28

Solution

Using a fixed inertial coordinate system, the velocity of flow at A and B and the mass flow rate can be obtained from Eq. 15–27. Since the density of water is constant, $Q_B = Q_A = Q$. Hence,

$$\frac{dm}{dt} = \rho_w Q = (1000\ \text{kg/m}^3)(0.2\ \text{m}^3/\text{s}) = 200\ \text{kg/s}$$

$$v_B = \frac{Q}{A_B} = \frac{0.2\ \text{m}^3/\text{s}}{\pi(0.05\ \text{m})^2} = 25.46\ \text{m/s} \downarrow$$

$$v_A = \frac{Q}{A_A} = \frac{0.2\ \text{m}^3/\text{s}}{\pi(0.1\ \text{m})^2} = 6.37\ \text{m/s} \rightarrow$$

Free-Body Diagram. As shown on the free-body diagram, Fig. 15–28*b*, the *fixed* connection at A exerts a resultant couple moment $\mathbf{M}_O$ and force components $\mathbf{F}_x$ and $\mathbf{F}_y$ on the elbow. Due to the static pressure of water in the pipe, the pressure force acting on the fluid at A is $F_A = p_A A_A$. Since $1\ \text{kPa} = 1000\ \text{N/m}^2$,

$$F_A = p_A A_A = [100(10^3)\ \text{N/m}^2][\pi(0.1\ \text{m})^2] = 3141.6\ \text{N}$$

There is no static pressure acting at B, since the water is discharged at atmospheric pressure; i.e., the pressure measured by a gauge at B is equal to zero, $p_B = 0$.

Equations of Steady Flow

$$\overset{+}{\rightarrow} \Sigma F_x = \frac{dm}{dt}(v_{Bx} - v_{Ax}); \quad -F_x + 3141.6\ \text{N} = 200\ \text{kg/s}(0 - 6.37\ \text{m/s})$$

$$F_x = 4.41\ \text{kN} \qquad \textit{Ans.}$$

$$+\uparrow \Sigma F_y = \frac{dm}{dt}(v_{By} - v_{Ay}); \quad -F_y - 20(9.81)\ \text{N} = 200\ \text{kg/s}(-25.46\ \text{m/s} - 0)$$

$$F_y = 4.90\ \text{kN} \qquad \textit{Ans.}$$

If moments are summed about point O, Fig. 15–28*b*, then $\mathbf{F}_x$, $\mathbf{F}_y$, and the static pressure $\mathbf{F}_A$ are eliminated, as well as the moment of momentum of the water entering at A, Fig. 15–28*a*. Hence,

$$\circlearrowleft + \Sigma M_O = \frac{dm}{dt}(d_{OB}v_B - d_{OA}v_A)$$

$$M_O + 20(9.81)\ \text{N}(0.125\ \text{m}) = 200\ \text{kg/s}[(0.3\ \text{m})(25.46\ \text{m/s}) - 0]$$

$$M_O = 1.50\ \text{kN}\cdot\text{m} \qquad \textit{Ans.}$$

EXAMPLE 15–17

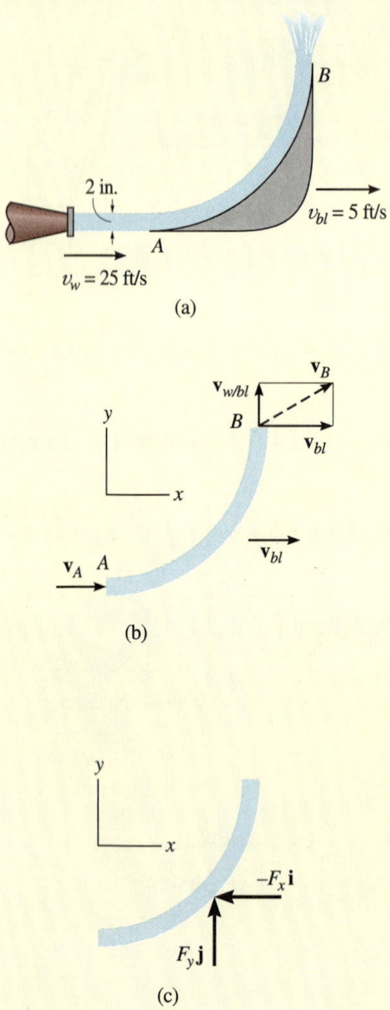

Fig. 15–29

A 2-in.-diameter water jet having a velocity of 25 ft/s impinges upon a single moving blade, Fig. 15–29*a*. If the blade is moving at 5 ft/s away from the jet, determine the horizontal and vertical components of force which the blade is exerting on the water. What power does the water generate on the blade? Water has a specific weight of $\gamma_w = 62.4 \text{ lb/ft}^3$.

Solution

Kinematic Diagram. From a fixed inertial coordinate system, Fig. 15–29*b*, the rate at which water enters the blade is

$$\mathbf{v}_A = \{25\mathbf{i}\} \text{ ft/s}$$

The *relative-flow velocity* of the water onto the blade is $\mathbf{v}_{w/bl} = \mathbf{v}_w - \mathbf{v}_{bl} = 25\mathbf{i} - 5\mathbf{i} = \{20\mathbf{i}\}$ ft/s. Since the blade is moving with a velocity of $\mathbf{v}_{bl} = \{5\mathbf{i}\}$ ft/s, the velocity of flow at B measured from x, y is the vector sum, shown in Fig. 15–29*b*. Here,

$$\begin{aligned}\mathbf{v}_B &= \mathbf{v}_{bl} + \mathbf{v}_{w/bl}\\ &= \{5\mathbf{i} + 20\mathbf{j}\} \text{ ft/s}\end{aligned}$$

Thus, the mass flow of water *onto* the blade that undergoes a momentum change is

$$\frac{dm}{dt} = \rho_w(v_{w/bl})A_A = \frac{62.4}{32.2}(20)\left[\pi\left(\frac{1}{12}\right)^2\right] = 0.846 \text{ slug/s}$$

Free-Body Diagram. The free-body diagram of a section of water acting on the blade is shown in Fig. 15–29*c*. The weight of the water will be neglected in the calculation, since this force will be small compared to the reactive components $\mathbf{F}_x$ and $\mathbf{F}_y$.

Equations of Steady Flow

$$\Sigma\mathbf{F} = \frac{dm}{dt}(\mathbf{v}_B - \mathbf{v}_A)$$

$$-F_x\mathbf{i} + F_y\mathbf{j} = 0.846(5\mathbf{i} + 20\mathbf{j} - 25\mathbf{i})$$

Equating the respective **i** and **j** components gives

$$F_x = 0.846(20) = 16.9 \text{ lb} \leftarrow \qquad \textit{Ans.}$$

$$F_y = 0.846(20) = 16.9 \text{ lb} \uparrow \qquad \textit{Ans.}$$

The water exerts equal but opposite forces on the blade.

Since the water force which causes the blade to move forward horizontally with a velocity of 5 ft/s is $F_x = 16.9$ lb, then from Eq. 14–10 the power is

$$P = \mathbf{F}\cdot\mathbf{v}; \qquad P = \frac{16.9 \text{ lb}(5 \text{ ft/s})}{550 \text{ hp/(ft}\cdot\text{lb/s)}} = 0.154 \text{ hp} \qquad \textit{Ans.}$$

*15.9 Propulsion with Variable Mass

In the previous section we considered the case in which a *constant* amount of mass dm enters and leaves a *"closed system."* There are, however, two other important cases involving mass flow, which are represented by a system that is either gaining or losing mass. In this section we will discuss each of these cases separately.

A System That Loses Mass. Consider a device such as a rocket which at an instant of time has a mass m and is moving forward with a velocity $\mathbf{v}$, Fig. 15–30*a*. At this same instant the device is expelling an amount of mass m_e with a mass flow velocity $\mathbf{v}_e$. For the analysis, the *"closed system"* includes *both the mass m of the device and the expelled mass m_e.* The impulse and momentum diagrams for the system are shown in Fig. 15–30*b*. During the time dt, the velocity of the device is increased from $\mathbf{v}$ to $\mathbf{v} + d\mathbf{v}$ since an amount of mass dm_e has been ejected and thereby gained in the exhaust. This increase in forward velocity, however, does not change the velocity $\mathbf{v}_e$ of the expelled mass, since this mass moves at a constant speed once it has been ejected. The impulses are created by $\Sigma \mathbf{F}_s$, which represents the resultant of all the external forces that *act on the system* in the direction of motion. This force resultant *does not include* the force which causes the device to move forward, since this force (called a *thrust*) is *internal to the system;* that is, the thrust acts with equal magnitude but opposite direction on the mass m of the device and the expelled exhaust mass m_e.* Applying the principle of impulse and momentum to the system, Fig. 15–30*b*, we have

$$(\overset{+}{\rightarrow}) \quad mv - m_e v_e + \Sigma F_s \, dt = (m - dm_e)(v + dv) - (m_e + dm_e)v_e$$

or

$$\Sigma F_s \, dt = -v \, dm_e + m \, dv - dm_e \, dv - v_e \, dm_e$$

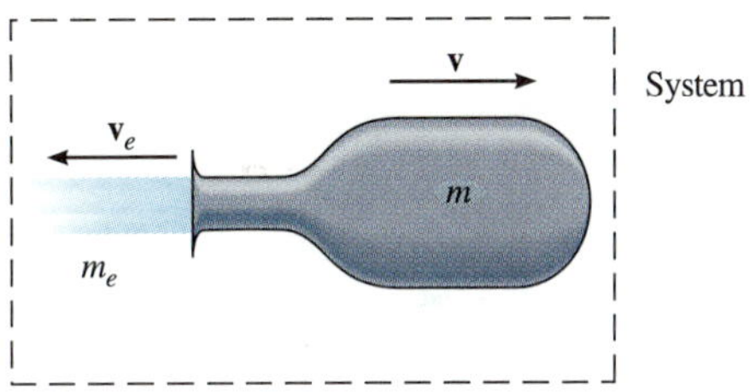

(a)

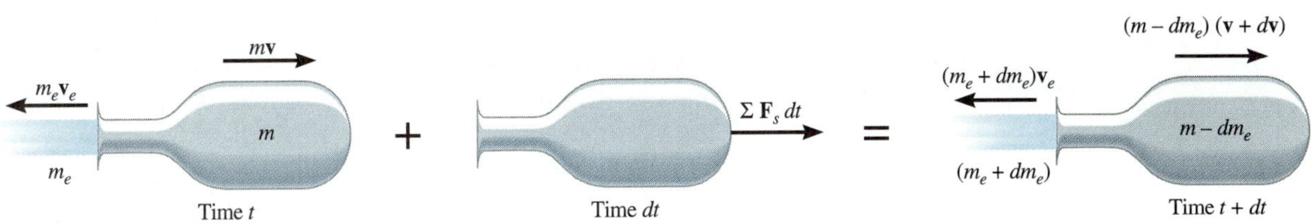

(b)

Fig. 15–30

*$\Sigma \mathbf{F}_s$ represents the external resultant force *acting on the system,* which is different from $\Sigma \mathbf{F}$, the resultant force acting only on the device.

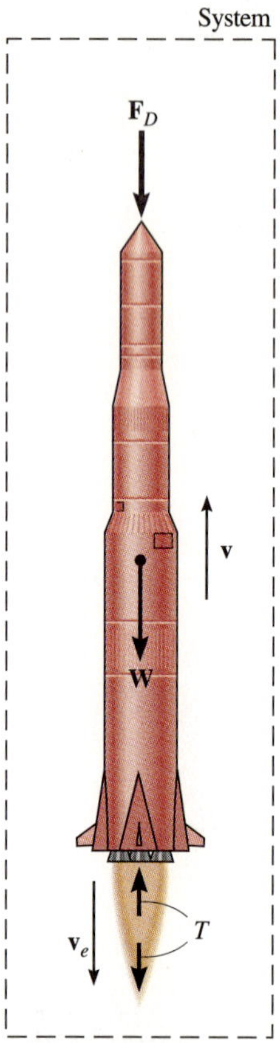

Fig. 15–31

Without loss of accuracy, the third term on the right side may be neglected since it is a "second-order" differential. Dividing by dt gives

$$\Sigma F_s = m\frac{dv}{dt} - (v + v_e)\frac{dm_e}{dt}$$

The relative velocity of the device as seen by an observer moving with the particles of the ejected mass is $v_{D/e} = (v + v_e)$, and so the final result can be written as

$$\Sigma F_s = m\frac{dv}{dt} - v_{D/e}\frac{dm_e}{dt} \tag{15–29}$$

Here the term dm_e/dt represents the rate at which mass is being ejected.

To illustrate an application of Eq. 15–29, consider the rocket shown in Fig. 15–31, which has a weight **W** and is moving upward against an atmospheric drag force $\mathbf{F}_D$. The system to be considered consists of the mass of the rocket and the mass of ejected gas m_e. Applying Eq. 15–29 to this system gives

$$(+\uparrow) \qquad -F_D - W = \frac{W}{g}\frac{dv}{dt} - v_{D/e}\frac{dm_e}{dt}$$

The last term of this equation represents the *thrust* **T** which the engine exhaust exerts on the rocket, Fig. 15–31. Recognizing that $dv/dt = a$, we may therefore write

$$(+\uparrow) \qquad T - F_D - W = \frac{W}{g}a$$

If a free-body diagram of the rocket is drawn, it becomes obvious that this equation represents an application of $\Sigma\mathbf{F} = m\mathbf{a}$ for the rocket.

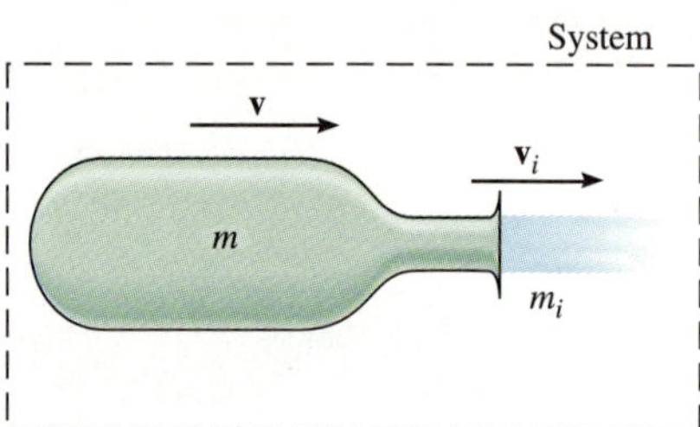

(a)

Fig. 15–32

A System That Gains Mass. A device such as a scoop or a shovel may gain mass as it moves forward. For example, the device shown in Fig. 15–32*a* has a mass m and is moving forward with a velocity **v**. At this instant, the device is collecting a particle stream of mass m_i. The flow velocity $\mathbf{v}_i$ of this injected mass is constant and independent of the velocity **v** such that $v > v_i$. The system to be considered includes both the mass of the device and the mass of the injected particles. The impulse and momentum diagrams for this system are shown in Fig. 15–32*b*. Along with an increase in mass dm_i gained by the device, there is an assumed increase in velocity $d\mathbf{v}$ during the time interval dt. This increase is caused by the impulse created by $\Sigma\mathbf{F}_s$, the resultant of all the external forces *acting on the system* in the direction of motion. The force summation

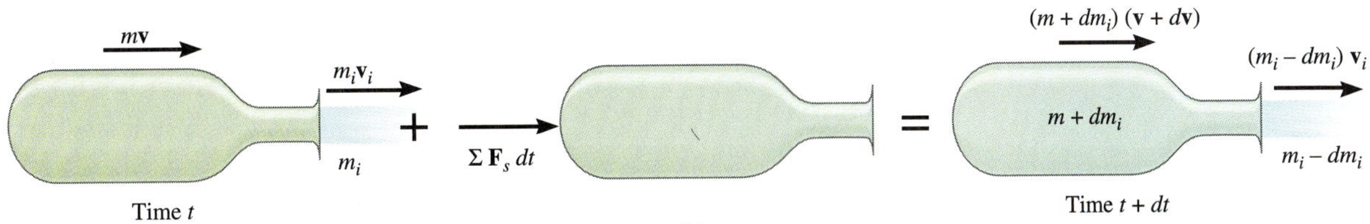

(b)

does not include the retarding force of the injected mass acting on the device. Why? Applying the principle of impulse and momentum to the system, we have

$$(\overset{+}{\rightarrow})\qquad mv + m_i v_i + \Sigma F_s\, dt = (m + dm_i)(v + dv) + (m_i - dm_i)v_i$$

Using the same procedure as in the previous case, we may write this equation as

$$\Sigma F_s = m\frac{dv}{dt} + (v - v_i)\frac{dm_i}{dt}$$

Since the relative velocity of the device as seen by an observer moving with the particles of the injected mass is $v_{D/i} = (v - v_i)$, the final result can be written as

$$\boxed{\Sigma F_s = m\frac{dv}{dt} + v_{D/i}\frac{dm_i}{dt}} \qquad (15\text{–}30)$$

where dm_i/dt is the rate of mass injected into the device. The last term in this equation represents the magnitude of force **R**, which the injected mass *exerts on the device.* Since $dv/dt = a$, Eq. 15–30 becomes

$$\Sigma F_s - R = ma$$

This is the application of $\Sigma\mathbf{F} = m\mathbf{a}$, Fig. 15–32*c*.

As in the case of steady flow, problems which are solved using Eqs. 15–29 and 15–30 should be accompanied by the necessary free-body diagram. With this diagram one can then determine ΣF_s *for the system* and isolate the force exerted on the device by the particle stream.

The scraper box behind this tractor represents a device that gains mass. If the tractor maintains a constant velocity v, then $dv/dt = 0$ and, because the soil is originally at rest, $v_{D/i} = v$. By Eq. 15–30, the horizontal towing force on the scraper box is then $T = 0 + v(dm/dt)$, where dm/dt is the rate of soil accumulation in the scraper box.

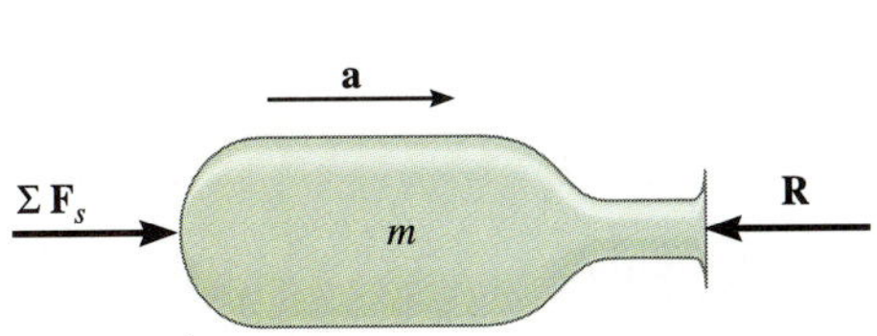

(c)

EXAMPLE 15-18

The initial combined mass of a rocket and its fuel is m_0. A total mass m_f of fuel is consumed at a constant rate of $dm_e/dt = c$ and expelled at a constant speed of u relative to the rocket. Determine the maximum velocity of the rocket, i.e., at the instant the fuel runs out. Neglect the change in the rocket's weight with altitude and the drag resistance of the air. The rocket is fired vertically from rest.

Solution

Since the rocket is losing mass as it moves upward, Eq. 15–29 can be used for the solution. The only *external force* acting on the *system* consisting of the rocket and a portion of the expelled mass is the weight **W**, Fig. 15–33. Hence,

$$+\uparrow \Sigma F_s = m\frac{dv}{dt} - v_{D/e}\frac{dm_e}{dt}; \qquad -W = m\frac{dv}{dt} - uc \tag{1}$$

The rocket's velocity is obtained by integrating this equation.

At any given instant t during the flight, the mass of the rocket can be expressed as $m = m_0 - (dm_e/dt)t = m_0 - ct$. Since $W = mg$, Eq. 1 becomes

$$-(m_0 - ct)g = (m_0 - ct)\frac{dv}{dt} - uc$$

Separating the variables and integrating, realizing that $v = 0$ at $t = 0$, we have

$$\int_0^v dv = \int_0^t \left(\frac{uc}{m_0 - ct} - g\right) dt$$

$$v = -u\ln(m_0 - ct) - gt\Big|_0^t = u\ln\left(\frac{m_0}{m_0 - ct}\right) - gt \tag{2}$$

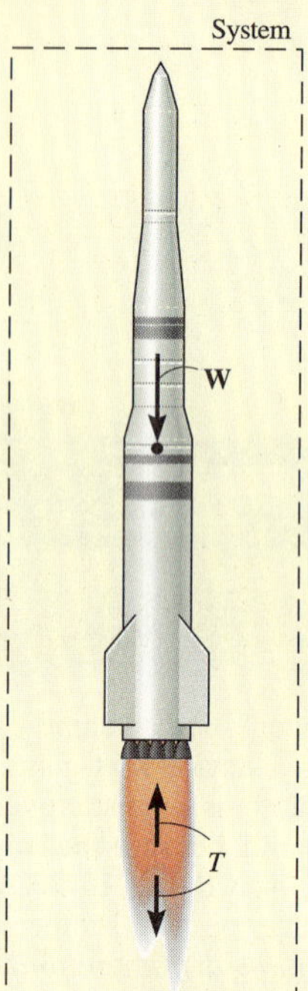

Fig. 15–33

Note that lift off requires the first term on the right to be greater than the second during the initial phase of motion. The time t' needed to consume all the fuel is given by

$$m_f = \left(\frac{dm_e}{dt}\right)t' = ct'$$

Hence,

$$t' = m_f/c$$

Substituting into Eq. 2 yields

$$v_{max} = u\ln\left(\frac{m_0}{m_0 - m_f}\right) - \frac{gm_f}{c} \qquad \textit{Ans.}$$

EXAMPLE 15-19

A chain of length l, Fig. 15–34*a*, has a mass m. Determine the magnitude of force **F** required to (*a*) raise the chain with a constant speed v_c, starting from rest when $y = 0$; and (*b*) lower the chain with a constant speed v_c, starting from rest when $y = l$.

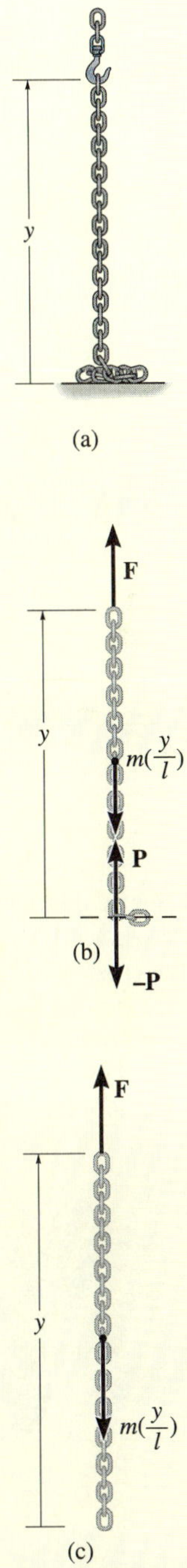

Fig. 15–34

Solution

Part (a). As the chain is raised, all the suspended links are given a sudden impulse downward by each added link which is lifted off the ground. Thus, the *suspended portion* of the chain may be considered as a device which is *gaining mass.* The system to be considered is the length of chain y which is suspended by **F** at any instant, including the next link which is about to be added but is still at rest, Fig. 15–34*b*. The forces acting on this system *exclude* the internal forces **P** and $-$**P**, which act between the added link and the suspended portion of the chain. Hence, $\Sigma F_s = F - mg(y/l)$.

To apply Eq. 15–30, it is also necessary to find the rate at which mass is being added to the system. The velocity $\mathbf{v}_c$ of the chain is equivalent to $\mathbf{v}_{D/i}$. Why? Since v_c is constant, $dv_c/dt = 0$ and $dy/dt = v_c$. Integrating, using the initial condition that $y = 0$ at $t = 0$, gives $y = v_c t$. Thus, the mass of the system at any instant is $m_s = m(y/l) = m(v_c t/l)$, and therefore the *rate* at which mass is *added* to the suspended chain is

$$\frac{dm_i}{dt} = m\left(\frac{v_c}{l}\right)$$

Applying Eq. 15–30 to the system, using this data, we have

$$+\uparrow \Sigma F_s = m\frac{dv_c}{dt} + v_{D/i}\frac{dm_i}{dt}$$

$$F - mg\left(\frac{y}{l}\right) = 0 + v_c m\left(\frac{v_c}{l}\right)$$

Hence,

$$F = (m/l)(gy + v_c^2) \qquad \textit{Ans.}$$

Part (b). When the chain is being lowered, the links which are expelled (given zero velocity) *do not* impart an impulse to the *remaining* suspended links. Why? Thus, the system in Part (*a*) cannot be considered. Instead, the equation of motion will be used to obtain the solution. At time t the portion of chain still off the floor is y. The free-body diagram for a suspended portion of the chain is shown in Fig. 15–34*c*. Thus,

$$+\uparrow \Sigma F = ma; \qquad F - mg\left(\frac{y}{l}\right) = 0$$

$$F = mg\left(\frac{y}{l}\right) \qquad \textit{Ans.}$$

Problems

15-111. The 150-lb fireman is holding a hose which has a nozzle diameter of 1 in. and hose diameter of 2 in. If the velocity of the water at discharge is 60 ft/s, determine the resultant normal and frictional force acting on the man's feet at the ground. Neglect the weight of the hose and the water within it. $\gamma_w = 62.4\ \text{lb/ft}^3$.

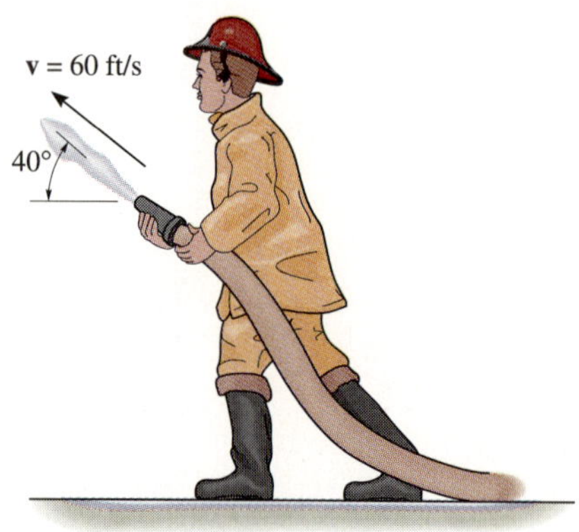

Prob. 15–111

***15-112.** The nozzle discharges water at a constant rate of 2 ft^3/s. The cross-sectional area of the nozzle at A is 4 in^2, and at B the cross-sectional area is 12 in^2. If the static gauge pressure due to the water at B is 2 lb/in^2, determine the magnitude of force which must be applied by the coupling at B to hold the nozzle in place. Neglect the weight of the nozzle and the water within it. $\gamma_w = 62.4\ \text{lb/ft}^3$.

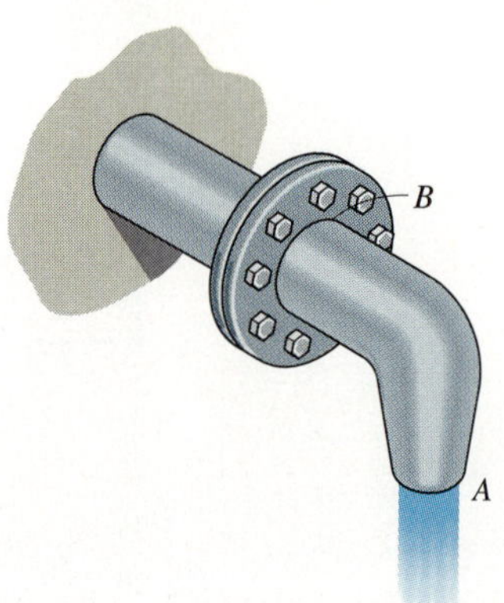

Prob. 15–112

15-113. When operating, the air-jet fan discharges air with a speed of $v_B = 20$ m/s into a slipstream having a diameter of 0.5 m. If air has a density of 1.22 kg/m^3, determine the horizontal and vertical components of reaction at C and the vertical reaction at each of the two wheels, D, when the fan is in operation. The fan and motor have a mass of 20 kg and a center of mass at G. Neglect the weight of the frame. Due to symmetry, both of the wheels support an equal load. Assume the air entering the fan at A is essentially at rest.

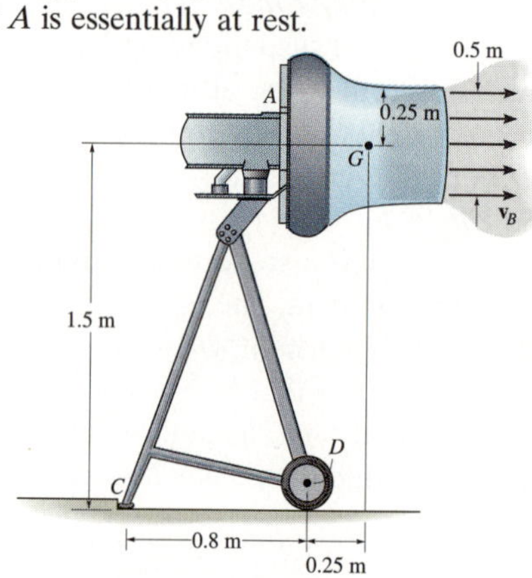

Prob. 15–113

15-114. A snowblower having a scoop S with a cross-sectional area of $A_S = 0.12\ \text{m}^2$ is pushed into snow with a speed of $v_S = 0.5$ m/s. The machine discharges the snow through a tube T that has a cross-sectional area of $A_T = 0.03\ \text{m}^2$ and is directed 60° from the horizontal. If the density of snow is $\rho_s = 104\ \text{kg/m}^3$, determine the horizontal force P required to push the blower forward, and the resultant frictional force F of the wheels on the ground, necessary to prevent the blower from moving sideways. The wheels roll freely.

Prob. 15–114

15-115. The nozzle has a diameter of 40 mm. If it discharges water uniformly with a downward velocity of 20 m/s against the fixed blade, determine the vertical force exerted by the water on the blade. $\rho_w = 1\ \text{Mg/m}^3$.

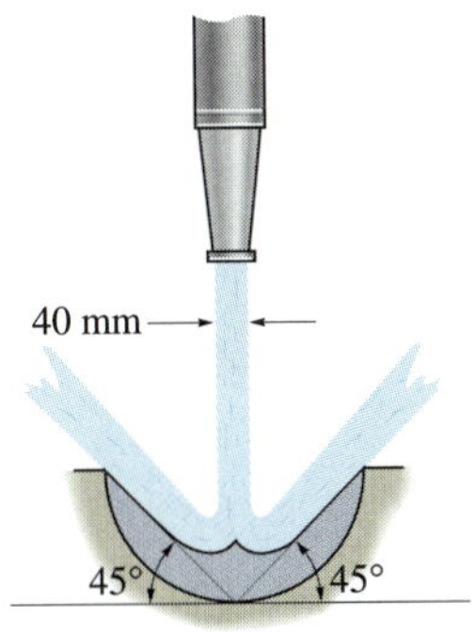

Prob. 15–115

***15-116.** A jet of water has a cross-sectional area of $5\ \text{in}^2$. If it strikes the fixed blade with a constant speed of 60 ft/s, determine the magnitude of force which the water exerts on the blade. $\gamma_w = 62.4\ \text{lb/ft}^3$.

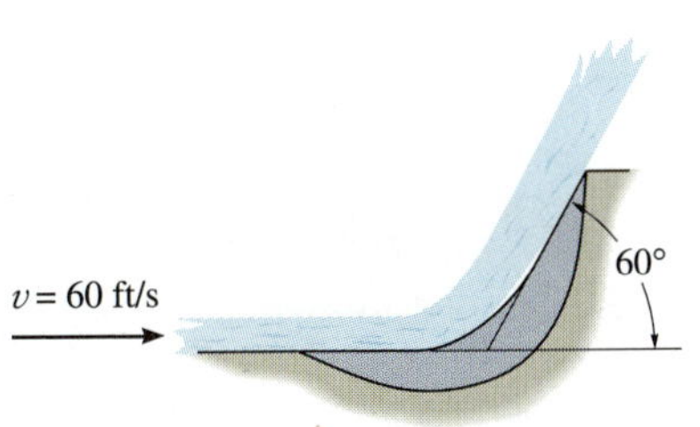

Prob. 15–116

15-117. The static pressure of water at C is $40\ \text{lb/in}^2$. If water flows out of the pipe at A and B with velocities $v_A = 12$ ft/s and $v_B = 25$ ft/s, determine the horizontal and vertical components of force exerted on the elbow at C necessary to hold the pipe assembly in equilibrium. Neglect the weight of water within the pipe and the weight of the pipe. The pipe has a diameter of 0.75 in. at C, and at A and B the diameter is 0.5 in. $\gamma_w = 62.4\ \text{lb/ft}^3$.

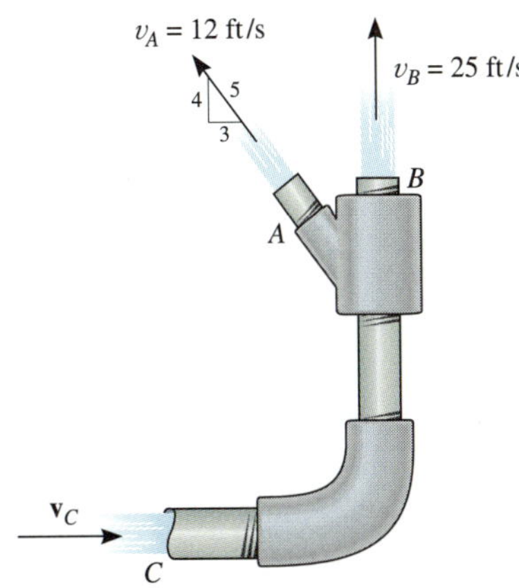

Prob. 15–117

15-118. Sand is deposited from a chute onto a conveyor belt which is moving at 0.5 m/s. If the sand is assumed to fall vertically onto the belt at A at the rate of 4 kg/s, determine the belt tension F_B to the right of A. The belt is free to move over the conveyor rollers and its tension to the left of A is $F_C = 400$ N.

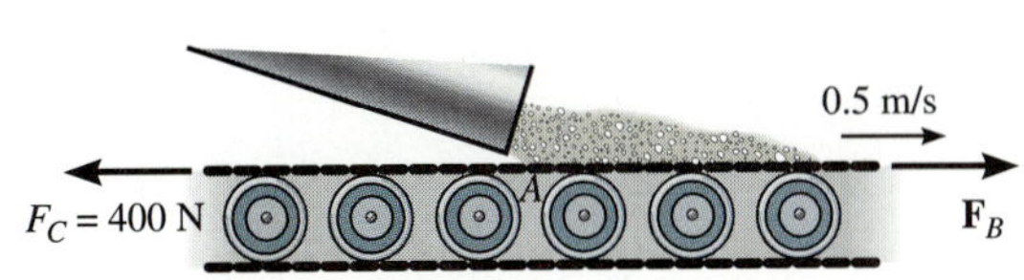

Prob. 15–118

15-119. The fireman sprays a 2-in.-diameter jet of water from a hose at the building. If the water is discharged at $0.80\ \text{ft}^3/\text{s}$, determine the elevation y to where the centerline of the water splashes on the wall. Also, compute the total normal reaction of the fireman's feet on the ground. He has a weight of 180 lb. Neglect the weight of the suspended hose and the water within it. $\gamma_w = 62.4\ \text{lb/ft}^3$.

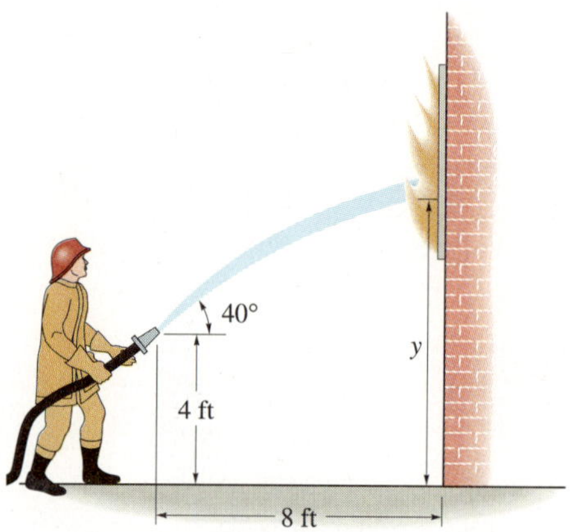

Prob. 15–119

***15-120.** The fountain shoots water in the direction shown. If the water is discharged at 30° from the horizontal, and the cross-sectional area of the water stream is approximately $2\ \text{in}^2$, determine the force it exerts on the concrete wall at B. $\gamma_w = 62.4\ \text{lb/ft}^3$.

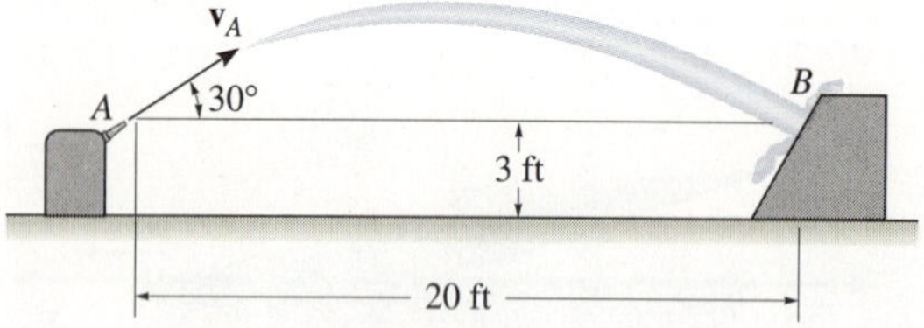

Prob. 15–120

15-121. The vane on a turbine is moving at 80 ft/s when a jet of water having a velocity of $v_A = 150$ ft/s strikes it. If the cross-sectional area of the jet is $1.5\ \text{in}^2$, and it is diverted as shown, determine the power developed by the water on the blade. $\gamma_w = 62.4\ \text{lb/ft}^3$.

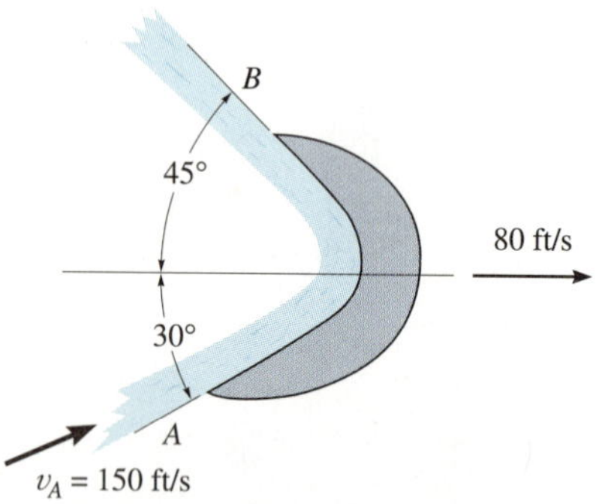

Prob. 15–121

15-122. A plow located on the front of a locomotive scoops up snow at the rate of $10\ \text{ft}^3/\text{s}$ and stores it in the train. If the locomotive is traveling at a constant speed of 12 ft/s, determine the resistance to motion caused by the shoveling. The specific weight of snow is $\gamma_s = 6\ \text{lb/ft}^3$.

15-123. The boat has a mass of 180 kg and is traveling forward on a river with a constant velocity of 70 km/h, measured *relative* to the river. The river is flowing in the opposite direction at 5 km/h. If a tube is placed in the water, as shown, and it collects 40 kg of water in the boat in 80 s, determine the horizontal thrust T on the tube that is required to overcome the resistance to the water collection. $\rho_w = 1\ \text{Mg/m}^3$.

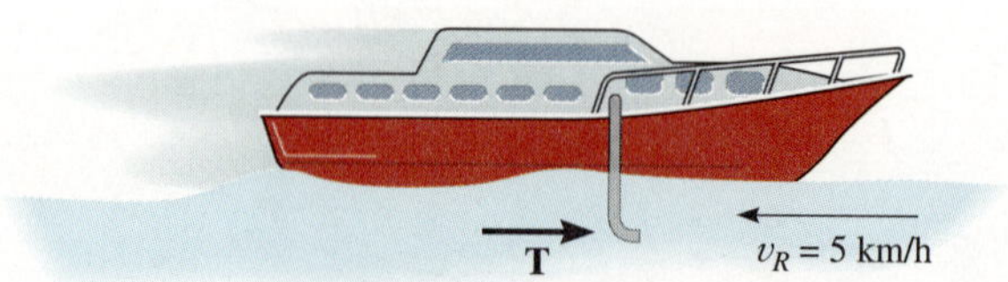

Prob. 15–123

*15-124. The second stage of the two-stage rocket weighs 2000 lb (empty) and is launched from the first stage with a velocity of 3000 mi/h. The fuel in the second stage weighs 1000 lb. If it is consumed at the rate of 50 lb/s and ejected with a relative velocity of 8000 ft/s, determine the acceleration of the second stage just after the engine is fired. What is the rocket's acceleration just before all the fuel is consumed? Neglect the effect of gravitation.

15-125. The missile weighs 40 000 lb. The constant thrust provided by the turbojet engine is $T = 15\,000$ lb. Additional thrust is provided by *two* rocket boosters B. The propellant in each booster is burned at a constant rate of 150 lb/s, with a relative exhaust velocity of 3000 ft/s. If the mass of the propellant lost by the turbojet engine can be neglected, determine the velocity of the missile after the 4-s burn time of the boosters. The initial velocity of the missile is 300 mi/h.

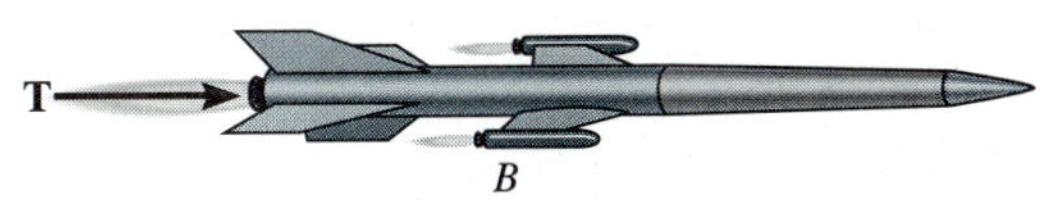

Prob. 15–125

15-126. The rocket car has a mass of 2 Mg (empty) and carries 120 kg of fuel. If the fuel is consumed at a constant rate of 6 kg/s and ejected from the car with a relative velocity of 800 m/s, determine the maximum speed attained by the car starting from rest. The drag resistance due to the atmosphere is $F_D = (6.8v^2)$ N, where v is the speed in m/s.

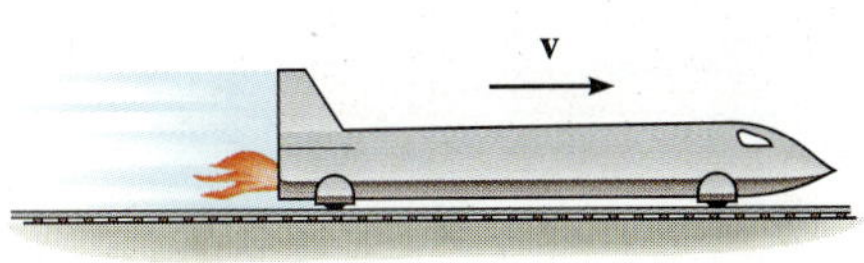

Prob. 15–126

15-127. The 10-Mg helicopter carries a bucket containing 500 kg of water, which is used to fight fires. If it hovers over the land in a fixed position and then releases 50 kg/s of water at 10 m/s, measured relative to the helicopter, determine the initial upward acceleration the helicopter experiences as the water is being released.

Prob. 15–127

*15-128. A rocket burns 2000 lb of fuel in 2 min. If the velocity of the gas is 6500 ft/s with respect to the rocket, determine the thrust of the rocket engine.

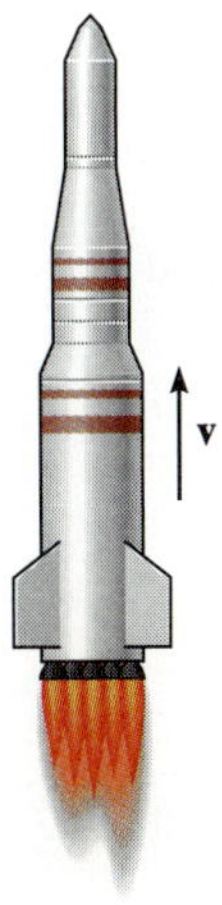

Prob. 15–128

15-129. A rocket has an empty weight of 500 lb and carries 300 lb of fuel. If the fuel is burned at the rate of 15 lb/s and ejected with a relative velocity of 4400 ft/s, determine the maximum speed attained by the rocket starting from rest. Neglect the effect of gravitation on the rocket.

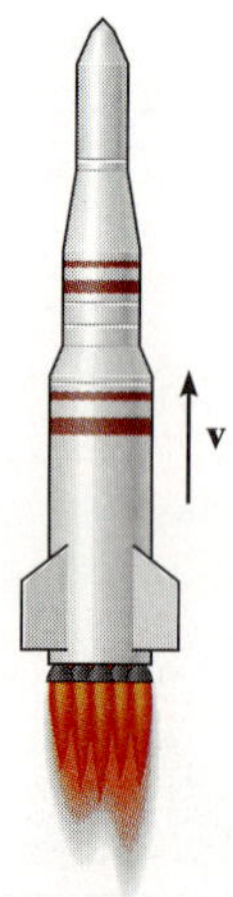

Prob. 15–129

15-130. The 12-Mg jet airplane has a constant speed of 950 km/h when it is flying along a horizontal straight line. Air enters the intake scoops S at the rate of 50 m^3/s. If the engine burns fuel at the rate of 0.4 kg/s and the gas (air and fuel) is exhausted relative to the plane with a speed of 450 m/s, determine the resultant drag force exerted on the plane by air resistance. Assume that air has a constant density of 1.22 kg/m^3. *Hint:* Since mass both enters and exits the plane, Eqs. 15–29 and 15–30 must be combined to yield

$$\Sigma F_s = m\frac{dv}{dt} - v_{D/e}\frac{dm_e}{dt} + v_{D/i}\frac{dm_i}{dt}$$

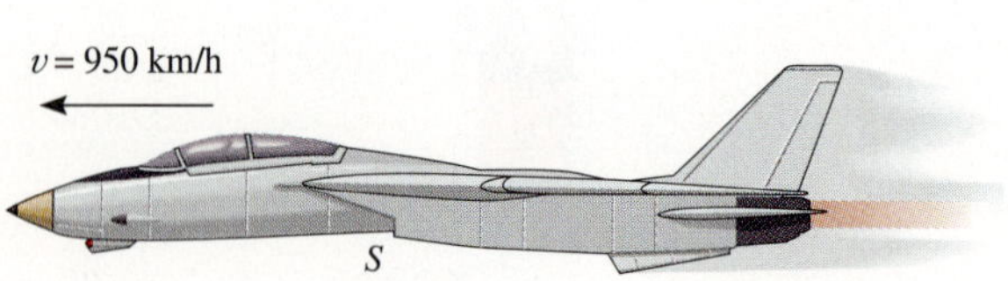

Prob. 15–130

15-131. The chain has a mass per unit length of m. Determine the magnitude of force, $F = F(t)$, which must be applied to the end of the chain to raise it with a constant speed v. Initially, the entire chain is at rest on the ground.

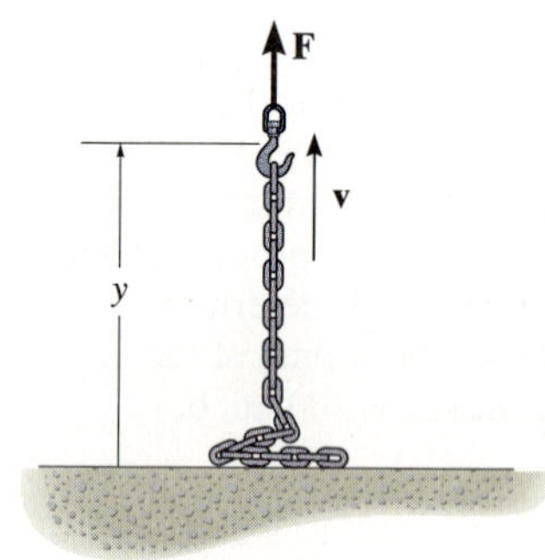

Prob. 15–131

***15-132.** The cart has a mass M and is filled with water that has a mass m_0. If a pump ejects the water through a nozzle having a cross-sectional area A at a constant rate of v_0 relative to the cart, determine the velocity of the cart as a function of time. What is the maximum speed developed by the cart assuming all the water can be pumped out? Neglect frictional resistance to forward motion. The density of the water is ρ.

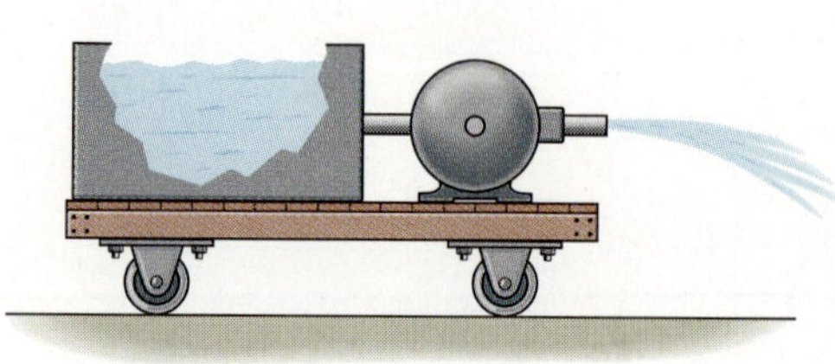

Prob. 15–132

15-133. The cart has a mass M and is filled with water that has a mass m_0. If a pump ejects the water through a nozzle having a cross-sectional area A at a constant rate of v_0 relative to the cart, determine the velocity of the cart as a function of time. What is the maximum speed developed by the cart assuming all the water can be pumped out? The frictional resistance to forward motion is F. The density of the water is ρ.

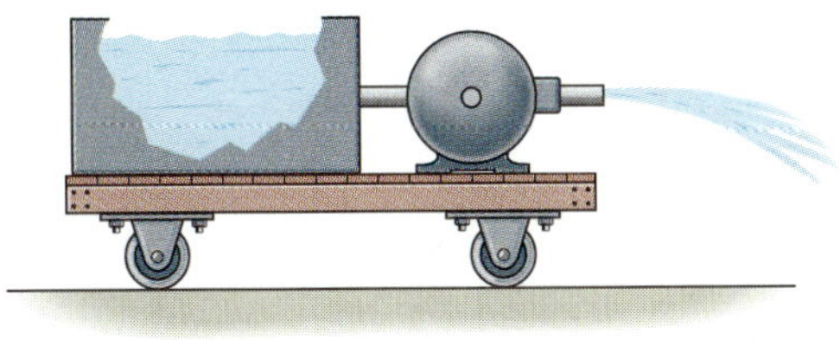

Prob. 15–133

15-134. The chain has a total length $L < d$ and a mass per unit length of m'. If a portion h of the chain is suspended over the table and released, determine the velocity of its end A as a function of its position y. Neglect friction.

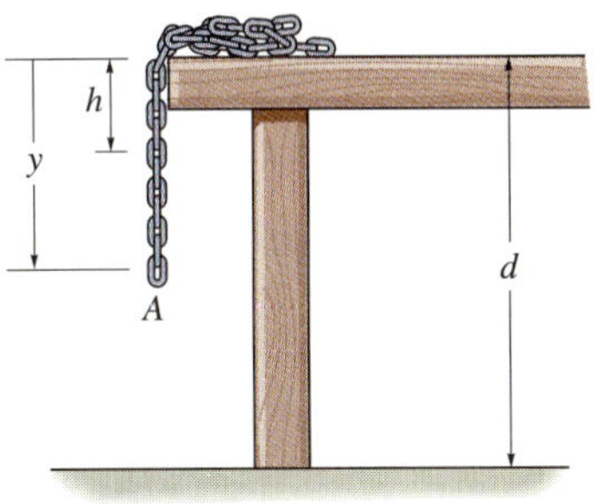

Prob. 15–134

Design Project

15–1D. DESIGN OF A CRANBERRY SELECTOR

The quality of a cranberry depends upon its firmness, which in turn is related to its bounce. Through experiment, it is found that berries that bounce to a height of $2.5 \leq h' \leq 3.25$ ft, when released from rest at a height of $h = 4$ ft, are appropriate for processing. Using this information, determine the berry's range of the allowable coefficient of restitution, and then design a manner in which good and bad berries can be separated. Submit a drawing of your design, and show calculations as to how the selection and collection of berries is made from your established geometry.

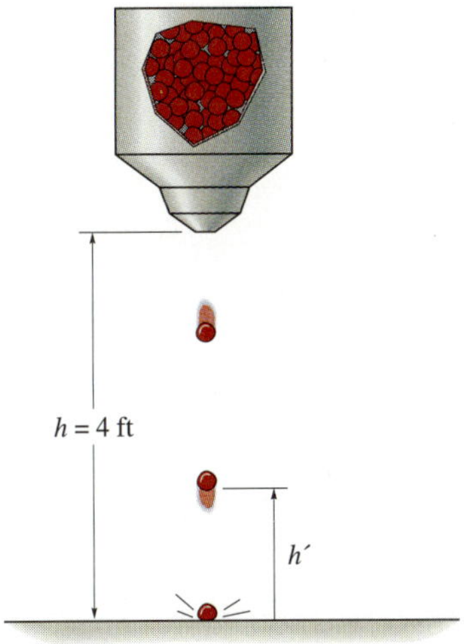

Prob. 15–1D

Review 1:

Kinematics and Kinetics of a Particle

The topics and problems presented in Chapters 12 through 15 have all been *categorized* in order to provide a *clear focus* for learning the various problem-solving principles involved. In engineering practice, however, it is most important to be able to *identify* an appropriate method for the solution of a particular problem. In this regard, one must fully understand the limitations and use of the equations of dynamics, and be able to recognize which equations and principles to use for the problem's solution. For these reasons, we will now summarize the equations and principles of particle dynamics and provide the opportunity for applying them to a variety of problems.

Kinematics. Problems in kinematics require a study only of the geometry of motion, and do not account for the forces causing the motion. When the equations of kinematics are applied, one should clearly establish a fixed origin and select an appropriate coordinate system used to define the position of the particle. Once the positive direction of each coordinate axis is established, then the directions of the components of position, velocity, and acceleration can be determined from the algebraic sign of their numerical quantities.

Rectilinear Motion

Variable Acceleration. If a mathematical (or graphical) relationship is established between *any two* of the *four* variables s, v, a, and t, then a *third* variable can be determined by solving one of the following equations which relates all three variables.

$$v = \frac{ds}{dt} \qquad a = \frac{dv}{dt} \qquad a\,ds = v\,dv$$

Constant Acceleration. Be *absolutely certain* that the acceleration is constant when using the following equations:

$$s = s_0 + v_0 t + \tfrac{1}{2}a_c t^2 \qquad v = v_0 + a_c t \qquad v^2 = v_0^2 + 2a_c(s - s_0)$$

Curvilinear Motion

x, y, z Coordinates. These coordinates are often used when the motion can be resolved into horizontal and vertical components. They are also useful for studying projectile motion since the acceleration of the projectile is *always* downward.

$$v_x = \dot{x} \qquad a_x = \dot{v}_x$$
$$v_y = \dot{y} \qquad a_y = \dot{v}_y$$
$$v_z = \dot{z} \qquad a_z = \dot{v}_z$$

n, t, b Coordinates. These coordinates are particularly advantageous for studying the particle's *acceleration* along a known path. This is because the t and n components of $\mathbf{a}$ represent the separate changes in the magnitude and direction of the velocity, respectively, and these components can be readily formulated.

$$v = \dot{s}$$

$$a_t = \dot{v} = v\frac{dv}{ds}$$

$$a_n = \frac{v^2}{\rho}$$

where

$$\rho = \left|\frac{[1 + (dy/dx)^2]^{3/2}}{d^2y/dx^2}\right|$$

when the path $y = f(x)$ is given.

r, θ, z Coordinates. These coordinates are used when data regarding the angular motion of the radial coordinate r is given to describe the particle's motion. Also, some paths of motion can conveniently be described using these coordinates.

$$v_r = \dot{r} \qquad a_r = \ddot{r} - r\dot{\theta}^2$$
$$v_\theta = r\dot{\theta} \qquad a_\theta = r\ddot{\theta} + 2\dot{r}\dot{\theta}$$
$$v_z = \dot{z} \qquad a_z = \ddot{z}$$

Relative Motion. If the origin of a *translating* coordinate system is established at particle A, then for particle B,

$$\mathbf{r}_B = \mathbf{r}_A + \mathbf{r}_{B/A}$$
$$\mathbf{v}_B = \mathbf{v}_A + \mathbf{v}_{B/A}$$
$$\mathbf{a}_B = \mathbf{a}_A + \mathbf{a}_{B/A}$$

Here the relative motion is measured by an observer fixed in the translating coordinate system.

Kinetics. Problems in kinetics involve the analysis of forces which cause the motion. When applying the equations of kinetics, it is absolutely necessary that measurements of the motion be made from an *inertial coordinate system,* i.e., one that does not rotate and is either fixed or translates with constant velocity. If a problem requires *simultaneous solution* of the equations of kinetics and kinematics, then it is important that the coordinate systems selected for writing each of the equations define the *positive directions* of the axes in the *same* manner.

Equations of Motion. These equations are used to solve for the particle's acceleration or the forces causing the motion. If they are used to determine a particle's position, velocity, or time of motion, then kinematics will also have to be considered in the solution. Before applying the equations of motion, *always draw a free-body diagram* to identify all the forces acting on the particle. Also, establish the direction of the particle's acceleration or its components. (A kinetic diagram may accompany the solution in order to graphically account for the $m\mathbf{a}$ vector.)

$$\Sigma F_x = ma_x \qquad \Sigma F_n = ma_n \qquad \Sigma F_r = ma_r$$
$$\Sigma F_y = ma_y \qquad \Sigma F_t = ma_t \qquad \Sigma F_\theta = ma_\theta$$
$$\Sigma F_z = ma_z \qquad \Sigma F_b = 0 \qquad \Sigma F_z = ma_z$$

Work and Energy. The equation of work and energy represents an integrated form of the tangential equation of motion, $\Sigma F_t = ma_t$, combined with kinematics ($a_t\,ds = v\,dv$). *It is used to solve problems involving force, velocity, and displacement.* Before applying this equation, *always draw a free-body diagram* in order to identify the forces which do work on the particle.

$$T_1 + \Sigma U_{1-2} = T_2$$

where

$$T = \tfrac{1}{2}mv^2 \qquad \text{(kinetic energy)}$$
$$U_F = \int_{s_1}^{s_2} F\cos\theta\,ds \qquad \text{(work of a variable force)}$$
$$U_{F_c} = F_c\cos\theta\,(s_2 - s_1) \qquad \text{(work of a constant force)}$$
$$U_W = -W\,\Delta y \qquad \text{(work of a weight)}$$
$$U_s = -\left(\tfrac{1}{2}ks_2^2 - \tfrac{1}{2}ks_1^2\right) \qquad \text{(work of an elastic spring)}$$

If the forces acting on the particle are *conservative forces,* i.e., those that *do not* cause a dissipation of energy, such as friction, then apply the conservation of energy equation. This equation is easier to use than the equation of work and energy since it applies only at *two points* on the path and *does not* require calculation of the work done by a force as the particle moves along the path.

$$T_1 + V_1 = T_2 + V_2$$

where

$$V_g = Wy \qquad \text{(gravitational potential energy)}$$
$$V_e = \tfrac{1}{2}ks^2 \qquad \text{(elastic potential energy)}$$

If the *power* developed by a force is to be determined, use

$$P = \frac{dU}{dt} = \mathbf{F}\cdot\mathbf{v}$$

where $\mathbf{v}$ is the velocity of a particle acted upon by the force $\mathbf{F}$.

Impulse and Momentum. The equation of *linear impulse and momentum* is an integrated form of the equation of motion, $\Sigma\mathbf{F} = m\mathbf{a}$, combined with kinematics ($\mathbf{a} = d\mathbf{v}/dt$). *It is used to solve problems involving force, velocity, and time.* Before applying this equation, one should *always draw the free-body diagram,* in order to identify all the forces that cause impulses on the particle. From the diagram the impulsive and nonimpulsive forces should be identified. Recall that the nonimpulsive forces can be neglected in the analysis during the time of impact. Also, establish the direction of the particle's velocity just before and just after the impulses are applied. As an alternative procedure, impulse and momentum diagrams may accompany the solution in order to graphically account for the terms in the equation.

$$m\mathbf{v}_1 + \Sigma\int_{t_1}^{t_2} \mathbf{F}\,dt = m\mathbf{v}_2$$

If several particles are involved in the problem, consider applying the *conservation of momentum* to the system in order to eliminate the internal impulses from the analysis. This can be done in a specified direction, provided no external impulses act on the particles in that direction.

$$\Sigma m\mathbf{v}_1 = \Sigma m\mathbf{v}_2$$

If the problem involves impact and the coefficient of restitution e is given, then apply the following equation.

$$e = \frac{(v_B)_2 - (v_A)_2}{(v_A)_1 - (v_B)_1} \qquad \text{(along line of impact)}$$

Remember that during impact the principle of work and energy cannot be used, since the particles deform and therefore the work due to the internal forces will be unknown. The principle of work and energy can be used, however, to determine the energy loss during the collision once the particle's initial and final velocities are determined.

The *principle of angular impulse and momentum* and the *conservation of angular momentum* may be applied about an axis in order to *eliminate* some of the unknown impulses acting on the particle. Investigation of the particle's free-body diagram (or the impulse diagram) will aid in choosing the axis for application.

$$(\mathbf{H}_O)_1 + \Sigma \int_{t_1}^{t_2} \mathbf{M}_O \, dt = (\mathbf{H}_O)_2$$

$$(\mathbf{H}_O)_1 = (\mathbf{H}_O)_2$$

The following problems provide an opportunity for applying the above concepts. They are presented in *random order* so that practice may be gained in identifying the various types of problems and developing the skills necessary for their solution.

Review Problems

R1-1. The block has a mass of 5 kg and is released from rest at A. It slides down the smooth plane onto the rough surface having a coefficient of kinetic friction of $\mu_k = 0.2$. Determine the total time of travel before the block stops sliding. Also, how far s does the block slide before stopping? Neglect the size of the block.

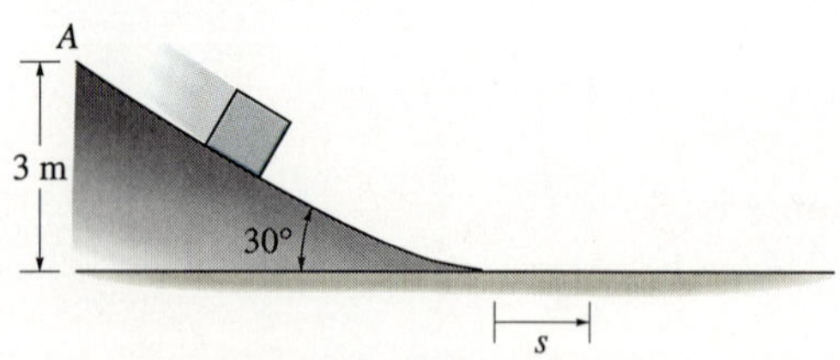

Prob. R1–1

R1-2. Cartons having a mass of 5 kg are required to move along the assembly line at a constant speed of 8 m/s. Determine the smallest radius of curvature, ρ, for the conveyor so the cartons do not slip. The coefficients of static and kinetic friction between a carton and the conveyor are $\mu_s = 0.7$ and $\mu_k = 0.5$, respectively.

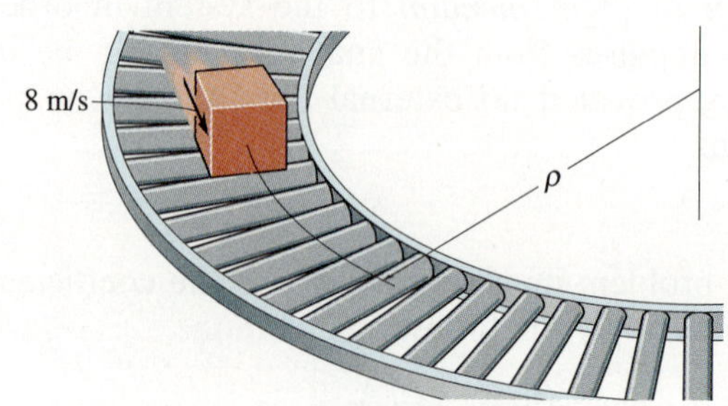

Prob. R1–2

R1-3. A small metal particle passes downward through a fluid medium while being subjected to the attraction of a magnetic field such that its position is observed to be $s = (15t^3 - 3t)$ mm, where t is in seconds. Determine (a) the particle's displacement from $t = 2$ s to $t = 4$ s, and (b) the velocity and acceleration of the particle when $t = 5$ s.

***R1-4.** The device shown is designed to produce the experience of weightlessness in a passenger when he reaches point A, $\theta = 90°$, along the path. If the passenger has a mass of 75 kg, determine the minimum speed he should have when he reaches A so that he does not exert a normal reaction on the seat. The chair is pin connected to the frame BC so that he is always seated in an upright position. During the motion his speed remains constant.

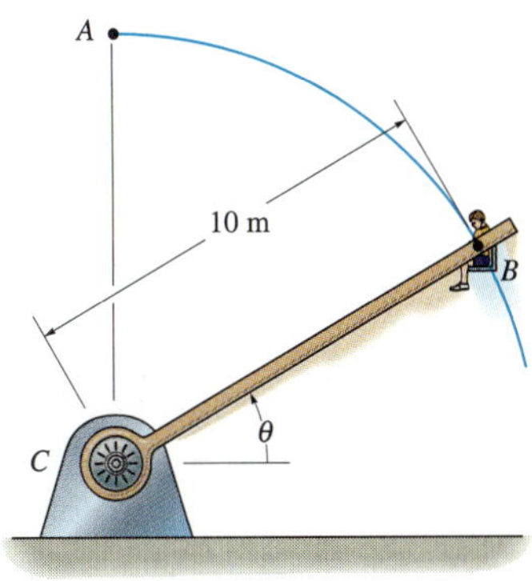

Prob. R1–4

R1-5. The passenger has a mass of 75 kg and always sits in an upright position on the chair. At the instant $\theta = 30°$, he has a speed of 5 m/s and an increase of speed of 2 m/s^2. Determine the horizontal and vertical forces that the chair exerts on him in order to produce this motion.

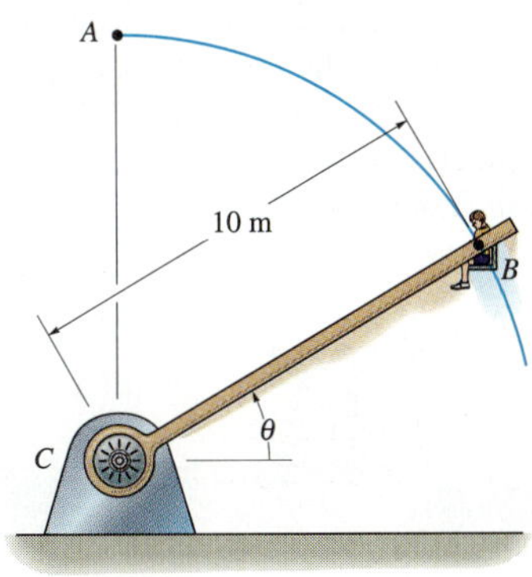

Prob. R1–5

R1-6. The 150-lb man lies against the cushion for which the coefficient of static friction is $\mu_s = 0.5$. Determine the resultant normal and frictional forces the cushion exerts on him if, due to rotation about the z axis, he has a constant speed $v = 20$ ft/s. Neglect the size of the man. Take $\theta = 60°$.

R1-7. The 150-lb man lies against the cushion for which the coefficient of static friction is $\mu_s = 0.5$. If he rotates about the z axis with a constant speed $v = 30$ ft/s, determine the smallest angle θ of the cushion at which he will begin to slip up the cushion.

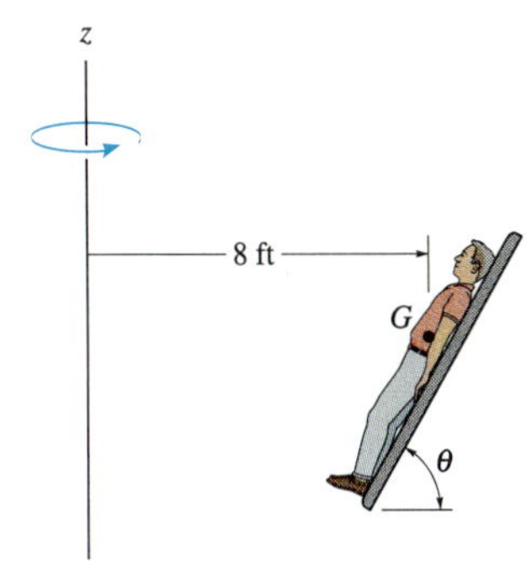

Probs. R1–6/7

***R1-8.** The baggage truck A has a mass of 800 kg and is used to pull each of the 300-kg cars. Determine the tension in the couplings at B and C if the tractive force $\mathbf{F}$ on the truck is $F = 480$ N. What is the speed of the truck when $t = 2$ s, starting from rest? The car wheels are free to roll. Neglect the mass of the wheels.

R1-9. The baggage truck A has a mass of 800 kg and is used to pull each of the 300-kg cars. If the tractive force $\mathbf{F}$ on the truck is $F = 480$ N, determine the initial acceleration of the truck. What is the acceleration of the truck if the coupling at C suddenly fails? The car wheels are free to roll. Neglect the mass of the wheels.

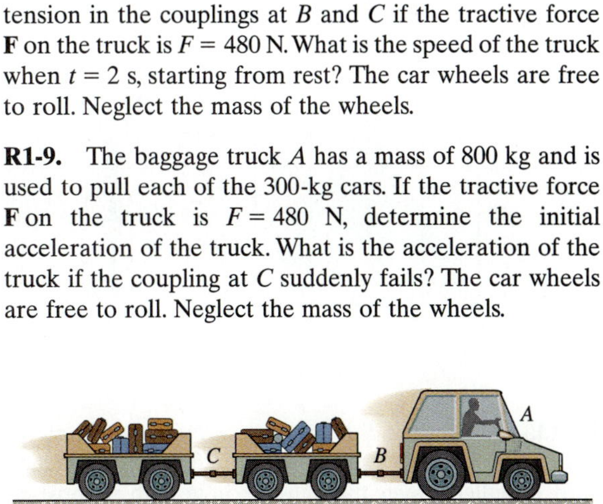

Probs. R1–8/9

R1-10. A car is traveling at 80 ft/s when the brakes are suddenly applied, causing a constant deceleration of 10 ft/s^2. Determine the time required to stop the car and the distance traveled before stopping.

R1-11. Determine the speed of block B if the end of the cable at C is pulled downward with a speed of 10 ft/s. What is the relative velocity of the block with respect to C?

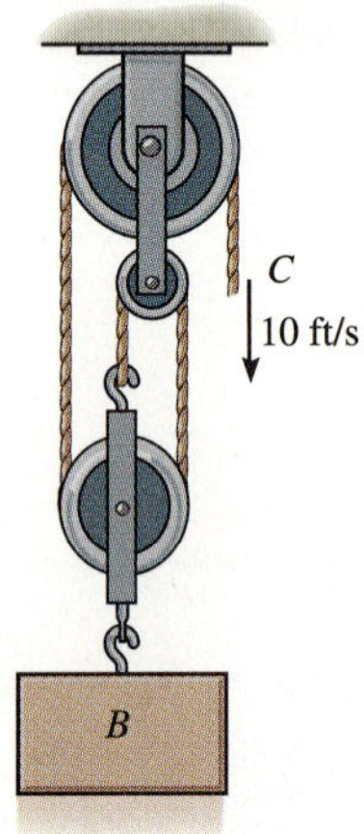

Prob. R1–11

***R1-12.** During operation the breaker hammer develops on the concrete surface a force which is indicated in the graph. To achieve this the 2-lb spike S is fired from rest into the surface at 200 ft/s. Determine the speed of the spike just after rebounding.

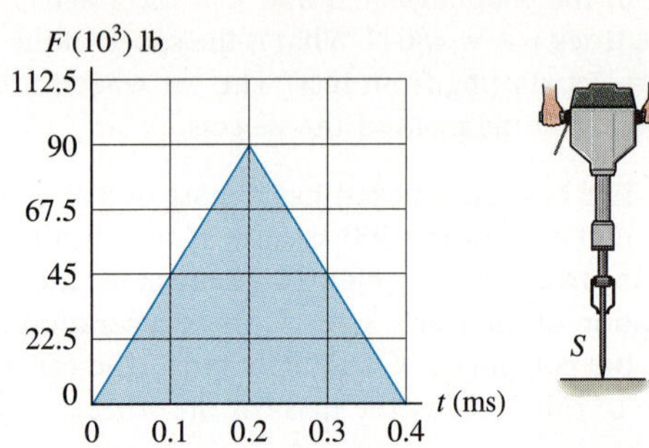

Prob. R1–12

R1-13. A 4-lb ball B is traveling around in a circle of radius $r_1 = 3$ ft with a speed $(v_B)_1 = 6$ ft/s. If the attached cord is pulled down through the hole with a constant speed $v_r = 2$ ft/s, determine the ball's speed at the instant $r_2 = 2$ ft. How much work has to be done to pull down the cord? Neglect friction and the size of the ball.

R1-14. A 4-lb ball B is traveling around in a circle of radius $r_1 = 3$ ft with a speed $(v_B)_1 = 6$ ft/s. If the attached cord is pulled down through the hole with a constant speed $v_r = 2$ ft/s, determine how much time is required for the ball to reach a speed of 12 ft/s. How far r_2 is the ball from the hole when this occurs? Neglect friction and the size of the ball.

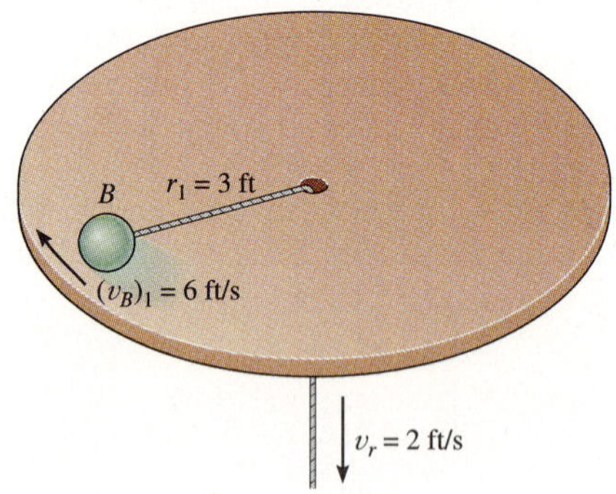

Probs. R1–13/14

R1-15. The block has a mass of 50 kg and rests on the surface of the cart having a mass of 75 kg. If the spring which is attached to the cart and not the block is compressed 0.2 m and the system is released from rest, determine the speed of the block after the spring becomes undeformed. Neglect the mass of the cart's wheels and the spring in the calculation. Also neglect friction. Take $k = 300$ N/m.

***R1-16.** The block has a mass of 50 kg and rests on the surface of the cart having a mass of 75 kg. If the spring which is attached to the cart and not the block is compressed 0.2 m and the system is released from rest, determine the speed of the block with respect to the cart after the spring becomes undeformed. Neglect the mass of the cart's wheels and the spring in the calculation. Also neglect friction. Take $k = 300$ N/m.

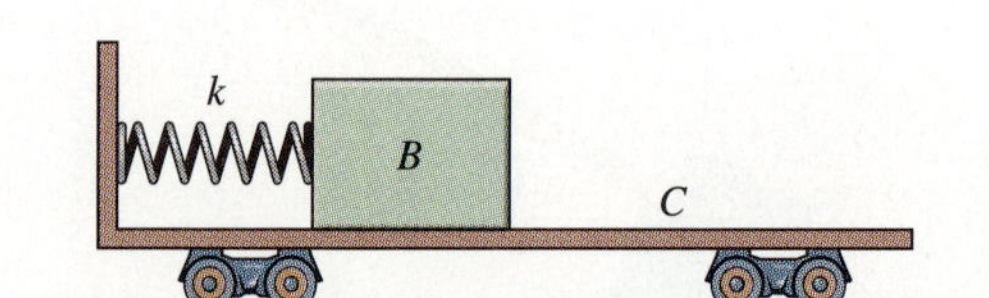

Probs. R1–15/16

R1-17. A ball is launched from point A at an angle of 30°. Determine the maximum and minimum speed v_A it can have so that it lands in the container.

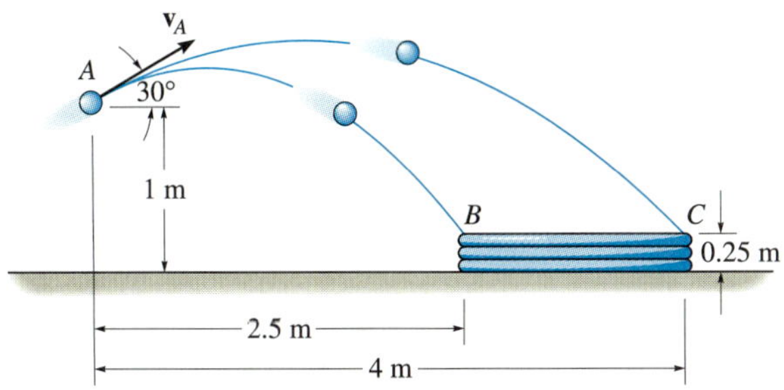

Prob. R1–17

R1-18. At the instant shown, cars A and B are traveling at speeds of 55 mi/h and 40 mi/h, respectively. If B is increasing its speed by 1200 mi/h^2, while A maintains a constant speed, determine the velocity and acceleration of B with respect to A. Car B moves along a curve having a radius of curvature of 0.5 mi.

R1-19. At the instant shown, cars A and B are traveling at speeds of 55 mi/h and 40 mi/h, respectively. If B is decreasing its speed at 1500 mi/h^2 while A is increasing its speed at 800 mi/h^2, determine the acceleration of B with respect to A. Car B moves along a curve having a radius of curvature of 0.75 mi.

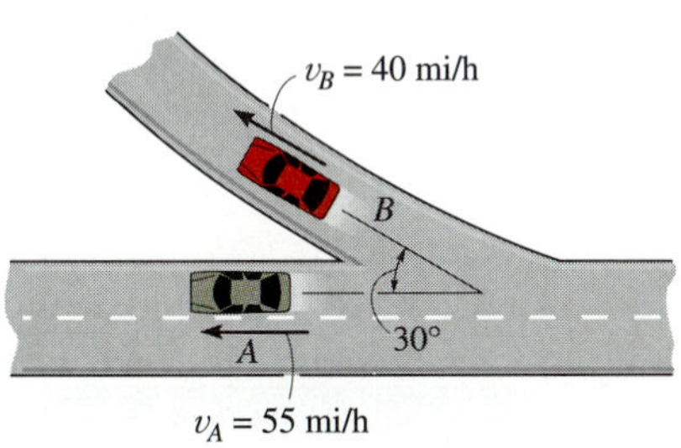

Probs. R1–18/19

***R1-20.** Four inelastic cables C are attached to a plate P and hold the 1-ft-long spring 0.25 ft in compression when *no weight* is on the plate. There is also an undeformed spring nested within this compressed spring. If the block, having a weight of 10 lb, is moving downward at $v = 4$ ft/s, when it is 2 ft above the plate, determine the maximum compression in each spring after it strikes the plate. Neglect the mass of the plate and springs and any energy lost in the collision.

R1-21. Four inelastic cables C are attached to a plate P and hold the 1-ft-long spring 0.25 ft in compression when *no weight* is on the plate. There is also a 0.5-ft long undeformed spring nested within this compressed spring. Determine the speed v of the 10-lb block when it is 2 ft above the plate, so that after it strikes the plate, it compresses the nested spring, having a stiffness of 50 lb/in., an amount of 0.20 ft. Neglect the mass of the plate and springs and any energy lost in the collision.

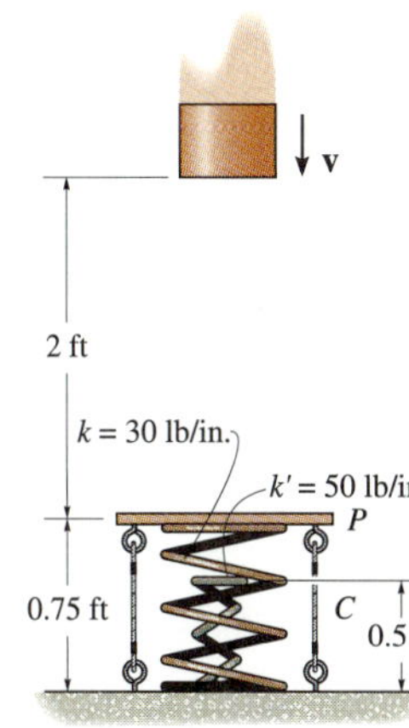

Probs. R1–20/21

R1-22. The 2-kg spool S fits loosely on the rotating inclined rod for which the coefficient of static friction is $\mu_s = 0.2$. If the spool is located 0.25 m from A, determine the minimum constant speed the spool can have so that it does not slip down the rod.

R1-23. The 2-kg spool S fits loosely on the inclined rod for which the coefficient of static friction is $\mu_s = 0.2$. If the spool is located 0.25 m from A, determine the maximum constant speed the spool can have so that it does not slip up the rod.

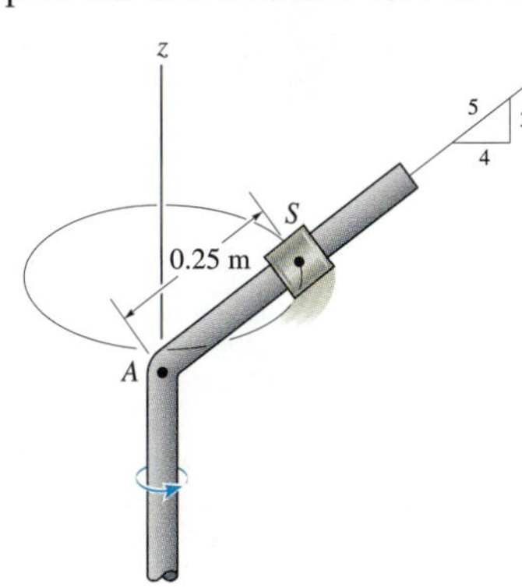

Probs. R1–22/23

***R1-24.** The winding drum D is drawing in the cable at an accelerated rate of 5 m/s^2. Determine the cable tension if the suspended crate has a mass of 800 kg.

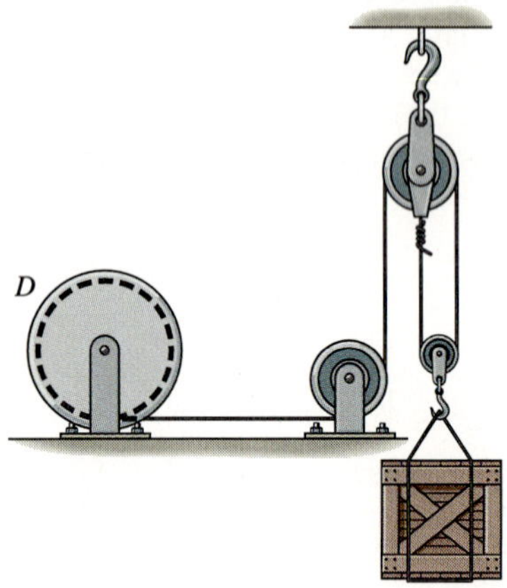

Prob. R1–24

R1-25. If the hoist H is moving upward at 6 ft/s, determine the speed at which the motor M must draw in the supporting cable.

R1-26. Starting from rest, the motor M can draw in the cable of the hoist H such that a point P on the cable has an acceleration of $a_{P/H} = (0.4t)$ ft/s^2, measured with respect to the hoist. Determine the velocity of the hoist when $t = 3$ s.

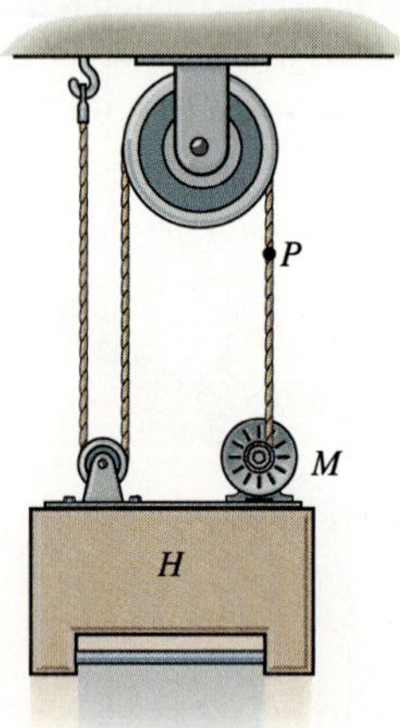

Probs. R1–25/26

R1-27. The winch delivers a horizontal towing force F to its cable at A which varies as shown in the graph. Determine the speed of the 70-kg block B when $t = 18$ s. Originally the block is moving upward at $v_1 = 3$ m/s.

***R1-28.** The winch delivers a horizontal towing force F to its cable at A which varies as shown in the graph. Determine the speed of the 40-kg block B when $t = 24$ s. Originally the block is resting on the ground.

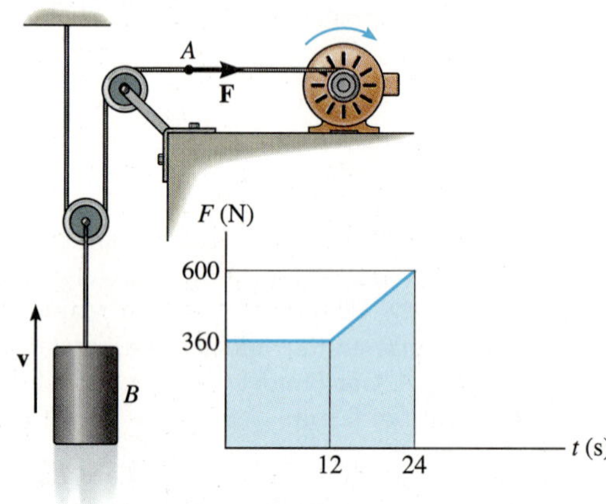

Probs. R1–27/28

R1-29. The motor pulls on the cable at A with a force $F = (30 + t^2)$ lb, where t is in seconds. If the 34-lb crate is originally at rest on the ground when $t = 0$, determine its speed when $t = 4$ s. Neglect the mass of the cable and pulleys. *Hint:* First find the time needed to begin lifting the crate.

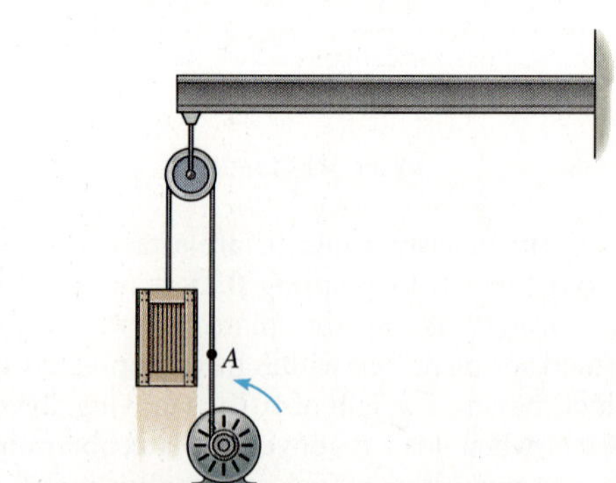

Prob. R1–29

R1-30. The motor pulls on the cable at A with a force $F = (e^{2t})$ lb, where t is in seconds. If the 34-lb crate is originally at rest on the ground when $t = 0$, determine the crate's velocity when $t = 2$ s. Neglect the mass of the cable and pulleys. *Hint:* First find the time needed to begin lifting the crate.

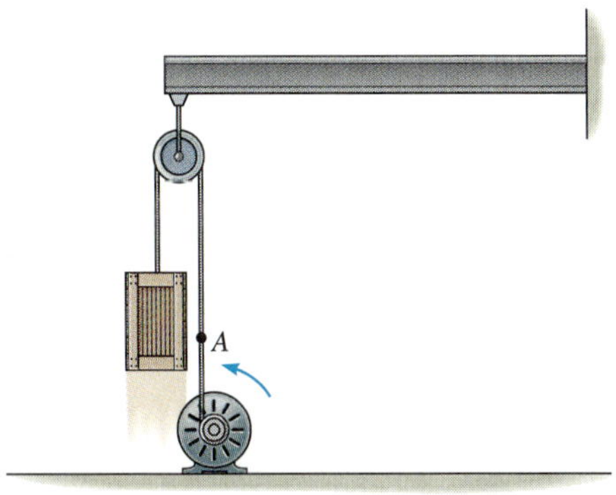

Prob. R1–30

R1-31. The collar has a mass of 2 kg and travels along the smooth *horizontal* rod defined by the equiangular spiral $r = (e^{\theta})$ m, where θ is in radians. Determine the tangential force F and the normal force N acting on the collar when $\theta = 45°$, if the force F maintains a constant angular motion $\dot{\theta} = 2$ rad/s.

***R1-32.** The collar has a mass of 2 kg and travels along the smooth *horizontal* rod defined by the equiangular spiral $r = (e^{\theta})$ m, where θ is in radians. Determine the tangential force F and the normal force N acting on the collar when $\theta = 90°$, if the force F maintains a constant angular motion $\dot{\theta} = 2$ rad/s.

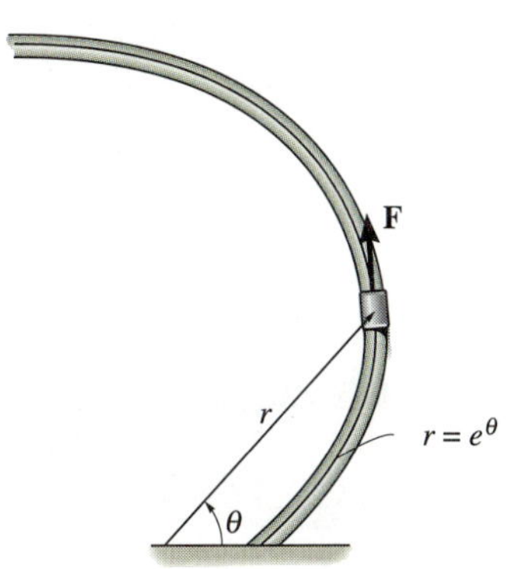

Probs. R1–31/32

R1-33. The acceleration of a particle along a straight line is defined by $a = (2t - 9)$ m/s^2, where t is in seconds. When $t = 0$, $s = 1$ m and $v = 10$ m/s. When $t = 9$ s, determine (a) the particle's position, (b) the total distance traveled, and (c) the velocity. Assume the positive direction is to the right.

R1-34. The 400-kg mine car is hoisted up the incline using the cable and motor M. For a short time, the force in the cable is $F = (3200t^2)$ N, where t is in seconds. If the car has an initial velocity $v_1 = 2$ m/s when $t = 0$, determine its velocity when $t = 2$ s.

R1-35. The 400-kg mine car is hoisted up the incline using the cable and motor M. For a short time, the force in the cable is $F = (3200t^2)$ N, where t is in seconds. If the car has an initial velocity $v_1 = 2$ m/s at $s = 0$ and $t = 0$, determine the distance it moves up the plane when $t = 2$ s.

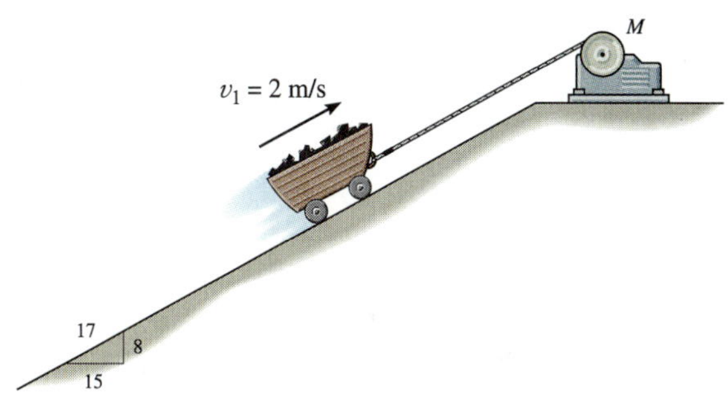

Probs. R1–34/35

***R1-36.** The rocket sled has a mass of 4 Mg and travels from rest along the smooth horizontal track such that it maintains a constant power output of 450 kW. Neglect the loss of fuel mass and air resistance, and determine how far it must travel to reach a speed of $v = 60$ m/s.

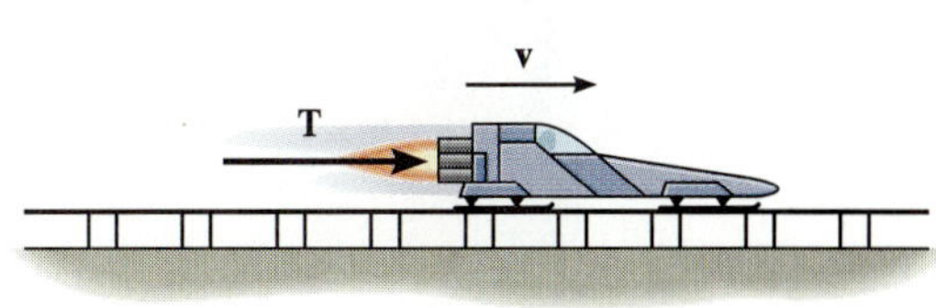

Prob. R1–36

R1-37. The collar has a mass of 20 kg and is supported on the smooth rod. The attached springs are undeformed when $d = 0.5$ m. Determine the speed of the collar after the applied force $F = 100$ N causes it to be displaced so that $d = 0.3$ m. When $d = 0.5$ m the collar is at rest.

R1-38. The collar has a mass of 20 kg and is supported on the smooth rod. The attached springs are both compressed 0.4 m when $d = 0.5$ m. Determine the speed of the collar after the applied force $F = 100$ N causes it to be displaced so that $d = 0.3$ m. When $d = 0.5$ m the collar is at rest.

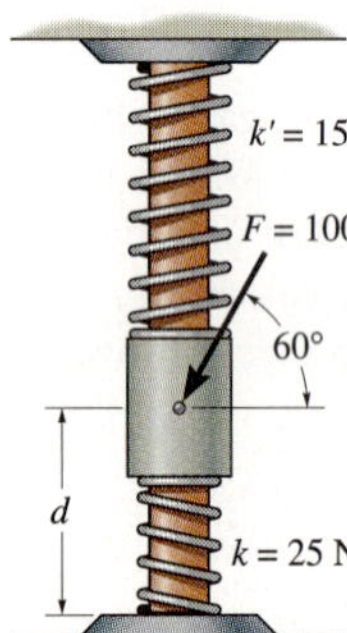

Probs. R1–37/38

R1-39. The assembly consists of two blocks A and B which have masses of 20 kg and 30 kg, respectively. Determine the speed of each block when B descends 1.5 m. The blocks are released from rest. Neglect the mass of the pulleys and cords.

***R1-40.** The assembly consists of two blocks A and B, which have masses of 20 kg and 30 kg, respectively. Determine the distance B must descend in order for A to achieve a speed of 3 m/s starting from rest.

Probs. R1–39/40

R1-41. Block A, having a mass m, is released from rest, falls a distance h and strikes the plate B having a mass $2m$. If the coefficient of restitution between A and B is e, determine the velocity of the plate just after collision. The spring has a stiffness k.

R1-42. Block A, having a mass of 2 kg, is released from rest, falls a distance $h = 0.5$ m, and strikes the plate B having a mass of 3 kg. If the coefficient of restitution between A and B is $e = 0.6$, determine the velocity of the block just after collision. The spring has a stiffness $k = 30$ N/m.

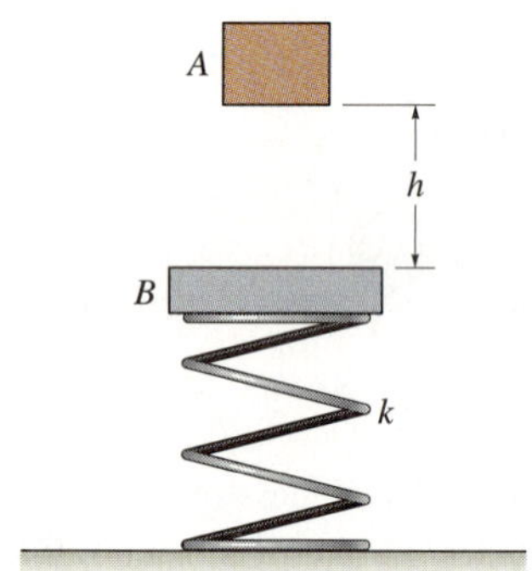

Probs. R1–41/42

R1-43. The cylindrical plug has a weight of 2 lb and it is free to move within the confines of the smooth pipe. The spring has a stiffness $k = 14$ lb/ft and when no motion occurs the distance $d = 0.5$ ft. Determine the force of the spring on the plug when the plug is at rest with respect to the pipe. The plug is traveling with a constant speed of 15 ft/s, which is caused by the rotation of the pipe about the vertical axis. Neglect the size of the plug.

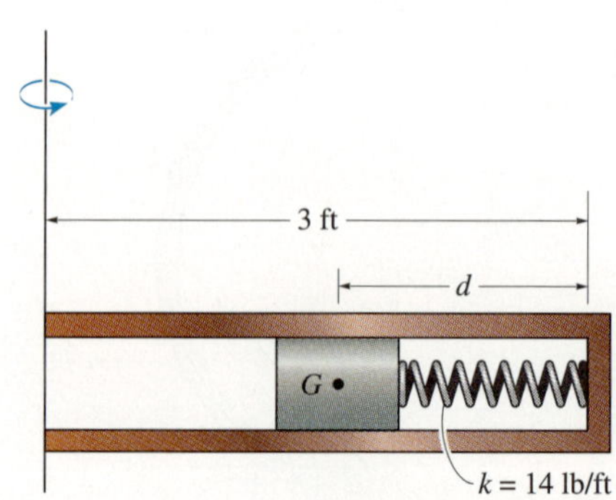

Prob. R1–43

*R1-44. A 20-g bullet is fired horizontally into the 300-g block which rests on the smooth surface. After the bullet becomes embedded into the block, the block moves to the right 0.3 m before momentarily coming to rest. Determine the speed $(v_B)_1$ of the bullet. The spring has a stiffness $k = 200$ N/m and is originally unstretched.

R1-45. The 20-g bullet is fired horizontally at $(v_B)_1 = 1200$ m/s into the 300-g block which rests on the smooth surface. Determine the distance the block moves to the right before momentarily coming to rest. The spring has a stiffness $k = 200$ N/m and is originally unstretched.

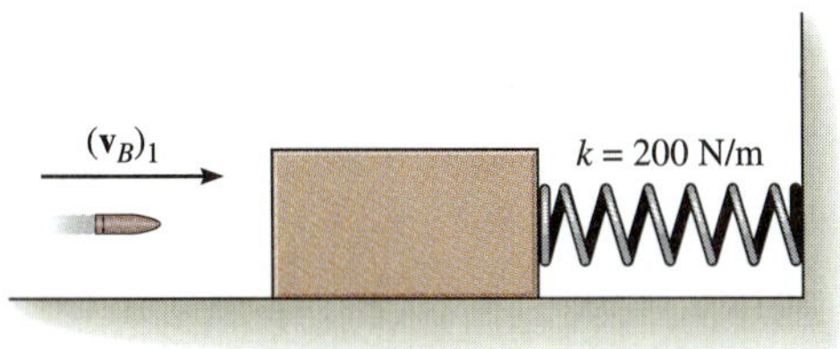

Probs. R1–44/45

R1-46. The collar has a weight of 8 lb. If it is pushed down so as to compress the spring 2 ft and then released from rest ($h = 0$), determine its speed when it is displaced $h = 4.5$ ft. The spring is not attached to the collar. Neglect friction.

R1-47. The collar has a weight of 8 lb. If it is released from rest at a height of $h = 2$ ft from the top of the uncompressed spring, determine the speed of the collar after it falls and compresses the spring 0.3 ft.

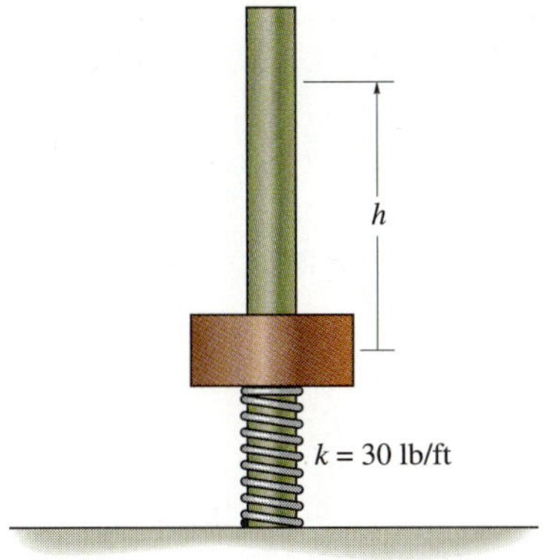

Probs. R1–46/47

*R1-48. The position of particles A and B are $\mathbf{r}_A = \{3t\mathbf{i} + 9t(2 - t)\mathbf{j}\}$ m and $\mathbf{r}_B = \{3(t^2 - 2t + 2)\mathbf{i} + 3(t - 2)\mathbf{j}\}$ m, respectively, where t is in seconds. Determine the point where the particles collide and their speeds just before the collision. How long does it take before the collision occurs?

R1-49. Determine the speed of the automobile if it has the acceleration shown and is traveling on a road which has a radius of curvature of $\rho = 50$ m. Also, what is the automobile's rate of increase in speed?

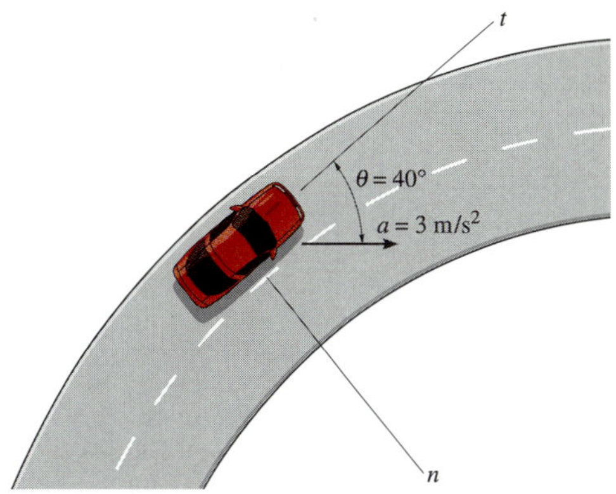

Prob. R1–49

R1-50. If block A of the pulley system is moving downward at 6 ft/s while block C is moving down at 18 ft/s, determine the relative velocity of block B with respect to C.

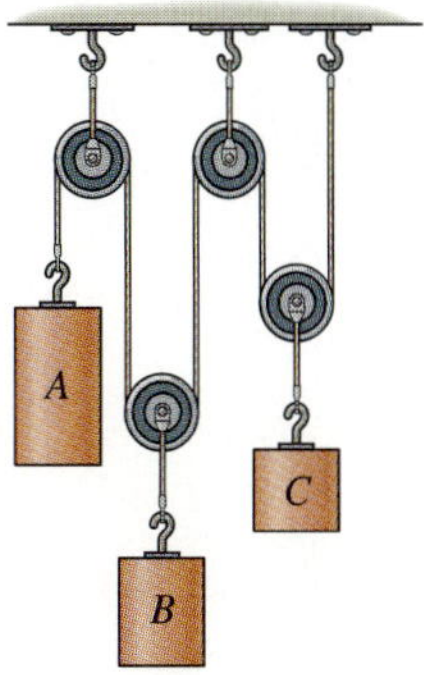

Prob. R1–50

The wind turbine rotates about a fixed axis with variable angular motion.

CHAPTER

16

Planar Kinematics of a Rigid Body

Chapter Objectives

- To classify the various types of rigid-body planar motion.
- To investigate rigid-body translation and show how to analyze motion about a fixed axis.
- To study planar motion using an absolute motion analysis.
- To provide a relative motion analysis of velocity and acceleration using a translating frame of reference.
- To show how to find the instantaneous center of zero velocity and determine the velocity of a point on a body using this method.
- To provide a relative motion analysis of velocity and acceleration using a rotating frame of reference.

16.1 Rigid-Body Motion

In this chapter, the planar kinematics of a rigid body will be discussed. This study is important for the design of gears, cams, and mechanisms used for many mechanical operations. Furthermore, once the kinematics of a rigid body is thoroughly understood, then it will be possible to apply the equations of motion, which relate the forces on the body to the body's motion.

When all the particles of a rigid body move along paths which are equidistant from a fixed plane, the body is said to undergo *planar motion.* There are three types of rigid body planar motion; in order of increasing complexity, they are

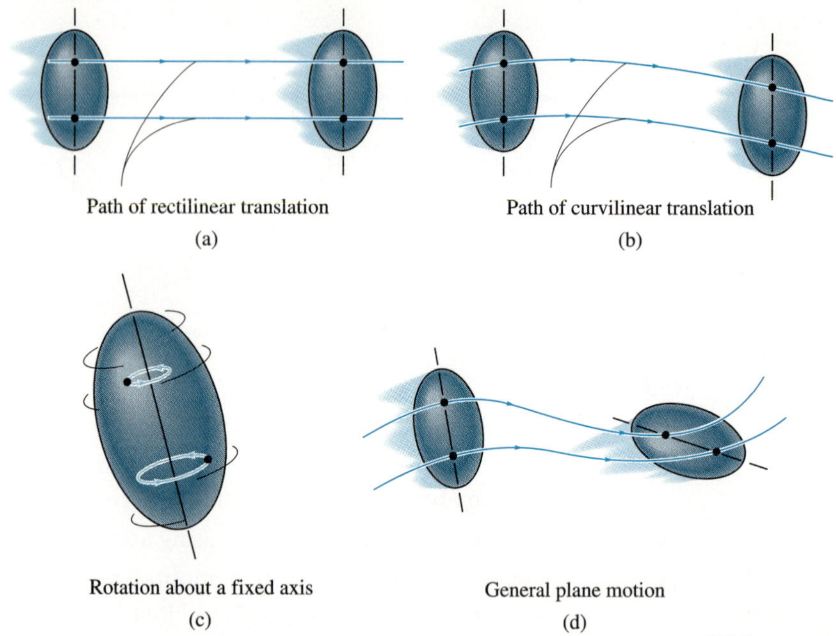

Fig. 16–1

1. *Translation.* This type of motion occurs if every line segment on the body remains parallel to its original direction during the motion. When the paths of motion for any two particles of the body are along equidistant straight lines the motion is called *rectilinear translation,* Fig. 16–1*a*. However, if the paths of motion are along curved lines which are equidistant the motion is called *curvilinear translation*, Fig. 16–1*b*.
2. *Rotation about a fixed axis.* When a rigid body rotates about a fixed axis, all the particles of the body, except those which lie on the axis of rotation, move along circular paths, Fig. 16–1*c*.
3. *General plane motion.* When a body is subjected to general plane motion, it undergoes a combination of translation *and* rotation, Fig. 16–1*d*. The translation occurs within a reference plane, and the rotation occurs about an axis perpendicular to the reference plane.

In the following sections we will consider each of these motions in detail. Examples of bodies undergoing these motions is shown in Fig. 16–2.

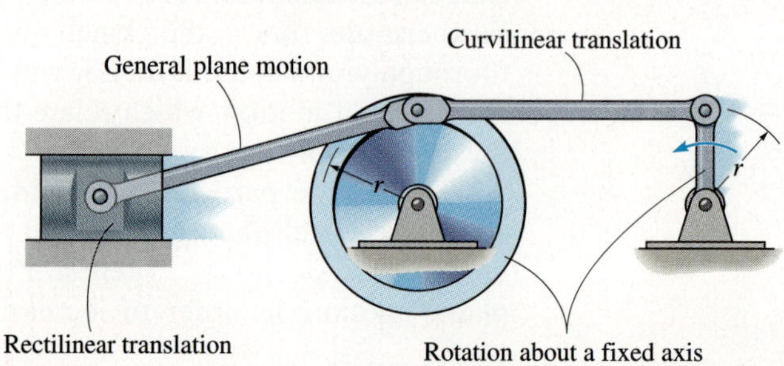

Fig. 16–2

16.2 Translation

Consider a rigid body which is subjected to either rectilinear or curvilinear translation in the x–y plane, Fig. 16–3.

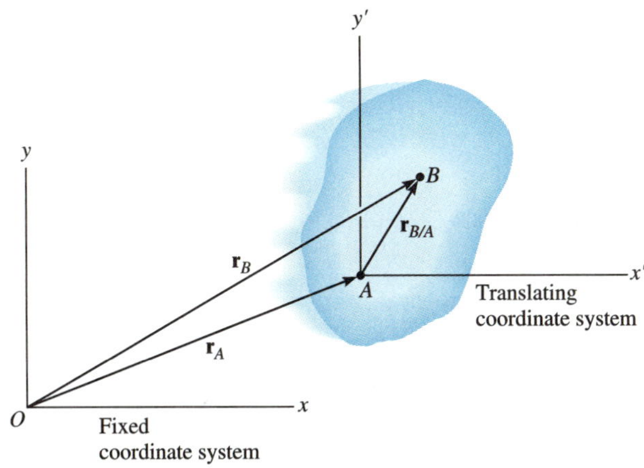

Fig. 16–3

Position. The locations of points A and B in the body are defined from the fixed x, y reference frame by using *position vectors* $\mathbf{r}_A$ and $\mathbf{r}_B$. The translating x', y' coordinate system is *fixed in the body* and has its origin located at A, hereafter referred to as the *base point.* The position of B with respect to A is denoted by the *relative-position vector* $\mathbf{r}_{B/A}$ ("$\mathbf{r}$ of B with respect to A"). By vector addition,

$$\mathbf{r}_B = \mathbf{r}_A + \mathbf{r}_{B/A}$$

Velocity. A relationship between the instantaneous velocities of A and B is obtained by taking the time derivative of the position equation, which yields $\mathbf{v}_B = \mathbf{v}_A + d\mathbf{r}_{B/A}/dt$. Here $\mathbf{v}_A$ and $\mathbf{v}_B$ denote *absolute velocities* since these vectors are measured from the x, y axes. The term $d\mathbf{r}_{B/A}/dt = \mathbf{0}$, since the *magnitude* of $\mathbf{r}_{B/A}$ is *constant* by definition of a rigid body, and because the body is translating the *direction* of $\mathbf{r}_{B/A}$ is *constant*. Therefore,

$$\mathbf{v}_B = \mathbf{v}_A$$

Acceleration. Taking the time derivative of the velocity equation yields a similar relationship between the instantaneous accelerations of A and B:

$$\mathbf{a}_B = \mathbf{a}_A$$

The above two equations indicate that *all points in a rigid body subjected to either rectilinear or curvilinear translation move with the same velocity and acceleration.* As a result, the kinematics of particle motion, discussed in Chapter 12, may also be used to specify the kinematics of points located in a translating rigid body.

Passengers on this amusement ride will be subjected to curvilinear translation since the vehicle moves in a circular path yet it always remains in an upright position.

16.3 Rotation About a Fixed Axis

When a body is rotating about a fixed axis, any point P located in the body travels along a *circular path.* To study this motion it is first necessary to discuss the angular motion of the body about the axis.

Angular Motion. A point is without dimension, and so it has no angular motion. *Only lines or bodies undergo angular motion.* For example, consider the body shown in Fig. 16–4*a* and the angular motion of a radial line r located within the shaded plane and directed from point O on the axis of rotation to point P.

Angular Position. At the instant shown, the *angular position* of r is defined by the angle θ, measured between a *fixed* reference line and r.

Angular Displacement. The change in the angular position, which can be measured as a differential $d\boldsymbol{\theta}$, is called the *angular displacement.** This vector has a *magnitude* of $d\theta$, measured in degrees, radians, or revolutions, where 1 rev = 2π rad. Since motion is about a *fixed axis,* the direction of $d\boldsymbol{\theta}$ is *always* along the axis. Specifically, the *direction* is determined by the right-hand rule; that is, the fingers of the right hand are curled with the sense of rotation, so that in this case the thumb, or $d\boldsymbol{\theta}$, points upward, Fig. 16–4*a*. In two dimensions, as shown by the top view of the shaded plane, Fig. 16–4*b*, both θ and $d\boldsymbol{\theta}$ are directed counterclockwise, and so the thumb points outward from the page.

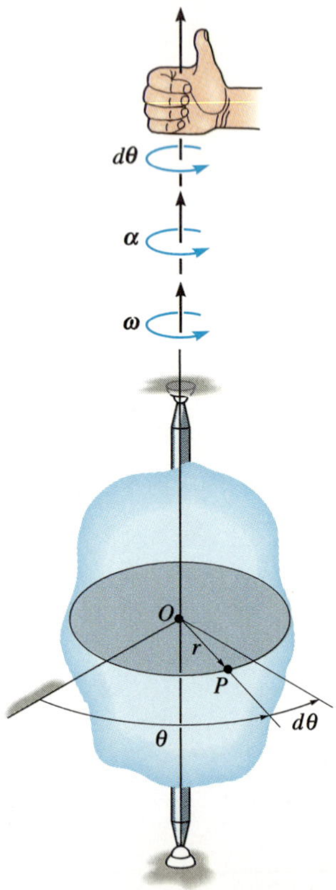

(a)

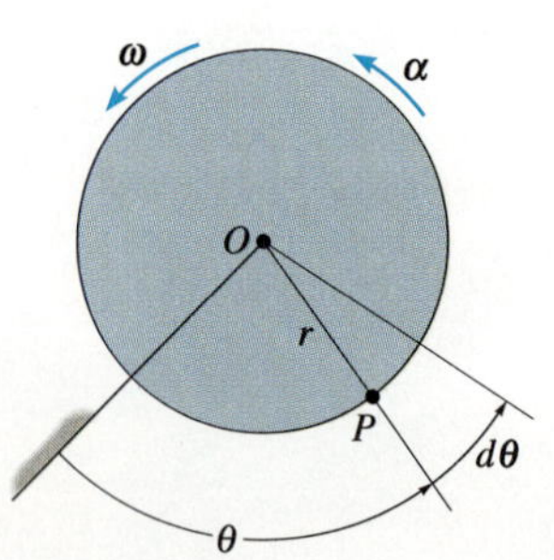

(b)

Fig. 16–4

Angular Velocity. The time rate of change in the angular position is called the *angular velocity* $\boldsymbol{\omega}$ (omega). Since $d\boldsymbol{\theta}$ occurs during an instant of time dt, then,

$$(\circlearrowleft +) \qquad \omega = \frac{d\theta}{dt} \qquad (16\text{–}1)$$

This vector has a *magnitude* which is often measured in rad/s. It is expressed here in scalar form since its *direction* is always along the axis of rotation, i.e., in the same direction as $d\boldsymbol{\theta}$, Fig. 16–4*a*. When indicating the angular motion in the shaded plane, Fig. 16–4*b*, we can refer to the sense of rotation as clockwise or counterclockwise. Here we have *arbitrarily* chosen counterclockwise rotations as *positive* and indicated this by the curl shown in parentheses next to Eq. 16–1. Realize, however, that the directional sense of $\boldsymbol{\omega}$ is actually outward from the page.

*It is shown in Sec. 20.1 that finite rotations or finite angular displacements are *not* vector quantities, although differential rotations $d\boldsymbol{\theta}$ are vectors.

Angular Acceleration. The *angular acceleration* $\boldsymbol{\alpha}$ (alpha) measures the time rate of change of the angular velocity. The *magnitude* of this vector may be written as

$$(\circlearrowleft+) \qquad \alpha = \frac{d\omega}{dt} \qquad (16\text{–}2)$$

Using Eq. 16–1, it is possible to express α as

$$(\circlearrowleft+) \qquad \alpha = \frac{d^2\theta}{dt^2} \qquad (16\text{–}3)$$

The line of action of $\boldsymbol{\alpha}$ is the same as that for $\boldsymbol{\omega}$, Fig. 16–4*a*; however, its sense of *direction* depends on whether $\boldsymbol{\omega}$ is increasing or decreasing. In particular, if $\boldsymbol{\omega}$ is decreasing, then $\boldsymbol{\alpha}$ is called an *angular deceleration* and it therefore has a sense of direction which is opposite to $\boldsymbol{\omega}$.

By eliminating dt from Eqs. 16–1 and 16–2, we obtain a differential relation between the angular acceleration, angular velocity, and angular displacement, namely,

$$(\circlearrowleft+) \qquad \alpha\, d\theta = \omega\, d\omega \qquad (16\text{–}4)$$

The similarity between the differential relations for angular motion and those developed for rectilinear motion of a particle ($v = ds/dt$, $a = dv/dt$, and $a\, ds = v\, dv$) should be apparent.

Constant Angular Acceleration. If the angular acceleration of the body is constant, $\boldsymbol{\alpha} = \boldsymbol{\alpha}_c$, then Eqs. 16–1, 16–2, and 16–4, when integrated, yield a set of formulas which relate the body's angular velocity, angular position, and time. These equations are similar to Eqs. 12–4 to 12–6 used for rectilinear motion. The results are

$$(\circlearrowleft+) \qquad \omega = \omega_0 + \alpha_c t \qquad (16\text{–}5)$$

$$(\circlearrowleft+) \qquad \theta = \theta_0 + \omega_0 t + \tfrac{1}{2}\alpha_c t^2 \qquad (16\text{–}6)$$

$$(\circlearrowleft+) \qquad \omega^2 = \omega_0^2 + 2\alpha_c(\theta - \theta_0) \qquad (16\text{–}7)$$

Constant Angular Acceleration

Here θ_0 and ω_0 are the initial values of the body's angular position and angular velocity, respectively.

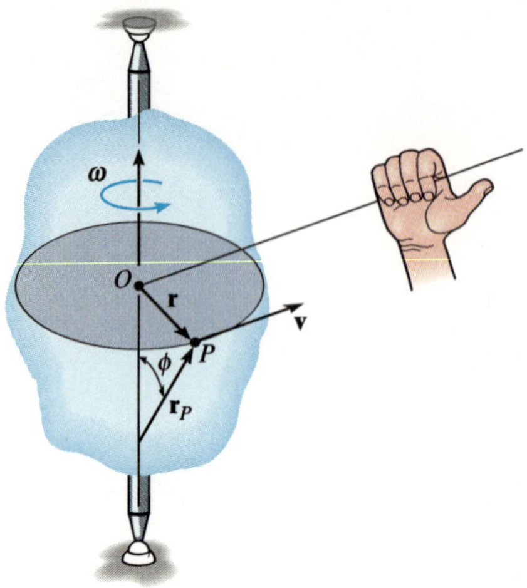

(c)

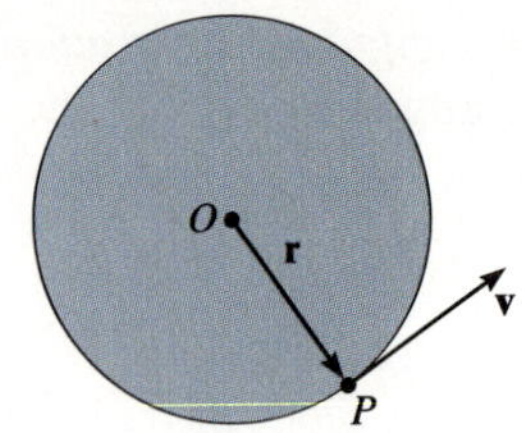

(d)

Fig. 16–4

Motion of Point P. As the rigid body in Fig. 16–4*c* rotates, point P travels along a *circular path* of radius r and center at point O. This path is contained within the shaded plane shown in top view, Fig. 16–4*d*.

Position. The position of P is defined by the position vector $\mathbf{r}$, which extends from O to P.

Velocity. The velocity of P has a magnitude which can be found from its polar coordinate components $v_r = \dot{r}$ and $v_\theta = r\dot{\theta}$, Eqs. 12–25. Since r is constant, the radial component $v_r = \dot{r} = 0$, and so $v = v_\theta = r\dot{\theta}$. Because $\omega = \dot{\theta}$, Eq. 16–1, the velocity is

$$v = \omega r \tag{16–8}$$

As shown in Figs. 16–4*c* and 16–4*d*, the *direction* of $\mathbf{v}$ is *tangent* to the circular path.

Both the magnitude and direction of $\mathbf{v}$ can also be accounted for by using the cross product of $\boldsymbol{\omega}$ and $\mathbf{r}_P$ (see Appendix C). Here, $\mathbf{r}_P$ is directed from *any point* on the axis of rotation to point P, Fig. 16–4*c*. We have

$$\mathbf{v} = \boldsymbol{\omega} \times \mathbf{r}_P \tag{16–9}$$

The order of the vectors in this formulation is important, since the cross product is not commutative, i.e., $\boldsymbol{\omega} \times \mathbf{r}_P \neq \mathbf{r}_P \times \boldsymbol{\omega}$. In this regard, notice in Fig. 16–4*c* how the correct direction of $\mathbf{v}$ is established by the right-hand rule. The fingers of the right hand are curled from $\boldsymbol{\omega}$ toward $\mathbf{r}_P$ ($\boldsymbol{\omega}$ "cross" $\mathbf{r}_P$). The thumb indicates the correct direction of $\mathbf{v}$, which is tangent to the path in the direction of motion. From Eq. C–8, the magnitude of $\mathbf{v}$ in Eq. 16–9 is $v = \omega r_P \sin \phi$, and since $r = r_P \sin \phi$, Fig. 16–4*c*, then $v = \omega r$, which agrees with Eq. 16–8. As a special case, the position vector $\mathbf{r}$ can be chosen for $\mathbf{r}_P$. Here $\mathbf{r}$ lies in the plane of motion and again the velocity of point P is

$$\mathbf{v} = \boldsymbol{\omega} \times \mathbf{r} \tag{16–10}$$

Acceleration. The acceleration of P can be expressed in terms of its normal and tangential components.* Since $a_t = dv/dt$ and $a_n = v^2/\rho$, where $\rho = r$, $v = \omega r$, and $\alpha = d\omega/dt$, we have

$$a_t = \alpha r \tag{16–11}$$

$$a_n = \omega^2 r \tag{16–12}$$

The *tangential component of acceleration,* Figs. 16–4*e* and 16–4*f*, represents the time rate of change in the velocity's magnitude. If the speed of P is increasing, then $\mathbf{a}_t$ acts in the same direction as $\mathbf{v}$; if the speed is decreasing, $\mathbf{a}_t$ acts in the opposite direction of $\mathbf{v}$; and finally, if the speed is constant, $\mathbf{a}_t$ is zero.

The *normal component of acceleration* represents the time rate of change in the velocity's direction. The *direction* of $\mathbf{a}_n$ is always toward O, the center of the circular path, Figs. 16–4*e* and 16–4*f*.

Like the velocity, the acceleration of point P may be expressed in terms of the vector cross product. Taking the time derivative of Eq. 16–9 we have

$$\mathbf{a} = \frac{d\mathbf{v}}{dt} = \frac{d\boldsymbol{\omega}}{dt} \times \mathbf{r}_P + \boldsymbol{\omega} \times \frac{d\mathbf{r}_P}{dt}$$

Recalling that $\boldsymbol{\alpha} = d\boldsymbol{\omega}/dt$, and using Eq. 16–9 ($d\mathbf{r}_P/dt = \mathbf{v} = \boldsymbol{\omega} \times \mathbf{r}_P$), yields

$$\mathbf{a} = \boldsymbol{\alpha} \times \mathbf{r}_P + \boldsymbol{\omega} \times (\boldsymbol{\omega} \times \mathbf{r}_P) \tag{16–13}$$

From the definition of the cross product, the first term on the right has a magnitude $a_t = \alpha r_P \sin\phi = \alpha r$, and by the right-hand rule, $\boldsymbol{\alpha} \times \mathbf{r}_P$ is in the direction of $\mathbf{a}_t$, Fig. 16–4*e*. Likewise, the second term has a magnitude $a_n = \omega^2 r_P \sin\phi = \omega^2 r$, and applying the right-hand rule twice, first to determine the result $\mathbf{v}_P = \boldsymbol{\omega} \times \mathbf{r}_P$ then $\boldsymbol{\omega} \times \mathbf{v}_P$, it can be seen that this result is in the same direction as $\mathbf{a}_n$, shown in Fig. 16–4*e*. Noting that this is also the *same* direction as $-\mathbf{r}$, which lies in the plane of motion, we can express $\mathbf{a}_n$ in a much simpler form as $\mathbf{a}_n = -\omega^2\mathbf{r}$. Hence, Eq. 16–12 can be identified by its two components as

$$\begin{aligned}\mathbf{a} &= \mathbf{a}_t + \mathbf{a}_n \\ &= \boldsymbol{\alpha} \times \mathbf{r} - \omega^2\mathbf{r}\end{aligned} \tag{16–14}$$

Since $\mathbf{a}_t$ and $\mathbf{a}_n$ are perpendicular to one another, if needed the magnitude of acceleration can be determined from the Pythagorean theorem; namely, $a = \sqrt{a_n^2 + a_t^2}$, Fig. 16–4*f*.

*Polar coordinates can also be used. Since $a_r = \ddot{r} - r\dot{\theta}^2$ and $a_\theta = r\ddot{\theta} + 2\dot{r}\dot{\theta}$, substituting $\dot{r} = \ddot{r} = 0$, $\dot{\theta} = \omega$, $\ddot{\theta} = \alpha$, we obtain Eqs. 16–11 and 16–12.

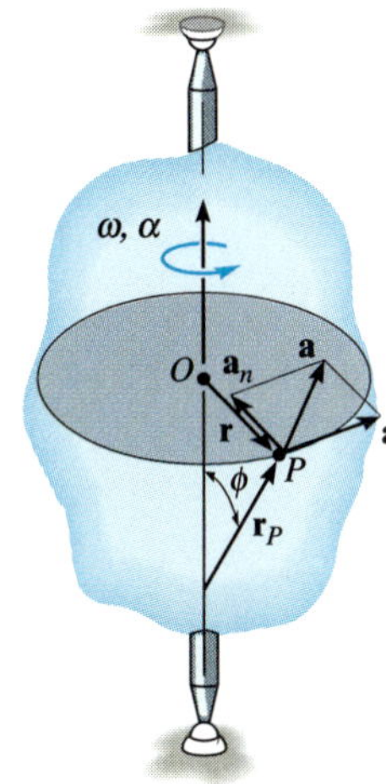

(e)

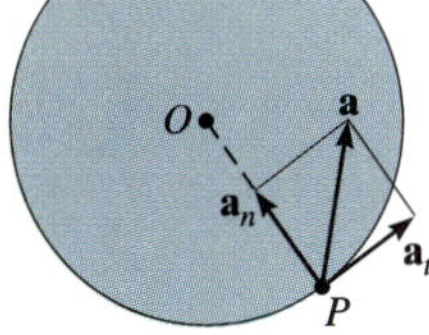

(f)

Fig. 16–4

The many gears used in the operation of a crane all rotate about fixed axes. Engineers must be able to relate their angular motions in order to properly design this gear system.

Important Points

- A body can undergo two types of translation. During rectilinear translation all points follow parallel straight-line paths, and during curvilinear translation the points follow curved paths that are the same shape and are equidistant from one another.
- All the points on a translating body move with the same velocity and acceleration.
- Points located on a body that rotates about a fixed axis follow circular paths.
- The relationship $\alpha\, d\theta = \omega\, d\omega$ is derived from $\alpha = d\omega/dt$ and $\omega = d\theta/dt$ by eliminating dt.
- Once the angular motions ω and α are known, the velocity and acceleration of any point on the body can be determined.
- The velocity always acts tangent to the path of motion.
- The acceleration has two components. The tangential acceleration measures the rate of change in the magnitude of the velocity and can be determined using $a_t = \alpha r$. The normal acceleration measures the rate of change in the direction of the velocity and can be determined from $a_n = \omega^2 r$.

Procedure for Analysis

The velocity and acceleration of a point located on a rigid body that is rotating about a fixed axis can be determined using the following procedure.

Angular Motion

- Establish the positive sense of direction along the axis of rotation and show it alongside each kinematic equation as it is applied.
- If a relationship is known between any *two* of the four variables α, ω, θ, and t, then a third variable can be obtained by using one of the following kinematic equations which relates all three variables.

$$\omega = \frac{d\theta}{dt} \qquad \alpha = \frac{d\omega}{dt} \qquad \alpha\, d\theta = \omega\, d\omega$$

- If the body's angular acceleration is *constant*, then the following equations can be used:

$$\omega = \omega_0 + \alpha_c t$$
$$\theta = \theta_0 + \omega_0 t + \tfrac{1}{2}\alpha_c t^2$$
$$\omega^2 = \omega_0^2 + 2\alpha_c(\theta - \theta_0)$$

- Once the solution is obtained, the sense of θ, ω, and α is determined from the algebraic signs of their numerical quantities.

Motion of P

- In most cases the velocity of P and its two components of acceleration can be determined from the scalar equations

$$v = \omega r$$
$$a_t = \alpha r$$
$$a_n = \omega^2 r$$

- If the geometry of the problem is difficult to visualize, the following vector equations should be used:

$$\mathbf{v} = \boldsymbol{\omega} \times \mathbf{r}_P = \boldsymbol{\omega} \times \mathbf{r}$$
$$\mathbf{a}_t = \boldsymbol{\alpha} \times \mathbf{r}_P = \boldsymbol{\alpha} \times \mathbf{r}$$
$$\mathbf{a}_n = \boldsymbol{\omega} \times (\boldsymbol{\omega} \times \mathbf{r}_P) = -\omega^2 \mathbf{r}$$

Here $\mathbf{r}_P$ is directed from any point on the axis of rotation to point P, whereas $\mathbf{r}$ lies in the plane of motion of P. Either of these vectors, along with $\boldsymbol{\omega}$ and $\boldsymbol{\alpha}$, should be expressed in terms of its $\mathbf{i}$, $\mathbf{j}$, $\mathbf{k}$ components, and, if necessary, the cross products determined by using a determinant expansion (see Eq. C–12).

EXAMPLE 16–1

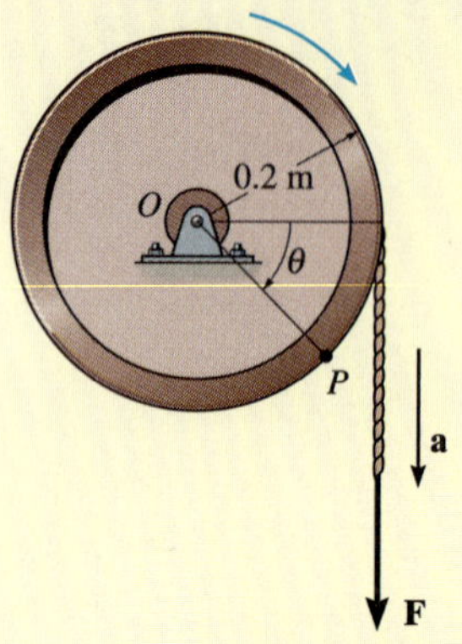

Fig. 16–5

A cord is wrapped around a wheel which is initially at rest as shown in Fig. 16–5. If a force is applied to the cord and gives it an acceleration $a = (4t)\ \text{m/s}^2$, where t is in seconds, determine as a function of time (*a*) the angular velocity of the wheel, and (*b*) the angular position of line OP in radians.

Solution

Part (a). The wheel is subjected to rotation about a fixed axis passing through point O. Thus, point P on the wheel has motion about a circular path, and the acceleration of this point has *both* tangential and normal components. The tangential component is $(a_P)_t = (4t)\ \text{m/s}^2$, since the cord is wrapped around the wheel and moves *tangent* to it. Hence the angular acceleration of the wheel is

$$(\curvearrowright +) \qquad (a_P)_t = \alpha r$$
$$(4t)\ \text{m/s}^2 = \alpha(0.2\ \text{m})$$
$$\alpha = 20t\ \text{rad/s}^2 \ \curvearrowright$$

Using this result, the wheel's angular velocity $\boldsymbol{\omega}$ can now be determined from $\alpha = d\omega/dt$, since this equation relates α, t, and ω. Integrating, with the initial condition that $\omega = 0$ at $t = 0$, yields

$$(\curvearrowright +) \qquad \alpha = \frac{d\omega}{dt} = (20t)\ \text{rad/s}^2$$
$$\int_0^{\omega} d\omega = \int_0^t 20t\ dt$$
$$\omega = 10t^2\ \text{rad/s} \ \curvearrowright \qquad \textit{Ans.}$$

Why not use Eq. 16–5 ($\omega = \omega_0 + \alpha_c t$) to obtain this result?

Part (b). Using this result, the angular position θ of OP can be found from $\omega = d\theta/dt$, since this equation relates θ, ω, and t. Integrating, with the initial condition $\theta = 0$ at $t = 0$, we have

$$(\curvearrowright +) \qquad \frac{d\theta}{dt} = \omega = (10t^2)\ \text{rad/s}$$
$$\int_0^{\theta} d\theta = \int_0^t 10t^2\ dt$$
$$\theta = 3.33\ t^3\ \text{rad} \qquad \textit{Ans.}$$

EXAMPLE 16–2

The motor shown in the photo is used to turn a wheel and attached blower contained within the housing. The details of the design are shown in Fig. 16–6*a*. If the pulley *A* connected to the motor begins rotating from rest with an angular acceleration of $\alpha_A = 2\ \text{rad/s}^2$, determine the magnitudes of the velocity and acceleration of point *P* on the wheel, after the wheel *B* has turned 10 revolutions. Assume the transmission belt does not slip on the pulley and wheel.

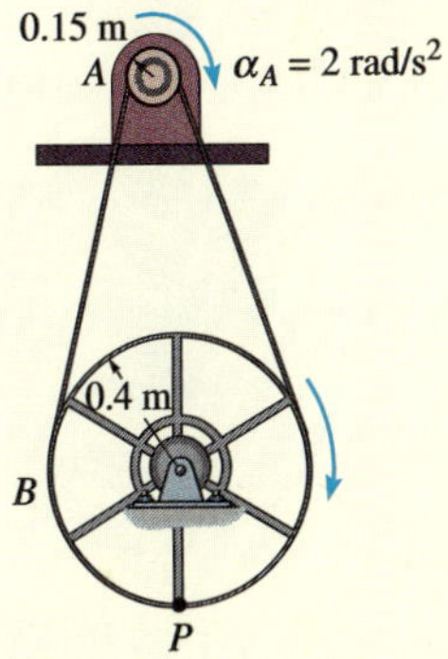

(a)

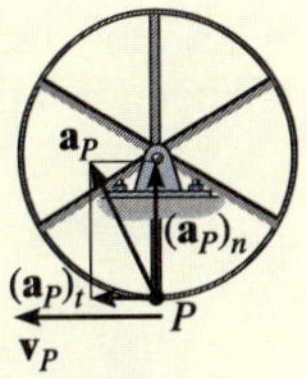

(b)

Fig. 16–6

Solution

Angular Motion. First we will convert the 10 revolutions to radians. Since there are 2π rad in one revolution, then

$$\theta_B = 10\ \text{rev}\left(\frac{2\pi\ \text{rad}}{1\ \text{rev}}\right) = 62.83\ \text{rad}$$

We can find the angular velocity of pulley *A* provided we first find its angular displacement. Since the belt does not slip, an equivalent length of belt *s* must be unraveled from both the pulley and wheel at all times. Thus,

$$s = \theta_A r_A = \theta_B r_B; \qquad \theta_A(0.15\ \text{m}) = 62.83(0.4\ \text{m})$$
$$\theta_A = 167.6\ \text{rad}$$

Since α_A is constant, the angular velocity of pulley *A* is therefore

$$(\circlearrowright +) \qquad \omega^2 = \omega_0^2 + 2\alpha_c(\theta - \theta_0)$$
$$\omega_A^2 = 0 + 2(2\ \text{rad/s}^2)(167.6\ \text{rad} - 0)$$
$$\omega_A = 18.31\ \text{rad/s}$$

The belt has the same speed and tangential component of acceleration as it passes over the pulley and wheel. Thus,

$$v = \omega_A r_A = \omega_B r_B; \quad 18.31\ \text{rad/s}(0.15\ \text{m}) = \omega_B(0.4\ \text{m})$$
$$\omega_B = 6.865\ \text{rad/s}$$

$$a_t = \alpha_A r_A = \alpha_B r_B; \quad 2\ \text{rad/s}^2(0.15\ \text{m}) = \alpha_B(0.4\ \text{m})$$
$$\alpha_B = 0.750\ \text{rad/s}^2$$

Motion of P. As shown on the kinematic diagram in Fig. 16–6*b*, we have

$$v_P = \omega_B r_B = 6.865\ \text{rad/s}(0.4\ \text{m}) = 2.75\ \text{m/s} \qquad \textit{Ans.}$$
$$(a_P)_t = \alpha_B r_B = 0.750\ \text{rad/s}^2(0.4\ \text{m}) = 0.3\ \text{m/s}^2$$
$$(a_P)_n = \omega_B^2 r_B = (6.865\ \text{rad/s})^2(0.4\ \text{m}) = 18.85\ \text{m/s}^2$$

Thus

$$a_P = \sqrt{(0.3)^2 + (18.85)^2} = 18.9\ \text{m/s}^2 \qquad \textit{Ans.}$$

Problems

16-1. A disk having a radius of 0.5 ft rotates with an initial angular velocity of 2 rad/s and has a constant angular acceleration of 1 rad/s^2. Determine the magnitudes of the velocity and acceleration of a point on the rim of the disk when $t = 2$ s.

16-2. The disk is originally rotating at $\omega_0 = 8$ rad/s. If it is subjected to a constant angular acceleration of $\alpha = 6$ rad/s^2, determine the magnitudes of the velocity and the n and t components of acceleration of point A at the instant $t = 0.5$ s.

16-3. The disk is originally rotating at $\omega_0 = 8$ rad/s. If it is subjected to a constant angular acceleration of $\alpha = 6$ rad/s^2, determine the magnitudes of the velocity and the n and t components of acceleration of point B just after the wheel undergoes 2 revolutions.

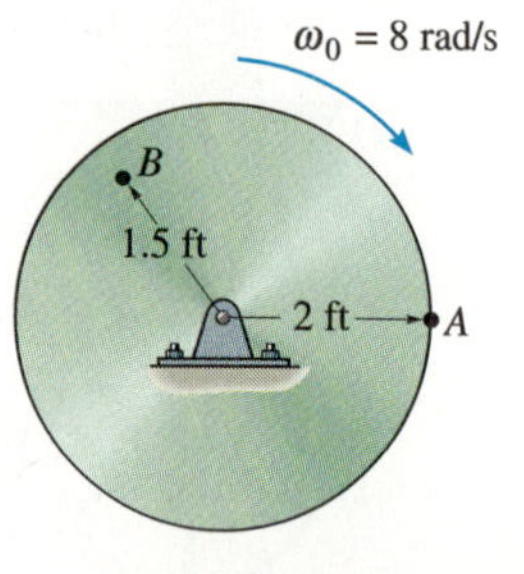

Probs. 16–2/3

***16-4.** Just after the fan is turned on, the motor gives the blade an angular acceleration $\alpha = (20e^{-0.6t})$ rad/s^2, where t is in seconds. Determine the speed of the tip P of one of the blades when $t = 3$ s. How many revolutions has the blade turned in 3 s? When $t = 0$ the blade is at rest.

Prob. 16–4

16-5. Due to an increase in power, the motor M rotates the shaft A with an angular acceleration of $\alpha = (0.06\theta^2)$ rad/s^2, where θ is in radians. If the shaft is initially turning at $\omega_0 = 50$ rad/s, determine the angular velocity of gear B after the shaft undergoes an angular displacement $\Delta\theta = 10$ rev.

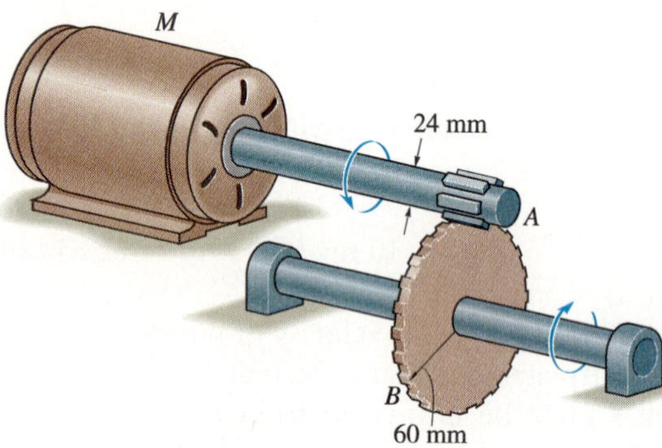

Prob. 16–5

16-6. The figure shows the internal gearing of a "spinner" used for drilling wells. With constant angular acceleration, the motor M rotates the shaft S to 100 rev/min in $t = 2$ s starting from rest. Determine the angular acceleration of the drill-pipe connection D and the number of revolutions it makes during the 2-s start up.

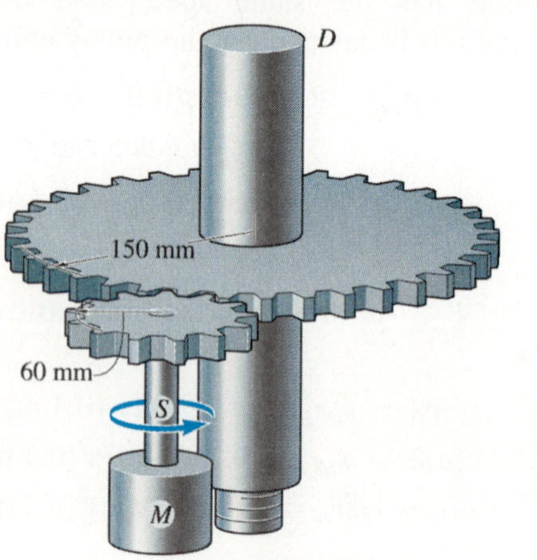

Prob. 16–6

16-7. If gear A starts from rest and has a constant angular acceleration of $\alpha_A = 2\ \text{rad/s}^2$, determine the time needed for gear B to attain an angular velocity of $\omega_B = 50\ \text{rad/s}$.

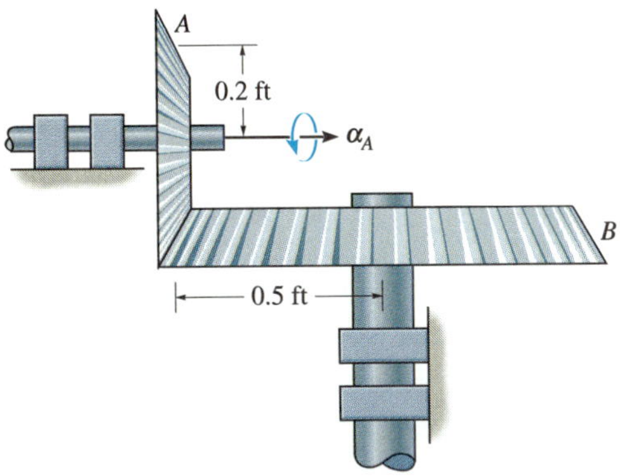

Prob. 16–7

***16-8.** If the armature A of the electric motor in the drill has a constant angular acceleration of $\alpha_A = 20\ \text{rad/s}^2$, determine its angular velocity and angular displacement when $t = 3$ s. The motor starts from rest.

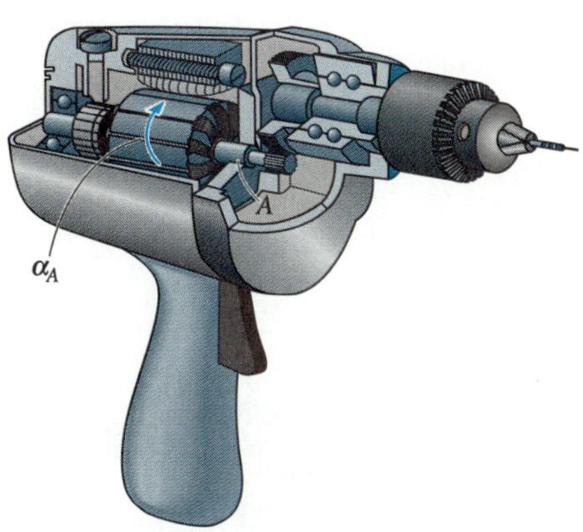

Prob. 16–8

16-9. The mechanism for a car window winder is shown in the figure. Here the handle turns the small cog C, which rotates the spur gear S, thereby rotating the fixed-connected lever AB which raises track D in which the window rests. The window is free to slide on the track. If the handle is wound at 0.5 rad/s, determine the speed of points A and E and the speed v_w of the window at the instant $\theta = 30°$.

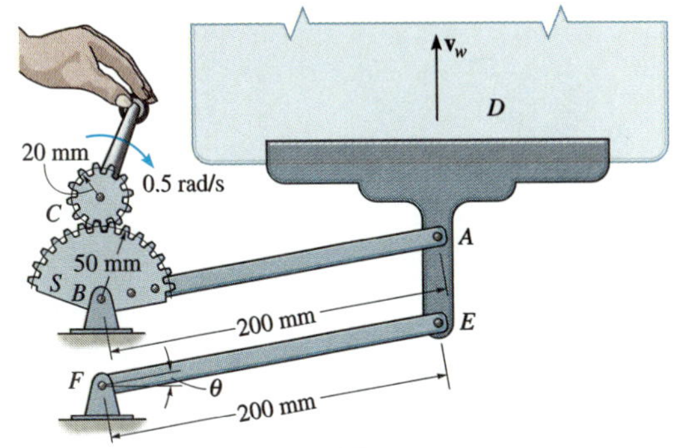

Prob. 16–9

16-10. The blade on the horizontal-axis windmill is turning with an angular velocity of $\omega_0 = 2\ \text{rad/s}$. Determine the distance point P on the tip of the blade has traveled if the blade attains an angular velocity of $\omega = 5\ \text{rad/s}$ in 3 s. The angular acceleration is constant. Also, what is the magnitude of the acceleration of this point when $t = 3$ s?

16-11. The blade on the horizontal-axis windmill is turning with an angular velocity of $\omega_0 = 2\ \text{rad/s}$. If it is given an angular acceleration of $\alpha_c = 0.6\ \text{rad/s}^2$, determine the angular velocity and the magnitude of acceleration of point P on the tip of the blade when $t = 3$ s.

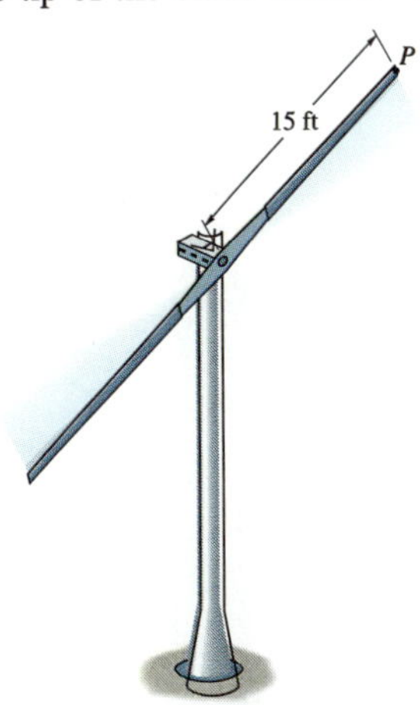

Probs. 16–10/11

***16-12.** When only two gears are in mesh, the driving gear A and the driven gear B will always turn in opposite directions. In order to get them to turn in the *same direction* an idler gear C is used. In the case shown, determine the angular velocity of gear B when $t = 5$ s, if gear A starts from rest and has an angular acceleration of $\alpha_A = (3t + 2)\ \text{rad/s}^2$, where t is in seconds.

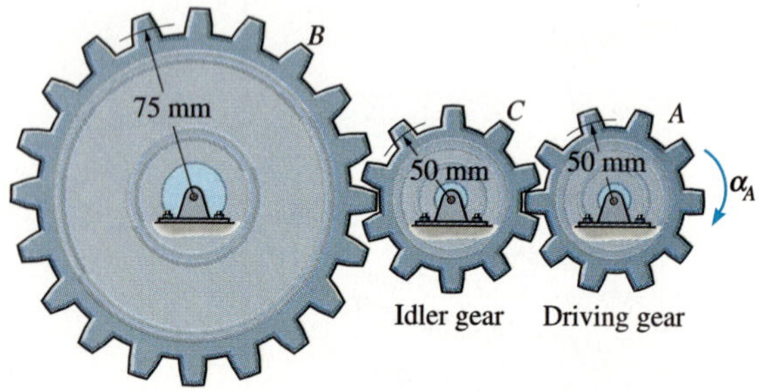

Prob. 16–12

16-13. The anemometer measures the speed of the wind due to the rotation of the three cups. If during a 3-s time period a wind gust causes the cups to have an angular velocity of $\omega = (2t^2 + 3)$ rad/s, where t is in seconds, determine (a) the speed of the cups when $t = 2$ s, (b) the total distance traveled by each cup during the 3-s time period, and (c) the angular acceleration of the cups when $t = 2$ s. Neglect the size of the cups for the calculation.

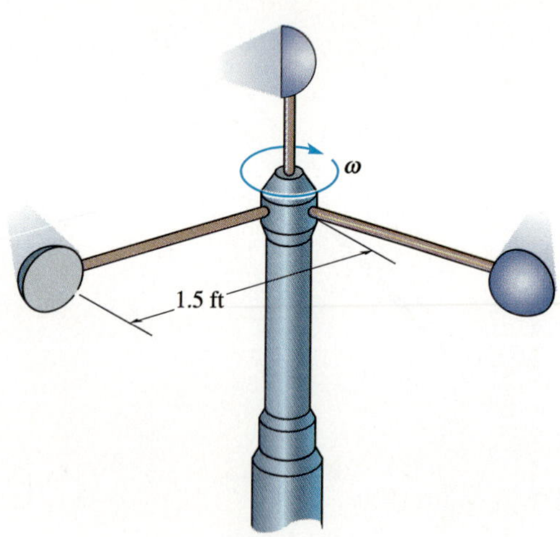

Prob. 16–13

16-14. The operation of reverse gear in an automotive transmission is shown. If the engine is turning shaft A at $\omega_A = 40$ rad/s, determine the rate of rotation of the drive shaft, ω_B. The radius of each gear is listed in the figure.

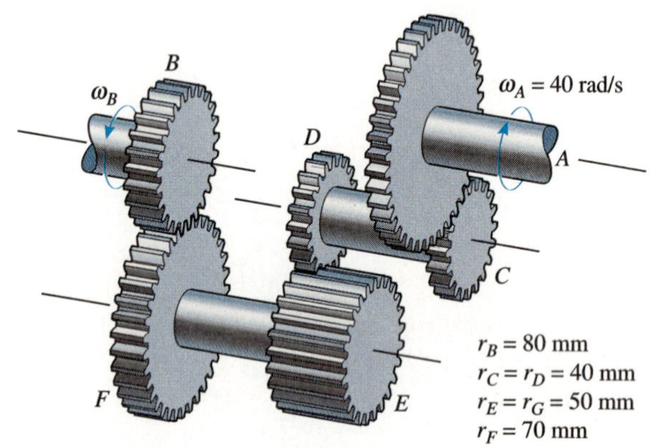

Prob. 16–14

16-15. The turntable T is driven by the frictional idler wheel A, which simultaneously bears against the inner rim of the turntable and the motor-shaft spindle B. Determine the required diameter d of the spindle if the motor turns it at 25 rad/s and it is required that the turntable rotate at 2 rad/s.

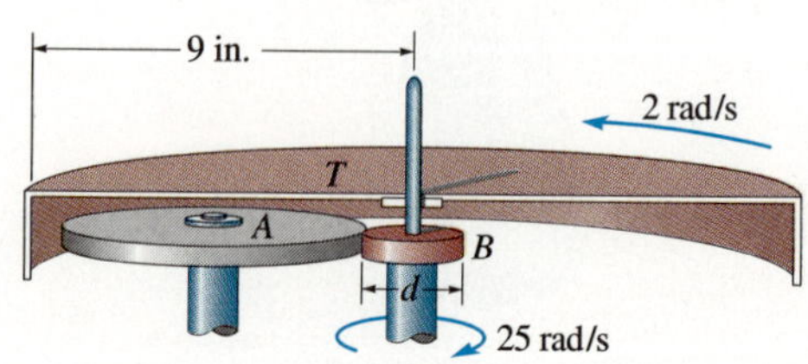

Prob. 16–15

***16-16.** The gear A on the drive shaft of the outboard motor has a radius $r_A = 0.5$ in. and the meshed pinion gear B on the propeller shaft has a radius $r_B = 1.2$ in. Determine the angular velocity of the propeller in $t = 1.5$ s if the drive shaft rotates with an angular acceleration $\alpha = (400t^3)$ rad/s^2, where t is in seconds. The propeller is originally at rest and the motor frame does not move.

16-17. For the outboard motor in Prob. 16-16, determine the magnitudes of the velocity and acceleration of a point P located on the tip of the propeller at the instant $t = 0.75$ s.

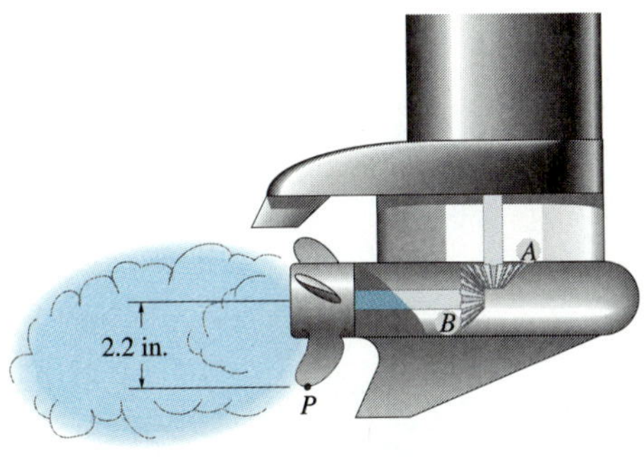

Probs. 16–16/17

16-18. For a short time a motor of the random-orbit sander drives the gear A with an angular velocity of $\omega_A = 40(t^3 + 6t)$ rad/s, where t is in seconds. This gear is connected to gear B, which is fixed connected to the shaft CD. The end of this shaft is connected to the eccentric spindle EF and pad P, which causes the pad to orbit around shaft CD at a radius of 15 mm. Determine the magnitudes of the velocity and the tangential and normal components of acceleration of the spindle EF when $t = 2$ s after starting from rest.

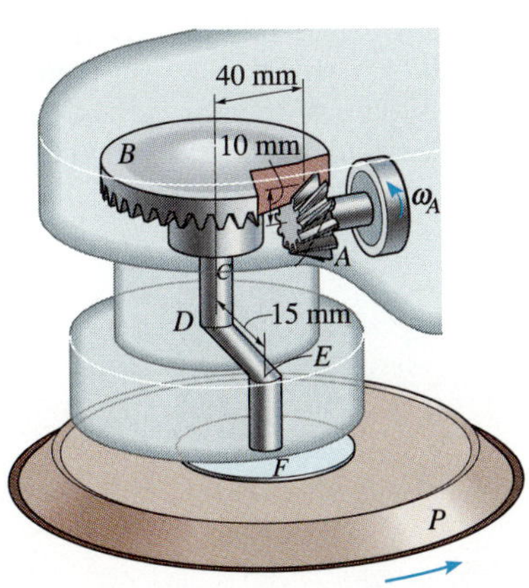

Prob. 16–18

16-19. For a short time the motor of the random-orbit sander drives the gear A with an angular velocity of $\omega_A = (5\theta^2)$ rad/s, where θ is in radians. This gear is connected to gear B, which is fixed connected to the shaft CD. The end of this shaft is connected to the eccentric spindle EF and pad P, which causes the pad to orbit around shaft CD at a radius of 15 mm. Determine the magnitudes of the velocity and the tangential and normal components of acceleration of the spindle EF when $\theta = 0.5$ revolutions starting from rest.

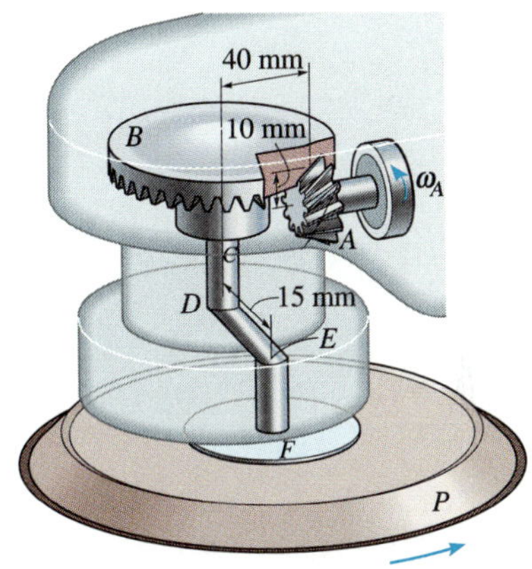

Prob. 16–19

***16-20.** The aerobic machine manufactured by Precor, Inc. transfers pedaling power to a flywheel F, which develops resistance using an AC-powered electromagnet. If the operator initially drives the pedals at 20 rev/min, and then begins an angular acceleration of 30 rev/min^2, determine the angular velocity of the flywheel when $t = 3$ s. Note that the pedal arm is fixed-connected to the chain wheel A, which in turn drives the sheave B using the fixed-connected clutch gear D. The poly-V belt wrapped around the sheave, then drives the pulley E and fixed-connected flywheel.

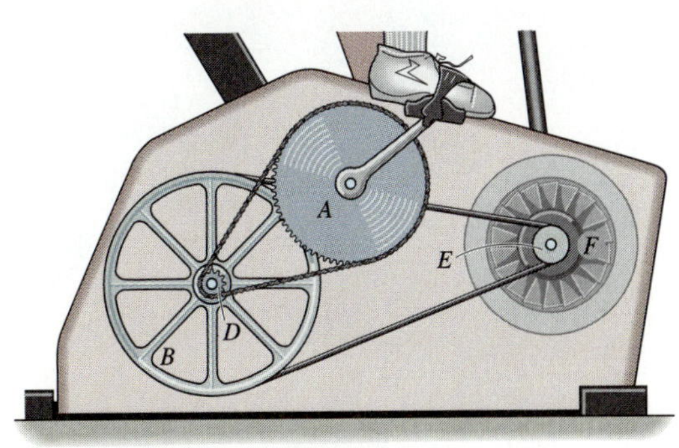

$r_A = 125$ mm $\quad r_B = 175$ mm
$r_D = 20$ mm $\quad r_E = 30$ mm

Prob. 16–20

16-21. The aerobic machine manufactured by Precor, Inc. transfers pedaling power to a flywheel F, which develops resistance using an AC-powered electromagnet. If the operator initially drives the pedals at 12 rev/min, and then begins an angular acceleration of 8 rev/min^2, determine the angular velocity of the flywheel after 2 revolutions of the pedal arm. Note that the pedal arm is fixed connected to the chain wheel A, which in turn drives the sheave B using the fixed-connected clutch gear D. The poly-V belt wrapped around the sheave, then drives the pulley E and fixed-connected flywheel.

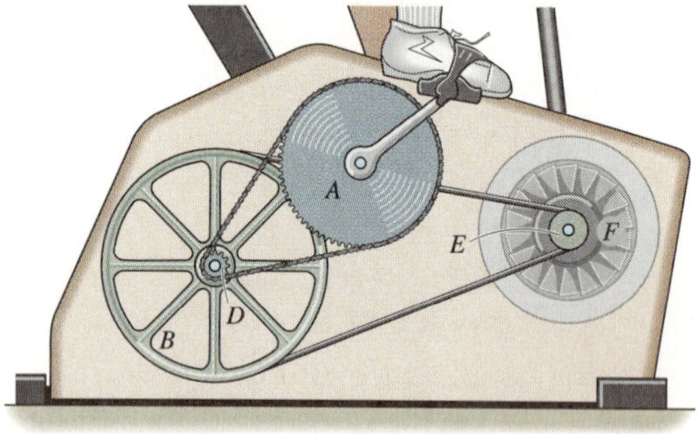

$r_A = 125$ mm $r_B = 175$ mm
$r_D = 20$ mm $r_E = 30$ mm

Prob. 16–21

16-22. The engine shaft S on the lawnmower rotates at a constant angular rate of 40 rad/s. Determine the magnitudes of the velocity and acceleration of point P on the blade and the distance P travels in 3 seconds. The shaft S is connected to the driver pulley A, and the motion is transmitted to the belt that passes over the idler pulleys at B and C and to the pulley at D. This pulley is connected to the blade and to another belt that drives the other blade.

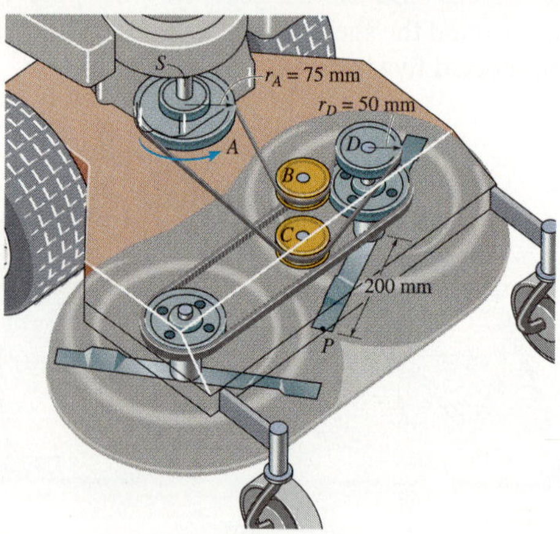

Prob. 16–22

16-23. The engine shaft S on the lawnmower is initially rotating with an angular velocity of 40 rad/s. If it experiences a constant angular acceleration of 3 rad/s^2 while the engine throttle is increased, determine the magnitudes of the velocity and the normal and tangential components of acceleration of point P on the blade when $t = 2$ s, and the distance P travels in 2 seconds. The shaft S is connected to the driver pulley A, and the motion is transmitted to the belt that passes over the idler pulleys at B and C and to the pulley at D. This pulley is connected to the blade and to another belt that drives the other blade.

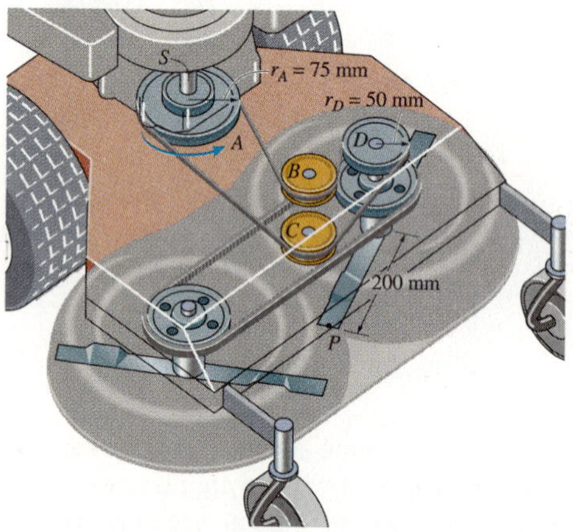

Prob. 16–23

***16-24.** The disk starts from rest and is given an angular acceleration $\alpha = (10\theta^{1/3})$ rad/s^2, where θ is in radians. Determine the angular velocity of the disk and its angular displacement when $t = 4$ s.

16-25. The disk starts from rest and is given an angular acceleration $\alpha = (10\theta^{1/3})$ rad/s^2, where θ is in radians. Determine the magnitudes of the normal and tangential components of acceleration of a point P on the rim of the disk when $t = 4$ s.

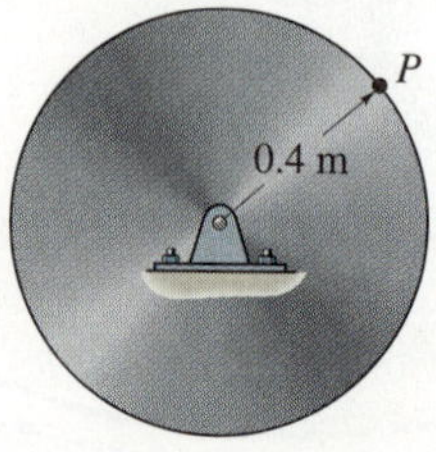

Probs. 16–24/25

16-26. Morse Industrial manufactures the speed reducer shown. If a motor drives the gear shaft S with an angular acceleration of $\alpha = (0.4e^t)$ rad/s^2, where t is in seconds, determine the angular velocity of shaft E when $t = 2$ s after starting from rest. The radius of each gear is listed in the figure. Note that gears B and C are fixed connected to the same shaft.

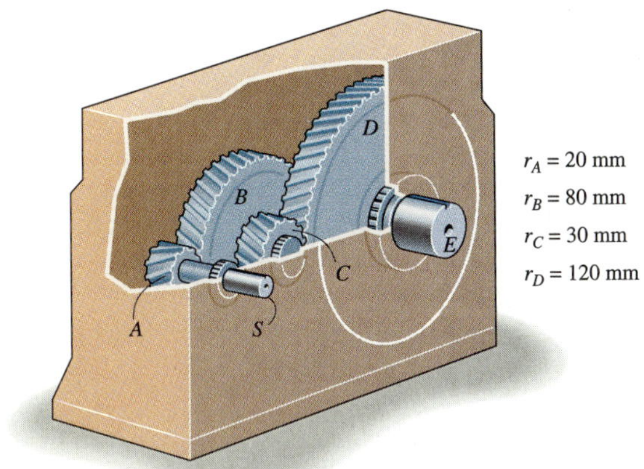

Prob. 16–26

16-27. Morse Industrial manufactures the speed reducer shown. If a motor drives the gear shaft S with an angular acceleration of $\alpha = (4\omega^{-3})$ rad/s^2, where ω is in rad/s, determine the angular velocity of shaft E when $t = 2$ s after starting from an angular velocity of 1 rad/s when $t = 0$. The radius of each gear is listed in the figure. Note that gears B and C are fixed connected to the same shaft.

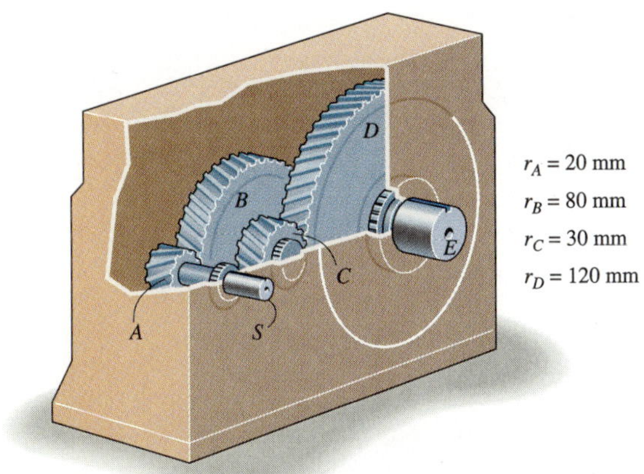

Prob. 16–27

***16-28.** The sphere starts from rest at $\theta = 0°$ and rotates with an angular acceleration of $\alpha = (4\theta)$ rad/s^2, where θ is in radians. Determine the magnitudes of the velocity and acceleration of point P on the sphere at the instant $\theta = 6$ rad.

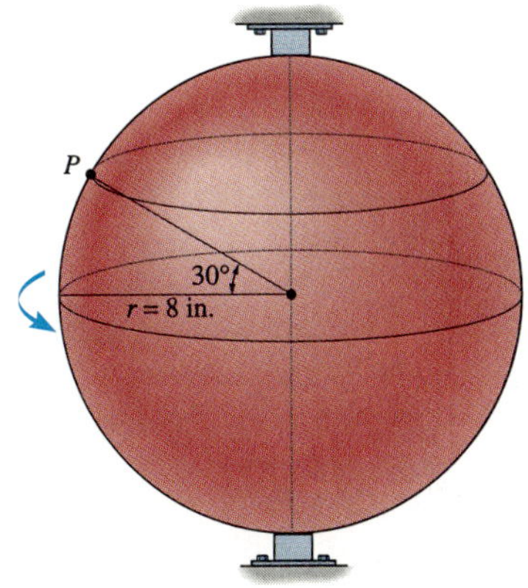

Prob. 16–28

16-29. At the instant shown, gear A is rotating with a constant angular velocity of $\omega_A = 6$ rad/s. Determine the largest angular velocity of gear B and the maximum speed of point C.

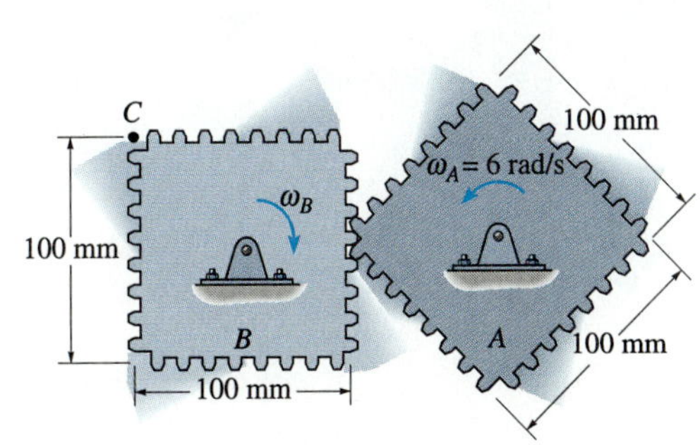

Prob. 16–29

16-30. If the rod starts from rest in the position shown and a motor drives it for a short time with an angular acceleration of $\alpha = (1.5e^t)$ rad/s^2, where t is in seconds, determine the magnitude of the angular velocity and the angular displacement of the rod when $t = 3$ s. Locate the point on the rod which has the greatest velocity and acceleration, and compute the magnitudes of the velocity and acceleration of this point when $t = 3$ s. The rod is defined by $z = 0.25 \sin(\pi y)$, where the argument for the sine is given in radians and y is in meters.

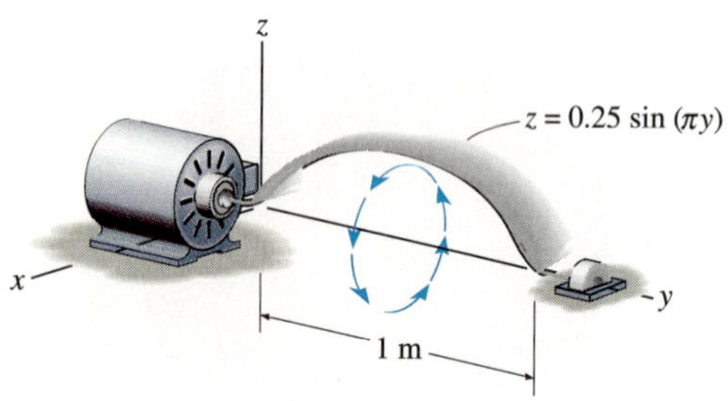

Prob. 16–30

16-31. A stamp S, located on the revolving drum, is used to label canisters. If the canisters are centered 200 mm apart on the conveyor, determine the radius r_A of the driving wheel A and the radius r_B of the conveyor belt drum so that for each revolution of the stamp it marks the top of a canister. How many canisters are marked per minute if the drum at B is rotating at $\omega_B = 0.2$ rad/s? Note that the driving belt is twisted as it passes between the wheels.

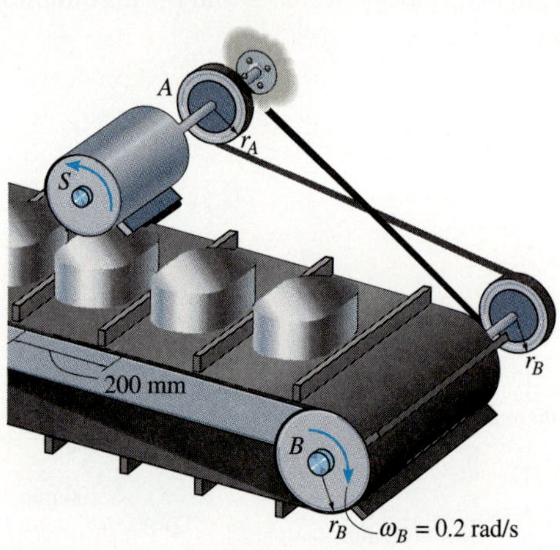

Prob. 16–31

***16-32.** A tape having a thickness s wraps around the wheel which is turning at a constant rate $\boldsymbol{\omega}$. Assuming the unwrapped portion of tape remains horizontal, determine the acceleration of point P of the unwrapped tape when the radius of the wrapped tape is r. *Hint:* Since $v_P = \omega r$, take the time derivative and note that $dr/dt = \omega(s/2\pi)$.

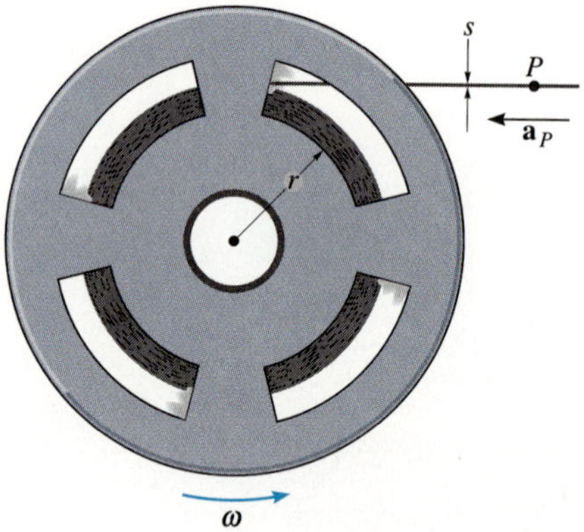

Prob. 16–32

*16.4 Absolute Motion Analysis

A body subjected to *general plane motion* undergoes a *simultaneous* translation and rotation. If the body is represented by a thin slab, the slab translates in the plane and rotates about an axis perpendicular to the plane. The motion can be completely specified by knowing *both* the angular rotation of a line fixed in the body and the motion of a point on the body. One way to define these motions is to use a rectilinear position coordinate s to locate the point along its path and an angular position coordinate θ to specify the orientation of the line. The two coordinates are then related using the geometry of the problem. By *direct application* of the time-differential equations $v = ds/dt$, $a = dv/dt$, $\omega = d\theta/dt$, and $\alpha = d\omega/dt$, the *motion* of the point and the *angular motion* of the line can then be related. In some cases, this procedure may also be used to relate the motions of one body to those of a connected body, or to study the motion of a body subjected to rotation about a fixed axis.

Procedure for Analysis

The velocity and acceleration of a point P undergoing rectilinear motion can be related to the angular velocity and angular acceleration of a line contained within a body using the following procedure.

Position Coordinate Equation

- Locate point P using a position coordinate s, which is measured from a *fixed origin* and is *directed along the straight-line path of motion* of point P.
- Measure from a fixed reference line the angular position θ of a line lying in the body.
- From the dimensions of the body, relate s to θ, $s = f(\theta)$, using geometry and/or trigonometry.

Time Derivatives

- Take the first derivative of $s = f(\theta)$ with respect to time to get a relationship between v and ω.
- Take the second time derivative to get a relationship between a and α.
- In each case the chain rule of calculus must be used when taking the derivatives of the position coordinate equation.

The dumping bin on the truck rotates about a fixed axis passing through the pin at A. It is operated by the extension of the hydraulic cylinder BC. The angular position of the bin can be specified using the angular position coordinate θ, and the position of point C on the bin is specified using the rectilinear coordinate s. Since a and b are fixed lengths, then the coordinates can be related by the cosine law, $s = \sqrt{a^2 + b^2 - 2ab\cos\theta}$. Using the chain rule, the time derivative of this equation relates the speed at which the hydraulic cylinder extends, to the angular velocity of the bin, i.e., $v = \frac{1}{2}(a^2 + b^2 - 2ab\cos\theta)^{-\frac{1}{2}}(2ab\sin\theta)\,\omega$.

EXAMPLE 16–3

At a given instant, the cylinder of radius r, shown in Fig. 16–7, has an angular velocity $\boldsymbol{\omega}$ and angular acceleration $\boldsymbol{\alpha}$. Determine the velocity and acceleration of its center G if the cylinder rolls without slipping.

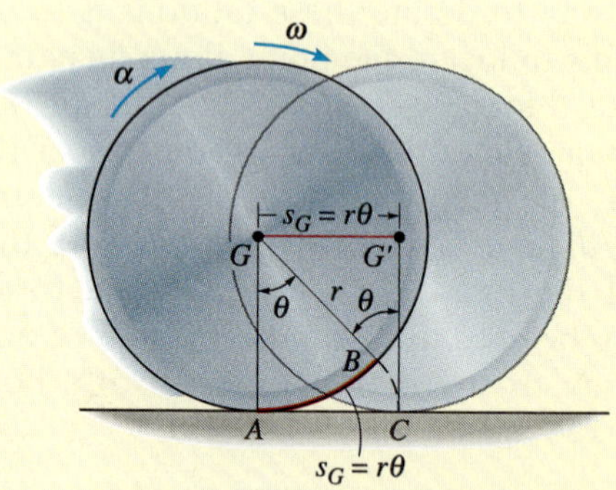

Fig. 16–7

Solution

Position Coordinate Equation. By inspection, point G moves *horizontally* to the right from G to G' as the cylinder rolls, Fig. 16–7. Consequently its new location G' will be specified by the *horizontal* position coordinate s_G, which is measured from the original position (G) of the cylinder's center. Notice also, that as the cylinder rolls (without slipping), points on its surface contact the ground such that the arc length AB of contact must be equal to the distance s_G. Consequently, the motion requires the radial line GB to rotate θ to the position $G'C$. Since the arc $AB = r\theta$, then G travels a distance

$$s_G = r\theta$$

Time Derivatives. Taking successive time derivatives of this equation, realizing that r is constant, $\omega = d\theta/dt$, and $\alpha = d\omega/dt$, gives the necessary relationships:

$$s_G = r\theta$$

$$v_G = r\omega \qquad \textit{Ans.}$$

$$a_G = r\alpha \qquad \textit{Ans.}$$

Remember that these relationships are valid only if the cylinder (disk, wheel, ball, etc.) rolls *without* slipping.

EXAMPLE 16-4

The end of rod R shown in Fig. 16–8 maintains contact with the cam by means of a spring. If the cam rotates about an axis through point O with an angular acceleration $\boldsymbol{\alpha}$ and angular velocity $\boldsymbol{\omega}$, determine the velocity and acceleration of the rod when the cam is in the arbitrary position θ.

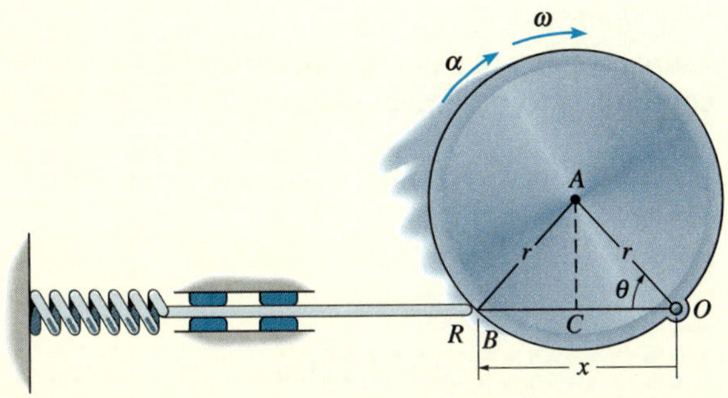

Fig. 16–8

Solution

Position Coordinate Equation. Coordinates θ and x are chosen in order to relate the *rotational motion* of the line segment OA on the cam to the *rectilinear motion* of the rod. These coordinates are measured from the *fixed point O* and may be related to each other using trigonometry. Since $OC = CB = r\cos\theta$, Fig. 16–8, then

$$x = 2r\cos\theta$$

Time Derivatives. Using the chain rule of calculus, we have

$$\frac{dx}{dt} = -2r(\sin\theta)\frac{d\theta}{dt}$$

$$v = -2r\,\omega\sin\theta \qquad \textit{Ans.}$$

$$\frac{dv}{dt} = -2r\left(\frac{d\omega}{dt}\right)\sin\theta - 2r\omega(\cos\theta)\frac{d\theta}{dt}$$

$$a = -2r(\alpha\sin\theta + \omega^2\cos\theta) \qquad \textit{Ans.}$$

The negative signs indicate that v and a are opposite to the direction of positive x.

EXAMPLE 16-5

The large window in Fig. 16-9 is opened using a hydraulic cylinder AB. If the cylinder extends at a constant rate of 0.5 m/s, determine the angular velocity and angular acceleration of the window at the instant $\theta = 30°$.

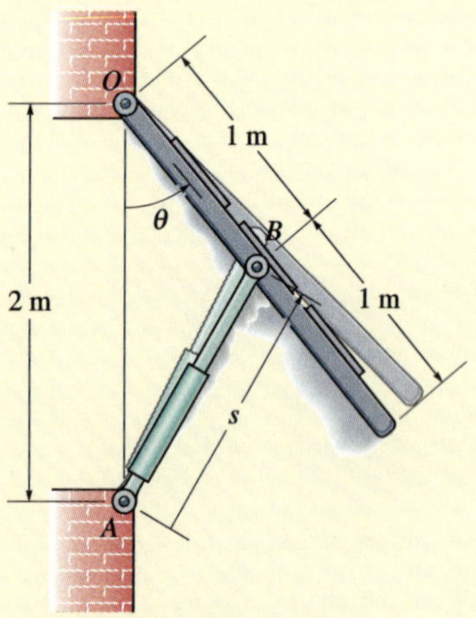

Fig. 16–9

Solution

Position Coordinate Equation. The angular motion of the window can be obtained using the coordinate θ, whereas the extension or motion *along the hydraulic cylinder* is defined using a coordinate s, which measures the length from the fixed point A to the moving point B. These coordinates can be related using the law of cosines, namely,

$$s^2 = (2\text{ m})^2 + (1\text{ m})^2 - 2(2\text{ m})(1\text{ m})\cos\theta$$

$$s^2 = 5 - 4\cos\theta \qquad (1)$$

When $\theta = 30°$,

$$s = 1.239\text{ m}$$

Time Derivatives. Taking the time derivatives of Eq. (1), we have

$$2s\frac{ds}{dt} = 0 - 4(-\sin\theta)\frac{d\theta}{dt}$$

$$s(v_s) = 2(\sin\theta)\omega \qquad (2)$$

Since $v_s = 0.5$ m/s, then at $\theta = 30°$,

$$(1.239\text{ m})(0.5\text{ m/s}) = 2\sin 30°\omega$$

$$\omega = 0.620\text{ rad/s} \qquad \textit{Ans.}$$

Taking the time derivative of Eq. (2) yields

$$\frac{ds}{dt}v_s + s\frac{dv_s}{dt} = 2(\cos\theta)\frac{d\theta}{dt}\omega + 2(\sin\theta)\frac{d\omega}{dt}$$

$$v_s^2 + sa_s = 2(\cos\theta)\omega^2 + 2(\sin\theta)\alpha$$

Since $a_s = dv_s/dt = 0$, then

$$(0.5\text{ m/s})^2 + 0 = 2\cos 30°(0.620\text{ rad/s})^2 + 2\sin 30°\alpha$$

$$\alpha = -0.415\text{ rad/s}^2 \qquad \textit{Ans.}$$

Because the result is negative, it indicates the window has an angular deceleration.

Problems

16-33. At the instant shown, $\theta = 60°$, and rod AB is subjected to a deceleration of $16\ \text{m/s}^2$ when the velocity is 10 m/s. Determine the angular velocity and angular acceleration of link CD at this instant.

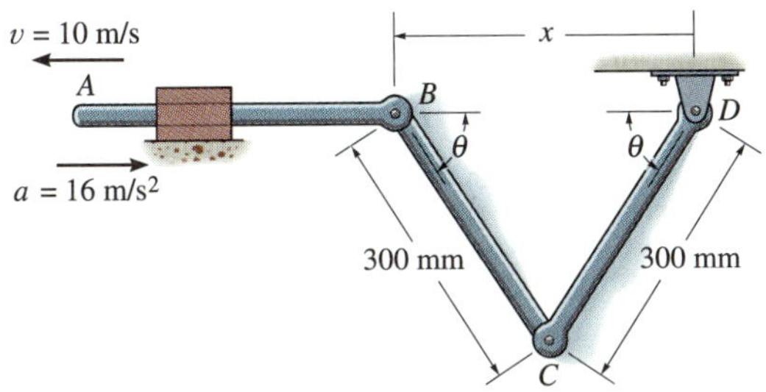

Prob. 16–33

16-34. The scaffold S is raised hydraulically by moving the roller at A toward the pin at B. If A is approaching B with a speed of 1.5 ft/s, determine the speed at which the platform is rising as a function of θ. The 4-ft links are pin-connected at their midpoint.

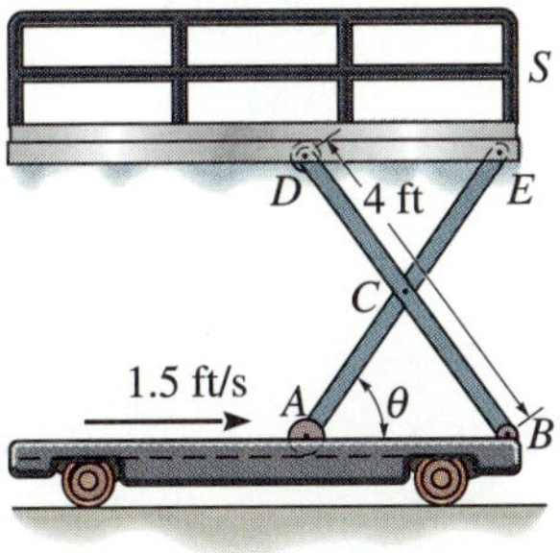

Prob. 16–34

16-35. The 2-m-long bar is confined to move in the horizontal and vertical slots A and B. If the velocity of the slider block at A is 8 m/s, determine the bar's angular velocity and the velocity of block B at the instant $\theta = 60°$.

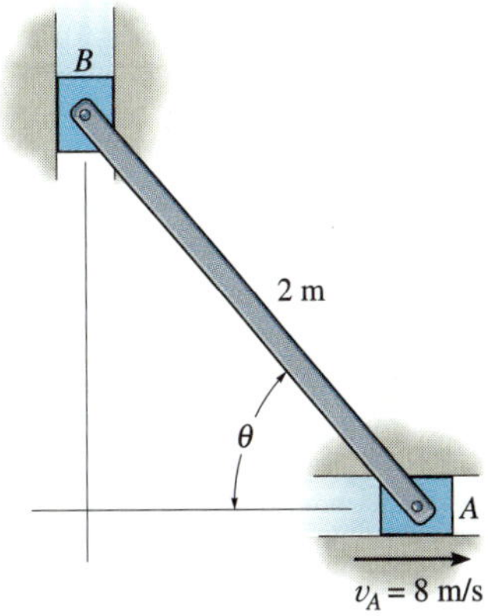

Prob. 16–35

***16-36.** Determine the angular velocity of rod AB when $\theta = 30°$. The shaft and the center of the roller C move forward at a constant rate $v = 5$ m/s.

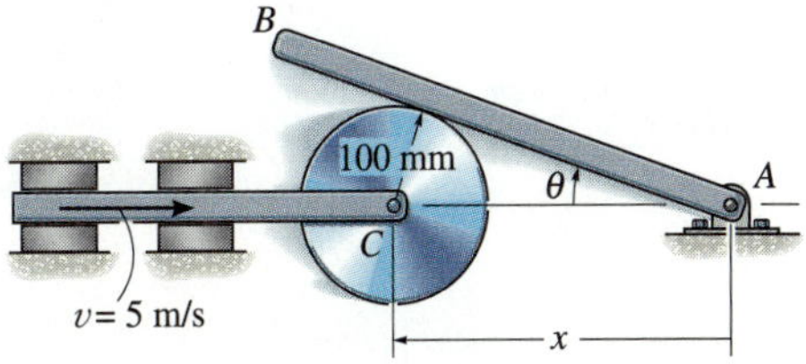

Prob. 16–36

16-37. The inclined plate moves to the left with a constant velocity **v**. Determine the angular velocity and angular acceleration of the slender rod of length l. The rod pivots about the step at C as it slides on the plate.

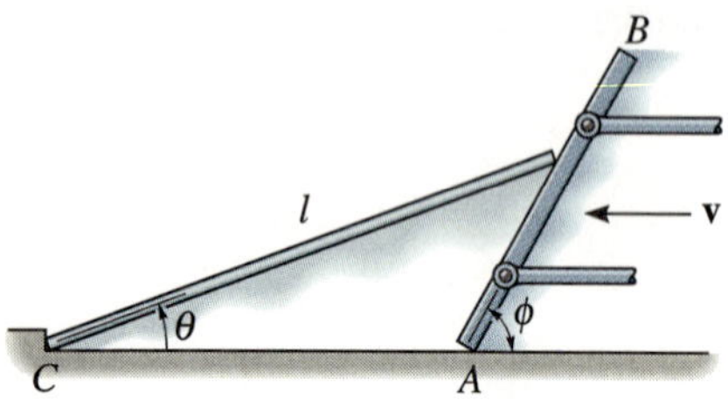

Prob. 16–37

16-38. The crankshaft AB is rotating at a constant angular velocity of $\omega = 150$ rad/s. Determine the velocity of the piston P at the instant $\theta = 30°$.

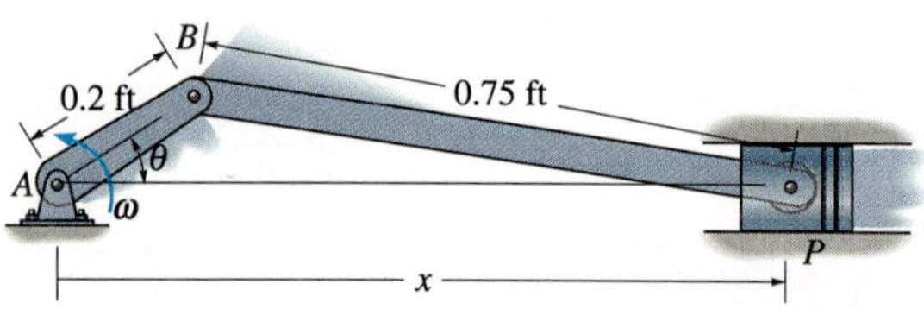

Prob. 16–38

16-39. Determine the velocity of the rod R for any angle θ of cam C as the cam rotates with a constant angular velocity $\boldsymbol{\omega}$. The pin connection at O does not cause an interference with the motion of plate A on C.

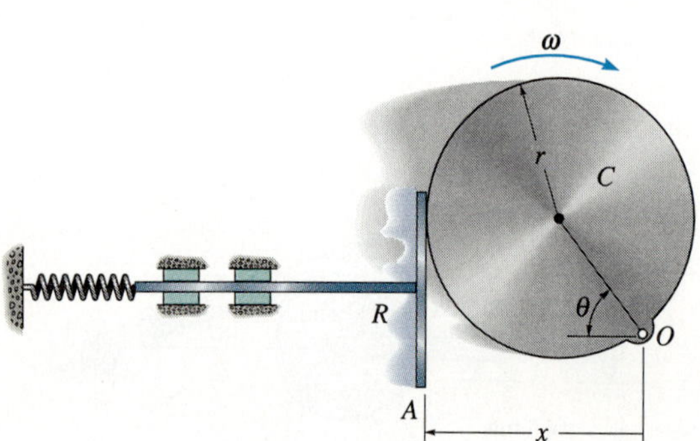

Prob. 16–39

***16-40.** Disk A rolls without slipping over the surface of the *fixed* cylinder B. Determine the angular velocity of A if its center C has a speed $v_C = 5$ m/s. How many revolutions will A have made about its center just after link DC completes one revolution?

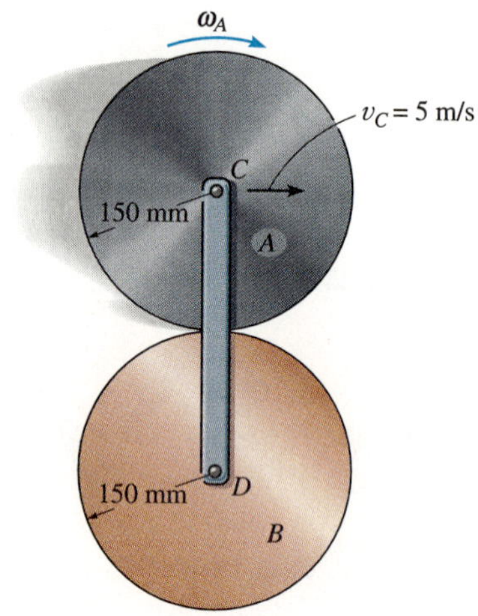

Prob. 16–40

16-41. Arm AB has an angular velocity of $\boldsymbol{\omega}$ and an angular acceleration of $\boldsymbol{\alpha}$. If no slipping occurs between the disk and the fixed curved surface, determine the angular velocity and angular acceleration of the disk.

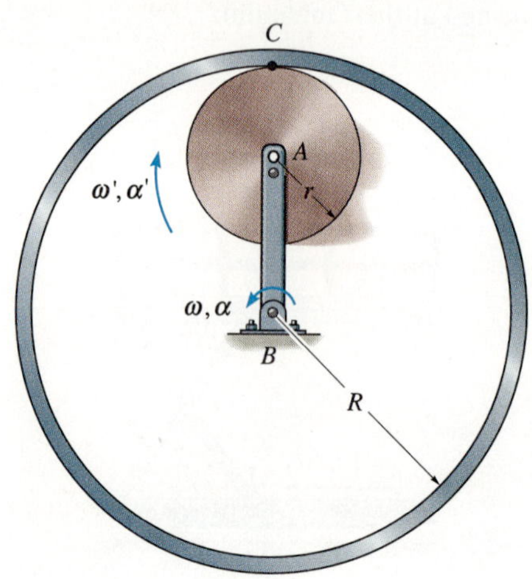

Prob. 16–41

16-42. Arm AB has an angular velocity of $\boldsymbol{\omega}$ and an angular acceleration of $\boldsymbol{\alpha}$. If no slipping occurs between the disk D and the fixed curved surface, determine the angular velocity and angular acceleration of the disk.

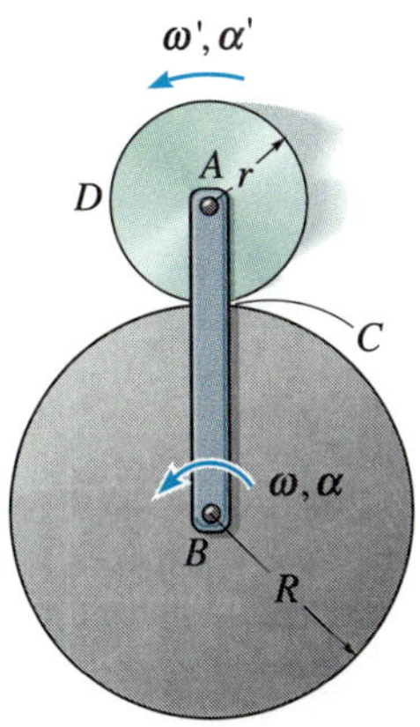

Prob. 16–42

■ **16-43.** The end A of the bar is moving downward along the slotted guide with a constant velocity $\mathbf{v}_A$. Determine the angular velocity $\boldsymbol{\omega}$ and angular acceleration $\boldsymbol{\alpha}$ of the bar as a function of its position y.

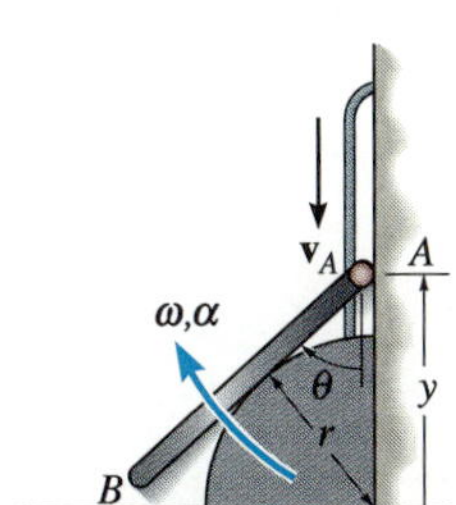

Prob. 16–43

***16-44.** The pins at A and B are confined to move in the vertical and horizontal tracks. If the slotted arm is causing A to move downward at $\mathbf{v}_A$, determine the velocity of B at the instant shown.

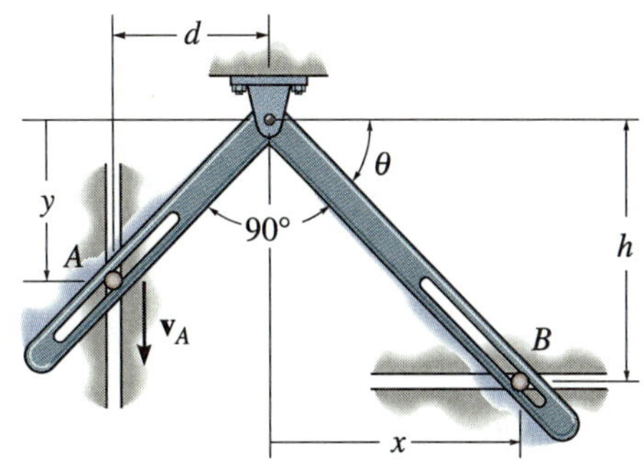

Prob. 16–44

16-45. The bar remains in contact with the floor and with point A. If point B moves to the right with a constant velocity $\mathbf{v}_B$, determine the angular velocity and angular acceleration of the bar as a function of x.

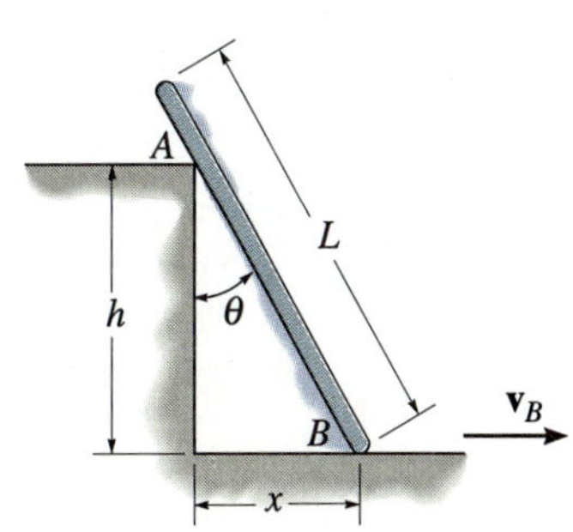

Prob. 16–45

16-46. The crate is transported on a platform which rests on rollers, each having a radius r. If the rollers do not slip, determine their angular velocity if the platform moves forward with a velocity $\mathbf{v}$.

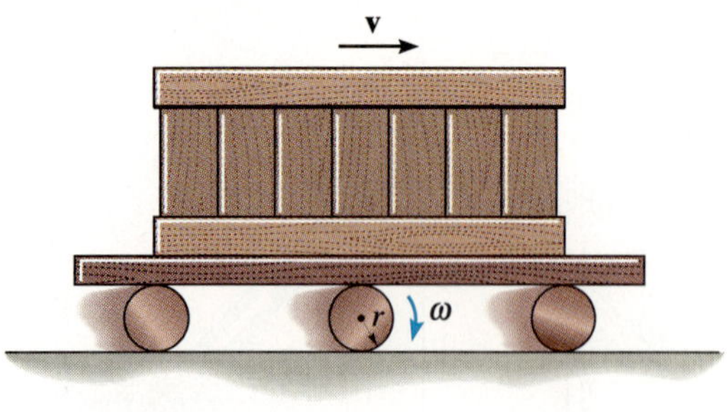

Prob. 16–46

16-47. The disk is rotating with an angular velocity of $\boldsymbol{\omega}$ and has an angular acceleration of $\boldsymbol{\alpha}$. Determine the velocity and acceleration of cylinder B. Neglect the size of the pulley at C.

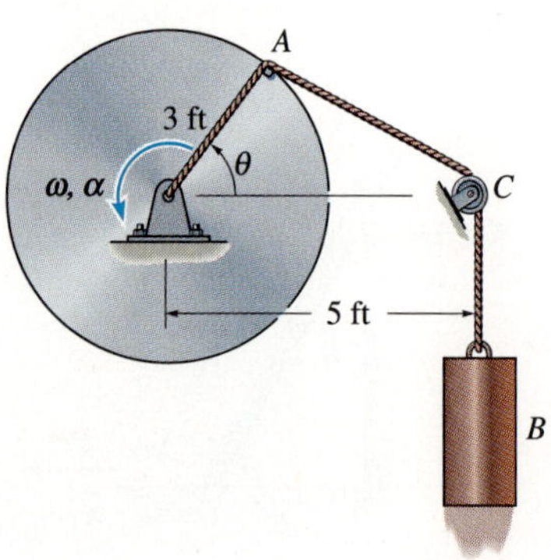

Prob. 16–47

***16-48.** When the bar is at the angle θ, the rod is rotating counter clockwise at $\boldsymbol{\omega}$ and has an angular acceleration of $\boldsymbol{\alpha}$. Determine the velocity and acceleration of the weight A at this instant. The cord is 20 ft long.

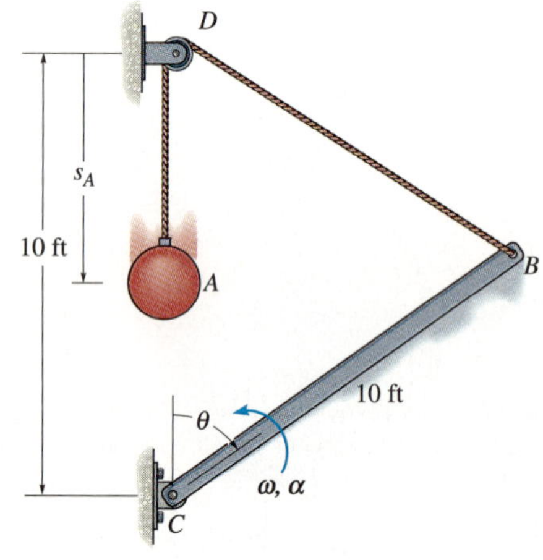

Prob. 16–48

■**16-49.** The crank AB has a constant angular velocity $\boldsymbol{\omega}$. Determine the velocity and acceleration of the slider at C as a function of θ. *Suggestion:* Use the x coordinate to express the motion of C and the ϕ coordinate for CB. $x = 0$ when $\phi = 0°$.

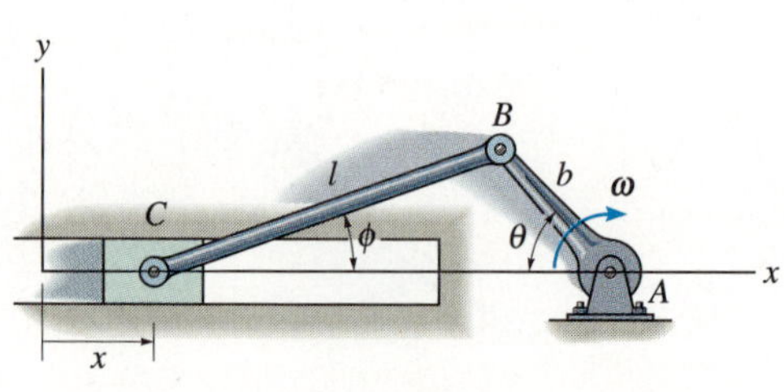

Prob. 16–49

16.5 Relative-Motion Analysis: Velocity

The general plane motion of a rigid body can be described as a *combination* of translation and rotation. To view these "component" motions *separately* we will use a *relative-motion analysis* involving two sets of coordinate axes. The x, y coordinate system is fixed and measures the *absolute* position of two points A and B on the body, Fig. 16–10*a*. The origin of the x', y' coordinate system will be attached to the selected "base point" A, which generally has a *known* motion. The axes of this coordinate system do not rotate with the body; rather they will only be allowed to *translate* with respect to the fixed frame.

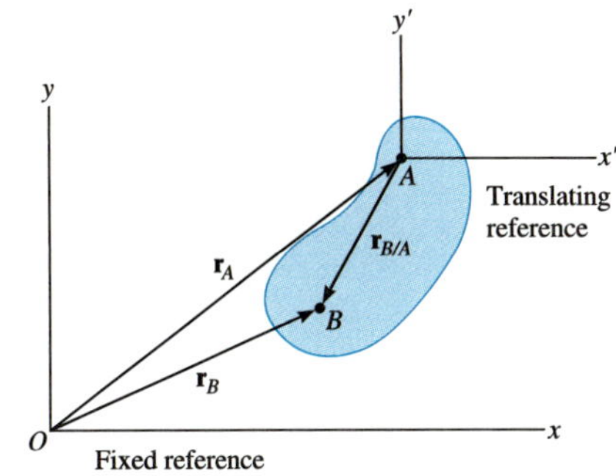

(a)

Fig. 16–10

Position. The position vector $\mathbf{r}_A$ in Fig. 16–10*a* specifies the location of the "base point" A, and the relative-position vector $\mathbf{r}_{B/A}$ locates point B with respect to point A. By vector addition, the *position* of B is then

$$\mathbf{r}_B = \mathbf{r}_A + \mathbf{r}_{B/A}$$

Displacement. During an instant of time dt, points A and B undergo displacements $d\mathbf{r}_A$ and $d\mathbf{r}_B$ as shown in Fig. 16–10*b*. If we consider the general plane motion by its component parts then the *entire body* first *translates* by an amount $d\mathbf{r}_A$ so that A, the base point, moves to its *final position* and point B moves to B', Fig. 16–10*c*. The body is then *rotated* about A by an amount $d\theta$ so that B' undergoes a *relative displacement* $d\mathbf{r}_{B/A}$ and thus moves to its final position B. Due to the rotation about A, $dr_{B/A} = r_{B/A}d\theta$, and the displacement of B is

$$d\mathbf{r}_B = d\mathbf{r}_A + d\mathbf{r}_{B/A}$$

due to rotation about A (for $d\mathbf{r}_{B/A}$)

due to translation of A (for $d\mathbf{r}_A$)

due to translation and rotation (for $d\mathbf{r}_B$)

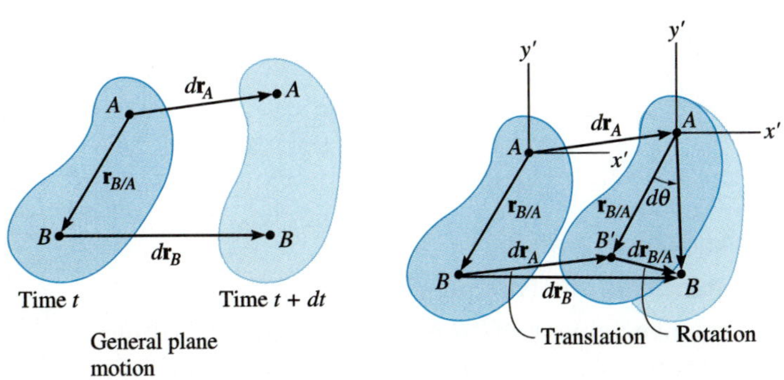

(b)

(c)

As the slider block A moves horizontally to the left with a velocity $\mathbf{v}_A$, it causes the link CB to rotate counterclockwise, such that $\mathbf{v}_B$ is directed tangent to its circular path, i.e., upward to the left. The connecting rod AB is subjected to general plane motion, and at the instant shown it has an angular velocity $\boldsymbol{\omega}$.

Velocity. To determine the relationship between the velocities of points A and B, it is necessary to take the time derivative of the position equation, or simply divide the displacement equation by dt. This yields

$$\frac{d\mathbf{r}_B}{dt} = \frac{d\mathbf{r}_A}{dt} + \frac{d\mathbf{r}_{B/A}}{dt}$$

The terms $d\mathbf{r}_B/dt = \mathbf{v}_B$ and $d\mathbf{r}_A/dt = \mathbf{v}_A$ are measured from the fixed x, y axes and represent the *absolute velocities* of points A and B, respectively. The magnitude of the third term is $r_{B/A}d\theta/dt = r_{B/A}\dot{\theta} = r_{B/A}\omega$, where ω is the angular velocity of the body at the instant considered. We will denote this term as the *relative velocity* $\mathbf{v}_{B/A}$, since it represents the velocity of B with respect to A as measured by an observer fixed to the translating x', y' axes. Since the body is rigid, realize that this observer only sees point B move along a *circular arc* that has a radius of curvature $r_{B/A}$. In other words, *the body appears to move as if it were rotating with an angular velocity* $\boldsymbol{\omega}$ *about the* z' *axis passing through* A. Consequently, $\mathbf{v}_{B/A}$ has a magnitude of $v_{B/A} = \omega r_{B/A}$ and a *direction* which is perpendicular to $\mathbf{r}_{B/A}$. We therefore have

$$\mathbf{v}_B = \mathbf{v}_A + \mathbf{v}_{B/A} \qquad (16\text{–}15)$$

where

$\mathbf{v}_B$ = velocity of point B

$\mathbf{v}_A$ = velocity of the base point A

$\mathbf{v}_{B/A}$ = relative velocity of "B with respect to A"
This relative motion is *circular*, the *magnitude* is $v_{B/A} = \omega r_{B/A}$ and the *direction* is perpendicular to $\mathbf{r}_{B/A}$.

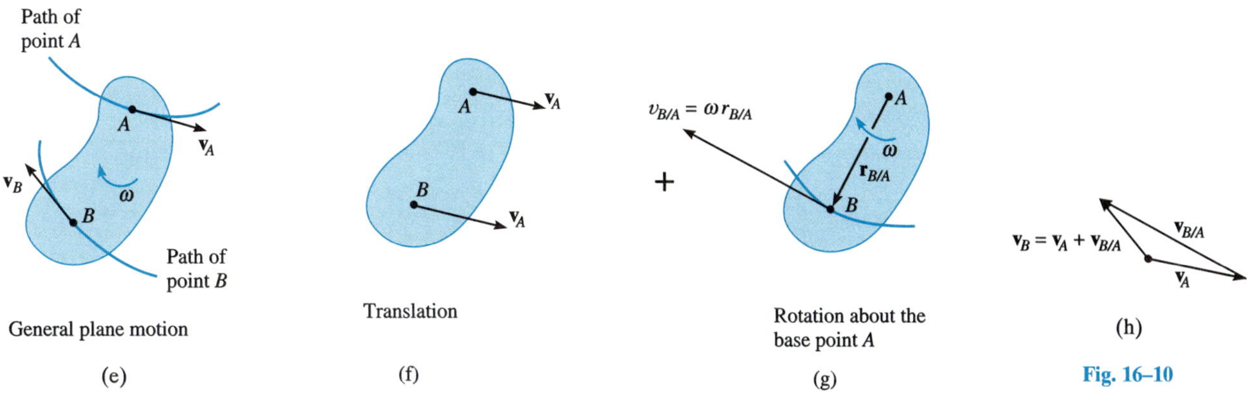

Fig. 16–10

Each of the three terms in Eq. 16–15 is represented graphically on the *kinematic diagrams* in Figs. 16–10*e*, 16–10*f*, and 16–10*g*. Here it is seen that the velocity of B, Fig. 16–10*e*, is determined by considering the entire body to translate with a velocity of $\mathbf{v}_A$, Fig. 16–10*f*, and rotate about A with an angular velocity $\boldsymbol{\omega}$, Fig. 16–10*g*. Vector addition of these two effects, applied to B, yields $\mathbf{v}_B$, as shown in Fig. 16–10*h*.

Since the relative velocity $\mathbf{v}_{B/A}$ represents the effect of *circular motion*, about A, this term can be expressed by the cross product $\mathbf{v}_{B/A} = \boldsymbol{\omega} \times \mathbf{r}_{B/A}$, Eq. 16–9. Hence, for application, we can also write Eq. 16–15 as

$$\mathbf{v}_B = \mathbf{v}_A + \boldsymbol{\omega} \times \mathbf{r}_{B/A} \qquad (16\text{–}16)$$

where

$\mathbf{v}_B$ = velocity of B
$\mathbf{v}_A$ = velocity of the base point A
$\boldsymbol{\omega}$ = angular velocity of the body
$\mathbf{r}_{B/A}$ = relative-position vector drawn from A to B

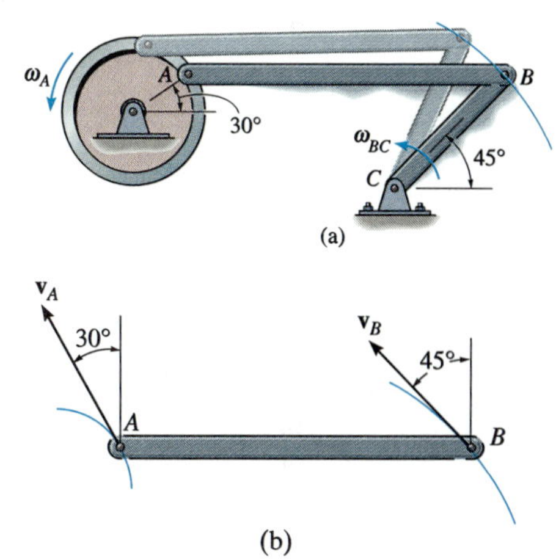

Fig. 16–11

The velocity equation 16–15 or 16–16 may be used in a practical manner to study the general plane motion of a rigid body which is either pin-connected to or in contact with other moving bodies. When applying this equation, points A and B should generally be selected as points on the body which are pin-connected to other bodies, or as points in contact with adjacent bodies which have a *known motion.* For example, both points A and B on link AB, Fig. 16–11*a*, have circular paths of motion since the wheel and link CB move in circular paths. The *directions* of $\mathbf{v}_A$ and $\mathbf{v}_B$ can therefore be established since they are always *tangent* to their paths of motion, Fig. 16–11*b*. In the case of the wheel in Fig. 16–12, which rolls without slipping, point A can be selected at the ground. Here A (momentarily) has zero velocity since the ground does not move. Furthermore, the center of the wheel, B, moves along a horizontal path so that $\mathbf{v}_B$ is horizontal.

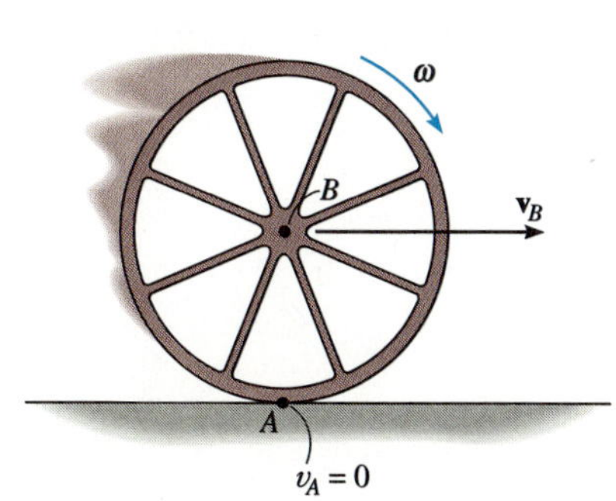

Fig. 16–12

Procedure for Analysis

The relative velocity equation can be applied either by using Cartesian vector analysis, or by writing the x and y scalar component equations directly. For application, it is suggested that the following procedure be used.

VECTOR ANALYSIS

Kinematic Diagram

- Establish the directions of the fixed x, y coordinates and draw a kinematic diagram of the body. Indicate on it the velocities $\mathbf{v}_A$, $\mathbf{v}_B$ of points A and B, the angular velocity $\boldsymbol{\omega}$, and the relative-position vector $\mathbf{r}_{B/A}$.
- If the magnitudes of $\mathbf{v}_A$, $\mathbf{v}_B$, or $\boldsymbol{\omega}$ are unknown, the sense of direction of these vectors may be assumed.

Velocity Equation

- To apply $\mathbf{v}_B = \mathbf{v}_A + \boldsymbol{\omega} \times \mathbf{r}_{B/A}$, express the vectors in Cartesian vector form and substitute them into the equation. Evaluate the cross product and then equate the respective $\mathbf{i}$ and $\mathbf{j}$ components to obtain two scalar equations.
- If the solution yields a *negative* answer for an *unknown* magnitude, it indicates the sense of direction of the vector is opposite to that shown on the kinematic diagram.

SCALAR ANALYSIS

Kinematic Diagram

- If the velocity equation is to be applied in scalar form, then the magnitude and direction of the relative velocity $\mathbf{v}_{B/A}$ must be established. Draw a kinematic diagram such as shown in Fig. 16–10g, which shows the relative motion. Since the body is considered to be "pinned" momentarily at the base point A, the *magnitude* is $v_{B/A} = \omega r_{B/A}$. The *sense of direction* of $\mathbf{v}_{B/A}$ is established from the diagram, such that $\mathbf{v}_{B/A}$ acts perpendicular to $\mathbf{r}_{B/A}$ in accordance with the rotational motion $\boldsymbol{\omega}$ of the body.*

Velocity Equation

- Write Eq. 16–15 in symbolic form, $\mathbf{v}_B = \mathbf{v}_A + \mathbf{v}_{B/A}$, and underneath each of the terms represent the vectors graphically by showing their magnitudes and directions. The scalar equations are determined from the x and y components of these vectors.

*The notation $\mathbf{v}_B = \mathbf{v}_A + \mathbf{v}_{B/A(\text{pin})}$ may be helpful in recalling that A is "pinned."

EXAMPLE 16–6

The link shown in Fig. 16–13*a* is guided by two blocks at A and B, which move in the fixed slots. If the velocity of A is 2 m/s downward, determine the velocity of B at the instant $\theta = 45^\circ$.

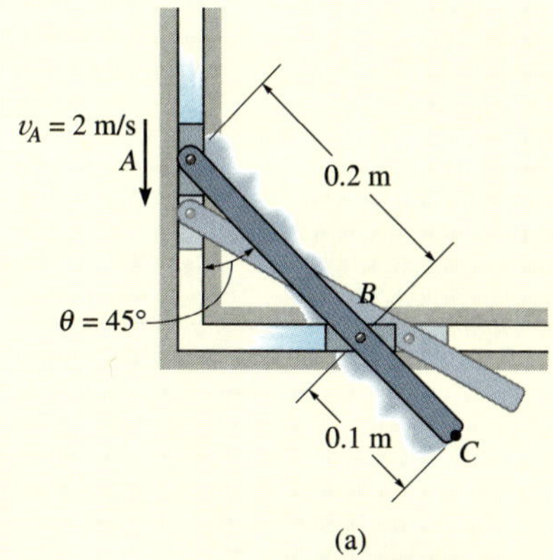

(a)

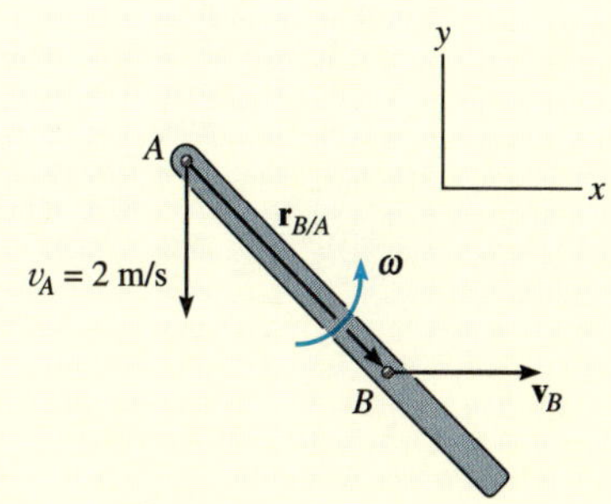

(b)

Fig. 16–13

Solution (VECTOR ANALYSIS)

Kinematic Diagram. Since points A and B are restricted to move along the fixed slots and $\mathbf{v}_A$ is directed downward, the velocity $\mathbf{v}_B$ must be directed horizontally to the right, Fig. 16–13*b*. This motion causes the link to rotate counterclockwise; that is, by the right-hand rule the angular velocity $\boldsymbol{\omega}$ is directed outward, perpendicular to the plane of motion. Knowing the magnitude and direction of $\mathbf{v}_A$ and the lines of action of $\mathbf{v}_B$ and $\boldsymbol{\omega}$, it is possible to apply the velocity equation $\mathbf{v}_B = \mathbf{v}_A + \boldsymbol{\omega} \times \mathbf{r}_{B/A}$ to points A and B in order to solve for the two unknown magnitudes v_B and ω. Since $\mathbf{r}_{B/A}$ is needed, it is also shown in Fig. 16–13*b*.

Velocity Equation. Expressing each of the vectors in Fig. 16–13*b* in terms of their **i, j, k** components and applying Eq. 16–16 to A, the base point, and B, we have

$$\mathbf{v}_B = \mathbf{v}_A + \boldsymbol{\omega} \times \mathbf{r}_{B/A}$$

$$v_B\mathbf{i} = -2\mathbf{j} + [\omega\mathbf{k} \times (0.2 \sin 45^\circ\mathbf{i} - 0.2 \cos 45^\circ\mathbf{j})]$$

$$v_B\mathbf{i} = -2\mathbf{j} + 0.2\omega \sin 45^\circ\mathbf{j} + 0.2\omega \cos 45^\circ\mathbf{i}$$

Equating the **i** and **j** components gives

$$v_B = 0.2\omega \cos 45^\circ \qquad 0 = -2 + 0.2\omega \sin 45^\circ$$

Thus,

$$\omega = 14.1 \text{ rad/s} \circlearrowleft$$

$$v_B = 2 \text{ m/s} \rightarrow \qquad \textit{Ans.}$$

Since both results are *positive*, the *directions* of $\mathbf{v}_B$ and $\boldsymbol{\omega}$ are indeed *correct* as shown in Fig. 16–13*b*. It should be emphasized that these results are *valid only* at the instant $\theta = 45^\circ$. A recalculation for $\theta = 44^\circ$ yields $v_B = 2.07$ m/s and $\omega = 14.4$ rad/s; whereas when $\theta = 46^\circ$, $v_B = 1.93$ m/s and $\omega = 13.9$ rad/s, etc.

Now that the velocity of a point (A) on the link and the angular velocity are *known*, the velocity of any other point on the link can be determined. As an exercise, see if you can apply Eq. 16–16 to points A and C or to points B and C and show that $v_C = 3.16$ m/s, directed at $\theta = 18.4^\circ$ up from the horizontal.

EXAMPLE 16–7

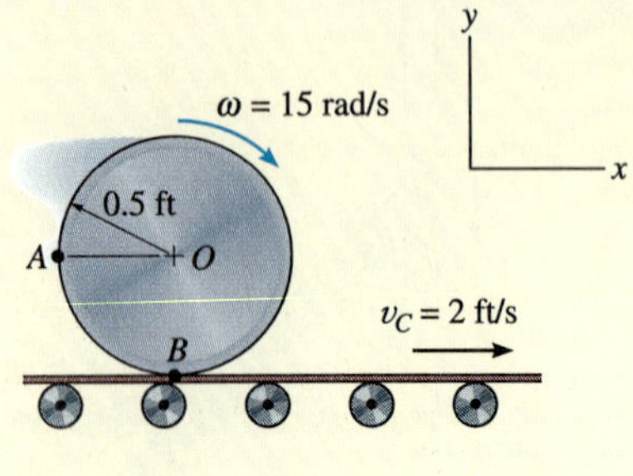

(a)

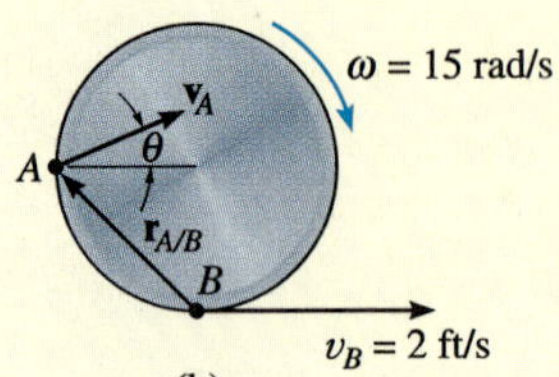

(b)

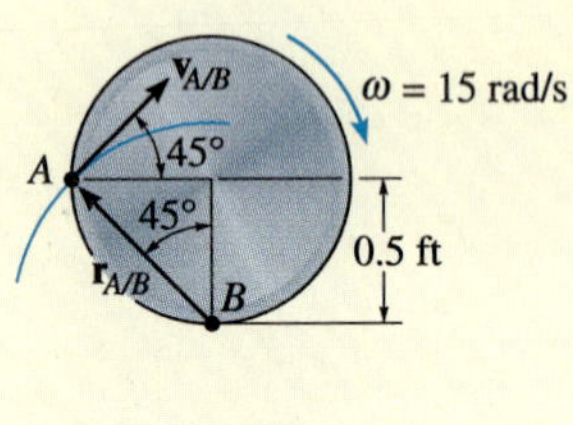

Relative motion

(c)

Fig. 16–14

The cylinder shown in Fig. 16–14*a* rolls without slipping on the surface of a conveyor belt which is moving at 2 ft/s. Determine the velocity of point A. The cylinder has a clockwise angular velocity $\omega = 15$ rad/s at the instant shown.

Solution I (Vector Analysis)

Kinematic Diagram. Since no slipping occurs, point B on the cylinder has the same velocity as the conveyor, Fig. 16–14*b*. Also, the angular velocity of the cylinder is known, so we can apply the velocity equation to B, the base point, and A to determine $\mathbf{v}_A$.

Velocity Equation

$$\mathbf{v}_A = \mathbf{v}_B + \boldsymbol{\omega} \times \mathbf{r}_{A/B}$$

$$(v_A)_x\mathbf{i} + (v_A)_y\mathbf{j} = 2\mathbf{i} + (-15\mathbf{k}) \times (-0.5\mathbf{i} + 0.5\mathbf{j})$$

$$(v_A)_x\mathbf{i} + (v_A)_y\mathbf{j} = 2\mathbf{i} + 7.50\mathbf{j} + 7.50\mathbf{i}$$

so that

$$(v_A)_x = 2 + 7.50 = 9.50 \text{ ft/s} \qquad (1)$$

$$(v_A)_y = 7.50 \text{ ft/s} \qquad (2)$$

Thus,

$$v_A = \sqrt{(9.50)^2 + (7.50)^2} = 12.1 \text{ ft/s} \qquad \textit{Ans.}$$

$$\theta = \tan^{-1}\frac{7.50}{9.50} = 38.3° \measuredangle^{\theta}_{\mathbf{v}_A} \qquad \textit{Ans.}$$

Solution II (Scalar Analysis)

As an alternative procedure, the scalar components of $\mathbf{v}_A = \mathbf{v}_B + \mathbf{v}_{A/B}$ can be obtained directly. From the kinematic diagram showing the relative "circular" motion $\mathbf{v}_{A/B}$, Fig. 16–14*c*, we have

$$v_{A/B} = \omega r_{A/B} = (15 \text{ rad/s})\left(\frac{0.5 \text{ ft}}{\cos 45°}\right) = 10.6 \text{ ft/s} \measuredangle^{45°}$$

Thus

$$\mathbf{v}_A = \mathbf{v}_B + \mathbf{v}_{A/B}$$

$$\begin{bmatrix}(v_A)_x \\ \rightarrow\end{bmatrix} + \begin{bmatrix}(v_A)_y \\ \uparrow\end{bmatrix} = \begin{bmatrix}2 \text{ ft/s} \\ \rightarrow\end{bmatrix} + \begin{bmatrix}10.6 \text{ ft/s} \\ \measuredangle^{45°}\end{bmatrix}$$

Equating the x and y components gives the same results as before, namely,

$$(\overset{+}{\rightarrow}) \qquad (v_A)_x = 2 + 10.6 \cos 45° = 9.50 \text{ ft/s}$$

$$(+\uparrow) \qquad (v_A)_y = 0 + 10.6 \sin 45° = 7.50 \text{ ft/s}$$

EXAMPLE 16–8

The collar C in Fig. 16–15a is moving downward with a velocity of 2 m/s. Determine the angular velocities of CB and AB at this instant.

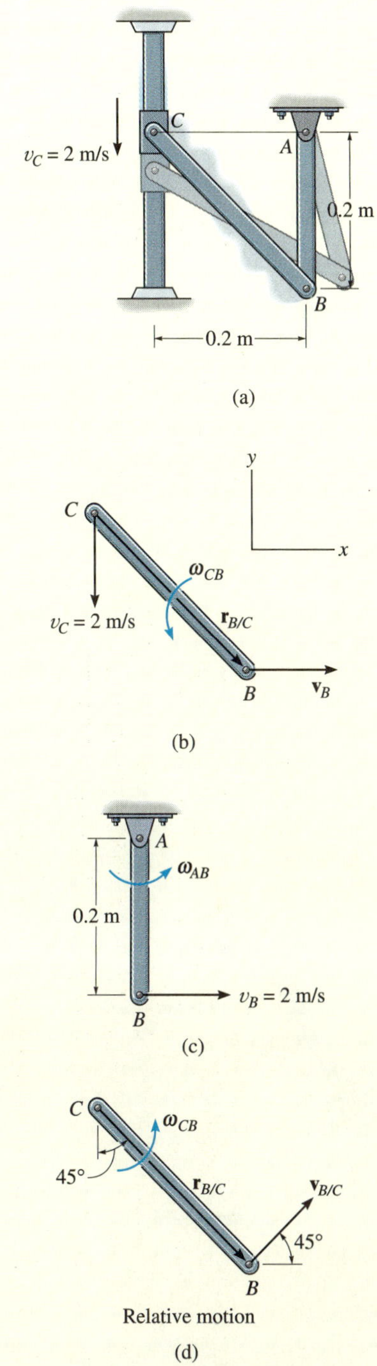

Fig. 16–15

Solution I (VECTOR ANALYSIS)

Kinematic Diagram. The downward motion of C causes B to move to the right. Also, CB and AB rotate counterclockwise. To solve, we will write the appropriate kinematic equation for each link.

Velocity Equation

Link CB (general plane motion): See Fig. 16–15b.

$$\mathbf{v}_B = \mathbf{v}_C + \boldsymbol{\omega}_{CB} \times \mathbf{r}_{B/C}$$

$$v_B\mathbf{i} = -2\mathbf{j} + \omega_{CB}\mathbf{k} \times (0.2\mathbf{i} - 0.2\mathbf{j})$$

$$v_B\mathbf{i} = -2\mathbf{j} + 0.2\omega_{CB}\mathbf{j} + 0.2\omega_{CB}\mathbf{i}$$

$$v_B = 0.2\omega_{CB} \qquad (1)$$

$$0 = -2 + 0.2\omega_{CB} \qquad (2)$$

$$\omega_{CB} = 10 \text{ rad/s} \circlearrowleft \qquad \textit{Ans.}$$

$$v_B = 2 \text{ m/s} \rightarrow$$

Link AB (rotation about a fixed axis): See Fig. 16–15c.

$$\mathbf{v}_B = \boldsymbol{\omega}_{AB} \times \mathbf{r}_B$$

$$2\mathbf{i} = \omega_{AB}\mathbf{k} \times (-0.2\mathbf{j})$$

$$2 = 0.2\omega_{AB}$$

$$\omega_{AB} = 10 \text{ rad/s} \circlearrowleft \qquad \textit{Ans.}$$

Solution II (SCALAR ANALYSIS)

The scalar component equations of $\mathbf{v}_B = \mathbf{v}_C + \mathbf{v}_{B/C}$ can be obtained directly. The kinematic diagram in Fig. 16–15d shows the relative "circular" motion $\mathbf{v}_{B/C}$. We have

$$\mathbf{v}_B = \mathbf{v}_C + \mathbf{v}_{B/C}$$

$$\begin{bmatrix} v_B \\ \rightarrow \end{bmatrix} = \begin{bmatrix} 2 \text{ m/s} \\ \downarrow \end{bmatrix} + \begin{bmatrix} \omega_{CB}(0.2\sqrt{2} \text{ m}) \\ \angle 45^\circ \end{bmatrix}$$

Resolving these vectors in the x and y directions yields

$$(\overset{+}{\rightarrow}) \qquad v_B = 0 + \omega_{CB}(0.2\sqrt{2}\cos 45^\circ)$$

$$(+\uparrow) \qquad 0 = -2 + \omega_{CB}(0.2\sqrt{2}\sin 45^\circ)$$

which is the same as Eqs. 1 and 2.

EXAMPLE 16–9

The bar AB of the linkage shown in Fig. 16–16a has a clockwise angular velocity of 30 rad/s when $\theta = 60°$. Determine the angular velocities of member BC and the wheel at this instant.

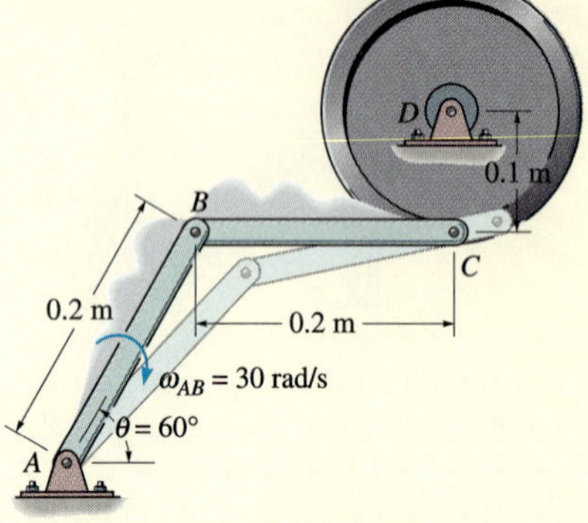

(a)

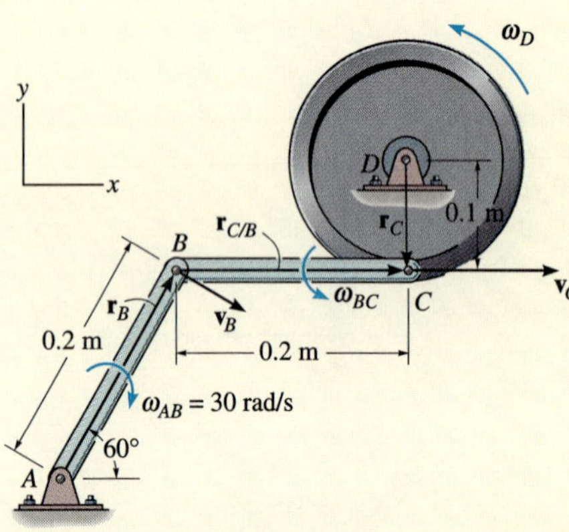

(b)

Fig. 16–16

Solution (VECTOR ANALYSIS)

Kinematic Diagram. By inspection, the velocities of points B and C are defined by the rotation of link AB and the wheel about their fixed axes. The position vectors and the angular velocity of each member are shown on the kinematic diagram in Fig. 16–16b. To solve, we will write the appropriate kinematic equation for each member.

Velocity Equation

Link AB (rotation about a fixed axis):

$$\mathbf{v}_B = \boldsymbol{\omega}_{AB} \times \mathbf{r}_B$$
$$= (-30\mathbf{k}) \times (0.2 \cos 60°\mathbf{i} + 0.2 \sin 60°\mathbf{j})$$
$$= \{5.20\mathbf{i} - 3.0\mathbf{j}\} \text{ m/s}$$

Link BC (general plane motion):

$$\mathbf{v}_C = \mathbf{v}_B + \boldsymbol{\omega}_{BC} \times \mathbf{r}_{C/B}$$
$$v_C\mathbf{i} = 5.20\mathbf{i} - 3.0\mathbf{j} + (\omega_{BC}\mathbf{k}) \times (0.2\mathbf{i})$$
$$v_C\mathbf{i} = 5.20\mathbf{i} + (0.2\omega_{BC} - 3.0)\mathbf{j}$$
$$v_C = 5.20 \text{ m/s}$$
$$0 = 0.2\omega_{BC} - 3.0$$
$$\omega_{BC} = 15 \text{ rad/s} \circlearrowleft \qquad \textit{Ans.}$$

Wheel (rotation about a fixed axis):

$$\mathbf{v}_C = \boldsymbol{\omega}_D \times \mathbf{r}_C$$
$$5.20\mathbf{i} = (\omega_D\mathbf{k}) \times (-0.1\mathbf{j})$$
$$5.20 = 0.1\omega_D$$
$$\omega_D = 52 \text{ rad/s} \circlearrowleft \qquad \textit{Ans.}$$

Note that, by inspection, Fig. 16–16a, $v_B = (0.2)(30) = 6$ m/s, $\measuredangle^{30°}$ and $\mathbf{v}_C$ is directed to the right. As an exercise, use this information and try to obtain ω_{BC} by applying $\mathbf{v}_C = \mathbf{v}_B + \mathbf{v}_{C/B}$ using scalar components.

Problems

16-50. If h and θ are known, and the speed of A and B is $v_A = v_B = v$, determine the angular velocity $\boldsymbol{\omega}$ of the body and the direction ϕ of $\mathbf{v}_B$.

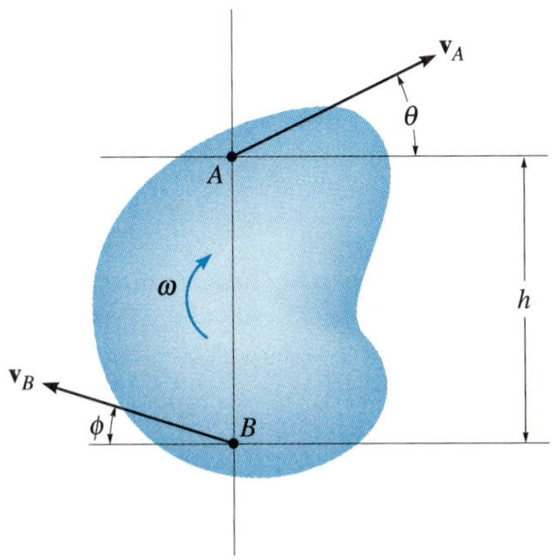

Prob. 16–50

16-51. The wheel is rotating with an angular velocity $\omega = 8$ rad/s. Determine the velocity of the collar A at the instant $\theta = 30°$and $\phi = 60°$. Also, sketch the location of bar AB when $\theta = 0°, 30°,$ and $60°$ to show its general plane motion.

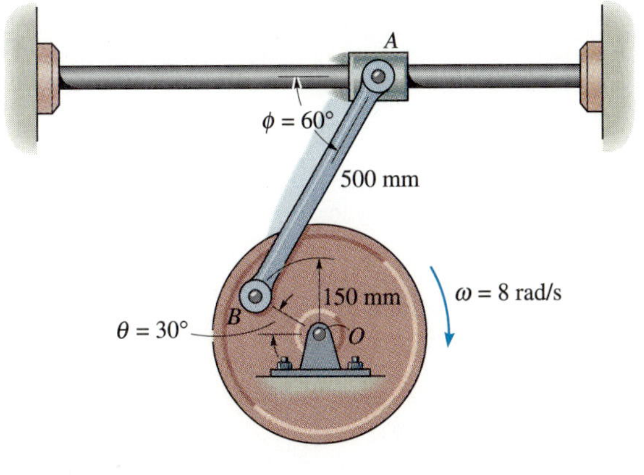

Prob. 16–51

***16-52.** The pinion gear A rolls on the fixed gear rack B with an angular velocity $\omega = 4$ rad/s. Determine the velocity of the gear rack C.

16-53. The pinion gear rolls on the gear racks. If B is moving to the right at 8 ft/s and C is moving to the left at 4 ft/s, determine the angular velocity of the pinion gear and the velocity of its center A.

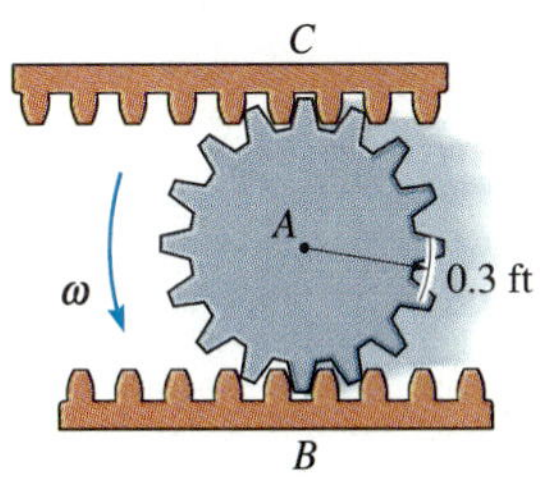

Probs. 16–52/53

16-54. The gear rests in a fixed horizontal rack. A cord is wrapped around the inner core of the gear so that it remains horizontally tangent to the inner core at A. If the cord is pulled to the right with a constant velocity of 2 ft/s, determine the velocity of the center of the gear, C.

16-55. Solve Prob. 16-54 assuming that the cord is wrapped around the gear in the opposite sense, so that the end of the cord remains horizontally tangent to the inner core at B and is pulled to the right at 2 ft/s.

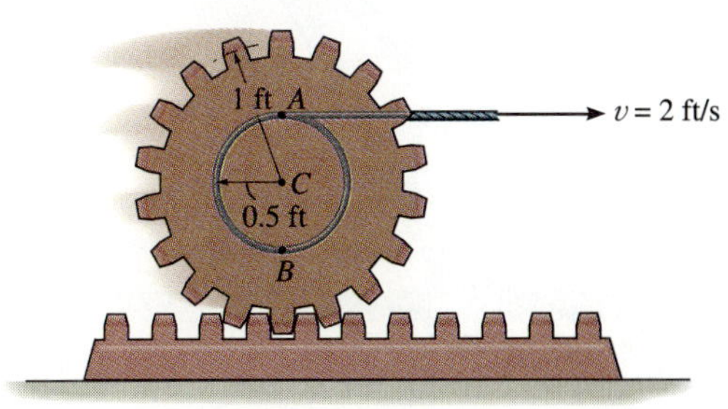

Probs. 16–54/55

***16-56.** A bowling ball is cast on the "alley" with a backspin of $\omega = 10$ rad/s while its center O has a forward velocity of $v_O = 8$ m/s. Determine the velocity of the contact point A touching the alley.

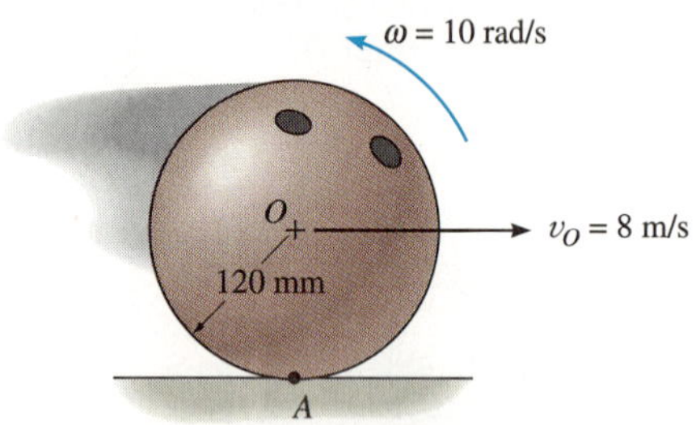

Prob. 16–56

16-57. Rod AB is rotating with an angular velocity $\omega_{AB} = 5$ rad/s. Determine the velocity of the collar C at the instant $\theta = 60°$ and $\phi = 45°$. Also, sketch the location of bar BC when $\theta = 30°$, $60°$ and $45°$ to show its general plane motion.

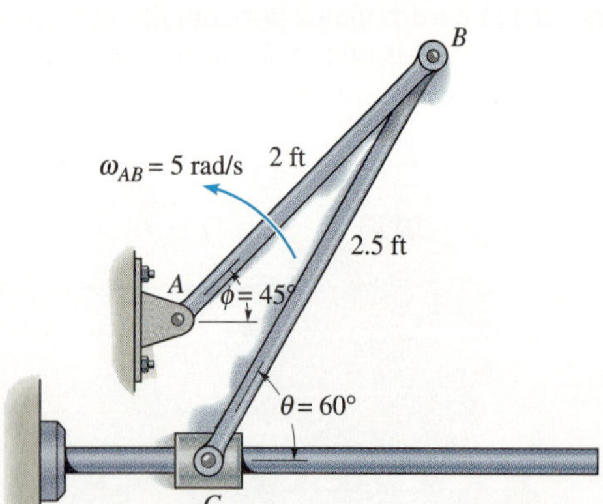

Prob. 16–57

16-58. The angular velocity of link AB is $\omega_{AB} = 4$ rad/s. Determine the velocity of the collar at C and the angular velocity of link CB at the instant $\theta = 60°$ and $\phi = 45°$. Link CB is horizontal at this instant. Also, sketch the location of link CB when $\theta = 30°$, $60°$, and $90°$ to show its general plane motion.

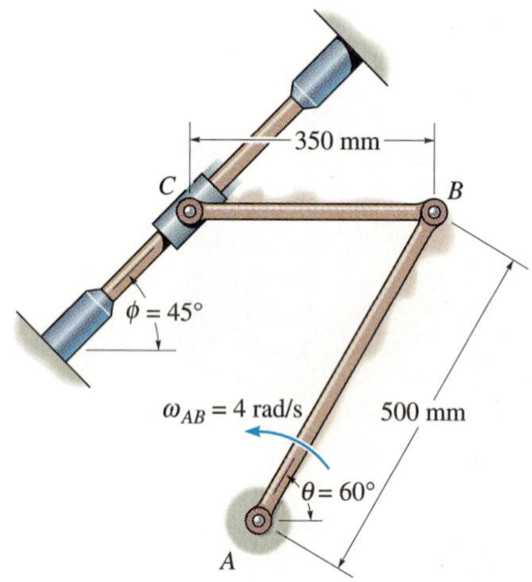

Prob. 16–58

16-59. The link AB has an angular velocity of 2 rad/s. Determine the velocity of block C at the instant $\theta = 45°$. Also, sketch the location of link BC when $\theta = 60°$, $45°$, and $30°$ to show its general plane motion.

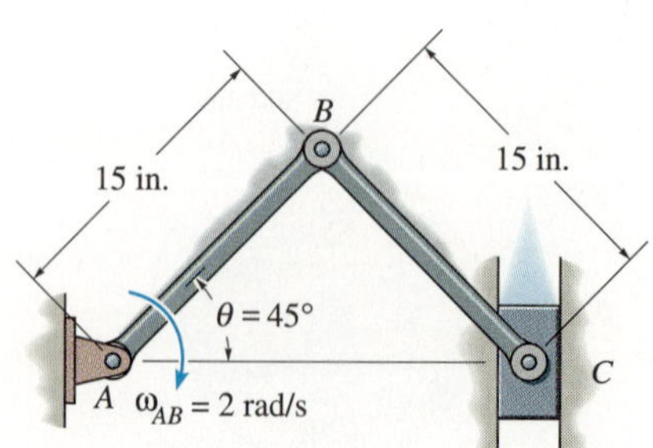

Prob. 16–59

***16-60.** If, at a given instant, point B has a downward velocity of $v_B = 3$ m/s, determine the velocity of point A at this instant. Notice that for this motion to occur, the wheel must slip at A.

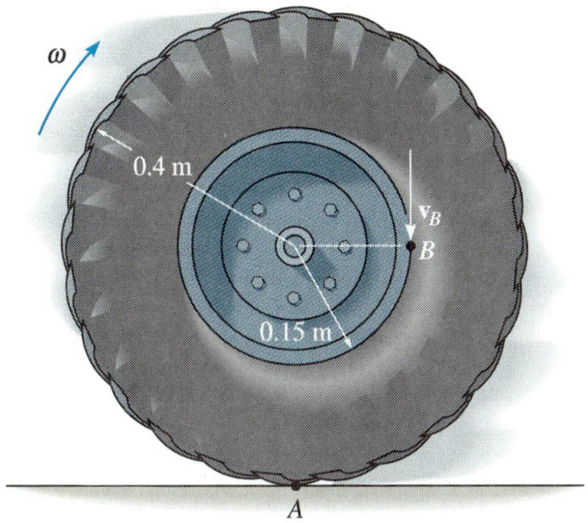

Prob. 16–60

16-61. The piston P is moving upward with a velocity of 300 in./s at the instant shown. Determine the angular velocity of the crankshaft AB at this instant.

16-62. Determine the velocity of the center of gravity G of the connecting rod at the instant shown. The piston is moving upward with a velocity of 300 in./s.

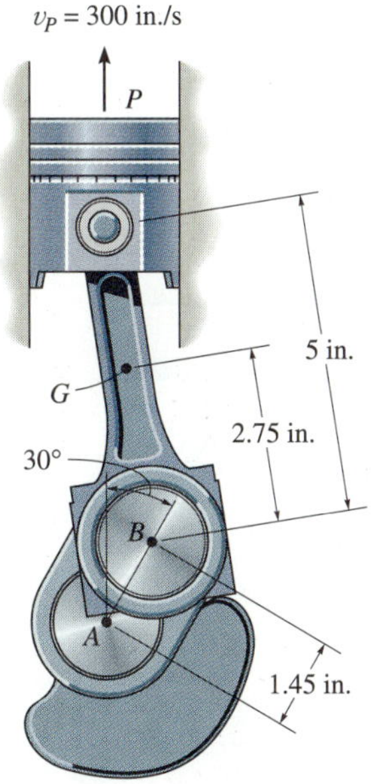

Probs. 16–61/62

16-63. The planetary gear system is used in an automatic transmission for an automobile. By locking or releasing certain gears, it has the advantage of operating the car at different speeds. Consider the case where the ring gear R is held fixed, $\omega_R = 0$, and the sun gear S is rotating at $\omega_S = 5$ rad/s. Determine the angular velocity of each of the planet gears P and shaft A.

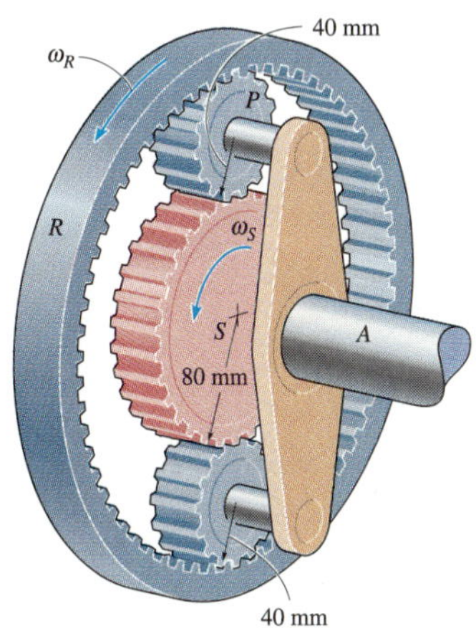

Prob. 16–63

***16-64.** The planetary gear system is used in an automatic transmission for an automobile. By locking or releasing certain gears, it has the advantage of operating the car at different speeds. Consider the case where the ring gear R is rotating at $\omega_R = 3$ rad/s, and the sun gear S is held fixed, $\omega_S = 0$. Determine the angular velocity of each of the planet gears P and shaft A.

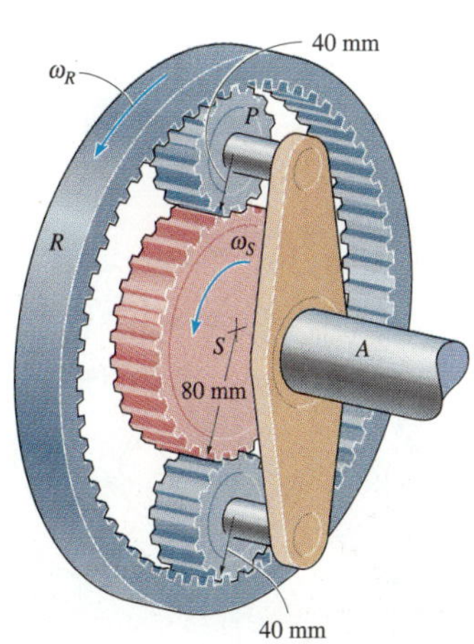

Prob. 16–64

16-65. If bar AB has an angular velocity $\omega_{AB} = 6$ rad/s, determine the velocity of the slider block C at the instant $\theta = 45°$ and $\phi = 30°$. Also, sketch the location of bar BC when $\theta = 30°, 45°$, and $60°$ to show its general plane motion.

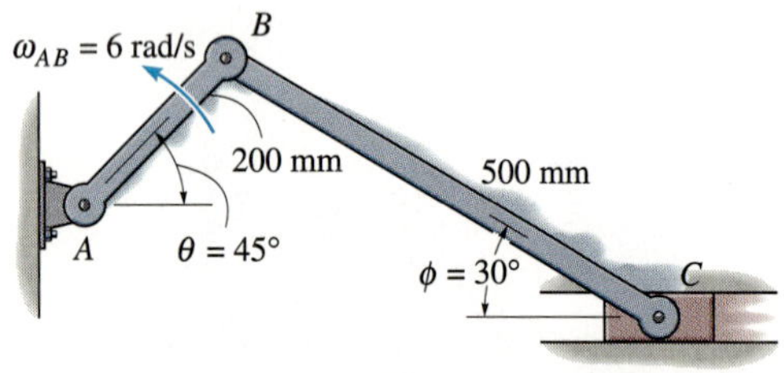

Prob. 16–65

16-66. The bicycle has a velocity $v = 4$ ft/s, and at the same instant the rear wheel has a clockwise angular velocity $\omega = 3$ rad/s, which causes it to slip at its contact point A. Determine the velocity of point A.

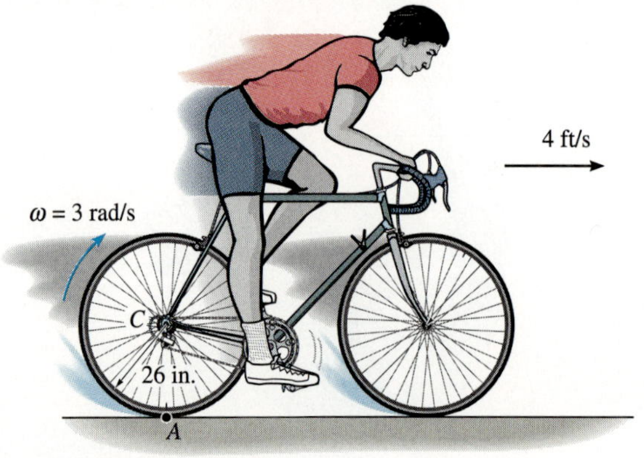

Prob. 16–66

16-67. If the angular velocity of link AB is $\omega_{AB} = 3$ rad/s, determine the velocity of the block at C and the angular velocity of the connecting link CB at the instant $\theta = 45°$ and $\phi = 30°$. Also, sketch the location of link BC when $\theta = 30°$, $45°$, and $60°$ to show its general plane motion.

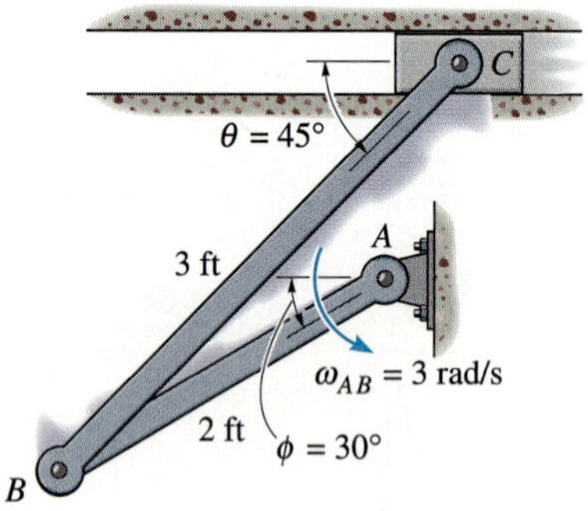

Prob. 16–67

***16-68.** If bar AB has an angular velocity $\omega_{AB} = 4$ rad/s, determine the velocity of the slider block C at the instant shown.

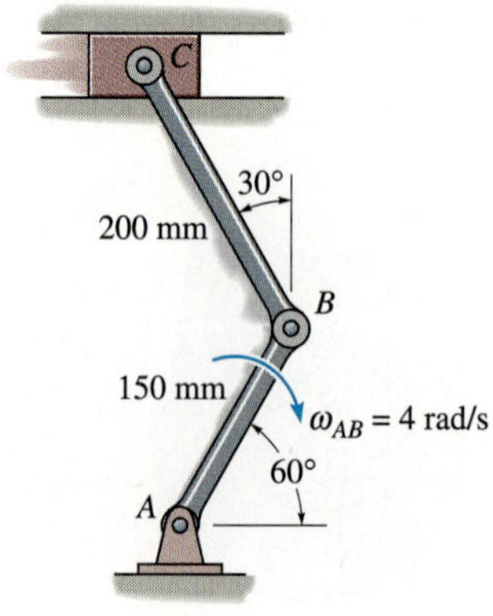

Prob. 16–68

16-69. At the instant shown, the truck is traveling to the right at 3 m/s, while the pipe is rolling counterclockwise at $\omega = 8$ rad/s without slipping at B. Determine the velocity of the pipe's center G.

16-70. At the instant shown, the truck is traveling to the right at 8 m/s. If the pipe does not slip at B, determine its angular velocity if its mass center G appears to an observer on the ground to remain stationary.

Probs. 16–69/70

16-71. The pinion gear A rolls on the fixed gear rack B with an angular velocity $\omega = 4$ rad/s. Determine the velocity of the gear rack C.

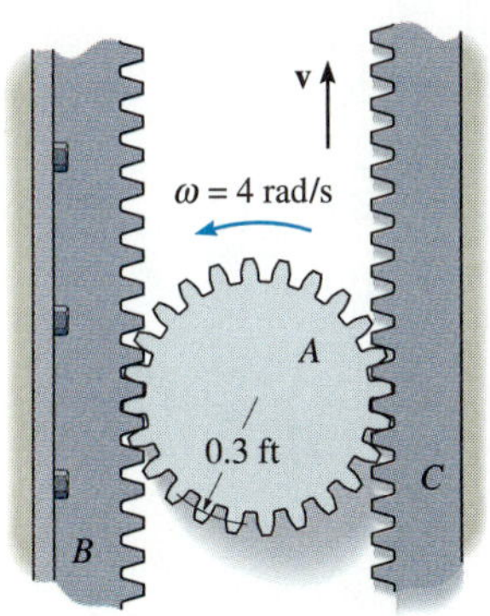

Prob. 16–71

***16-72.** When the crank on the Chinese windlass is turning, the rope on shaft A unwinds while that on shaft B winds up. Determine the speed at which the block D lowers if the crank is turning with an angular velocity $\omega = 4$ rad/s. What is the angular velocity of the pulley at C? The rope segments on each side of the pulley are both parallel and vertical, and the rope does not slip on the pulley.

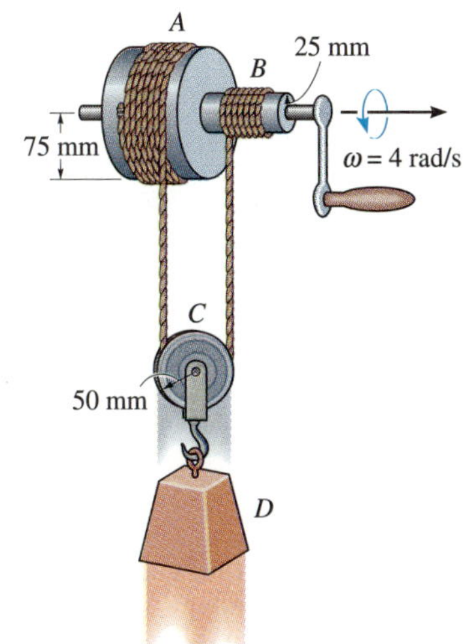

Prob. 16–72

16-73. The cylinder B rolls on the *fixed cylinder A* without slipping. If the connected bar CD is rotating with an angular velocity of $\omega_{CD} = 5$ rad/s, determine the angular velocity of cylinder B.

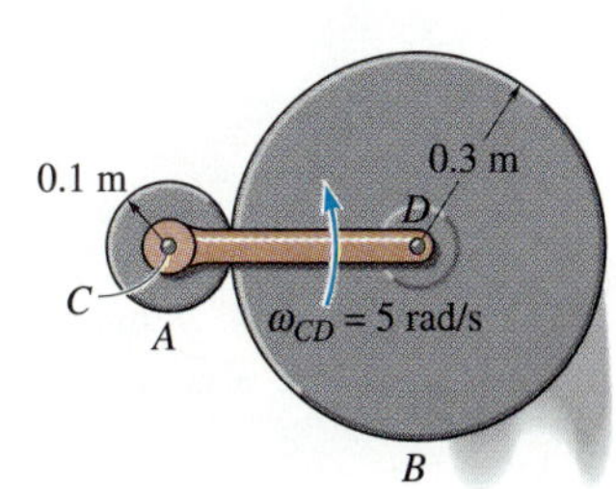

Prob. 16–73

16-74. The slider mechanism is used to increase the stroke of travel of one slider with respect to that of another. As shown, when the slider A is moving forward, the attached pinion F rolls on the *fixed* rack D, forcing slider C to move forward. This in turn causes the attached pinion G to roll on the *fixed* rack E, thereby moving slider B. If A has a velocity of $v_A = 4$ ft/s at the instant shown, determine the velocity of B. $r = 0.2$ ft.

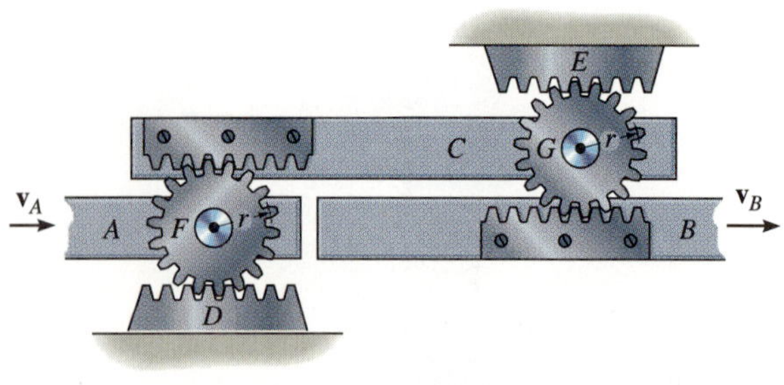

Prob. 16–74

16-75. The epicyclic gear train consists of the sun gear A which is in mesh with the planet gear B. This gear has an inner hub C which is fixed to B and in mesh with the fixed ring gear R. If the connecting link DE attached to B and C is rotating at $\omega_{DE} = 18$ rad/s, determine the angular velocities of the planet and sun gears.

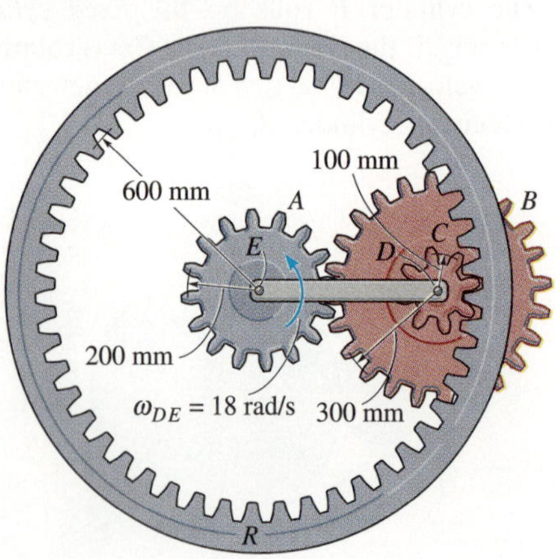

Prob. 16–75

***16-76.** If link AB has an angular velocity of $\omega_{AB} = 4$ rad/s at the instant shown, determine the velocity of the slider block E at this instant. Also, identify the type of motion of each of the four links.

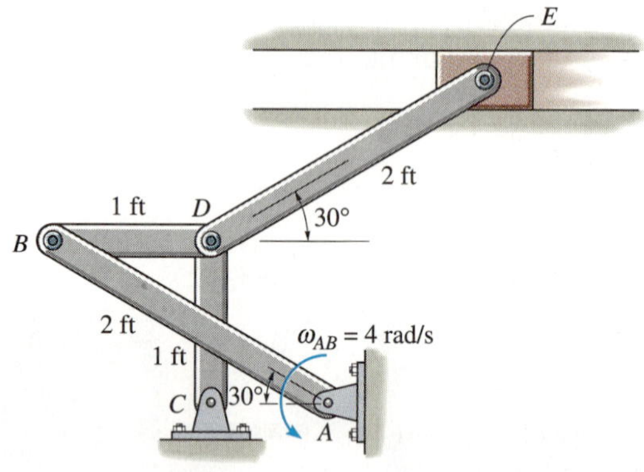

Prob. 16–76

16-77. The gauge is used to indicate the safe load acting at the end of the boom, B, when it is in any angular position. It consists of a fixed dial plate D and an indicator arm ACE which is pinned to the plate at C and to a short link EF. If the boom is pin-connected to the trunk frame at G and is rotating downward at $\omega_B = 4$ rad/s, determine the velocity of the dial pointer A at the instant shown, i.e., when EF and AC are in the vertical position.

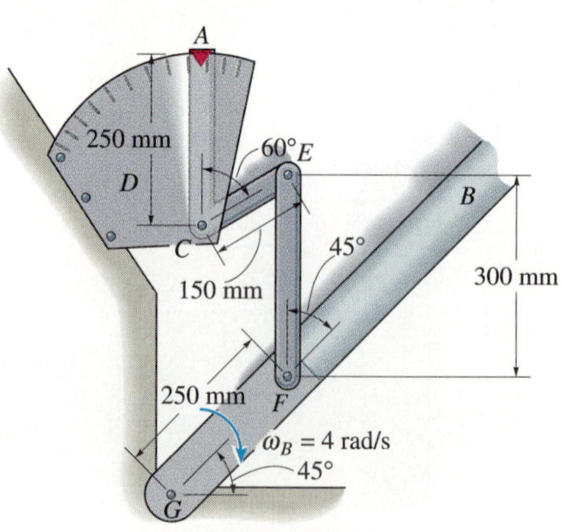

Prob. 16–77

16.6 Instantaneous Center of Zero Velocity

The velocity of any point B located on a rigid body can be obtained in a very direct way if one chooses the base point A to be a point that has *zero velocity* at the instant considered. In this case, $\mathbf{v}_A = \mathbf{0}$, and therefore the velocity equation, $\mathbf{v}_B = \mathbf{v}_A + \boldsymbol{\omega} \times \mathbf{r}_{B/A}$, becomes $\mathbf{v}_B = \boldsymbol{\omega} \times \mathbf{r}_{B/A}$. For a body having general plane motion, point A so chosen is called the *instantaneous center of zero velocity (IC),* and it lies on the *instantaneous axis of zero velocity.* This axis is always perpendicular to the plane of motion, and the intersection of the axis with this plane defines the location of the IC. Since point A is coincident with the IC, then $\mathbf{v}_B = \boldsymbol{\omega} \times \mathbf{r}_{B/IC}$ and so point B moves momentarily about the IC in a *circular path;* in other words, the body appears to rotate about the instantaneous axis. The *magnitude* of $\mathbf{v}_B$ is simply $v_B = \omega r_{B/IC}$, where ω is the angular velocity of the body. Due to the circular motion, the *direction* of $\mathbf{v}_B$ must always be *perpendicular* to $\mathbf{r}_{B/IC}$.

For example, consider the wheel in Fig. 16–17*a*. If it rolls *without slipping,* then the point of *contact* with the ground has *zero velocity.* Hence this point represents the IC for the wheel, Fig. 16–17*b*. If it is imagined that the wheel is momentarily pinned at this point, the velocities of points A, B, O, and so on, can be found using $v = \omega r$. Here the radial distances $r_{A/IC}$, $r_{B/IC}$, and $r_{O/IC}$, shown in Fig. 16–17*b*, must be determined from the geometry of the wheel.

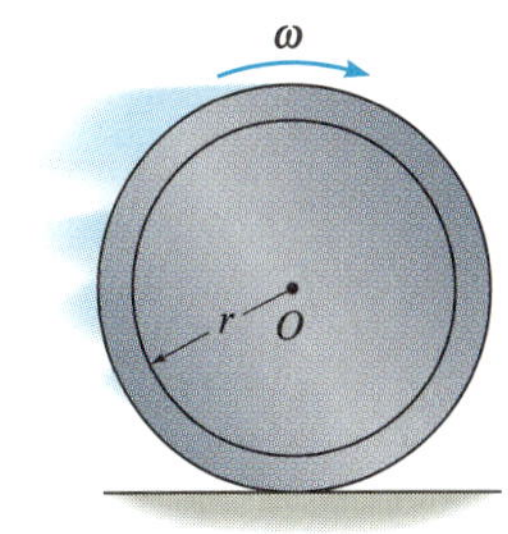

(a)

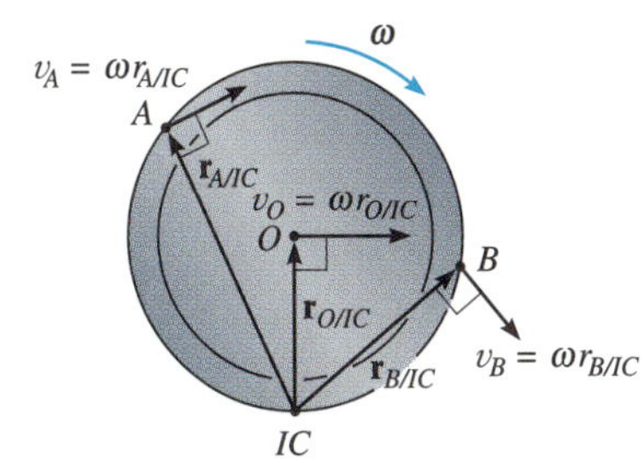

(b)

Fig. 16–17

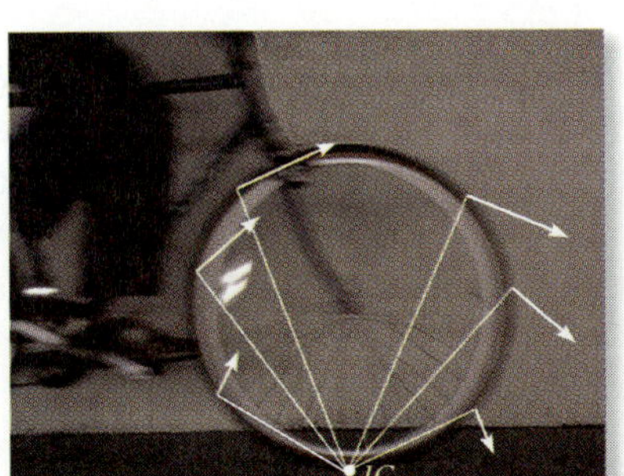

The IC for this bicycle wheel is at the ground. There the spokes are somewhat visible, whereas at the top of the wheel they become blurred. Note also how points on the side portions of the wheel move as shown by their velocities.

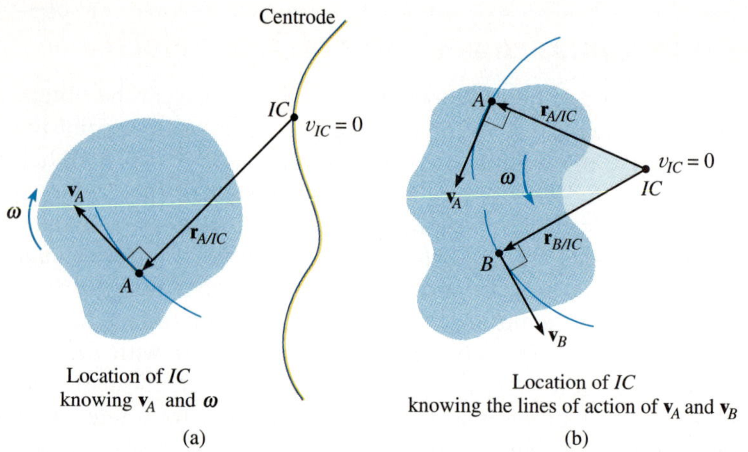

Fig. 16–18

Location of the IC. To locate the *IC* we can use the fact that the *velocity* of a point on the body is *always perpendicular* to the *relative-position vector* extending from the *IC* to the point. Several possibilities exist:

- *Given the velocity* $\mathbf{v}_A$ *of a point A on the body, and the angular velocity* $\boldsymbol{\omega}$ *of the body,* Fig. 16–18*a*. In this case, the *IC* is located along the line drawn perpendicular to $\mathbf{v}_A$ at *A*, such that the distance from *A* to the *IC* is $r_{A/IC} = v_A/\omega$. Note that the *IC* lies up and to the right of *A* since $\mathbf{v}_A$ must cause a clockwise angular velocity $\boldsymbol{\omega}$ about the *IC*.

- *Given the lines of action of two nonparallel velocities* $\mathbf{v}_A$ and $\mathbf{v}_B$, Fig. 16–18*b*. Construct at points *A* and *B* line segments that are perpendicular to $\mathbf{v}_A$ and $\mathbf{v}_B$. Extending these perpendiculars to their *point of intersection* as shown locates the *IC* at the instant considered.

- *Given the magnitude and direction of two parallel velocities* $\mathbf{v}_A$ *and* $\mathbf{v}_B$. Here the location of the *IC* is determined by proportional triangles. Examples are shown in Fig. 16–18*c* and *d*. In both cases $r_{A/IC} = v_A/\omega$ and $r_{B/IC} = v_B/\omega$. If *d* is a known distance between points *A* and *B*, then in Fig. 16–18*c*, $r_{A/IC} + r_{B/IC} = d$ and in Fig. 16–18*d*, $r_{B/IC} - r_{A/IC} = d$. As a special case, note that if the body is *translating*, $\mathbf{v}_A = \mathbf{v}_B$ then the *IC* would be located at infinity, in which case $r_{A/IC} = r_{B/IC} \rightarrow \infty$. This being the case, $\omega = (v_A/r_{A/IC}) = (v_B/r_{B/IC}) \rightarrow 0$, as expected.

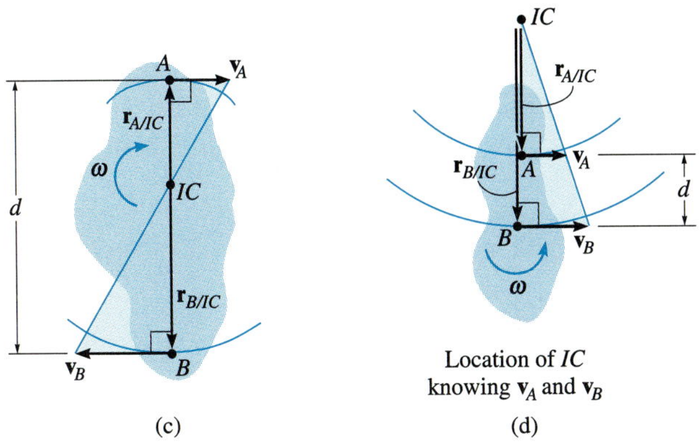

Fig. 16–18

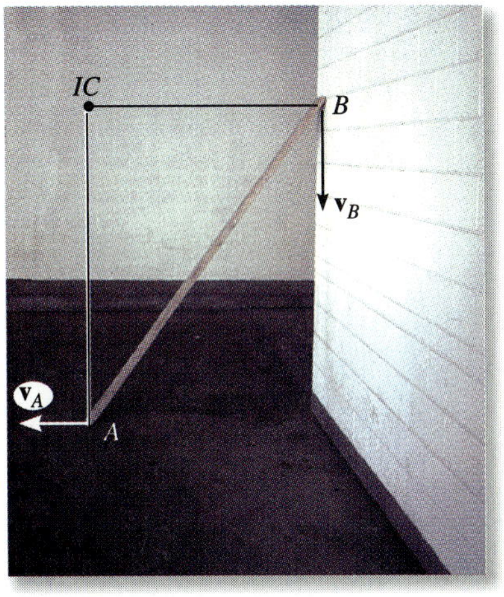

As the board slides downward to the left it is subjected to general plane motion. Since the directions of the velocities of its ends A and B are known, the IC is located as shown. At this instant the board will momentarily rotate about this point. Draw the board in several other positions, establish the IC for each case and sketch the centrode.

Realize that the point chosen as the instantaneous center of zero velocity for the body *can only be used for an instant of time* since the body changes its position from one instant to the next. The locus of points which define the location of the IC during the body's motion is called a *centrode,* Fig. 16–18*a*, and so each point on the centrode acts as the IC for the body only for an instant.

Although the IC may be conveniently used to determine the velocity of any point in a body, it generally *does not have zero acceleration* and therefore it *should not* be used for finding the accelerations of points in a body.

Procedure for Analysis

The velocity of a point on a body which is subjected to general plane motion can be determined with reference to its instantaneous center of zero velocity provided the location of the IC is first established using one of the three methods described above.

- As shown on the kinematic diagram in Fig. 16–19, the body is imagined as "extended and pinned" at the IC such that, at the instant considered, it rotates about this pin with its angular velocity $\boldsymbol{\omega}$.
- The *magnitude* of velocity for each of the arbitrary points A, B, and C on the body can be determined by using the equation $v = \omega r$, where r is the radial line drawn from the IC to each point.
- The line of action of each velocity vector $\mathbf{v}$ is *perpendicular* to its associated radial line $\mathbf{r}$, and the velocity has a *sense of direction* which tends to move the point in a manner consistent with the angular rotation $\boldsymbol{\omega}$ of the radial line, Fig. 16–19.

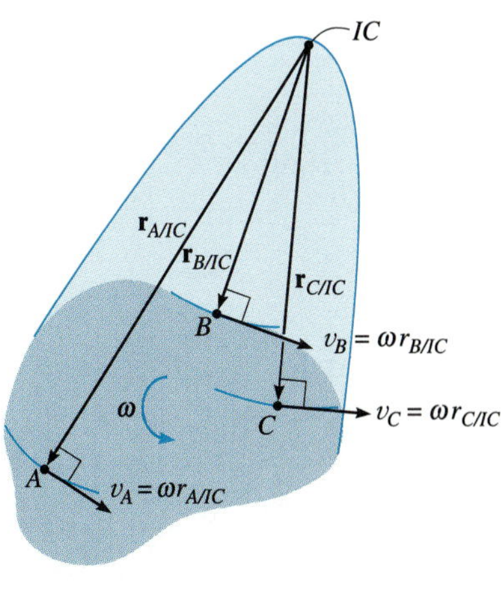

Fig. 16–19

EXAMPLE 16–10

Show how to determine the location of the instantaneous center of zero velocity for (*a*) member *BC* shown in Fig. 16–20*a*; and (*b*) the link *CB* shown in Fig. 16–20*b*.

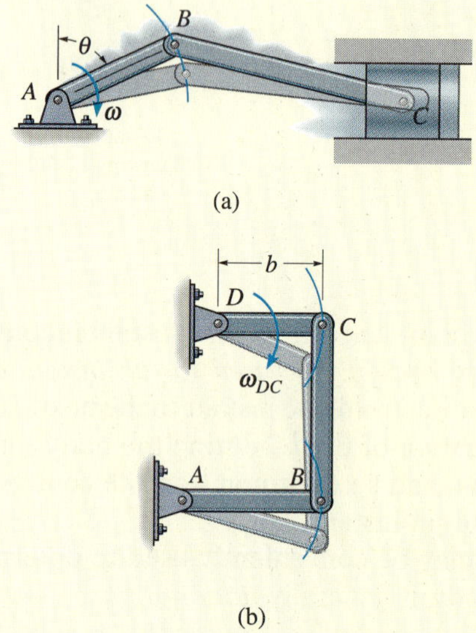

Fig. 16–20

Solution

Part (a). As shown in Fig. 16–20*a*, point *B* has a velocity $\mathbf{v}_B$, which is caused by the clockwise rotation of link *AB*. Point *B* moves in a circular path such that $\mathbf{v}_B$ is perpendicular to *AB*, and so it acts at an angle θ from the horizontal as shown in Fig. 16–20*c*. The motion of point *B* causes the piston to move forward *horizontally* with a velocity $\mathbf{v}_C$. When lines are drawn perpendicular to $\mathbf{v}_B$ and $\mathbf{v}_C$, Fig. 16–20*c*, they intersect at the *IC*.

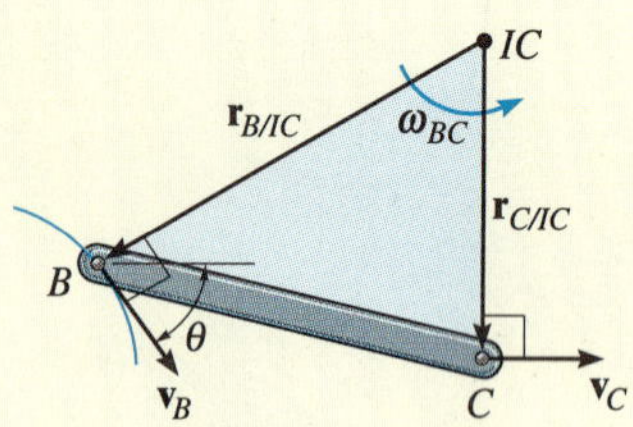

(c)

Part (b). Points *B* and *C* follow circular paths of motion since rods *AB* and *DC* are each subjected to rotation about a fixed axis, Fig. 16–20*b*. Since the velocity is always tangent to the path, at the instant considered, $\mathbf{v}_C$ on rod *DC* and $\mathbf{v}_B$ on rod *AB* are both directed vertically downward, along the axis of link *CB*, Fig. 16–20*d*. Radial lines drawn perpendicular to these two velocities form parallel lines which intersect at "infinity;" i.e., $r_{C/IC} \rightarrow \infty$ and $r_{B/IC} \rightarrow \infty$. Thus, $\omega_{CB} = (v_C/r_{C/IC}) = \rightarrow 0$. As a result, rod *CB* momentarily *translates.* An instant later, however, *CB* will move to a tilted position, causing the instantaneous center to move to some finite location.

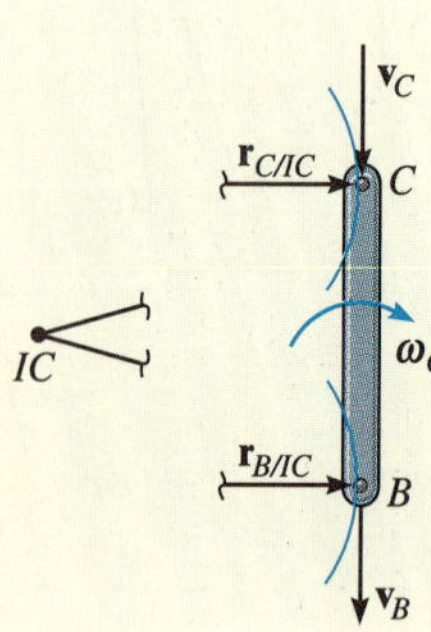

(d)

EXAMPLE 16–11

Block D shown in Fig. 16–21a moves with a speed of 3 m/s. Determine the angular velocities of links BD and AB, at the instant shown.

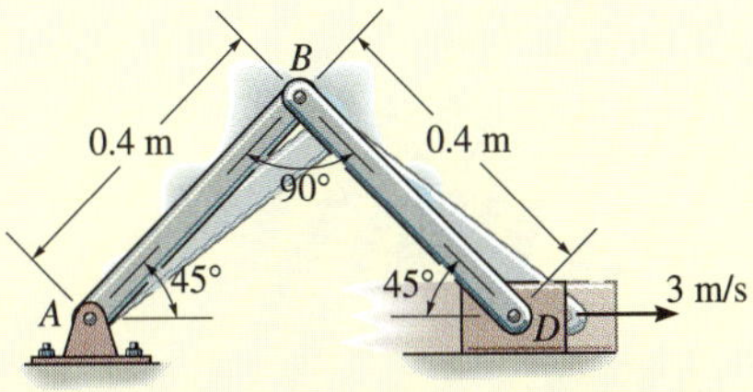

(a)

Fig. 16–21

Solution

As D moves to the right, it causes arm AB to rotate clockwise about point A. Hence, $\mathbf{v}_B$ is directed perpendicular to AB. The instantaneous center of zero velocity for BD is located at the intersection of the line segments drawn perpendicular to $\mathbf{v}_B$ and $\mathbf{v}_D$, Fig. 16–21b. From the geometry,

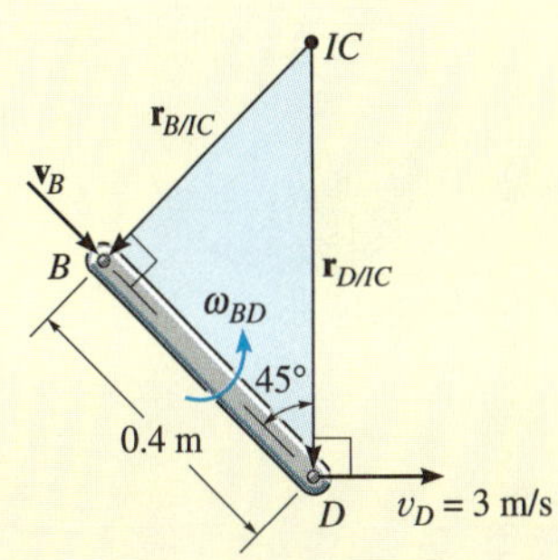

(b)

$$r_{B/IC} = 0.4 \tan 45° \text{ m} = 0.4 \text{ m}$$

$$r_{D/IC} = \frac{0.4 \text{ m}}{\cos 45°} = 0.566 \text{ m}$$

Since the magnitude of $\mathbf{v}_D$ is known, the angular velocity of link BD is

$$\omega_{BD} = \frac{v_D}{r_{D/IC}} = \frac{3 \text{ m/s}}{0.566 \text{ m}} = 5.30 \text{ rad/s} \circlearrowleft \qquad \textit{Ans.}$$

The velocity of B is therefore

$$v_B = \omega_{BD}(r_{B/IC}) = 5.30 \text{ rad/s}(0.4 \text{ m}) = 2.12 \text{ m/s} \searrow^{45°}$$

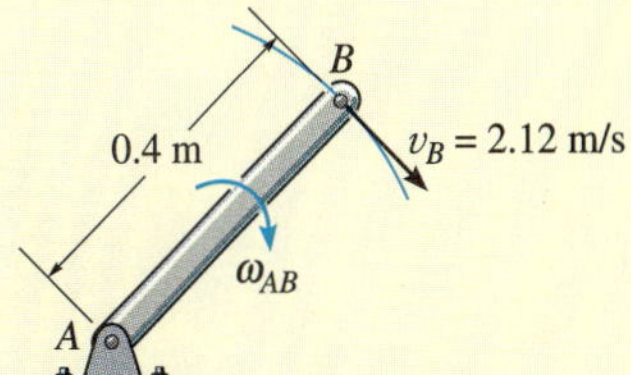

(c)

From Fig. 16–21c, the angular velocity of AB is

$$\omega_{AB} = \frac{v_B}{r_{B/A}} = \frac{2.12 \text{ m/s}}{0.4 \text{ m}} = 5.30 \text{ rad/s} \circlearrowright \qquad \textit{Ans.}$$

EXAMPLE 16–12

The cylinder shown in Fig. 16–22*a* rolls without slipping between the two moving plates *E* and *D*. Determine the angular velocity of the cylinder and the velocity of its center *C* at the instant shown.

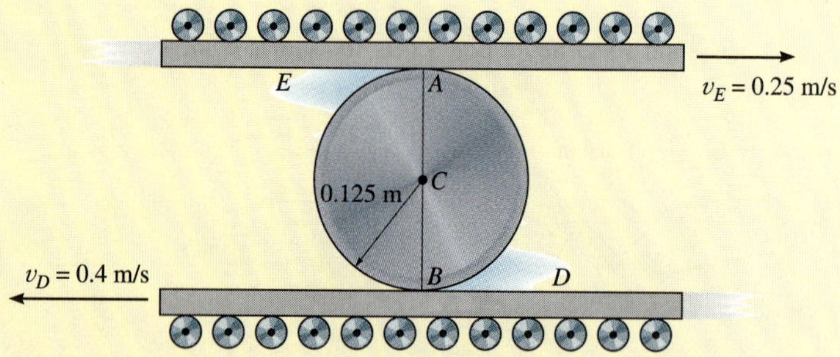

(a)

Fig. 16–22

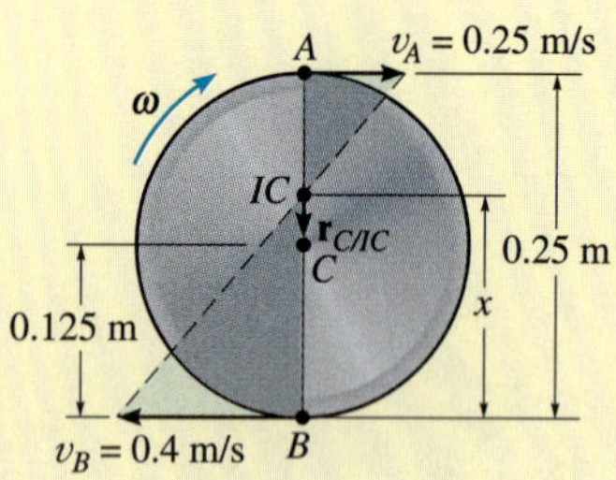

(b)

Solution

Since no slipping occurs, the contact points *A* and *B* on the cylinder have the same velocities as the plates *E* and *D*, respectively. Furthermore, the velocities $\mathbf{v}_A$ and $\mathbf{v}_B$ are *parallel*, so that by the proportionality of right triangles the *IC* is located at a point on line *AB*, Fig. 16–22*b*. Assuming this point to be a distance *x* from *B*, we have

$$v_B = \omega x; \qquad 0.4\text{ m/s} = \omega x$$

$$v_A = \omega(0.25\text{ m} - x); \qquad 0.25\text{ m/s} = \omega(0.25\text{ m} - x)$$

Dividing one equation into the other eliminates ω and yields

$$0.4(0.25 - x) = 0.25x$$

$$x = \frac{0.1}{0.65} = 0.154\text{ m}$$

Hence, the angular velocity of the cylinder is

$$\omega = \frac{v_B}{x} = \frac{0.4\text{ m/s}}{0.154\text{ m}} = 2.60\text{ rad/s} \circlearrowright \qquad \textit{Ans.}$$

The velocity of point *C* is therefore

$$v_C = \omega r_{C/IC} = 2.60\text{ rad/s}(0.154\text{ m} - 0.125\text{ m})$$

$$= 0.0750\text{ m/s} \leftarrow \qquad \textit{Ans.}$$

Problems

16-78. Solve Prob. 16-51 using the method of instantaneous center of zero velocity.

16-79. Solve Prob. 16-54 using the method of instantaneous center of zero velocity.

***16-80.** Solve Prob. 16-56 using the method of instantaneous center of zero velocity.

16-81. Solve Prob. 16-58 using the method of instantaneous center of zero velocity.

16-82. Solve Prob. 16-60 using the method of instantaneous center of zero velocity.

16-83. Solve Prob. 16-63 using the method of instantaneous center of zero velocity.

***16-84.** Solve Prob. 16-68 using the method of instantaneous center of zero velocity.

16-85. Solve Prob. 16-66 using the method of instantaneous center of zero velocity.

16-86. The instantaneous center of zero velocity for the body is located at point *IC* (0.5 m, 2 m). If the body has an angular velocity of 4 rad/s, as shown, determine the velocity of *B* with respect to *A*.

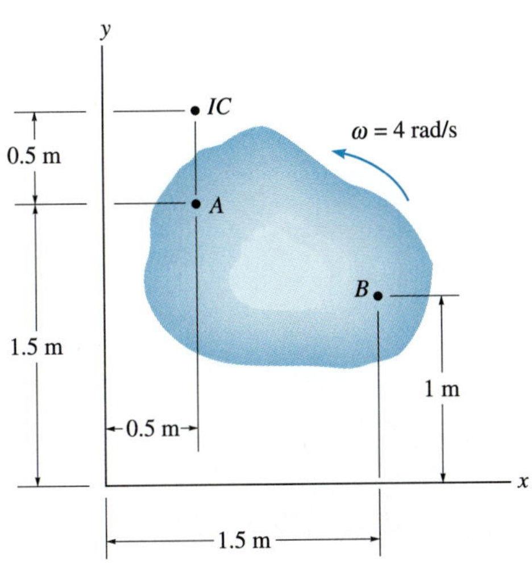

Prob. 16–86

16-87. The slider block *C* is moving 4 ft/s up the incline. Determine the angular velocities of links *AB* and *BC* and the velocity of point *B* at the instant shown.

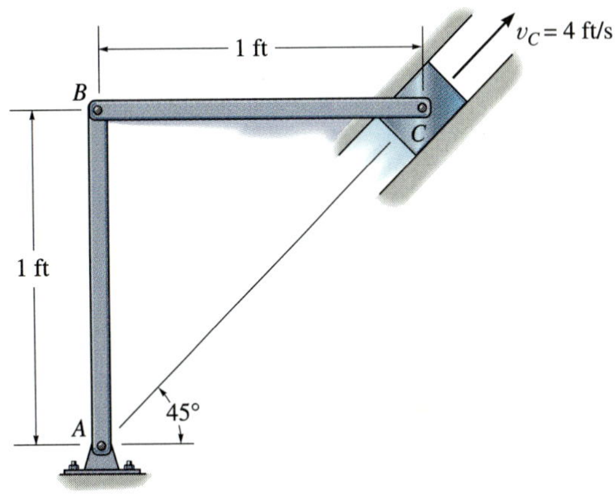

Prob. 16–87

***16-88.** As the cord unravels from the wheel's inner hub, the wheel is rotating at $\omega = 2$ rad/s at the instant shown. Determine the velocities of points *A* and *B*.

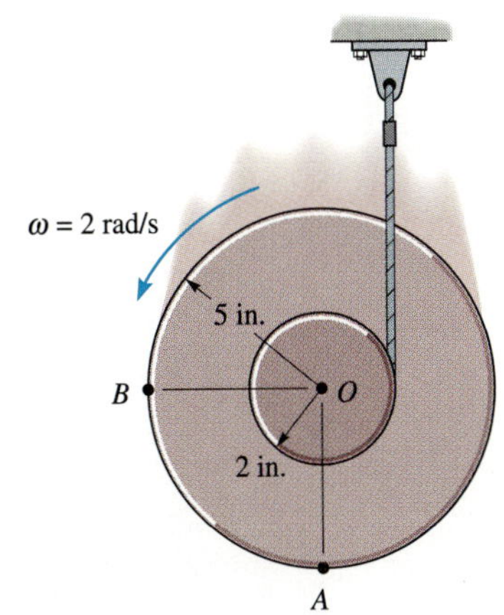

Prob. 16–88

16-89. The wheel rolls on its hub without slipping on the horizontal surface. If the velocity of the center of the wheel is $v_C = 2$ ft/s to the right, determine the velocities of points A and B at the instant shown.

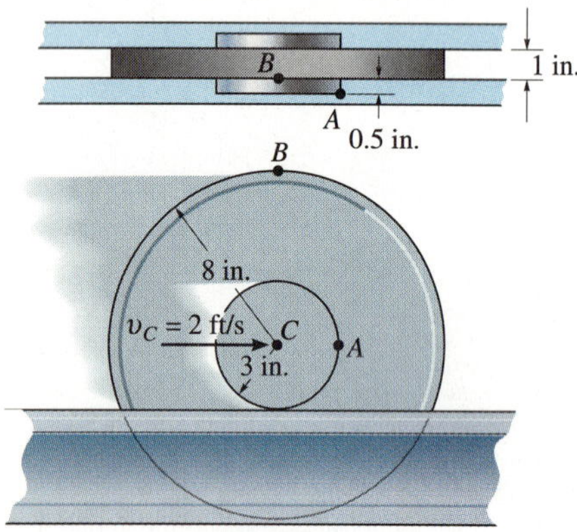

Prob. 16–89

16-90. If link CD has an angular velocity of $\omega_{CD} = 6$ rad/s, determine the velocity of point E on link BC and the angular velocity of link AB at the instant shown.

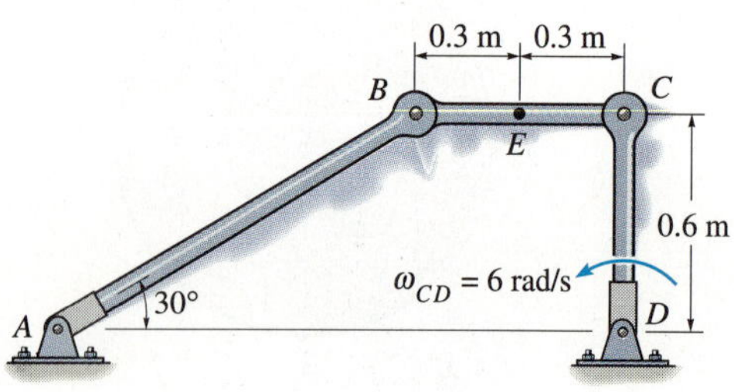

Prob. 16–90

16-91. The disk of radius r is confined to roll without slipping at A and B. If the plates have the velocities shown, determine the angular velocity of the disk.

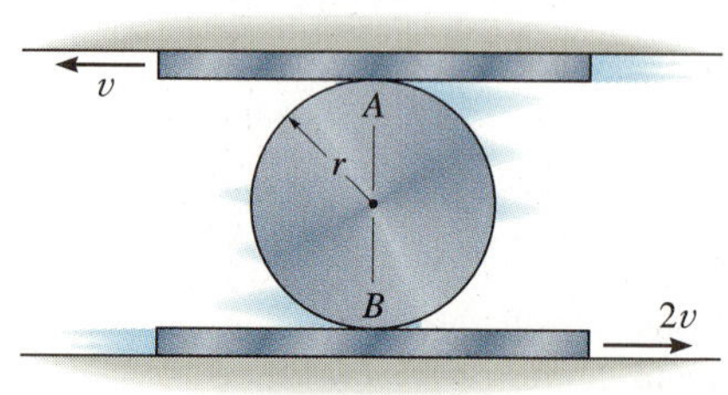

Prob. 16–91

***16-92.** Show that if the rim of the wheel and its hub maintain contact with the three tracks as the wheel rolls, it is necessary that slipping occurs at the hub A if no slipping occurs at B. Under these conditions, what is the speed at A if the wheel has an angular velocity $\boldsymbol{\omega}$?

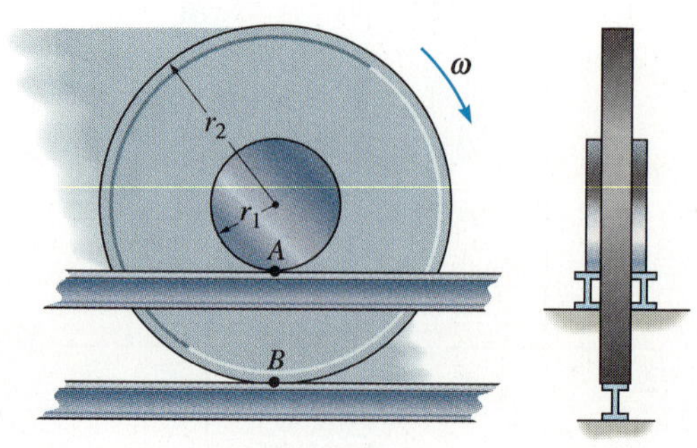

Prob. 16–92

16-93. In an automobile transmission the planet pinions A and B rotate on shafts that are mounted on the planet-pinion carrier CD. As shown, CD is attached to a shaft at E which is aligned with the center of the *fixed* sun gear S. This shaft is not attached to the sun gear. If CD is rotating at $\omega_{CD} = 8$ rad/s, determine the angular velocity of the ring gear R.

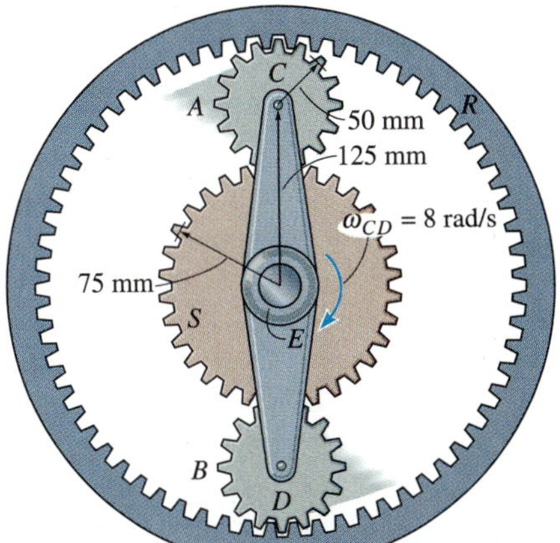

Prob. 16–93

16-94. If the hub gear H and ring gear R have angular velocities $\omega_H = 5$ rad/s and $\omega_R = 20$ rad/s, respectively, determine the angular velocity ω_S of the spur gear S and the angular velocity of its attached arm OA.

16-95. If the hub gear H has an angular velocity $\omega_H = 5$ rad/s, determine the angular velocity of the ring gear R so that the arm OA attached to the spur gear S remains stationary ($\omega_{OA} = 0$). What is the angular velocity of the spur gear?

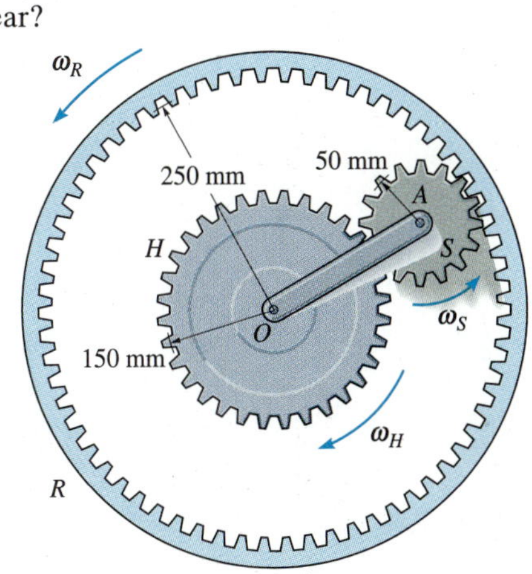

Probs. 16–94/95

***16-96.** The planet gear A is pin connected to the end of the link BC. If the link rotates about the fixed point B at 4 rad/s, determine the angular velocity of the ring gear R. The sun gear D is fixed from rotating.

16-97. Solve Prob. 16-96 if the sun gear D is rotating clockwise at $\omega_D = 5$ rad/s while link BC rotates counterclockwise at $\omega_{BC} = 4$ rad/s.

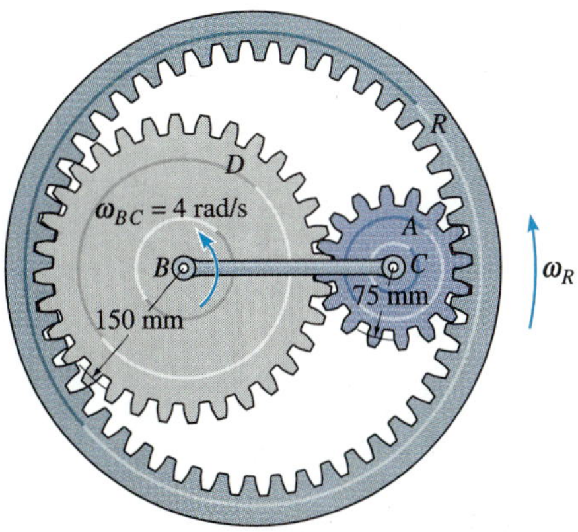

Probs. 16–96/97

16-98. The mechanism used in a marine engine consists of a single crank AB and two connecting rods BC and BD. Determine the velocity of the piston at C the instant the crank is in the position shown and has an angular velocity of 5 rad/s.

16-99. The mechanism used in a marine engine consists of a single crank AB and two connecting rods BC and BD. Determine the velocity of the piston at D the instant the crank is in the position shown and has an angular velocity of 5 rad/s.

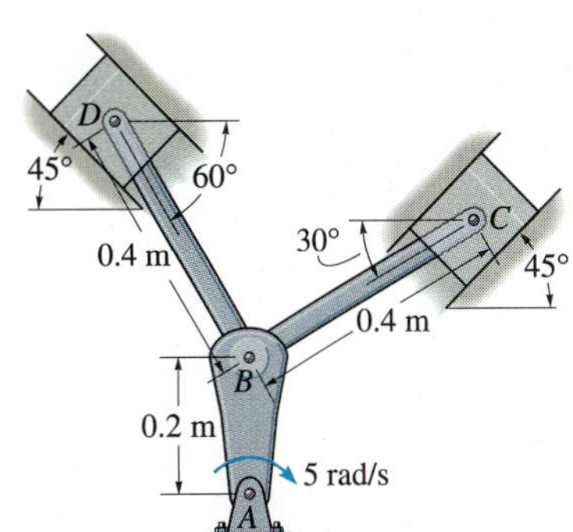

Probs. 16–98/99

***16-100.** The square plate is confined within the slots at A and B. When $\theta = 30°$, point A is moving at $v_A = 8$ m/s. Determine the velocity of point C at this instant.

16-101. The square plate is confined within the slots at A and B. When $\theta = 30°$, point A is moving at $v_A = 8$ m/s. Determine the velocity of point D at this instant.

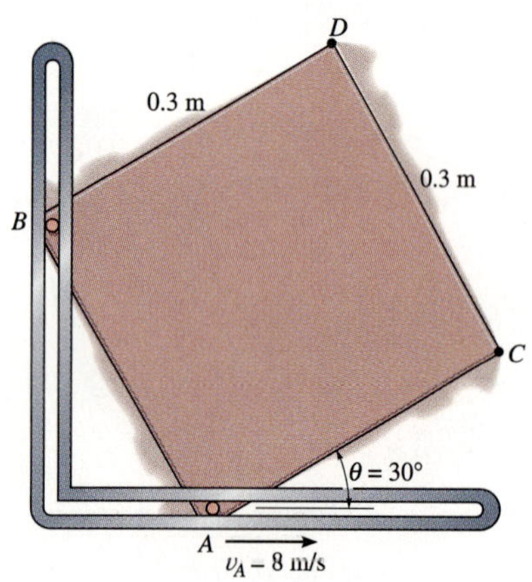

Probs. 16–100/101

16-102. If the slider block A is moving to the right at $v_A = 8$ ft/s, determine the velocities of blocks B and C at the instant shown.

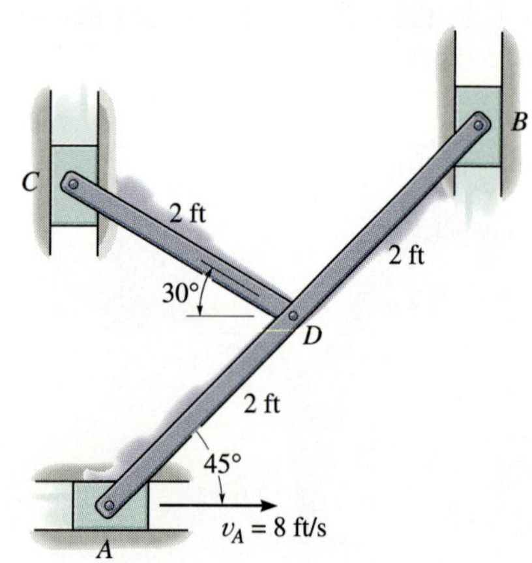

Prob. 16–102

16-103. The crankshaft AB rotates at $\omega_{AB} = 50$ rad/s about the fixed axis through point A, and the disk at C is held fixed in its support at E. Determine the angular velocity of rod CD at the instant shown.

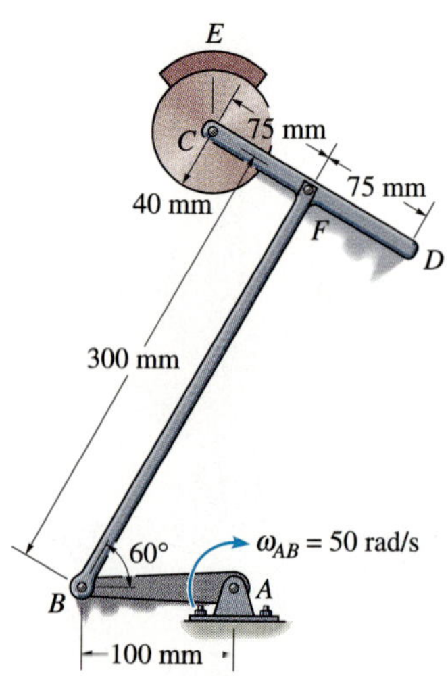

Prob. 16–103

***16-104.** The mechanism shown is used in a riveting machine. It consists of a driving piston A, three members, and a riveter which is attached to the slider block D. Determine the velocity of D at the instant shown, when the piston at A is traveling at $v_A = 30$ m/s.

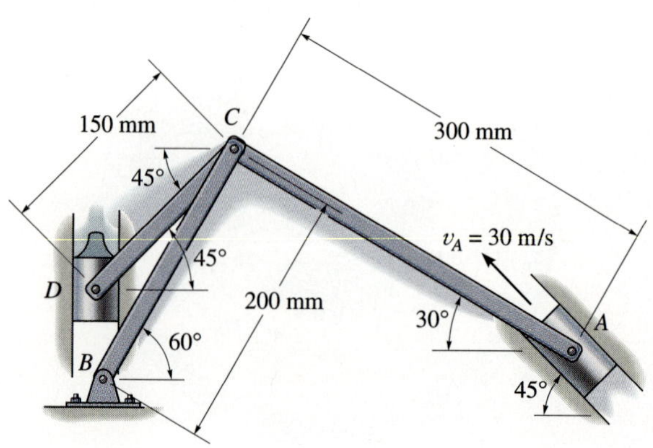

Prob. 16–104

16.7 Relative-Motion Analysis: Acceleration

An equation that relates the accelerations of two points on a rigid body subjected to general plane motion may be determined by differentiating the velocity equation $\mathbf{v}_B = \mathbf{v}_A + \mathbf{v}_{B/A}$ with respect to time. This yields

$$\frac{d\mathbf{v}_B}{dt} = \frac{d\mathbf{v}_A}{dt} + \frac{d\mathbf{v}_{B/A}}{dt}$$

The terms $d\mathbf{v}_B/dt = \mathbf{a}_B$ and $d\mathbf{v}_A/dt = \mathbf{a}_A$ are measured from a set of *fixed x, y axes* and represent the *absolute accelerations* of points B and A. The last term represents the acceleration of B with respect to A as measured by an observer fixed to translating x', y' axes which have their origin at the base point A. In Sec. 16.5 it was shown that to this observer point B appears to move along a *circular arc* that has a radius of curvature $r_{B/A}$. Consequently, $\mathbf{a}_{B/A}$ can be expressed in terms of its tangential and normal components of motion; i.e., $\mathbf{a}_{B/A} = (\mathbf{a}_{B/A})_t + (\mathbf{a}_{B/A})_n$, where $(a_{B/A})_t = \alpha r_{B/A}$ and $(a_{B/A})_n = \omega^2 r_{B/A}$. Hence, the relative-acceleration equation can be written in the form

$$\mathbf{a}_B = \mathbf{a}_A + (\mathbf{a}_{B/A})_t + (\mathbf{a}_{B/A})_n \qquad (16\text{–}17)$$

where

$\mathbf{a}_B$ = acceleration of point B

$\mathbf{a}_A$ = acceleration of point A

$(\mathbf{a}_{B/A})_t$ = relative tangential acceleration component of "B with respect to A." The *magnitude* is $(a_{B/A})_t = \alpha r_{B/A}$, and the *direction* is perpendicular to $\mathbf{r}_{B/A}$.

$(\mathbf{a}_{B/A})_n$ = relative normal acceleration component of "B with respect to A." The *magnitude* is $(a_{B/A})_n = \omega^2 r_{B/A}$, and the *direction* is always from B towards A.

Each of the four terms in Eq. 16–17 is represented graphically on the *kinematic diagrams* shown in Fig. 16–23. Here it is seen that at a given instant the acceleration of B, Fig. 16–23*a*, is determined by considering the body to translate with an acceleration $\mathbf{a}_A$, Fig. 16–23*b*, and simultaneously rotate about the base point A with an instantaneous angular velocity $\boldsymbol{\omega}$ and angular acceleration $\boldsymbol{\alpha}$, Fig. 16–23*c*. Vector addition of these two effects, applied to B, yields $\mathbf{a}_B$, as shown in Fig. 16–23*d*. It should be noted from Fig. 16–23*a* that since points A and B move along *curved paths*, the accelerations of these points will have *both tangential and normal components.* (Recall that the acceleration of a point is *tangent to the path only* when the path is *rectilinear* or when it is an inflection point on a curve.)

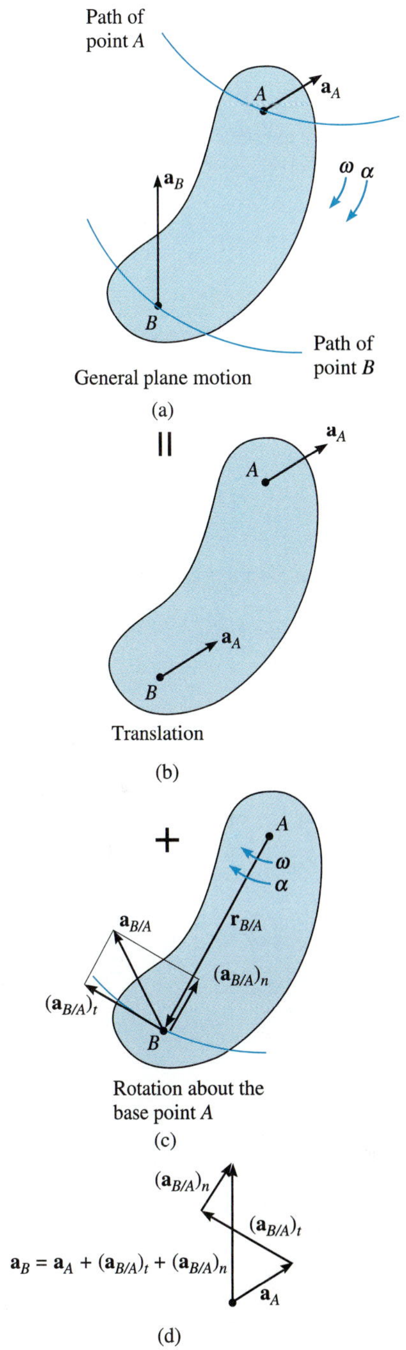

Fig. 16–23

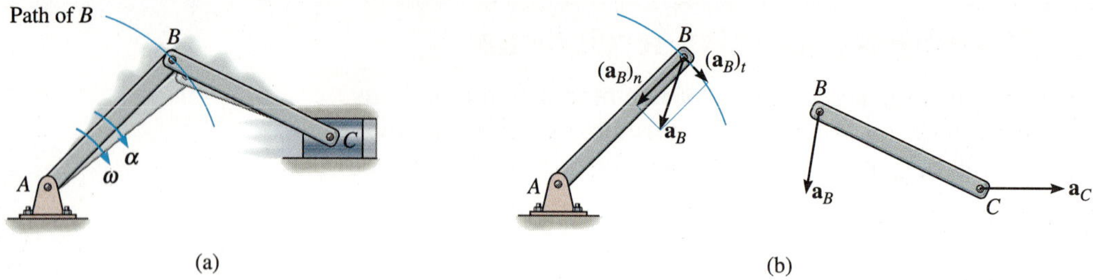

Fig. 16–24

Since the relative-acceleration components represent the effect of *circular motion* observed from translating axes having their origin at the base point A, these terms can be expressed as $(\mathbf{a}_{B/A})_t = \boldsymbol{\alpha} \times \mathbf{r}_{B/A}$ and $(\mathbf{a}_{B/A})_n = -\omega^2 \mathbf{r}_{B/A}$, Eq. 16–14. Hence, Eq. 16–17 becomes

$$\mathbf{a}_B = \mathbf{a}_A + \boldsymbol{\alpha} \times \mathbf{r}_{B/A} - \omega^2 \mathbf{r}_{B/A} \tag{16–18}$$

where

$\mathbf{a}_B$ = acceleration of point B

$\mathbf{a}_A$ = acceleration of the base point A

$\boldsymbol{\alpha}$ = angular acceleration of the body

$\boldsymbol{\omega}$ = angular velocity of the body

$\mathbf{r}_{B/A}$ = relative-position vector drawn from A to B

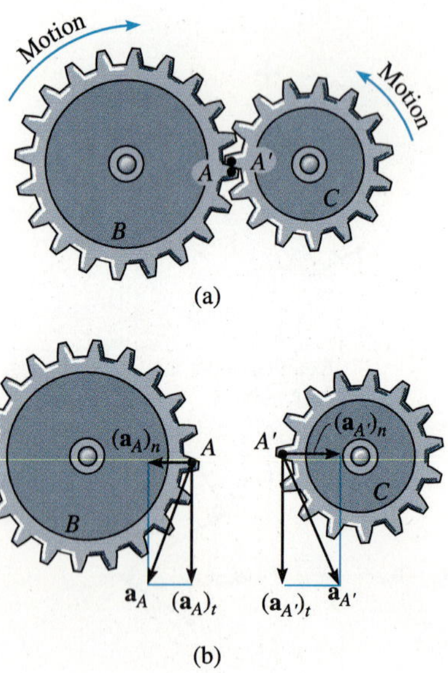

Fig. 16–25

If Eq. 16–17 or 16–18 is applied in a practical manner to study the accelerated motion of a rigid body which is pin connected to two other bodies, it should be realized that points which are *coincident at the pin* move with the *same acceleration,* since the path of motion over which they travel is the *same.* For example, point B lying on either rod AB or BC of the crank mechanism shown in Fig. 16–24*a* has the same acceleration, since the rods are pin connected at B. Here the motion of B is along a *curved path,* so that $\mathbf{a}_B$ can be expressed in terms of its tangential and normal components. At the other end of rod BC point C moves along a *straight-lined path,* which is defined by the piston. Hence, $\mathbf{a}_C$ is horizontal, Fig. 16–24*b*.

If two bodies contact one another *without slipping,* and the *points in contact* move along *different paths,* the *tangential components* of acceleration of the points will be the *same;* however, the *normal components* will *not* be the same. For example, consider the two meshed gears in Fig. 16–25*a*. Point A is located on gear B and a coincident point A' is located on gear C. Due to the rotational motion, $(\mathbf{a}_A)_t = (\mathbf{a}_{A'})_t$; however, since both points follow different curved paths, $(\mathbf{a}_A)_n \neq (\mathbf{a}_{A'})_n$ and therefore $\mathbf{a}_A \neq \mathbf{a}_{A'}$, Fig. 16–25*b*.

Procedure for Analysis

The relative acceleration equation can be applied between any two points A and B on a body either by using a Cartesian vector analysis, or by writing the x and y scalar component equations directly.

Velocity Analysis

- Determine the angular velocity $\boldsymbol{\omega}$ of the body by using a velocity analysis as discussed in Sec. 16.5 or 16.6. Also, determine the velocities $\mathbf{v}_A$ and $\mathbf{v}_B$ of points A and B *if these points move along curved paths.*

VECTOR ANALYSIS

Kinematic Diagram

- Establish the directions of the fixed x, y coordinates and draw the kinematic diagram of the body. Indicate on it $\mathbf{a}_A$, $\mathbf{a}_B$, $\boldsymbol{\omega}$, $\boldsymbol{\alpha}$, and $\mathbf{r}_{B/A}$.
- If points A and B move along *curved paths,* then their accelerations should be indicated in terms of their tangential and normal components, i.e., $\mathbf{a}_A = (\mathbf{a}_A)_t + (\mathbf{a}_A)_n$ and $\mathbf{a}_B = (\mathbf{a}_B)_t + (\mathbf{a}_B)_n$.

Acceleration Equation

- To apply $\mathbf{a}_B = \mathbf{a}_A + \boldsymbol{\alpha} \times \mathbf{r}_{B/A} - \omega^2 \mathbf{r}_{B/A}$ express the vectors in Cartesian vector form and substitute them into the equation. Evaluate the cross product and then equate the respective $\mathbf{i}$ and $\mathbf{j}$ components to obtain two scalar equations.
- If the solution yields a *negative* answer for an *unknown* magnitude, it indicates the sense of direction of the vector is opposite to that shown on the kinematic diagram.

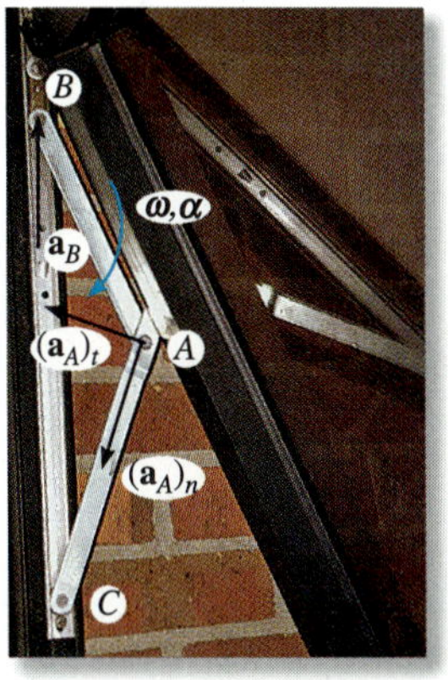

The mechanism for a window is shown in the photo. Here CA rotates about a fixed axis through C, and AB undergoes general plane motion. Since point A moves along a curved path it has two components of acceleration, whereas point B moves along a straight track and thus its direction is specified.

SCALAR ANALYSIS

Kinematic Diagram

- If the equation $\mathbf{a}_B = \mathbf{a}_A + (\mathbf{a}_{B/A})_t - (\mathbf{a}_{B/A})_n$ is applied, then the magnitudes and directions of the relative-acceleration components $(\mathbf{a}_{B/A})_t$ and $(\mathbf{a}_{B/A})_n$ must be established. To do this draw a kinematic diagram such as shown in Fig. 16–23*c*. Since the body is considered to be momentarily "pinned" at the base point A, the *magnitudes* are $(a_{B/A})_t = \alpha r_{B/A}$ and $(a_{B/A})_n = \omega^2 r_{B/A}$. Their *sense of direction* is established from the diagram such that $(\mathbf{a}_{B/A})_t$ acts perpendicular to $\mathbf{r}_{B/A}$, in accordance with the rotational motion $\boldsymbol{\alpha}$ of the body, and $(\mathbf{a}_{B/A})_n$ is directed from B towards A.*

Acceleration Equation

- Represent the vectors in $\mathbf{a}_B = \mathbf{a}_A + (\mathbf{a}_{B/A})_t + (\mathbf{a}_{B/A})_n$ graphically by showing their magnitudes and directions underneath each term. The scalar equations are determined from the x and y components of these vectors.

*Perhaps the notation $\mathbf{a}_B = \mathbf{a}_A + (\mathbf{a}_{B/A(\text{pin})})_t + (\mathbf{a}_{B/A(\text{pin})})_n$ may be helpful in recalling that A is pinned.

EXAMPLE 16–13

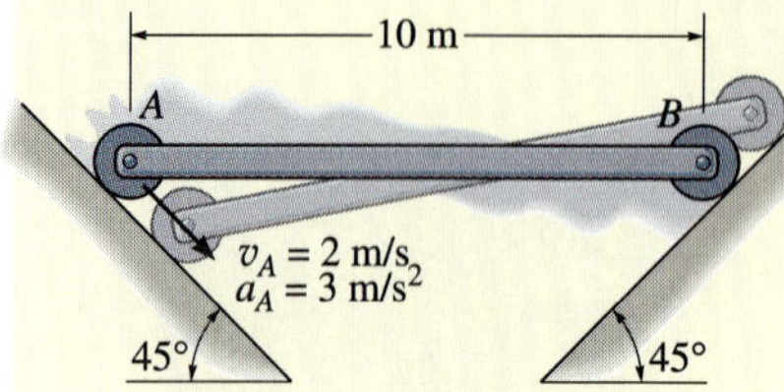

(a)

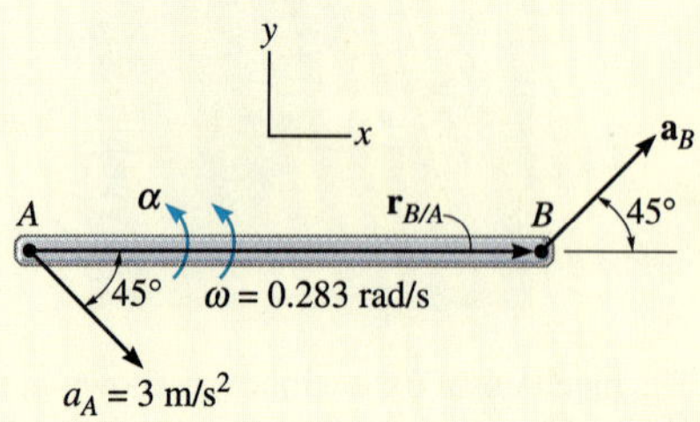

(b)

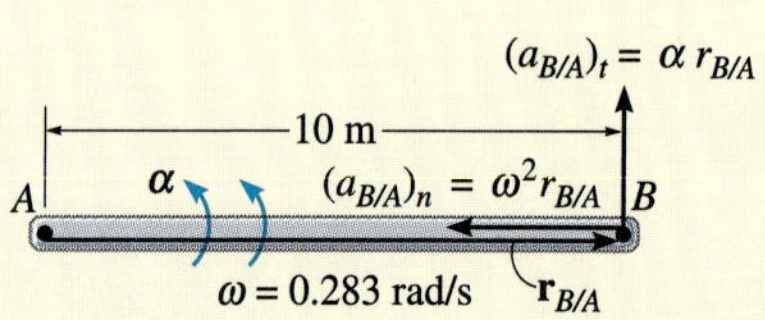

(c)

Fig. 16–26

The rod AB shown in Fig. 16–26a is confined to move along the inclined planes at A and B. If point A has an acceleration of 3 m/s² and a velocity of 2 m/s, both directed down the plane at the instant the rod becomes horizontal, determine the angular acceleration of the rod at this instant.

Solution I (VECTOR ANALYSIS)

We will apply the acceleration equation to points A and B on the rod. To do so it is first necessary to determine the angular velocity of the rod. Show that it is $\omega = 0.283$ rad/s ↺ using either the velocity equation or the method of instantaneous centers.

Kinematic Diagram. Since points A and B both move along straight-line paths, they have *no* components of acceleration normal to the paths. There are two unknowns in Fig. 16–26b, namely, a_B and α.

Acceleration Equation. Applying Eq. 16–18 to points A and B on the rod, and expressing each of the vectors in Cartesian vector form, we have

$$\mathbf{a}_B = \mathbf{a}_A + \boldsymbol{\alpha} \times \mathbf{r}_{B/A} - \omega^2 \mathbf{r}_{B/A}$$

$$a_B \cos 45°\mathbf{i} + a_B \sin 45°\mathbf{j} = 3 \cos 45°\mathbf{i} - 3 \sin 45°\mathbf{j} + (\alpha\mathbf{k}) \times (10\mathbf{i}) - (0.283)^2(10\mathbf{i})$$

Carrying out the cross product and equating the **i** and **j** components yields

$$a_B \cos 45° = 3 \cos 45° - (0.283)^2(10) \qquad (1)$$

$$a_B \sin 45° = -3 \sin 45° + \alpha(10) \qquad (2)$$

Solving, we have

$$a_B = 1.87 \text{ m/s}^2 \measuredangle^{45°}$$

$$\alpha = 0.344 \text{ rad/s}^2 ↺ \qquad \textit{Ans.}$$

Solution II (SCALAR ANALYSIS)

As an alternative procedure, the scalar component equations 1 and 2 can be obtained directly. From the kinematic diagram, showing the relative-acceleration components $(\mathbf{a}_{B/A})_t$ and $(\mathbf{a}_{B/A})_n$, Fig. 16–26c, we have

$$\mathbf{a}_B = \mathbf{a}_A + (\mathbf{a}_{B/A})_t + (\mathbf{a}_{B/A})_n$$

$$\begin{bmatrix} a_B \\ \measuredangle^{45°} \end{bmatrix} = \begin{bmatrix} 3 \text{ m/s}^2 \\ \searrow 45° \end{bmatrix} + \begin{bmatrix} \alpha(10 \text{ m}) \\ \uparrow \end{bmatrix} + \begin{bmatrix} (0.283 \text{ rad/s})^2(10 \text{ m}) \\ \leftarrow \end{bmatrix}$$

Equating the x and y components yields Eqs. 1 and 2, and the solution proceeds as before.

EXAMPLE 16-14

At a given instant, the cylinder of radius r, shown in Fig. 16–27a, has an angular velocity $\boldsymbol{\omega}$ and angular acceleration $\boldsymbol{\alpha}$. Determine the velocity and acceleration of its center G if it rolls without slipping.

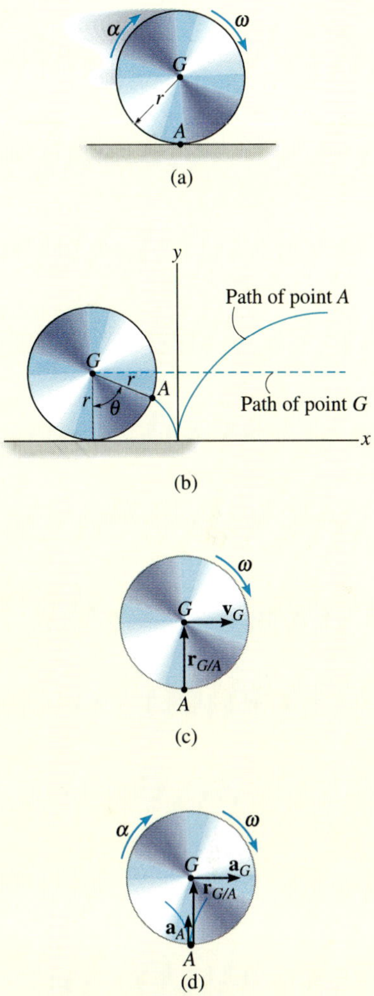

Fig. 16–27

Solution (Vector Analysis)

As the cylinder rolls, point G moves along a straight line, and point A, located on the rim of the cylinder, moves along a curved path called a *cycloid,* Fig. 16–27b. We will apply the velocity and acceleration equations to these two points.

Velocity Analysis. Since no slipping occurs, at the instant A contacts the ground, $\mathbf{v}_A = \mathbf{0}$. Thus, from the kinematic diagram in Fig. 16–27c we have

$$\mathbf{v}_G = \mathbf{v}_A + \boldsymbol{\omega} \times \mathbf{r}_{G/A}$$

$$v_G\mathbf{i} = \mathbf{0} + (-\omega\mathbf{k}) \times (r\mathbf{j})$$

$$v_G = \omega r \qquad (1) \qquad \textit{Ans.}$$

This same result can also be obtained directly by noting that point A represents the instantaneous center of zero velocity.

Kinematic Diagram. The acceleration of point G is horizontal since it moves along a *straight-line path. Just before* point A touches the ground, its velocity is directed downward along the y axis, Fig. 16–27b, and just after contact, its velocity is directed *upward.* For this reason, point A begins to accelerate upward when it leaves the ground at A, Fig. 16–27d. The magnitudes of $\mathbf{a}_A$ and $\mathbf{a}_G$ are unknown.

Acceleration Equation

$$\mathbf{a}_G = \mathbf{a}_A + \boldsymbol{\alpha} \times \mathbf{r}_{G/A} - \omega^2\mathbf{r}_{G/A}$$

$$a_G\mathbf{i} = a_A\mathbf{j} + (-\alpha\mathbf{k}) \times (r\mathbf{j}) - \omega^2(r\mathbf{j})$$

Evaluating the cross product and equating the $\mathbf{i}$ and $\mathbf{j}$ components yields

$$a_G = \alpha r \qquad (2) \qquad \textit{Ans.}$$

$$a_A = \omega^2 r \qquad (3)$$

These important results, that $v_G = \omega r$ and $a_G = \alpha r$, were also obtained in Example 16–3. They apply to any circular object, such as a ball, pulley, disk, etc., that rolls *without* slipping. Also, the fact that $a_A = \omega^2 r$ indicates that the instantaneous center of zero velocity, point A, *is not* a point of zero acceleration.

EXAMPLE 16–15

The ball rolls without slipping and has the angular motion shown in Fig. 16–28*a*. Determine the accelerations of point *B* and point *A* at this instant.

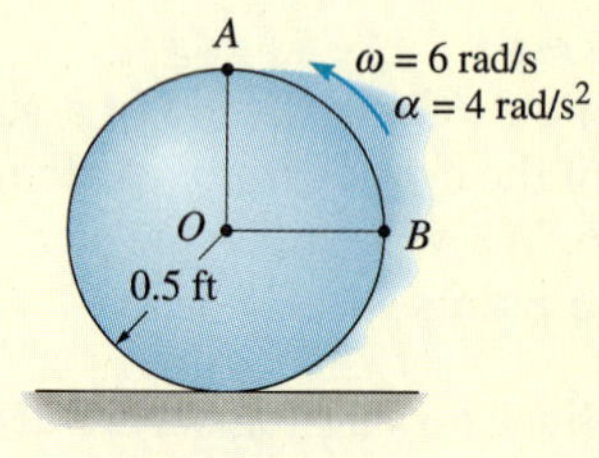

(a)

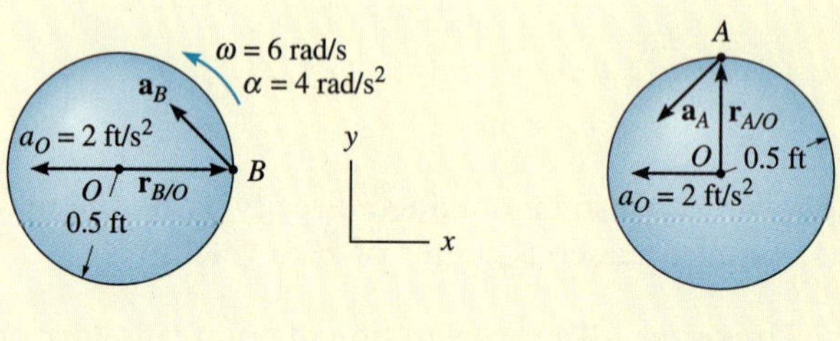

(b) (c)

Fig. 16–28

Solution (VECTOR ANALYSIS)

Kinematic Diagram. Using the results of the previous example, the center of the ball has an acceleration of $a_O = \alpha r = (4 \text{ rad/s}^2)(0.5 \text{ ft}) = 2 \text{ ft/s}^2$. We will apply the acceleration equation to points O and B and points O and A.

Acceleration Equation

For point B, Fig. 16–28*b*,

$$\mathbf{a}_B = \mathbf{a}_O + \boldsymbol{\alpha} \times \mathbf{r}_{B/O} - \omega^2 \mathbf{r}_{B/O}$$

$$\mathbf{a}_B = -2\mathbf{i} + (4\mathbf{k}) \times (0.5\mathbf{i}) - (6)^2(0.5\mathbf{i})$$

$$\mathbf{a}_B = \{-20\mathbf{i} + 2\mathbf{j}\} \text{ ft/s}^2 \qquad \textit{Ans.}$$

For point A, Fig. 16–28*c*,

$$\mathbf{a}_A = \mathbf{a}_O + \boldsymbol{\alpha} \times \mathbf{r}_{A/O} - \omega^2 \mathbf{r}_{A/O}$$

$$\mathbf{a}_A = -2\mathbf{i} + (4\mathbf{k}) \times (0.5\mathbf{j}) - (6)^2(0.5\mathbf{j})$$

$$\mathbf{a}_A = \{-4\mathbf{i} - 18\mathbf{j}\} \text{ ft/s}^2 \qquad \textit{Ans.}$$

EXAMPLE 16-16

The spool shown in Fig. 16–29*a* unravels from the cord, such that at the instant shown it has an angular velocity of 3 rad/s and an angular acceleration of 4 rad/s². Determine the acceleration of point *B*.

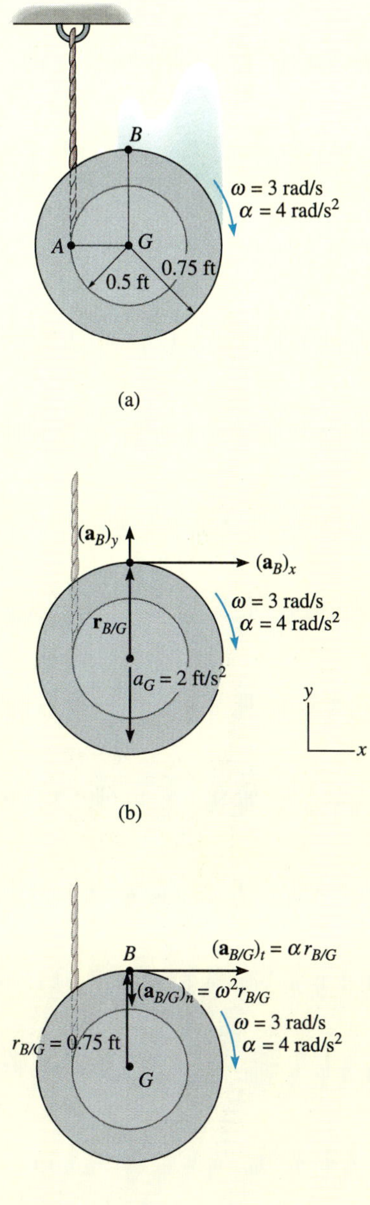

Fig. 16–29

Solution I (VECTOR ANALYSIS)

The spool "appears" to be rolling downward without slipping at point *A*. Therefore, we can use the results of Example 16–14 to determine the acceleration of point *G*, i.e.,

$$a_G = \alpha r = 4 \text{ rad/s}^2 (0.5 \text{ ft}) = 2 \text{ ft/s}^2$$

We will apply the acceleration equation at points *G* and *B*.

Kinematic Diagram. Point *B* moves along a *curved path* having an *unknown* radius of curvature.* Its acceleration will be represented by its unknown *x* and *y* components as shown in Fig. 16–29*b*.

Acceleration Equation

$$\mathbf{a}_B = \mathbf{a}_G + \boldsymbol{\alpha} \times \mathbf{r}_{B/G} - \omega^2 \mathbf{r}_{B/G}$$

$$(a_B)_x\mathbf{i} + (a_B)_y\mathbf{j} = -2\mathbf{j} + (-4\mathbf{k}) \times (0.75\mathbf{j}) - (3)^2(0.75\mathbf{j})$$

Equating the **i** and **j** terms, the component equations are

$$(a_B)_x = 4(0.75) = 3 \text{ ft/s}^2 \rightarrow \qquad (1)$$

$$(a_B)_y = -2 - 6.75 = -8.75 \text{ ft/s}^2 = 8.75 \text{ ft/s}^2 \downarrow \qquad (2)$$

The magnitude and direction of $\mathbf{a}_B$ are therefore

$$a_B = \sqrt{(3)^2 + (8.75)^2} = 9.25 \text{ ft/s}^2 \qquad \textit{Ans.}$$

$$\theta = \tan^{-1}\frac{8.75}{3} = 71.1^\circ \qquad \textit{Ans.}$$

Solution II (SCALAR ANALYSIS)

This problem may be solved by writing the scalar component equations directly. The kinematic diagram in Fig. 16–29*c* shows the relative-acceleration components $(\mathbf{a}_{B/G})_t$ and $(\mathbf{a}_{B/G})_n$. Thus,

$$\mathbf{a}_B = \mathbf{a}_G + (\mathbf{a}_{B/G})_t + (\mathbf{a}_{B/G})_n$$

$$\begin{bmatrix}(a_B)_x \\ \rightarrow\end{bmatrix} + \begin{bmatrix}(a_B)_y \\ \uparrow\end{bmatrix}$$

$$= \begin{bmatrix}2 \text{ ft/s}^2 \\ \downarrow\end{bmatrix} + \begin{bmatrix}4 \text{ rad/s}^2(0.75 \text{ ft}) \\ \rightarrow\end{bmatrix} + \begin{bmatrix}(3 \text{ rad/s})^2(0.75 \text{ ft}) \\ \downarrow\end{bmatrix}$$

The *x* and *y* components yield Eqs. 1 and 2 above.

*Realize that the path's radius of curvature ρ is *not* equal to the radius of the spool since the spool is *not* rotating about point *G*. Furthermore, ρ is *not* defined as the distance from *A* (*IC*) to *B*, since the location of the *IC* depends only on the velocity of a point and *not* the geometry of its path.

EXAMPLE 16–17

The collar C in Fig. 16–30a is moving downward with an acceleration of 1 m/s^2. At the instant shown, it has a speed of 2 m/s which gives links CB and AB an angular velocity $\omega_{AB} = \omega_{CB} = 10$ rad/s. (See Example 16–8.) Determine the angular accelerations of CB and AB at this instant.

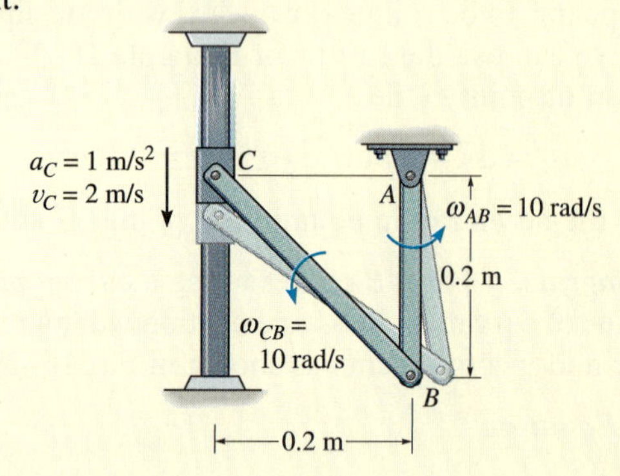

(a)

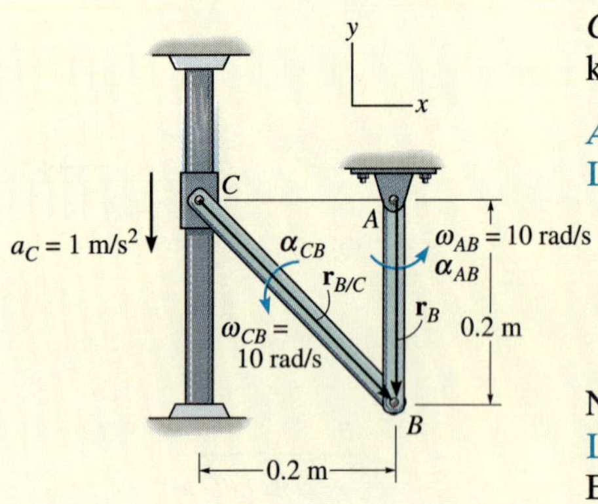

(b)

Fig. 16–30

Solution (VECTOR ANALYSIS)

Kinematic Diagram. The kinematic diagrams of *both* links AB and CB are shown in Fig. 16–30b. To solve, we will apply the appropriate kinematic equation to each link.

Acceleration Equation

Link AB (rotation about a fixed axis):

$$\mathbf{a}_B = \boldsymbol{\alpha}_{AB} \times \mathbf{r}_B - \omega_{AB}^2 \mathbf{r}_B$$
$$\mathbf{a}_B = (\alpha_{AB}\mathbf{k}) \times (-0.2\mathbf{j}) - (10)^2(-0.2\mathbf{j})$$
$$\mathbf{a}_B = 0.2\alpha_{AB}\mathbf{i} + 20\mathbf{j}$$

Note that $\mathbf{a}_B$ has two components since it moves along a *curved path.* Link BC (general plane motion): Using the result for $\mathbf{a}_B$ and applying Eq. 16–18, we have

$$\mathbf{a}_B = \mathbf{a}_C + \boldsymbol{\alpha}_{CB} \times \mathbf{r}_{B/C} - \omega_{CB}^2 \mathbf{r}_{B/C}$$
$$0.2\alpha_{AB}\mathbf{i} + 20\mathbf{j} = -1\mathbf{j} + (\alpha_{CB}\mathbf{k}) \times (0.2\mathbf{i} - 0.2\mathbf{j}) - (10)^2(0.2\mathbf{i} - 0.2\mathbf{j})$$
$$0.2\alpha_{AB}\mathbf{i} + 20\mathbf{j} = -1\mathbf{j} + 0.2\alpha_{CB}\mathbf{j} + 0.2\alpha_{CB}\mathbf{i} - 20\mathbf{i} + 20\mathbf{j}$$

Thus,

$$0.2\alpha_{AB} = 0.2\alpha_{CB} - 20$$
$$20 = -1 + 0.2\alpha_{CB} + 20$$

Solving,

$$\alpha_{CB} = 5 \text{ rad/s}^2 \circlearrowleft \qquad \textit{Ans.}$$
$$\alpha_{AB} = -95 \text{ rad/s}^2 = 95 \text{ rad/s}^2 \circlearrowright \qquad \textit{Ans.}$$

EXAMPLE 16–18

The crankshaft AB of an engine turns with a clockwise angular acceleration of 20 rad/s^2, Fig. 16–31a. Determine the acceleration of the piston at the instant AB is in the position shown. At this instant $\omega_{AB} = 10$ rad/s and $\omega_{BC} = 2.43$ rad/s.

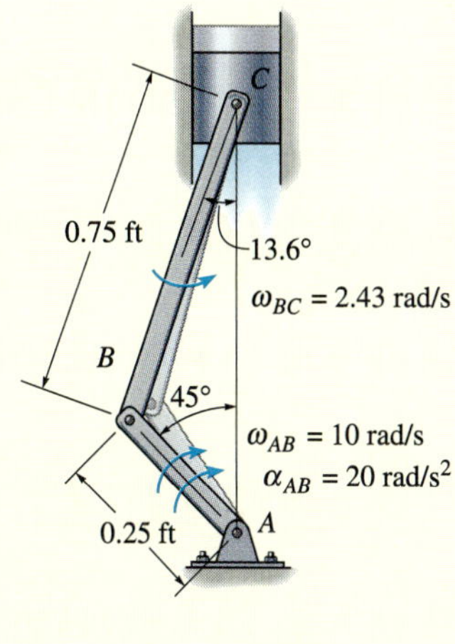

(a)

Solution (VECTOR ANALYSIS)

Kinematic Diagram. The kinematic diagrams for both AB and BC are shown in Fig. 16–31b. Here $\mathbf{a}_C$ is vertical since C moves along a straight-line path.

Acceleration Equation. Expressing each of the position vectors in Cartesian vector form

$$\mathbf{r}_B = \{-0.25 \sin 45°\mathbf{i} + 0.25 \cos 45°\mathbf{j}\} \text{ ft} = \{-0.177\mathbf{i} + 0.177\mathbf{j}\} \text{ ft}$$
$$\mathbf{r}_{C/B} = \{0.75 \sin 13.6°\mathbf{i} + 0.75 \cos 13.6°\mathbf{j}\} \text{ ft} = \{0.176\mathbf{i} + 0.729\mathbf{j}\} \text{ ft}$$

Crankshaft AB (rotation about a fixed axis):

$$\mathbf{a}_B = \boldsymbol{\alpha}_{AB} \times \mathbf{r}_B - \omega_{AB}^2\mathbf{r}_B$$
$$= (-20\mathbf{k}) \times (-0.177\mathbf{i} + 0.177\mathbf{j}) - (10)^2(-0.177\mathbf{i} + 0.177\mathbf{j})$$
$$= \{21.21\mathbf{i} - 14.14\mathbf{j}\} \text{ ft/s}^2$$

Connecting rod BC (general plane motion): Using the result for $\mathbf{a}_B$ and noting that $\mathbf{a}_C$ is in the vertical direction, we have

$$\mathbf{a}_C = \mathbf{a}_B + \boldsymbol{\alpha}_{BC} \times \mathbf{r}_{C/B} - \omega_{BC}^2\mathbf{r}_{C/B}$$
$$a_C\mathbf{j} = 21.21\mathbf{i} - 14.14\mathbf{j} + (\alpha_{BC}\mathbf{k}) \times (0.176\mathbf{i} + 0.729\mathbf{j}) - (2.43)^2(0.176\mathbf{i} + 0.729\mathbf{j})$$
$$a_C\mathbf{j} = 21.21\mathbf{i} - 14.14\mathbf{j} + 0.176\alpha_{BC}\mathbf{j} - 0.729\alpha_{BC}\mathbf{i} - 1.04\mathbf{i} - 4.30\mathbf{j}$$
$$0 = 20.17 - 0.729\alpha_{BC}$$
$$a_C = 0.176\alpha_{BC} - 18.45$$

Solving yields

$$\alpha_{BC} = 27.7 \text{ rad/s}^2 \circlearrowleft$$
$$a_C = -13.6 \text{ ft/s}^2 \qquad \textit{Ans.}$$

Since the piston is moving upward, the negative sign for a_C indicates that the piston is decelerating, i.e., $\mathbf{a}_C = \{-13.6\mathbf{j}\}$ ft/s^2. This causes the speed of the piston to decrease until AB becomes vertical, at which time the piston is momentarily at rest.

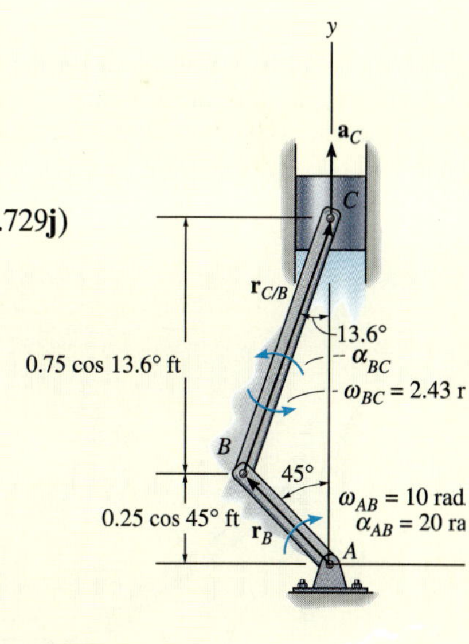

(b)

Fig. 16–31

Problems

16-105. At a given instant A has the motion shown. Determine the acceleration of B and the angular acceleration of the bar at this same instant.

16-106. At a given instant A has the motion shown. Determine the acceleration of point C at this same instant.

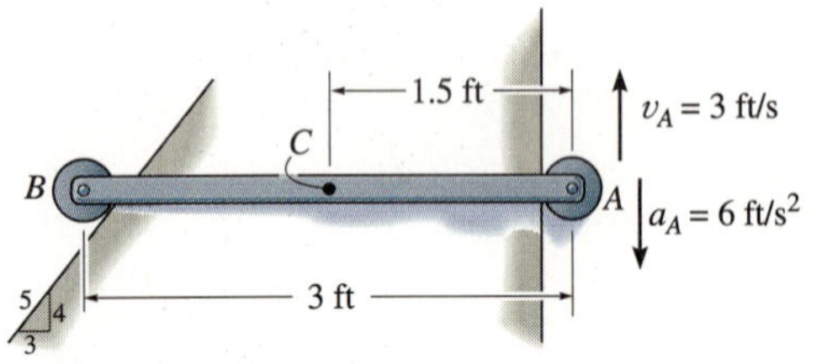

Probs. 16–105/106

***16-108.** At a given instant, the slider block A has the velocity and deceleration shown. Determine the acceleration of block B and the angular acceleration of the link at this instant.

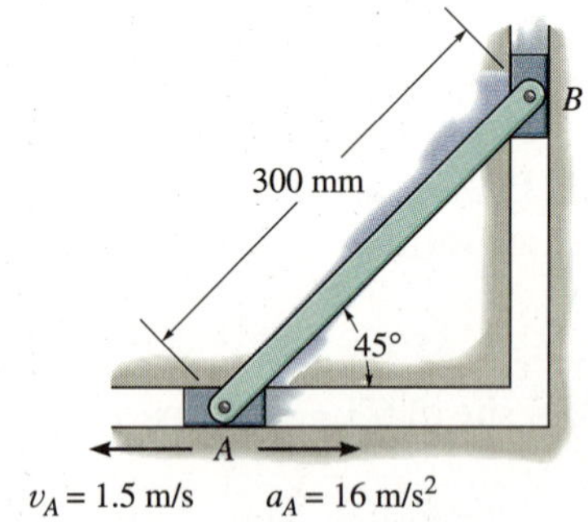

Prob. 16–108

16-107. The 10-ft rod slides down the inclined plane, such that when it is at B it has the motion shown. Determine the velocity and acceleration of A at this instant.

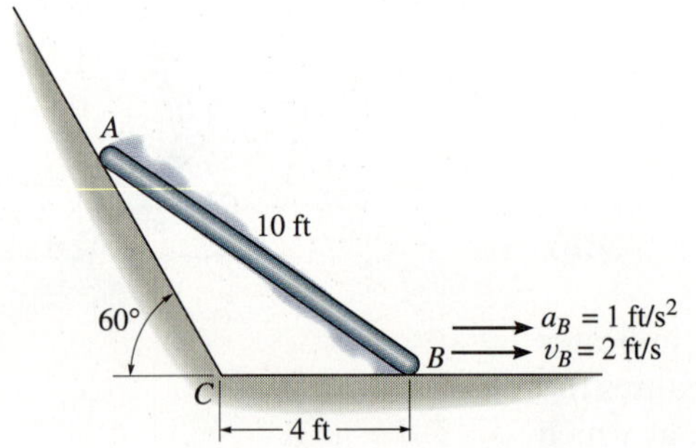

Prob. 16–107

16-109. Determine the angular acceleration of link AB at the instant $\theta = 90°$ if the collar C has a velocity of $v_C = 4$ ft/s and deceleration of $a_C = 3$ ft/s² as shown.

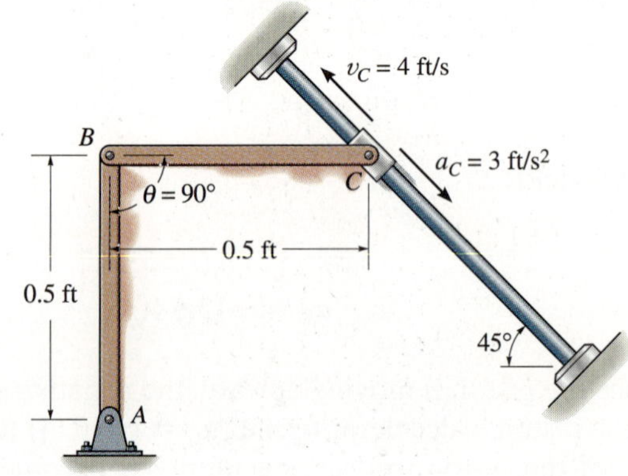

Prob. 16–109

16-110. At a given instant the slider block B is moving to the right with the motion shown. Determine the angular acceleration of link AB and the acceleration of point A at this instant.

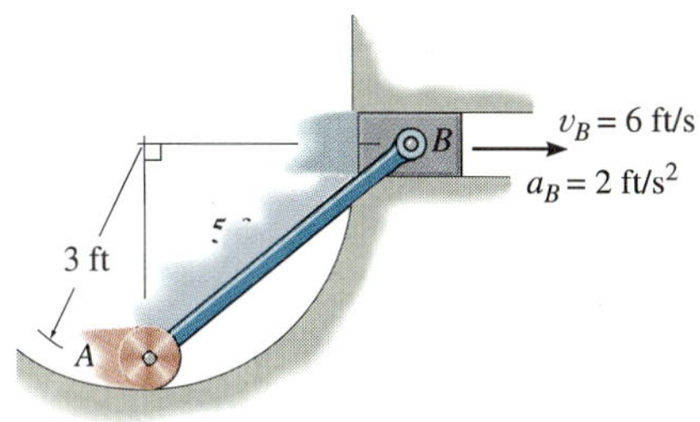

Prob. 16–110

16-111. The rod is confined to move along the path due to the pins at its ends. At the instant shown, point A has the motion shown. Determine the velocity and acceleration of point B at this instant.

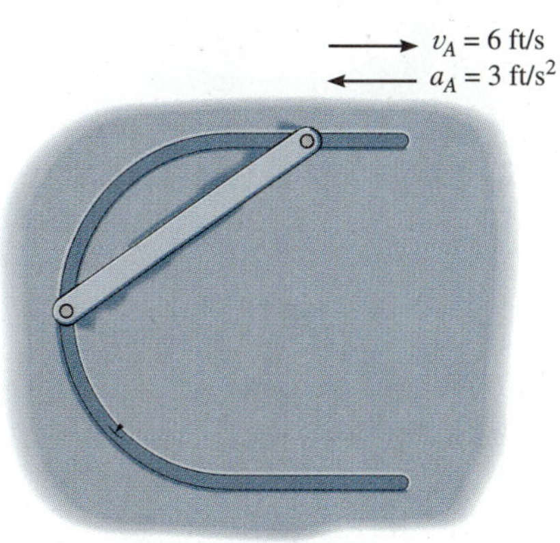

Prob. 16–111

***16-112.** The closure is manufactured by the LCN Company and is used to control the restricted motion of a heavy door. If the door to which is it connected has an angular acceleration of 3 rad/s^2, determine the angular acceleration of links BC and CD. Originally the door is not rotating but is hinged at A.

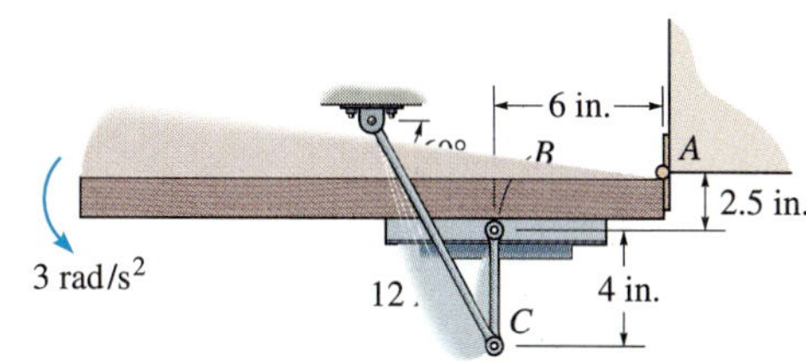

Prob. 16–112

16-113. The disk is moving to the left such that it has an angular acceleration $\alpha = 8$ rad/s^2 and angular velocity $\omega = 3$ rad/s at the instant shown. If it does not slip at A, determine the acceleration of point B.

16-114. The disk is moving to the left such that it has an angular acceleration $\alpha = 8$ rad/s^2 and angular velocity $\omega = 3$ rad/s at the instant shown. If it does not slip at A, determine the acceleration of point D.

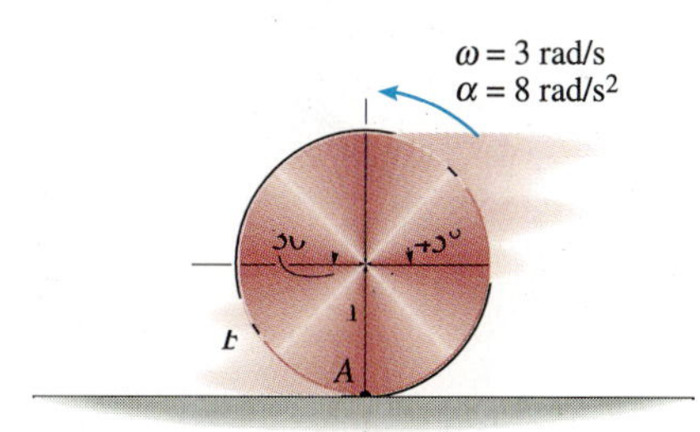

Probs. 16–113/114

16-115. The hoop is cast on the rough surface such that it has an angular velocity $\omega = 4$ rad/s and an angular deceleration $\alpha = 5\ \text{rad/s}^2$. Also, its center has a velocity $v_O = 5$ m/s and a deceleration $a_O = 2\ \text{m/s}^2$. Determine the acceleration of point A at this instant.

***16-116.** The hoop is cast on the rough surface such that it has an angular velocity $\omega = 4$ rad/s and an angular deceleration $\alpha = 5\ \text{rad/s}^2$. Also, its center has a velocity of $v_O = 5$ m/s and a deceleration $a_O = 2\ \text{m/s}^2$. Determine the acceleration of point B at this instant.

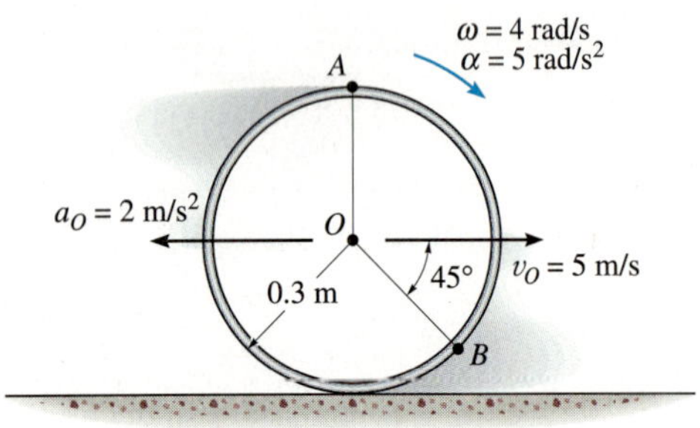

Probs. 16–115/116

16-117. Rod AB has the angular motion shown. Determine the acceleration of block C at this instant.

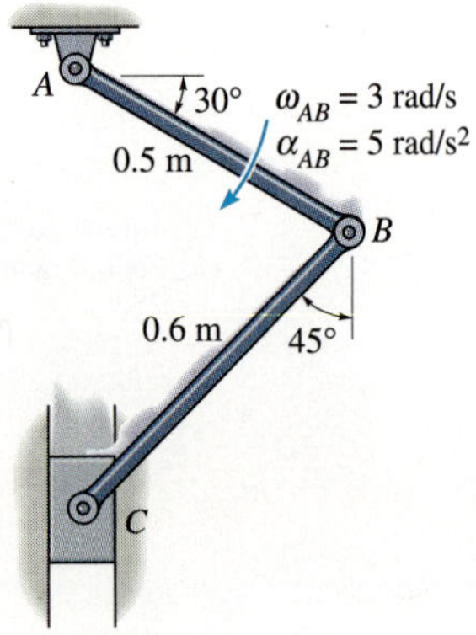

Prob. 16–117

16-118. The flywheel rotates with an angular velocity $\omega = 2$ rad/s and an angular acceleration $\alpha = 6\ \text{rad/s}^2$. Determine the angular accelerations of links AB and BC at this instant.

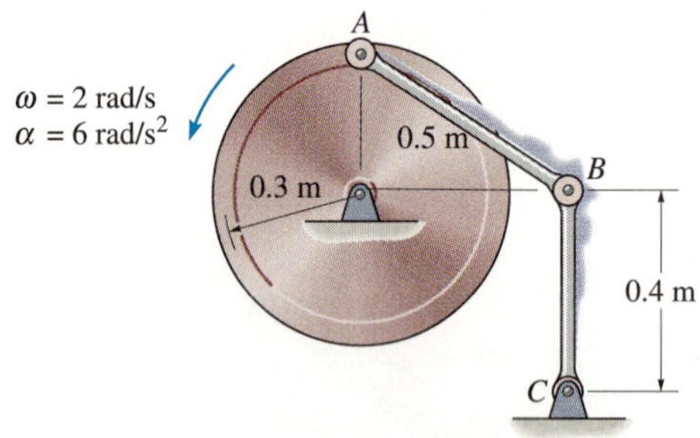

Prob. 16–118

16-119. The ends of the bar AB are confined to move along the paths shown. At a given instant, A has a velocity of 8 ft/s and an acceleration of 3 ft/s^2. Determine the angular velocity and angular acceleration of AB at this instant.

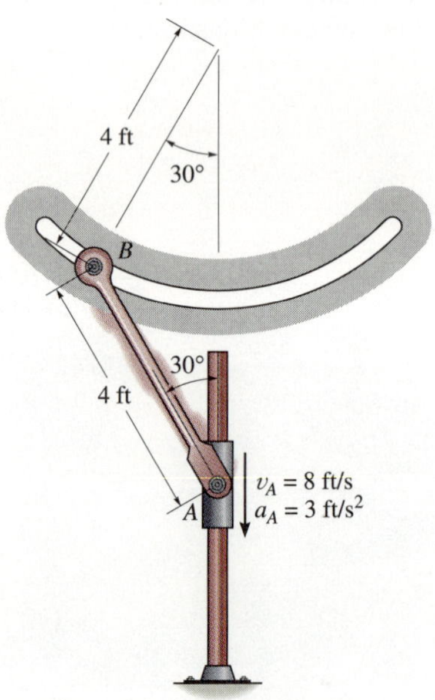

Prob. 16–119

***16-120.** Rod AB has the angular motion shown. Determine the acceleration of the collar C at this instant.

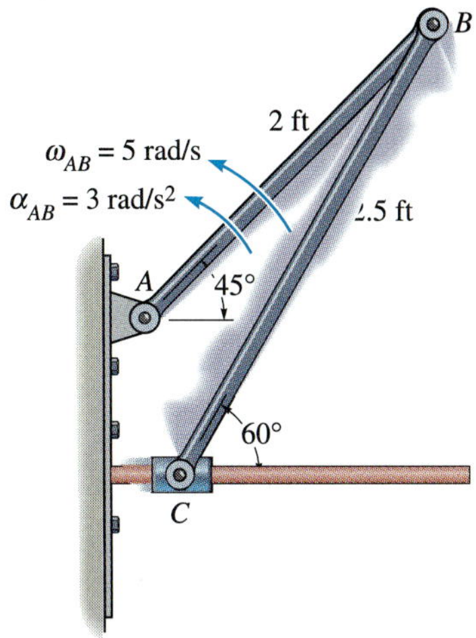

Prob. 16–120

16-121. At the given instant member AB has the angular motions shown. Determine the velocity and acceleration of the slider block C at this instant.

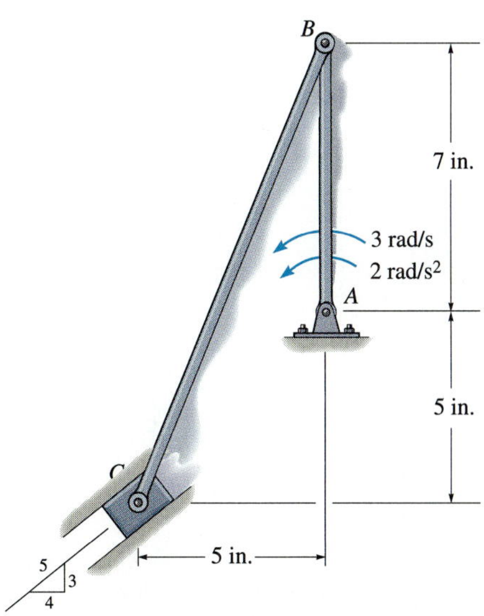

Prob. 16–121

16-122. A cord is wrapped around the inner spool of the gear. If it is pulled with a constant velocity **v**, determine the velocities and accelerations of points A and B. The gear rolls on the fixed gear rack.

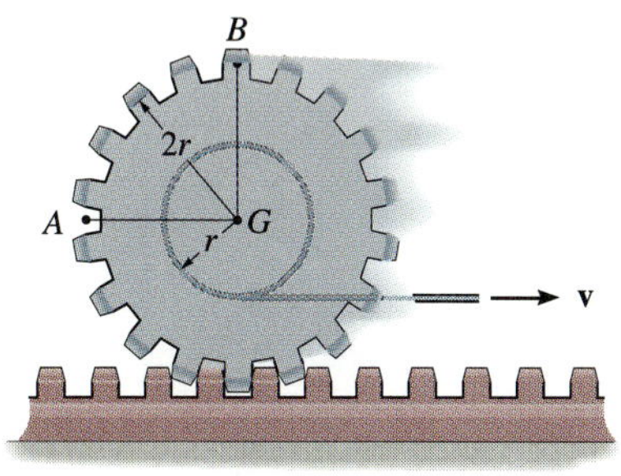

Prob. 16–122

16-123. At a given instant the wheel is rotating with the angular velocity and angular acceleration shown. Determine the acceleration of block B at this instant.

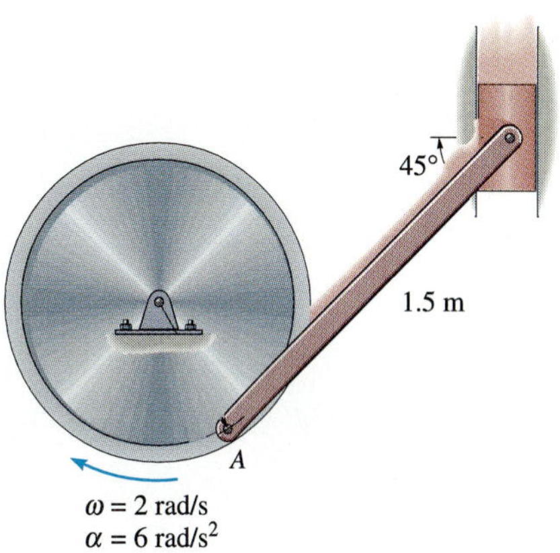

Prob. 16–123

***16-124.** As the cord unravels from the cylinder, the cylinder has an angular acceleration of $\alpha = 4 \text{ rad/s}^2$ and an angular velocity of $\omega = 2$ rad/s at the instant shown. Determine the accelerations of points A and B at this instant.

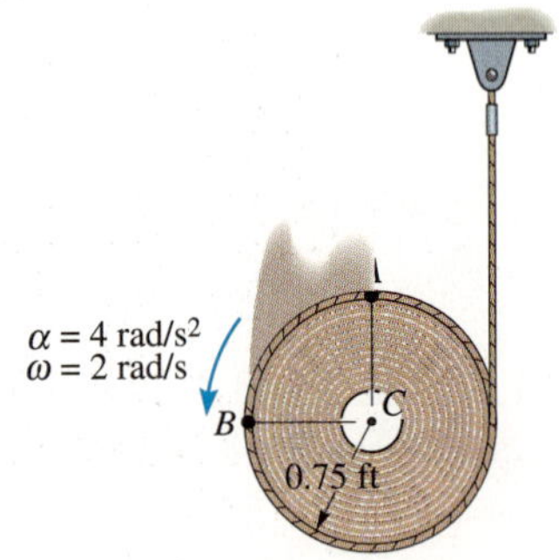

Prob. 16–124

16-125. The disk rolls without slipping such that it has an angular acceleration of $\alpha = 4 \text{ rad/s}^2$ and angular velocity of $\omega = 2$ rad/s at the instant shown. Determine the accelerations of points A and B on the link and the link's angular acceleration at this instant. Assume point A lies on the periphery of the disk, 150 mm from C.

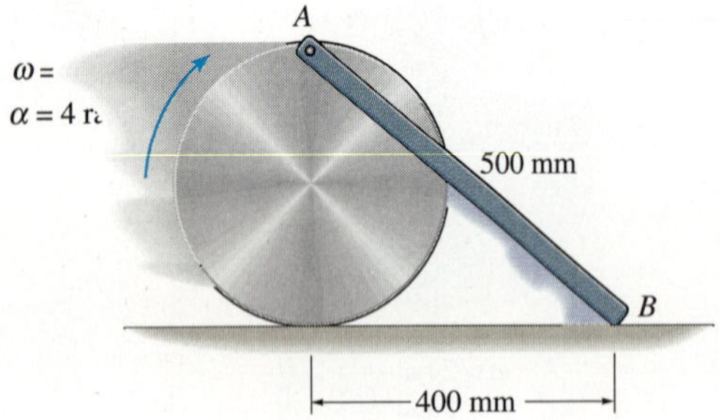

Prob. 16–125

16-126. At a given instant, the gear has the angular motion shown. Determine the accelerations of points A and B on the link and the link's angular acceleration at this instant.

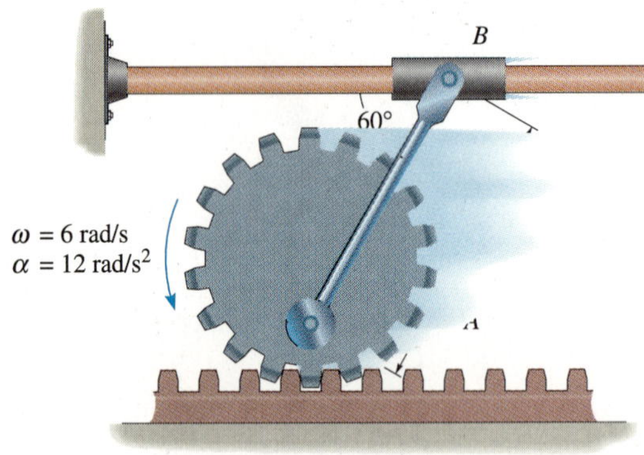

Prob. 16–126

16-127. Determine the angular acceleration of link AB if link CD has the angular velocity and angular deceleration shown.

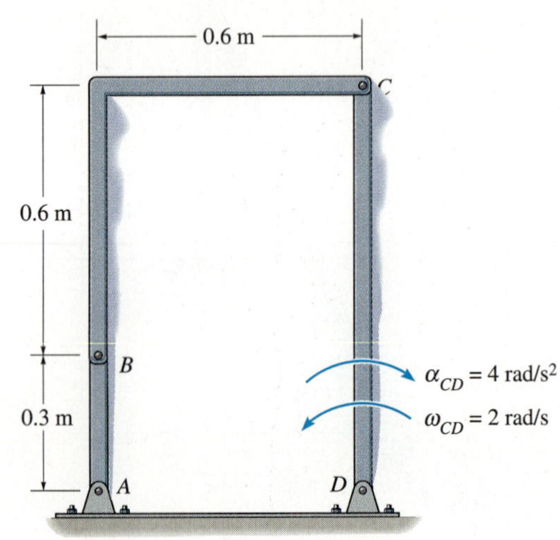

Prob. 16–127

***16-128.** The gear at A is subjected to the angular motion shown. Determine the angular velocity and angular acceleration of link CD at this instant.

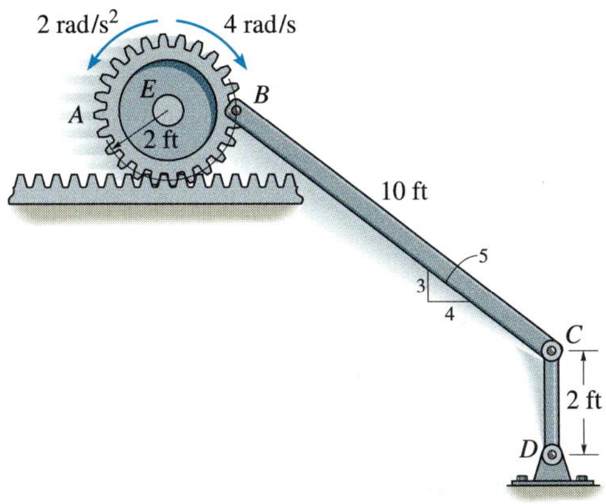

Prob. 16–128

16-129. Determine the angular velocity and the angular acceleration of the plate CD of the stone-crushing mechanism at the instant AB is horizontal. At this instant $\theta = 30°$ and $\phi = 90°$. The driving link AB is turning with a constant angular velocity of $\omega_{AB} = 4$ rad/s.

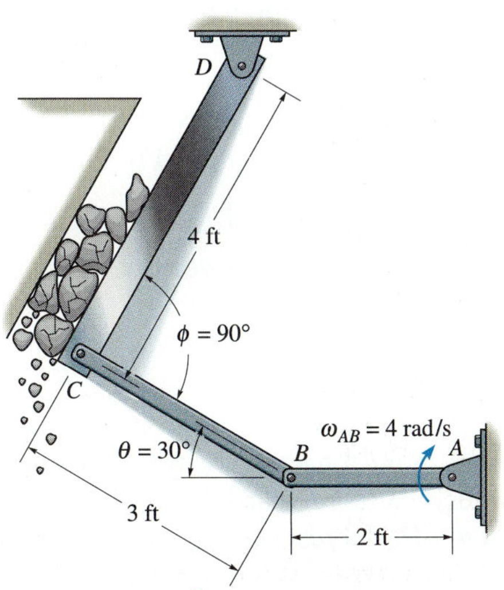

Prob. 16–129

16-130. The mechanism produces intermittent motion of link AB. If the sprocket S is turning with an angular acceleration $\alpha_S = 2$ rad/s^2 and has an angular velocity $\omega_S = 6$ rad/s at the instant shown, determine the angular velocity and angular acceleration of link AB at this instant. The sprocket S is mounted on a shaft which is *separate* from a collinear shaft attached to AB at A. The pin at C is attached to one of the chain links such that it moves vertically downward.

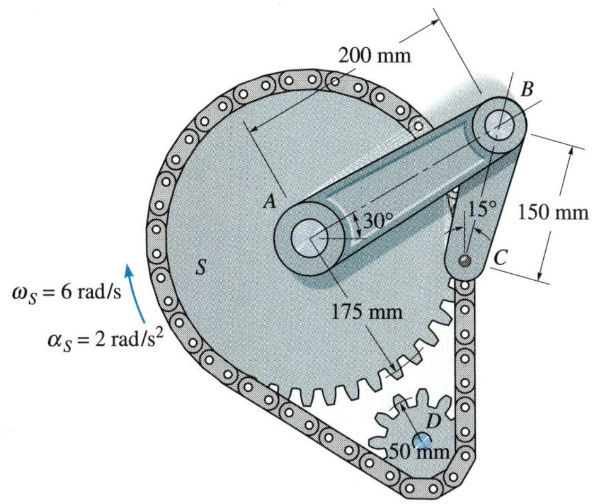

Prob. 16–130

16.8 Relative-Motion Analysis Using Rotating Axes

In the previous sections the relative-motion analysis for velocity and acceleration was described using a translating coordinate system. This type of analysis is useful for determining the motion of points on the *same* rigid body, or the motion of points located on several pin-connected rigid bodies. In some problems, however, rigid bodies (mechanisms) are constructed such that *sliding* will occur at their connections. The kinematic analysis for such cases is best performed if the motion is analyzed using a coordinate system which both *translates* and *rotates*. Furthermore, this frame of reference is useful for analyzing the motions of two points on a mechanism which are *not* located in the *same* rigid body and for specifying the kinematics of particle motion when the particle is moving along a rotating path.

In the following analysis two equations are developed which relate the velocity and acceleration of two points, one of which is the origin of a moving frame of reference subjected to both a translation and a rotation in the plane.* Due to the generality in the derivation which follows, these two points may represent either two particles moving independently of one another or two points located on the same (or different) rigid bodies.

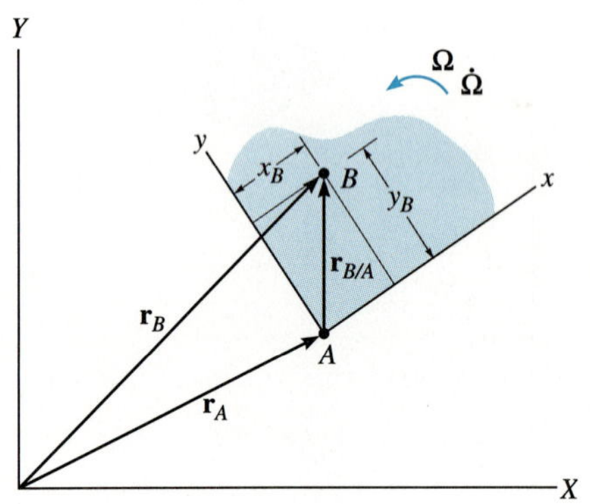

(a)

Fig. 16–32

Position. Consider the two points A and B shown in Fig. 16–32*a*. Their location is specified by the position vectors $\mathbf{r}_A$ and $\mathbf{r}_B$, which are measured from the fixed X, Y, Z coordinate system. As shown in the figure, the "base point" A represents the origin of the x, y, z coordinate system, which is assumed to be both translating and rotating with respect to the X, Y, Z system. The position of B with respect to A is specified by the relative-position vector $\mathbf{r}_{B/A}$. The components of this vector may be expressed either in terms of unit vectors along the X, Y axes, i.e., $\mathbf{I}$ and $\mathbf{J}$, or by unit vectors along the x, y axes, i.e., $\mathbf{i}$ and $\mathbf{j}$. For the development which follows, $\mathbf{r}_{B/A}$ will be measured relative to the moving x, y frame of reference. Thus, if B has coordinates (x_B, y_B), Fig. 16–32*a*, then

$$\mathbf{r}_{B/A} = x_B\mathbf{i} + y_B\mathbf{j}$$

Using vector addition, the three position vectors in Fig. 16–32*a* are related by the equation

$$\mathbf{r}_B = \mathbf{r}_A + \mathbf{r}_{B/A} \tag{16–19}$$

At the instant considered, point A has a velocity $\mathbf{v}_A$ and an acceleration $\mathbf{a}_A$, while the angular velocity and angular acceleration of the x, y axes are $\boldsymbol{\Omega}$ (omega) and $\dot{\boldsymbol{\Omega}} = d\boldsymbol{\Omega}/dt$, respectively. All these vectors are measured from the X, Y, Z frame of reference, although they may be

*The more general, three-dimensional motion of the points is developed in Sec. 20.4.

expressed in terms of either **I, J, K** or **i, j, k** components. Since planar motion is specified, by the right-hand rule $\mathbf{\Omega}$ and $\dot{\mathbf{\Omega}}$ are always directed *perpendicular* to the reference plane of motion, whereas $\mathbf{v}_A$ and $\mathbf{a}_A$ lie in this plane.

Velocity. The velocity of point B is determined by taking the time derivative of Eq. 16–19, which yields

$$\mathbf{v}_B = \mathbf{v}_A + \frac{d\mathbf{r}_{B/A}}{dt} \qquad (16\text{–}20)$$

The last term in this equation is evaluated as follows:

$$\begin{aligned}\frac{d\mathbf{r}_{B/A}}{dt} &= \frac{d}{dt}(x_B\mathbf{i} + y_B\mathbf{j}) \\ &= \frac{dx_B}{dt}\mathbf{i} + x_B\frac{d\mathbf{i}}{dt} + \frac{dy_B}{dt}\mathbf{j} + y_B\frac{d\mathbf{j}}{dt} \\ &= \left(\frac{dx_B}{dt}\mathbf{i} + \frac{dy_B}{dt}\mathbf{j}\right) + \left(x_B\frac{d\mathbf{i}}{dt} + y_B\frac{d\mathbf{j}}{dt}\right) \qquad (16\text{–}21)\end{aligned}$$

The two terms in the first set of parentheses represent the components of velocity of point B as measured by an observer attached to the moving x, y, z coordinate system. These terms will be denoted by vector $(\mathbf{v}_{B/A})_{xyz}$. In the second set of parentheses the instantaneous time rate of change of the unit vectors **i** and **j** is measured by an observer located in the fixed X, Y, Z coordinate system. These changes, $d\mathbf{i}$ and $d\mathbf{j}$, are due *only* to the instantaneous *rotation* $d\theta$ of the x, y, z axes, causing **i** to become $\mathbf{i}' = \mathbf{i} + d\mathbf{i}$ and **j** to become $\mathbf{j}' = \mathbf{j} + d\mathbf{j}$, Fig. 16–32*b*. As shown, the *magnitudes* of both $d\mathbf{i}$ and $d\mathbf{j}$ equal 1 $(d\theta)$, since $i = i' = j = j' = 1$. The *direction* of $d\mathbf{i}$ is defined by $+\mathbf{j}$, since $d\mathbf{i}$ is tangent to the path described by the arrowhead of **i** in the limit as $\Delta t \to dt$. Likewise, $d\mathbf{j}$ acts in the $-\mathbf{i}$ direction, Fig. 16–32*b*. Hence,

$$\frac{d\mathbf{i}}{dt} = \frac{d\theta}{dt}(\mathbf{j}) = \Omega\mathbf{j} \qquad \frac{d\mathbf{j}}{dt} = \frac{d\theta}{dt}(-\mathbf{i}) = -\Omega\mathbf{i}$$

Viewing the axes in three dimensions, Fig. 16–32*c*, and noting that $\mathbf{\Omega} = \Omega\mathbf{k}$, we can express the above derivatives in terms of the cross product as

$$\frac{d\mathbf{i}}{dt} = \mathbf{\Omega} \times \mathbf{i} \qquad \frac{d\mathbf{j}}{dt} = \mathbf{\Omega} \times \mathbf{j} \qquad (16\text{–}22)$$

Substituting these results into Eq. 16–21 and using the distributive property of the vector cross product, we obtain

$$\frac{d\mathbf{r}_{B/A}}{dt} = (\mathbf{v}_{B/A})_{xyz} + \mathbf{\Omega} \times (x_B\mathbf{i} + y_B\mathbf{j}) = (\mathbf{v}_{B/A})_{xyz} + \mathbf{\Omega} \times \mathbf{r}_{B/A} \qquad (16\text{–}23)$$

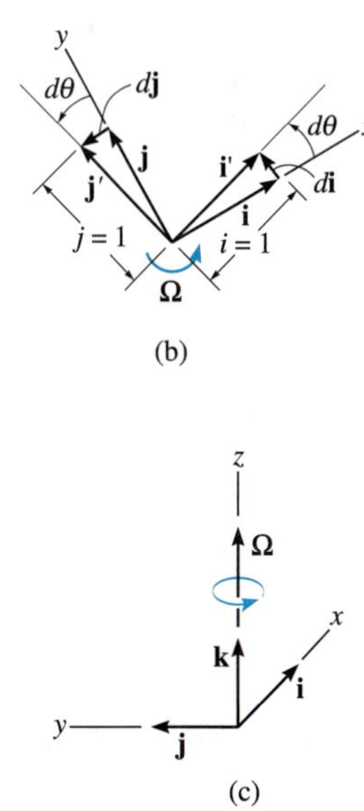

Fig. 16–32

Hence, Eq. 16–20 becomes

$$\mathbf{v}_B = \mathbf{v}_A + \boldsymbol{\Omega} \times \mathbf{r}_{B/A} + (\mathbf{v}_{B/A})_{xyz} \quad (16\text{–}24)$$

where

$\mathbf{v}_B$ = velocity of B, measured from the X, Y, Z reference

$\mathbf{v}_A$ = velocity of the origin A of the x, y, z reference, measured from the X, Y, Z reference

$(\mathbf{v}_{B/A})_{xyz}$ = relative velocity of "B with respect to A," as measured by an observer attached to the *rotating* x, y, z reference

$\boldsymbol{\Omega}$ = angular velocity of the x, y, z reference, measured from the X, Y, Z reference

$\mathbf{r}_{B/A}$ = relative position of "B with respect to A"

Comparing Eq. 16–24 with Eq. 16–16 ($\mathbf{v}_B = \mathbf{v}_A + \boldsymbol{\Omega} \times \mathbf{r}_{B/A}$), which is valid for a translating frame of reference, it can be seen that the only difference between the equations is represented by the term $(\mathbf{v}_{B/A})_{xyz}$.

When applying Eq. 16–24 it is often useful to understand what each of the terms represents. In order of appearance, they are as follows:

$\mathbf{v}_B$	absolute velocity of B	motion of B observed from the X, Y, Z frame
	(equals)	
$\mathbf{v}_A$	absolute velocity of origin of x, y, z frame	motion of x, y, z frame observed from the X, Y, Z frame
	(plus)	
$\boldsymbol{\Omega} \times \mathbf{r}_{B/A}$	angular velocity effect caused by rotation of x, y, z frame	
	(plus)	
$(\mathbf{v}_{B/A})_{xyz}$	relative velocity of B with respect to A	motion of B observed from the x, y, z frame

Acceleration. The acceleration of B, observed from the X, Y, Z coordinate system, may be expressed in terms of its motion measured with respect to the rotating or moving system of coordinates by taking the time derivative of Eq. 16–24, i.e.,

$$\frac{d\mathbf{v}_B}{dt} = \frac{d\mathbf{v}_A}{dt} + \frac{d\boldsymbol{\Omega}}{dt}\times\mathbf{r}_{B/A} + \boldsymbol{\Omega}\times\frac{d\mathbf{r}_{B/A}}{dt} + \frac{d(\mathbf{v}_{B/A})_{xyz}}{dt}$$

$$\mathbf{a}_B = \mathbf{a}_A + \dot{\boldsymbol{\Omega}}\times\mathbf{r}_{B/A} + \boldsymbol{\Omega}\times\frac{d\mathbf{r}_{B/A}}{dt} + \frac{d(\mathbf{v}_{B/A})_{xyz}}{dt} \qquad (16\text{–}25)$$

Here $\dot{\boldsymbol{\Omega}} = d\boldsymbol{\Omega}/dt$ is the angular acceleration of the x, y, z coordinate system. For planar motion $\boldsymbol{\Omega}$ is always perpendicular to the plane of motion, and therefore $\dot{\boldsymbol{\Omega}}$ measures *only the change in magnitude* of $\boldsymbol{\Omega}$. The derivative $d\mathbf{r}_{B/A}/dt$ in Eq. 16–25 is defined by Eq. 16–23, so that

$$\boldsymbol{\Omega}\times\frac{d\mathbf{r}_{B/A}}{dt} = \boldsymbol{\Omega}\times(\mathbf{v}_{B/A})_{xyz} + \boldsymbol{\Omega}\times(\boldsymbol{\Omega}\times\mathbf{r}_{B/A}) \qquad (16\text{–}26)$$

Finding the time derivative of $(\mathbf{v}_{B/A})_{xyz} = (v_{B/A})_x\mathbf{i} + (v_{B/A})_y\mathbf{j}$,

$$\frac{d(\mathbf{v}_{B/A})_{xyz}}{dt} = \left[\frac{d(v_{B/A})_x}{dt}\mathbf{i} + \frac{d(v_{B/A})_y}{dt}\mathbf{j}\right] + \left[(v_{B/A})_x\frac{d\mathbf{i}}{dt} + (v_{B/A})_y\frac{d\mathbf{j}}{dt}\right]$$

The two terms in the first set of brackets represent the components of acceleration of point B as measured by an observer attached to the moving coordinate system. These terms will be denoted by $(\mathbf{a}_{B/A})_{xyz}$. The terms in the second set of brackets can be simplified using Eqs. 16–22.

$$\frac{d(\mathbf{v}_{B/A})_{xyz}}{dt} = (\mathbf{a}_{B/A})_{xyz} + \boldsymbol{\Omega}\times(\mathbf{v}_{B/A})_{xyz}$$

Substituting this and Eq. 16–26 into Eq. 16–25 and rearranging terms,

$$\mathbf{a}_B = \mathbf{a}_A + \dot{\boldsymbol{\Omega}}\times\mathbf{r}_{B/A} + \boldsymbol{\Omega}\times(\boldsymbol{\Omega}\times\mathbf{r}_{B/A}) + 2\boldsymbol{\Omega}\times(\mathbf{v}_{B/A})_{xyz} + (\mathbf{a}_{B/A})_{xyz} \qquad (16\text{–}27)$$

where

$\mathbf{a}_B$ = acceleration of B, measured from the X, Y, Z reference

$\mathbf{a}_A$ = acceleration of the origin A of the x, y, z reference, measured from the X, Y, Z reference

$(\mathbf{a}_{B/A})_{xyz}$, $(\mathbf{v}_{B/A})_{xyz}$ = relative acceleration and relative velocity of "B with respect to A," as measured by an observer attached to the *rotating* x, y, z reference

$\dot{\boldsymbol{\Omega}}$, $\boldsymbol{\Omega}$ = angular acceleration and angular velocity of the x, y, z reference, measured from the X, Y, Z reference

$\mathbf{r}_{B/A}$ = relative position of "B with respect to A"

If Eq. 16–27 is compared with Eq. 16–18, written in the form $\mathbf{a}_B = \mathbf{a}_A + \dot{\mathbf{\Omega}} \times \mathbf{r}_{B/A} + \mathbf{\Omega} \times (\mathbf{\Omega} \times \mathbf{r}_{B/A})$, which is valid for a translating frame of reference, it can be seen that the difference between the equations is represented by the terms $2\mathbf{\Omega} \times (\mathbf{v}_{B/A})_{xyz}$ and $(\mathbf{a}_{B/A})_{xyz}$. In particular, $2\mathbf{\Omega} \times (\mathbf{v}_{B/A})_{xyz}$ is called the *Coriolis acceleration,* named after the French engineer G. C. Coriolis, who was the first to determine it. This term represents the difference in the acceleration of B as measured from nonrotating and rotating x, y, z axes. As indicated by the vector cross product, the Coriolis acceleration will *always* be perpendicular to both $\mathbf{\Omega}$ and $(\mathbf{v}_{B/A})_{xyz}$. It is an important component of the acceleration which must be considered whenever rotating reference frames are used. This often occurs, for example, when studying the accelerations and forces which act on rockets, long-range projectiles, or other bodies having motions whose measurements are significantly affected by the rotation of the earth.

The following interpretation of the terms in Eq. 16–27 may be useful when applying this equation to the solution of problems.

$\mathbf{a}_B$	{absolute acceleration of B	motion of B observed from the X, Y, Z frame
	(equals)	
$\mathbf{a}_A$	absolute acceleration of origin of x, y, z frame	
	(plus)	
$\dot{\mathbf{\Omega}} \times \mathbf{r}_{B/A}$	angular acceleration effect caused by rotation of x, y, z frame	motion of x, y, z frame observed from the X, Y, Z frame
	(plus)	
$\mathbf{\Omega} \times (\mathbf{\Omega} \times \mathbf{r}_{B/A})$	angular velocity effect caused by rotation of x, y, z frame	
	(plus)	
$2\mathbf{\Omega} \times (\mathbf{v}_{B/A})_{xyz}$	combined effect of B moving relative to x, y, z coordinates and rotation of x, y, z frame	interacting motion
	(plus)	
$(\mathbf{a}_{B/A})_{xyz}$	relative acceleration of B with respect to A	motion of B observed from the x, y, z frame

Procedure for Analysis

Equations 16–24 and 16–27 can be applied to the solution of problems involving the planar motion of particles or rigid bodies using the following procedure.

Coordinate Axes

- Choose an appropriate location for the origin and proper orientation of the axes for both the X, Y, Z and moving x, y, z reference frames.
- Most often solutions are easily obtained if at the instant considered:

  (1) the origins are coincident
  (2) the corresponding axes are collinear
  (3) the corresponding axes are parallel
- The moving frame should be selected fixed to the body or device along which the relative motion occurs.

Kinematic Equations

- After defining the origin A of the moving reference and specifying the moving point B, Eqs. 16–24 and 16–27 should be written in symbolic form

$$\mathbf{v}_B = \mathbf{v}_A + \mathbf{\Omega} \times \mathbf{r}_{B/A} + (\mathbf{v}_{B/A})_{xyz}$$

$$\mathbf{a}_B = \mathbf{a}_A + \dot{\mathbf{\Omega}} \times \mathbf{r}_{B/A} + \mathbf{\Omega} \times (\mathbf{\Omega} \times \mathbf{r}_{B/A}) + 2\mathbf{\Omega} \times (\mathbf{v}_{B/A})_{xyz} + (\mathbf{a}_{B/A})_{xyz}$$

- The Cartesian components of all these vectors may be expressed along either the X, Y, Z axes or the x, y, z axes. The choice is arbitrary provided a consistent set of unit vectors is used.
- Motion of the moving reference, is expressed by $\mathbf{v}_A$, $\mathbf{a}_A$, $\mathbf{\Omega}$, and $\dot{\mathbf{\Omega}}$; and motion of B with respect to the moving reference, is expressed by $\mathbf{r}_{B/A}$, $(\mathbf{v}_{B/A})_{xyz}$, and $(\mathbf{a}_{B/A})_{xyz}$.

The rotation of the dumping bin of the truck about point C is operated by the extension of the hydraulic cylinder AB. To determine the rotation of the bin due to this extension, we can use the equations of relative motion and fix the x,y axes to the cylinder so that the relative motion of the cylinder's extension occurs along the x axis.

EXAMPLE 16–19

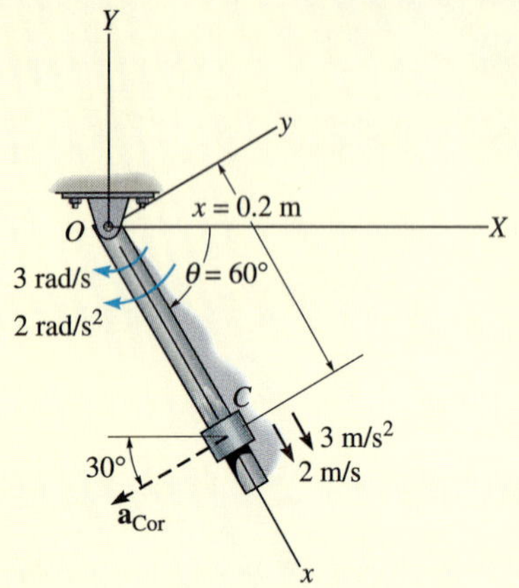

Fig. 16–33

At the instant $\theta = 60°$, the rod in Fig. 16–33 has an angular velocity of 3 rad/s and an angular acceleration of 2 rad/s^2. At this same instant, the collar C is traveling outward along the rod such that when $x =$ 0.2 m the velocity is 2 m/s and the acceleration is 3 m/s^2, both measured relative to the rod. Determine the Coriolis acceleration and the velocity and acceleration of the collar at this instant.

Solution

Coordinate Axes. The origin of both coordinate systems is located at point O, Fig. 16–33. Since motion of the collar is reported relative to the rod, the moving x, y, z frame of reference is *attached* to the rod.

Kinematic Equations

$$\mathbf{v}_C = \mathbf{v}_O + \boldsymbol{\Omega} \times \mathbf{r}_{C/O} + (\mathbf{v}_{C/O})_{xyz} \tag{1}$$

$$\mathbf{a}_C = \mathbf{a}_O + \dot{\boldsymbol{\Omega}} \times \mathbf{r}_{C/O} + \boldsymbol{\Omega} \times (\boldsymbol{\Omega} \times \mathbf{r}_{C/O}) + 2\boldsymbol{\Omega} \times (\mathbf{v}_{C/O})_{xyz} + (\mathbf{a}_{C/O})_{xyz} \tag{2}$$

It will be simpler to express the data in terms of **i, j, k** component vectors rather than **I, J, K** components. Hence,

Motion of moving reference	*Motion of C with respect to moving reference*
$\mathbf{v}_O = \mathbf{0}$	$\mathbf{r}_{C/O} = \{0.2\mathbf{i}\}$ m
$\mathbf{a}_O = \mathbf{0}$	$(\mathbf{v}_{C/O})_{xyz} = \{2\mathbf{i}\}$ m/s
$\boldsymbol{\Omega} = \{-3\mathbf{k}\}$ rad/s	$(\mathbf{a}_{C/O})_{xyz} = \{3\mathbf{i}\}$ m/s^2
$\dot{\boldsymbol{\Omega}} = \{-2\mathbf{k}\}$ rad/s^2	

From Eq. 2, the Coriolis acceleration is defined as

$$\mathbf{a}_{Cor} = 2\boldsymbol{\Omega} \times (\mathbf{v}_{C/O})_{xyz} = 2(-3\mathbf{k}) \times (2\mathbf{i}) = \{-12\mathbf{j}\} \text{ m/s}^2 \qquad \textit{Ans.}$$

This vector is shown dashed in Fig. 16–33. If desired, it may be resolved into **I, J** components acting along the X and Y axes, respectively.

The velocity and acceleration of the collar are determined by substituting the data into Eqs. 1 and 2 and evaluating the cross products, which yields

$$\begin{aligned}\mathbf{v}_C &= \mathbf{v}_O + \boldsymbol{\Omega} \times \mathbf{r}_{C/O} + (\mathbf{v}_{C/O})_{xyz} \\ &= \mathbf{0} + (-3\mathbf{k}) \times (0.2\mathbf{i}) + 2\mathbf{i} \\ &= \{2\mathbf{i} - 0.6\mathbf{j}\} \text{ m/s} \qquad \textit{Ans.}\end{aligned}$$

$$\begin{aligned}\mathbf{a}_C &= \mathbf{a}_O + \dot{\boldsymbol{\Omega}} \times \mathbf{r}_{C/O} + \boldsymbol{\Omega} \times (\boldsymbol{\Omega} \times \mathbf{r}_{C/O}) + 2\boldsymbol{\Omega} \times (\mathbf{v}_{C/O})_{xyz} + (\mathbf{a}_{C/O})_{xyz} \\ &= \mathbf{0} + (-2\mathbf{k}) \times (0.2\mathbf{i}) + (-3\mathbf{k}) \times [(-3\mathbf{k}) \times (0.2\mathbf{i})] + 2(-3\mathbf{k}) \times (2\mathbf{i}) + 3\mathbf{i} \\ &= \mathbf{0} - 0.4\mathbf{j} - 1.80\mathbf{i} - 12\mathbf{j} + 3\mathbf{i} \\ &= \{1.20\mathbf{i} - 12.4\mathbf{j}\} \text{ m/s}^2 \qquad \textit{Ans.}\end{aligned}$$

EXAMPLE 16-20

The rod AB, shown in Fig. 16–34, rotates clockwise such that it has an angular velocity $\omega_{AB} = 3$ rad/s and angular acceleration $\alpha_{AB} = 4$ rad/s^2 when $\theta = 45°$. Determine the angular motion of rod DE at this instant. The collar at C is pin connected to AB and slides over rod DE.

Solution

Coordinate Axes. The origin of both the fixed and moving frames of reference is located at D, Fig. 16–34. Furthermore, the x, y, z reference is attached to and rotates with rod DE so that the relative motion of the collar is easy to follow.

Kinematic Equations

$$\mathbf{v}_C = \mathbf{v}_D + \boldsymbol{\Omega} \times \mathbf{r}_{C/D} + (\mathbf{v}_{C/D})_{xyz} \quad (1)$$

$$\mathbf{a}_C = \mathbf{a}_D + \dot{\boldsymbol{\Omega}} \times \mathbf{r}_{C/D} + \boldsymbol{\Omega} \times (\boldsymbol{\Omega} \times \mathbf{r}_{C/D}) + 2\boldsymbol{\Omega} \times (\mathbf{v}_{C/D})_{xyz} + (\mathbf{a}_{C/D})_{xyz} \quad (2)$$

All vectors will be expressed in terms of **i, j, k** components.

Motion of moving reference	*Motion of C with respect to moving reference*
$\mathbf{v}_D = \mathbf{0}$	$\mathbf{r}_{C/D} = \{0.4\mathbf{i}\}$ m
$\mathbf{a}_D = \mathbf{0}$	$(\mathbf{v}_{C/D})_{xyz} = (v_{C/D})_{xyz}\mathbf{i}$
$\boldsymbol{\Omega} = -\omega_{DE}\mathbf{k}$	$(\mathbf{a}_{C/D})_{xyz} = (a_{C/D})_{xyz}\mathbf{i}$
$\dot{\boldsymbol{\Omega}} = -\alpha_{DE}\mathbf{k}$	

Motion of C: Since the collar moves along a *circular path,* its velocity and acceleration can be determined using Eqs. 16–9 and 16–14.

$$\mathbf{v}_C = \boldsymbol{\omega}_{AB} \times \mathbf{r}_{C/A} = (-3\mathbf{k}) \times (0.4\mathbf{i} + 0.4\mathbf{j}) = \{1.2\mathbf{i} - 1.2\mathbf{j}\} \text{ m/s}$$

$$\mathbf{a}_C = \boldsymbol{\alpha}_{AB} \times \mathbf{r}_{C/A} - \omega_{AB}^2\mathbf{r}_{C/A}$$
$$= (-4\mathbf{k}) \times (0.4\mathbf{i} + 0.4\mathbf{j}) - (3)^2(0.4\mathbf{i} + 0.4\mathbf{j}) = \{-2\mathbf{i} - 5.2\mathbf{j}\} \text{ m/s}^2$$

Substituting the data into Eqs. 1 and 2, we have

$$\mathbf{v}_C = \mathbf{v}_D + \boldsymbol{\Omega} \times \mathbf{r}_{C/D} + (\mathbf{v}_{C/D})_{xyz}$$
$$1.2\mathbf{i} - 1.2\mathbf{j} = \mathbf{0} + (-\omega_{DE}\mathbf{k}) \times (0.4\mathbf{i}) + (v_{C/D})_{xyz}\mathbf{i}$$
$$1.2\mathbf{i} - 1.2\mathbf{j} = \mathbf{0} - 0.4\omega_{DE}\mathbf{j} + (v_{C/D})_{xyz}\mathbf{i}$$
$$(v_{C/D})_{xyz} = 1.2 \text{ m/s}$$
$$\omega_{DE} = 3 \text{ rad/s} \circlearrowright \qquad \textit{Ans.}$$

$$\mathbf{a}_C = \mathbf{a}_D + \dot{\boldsymbol{\Omega}} \times \mathbf{r}_{C/D} + \boldsymbol{\Omega} \times (\boldsymbol{\Omega} \times \mathbf{r}_{C/D}) + 2\boldsymbol{\Omega} \times (\mathbf{v}_{C/D})_{xyz} + (\mathbf{a}_{C/D})_{xyz}$$
$$-2\mathbf{i} - 5.2\mathbf{j} = \mathbf{0} + (-\alpha_{DE}\mathbf{k}) \times (0.4\mathbf{i}) + (-3\mathbf{k}) \times [(-3\mathbf{k}) \times (0.4\mathbf{i})]$$
$$+ 2(-3\mathbf{k}) \times (1.2\mathbf{i}) + (a_{C/D})_{xyz}\mathbf{i}$$
$$-2\mathbf{i} - 5.2\mathbf{j} = -0.4\alpha_{DE}\mathbf{j} - 3.6\mathbf{i} - 7.2\mathbf{j} + (a_{C/D})_{xyz}\mathbf{i}$$
$$(a_{C/D})_{xyz} = 1.6 \text{ m/s}^2$$
$$\alpha_{DE} = -5 \text{ rad/s}^2 = 5 \text{ rad/s}^2 \circlearrowleft \qquad \textit{Ans.}$$

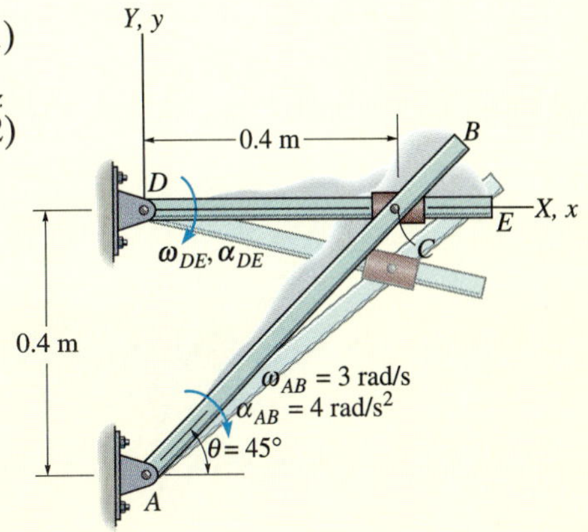

Fig. 16–34

EXAMPLE 16–21

Two planes A and B are flying at the same elevation and have the motions shown in Fig. 16–35. Determine the velocity and acceleration of A as measured by the pilot of B.

Solution

Coordinate Axes. Since the relative motion of A with respect to the pilot in B is being sought, the x, y, z axes are attached to plane B, Fig. 16–35. At the *instant* considered, the origin B coincides with the origin of the fixed X, Y, Z frame.

Kinematic Equations

$$\mathbf{v}_A = \mathbf{v}_B + \mathbf{\Omega} \times \mathbf{r}_{A/B} + (\mathbf{v}_{A/B})_{xyz} \tag{1}$$

$$\mathbf{a}_A = \mathbf{a}_B + \dot{\mathbf{\Omega}} \times \mathbf{r}_{A/B} + \mathbf{\Omega} \times (\mathbf{\Omega} \times \mathbf{r}_{A/B}) + 2\mathbf{\Omega} \times (\mathbf{v}_{A/B})_{xyz} + (\mathbf{a}_{A/B})_{xyz} \tag{2}$$

Motion of moving reference:

$$\mathbf{v}_B = \{600\mathbf{j}\} \text{ km/h}$$

$$(a_B)_n = \frac{v_B^2}{\rho} = \frac{(600)^2}{400} = 900 \text{ km/h}^2$$

$$\mathbf{a}_B = (\mathbf{a}_B)_n + (\mathbf{a}_B)_t = \{900\mathbf{i} - 100\mathbf{j}\} \text{ km/h}^2$$

$$\Omega = \frac{v_B}{\rho} = \frac{600 \text{ km/h}}{400 \text{ km}} = 1.5 \text{ rad/h} \circlearrowright \qquad \mathbf{\Omega} = \{-1.5\mathbf{k}\} \text{ rad/h}$$

$$\dot{\Omega} = \frac{(a_B)_t}{\rho} = \frac{100 \text{ km/h}^2}{400 \text{ km}} = 0.25 \text{ rad/h}^2 \circlearrowleft \qquad \dot{\mathbf{\Omega}} = \{0.25\mathbf{k}\} \text{ rad/h}^2$$

Motion of A with respect to moving reference:

$$\mathbf{r}_{A/B} = \{-4\mathbf{i}\} \text{ km} \qquad (\mathbf{v}_{A/B})_{xyz} = ? \qquad (\mathbf{a}_{A/B})_{xyz} = ?$$

Substituting the data into Eqs. 1 and 2, realizing that $\mathbf{v}_A = \{700\mathbf{j}\}$ km/h and $\mathbf{a}_A = \{50\mathbf{j}\}$ km/h^2, we have

$$\mathbf{v}_A = \mathbf{v}_B + \mathbf{\Omega} \times \mathbf{r}_{A/B} + (\mathbf{v}_{A/B})_{xyz}$$

$$700\mathbf{j} = 600\mathbf{j} + (-1.5\mathbf{k}) \times (-4\mathbf{i}) + (\mathbf{v}_{A/B})_{xyz}$$

$$(\mathbf{v}_{A/B})_{xyz} = \{94\mathbf{j}\} \text{ km/h} \qquad \textit{Ans.}$$

$$\mathbf{a}_A = \mathbf{a}_B + \dot{\mathbf{\Omega}} \times \mathbf{r}_{A/B} + \mathbf{\Omega} \times (\mathbf{\Omega} \times \mathbf{r}_{A/B}) + 2\mathbf{\Omega} \times (\mathbf{v}_{A/B})_{xyz} + (\mathbf{a}_{A/B})_{xyz}$$

$$50\mathbf{j} = (900\mathbf{i} - 100\mathbf{j}) + (0.25\mathbf{k}) \times (-4\mathbf{i}) + (-1.5\mathbf{k}) \times [(-1.5\mathbf{k}) \times (-4\mathbf{i})] + 2(-1.5\mathbf{k}) \times (94\mathbf{j}) + (\mathbf{a}_{A/B})_{xyz}$$

$$(\mathbf{a}_{A/B})_{xyz} = \{-1191\mathbf{i} + 151\mathbf{j}\} \text{ km/h}^2 \qquad \textit{Ans.}$$

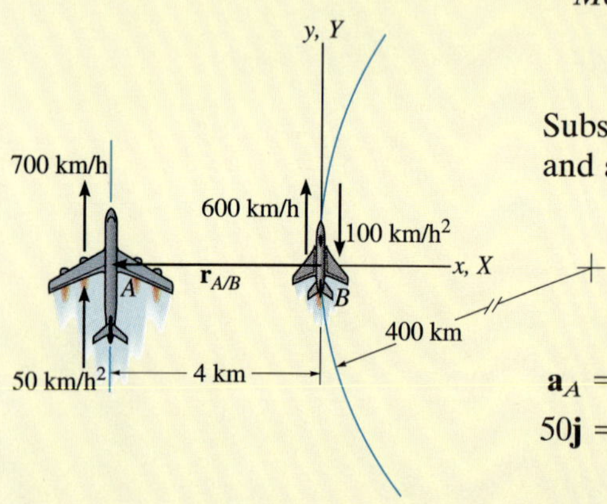

Fig. 16–35

The solution of this problem should be compared with that of Example 12–26.

Problems

16-131. At the instant shown, ball B is rolling along the slot in the disk with a velocity of 600 mm/s and an acceleration of 150 mm/s^2, both measured relative to the disk and directed away from O. If at the same instant the disk has the angular velocity and angular acceleration shown, determine the velocity and acceleration of the ball at this instant.

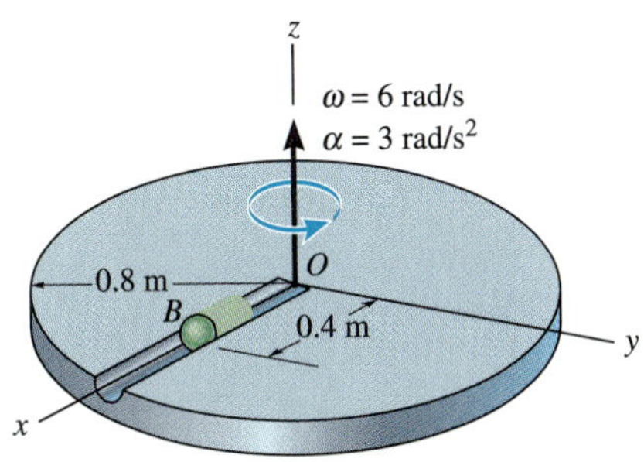

Prob. 16–131

***16-132.** The ball B of negligible size rolls through the tube such that at the instant shown it has a velocity of 5 ft/s and an acceleration of 3 ft/s^2, measured relative to the tube. If the tube has an angular velocity of $\omega = 3$ rad/s and an angular acceleration of $\alpha = 5$ rad/s^2 at this same instant, determine the velocity and acceleration of the ball.

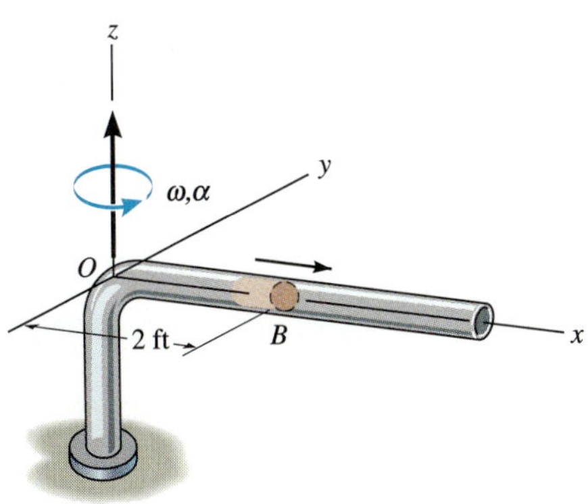

Prob. 16–132

16-133. The man stands on the platform at O and runs out toward the edge such that when he is at A, $y = 5$ ft, his mass center has a velocity of 2 ft/s and an acceleration of 3 ft/s^2, both measured with respect to the platform and directed along the y axis. If the platform has the angular motions shown, determine the velocity and acceleration of his mass center at this instant.

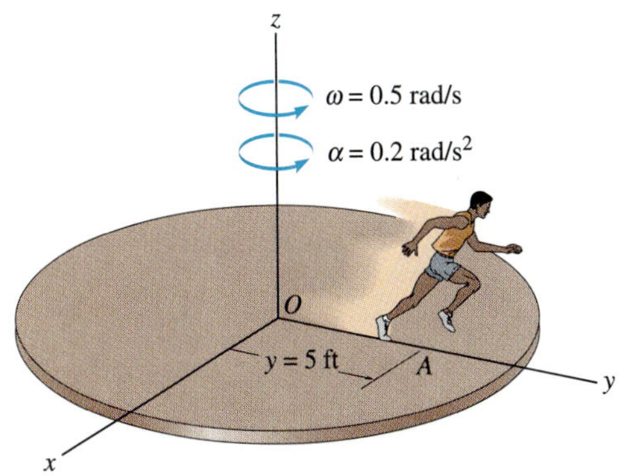

Prob. 16–133

16-134. The dumpster pivots about C and is operated by the hydraulic cylinder AB. If the cylinder is extending at a constant rate of 0.5 ft/s, determine the angular velocity $\boldsymbol{\omega}$ of the container at the instant it becomes horizontal as shown.

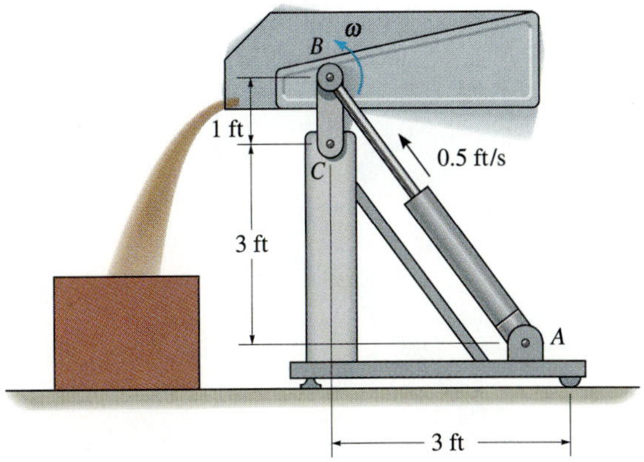

Prob. 16–134

16-135. Rod AB rotates counterclockwise with a constant angular velocity of 2 rad/s. Determine the velocity and acceleration of point C located on the double collar when $\theta = 45°$. The double collar consists of two pin-connected sliders which are constrained to move along the circular shaft and the rod AB.

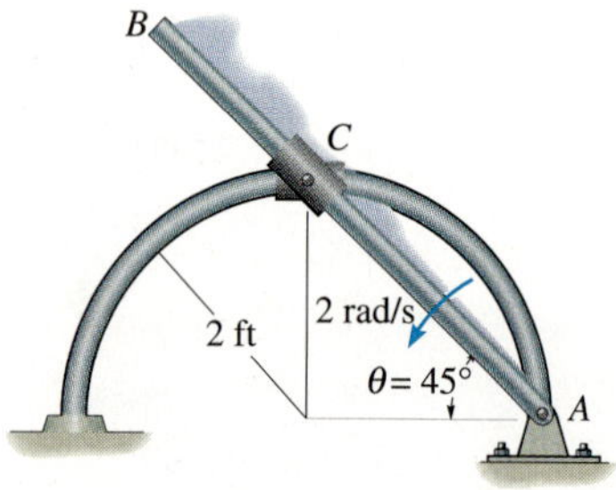

Prob. 16–135

16-137. The double collar E consists of two pin-connected collars that slide freely over rods AB and CD. At the instant shown, rod AB has a clockwise angular velocity of 5 rad/s and an angular acceleration of 3 rad/s². Determine the angular velocity and angular acceleration of rod CD at this instant.

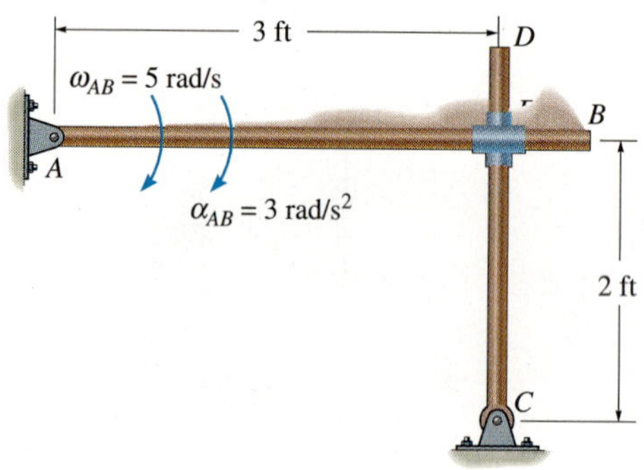

Prob. 16–137

***16-136.** The collar E is attached to, and pivots about, rod AB while it slides on rod CD. If rod AB has an angular velocity of 6 rad/s and an angular acceleration of 1 rad/s², both acting clockwise, determine the angular velocity and the angular acceleration of rod CD at the instant shown.

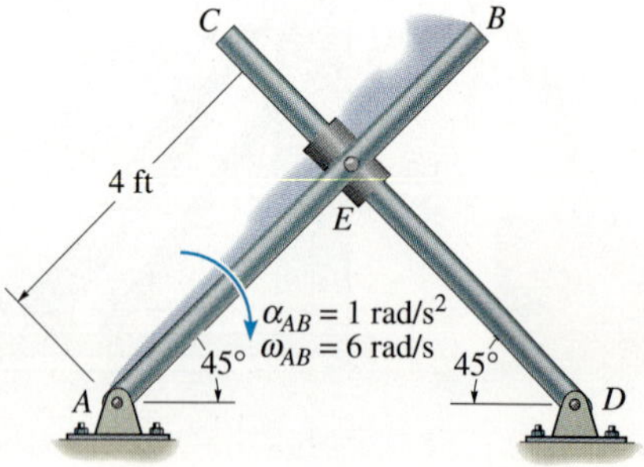

Prob. 16–136

16-138. At the instant $\theta = 45°$, link DC has an angular velocity of $\omega_{DC} = 4$ rad/s and an angular acceleration of $\alpha_{DC} = 2$ rad/s². Determine the angular velocity and angular acceleration of rod AB at this instant. The collar at C is pin connected to DC and slides over AB.

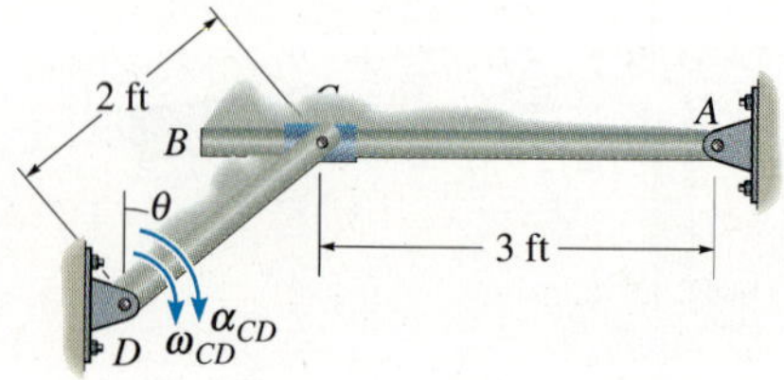

Prob. 16–138

16-139. At a given instant, rod AB has the angular motions shown. Determine the angular velocity and angular acceleration of rod CD at this instant. There is a collar at C.

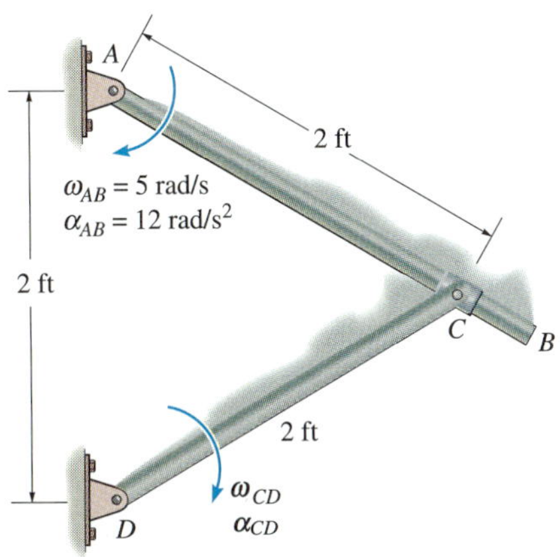

Prob. 16–139

*16-140. The disk rolls without slipping and at a given instant has the angular motion shown. Determine the angular velocity and angular acceleration of the slotted link BC at this instant. The peg at A is fixed to the disk.

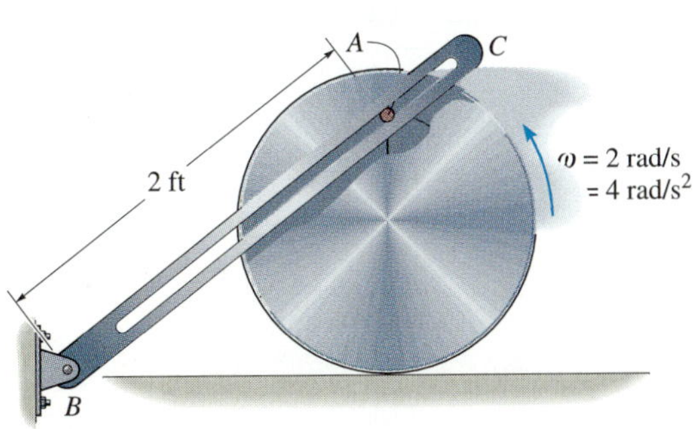

Prob. 16–140

16-141. The disk rotates with the angular motion shown. Determine the angular velocity and angular acceleration of the slotted link AC at this instant. The peg at B is fixed to the disk.

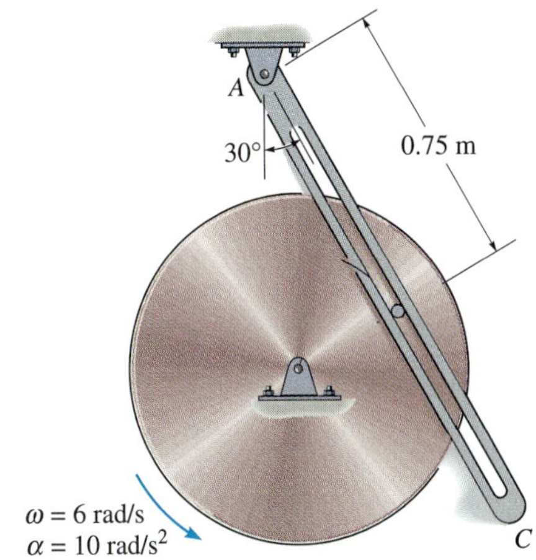

Prob. 16–141

16-142. If the slider block C is fixed to the disk that has a constant counterclockwise angular velocity of 4 rad/s, determine the angular velocity and angular acceleration of the slotted arm AB at the instant shown.

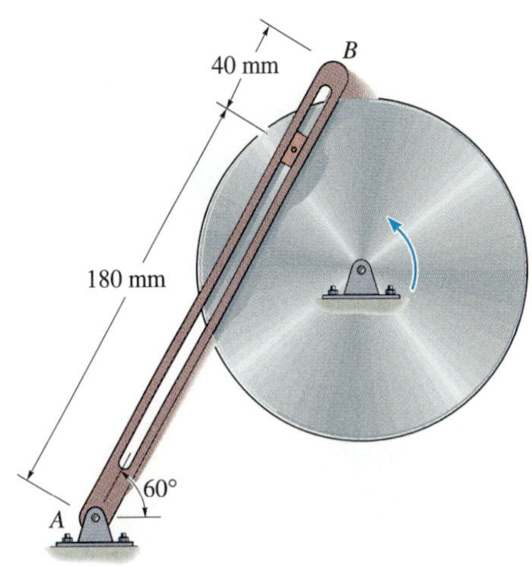

Prob. 16–142

16-143. The two-link mechanism serves to amplify angular motion. Link AB has a pin at B which is confined to move within the slot of link CD. If at the instant shown, AB (input) has an angular velocity of $\omega_{AB} = 2.5$ rad/s and an angular acceleration of $\alpha_{AB} = 3$ rad/s², determine the angular velocity and angular acceleration of CD (output) at this instant.

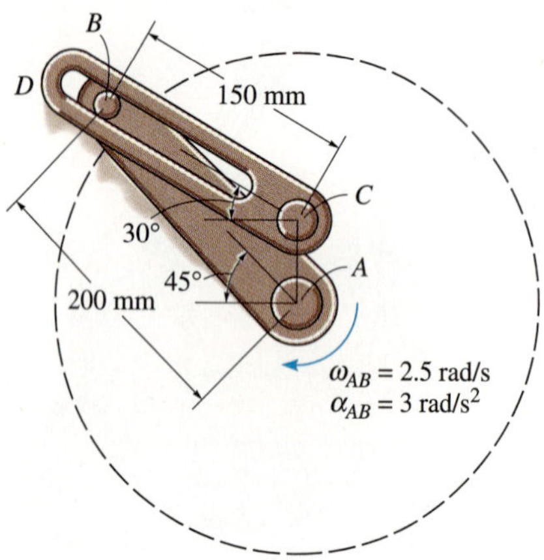

Prob. 16–143

***16-144.** The block B of the quick-return mechanism is confined to move within the slot in member CD. If AB is rotating at a constant rate of $\omega_{AB} = 3$ rad/s, determine the angular velocity and angular acceleration of member CD at the instant shown.

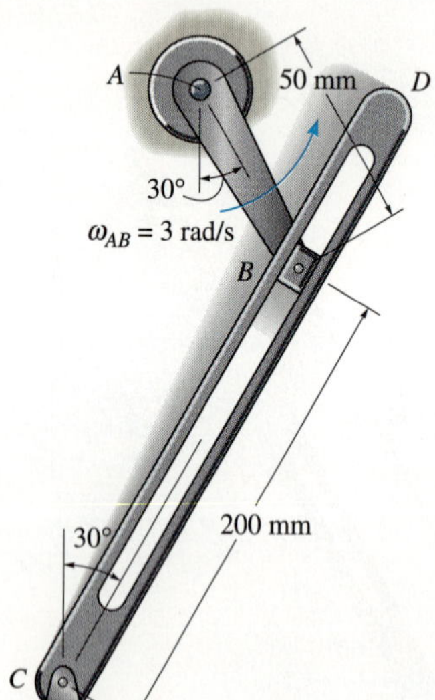

Prob. 16–144

16-145. The quick-return mechanism consists of a crank AB, slider block B, and slotted link CD. If the crank has the angular motion shown, determine the angular motion of the slotted link at this instant.

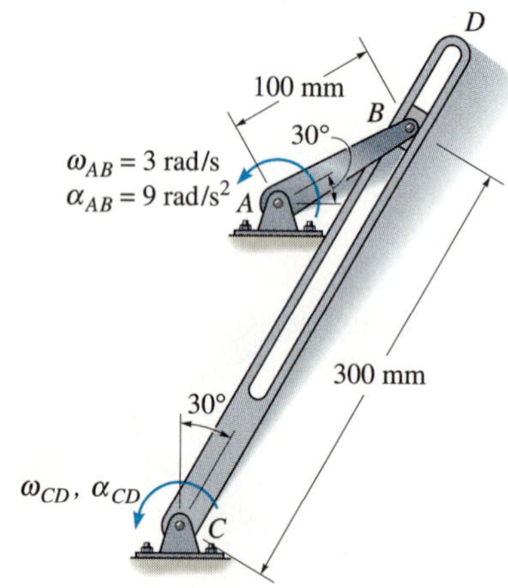

Prob. 16–145

16-146. Particles B and A move along the parabolic and circular paths, respectively. If B has a velocity of 7 m/s in the direction shown and its speed is increasing at 4 m/s², while A has a velocity of 8 m/s in the direction shown and its speed is decreasing at 6 m/s², determine the relative velocity and relative acceleration of B with respect to A.

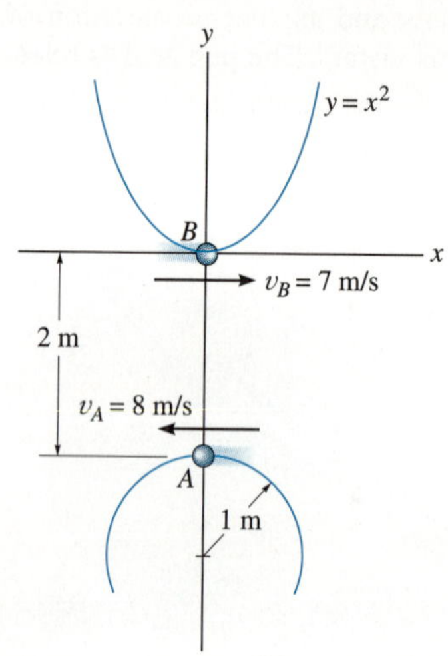

Prob. 16–146

16-147. The Geneva mechanism is used in a packaging system to convert constant angular motion into intermittent angular motion. The star wheel A makes one sixth of a revolution for each full revolution of the driving wheel B and the attached guide C. To do this, pin P, which is attached to B, slides into one of the radial slots of A, thereby turning wheel A, and then exits the slot. If B has a constant angular velocity of $\omega_B = 4$ rad/s, determine $\boldsymbol{\omega}_A$ and $\boldsymbol{\alpha}_A$ of wheel A at the instant $\theta = 30°$ as shown.

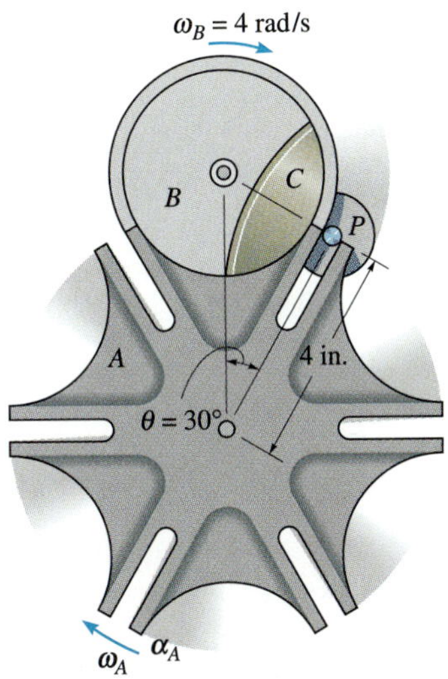

Prob. 16–147

***16-148.** The combination hinge and support for a lid uses the mechanism shown. If the lid is given an angular velocity of 4 rad/s when it is in the position shown, determine the angular velocity of link DC at this instant. Note that the end E of member FBE slides within the grooved bracket on the lid.

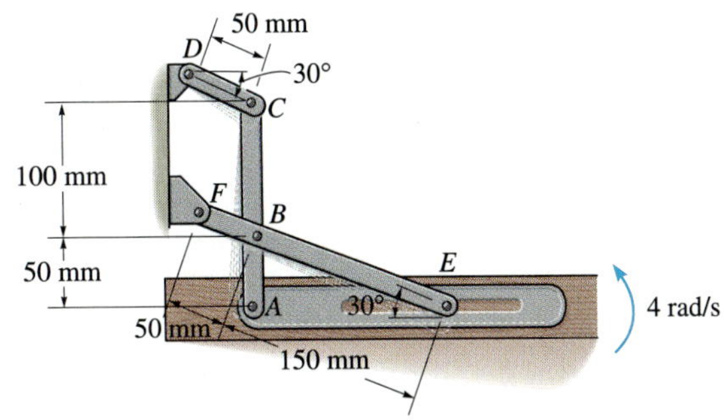

Prob. 16–148

Design Projects

16–1D. DESIGN OF A BELT TRANSMISSION SYSTEM

The wheel A is used in a textile mill and must be rotated counterclockwise at 4 rad/s. This can be done using a motor which is mounted on the platform at the location shown. If the shaft B on the motor can rotate clockwise 50 rad/s, design a method for transmitting the rotation from B to A. Use a series of belts and pulleys as a basis for your design. A belt drives the rotation of wheel A by wrapping around its outer surface, and a pulley can be attached to the shaft of the motor as well as anywhere else. Do not let the length of any belt be longer than 6 ft. Submit a drawing of your design and the calculations of the kinematics. Also, determine the total cost of materials if any belt costs \$2.50, and any pulley costs \$2r, where r is the radius of the pulley in inches.

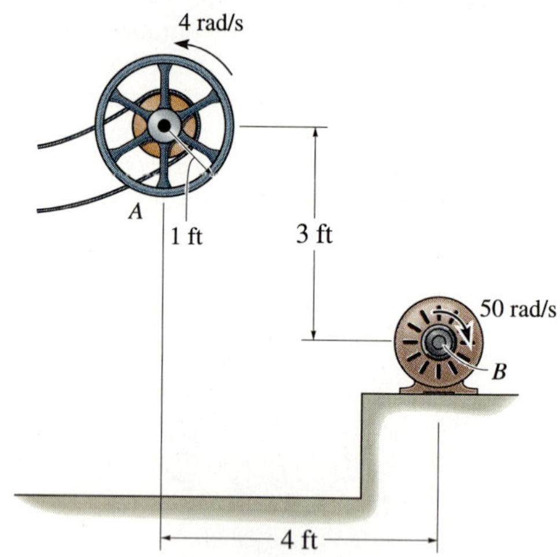

Prob. 16–1D

16–2D. DESIGN OF AN OSCILLATING LINK MECHANISM

The operation of a sewing machine requires the 200-mm-long bar to oscillate back-and-forth through an angle of 60° every 0.2 seconds. A motor having a drive shaft which turns at 40 rad/s is available to provide the necessary power. Specify the location of the motor and design a mechanism required to perform the motion. Submit a drawing of your design, showing the placement of the motor, and compute the velocity and acceleration of the end A of the link as a function of its angle of rotation $0° \leq \theta \leq 60°$.

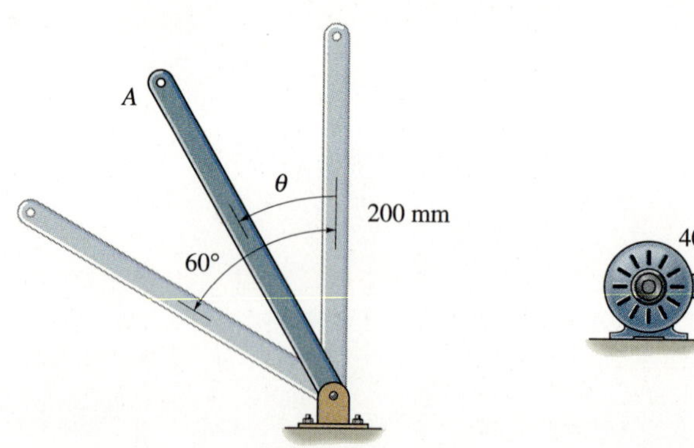

Prob. 16–2D

16–3D. DESIGN OF A RETRACTABLE AIRCRAFT LANDING-GEAR MECHANISM

The nose wheel of a small plane is attached to member AB, which is pinned to the aircraft frame at B. Design a mechanism that will allow the wheel to be fully retracted forward; i.e., rotated clockwise 90°, in $t \leq 4$ seconds. Use a hydraulic cylinder which has a closed length of 1.25 ft and, if needed, a fully extended length of 2 ft. Make sure that your design holds the wheel in a stable position when the wheel is on the ground. Show plots of the angular velocity and angular acceleration of AB versus its angular position $0° \leq \theta \leq 90°$.

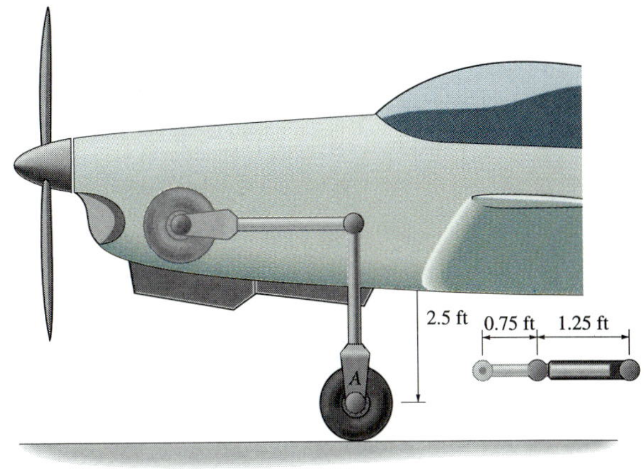

Prob. 16–3D

16–4D. DESIGN OF A SAW LINK MECHANISM

The saw blade in a lumber mill is required to remain in the horizontal position and undergo a complete back-and-forth motion in 2 seconds. An electric motor, having a shaft rotation of 50 rad/s, is available to power the saw and can be located anywhere. Design a mechanism that will transfer the rotation of the motor's shaft to the saw blade. Submit drawings of your design and calculations of the kinematics of the saw blade. Include a plot of the velocity and acceleration of the saw blade as a function of it horizontal position. Note that to cut through the log the blade must be allowed to move freely downward as well as back and forth.

Prob. 16–4D

The forces acting on this dragster as it begins to accelerate are quite severe and must be accounted for in the design of its structure.

CHAPTER 17

Planar Kinetics of a Rigid Body: Force and Acceleration

Chapter Objectives

- To introduce the methods used to determine the mass moment of inertia of a body.
- To develop the planar kinetic equations of motion for a symmetric rigid body.
- To discuss applications of these equations to bodies undergoing translation, rotation about a fixed axis, and general plane motion.

17.1 Moment of Inertia

Since a body has a definite size and shape, an applied nonconcurrent force system may cause the body to both translate and rotate. The translational aspects of the motion were studied in Chapter 13 and are governed by the equation $\mathbf{F} = m\mathbf{a}$. It will be shown in Sec. 17.2 that the rotational aspects, caused by a moment $\mathbf{M}$, are governed by an equation of the form $\mathbf{M} = I\boldsymbol{\alpha}$. The symbol I in this equation is termed the moment of inertia. By comparison, the *moment of inertia* is a measure of the resistance of a body to *angular acceleration* ($\mathbf{M} = I\boldsymbol{\alpha}$) in the same way that *mass* is a measure of the body's resistance to *acceleration* ($\mathbf{F} = m\mathbf{a}$).

The flywheel on the engine of this tractor has a large moment of inertia about its axis of rotation. Once it is set into motion, it will be difficult to stop, and this in turn will prevent the engine from stalling and instead it will allow it to maintain a constant power.

We define the *moment of inertia* as the integral of the "second moment" about an axis of all the elements of mass dm which compose the body.* For example, the body's moment of inertia about the z axis in Fig. 17–1 is

$$I = \int_m r^2\,dm \tag{17–1}$$

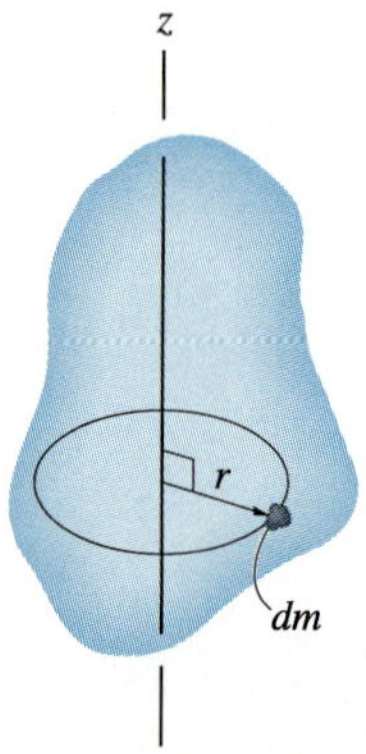

Fig. 17–1

Here the "moment arm" r is the perpendicular distance from the z axis to the arbitrary element dm. Since the formulation involves r, the value of I is different for each axis about which it is computed. In the study of planar kinetics, the axis which is generally chosen for analysis passes through the body's mass center G and is always perpendicular to the plane of motion. The moment of inertia computed about this axis will be denoted as I_G. Realize that because r is squared in Eq. 17–1, the mass moment of inertia is always a *positive* quantity. Common units used for its measurement are $\text{kg}\cdot\text{m}^2$ or $\text{slug}\cdot\text{ft}^2$.

If the body consists of material having a variable density, $\rho = \rho\,(x, y, z)$, the elemental mass dm of the body may be expressed in terms of its density and volume as $dm = \rho\,dV$. Substituting dm into Eq. 17–1, the body's moment of inertia is then computed using *volume elements* for integration; i.e.,

$$I = \int_V r^2 \rho\,dV \tag{17–2}$$

*Another property of the body, which measures the symmetry of the body's mass with respect to a coordinate system, is the product of inertia. This property applies to the three-dimensional motion of a body and will be discussed in Chapter 21.

In the special case of ρ being a *constant,* this term may be factored out of the integral, and the integration is then purely a function of geometry,

$$I = \rho \int_V r^2 \, dV \qquad (17\text{–}3)$$

When the elemental volume chosen for integration has infinitesimal dimensions in all three directions, e.g., $dV = dx\, dy\, dz$, Fig. 17–2*a*, the moment of inertia of the body must be determined using "triple integration." The integration process can, however, be simplified to a *single integration* provided the chosen elemental volume has a differential size or thickness in only *one direction.* Shell or disk elements are often used for this purpose.

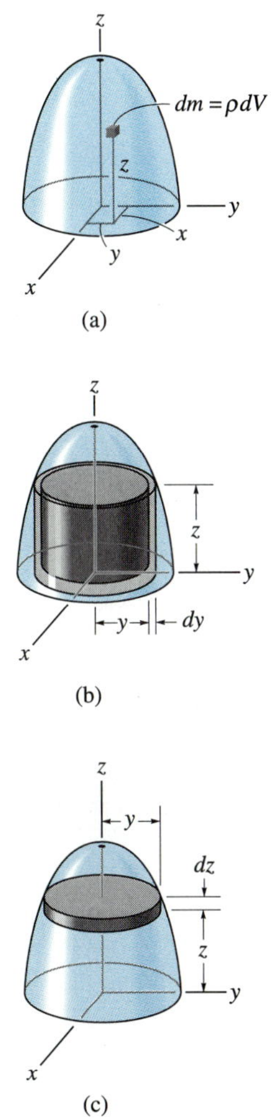

Fig. 17–2

Procedure for Analysis

For integration, we will consider only symmetric bodies having surfaces which are generated by revolving a curve about an axis. An example of such a body which is generated about the z axis is shown in Fig. 17–2*a*. Two types of differential elements can be chosen.

Shell Element

- If a *shell element* having a height z, radius $r = y$, and thickness dy is chosen for integration, Fig. 17–2*b*, then the volume is $dV = (2\pi y)(z)\, dy$.
- This element may be used in Eq. 17–2 or 17–3 for determining the moment of inertia I_z of the body about the z axis, since the *entire element,* due to its "thinness," lies at the *same* perpendicular distance $r = y$ from the z axis (see Example 17–1).

Disk Element

- If a disk element having a radius y and a thickness dz is chosen for integration, Fig. 17–2*c*, then the volume is $dV = (\pi y^2)\, dz$.
- This element is *finite* in the radial direction, and consequently its parts *do not* all lie at the *same radial distance r* from the z axis. As a result, Eq. 17–2 or 17–3 *cannot* be used to determine I_z directly. Instead, to perform the integration it is first necessary to determine the moment of inertia *of the element* about the z axis and then integrate this result (see Example 17–2).

EXAMPLE 17–1

Determine the moment of inertia of the cylinder shown in Fig. 17–3*a* about the z axis. The density of the material, ρ, is constant.

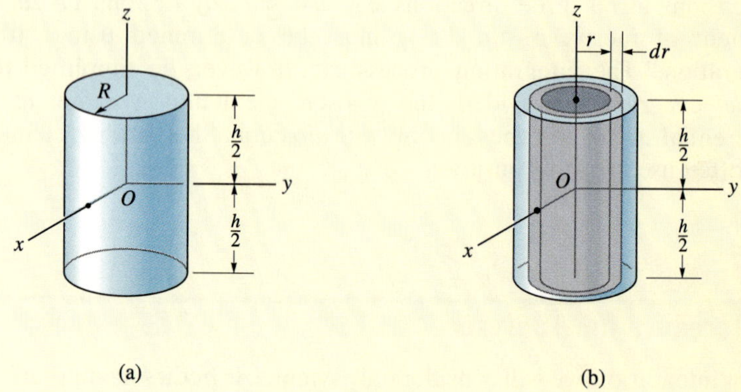

Fig. 17–3

Solution

Shell Element. This problem may be solved using the *shell element* in Fig. 17–3*b* and single integration. The volume of the element is $dV = (2\pi r)(h)\,dr$, so that its mass is $dm = \rho\,dV = \rho(2\pi hr\,dr)$. Since the *entire element* lies at the same distance r from the z axis, the moment of inertia *of the element* is

$$dI_z = r^2\,dm = \rho 2\pi h r^3\,dr$$

Integrating over the entire region of the cylinder yields

$$I_z = \int_m r^2\,dm = \rho 2\pi h \int_0^R r^3\,dr = \frac{\rho\pi}{2} R^4 h$$

The mass of the cylinder is

$$m = \int_m dm = \rho 2\pi h \int_0^R r\,dr = \rho\pi h R^2$$

so that

$$I_z = \frac{1}{2} m R^2 \qquad \textit{Ans.}$$

EXAMPLE 17-2

A solid is formed by revolving the shaded area shown in Fig. 17–4*a* about the y axis. If the density of the material is 5 slug/ft^3, determine the moment of inertia about the y axis.

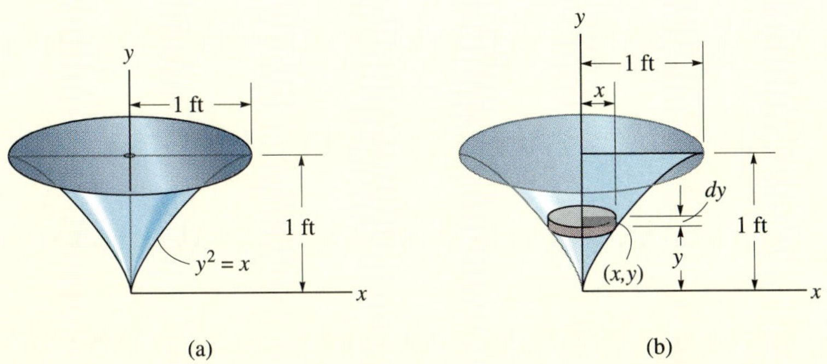

Fig. 17–4

Solution

Disk Element. The moment of inertia will be computed using a *disk element,* as shown in Fig. 17–4*b*. Here the element intersects the curve at the arbitrary point (x, y) and has a mass

$$dm = \rho\, dV = \rho(\pi x^2)\, dy$$

Although all portions of the element are *not* located at the same distance from the y axis, it is still possible to determine the moment of inertia dI_y *of the element* about the y axis. In the preceding example it was shown that the moment of inertia of a cylinder about its longitudinal axis is $I = \frac{1}{2}mR^2$, where m and R are the mass and radius of the cylinder. Since the height of the cylinder is not involved in this formula, the cylinder itself can be thought of as a disk. Thus, for the disk element in Fig. 17–4*b*, we have

$$dI_y = \tfrac{1}{2}(dm)x^2 = \tfrac{1}{2}[\rho(\pi x^2)\, dy]x^2$$

Substituting $x = y^2$, $\rho = 5$ slug/ft^3, and integrating with respect to y, from $y = 0$ to $y = 1$ ft, yields the moment of inertia for the entire solid.

$$I_y = \frac{\pi(5)}{2}\int_0^1 x^4\, dy = \frac{\pi(5)}{2}\int_0^1 y^8\, dy = 0.873 \text{ slug}\cdot\text{ft}^2 \qquad \textit{Ans.}$$

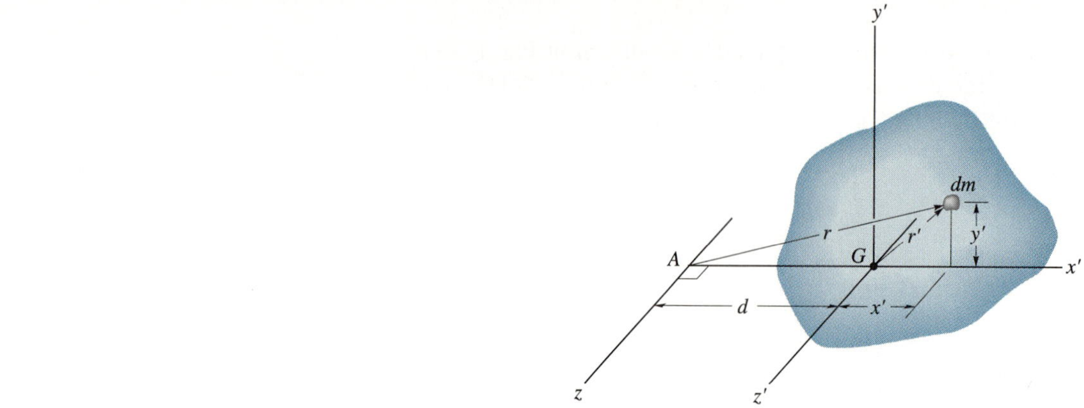

Fig. 17–5

Parallel-Axis Theorem. If the moment of inertia of the body about an axis passing through the body's mass center is known, then the moment of inertia about any other *parallel axis* may be determined by using the *parallel-axis theorem.* This theorem can be derived by considering the body shown in Fig. 17–5. The z' axis passes through the mass center G, whereas the corresponding *parallel z axis* lies at a constant distance d away. Selecting the differential element of mass dm, which is located at point (x', y'), and using the Pythagorean theorem, $r^2 = (d + x')^2 + y'^2$, we can express the moment of inertia of the body about the z axis as

$$I = \int_m r^2\,dm = \int_m [(d + x')^2 + y'^2]\,dm$$

$$= \int_m (x'^2 + y'^2)\,dm + 2d\int_m x'\,dm + d^2\int_m dm$$

Since $r'^2 = x'^2 + y'^2$, the first integral represents I_G. The second integral equals *zero,* since the z' axis passes through the body's mass center, i.e., $\int x'\,dm = \bar{x}'\int dm = 0$ since $\bar{x}' = 0$. Finally, the third integral represents the total mass m of the body. Hence, the moment of inertia about the z

axis can be written as

$$I = I_G + md^2 \tag{17–4}$$

where

I_G = moment of inertia about the z' axis passing through the mass center G

m = mass of the body

d = perpendicular distance between the parallel axes

Radius of Gyration. Occasionally, the moment of inertia of a body about a specified axis is reported in handbooks using the *radius of gyration, k*. This value has units of length, and when it and the body's mass m are known, the body's moment of inertia is determined from the equation

$$I = mk^2 \quad \text{or} \quad k = \sqrt{\frac{I}{m}} \tag{17–5}$$

Note the *similarity* between the definition of k in this formula and r in the equation $dI = r^2\, dm$, which defines the moment of inertia of an elemental mass dm of the body about an axis.

Composite Bodies. If a body is constructed of a number of simple shapes such as disks, spheres, and rods, the moment of inertia of the body about any axis z can be determined by adding algebraically the moments of inertia of all the composite shapes computed about the z axis. Algebraic addition is necessary since a composite part must be considered as a negative quantity if it has already been counted as a piece of another part—for example, a "hole" subtracted from a solid plate. The parallel-axis theorem is needed for the calculations if the center of mass of each composite part does not lie on the z axis. For the calculation, then, $I = \Sigma(I_G + md^2)$. Here I_G for each of the composite parts is computed by integration or can be determined from a table, such as the one given on the inside back cover of this book.

EXAMPLE 17–3

If the plate shown in Fig. 17–6*a* has a density of 8000 kg/m³ and a thickness of 10 mm, determine its moment of inertia about an axis directed perpendicular to the page and passing through point *O*.

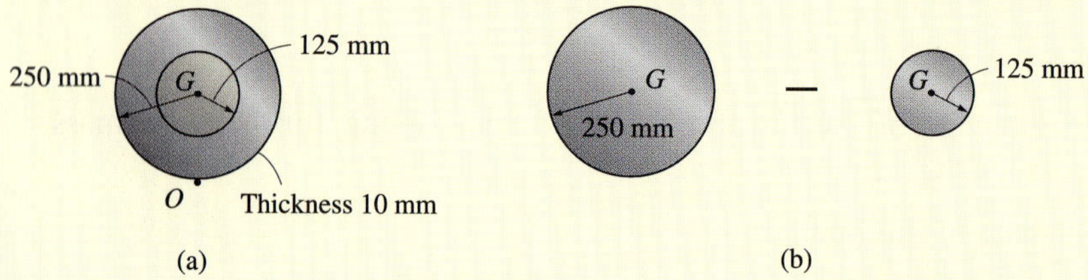

Fig. 17–6

Solution

The plate consists of two composite parts, the 250-mm-radius disk *minus* a 125-mm-radius disk, Fig. 17–6*b*. The moment of inertia about *O* can be determined by computing the moment of inertia of each of these parts about *O* and then adding the results *algebraically*. The calculations are performed by using the parallel-axis theorem in conjunction with the data listed in the table on the inside back cover.

Disk. The moment of inertia of a disk about the centroidal axis perpendicular to the plane of the disk is $I_G = \frac{1}{2}mr^2$. The mass center of the disk is located at a distance of 0.25 m from point *O*. Thus,

$$m_d = \rho_d V_d = 8000\ \text{kg/m}^3[\pi(0.25\ \text{m})^2(0.01\ \text{m})] = 15.71\ \text{kg}$$

$$\begin{aligned}(I_d)_O &= \tfrac{1}{2}m_d r_d^2 + m_d d^2\\ &= \frac{1}{2}(15.71\ \text{kg})(0.25\ \text{m})^2 + (15.71\ \text{kg})(0.25\ \text{m})^2\\ &= 1.473\ \text{kg}\cdot\text{m}^2\end{aligned}$$

Hole. For the 125-mm-radius disk (hole), we have

$$m_h = \rho_h V_h = 8000\ \text{kg/m}^3[\pi(0.125\ \text{m})^2(0.01\ \text{m})] = 3.93\ \text{kg}$$

$$\begin{aligned}(I_h)_O &= \tfrac{1}{2}m_h r_h^2 + m_h d^2\\ &= \frac{1}{2}(3.93\ \text{kg})(0.125\ \text{m})^2 + (3.93\ \text{kg})(0.25\ \text{m})^2\\ &= 0.276\ \text{kg}\cdot\text{m}^2\end{aligned}$$

The moment of inertia of the plate about point *O* is therefore

$$\begin{aligned}I_O &= (I_d)_O - (I_h)_O\\ &= 1.473\ \text{kg}\cdot\text{m}^2 - 0.276\ \text{kg}\cdot\text{m}^2\\ &= 1.20\ \text{kg}\cdot\text{m}^2\end{aligned}$$

Ans.

EXAMPLE 17–4

The pendulum in Fig. 17–7 is suspended from point O and consists of two thin rods, each having a weight of 10 lb. Determine the pendulum's moment of inertia about an axis passing through (*a*) the pin at O, and (*b*) the mass center G of the pendulum.

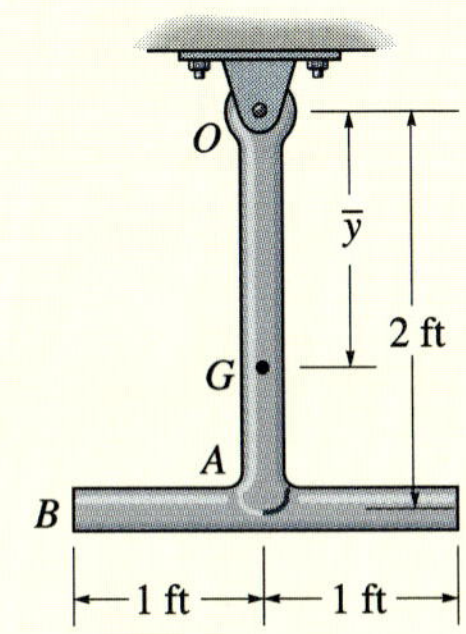

Fig. 17–7

Solution

Part (a). Using the table on the inside back cover, the moment of inertia of rod OA about an axis perpendicular to the page and passing through the end point O of the rod is $I_O = \frac{1}{3}ml^2$. Hence,

$$(I_{OA})_O = \frac{1}{3}ml^2 = \frac{1}{3}\left(\frac{10\text{ lb}}{32.2\text{ ft/s}^2}\right)(2\text{ ft})^2 = 0.414\text{ slug}\cdot\text{ft}^2$$

This same value is obtained using $I_G = \frac{1}{12}ml^2$ and the parallel-axis theorem.

$$(I_{OA})_O = \frac{1}{12}ml^2 + md^2 = \frac{1}{12}\left(\frac{10\text{ lb}}{32.2\text{ ft/s}^2}\right)(2\text{ ft})^2 + \left(\frac{10\text{ lb}}{32.2\text{ ft/s}^2}\right)(1\text{ ft})^2$$
$$= 0.414\text{ slug}\cdot\text{ft}^2$$

For rod BC we have

$$(I_{BC})_O = \frac{1}{12}ml^2 + md^2 = \frac{1}{12}\left(\frac{10\text{ lb}}{32.2\text{ ft/s}^2}\right)(2\text{ ft})^2 + \left(\frac{10\text{ lb}}{32.2\text{ ft/s}^2}\right)(2\text{ ft})^2$$
$$= 1.346\text{ slug}\cdot\text{ft}^2$$

The moment of inertia of the pendulum about O is therefore

$$I_O = 0.414 + 1.346 = 1.76\text{ slug}\cdot\text{ft}^2 \qquad \textit{Ans.}$$

Part (b). The mass center G will be located relative to the pin at O. Assuming this distance to be $\bar{y}$, Fig. 17–7, and using the formula for determining the mass center, we have

$$\bar{y} = \frac{\Sigma\tilde{y}m}{\Sigma m} = \frac{1(10/32.2) + 2(10/32.2)}{(10/32.2) + (10/32.2)} = 1.50\text{ ft}$$

The moment of inertia I_G may be computed in the same manner as I_O, which requires successive applications of the parallel-axis theorem to transfer the moments of inertia of rods OA and BC to G. A more direct solution, however, involves using the result for I_O, i.e.,

$$I_O = I_G + md^2; \qquad 1.76\text{ slug}\cdot\text{ft}^2 = I_G + \left(\frac{20\text{ lb}}{32.2\text{ ft/s}^2}\right)(1.50\text{ ft})^2$$
$$I_G = 0.362\text{ slug}\cdot\text{ft}^2 \qquad \textit{Ans.}$$

Problems

17-1. Determine the moment of inertia I_y for the slender rod. The rod's density ρ and cross-sectional area A are constant. Express the result in terms of the rod's total mass m.

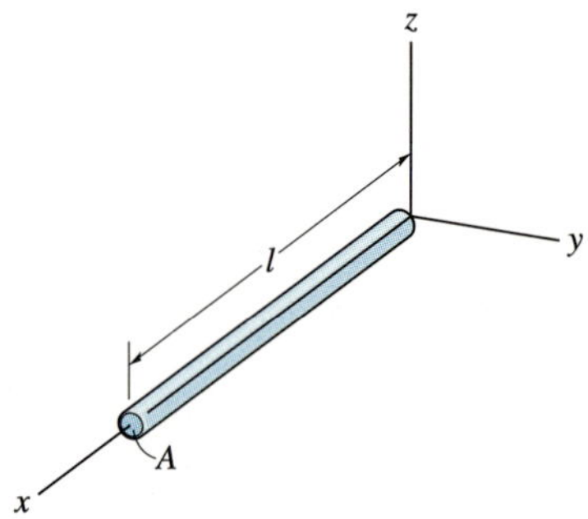

Prob. 17–1

17-2. Determine the moment of inertia of the thin ring about the z axis. The ring has a mass m.

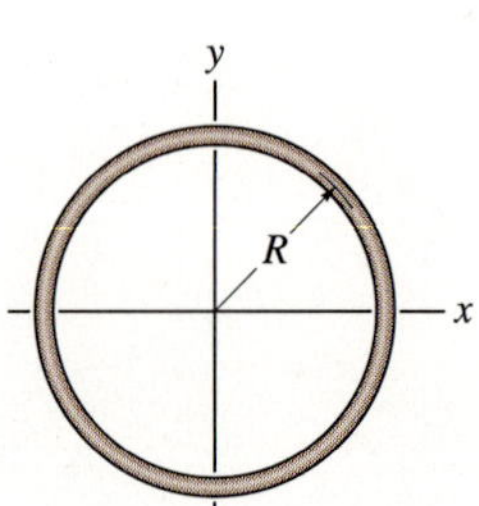

Prob. 17–2

17-3. The right circular cone is formed by revolving the shaded area around the x axis. Determine the moment of inertia I_x and express the result in terms of the total mass m of the cone. The cone has a constant density ρ.

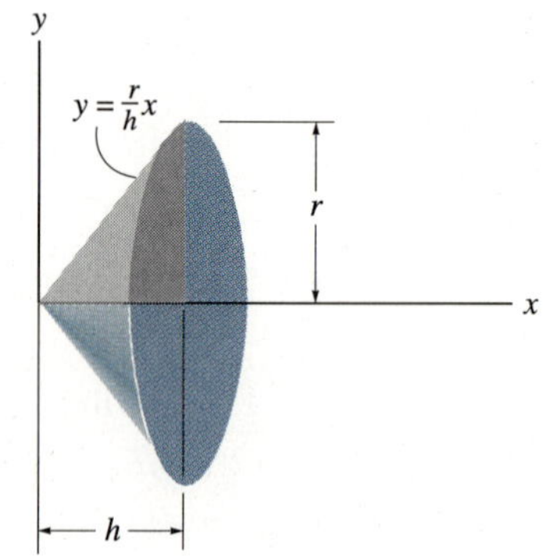

Prob. 17–3

***17-4.** The paraboloid is formed by revolving the shaded area around the x axis. Determine the radius of gyration k_x. The density of the material is $\rho = 5\ \text{Mg/m}^3$.

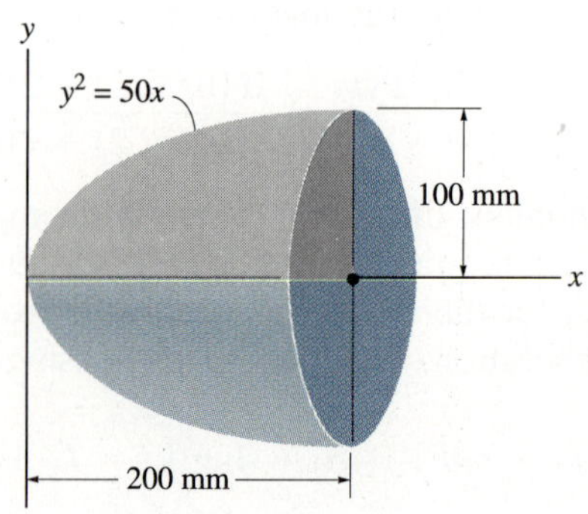

Prob. 17–4

17-5. The solid is formed by revolving the shaded area around the y axis. Determine the radius of gyration k_y. The specific weight of the material is $\gamma = 380\ \text{lb/ft}^3$.

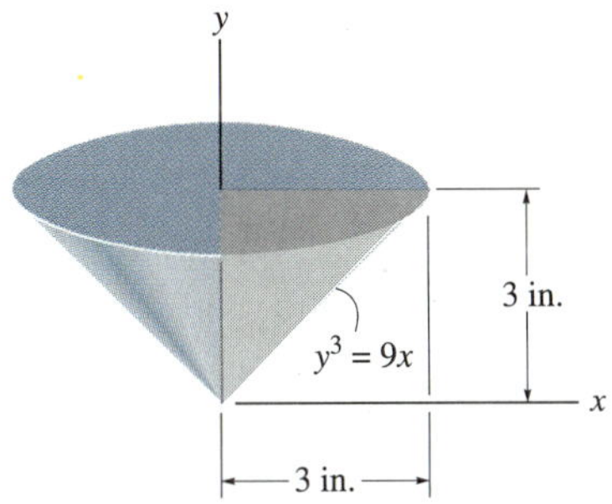

Prob. 17–5

17-6. The sphere is formed by revolving the shaded area around the x axis. Determine the moment of inertia I_x and express the result in terms of the total mass m of the sphere. The material has a constant density ρ.

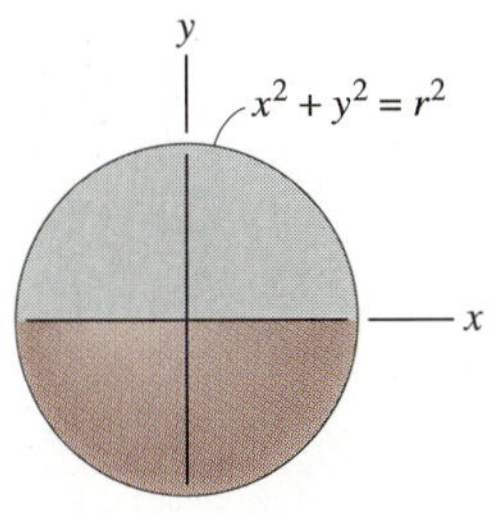

Prob. 17–6

17-7. The paraboloid is formed by revolving the shaded area around the x axis. Determine the moment of inertia with respect to the x axis and express the result in terms of the mass m of the paraboloid. The material has a constant density ρ.

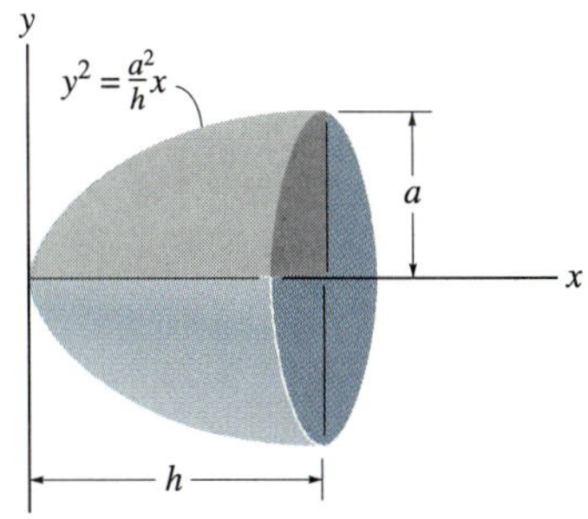

Prob. 17–7

***17-8.** The hemisphere is formed by rotating the shaded area about the y axis. Determine the moment of inertia I_y and express the result in terms of the total mass m of the hemisphere. The material has a constant density ρ.

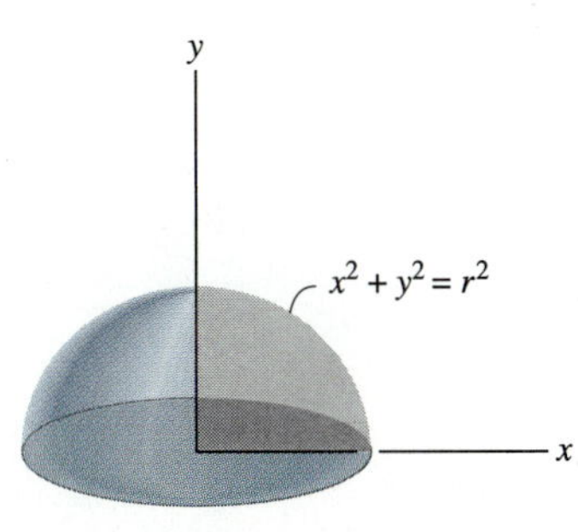

Prob. 17–8

17-9. The concrete shape is formed by rotating the shaded area about the y axis. Determine the moment of inertia I_y. The specific weight of concrete is $\gamma = 150\ \text{lb/ft}^3$.

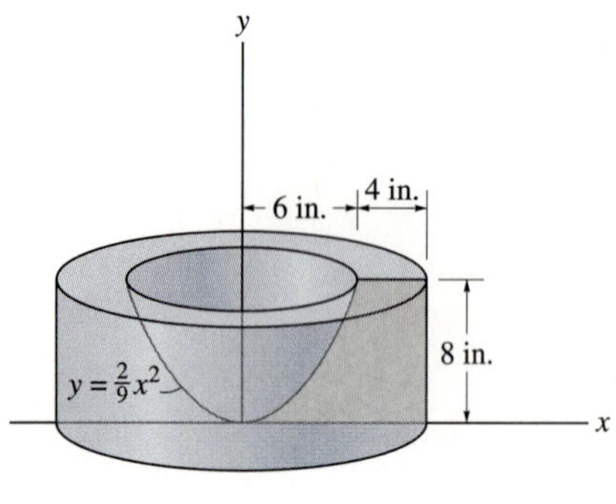

Prob. 17–9

17-10. The frustum is formed by rotating the shaded area around the x axis. Determine the moment of inertia I_x and express the result in terms of the total mass m of the frustum. The frustum has a constant density.

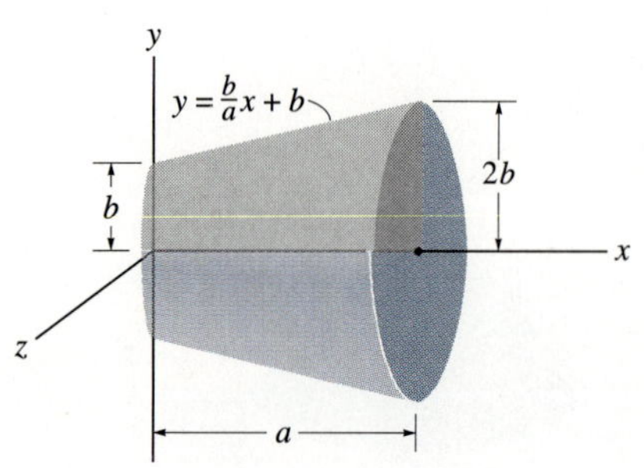

Prob. 17–10

17-11. Determine the moment of inertia of the homogeneous pyramid of mass m with respect to the z axis. The density of the material is ρ. *Suggestion:* Use a rectangular plate element having a volume of $dV = (2x)(2y)dz$.

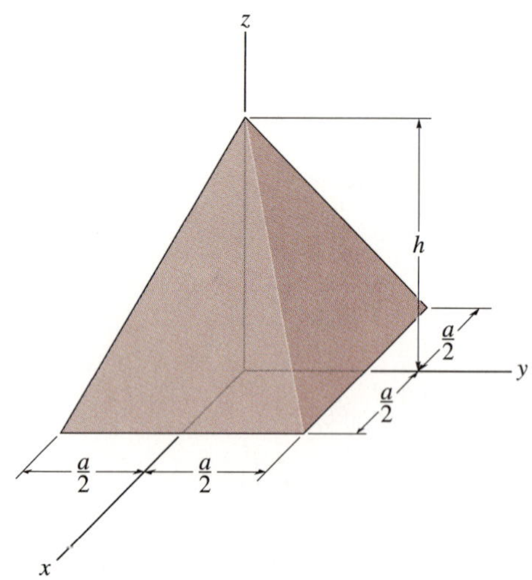

Prob. 17–11

***17-12.** Determine the moment of inertia of the thin plate about an axis perpendicular to the page and passing through the pin at O. The plate has a hole in its center. Its thickness is 50 mm and the material has a density of $\rho = 50\ \text{kg/m}^3$.

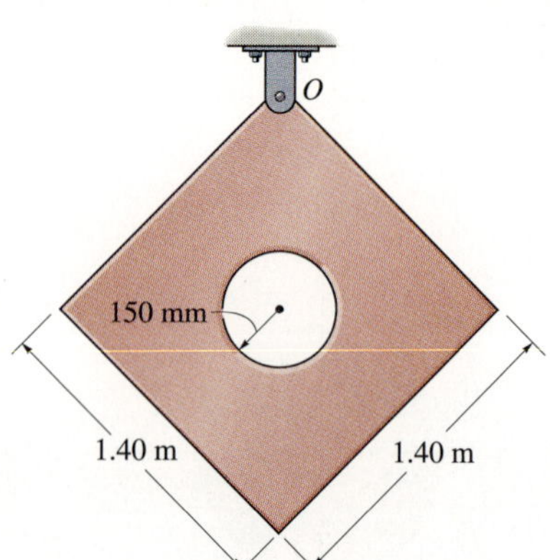

Prob. 17–12

17-13. Determine the moment of inertia of the assembly about an axis which is perpendicular to the page and passes through the center of mass G. The material has a specific weight of $\gamma = 90\ \text{lb/ft}^3$.

17-14. Determine the moment of inertia of the assembly about an axis which is perpendicular to the page and passes through point O. The material has a specific weight of $\gamma = 90\ \text{lb/ft}^3$.

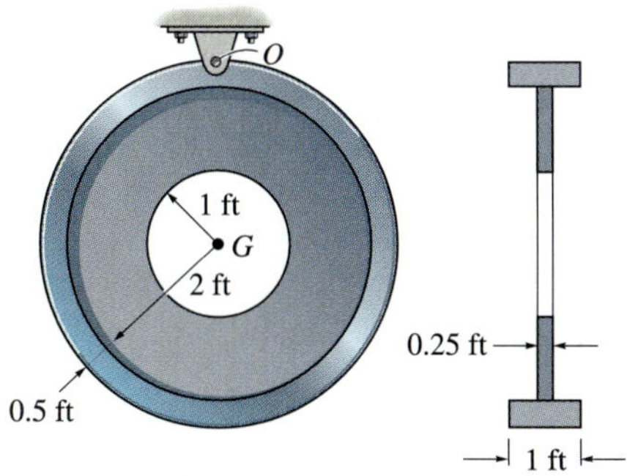

Probs. 17–13/14

***17-16.** The pendulum consists of the 3-kg slender rod and the 5-kg thin plate. Determine the location $\bar{y}$ of the center of mass G of the pendulum; then calculate the moment of inertia of the pendulum about an axis perpendicular to the page and passing through G.

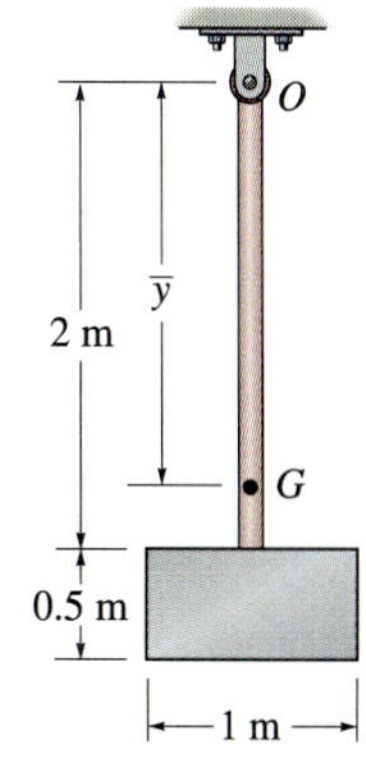

Prob. 17–16

17-15. The wheel consists of a thin ring having a mass of 10 kg and four spokes made from slender rods and each having a mass of 2 kg. Determine the wheel's moment of inertia about an axis perpendicular to the page and passing through point A.

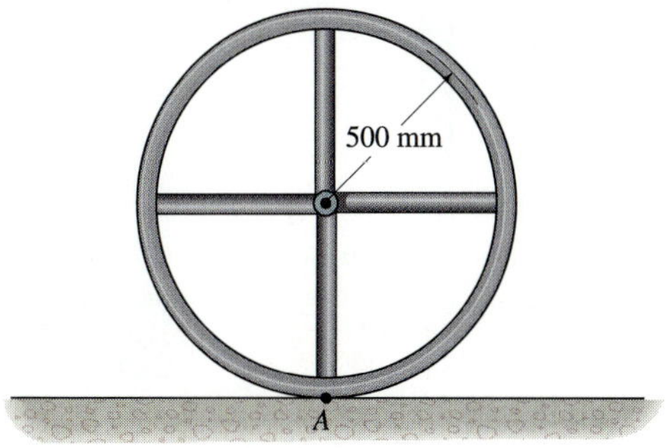

Prob. 17–15

17-17. Each of the three rods has a mass m. Determine the moment of inertia of the assembly about an axis which is perpendicular to the page and passes through the center point O.

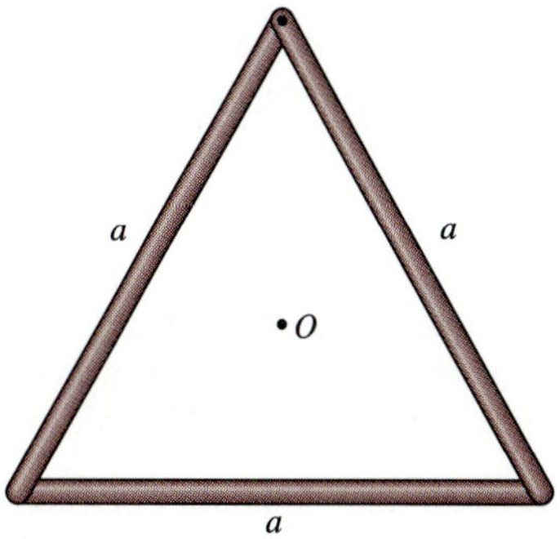

Prob. 17–17

17-18. The slender rods have a weight of 3 lb/ft. Determine the moment of inertia of the assembly about an axis perpendicular to the page and passing through point A.

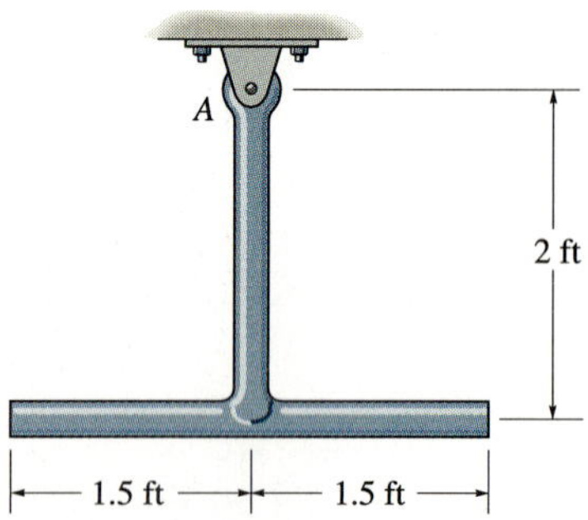

Prob. 17–18

17-19. The pendulum consists of a plate having a weight of 12 lb and a slender rod having a weight of 4 lb. Determine the radius of gyration of the pendulum about an axis perpendicular to the page and passing through point O.

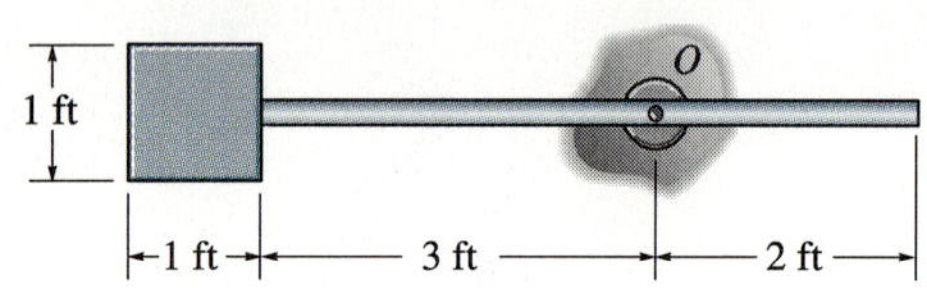

Prob. 17–19

***17-20.** Determine the moment of inertia of the overhung crank about the x axis. The material is steel for which the density is $\rho = 7.85\ \text{Mg/m}^3$.

17-21. Determine the moment of inertia of the overhung crank about the x' axis. The material is steel for which the density is $\rho = 7.85\ \text{Mg/m}^3$.

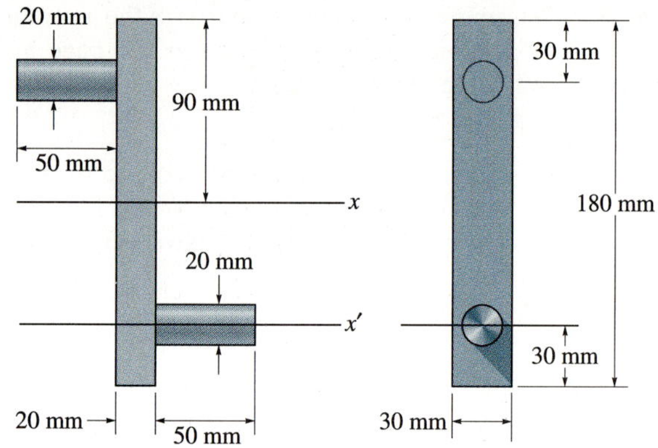

Probs. 17–20/21

17-22. Determine the moment of inertia of the solid steel assembly about the x axis. Steel has a specific weight of $\gamma_{st} = 490\ \text{lb/ft}^3$.

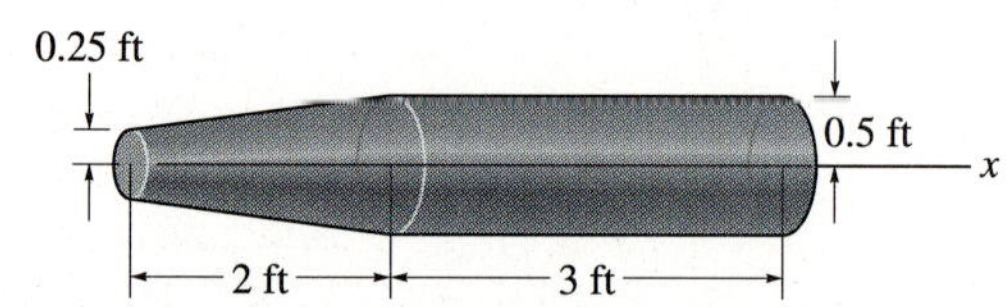

Prob. 17–22

17-23. Determine the moment of inertia of the center crank about the x axis. The material is steel having a specific weight of $\gamma_{st} = 490\ \text{lb/ft}^3$.

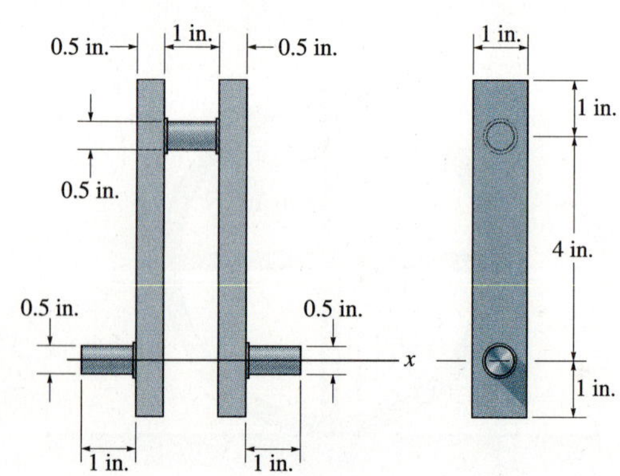

Prob. 17–23

17.2 Planar Kinetic Equations of Motion

In the following analysis we will limit our study of planar kinetics to rigid bodies which, along with their loadings, are considered to be *symmetrical* with respect to a fixed reference plane.* In this case the path of motion of each particle of the body is a plane curve parallel to a fixed reference plane. Since the motion of the body may be viewed within the reference plane, all the forces (and couple moments) acting on the body can then be projected onto the plane. An example of an arbitrary body of this type is shown in Fig. 17–8*a*. Here the *inertial frame of reference x, y, z* has its origin *coincident* with the arbitrary point P in the body. By definition, *these axes do not rotate and are either fixed or translate with constant velocity.*

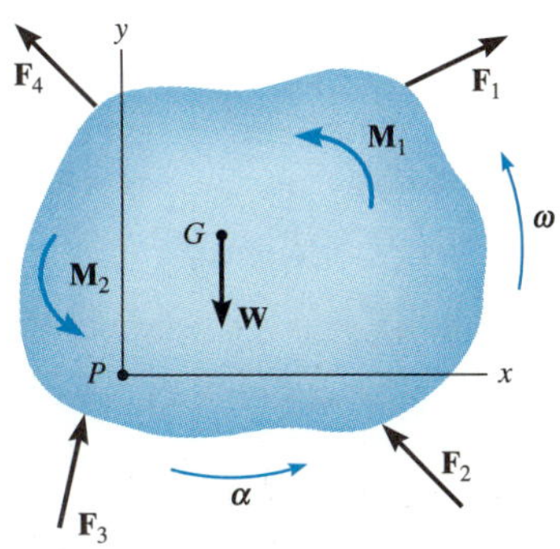

(a)

Fig. 17–8

Equation of Translational Motion. The external forces shown on the body in Fig. 17–8*a* represent the effect of gravitational, electrical, magnetic, or contact forces between adjacent bodies. Since this force system has been considered previously in Sec. 13.3 for the analysis of a system of particles, the resulting Eq. 13–6 may be used here, in which case

$$\Sigma \mathbf{F} = m\mathbf{a}_G$$

This equation is referred to as the *translational equation of motion* for the mass center of a rigid body. It states that *the sum of all the external forces acting on the body is equal to the body's mass times the acceleration of its mass center G.*

For motion of the body in the x–y plane, the translational equation of motion may be written in the form of two independent scalar equations, namely,

$$\Sigma F_x = m(a_G)_x$$
$$\Sigma F_y = m(a_G)_y$$

*By doing this, the rotational equation of motion reduces to a rather simplified form. The more general case of body shape and loading is considered in Chapter 21.

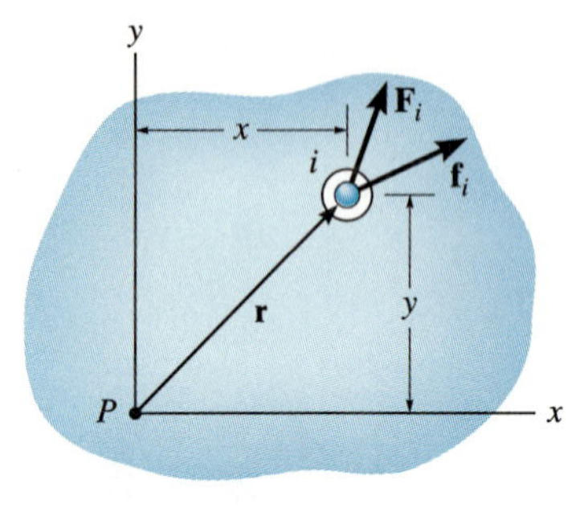

Particle free-body diagram

(b)

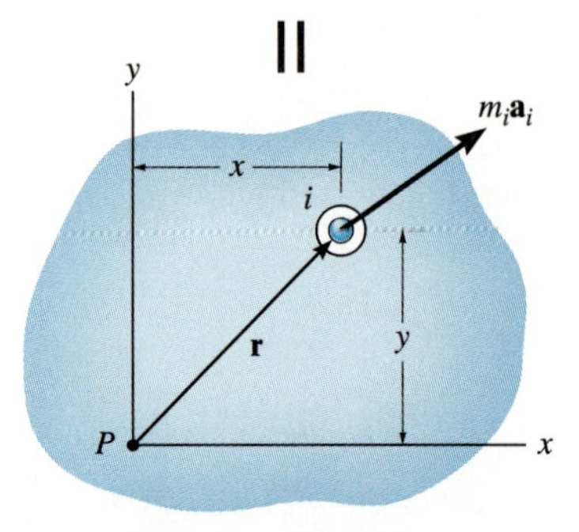

Particle kinetic diagram

(c)

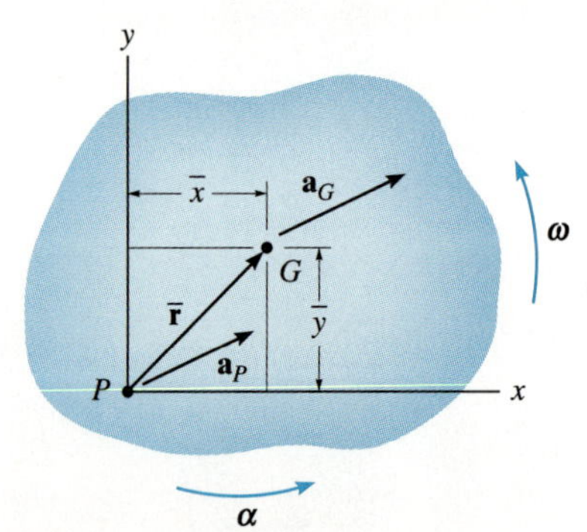

(d)

Fig. 17–8

Equation of Rotational Motion. We will now determine the effects caused by the moments of the external force system computed about an axis perpendicular to the plane of motion (the z axis) and passing through point P. As shown on the free-body diagram of the ith particle, Fig. 17–8b, $\mathbf{F}_i$ represents the *resultant external force* acting on the particle, and $\mathbf{f}_i$ is the *resultant of the internal forces* caused by interactions with adjacent particles. If the particle has a mass m_i and at the instant considered its acceleration is $\mathbf{a}_i$, then the kinetic diagram is constructed as shown in Fig. 17–8c. If moments of the forces acting on the particle are summed about point P, we require

$$\mathbf{r} \times \mathbf{F}_i + \mathbf{r} \times \mathbf{f}_i = \mathbf{r} \times m_i\mathbf{a}_i$$

or

$$(\mathbf{M}_P)_i = \mathbf{r} \times m_i\mathbf{a}_i$$

The moments about P can be expressed in terms of the acceleration of point P, Fig. 17–8d. If the body has an angular acceleration $\boldsymbol{\alpha}$ and angular velocity $\boldsymbol{\omega}$, then using Eq. 16–18 we have

$$\begin{aligned}(\mathbf{M}_P)_i &= m_i\mathbf{r} \times (\mathbf{a}_P + \boldsymbol{\alpha} \times \mathbf{r} - \omega^2\mathbf{r}) \\ &= m_i[\mathbf{r} \times \mathbf{a}_P + \mathbf{r} \times (\boldsymbol{\alpha} \times \mathbf{r}) - \omega^2(\mathbf{r} \times \mathbf{r})]\end{aligned}$$

The last term is zero, since $\mathbf{r} \times \mathbf{r} = \mathbf{0}$. Expressing the vectors with Cartesian components and carrying out the cross-product operations yields

$$\begin{aligned}(M_P)_i\,\mathbf{k} &= m_i\{(x\mathbf{i} + y\mathbf{j}) \times [(a_P)_x\mathbf{i} + (a_P)_y\mathbf{j}] \\ &\quad + (x\mathbf{i} + y\mathbf{j}) \times [\alpha\mathbf{k} \times (x\mathbf{i} + y\mathbf{j})]\} \\ (M_P)_i\mathbf{k} &= m_i[-y(a_P)_x + x(a_P)_y + \alpha x^2 + \alpha y^2]\mathbf{k} \\ \circlearrowleft (M_P)_i &= m_i[-y(a_P)_x + x(a_P)_y + \alpha r^2]\end{aligned}$$

Letting $m_i \rightarrow dm$ and integrating with respect to the entire mass m of the body, we obtain the resultant moment equation

$$\circlearrowleft \Sigma M_P = -\left(\int_m y\,dm\right)(a_P)_x + \left(\int_m x\,dm\right)(a_P)_y + \left(\int_m r^2\,dm\right)\alpha$$

Here ΣM_P represents only the moment of the *external forces* acting on the body about point P. The resultant moment of the internal forces is zero, since for the entire body these forces occur in equal and opposite collinear pairs and thus the moment of each pair of forces about P cancels. The integrals in the first and second terms on the right are used to locate the body's center of mass G with respect to P, since $\bar{y}m = \int y\,dm$ and $\bar{x}m = \int x\,dm$, Fig. 17–8d. Also, the last integral represents the body's moment of inertia computed about the z axis, i.e., $I_P = \int r^2\,dm$. Thus,

$$\circlearrowleft \Sigma M_P = -\bar{y}m(a_P)_x + \bar{x}m(a_P)_y + I_P\alpha \qquad (17\text{–}6)$$

It is possible to reduce this equation to a simpler form if point P coincides with the mass center G for the body. If this is the case, then $\bar{x} = \bar{y} = 0$, and therefore*

$$\Sigma M_G = I_G\alpha \qquad (17\text{–}7)$$

This rotational equation of motion states that the sum of the moments of all the external forces computed about the body's mass center G is equal to the product of the moment of inertia of the body about an axis passing through G and the body's angular acceleration.

Equation 17–6 can also be rewritten in terms of the x and y components of $\mathbf{a}_G$ and the body's moment of inertia I_G. If point G is located at point $(\bar{x}, \bar{y})$, Fig. 17–8*d*, then by the parallel-axis theorem, $I_P = I_G + m(\bar{x}^2 + \bar{y}^2)$. Substituting into Eq. 17–6 and rearranging terms, we get

$$\circlearrowright \Sigma M_P = \bar{y}m[-(a_P)_x + \bar{y}\alpha] + \bar{x}m[(a_P)_y + \bar{x}\alpha] + I_G\alpha \qquad (17\text{–}8)$$

From the kinematic diagram of Fig. 17–8*d*, $\mathbf{a}_P$ can be expressed in terms of $\mathbf{a}_G$ as

$$\mathbf{a}_G = \mathbf{a}_P + \boldsymbol{\alpha} \times \bar{\mathbf{r}} - \omega^2\bar{\mathbf{r}}$$

$$(a_G)_x\mathbf{i} + (a_G)_y\mathbf{j} = (a_P)_x\mathbf{i} + (a_P)_y\mathbf{j} + \alpha\mathbf{k} \times (\bar{x}\mathbf{i} + \bar{y}\mathbf{j}) - \omega^2(\bar{x}\mathbf{i} + \bar{y}\mathbf{j})$$

Carrying out the cross product and equating the respective $\mathbf{i}$ and $\mathbf{j}$ components yields the two scalar equations

$$(a_G)_x = (a_P)_x - \bar{y}\alpha - \bar{x}\omega^2$$
$$(a_G)_y = (a_P)_y + \bar{x}\alpha - \bar{y}\omega^2$$

From these equations, $[-(a_P)_x + \bar{y}\alpha] = [-(a_G)_x - \bar{x}\omega^2]$ and $[(a_P)_y + \bar{x}\alpha] = [(a_G)_y + \bar{y}\omega^2]$. Substituting these results into Eq. 17–8 and simplifying gives

$$\circlearrowright \Sigma M_P = -\bar{y}m(a_G)_x + \bar{x}m(a_G)_y + I_G\alpha \qquad (17\text{–}9)$$

This important result indicates that when moments of the external forces shown on the free-body diagram are summed about point P, Fig. 17–8e, they are equivalent to the sum of the "kinetic moments" of the components of m$\mathbf{a}_G$ *about P plus the "kinetic moment" of* $I_G\boldsymbol{\alpha}$*, Fig. 17–8f.* In other words, when the "kinetic moments," $\Sigma(\mathcal{M}_k)_P$, are computed, Fig. 17–8*f*, the vectors $m(\mathbf{a}_G)_x$ and $m(\mathbf{a}_G)_y$ are treated as sliding vectors; that is, they can act at *any point along their line of action.* In a similar manner, $I_G\boldsymbol{\alpha}$ can be treated as a free vector and can therefore act at *any point. It is important to keep in mind that m*$\mathbf{a}_G$ *and* $I_G\boldsymbol{\alpha}$ *are not the same as a force or a couple moment. Instead, they are caused by the external effects of forces and couple moments acting on the body.* With this in mind we can therefore write Eq. 17–9 in a more general form as

$$\Sigma M_P = \Sigma(\mathcal{M}_k)_P \qquad (17\text{–}10)$$

*It also reduces to this same simple form $\Sigma M_P = I_P\alpha$ if point P is a *fixed point* (see Eq. 17–16) or the acceleration of point P is directed along the line PG.

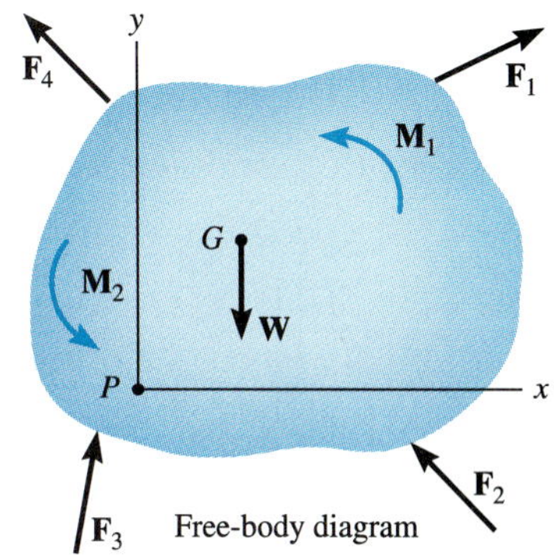

(e)

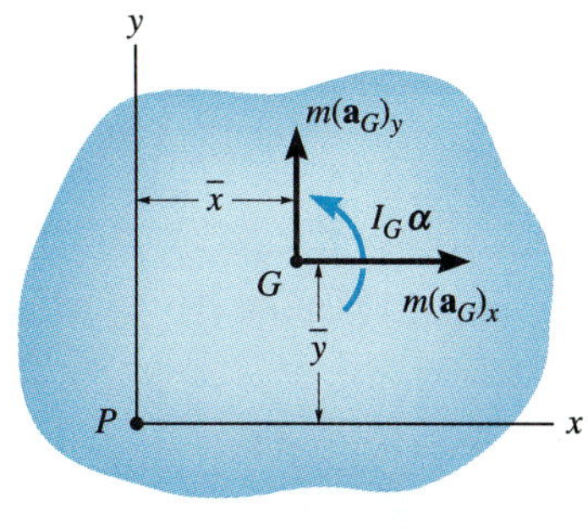

(f)

Fig. 17–8

General Application of the Equations of Motion. To summarize this analysis, *three* independent scalar equations may be written to describe the general plane motion of a symmetrical rigid body.

$$\Sigma F_x = m(a_G)_x$$
$$\Sigma F_y = m(a_G)_y$$
$$\Sigma M_G = I_G\alpha \quad \text{or} \quad \Sigma M_P = \Sigma(\mathcal{M}_k)_P \qquad (17\text{–}11)$$

When applying these equations, one should *always* draw a free-body diagram, Fig. 17–8*e*, in order to account for the terms involved in ΣF_x, ΣF_y, ΣM_G, or ΣM_P. In some problems it may also be helpful to draw the *kinetic diagram* for the body. This diagram graphically accounts for the terms $m(\mathbf{a}_G)_x$, $m(\mathbf{a}_G)_y$, and $I_G\boldsymbol{\alpha}$, and it is especially convenient when used to determine the components of $m\mathbf{a}_G$ and the moment terms in $\Sigma(\mathcal{M}_k)_P$.*

17.3 Equations of Motion: Translation

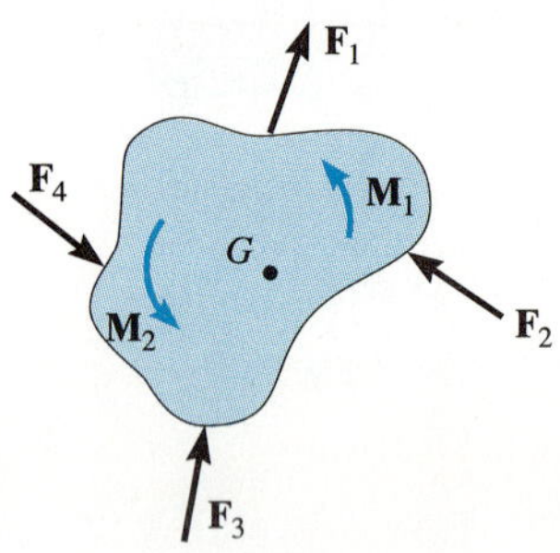

(a)

Fig. 17–9

When a rigid body undergoes a *translation,* Fig. 17–9*a*, all the particles of the body have the *same acceleration,* so that $\mathbf{a}_G = \mathbf{a}$. Furthermore, $\boldsymbol{\alpha} = \mathbf{0}$, in which case the rotational equation of motion applied at point G reduces to a simplified form, namely, $\Sigma M_G = 0$. Application of this and the translational equations of motion will now be discussed for each of the two types of translation.

Rectilinear Translation. When a body is subjected to *rectilinear translation,* all the particles of the body (slab) travel along parallel straight-line paths. The free-body and kinetic diagrams are shown in Fig. 17–9*b*. Since $I_G\boldsymbol{\alpha} = \mathbf{0}$, only $m\mathbf{a}_G$ is shown on the kinetic diagram. Hence, the equations of motion which apply in this case become

$$\Sigma F_x = m(a_G)_x$$
$$\Sigma F_y = m(a_G)_y \qquad (17\text{–}12)$$
$$\Sigma M_G = 0$$

The last equation requires that the sum of the moments of all the external forces (and couple moments) computed about the body's center of mass be equal to zero. It is possible, of course, to sum moments about other points on or off the body, in which case the moment of $m\mathbf{a}_G$ must be

*For this reason, the kinetic diagram will be used in the solution of an example problem whenever $\Sigma M_P = \Sigma(\mathcal{M}_k)_P$ is applied.

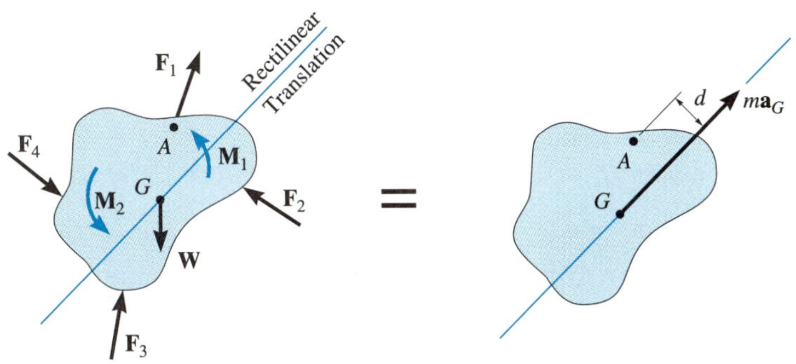

(b)

Fig. 17–9

taken into account. For example, if point A is chosen, which lies at a perpendicular distance d from the line of action of $m\mathbf{a}_G$, the following moment equation applies:

$$\circlearrowright+\Sigma M_A = \Sigma(\mathcal{M}_k)_A; \qquad \Sigma M_A = (ma_G)d$$

Here the sum of moments of the external forces and couple moments about A (ΣM_A, free-body diagram) equals the moment of $m\mathbf{a}_G$ about A ($\Sigma(\mathcal{M}_k)_A$, kinetic diagram).

Curvilinear Translation. When a rigid body is subjected to *curvilinear translation,* all the particles of the body travel along *parallel curved paths.* For analysis, it is often convenient to use an inertial coordinate system having an origin which is coincident with the body's mass center at the instant considered, and axes which are oriented in the normal and tangential directions to the path of motion, Fig. 17–9c. The three scalar equations of motion are then

$$\begin{aligned} \Sigma F_n &= m(a_G)_n \\ \Sigma F_t &= m(a_G)_t \\ \Sigma M_G &= 0 \end{aligned} \qquad (17\text{–}13)$$

Here $(a_G)_t$ and $(a_G)_n$ represent, respectively, the magnitudes of the tangential and normal components of acceleration of point G.

If the moment equation $\Sigma M_G = 0$ is replaced by a moment summation about the arbitrary point B, Fig. 17–9c, it is necessary to account for the moments, $\Sigma(\mathcal{M}_k)_B$, of the two components $m(\mathbf{a}_G)_n$ and $m(\mathbf{a}_G)_t$ about this point. From the kinetic diagram, h and e represent the perpendicular distances (or "moment arms") from B to the lines of action of the components. The required moment equation therefore becomes

$$\circlearrowright+\Sigma M_B = \Sigma(\mathcal{M}_k)_B; \quad \Sigma M_B = e[m(a_G)_t] - h[m(a_G)_n]$$

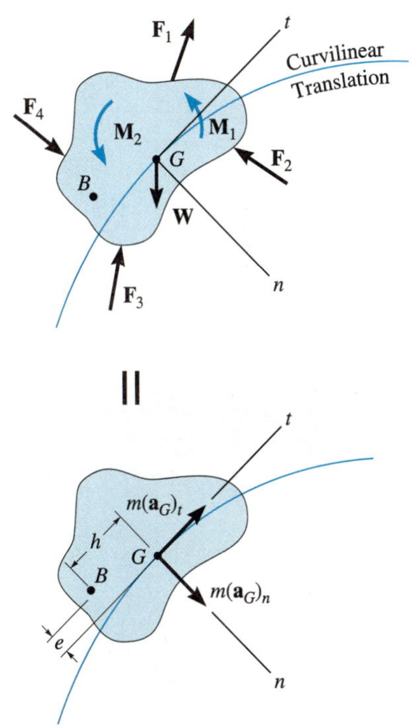

(c)

Fig. 17–9

Procedure for Analysis

Kinetic problems involving rigid-body *translation* can be solved using the following procedure.

Free-Body Diagram

- Establish the x, y or n, t inertial coordinate system and draw the free-body diagram in order to account for all the external forces and couple moments that act on the body.
- The direction and sense of the acceleration of the body's mass center $\mathbf{a}_G$ should be established.
- Identify the unknowns in the problem.
- If it is decided that the rotational equation of motion $\Sigma M_P = \Sigma(\mathcal{M}_k)_P$ is to be used in the solution, then consider drawing the kinetic diagram, since it graphically accounts for the components $m(\mathbf{a}_G)_x$, $m(\mathbf{a}_G)_y$ or $m(\mathbf{a}_G)_t$, $m(\mathbf{a}_G)_n$ and is therefore convenient for "visualizing" the terms needed in the moment sum $\Sigma(\mathcal{M}_k)_P$.

The free-body and kinetic diagrams for this boat and trailer are first established in order to apply the equations of motion. Here the forces on the free-body diagram cause the effect shown on the kinetic diagram. If moments are summed about the mass center, G, then $\Sigma M_G = 0$. However, if moments are summed about point B then $\curvearrowleft + \Sigma M_B = ma_G(d)$.

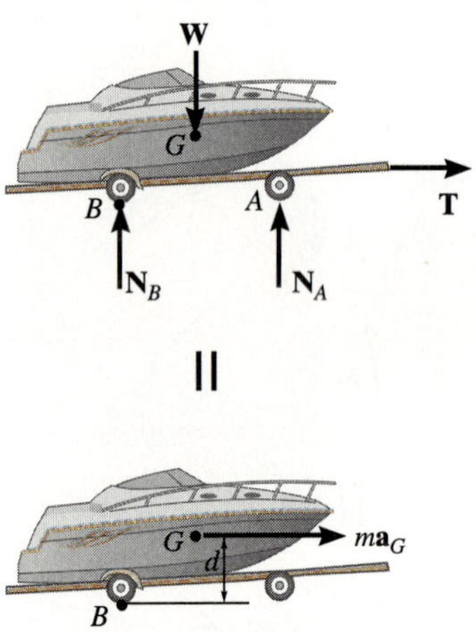

Equations of Motion

- Apply the three equations of motion in accordance with the established sign convention.
- To simplify the analysis, the moment equation $\Sigma M_G = 0$ can be replaced by the more general equation $\Sigma M_P = \Sigma(\mathcal{M}_k)_P$, where point P is usually located at the intersection of the lines of action of as many unknown forces as possible.
- If the body is in contact with a *rough surface* and slipping occurs, use the frictional equation $F = \mu_k N$. Remember, $\mathbf{F}$ always acts on the body so as to oppose the motion of the body relative to the surface it contacts.

Kinematics

- Use kinematics if the velocity and position of the body are to be determined.
- For *rectilinear translation* with *variable acceleration,* use $a_G = dv_G/dt$, $a_G\,ds_G = v_G\,dv_G$, $v_G = ds_G/dt$
- For *rectilinear translation* with *constant acceleration,* use $v_G = (v_G)_0 + a_G t$, $v_G^2 = (v_G)_0^2 + 2a_G[s_G - (s_G)_0]$, $s_G = (s_G)_0 + (v_G)_0 t + \frac{1}{2}a_G t^2$.
- For *curvilinear translation,* use $(a_G)_n = v_G^2/\rho = \omega^2\rho$, $(a_G)_t = dv_G/dt$, $(a_G)_t\,ds_G = v_G\,dv_G$, $(a_G)_t = \alpha\rho$.

EXAMPLE 17–5

The car shown in Fig. 17–10*a* has a mass of 2 Mg and a center of mass at G. Determine the car's acceleration if the "driving" wheels in the back are always slipping, whereas the front wheels freely rotate. Neglect the mass of the wheels. The coefficient of kinetic friction between the wheels and the road is $\mu_k = 0.25$.

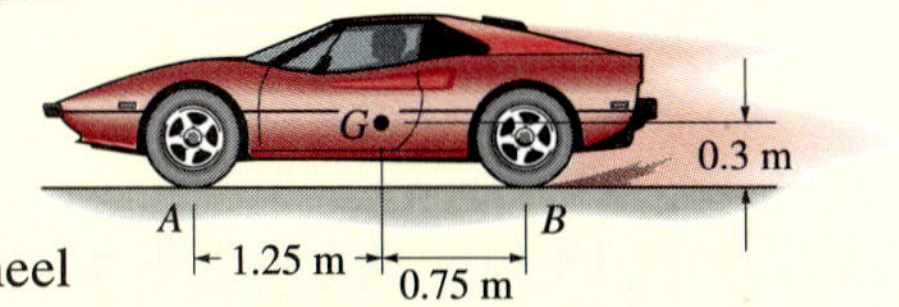

(a)

Solution I

Free-Body Diagram. As shown in Fig. 17–10*b*, the rear-wheel frictional force $\mathbf{F}_B$ pushes the car forward, and since *slipping occurs,* $F_B = 0.25N_B$. The frictional forces acting on the *front wheels* are *zero,* since these wheels have negligible mass.* There are three unknowns in the problem, N_A, N_B, and a_G. Here we will sum moments about the mass center. The car (point G) is assumed to accelerate to the left, i.e., in the negative x direction, Fig. 17–10*b*.

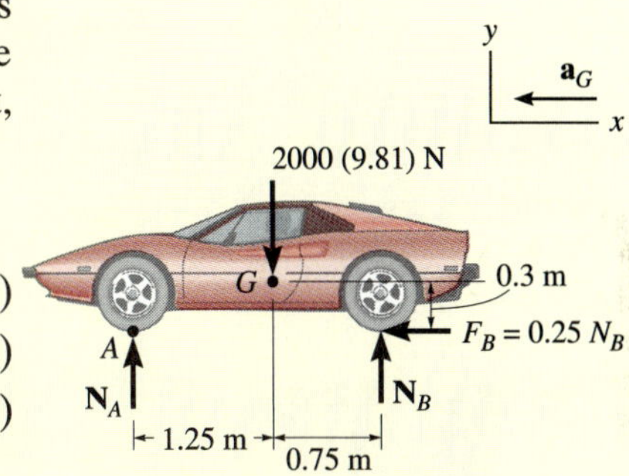

(b)

Equations of Motion

$$\overset{+}{\rightarrow}\Sigma F_x = m(a_G)_x; \qquad -0.25N_B = -(2000\text{ kg})a_G \qquad (1)$$

$$+\uparrow \Sigma F_y = m(a_G)_y; \qquad N_A + N_B - 2000(9.81)\text{ N} = 0 \qquad (2)$$

$$\circlearrowright + \Sigma M_G = 0; \quad -N_A(1.25\text{ m}) - 0.25N_B(0.3\text{ m}) + N_B(0.75\text{ m}) = 0 \quad (3)$$

Solving,

$$a_G = 1.59\text{ m/s}^2 \leftarrow \qquad \textit{Ans.}$$

$$N_A = 6.88\text{ kN}$$

$$N_B = 12.7\text{ kN}$$

Solution II

Free-Body and Kinetic Diagrams. If the "moment" equation is applied about point A, then the unknown N_A will be eliminated from the equation. To "visualize" the moment of $m\mathbf{a}_G$ about A, we will include the kinetic diagram as part of the analysis, Fig. 17–10*c*.

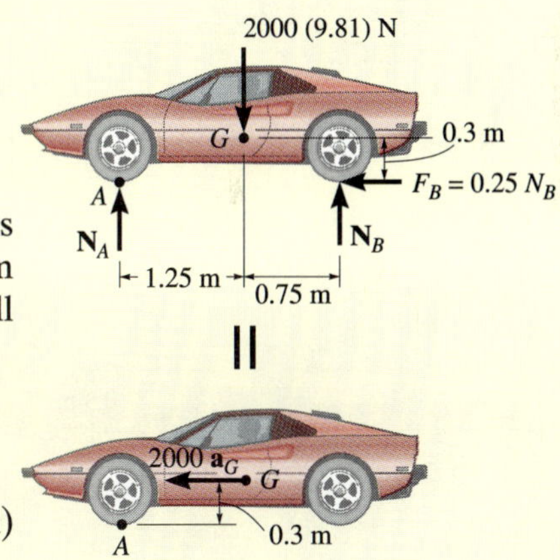

(c)

Fig. 17–10

Equation of Motion. We require

$$\circlearrowright + \Sigma M_A = \Sigma(\mathcal{M}_k)_A; \qquad N_B(2\text{ m}) - 2000(9.81)\text{ N}(1.25\text{ m}) = (2000\text{ kg})a_G(0.3\text{ m})$$

Solving this and Eq. 1 for a_G leads to a simpler solution than that obtained from Eqs. 1 to 3.

*If the mass of the front wheels were to be included in the analysis, the frictional force acting at A would be *directed to the right* to create the necessary counterclockwise rotation of the wheels. The problem solution for this case would be more involved since a general-plane-motion analysis of the wheels would have to be considered (see Sec. 17.5).

EXAMPLE 17–6

The motorcycle shown in Fig. 17–11*a* has a mass of 125 kg and a center of mass at G_1, while the rider has a mass of 75 kg and a center of mass at G_2. Determine the minimum coefficient of static friction between the wheels and the pavement in order for the rider to do a "wheely," i.e., lift the front wheel off the ground as shown in the photo. What acceleration is necessary to do this? Neglect the mass of the wheels and assume that the front wheel is free to roll.

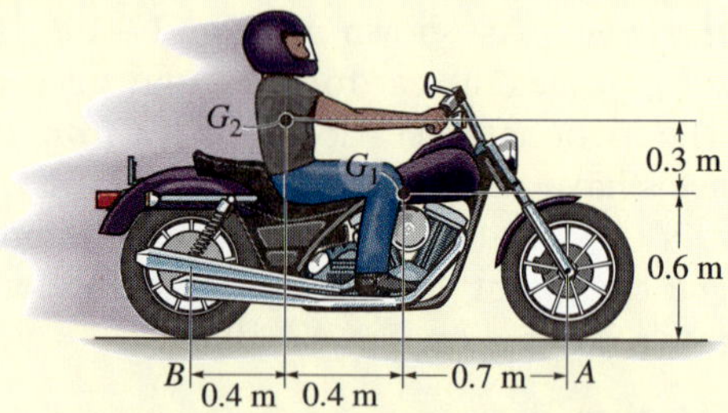

(a)

Solution

Free-Body and Kinetic Diagrams. In this problem we will consider both the motorcycle and the rider as the "system" to be analyzed. It is possible first to determine the location of the center of mass for this "system" by using the equations $\bar{x} = \Sigma\tilde{x}m/\Sigma m$ and $\bar{y} = \Sigma\tilde{y}m/\Sigma m$. Here, however, we will consider the separate weight and mass of each of its *component parts* as shown on the free-body and kinetic diagrams, Fig. 17–11*b*. Both parts move with the *same* acceleration and we have assumed that the front wheel is *about* to leave the ground, so that the normal reaction $N_A \approx 0$. The three unknowns in the problem are N_B, F_B, and a_G.

Equations of Motion

$$\xrightarrow{+}\Sigma F_x = m(a_G)_x; \quad F_B = (75\text{ kg} + 125\text{ kg})a_G \qquad (1)$$

$$+\uparrow \Sigma F_y = m(a_G)_y; \quad N_B - 735.75\text{ N} - 1226.25\text{ N} = 0 \qquad (2)$$

$$\circlearrowleft + \Sigma M_B = \Sigma(\mathcal{M}_k)_B; \; -(735.75\text{ N})(0.4\text{ m}) - (1226.25\text{ N})(0.8\text{ m}) = -(75\text{ kg } a_G)(0.9\text{ m}) - (125\text{ kg } a_G)(0.6\text{ m})$$

Solving,

$$a_G = 8.94\text{ m/s}^2 \rightarrow \qquad \textit{Ans.}$$

$$N_B = 1962\text{ N}$$

$$F_B = 1790\text{ N}$$

Thus the minimum coefficient of static friction is

$$(\mu_s)_{min} = \frac{F_B}{N_B} = \frac{1790\text{ N}}{1962\text{ N}} = 0.912 \qquad \textit{Ans.}$$

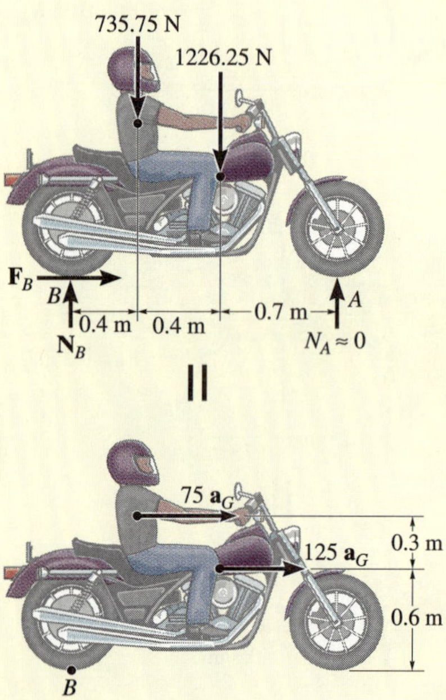

(b)

Fig. 17–11

EXAMPLE 17–7

A uniform 50-kg crate rests on a horizontal surface for which the coefficient of kinetic friction is $\mu_k = 0.2$. Determine the crate's acceleration if a force of $P = 600$ N is applied to the crate as shown in Fig. 17–12*a*.

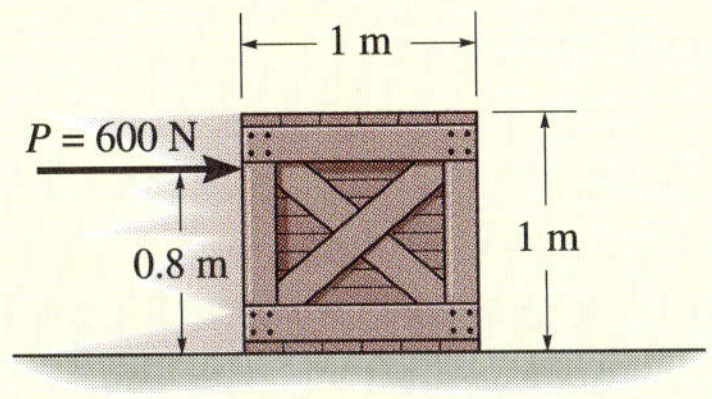

(a)

Solution

Free-Body Diagram. The force **P** can cause the crate either to slide or to tip over. As shown in Fig. 17–12*b*, it is assumed that the crate slides, so that $F = \mu_k N_C = 0.2N_C$. Also, the resultant normal force $\mathbf{N}_C$ acts at O, a distance x (where $0 < x \leq 0.5$ m) from the crate's center line.* The three unknowns are N_C, x, and a_G.

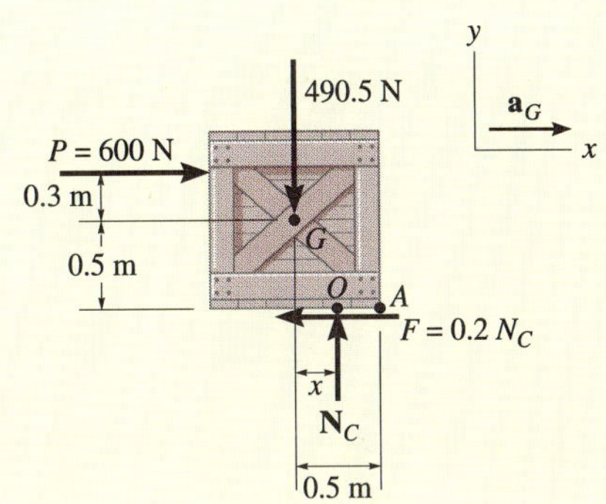

(b)

Fig. 17–12

Equations of Motion

$$\overset{+}{\rightarrow}\Sigma F_x = m(a_G)_x; \qquad 600\text{ N} - 0.2N_C = (50\text{ kg})\,a_G \qquad (1)$$

$$+\uparrow \Sigma F_y = m(a_G)_y; \qquad N_C - 490.5\text{ N} = 0 \qquad (2)$$

$$\circlearrowright + \Sigma M_G = 0; \qquad -600\text{ N}(0.3\text{ m}) + N_C(x) - 0.2N_C(0.5\text{ m}) = 0 \qquad (3)$$

Solving, we obtain

$$N_C = 490\text{ N}$$

$$x = 0.467\text{ m}$$

$$a_G = 10.0\text{ m/s}^2 \rightarrow \qquad \textit{Ans.}$$

Since $x = 0.467\ \text{m} < 0.5$ m, indeed the crate slides as originally assumed. If the solution had given a value of $x > 0.5$ m, the problem would have to be reworked with the assumption that tipping occurred. If this were the case, $\mathbf{N}_C$ would act at the *corner point A* and $F \leq 0.2N_C$.

*The line of action of $\mathbf{N}_C$ does not necessarily pass through the mass center G ($x = 0$), since $\mathbf{N}_C$ must counteract the tendency for tipping caused by **P**. See Sec. 8.1 of *Engineering Mechanics: Statics*.

EXAMPLE 17-8

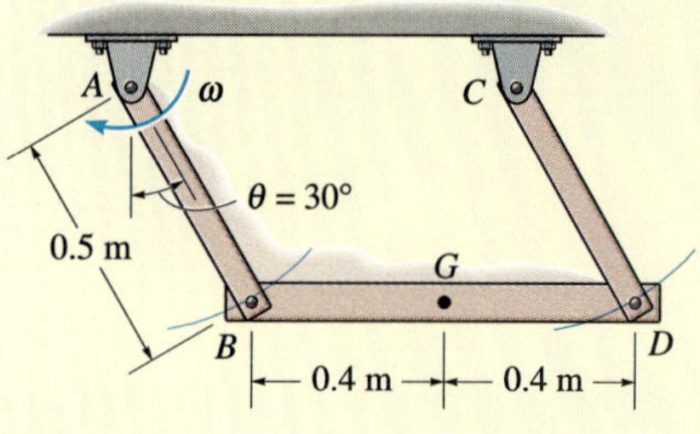

(a)

The 100-kg beam BD shown in Fig. 17–13a is supported by two rods having negligible mass. Determine the force created in each rod if at the instant $\theta = 30°$ the rods are both swinging freely with an angular velocity of $\omega = 6$ rad/s.

Solution

Free-Body Diagram. The beam moves with *curvilinear translation* since points B and D and the center of mass G all move along circular paths, each path having the same radius of 0.5 m. Using normal and tangential coordinates, the free-body diagram for the beam is shown in Fig. 17–13b. Because of the *translation*, G has the *same* motion as the pin at B, which is connected to both the rod and the beam. By studying the angular motion of rod AB, Fig. 17–13c, note that the tangential component of acceleration acts downward to the left due to the clockwise direction of $\boldsymbol{\alpha}$. Furthermore, the normal component of acceleration is *always* directed toward the center of curvature (toward point A for rod AB). Since the angular velocity of AB is 6 rad/s, then

$$(a_G)_n = \omega^2 r = (6\text{ rad/s})^2(0.5\text{ m}) = 18\text{ m/s}^2$$

The three unknowns are T_B, T_D, and $(a_G)_t$. The directions of $(\mathbf{a}_G)_n$ and $(\mathbf{a}_G)_t$ have been established, and are indicated on the coordinate axes.

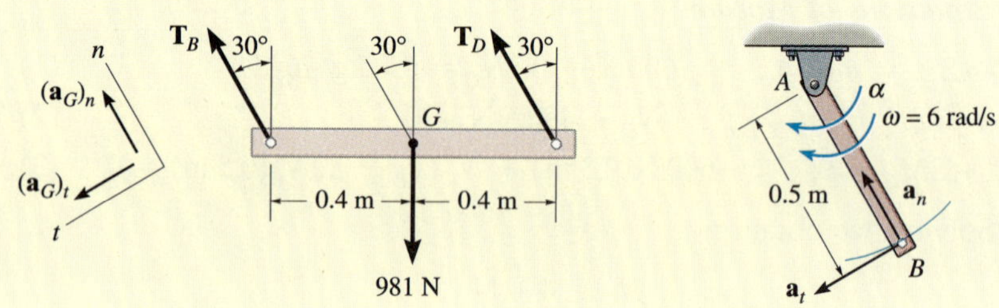

(b) (c)

Fig. 17–13

Equations of Motion

$$+\nwarrow\Sigma F_n = m(a_G)_n;\; T_B + T_D - 981\cos 30°\text{ N} = 100\text{ kg}(18\text{ m/s}^2) \quad (1)$$

$$+\swarrow\Sigma F_t = m(a_G)_t;\quad 981\sin 30° = 100\text{ kg}(a_G)_t \quad (2)$$

$$\circlearrowright+\Sigma M_G = 0;\quad -(T_B\cos 30°)(0.4\text{ m}) + (T_D\cos 30°)(0.4\text{ m}) = 0 \quad (3)$$

Simultaneous solution of these three equations gives

$$T_B = T_D = 1.32\text{ kN} \;\angle^{30°} \qquad \textit{Ans.}$$

$$(a_G)_t = 4.90\text{ m/s}^2$$

Problems

***17-24.** The 4-Mg canister contains nuclear waste material encased in concrete. If the mass of the spreader beam *BD* is 50 kg, determine the force in each of the links *AB*, *CD*, *EF*, and *GH* when the system is lifted with an acceleration of $a = 2\ \text{m/s}^2$ for a short period of time.

17-25. The 4-Mg canister contains nuclear waste material encased in concrete. If the mass of the spreader beam *BD* is 50 kg, determine the largest vertical acceleration **a** of the system so that each of the links *AB* and *CD* are not subjected to a force greater than 30 kN and links *EF* and *GH* are not subjected to a force greater than 34 kN.

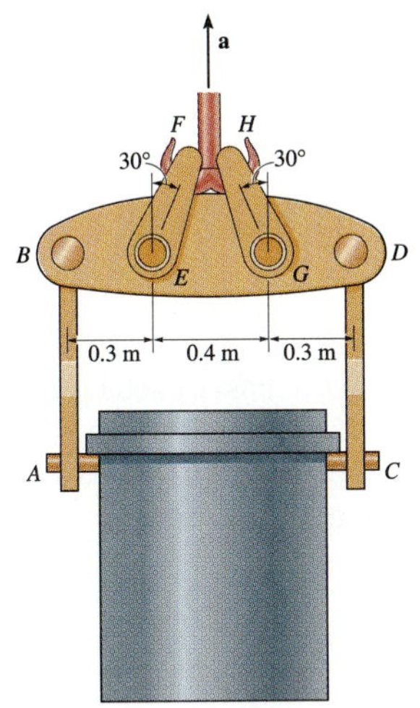

Probs. 17–24/25

17-26. The machine has a mass of 1.5 Mg and rests on the bed of the truck and on the smooth surface at *B*. If it does not slip at *A*, determine the maximum acceleration of the truck so that the machine will not move relative to the truck. Also, what are the horizontal and vertical components of reaction at *A* when this occurs?

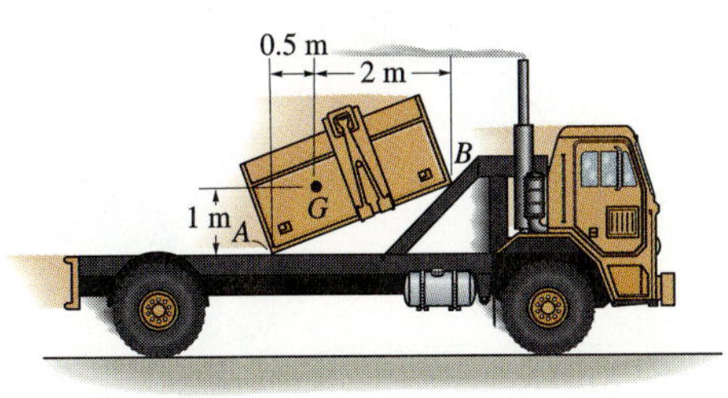

Prob. 17–26

17-27. The fork lift has a boom with a mass of 800 kg and a mass center at *G*. If the vertical acceleration of the boom is $4\ \text{m/s}^2$, determine the horizontal and vertical reactions at the pin *A* and on the short link *BC* when the 1.25-Mg load is lifted.

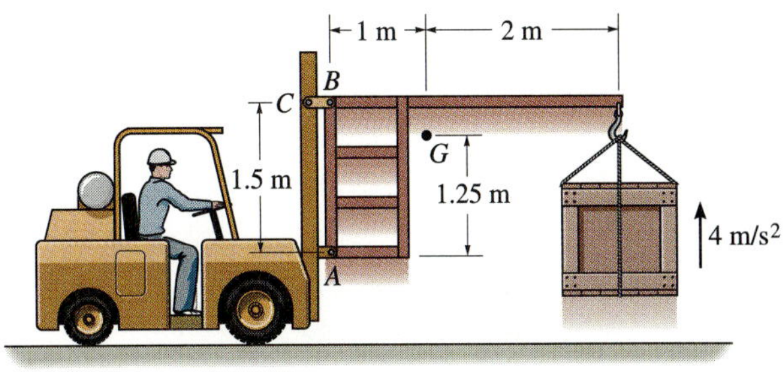

Prob. 17–27

***17-28.** The pipe has a mass of 460 kg and is held in place on the truck bed using the two boards *A* and *B*. Determine the greatest acceleration of the truck so that the pipe begins to lose contact at *A* and the bed of the truck and starts to pivot about *B*. Assume board *B* will not slip on the bed of the truck, and the pipe is smooth. Also, what force does board *B* exert on the pipe during the acceleration?

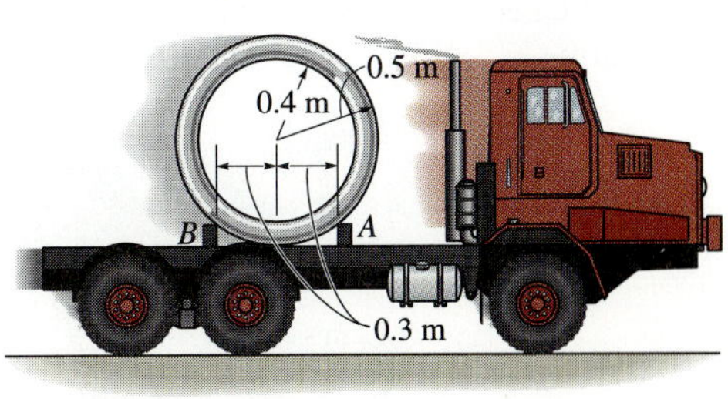

Prob. 17–28

17-29. The lift truck has a mass of 70 kg and mass center at G. If it lifts the 120-kg spool with an acceleration of $3\ m/s^2$, determine the reactions of each of the four wheels on the ground. The loading is symmetric. Neglect the mass of the movable arm CD.

17-30. The lift truck has a mass of 70 kg and mass center at G. Determine the largest upward acceleration of the 120-kg spool so that no reaction of the wheels on the ground exceeds 600 N.

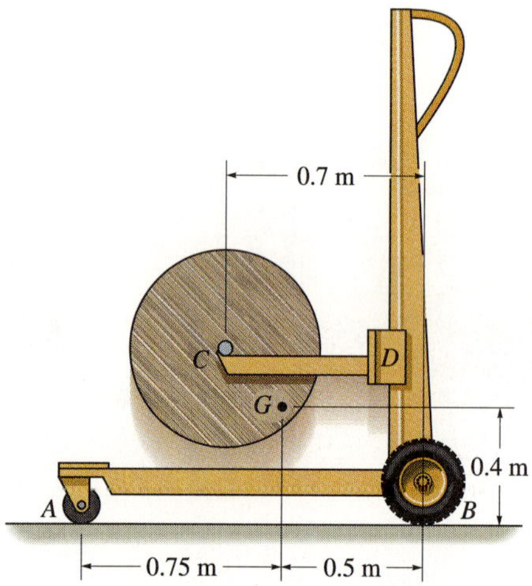

Probs. 17–29/30

17-31. Determine the greatest possible acceleration of the 975-kg race car so that its front tires do not leave the ground or the tires slip on the track. The coefficients of static and kinetic friction are $\mu_s = 0.8$ and $\mu_k = 0.6$, respectively. Neglect the mass of the tires. The car has rear-wheel drive and the front tires are free to roll.

***17-32.** Determine the greatest possible acceleration of the 975-kg race car so that its front wheels do not leave the ground or the tires slip on the track. The coefficients of static and kinetic friction are $\mu_s = 0.8$ and $\mu_k = 0.6$, respectively. Neglect the mass of the tires. The car has four-wheel drive.

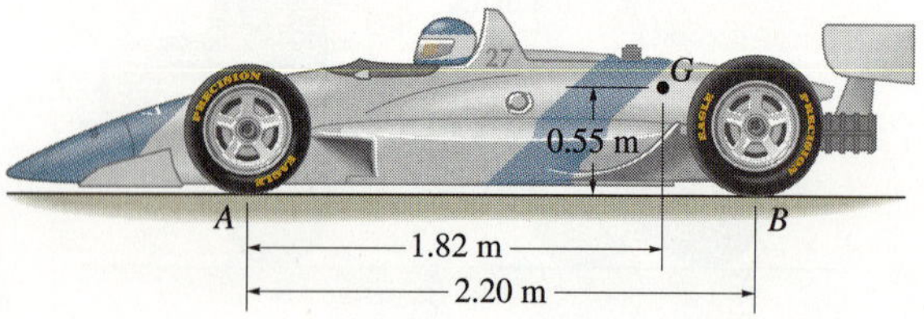

Probs. 17–31/32

17-33. The 1.6-Mg car shown has been "raked" by increasing the height of its center of mass to $h = 0.2$ m. This was done by raising the springs on the rear axle. If the coefficient of kinetic friction between the rear wheels and the ground is $\mu_k = 0.3$, show that the car can accelerate slightly faster than its counterpart for which $h = 0$. Neglect the mass of the wheels and driver and assume the front wheels at B are free to roll while the rear wheels slip.

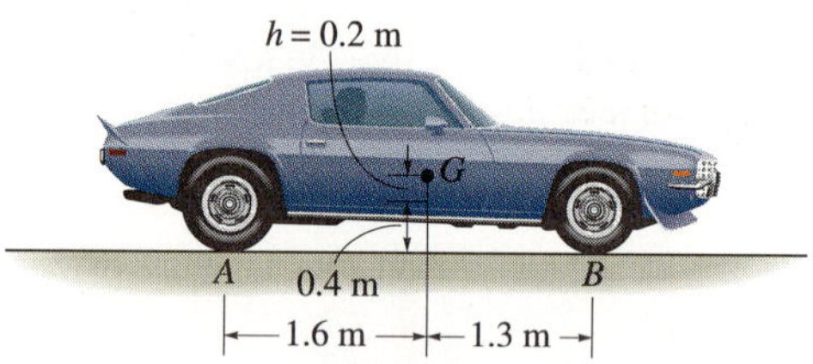

Prob. 17–33

17-34. The pipe has a mass of 800 kg and is being towed behind the truck. If the acceleration of the truck is $a_t = 0.5\ m/s^2$, determine the angle θ and the tension in the cable. The coefficient of kinetic friction between the pipe and the ground is $\mu_k = 0.1$.

17-35. The pipe has a mass of 800 kg and is being towed behind a truck. If the angle $\theta = 30°$, determine the acceleration of the truck and the tension in the cable. The coefficient of kinetic friction between the pipe and the ground is $\mu_k = 0.1$.

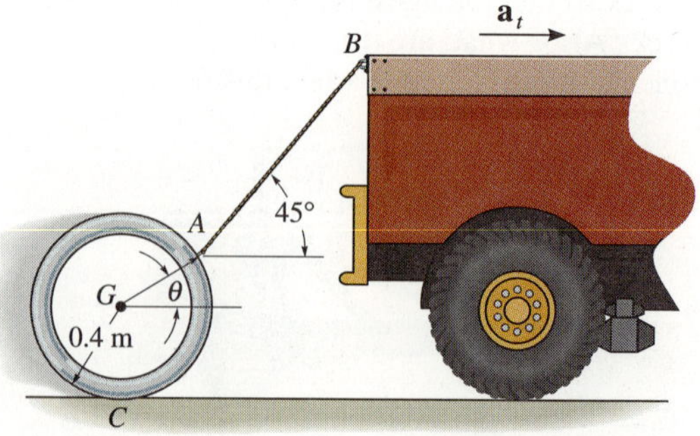

Probs. 17–34/35

*17-36. The pipe has a length of 3 m and a mass of 500 kg. It is attached to the back of the truck using a 0.6-m-long chain AB. If the coefficient of kinetic friction at C is $\mu_k = 0.4$, determine the acceleration of the truck if the angle $\theta = 10°$ with the road as shown.

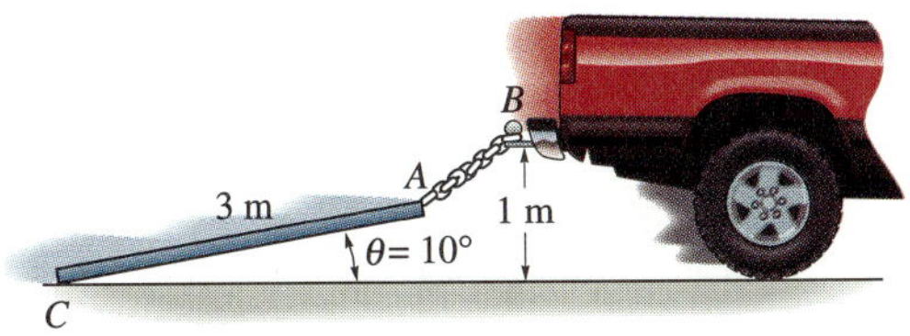

Prob. 17–36

17-37. Block A weighs 50 lb and the platform weighs 10 lb. If $P = 100$ lb, determine the normal force exerted by block A on B. Neglect the weight of the pulleys and bars of the triangular frame.

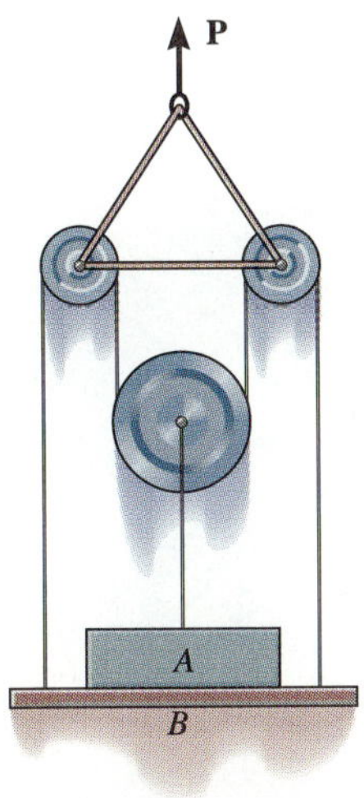

Prob. 17–37

17-38. The sports car has a mass of 1.5 Mg and a center of mass at G. Determine the shortest time it takes for it to reach a speed of 80 km/h, starting from rest, if the engine only drives the rear wheels, whereas the front wheels are free rolling. The coefficient of static friction between the wheels and the road is $\mu_s = 0.2$. Neglect the mass of the wheels for the calculation. If driving power could be supplied to all four wheels, what would be the shortest time for the car to reach a speed of 80 km/h?

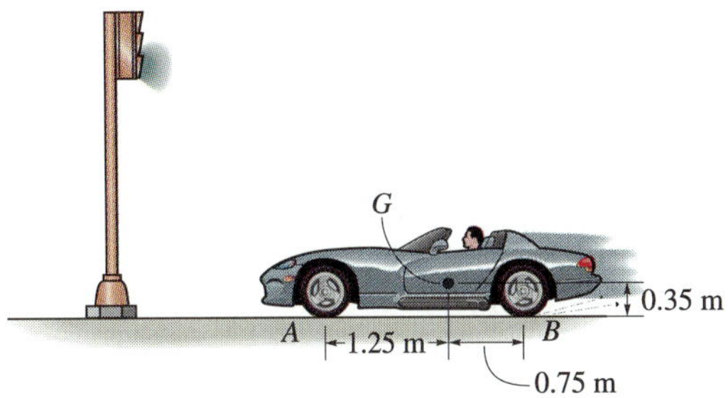

Prob. 17–38

17-39. The "muscle car" is designed to do a "wheeley," i.e., to be able to lift its front wheels off the ground in the manner shown when it accelerates. If the 1.35-Mg car has a center of mass at G, determine the minimum torque that must be developed at both rear wheels in order to do this. Also, what is the smallest necessary coefficient of static friction assuming the thick-walled rear wheels do not slip on the pavement? Neglect the mass of the wheels.

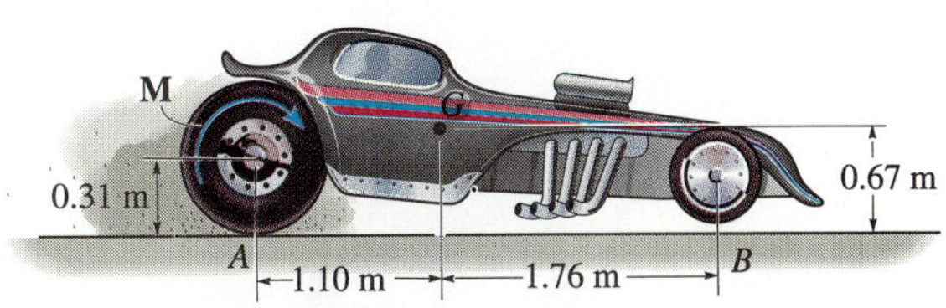

Prob. 17–39

***17-40.** The crate is uniform and weighs 200 lb. If the coefficient of static friction between it and the truck is $\mu_s = 0.3$, determine the shortest distance s in which the truck can stop without causing the crate to tip or slide. The truck is traveling at $v = 20$ ft/s.

Prob. 17–40

17-41. The crate of mass m is supported on a cart of negligible mass. Determine the maximum force P that can be applied a distance d from the cart bottom without causing the crate to tip on the cart.

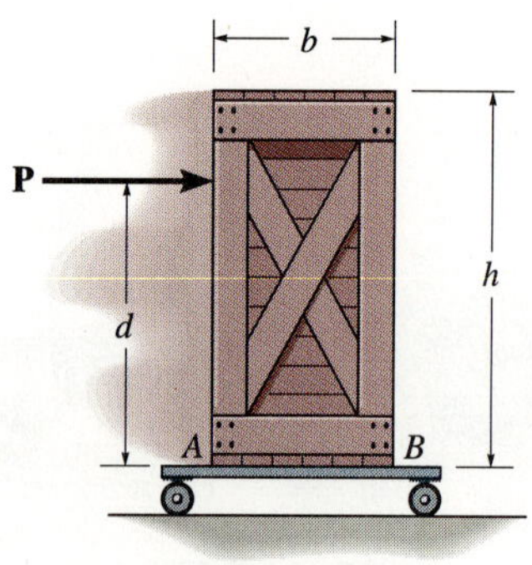

Prob. 17–41

17-42. The uniform crate has a mass m and rests on a rough pallet for which the coefficient of static friction between the crate and pallet is μ_s. If the pallet is given an acceleration of a_p, show that the crate will tip and slip at the same time provided $\mu_s = b/h$.

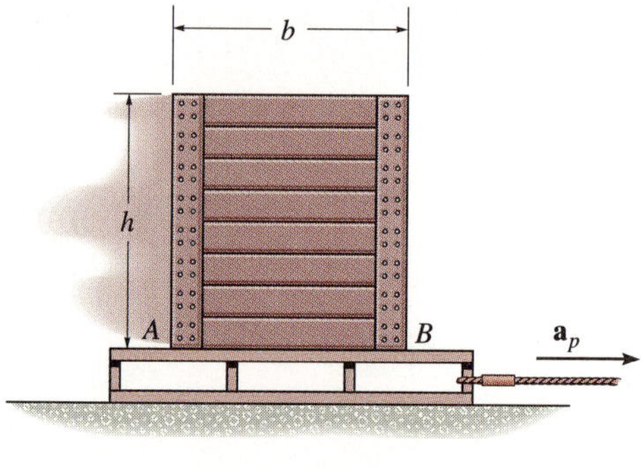

Prob. 17–42

17-43. The bicycle and rider have a mass of 80 kg with center of mass located at G. If the coefficient of kinetic friction at the rear tire is $\mu_B = 0.8$, determine the normal reactions at the tires A and B, and the deceleration of the rider, when the rear wheel locks for braking. What is the normal reaction at the rear wheel when the bicycle is traveling at constant velocity and the brakes are not applied? Neglect the mass of the wheels.

***17-44.** The bicycle and rider have a mass of 80 kg with center of mass located at G. Determine the minimum coefficient of kinetic friction between the road and the wheels so that the rear wheel B starts to lift off the ground when the rider applies the brakes to the front wheel. Neglect the mass of the wheels.

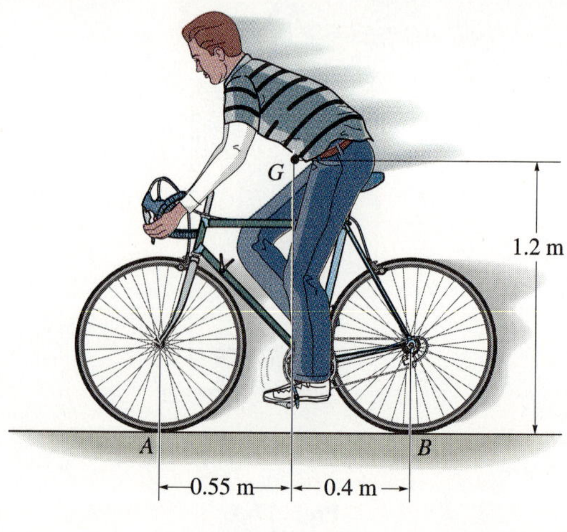

Probs. 17–43/44

17-45. The dresser has a weight of 80 lb and is pushed along the floor. If the coefficient of static friction at A and B is $\mu_s = 0.3$ and the coefficient of kinetic friction is $\mu_k = 0.2$, determine the smallest horizontal force P needed to cause motion. If this force is increased slightly, determine the acceleration of the dresser. Also, what are the normal reactions at A and B when it begins to move?

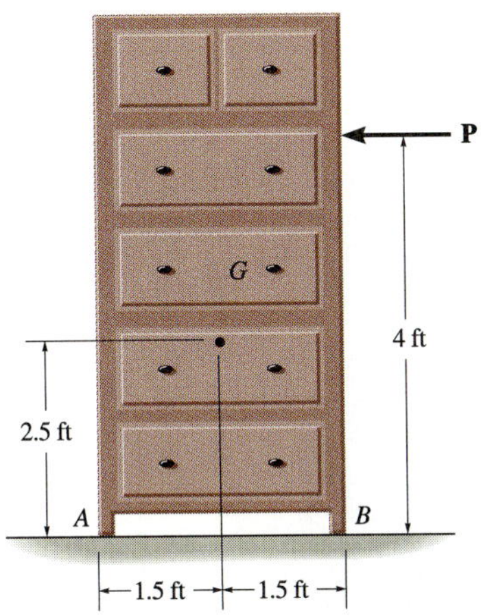

Prob. 17–45

17-46. The dresser has a weight of 80 lb and is pushed along the floor. If the coefficient of static friction at A and B is $\mu_s = 0.3$ and the coefficient of kinetic friction is $\mu_k = 0.2$, determine the maximum horizontal force P that can be applied without causing the dresser to tip over.

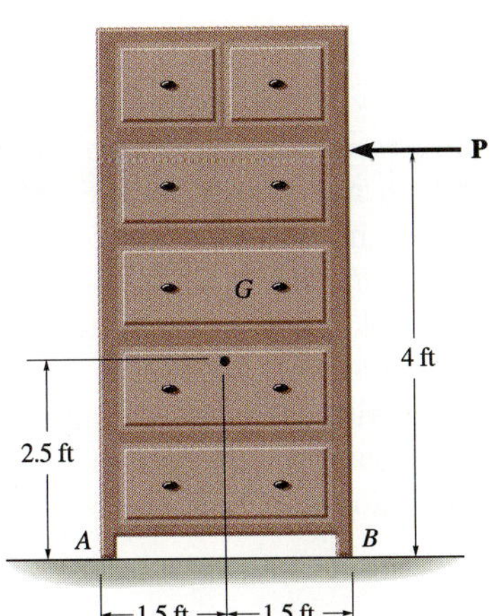

Prob. 17–46

17-47. The handcart has a mass of 200 kg and center of mass at G. Determine the normal reactions at each of the two wheels at A and the two wheels at B if a force of $P = 50$ N is applied to the handle. Neglect the mass of the wheels.

***17-48.** The handcart has a mass of 200 kg and center of mass at G. Determine the magnitude of the largest force **P** that can be applied to the handle so that the wheels at A or B continue to maintain contact with the ground. Neglect the mass of the wheels.

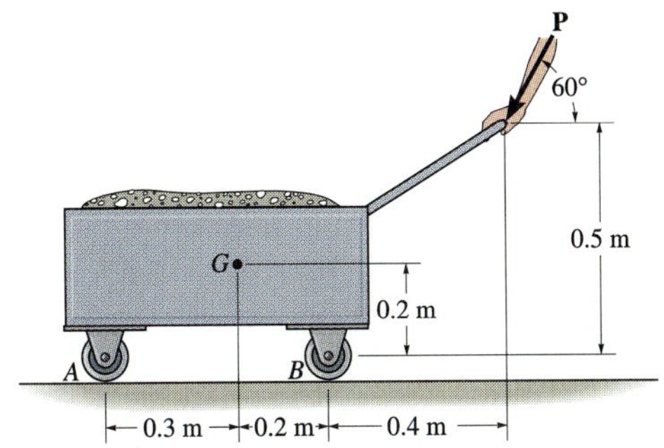

Probs. 17–47/48

17-49. The 50-kg uniform crate rests on the platform for which the coefficient of static friction is $\mu_s = 0.5$. If the supporting links have an angular velocity $\omega = 1$ rad/s, determine the greatest angular acceleration α they can have so that the crate does not slip or tip at the instant $\theta = 30°$ as shown.

17-50. The 50-kg uniform crate rests on the platform for which the coefficient of static friction is $\mu_s = 0.5$. If at the instant $\theta = 30°$ the supporting links have an angular velocity $\omega = 1$ rad/s and angular acceleration $\alpha = 0.5$ rad/s^2, determine the friction force on the crate.

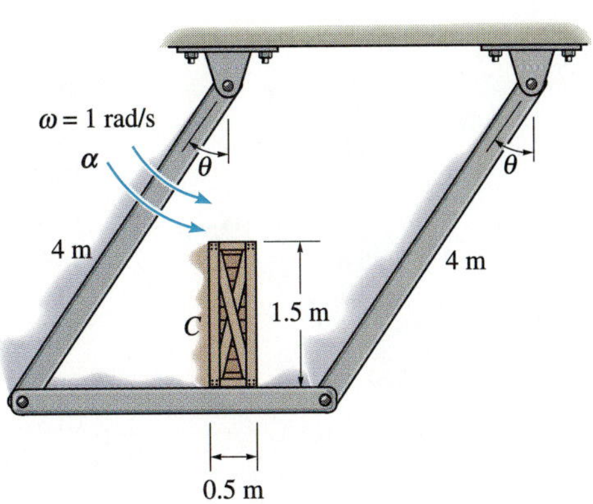

Probs. 17–49/50

17-51. The two 3-lb rods EF and HI are fixed (welded) to the link AC at E. Determine the normal force N_E, shear force V_E, and moment M_E, which the bar AC exerts on FE at E if at the instant $\theta = 30°$ link AB has an angular velocity $\omega = 5$ rad/s and an angular acceleration $\alpha = 8$ rad/s^2 as shown.

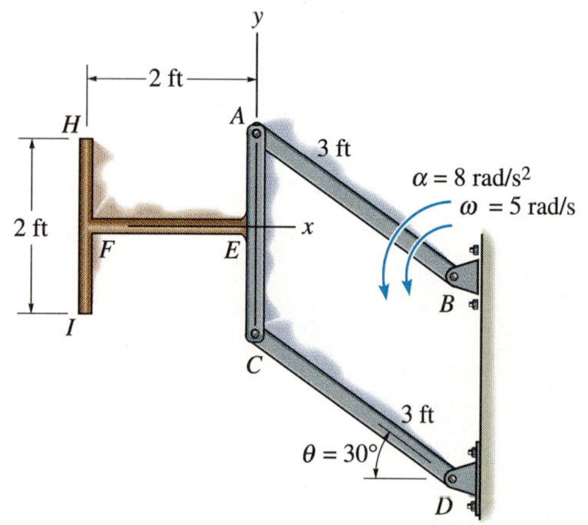

Prob. 17–51

***17-52.** The arm BDE of the industrial robot manufactured by Cincinnati Milacron is activated by applying the torque of $M = 50$ N·m to link CD. Determine the reactions at the pins B and D when the links are in the position shown and have an angular velocity of 2 rad/s. The uniform arm BDE has a mass of 10 kg and a center of mass at G_1. The container held in its grip at E has a mass of 12 kg and center of mass at G_2. Neglect the mass of links AB and CD.

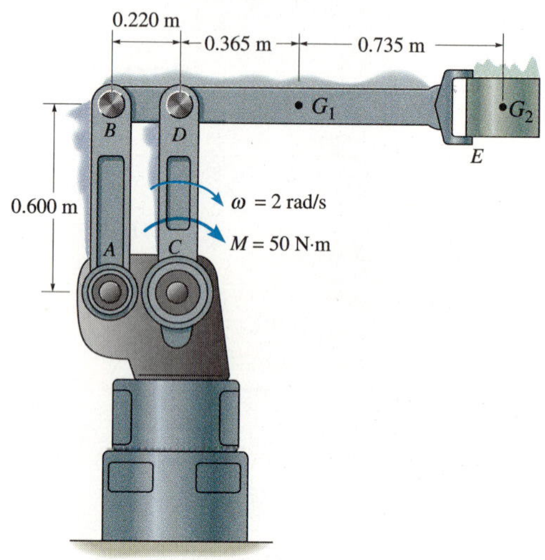

Prob. 17–52

17.4 Equations of Motion: Rotation About a Fixed Axis

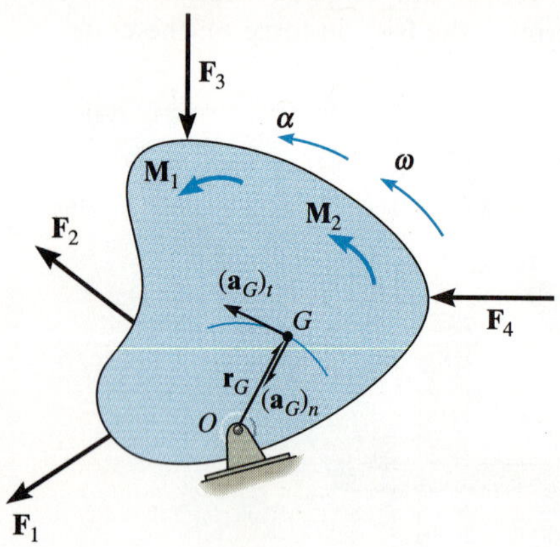

(a)

Consider the rigid body (or slab) shown in Fig. 17–14a, which is constrained to rotate in the vertical plane about a fixed axis perpendicular to the page and passing through the pin at O. The angular velocity and angular acceleration are caused by the external force and couple moment system acting on the body. Because the body's center of mass G moves in a *circular path,* the acceleration of this point is represented by its tangential and normal components. The *tangential component of acceleration* has a *magnitude* of $(a_G)_t = \alpha r_G$ and must act in a *direction* which is *consistent* with the body's angular acceleration $\boldsymbol{\alpha}$. The *magnitude* of the *normal component of acceleration* is $(a_G)_n = \omega^2 r_G$. This component is *always directed* from point G to O, regardless of the direction of $\boldsymbol{\omega}$.

The free-body and kinetic diagrams for the body are shown in Fig. 17–14*b*. The weight of the body, $W = mg$, and the pin reaction $\mathbf{F}_O$ are included on the free-body diagram since they represent external forces acting on the body. The two components $m(\mathbf{a}_G)_t$ and $m(\mathbf{a}_G)_n$, shown on the kinetic diagram, are associated with the tangential and normal acceleration components of the body's mass center. These vectors act in the same *direction* as the acceleration components and have *magnitudes* of $m(a_G)_t$ and $m(a_G)_n$. The $I_G\boldsymbol{\alpha}$ vector acts in the same *direction* as $\boldsymbol{\alpha}$ and has a *magnitude* of $I_G\alpha$, where I_G is the body's moment of inertia calculated about an axis which is perpendicular to the page and passing through G. From the derivation given in Sec. 17.2, the equations of motion which apply to the body may be written in the form

$$\begin{aligned} \Sigma F_n &= m(a_G)_n = m\omega^2 r_G \\ \Sigma F_t &= m(a_G)_t = m\alpha r_G \\ \Sigma M_G &= I_G\alpha \end{aligned} \qquad (17\text{–}14)$$

The moment equation may be replaced by a moment summation about any arbitrary point P on or off the body provided one accounts for the moments $\Sigma(\mathcal{M}_k)_P$ produced by $I_G\boldsymbol{\alpha}$, $m(\mathbf{a}_G)_t$, and $m(\mathbf{a}_G)_n$ about the point. In many problems it is convenient to sum moments about the pin at O in order to eliminate the *unknown* force $\mathbf{F}_O$. From the kinetic diagram, Fig. 17–14*b*, this requires

$$\circlearrowleft+\Sigma M_O = \Sigma(\mathcal{M}_k)_O; \qquad \Sigma M_O = r_G m(a_G)_t + I_G\alpha \qquad (17\text{–}15)$$

Note that the moment of $m(\mathbf{a}_G)_n$ is not included in the summation since the line of action of this vector passes through O. Substituting $(a_G)_t = r_G\alpha$, we may rewrite the above equation as $\circlearrowleft+\Sigma M_O = (I_G + mr_G^2)\alpha$. From the parallel-axis theorem, $I_O = I_G + md^2$, and therefore the term in parentheses represents the *moment of inertia of the body about the fixed axis of rotation passing through O.** Consequently, we can write the three equations of motion for the body as

$$\begin{aligned} \Sigma F_n &= m(a_G)_n = m\omega^2 r_G \\ \Sigma F_t &= m(a_G)_t = m\alpha r_G \\ \Sigma M_O &= I_O\alpha \end{aligned} \qquad (17\text{–}16)$$

For applications, one should remember that "$I_O\alpha$" accounts for the "moment" of *both* $m(\mathbf{a}_G)_t$ *and* $I_G\boldsymbol{\alpha}$ about point O, Fig. 17–14*b*. In other words, $\Sigma M_O = \Sigma(\mathcal{M}_k)_O = I_O\alpha$, as indicated by Eqs. 17–15 and 17–16.

*The result $\Sigma M_O = I_O\alpha$ can also be obtained *directly* from Eq. 17–6 by selecting point P to coincide with O, realizing that $(a_P)_x = (a_P)_y = 0$.

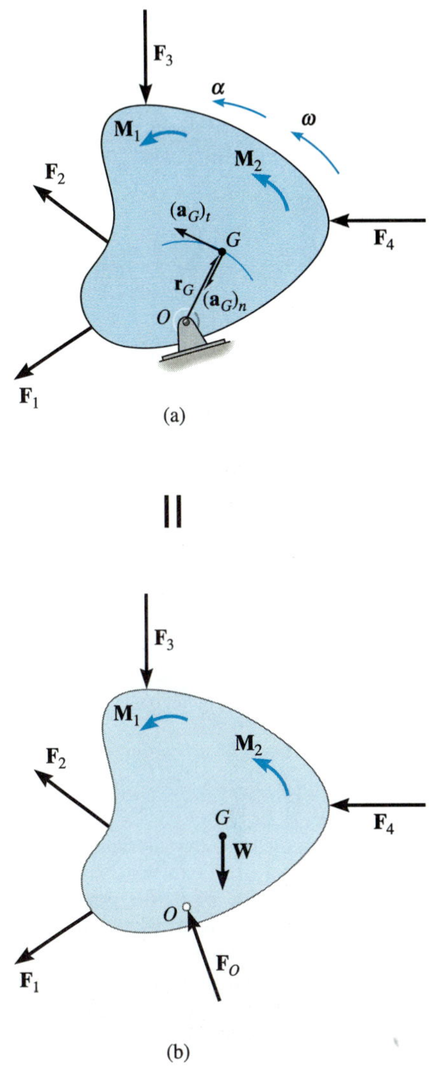

Fig. 17–14

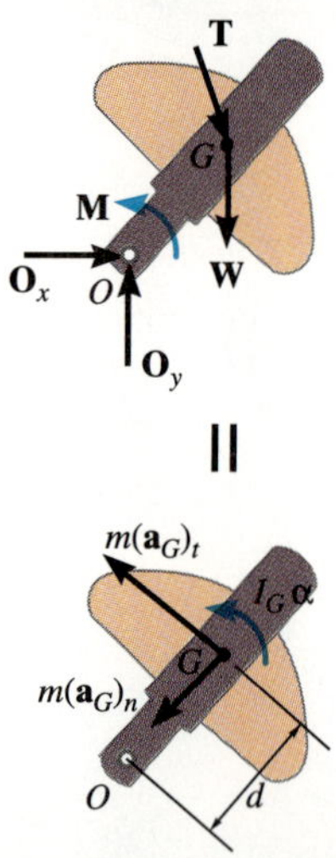

The crank on the oil-pumping rig undergoes rotation about a fixed axis which is caused by a driving torque **M** of the motor. The loadings shown on the free-body diagram cause the effects shown on the kinetic diagram. If moments are summed about the mass center, G, then $\Sigma M_G = I_G\alpha$. However, if moments are summed about point O, noting that $(a_G)_t = \alpha d$, then $\circlearrowright + \Sigma M_O = I_G\alpha + m(a_G)_t d + m(a_G)_n(0) = (I_G + md^2)\alpha = I_O\alpha$.

Procedure for Analysis

Kinetic problems which involve the rotation of a body about a fixed axis can be solved using the following procedure.

Free-Body Diagram

- Establish the inertial x, y or n, t coordinate system and specify the direction and sense of the accelerations $(\mathbf{a}_G)_n$ and $(\mathbf{a}_G)_t$ and the angular acceleration $\boldsymbol{\alpha}$ of the body. Recall that $(\mathbf{a}_G)_t$ must act in a direction which is in accordance with $\boldsymbol{\alpha}$, whereas $(\mathbf{a}_G)_n$ always acts toward the axis of rotation, point O.
- Draw the free-body diagram to account for all the external forces and couple moments that act on the body.
- Compute the moment of inertia I_G or I_O.
- Identify the unknowns in the problem.
- If it is decided that the rotational equation of motion $\Sigma M_P = \Sigma(\mathcal{M}_k)_P$ is to be used, i.e., P is a point other than G or O, then consider drawing the kinetic diagram in order to help "visualize" the "moments" developed by the components $m(\mathbf{a}_G)_n$, $m(\mathbf{a}_G)_t$, and $I_G\boldsymbol{\alpha}$ when writing the terms for the moment sum $\Sigma(\mathcal{M}_k)_P$.

Equations of Motion

- Apply the three equations of motion in accordance with the established sign convention.
- If moments are summed about the body's mass center, G, then $\Sigma M_G = I_G\alpha$, since $(m\mathbf{a}_G)_t$ and $(m\mathbf{a}_G)_n$ create no moment about G.
- If moments are summed about the pin support O on the axis of rotation, then $(m\mathbf{a}_G)_n$ creates no moment about G, and it can be shown that $\Sigma M_O = I_O\alpha$.

Kinematics

- Use kinematics if a complete solution cannot be obtained strictly from the equations of motion.
- If the *angular acceleration is variable,* use

$$\alpha = \frac{d\omega}{dt} \qquad \alpha\, d\theta = \omega\, d\omega \qquad \omega = \frac{d\theta}{dt}$$

- If the *angular acceleration is constant,* use

$$\omega = \omega_0 + \alpha_c t$$
$$\theta = \theta_0 + \omega_0 t + \tfrac{1}{2}\alpha_c t^2$$
$$\omega^2 = \omega_0^2 + 2\alpha_c(\theta - \theta_0)$$

EXAMPLE 17–9

The 30-kg uniform disk shown in Fig. 17–15*a* is pin supported at its center. If it starts from rest, determine the number of revolutions it must make to attain an angular velocity of 20 rad/s. Also, what are the reactions at the pin? The disk is acted upon by a constant force $F = 10$ N, which is applied to a cord wrapped around its periphery, and a constant couple moment $M = 5$ N·m. Neglect the mass of the cord in the calculation.

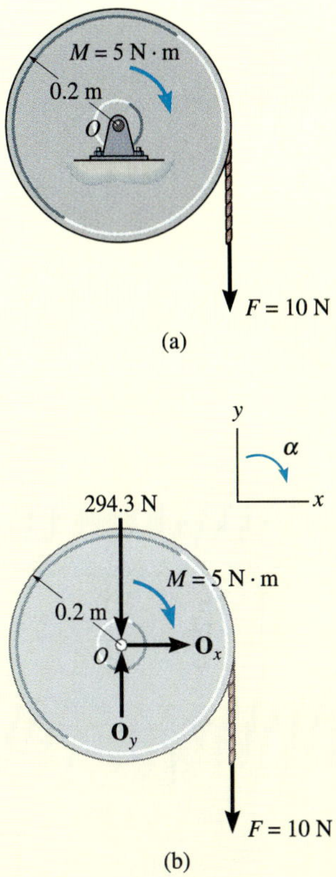

Fig. 17–15

Solution

Free-Body Diagram. Fig. 17–15*b*. Note that the mass center is not subjected to an acceleration; however, the disk has a clockwise angular acceleration.

The moment of inertia of the disk about the pin is

$$I_O = \tfrac{1}{2}mr^2 = \frac{1}{2}(30\text{ kg})(0.2\text{ m})^2 = 0.6\text{ kg}\cdot\text{m}^2$$

The three unknowns are O_x, O_y, and α.

Equations of Motion

$$\overset{+}{\rightarrow}\Sigma F_x = m(a_G)_x; \qquad O_x = 0 \qquad \textit{Ans.}$$

$$+\uparrow \Sigma F_y = m(a_G)_y; \quad O_y - 294.3\text{ N} - 10\text{ N} = 0$$

$$O_y = 304\text{ N} \qquad \textit{Ans.}$$

$$\circlearrowright + \Sigma M_O = I_O\alpha; \quad -10\text{ N}(0.2\text{ m}) - 5\text{ N}\cdot\text{m} = -(0.6\text{ kg}\cdot\text{m}^2)\alpha$$

$$\alpha = 11.7\text{ rad/s}^2 \circlearrowright$$

Kinematics. Since α is constant and is clockwise, the number of radians the disk must turn to obtain a clockwise angular velocity of 20 rad/s is

$$\circlearrowright + \qquad \omega^2 = \omega_0^2 + 2\alpha_c(\theta - \theta_0)$$

$$(-20\text{ rad/s})^2 = 0 + 2(-11.7\text{ rad/s}^2)(\theta - 0)$$

$$\theta = -17.1\text{ rad} = 17.1\text{ rad} \circlearrowright$$

Hence,

$$\theta = 17.1\text{ rad}\left(\frac{1\text{ rev}}{2\pi\text{ rad}}\right) = 2.73\text{ rev} \circlearrowright \qquad \textit{Ans.}$$

EXAMPLE 17–10

The 20-kg slender rod shown in Fig. 17–16*a* is rotating in the vertical plane, and at the instant shown it has an angular velocity of $\omega = 5$ rad/s. Determine the rod's angular acceleration and the horizontal and vertical components of reaction at the pin at this instant.

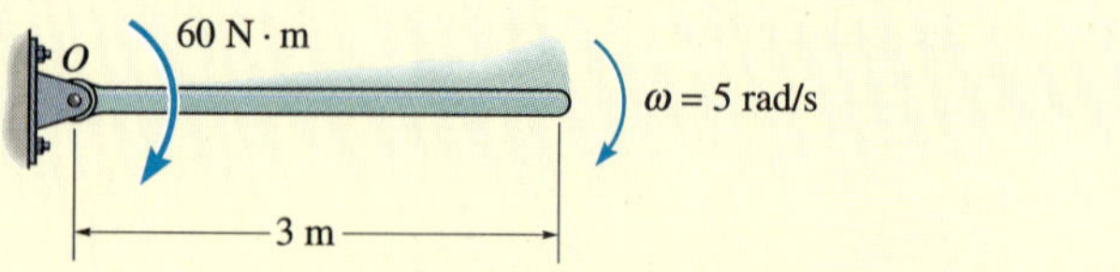

(a)

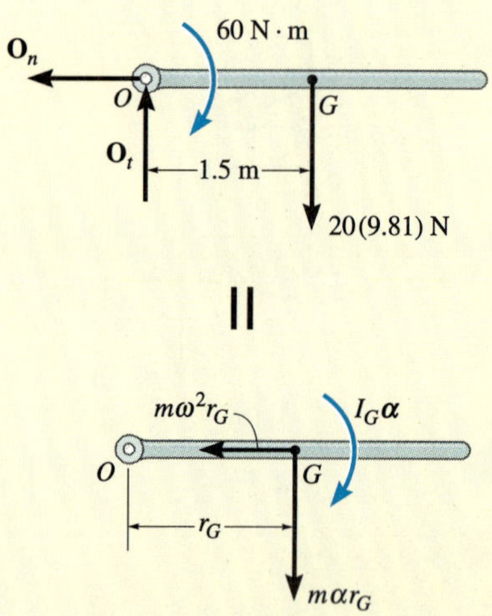

(b)

Fig. 17–16

Solution

Free-Body and Kinetic Diagrams. Fig. 17–16*b*. As shown on the kinetic diagram, point G moves in a circular path and so has two components of acceleration. It is important that the tangential component $a_t = \alpha r_G$ act downward since it must be in accordance with the angular acceleration $\boldsymbol{\alpha}$ of the rod. The three unknowns are O_n, O_t, and α.

Equations of Motion

$$\overset{+}{\leftarrow} \Sigma F_n = m\omega^2 r_G; \qquad O_n = (20\text{ kg})(5\text{ rad/s})^2(1.5\text{ m})$$

$$+\downarrow \Sigma F_t = m\alpha r_G; \qquad -O_t + 20(9.81)\text{ N} = (20\text{ kg})(\alpha)(1.5\text{ m})$$

$$\circlearrowright + \Sigma M_G = I_G\alpha; \quad O_t(1.5\text{ m}) + 60\text{ N}\cdot\text{m} = [\tfrac{1}{12}(20\text{ kg})(3\text{ m})^2]\alpha$$

Solving

$$O_n = 750\text{ N} \qquad O_t = 19.0\text{ N} \qquad \alpha = 5.90\text{ rad/s}^2 \qquad \textit{Ans.}$$

A more direct solution to this problem would be to sum moments about point O to eliminate $\mathbf{O}_n$ and $\mathbf{O}_t$ and obtain a *direct solution* for α. Here,

$$\circlearrowright + \Sigma M_O = \Sigma(\mathcal{M}_k)_O; \qquad 60\text{ N}\cdot\text{m} + 20(9.81)\text{ N}(1.5\text{ m}) =$$
$$[\tfrac{1}{12}(20\text{ kg})(3\text{ m})^2]\alpha + [20\text{ kg}(\alpha)(1.5\text{ m})](1.5\text{ m})$$
$$\alpha = 5.90\text{ rad/s}^2 \qquad \textit{Ans.}$$

Also, since $I_O = \frac{1}{3}ml^2$ for a slender rod, we can apply

$$\circlearrowright + \Sigma M_O = I_O\alpha; \quad 60\text{ N}\cdot\text{m} + 20(9.81)\text{ N}(1.5\text{ m}) = [\tfrac{1}{3}(20\text{ kg})(3\text{ m})^2]\alpha$$
$$\alpha = 5.90\text{ rad/s}^2 \qquad \textit{Ans.}$$

By comparison, the last equation provides the simplest solution for α and *does not* require use of the kinetic diagram.

EXAMPLE 17–11

The drum shown in Fig. 17–17*a* has a mass of 60 kg and a radius of gyration $k_O = 0.25$ m. A cord of negligible mass is wrapped around the periphery of the drum and attached to a block having a mass of 20 kg. If the block is released, determine the drum's angular acceleration.

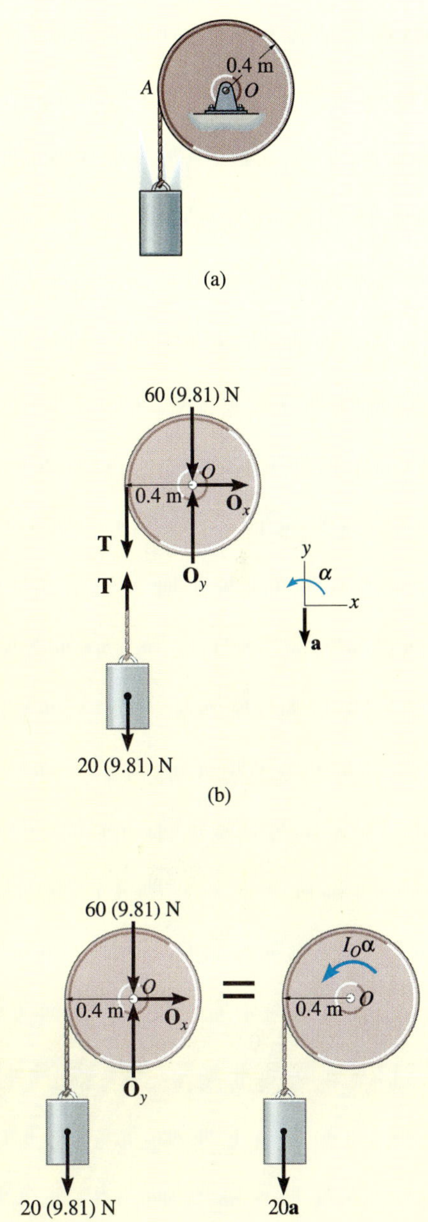

Fig. 17–17

Solution I

Free-Body Diagram. Here we will consider the drum and block separately, Fig. 17–17*b*. Assuming the block accelerates *downward* at **a**, it creates a *counterclockwise* angular acceleration $\boldsymbol{\alpha}$ of the drum.

The moment of inertia of the drum is

$$I_O = mk_O^2 = (60\text{ kg})(0.25\text{ m})^2 = 3.75\text{ kg}\cdot\text{m}^2$$

There are five unknowns, namely O_x, O_y, T, a, and α.

Equations of Motion. Applying the translational equations of motion $\Sigma F_x = m(a_G)_x$ and $\Sigma F_y = m(a_G)_y$ to the drum is of no consequence to the solution, since these equations involve the unknowns O_x and O_y. Thus, for the drum and block, respectively,

$$\circlearrowleft\!+\Sigma M_O = I_O\alpha; \qquad T(0.4\text{ m}) = (3.75\text{ kg}\cdot\text{m}^2)\alpha \qquad (1)$$

$$+\uparrow \Sigma F_y = m(a_G)_y; \quad -20(9.81)\text{ N} + T = -20a \qquad (2)$$

Kinematics. Since the point of contact A between the cord and drum has a tangential component of acceleration **a**, Fig. 17–17*a*, then

$$\circlearrowleft\!+a = \alpha r; \qquad a = \alpha(0.4) \qquad (3)$$

Solving the above equations,

$$T = 106\text{ N}$$
$$a = 4.52\text{ m/s}^2$$
$$\alpha = 11.3\text{ rad/s}^2 \circlearrowleft \qquad \textit{Ans.}$$

Solution II

Free-Body and Kinetic Diagrams. The cable tension T can be eliminated from the analysis by considering the drum and block as a *single system*, Fig. 17–17*c*. The kinetic diagram is shown since moments will be summed about point O.

Equations of Motion. Using Eq. 3 and applying the moment equation about O to eliminate the unknowns O_x and O_y, we have

$$\circlearrowleft\!+\Sigma M_O = \Sigma(\mathcal{M}_k)_O; \qquad 20(9.81)\text{ N}(0.4\text{ m}) =$$
$$(3.75\text{ kg}\cdot\text{m}^2)\alpha + [20\text{ kg}(0.4\text{ m }\alpha)](0.4\text{ m})$$
$$\alpha = 11.3\text{ rad/s}^2 \circlearrowleft \qquad \textit{Ans.}$$

Note: If the block were *removed* and a force of 20(9.81) N were applied to the cord, show that $\alpha = 20.9\text{ rad/s}^2$ and explain the reason for the difference in the results.

EXAMPLE 17–12

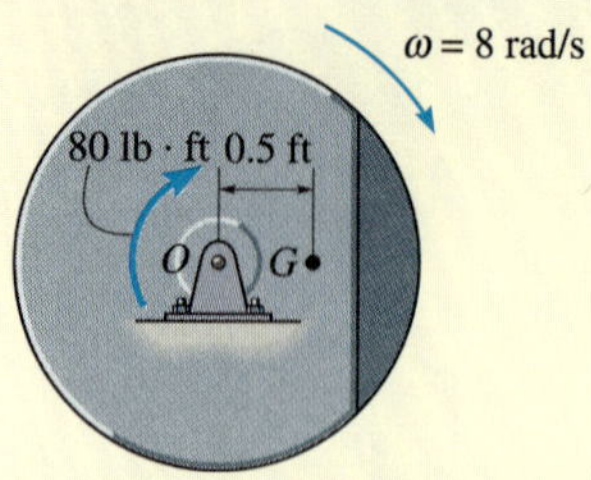

(a)

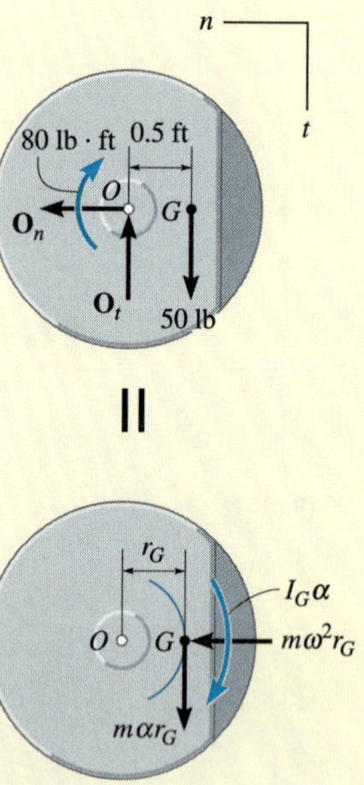

(b)

Fig. 17–18

The unbalanced 50-lb flywheel shown in Fig. 17–18*a* has a radius of gyration of $k_G = 0.6$ ft about an axis passing through its mass center G. If it has a clockwise angular velocity of 8 rad/s at the instant shown, determine the horizontal and vertical components of reaction at the pin O.

Solution

Free-Body and Kinetic Diagrams. Since G moves in a circular path, it will have both normal and tangential components of acceleration. Also, since α, which is caused by the flywheel's weight, acts clockwise, the tangential component of acceleration will act downward. Why? The vectors $m(a_G)_t = m\alpha r_G$, $m(a_G)_n = m\omega^2 r_G$, and $I_G\alpha$ are shown on the kinematic diagram in Fig. 17–18*b*. Here, the moment of inertia of the flywheel about its mass center is determined from the radius of gyration and the flywheel's mass; i.e., $I_G = mk_G^2 = (50\text{ lb}/32.2\text{ ft/s}^2)(0.6\text{ ft})^2 = 0.559\text{ slug}\cdot\text{ft}^2$.

The three unknowns are O_x, O_y, and α.

Equations of Motion

$$\overset{+}{\leftarrow}\Sigma F_n = m\omega^2 r_G; \qquad O_n = \left(\frac{50\text{ lb}}{32.2\text{ ft/s}^2}\right)(8\text{ rad/s})^2(0.5\text{ ft}) \tag{1}$$

$$+\downarrow \Sigma F_t = m\alpha r_G; \quad -O_t + 50\text{ lb} = \left(\frac{50\text{ lb}}{32.2\text{ ft/s}^2}\right)(\alpha)(0.5\text{ ft}) \tag{2}$$

$$\curvearrowright\!+\Sigma M_G = I_G\alpha; \qquad 80\text{ lb}\cdot\text{ft} + O_t(0.5\text{ ft}) = (0.559\text{ slug}\cdot\text{ft}^2)\alpha \tag{3}$$

Solving, $\alpha = 111\text{ rad/s}^2 \qquad O_n = 49.7\text{ lb} \qquad O_t = -36.1\text{ lb}$ *Ans.*

Moments can also be summed about point O in order to eliminate $\mathbf{O}_n$ and $\mathbf{O}_t$ and thereby obtain a *direct solution* for $\boldsymbol{\alpha}$, Fig. 17–18*b*. This can be done in one of *two* ways, i.e., by using either $\Sigma M_O = \Sigma(\mathcal{M}_k)_O$ or $\Sigma M_O = I_O\alpha$. If the first of these equations is applied, we have

$$\curvearrowright\!+\Sigma M_O = \Sigma(\mathcal{M}_k)_O; \quad 80\text{ lb}\cdot\text{ft} + 50\text{ lb}(0.5\text{ ft}) =$$

$$(0.559\text{ slug}\cdot\text{ft}^2)\alpha + \left[\left(\frac{50\text{ lb}}{32.2\text{ ft/s}^2}\right)\alpha(0.5\text{ ft})\right](0.5\text{ ft})$$

$$105 = 0.947\alpha \tag{4}$$

If $\Sigma M_O = I_O\alpha$ is applied, then by the parallel-axis theorem the moment of inertia of the flywheel about O is

$$I_O = I_G + mr_G^2 = 0.559 + \left(\frac{50}{32.2}\right)(0.5)^2 = 0.947\text{ slug}\cdot\text{ft}^2$$

Hence, from the free-body diagram, Fig. 17–18*b*, we require

$$\curvearrowright\!+\Sigma M_O = I_O\alpha; \quad 80\text{ lb}\cdot\text{ft} + 50\text{ lb}(0.5\text{ ft}) = (0.947\text{ slug}\cdot\text{ft}^2)\alpha$$

which is the same as Eq. 4. Solving for α and substituting into Eq. 2 yields the answer for O_t obtained previously.

EXAMPLE 17–13

The slender rod shown in Fig. 17–19*a* has a mass m and length l and is released from rest when $\theta = 0°$. Determine the horizontal and vertical components of force which the pin at A exerts on the rod at the instant $\theta = 90°$.

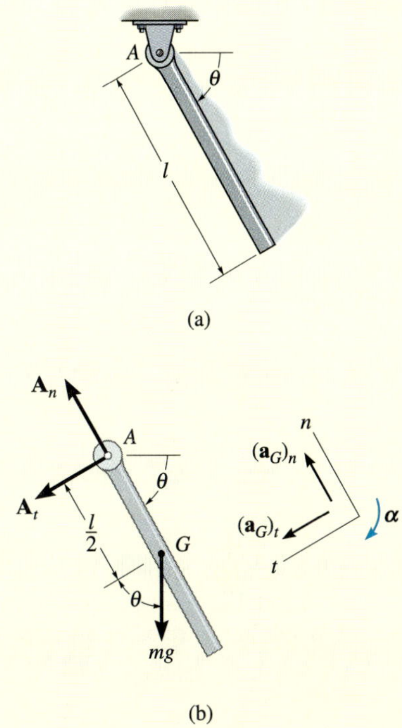

Fig. 17–19

Solution

Free-Body Diagram. The free-body diagram for the rod is shown when the rod is in the general position θ, Fig. 17–19*b*. For convenience, the force components at A are shown acting in the n and t directions. Note that $\boldsymbol{\alpha}$ acts clockwise.

The moment of inertia of the rod about point A is $I_A = \frac{1}{3}ml^2$.

Equations of Motion. Moments will be summed about A in order to eliminate the reactive forces there.*

$$+\nwarrow \Sigma F_n = m\omega^2 r_G; \quad A_n - mg\sin\theta = m\omega^2(l/2) \quad (1)$$

$$+\swarrow \Sigma F_t = m\alpha r_G; \quad A_t + mg\cos\theta = m\alpha(l/2) \quad (2)$$

$$\curvearrowright + \Sigma M_A = I_A\alpha; \quad mg\cos\theta(l/2) = \left(\tfrac{1}{3}ml^2\right)\alpha \quad (3)$$

Kinematics. For a given angle θ there are four unknowns in the above three equations: A_n, A_t, ω, and α. As shown by Eq. 3, α is *not constant;* rather, it depends on the position θ of the rod. The necessary fourth equation is obtained using kinematics, where α and ω can be related to θ by the equation

$$(\curvearrowright +) \qquad \omega\, d\omega = \alpha\, d\theta \qquad (4)$$

Note that the positive clockwise direction for this equation *agrees* with that of Eq. 3. This is important since we are seeking a simultaneous solution.

In order to solve for ω at $\theta = 90°$, eliminate α from Eqs. 3 and 4, which yields

$$\omega\, d\omega = (1.5\, g/l)\cos\theta\, d\theta$$

Since $\omega = 0$ at $\theta = 0°$, we have

$$\int_0^{\omega} \omega\, d\omega = (1.5\, g/l)\int_{0°}^{90°} \cos\theta\, d\theta$$

$$\omega^2 = 3\, g/l$$

Substituting this value into Eq. 1 with $\theta = 90°$ and solving Eqs. 1 to 3 yields

$$\alpha = 0 \qquad A_t = 0 \qquad A_n = 2.5mg \qquad \textit{Ans.}$$

*If $\Sigma M_A = \Sigma(\mathcal{M}_k)_A$ is used, one must account for the moments of $I_G\boldsymbol{\alpha}$ and $m(\mathbf{a}_G)_t$ about A. Here, however, we have used $\Sigma M_A = I_A\alpha$.

Problems

17-53. The 10-lb rod is pin connected to its support at A and has an angular velocity $\omega = 4$ rad/s when it is in the horizontal position shown. Determine its angular acceleration and the horizontal and vertical components of reaction which the pin exerts on the rod at this instant.

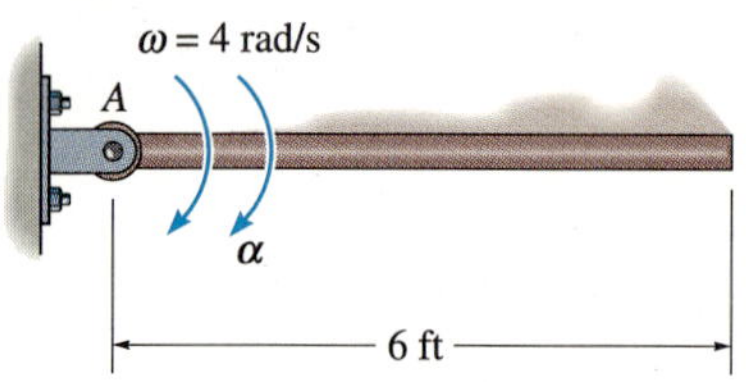

Prob. 17–53

17-54. The pendulum consists of a 20-lb sphere and a 5-lb slender rod. Determine the reaction at the pin O just after the pendulum is released from the position shown.

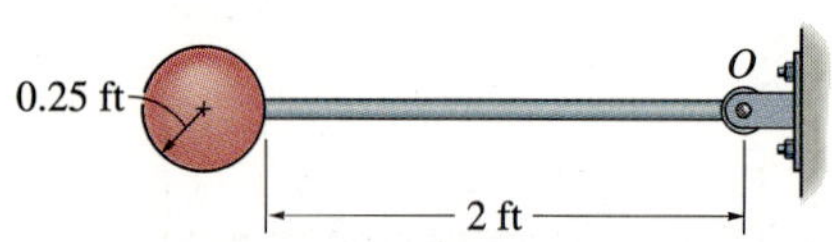

Prob. 17–54

17-55. The fan blade has a mass of 2 kg and a moment of inertia $I_O = 0.18\ \text{kg}\cdot\text{m}^2$ about an axis passing through its center O. If it is subjected to a moment of $M = 3(1 - e^{-0.2t})$ N·m, where t is in seconds, determine its angular velocity when $t = 4$ s starting from rest.

Prob. 17–55

***17-56.** The pendulum consists of a 15-lb disk and a 10-lb slender rod. Determine the horizontal and vertical components of reaction that the pin O exerts on the rod just as it passes the horizontal position, at which time its angular velocity is $\omega = 8$ rad/s.

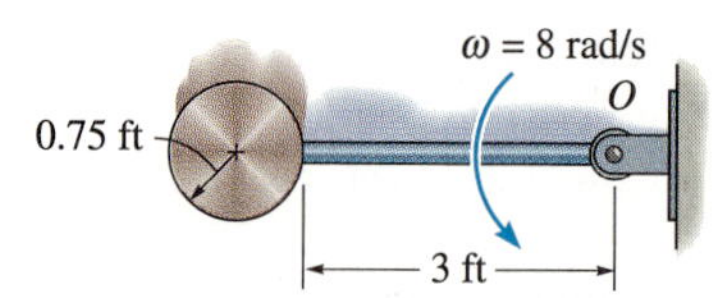

Prob. 17–56

17-57. The spool is supported on small rollers at A and B. Determine the constant force P that must be applied to the cable in order to unwind 8 m of cable in 4 s starting from rest. Also calculate the normal forces at A and B during this time. The spool has a mass of 60 kg and a radius of gyration $k_O = 0.65$ m. For the calculation neglect the mass of the cable and the mass of the rollers at A and B.

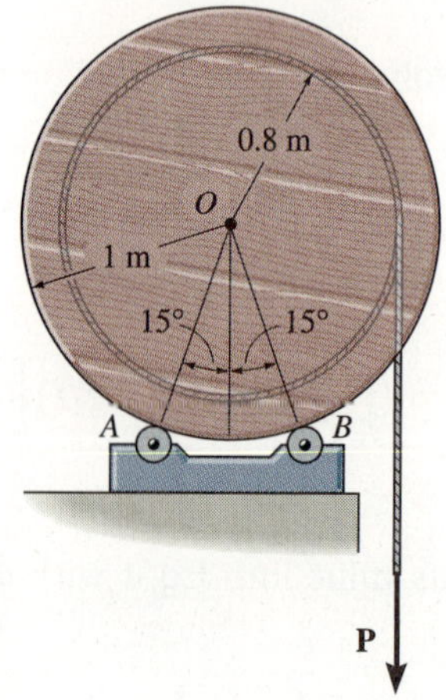

Prob. 17–57

17-58. A cord is wrapped around the inner core of a spool. If the cord is pulled with a constant tension of 30 lb and the spool is originally at rest, determine the spool's angular velocity when $s = 8$ ft of cord has unwound. Neglect the weight of the 8-ft portion of cord. The spool and the entire cord have a total weight of 400 lb, and the radius of gyration about the axle A is $k_A = 1.30$ ft.

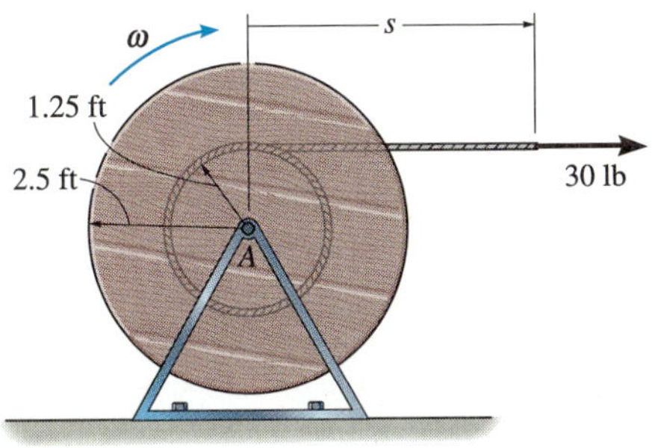

Prob. 17–58

17-59. A motor supplies a constant torque $M = 2$ N·m to a 50-mm-diameter shaft O connected to the center of the 30-kg flywheel. The resultant bearing friction **F**, which the pin exerts on the shaft, acts tangent to the supporting shaft and has a magnitude of 50 N. Determine how long the torque must be applied to the shaft to increase the flywheel's rotational speed from 4 rad/s to 15 rad/s. The flywheel has a radius of gyration $k_O = 0.15$ m about its center.

***17-60.** If the motor in Prob. 17-59 is disengaged from the shaft once the flywheel is rotating at 15 rad/s, so that $M = 0$, determine how long it will take before the resultant bearing frictional force $F = 50$ N stops the flywheel from rotating.

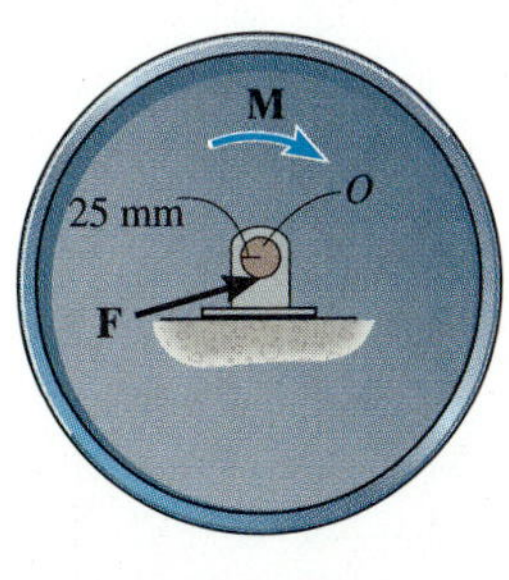

Probs. 17–59/60

17-61. The pendulum consists of a uniform 5-kg plate and a 2-kg slender rod. Determine the horizontal and vertical components of reaction that the pin O exerts on the rod at the instant $\theta = 30°$, at which time its angular velocity is $\omega = 3$ rad/s.

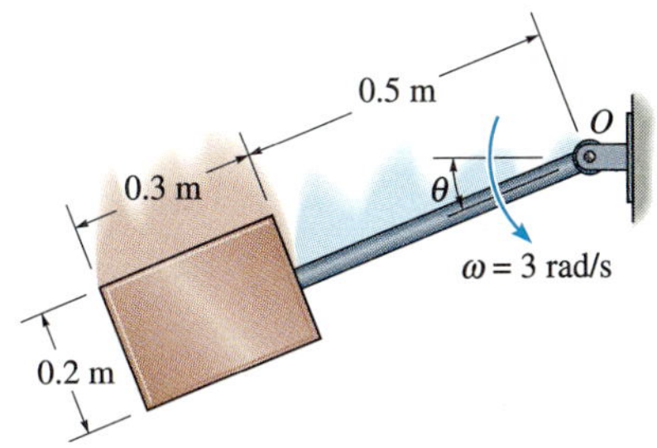

Prob. 17–61

17-62. The cylinder has a radius r and mass m and rests in the trough for which the coefficient of kinetic friction at A and B is μ_k. If a horizontal force **P** is applied to the cylinder, determine the cylinder's angular acceleration when it begins to spin.

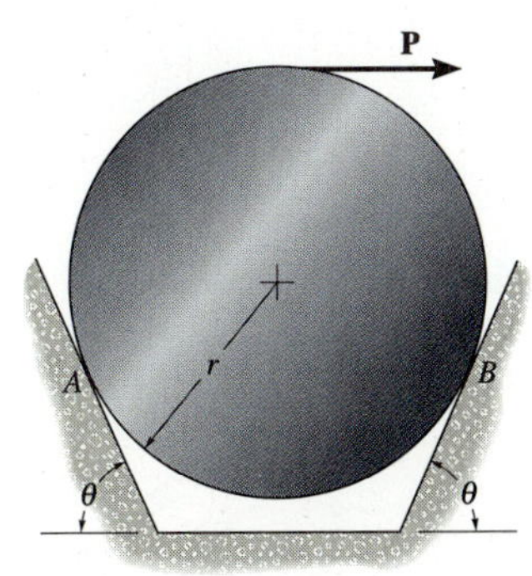

Prob. 17–62

17-63. The uniform slender rod has a mass of 5 kg. If the cord at A is cut, determine the reaction at the pin O, (a) when the rod is still in the horizontal position, and (b) when the rod swings to the vertical position.

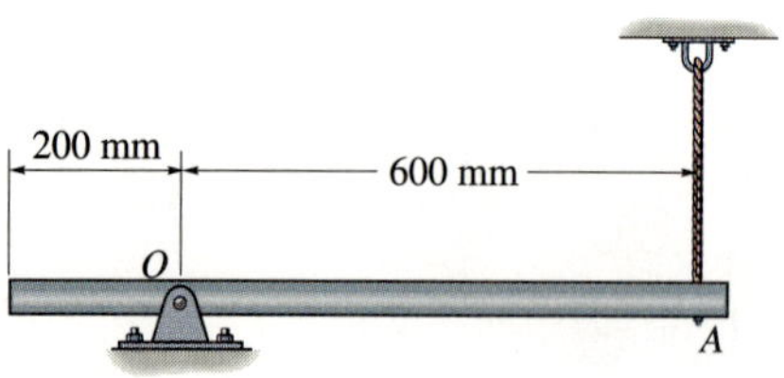

Prob. 17–63

17-65. The kinetic diagram representing the general rotational motion of a rigid body about a fixed axis at O is shown in the figure. Show that $I_G\boldsymbol{\alpha}$ may be eliminated by moving the vectors $m(\mathbf{a}_G)_t$ and $m(\mathbf{a}_G)_n$ to point P, located a distance $r_{GP} = k_G^2/r_{OG}$ from the center of mass G of the body. Here k_G represents the radius of gyration of the body about G. The point P is called the *center of percussion* of the body.

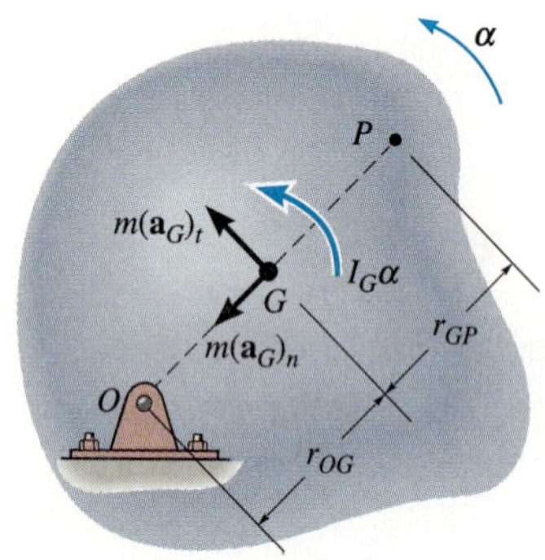

Prob. 17–65

***17-64.** The bar has a mass m and length l. If it is released from rest from the position $\theta = 30°$, determine its angular acceleration and the horizontal and vertical components of reaction at the pin O.

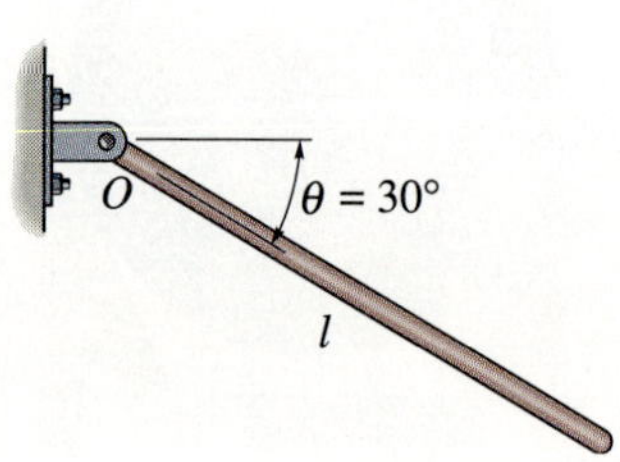

Prob. 17–64

17-66. Determine the position r_P of the center of percussion P of the 10-lb slender bar. (See Prob. 17-65.) What is the horizontal force A_x at the pin when the bar is struck at P with a force of $F = 20$ lb?

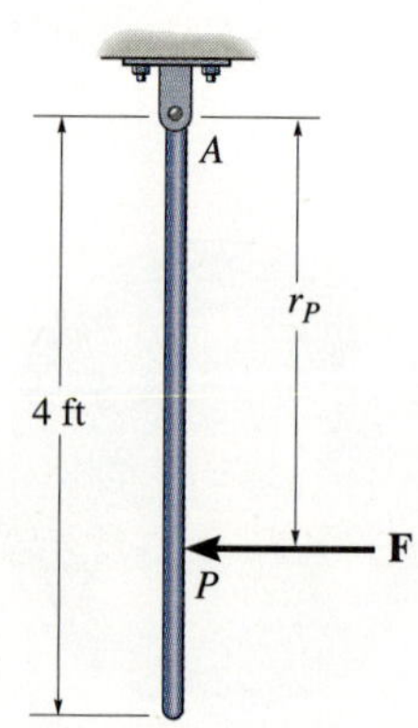

Prob. 17–66

17-67. In order to experimentally determine the moment of inertia I_G of a 4-kg connecting rod, the rod is suspended horizontally at A by a cord and at B by a bearing and piezoelectric sensor, an instrument used for measuring force. Under these equilibrium conditions, the force at B is measured as 14.6 N. If, at the instant the cord is released, the reaction at B is measured as 9.3 N, determine the value of I_G. The support at B does not move when the measurement is taken. For the calculation, the horizontal location of G must be determined.

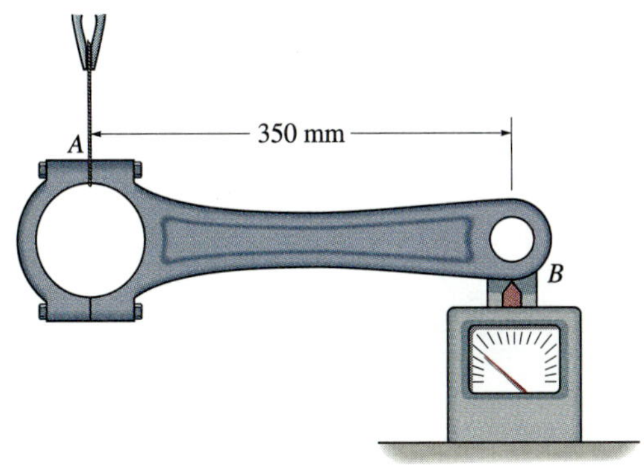

Prob. 17–67

*17-68. The 4-kg slender rod is supported horizontally by a spring at A and a cord at B. Determine the angular acceleration of the rod and the acceleration of the rod's mass center at the instant the cord at B is cut. *Hint:* The stiffness of the spring is not needed for the calculation.

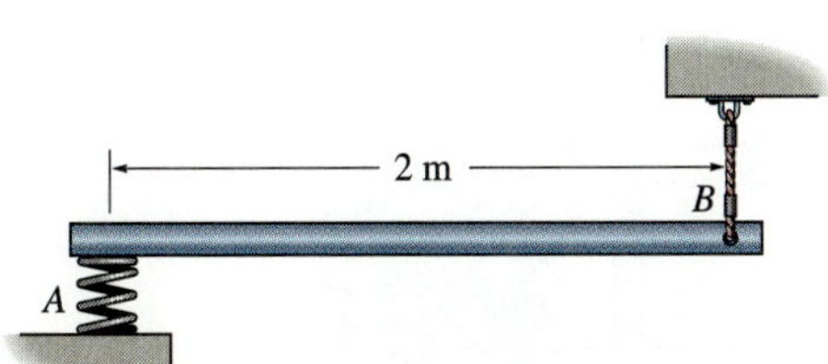

Prob. 17–68

17-69. The 10-lb disk D is subjected to a counterclockwise moment of $M = (10t)$ lb·ft, where t is in seconds. Determine the angular velocity of the disk 2 s after the moment is applied. Due to the spring the plate P exerts a constant force of 100 lb on the disk. The coefficients of static and kinetic friction between the disk and the plate are $\mu_s = 0.3$ and $\mu_k = 0.2$, respectively. *Hint:* First find the time needed to start the disk rotating.

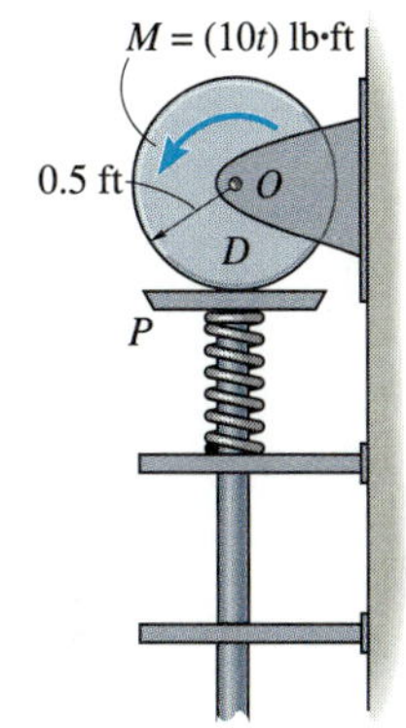

Prob. 17–69

17-70. The furnace cover has a mass of 20 kg and a radius of gyration $k_G = 0.25$ m about its mass center G. If an operator applies a force $F = 120$ N to the handle in order to open the cover, determine the cover's initial angular acceleration and the horizontal and vertical components of reaction which the pin at A exerts on the cover at the instant the cover begins to open. Neglect the mass of the handle BAC in the calculation.

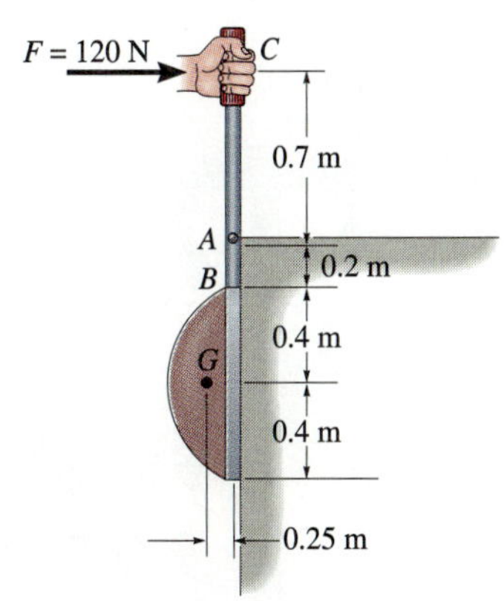

Prob. 17–70

17-71. The variable-reluctance motor is often used for appliances, pumps, and blowers. By applying a current through the stator S, an electromagnetic field is created that "pulls in" the nearest rotor poles. The result of this is to create a torque of 4 N·m about the bearing at A. If the rotor is made from iron and has a 3-kg cylindrical core of 50-mm diameter and eight extended slender rods, each having a mass of 1 kg and 100-mm length, determine its angular velocity in 5 seconds starting from rest.

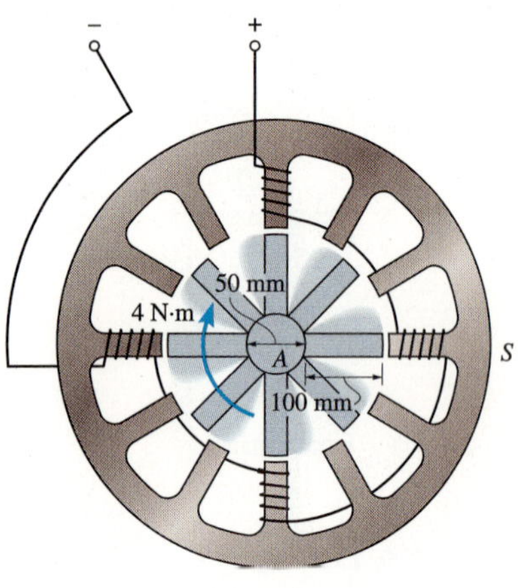

Prob. 17–71

***17-72.** The variable-reluctance motor is often used for appliances, pumps, and blowers. By applying a current through the stator S, an electromagnetic field is created that "pulls in" the nearest rotor poles. The result of this is to create a torque of 4 N·m about the bearing at A. If the rotor is made from iron and has a 3-kg cylindrical core and eight extended slender rods, each having a mass of 1 kg, determine its angular velocity at the instant the rotor has undergone 15 revolutions, starting from rest.

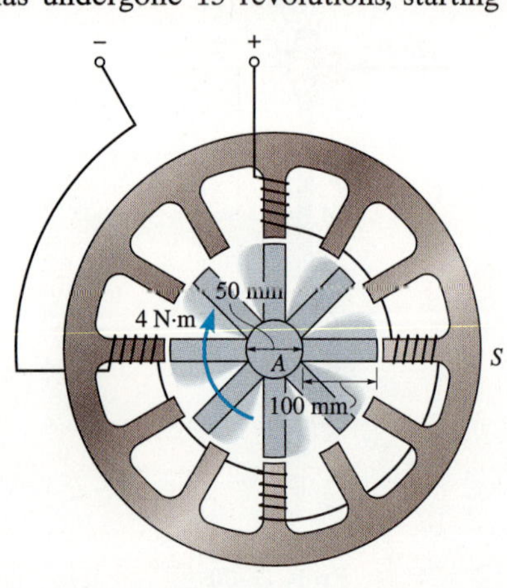

Probs. 17–72

17-73. The disk has a mass of 20 kg and is originally spinning at the end of the strut with an angular velocity of $\omega = 60$ rad/s. If it is then placed against the wall, for which the coefficient of kinetic friction is $\mu_k = 0.3$, determine the time required for the motion to stop. What is the force in strut BC during this time?

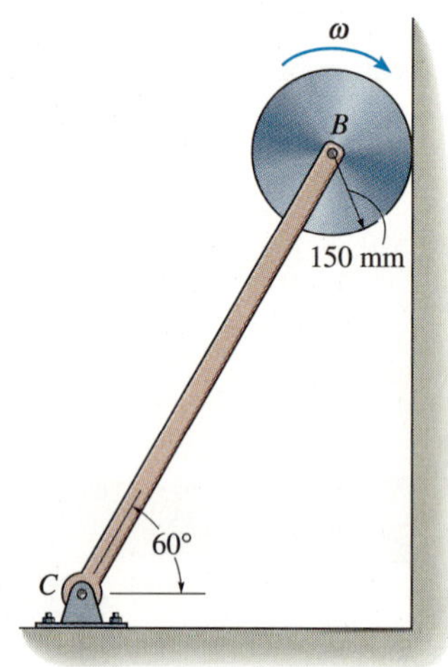

Prob. 17–73

17-74. The relay switch consists of an electromagnet E and a 20-g armature AB (slender bar) which is pinned at A and lies in the vertical plane. When the current is turned off, the armature is held open against the smooth stop at B by the spring CD, which exerts an upward vertical force $F_s = 0.85$ N on the armature at C. When the current is turned on, the electromagnet attracts the armature at E with a vertical force $F = 0.8$ N. Determine the initial angular acceleration of the armature when the contact BF begins to close.

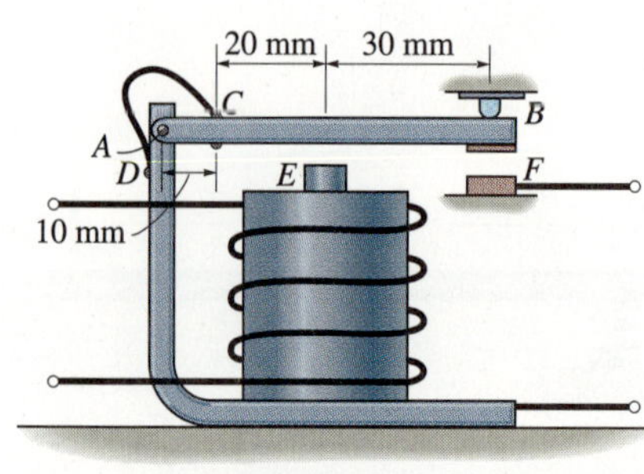

Prob. 17–74

17-75. The rod has a length L and mass m. If it is released from rest when $\theta \approx 0^\circ$, determine its angular velocity as a function of θ. Also, express the horizontal and vertical components of reaction at the pin O as a function of θ.

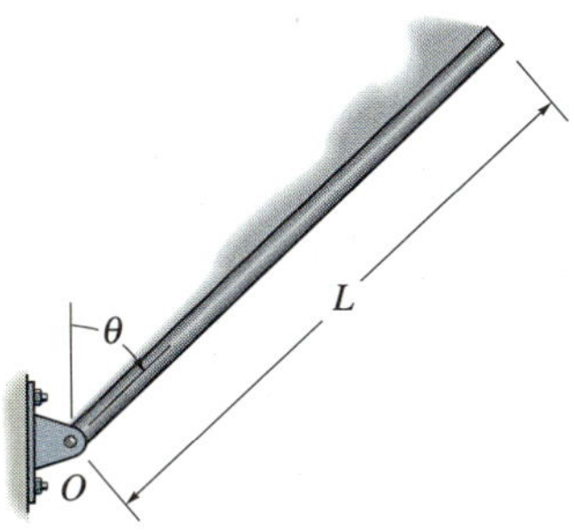

Prob. 17–75

***17-76.** The lightweight turbine consists of a rotor which is powered from a torque appled at its center. At the instant the rotor is horizontal it has an angular velocity of 15 rad/s and an angular acceleration of 8 rad/s^2. Determine the internal normal force, shear force, and moment at a section through A. Assume the rotor is a 50-m-long slender rod, having a mass of 3 kg/m.

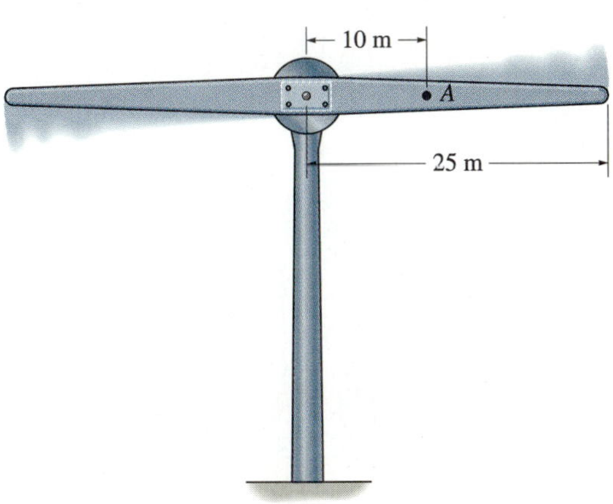

Prob. 17–76

17-77. The bar has a weight per length of w. If it is rotating in the vertical plane at a constant rate $\boldsymbol{\omega}$ about point O, determine the internal normal force, shear force, and moment as a function of x and θ.

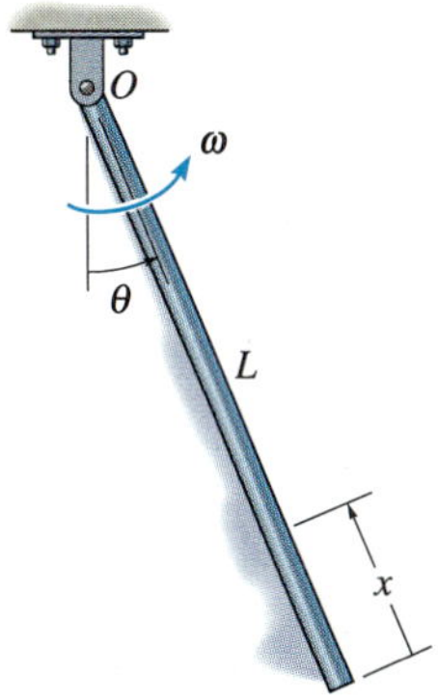

Prob. 17–77

17-78. Disk A has a weight of 5 lb and disk B has a weight of 10 lb. If no slipping occurs between them, determine the couple moment **M** which must be applied to disk A to give it an angular acceleration of 4 rad/s^2.

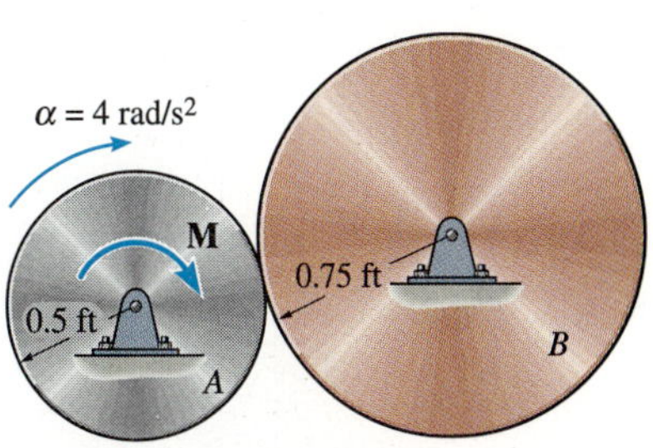

Prob. 17–78

17-79. Determine the force $\mathbf{T}_A$ which must be applied to the cable at A in order to give the 10-kg block an upward acceleration of 200 mm/s^2. Assume that the cable does not slip over the surface of the 20-kg disk. Determine the tension in the vertical segment of the cord that supports the block and explain why this tension is different from that at A. The disk is pinned at its center C and is free to rotate. Neglect the mass of the cable.

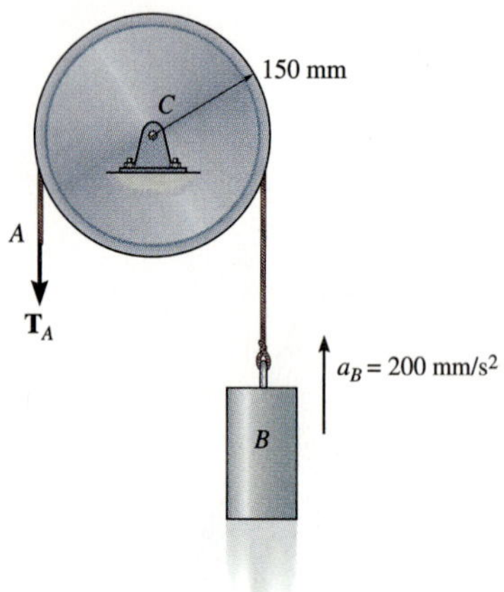

Prob. 17–79

***17-80.** The cord is wrapped around the inner core of the spool. If a 5-lb block B is suspended from the cord and released from rest, determine the spool's angular velocity when $t = 3$ s. Neglect the mass of the cord. The spool has a weight of 180 lb and the radius of gyration about the axle A is $k_A = 1.25$ ft. Solve the problem in two ways, first by considering the "system" consisting of the block and spool, and then by considering the block and spool separately.

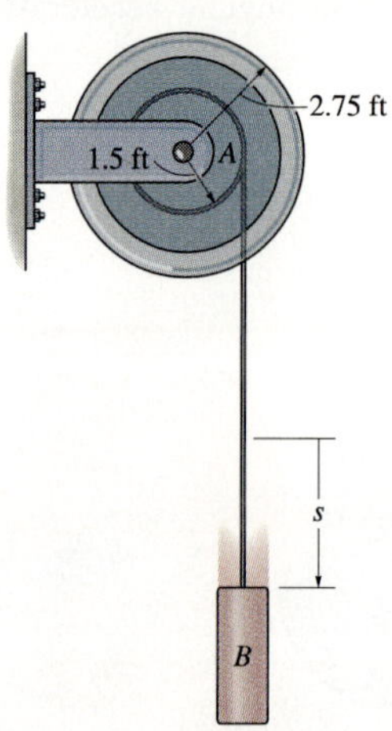

Prob. 17–80

17-81. A cord having negligible mass is wrapped over the 15-lb disk at A and passes over the 5-lb disk at B. If a 3-lb block C is attached to its end and released from rest, determine the speed of the block after it descends 3 ft. Also, what is the tension in the horizontal and vertical segments of the cord? Assume no slipping of the cord over the disk at B. Neglect friction at the pins D and E.

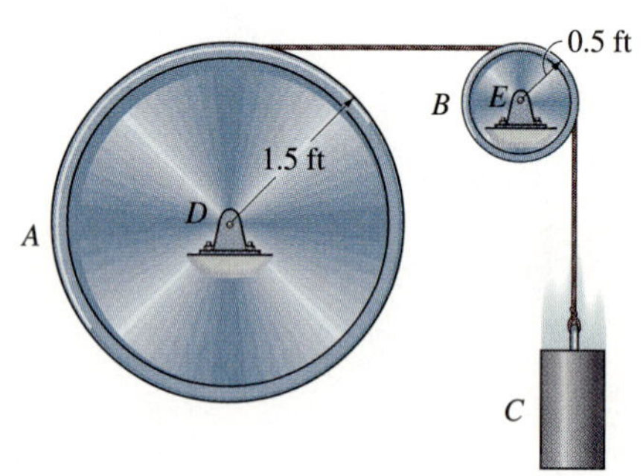

Prob. 17–81

17-82. The two blocks A and B have a mass of m_A and m_B, respectively, where $m_B > m_A$. If the pulley can be treated as a disk of mass M, determine the acceleration of block A. Neglect the mass of the cord and any slipping on the pulley.

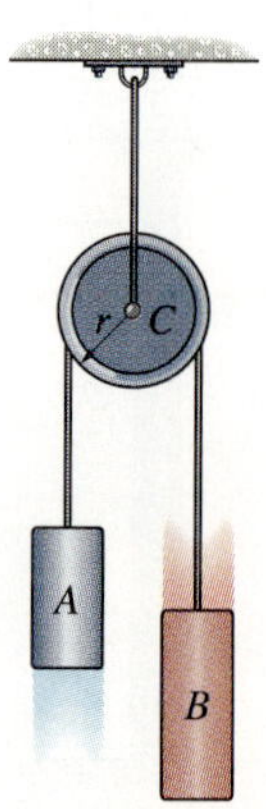

Prob. 17–82

17-83. Block A has a mass m and rests on a surface having a coefficient of kinetic friction μ_k. The cord attached to A passes over a pulley at C and is attached to a block B having a mass $2m$. If B is released, determine the acceleration of A. Assume that the cord does not slip over the pulley. The pulley can be approximated as a thin disk of radius r and mass $\frac{1}{4}m$. Neglect the mass of the cord.

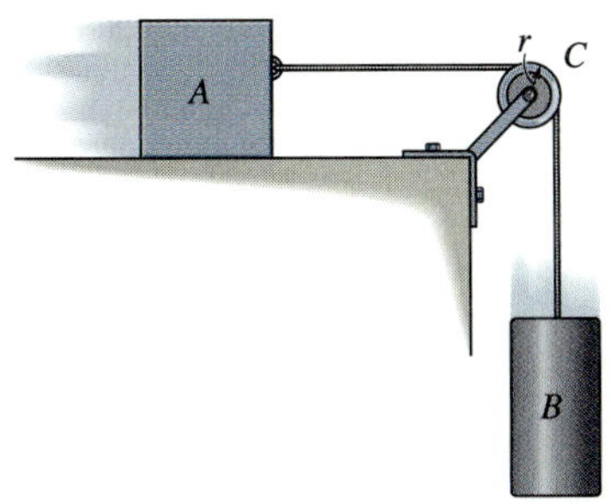

Prob. 17–83

*__17-84.__ The slender rod of mass m is released from rest when $\theta = 45°$. At the same instant ball B having the same mass m is released. Will B or the end A of the rod have the greatest speed when they pass the horizontal ($\theta = 0°$)? What is the difference in their speeds?

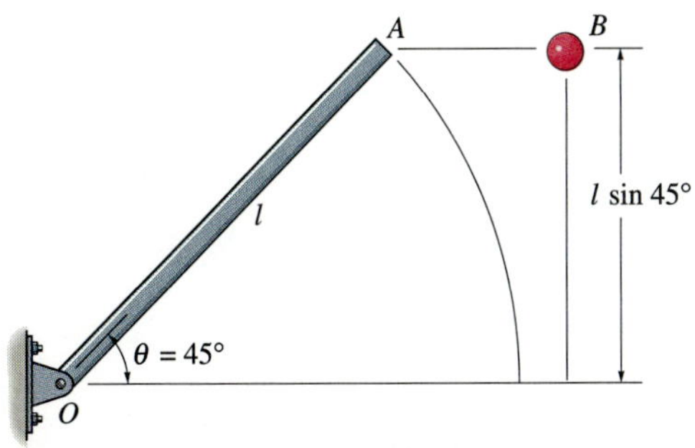

Prob. 17–84

17-85. The "Catherine wheel" is a firework that consists of a coiled tube of powder which is pinned at its center. If the powder burns at a constant rate of 20 g/s such that the exhaust gases always exert a force having a constant magnitude of 0.3 N, directed tangent to the wheel, determine the angular velocity of the wheel when 75% of the mass is burned off. Initially, the wheel is at rest and has a mass of 100 g and a radius of $r = 75$ mm. For the calculation, consider the wheel to always be a thin disk.

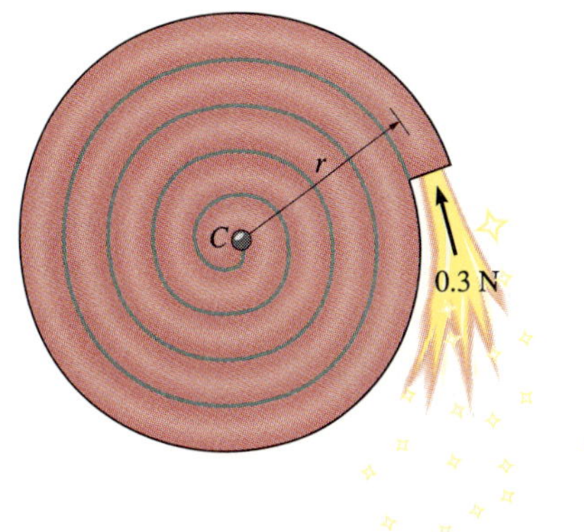

Prob. 17–85

17-86. The drum has a weight of 50 lb and a radius of gyration $k_A = 0.4$ ft. A 35-ft-long chain having a weight of 2 lb/ft is wrapped around the outer surface of the drum so that a chain length of $s = 3$ ft is suspended as shown. If the drum is originally at rest, determine its angular velocity after the end B has descended $s = 13$ ft. Neglect the thickness of the chain.

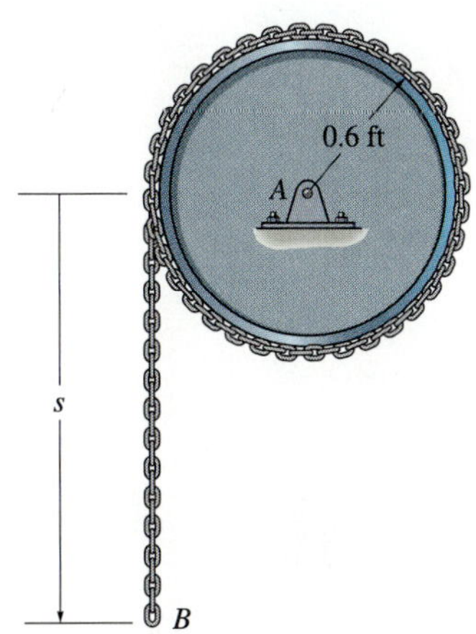

Prob. 17–86

17.5 Equations of Motion: General Plane Motion

The rigid body (or slab) shown in Fig. 17–20*a* is subjected to general plane motion caused by the externally applied force and couple-moment system. The free-body and kinetic diagrams for the body are shown in Fig. 17–20*b*. If an x and y inertial coordinate system is chosen as shown, the three equations of motion may be written as

$$\begin{aligned} \Sigma F_x &= m(a_G)_x \\ \Sigma F_y &= m(a_G)_y \\ \Sigma M_G &= I_G \alpha \end{aligned} \qquad (17\text{–}17)$$

In some problems it may be convenient to sum moments about some point P other than G. This is usually done in order to eliminate unknown forces from the moment summation. When used in this more general sense, the three equations of motion become

$$\begin{aligned} \Sigma F_x &= m(a_G)_x \\ \Sigma F_y &= m(a_G)_y \\ \Sigma M_P &= \Sigma(\mathcal{M}_k)_P \end{aligned} \qquad (17\text{–}18)$$

Here $\Sigma(\mathcal{M}_k)_P$ represents the moment sum of $I_G\boldsymbol{\alpha}$ and $m\mathbf{a}_G$ (or its components) about P as determined by the data on the kinetic diagram.

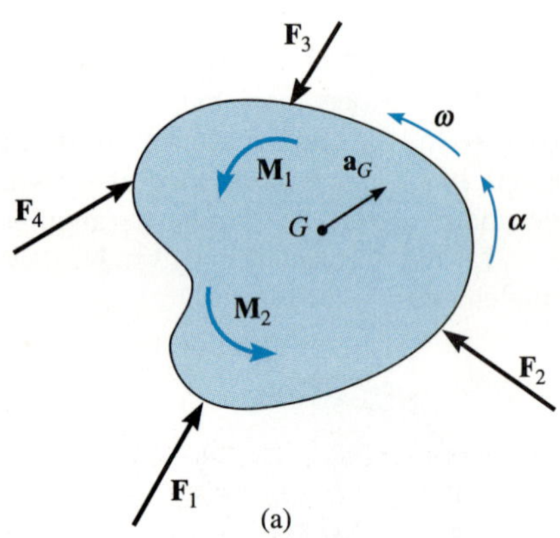

(a)

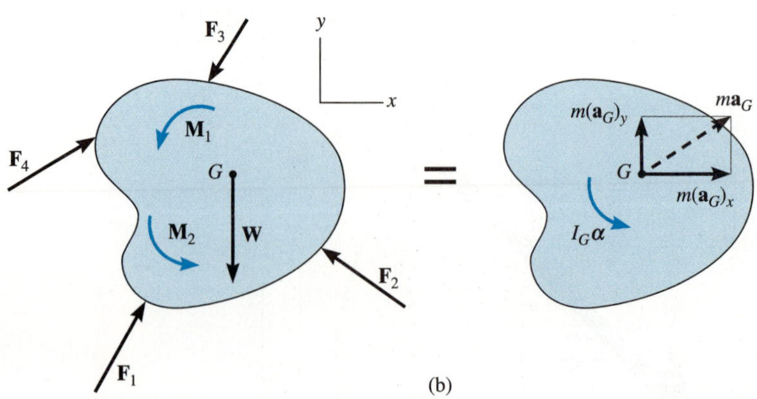

(b)

Fig. 17–20

Frictional Rolling Problems. There is a class of planar kinetics problems which deserves special mention. These problems involve wheels, cylinders, or bodies of similar shape, which roll on a *rough* plane surface. Because of the applied loadings, it may not be known if the body *rolls without slipping,* or if it *slides as it rolls.* For example, consider the homogeneous disk shown in Fig. 17–21*a*, which has a mass m and is subjected to a known horizontal force **P**. The free-body diagram is shown in Fig. 17–21*b*. Since $\mathbf{a}_G$ is directed to the right and $\boldsymbol{\alpha}$ is clockwise, we have

$$\xrightarrow{+}\Sigma F_x = m(a_G)_x; \qquad P - F = ma_G \qquad (17\text{–}19)$$

$$+\uparrow \Sigma F_y = m(a_G)_y; \qquad N - mg = 0 \qquad (17\text{–}20)$$

$$\curvearrowright\!+\Sigma M_G = I_G\alpha; \qquad Fr = I_G\alpha \qquad (17\text{–}21)$$

A fourth equation is needed since these *three equations* contain *four unknowns: F, N,* α, and a_G.

No Slipping. If the frictional force **F** is great enough to allow the disk to roll *without slipping,* then a_G may be related to α by the *kinematic equation,**

$$(\curvearrowright\!+) \qquad a_G = \alpha r \qquad (17\text{–}22)$$

The four unknowns are determined by *solving simultaneously* Eqs 17–19 to 17–22. When the solution is obtained, the assumption of no slipping must be *checked.* Recall that no slipping occurs provided $F \le \mu_s N$, where μ_s is the coefficient of static friction. If the inequality is satisfied, the problem is solved. However, if $F > \mu_s N$, the problem must be *reworked,* since then the disk slips as it rolls.

Slipping. In the case of slipping, α and a_G are *independent of one another* so that Eq. 17–22 does not apply. Instead, the magnitude of the frictional force is related to the magnitude of the normal force using the coefficient of kinetic friction μ_k, i.e.,

$$F = \mu_k N \qquad (17\text{–}23)$$

In this case Eqs. 17–19 to 17–21 and 17–23 are used for the solution. It is important to keep in mind that whenever Eq. 17–22 or 17–23 is applied, it is necessary to have consistency in the directional sense of the vectors. In the case of Eq. 17–22, $\mathbf{a}_G$ must be directed to the right when $\boldsymbol{\alpha}$ is clockwise, since the rolling motion requires it. And in Eq. 17–23, **F** must be directed to the left to prevent the assumed slipping motion to the right, Fig. 17–21*b*. On the other hand, if these equations are *not used* for the solution, these vectors can have *any* assumed directional sense. Then if the calculated numerical value of these quantities is negative, the vectors act in their opposite sense of direction. Examples 17–15 and 17–16 illustrate these concepts numerically.

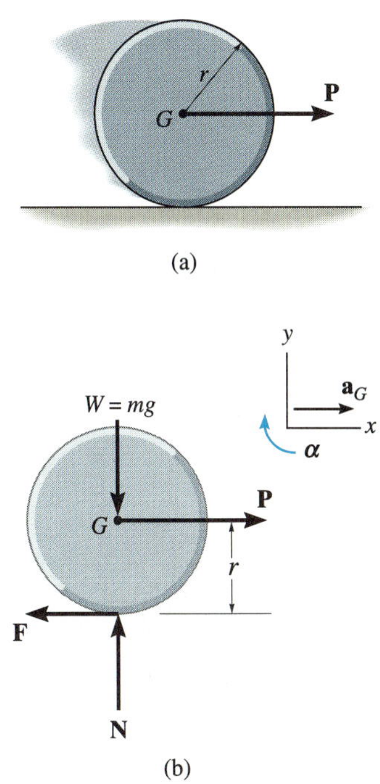

Fig. 17–21

*See Example 16–3 or 16–14.

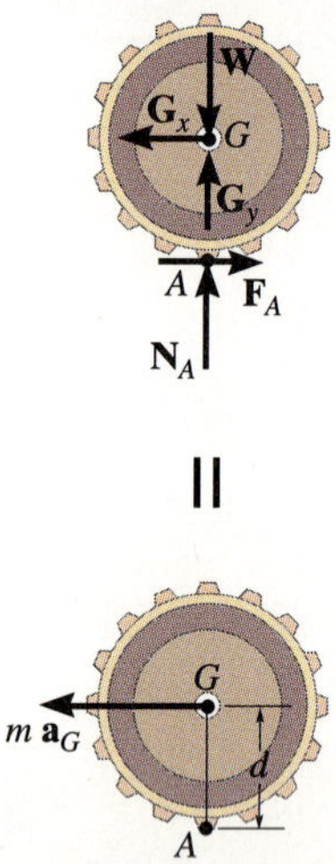

As the soil compactor accelerates forward, the roller has general plane motion. The forces shown on the roller's free-body diagram cause the effects shown on the kinetic diagram. If moments are summed about the mass center, G, then $\Sigma M_G = I_G\alpha$. However, if moments are summed about point A then $\circlearrowright + \Sigma M_A = I_G\alpha + (ma_G)d$.

Procedure for Analysis

Kinetic problems involving general plane motion of a rigid body can be solved using the following procedure.

Free-Body Diagram

- Establish the x, y inertial coordinate system and draw the free-body diagram for the body.
- Specify the direction and sense of the acceleration of the mass center, $\mathbf{a}_G$, and the angular acceleration $\boldsymbol{\alpha}$ of the body.
- Compute the moment of inertia I_G.
- Identify the unknowns in the problem.
- If it is decided that the rotational equation of motion $\Sigma M_P = \Sigma(\mathcal{M}_k)_P$ is to be used, then consider drawing the kinetic diagram in order to help "visualize" the "moments" developed by the components $m(\mathbf{a}_G)_x$, $m(\mathbf{a}_G)_y$, and $I_G\boldsymbol{\alpha}$ when writing the terms in the moment sum $\Sigma(\mathcal{M}_k)_P$.

Equations of Motion

- Apply the three equations of motion in accordance with the established sign convention.
- When friction is present, there is the possibility for motion with no slipping or tipping. Each possibility for motion should be considered.

Kinematics

- Use kinematics if a complete solution cannot be obtained strictly from the equations of motion.
- If the body's motion is *constrained* due to its supports, additional equations may be obtained by using $\mathbf{a}_B = \mathbf{a}_A + \mathbf{a}_{B/A}$, which relates the accelerations of any two points A and B on the body.
- When a wheel, disk, cylinder, or ball *rolls without slipping,* then $a_G = \alpha r$.

EXAMPLE 17–14

The spool in Fig. 17–22a has a mass of 8 kg and a radius of gyration of $k_G = 0.35$ m. If cords of negligible mass are wrapped around its inner hub and outer rim as shown, determine the spool's angular acceleration.

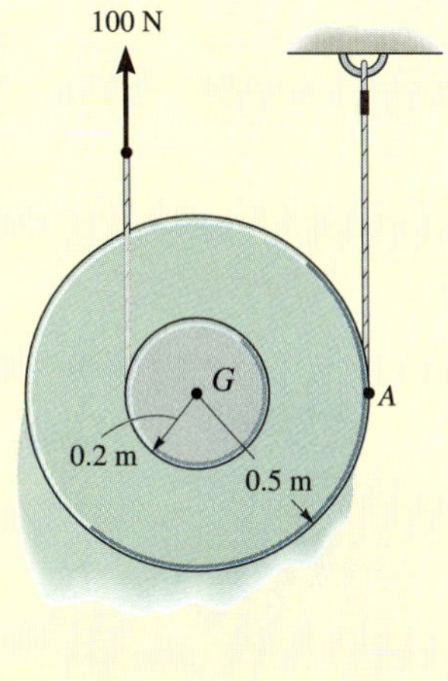

(a)

Solution I

Free-Body Diagram. Fig. 17–22b. The 100-N force causes $\mathbf{a}_G$ to act upward. Also, $\boldsymbol{\alpha}$ acts clockwise, since the spool winds around the cord at A.

There are three unknowns T, a_G, and α. The moment of inertia of the spool about its mass center is

$$I_G = mk_G^2 = 8\text{ kg}(0.35\text{ m})^2 = 0.980\text{ kg}\cdot\text{m}^2$$

Equations of Motion

$$+\uparrow \Sigma F_y = m(a_G)_y; \quad T + 100\text{ N} - 78.48\text{ N} = (8\text{ kg})a_G \qquad (1)$$

$$\circlearrowleft + \Sigma M_G = I_G\alpha; \quad 100\text{ N}(0.2\text{ m}) - T(0.5\text{ m}) = (0.980\text{ kg}\cdot\text{m}^2)\alpha \qquad (2)$$

Kinematics. A complete solution is obtained if kinematics is used to relate a_G to α. In this case the spool "rolls without slipping" on the cord at A. Hence, we can use the results of Example 16–3 or 16–14, so that

$$(\circlearrowleft+)\, a_G = \alpha r; \qquad a_G = 0.5\alpha \qquad (3)$$

Solving Eqs. 1 to 3, we have

$$\alpha = 10.3\text{ rad/s}^2 \qquad \textit{Ans.}$$

$$a_G = 5.16\text{ m/s}^2$$

$$T = 19.8\text{ N}$$

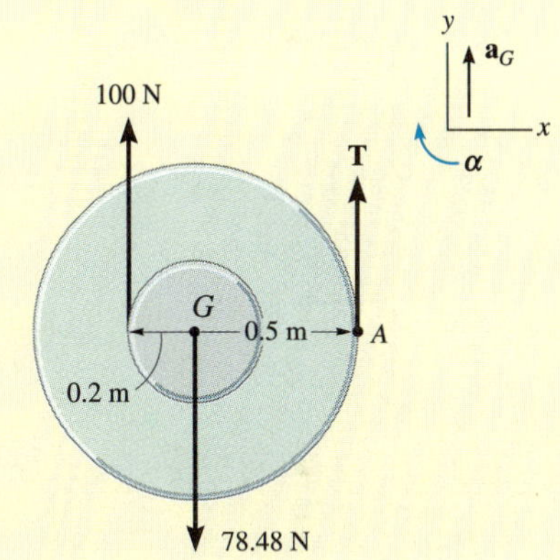

(b)

Solution II

Equations of Motion. We can eliminate the unknown T by summing moments about point A. From the free-body and kinetic diagrams Figs. 17–22b and 17–22c, we have

$$\circlearrowleft + \Sigma M_A = \Sigma(\mathcal{M}_k)_A; \quad 100\text{ N}(0.7\text{ m}) - 78.48\text{ N}(0.5\text{ m})$$
$$= (0.980\text{ kg}\cdot\text{m}^2)\alpha + [(8\text{ kg})a_G](0.5\text{ m})$$

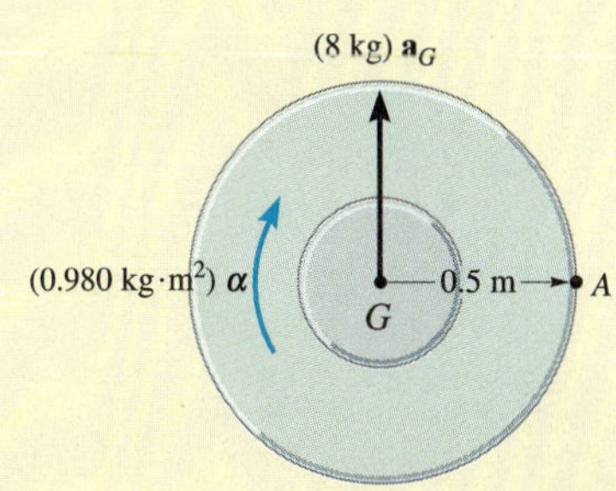

(c)

Using Eq. (3),

$$\alpha = 10.3\text{ rad/s}^2 \qquad \textit{Ans.}$$

Fig. 17–22

EXAMPLE 17–15

The 50-lb wheel shown in Fig. 17–23*a* has a radius of gyration $k_G = 0.70$ ft. If a 35-lb·ft couple moment is applied to the wheel, determine the acceleration of its mass center G. The coefficients of static and kinetic friction between the wheel and the plane at A are $\mu_s = 0.3$ and $\mu_k = 0.25$, respectively.

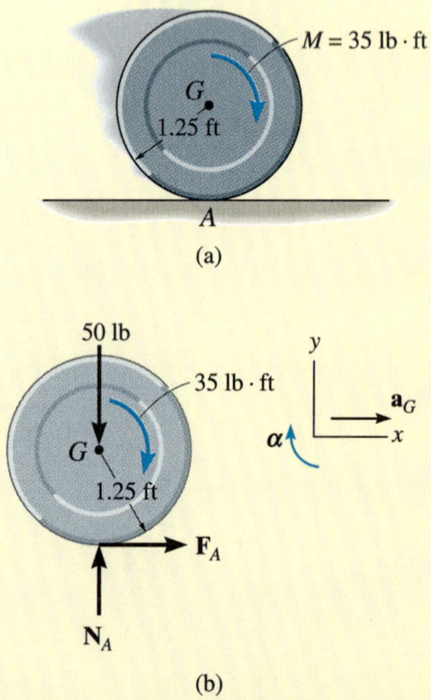

Fig. 17–23

Solution

Free-Body Diagram. By inspection of Fig. 17–23*b*, it is seen that the couple moment causes the wheel to have a clockwise angular acceleration of $\boldsymbol{\alpha}$. As a result, the acceleration of the mass center, $\mathbf{a}_G$, is directed to the right. The moment of inertia is

$$I_G = mk_G^2 = \frac{50 \text{ lb}}{32.2 \text{ ft/s}^2}(0.70 \text{ ft})^2 = 0.761 \text{ slug}\cdot\text{ft}^2$$

The unknowns are N_A, F_A, a_G, and α.

Equations of Motion

$$\overset{+}{\rightarrow}\Sigma F_x = m(a_G)_x; \qquad F_A = \frac{50 \text{ lb}}{32.2 \text{ ft/s}^2} a_G \qquad (1)$$

$$+\uparrow \Sigma F_y = m(a_G)_y; \qquad N_A - 50 \text{ lb} = 0 \qquad (2)$$

$$\curvearrowright+\Sigma M_G = I_G\alpha; \qquad 35 \text{ lb}\cdot\text{ft} - 1.25 \text{ ft}(F_A) = (0.761 \text{ slug}\cdot\text{ft}^2)\alpha \qquad (3)$$

A fourth equation is needed for a complete solution.

Kinematics (No Slipping). If this assumption is made, then

$$(\curvearrowright+) \qquad a_G = (1.25 \text{ ft})\alpha \qquad (4)$$

Solving Eqs. 1 to 4,

$$N_A = 50.0 \text{ lb} \qquad F_A = 21.3 \text{ lb}$$
$$\alpha = 11.0 \text{ rad/s}^2 \qquad a_G = 13.7 \text{ ft/s}^2$$

The original assumption of no slipping requires $F_A \le \mu_s N_A$. However, since 21.3 lb > 0.3(50 lb) = 15 lb, the wheel slips as it rolls.

(Slipping). Equation 4 is not valid. Instead, it is necessary that $\mathbf{F}_A$ acts to the right. Why? Also, $F_A = \mu_k N_A$, or

$$F_A = 0.25N_A \qquad (5)$$

Solving Eqs. 1 to 3 and 5 yields

$$N_A = 50.0 \text{ lb} \qquad F_A = 12.5 \text{ lb}$$
$$\alpha = 25.5 \text{ rad/s}^2$$
$$a_G = 8.05 \text{ ft/s}^2 \rightarrow \qquad \textit{Ans.}$$

EXAMPLE 17–16

The uniform slender pole shown in Fig. 17–24*a* has a mass of 100 kg and a moment of inertia $I_G = 75\ \text{kg}\cdot\text{m}^2$. If the coefficients of static and kinetic friction between the end of the pole and the surface are $\mu_s = 0.3$ and $\mu_k = 0.25$, respectively, determine the pole's angular acceleration at the instant the 400-N horizontal force is applied. The pole is originally at rest.

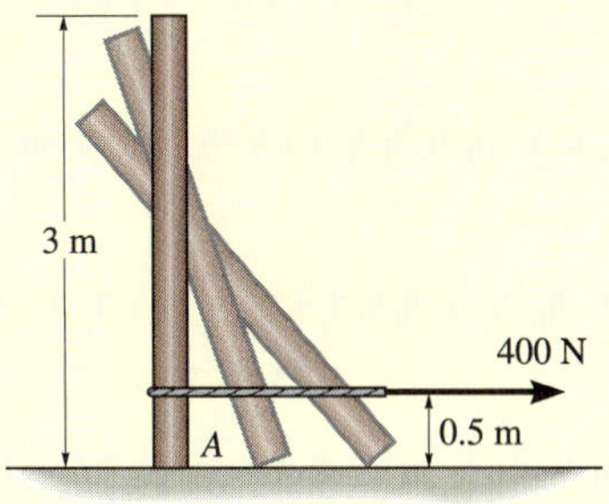

(a)

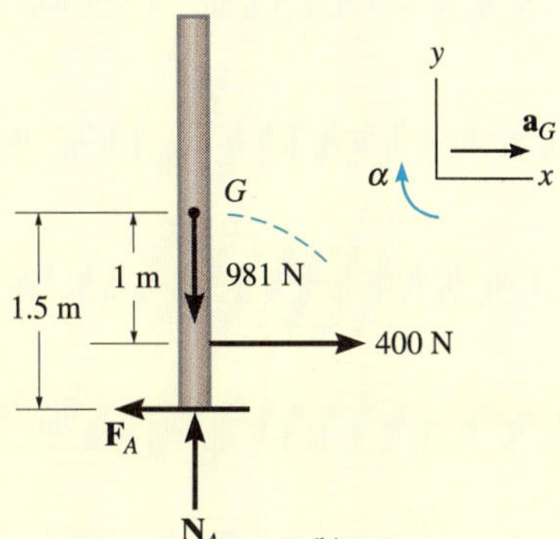

(b)

Fig. 17–24

Solution

Free-Body Diagram. Figure 17–24*b*. The path of motion of the mass center G will be along an unknown curved path having a radius of curvature ρ, which is initially parallel to the y axis. There is no normal or y component of acceleration since the pole is originally at rest, i.e., $\mathbf{v}_G = \mathbf{0}$, so that $(a_G)_y = v_G^2/\rho = 0$. We will assume the mass center accelerates to the right and that the pole has a clockwise angular acceleration of $\boldsymbol{\alpha}$. The unknowns are N_A, F_A, a_G, and α.

Equations of Motion

$$\xrightarrow{+} \Sigma F_x = m(a_G)_x; \qquad 400\text{ N} - F_A = (100\text{ kg})a_G \qquad (1)$$

$$+\uparrow \Sigma F_y = m(a_G)_y; \qquad N_A - 981\text{ N} = 0 \qquad (2)$$

$$\curvearrowright + \Sigma M_G = I_G\alpha; \qquad F_A(1.5\text{ m}) - 400\text{ N}(1\text{ m}) = (75\text{ kg}\cdot\text{m}^2)\alpha \qquad (3)$$

A fourth equation is needed for a complete solution.

Kinematics (No Slipping). In this case point A acts as a "pivot" so that indeed, if α is clockwise, then a_G is directed to the right.

$$\curvearrowright + a_G = \alpha r_{AG}; \qquad a_G = (1.5\text{ m})\alpha \qquad (4)$$

Solving Eqs. 1 to 4 yields

$$N_A = 981\text{ N} \qquad F_A = 300\text{ N}$$
$$a_G = 1\text{ m/s}^2 \qquad \alpha = 0.667\text{ rad/s}^2$$

Testing the original assumption of no slipping requires $F_A \le \mu_s N_A$. However, $300\text{ N} > 0.3(981\text{ N}) = 294\text{ N}$. (Slips at A.)

(Slipping). For this case Eq. 4 does *not* apply. Instead the frictional equation $F_A = \mu_k N_A$ is used. Hence,

$$F_A = 0.25N_A \qquad (5)$$

Solving Eqs. 1 to 3 and 5 simultaneously yields

$$N_A = 981\text{ N} \qquad F_A = 245\text{ N} \qquad a_G = 1.55\text{ m/s}^2$$
$$\alpha = -0.428\text{ rad/s}^2 = 0.428\text{ rad/s}^2 \curvearrowleft \qquad \textit{Ans.}$$

EXAMPLE 17-17

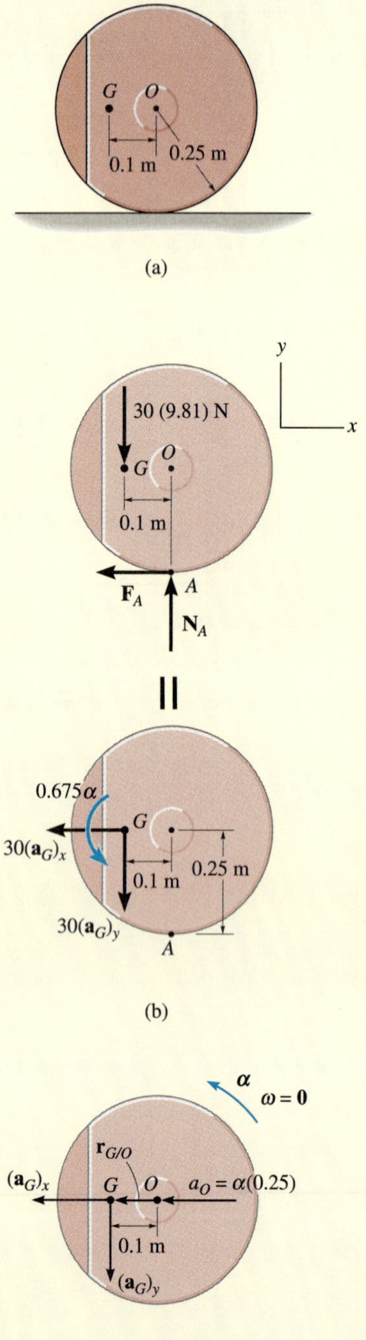

Fig. 17–25

The 30-kg wheel shown in Fig. 17–25*a* has a mass center at G and a radius of gyration $k_G = 0.15$ m. If the wheel is originally at rest and released from the position shown, determine its angular acceleration. No slipping occurs.

Solution

Free-Body and Kinetic Diagrams. The two unknowns $\mathbf{F}_A$ and $\mathbf{N}_A$ shown on the free-body diagram, Fig. 17–25*b*, can be eliminated from the analysis by summing moments about point A. The kinetic diagram accompanies the solution in order to illustrate application of $\Sigma(\mathcal{M}_k)_A$. Since point G moves along a curved path, the two components $m(\mathbf{a}_G)_x$ and $m(\mathbf{a}_G)_y$ are shown on the kinetic diagram, Fig. 17–25*b*.

The moment of inertia is

$$I_G = mk_G^2 = 30(0.15)^2 = 0.675 \text{ kg}\cdot\text{m}^2$$

There are five unknowns, N_A, F_A, $(a_G)_x$, $(a_G)_y$, and α.

Equation of Motion. Applying the rotational equation of motion about point A, to eliminate N_A, and F_A, we have

$$\circlearrowleft + \Sigma M_A = \Sigma(\mathcal{M}_k)_A; \qquad 30(9.81)\text{ N}(0.1\text{ m}) =$$
$$(0.675 \text{ kg}\cdot\text{m}^2)\alpha + (30\text{ kg})(a_G)_x(0.25\text{ m}) + (30\text{ kg})(a_G)_y(0.1\text{ m}) \quad (1)$$

There are three unknowns in this equation: $(a_G)_x$, $(a_G)_y$, and α.

Kinematics. Using kinematics, $(a_G)_x$, $(a_G)_y$ will be related to α. As shown in Fig. 17–25*c*, these vectors must have the same directional sense as the corresponding vectors on the kinetic diagram since we are seeking a simultaneous solution with Eq. 1. Since no slipping occurs, $a_O = \alpha r = \alpha(0.25\text{ m})$, directed to the left, Fig. 17–25*c*. Also, $\omega = 0$, since the wheel is originally at rest. Applying the acceleration equation to point O (base point) and point G, we have

$$\mathbf{a}_G = \mathbf{a}_O + \boldsymbol{\alpha} \times \mathbf{r}_{G/O} - \omega^2\mathbf{r}_{G/O}$$
$$-(a_G)_x\mathbf{i} - (a_G)_y\mathbf{j} = -\alpha(0.25)\mathbf{i} + (\alpha\mathbf{k}) \times (-0.1\mathbf{i}) - \mathbf{0}$$

Expanding and equating the respective **i** and **j** components, we have

$$(a_G)_x = \alpha(0.25) \quad (2)$$
$$(a_G)_y = \alpha(0.1) \quad (3)$$

Solving Eqs. 1 to 3 yields

$$\alpha = 10.3 \text{ rad/s}^2 \circlearrowleft \qquad \textit{Ans.}$$
$$(a_G)_x = 2.58 \text{ m/s}^2$$
$$(a_G)_y = 1.03 \text{ m/s}^2\text{\'[pf}$$

As an exercise, show that $F_A = 77.4$ N and $N_A = 263$ N.

Problems

17-87. If the disk in Fig. 17-21*a rolls without slipping,* show that when moments are summed about the instantaneous center of zero velocity, *IC*, it is possible to use the moment equation $\Sigma M_{IC} = I_{IC}\alpha$, where I_{IC} represents the moment of inertia of the disk calculated about the instantaneous axis of zero velocity.

***17-88.** The wheel has a weight of 30 lb and a radius of gyration of $k_G = 0.6$ ft. If the coefficients of static and kinetic friction between the wheel and the plane are $\mu_s = 0.2$ and $\mu_k = 0.15$, determine the wheel's angular acceleration as it rolls down the incline. Set $\theta = 12°$.

17-89. The wheel has a weight of 30 lb and a radius of gyration of $k_G = 0.6$ ft. If the coefficients of static and kinetic friction between the wheel and the plane are $\mu_s = 0.2$ and $\mu_k = 0.15$, determine the maximum angle θ of the inclined plane so that the wheel rolls without slipping.

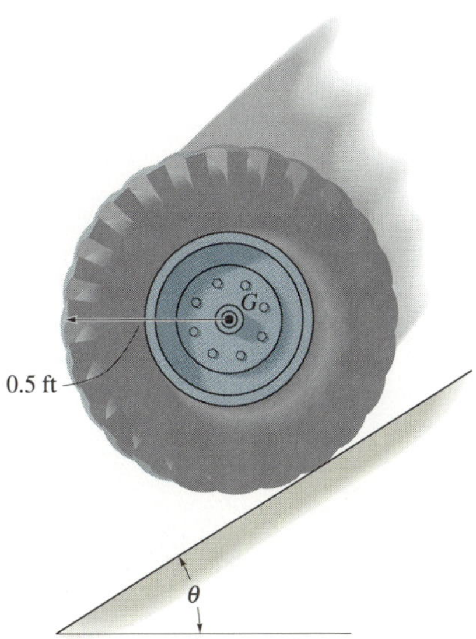

Probs. 17–88/89

■**17-90.** A rocket *CD*, having a mass of 20 Mg with center of mass at *G*, is located in deep space so that the effect of gravitation (weight) can be neglected. The smaller rockets *A* and *B* each have a mass of 4 Mg and center of mass at G_A and G_B, respectively. If these rockets travel in a straight line and exert a constant thrust of $T = 7$ kN perpendicular to *CD*, determine the angular acceleration of *CD* and the accelerations of rockets *A* and *B*. Assume that *CD* is initially at rest and that its radius of gyration about an axis passing through *G* and perpendicular to the plane of motion is $k_G = 4.60$ m.

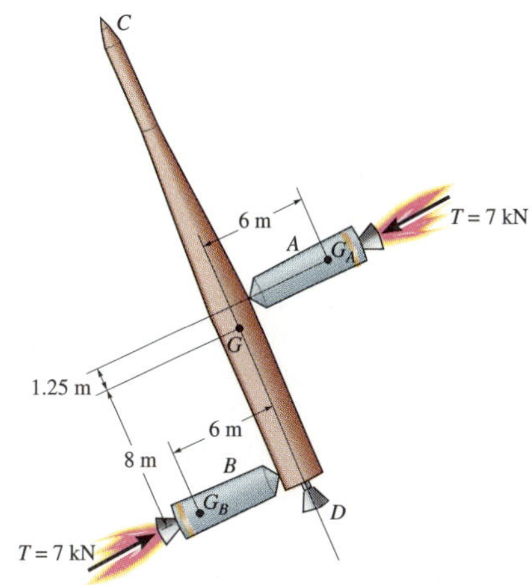

Prob. 17–90

17-91. The trailer has a mass of 580 kg and a mass center at *G*, whereas the spool has a mass of 200 kg, mass center at *O*, and a radius of gyration about an axis passing through *O* of $k_O = 0.45$ m. If a force of 60 N is applied to the cable, determine the angular acceleration of the spool and the acceleration of the trailer. The wheels have negligible mass and are free to roll.

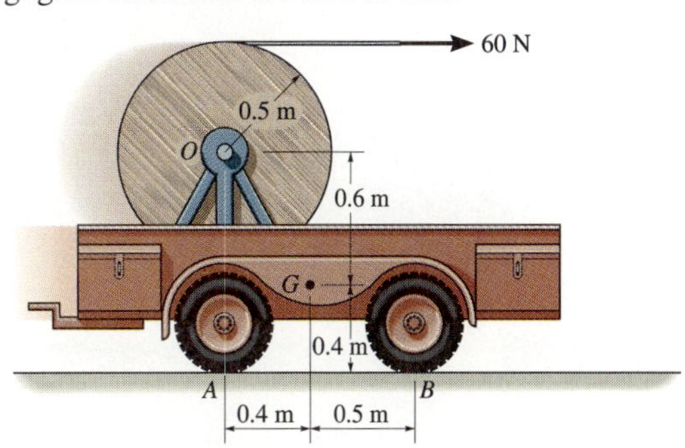

Prob. 17–91

***17-92.** The spool and wire wrapped around its core have a mass of 20 kg and a centroidal radius of gyration $k_G = 250$ mm. If the coefficient of kinetic friction at the ground is $\mu_k = 0.1$, determine the angular acceleration of the spool when the 30-N·m couple is applied.

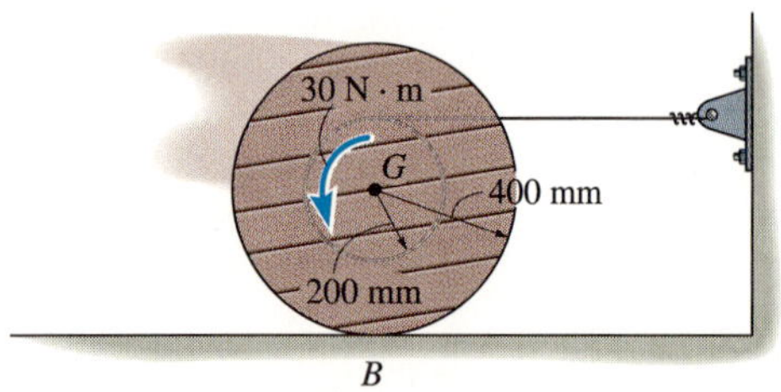

Prob. 17–92

17-93. The lawn roller has a mass of 80 kg and a radius of gyration $k_G = 0.175$ m. If it is pushed forward with a force of 200 N when the handle is at 45°, determine its angular acceleration. The coefficients of static and kinetic friction between the ground and the roller are $\mu_s = 0.12$ and $\mu_k = 0.1$, respectively.

17-94. Solve Prob. 17-93 if $\mu_s = 0.6$ and $\mu_k = 0.45$.

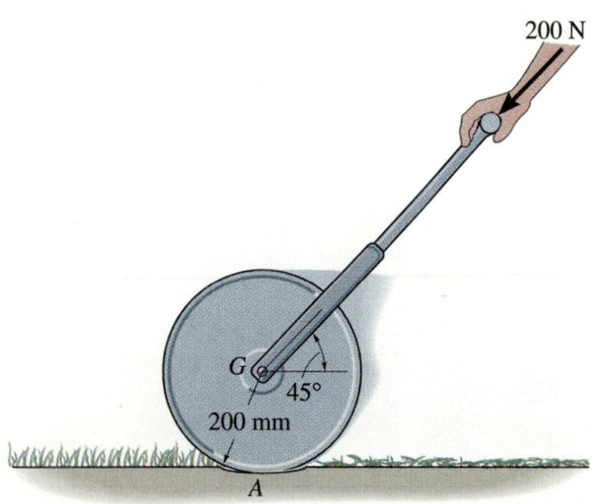

Probs. 17–93/94

17-95. The spool has a mass of 100 kg and a radius of gyration of $k_G = 0.3$ m. If the coefficients of static and kinetic friction at A are $\mu_s = 0.2$ and $\mu_k = 0.15$, respectively, determine the angular acceleration of the spool if $P = 50$ N.

***17-96.** Solve Prob. 17-95 if the cord and force $P = 50$ N are directed vertically upwards.

17-97. The spool has a mass of 100 kg and a radius of gyration $k_G = 0.3$ m. If the coefficients of static and kinetic friction at A are $\mu_s = 0.2$ and $\mu_k = 0.15$, respectively, determine the angular acceleration of the spool if $P = 600$ N.

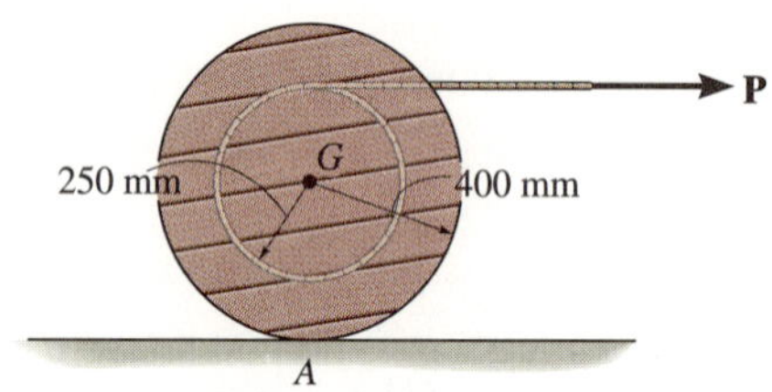

Probs. 17–95/96/97

17-98. The spool has a mass of 75 kg and a radius of gyration $k_G = 0.380$ m. It rests on the inclined surface for which the coefficient of kinetic friction is $\mu_k = 0.15$. If the spool is released from rest and slips at A, determine the initial tension in the cord and the angular acceleration of the spool.

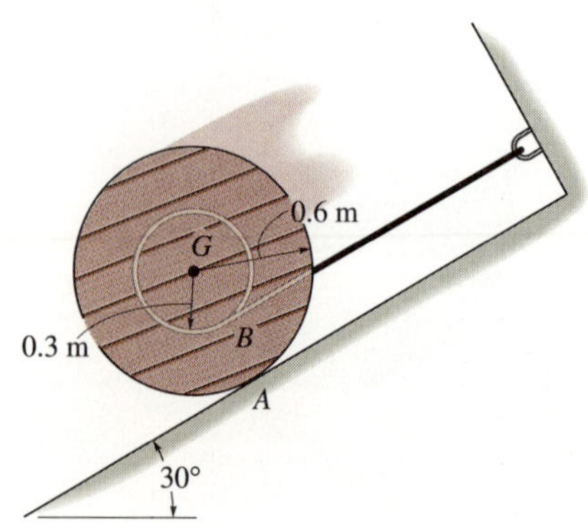

Prob. 17–98

17-99. The wheel has a mass of 80 kg and a radius of gyration $k_G = 0.25$ m. If it is subjected to a couple moment of $M = 50$ N·m, determine its angular acceleration. The coefficients of static and kinetic friction between the ground and the wheel are $\mu_s = 0.2$ and $\mu_k = 0.15$, respectively.

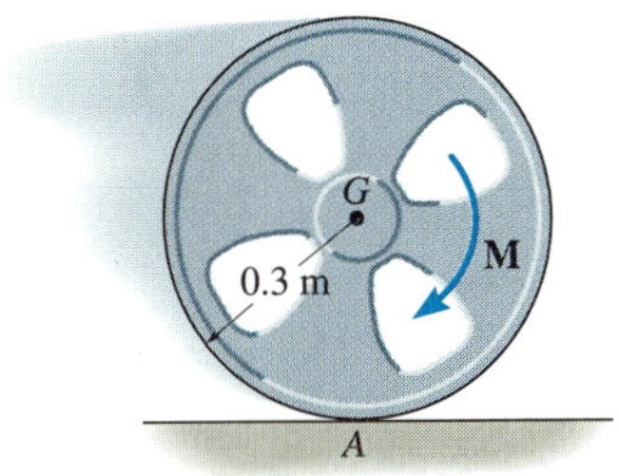

Prob. 17–99

*17-100. The truck carries the spool which has a weight of 500 lb and a radius of gyration of $k_G = 2$ ft. Determine the angular acceleration of the spool if it is not tied down on the truck and the truck begins to accelerate at 3 ft/s^2. Assume the spool does not slip on the bed of the truck.

17-101. The truck carries the spool which has a weight of 200 lb and a radius of gyration of $k_G = 2$ ft. Determine the angular acceleration of the spool if it is not tied down on the truck and the truck begins to accelerate at 5 ft/s^2. The coefficients of static and kinetic friction between the spool and the truck bed are $\mu_s = 0.15$ and $\mu_k = 0.1$, respectively.

Probs. 17–100/101

17-102. The uniform beam has a weight W. If it is originally at rest while being supported at A and B by cables, determine the tension in cable A if cable B suddenly fails. Assume the beam is a slender rod.

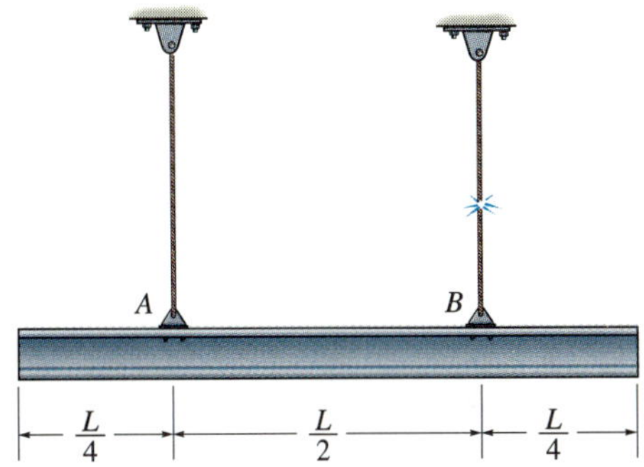

Prob. 17–102

17-103. The slender 150-lb bar is supported by two cords AB and AC. If cord AC suddenly breaks, determine the initial angular acceleration of the bar and the tension in cord AB.

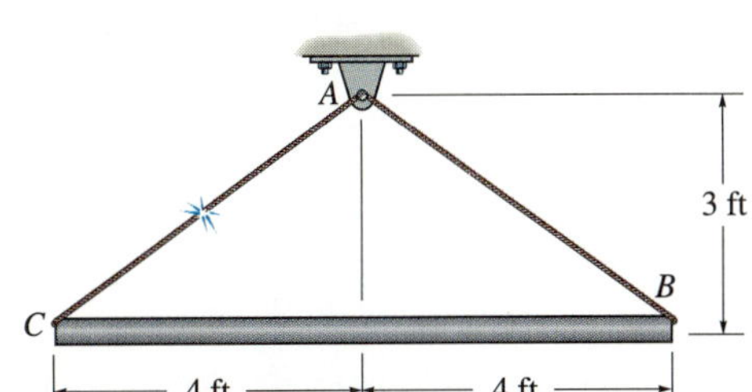

Prob. 17–103

***17-104.** A long strip of paper is wrapped into two rolls, each having a mass of 8 kg. Roll A is pin supported about its center whereas roll B is not centrally supported. If B is brought into contact with A and released from rest, determine the initial tension in the paper between the rolls and the angular acceleration of each roll. For the calculation, assume the rolls to be approximated by cylinders.

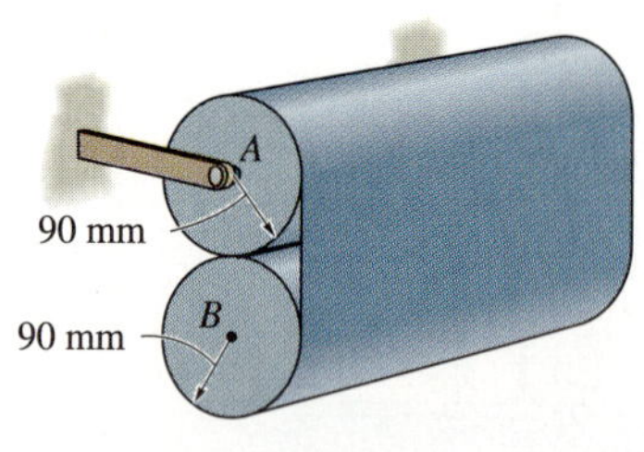

Prob. 17–104

17-105. The 20-kg canister has a radius of gyration about its center of mass G of $k_G = 0.4$ m. If it is subjected to a horizontal force of $F = 30$ N, determine the initial angular acceleration of the canister and the tension in the supporting cable AB.

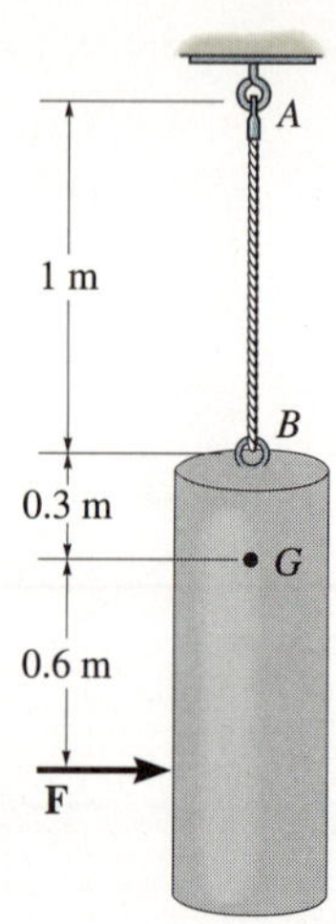

Prob. 17–105

17-106. A woman sits in a rigid position in the middle of the swing. The combined weight of the woman and swing is 180 lb and the radius of gyration about the center of mass G is $k_G = 2.5$ ft. If a man pushes on the swing with a horizontal force $F = 20$ lb as shown, determine the initial angular acceleration and the tension in each of the two supporting chains AB. During the motion, assume that the chain segment CAD remains rigid. The swing is originally at rest.

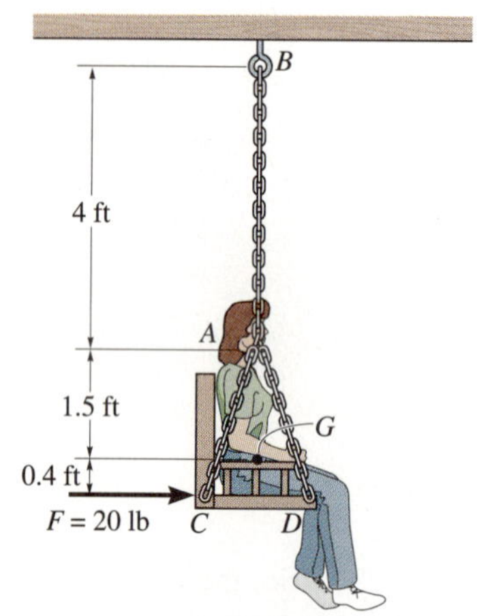

Prob. 17–106

17-107. The 16-lb bowling ball is cast horizontally onto a lane such that initially $\omega = 0$ and its mass center has a velocity $v = 8$ ft/s. If the coefficient of kinetic friction between the lane and the ball is $\mu_k = 0.12$, determine the distance the ball travels before it rolls without slipping. For the calculation, neglect the finger holes in the ball and assume the ball has a uniform density.

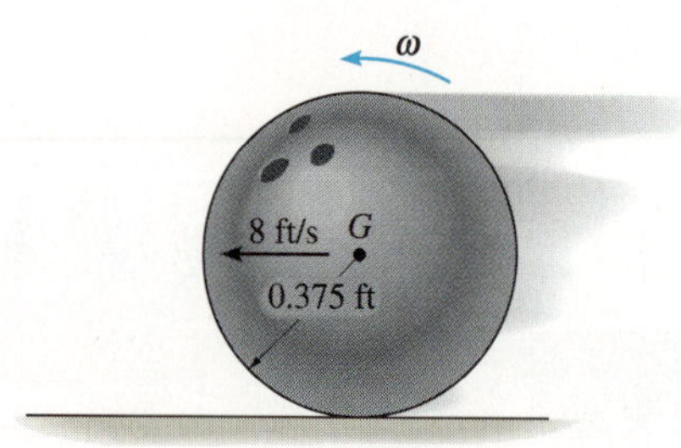

Prob. 17–107

***17-108.** By pressing down with the finger at B, a thin ring having a mass m is given an initial velocity $\mathbf{v}_0$ and a backspin $\boldsymbol{\omega}_0$ when the finger is released. If the coefficient of kinetic friction between the table and the ring is μ_k, determine the distance the ring travels forward before backspinning stops.

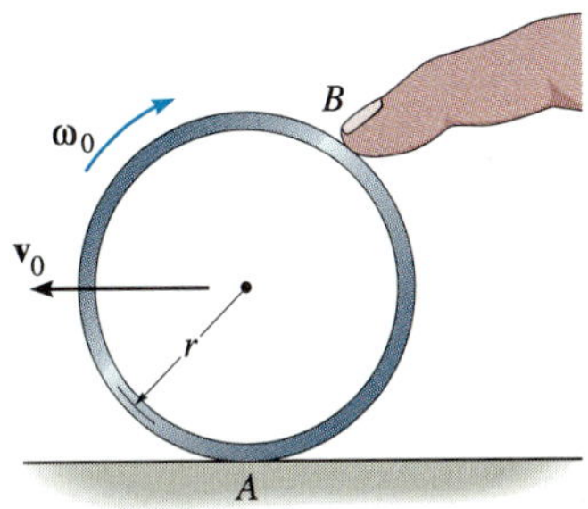

Prob. 17–108

17-109. A girl sits snugly inside a large tire such that both the girl and tire have a total weight of 185 lb, a center of mass at G, and a radius of gyration $k_G = 1.65$ ft about G. If the tire rolls freely down the incline, determine the normal and frictional forces it exerts on the ground when it is in the position shown and has an angular velocity of 6 rad/s. Assume that the tire does not slip as it rolls.

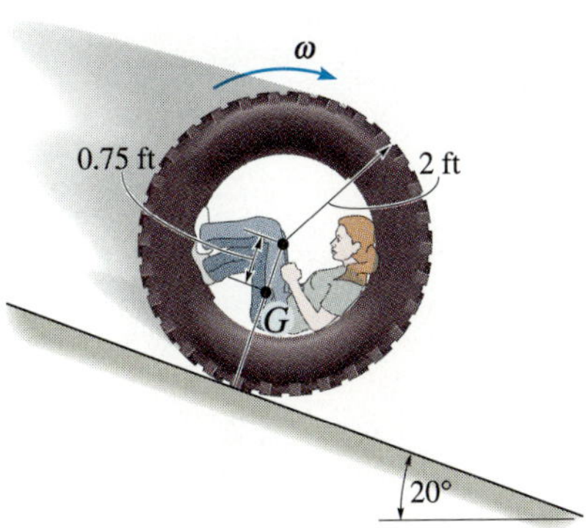

Prob. 17–109

17-110. Wheel C has a mass of 60 kg and a radius of gyration of 0.4 m, whereas wheel D has a mass of 40 kg and a radius of gyration of 0.35 m. Determine the angular acceleration of each wheel at the instant shown. Neglect the mass of the link and assume that the assembly does not slip on the plane.

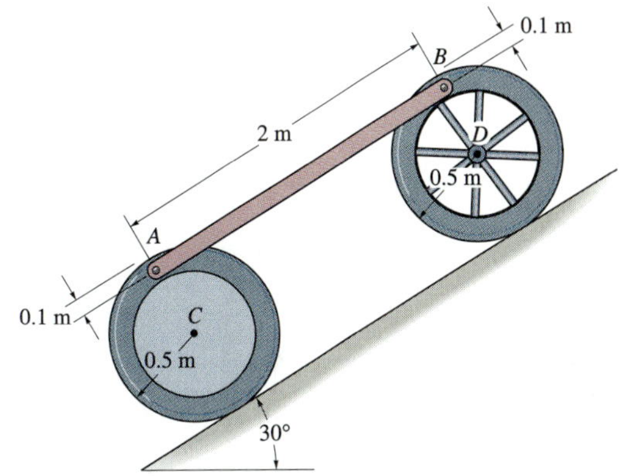

Prob. 17–110

■**17-111.** The assembly consists of an 8-kg disk and a 10-kg bar which is pin connected to the disk. If the system is released from rest, determine the angular acceleration of the disk. The coefficients of static and kinetic friction between the disk and the inclined plane are $\mu_s = 0.6$ and $\mu_k = 0.4$, respectively. Neglect friction at B.

***17-112.** Solve Prob. 17-111 if the bar is removed. The coefficients of static and kinetic friction between the disk and inclined plane are $\mu_s = 0.15$ and $\mu_k = 0.1$, respectively.

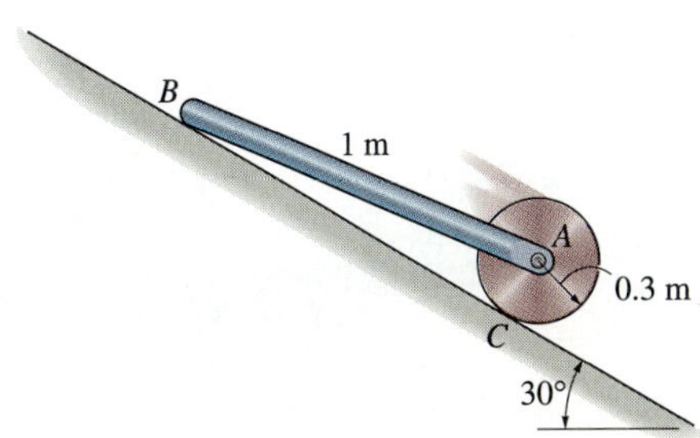

Probs. 17–111/112

17-113. The 20-kg disk A is attached to the 10-kg block B using the cable and pulley system shown. If the disk rolls without slipping, determine its angular acceleration and the acceleration of the block when they are released. Also, what is the tension in the cable? Neglect the mass of the pulleys.

17-114. Determine the minimum coefficient of static friction between the disk and the surface in Prob. 17-113 so that the disk will roll without slipping. Neglect the mass of the pulleys.

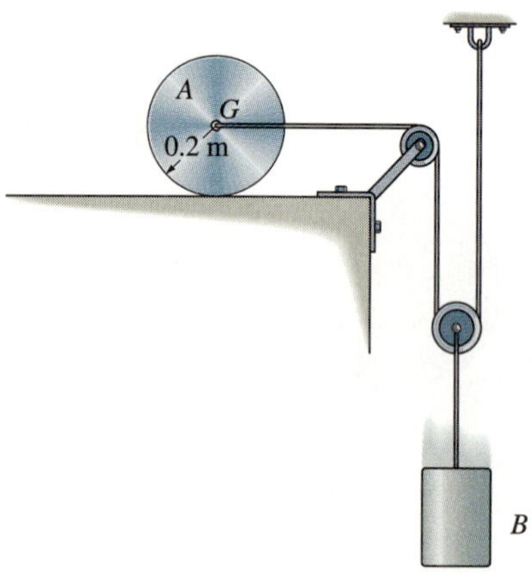

Probs. 17–113/114

17-115. A "lifted" truck can become a road hazard since the bumper is high enough to ride up a standard car in the event the car is rear-ended. As a model of this case consider the truck to have a mass of 2.70 Mg, a mass center G, and a radius of gyration about G of $k_G = 1.45$ m. Determine the horizontal and vertical components of acceleration of the mass center G, and the angular acceleration of the truck, at the moment its front wheels at C have just left the ground and its smooth front bumper begins to ride up the back of the stopped car so that point B has a velocity of $v_B = 8$ m/s at 20° from the horizontal. Assume the wheels are free to roll, and neglect the size of the wheels and the deformation of the material.

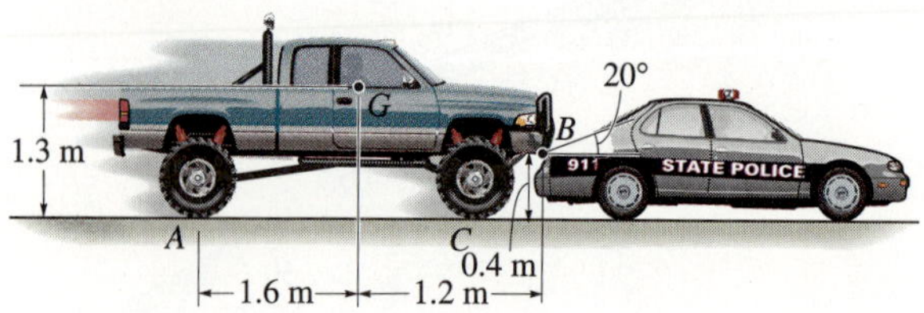

Prob. 17–115

***17-116.** The solid ball of radius r and mass m rolls without slipping down the 60° trough. Determine its angular acceleration.

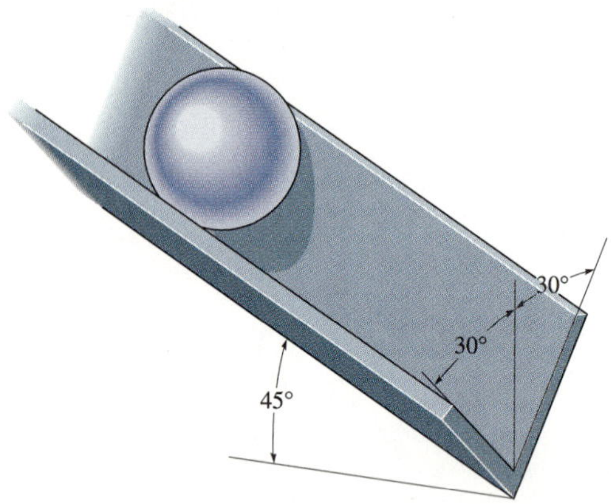

Prob. 17–116

Design Projects

17–1D. DESIGN OF A DYNAMOMETER

In order to test the dynamic strength of cables, an instrument called a *dynamometer* must be used that will measure the tension in a cable when it hoists a very heavy object with accelerated motion. Design such an instrument, based on the use of single or multiple springs so that it can be used on the cable supporting the 300-kg pipe that is given an upward acceleration of $2\ \text{m/s}^2$. Submit a drawing and explain how your dynamometer operates.

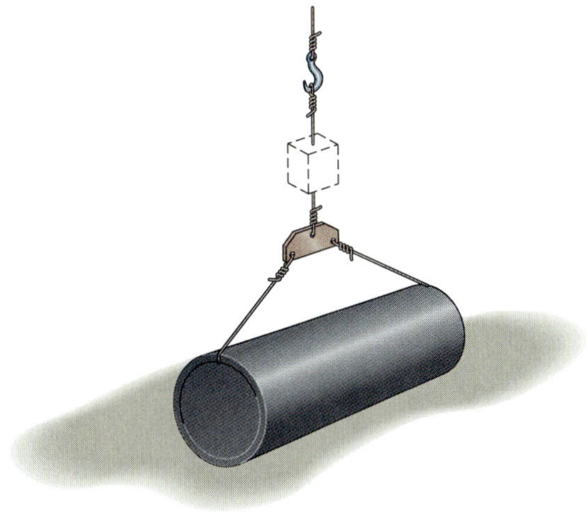

Prob. 17–1D

17–2D. DESIGN OF A SMALL ELEVATOR BRAKE

A small household elevator is operated using a hoist. For safety purposes it is necessary to install a braking mechanism which will automatically engage in case the cable fails during operation. Design the braking mechanism using steel members and springs. The elevator and its contents are assumed to have a mass of 300 kg, and it travels at 2.5 m/s. The maximum allowable deceleration to stop the motion is to be $4\ \text{m/s}^2$. Assume the coefficient of kinetic friction between any steel members and the walls of the elevator shaft is $\mu_k = 0.3$. The gap between the elevator frame and each wall of the shaft is 50 mm. Submit a scale drawing of your design along with a force analysis to show that your design will arrest the motion as required. Discuss the safety and reliability of the mechanism.

Prob. 17–2D

17–3D. SAFETY PERFORMANCE OF A BICYCLE

One of the most common accidents one can have on a bicycle is to flip over the handle bars. Obtain the necessary measurements of a standard-size bicycle and its mass and center of mass. Consider yourself as the rider, with center of mass at your navel. Perform an experiment to determine the coefficient of kinetic friction between the wheels and the pavement. With this data, calculate the possibility of flipping over when (a) only the rear brakes are applied, (b) only the front brakes are applied, and (c) both front and rear brakes are applied simultaneously. What effect does the height of the seat have on these results? Suggest a way to improve the bicycle's design, and write a report on the safety of cycling based on this analysis.

Prob. 17–3D

The principle of work and energy plays an important role in the motion of the draw works used to lift pipe on this drilling rig.

CHAPTER 18

Planar Kinetics of a Rigid Body: Work and Energy

Chapter Objectives

- To develop formulations for the kinetic energy of a body, and define the various ways a force and couple do work.
- To apply the principle of work and energy to solve rigid-body planar kinetic problems that involve force, velocity, and displacement.
- To show how the conservation of energy can be used to solve rigid-body planar kinetic problems.

18.1 Kinetic Energy

In this chapter we will apply work and energy methods to problems involving force, velocity, and displacement related to the planar motion of a rigid body. Before doing this, however, it will first be necessary to develop a means of obtaining the body's kinetic energy when the body is subjected to translation, rotation about a fixed axis, or general plane motion.

To do this we will consider the rigid body shown in Fig. 18–1, which is represented here by a *slab* moving in the inertial x–y reference plane. An arbitrary ith particle of the body, having a mass dm, is located at r from the arbitrary point P. If at the *instant* shown the particle has a velocity $\mathbf{v}_i$, then the particle's kinetic energy is $T_i = \frac{1}{2} dm\, v_i^2$. The kinetic

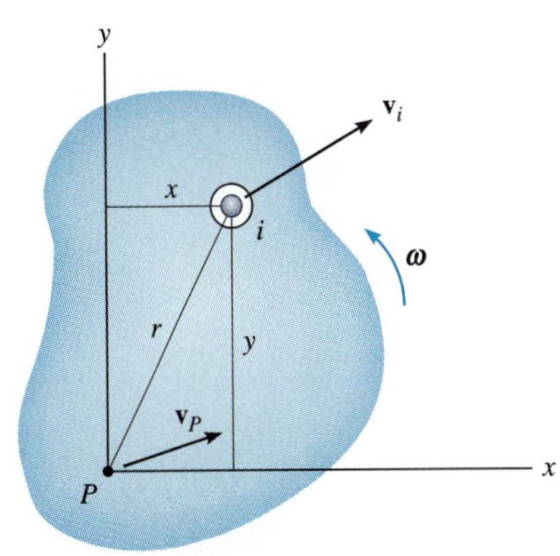

Fig. 18–1

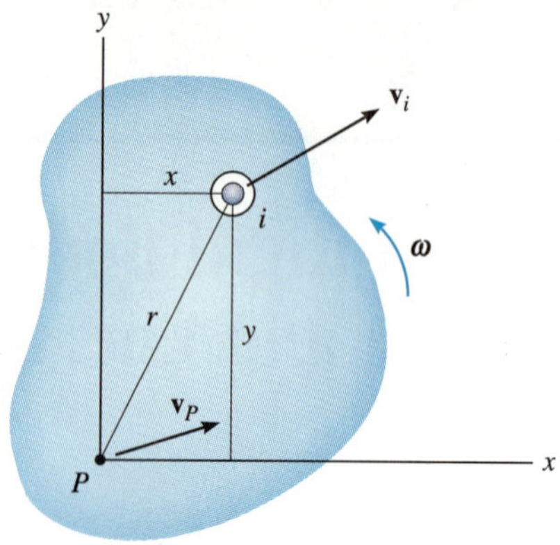

Fig. 18–1

energy of the entire body is determined by writing similar expressions for each particle of the body and integrating the results, i.e.,

$$T = \frac{1}{2}\int_m dm\, v_i^2$$

This equation may also be expressed in terms of the velocity of point P. If the body has an angular velocity $\boldsymbol{\omega}$, then from Fig. 18–1 we have

$$\begin{aligned}\mathbf{v}_i &= \mathbf{v}_P + \mathbf{v}_{i/P}\\ &= (v_P)_x\mathbf{i} + (v_P)_y\mathbf{j} + \omega\mathbf{k}\times(x\mathbf{i} + y\mathbf{j})\\ &= [(v_P)_x - \omega y]\mathbf{i} + [(v_P)_y + \omega x]\mathbf{j}\end{aligned}$$

The square of the magnitude of $\mathbf{v}_i$ is thus

$$\begin{aligned}\mathbf{v}_i\cdot\mathbf{v}_i = v_i^2 &= [(v_P)_x - \omega y]^2 + [(v_P)_y + \omega x]^2\\ &= (v_P)_x^2 - 2(v_P)_x\omega y + \omega^2 y^2 + (v_P)_y^2 + 2(v_P)_y\omega x + \omega^2 x^2\\ &= v_P^2 - 2(v_P)_x\omega y + 2(v_P)_y\omega x + \omega^2 r^2\end{aligned}$$

Substituting into the equation of kinetic energy yields

$$T = \tfrac{1}{2}\left(\int_m dm\right)v_P^2 - (v_P)_x\omega\left(\int_m y\,dm\right) + (v_P)_y\omega\left(\int_m x\,dm\right) + \tfrac{1}{2}\omega^2\left(\int_m r^2\,dm\right)$$

The first integral on the right represents the entire mass m of the body. Since $\bar{y}m = \int y\,dm$ and $\bar{x}m = \int x\,dm$, the second and third integrals locate the body's center of mass G with respect to P. The last integral represents the body's moment of inertia I_P, computed about the z axis passing through point P. Thus,

$$T = \tfrac{1}{2}mv_P^2 - (v_P)_x\,\omega\bar{y}m + (v_P)_y\,\omega\bar{x}m + \tfrac{1}{2}I_P\,\omega^2 \qquad (18\text{–}1)$$

As a special case, if point P coincides with the mass center G for the body, then $\bar{y} = \bar{x} = 0$, and therefore

$$T = \tfrac{1}{2}mv_G^2 + \tfrac{1}{2}I_G\omega^2 \qquad (18\text{–}2)$$

Here I_G is the moment of inertia for the body about an axis which is perpendicular to the plane of motion and passes through the mass center. Both terms on the right side are *always positive,* since the velocities are squared. Furthermore, it may be verified that these terms have units of length times force, common units being m·N or ft·lb. Recall, however, that in the SI system the unit of energy is the joule (J), where 1 J = 1 m·N.

Translation. When a rigid body of mass m is subjected to either rectilinear or curvilinear *translation,* the kinetic energy due to rotation is zero, since $\boldsymbol{\omega} = \mathbf{0}$. From Eq. 18–2, the kinetic energy of the body is therefore

$$T = \frac{1}{2} m v_G^2 \qquad (18\text{–}3)$$

where v_G is the magnitude of the translational velocity $\mathbf{v}$ at the instant considered, Fig. 18–2.

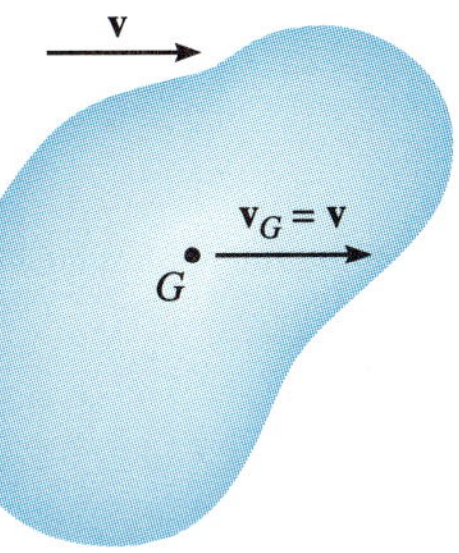

Translation

Fig. 18–2

Rotation About a Fixed Axis. When a rigid body is *rotating about a fixed axis* passing through point O, Fig. 18–3, the body has both *translational* and *rotational* kinetic energy as defined by Eq. 18–2, i.e.,

$$T = \frac{1}{2} m v_G^2 + \frac{1}{2} I_G \omega^2 \qquad (18\text{–}4)$$

The body's kinetic energy may also be formulated by noting that $v_G = r_G \omega$, in which case $T = \frac{1}{2}(I_G + m r_G^2)\omega^2$. By the parallel-axis theorem, the terms inside the parentheses represent the moment of inertia I_O of the body about an axis perpendicular to the plane of motion and passing through point O. Hence,*

$$T = \frac{1}{2} I_O \omega^2 \qquad (18\text{–}5)$$

From the derivation, this equation will give the same result as Eq. 18–4, since it accounts for *both* the translational and rotational kinetic energies of the body.

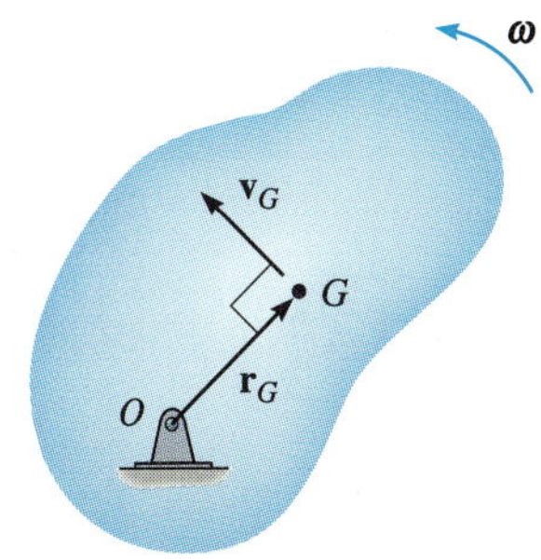

Rotation About a Fixed Axis

Fig. 18–3

General Plane Motion. When a rigid body is subjected to general plane motion, Fig. 18–4, it has an angular velocity $\boldsymbol{\omega}$ and its mass center has a velocity $\mathbf{v}_G$. Hence, the kinetic energy is defined by Eq. 18–2, i.e.,

$$T = \frac{1}{2} m v_G^2 + \frac{1}{2} I_G \omega^2 \qquad (18\text{–}6)$$

Here it is seen that the total kinetic energy of the body consists of the *scalar* sum of the body's *translational* kinetic energy, $\frac{1}{2} m v_G^2$, and *rotational* kinetic energy about its mass center, $\frac{1}{2} I_G \omega^2$.

Because energy is a scalar quantity, the total kinetic energy for a system of *connected* rigid bodies is the sum of the kinetic energies of all its moving parts. Depending on the type of motion, the kinetic energy of *each body* is found by applying Eq. 18–2 or the alternative forms mentioned above.

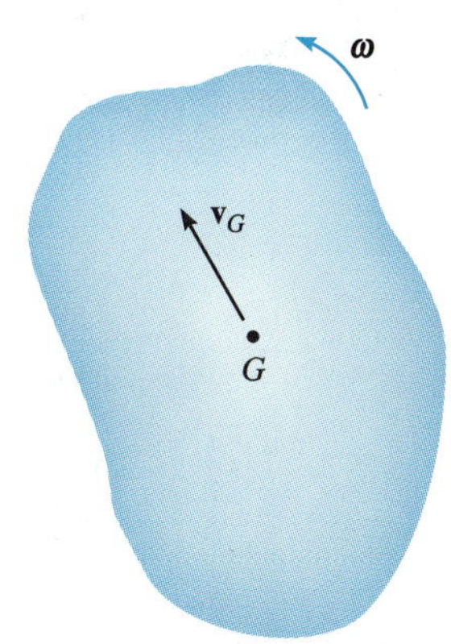

General Plane Motion

Fig. 18–4

*The similarity between this derivation and that of $\Sigma M_O = I_O \alpha$, Eq. 17–16, should be noted. Also note that the same result can be obtained directly from Eq. 18–1 by selecting point P at O, realizing that $\mathbf{v}_O = \mathbf{0}$.

The total kinetic energy of this soil compactor consists of the kinetic energy of the body or frame of the machine due to its translation, and the translational and rotational kinetic energies of the roller and the wheels due to their general plane motion. Here we exclude the additional kinetic energy developed by the moving parts of the engine and drive train.

EXAMPLE 18–1

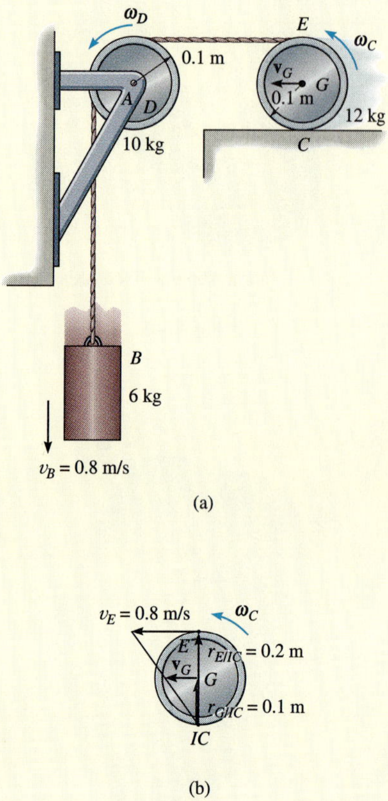

Fig. 18–5

The system of three elements shown in Fig. 18–5*a* consists of a 6-kg block *B*, a 10-kg disk *D*, and a 12-kg cylinder *C*. If no slipping occurs, determine the total kinetic energy of the system at the instant shown.

Solution

In order to compute the kinetic energy of the disk and cylinder, it is first necessary to determine ω_D, ω_C, and v_G, Fig. 18–5*a*. From the *kinematics* of the disk,

$$v_B = r_D\omega_D; \qquad 0.8\text{ m/s} = (0.1\text{ m})\omega_D \qquad \omega_D = 8\text{ rad/s}$$

Since the cylinder rolls without slipping, the instantaneous center of zero velocity is at the point of contact with the ground, Fig. 18–5*b*, hence,

$$v_E = r_{E/IC}\omega_C; \qquad 0.8\text{ m/s} = (0.2\text{ m})\omega_C \qquad \omega_C = 4\text{ rad/s}$$

$$v_G = r_{G/IC}\omega_C; \qquad v_G = (0.1\text{ m})(4\text{ rad/s}) = 0.4\text{ m/s}$$

Block

$$T_B = \tfrac{1}{2}m_B v_B^2 = \tfrac{1}{2}(6\text{ kg})(0.8\text{ m/s})^2 = 1.92\text{ J}$$

Disk

$$T_D = \tfrac{1}{2}I_D\omega_D^2 = \tfrac{1}{2}\left(\tfrac{1}{2}m_D r_D^2\right)\omega_D^2$$
$$= \tfrac{1}{2}\left[\tfrac{1}{2}(10\text{ kg})(0.1\text{ m})^2\right](8\text{ rad/s})^2 = 1.60\text{ J}$$

Cylinder

$$T_C = \tfrac{1}{2}mv_G^2 + \tfrac{1}{2}I_G\omega_C^2 = \tfrac{1}{2}mv_G^2 + \tfrac{1}{2}\left(\tfrac{1}{2}m_C r_C^2\right)\omega_C^2$$
$$= \tfrac{1}{2}(12\text{ kg})(0.4\text{ m/s})^2 + \tfrac{1}{2}\left[\tfrac{1}{2}(12\text{ kg})(0.1\text{ m})^2\right](4\text{ rad/s})^2 = 1.44\text{ J}$$

The total kinetic energy of the system is therefore

$$T = T_B + T_D + T_C$$
$$= 1.92\text{ J} + 1.60\text{ J} + 1.44\text{ J} = 4.96\text{ J} \qquad \textit{Ans.}$$

18.2 The Work of a Force

Several types of forces are often encountered in planar kinetics problems involving a rigid body. The work of each of these forces has been presented in Sec. 14.1 and is listed below as a summary.

Work of a Variable Force. If an external force **F** acts on a rigid body, the work done by the force when it moves along the path s, Fig. 18–6, is defined as

$$U_F = \int_s F \cos \theta \, ds \qquad (18\text{–}7)$$

Here θ is the angle between the "tails" of the force vector and the differential displacement. In general, the integration must account for the variation of the force's direction and magnitude.

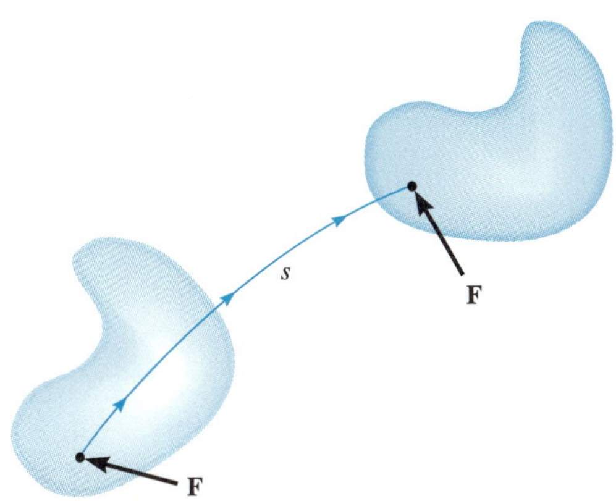

Fig. 18–6

Work of a Constant Force. If an external force $\mathbf{F}_c$ acts on a rigid body, Fig. 18–7, and maintains a constant magnitude F_c and constant direction θ, while the body undergoes a translation s, Eq. 18–7 can be integrated so that the work becomes

$$U_{F_c} = (F_c \cos \theta)s \qquad (18\text{–}8)$$

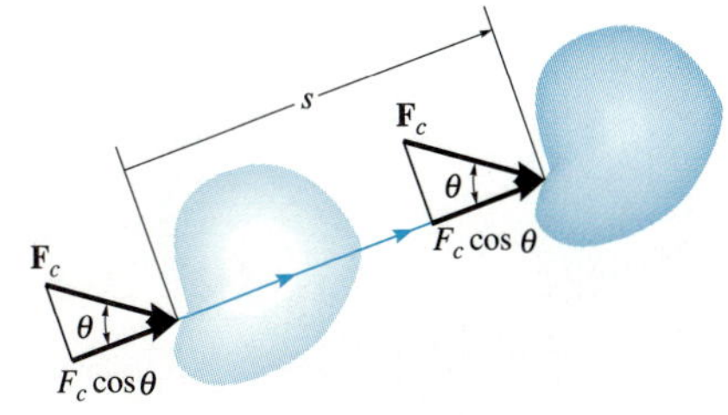

Fig. 18–7

Here F_c cos θ represents the magnitude of the component of force in the direction of displacement.

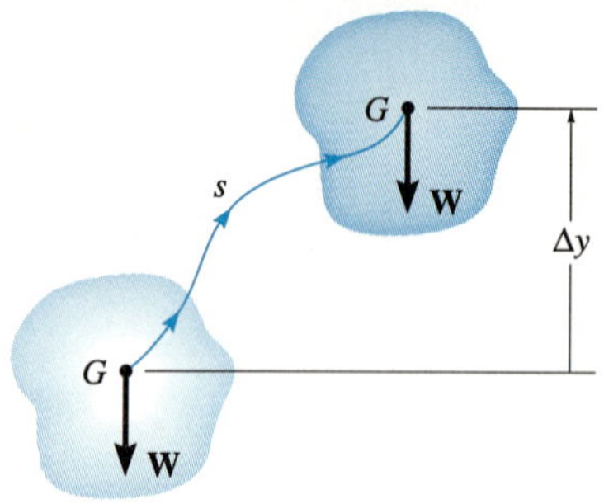

Fig. 18–8

Work of a Weight. The weight of a body does work only when the body's center of mass G undergoes a *vertical displacement* Δy. If this displacement is *upward,* Fig. 18–8, the work is negative, since the weight and displacement are in opposite directions.

$$U_W = -W\,\Delta y \tag{18–9}$$

Likewise, if the displacement is *downward* $(-\Delta y)$ the work becomes *positive.* In both cases the elevation change is considered to be small so that **W**, which is caused by gravitation, is constant.

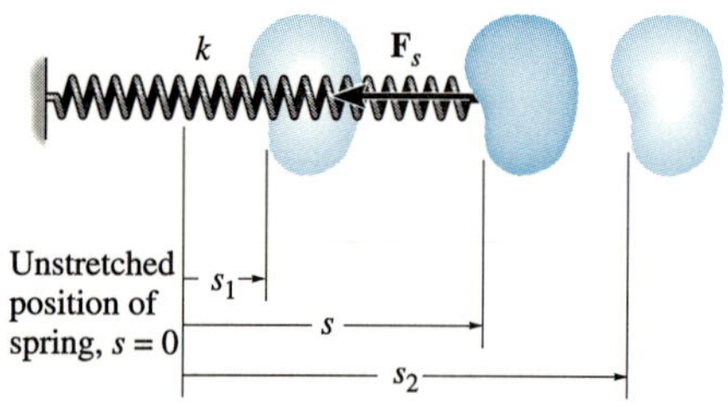

Fig. 18–9

Work of a Spring Force. If a linear elastic spring is attached to a body, the spring force $F_s = ks$ *acting on the body* does work when the spring either stretches or compresses from s_1 to a *further* position s_2. In both cases the work will be *negative* since the *displacement of the body* is in the opposite direction to the force, Fig. 18–9. The work done is

$$U_s = -\left(\tfrac{1}{2}ks_2^2 - \tfrac{1}{2}ks_1^2\right) \tag{18–10}$$

where $|s_2| > |s_1|$.

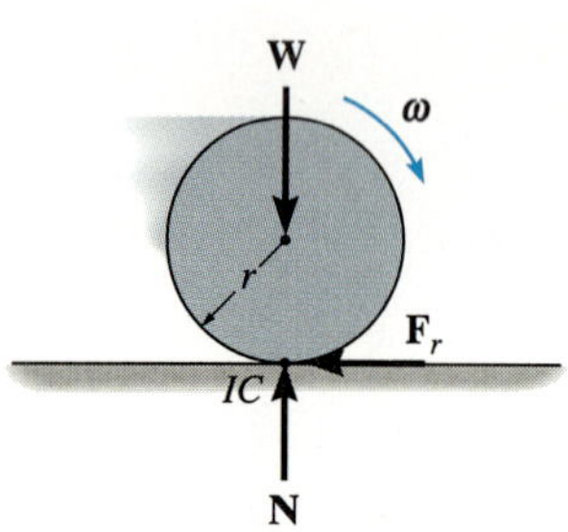

Fig. 18–10

Forces That Do No Work. There are some external forces that do no work when the body is displaced. These forces can act either at *fixed points* on the body, or they can have a direction *perpendicular to their displacement.* Examples include the reactions at a pin support about which a body rotates, the normal reaction acting on a body that moves along a fixed surface, and the weight of a body when the center of gravity of the body moves in a *horizontal plane,* Fig. 18–10. A rolling resistance force $\mathbf{F}_r$ acting on a round body as it *rolls without slipping* over a rough surface also does no work, Fig. 18–10.* This is because, during any *instant of time dt,* $\mathbf{F}_r$ acts at a point on the body which has *zero velocity* (instantaneous center, *IC*), and so the work done by the force on the point is zero. In other words, the point is not displaced in the direction of the force during this instant. Since $\mathbf{F}_r$ contacts successive points for only an instant, the work of $\mathbf{F}_r$ will be zero.

*The work done by the frictional force *when the body slips* has been discussed in Sec. 14.3.

18.3 The Work of a Couple

When a body subjected to a couple undergoes general plane motion, the two couple forces do work *only* when the body undergoes a *rotation.* To show this, consider the body in Fig. 18–11*a*, which is subjected to a couple moment $M = Fr$. Any general differential displacement of the body can be considered as a translation plus rotation. When the body *translates*, such that the *component of displacement* along the line of action of the forces is ds_t, Fig. 18–11*b*, clearly the "positive" work of one force *cancels* the "negative" work of the other. If the body undergoes a differential rotation $d\theta$ about an axis which is perpendicular to the plane of the couple and intersects the plane at point O, Fig. 18–11*c*, then each force undergoes a displacement $ds_\theta = (r/2)\,d\theta$ in the direction of the force. Hence, the total work done is

$$dU_M = F\left(\frac{r}{2}d\theta\right) + F\left(\frac{r}{2}d\theta\right) = (Fr)\,d\theta$$
$$= M\,d\theta$$

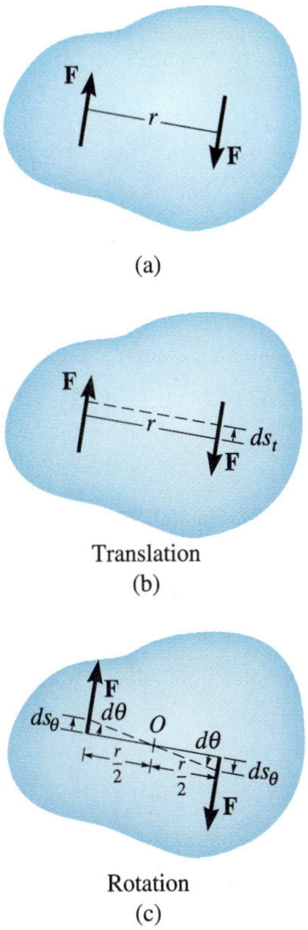

Fig. 18–11

Here the line of action of $d\theta$ is parallel to the line of action of M. This is *always the case for general plane motion,* since **M** and $d\boldsymbol{\theta}$ are perpendicular to the plane of motion. Furthermore, the resultant work is *positive* when **M** and $d\boldsymbol{\theta}$ have the *same sense of direction* and *negative* if these vectors have an *opposite sense of direction.*

When the body rotates in the plane through a finite angle θ measured in radians, from θ_1 to θ_2, the work of a couple is

$$U_M = \int_{\theta_1}^{\theta_2} M\,d\theta \qquad (18\text{–}11)$$

If the couple moment **M** has a *constant magnitude,* then

$$U_M = M(\theta_2 - \theta_1) \qquad (18\text{–}12)$$

Here the work is *positive* provided **M** and $(\boldsymbol{\theta}_2 - \boldsymbol{\theta}_1)$ are in the same direction.

EXAMPLE 18–2

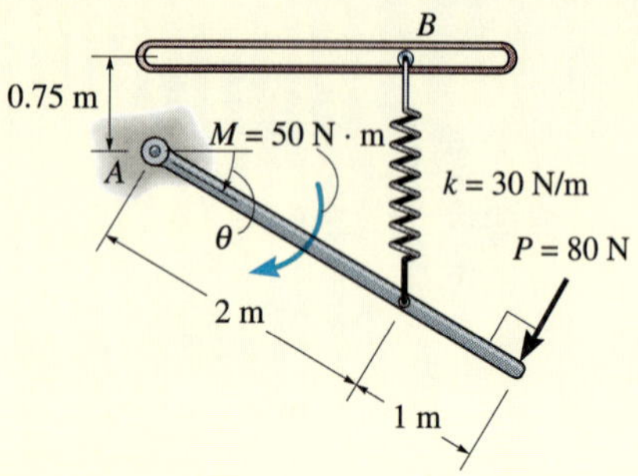

(a)

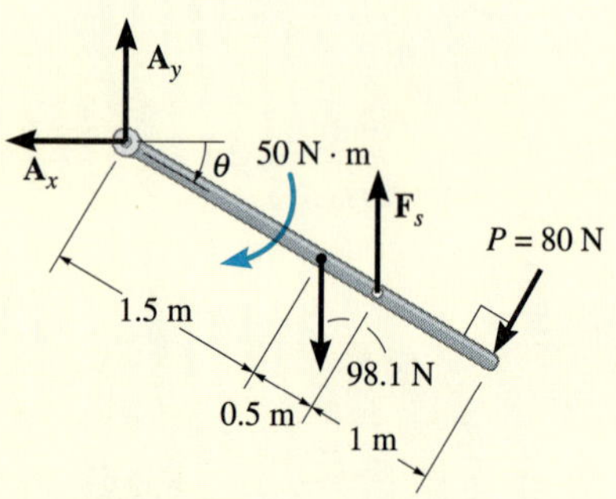

(b)

Fig. 18–12

The bar shown in Fig. 18–12*a* has a mass of 10 kg and is subjected to a couple moment of $M = 50$ N·m and a force of $P = 80$ N, which is always applied perpendicular to the end of the bar. Also, the spring has an unstretched length of 0.5 m and remains in the vertical position due to the roller guide at B. Determine the total work done by all the forces acting on the bar when it has rotated downward from $\theta = 0°$ to $\theta = 90°$.

Solution

First the free-body diagram of the bar is drawn in order to account for all the forces that act on it, Fig. 18–12*b*.

Weight W. Since the weight $10(9.81)\text{ N} = 98.1\text{ N}$ is displaced downward 1.5 m, the work is

$$U_W = 98.1\text{ N}(1.5\text{ m}) = 147.2\text{ J}$$

Why is the work positive?

Couple Moment M. The couple moment rotates through an angle of $\theta = \pi/2$ rad. Hence

$$U_M = 50\text{ N}\cdot\text{m}(\pi/2) = 78.5\text{ J}$$

Spring Force F_s. When $\theta = 0°$ the spring is stretched $(0.75\text{ m} - 0.5\text{ m}) = 0.25$ m, and when $\theta = 90°$, the stretch is $(2\text{ m} + 0.75\text{ m}) - 0.5\text{ m} = 2.25$ m. Thus

$$U_s = -\left[\tfrac{1}{2}(30\text{ N/m})(2.25\text{ m})^2 - \tfrac{1}{2}(30\text{ N/m})(0.25\text{ m})^2\right] = -75.0\text{ J}$$

By inspection the spring does negative work on the bar since F_s acts in the opposite direction to displacement. This checks with the result.

Force P. As the bar moves downward, the force is displaced through a distance of $(\pi/2)(3\text{ m}) = 4.712$ m. The work is positive. Why?

$$U_P = 80\text{ N}(4.712\text{ m}) = 377.0\text{ J}$$

Pin Reactions. Forces $\mathbf{A}_x$ and $\mathbf{A}_y$ do no work since they are not displaced.

Total Work. The work of all the forces when the bar is displaced is thus

$$U = 147.2\text{ J} + 78.5\text{ J} - 75.0\text{ J} + 377.0\text{ J} = 528\text{ J} \qquad \textit{Ans.}$$

18.4 Principle of Work and Energy

By applying the principle of work and energy developed in Sec. 14.2 to each of the particles of a rigid body and adding the results algebraically, since energy is a scalar, the principle of work and energy for a rigid body becomes

$$T_1 + \Sigma U_{1-2} = T_2 \tag{18–13}$$

This equation states that the body's initial translational *and* rotational kinetic energy, plus the work done by all the external forces and couple moments acting on the body as the body moves from its initial to its final position, is equal to the body's final translational *and* rotational kinetic energy. Note that the work of the body's *internal forces* does not have to be considered since the body is rigid. These forces occur in equal but opposite collinear pairs, so that when the body moves, the work of one force cancels that of its counterpart. Furthermore, since the body is rigid, *no relative movement* between these forces occurs, so that no internal work is done.

When several rigid bodies are pin connected, connected by inextensible cables, or in mesh with one another, Eq. 18–13 may be applied to the entire system of connected bodies. In all these cases the internal forces, which hold the various members together, do no work and hence are eliminated from the analysis.

The work of the torque or moment developed by the driving gears on the two motors is transformed into kinetic energy of rotation of the drum of the mixer.

Procedure for Analysis

The principle of work and energy is used to solve kinetic problems that involve *velocity, force,* and *displacement,* since these terms are involved in the formulation. For application, it is suggested that the following procedure be used.

Kinetic Energy (Kinematic Diagrams)

- The kinetic energy of a body is made up of two parts. Kinetic energy of translation is referenced to the velocity of the mass center, $T = \frac{1}{2}mv_G^2$, and kinetic energy of rotation is determined from knowing the moment of inertia about the mass center, $T = \frac{1}{2}I_G\omega^2$. In the special case of rotation about a fixed axis, these two kinetic energies are combined and can be expressed as $T = \frac{1}{2}I_O\omega^2$, where I_O is the moment of inertia about the axis of rotation.
- *Kinematic diagrams* for velocity may be useful for determining v_G and ω or for establishing a *relationship* between v_G and ω.*

Work (Free-Body Diagram)

- Draw a free-body diagram of the body when it is located at an intermediate point along the path in order to account for all the forces and couple moments which do work on the body as it moves along the path.
- A force does work when it moves through a displacement in the direction of the force.
- Forces that are functions of displacement must be integrated to obtain the work. Graphically, the work is equal to the area under the force–displacement curve.
- The work of a weight is the product of its magnitude and the vertical displacement, $U_W = Wy$. It is positive when the weight moves downwards.
- The work of a spring is of the form $U_s = \frac{1}{2}ks^2$, where k is the spring stiffness and s is the stretch or compression of the spring.
- The work of a couple is the product of the couple moment and the angle in radians through which it rotates.
- Since *algebraic addition* of the work terms is required, it is important that the proper sign of each term be specified. Specifically, work is *positive* when the force (couple moment) is in the *same direction* as its displacement (rotation); otherwise, it is negative.

Principle of Work and Energy

- Apply the principle of work and energy, $T_1 + \Sigma U_{1-2} = T_2$. Since this is a scalar equation, it can be used to solve for only one unknown when it is applied to a single rigid body.

*A brief review of Secs. 16.5 to 16.7 may prove helpful in solving problems, since computations for kinetic energy require a kinematic analysis of velocity.

EXAMPLE 18-3

The 30-kg disk shown in Fig. 18–13*a* is pin supported at its center. Determine the number of revolutions it must make to attain an angular velocity of 20 rad/s starting from rest. It is acted upon by a constant force $F = 10$ N, which is applied to a cord wrapped around its periphery, and a constant couple moment $M = 5$ N·m. Neglect the mass of the cord in the calculation.

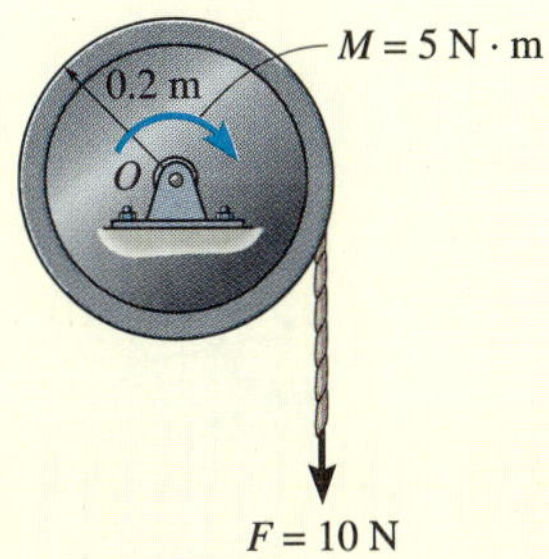

(a)

Solution

Kinetic Energy. Since the disk rotates about a fixed axis, the kinetic energy can be computed using $T = \frac{1}{2}I_O\omega^2$, where the moment of inertia is $I_O = \frac{1}{2}mr^2$. Initially, the disk is at rest, so that

$$T_1 = 0$$

$$T_2 = \tfrac{1}{2}I_O\omega_2^2 = \tfrac{1}{2}\left[\tfrac{1}{2}(30\text{ kg})(0.2\text{ m})^2\right](20\text{ rad/s})^2 = 120\text{ J}$$

Work (Free-Body Diagram). As shown in Fig. 18–13*b*, the pin reactions $\mathbf{O}_x$ and $\mathbf{O}_y$ and the weight (294.3 N) do no work, since they are not displaced. The *couple moment,* having a constant magnitude, does positive work $U_M = M\theta$ as the disk *rotates* through a clockwise angle of θ rad, and the *constant force* **F** does positive work $U_{F_c} = Fs$ as the cord *moves* downward $s = \theta r = \theta(0.2\text{ m})$.

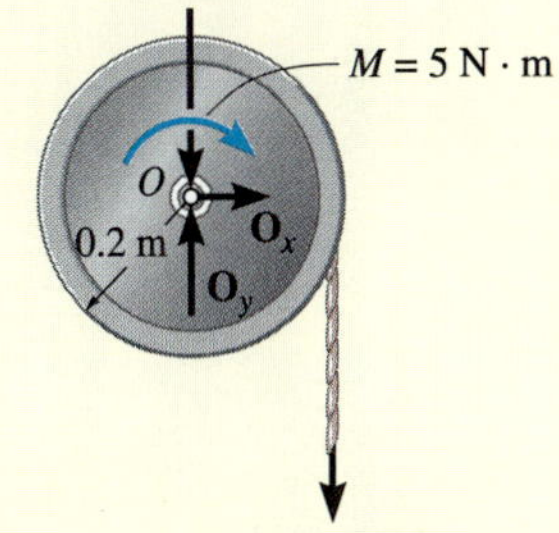

(b)

Fig. 18–13

Principle of Work and Energy

$$\{T_1\} + \{\Sigma U_{1-2}\} = \{T_2\}$$

$$\{T_1\} + \{M\theta + Fs\} = \{T_2\}$$

$$\{0\} + \{(5\text{ N}\cdot\text{m})\theta + (10\text{ N})\theta(0.2\text{ m})\} = \{120\text{ J}\}$$

$$\theta = 17.1\text{ rad} = 17.1\text{ rad}\left(\frac{1\text{ rev}}{2\pi\text{ rad}}\right) = 2.73\text{ rev} \qquad \textit{Ans.}$$

This problem has also been solved in Example 17–9. Compare the two methods of solution and note that since force, velocity, and displacement θ are involved, a work–energy approach yields a more direct solution.

EXAMPLE 18–4

The 700-kg pipe is equally suspended from the two tines of the fork lift shown in the photo. It is undergoing a swinging motion such that when $\theta = 30°$ it is momentarily at rest. Determine the normal and frictional forces acting on each tine which are needed to support the pipe at the instant $\theta = 0°$. Measurements of the pipe and the suspender are shown in Fig. 18–14*a*. Neglect the mass of the suspender and the thickness of the pipe.

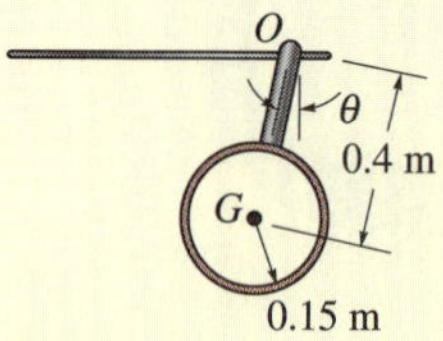

(a)

Fig. 18–14

Solution

We must use the equations of motion to find the forces on the tines since these forces do no work. Before doing this, however, we will apply the principle of work and energy to determine the angular velocity of the pipe when $\theta = 0°$.

Kinetic Energy (Kinematic Diagram). Since the pipe is originally at rest, then

$$T_1 = 0$$

The final kinetic energy may be computed with reference to either the fixed point O or the center of mass G. For the calculation we will consider the pipe to be a thin ring so that $I_G = mr^2$. If point G is considered, we have

$$\begin{aligned} T_2 &= \tfrac{1}{2}m(v_G)_2^2 + \tfrac{1}{2}I_G\omega_2^2 \\ &= \tfrac{1}{2}(700\text{ kg})[(0.4\text{ m})\omega_2]^2 + \tfrac{1}{2}[700\text{ kg}(0.15\text{ m})^2]\omega_2^2 \\ &= 63.875\omega_2^2 \end{aligned}$$

If point O is considered then the parallel-axis theorem must be used to determine I_O. Hence,

$$\begin{aligned} T_2 = \tfrac{1}{2}I_O\omega_2^2 &= \tfrac{1}{2}[700\text{ kg}(0.15\text{ m})^2 + 700\text{ kg}(0.4\text{ m})^2]\omega_2^2 \\ &= 63.875\omega_2^2 \end{aligned}$$

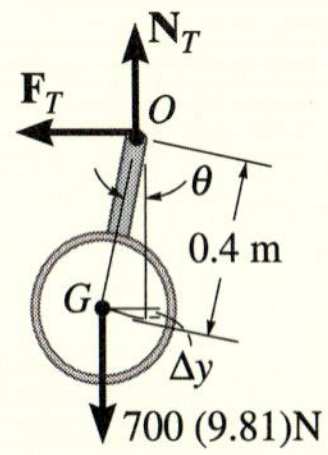

(b)

Work (Free-Body Diagram). Fig. 18–14*b*. The normal and frictional forces on the tines do no work since they do not move as the pipe swings. The weight, centered at G, does positive work since the weight moves downward through a vertical distance $\Delta y = 0.4\text{ m} - 0.4\cos 30^\circ\text{ m} = 0.05359\text{ m}$.

Principle of Work and Energy

$$\{T_1\} + \{\Sigma U_{1-2}\} = \{T_2\}$$

$$\{0\} + \{700(9.81)\text{ N}(0.05359\text{ m})\} = \{63.875\omega_2^2\}$$

$$\omega_2 = 2.40\text{ rad/s}$$

Equations of Motion. Referring to the free-body and kinetic diagrams shown in Fig. 18–14*c*, and using the result for ω_2, we have

$\overset{+}{\leftarrow}\Sigma F_t = m(a_G)_t;\qquad F_T = 700(a_G)_t$

$+\uparrow\Sigma F_n = m(a_G)_n;\quad N_T - 700(9.81)\text{ N} = 700\text{ kg}(2.40\text{ rad/s})^2(0.4\text{ m})$

$\circlearrowleft+\Sigma M_O = I_O\alpha;\quad 0 = [700\text{ kg}(0.15\text{ m})^2 + 700\text{ kg}(0.4\text{ m})^2]\alpha$

Since $(a_G)_t = 0.4\alpha$, then

$$\alpha = 0,\ (a_G)_t = 0$$

$$F_T = 0$$

$$N_T = 8.48\text{ kN}$$

There are two tines used to support the load, therefore

$$F_T' = 0 \qquad \textit{Ans.}$$

$$N_T' = \frac{8.48\text{ kN}}{2} = 4.24\text{ kN} \qquad \textit{Ans.}$$

Due to the swinging motion the tines are subjected to a *greater* normal force than would be the case if the load was static, in which case $N_T' = 700(9.81)\text{N}/2 = 3.43\text{ kN}$.

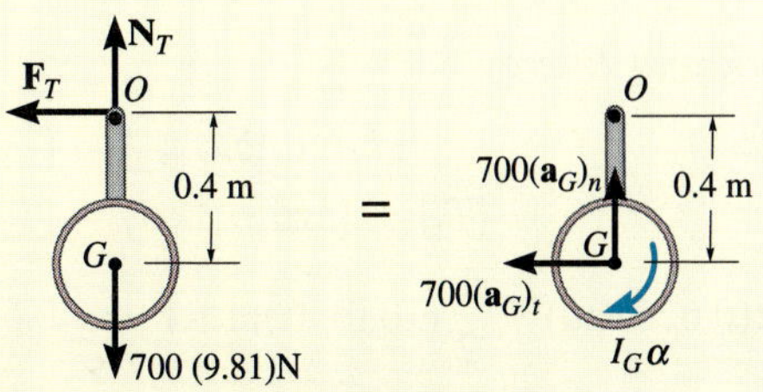

(c)

EXAMPLE 18–5

The wheel shown in Fig. 18–15*a* weighs 40 lb and has a radius of gyration $k_G = 0.6$ ft about its mass center G. If it is subjected to a clockwise couple moment of 15 lb·ft and rolls from rest without slipping, determine its angular velocity after its center G moves 0.5 ft. The spring has a stiffness $k = 10$ lb/ft and is initially unstretched when the couple moment is applied.

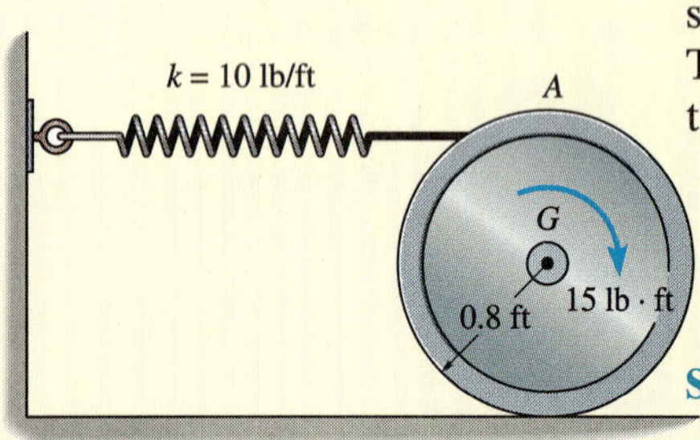

(a)

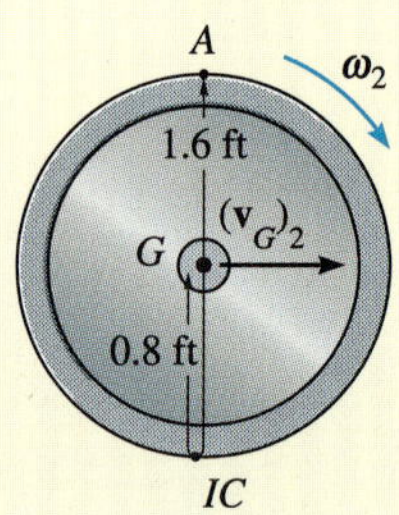

(b)

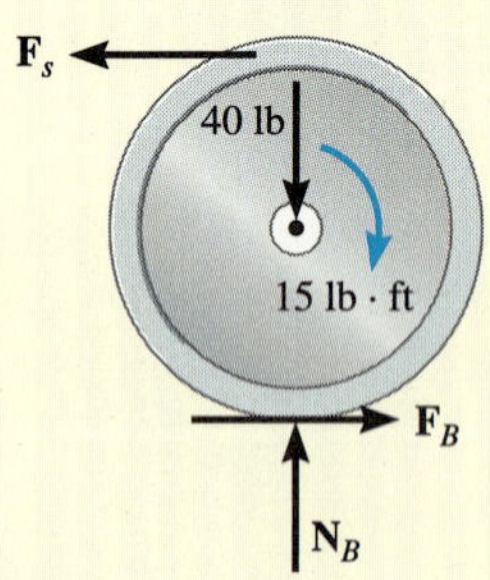

(c)

Fig. 18–15

Solution

Kinetic Energy (Kinematic Diagram). Since the wheel is initially at rest,

$$T_1 = 0$$

The kinematic diagram of the wheel when it is in the final position is shown in Fig. 18–15*b*. Hence, the final kinetic energy is

$$T_2 = \tfrac{1}{2}m(v_G)_2^2 + \tfrac{1}{2}I_G\omega_2^2$$

$$= \frac{1}{2}\left(\frac{40\text{ lb}}{32.2\text{ ft/s}^2}\right)(v_G)_2^2 + \frac{1}{2}\left[\frac{40\text{ lb}}{32.2\text{ ft/s}^2}(0.6\text{ ft})^2\right]\omega_2^2$$

The velocity of the mass center can be related to the angular velocity from the instantaneous center of zero velocity (IC), i.e., $(v_G)_2 = 0.8\omega_2$. Substituting into the above equation and simplifying, we have

$$T_2 = 0.621\omega_2^2$$

Work (Free-Body Diagram). As shown in Fig. 18–15*c*, only the spring force $\mathbf{F}_s$ and the couple moment do work. The normal force does not move along its line of action and the frictional force does *no work,* since the wheel does not slip as it rolls.

The work of $\mathbf{F}_s$ may be computed using $U_s = -\tfrac{1}{2}ks^2$. Here the work is negative since $\mathbf{F}_s$ is in the opposite direction to displacement. Since the wheel does not slip when the center G moves 0.5 ft, then the wheel rotates $\theta = s_G/r_{G/IC} = 0.5\text{ ft}/0.8\text{ ft} = 0.625$ rad, Fig. 18–15*b*. Hence, the spring stretches $s_A = \theta r_{A/IC} = 0.625\text{ rad}(1.6\text{ ft}) = 1$ ft.

Principle of Work and Energy

$$\{T_1\} + \{\Sigma U_{1-2}\} = \{T_2\}$$

$$\{T_1\} + \{M\theta - \tfrac{1}{2}ks^2\} = \{T_2\}$$

$$\{0\} + \{15\text{ lb}\cdot\text{ft}(0.625\text{ rad}) - \frac{1}{2}(10\text{ lb/ft})(1\text{ ft})^2\} = \{0.621\omega_2^2\text{ft}\cdot\text{lb}\}$$

$$\omega_2 = 2.65\text{ rad/s} \circlearrowright$$ *Ans.*

EXAMPLE 18–6

The 10-kg rod shown in Fig. 18–16a is constrained so that its ends move along the grooved slots. The rod is initially at rest when $\theta = 0°$. If the slider block at B is acted upon by a horizontal force $P = 50$ N, determine the angular velocity of the rod at the instant $\theta = 45°$. Neglect friction and the mass of blocks A and B.

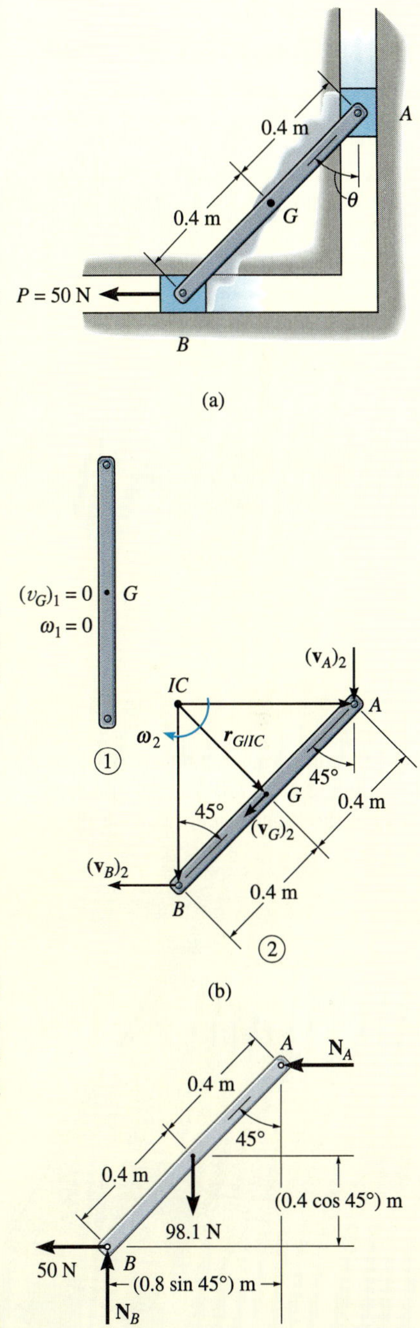

Fig. 18–16

Solution

Why can the principle of work and energy be used to solve this problem?

Kinetic Energy (Kinematic Diagrams). Two kinematic diagrams of the rod, when it is in the initial position 1 and final position 2, are shown in Fig. 18–16b. When the rod is in position 1, $T_1 = 0$ since $(\mathbf{v}_G)_1 = \boldsymbol{\omega}_1 = \mathbf{0}$. In position 2 the angular velocity is $\boldsymbol{\omega}_2$ and the velocity of the mass center is $(\mathbf{v}_G)_2$. Hence, the kinetic energy is

$$\begin{aligned} T_2 &= \tfrac{1}{2}m(v_G)_2^2 + \tfrac{1}{2}I_G\omega_2^2 \\ &= \tfrac{1}{2}(10\text{ kg})(v_G)_2^2 + \tfrac{1}{2}[\tfrac{1}{12}(10\text{ kg})(0.8\text{ m})^2]\omega_2^2 \\ &= 5(v_G)_2^2 + 0.267(\omega_2)^2 \end{aligned}$$

The two unknowns $(v_G)_2$ and ω_2 may be related from the instantaneous center of zero velocity for the rod, Fig. 18–16b. It is seen that as A moves downward with a velocity $(\mathbf{v}_A)_2$, B moves horizontally to the left with a velocity $(\mathbf{v}_B)_2$. Knowing these directions, the IC is determined as shown in the figure. Hence,

$$\begin{aligned} (v_G)_2 &= r_{G/IC}\,\omega_2 = (0.4 \tan 45°\text{ m})\omega_2 \\ &= 0.4\omega_2 \end{aligned}$$

Therefore,

$$T_2 = 0.8\omega_2^2 + 0.267\omega_2^2 = 1.067\omega_2^2$$

Work (Free-Body Diagram). Fig. 18–16c. The normal forces $\mathbf{N}_A$ and $\mathbf{N}_B$ do no work as the rod is displaced. Why? The 98.1-N weight is displaced a vertical distance of $\Delta y = (0.4 - 0.4 \cos 45°)$ m; whereas the 50-N force moves a horizontal distance of $s = (0.8 \sin 45°)$ m. Both of these forces do positive work. Why?

Principle of Work and Energy

$$\{T_1\} + \{\Sigma U_{1-2}\} = \{T_2\}$$

$$\{T_1\} + \{W\Delta y + Ps\} = \{T_2\}$$

$$\{0\} + \{98.1\text{ N}(0.4\text{ m} - 0.4 \cos 45°\text{ m}) + 50\text{ N}(0.8 \sin 45°\text{ m})\} = \{1.067\omega_2^2\text{ J}\}$$

Solving for ω_2 gives

$$\omega_2 = 6.11\text{ rad/s} \circlearrowright \qquad \textit{Ans.}$$

Problems

18-1. At a given instant the body of mass m has an angular velocity $\boldsymbol{\omega}$ and its mass center has a velocity $\mathbf{v}_G$. Show that its kinetic energy can be represented as $T = \frac{1}{2}I_{IC}\omega^2$, where I_{IC} is the moment of inertia of the body computed about the instantaneous axis of zero velocity, located a distance $r_{G/IC}$ from the mass center as shown.

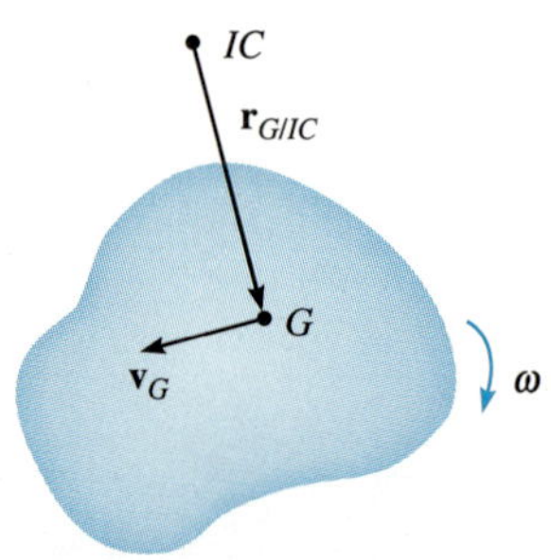

Prob. 18–1

18-2. The uniform rectangular plate weighs 30 lb. If the plate is pinned at A and has an angular velocity of 3 rad/s, determine the kinetic energy of the plate.

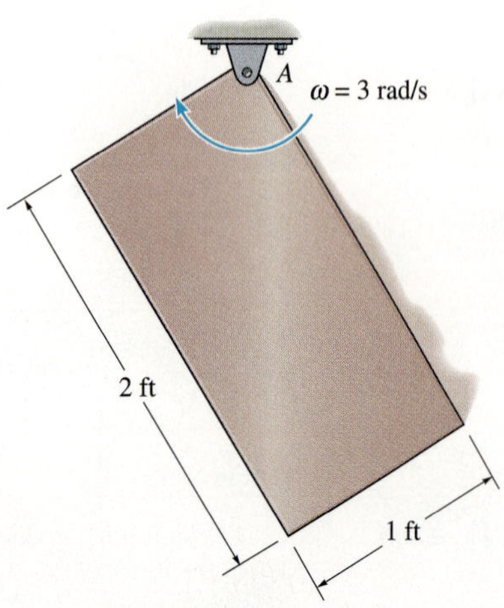

Prob. 18–2

18-3. Determine the kinetic energy of the system of three links. Links AB and CD each weigh 10 lb, and link BC weighs 20 lb.

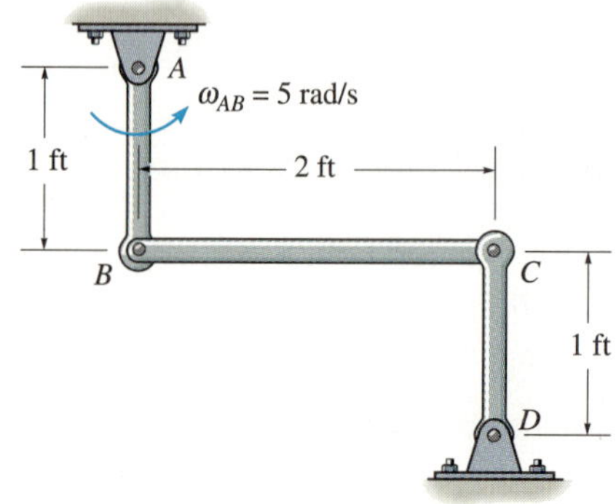

Prob. 18–3

***18-4.** The double pulley consists of two parts that are attached to one another. It has a weight of 50 lb and a centroidal radius of gyration of $k_O = 0.6$ ft and is turning with an angular velocity of 20 rad/s clockwise. Determine the kinetic energy of the system. Assume that neither cable slips on the pulley.

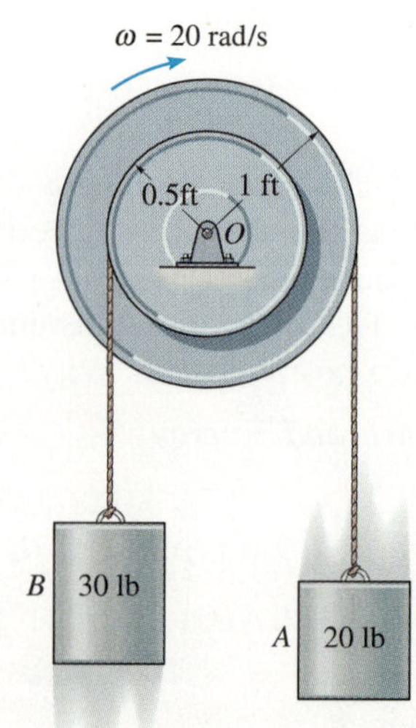

Prob. 18–4

18-5. The mechanism consists of two rods, AB and BC, which weigh 10 lb and 20 lb, respectively, and a 4-lb block at C. Determine the kinetic energy of the system at the instant shown, when the block is moving at 3 ft/s.

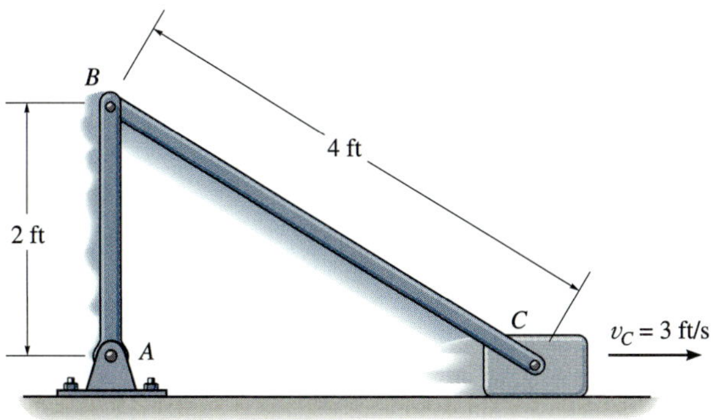

Prob. 18–5

18-6. Solve Prob. 17-58 using the principle of work and energy.

18-7. Solve Prob. 17-81 using the principle of work and energy.

***18-8.** Solve Prob. 17-72 using the principle of work and energy.

18-9. A force of $P = 20$ N is applied to the cable, which causes the 175-kg reel to turn since it is resting on the two rollers A and B of the dispenser. Determine the angular velocity of the reel after it has made two revolutions starting from rest. Neglect the mass of the rollers and the mass of the cable. The radius of gyration of the reel about its center axis is $k_G = 0.42$ m.

18-10. A force of $P = 20$ N is applied to the cable, which causes the 175-kg reel to turn without slipping on the two rollers A and B of the dispenser. Determine the angular velocity of the reel after it has made two revolutions starting from rest. Neglect the mass of the cable. Each roller can be considered as an 18-kg cylinder, having a radius of 0.1 m. The radius of gyration of the reel about its center axis is $k_G = 0.42$ m.

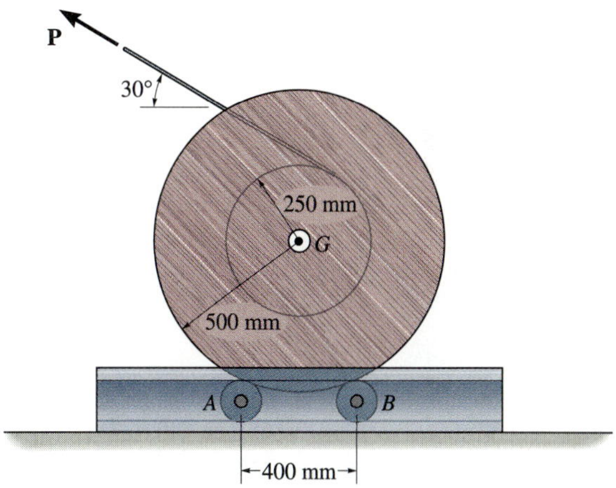

Probs. 18–9/10

18-11. The rotary screen S is used to wash limestone. When empty it has a mass of 800 kg and a radius of gyration of $k_G = 1.75$ m. Rotation is achieved by applying a torque of $M = 280$ N·m about the drive wheel A. If no slipping occurs at A and the supporting wheel at B is free to roll, determine the angular velocity of the screen after it has rotated 5 revolutions. Neglect the mass of A and B.

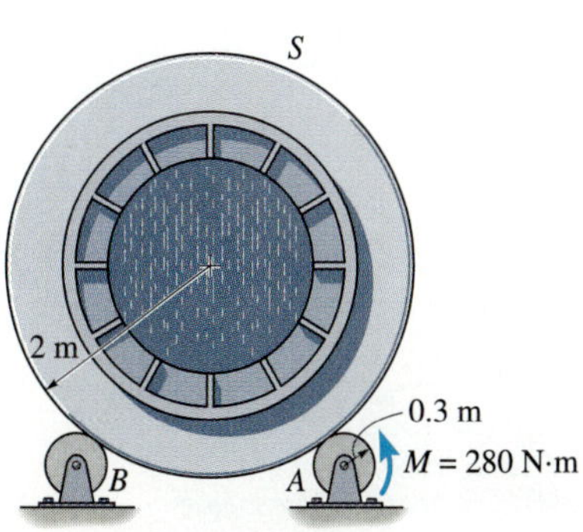

Prob. 18–11

***18-12.** The spool of cable, originally at rest, has a mass of 200 kg and a radius of gyration of $k_G = 325$ mm. If the spool rests on two small rollers A and B and a constant horizontal force of $P = 400$ N is applied to the end of the cable, determine the angular velocity of the spool when 8 m of cable has been unwound. Neglect friction and the mass of the rollers and unwound cable.

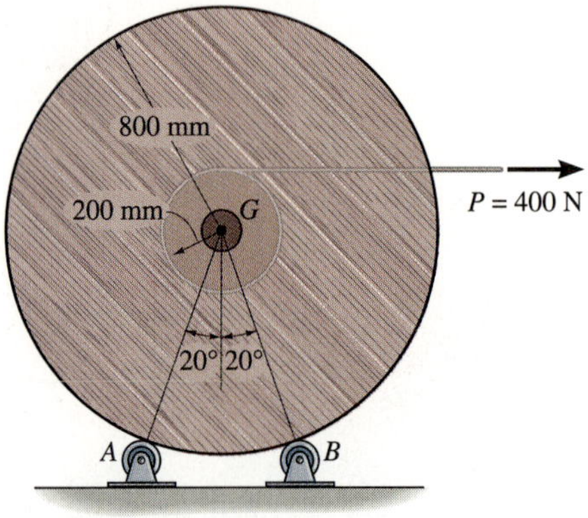

Prob. 18–12

18-13. The pendulum of the Charpy impact machine has a mass of 50 kg and a radius of gyration of $k_A = 1.75$ m. If it is released from rest when $\theta = 0°$, determine its angular velocity just before it strikes the specimen S, $\theta = 90°$.

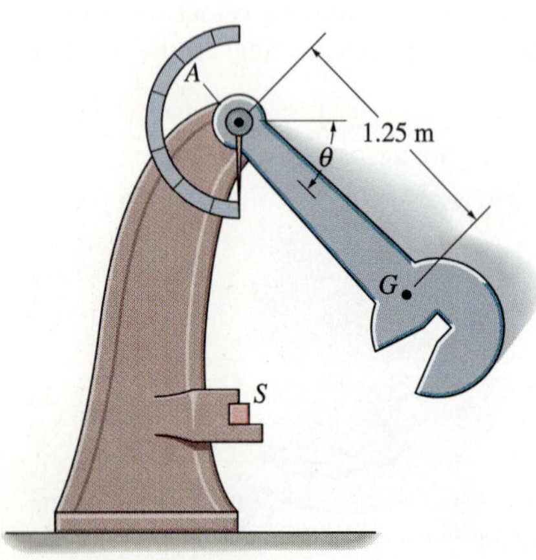

Prob. 18–13

18-14. The 1500-lb cement bucket is hoisted using a motor that supplies a torque of $M = 2000$ lb·ft to the axle of the wheel. If the wheel has a weight of 115 lb and a radius of gyration about O of $k_O = 0.95$ ft, determine the speed of the bucket when it has been hoisted 10 ft starting from rest.

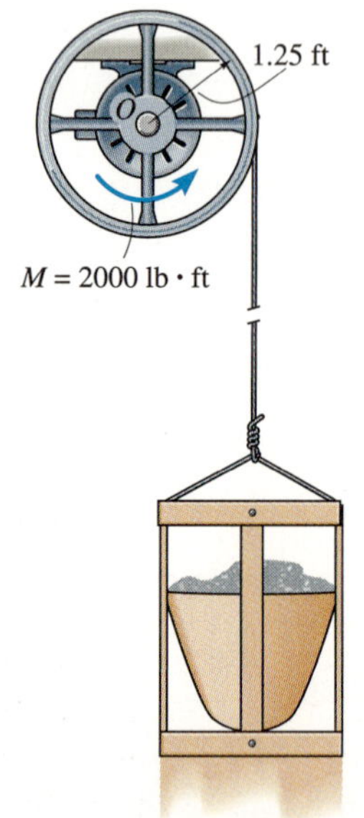

Prob. 18–14

18-15. The 10-kg pulley has a radius of gyration about O of $k_O = 0.21$ m. If a motor M supplies a force to the cable of $P = 800(3 - 2e^{-x})$ N, where x is the amount of cable wound up in meters, determine the speed of the 50-kg crate when it has been hoisted 2 m starting from rest. Neglect the mass of the cable and assume the cable does not slip on the pulley.

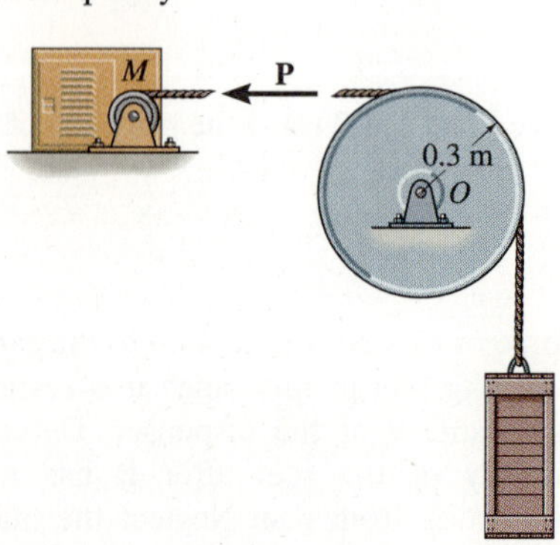

Prob. 18–15

***18-16.** The drum has a mass of 50 kg and a radius of gyration about the pin at O of $k_O = 0.23$ m. Starting from rest, the suspended 15-kg block B is allowed to fall 3 m without applying the brake ACD. Determine the speed of the block at this instant. If the coefficient of kinetic friction at the brake pad C is $\mu_k = 0.5$, determine the force P that must be applied at the brake handle which will then stop the block after it descends *another* 3 m. Neglect the thickness of the handle.

18-17. The drum has a mass of 50 kg and a radius of gyration about the pin at O of $k_O = 0.23$ m. If the 15-kg block is moving downward at 3 m/s, and a force of $P = 100$ N is applied to the brake arm, determine how far the block descends from the instant the brake is applied until it stops. Neglect the thickness of the handle. The coefficient of kinetic friction at the brake pad is $\mu_k = 0.5$.

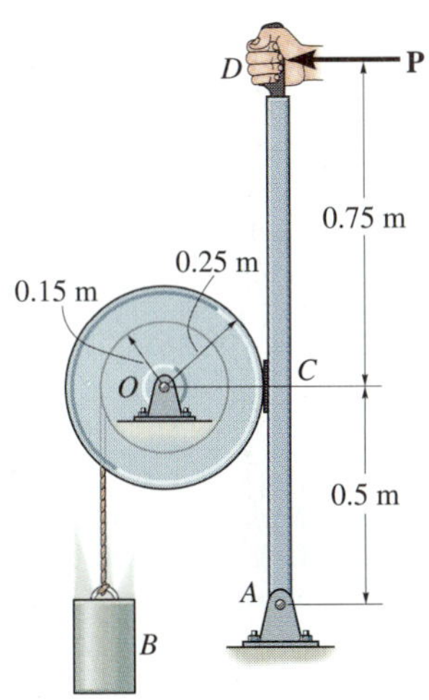

Probs. 18–16/17

18-18. The elevator car E has a mass of 1.80 Mg and the counterweight C has a mass of 2.30 Mg. If a motor turns the driving sheave A with a constant torque of $M = 100$ N·m, determine the speed of the elevator when it has ascended 10 m starting from rest. Each sheave A and B has a mass of 150 kg and a radius of gyration of $k = 0.2$ m about its mass center or pinned axis. Neglect the mass of the cable and assume the cable does not slip on the sheaves.

18-19. The elevator car E has a mass of 1.80 Mg and the counterweight C has a mass of 2.30 Mg. If a motor turns the driving sheave A with a torque of $M = (0.06\theta^2 + 7.5)$ N·m, where θ is in radians, determine the speed of the elevator when it has ascended 12 m starting from rest. Each sheave A and B has a mass of 150 kg and a radius of gyration of $k = 0.2$ m about its mass center or pinned axis. Neglect the mass of the cable and assume the cable does not slip on the sheaves.

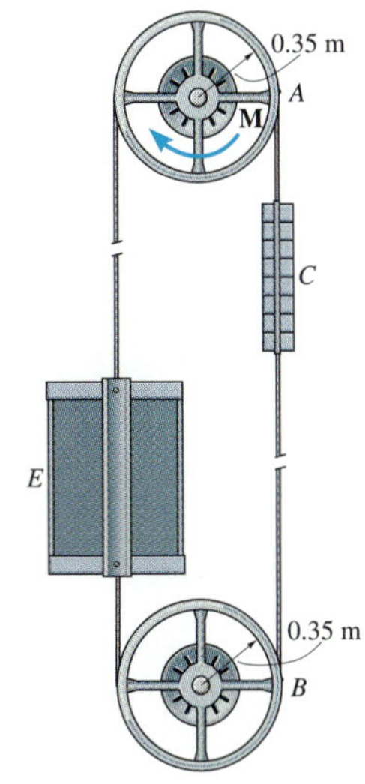

Probs. 18–18/19

***18-20.** The wheel has a mass of 100 kg and a radius of gyration $k_O = 0.2$ m. A motor supplies a torque $M = (40\theta + 900)$ N·m, where θ is in radians, about the drive shaft at O. Determine the speed of the loading car, which has a mass of 300 kg, after it travels $s = 4$ m. Initially the car is at rest when $s = 0$ and $\theta = 0°$. Neglect the mass of the attached cable and the mass of the car's wheels.

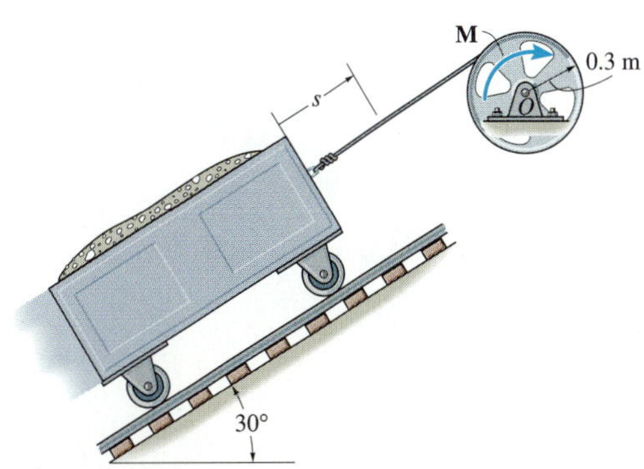

Prob. 18–20

18-21. A man having a weight of 180 lb sits in a chair of the Ferris wheel, which, excluding the man, has a weight of 15 000 lb and a radius of gyration $k_O = 37$ ft. If a torque $M = 80(10^3)$ lb·ft is applied about O, determine the angular velocity of the wheel after it has rotated 180°. Neglect the weight of the chairs and note that the man remains in an upright position as the wheel rotates. The wheel starts from rest in the position shown.

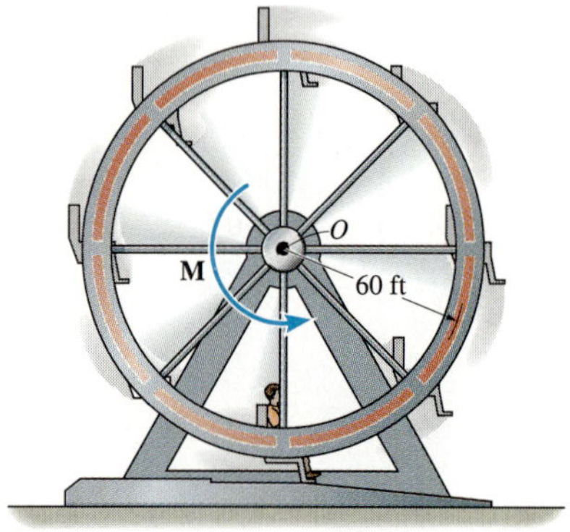

Prob. 18–21

18-22. The 20-kg disk is originally at rest, and the spring holds it in equilibrium. A couple moment of $M = 30$ N·m is then applied to the disk as shown. Determine its angular velocity at the instant its mass center G has moved 0.8 m down along the inclined plane. The disk rolls without slipping.

18-23. The 20-kg disk is originally at rest, and the spring holds it in equilibrium. A couple moment of $M = 30$ N·m is then applied to the disk as shown. Determine how far the center of mass of the disk travels down along the incline, measured from the equilibrium position, before it stops. The disk rolls without slipping.

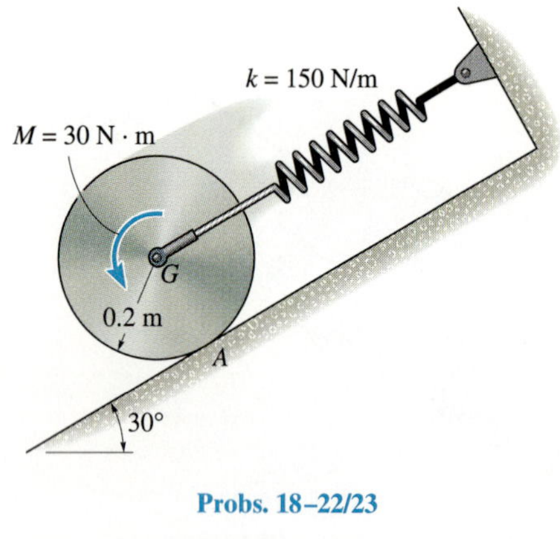

Probs. 18–22/23

***18-24.** The spool has a mass of 60 kg and a radius of gyration $k_G = 0.3$ m. If it is released from rest, determine how far its center descends down the smooth plane before it attains an angular velocity of $\omega = 6$ rad/s. Neglect friction and the mass of the cord which is wound around the central core.

18-25. Solve Prob. 18-24 if the coefficient of kinetic friction between the spool and plane at A is $\mu_k = 0.2$.

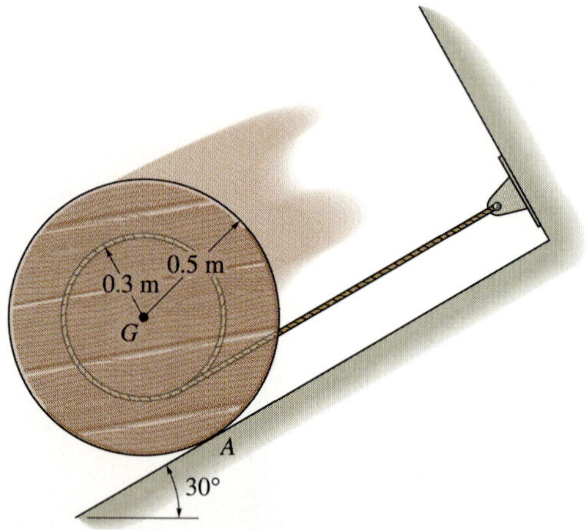

Probs. 18–24/25

18-26. The spool has a weight of 500 lb and a radius of gyration of $k_G = 1.75$ ft. A horizontal force of $P = 15$ lb is applied to a cable wrapped around its inner core. If the spool is originally at rest, determine its angular velocity after the mass center G has moved 6 ft to the left. The spool rolls without slipping. Neglect the mass of the cable.

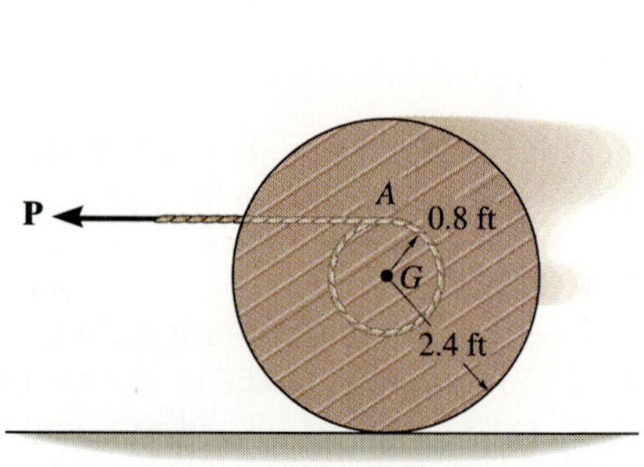

Prob. 18–26

18-27. The gear has a weight of 15 lb and a radius of gyration $k_G = 0.375$ ft. If the spring is unstretched when the torque $M = 6$ lb·ft is applied, determine the gear's angular velocity after its mass center G has moved to the left 2 ft.

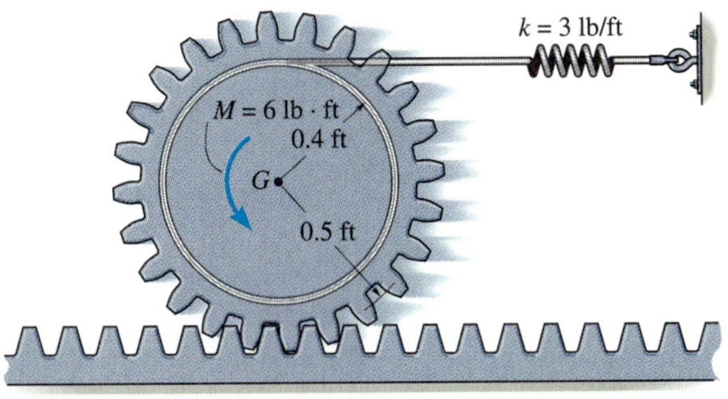

Prob. 18–27

***18-28.** A man having a weight of 150 lb crouches down on the end of a diving board as shown. In this position the radius of gyration about his center of gravity is $k_G = 1.2$ ft. While holding this position at $\theta = 0°$, he rotates about his toes at A until he loses contact with the board when $\theta = 90°$. If he remains rigid, determine approximately how many revolutions he then makes before striking the water after falling 30 ft.

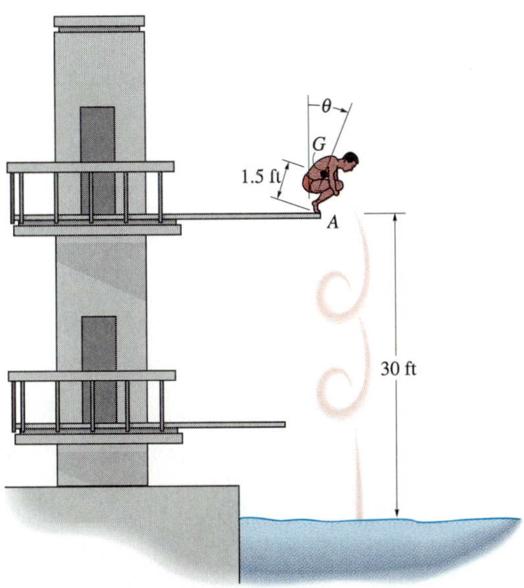

Prob. 18–28

18-29 The two 2-kg gears A and B are attached to the ends of a 3-kg slender bar. The gears roll within the fixed ring gear C, which lies in the horizontal plane. If a 10-N·m torque is applied to the center of the bar as shown, determine the number of revolutions the bar must rotate starting from rest in order for it to have an angular velocity of $\omega_{AB} = 20$ rad/s. For the calculation, assume the gears can be approximated by thin disks. What is the result if the gears lie in the vertical plane?

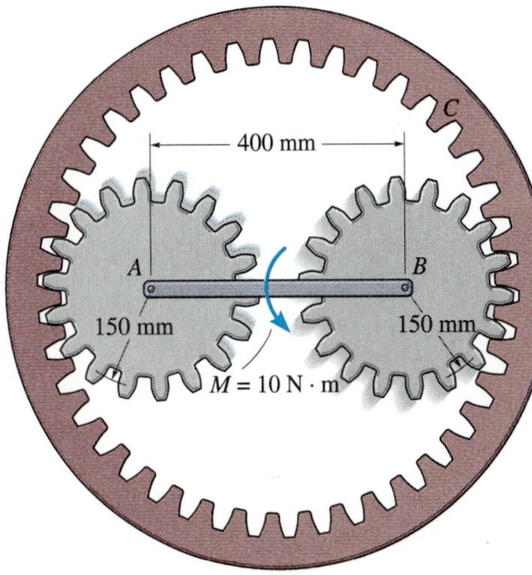

Prob. 18–29

18-30. The assembly consists of two 15-lb slender rods and a 20-lb disk. If the spring is unstretched when $\theta = 45°$ and the assembly is released from rest at this position, determine the angular velocity of rod AB at the instant $\theta = 0°$. The disk rolls without slipping.

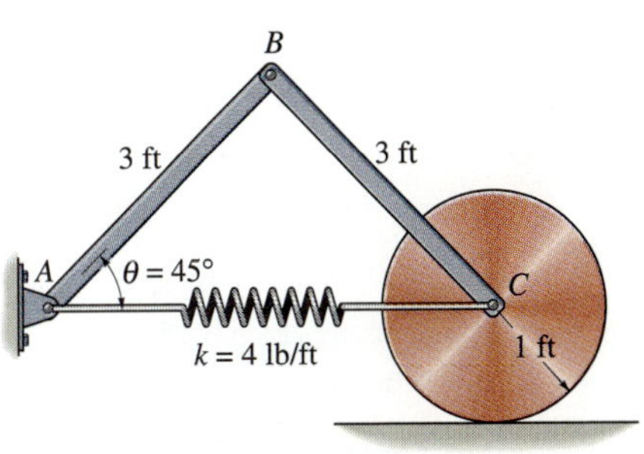

Prob. 18–30

18-31. The 100-lb block is transported a short distance by using two cylindrical rollers, each having a weight of 35 lb. If a horizontal force $P = 25$ lb is applied to the block, determine the block's speed after it has been displaced 2 ft to the left. Originally the block is at rest. No slipping occurs.

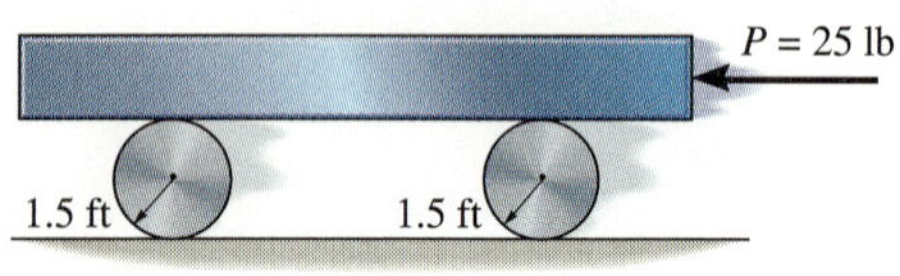

Prob. 18–31

18-33. The 10-lb sphere starts from rest at $\theta = 0°$ and rolls without slipping down the cylindrical surface which has a radius of 10 ft. Determine the speed of the sphere's center of mass at the instant $\theta = 45°$.

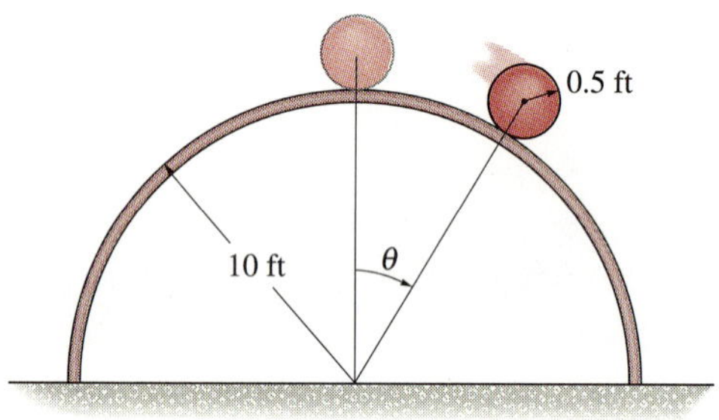

Prob. 18–33

***18-32.** The linkage consists of two 8-lb rods AB and CD and a 10-lb rod AD. When $\theta = 0°$, rod AB is rotating with an angular velocity $\omega_{AB} = 2$ rad/s. If at this instant rod CD is subjected to a couple moment $M = 15$ lb·ft and rod AD is subjected to a horizontal force $P = 20$ lb as shown, determine ω_{AB} at the instant $\theta = 90°$.

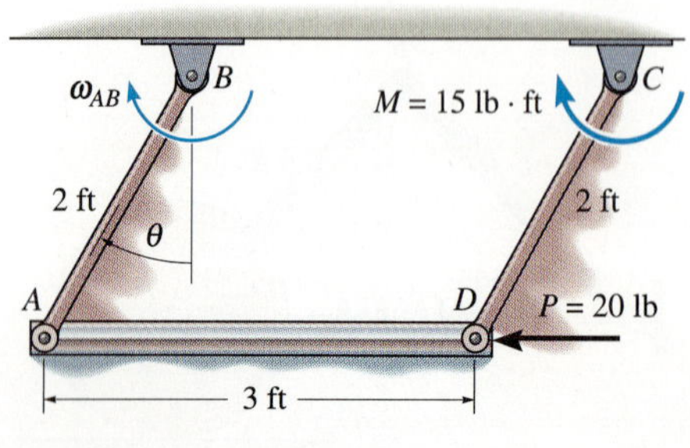

Prob. 18–32

18-34. The slender beam having a weight of 150 lb is supported by two cables. If the cable at end B is cut so that the beam is released from rest when $\theta = 30°$, determine the speed at which end A strikes the wall. Neglect friction at B.

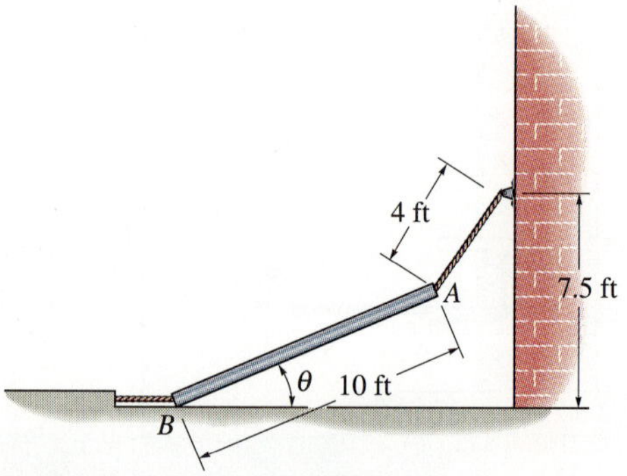

Prob. 18–34

18.5 Conservation of Energy

When a force system acting on a rigid body consists only of *conservative forces,* the conservation of energy theorem may be used to solve a problem which otherwise would be solved using the principle of work and energy. This theorem is often easier to apply since the work of a conservative force is *independent of the path* and depends only on the initial and final positions of the body. It was shown in Sec. 14.5 that the work of a conservative force may be expressed as the difference in the body's potential energy measured from an arbitrarily selected reference or datum.

Gravitational Potential Energy. Since the total weight of a body can be considered concentrated at its center of gravity, the *gravitational potential energy* of the body is determined by knowing the height of the body's center of gravity above or below a horizontal datum. Measuring y_G as *positive upward,* the gravitational potential energy of the body is thus

$$V_g = Wy_G \qquad (18\text{–}14)$$

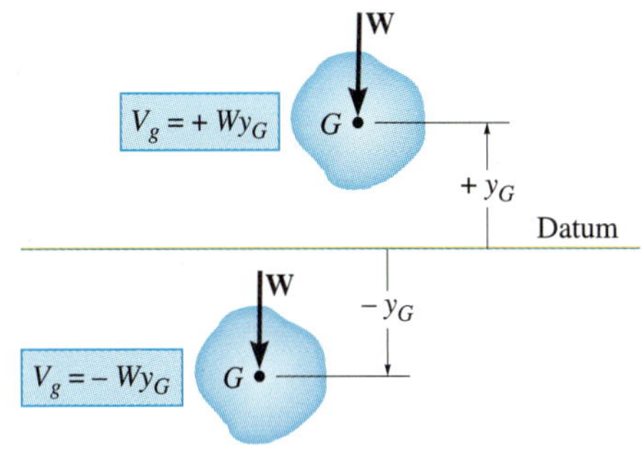

Gravitational potential energy

Fig. 18–17

Here the potential energy is *positive* when y_G is positive, since the weight has the ability to do *positive work* when the body is moved back to the datum, Fig. 18–17. Likewise, if the body is located *below* the datum $(-y_G)$, the gravitational potential energy is *negative,* since the weight does *negative work* when the body is returned to the datum.

Elastic Potential Energy. The force developed by an elastic spring is also a conservative force. The *elastic potential energy* which a spring imparts to an attached body when the spring is elongated or compressed from an initial undeformed position ($s = 0$) to a final position s, Fig. 18–18, is

$$V_e = +\tfrac{1}{2}ks^2 \qquad (18\text{–}15)$$

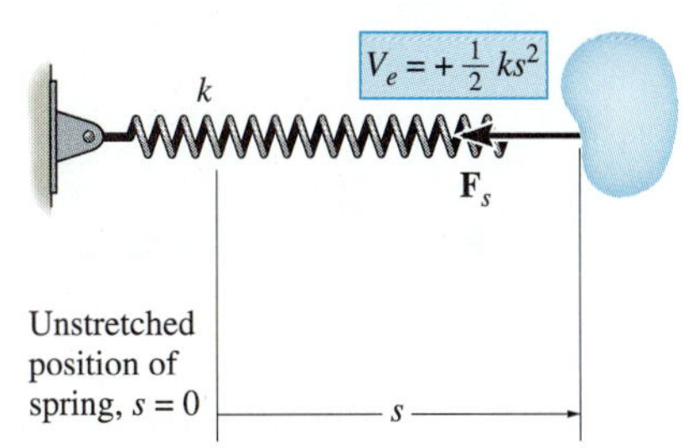

Elastic potential energy

Fig. 18–18

In the deformed position, the spring force acting *on the body* always has the capacity for doing positive work when the spring is returned back to its original undeformed position (see Sec. 14.5).

Conservation of Energy. In general, if a body is subjected to both gravitational and elastic forces, the total *potential energy* is expressed as a potential function V represented as the algebraic sum

$$V = V_g + V_e \tag{18–16}$$

Here measurement of V depends on the location of the body with respect to the selected datum.

Realizing that the work of conservative forces can be written as a difference in their potential energies, i.e., $(\Sigma U_{1-2})_{\text{cons}} = V_1 - V_2$, Eq. 14–16, we can rewrite the principle of work and energy for a rigid body as

$$T_1 + V_1 + (\Sigma U_{1-2})_{\text{noncons}} = T_2 + V_2 \tag{18–17}$$

Here $(\Sigma U_{1-2})_{\text{noncons}}$ represents the work of the nonconservative forces such as friction. If this term is zero, then

$$T_1 + V_1 = T_2 + V_2 \tag{18–18}$$

The torsional springs located at the top of the garage door wind up as the door is lowered. When the door is raised, the potential energy stored in the springs is then transferred into gravitational potential energy of the door's weight, thereby making it easy to open.

This equation is referred to as the conservation of mechanical energy. It states that the *sum* of the potential and kinetic energies of the body remains *constant* when the body moves from one position to another. It also applies to a system of smooth, pin-connected rigid bodies, bodies connected by inextensible cords, and bodies in mesh with other bodies. In all these cases the forces acting at the points of contact are *eliminated* from the analysis, since they occur in equal but opposite collinear pairs and each pair of forces moves through an equal distance when the system undergoes a displacement.

It is important to remember that only problems involving conservative force systems may be solved by using Eq. 18–18. As stated in Sec. 14.5, friction or other drag-resistant forces, which depend on velocity or acceleration, are nonconservative. The work of such forces is transformed into thermal energy used to heat up the surfaces of contact, and consequently this energy is dissipated into the surroundings and may not be recovered. Therefore, problems involving frictional forces can be solved by using either the principle of work and energy written in the form of Eq. 18–17, if it applies, or the equations of motion.

Procedure for Analysis

The conservation of energy equation is used to solve problems involving *velocity, displacement,* and *conservative force systems.* For application it is suggested that the following procedure be used.

Potential Energy

- Draw two diagrams showing the body located at its initial and final positions along the path.
- If the center of gravity, G, is subjected to a *vertical displacement,* establish a fixed horizontal datum from which to measure the body's gravitational potential energy V_g.
- Data pertaining to the elevation y_G of the body's center of gravity from the datum and the extension or compression of any connecting springs can be determined from the problem geometry and listed on the two diagrams.
- Recall that the potential energy $V = V_g + V_e$. Here $V_g = Wy_G$, which can be positive or negative, and $V_e = \frac{1}{2}ks^2$, which is always positive.

Kinetic Energy

- The kinetic energy of the body consists of two parts, namely translational kinetic energy, $T = \frac{1}{2}mv_G^2$, and rotational kinetic energy, $T = \frac{1}{2}I_G\omega^2$.
- Kinematic diagrams for velocity may be useful for determining v_G and ω or for establishing a *relationship* between these quantities.

Conservation of Energy

- Apply the conservation of energy equation $T_1 + V_1 = T_2 + V_2$.

EXAMPLE 18–7

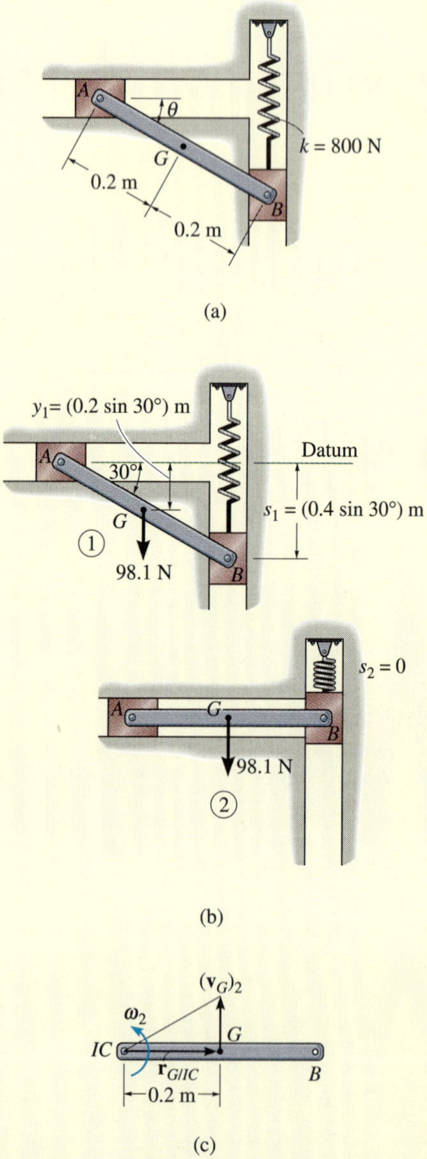

Fig. 18–19

The 10-kg rod AB shown in Fig. 18–19a is confined so that its ends move in the horizontal and vertical slots. The spring has a stiffness of $k = 800$ N/m and is unstretched when $\theta = 0°$. Determine the angular velocity of AB when $\theta = 0°$, if the rod is released from rest when $\theta = 30°$. Neglect the mass of the slider blocks.

Solution

Potential Energy. The two diagrams of the rod, when it is located at its initial and final positions, are shown in Fig. 18–19b. The datum, used to measure the gravitational potential energy, is placed in line with the rod when $\theta = 0°$.

When the rod is in position 1, the center of gravity G is located *below the datum* so that the gravitational potential energy is *negative*. Furthermore, (positive) elastic potential energy is stored in the spring, since it is stretched a distance of $s_1 = (0.4 \sin 30°)$ m. Thus,

$$V_1 = -Wy_1 + \tfrac{1}{2}ks_1^2$$
$$= -98.1\text{ N}(0.2 \sin 30°\text{ m}) + \tfrac{1}{2}(800\text{ N/m})(0.4 \sin 30°\text{ m})^2 = 6.19\text{ J}$$

When the rod is in position 2, the potential energy of the rod is zero, since the spring is unstretched, $s_2 = 0$, and the center of gravity G is located at the datum. Thus,

$$V_2 = 0$$

Kinetic Energy. The rod is released from rest from position 1, thus $(\mathbf{v}_G)_1 = \mathbf{0}$ and $\boldsymbol{\omega}_1 = \mathbf{0}$, and

$$T_1 = 0$$

In position 2, the angular velocity is $\boldsymbol{\omega}_2$ and the rod's mass center has a velocity of $(\mathbf{v}_G)_2$. Thus,

$$T_2 = \tfrac{1}{2}m(v_G)_2^2 + \tfrac{1}{2}I_G\omega_2^2$$
$$= \tfrac{1}{2}(10\text{ kg})(v_G)_2^2 + \tfrac{1}{2}[\tfrac{1}{12}(10\text{ kg})(0.4\text{ m})^2]\omega_2^2$$

Using *kinematics*, $(\mathbf{v}_G)_2$ can be related to $\boldsymbol{\omega}_2$ as shown in Fig. 18–19c. At the instant considered, the instantaneous center of zero velocity (IC) for the rod is at point A; hence, $(v_G)_2 = (r_{G/IC})\omega_2 = (0.2)\omega_2$. Substituting into the above expression and simplifying, we get

$$T_2 = 0.267\omega_2^2$$

Conservation of Energy

$$\{T_1\} + \{V_1\} = \{T_2\} + \{V_2\}$$
$$\{0\} + \{6.19\} = \{0.267\omega_2^2\} + \{0\}$$
$$\omega_2 = 4.82\text{ rad/s} \circlearrowleft \qquad \textit{Ans.}$$

EXAMPLE 18-8

The disk shown in Fig. 18–20*a* has a weight of 30 lb, a radius of gyration of $k_G = 0.6$ ft and is attached to a spring which has a stiffness $k = 2$ lb/ft and an unstretched length of 1 ft. If the disk is released from rest in the position shown and rolls without slipping, determine its angular velocity at the instant B moves 3 ft to the left.

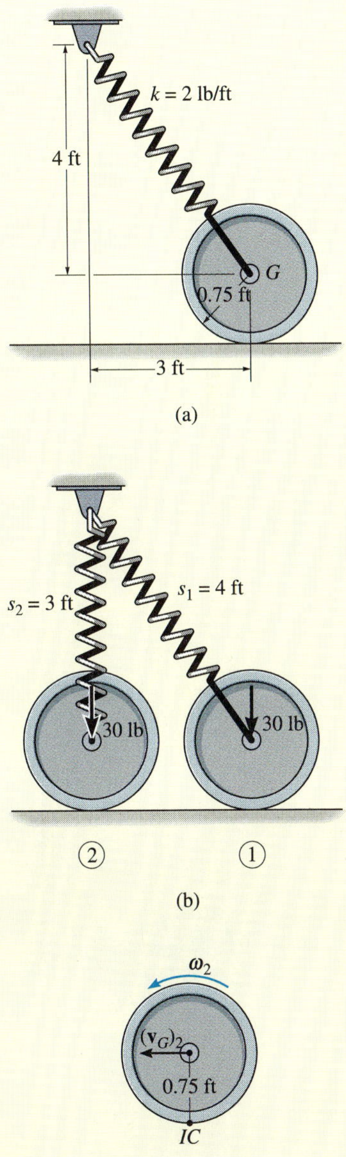

Fig. 18–20

Solution

Potential Energy. Two diagrams of the disk, when it is located in its initial and final positions, are shown in Fig. 18–20*b*. A gravitational datum is not needed here since the weight is not displaced vertically. From the problem geometry the spring is stretched $s_1 = (\sqrt{3^2 + 4^2} - 1) = 4$ ft and $s_2 = (4 - 1) = 3$ ft in the initial and final positions, respectively. Hence,

$$V_1 = \tfrac{1}{2}ks_1^2 = \tfrac{1}{2}(2\text{ lb/ft})(4\text{ ft})^2 = 16\text{ J}$$
$$V_2 = \tfrac{1}{2}ks_2^2 = \tfrac{1}{2}(2\text{ lb/ft})(3\text{ ft})^2 = 9\text{ J}$$

Kinetic Energy. The disk is released from rest so that $(\mathbf{v}_G)_1 = \mathbf{0}$, $\boldsymbol{\omega}_1 = \mathbf{0}$, and

$$T_1 = 0$$

In the final position,

$$T_2 = \frac{1}{2}m(v_G)_2^2 + \frac{1}{2}I_G\omega_2^2$$
$$= \frac{1}{2}\left(\frac{30\text{ lb}}{32.2\text{ ft/s}^2}\right)(v_G)_2^2 + \frac{1}{2}\left[\left(\frac{30\text{ lb}}{32.2\text{ ft/s}^2}\right)(0.6\text{ ft})^2\right]\omega_2^2$$

Since the disk rolls without slipping, $(\mathbf{v}_G)_2$ can be related to $\boldsymbol{\omega}_2$ from the instantaneous center of zero velocity, Fig. 18–20*c*, i.e., $(v_G)_2 = (0.75\text{ ft})\omega_2$. Substituting and simplifying yields

$$T_2 = 0.430\,\omega_2^2$$

Conservation of Energy

$$\{T_1\} + \{V_1\} = \{T_2\} + \{V_2\}$$
$$\{0\} + \{16\} = \{0.430\,\omega_2^2\} + \{9\}$$
$$\omega_2 = 4.04\text{ rad/s} \circlearrowleft \qquad \textit{Ans.}$$

EXAMPLE 18-9

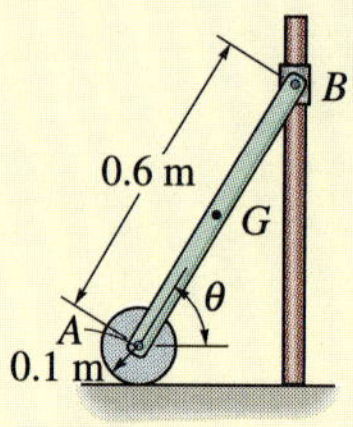

(a)

The 10-kg homogeneous disk shown in Fig. 18–21*a* is attached to a uniform 5-kg rod *AB*. If the assembly is released from rest when $\theta = 60°$, determine the angular velocity of the rod when $\theta = 0°$. Assume that the disk rolls without slipping. Neglect friction along the guide and the mass of the collar at *B*.

Solution

Potential Energy. Two diagrams for the rod and disk, when they are located at their initial and final positions, are shown in Fig. 18–21*b*. For convenience the datum passes through point *A*.

When the system is in position 1, the rod's weight has positive potential energy. Thus,

$$V_1 = W_R y_1 = 49.05\text{ N}(0.3 \sin 60° \text{ m}) = 12.74 \text{ J}$$

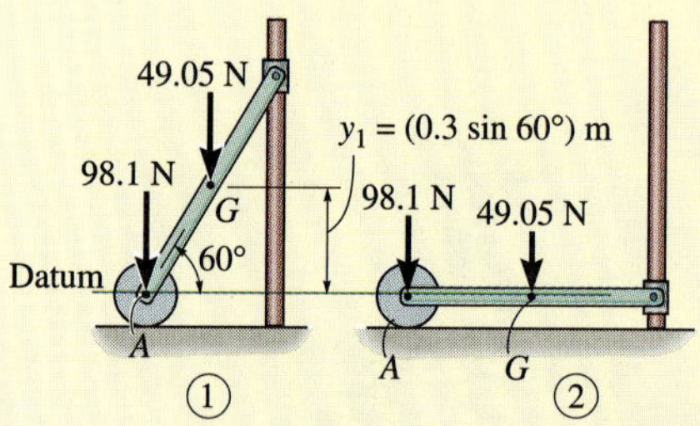

(b)

When the system is in position 2, both the weight of the rod and the weight of the disk have zero potential energy. Why? Thus,

$$V_2 = 0$$

Kinetic Energy. Since the entire system is at rest at the initial position,

$$T_1 = 0$$

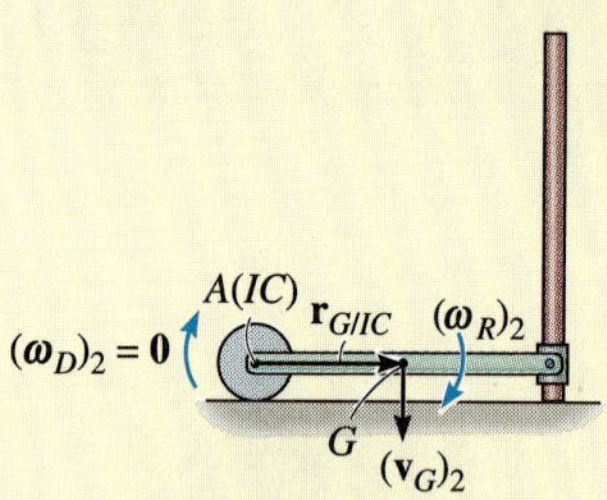

(c)

Fig. 18–21

In the final position the rod has an angular velocity $(\boldsymbol{\omega}_R)_2$ and its mass center has a velocity $(\mathbf{v}_G)_2$, Fig. 18–21*c*. Since the rod is *fully extended* in this position, the disk is momentarily at rest, so $(\boldsymbol{\omega}_D)_2 = \mathbf{0}$ and $(\mathbf{v}_A)_2 = \mathbf{0}$. For the rod $(\mathbf{v}_G)_2$ can be related to $(\boldsymbol{\omega}_R)_2$ from the instantaneous center of zero velocity, which is located at point *A*, Fig. 18–21*c*. Hence, $(v_G)_2 = r_{G/IC}(\omega_R)_2$ or $(v_G)_2 = 0.3(\omega_R)_2$. Thus,

$$T_2 = \frac{1}{2}m_R(v_G)_2^2 + \frac{1}{2}I_G(\omega_R)_2^2 + \frac{1}{2}m_D(v_A)_2^2 + \frac{1}{2}I_A(\omega_D)_2^2$$

$$= \frac{1}{2}(5\text{ kg})[0.3\text{ m}(\omega_R)_2]^2 + \frac{1}{2}\left[\frac{1}{12}(5\text{ kg})(0.6\text{ m})^2\right](\omega_R)_2^2 + 0 + 0$$

$$= 0.3(\omega_R)_2^2$$

Conservation of Energy

$$\{T_1\} + \{V_1\} = \{T_2\} + \{V_2\}$$

$$\{0\} + \{12.74\} = \{0.3(\omega_R)_2^2\} + \{0\}$$

$$(\omega_R)_2 = 6.52 \text{ rad/s} \circlearrowright \qquad \textit{Ans.}$$

Problems

18-35. Solve Prob. 18-13 using the conservation of energy equation.

***18-36.** Solve Prob. 18-24 using the conservation of energy equation.

18-37. Solve Prob. 18-30 using the conservation of energy equation.

18-38. Solve Prob. 18-33 using the conservation of energy equation.

18-39. Solve Prob. 18-34 using the conservation of energy equation.

***18-40.** Solve Prob. 18-28 using the conservation of energy equation.

18-41. The 15-lb disk is rotating about pin A in the vertical plane with an angular velocity $\omega = 2$ rad/s when $\theta = 0°$. Determine its angular velocity at the instant $\theta = 90°$. Also, compute the horizontal and vertical components of reaction at A at this instant.

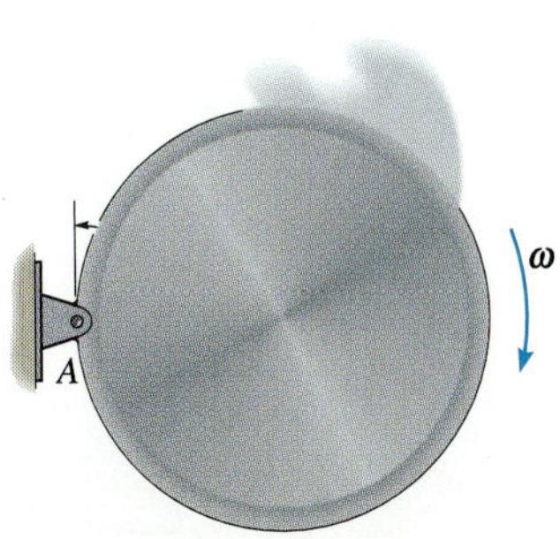

Prob. 18–41

18-42. The link is rotating about point O in the vertical plane with an angular velocity $\omega = 5$ rad/s when $\theta = 0°$. If it has a mass of 20 kg, a mass center at G, and a radius of gyration $k_G = 800$ mm, determine its angular velocity at the instant $\theta = 90°$.

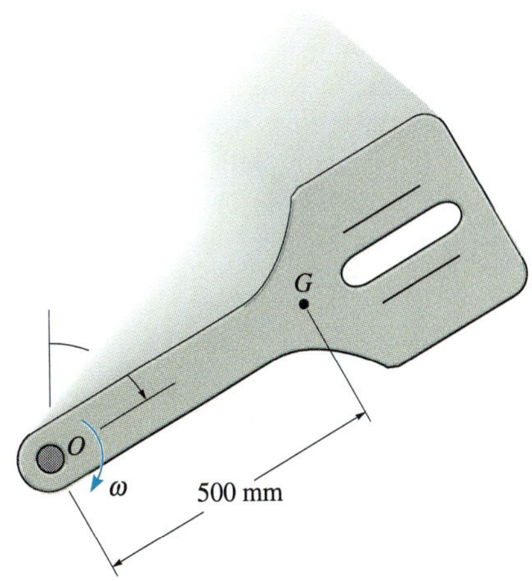

Prob. 18–42

18-43. The overhead door BC is pushed slightly from its open position and then rotates downward about the pin at A. Determine its angular velocity just before its end B strikes the floor. Assume the door is a thin plate having a mass of 180 kg and length of 6 m. Neglect the mass of the supporting frame AB and AC.

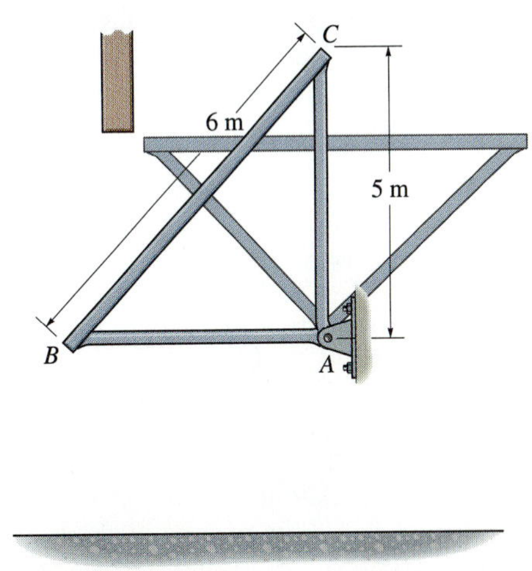

Prob. 18–43

***18-44.** The uniform garage door has a mass of 150 kg and is guided along tracks at its ends. Lifting is done using the two springs, each of which is attached to the anchor bracket at A and to the counterbalance shaft at B and C. As the door is raised, the springs begin to unwind from the shaft, thereby assisting the lift. If each spring provides a torsional moment of $M = (0.7\theta)$ N·m, where θ is in radians, determine the angle θ_0 at which both the left-wound and right-wound spring should be attached so that the door is completely balanced by the springs, i.e., when the door is in the vertical position and is given a slight force upwards, the springs will lift it along the side tracks to the horizontal plane with no final angular velocity. *Note:* The elastic potential energy of the torsional spring is $V_e = \frac{1}{2}k\theta^2$, where $M = k\theta$ and $k = 0.7$ N·m/rad.

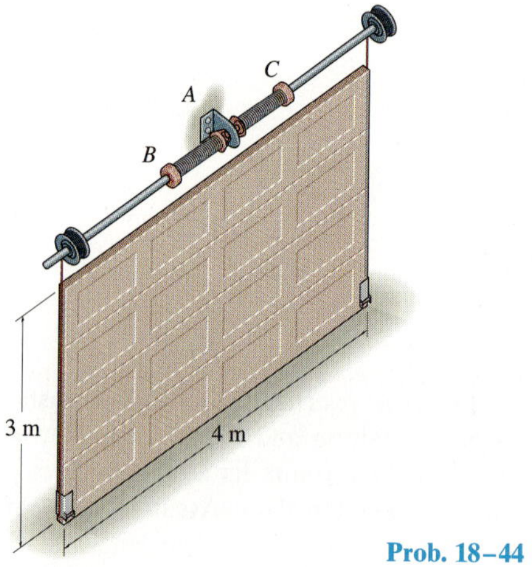

Prob. 18–44

18-45. The 80-lb cylinder is attached to the 10-lb slender rod which is pinned at point A. At the instant $\theta = 30°$ the rod has an angular velocity of $\omega_0 = 1$ rad/s as shown. Determine the angle θ to which the rod swings before it momentarily stops.

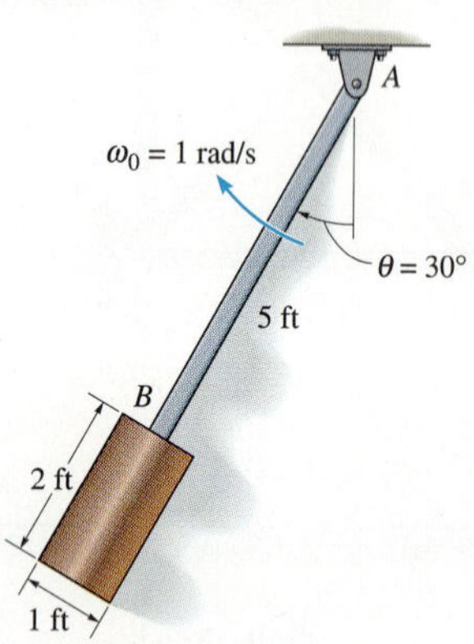

Prob. 18–45

18-46. An automobile tire has a mass of 7 kg and radius of gyration $k_G = 0.3$ m. If it is released from rest at A on the incline, determine its angular velocity when it reaches the horizontal plane. The tire rolls without slipping.

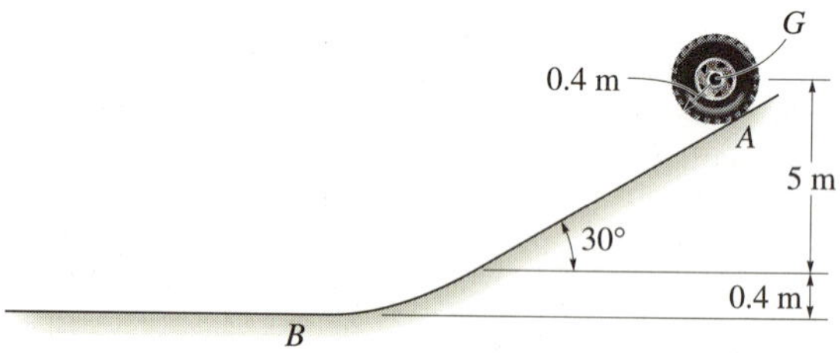

Prob. 18–46

18-47. The disk has a mass of 5 kg, a center of gravity at G, and a radius of gyration of $k_O = 80$ mm. A cord is placed over the rim of the disk, and the attached blocks A and B are released from rest when the disk is in the position shown. If the cord does not slip on the rim, determine the angular velocity of the disk when it has rotated 180°. Blocks A and B have a mass of 4 kg and 10 kg, respectively.

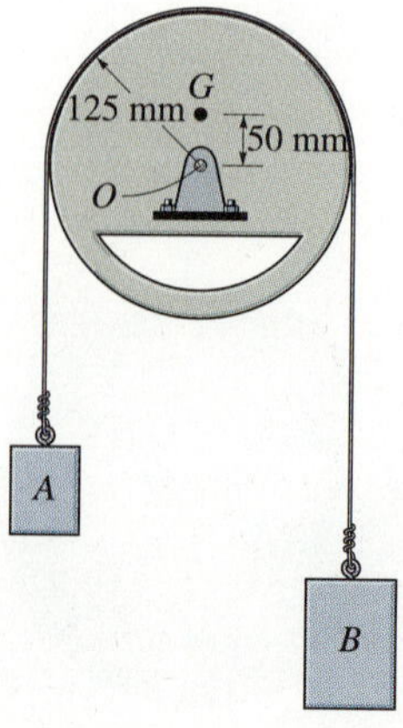

Prob. 18–47

*18-48. At the instant the spring becomes undeformed, the center of the 40-kg disk has a speed of 4 m/s. From this point determine the distance d the disk moves down the plane before momentarily stopping. The disk rolls without slipping.

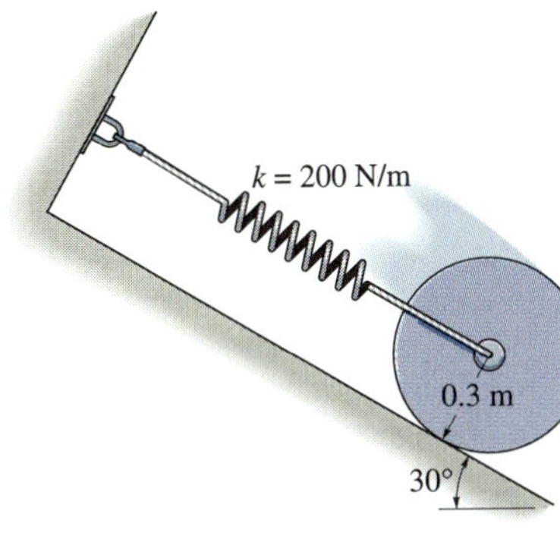

Prob. 18–48

18-49. The 15-kg semicircular segment is released from rest in the position shown. Determine the velocity of point A when it has rotated counterclockwise 90°. Assume that the segment rolls without slipping on the surface. The moment of inertia about its mass center is $I_G = 0.25\ \text{kg}\cdot\text{m}^2$.

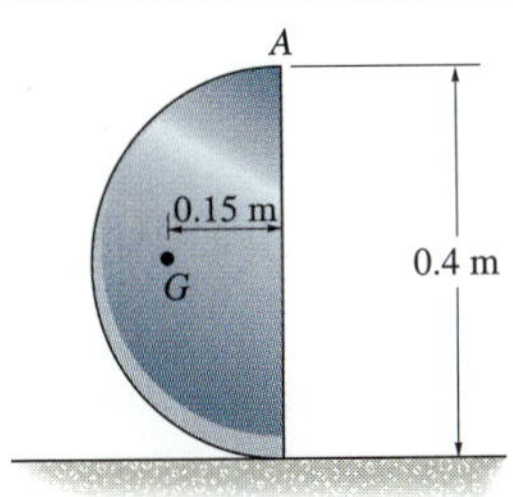

Prob. 18–49

18-50. The pendulum consists of a rod AB which weighs 10 lb and a disk which weighs 20 lb. The disk maintains a *constant* counterclockwise angular velocity of 2 rad/s *relative to the rod.* Determine the angular velocity of the rod when $\theta = 90°$ if it is released from rest when $\theta = 0°$.

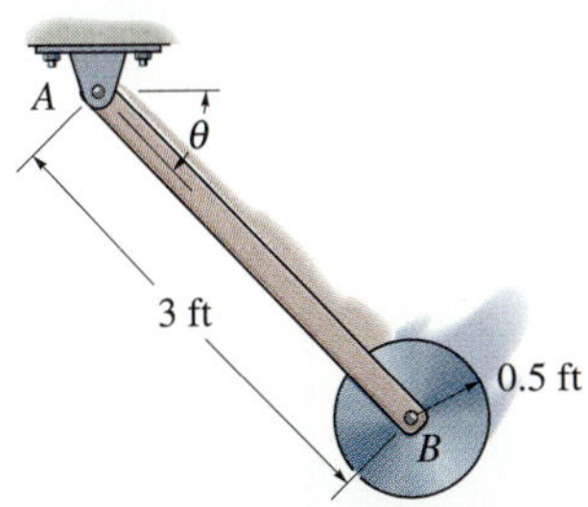

Prob. 18–50

18-51. The uniform 150-lb stone (rectangular block) is being turned over on its side by pulling the vertical cable *slowly* upward until the stone begins to tip. If it then falls freely $(\mathbf{T} = \mathbf{0})$ from an essentially balanced at-rest position, determine the speed at which the corner A strikes the pad at B. The stone does not slip at its corner C as it falls.

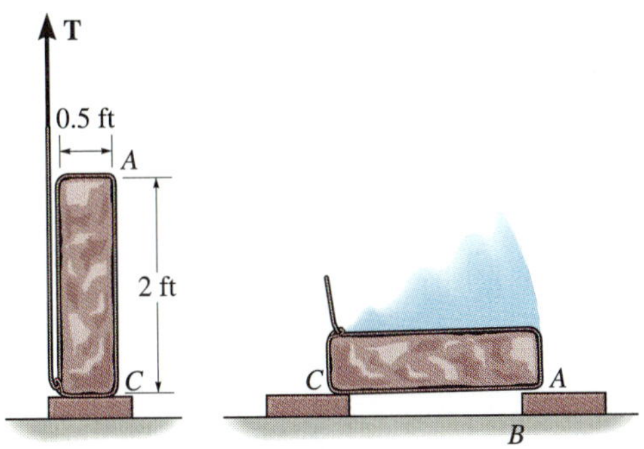

Prob. 18–51

*18-52. The system consists of a 20-lb disk A, 4-lb slender rod BC, and a 1-lb smooth collar C. If the disk rolls without slipping, determine the velocity of the collar at the instant the rod becomes horizontal, i.e., $\theta = 0°$. The system is released from rest when $\theta = 45°$.

18-53. The system consists of a 20-lb disk A, 4-lb slender rod BC, and a 1-lb smooth collar C. If the disk rolls without slipping, determine the velocity of the collar at the instant $\theta = 30°$. The system is released from rest when $\theta = 45°$.

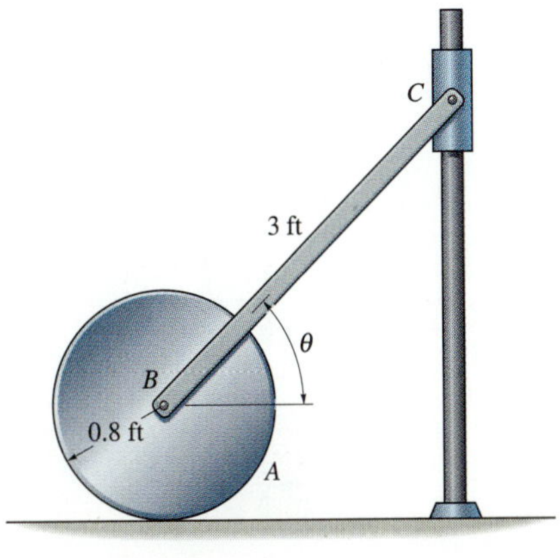

Probs. 18–52/53

18-54. A chain that has a negligible mass is draped over the sprocket which has a mass of 2 kg and a radius of gyration of $k_O = 50$ mm. If the 4-kg block A is released from rest in the position $s = 1$ m, determine the angular velocity of the sprocket at the instant $s = 2$ m.

18-55. Solve Prob. 18-54 if the chain has a mass of 0.8 kg/m. For the calculation neglect the portion of the chain that wraps over the sprocket.

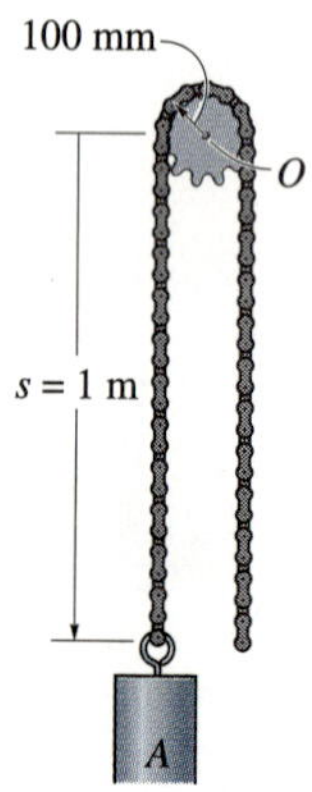

Probs. 18–54/55

***18-56.** Pulley A has a weight of 30 lb and a centroidal radius of gyration $k_B = 0.6$ ft. Determine the speed of the 20-lb crate C at the instant $s = 10$ ft. Initially, the crate is released from rest when $s = 5$ ft. The pulley at P "rolls" downward on the cord without slipping. For the calculation, neglect the mass of this pulley and the cord as it unwinds from the inner and outer hubs of pulley A.

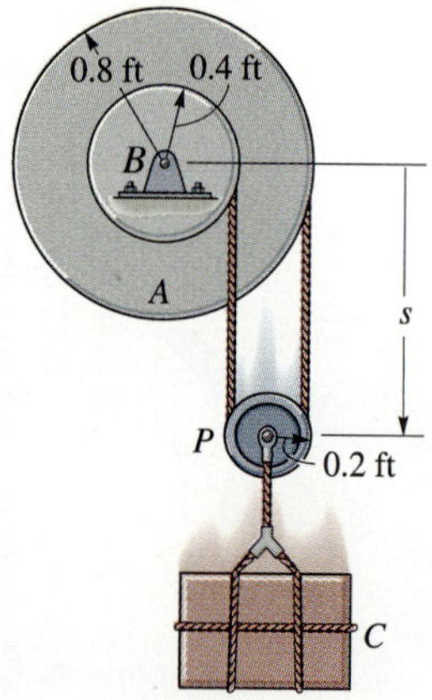

Prob. 18–56

18-57. At the instant shown, the 50-lb bar is rotating downwards at 2 rad/s. The spring attached to its end always remains vertical due to the roller guide at C. If the spring has an unstretched length of 2 ft and a stiffness of $k = 6$ lb/ft, determine the angular velocity of the bar the instant it has rotated downward 30° below the horizontal.

18-58. At the instant shown, the 50-lb bar is rotating downwards at 2 rad/s. The spring attached to its end always remains vertical due to the roller guide at C. If the spring has an unstretched length of 2 ft and a stiffness of $k = 12$ lb/ft, determine the angle θ, measured below the horizontal, to which the bar rotates before it momentarily stops.

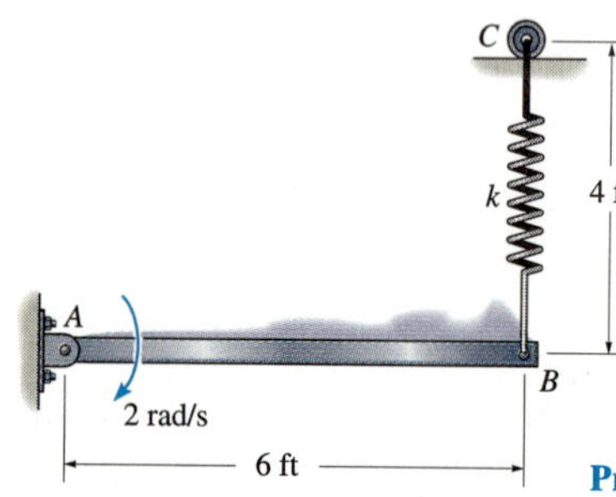

Probs. 18–57/58

18-59. The uniform window shade AB has a total weight of 0.4 lb. When it is released, it winds up around the spring-loaded core O. Motion is caused by a spring within the core, which is coiled so that it exerts a torque $M = 0.3(10^{-3})\theta$ lb·ft, where θ is in radians, on the core. If the shade is released from rest, determine the angular velocity of the core at the instant the shade is completely rolled up, i.e., after 12 revolutions. When this occurs, the spring becomes uncoiled and the radius of gyration of the shade about the axle at O is $k_O = 0.9$ in. *Note:* The elastic potential energy of the torsional spring is $V_e = \frac{1}{2}k\theta^2$, where $M = k\theta$ and $k = 0.3(10^{-3})$ lb·ft/rad.

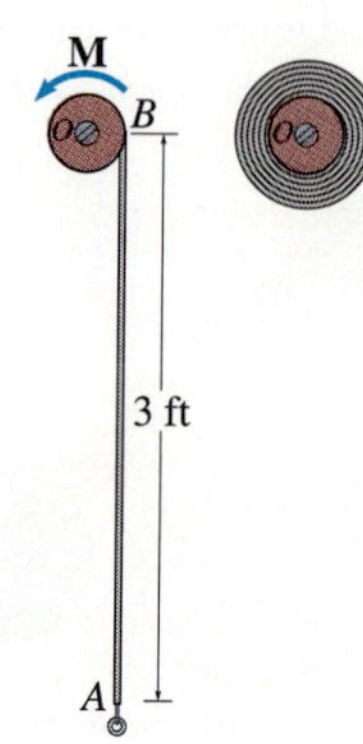

Prob. 18–59

*18-60. The garage door CD has a mass of 50 kg and can be treated as a thin plate. If it is released from rest from the horizontal position shown, determine its angular velocity at the instant the two side bars AC become horizontal. Each of the two side springs has a stiffness of $k = 30$ N/m and is originally unstretched. Neglect the mass of the side bars AC.

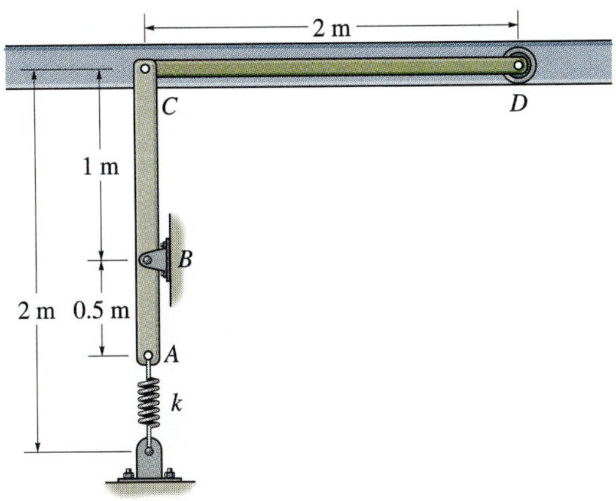

Prob. 18–60

18-61. The garage door CD has a mass of 50 kg and can be treated as a thin plate. Determine the required initial unstretched length of each of the two side springs when the door is open, so that when it falls freely it comes to rest when it reaches the fully closed position, i.e., when AC rotates 180°. Each of the two side springs has a stiffness of $k = 350$ N/m. Neglect the mass of the side bars AC.

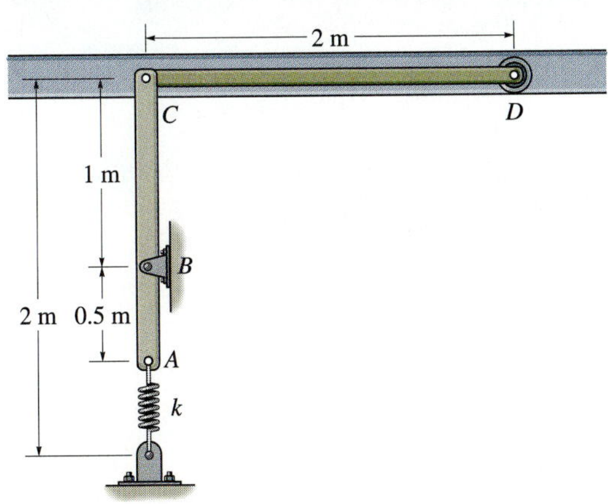

Prob. 18–61

18-62. The uniform 80-lb garage door is guided at its ends on the track. Determine the required initial stretch in the spring when the door is open, $\theta = 0°$, so that when it falls freely it comes to rest when it just reaches the fully closed position, $\theta = 90°$. Assume the door can be treated as a thin plate, and there is a spring and pulley system on each of the two sides of the door.

18-63. The uniform 80-lb garage door is guided at its ends on the track. If it is released from rest at $\theta = 0°$, determine the door's angular velocity at the instant $\theta = 30°$. The spring is originally stretched 1 ft when the door is held open, $\theta = 0°$. Assume the door can be treated as a thin plate, and there is a spring and pulley system on each of the two sides of the door.

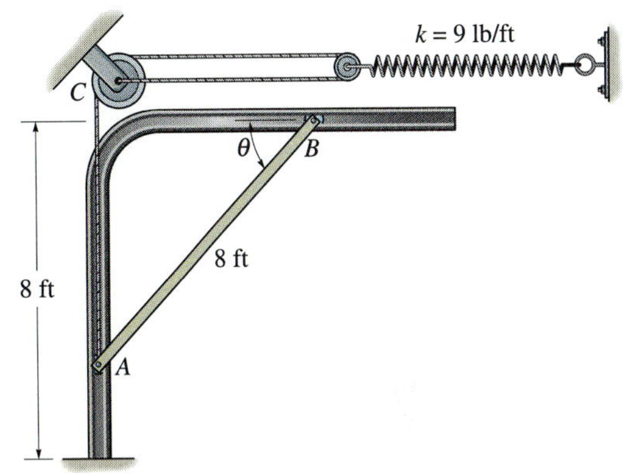

Probs. 18–62/63

*18-64. The 500-g rod AB rests along the smooth inner surface of a hemispherical bowl. If the rod is released from rest from the position shown, determine its angular velocity at the instant it swings downward and becomes horizontal.

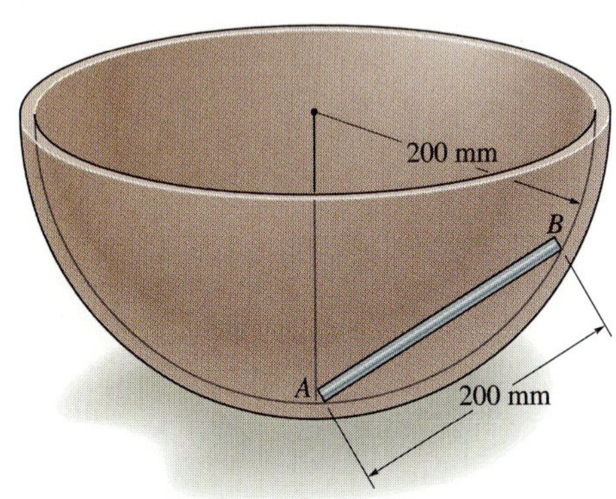

Prob. 18–64

The launching of this weather satellite requires application of impulse and momentum principles to accurately predict its orbital angular motion and proper orientation.

CHAPTER 19

Planar Kinetics of a Rigid Body: Impulse and Momentum

Chapter Objectives

- To develop formulations for the linear and angular momentum of a body.
- To apply the principles of linear and angular impulse and momentum to solve rigid-body planar kinetic problems that involve force, velocity, and time.
- To discuss application of the conservation of momentum.
- To analyze the mechanics of eccentric impact.

19.1 Linear and Angular Momentum

In this chapter we will use the principles of linear and angular impulse and momentum to solve problems involving force, velocity, and time as related to the planar motion of a rigid body. Before doing this, however, it will first be necessary to formalize the methods for obtaining the body's linear and angular momentum when the body is subjected to translation, rotation about a fixed axis, and general plane motion. Here we will assume the body is symmetric with respect to an inertial x–y reference plane.

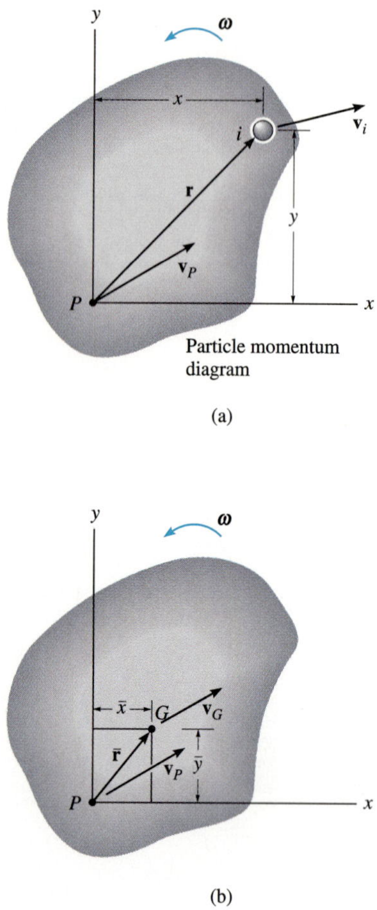

Fig. 19–1

Linear Momentum. The linear momentum of a rigid body is determined by summing vectorially the linear momenta of all the particles of the body, i.e., $\mathbf{L} = \Sigma m_i \mathbf{v}_i$. Since $\Sigma m_i \mathbf{v}_i = m\mathbf{v}_G$ (see Sec. 15.2) we can also write

$$\mathbf{L} = m\mathbf{v}_G \tag{19–1}$$

This equation states that the body's linear momentum is a vector quantity having a *magnitude* mv_G, which is commonly measured in units of kg·m/s or slug·ft/s, and a *direction* defined by $\mathbf{v}_G$, the velocity of the body's mass center.

Angular Momentum. Consider the body in Fig. 19–1*a*, which is subjected to general plane motion. At the instant shown, the arbitrary point *P* has a velocity $\mathbf{v}_P$, and the body has an angular velocity $\boldsymbol{\omega}$. If the velocity of the *i*th particle of the body is to be determined, Fig. 19–1*a*, then

$$\mathbf{v}_i = \mathbf{v}_P + \mathbf{v}_{i/P} = \mathbf{v}_P + \boldsymbol{\omega} \times \mathbf{r}$$

The angular momentum of particle *i* about point *P* is equal to the "moment" of the particle's linear momentum about *P*, Fig. 19–1*a*. Thus,

$$(\mathbf{H}_P)_i = \mathbf{r} \times m_i \mathbf{v}_i$$

Expressing $\mathbf{v}_i$ in terms of $\mathbf{v}_P$ and using Cartesian vectors, we have

$$(H_P)_i \mathbf{k} = m_i(x\mathbf{i} + y\mathbf{j}) \times [(v_P)_x\mathbf{i} + (v_P)_y\mathbf{j} + \omega\mathbf{k} \times (x\mathbf{i} + y\mathbf{j})]$$
$$(H_P)_i = -m_i\, y(v_P)_x + m_i\, x(v_P)_y + m_i\, \omega r^2$$

Letting $m_i \rightarrow dm$ and integrating over the entire mass *m* of the body, we obtain

$$H_P = -\left(\int_m y\, dm\right)(v_P)_x + \left(\int_m x\, dm\right)(v_P)_y + \left(\int_m r^2\, dm\right)\omega$$

Here H_P represents the angular momentum of the body about an axis (the *z* axis) perpendicular to the plane of motion and passing through point *P*. Since $\bar{y}m = \int y\, dm$ and $\bar{x}m = \int x\, dm$, the integrals for the first and second terms on the right are used to locate the body's center of mass *G* with respect to *P*, Fig. 19-1*b*. Also, the last integral represents the body's moment of inertia computed about the *z* axis, i.e., $I_P = \int r^2\, dm$. Thus,

$$H_P = -\bar{y}m(v_P)_x + \bar{x}m(v_P)_y + I_P\omega \tag{19–2}$$

This equation reduces to a simpler form if point *P* coincides with the mass center *G* for the body,* in which case $\bar{x} = \bar{y} = 0$. Hence,

*It *also* reduces to the same simple form, $H_P = I_P\omega$, if point *P* is a *fixed point* (see Eq. 19–9) or the velocity of point *P* is directed along the line *PG*.

$$H_G = I_G\omega \tag{19–3}$$

This equation states that the angular momentum of the body computed about G is equal to the product of the moment of inertia of the body about an axis passing through G and the body's angular velocity. It is important to realize that $\mathbf{H}_G$ is a vector quantity having a *magnitude* $I_G\omega$, which is commonly measured in units of $kg\cdot m^2/s$ or $slug\cdot ft^2/s$, and a *direction* defined by $\boldsymbol{\omega}$, which is always perpendicular to the plane of motion.

Equation 19–2 can also be rewritten in terms of the x and y components of the velocity of the body's mass center, $(v_G)_x$ and $(v_G)_y$, and the body's moment of inertia I_G. Since point G is located at coordinates $(\bar{x}, \bar{y})$, then by the parallel-axis theorem, $I_P = I_G + m(\bar{x}^2 + \bar{y}^2)$. Substituting into Eq. 19–2 and rearranging terms, we have

$$H_P = \bar{y}m[-(v_P)_x + \bar{y}\omega] + \bar{x}m[(v_P)_y + \bar{x}\omega] + I_G\omega \tag{19–4}$$

From the kinematic diagram of Fig. 19–1*b*, $\mathbf{v}_G$ can be expressed in terms of $\mathbf{v}_P$ as

$$\mathbf{v}_G = \mathbf{v}_P + \boldsymbol{\omega} \times \bar{\mathbf{r}}$$
$$(v_G)_x\mathbf{i} + (v_G)_y\mathbf{j} = (v_P)_x\mathbf{i} + (v_P)_y\mathbf{j} + \omega\mathbf{k} \times (\bar{x}\mathbf{i} + \bar{y}\mathbf{j})$$

Carrying out the cross product and equating the respective $\mathbf{i}$ and $\mathbf{j}$ components yields the two scalar equations

$$(v_G)_x = (v_P)_x - \bar{y}\omega$$
$$(v_G)_y = (v_P)_y + \bar{x}\omega$$

Substituting these results into Eq. 19–4 yields

$$H_P = -\bar{y}m(v_G)_x + \bar{x}m(v_G)_y + I_G\omega \tag{19–5}$$

With reference to Fig. 19–1*c*, *this result indicates that when the angular momentum of the body is computed about point P, it is equivalent to the moment of the linear momentum $m\mathbf{v}_G$ or its components $m(\mathbf{v}_G)_x$ and $m(\mathbf{v}_G)_y$ about P plus the angular momentum $I_G\boldsymbol{\omega}$.* Note that since $\boldsymbol{\omega}$ is a free vector, $\mathbf{H}_G$ *can act at any point on the body* provided it preserves its same magnitude and direction. Furthermore, since angular momentum is equal to the "moment" of the linear momentum, the *line of action of* $\mathbf{L}$ *must pass through the body's mass center G* in order to preserve the correct magnitude of $\mathbf{H}_P$ when "moments" are computed about *P*, Fig. 19–1*c*. As a result of this analysis, we will now consider three types of motion.

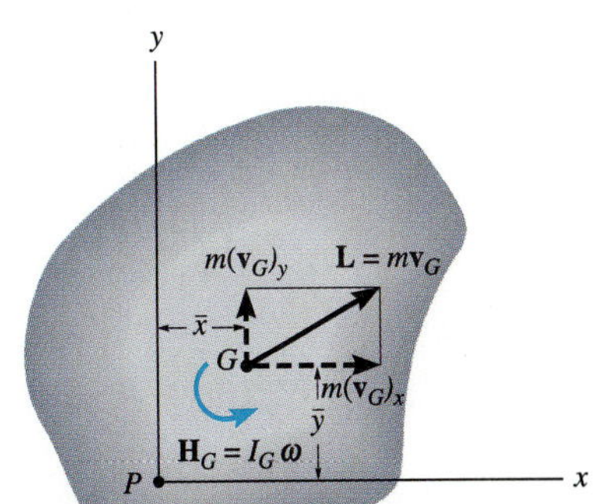

(c)

Fig. 19–1

Translation. When a rigid body of mass m is subjected to either rectilinear or curvilinear *translation,* Fig. 19–2*a*, its mass center has a velocity of $\mathbf{v}_G = \mathbf{v}$ and $\boldsymbol{\omega} = \mathbf{0}$ for the body. Hence, the linear momentum and the angular momentum computed about G become

$$L = mv_G \qquad H_G = 0 \tag{19–6}$$

If the angular momentum is computed about any other point A on or off the body, Fig. 19–2*a*, the "moment" of the linear momentum **L** must be computed about the point. Since d is the "moment arm" as shown in the figure, then in accordance with Eq. 19–5, $H_A = (d)(mv_G)\ \circlearrowleft$.

Rotation About a Fixed Axis. When a rigid body is *rotating about a fixed axis* passing through point O, Fig. 19–2*b*, the linear momentum and the angular momentum computed about G are

$$L = mv_G \qquad H_G = I_G\omega \tag{19–7}$$

It is sometimes convenient to compute the angular momentum of the body about point O. In this case it is necessary to account for the "moment" of *both* **L** and $\mathbf{H}_G$ about O. Noting that **L** (or $\mathbf{v}_G$) is always *perpendicular to* $\mathbf{r}_G$, we have

$$\circlearrowleft + \qquad H_O = I_G\omega + r_G(mv_G) \tag{19–8}$$

This equation may be *simplified* by first substituting $v_G = r_G\omega$, in which case $H_O = (I_G + mr_G^2)\omega$, and, by the parallel-axis theorem, noting that the terms inside the parentheses represent the moment of inertia I_O of the body about an axis perpendicular to the plane of motion and passing through point O. Hence,*

$$H_O = I_O\omega \tag{19–9}$$

For the calculation, then, either Eq. 19–8 or 19–9 can be used.

*The similarity between this derivation and that of Eq. 17–16 ($\Sigma M_O = I_O\alpha$) and Eq. 18–5 ($T = \frac{1}{2}I_O\omega^2$) should be noted. Also note that the same result can be obtained from Eq. 19–2 by selecting point P at O, realizing that $(v_O)_x = (v_O)_y = 0$.

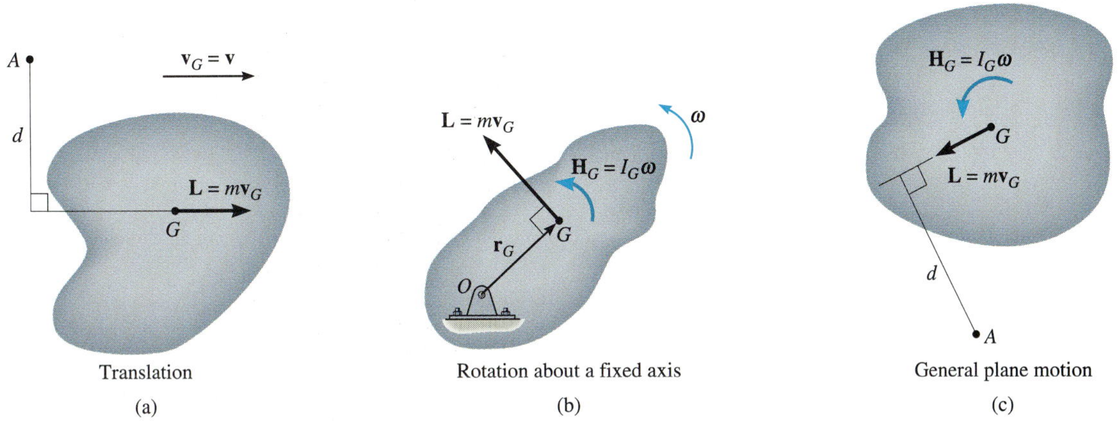

Fig. 19–2

General Plane Motion. When a rigid body is subjected to general plane motion, Fig. 19–2*c*, the linear momentum and the angular momentum computed about *G* become

$$L = mv_G$$
$$H_G = I_G\omega \qquad (19\text{–}10)$$

If the angular momentum is computed about a point *A* located either on or off the body, Fig. 19–2*c*, it is necessary to find the moments of *both* **L** and $\mathbf{H}_G$ about this point. In this case,

$$\circlearrowright + \qquad H_A = I_G\omega + (d)(mv_G)$$

Here *d* is the moment arm, as shown in the figure.

As this pendulum swings downward, its angular momentum about point *O* can be determined by computing the moment of $I_G\boldsymbol{\omega}$ and $m\mathbf{v}_G$ about *O*. This is $H_O = I_G\omega + (mv_G)d$. Since $v_G = \omega d$, then $H_O = I_G\omega + m(\omega d)d = (I_G + md^2)\omega = I_O\omega$.

EXAMPLE 19–1

At a given instant the 10-kg disk and 5-kg bar have the motions shown in Fig. 19–3*a*. Determine their angular momenta about point G and about point B for the disk and about G and about the IC for the bar at this instant.

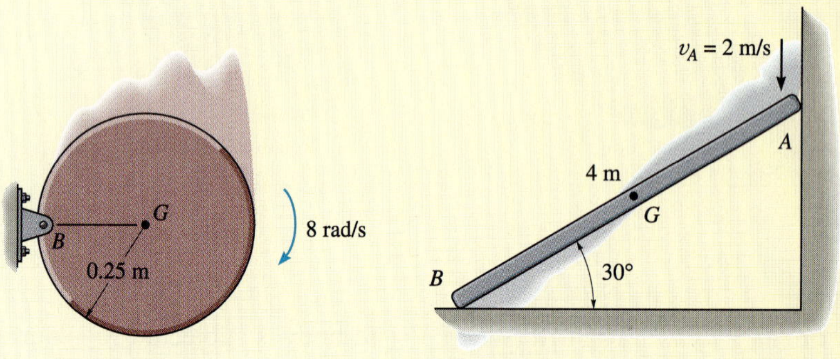

(a)

Solution

Disk. Since the disk is *rotating about a fixed axis* (through point B), then $v_G = (8\text{ rad/s})(0.25\text{ m}) = 2\text{ m/s}$, Fig. 19–3*b*. Hence

$$\circlearrowleft\!+ H_G = I_G\omega = [\tfrac{1}{2}(10\text{ kg})(0.25\text{ m})^2](8\text{ rad/s}) = 2.50\text{ kg}\cdot\text{m}^2/\text{s} \qquad \textit{Ans.}$$

$$\circlearrowleft\!+ H_B = I_G\omega + (mv_G)r_G = 2.50\text{ kg}\cdot\text{m}^2/\text{s} + (10\text{ kg})(2\text{ m/s})(0.25\text{ m})$$
$$= 7.50\text{ kg}\cdot\text{m}^2/\text{s} \circlearrowright \qquad \textit{Ans.}$$

Also, from the table on the inside back cover, $I_B = (3/2)mr^2$, so that

$$\circlearrowleft\!+ H_B = I_B\omega = [\tfrac{3}{2}(10\text{ kg})(0.25\text{ m})^2](8\text{ rad/s}) = 7.50\text{ kg}\cdot\text{m}^2/\text{s} \circlearrowright \qquad \textit{Ans.}$$

Bar. The bar undergoes *general plane motion.* The IC is established in Fig. 19–3*c*, so that $\omega = (2\text{ m/s})/(3.464\text{ m}) = 0.5774\text{ rad/s}$ and $v_G = (0.5774\text{ rad/s})(2\text{ m}) = 1.155\text{ m/s}$. Thus,

$$\circlearrowleft\!+ H_G = I_G\omega = [\tfrac{1}{12}(5\text{ kg})(4\text{ m})^2](0.5774\text{ rad/s}) = 3.85\text{ kg}\cdot\text{m}^2/\text{s} \circlearrowright \qquad \textit{Ans.}$$

Moments of $I_G\omega$ and mv_G about the IC yield

$$\circlearrowleft\!+ H_{IC} = I_G\omega + d(mv_G) = 3.85\text{ kg}\cdot\text{m}^2/\text{s} + (2\text{ m})(5\text{ kg})(1.155\text{ m/s})$$
$$= 15.4\text{ kg}\cdot\text{m}^2/\text{s} \circlearrowright \qquad \textit{Ans.}$$

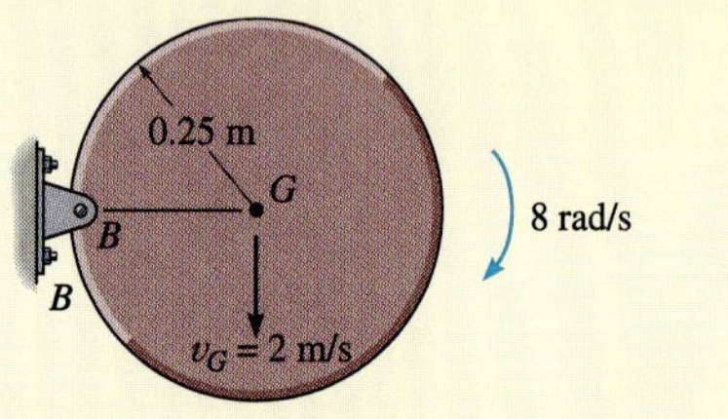

(b)

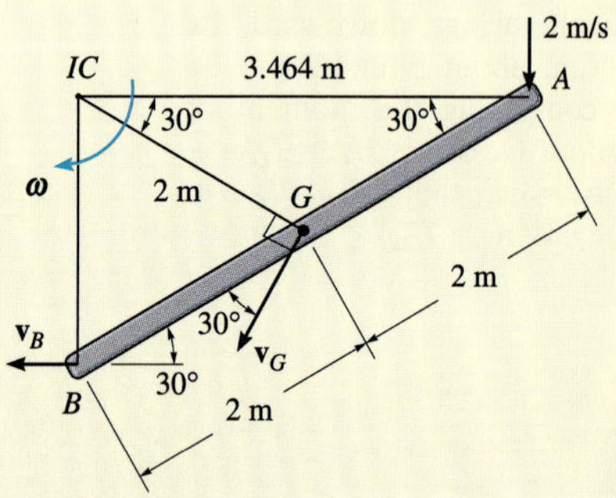

(c)

Fig. 19–3

19.2 Principle of Impulse and Momentum

Like the case for particle motion, the principle of impulse and momentum for a rigid body is developed by *combining* the equation of motion with kinematics. The resulting equation will allow a *direct solution to problems involving force, velocity, and time.*

Principle of Linear Impulse and Momentum. The equation of translational motion for a rigid body can be written as $\Sigma \mathbf{F} = m\mathbf{a}_G = m\,(d\mathbf{v}_G/dt)$. Since the mass of the body is constant,

$$\Sigma \mathbf{F} = \frac{d}{dt}(m\mathbf{v}_G)$$

Multiplying both sides by dt and integrating from $t = t_1$, $\mathbf{v}_G = (\mathbf{v}_G)_1$ to $t = t_2$, $\mathbf{v}_G = (\mathbf{v}_G)_2$ yields

$$\Sigma \int_{t_1}^{t_2} \mathbf{F}\,dt = m(\mathbf{v}_G)_2 - m(\mathbf{v}_G)_1 \qquad (19\text{–}11)$$

This equation is referred to as the *principle of linear impulse and momentum.* It states that the sum of all the impulses created by the *external force system* which acts on the body during the time interval t_1 to t_2 is equal to the change in the linear momentum of the body during the time interval, Fig. 19–4.

Principle of Angular Impulse and Momentum. If the body has *general plane motion* we can write $\Sigma M_G = I_G\alpha = I_G(d\omega/dt)$. Since the moment of inertia is constant,

$$\Sigma M_G = \frac{d}{dt}(I_G\omega)$$

Multiplying both sides by dt and integrating from $t = t_1$, $\omega = \omega_1$ to $t = t_2$, $\omega = \omega_2$ gives

$$\Sigma \int_{t_1}^{t_2} M_G\,dt = I_G\omega_2 - I_G\omega_1 \qquad (19\text{–}12)$$

In a similar manner, for *rotation about a fixed axis* passing through point O, Eq. 17–16 ($\Sigma M_O = I_O\alpha$) when integrated becomes

$$\Sigma \int_{t_1}^{t_2} M_O\,dt = I_O\omega_2 - I_O\omega_1 \qquad (19\text{–}13)$$

Equations 19–12 and 19–13 are referred to as the *principle of angular impulse and momentum.* Both equations state that the sum of the angular impulses acting on the body during the time interval t_1 to t_2 is equal to the change in the body's angular momentum during this time interval. In particular, the angular impulse considered is determined by integrating the moments about point G or O of all the external forces and couple moments applied to the body.

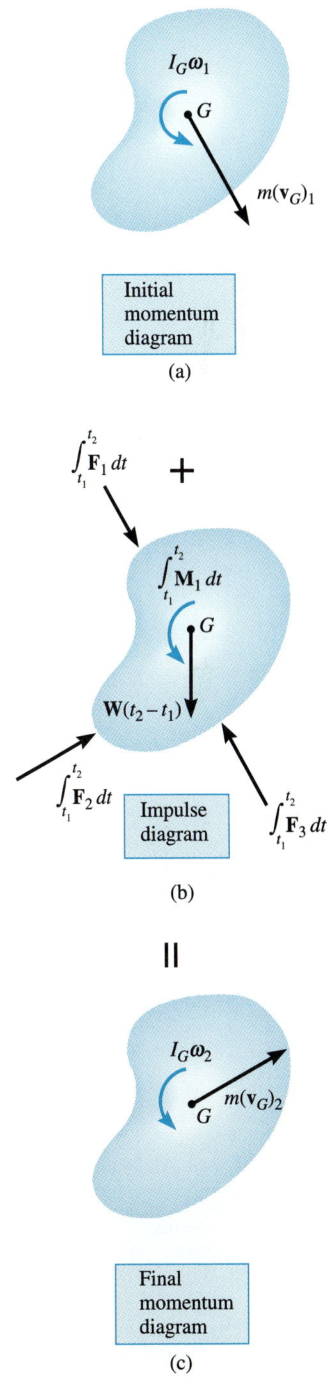

Fig. 19–4

To summarize the preceding concepts, if motion is occurring in the x–y plane, using impulse and momentum principles the following *three scalar equations* may be written which describe the *planar motion* of the body:

$$m(v_{Gx})_1 + \Sigma \int_{t_1}^{t_2} F_x\,dt = m(v_{Gx})_2$$
$$m(v_{Gy})_1 + \Sigma \int_{t_1}^{t_2} F_y\,dt = m(v_{Gy})_2 \qquad (19\text{–}14)$$
$$I_G\omega_1 + \Sigma \int_{t_1}^{t_2} M_G\,dt = I_G\omega_2$$

The first two of these equations represent the principle of linear impulse and momentum in the x–y plane, Eq. 19–11, and the third equation represents the principle of angular impulse and momentum about the z axis, which passes through the body's mass center G, Eq. 19–12.

The terms in Eqs. 19–14 can be graphically accounted for by drawing a set of impulse and momentum diagrams for the body, Fig. 19–4. Note that the linear momenta $m\mathbf{v}_G$ are applied at the body's mass center, Figs. 19–4*a* and 19–4*c*; whereas the angular momenta $I_G\boldsymbol{\omega}$ are free vectors, and therefore, like a couple moment, they may be applied at any point on the body. When the impulse diagram is constructed, Fig. 19–4*b*, the forces **F** and moment **M** vary with time, and are indicated by the integrals. However, if **F** and **M** are *constant* from t_1 to t_2, integration of the impulses yields $\mathbf{F}(t_2 - t_1)$ and $\mathbf{M}(t_2 - t_1)$, respectively. Such is the case for the body's weight **W**, Fig. 19–4*b*.

Equations 19–14 may also be applied to an entire system of connected bodies rather than to each body separately. Doing this eliminates the need to include reactive impulses which occur at the connections since they are *internal* to the system. The resultant equations may be written in symbolic form as

$$\left(\Sigma \begin{matrix}\text{syst. linear}\\ \text{momentum}\end{matrix}\right)_{x1} + \left(\Sigma \begin{matrix}\text{syst. linear}\\ \text{impulse}\end{matrix}\right)_{x(1-2)} = \left(\Sigma \begin{matrix}\text{syst. linear}\\ \text{momentum}\end{matrix}\right)_{x2}$$
$$\left(\Sigma \begin{matrix}\text{syst. linear}\\ \text{momentum}\end{matrix}\right)_{y1} + \left(\Sigma \begin{matrix}\text{syst. linear}\\ \text{impulse}\end{matrix}\right)_{y(1-2)} = \left(\Sigma \begin{matrix}\text{syst. linear}\\ \text{momentum}\end{matrix}\right)_{y2}$$
$$\left(\Sigma \begin{matrix}\text{syst. angular}\\ \text{momentum}\end{matrix}\right)_{O1} + \left(\Sigma \begin{matrix}\text{syst. angular}\\ \text{impulse}\end{matrix}\right)_{O(1-2)} = \left(\Sigma \begin{matrix}\text{syst. angular}\\ \text{momentum}\end{matrix}\right)_{O2}$$

(19–15)

As indicated, the system's angular momentum and angular impulse must be computed with respect to the *same fixed reference point O* for all the bodies of the system.

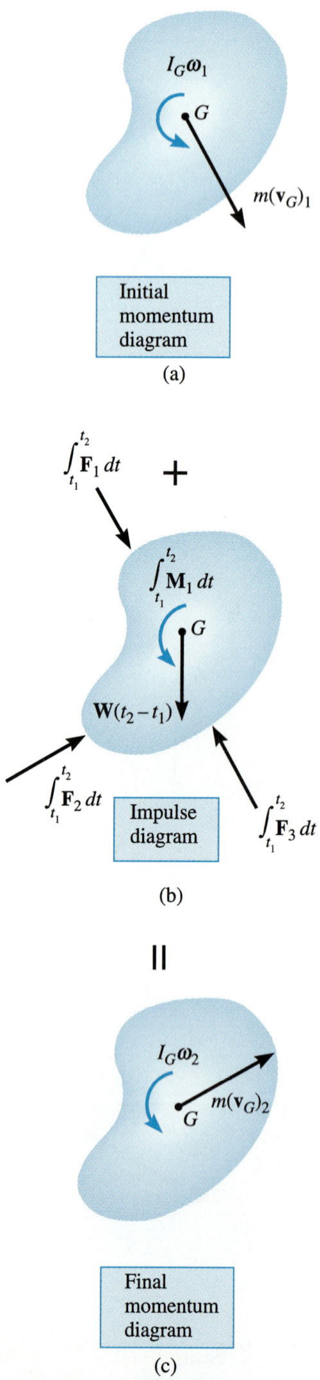

Fig. 19–4

Procedure for Analysis

Impulse and momentum principles are used to solve kinetic problems that involve *velocity, force,* and *time* since these terms are involved in the formulation.

Free-Body Diagram

- Establish the x, y, z inertial frame of reference and draw the free-body diagram in order to account for all the forces and couple moments that produce impulses on the body.
- The direction and sense of the initial and final velocity of the body's mass center, $\mathbf{v}_G$, and the body's angular velocity $\boldsymbol{\omega}$ should be established. If any of these motions is unknown, assume that the sense of its components is in the direction of the positive inertial coordinates.
- Compute the moment of inertia I_G or I_O.
- As an alternative procedure, draw the impulse and momentum diagrams for the body or system of bodies. Each of these diagrams represents an outlined shape of the body which graphically accounts for the data required for each of the three terms in Eqs. 19–14 or 19–15, Fig. 19–4. These diagrams are particularly helpful in order to visualize the "moment" terms used in the principle of angular impulse and momentum, if application is about a point other than the body's mass center G or a fixed point O.

Principle of Impulse and Momentum

- Apply the three scalar equations of impulse and momentum.
- The angular momentum of a rigid body rotating about a fixed axis is the moment of $m\mathbf{v}_G$ plus $I_G\boldsymbol{\omega}$ about the axis. This can be shown to be equal to $H_O = I_O\boldsymbol{\omega}$, where I_O is the moment of inertia of the body about the axis.
- All the forces acting on the body's free-body diagram will create an impulse; however, some of these forces will do no work.
- Forces that are functions of time must be integrated to obtain the impulse. Graphically, the impulse is equal to the area under the force–time curve.
- The principle of angular impulse and momentum is often used to eliminate unknown impulsive forces that are parallel or pass through a common axis, since the moment of these forces is zero about this axis.

Kinematics

- If more than three equations are needed for a complete solution, it may be possible to relate the velocity of the body's mass center to the body's angular velocity using *kinematics.* If the motion appears to be complicated, kinematic (velocity) diagrams may be helpful in obtaining the necessary relation.

EXAMPLE 19–2

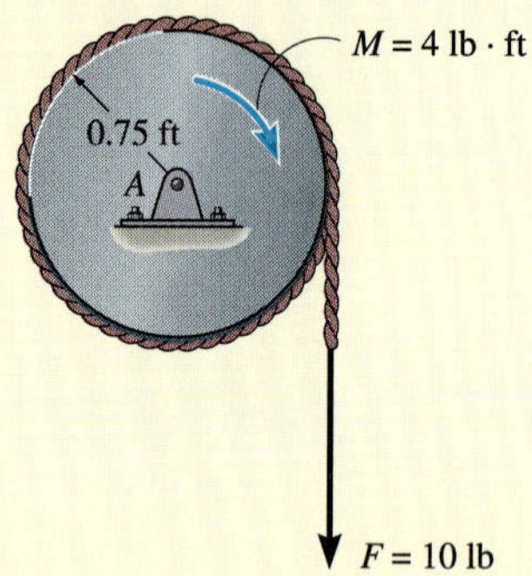

(a)

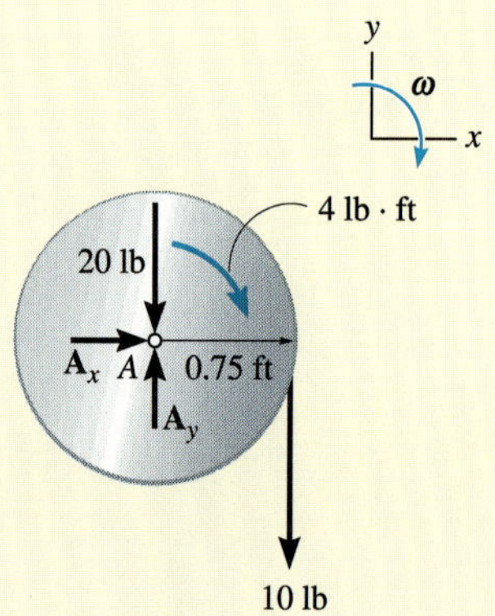

(b)

Fig. 19–5

The 20-lb disk shown in Fig. 19–5*a* is assumed to be uniform and is pin supported at its center. If it is acted upon by a constant couple moment of 4 lb·ft and a force of 10 lb which is applied to a cord wrapped around its periphery, determine the angular velocity of the disk two seconds after starting from rest. Also, what are the force components of reaction at the pin?

Solution

Since angular velocity, force, and time are involved in the problems we will apply the principles of impulse and momentum to the solution.

Free-Body Diagram Fig. 19–5*b*. The disk's mass center does not move; however, the loading causes the disk to rotate clockwise.

The moment of inertia of the disk about its fixed axis of rotation is

$$I_A = \frac{1}{2}mr^2 = \frac{1}{2}\left(\frac{20\text{ lb}}{32.2\text{ ft/s}^2}\right)(0.75\text{ ft})^2 = 0.175\text{ slug}\cdot\text{ft}^2$$

Principle of Impulse and Momentum

$$(\overset{+}{\rightarrow}) \qquad m(v_{Ax})_1 + \Sigma\int_{t_1}^{t_2} F_x\,dt = m(v_{Ax})_2$$

$$0 + A_x(2\text{ s}) = 0$$

$$(+\uparrow) \qquad m(v_{Ay})_1 + \Sigma\int_{t_1}^{t_2} F_y\,dt = m(v_{Ay})_2$$

$$0 + A_y(2\text{ s}) - 20\text{ lb}(2\text{ s}) - 10\text{ lb}(2\text{ s}) = 0$$

$$(\curvearrowright +) \qquad I_A\omega_1 + \Sigma\int_{t_1}^{t_2} M_A\,dt = I_A\omega_2$$

$$0 + 4\text{ lb}\cdot\text{ft}(2\text{ s}) + [10\text{ lb}(2\text{ s})](0.75\text{ ft}) = 0.175\omega_2$$

Solving these equations yields

$$A_x = 0 \qquad \textit{Ans.}$$

$$A_y = 30\text{ lb} \qquad \textit{Ans.}$$

$$\omega_2 = 132\text{ rad/s} \curvearrowright \qquad \textit{Ans.}$$

EXAMPLE 19–3

The 100-kg spool shown in Fig. 19–6*a* has a radius of gyration $k_G = 0.35$ m. A cable is wrapped around the central hub of the spool, and a horizontal force having a variable magnitude of $P = (t + 10)$ N is applied, where t is in seconds. If the spool is initially at rest, determine its angular velocity in 5 s. Assume that the spool rolls without slipping at A.

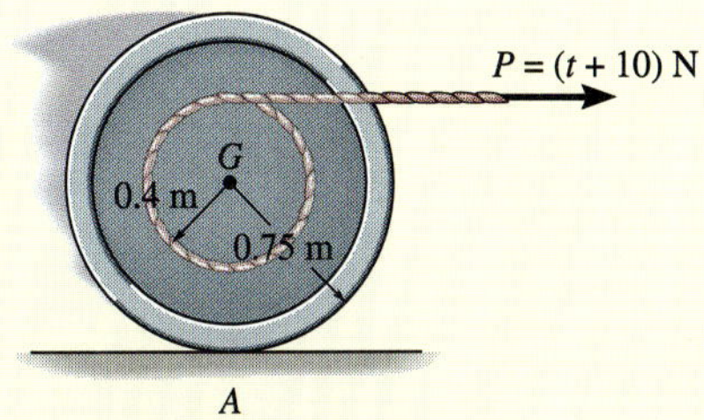

(a)

Solution

Free-Body Diagram. By inspection of the free-body diagram, Fig. 19–6*b*, the *variable* force P will cause the friction force F_A to be variable, and thus the impulses created by both P and F_A must be determined by integration. The force **P** causes the mass center to have a velocity $\mathbf{v}_G$ to the right, and the spool has a clockwise angular velocity $\boldsymbol{\omega}$.

The moment of inertia of the spool about its mass center is

$$I_G = mk_G^2 = 100\text{ kg}(0.35\text{ m})^2 = 12.25\text{ kg}\cdot\text{m}^2$$

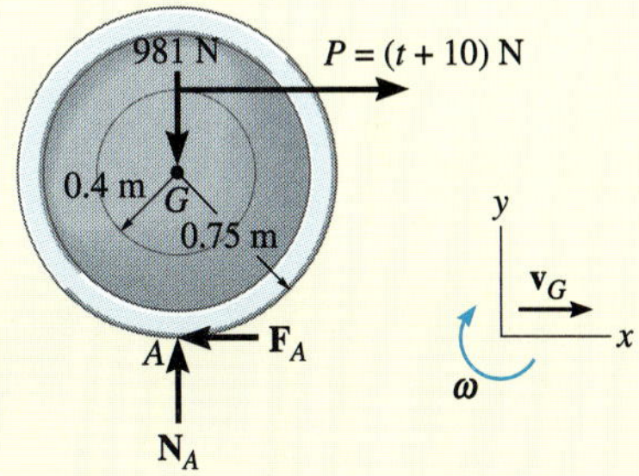

(b)

Principle of Impulse and Momentum

$$(\overset{+}{\rightarrow}) \qquad m(v_G)_1 + \Sigma\int F_x\,dt = m(v_G)_2$$

$$0 + \int_0^{5\text{ s}} (t + 10)\text{ N}\,dt - \int F_A\,dt = 100\text{ kg}(v_G)_2$$

$$62.5 - \int F_A\,dt = 100(v_G)_2 \qquad (1)$$

$$(\curvearrowleft +) \qquad I_G\omega_1 + \Sigma\int M_G\,dt = I_G\omega_2$$

$$0 + \left[\int_0^{5\text{ s}} (t + 10)\text{ N}\,dt\right](0.4\text{ m}) + \left(\int F_A\,dt\right)(0.75\text{ m}) = (12.25\text{ kg}\cdot\text{m}^2)\,\omega_2$$

$$25 + \left(\int F_A\,dt\right)(0.75) = 12.25\,\omega_2 \qquad (2)$$

Kinematics. Since the spool does not slip, the instantaneous center of zero velocity is at point A, Fig. 19–6*c*. Hence, the velocity of G can be expressed in terms of the spool's angular velocity as $(v_G)_2 = (0.75\text{ m})\omega_2$. Substituting this into Eq. 1 and eliminating the unknown impulse $\int F_A\,dt$ between Eqs. 1 and 2, we obtain

$$\omega_2 = 1.05\text{ rad/s} \curvearrowright \qquad \textit{Ans.}$$

Note: A more direct solution can be obtained by applying the principle of angular impulse and momentum about point A. As an exercise, do this and show that one obtains the same result.

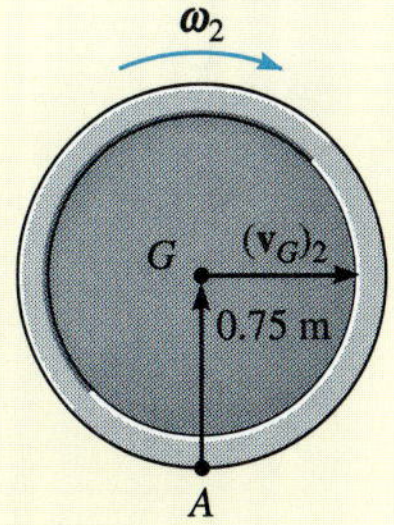

(c)

Fig. 19–6

EXAMPLE 19–4

The block shown in Fig. 19–7*a* has a mass of 6 kg. It is attached to a cord which is wrapped around the periphery of a 20-kg disk that has a moment of inertia $I_A = 0.40\ \text{kg}\cdot\text{m}^2$. If the block is initially moving downward with a speed of 2 m/s, determine its speed in 3 s. Neglect the mass of the cord in the calculation.

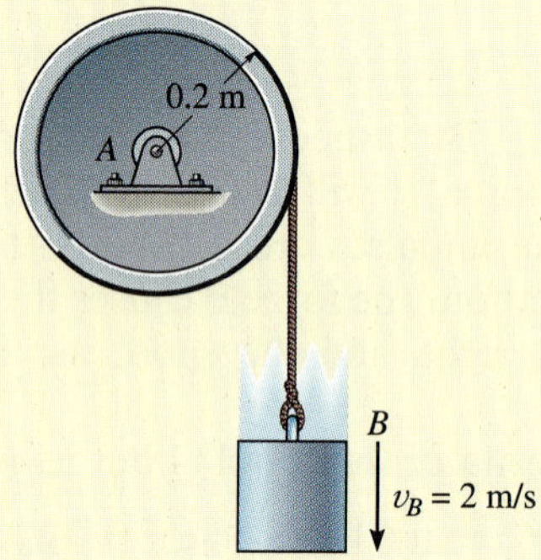

(a)

Solution I

Free-Body Diagram. The free-body diagrams of the block and disk are shown in Fig. 19–7*b*. All the forces are *constant* since the weight of the block causes the motion. The downward motion of the block, $\mathbf{v}_B$, causes $\boldsymbol{\omega}$ of the disk to be clockwise.

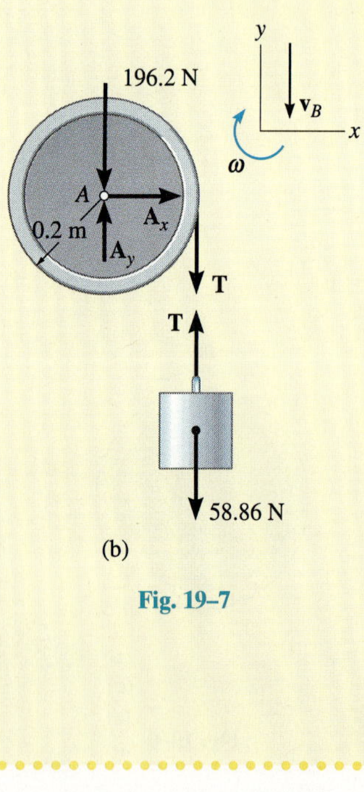

(b)

Fig. 19–7

Principle of Impulse and Momentum. We can eliminate $\mathbf{A}_x$ and $\mathbf{A}_y$ from the analysis by applying the principle of angular impulse and momentum about point *A*. Hence

Disk

$(\curvearrowright +)$
$$I_A\omega_1 + \Sigma \int M_A\, dt = I_A\omega_2$$

$$0.40\ \text{kg}\cdot\text{m}^2(\omega_1) + T(3\ \text{s})(0.2\ \text{m}) = (0.4\ \text{kg}\cdot\text{m}^2)\omega_2$$

Block

$(+\uparrow)$
$$m_B(v_B)_1 + \Sigma \int F_y\, dt = m_B(v_B)_2$$

$$-6\ \text{kg}(2\ \text{m/s}) + T(3\ \text{s}) - 58.86\ \text{N}(3\ \text{s}) = -6\ \text{kg}(v_B)_2$$

Kinematics. Since $\omega = v_B/r$, then $\omega_1 = (2\ \text{m/s})/(0.2\ \text{m}) = 10\ \text{rad/s}$ and $\omega_2 = (v_B)_2/0.2\ \text{m} = 5(v_B)_2$. Substituting and solving the equations simultaneously for $(v_B)_2$ yields

$$(v_B)_2 = 13.0\ \text{m/s} \downarrow \qquad \textit{Ans.}$$

Solution II

Impulse and Momentum Diagrams. We can obtain $(v_B)_2$ *directly* by considering the *system* consisting of the block, the cord, and the disk. The impulse and momentum diagrams have been drawn to clarify application of the principle of angular impulse and momentum about point A, Fig. 19–7*c*.

Principle of Angular Impulse and Momentum. Realizing that $\omega_1 =$ 10 rad/s and $\omega_2 = 5(v_B)_2$, we have

$$(\circlearrowright+)\quad \left(\Sigma \begin{matrix}\text{syst. angular}\\ \text{momentum}\end{matrix}\right)_{A1} + \left(\Sigma \begin{matrix}\text{syst. angular}\\ \text{impulse}\end{matrix}\right)_{A(1-2)} = \left(\Sigma \begin{matrix}\text{syst. angular}\\ \text{momentum}\end{matrix}\right)_{A2}$$

$$6\text{ kg}(2\text{ m/s})(0.2\text{ m}) + 0.4\text{ kg}\cdot\text{m}^2(10\text{ rad/s}) + 58.86\text{ N}(3\text{ s})(0.2\text{ m})$$
$$= 6\text{ kg}(v_B)_2(0.2\text{ m}) + 0.40\text{ kg}\cdot\text{m}^2[5(v_B)_2(0.2\text{ m})]$$
$$(v_B)_2 = 13.0\text{ m/s}\downarrow \qquad \textit{Ans.}$$

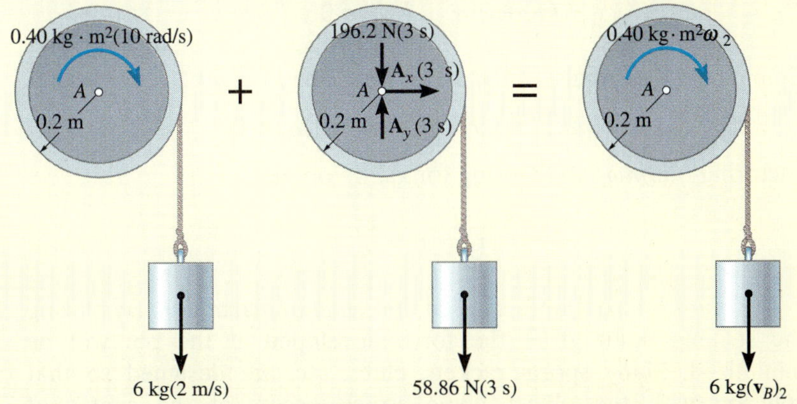

(c)

Fig. 19–7

EXAMPLE 19–5

The Charpy impact test is used in materials testing to determine the energy absorption characteristics of a material during impact. The test is performed using the pendulum shown in Fig. 19–8*a*, which has a mass m, mass center at G, and a radius of gyration k_G about G. Determine the distance r_P from the pin at A to the point P where the impact with the specimen S should occur so that the horizontal force at the pin is essentially zero during the impact. For the calculation, assume the specimen absorbs all the pendulum's kinetic energy gained during the time it falls and thereby stops the pendulum from swinging when $\theta = 0°$.

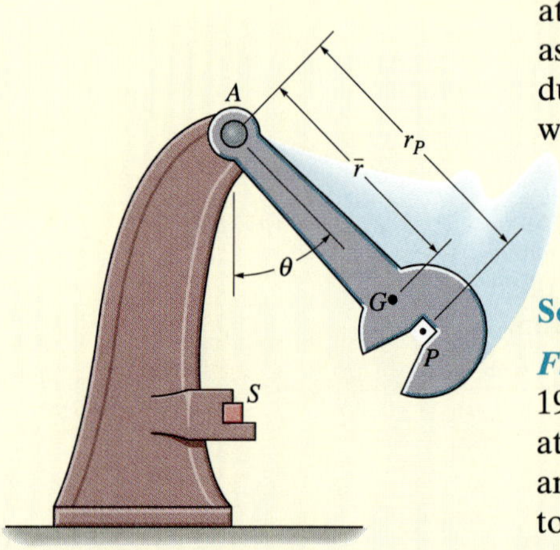

(a)

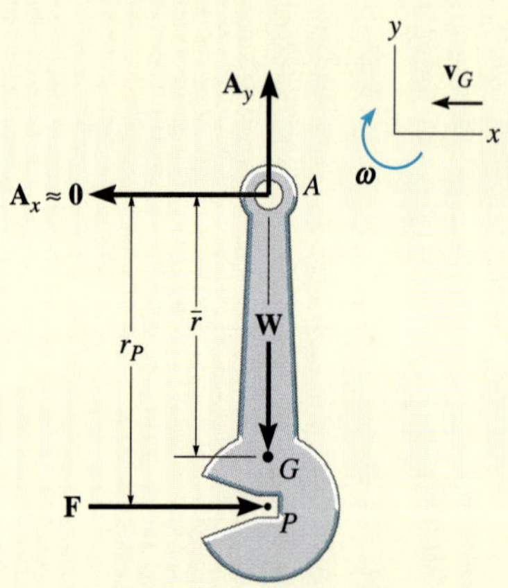

(b)

Fig. 19–8

Solution

Free-Body Diagram. As shown on the free-body diagram, Fig. 19–8*b*, the conditions of the problem require the horizontal impulse at A to be zero. Just before impact, the pendulum has a clockwise angular velocity $\boldsymbol{\omega}_1$, and the mass center of the pendulum is moving to the left at $(v_G)_1 = \bar{r}\omega_1$.

Principle of Impulse and Momentum. We will apply the principle of angular impulse and momentum about point A. Thus,

$$(\circlearrowright+) \qquad I_A\omega_1 + \Sigma M_A\,dt = I_A\omega_2$$

$$I_A\omega_1 - \left(\int F\,dt\right)r_P = 0$$

$$(\xrightarrow{+}) \qquad m(v_G)_1 + \Sigma F\,dt = m(v_G)_2$$

$$-m(\bar{r}\omega_1) + \int F\,dt = 0$$

Eliminating the impulse $\int F\,dt$ and substituting $I_A = mk_G^2 + m\bar{r}^2$ yields

$$[mk_G^2 + m\bar{r}^2]\omega_1 - m(\bar{r}\omega_1)r_P = 0$$

Factoring out $m\omega_1$ and solving for r_P, we obtain

$$r_P = \bar{r} + \frac{k_G^2}{\bar{r}} \qquad \textit{Ans.}$$

The point P, so defined, is called the *center of percussion*. By placing the striking point at P, the force developed at the pin will be minimized. Many sports rackets, clubs, etc. are designed so that collision with the object being struck occurs at the center of percussion. As a consequence, no "sting" or little sensation occurs in the hand of the player. (Also see Probs. 17–65 and 19–1.)

Problems

19-1. The rigid body (slab) has a mass m and is rotating with an angular velocity $\boldsymbol{\omega}$ about an axis passing through the fixed point O. Show that the momenta of all the particles composing the body can be represented by a single vector having a magnitude mv_G and acting through point P, called the *center of percussion,* which lies at a distance $r_{P/G} = k_G^2/r_{G/O}$ from the mass center G. Here k_G is the radius of gyration of the body, computed about an axis perpendicular to the plane of motion and passing through G.

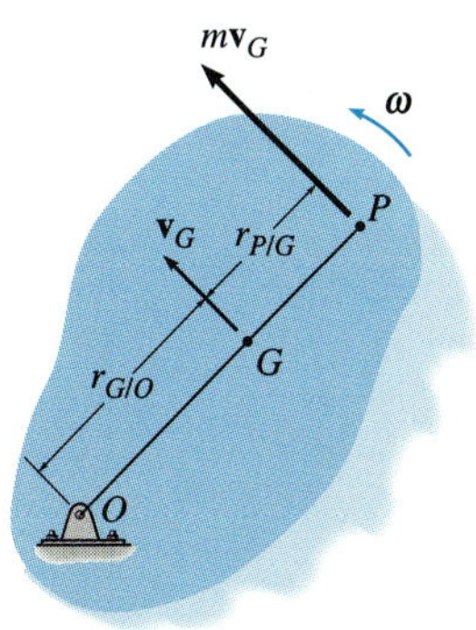

Prob. 19–1

19-2. At a given instant, the body has a linear momentum $\mathbf{L} = m\mathbf{v}_G$ and an angular momentum $\mathbf{H}_G = I_G\boldsymbol{\omega}$ computed about its mass center. Show that the angular momentum of the body computed about the instantaneous center of zero velocity IC can be expressed as $\mathbf{H}_{IC} = I_{IC}\boldsymbol{\omega}$, where I_{IC} represents the body's moment of inertia computed about the instantaneous axis of zero velocity. As shown, the IC is located at a distance $r_{G/IC}$ away from the mass center G.

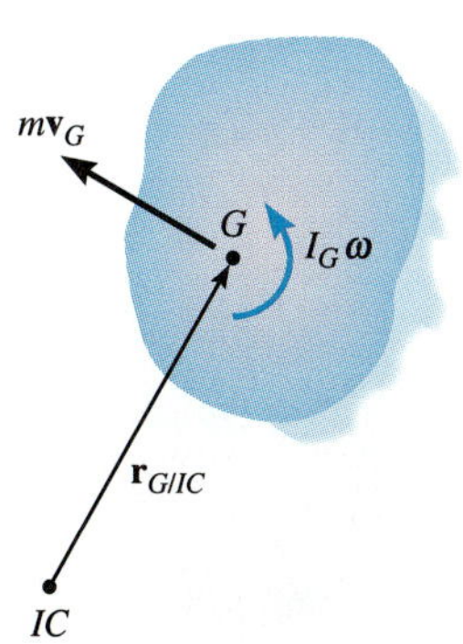

Prob. 19–2

19-3. Show that if a slab is rotating about a fixed axis perpendicular to the slab and passing through its mass center G, the angular momentum is the same when computed about any other point P on the slab.

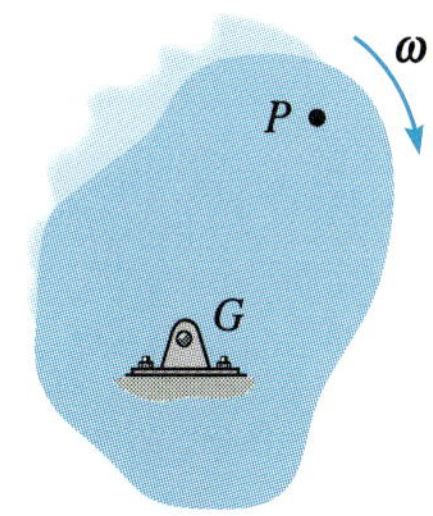

Prob. 19–3

***19-4.** Gear A rotates along the inside of the circular gear rack R. If A has a weight of 4 lb and a radius of gyration of $k_B = 0.5$ ft, determine its angular momentum about point C when $\omega_{CB} = 30$ rad/s and (a) $\omega_R = 0$, (b) $\omega_R = 20$ rad/s.

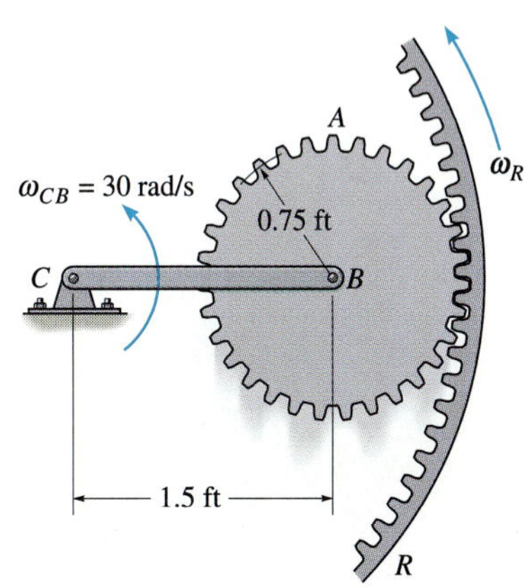

Prob. 19–4

19-5. Solve Prob. 17-55 using the principle of impulse and momentum.

19-6. Solve Prob. 17-59 using the principle of impulse and momentum.

19-7. Solve Prob. 17-69 using the principle of impulse and momentum.

***19-8.** Solve Prob. 17-80 using the principle of impulse and momentum.

19-9. Solve Prob. 17-73 using the principle of impulse and momentum.

19-10. A flywheel has a mass of 60 kg and a radius of gyration of $k_G = 150$ mm about an axis of rotation passing through its mass center. If a motor supplies a clockwise torque having a magnitude of $M = (5t)$ N·m, where t is in seconds, determine the flywheel's angular velocity in $t = 3$ s. Initially the flywheel is rotating clockwise at $\omega_1 = 2$ rad/s.

19-11. The pilot of a crippled F-15 fighter was able to control his plane by throttling the two engines. If the plane has a weight of 17 000 lb and a radius of gyration of $k_G = 4.7$ ft about the mass center G, determine the angular velocity of the plane and the velocity of its mass center G in $t = 5$ s if the thrust in each engine is altered to $T_1 = 5000$ lb and $T_2 = 800$ lb as shown. Originally the plane is flying straight at 1200 ft/s. Neglect the effects of drag and the loss of fuel.

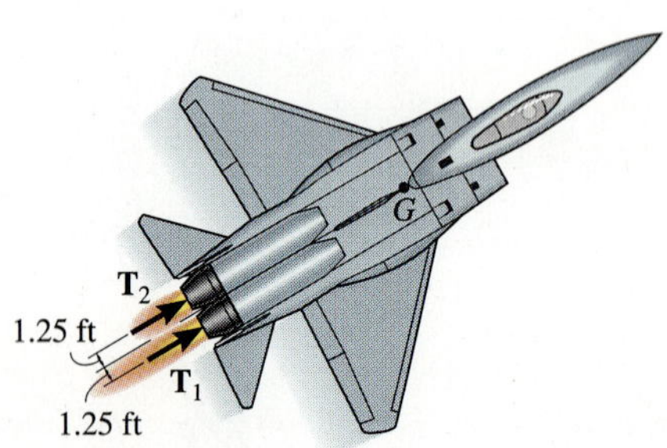

Prob. 19–11

***19-12.** The disk has a weight of 10 lb and is pinned at its center O. If a vertical force of $P = 2$ lb is applied to the cord wrapped around its outer rim, determine the angular velocity of the disk in four seconds starting from rest. Neglect the mass of the cord.

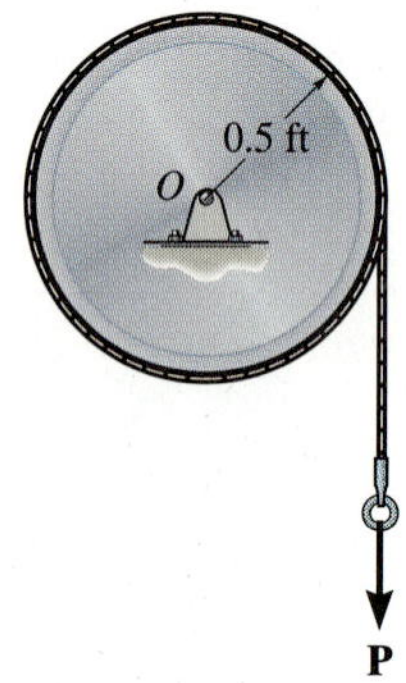

Prob. 19–12

19-13. A wire of negligible mass is wrapped around the outer surface of the 2-kg disk. If the disk is released from rest, determine its angular velocity in 3 s.

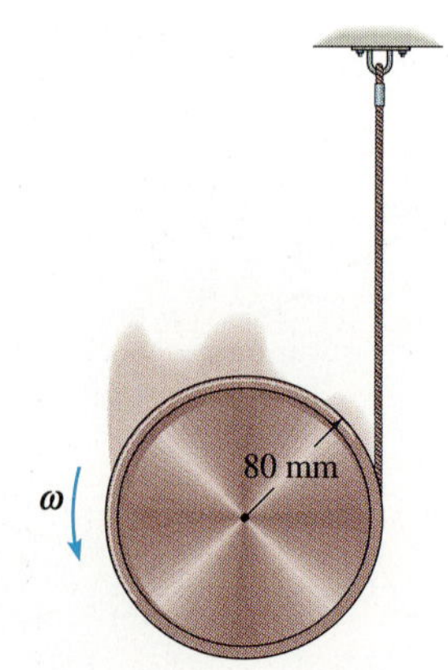

Prob. 19–13

19-14. The 4-kg slender rod rests on a smooth floor. If it is kicked so as to receive a horizontal impulse $I = 8\ \text{N}\cdot\text{s}$ at point A as shown, determine its angular velocity and the speed of its mass center.

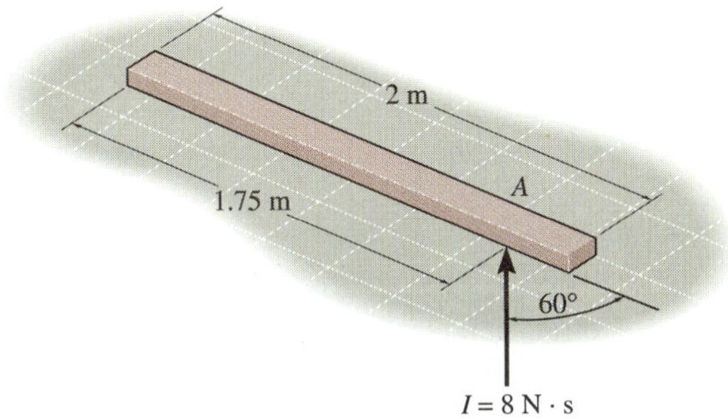

Prob. 19–14

19-15. The impact wrench consists of a slender 1-kg rod AB which is 580 mm long, and cylindrical end weights at A and B that each have a diameter of 20 mm and a mass of 1 kg. This assembly is free to turn about the handle and socket, which are attached to the lug nut on the wheel of a car. If the rod AB is given an angular velocity of 4 rad/s and it strikes the bracket C on the handle without rebounding, determine the angular impulse imparted to the lug nut.

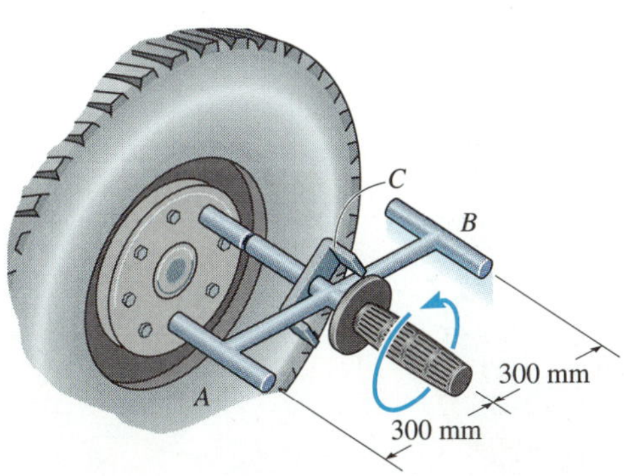

Prob. 19–15

***19-16.** The space shuttle is located in "deep space," where the effects of gravity can be neglected. It has a mass of 120 Mg, a center of mass at G, and a radius of gyration $(k_G)_x = 14$ m about the x axis. It is originally traveling forward at $v = 3$ km/s when the pilot turns on the engine at A, creating a thrust $T = 600(1 - e^{-0.3t})$ kN, where t is in seconds. Determine the shuttle's angular velocity 2 s later.

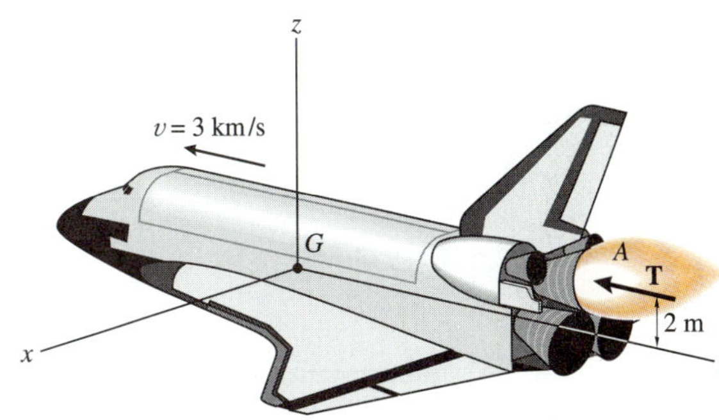

Prob. 19–16

19-17. A cord of negligible mass is wrapped around the outer surface of the 50-lb cylinder and its end is subjected to a constant horizontal force of $P = 2$ lb. If the cylinder rolls without slipping at A, determine its angular velocity in 4 s starting from rest. Neglect the thickness of the cord.

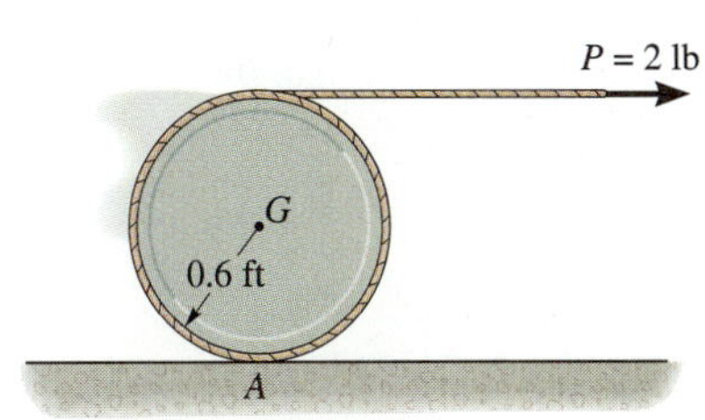

Prob. 19–17

19-18. The two gears A and B have weights and radii of gyration of $W_A = 15$ lb, $k_A = 0.5$ ft and $W_B = 10$ lb, $k_B = 0.35$ ft, respectively. If a motor transmits a couple moment to gear B of $M = 2(1 - e^{-0.5t})$ lb·ft, where t is in seconds, determine the angular velocity of gear A in $t = 5$ s, starting from rest.

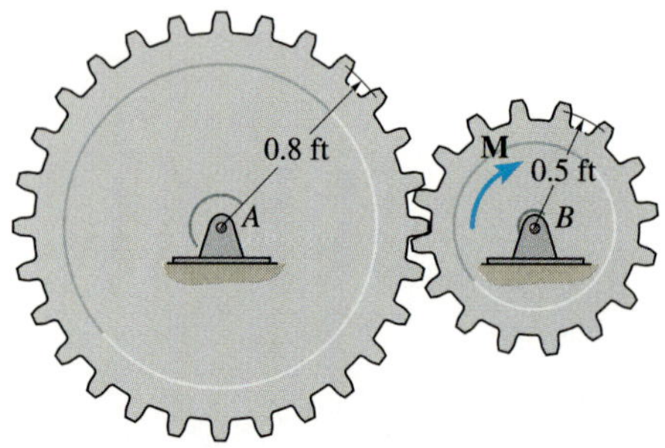

Prob. 19–18

19-19. A motor transmits a torque of $M = 0.05$ N·m to the center of gear A. Determine the angular velocity of each of the three (equal) smaller gears in 2 s starting from rest. The smaller gears (B) are pinned at their centers, and the masses and centroidal radii of gyration of the gears are given in the figure.

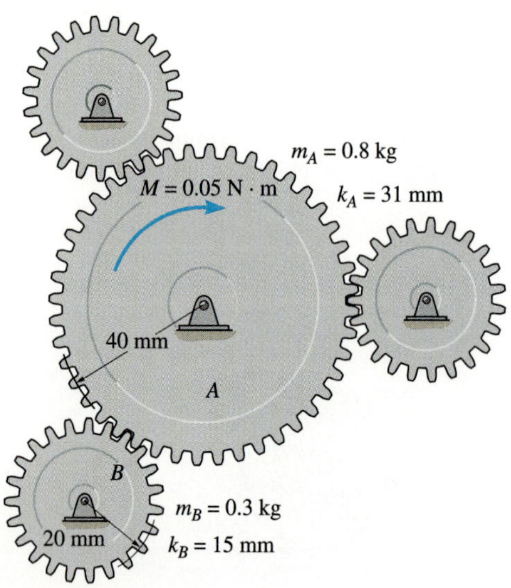

Prob. 19–19

***19-20.** The drum of mass m, radius r, and radius of gyration k_O rolls along an inclined plane for which the coefficient of static friction is μ_s. If the drum is released from rest, determine the maximum angle θ for the incline so that it rolls without slipping.

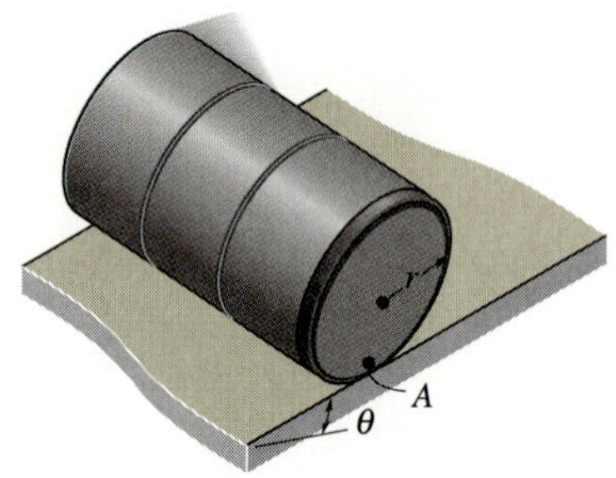

Prob. 19–20

19-21. Spool B is at rest and spool A is rotating at 6 rad/s when the slack in the cord connecting them is taken up. Determine the angular velocity of each spool immediately after the cord is jerked tight by the spinning of spool A. The weights and radii of gyration of A and B are $W_A = 30$ lb, $k_A = 0.8$ ft and $W_B = 15$ lb, $k_B = 0.6$ ft, respectively.

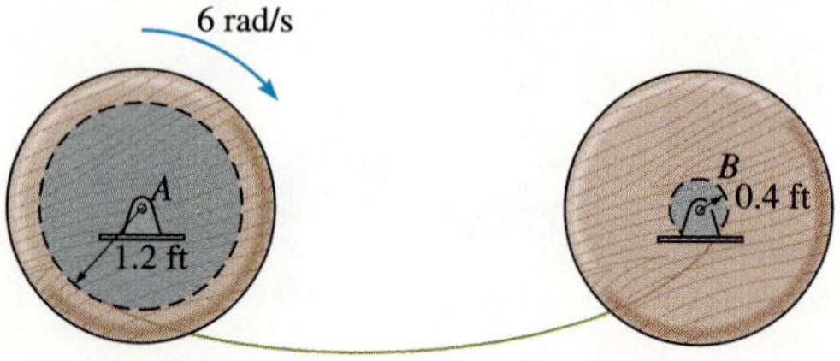

Prob. 19–21

19-22. A 4-kg disk A is mounted on arm BC, which has a negligible mass. If a torque of $M = (5e^{0.5t})$ N·m, where t is in seconds, is applied to the arm at C, determine the angular velocity of BC in 2 s starting from rest. Solve the problem assuming that (a) the disk is set in a smooth bearing at B so that it rotates with curvilinear translation, (b) the disk is fixed to the shaft BC, and (c) the disk is given an initial freely spinning angular velocity of $\boldsymbol{\omega}_D = \{-80\mathbf{k}\}$ rad/s prior to application of the torque.

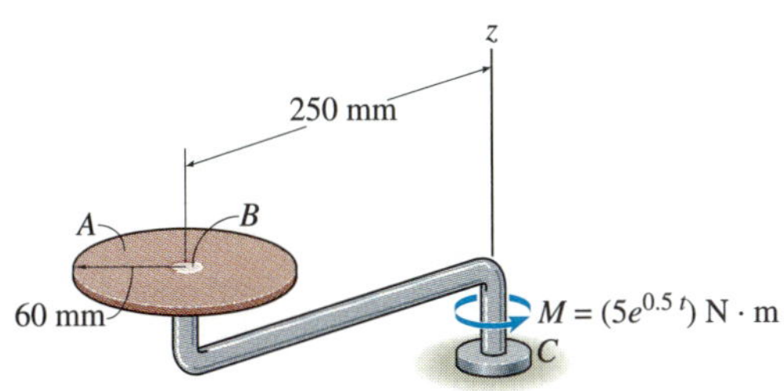

Prob. 19–22

19-23. If the hoop has a weight W and radius r and is thrown onto a *rough surface* with a velocity $\mathbf{v}_G$ parallel to the surface, determine the amount of backspin, ω_0, it must be given so that it stops spinning at the same instant that its forward velocity is zero. It is not necessary to know the coefficient of kinetic friction at A for the calculation.

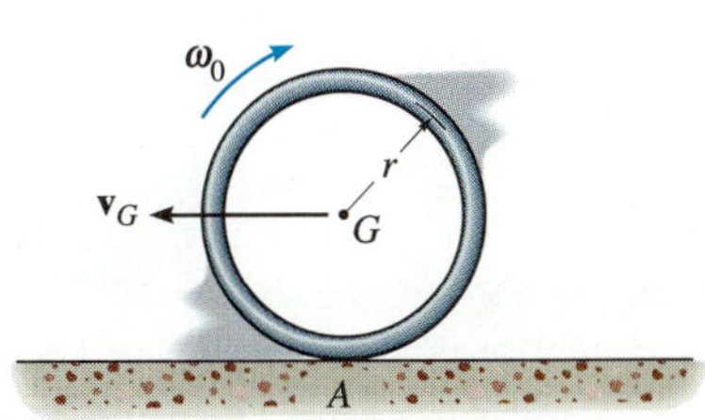

Prob. 19–23

***19-24.** For safety reasons, the 20-kg supporting leg of a sign is designed to break away with negligible resistance at B when the leg is subjected to the impact of a car. Assuming that the leg is pin supported at A and approximates a thin rod, determine the impulse the car bumper exerts on it, if after the impact the leg appears to rotate upward to an angle of $\theta_{max} = 150°$.

Prob. 19–24

19-25. The slender rod has a mass m and is suspended at its end A by a cord. If the rod receives a horizontal blow giving it an impulse $\mathbf{I}$ at its bottom B, determine the location y of the point P about which the rod appears to rotate during the impact.

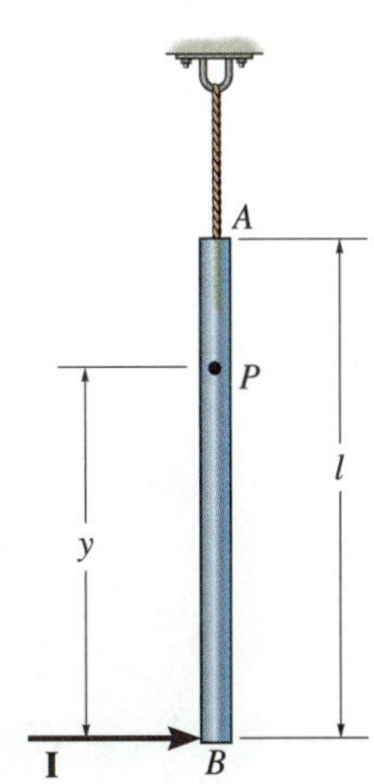

Prob. 19–25

19-26. A thin rod having a mass of 4 kg is balanced vertically as shown. Determine the height h at which it can be struck with a horizontal force **F** and not slip on the floor. This requires that the frictional force at A be essentially zero.

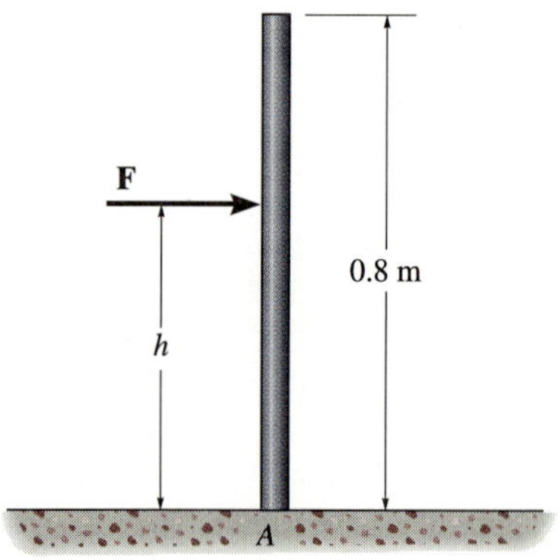

Prob. 19–26

19-27. Determine the height h of the bumper of the pool table, so that when the pool ball of mass m strikes it, no frictional force will be developed between the ball and the table at A. Assume the bumper exerts only a horizontal force on the ball.

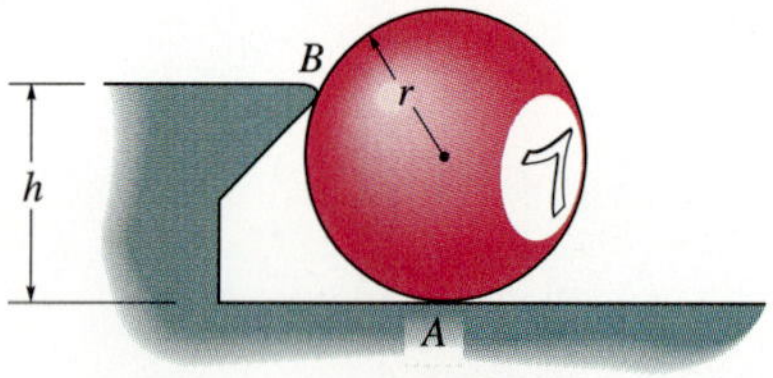

Prob. 19–27

***19-28.** If the ball has a weight W and radius r and is thrown onto a *rough surface* with a velocity $\mathbf{v}_0$ parallel to the surface, determine the amount of backspin, ω_0, it must be given so that it stops spinning at the same instant that its forward velocity is zero. It is not necessary to know the coefficient of friction at A for the calculation.

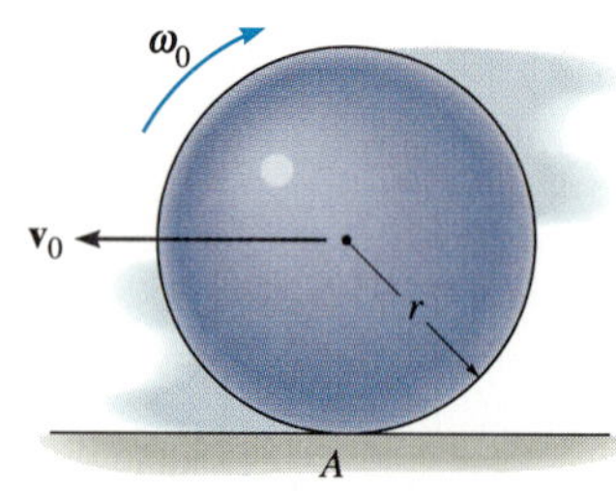

Prob. 19–28

19-29. The frame of the roller has a mass of 5.5 Mg and a center of mass at G. The roller has a mass of 2 Mg and a radius of gyration about its mass center of $k_A = 0.45$ m. If a torque of $M = 600$ N·m is applied to the rear wheels, determine the speed of the compactor in $t = 4$ s, starting from rest. No slipping occurs. Neglect the mass of the driving wheels.

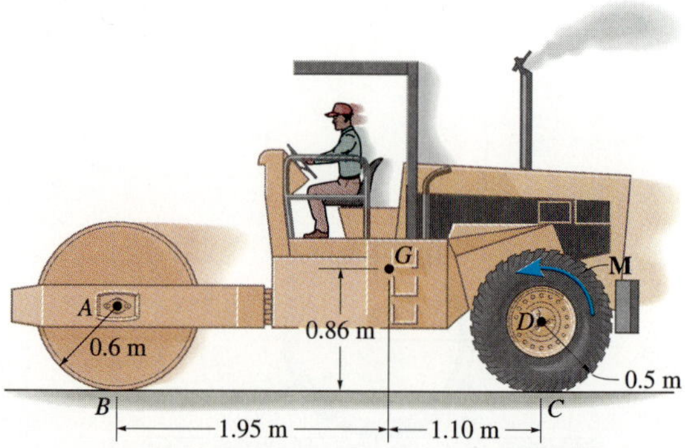

Prob. 19–29

19-30. The car strikes the side of a light pole, which is designed to break away from its base with negligible resistance. From a video taken of the collision it is observed that the pole was given an angular velocity of 60 rad/s when AC was vertical. The pole has a mass of 175 kg, a center of mass at G, and a radius of gyration about an axis perpendicular to the plane of the pole assembly and passing through G of $k_G = 2.25$ m. Determine the horizontal impulse which the car exerts on the pole while AC is essentially vertical.

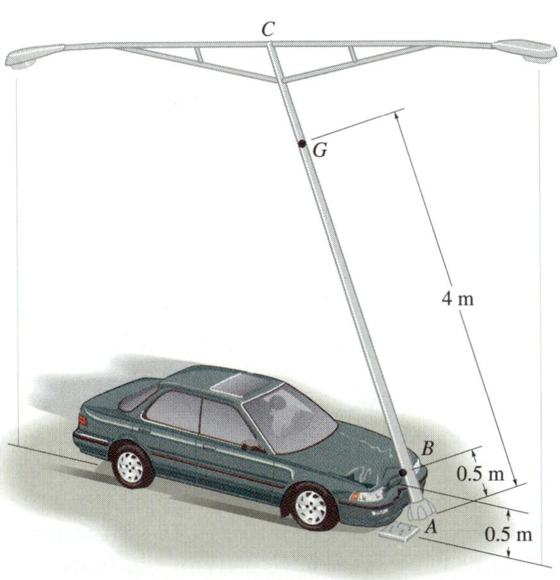

Prob. 19–30

19-31. The double pulley consists of two wheels which are attached to one another and turn at the same rate. The pulley has a mass of 15 kg and a radius of gyration of $k_O = 110$ mm. If the block at A has a mass of 40 kg and the container at B has a mass of 85 kg, including its contents, determine the speed of the container when $t = 3$ s after it is released from rest.

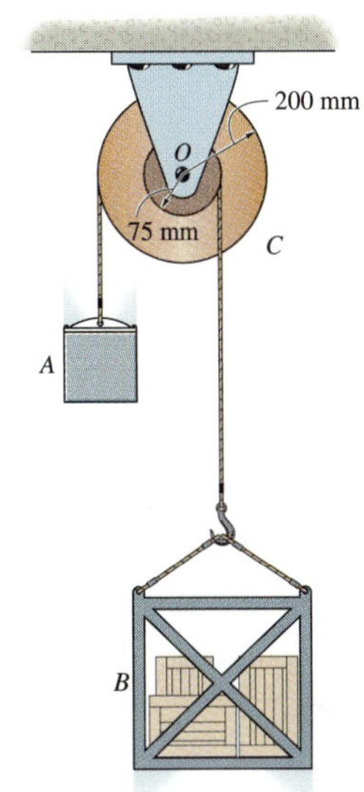

Prob. 19–31

***19-32.** The two rods each have a mass m and a length l, and lie on the smooth horizontal plane. If an impulse **I** is applied at an angle of 45° to one of the rods at mid length as shown, determine the angular velocity of each rod just after the impulse. The rods are pin connected at B.

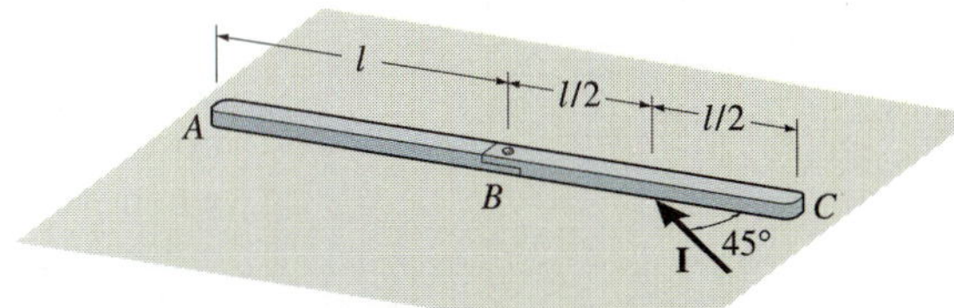

Prob. 19–32

19.3 Conservation of Momentum

Conservation of Linear Momentum. If the sum of all the *linear impulses* acting on a system of connected rigid bodies is *zero,* the linear momentum of the system is constant, or conserved. Consequently, the first two of Eqs. 19–15 reduce to the form

$$\left(\Sigma \begin{matrix}\text{syst. linear}\\ \text{momentum}\end{matrix}\right)_1 = \left(\Sigma \begin{matrix}\text{syst. linear}\\ \text{momentum}\end{matrix}\right)_2 \tag{19–16}$$

This equation is referred to as the *conservation of linear momentum.*

Without inducing appreciable errors in the calculations, it may be possible to apply Eq. 19–16 in a specified direction for which the linear impulses are small or *nonimpulsive.* Specifically, nonimpulsive forces occur when small forces act over very short periods of time. Typical examples include the force of a slightly deformed spring, the initial contact with soft ground, and in some cases the weight of the body.

Conservation of Angular Momentum. The angular momentum of a system of connected rigid bodies is conserved about the system's center of mass G, or a fixed point O, when the sum of all the angular impulses created by the external forces acting on the system is zero or appreciably small (nonimpulsive) when computed about these points. The third of Eqs. 19–15 then becomes

$$\left(\Sigma \begin{matrix}\text{syst. angular}\\ \text{momentum}\end{matrix}\right)_{O1} = \left(\Sigma \begin{matrix}\text{syst. angular}\\ \text{momentum}\end{matrix}\right)_{O2} \tag{19–17}$$

This equation is referred to as the *conservation of angular momentum.* In the case of a single rigid body, Eq. 19–17 applied to point G becomes $(I_G\omega)_1 = (I_G\omega)_2$. To illustrate an application of this equation, consider a swimmer who executes a somersault after jumping off a diving board. By tucking his arms and legs in close to his chest, he *decreases* his body's moment of inertia and thus *increases* his angular velocity ($I_G\omega$ must be constant). If he straightens out just before entering the water, his body's moment of inertia is *increased,* and his angular velocity *decreases.* Since the weight of his body creates a linear impulse during the time of motion, this example also illustrates how the angular momentum of a body can be conserved and yet the linear momentum is *not.* Such cases occur whenever the external forces creating the linear impulse pass through either the center of mass of the body or a fixed axis of rotation.

Provided the initial linear or angular velocity of the body is known, the conservation of linear or angular momentum is used to determine the respective final linear or angular velocity of the body *just after* the time period considered. Furthermore, by applying these equations to a *system* of bodies, the internal impulses acting within the system, which may be unknown, are eliminated from the analysis, since they occur in equal but opposite collinear pairs. If it is necessary to determine an *internal impulsive force* acting on only one body of a system of connected bodies, the body must be *isolated* (free-body diagram) and the principle of linear or angular impulse and momentum must be applied *to the body*. After the impulse $\int F\,dt$ is calculated, then, provided the time Δt for which the impulse acts is known, the *average impulsive force* F_{avg} can be determined from $F_{avg} = (\int F\,dt)/\Delta t$.

Procedure for Analysis

The conservation of linear or angular momentum should be applied using the following procedure.

Free-Body Diagram

- Establish the x, y inertial frame of reference and draw the free-body diagram for the body or system of bodies during the time of impact. From this diagram classify each of the applied forces as being either "impulsive" or "nonimpulsive."
- By inspection of the free-body diagram, the *conservation of linear momentum* applies in a given direction when *no* external impulsive forces act on the body or system in that direction; whereas the *conservation of angular momentum* applies about a fixed point O or at the mass center G of a body or system of bodies when all the external impulsive forces acting on the body or system create zero moment (or zero angular impulse) about O or G.
- As an alternative procedure, draw the impulse and momentum diagrams for the body or system of bodies. These diagrams are particularly helpful in order to visualize the "moment" terms used in the conservation of angular momentum equation, when it has been decided that angular momenta are to be computed about a point other than the body's mass center G.

Conservation of Momentum

- Apply the conservation of linear or angular momentum in the appropriate directions.

Kinematics

- If the motion appears to be complicated, kinematic (velocity) diagrams may be helpful in obtaining the necessary kinematic relations.

EXAMPLE 19-6

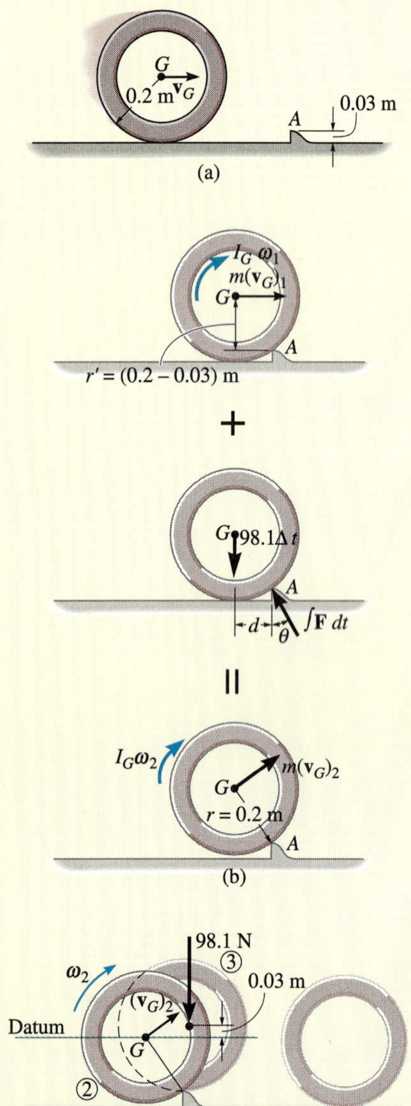

Fig. 19–9

The 10-kg wheel shown in Fig. 19–9*a* has a moment of inertia $I_G = 0.156$ kg·m^2. Assuming that the wheel does not slip or rebound, determine the minimum velocity $\mathbf{v}_G$ it must have to just roll over the obstruction at A.

Solution

Impulse and Momentum Diagrams. Since no slipping or rebounding occurs, the wheel essentially *pivots* about point A during contact. This condition is shown in Fig. 19–9*b*, which indicates, respectively, the momentum of the wheel *just before impact,* the impulses given to the wheel *during impact,* and the momentum of the wheel *just after impact.* Only two impulses (forces) act on the wheel. By comparison, the force at A is much greater than that of the weight, and since the time of impact is very short, the weight can be considered nonimpulsive. The impulsive force $\mathbf{F}$ at A has both an unknown magnitude and an unknown direction θ. To eliminate this force from the analysis, note that angular momentum about A is essentially *conserved* since $(98.1\Delta t)d \approx 0$.

Conservation of Angular Momentum. With reference to Fig. 19–9*b*,

$$(\circlearrowleft +) \qquad (H_A)_1 = (H_A)_2$$

$$r'm(v_G)_1 + I_G\omega_1 = rm(v_G)_2 + I_G\omega_2$$

$$(0.2 \text{ m} - 0.03 \text{ m})(10 \text{ kg})(v_G)_1 + (0.156 \text{ kg}\cdot\text{m}^2)(\omega_1) = (0.2 \text{ m})(10 \text{ kg})(v_G)_2 + (0.156 \text{ kg}\cdot\text{m}^2)(\omega_2)$$

Kinematics. Since no slipping occurs, in general $\omega = v_G/r = v_G/0.2 \text{ m} = 5v_G$. Substituting this into the above equation and simplifying yields

$$(v_G)_2 = 0.892(v_G)_1 \qquad (1)$$

Conservation of Energy.* In order to roll over the obstruction, the wheel must pass position 3 shown in Fig. 19–9*c*. Hence, if $(v_G)_2$ [or $(v_G)_1$] is to be a minimum, it is necessary that the kinetic energy of the wheel at position 2 be equal to the potential energy at position 3. Constructing the datum through the center of gravity, as shown in the figure, and applying the conservation of energy equation, we have

$$\{T_2\} + \{V_2\} = \{T_3\} + \{V_3\}$$

$$\{\tfrac{1}{2}(10 \text{ kg})(v_G)_2^2 + \tfrac{1}{2}(0.156 \text{ kg}\cdot\text{m}^2)\omega_2^2\} + \{0\} = \{0\} + \{(98.1 \text{ N})(0.03 \text{ m})\}$$

Substituting $\omega_2 = 5(v_G)_2$ and Eq. 1 into this equation, and solving,

$$(v_G)_1 = 0.729 \text{ m/s} \rightarrow \qquad \textit{Ans.}$$

*This principle *does not* apply during *impact,* since energy is *lost* during the collision; however, just after impact, position 2, it can be used.

EXAMPLE 19–7

The 5-kg slender rod shown in Fig. 19–10*a* is pinned at *O* and is initially at rest. If a 4-g bullet is fired into the rod with a velocity of 400 m/s, as shown in the figure, determine the angular velocity of the rod just after the bullet becomes embedded in it.

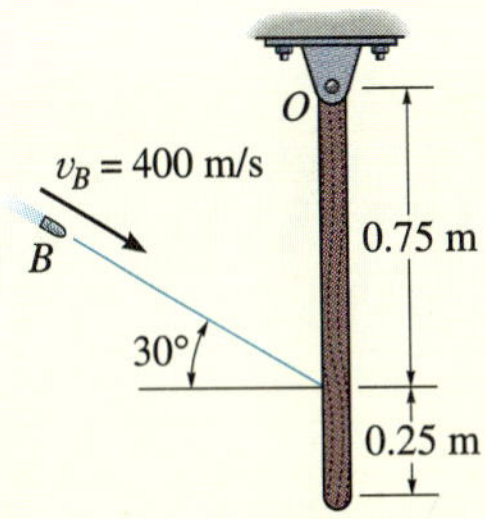

(a)

Solution

Impulse and Momentum Diagrams. The impulse which the bullet exerts on the rod can be eliminated from the analysis, and the angular velocity of the rod just after impact can be determined by considering the bullet and rod as a single system. To clarify the principles involved, the impulse and momentum diagrams are shown in Fig. 19–10*b*. The momentum diagrams are drawn *just before and just after impact.* During impact, the bullet and rod exchange equal but *opposite internal impulses* at *A*. As shown on the impulse diagram, the impulses that are external to the system are due to the reactions at *O* and the weights of the bullet and rod. Since the time of impact, Δt, is very short, the rod moves only a slight amount, and so the "moments" of the weight impulses about point *O* are essentially zero. Therefore angular momentum is conserved about this point.

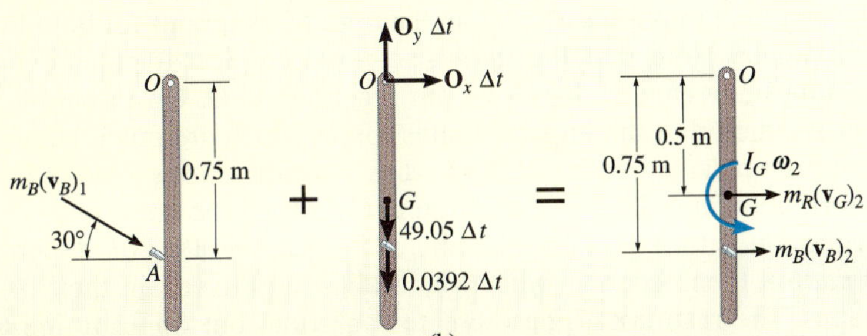

(b)

Conservation of Angular Momentum. From Fig. 19–10*b*, we have

$$(\circlearrowright +) \qquad \Sigma(H_O)_1 = \Sigma(H_O)_2$$

$$m_B(v_B)_1 \cos 30°(0.75\text{ m}) = m_B(v_B)_2(0.75\text{ m}) + m_R(v_G)_2(0.5\text{ m}) + I_G\omega_2$$

$$(0.004\text{ kg})(400 \cos 30°\text{ m/s})(0.75\text{ m}) =$$

$$(0.004\text{ kg})(v_B)_2(0.75\text{ m}) + (5\text{ kg})(v_G)_2(0.5\text{ m}) + [\tfrac{1}{12}(5\text{ kg})(1\text{ m})^2]\omega_2$$

or

$$1.039 = 0.003(v_B)_2 + 2.50(v_G)_2 + 0.417\omega_2 \qquad (1)$$

Kinematics. Since the rod is pinned at *O*, from Fig. 19–10*c* we have

$$(v_G)_2 = (0.5\text{ m})\omega_2 \qquad (v_B)_2 = (0.75\text{ m})\omega_2$$

Substituting into Eq. 1 and solving yields

$$\omega_2 = 0.623\text{ rad/s} \circlearrowright \qquad \textit{Ans.}$$

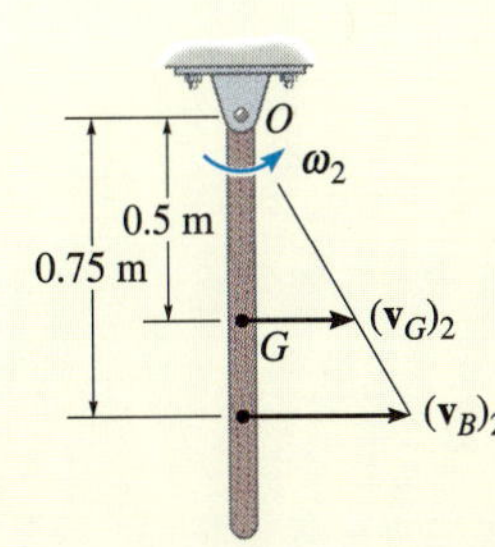

(c)

Fig. 19–10

*19.4 Eccentric Impact

An example of eccentric impact occurring between this bowling ball and pin.

The concepts involving central and oblique impact of particles have been presented in Sec. 15.4. We will now expand this treatment and discuss the eccentric impact of two bodies. *Eccentric impact* occurs when the line connecting the *mass centers* of the two bodies *does not* coincide with the line of impact.* This type of impact often occurs when one or both of the bodies are constrained to rotate about a fixed axis. Consider, for example, the collision at C between the two bodies A and B, shown in Fig. 19–11a. It is assumed that just before collision B is rotating counterclockwise with an angular velocity $(\boldsymbol{\omega}_B)_1$, and the velocity of the contact point C located on A is $(\mathbf{u}_A)_1$. Kinematic diagrams for both bodies just before collision are shown in Fig. 19–11b. Provided the bodies are smooth, the *impulsive forces* they exert on each other *are directed along the line of impact.* Hence, the component of velocity of point C on body B, which is directed along the line of impact, is $(v_B)_1 = (\omega_B)_1 r$, Fig. 19–11b. Likewise, on body A the component of velocity $(\mathbf{u}_A)_1$ along the line of impact is $(\mathbf{v}_A)_1$. In order for a collision to occur, $(v_A)_1 > (v_B)_1$.

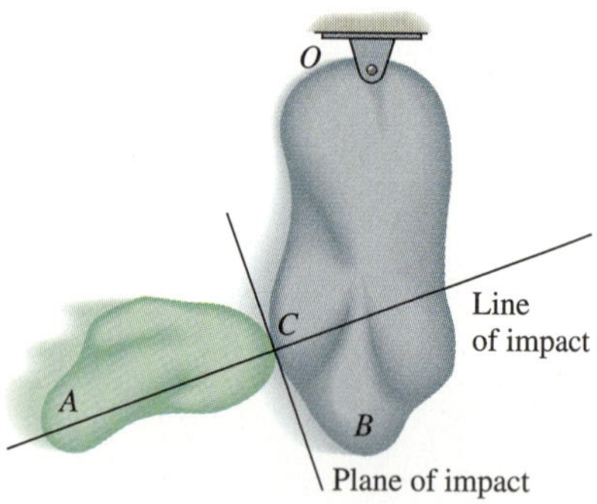

(a)

Fig. 19–11

During the impact an equal but opposite impulsive force $\mathbf{P}$ is exerted between the bodies which *deforms* their shapes at the point of contact. The resulting impulse is shown on the impulse diagrams for both bodies, Fig. 19–11c. Note that the impulsive force created at point C on the rotating body creates impulsive pin reactions at O. On these diagrams it is assumed that the impact creates forces which are much larger than the nonimpulsive weights of the bodies, which are not shown. When the deformation at point C is a maximum, C on both the bodies moves with a common velocity $\mathbf{v}$ along the line of impact, Fig. 19–11d. A period of *restitution* then occurs in which the bodies tend to regain their original shapes. The restitution phase creates an equal but opposite impulsive force $\mathbf{R}$ acting between the bodies as shown on the impulse diagram, Fig. 19–11e. After restitution the bodies move apart such that point C on body B has a velocity $(\mathbf{v}_B)_2$ and point C on body A has a velocity $(\mathbf{u}_A)_2$, Fig. 19–11f, where $(v_B)_2 > (v_A)_2$.

In general, a problem involving the impact of two bodies requires determining the *two unknowns* $(v_A)_2$ and $(v_B)_2$, assuming $(v_A)_1$ and $(v_B)_1$ are known (or can be determined using kinematics, energy methods, the equations of motion, etc.). To solve this problem, two equations must be written. The *first equation* generally involves application of *the conservation of angular momentum to the two bodies.* In the case of both bodies A and B, we can state that angular momentum is conserved about point O since the impulses at C are internal to the system and the impulses at O create zero moment (or zero angular impulse) about point

*When these lines coincide, central impact occurs and the problem can be analyzed as discussed in Sec. 15.4.

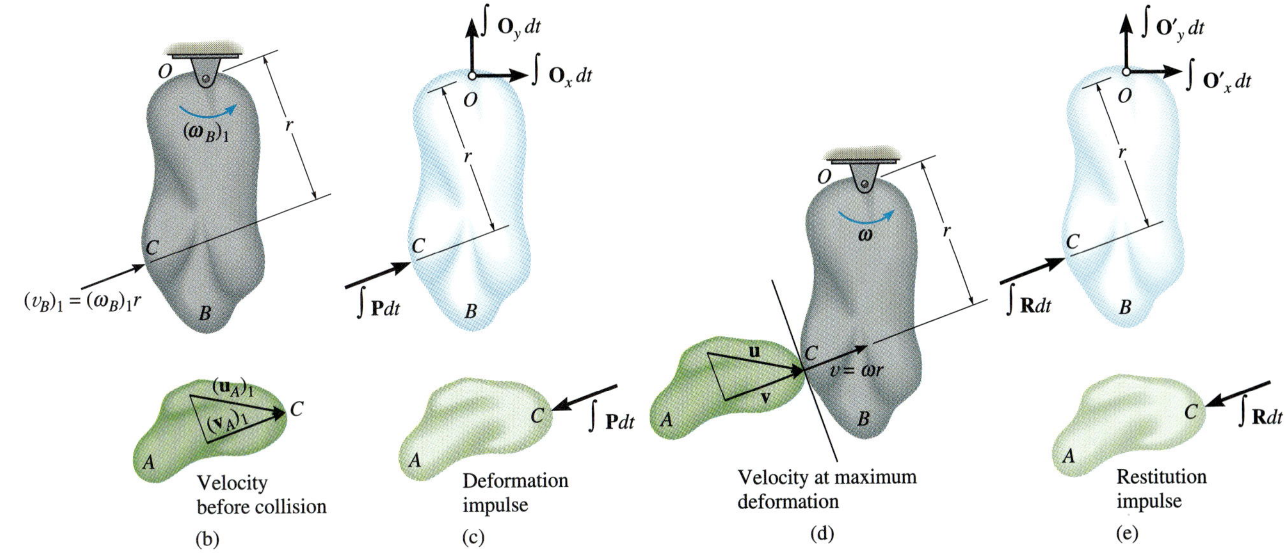

O. The *second equation* is obtained using the definition of the *coefficient of restitution, e,* which is a ratio of the restitution impulse to the deformation impulse. To establish a useful form of this equation we must first apply the principle of angular impulse and momentum about point O to bodies B and A separately. Combining the results, we then obtain the necessary equation. Proceeding in this manner, the principle of impulse and momentum applied to body B from the time just before the collision to the instant of maximum deformation, Figs. 19–11*b*, 19–11*c*, and 19–11*d*, becomes

$$(\circlearrowright+) \qquad I_O(\omega_B)_1 + r\int P\,dt = I_O\omega \qquad (19\text{–}18)$$

Here I_O is the moment of inertia of body B about point O. Similarly, applying the principle of angular impulse and momentum from the instant of maximum deformation to the time just after the impact, Figs. 19–11*d*, 19–11*e*, and 19–11*f*, yields

$$(\circlearrowright+) \qquad I_O\omega + r\int R\,dt = I_O(\omega_B)_2 \qquad (19\text{–}19)$$

Solving Eqs. 19–18 and 19–19 for $\int P\,dt$ and $\int R\,dt$, respectively, and formulating e, we have

$$e = \frac{\int R\,dt}{\int P\,dt} = \frac{r(\omega_B)_2 - r\omega}{r\omega - r(\omega_B)_1} = \frac{(v_B)_2 - v}{v - (v_B)_1}$$

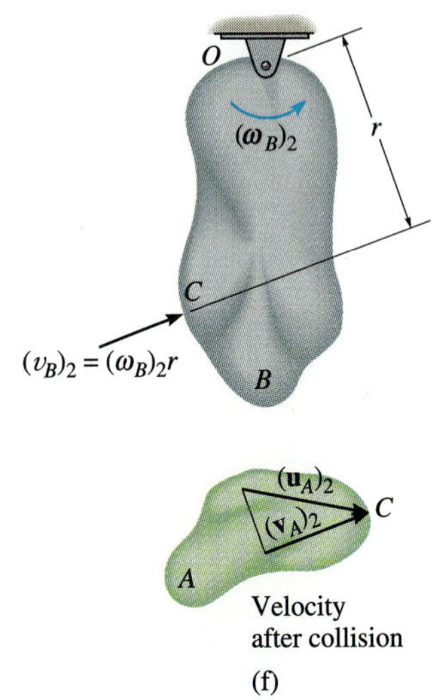

Fig. 19–11

In the same manner, we may write an equation which relates the magnitudes of velocity $(v_A)_1$ and $(v_A)_2$ of body A. The result is

$$e = \frac{v - (v_A)_2}{(v_A)_1 - v}$$

Combining the above equations by eliminating the common velocity v yields the desired result, i.e.,

$$(+\nearrow) \qquad e = \frac{(v_B)_2 - (v_A)_2}{(v_A)_1 - (v_B)_1} \qquad (19\text{–}20)$$

This equation is identical to Eq. 15–11, which was derived for the central impact between two particles. Equation 19–20 states that the coefficient of restitution is equal to the ratio of the relative velocity of *separation* of the points of contact (C) *just after impact* to the relative velocity at which the points *approach* one another *just* before impact. In deriving this equation, we assumed that the points of contact for both bodies move up and to the right *both* before and after impact. If motion of any one of the contacting points occurs down and to the left, the velocity of this point is considered a negative quantity in Eq. 19–20.

As stated previously, when Eq. 19–20 is used in conjunction with the conservation of angular momentum for the bodies, it provides a useful means of obtaining the velocities of two colliding bodies just after collision.

During collision the columns on many highway signs are intended to break out of their supports and easily collapse at their joints as shown by the slotted connections at their base and the breaks at the column's midsection. The mechanics of eccentric impact is used in the design of these structures.

EXAMPLE 19–8

The 10-lb slender rod is suspended from the pin at A, Fig. 19–12a. If a 2-lb ball B is thrown at the rod and strikes its center with a horizontal velocity of 30 ft/s, determine the angular velocity of the rod just after impact. The coefficient of restitution is $e = 0.4$.

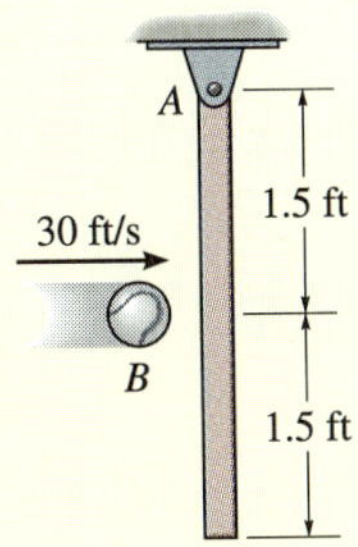

(a)

Solution

Conservation of Angular Momentum. Consider the ball and rod as a system, Fig. 19–12b. Angular momentum is conserved about point A since the impulsive force between the rod and ball is *internal*. Also, the *weights* of the ball and rod are *nonimpulsive*. Noting the directions of the velocities of the ball and rod just after impact as shown on the kinematic diagram, Fig. 19–12c, we require

$$(\circlearrowright +) \qquad (H_A)_1 = (H_A)_2$$

$$m_B(v_B)_1(1.5\text{ ft}) = m_B(v_B)_2(1.5\text{ ft}) + m_R(v_G)_2(1.5\text{ ft}) + I_G\omega_2$$

$$\left(\frac{2\text{ lb}}{32.2\text{ ft/s}^2}\right)(30\text{ ft/s})(1.5\text{ ft}) = \left(\frac{2\text{ lb}}{32.2\text{ ft/s}^2}\right)(v_B)_2(1.5\text{ ft}) + \left(\frac{10\text{ lb}}{32.2\text{ ft/s}^2}\right)(v_G)_2(1.5\text{ ft}) + \left[\frac{1}{12}\left(\frac{10\text{ lb}}{32.2\text{ ft/s}^2}\right)(3\text{ ft})^2\right]\omega_2$$

Since $(v_G)_2 = 1.5\omega_2$ then

$$2.795 = 0.09317(v_B)_2 + 0.9317\omega_2 \qquad (1)$$

Coefficient of Restitution. With reference to Fig. 19–12c, we have

$$(\overset{+}{\rightarrow}) \qquad e = \frac{(v_G)_2 - (v_B)_2}{(v_B)_1 - (v_G)_1} \qquad 0.4 = \frac{(1.5\text{ ft})\omega_2 - (v_B)_2}{30\text{ ft/s} - 0}$$

$$12.0 = 1.5\omega_2 - (v_B)_2$$

Solving,

$$(v_B)_2 = -6.52\text{ ft/s} = 6.52\text{ ft/s} \leftarrow$$

$$\omega_2 = 3.65\text{ rad/s} \circlearrowleft \qquad \textit{Ans.}$$

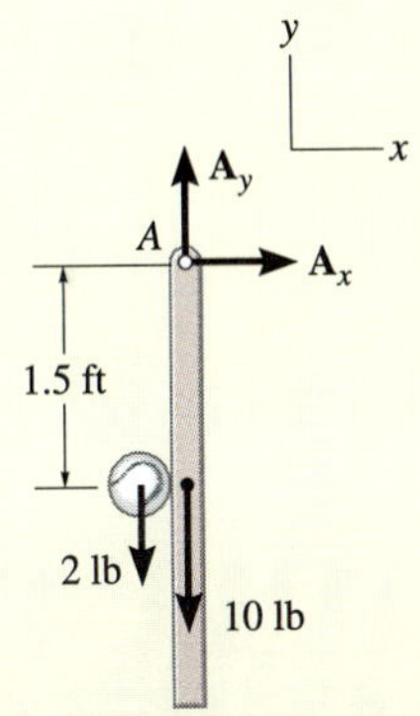

(b)

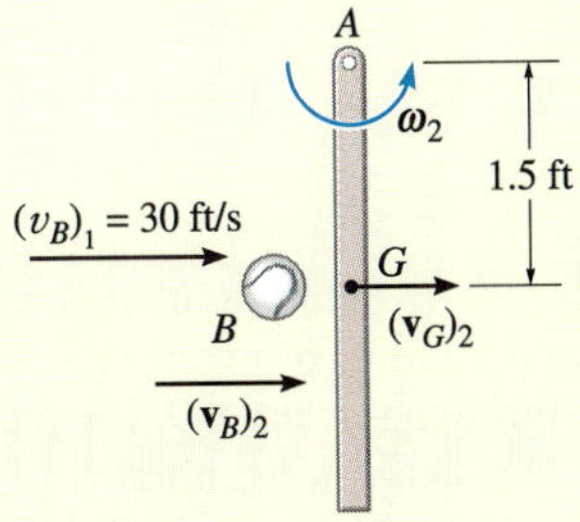

(c)

Fig. 19–12

Problems

19-33. Two wheels A and B have masses m_A and m_B, and radii of gyration about their central vertical axes of k_A and k_B, respectively. If they are freely rotating in the same direction at $\boldsymbol{\omega}_A$ and $\boldsymbol{\omega}_B$ about the same vertical axis, determine their common angular velocity after they are brought into contact and slipping between them stops.

19-34. A horizontal circular platform has a weight of 300 lb and a radius of gyration $k_z = 8$ ft about the z axis passing through its center O. The platform is free to rotate about the z axis and is initially at rest. A man having a weight of 150 lb begins to run along the edge in a circular path of radius 10 ft. If he has a speed of 4 ft/s and maintains this speed relative to the platform, determine the angular velocity of the platform. Neglect friction.

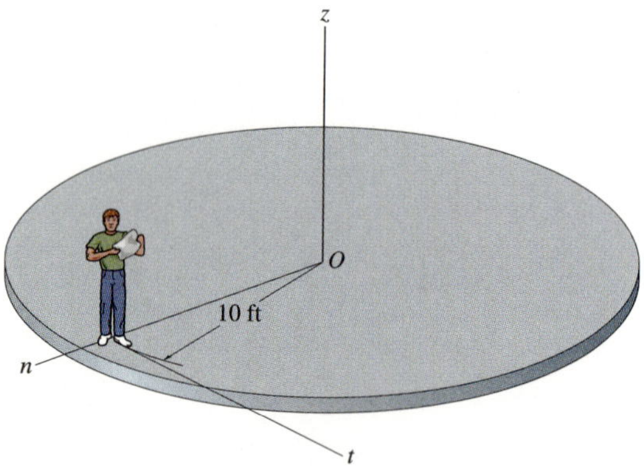

Prob. 19–35

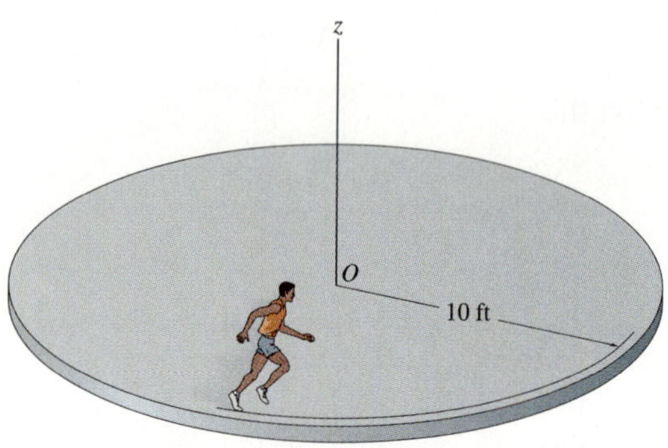

Prob. 19–34

19-35. A horizontal circular platform has a weight of 300 lb and a radius of gyration $k_z = 8$ ft about the z axis passing through its center O. The platform is free to rotate about the z axis and is initially at rest. A man having a weight of 150 lb throws a 15-lb block off the edge of the platform with a horizontal velocity of 5 ft/s, *measured relative to the platform.* Determine the angular velocity of the platform if the block is thrown (a) tangent to the platform, along the $+t$ axis, and (b) outward along a radial line, or $+n$ axis. Neglect the size of the man.

***19-36.** The Hubble Space Telescope is powered by two solar panels as shown. The body of the telescope has a mass of 11 Mg and radii of gyration $k_x = 1.64$ m and $k_y = 3.85$ m, whereas the solar panels can be considered as thin plates, each having a mass of 54 kg. Due to an internal drive, the panels are given an angular velocity of $\{0.6\mathbf{j}\}$ rad/s, measured relative to the telescope. Determine the angular velocity of the telescope due to the rotation of the panels. Prior to rotating the panels, the telescope was originally traveling at $\mathbf{v}_G = \{-400\mathbf{i} + 250\mathbf{j} + 175\mathbf{k}\}$ m/s. Neglect its orbital rotation.

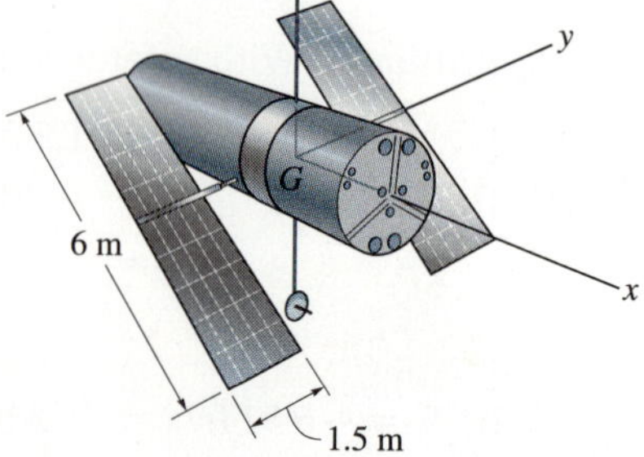

Prob. 19–36

19-37. The platform swing consists of a 200-lb flat plate suspended by four rods of negligible weight. When the swing is at rest, the 150-lb man jumps off the platform when his center of gravity G is 10 ft from the pin at A. This is done with a horizontal velocity of 5 ft/s, measured relative to the swing at the level of G. Determine the angular velocity he imparts to the swing just after jumping off.

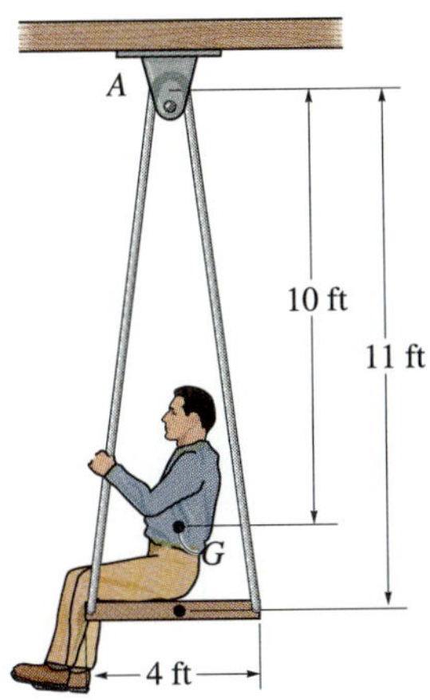

Prob. 19–37

19-38. A man has a moment of inertia I_z about the z axis. He is originally at rest and standing on a small platform which can turn freely. If he is handed a wheel which is rotating at $\boldsymbol{\omega}$ and has a moment of inertia I about its spinning axis, determine his angular velocity if (a) he holds the wheel upright as shown, (b) turns the wheel out, $\theta = 90°$, and (c) turns the wheel downward, $\theta = 180°$.

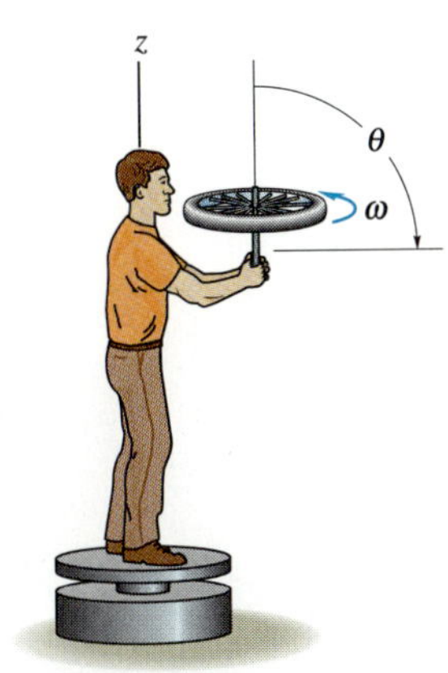

Prob. 19–38

19-39. A man has a moment of inertia I_z about the z axis. He is originally at rest and standing on a small platform which can turn freely. If he is handed a wheel when it is at *rest* and he starts it spinning with an angular velocity $\boldsymbol{\omega}$, determine his angular velocity if (a) he holds the wheel upright as shown, (b) turns the wheel out, $\theta = 90°$, and (c) turns the wheel downward, $\theta = 180°$.

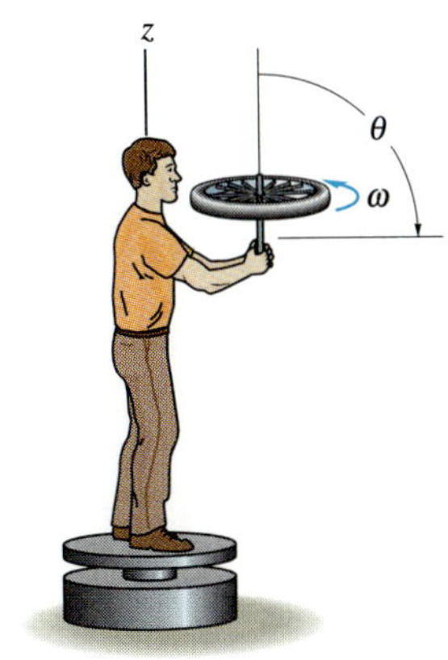

Prob. 19–39

***19-40.** A 7-g bullet having a velocity of 800 m/s is fired into the edge of the 5-kg disk as shown. Determine the angular velocity of the disk just after the bullet becomes embedded in it. Also, calculate how far θ the disk will swing until it momentarily stops. The disk is originally at rest.

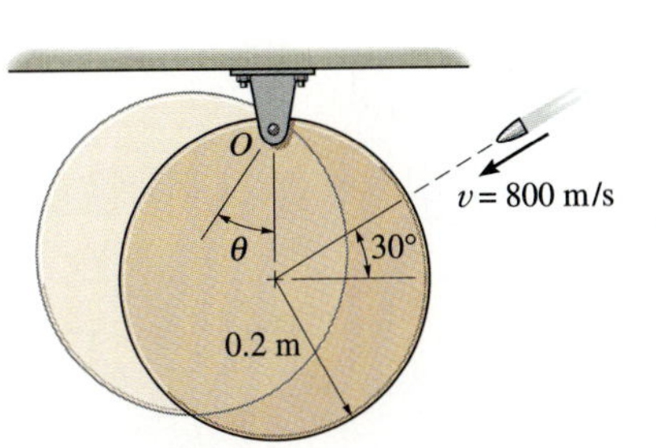

Prob. 19–40

19-41. The pendulum consists of a slender 2-kg rod AB and 5-kg disk. It is released from rest without rotating. When it falls 0.3 m, the end A strikes the hook S, which provides a permanent connection. Determine the angular velocity of the pendulum after it has rotated 90°. Treat the pendulum's weight during impact as a nonimpulsive force.

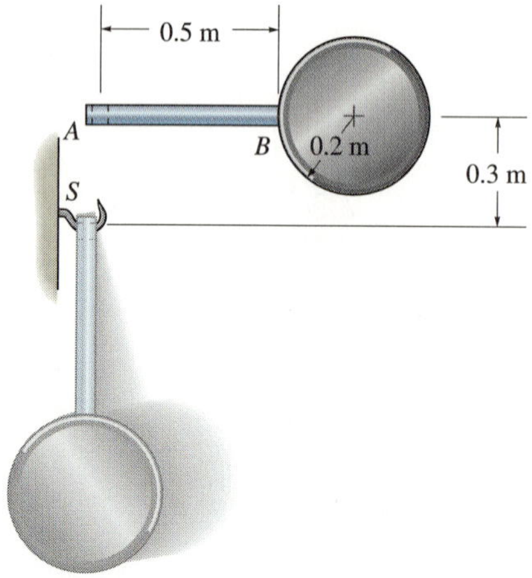

Prob. 19–41

19-42. A thin rod of mass m has an angular velocity $\boldsymbol{\omega}_0$ while rotating on a smooth surface. Determine its new angular velocity just after its end strikes and hooks onto the peg and the rod starts to rotate about P without rebounding. Solve the problem (a) using the parameters given, (b) setting $m = 2$ kg, $\omega_0 = 4$ rad/s, $l = 1.5$ m.

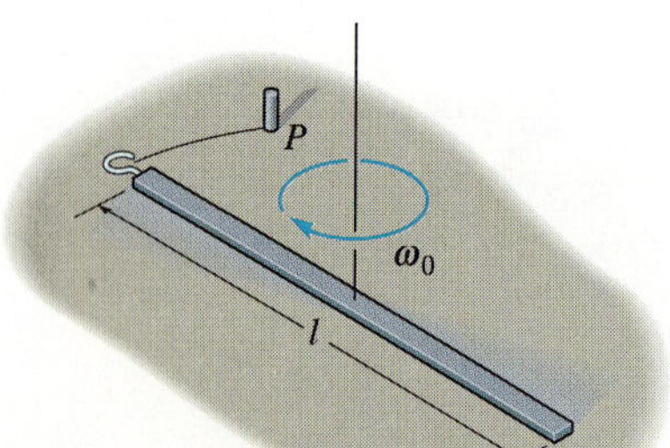

Prob. 19–42

19-43. The square plate has a weight W and is rotating on the smooth surface with a constant angular velocity $\boldsymbol{\omega}_0$. Determine the new angular velocity of the plate just after its corner strikes the peg P and the plate starts to rotate about P without rebounding.

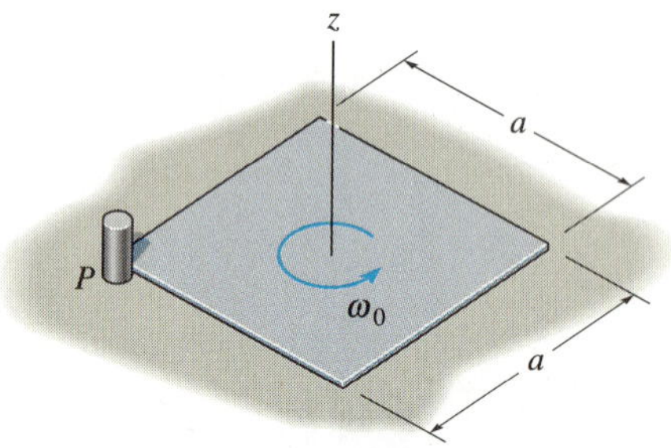

Prob. 19–43

***19-44.** A ball having a mass of 8 kg and initial speed of $v_1 = 0.2$ m/s rolls over a 30-mm-long depression. Assuming that the ball rolls off the edges of contact, first A, then B, determine its final velocity $\mathbf{v}_2$ when it reaches the other side.

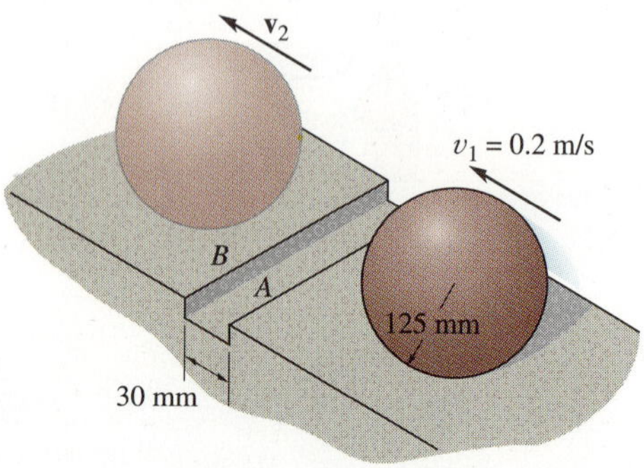

Prob. 19–44

19-45. A thin ring having a mass of 15 kg strikes the 20-mm-high step. Determine the largest angular velocity ω_1 the ring can have so that it will not rebound off the step at A when it strikes it.

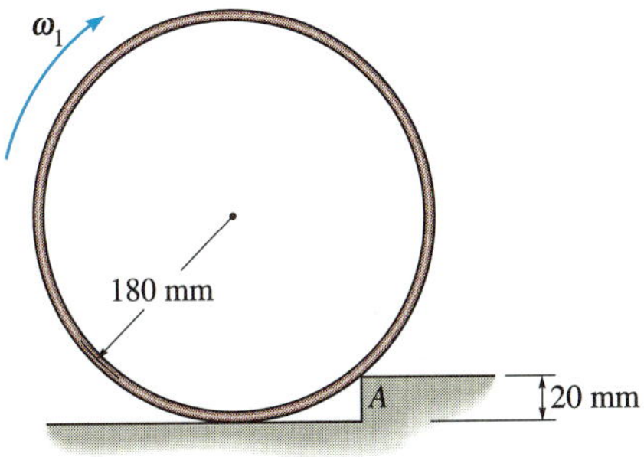

Prob. 19–45

19-46. The solid ball of mass m is dropped with a velocity $\mathbf{v}_1$ onto the edge of the rough step. If it rebounds horizontally off the step with a velocity $\mathbf{v}_2$, determine the angle θ at which contact occurs. Assume no slipping when the ball strikes the step. The coefficient of restitution is e.

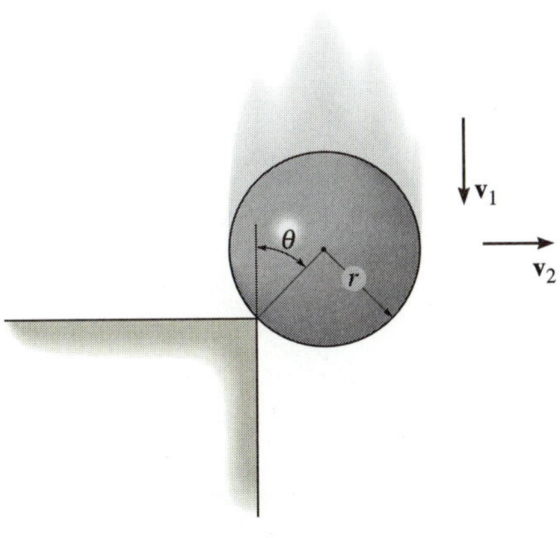

Prob. 19–46

19-47. Tests of impact on the fixed crash dummy are conducted using the 300-lb ram that is released from rest at $\theta = 30°$, and allowed to fall and strike the dummy at $\theta = 90°$. If the coefficient of restitution between the dummy and the ram is $e = 0.4$, determine the angle θ to which the ram will rebound before momentarily coming to rest.

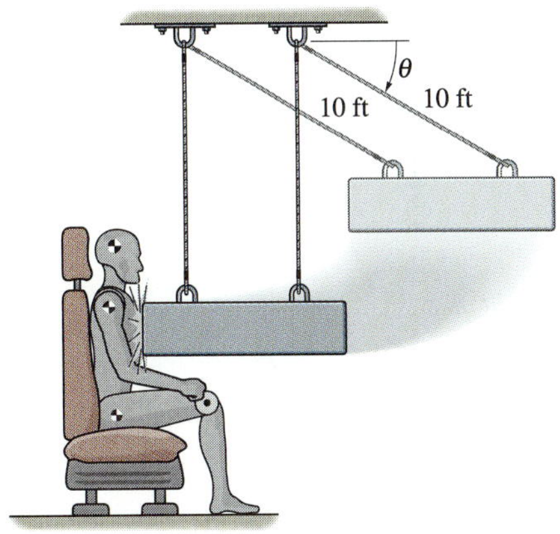

Prob. 19–47

***19-48.** The disk has a mass of 15 kg. If it is released from rest when $\theta = 30°$, determine the maximum angle θ of rebound after it collides with the wall. The coefficient of restitution between the disk and the wall is $e = 0.6$. When $\theta = 0°$, the disk hangs such that it just touches the wall. Neglect friction at the pin C.

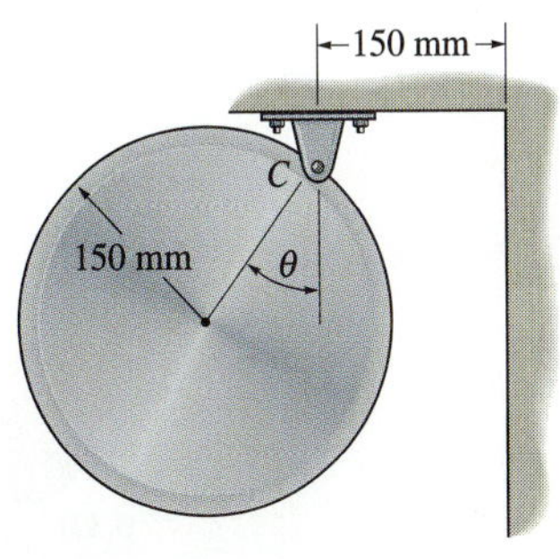

Prob. 19–48

19-49. The 6-lb slender rod AB is released from rest when it is in the *horizontal position* so that it begins to rotate clockwise. A 1-lb ball is thrown at the rod with a velocity $v = 50$ ft/s. The ball strikes the rod at C at the instant the rod is in the vertical position as shown. Determine the angular velocity of the rod just after the impact. Take $e = 0.7$.

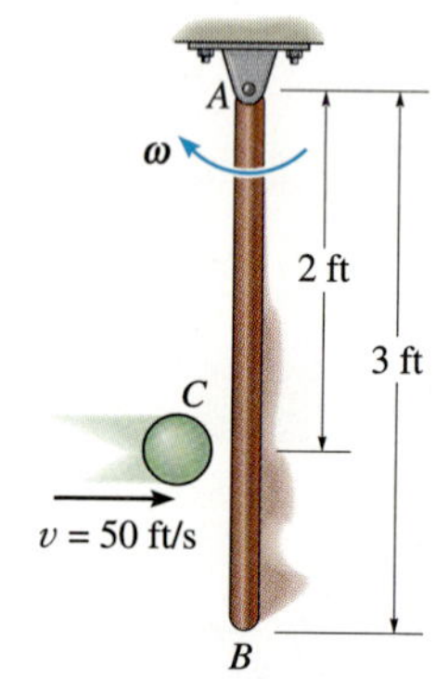

Prob. 19–49

19-50. The pendulum consists of a 10-lb solid ball and 4-lb rod. If it is released from rest when $\theta_0 = 0°$, determine the angle θ_1 of rebound after the ball strikes the wall and the pendulum swings up to the point of momentary rest. Take $e = 0.6$.

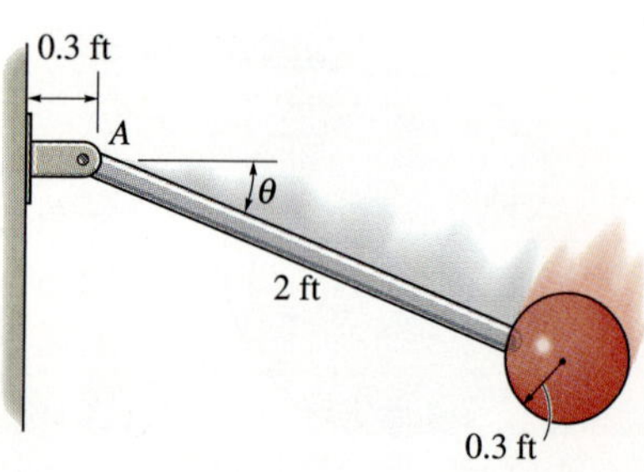

Prob. 19–50

19-51. The two disks each weigh 10 lb. If they are released from rest when $\theta = 30°$, determine the maximum angle θ after they collide and rebound from each other. The coefficient of restitution is $e = 0.75$.

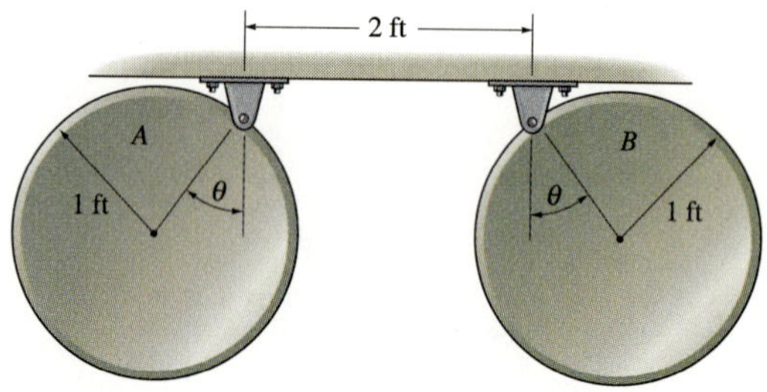

Prob. 19–51

***19-52.** The pendulum consists of a 10-lb sphere and 4-lb rod. If it is released from rest when $\theta_0 = 90°$, determine the angle θ_1 of rebound after the sphere strikes the floor. Take $e = 0.8$.

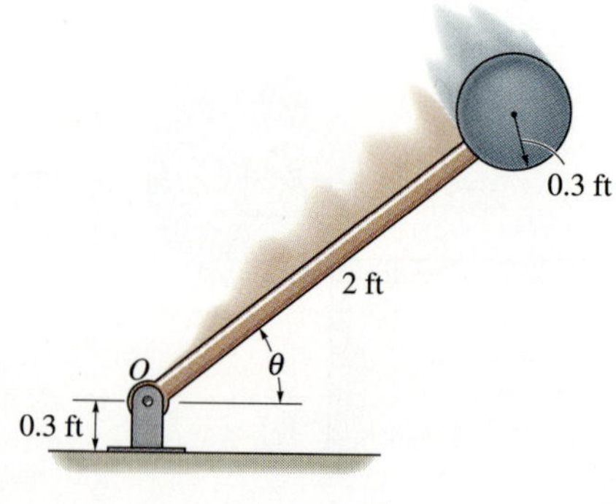

Prob. 19–52

19-53. The plank has a weight of 30 lb, center of gravity at G, and it rests on the two sawhorses at A and B. If the end D is raised 2 ft above the top of the sawhorses and is released from rest, determine how high end C will rise from the top of the sawhorses after the plank falls so that it rotates clockwise about A, strikes and pivots on the sawhorse at B, and rotates clockwise off the sawhorse at A.

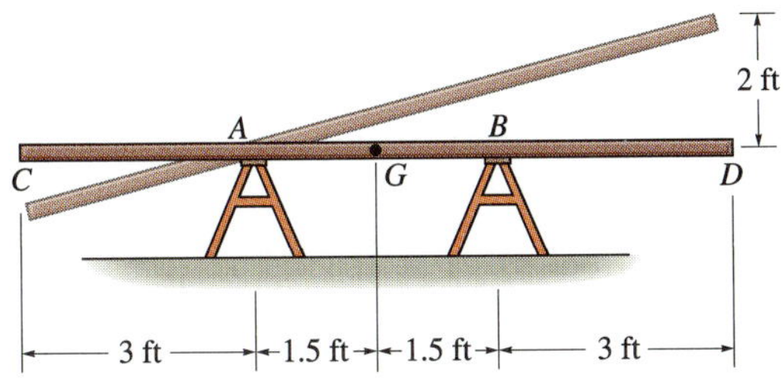

Prob. 19–53

19-55. The 15-lb rod AB is released from rest in the vertical position. If the coefficient of restitution between the floor and the cushion at B is $e = 0.7$, determine how high the end of the rod rebounds after impact with the floor.

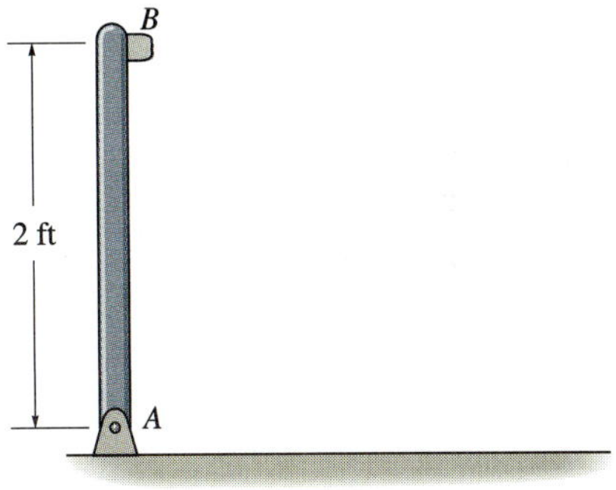

Prob. 19–55

19-54. The disk has a mass m and radius r. If it strikes the rough step having a height $\frac{1}{8}r$ as shown, determine the largest angular velocity ω_1 the disk can have and not rebound off the step when it strikes it.

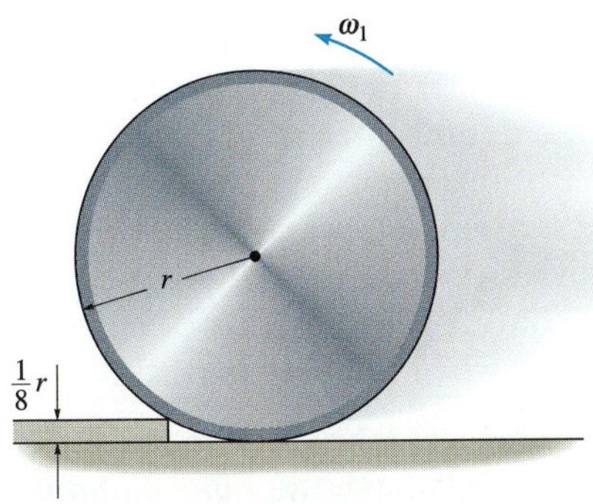

Prob. 19–54

***19-56.** A solid ball with a mass m is thrown on the ground such that at the instant of contact it has an angular velocity $\boldsymbol{\omega}_1$ and velocity components $(\mathbf{v}_G)_{x1}$ and $(\mathbf{v}_G)_{y1}$ as shown. If the ground is rough so no slipping occurs, determine the components of the velocity of its mass center just after impact. The coefficient of restitution is e.

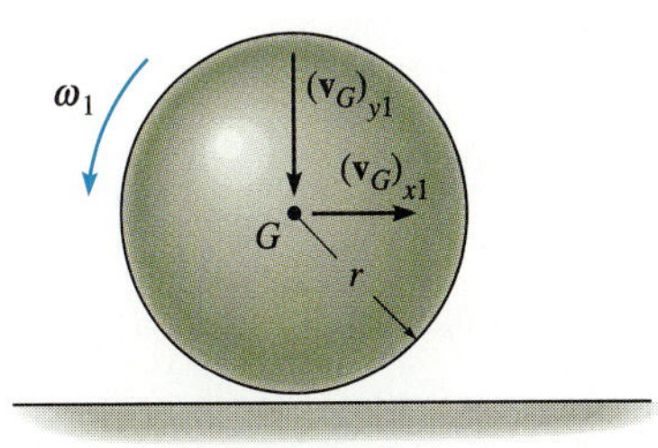

Prob. 19–56

Review 2:

Planar Kinematics and Kinetics of a Rigid Body

Having presented the various topics in planar kinematics and kinetics in Chapters 16 through 19, we will now summarize these principles and provide an opportunity for applying them to the solution of various types of problems.

Kinematics. Here we are interested in studying the geometry of motion, without concern for the forces which cause the motion. Before solving a planar kinematics problem, it is *first* necessary to *classify the motion* as being either rectilinear or curvilinear translation, rotation about a fixed axis, or general plane motion. In particular, problems involving general plane motion can be solved either with reference to a fixed axis (absolute motion analysis) or using translating or rotating frames of reference (relative motion analysis). The choice generally depends upon the type of constraints and the problem's geometry. In all cases, application of the necessary equations may be clarified by drawing a kinematic diagram. Remember that the *velocity* of a point is always *tangent* to its path of motion, and the *acceleration* of a point can have *components* in the *n–t* directions when the path is *curved.*

Translation. When the body moves with rectilinear or curvilinear translation, *all* the points on the body have the *same motion.*

$$\mathbf{v}_B = \mathbf{v}_A \qquad \mathbf{a}_B = \mathbf{a}_A$$

Rotation About a Fixed Axis

Angular Motion

Variable Angular Acceleration. Provided a mathematical relationship is given between *any two* of the *four* variables θ, ω, α, and t, then a *third* variable can be determined by solving one of the following equations which relate all three variables.

$$\omega = \frac{d\theta}{dt} \qquad \alpha = \frac{d\omega}{dt} \qquad \alpha\, d\theta = \omega\, d\omega$$

Constant Angular Acceleration. The following equations apply when it is *absolutely certain* that the angular acceleration is constant.

$$\theta = \theta_0 + \omega_0 t + \tfrac{1}{2}\alpha_c t^2 \qquad \omega = \omega_0 + \alpha_c t \qquad \omega^2 = \omega_0^2 + 2\alpha_c(\theta - \theta_0)$$

Motion of Point P

Once $\boldsymbol{\omega}$ and $\boldsymbol{\alpha}$ have been determined, then the circular motion of point P can be specified using the following scalar or vector equations.

$$v = \omega r \qquad \mathbf{v} = \boldsymbol{\omega} \times \mathbf{r}$$

$$a_t = \alpha r \quad a_n = \omega^2 r \qquad \mathbf{a} = \boldsymbol{\alpha} \times \mathbf{r} + \boldsymbol{\omega} \times (\boldsymbol{\omega} \times \mathbf{r})$$

General Plane Motion—Relative-Motion Analysis. Recall that when *translating axes* are placed at the "base point" A, the *relative motion* of point B with respect to A is simply *circular motion of B about A*. The following equations apply to two points A and B located on the *same* rigid body.

$$\mathbf{v}_B = \mathbf{v}_A + \mathbf{v}_{B/A} = \mathbf{v}_A + (\boldsymbol{\omega} \times \mathbf{r}_{B/A})$$

$$\mathbf{a}_B = \mathbf{a}_A + \mathbf{a}_{B/A} = \mathbf{a}_A + \boldsymbol{\alpha} \times \mathbf{r}_{B/A} + \boldsymbol{\omega} \times (\boldsymbol{\omega} \times \mathbf{r}_{B/A})$$

Rotating and translating axes are often used to analyze the motion of rigid bodies which are connected together by collars or slider blocks.

$$\mathbf{v}_B = \mathbf{v}_A + \boldsymbol{\Omega} \times \mathbf{r}_{B/A} + (\mathbf{v}_{B/A})_{xyz}$$

$$\mathbf{a}_B = \mathbf{a}_A + \dot{\boldsymbol{\Omega}} \times \mathbf{r}_{B/A} + \boldsymbol{\Omega} \times (\boldsymbol{\Omega} \times \mathbf{r}_{B/A}) + 2\boldsymbol{\Omega} \times (\mathbf{v}_{B/A})_{xyz} + (\mathbf{a}_{B/A})_{xyz}$$

Kinetics. To analyze the forces which cause the motion we must use the principles of kinetics. When applying the necessary equations, it is important to first establish the inertial coordinate system and define the positive directions of the axes. The *directions* should be the *same* as those selected when writing any equations of kinematics provided *simultaneous solution* of equations becomes necessary.

Equations of Motion. These equations are used to determine accelerated motions or forces causing the motion. If used to determine position, velocity, or time of motion, then kinematics will have to be considered for part of the solution. Before applying the equations of motion, *always draw a free-body diagram* in order to identify all the forces acting on the body. Also, establish the directions of the acceleration of the mass center and the angular acceleration of the body. (A kinetic diagram may also be drawn in order to represent $m\mathbf{a}_G$ and $I_G\boldsymbol{\alpha}$ graphically. This diagram is particularly convenient for resolving $m\mathbf{a}_G$ into components and for identifying the terms in the moment sum $\Sigma(\mathcal{M}_k)_P$.)

The three equations of motion are

$$\Sigma F_x = m(a_G)_x$$
$$\Sigma F_y = m(a_G)_y$$
$$\Sigma M_G = I_G\alpha \quad \text{or} \quad \Sigma M_P = \Sigma(\mathcal{M}_k)_P$$

In particular, if the body is *rotating about a fixed axis,* moments may also be summed about point O on the axis, in which case

$$\Sigma M_O = \Sigma(\mathcal{M}_k)_O = I_O\alpha$$

Work and Energy. *The equation of work and energy is used to solve problems involving force, velocity, and displacement.* Before applying this equation, *always draw a free-body diagram* of the body in order to identify the forces which do work. Recall that the kinetic energy of the body is due to translational motion of the mass center, $\mathbf{v}_G$, *and* rotational motion of the body, $\boldsymbol{\omega}$.

$$T_1 + \Sigma U_{1-2} = T_2$$

where

$$T = \tfrac{1}{2}mv_G^2 + \tfrac{1}{2}I_G\omega^2$$

$$U_F = \int F\cos\theta\, ds \qquad \text{(variable force)}$$
$$U_{F_c} = F_c\cos\theta(s_2 - s_1) \qquad \text{(constant force)}$$
$$U_W = -W\,\Delta y \qquad \text{(weight)}$$
$$U_s = -(\tfrac{1}{2}ks_2^2 - \tfrac{1}{2}ks_1^2) \qquad \text{(spring)}$$
$$U_M = M\theta \qquad \text{(constant couple moment)}$$

If the forces acting on the body are *conservative forces,* then apply the *conservation of energy equation.* This equation is easier to use than the equation of work and energy, since it applies only at *two points* on the path and *does not* require calculation of the work done by a force as the body moves along the path.

$$T_1 + V_1 = T_2 + V_2$$

where

$$V_g = Wy \qquad \text{(gravitational potential energy)}$$
$$V_e = \tfrac{1}{2}ks^2 \qquad \text{(elastic potential energy)}$$

Impulse and Momentum. *The principles of linear and angular impulse and momentum are used to solve problems involving force, velocity, and time.* Before applying the equations, *draw a free-body diagram* in order to identify all the forces which cause linear and angular impulses on the body. Also, establish the directions of the velocity of the mass center and the angular velocity of the body just before and just after the impulses are applied. (As an alternative procedure, the impulse and momentum diagrams may accompany the solution in order to graphically account for the terms in the equations. These diagrams are particularly advantageous when computing the angular impulses and angular momenta about a point other than the body's mass center.)

$$m(\mathbf{v}_G)_1 + \Sigma \int \mathbf{F}\,dt = m(\mathbf{v}_G)_2$$

$$(\mathbf{H}_G)_1 + \Sigma \int \mathbf{M}_G\,dt = (\mathbf{H}_G)_2$$

or

$$(\mathbf{H}_O)_1 + \Sigma \int \mathbf{M}_O\,dt = (\mathbf{H}_O)_2$$

Conservation of Momentum. If nonimpulsive forces or no impulsive forces act on the body in a particular direction, or if the motions of several bodies are involved in the problem, then consider applying the conservation of linear or angular momentum for the solution. Investigation of the free-body diagram (or the impulse diagram) will aid in determining the directions for which the impulsive forces are zero, or axes about which the impulsive forces cause zero angular momentum. For these cases,

$$m(\mathbf{v}_G)_1 = m(\mathbf{v}_G)_2$$
$$(\mathbf{H}_O)_1 = (\mathbf{H}_O)_2$$

The problems that follow involve application of all the above concepts. They are presented in *random order* so that practice may be gained at identifying the various types of problems and developing the skills necessary for their solution.

Review Problems

R2-1. An automobile transmission consists of the planetary gear system shown. If the ring gear R is held fixed so that $\omega_R = 0$, and the shaft s, which is fixed to the sun gear S, is rotating at 20 rad/s, determine the angular velocity of each planet gear P and the angular velocity of the connecting rack D, which is free to turn about the center shaft s.

R2-2. An automobile transmission consists of the planetary gear system shown. If the ring gear R is rotating at $\omega_R = 2$ rad/s, and the shaft s, which is fixed to the sun gear S, is rotating at 20 rad/s, determine the angular velocity of each planet gear P and the angular velocity of the connecting rack D, which is free to turn about the center shaft s.

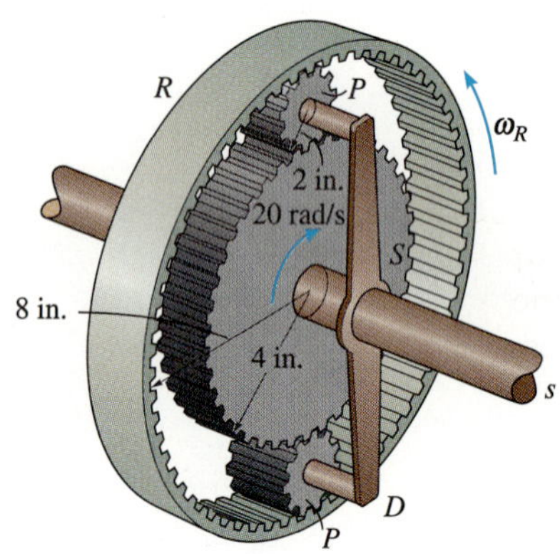

Probs. R2–1/2

R2-3. The 6-lb slender rod AB is released from rest when it is in the *horizontal position* so that it begins to rotate clockwise. A 1-lb ball is thrown at the rod with a velocity $v = 50$ ft/s. The ball strikes the rod at C at the instant the rod is in the vertical position as shown. Determine the angular velocity of the rod just after the impact. Take $e = 0.7$ and $d = 2$ ft.

***R2-4.** The 6-lb slender rod AB is originally at rest, suspended in the vertical position. A 1-lb ball is thrown at the rod with a velocity $v = 50$ ft/s and strikes the rod at C. Determine the angular velocity of the rod just after the impact. Take $e = 0.7$ and $d = 2$ ft.

R2-5. The 6-lb slender rod is originally at rest, suspended in the vertical position. Determine the distance d where the 1-lb ball, traveling at $v = 50$ ft/s, should strike the rod so that it does not create a horizontal impulse at A. What is the rod's angular velocity just after the impact? Take $e = 0.5$.

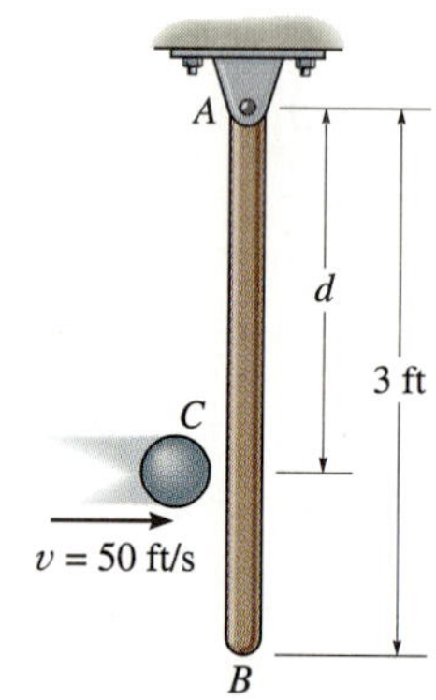

Probs. R2–3/4/5

R2-6. At a given instant, the wheel is rotating with the angular motions shown. Determine the acceleration of the collar at A at this instant.

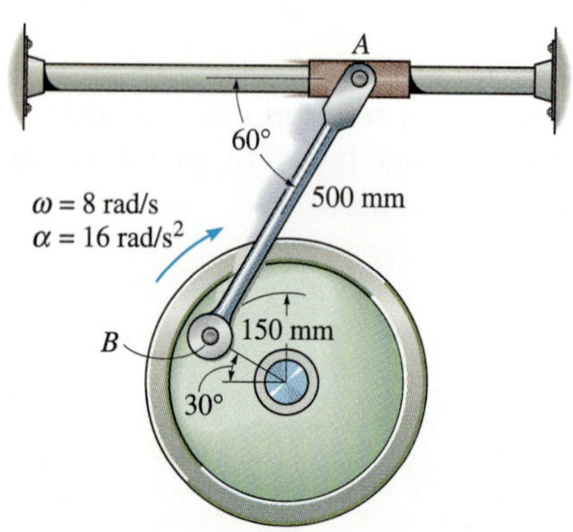

Prob. R2–6

R2-7. The transmission gear fixed onto the frame of an electric train turns at a constant rate of $\omega_t = 30\ \text{rad/s}$. This gear is in mesh with the gear that is fixed to the axle of the engine. Determine the velocity of the train, assuming the wheels do not slip on the track.

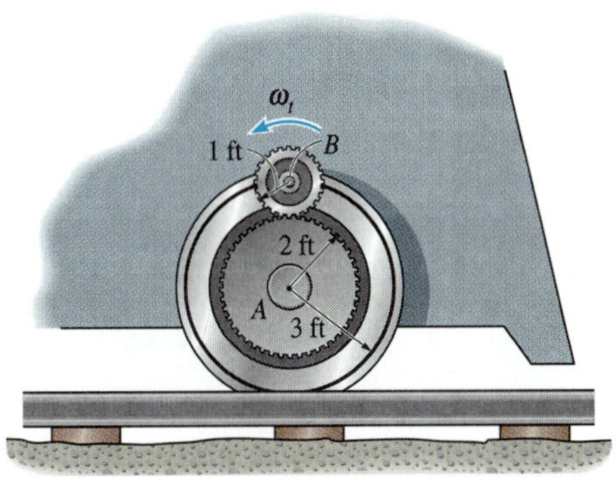

Prob. R2–7

***R2-8.** The 50-kg cylinder has an angular velocity of 30 rad/s when it is brought into contact with the surface at C. If the coefficient of kinetic friction is $\mu_k = 0.2$, determine how long it takes for the cylinder to stop spinning. What force is developed in the link AB during this time? The axis of the cylinder is connected to *two* symmetrical links. (Only AB is shown.) For the computation, neglect the weight of the links.

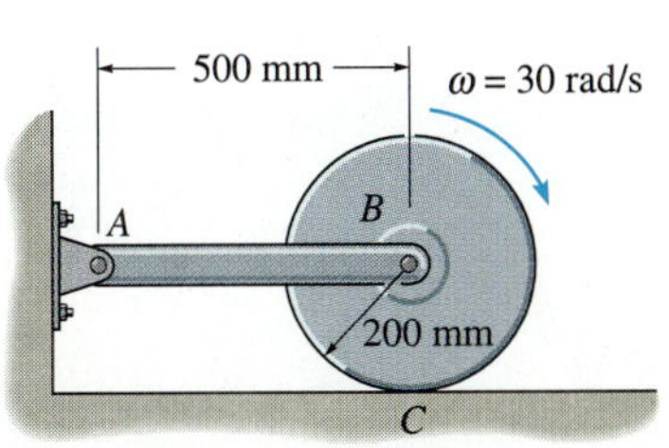

Prob. R2–8

R2-9. The gear rack has a mass of 6 kg, and the gears each have a mass of 4 kg and a radius of gyration of $k = 30$ mm. If the rack is originally moving downward at 2 m/s, when $s = 0$, determine the speed of the rack when $s = 600$ mm. The gears are free to turn about their centers, A and B.

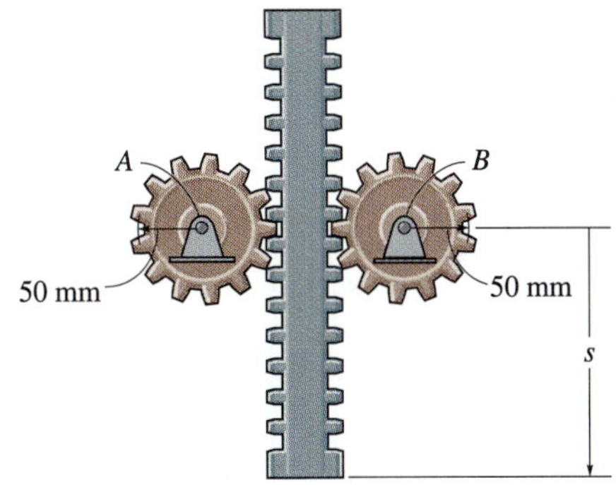

Prob. R2–9

R2-10. The gear has a mass of 2 kg and a radius of gyration $k_A = 0.15$ m. The connecting link (slender rod) and slider block at B have a mass of 4 kg and 1 kg, respectively. If the gear has an angular velocity $\omega = 8$ rad/s at the instant $\theta = 45°$, determine the gear's angular velocity when $\theta = 0°$.

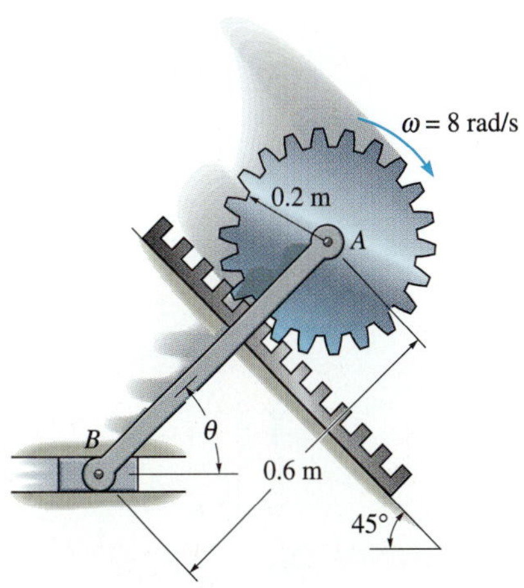

Prob. R2–10

R2-11. The rotation of link AB creates an oscillating movement of gear F. If AB has an angular velocity of $\omega_{AB} = 6$ rad/s, determine the angular velocity of gear F at the instant shown. Gear E is rigidly attached to arm CD and pinned at D to a fixed point.

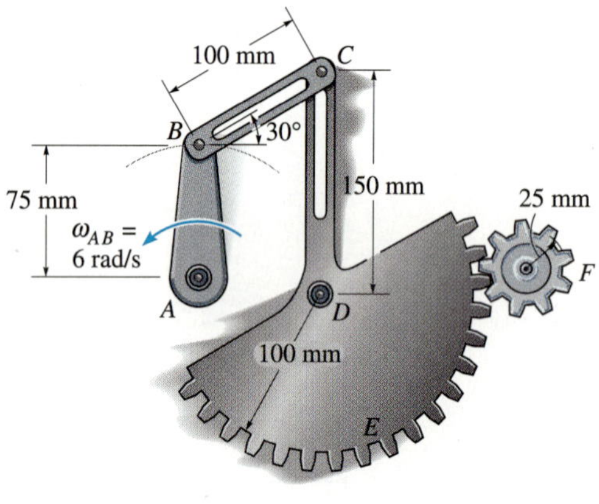

Prob. R2–11

***R2-12.** The revolving door consists of four doors which are attached to an axle AB. Each door can be assumed to be a 50-lb thin plate. Friction at the axle contributes a moment of 2 lb·ft which resists the rotation of the doors. If a woman passes through one door by always pushing with a force $P = 15$ lb perpendicular to the plane of the door as shown, determine the door's angular velocity after it has rotated 90°. The doors are originally at rest.

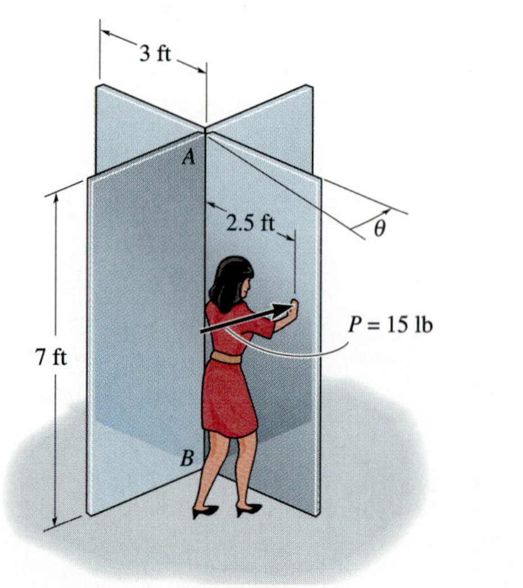

Prob. R2–12

R2-13. The 10-lb cylinder rests on the 20-lb dolly. If the system is released from rest, determine the angular velocity of the cylinder in 2 s. The cylinder does not slip on the dolly. Neglect the mass of the wheels on the dolly.

R2-14. Solve Prob. R2-13 if the coefficients of static and kinetic friction between the cylinder and the dolly are $\mu_s = 0.3$ and $\mu_k = 0.2$, respectively.

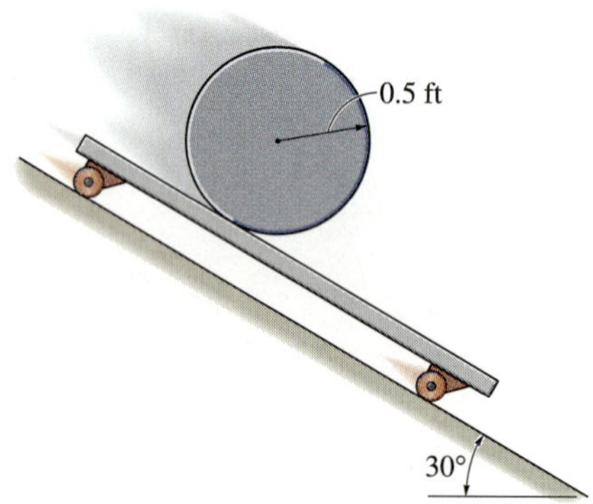

Probs. R2–13/14

R2-15. Gears C and H each have a weight of 0.4 lb and a radius of gyration about their mass center of $(k_H)_B = (k_C)_A = 2$ in. The uniform link AB has a weight of 0.2 lb and a radius of gyration of $(k_{AB})_A = 3$ in., whereas link DE has a weight of 0.15 lb and a radius of gyration of $(k_{DE})_B = 4.5$ in. If a couple moment of $M = 3$ lb·ft is applied to link AB and the assembly is originally at rest, determine the angular velocity of link DE when link AB has rotated 360°. Gear C is fixed from rotating and motion occurs in the horizontal plane. Also, gear H and link DE rotate together about the same shaft at B.

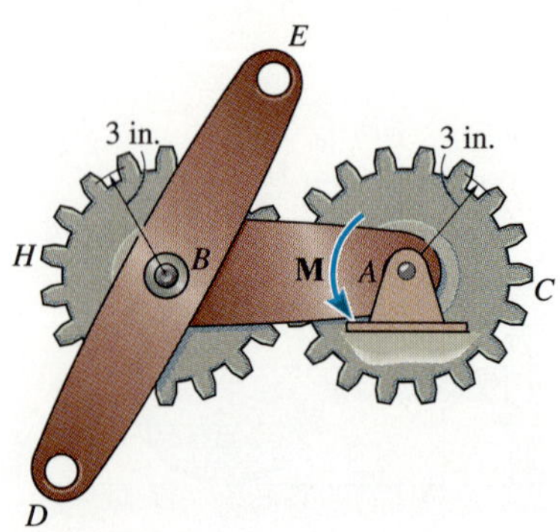

Prob. R2–15

***R2-16.** The inner hub of the roller bearing is rotating with an angular velocity of $\omega_i = 6$ rad/s, while the outer hub is rotating in the opposite direction at $\omega_o = 4$ rad/s. Determine the angular velocity of each of the rollers if they roll on the hubs without slipping.

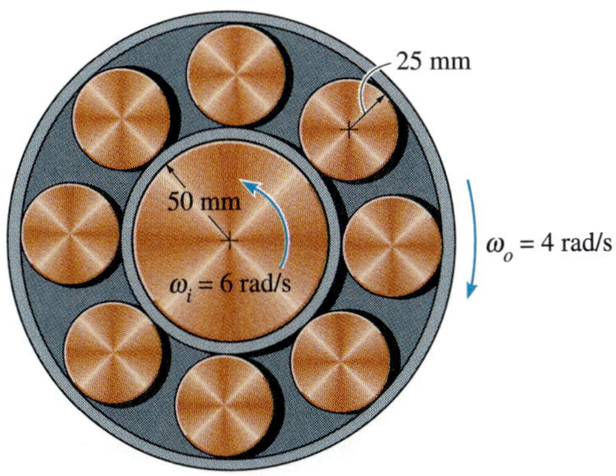

Prob. R2–16

R2-17. The hoop (thin ring) has a mass of 5 kg and is released down the inclined plane such that it has a backspin $\omega = 8$ rad/s and its center has a velocity $v_G = 3$ m/s as shown. If the coefficient of kinetic friction between the hoop and the plane is $\mu_k = 0.6$, determine how long the hoop rolls before it stops slipping.

R2-18. The hoop (thin ring) has a mass of 5 kg and is released down the inclined plane such that it has a backspin $\omega = 8$ rad/s and its center has a velocity $v_G = 3$ m/s as shown. If the coefficient of kinetic friction between the hoop and the plane is $\mu_k = 0.6$, determine the hoop's angular velocity in 1 s.

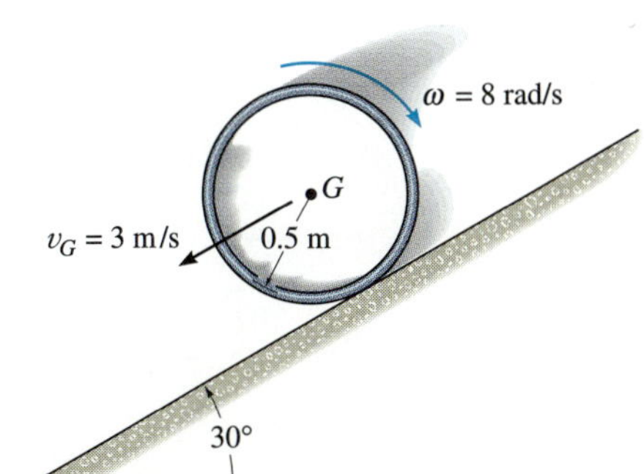

Probs. R2–17/18

R2-19. Determine the angular velocity of rod CD at the instant $\theta = 30°$. Rod AB moves to the left at a constant rate $v_{AB} = 5$ m/s.

***R2-20.** Determine the angular acceleration of rod CD at the instant $\theta = 30°$. Rod AB has zero velocity, i.e., $v_{AB} = 0$, and an acceleration of $a_{AB} = 2\ \text{m/s}^2$ to the right when $\theta = 30°$.

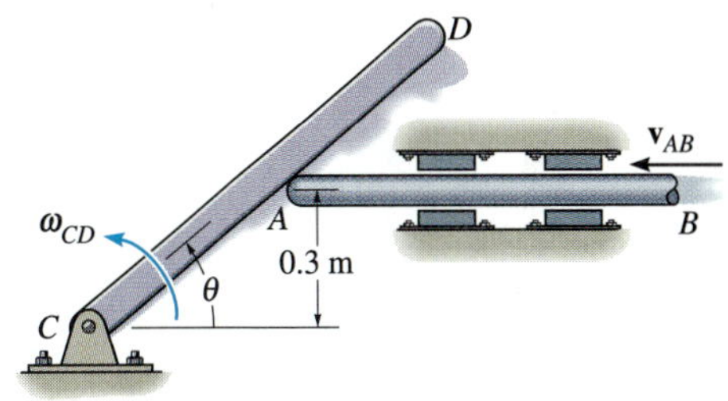

Probs. R2–19/20

R2-21. The slender 15-kg bar is initially at rest and standing in the vertical position when the bottom end A is displaced slightly to the right. If the track in which it moves is smooth, determine the speed at which end A strikes the corner D. The bar is constrained to move in the vertical plane. Neglect the mass of the cord BC.

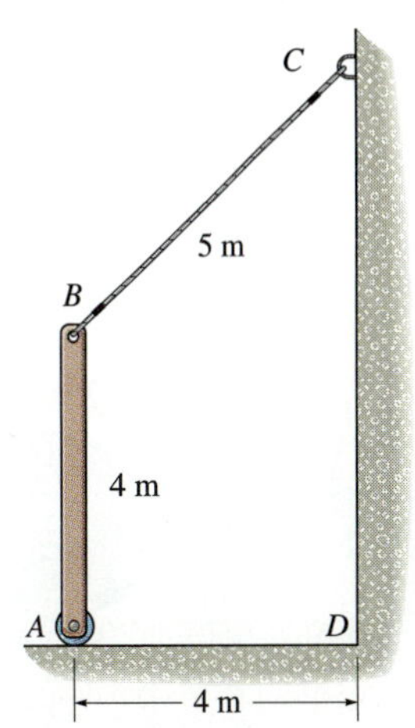

Prob. R2–21

R2-22. The four-wheeler has a weight of 335 lb and center of gravity at G_1, whereas the rider has a weight of 150 lb and center of gravity at G_2. If the engine can develop enough torque to cause the wheels to slip, determine the largest coefficient of static friction between the rear wheels and the ground so that the vehicle will accelerate without tipping over. What is this maximum acceleration? In order to *increase* the acceleration, should the rider crouch down or sit up straight from the position shown? Explain. The front wheels are free to roll. Neglect the mass of the wheels in the calculation.

Prob. R2–22

R2-23. The four-wheeler has a weight of 335 lb and center of gravity at G_1, whereas the rider has a weight of 150 lb and center of gravity at G_2. If the coefficient of static friction between the rear wheels and the ground is $\mu_s = 0.3$, determine the greatest acceleration the vehicle can have. The front wheels are free to roll. Neglect the mass of the wheels in the calculation.

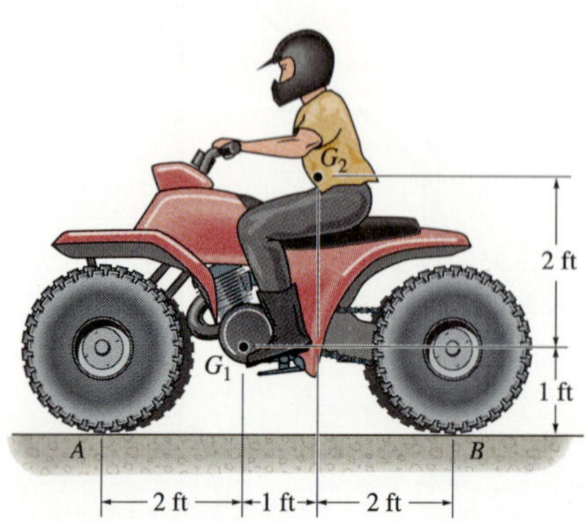

Prob. R2–23

***R2-24.** The pavement roller is traveling down the incline at $v_1 = 5$ ft/s when the motor is disengaged. Determine the speed of the roller when it has traveled 20 ft down the plane. The body of the roller, excluding the rollers, has a weight of 8000 lb and a center of gravity at G. Each of the two rear rollers weighs 400 lb and has a radius of gyration of $k_A = 3.3$ ft. The front roller has a weight of 800 lb and a radius of gyration of $k_B = 1.8$ ft. The rollers do not slip as they rotate.

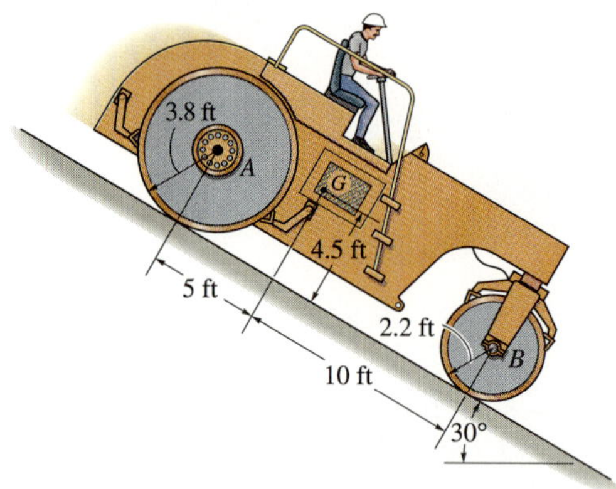

Prob. R2–24

R2-25. The cylinder B rolls on the fixed cylinder A without slipping. If bar CD is rotating with an angular velocity $\omega_{CD} = 5$ rad/s, determine the angular velocity of cylinder B. Point C is a fixed point.

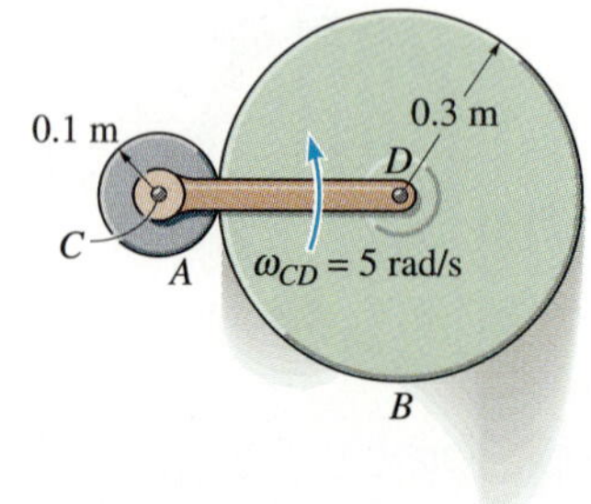

Prob. R2–25

R2-26. The disk has a mass M and a radius R. If a block of mass m is attached to the cord, determine the angular acceleration of the disk when the block is released from rest. Also, what is the distance the block falls from rest in the time t?

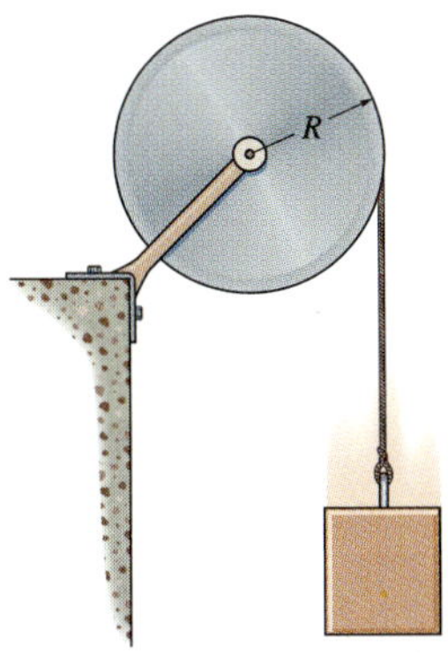

Prob. R2–26

R2-27. The tub of the mixer has a weight of 70 lb and a radius of gyration $k_G = 1.3$ ft about its center of gravity. If a constant torque $M = 60$ lb·ft is applied to the dumping wheel, determine the angular velocity of the tub when it has rotated $\theta = 90°$. Originally the tub is at rest when $\theta = 0°$.

***R2-28.** Solve Prob. R2-27 if the applied torque is $M = (50\theta)$ lb·ft, where θ is in radians.

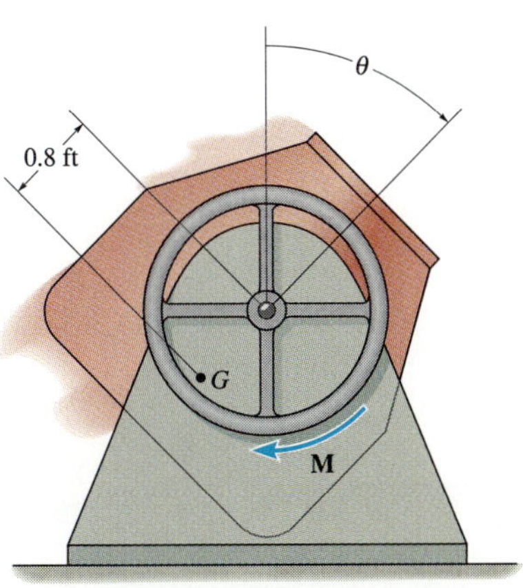

Probs. R2–27/28

R2-29. The spool has a weight of 30 lb and a radius of gyration $k_O = 0.45$ ft. A cord is wrapped around the spool's inner hub and its end subjected to a horizontal force $P = 5$ lb. Determine the spool's angular velocity in 4 s starting from rest. Assume the spool rolls without slipping.

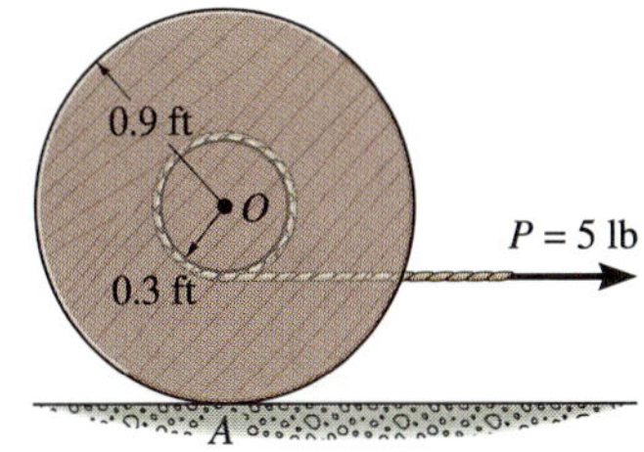

Prob. R2–29

R2-30. The slender rod of length L and mass m is released from rest when $\theta = 0°$. Determine, as a function of θ, the normal and frictional forces which are exerted on the ledge at A as the rod falls downward. At what angle θ does it begin to slip if the coefficient of static friction at A is μ_s?

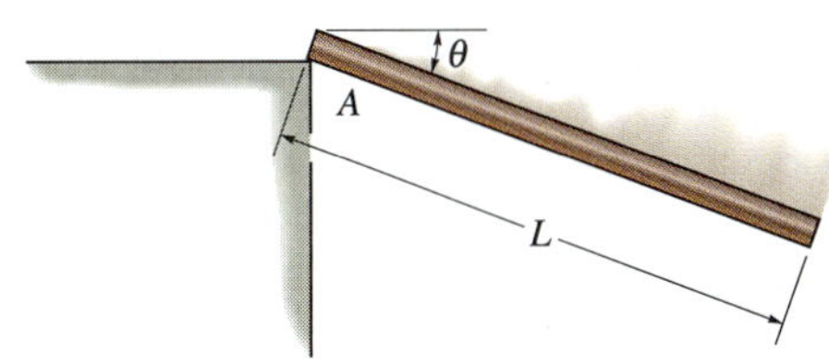

Prob. R2–30

R2-31. A sphere and cylinder are released from rest on the ramp at $t = 0$. If each has a mass m and a radius r, determine their angular velocities at time t. Assume no slipping occurs.

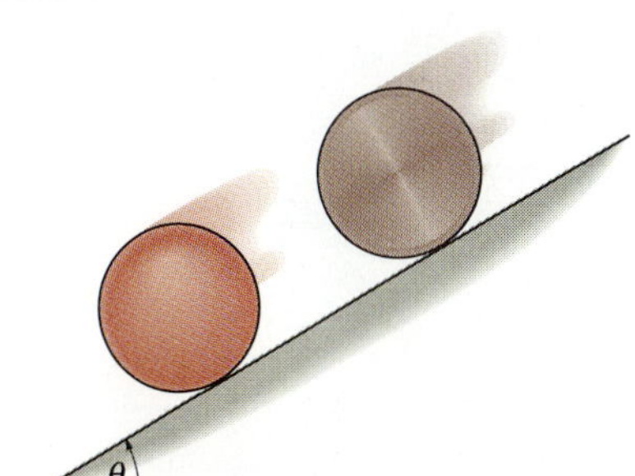

Prob. R2–31

***R2-32.** At a given instant, link AB has an angular acceleration $\alpha_{AB} = 12$ rad/s^2 and an angular velocity $\omega_{AB} = 4$ rad/s. Determine the angular velocity and angular acceleration of link CD at this instant.

R2-33. At a given instant, link CD has an angular acceleration $\alpha_{CD} = 5$ rad/s^2 and angular velocity $\omega_{CD} = 2$ rad/s. Determine the angular velocity and angular acceleration of link AB at this instant.

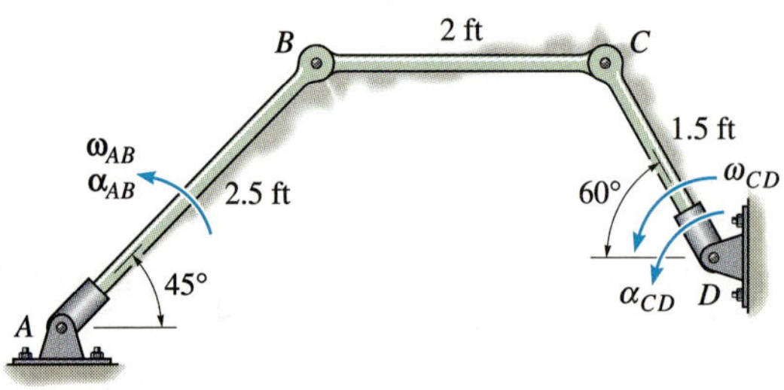

Probs. R2–32/33

R2-34. The spool and the wire wrapped around its core have a mass of 50 kg and a centroidal radius of gyration of $k_G = 235$ mm. If the coefficient of kinetic friction at the surface is $\mu_k = 0.15$, determine the angular acceleration of the spool after it is released from rest.

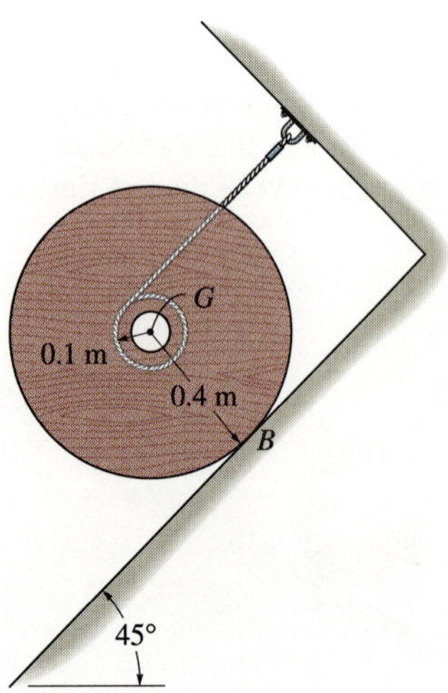

Prob. R2–34

R2-35. The bar is confined to move along the vertical and inclined planes. If the velocity of the roller at A is $v_A = 6$ ft/s when $\theta = 45°$, determine the bar's angular velocity and the velocity of B at this instant.

***R2-36.** The bar is confined to move along the vertical and inclined planes. If the roller at A has a constant velocity of $v_A = 6$ ft/s, determine the bar's angular acceleration and the acceleration of B when $\theta = 45°$.

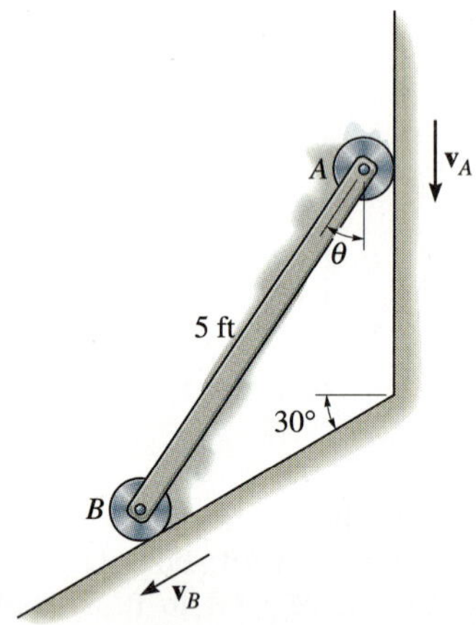

Probs. R2–35/36

R2-37. The 15-lb cylinder is initially at rest on a 5-lb plate. If a couple moment of $M = 40$ lb·ft is applied to the cylinder, determine the angular acceleration of the cylinder and the time needed for the end B of the plate to travel 3 ft and strike the wall. Assume the cylinder does not slip on the plate, and neglect the mass of the rollers under the plate.

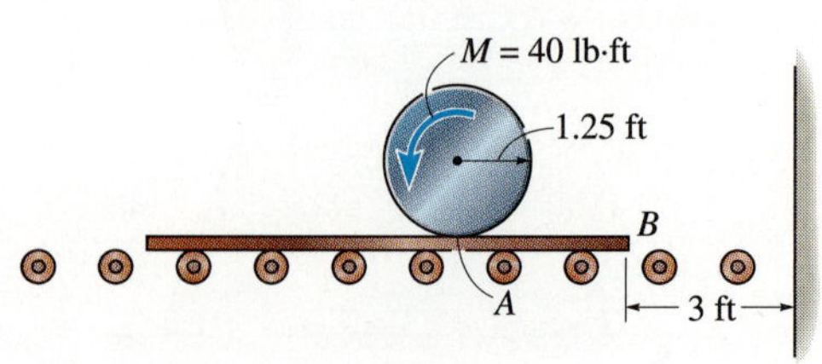

Prob. R2–37

R2-38. Each gear has a mass of 2 kg and a radius of gyration about its pinned mass centers A and B of k_g = 40 mm. Each link has a mass of 2 kg and a radius of gyration about its pinned ends A and B of $k_l = 50$ mm. If originally the spring is unstretched when the couple moment $M = 20$ N·m is applied to link AC, determine the angular velocities of the links at the instant link AC rotates $\theta = 45°$. Each gear and link is connected together and rotates about the fixed pins A and B.

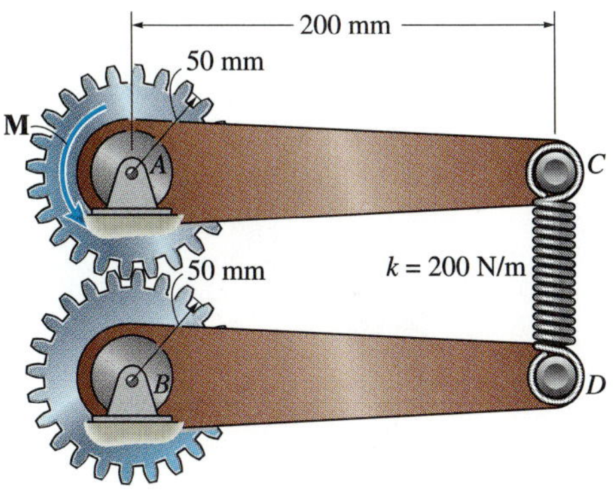

Prob. R2–38

R2-39. The 5-lb rod AB supports the 3-lb disk at its end. If the disk is given an angular velocity $\omega_D = 8$ rad/s while the rod is held stationary and then released, determine the angular velocity of the rod after the disk has stopped spinning relative to the rod due to frictional resistance at the bearing A. Motion is in the *horizontal plane.* Neglect friction at the fixed bearing B.

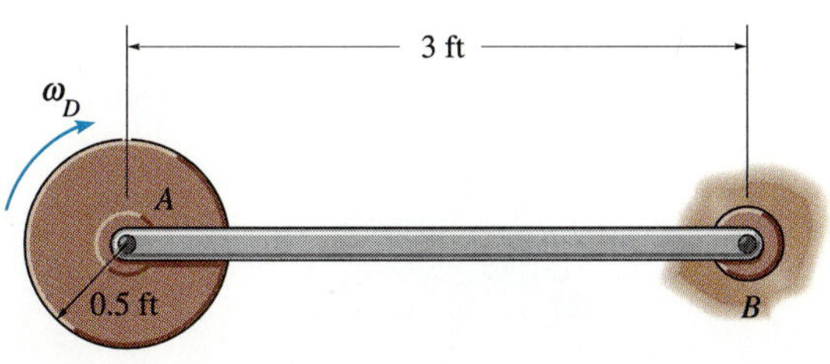

Prob. R2–39

***R2-40.** If link AB is rotating at $\omega_{AB} = 3$ rad/s, determine the angular velocity of link CD at the instant shown.

R2-41. If link CD is rotating at $\omega_{CD} = 5$ rad/s, determine the angular velocity of link AB at the instant shown.

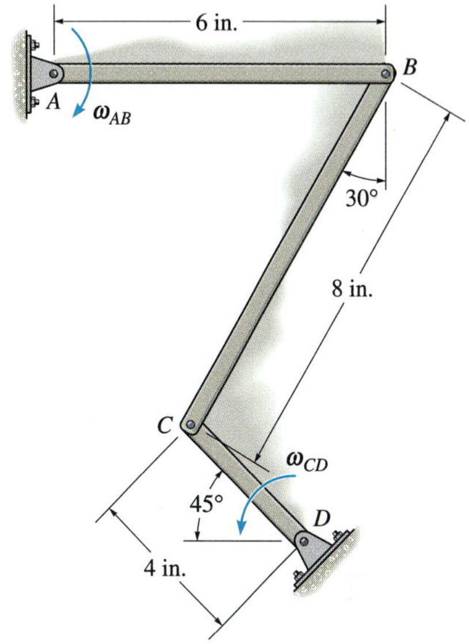

Probs. R2–40/41

R2-42. The 15-kg disk is pinned at O and is initially at rest. If a 10-g bullet is fired into the disk with a velocity of 200 m/s, as shown, determine the maximum angle θ to which the disk swings. The bullet becomes embedded in the disk.

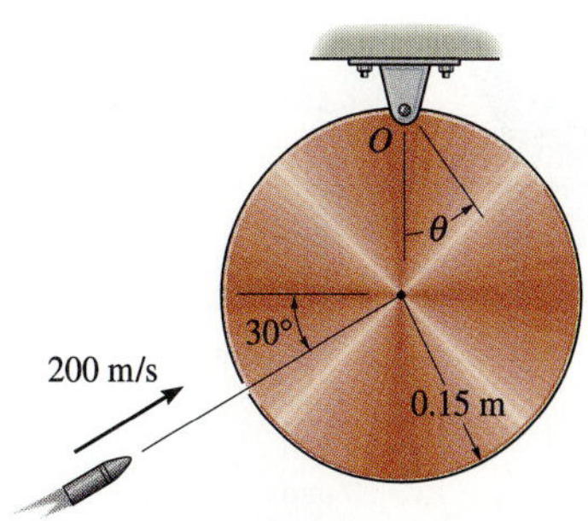

Prob. R2–42

R2-43. The disk is rotating at a constant rate of 4 rad/s as it falls freely, its center G having an acceleration of 32.2 ft/s^2. Determine the accelerations of points A and B on the rim of the disk at the instant shown.

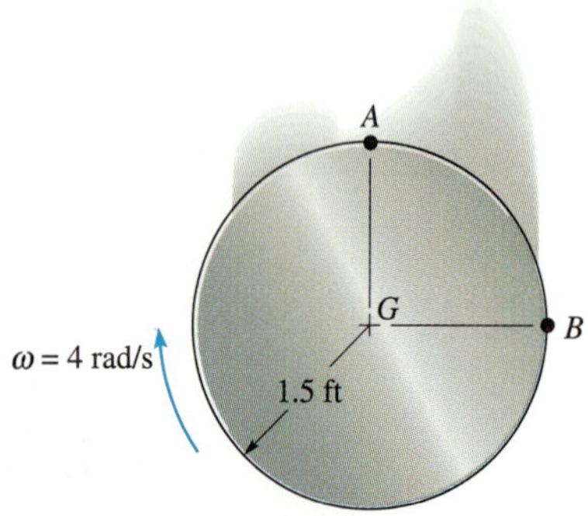

Prob. R2–43

***R2-44.** The operation of "reverse" for a three-speed automotive transmission is illustrated schematically in the figure. If the crankshaft G is turning with an angular speed of 60 rad/s, determine the angular speed of the drive shaft H. Each of the gears rotates about a fixed axis. Note that gears A and B, C and D, E and F are in mesh. The radius of each of these gears is reported in the figure.

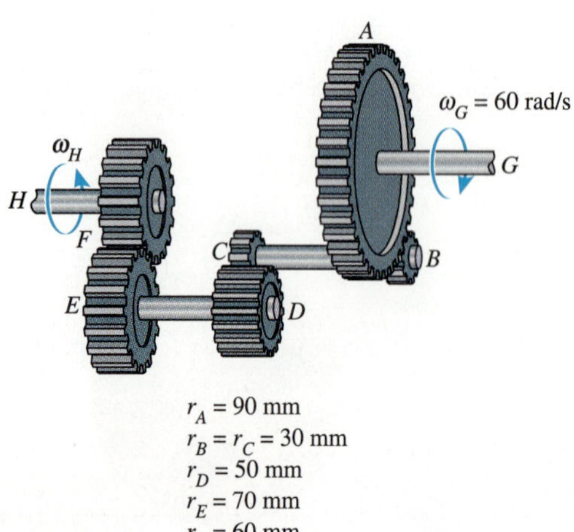

Prob. R2–44

R2-45. Shown is the internal gearing of a "spinner" used for drilling wells. With constant angular acceleration, the motor M rotates the shaft S to 100 rev/min in $t = 2$ s starting from rest. Determine the angular acceleration of the drill-pipe connection D and the number of revolutions it makes during the 2-s startup.

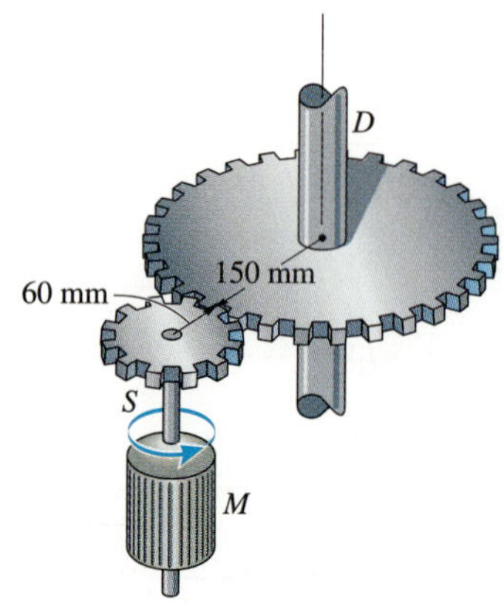

Prob. R2–45

R2-46. The plate weighs 40 lb and is supported by a roller at A. If a horizontal force of $F = 70$ lb is suddenly applied to the roller, determine the acceleration of the center of the roller at the instant the force is applied. The plate has a moment of inertia about its center of mass of $I_G = 0.414 \text{ slug}\cdot\text{ft}^2$. Neglect the weight and the size d of the roller.

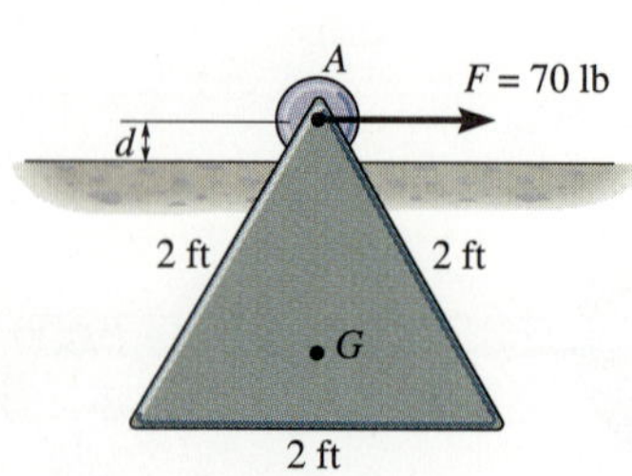

Prob. R2–46

R2-47. The 15-kg cylinder is rotating with an angular velocity of $\omega = 40$ rad/s. If a force $F = 6$ N is applied to link AB, as shown, determine the time needed to stop the rotation. The coefficient of kinetic friction between AB and the cylinder is $\mu_k = 0.4$.

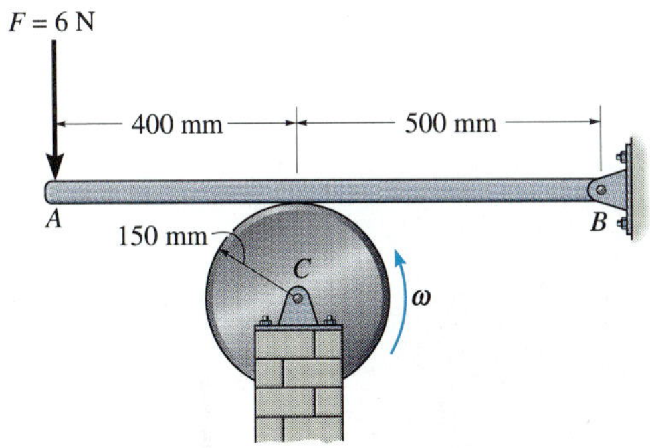

Prob. R2–47

***R2-48.** If link AB is rotating at $\omega_{AB} = 6$ rad/s, determine the angular velocities of links BC and CD at the instant shown.

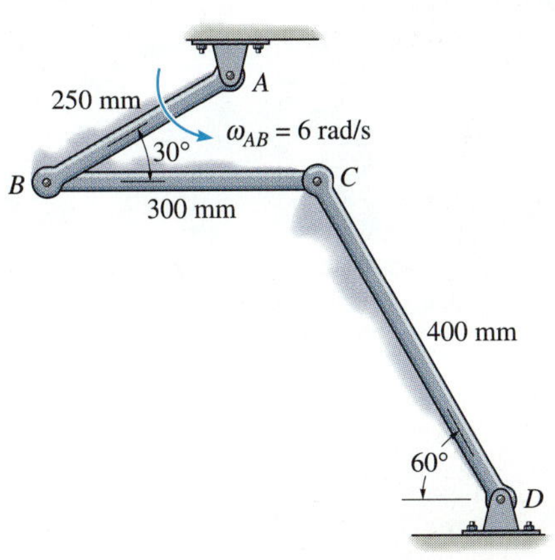

Prob. R2–48

R2-49. The wheel has a mass of 25 kg and a radius of gyration $k_B = 0.15$ m. It is originally spinning at $\omega_1 = 40$ rad/s. If it is placed on the ground, for which the coefficient of kinetic friction is $\mu_k = 0.5$, determine the time required for the motion to stop. What are the horizontal and vertical components of reaction which the pin at A exerts on AB during this time? Neglect the mass of AB.

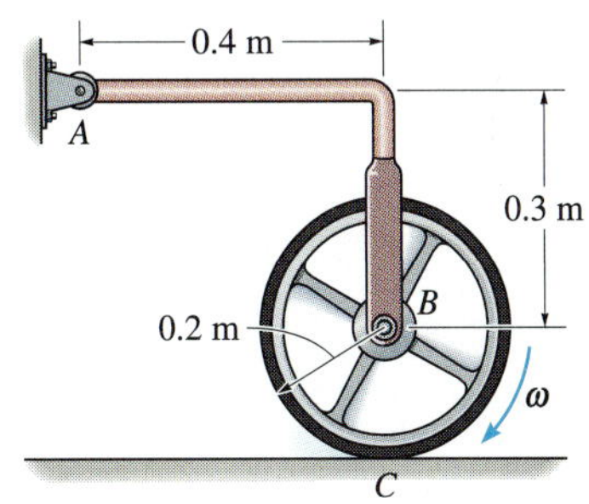

Prob. R2–49

R2-50. At the *start* of take-off, the propeller on the 2-Mg plane exerts a horizontal thrust of 600 N on the plane. Determine the plane's acceleration and the vertical reactions at the nose wheel A and each of the *two* wing wheels B. Neglect the lifting force of the wings since the plane is originally at rest. The mass center is at G.

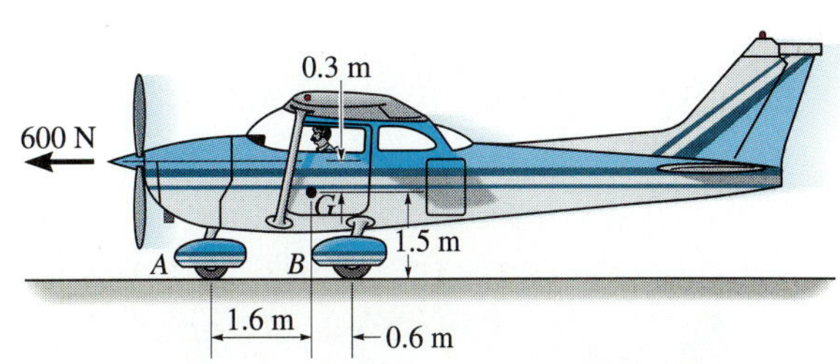

Prob. R2–50

The three-dimensional motion of these industrial robots used in the manufacturing of automobiles must be accurately specified.

CHAPTER 20

Three-Dimensional Kinematics of a Rigid Body

Chapter Objectives

- To analyze the kinematics of a body subjected to rotation about a fixed axis and general plane motion.
- To provide a relative-motion analysis of a rigid body using translating and rotating axes.

*20.1 Rotation About a Fixed Point

When a rigid body rotates about a fixed point, the distance r from the point to a particle P located on the body is the *same* for *any position* of the body. Thus, the path of motion for the particle lies on the *surface of a sphere* having a radius r and centered at the fixed point. Since motion along this path occurs only from a series of rotations made during a finite time interval, we will first develop a familiarity with some of the properties of rotational displacements.

The boom can rotate up and down, and because it is hinged to a point on the vertical axis about which it turns it is subjected to rotation about a fixed point.

Euler's Theorem. Euler's theorem states that two "component" rotations about different axes passing through a point are equivalent to a single resultant rotation about an axis passing through the point. If more than two rotations are applied, they can be combined into pairs, and each pair can be further reduced to combine into one rotation.

Finite Rotations. If the component rotations used in Euler's theorem are *finite,* it is important that the *order* in which they are applied be maintained. This is because finite rotations do *not* obey the law of vector addition, and hence they cannot be classified as vector quantities. To show this, consider the two finite rotations $\boldsymbol{\theta}_1 + \boldsymbol{\theta}_2$ applied to the block in Fig. 20–1*a*. Each rotation has a magnitude of 90° and a direction defined by the right-hand rule, as indicated by the arrow. The resultant orientation of the block is shown at the right. When these two rotations are applied in the order $\boldsymbol{\theta}_2 + \boldsymbol{\theta}_1$, as shown in Fig. 20–1*b*, the resultant position of the block is *not* the same as it is in Fig. 20–1*a*. Consequently, *finite rotations* do not obey the commutative law of addition ($\boldsymbol{\theta}_1 + \boldsymbol{\theta}_2 \neq \boldsymbol{\theta}_2 + \boldsymbol{\theta}_1$), and therefore *they cannot be classified as vectors.* If smaller, yet finite, rotations had been used to illustrate this point, e.g., 10° instead of 90°, the *resultant* orientation of the block after each combination of rotations would also be different; however, in this case, the difference is only a small amount.

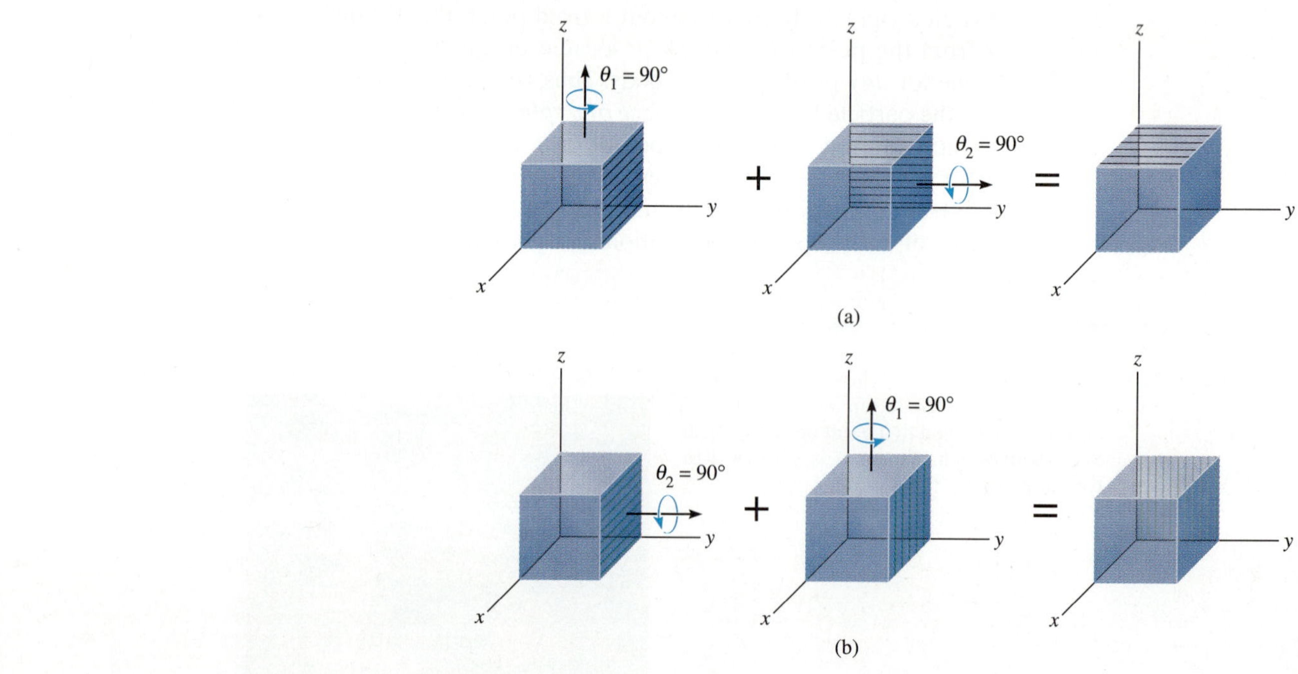

Fig. 20–1

Infinitesimal Rotations. When defining the angular motions of a body subjected to three-dimensional motion, only rotations which are *infinitesimally small* will be considered. *Such rotations may be classified as vectors, since they can be added vectorially in any manner.* To show this, let us for purposes of simplicity consider the rigid body itself to be a sphere which is allowed to rotate about its central fixed point O, Fig. 20–2a. If we impose two infinitesimal rotations $d\boldsymbol{\theta}_1 + d\boldsymbol{\theta}_2$ on the body, it is seen that point P moves along the path $d\boldsymbol{\theta}_1 \times \mathbf{r} + d\boldsymbol{\theta}_2 \times \mathbf{r}$ and ends up at P'. Had the two successive rotations occurred in the order $d\boldsymbol{\theta}_2 + d\boldsymbol{\theta}_1$, then the resultant displacements of P would have been $d\boldsymbol{\theta}_2 \times \mathbf{r} + d\boldsymbol{\theta}_1 \times \mathbf{r}$. Since the vector cross product obeys the distributive law, by comparison $(d\boldsymbol{\theta}_1 + d\boldsymbol{\theta}_2) \times \mathbf{r} = (d\boldsymbol{\theta}_2 + d\boldsymbol{\theta}_1) \times \mathbf{r}$. Here infinitesimal rotations $d\boldsymbol{\theta}$ are vectors, since these quantities have both a magnitude and direction for which the order of (vector) addition is not important, i.e., $d\boldsymbol{\theta}_1 + d\boldsymbol{\theta}_2 = d\boldsymbol{\theta}_2 + d\boldsymbol{\theta}_1$. Furthermore, as shown in Fig. 20–2a, the two "component" rotations $d\boldsymbol{\theta}_1$ and $d\boldsymbol{\theta}_2$ are equivalent to a single resultant rotation $d\boldsymbol{\theta} = d\boldsymbol{\theta}_1 + d\boldsymbol{\theta}_2$, a consequence of Euler's theorem.

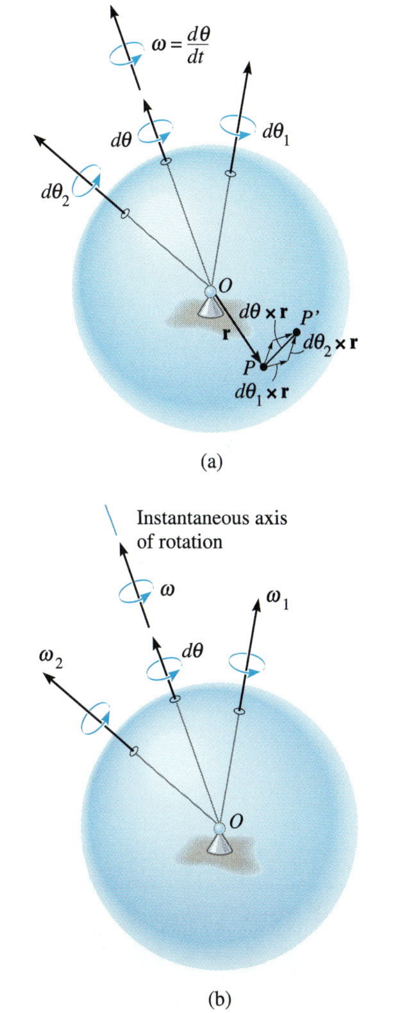

Fig. 20–2

Angular Velocity. If the body is subjected to an angular rotation $d\boldsymbol{\theta}$ about a fixed point, the angular velocity of the body is defined by the time derivative,

$$\boldsymbol{\omega} = \dot{\boldsymbol{\theta}} \tag{20–1}$$

The line specifying the direction of $\boldsymbol{\omega}$, which is collinear with $d\boldsymbol{\theta}$, is referred to as the *instantaneous axis of rotation*, Fig. 20–2b. In general, this axis changes direction during each instant of time. Since $d\boldsymbol{\theta}$ is a vector quantity, so too is $\boldsymbol{\omega}$, and it follows from vector addition that if the body is subjected to two component angular motions, $\boldsymbol{\omega}_1 = \dot{\boldsymbol{\theta}}_1$ and $\boldsymbol{\omega}_2 = \dot{\boldsymbol{\theta}}_2$, the resultant angular velocity is $\boldsymbol{\omega} = \boldsymbol{\omega}_1 + \boldsymbol{\omega}_2$.

Angular Acceleration. The body's angular acceleration is determined from the time derivative of the angular velocity, i.e.,

$$\boldsymbol{\alpha} = \dot{\boldsymbol{\omega}} \tag{20–2}$$

For motion about a fixed point, $\boldsymbol{\alpha}$ must account for a change in *both* the magnitude and direction of $\boldsymbol{\omega}$, so that, in general, $\boldsymbol{\alpha}$ is not directed along the instantaneous axis of rotation, Fig. 20–3.

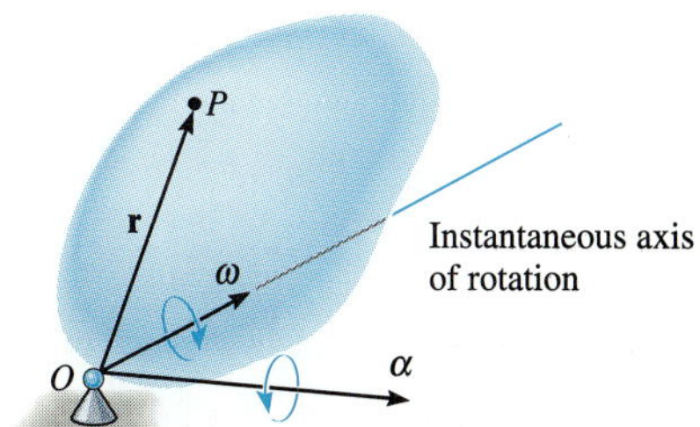

Fig. 20–3

As the direction of the instantaneous axis of rotation (or the line of action of $\boldsymbol{\omega}$) changes in space, the locus of points defined by the axis generates a fixed *space cone.* If the change in this axis is viewed with respect to the rotating body, the locus of the axis generates a *body cone,*

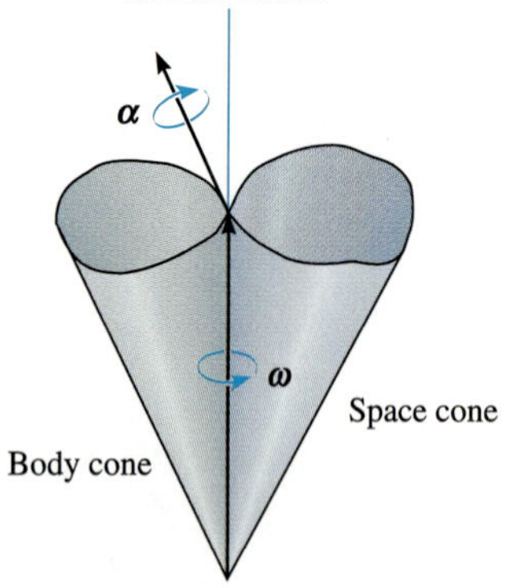

Fig. 20–4

Fig. 20–4. At any given instant, these cones are tangent along the instantaneous axis of rotation, and when the body is in motion, the body cone appears to roll either on the inside or the outside surface of the fixed space cone. Provided the paths defined by the open ends of the cones are described by the head of the $\boldsymbol{\omega}$ vector, $\boldsymbol{\alpha}$ must act tangent to these paths at any given instant, since the time rate of change of $\boldsymbol{\omega}$ is equal to $\boldsymbol{\alpha}$, Fig. 20–4.

Velocity. Once $\boldsymbol{\omega}$ is specified, the velocity of any point P on a body rotating about a fixed point can be determined using the same methods as for a body rotating about a fixed axis (Sec. 16.3). Hence, by the cross product,

$$\mathbf{v} = \boldsymbol{\omega} \times \mathbf{r} \tag{20–3}$$

Here $\mathbf{r}$ defines the position of P measured from the fixed point O, Fig. 20–3.

Acceleration. If $\boldsymbol{\omega}$ and $\boldsymbol{\alpha}$ are known at a given instant, the acceleration of any point P on the body can be obtained by time differentiation of Eq. 20–3, which yields

$$\mathbf{a} = \boldsymbol{\alpha} \times \mathbf{r} + \boldsymbol{\omega} \times (\boldsymbol{\omega} \times \mathbf{r}) \tag{20–4}$$

The form of this equation is the same as that developed in Sec. 16.3, which defines the acceleration of a point located on a body subjected to rotation about a fixed axis.

*20.2 The Time Derivative of a Vector Measured from Either a Fixed or Translating-Rotating System

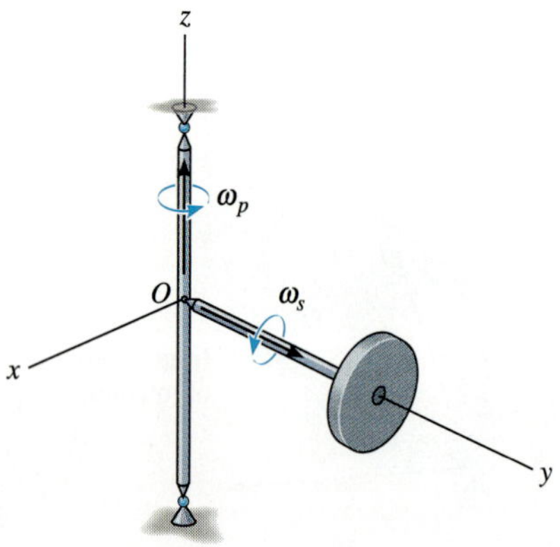

Fig. 20–5

In many types of problems involving the motion of a body about a fixed point, the angular velocity $\boldsymbol{\omega}$ is specified in terms of its component angular motions. For example, the disk in Fig. 20–5 spins about the horizontal y axis at $\boldsymbol{\omega}_s$ while it rotates or precesses about the vertical z axis at $\boldsymbol{\omega}_p$. Therefore, its resultant angular velocity is $\boldsymbol{\omega} = \boldsymbol{\omega}_s + \boldsymbol{\omega}_p$. If the angular acceleration $\boldsymbol{\alpha} = \dot{\boldsymbol{\omega}}$ of such a body is to be determined, it is sometimes easier to compute the time derivative of $\boldsymbol{\omega}$ by using a coordinate system which has a *rotation* defined by one or more of the components of $\boldsymbol{\omega}$.* For this reason, and for other uses later, an equation will presently be derived that relates the time derivative of any vector $\mathbf{A}$ defined from a translating-rotating reference to its time derivative defined from a fixed reference.

*In the case of the spinning disk, Fig. 20–5, the x, y, z axes may be given an angular velocity of $\boldsymbol{\omega}_p$.

Consider the x, y, z axes of the moving frame of reference to have an angular velocity $\mathbf{\Omega}$ which is measured from the fixed X, Y, Z axes, Fig. 20–6a. In the following discussion, it will be convenient to express vector $\mathbf{A}$ in terms of its $\mathbf{i}$, $\mathbf{j}$, $\mathbf{k}$ components, which define the directions of the moving axes. Hence,

$$\mathbf{A} = A_x\mathbf{i} + A_y\mathbf{j} + A_z\mathbf{k}$$

In general, the time derivative of $\mathbf{A}$ must account for the change in both the vector's magnitude and direction. However, if this derivative is taken *with respect to the moving frame of reference,* only a change in the magnitudes of the components of $\mathbf{A}$ must be accounted for, since the directions of the components do not change with respect to the moving reference. Hence,

$$(\dot{\mathbf{A}})_{xyz} = \dot{A}_x\mathbf{i} + \dot{A}_y\mathbf{j} + \dot{A}_z\mathbf{k} \tag{20–5}$$

When the time derivative of $\mathbf{A}$ is taken *with respect to the fixed frame of reference,* the *directions* of $\mathbf{i}$, $\mathbf{j}$, and $\mathbf{k}$ change only on account of the *rotation* $\mathbf{\Omega}$ of the axes and not their translation. Hence, in general,

$$\dot{\mathbf{A}} = \dot{A}_x\mathbf{i} + \dot{A}_y\mathbf{j} + \dot{A}_z\mathbf{k} + A_x\dot{\mathbf{i}} + A_y\dot{\mathbf{j}} + A_z\dot{\mathbf{k}}$$

The time derivatives of the unit vectors will now be considered. For example, $\dot{\mathbf{i}} = d\mathbf{i}/dt$ represents only a change in the *direction* of $\mathbf{i}$ with respect to time, since $\mathbf{i}$ has a fixed magnitude of 1 unit. As shown in Fig. 20–6b, the change, $d\mathbf{i}$, is *tangent to the path* described by the arrowhead of $\mathbf{i}$ as $\mathbf{i}$ moves due to the rotation $\mathbf{\Omega}$. Accounting for both the magnitude and direction of $d\mathbf{i}$, we can therefore define $\dot{\mathbf{i}}$ using the cross product, $\dot{\mathbf{i}} = \mathbf{\Omega} \times \mathbf{i}$. In general,

$$\dot{\mathbf{i}} = \mathbf{\Omega} \times \mathbf{i} \qquad \dot{\mathbf{j}} = \mathbf{\Omega} \times \mathbf{j} \qquad \dot{\mathbf{k}} = \mathbf{\Omega} \times \mathbf{k}$$

These formulations were also developed in Sec. 16.8, regarding planar motion of the axes. Substituting the results into the above equation and using Eq. 20–5 yields

$$\dot{\mathbf{A}} = (\dot{\mathbf{A}})_{xyz} + \mathbf{\Omega} \times \mathbf{A} \tag{20–6}$$

This result is rather important, and it will be used throughout Sec. 20.4 and Chapter 21. It states that the time derivative of *any vector* $\mathbf{A}$ as observed from the fixed X, Y, Z frame of reference is equal to the time rate of change of $\mathbf{A}$ as observed from the x, y, z translating-rotating frame of reference, Eq. 20–5, plus $\mathbf{\Omega} \times \mathbf{A}$, the change of $\mathbf{A}$ caused by the rotation of the x, y, z frame. As a result, Eq. 20–6 should always be used whenever $\mathbf{\Omega}$ produces a change in the direction of $\mathbf{A}$ as seen from the X, Y, Z reference. If this change does not occur, i.e., $\mathbf{\Omega} = \mathbf{0}$, then $\dot{\mathbf{A}} = (\dot{\mathbf{A}})_{xyz}$, and so the time rate of change of $\mathbf{A}$ as observed from both coordinate systems will be the *same.*

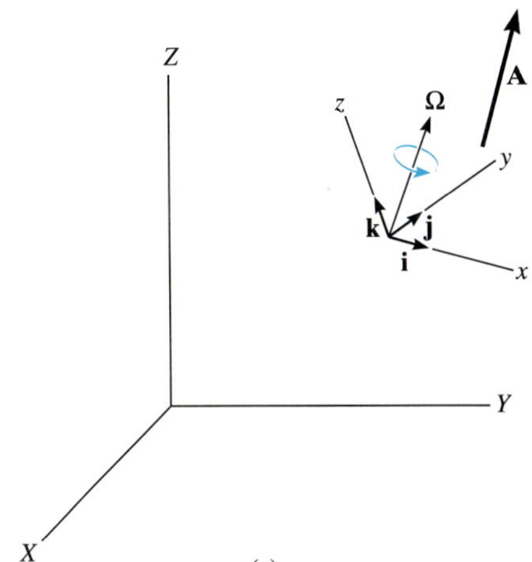

(a)

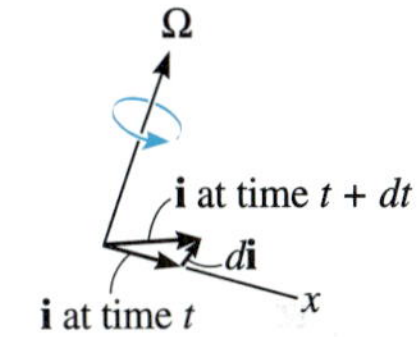

(b)

Fig. 20–6

EXAMPLE 20–1

The disk shown in Fig. 20–7*a* is spinning about its horizontal axis with a constant angular velocity $\omega_s = 3$ rad/s, while the horizontal platform on which the disk is mounted is rotating about the vertical axis at a constant rate $\omega_p = 1$ rad/s. Determine the angular acceleration of the disk and the velocity and acceleration of point *A* on the disk when it is in the position shown.

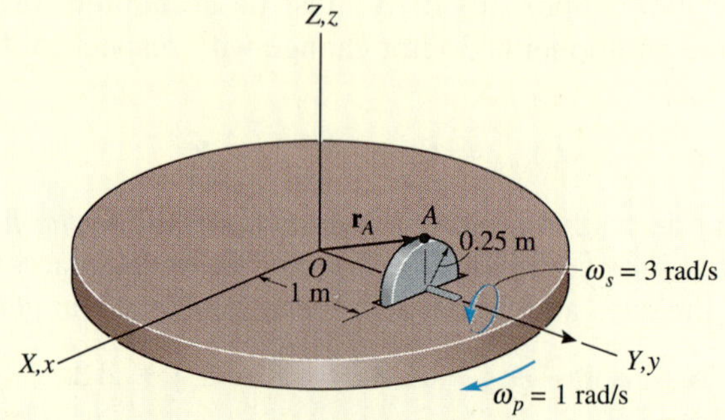

(a)

Fig. 20–7

Solution

Point *O* represents a fixed point of rotation for the disk if one considers a hypothetical extension of the disk to this point. To determine the velocity and acceleration of point *A*, it is first necessary to determine the resultant angular velocity $\boldsymbol{\omega}$ and angular acceleration $\boldsymbol{\alpha}$ of the disk, since these vectors are used in Eqs. 20–3 and 20–4.

Angular Velocity. The angular velocity, which is measured from *X*, *Y*, *Z*, is simply the vector addition of the two component motions. Thus,

$$\boldsymbol{\omega} = \boldsymbol{\omega}_s + \boldsymbol{\omega}_p = \{3\mathbf{j} - 1\mathbf{k}\} \text{ rad/s}$$

At first glance, it may not appear that the disk is actually rotating with this angular velocity, since it is generally more difficult to imagine the resultant of angular motions in comparison with linear motions. To further understand the angular motion, consider the disk as being replaced by a cone (a body cone), which is rolling over the stationary space cone, Fig. 20–7*b*. The instantaneous axis of rotation is along the line of contact of the cones. This axis defines the direction of the resultant $\boldsymbol{\omega}$, which has components $\boldsymbol{\omega}_s$ and $\boldsymbol{\omega}_p$.

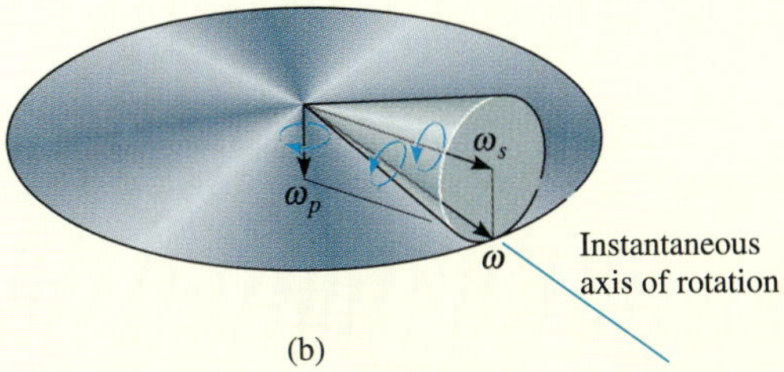

(b)

Angular Acceleration. Since the magnitude of $\boldsymbol{\omega}$ is constant, only a change in its direction, as seen from a fixed reference, creates the angular acceleration $\boldsymbol{\alpha}$ of the disk. One way to obtain $\boldsymbol{\alpha}$ is to compute the time derivative of *each of the two components* of $\boldsymbol{\omega}$ using Eq. 20–6. At the instant shown in Fig. 20–7*a*, imagine the fixed X, Y, Z and a rotating x, y, z frame to be coincident. If the rotating x, y, z frame is chosen to have an angular velocity of $\boldsymbol{\Omega} = \boldsymbol{\omega}_p = \{-1\mathbf{k}\}$ rad/s, then $\boldsymbol{\omega}_s$ will *always* be directed along the y (not Y) axis, and the time rate of change of $\boldsymbol{\omega}_s$ *as seen from* x, y, z *is zero;* i.e., $(\dot{\boldsymbol{\omega}}_s)_{xyz} = \mathbf{0}$ (the magnitude and direction of $\boldsymbol{\omega}_s$ is constant). Thus, by Eq. 20–6,

$$\dot{\boldsymbol{\omega}}_s = (\dot{\boldsymbol{\omega}}_s)_{xyz} + \boldsymbol{\omega}_p \times \boldsymbol{\omega}_s = \mathbf{0} + (-1\mathbf{k}) \times (3\mathbf{j}) = \{3\mathbf{i}\}\ \text{rad/s}^2$$

By the same choice of axes rotation, $\boldsymbol{\Omega} = \boldsymbol{\omega}_p$, or even with $\boldsymbol{\Omega} = \mathbf{0}$, the time derivative $(\dot{\boldsymbol{\omega}}_p)_{xyz} = \mathbf{0}$, since $\boldsymbol{\omega}_p$ has a constant magnitude and direction. Hence,

$$\dot{\boldsymbol{\omega}}_p = (\dot{\boldsymbol{\omega}}_p)_{xyz} + \boldsymbol{\omega}_p \times \boldsymbol{\omega}_p = \mathbf{0} + \mathbf{0} = \mathbf{0}$$

The angular acceleration of the disk is therefore

$$\boldsymbol{\alpha} = \dot{\boldsymbol{\omega}} = \dot{\boldsymbol{\omega}}_s + \dot{\boldsymbol{\omega}}_p = \{3\mathbf{i}\}\ \text{rad/s}^2 \qquad \textit{Ans.}$$

Velocity and Acceleration. Since $\boldsymbol{\omega}$ and $\boldsymbol{\alpha}$ have been determined, the velocity and acceleration of point A can be computed using Eqs. 20–3 and 20–4. Realizing that $\mathbf{r}_A = \{1\mathbf{j} + 0.25\mathbf{k}\}$ m, Fig. 20–7*a*, we have

$$\mathbf{v}_A = \boldsymbol{\omega} \times \mathbf{r}_A = (3\mathbf{j} - 1\mathbf{k}) \times (1\mathbf{j} + 0.25\mathbf{k}) = \{1.75\mathbf{i}\}\ \text{m/s} \qquad \textit{Ans.}$$

$$\begin{aligned}\mathbf{a}_A &= \boldsymbol{\alpha} \times \mathbf{r}_A + \boldsymbol{\omega} \times (\boldsymbol{\omega} \times \mathbf{r}_A)\\ &= (3\mathbf{i}) \times (1\mathbf{j} + 0.25\mathbf{k}) + (3\mathbf{j} - 1\mathbf{k}) \times [(3\mathbf{j} - 1\mathbf{k}) \times (1\mathbf{j} + 0.25\mathbf{k})]\\ &= \{-2.50\mathbf{j} - 2.25\mathbf{k}\}\ \text{m/s}^2 \qquad \textit{Ans.}\end{aligned}$$

EXAMPLE 20–2

At the instant $\theta = 60°$, the gyrotop in Fig. 20–8 has three components of angular motion directed as shown and having magnitudes defined as:

spin: $\omega_s = 10$ rad/s, increasing at the rate of 6 rad/s^2

nutation: $\omega_n = 3$ rad/s, increasing at the rate of 2 rad/s^2

precession: $\omega_p = 5$ rad/s, increasing at the rate of 4 rad/s^2

Determine the angular velocity and angular acceleration of the top.

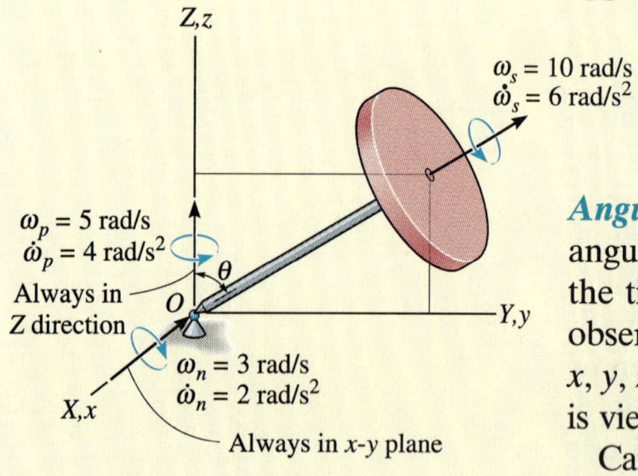

Fig. 20–8

Solution

Angular Velocity. The top is rotating about the fixed point O. If the fixed and rotating frames are coincident at the instant shown, then the angular velocity can be expressed in terms of **i, j, k** components, appropriate to the x, y, z frame; i.e.,

$$\begin{aligned}\boldsymbol{\omega} &= -\omega_n\mathbf{i} + \omega_s \sin\theta\mathbf{j} + (\omega_p + \omega_s\cos\theta)\mathbf{k}\\ &= -3\mathbf{i} + 10\sin 60°\mathbf{j} + (5 + 10\cos 60°)\mathbf{k}\\ &= \{-3\mathbf{i} + 8.66\mathbf{j} + 10\mathbf{k}\}\text{ rad/s}\end{aligned}$$ *Ans.*

Angular Acceleration. As in the solution of Example 20–1, the angular acceleration $\boldsymbol{\alpha}$ will be determined by investigating separately the time rate of change of *each of the angular velocity components* as observed from the fixed X, Y, Z reference. We will choose an $\boldsymbol{\Omega}$ for the x, y, z reference so that the component of $\boldsymbol{\omega}$ which is being considered is viewed as having a *constant direction* when observed from x, y, z.

Careful examination of the motion of the top reveals that $\boldsymbol{\omega}_s$ has a *constant direction* relative to x, y, z if these axes rotate at $\boldsymbol{\Omega} = \boldsymbol{\omega}_n + \boldsymbol{\omega}_p$. Thus,

$$\begin{aligned}\dot{\boldsymbol{\omega}}_s &= (\dot{\boldsymbol{\omega}}_s)_{xyz} + (\boldsymbol{\omega}_n + \boldsymbol{\omega}_p)\times\boldsymbol{\omega}_s\\ &= (6\sin 60°\mathbf{j} + 6\cos 60°\mathbf{k}) + (-3\mathbf{i} + 5\mathbf{k})\times(10\sin 60°\mathbf{j} + 10\cos 60°\mathbf{k})\\ &= \{-43.30\mathbf{i} + 20.20\mathbf{j} - 22.98\mathbf{k}\}\text{ rad/s}^2\end{aligned}$$

Since $\boldsymbol{\omega}_n$ *always* lies in the fixed X–Y plane, this vector has a *constant direction* if the motion is viewed from axes x, y, z having a rotation of $\boldsymbol{\Omega} = \boldsymbol{\omega}_p$ (not $\boldsymbol{\Omega} = \boldsymbol{\omega}_s + \boldsymbol{\omega}_p$). Thus,

$$\dot{\boldsymbol{\omega}}_n = (\dot{\boldsymbol{\omega}}_n)_{xyz} + \boldsymbol{\omega}_p\times\boldsymbol{\omega}_n = -2\mathbf{i} + (5\mathbf{k})\times(-3\mathbf{i}) = \{-2\mathbf{i} - 15\mathbf{j}\}\text{ rad/s}^2$$

Finally, the component $\boldsymbol{\omega}_p$ is *always directed* along the Z axis so that here it is not necessary to think of x, y, z as rotating, i.e., $\boldsymbol{\Omega} = \mathbf{0}$. Expressing the data in terms of the **i, j, k** components, we therefore have

$$\dot{\boldsymbol{\omega}}_p = (\dot{\boldsymbol{\omega}}_p)_{xyz} + \mathbf{0}\times\boldsymbol{\omega}_p = \{4\mathbf{k}\}\text{ rad/s}^2$$

Thus, the angular acceleration of the top is

$$\boldsymbol{\alpha} = \dot{\boldsymbol{\omega}}_s + \dot{\boldsymbol{\omega}}_n + \dot{\boldsymbol{\omega}}_p = \{-45.3\mathbf{i} + 5.20\mathbf{j} - 19.0\mathbf{k}\}\text{ rad/s}^2$$ *Ans.*

*20.3 General Motion

Shown in Fig. 20–9 is a rigid body subjected to general motion in three dimensions for which the angular velocity is $\boldsymbol{\omega}$ and the angular acceleration is $\boldsymbol{\alpha}$. If point A has a known motion of $\mathbf{v}_A$ and $\mathbf{a}_A$, the motion of any other point B may be determined by using a relative-motion analysis. In this section a *translating coordinate system* will be used to define the relative motion, and in the next section a reference that is both rotating and translating will be considered.

If the origin of the translating coordinate system x, y, z ($\boldsymbol{\Omega} = \mathbf{0}$) is located at the "base point" A, then, at the instant shown, the motion of the body may be regarded as the sum of an instantaneous translation of the body having a motion of $\mathbf{v}_A$ and $\mathbf{a}_A$ and a rotation of the body about an instantaneous axis passing through the base point. Since the body is rigid, the motion of point B measured by an observer located at A is the same as *motion of the body about a fixed point.* This relative motion occurs about the instantaneous axis of rotation and is defined by $\mathbf{v}_{B/A} = \boldsymbol{\omega} \times \mathbf{r}_{B/A}$, Eq. 20–3, and $\mathbf{a}_{B/A} = \boldsymbol{\alpha} \times \mathbf{r}_{B/A} + \boldsymbol{\omega} \times (\boldsymbol{\omega} \times \mathbf{r}_{B/A})$, Eq. 20–4. For translating axes the relative motions are related to absolute motions by $\mathbf{v}_B = \mathbf{v}_A + \mathbf{v}_{B/A}$ and $\mathbf{a}_B = \mathbf{a}_A + \mathbf{a}_{B/A}$, Eqs. 16–15 and 16–17, so that the absolute velocity and acceleration of point B can be determined from the equations

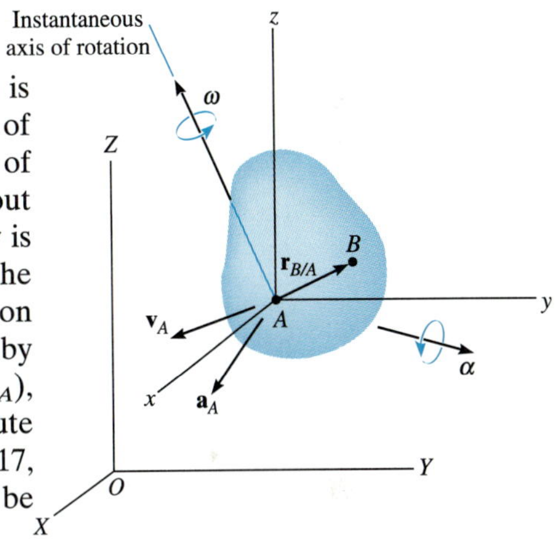

Fig. 20–9

$$\mathbf{v}_B = \mathbf{v}_A + \boldsymbol{\omega} \times \mathbf{r}_{B/A} \tag{20–7}$$

and

$$\mathbf{a}_B = \mathbf{a}_A + \boldsymbol{\alpha} \times \mathbf{r}_{B/A} + \boldsymbol{\omega} \times (\boldsymbol{\omega} \times \mathbf{r}_{B/A}) \tag{20–8}$$

These two equations are identical to those describing the general plane motion of a rigid body, Eqs. 16–16 and 16–18. However, difficulty in application arises for three-dimensional motion, because $\boldsymbol{\alpha}$ measures the change in *both* the magnitude and direction of $\boldsymbol{\omega}$. (Recall that, for general plane motion, $\boldsymbol{\alpha}$ and $\boldsymbol{\omega}$ are always parallel or perpendicular to the plane of motion, and therefore $\boldsymbol{\alpha}$ measures only a change in the magnitude of $\boldsymbol{\omega}$.) In some problems the constraints or connections of a body will require that the directions of the angular motions or displacement paths of points on the body be defined. As illustrated in the following example, this information is useful for obtaining some of the terms in the above equations.

EXAMPLE 20–3

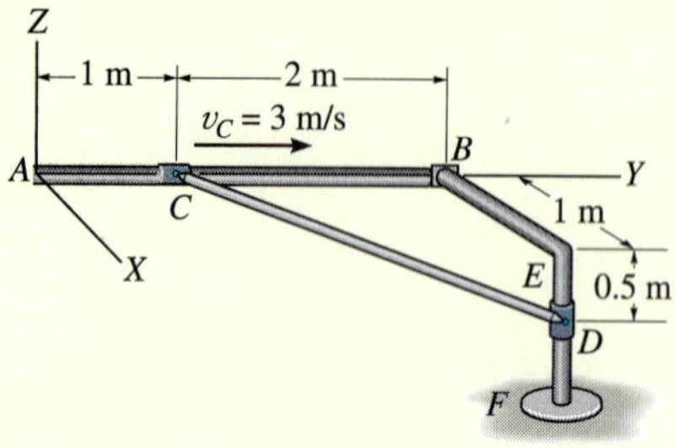

(a)

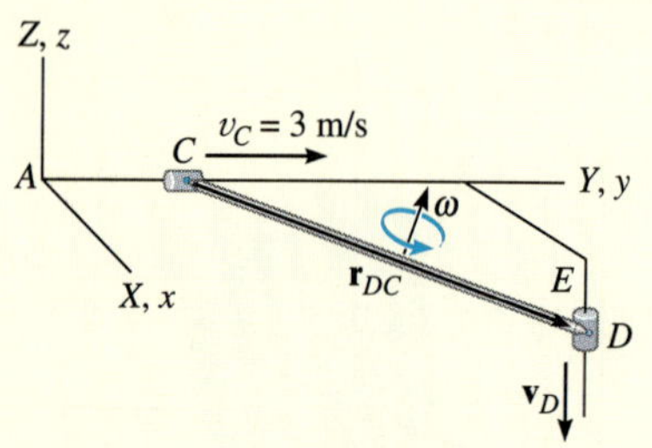

(b)

Fig. 20–10

One end of the rigid bar *CD* shown in Fig. 20–10*a* slides along the horizontal member *AB*, and the other end slides along the vertical member *EF*. If the collar at *C* is moving towards *B* at a speed of 3 m/s, determine the velocity of the collar at *D* and the angular velocity of the bar at the instant shown. The bar is connected to the collars at its end points by ball-and-socket joints.

Solution

Bar *CD* is subjected to general motion. Why? The velocity of point *D* on the bar may be related to the velocity of point *C* by the equation

$$\mathbf{v}_D = \mathbf{v}_C + \boldsymbol{\omega} \times \mathbf{r}_{D/C}$$

The fixed and translating frames of reference are assumed to coincide at the instant considered, Fig. 20–10*b*. We have

$$\mathbf{v}_D = -v_D\mathbf{k} \qquad \mathbf{v}_C = \{3\mathbf{j}\}\text{ m/s}$$
$$\mathbf{r}_{D/C} = \{1\mathbf{i} + 2\mathbf{j} - 0.5\mathbf{k}\}\text{ m} \qquad \boldsymbol{\omega} = \omega_x\mathbf{i} + \omega_y\mathbf{j} + \omega_z\mathbf{k}$$

Substituting these quantities into the above equation gives

$$-v_D\mathbf{k} = 3\mathbf{j} + \begin{vmatrix} \mathbf{i} & \mathbf{j} & \mathbf{k} \\ \omega_x & \omega_y & \omega_z \\ 1 & 2 & -0.5 \end{vmatrix}$$

Expanding and equating the respective **i**, **j**, **k** components yields

$$-0.5\omega_y - 2\omega_z = 0 \qquad (1)$$
$$0.5\omega_x + 1\omega_z + 3 = 0 \qquad (2)$$
$$2\omega_x - 1\omega_y + v_D = 0 \qquad (3)$$

These equations contain four unknowns.* A fourth equation can be written if the direction of $\boldsymbol{\omega}$ is specified. In particular, any component of $\boldsymbol{\omega}$ acting along the bar's axis has no effect on moving the collars. This is because the bar is *free to rotate* about its axis. Therefore, if $\boldsymbol{\omega}$ is specified as acting *perpendicular* to the axis of the bar, then $\boldsymbol{\omega}$ must have a unique magnitude to satisfy the above equations. Perpendicularity is guaranteed provided the dot product of $\boldsymbol{\omega}$ and $\mathbf{r}_{D/C}$ is zero (see Eq. C–14 of Appendix C). Hence,

$$\boldsymbol{\omega}\cdot\mathbf{r}_{D/C} = (\omega_x\mathbf{i} + \omega_y\mathbf{j} + \omega_z\mathbf{k})\cdot(1\mathbf{i} + 2\mathbf{j} - 0.5\mathbf{k}) = 0$$
$$1\omega_x + 2\omega_y - 0.5\omega_z = 0 \qquad (4)$$

Solving Eqs. 1 through 4 simultaneously yields

$$\omega_x = -4.86\text{ rad/s} \quad \omega_y = 2.29\text{ rad/s} \quad \omega_z = -0.571\text{ rad/s} \qquad \textit{Ans.}$$
$$v_D = 12.0\text{ m/s}\downarrow \qquad \textit{Ans.}$$

*Although this is the case, the magnitude of $\mathbf{v}_D$ can be obtained. For example, solve Eqs. 1 and 2 for ω_y and ω_x in terms of ω_z and substitute into Eq. 3. It will be noted that ω_z will *cancel out,* which will allow a solution for v_D.

Problems

20-1. The anemometer located on the ship at A is spinning about its own axis at a rate ω_s, while the ship is rolling about the x axis at the rate ω_x and about the y axis at the rate ω_y. Determine the angular velocity and angular acceleration of the anemometer at the instant the ship is level as shown. Assume that the magnitudes of all components of angular velocity are constant and that the rolling motion caused by the sea is independent in the x and y directions.

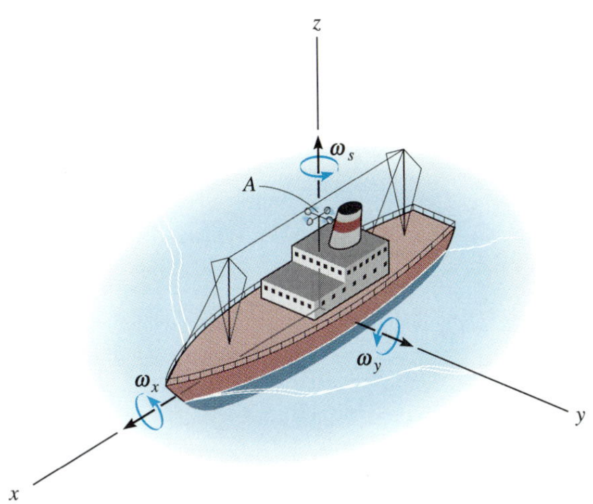

Prob. 20–1

20-2. The motion of the top is such that at the instant shown it is rotating about the z axis at $\omega_1 = 0.6$ rad/s, while it is spinning at $\omega_2 = 8$ rad/s. Determine the angular velocity and angular acceleration of the top at this instant. Express the result as a Cartesian vector.

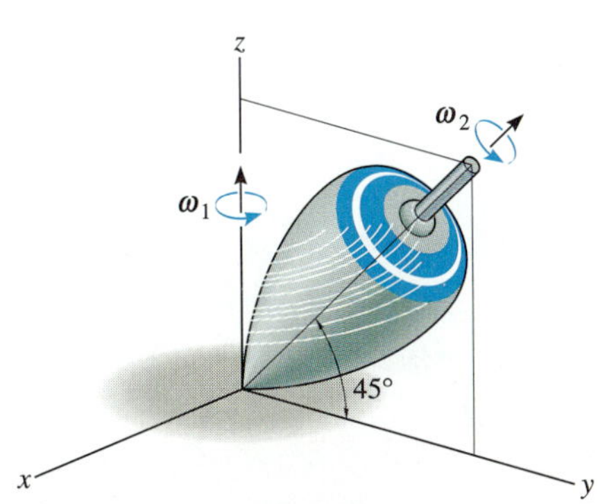

Prob. 20–2

20-3. At the instant shown, the radar dish is rotating about the z axis at $\omega_1 = 3$ rad/s, which is increasing at 4 rad/s^2. Also, at this instant the angle of tilt $\phi = 30°$, and $\dot{\phi} = 2$ rad/s, $\ddot{\phi} = 6$ rad/s^2. Determine the angular velocity and angular acceleration of the dish at this instant.

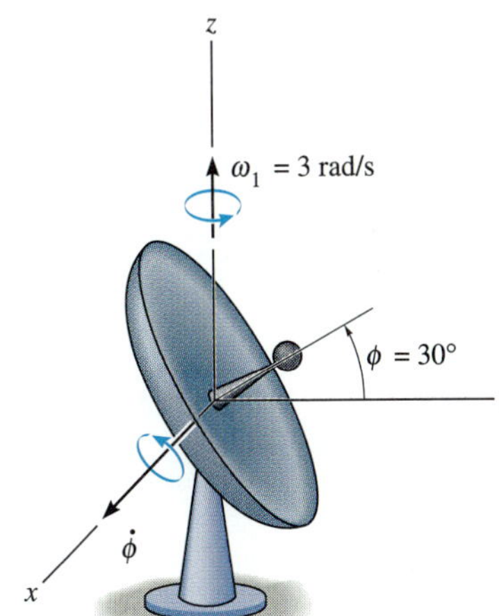

Prob. 20–3

***20-4.** The antenna is following the motion of a jet plane. At the instant $\theta = 25°$ and $\phi = 75°$, the constant angular rates of change are $\dot{\theta} = 0.4$ rad/s and $\dot{\phi} = 0.6$ rad/s. Determine the velocity and acceleration of the signal horn A at this instant. The distance OA is 0.8 m.

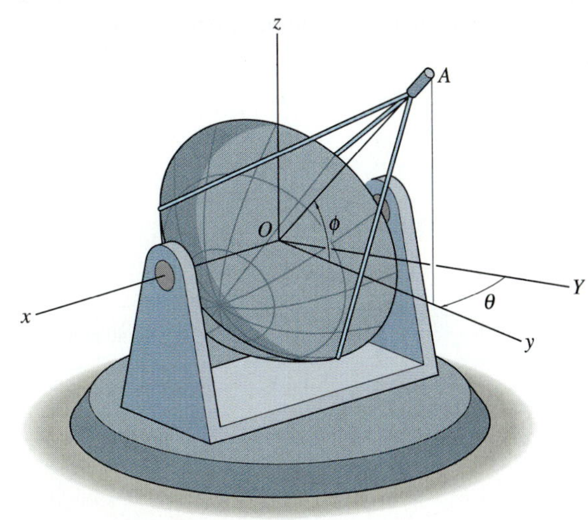

Prob. 20–4

20-5. The fan is mounted on a swivel support such that at the instant shown it is rotating about the z axis at $\omega_1 = 0.8$ rad/s, which is increasing at 12 rad/s^2. The blade is spinning at $\omega_2 = 16$ rad/s, which is decreasing at 2 rad/s^2. Determine the angular velocity and angular acceleration of the blade at this instant.

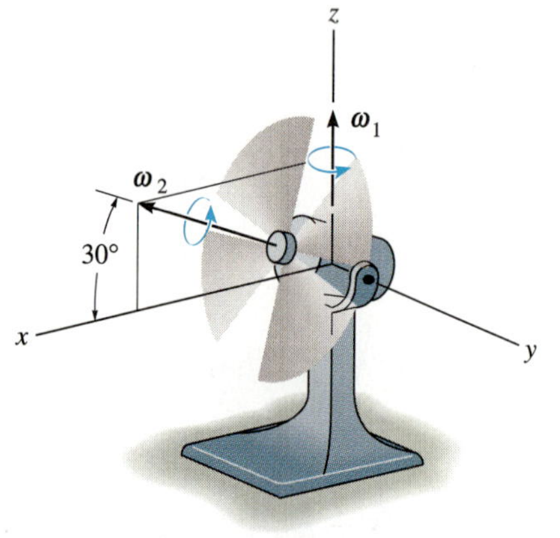

Prob. 20–5

20-6. Gear B is connected to the rotating shaft, while the plate gear A is fixed. If the shaft is turning at a constant rate of $\omega_z = 10$ rad/s about the z axis, determine the magnitudes of the angular velocity and the angular acceleration of gear B. Also, determine the magnitudes of the velocity and acceleration of point P.

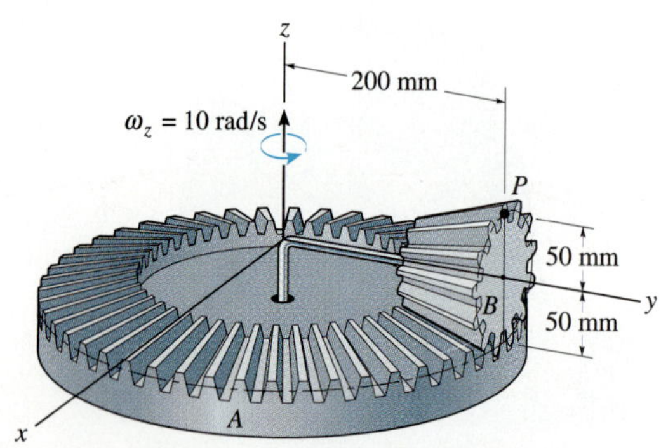

Prob. 20–6

20-7. Gears A and B are fixed, while gears C and D are free to rotate on the shaft S. If the shaft is turning about the z axis at a constant rate of $\omega_1 = 4$ rad/s, determine the angular velocity and angular acceleration of gear C.

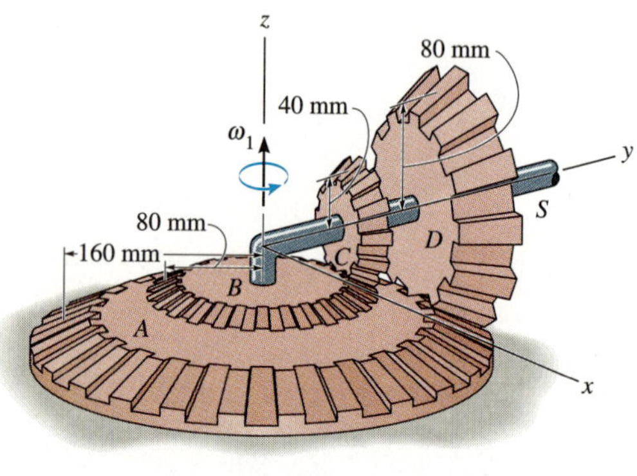

Prob. 20–7

***20-8.** The propeller of an airplane is rotating at a constant speed $\omega_s\mathbf{i}$, while the plane is undergoing a turn at a constant rate ω_t. Determine the angular acceleration of the propeller if (a) the turn is horizontal, i.e., $\omega_t\mathbf{k}$, and (b) the turn is vertical, downward, i.e., $\omega_t\mathbf{j}$.

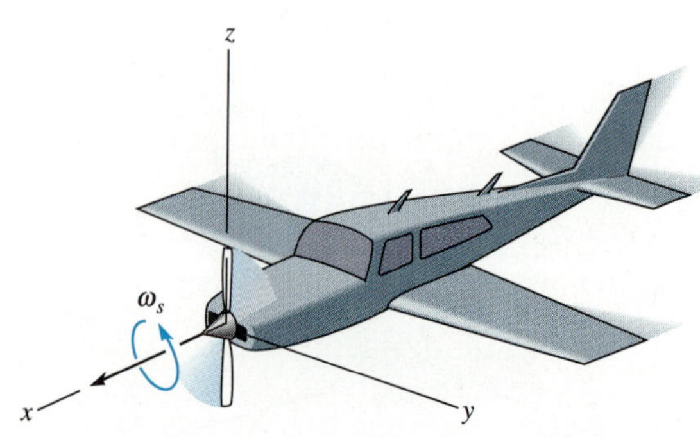

Prob. 20–8

20-9. The conical spool rolls on the plane without slipping. If the axle has an angular velocity of $\omega_1 = 3$ rad/s and an angular acceleration of $\alpha_1 = 2$ rad/s^2 at the instant shown, determine the angular velocity and angular acceleration of the spool at this instant.

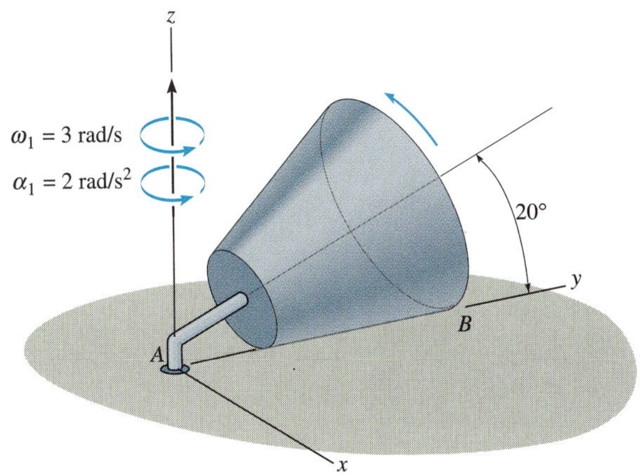

Prob. 20–9

20-10. If the top gear B is rotating at a constant rate of $\boldsymbol{\omega}$, determine the angular velocity of gear A, which is free to turn about the shaft and rolls on the bottom fixed gear C.

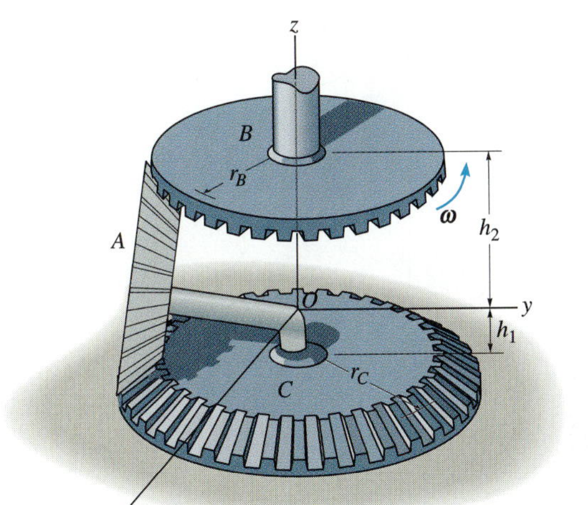

Prob. 20–10

20-11. The telescope is mounted on the frame F that allows it to be directed to any point in the sky. At the instant shown, the frame has an angular acceleration of $\alpha_{y'} = 0.2$ rad/s^2 and an angular velocity of $\omega_{y'} = 0.3$ rad/s about the y' axis, and $\ddot{\theta} = 0.5$ rad/s^2 while $\dot{\theta} = 0.4$ rad/s. Determine the velocity and acceleration of the observing capsule at C at this instant when $\theta = 30°$.

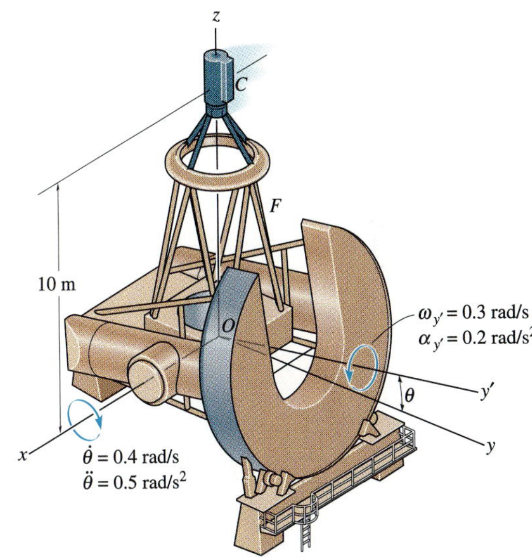

Prob. 20–11

***20-12.** If the plate gears A and B are rotating with the angular velocities shown, determine the angular velocity of gear C about the shaft DE. What is the angular velocity of DE about the y axis?

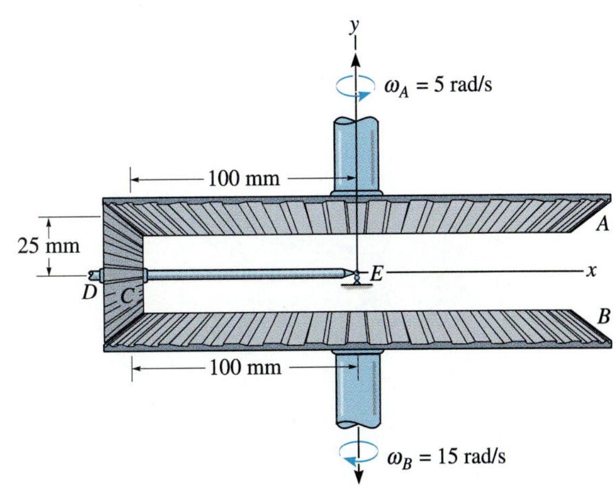

Prob. 20–12

20-13. The right circular cone rotates about the z axis at a constant rate of $\omega_1 = 4$ rad/s without slipping on the horizontal plane. Determine the magnitudes of the velocity and acceleration of points B and C.

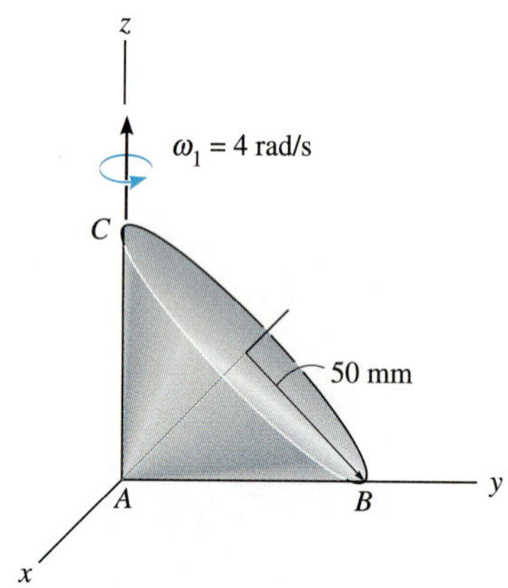

Prob. 20–13

20-14. The tower crane is rotating about the z axis at a constant rate $\omega_1 = 0.25$ rad/s, while the boom OA is rotating downward at a constant rate $\omega_2 = 0.4$ rad/s. Determine the velocity and acceleration of point A located at the top of the boom at the instant shown.

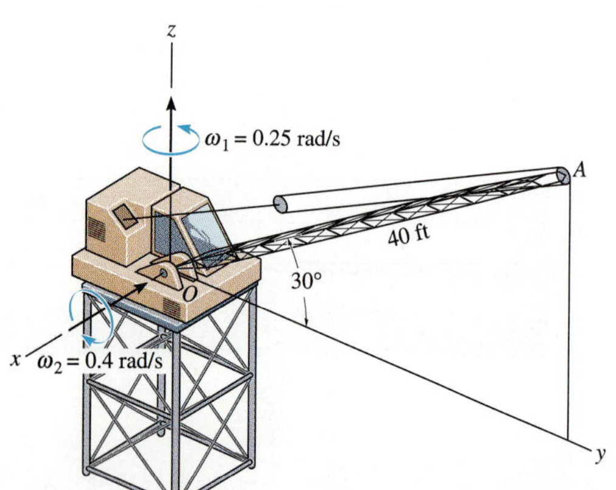

Prob. 20–14

20-15. At the instant shown, the tower crane is rotating about the z axis with an angular velocity $\omega_1 = 0.25$ rad/s, which is increasing at 0.6 rad/s^2. The boom OA is rotating downward with an angular velocity $\omega_2 = 0.4$ rad/s, which is increasing at 0.8 rad/s^2. Determine the velocity and acceleration of point A located at the top of the boom at this instant.

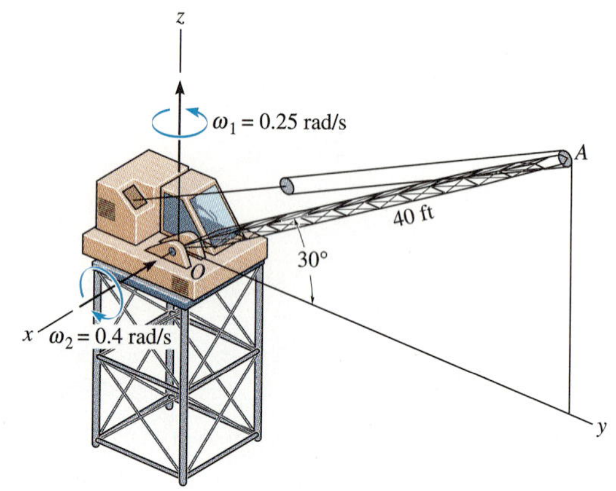

Prob. 20–15

***20-16.** The construction boom OA is rotating about the z axis with a constant angular velocity of $\omega_1 = 0.15$ rad/s, while it is rotating downward with a constant angular velocity of $\omega_2 = 0.2$ rad/s. Determine the velocity and acceleration of point A located at the tip of the boom at the instant shown.

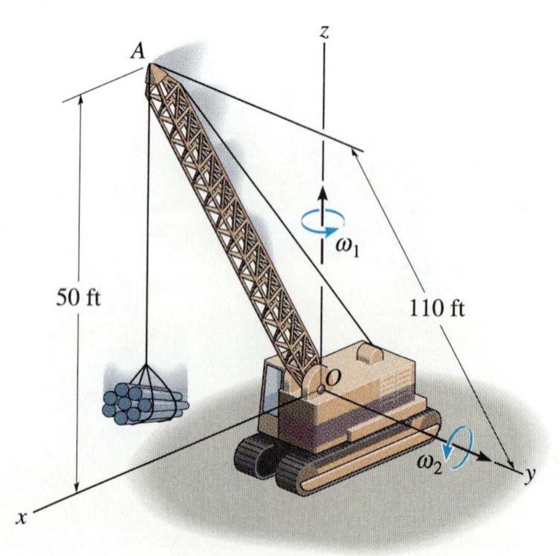

Prob. 20–16

20-17. The differential of an automobile allows the two rear wheels to rotate at different speeds when the automobile travels along a curve. For operation, the rear axles are attached to the wheels at one end and have beveled gears A and B on their other ends. The differential case D is placed over the left axle but can rotate about C independent of the axle. The case supports a pinion gear E on a shaft, which meshes with gears A and B. Finally, a ring gear G is *fixed* to the differential case so that the case rotates with the ring gear when the latter is driven by the drive pinion H. This gear, like the differential case, is free to rotate about the left wheel axle. If the drive pinion is turning at $\omega_H = 100$ rad/s and the pinion gear E is spinning about its shaft at $\omega_E = 30$ rad/s, determine the angular velocity, ω_A and ω_B, of each axle.

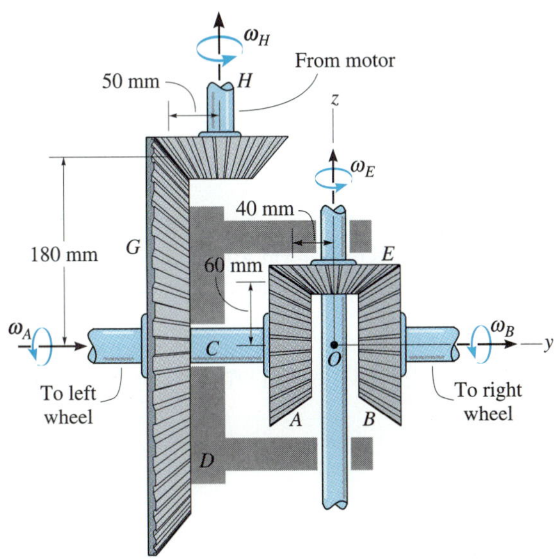

Prob. 20–17

20-18. The rod AB is attached to collars at its ends by ball-and-socket joints. If collar A has a velocity $v_A = 15$ ft/s at the instant shown, determine the velocity of collar B.

20-19. The rod AB is attached to collars at its ends by ball-and-socket joints. If collar A has an acceleration of $a_A = 2$ ft/s^2 at the instant shown, determine the acceleration of collar B.

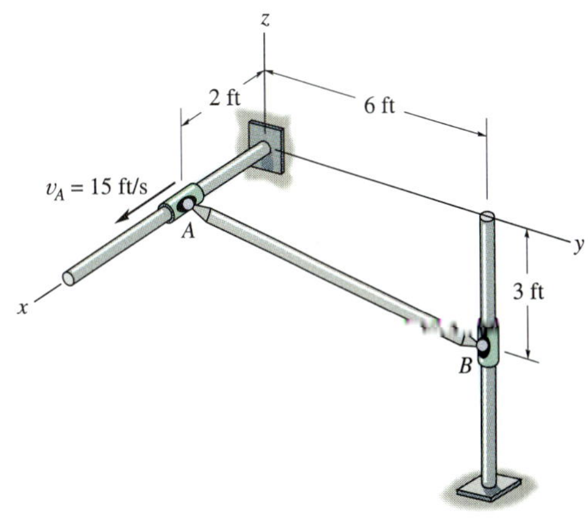

Probs. 20–18/19

***20-20.** If the rod is attached with ball-and-socket joints to smooth collars A and B at its end points, determine the speed of B at the instant shown if A is moving downward at a constant speed of $v_A = 8$ ft/s. Also, determine the angular velocity of the rod if it is directed perpendicular to the axis of the rod.

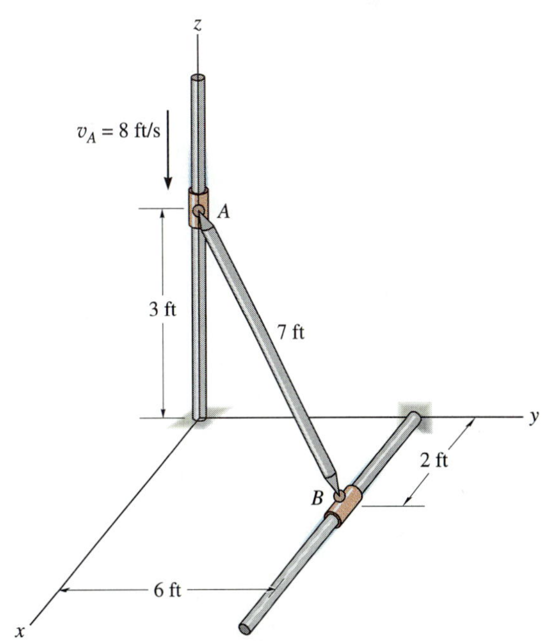

Prob. 20–20

20-21. If the collar at A is moving downward with an acceleration $\mathbf{a}_A = \{-5\mathbf{k}\}$ ft/s^2, at the instant its speed is $v_A = 8$ ft/s, determine the acceleration of the collar at B at this instant.

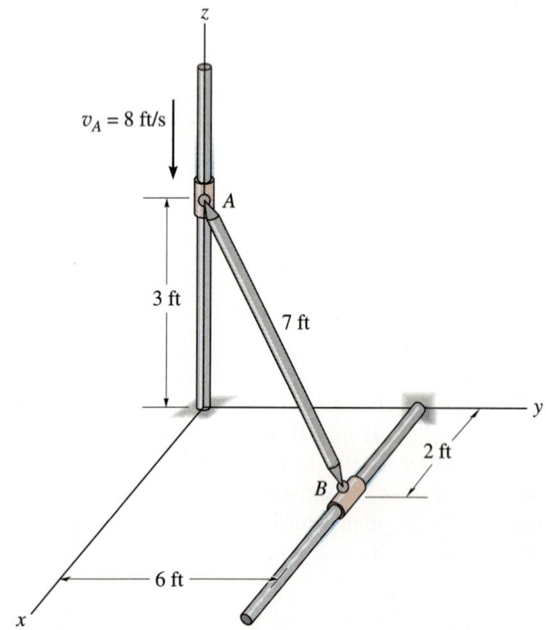

Prob. 20–21

20-22. Rod AB is attached to a disk and a collar by ball-and-socket joints. If the disk is rotating at a constant angular velocity $\boldsymbol{\omega} = \{2\mathbf{i}\}$ rad/s, determine the velocity and acceleration of the collar at A at the instant shown. Assume the angular velocity is directed perpendicular to the rod.

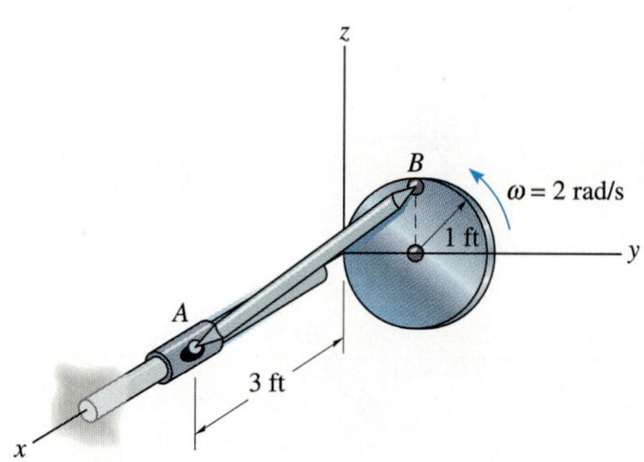

Prob. 20–22

20-23. Rod AB is attached to a disk and a collar by ball-and-socket joints. If the disk is rotating with an angular acceleration $\boldsymbol{\alpha} = \{4\mathbf{i}\}$ rad/s^2, and at the instant shown has an angular velocity $\boldsymbol{\omega} = \{2\mathbf{i}\}$ rad/s, determine the velocity and acceleration of the collar at A at the instant shown.

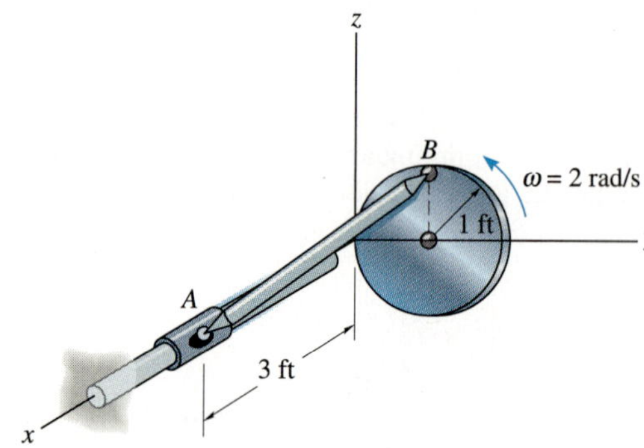

Prob. 20–23

***20-24.** Disk A is rotating at a constant angular velocity of 10 rad/s. If rod BC is joined to the disk and a collar by ball-and-socket joints, determine the velocity of collar B at the instant shown. Also, what is the rod's angular velocity $\boldsymbol{\omega}_{BC}$ if it is directed perpendicular to the axis of the rod?

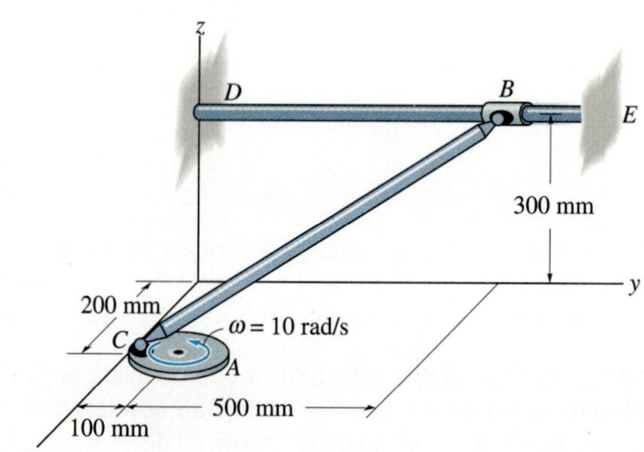

Prob. 20–24

*20-25. Solve Prob. 20-24 if the connection at B consists of a pin as shown in the figure below, rather than a ball-and-socket joint. *Hint:* The constraint allows rotation of the rod both along bar DE ($\mathbf{j}$ direction) and along the axis of the pin ($\mathbf{n}$ direction). Since there is no rotational component in the $\mathbf{u}$ direction, i.e., perpendicular to $\mathbf{n}$ and $\mathbf{j}$ where $\mathbf{u} = \mathbf{j} \times \mathbf{n}$, an additional equation for solution can be obtained from $\boldsymbol{\omega} \cdot \mathbf{u} = 0$. The vector $\mathbf{n}$ is in the same direction as $\mathbf{r}_{B/C} \times \mathbf{r}_{D/C}$.

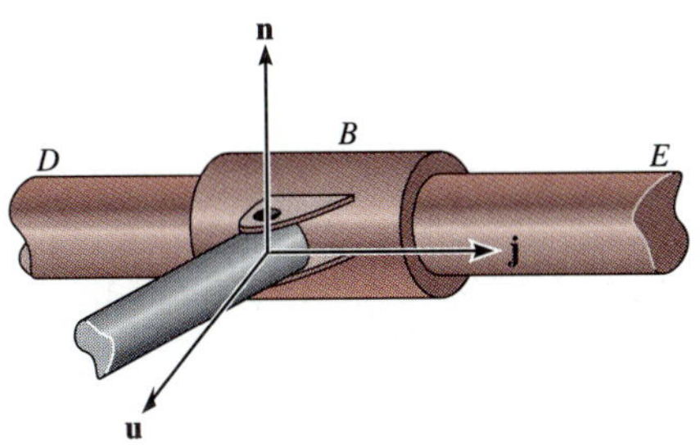

Prob. 20–25

20-26. Rod AB is attached to the rotating arm using ball-and-socket joints. If AC is rotating with a constant angular velocity of 8 rad/s about the pin at C, determine the angular velocity of link BD at the instant shown.

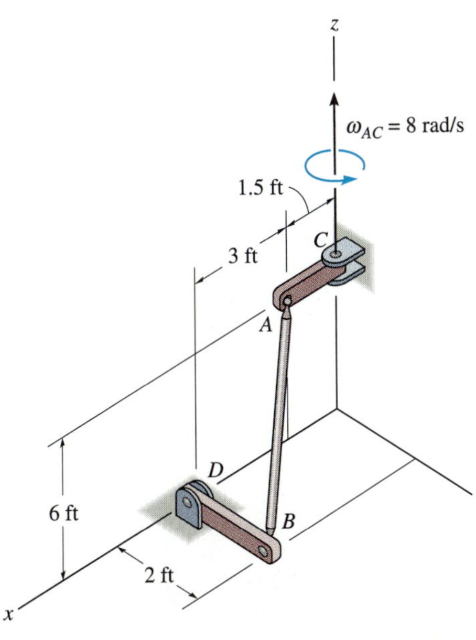

Prob. 20–26

20-27. Rod AB is attached to collars at its ends by ball-and-socket joints. If collar A moves upward with a velocity of 8 ft/s, determine the angular velocity of the rod and the speed of collar B at the instant shown. Assume that the rod's angular velocity is directed perpendicular to the rod.

*20-28. Rod AB is attached to collars at its ends by ball-and-socket joints. If collar A moves upward with an acceleration of $a_A = 4\ \text{ft/s}^2$, determine the angular acceleration of rod AB and the magnitude of acceleration of collar B. Assume that the rod's angular acceleration is directed perpendicular to the rod.

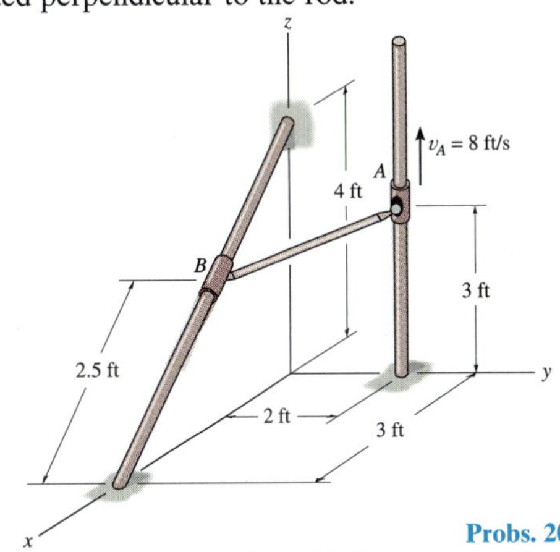

Probs. 20–27/28

20-29. The triangular plate ABC is supported at A by a ball-and-socket joint and at C by the x–z plane. The side AB lies in the x–y plane. At the instant $\theta = 60°$, $\dot{\theta} = 2$ rad/s and point C has the coordinates shown. Determine the angular velocity of the plate and the velocity of point C at this instant.

20-30. The triangular plate ABC is supported at A by a ball-and-socket joint and at C by the x–z plane. The side AB lies in the x–y plane. At the instant $\theta = 60°$, $\dot{\theta} = 2$ rad/s, $\ddot{\theta} = 3\ \text{rad/s}^2$, and point C has the coordinates shown. Determine the angular acceleration of the plate and the acceleration of point C at this instant.

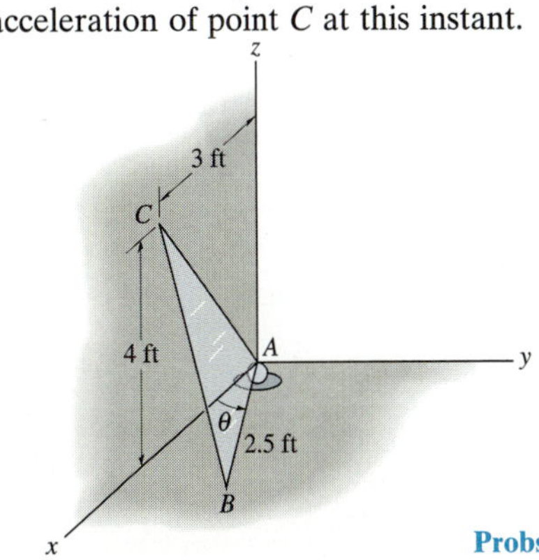

Probs. 20–29/30

★20.4 Relative-Motion Analysis Using Translating and Rotating Axes

The most general way to analyze the three-dimensional motion of a rigid body requires the use of x, y, z axes that both translate and rotate relative to a second frame X, Y, Z. This analysis will also provide a means for determining the motions of two points A and B located on separate members of a mechanism, and for determining the relative motion of one particle with respect to another when one or both particles are moving along *rotating paths.*

As shown in Fig. 20–11, the locations of points A and B are specified relative to the X, Y, Z frame of reference by position vectors $\mathbf{r}_A$ and $\mathbf{r}_B$. The base point A represents the origin of the x, y, z coordinate system, which is translating and rotating with respect to X, Y, Z. At the instant considered, the velocity and acceleration of point A are $\mathbf{v}_A$ and $\mathbf{a}_A$, respectively, and the angular velocity and angular acceleration of the x, y, z axes are $\mathbf{\Omega}$ and $\dot{\mathbf{\Omega}} = d\mathbf{\Omega}/dt$, respectively. All these vectors are *measured* with respect to the X, Y, Z frame of reference, although they may be expressed in Cartesian component form along either set of axes.

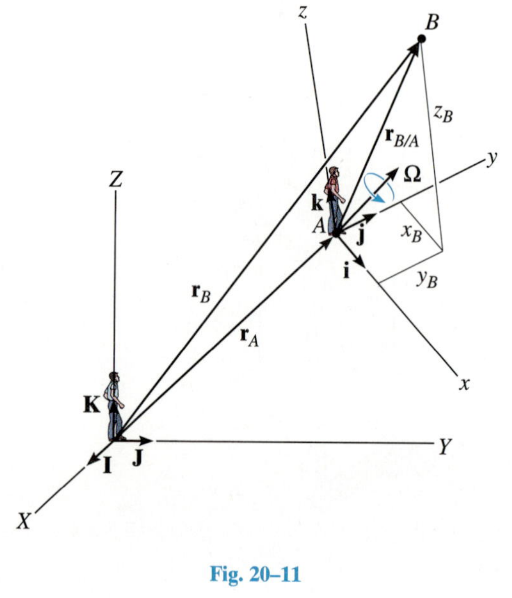

Fig. 20–11

Position. If the position of "B with respect to A" is specified by the *relative-position vector* $\mathbf{r}_{B/A}$, Fig. 20–11, then, by vector addition,

$$\mathbf{r}_B = \mathbf{r}_A + \mathbf{r}_{B/A} \tag{20–9}$$

where

$\mathbf{r}_B$ = position of B
$\mathbf{r}_A$ = position of the origin A
$\mathbf{r}_{B/A}$ = relative position of "B with respect to A"

Velocity. The velocity of point B measured from X, Y, Z is determined by taking the time derivative of Eq. 20–9, which yields

$$\dot{\mathbf{r}}_B = \dot{\mathbf{r}}_A + \dot{\mathbf{r}}_{B/A}$$

The first two terms represent $\mathbf{v}_B$ and $\mathbf{v}_A$. The last term is evaluated by applying Eq. 20–6, since $\mathbf{r}_{B/A}$ is measured between two points in a rotating reference. Hence,

$$\dot{\mathbf{r}}_{B/A} = (\dot{\mathbf{r}}_{B/A})_{xyz} + \boldsymbol{\Omega} \times \mathbf{r}_{B/A} = (\mathbf{v}_{B/A})_{xyz} + \boldsymbol{\Omega} \times \mathbf{r}_{B/A} \tag{20–10}$$

Here $(\mathbf{v}_{B/A})_{xyz}$ is the relative velocity of B with respect to A measured from x, y, z. Thus,

$$\mathbf{v}_B = \mathbf{v}_A + \boldsymbol{\Omega} \times \mathbf{r}_{B/A} + (\mathbf{v}_{B/A})_{xyz} \tag{20–11}$$

where

$\mathbf{v}_B$ = velocity of B
$\mathbf{v}_A$ = velocity of the origin A of the x, y, z frame of reference
$(\mathbf{v}_{B/A})_{xyz}$ = relative velocity of "B with respect to A" as measured by an observer attached to the rotating x, y, z frame of reference
$\boldsymbol{\Omega}$ = angular velocity of the x, y, z frame of reference
$\mathbf{r}_{B/A}$ = relative position of "B with respect to A"

Acceleration. The acceleration of point B measured from X, Y, Z is determined by taking the time derivative of Eq. 20–11, which yields

$$\dot{\mathbf{v}}_B = \dot{\mathbf{v}}_A + \dot{\mathbf{\Omega}} \times \mathbf{r}_{B/A} + \mathbf{\Omega} \times \dot{\mathbf{r}}_{B/A} + \frac{d}{dt}(\mathbf{v}_{B/A})_{xyz}$$

The time derivatives defined in the first and second terms represent $\mathbf{a}_B$ and $\mathbf{a}_A$ respectively. The fourth term is evaluated using Eq. 20–10, and the last term is evaluated by applying Eq. 20–6, which yields

$$\begin{aligned}\frac{d}{dt}(\mathbf{v}_{B/A})_{xyz} &= (\dot{\mathbf{v}}_{B/A})_{xyz} + \mathbf{\Omega} \times (\mathbf{v}_{B/A})_{xyz}\\ &= (\mathbf{a}_{B/A})_{xyz} + \mathbf{\Omega} \times (\mathbf{v}_{B/A})_{xyz}\end{aligned}$$

Here $(\mathbf{a}_{B/A})_{xyz}$ is the relative acceleration of B with respect to A measured from x, y, z. Substituting this result and Eq. 20–10 into the above equation and simplifying, we have

$$\mathbf{a}_B = \mathbf{a}_A + \dot{\mathbf{\Omega}} \times \mathbf{r}_{B/A} + \mathbf{\Omega} \times (\mathbf{\Omega} \times \mathbf{r}_{B/A}) + 2\mathbf{\Omega} \times (\mathbf{v}_{B/A})_{xyz} + (\mathbf{a}_{B/A})_{xyz} \qquad (20\text{–}12)$$

where

$\mathbf{a}_B$ = acceleration of B

$\mathbf{a}_A$ = acceleration of the origin A of the x, y, z frame of reference

$(\mathbf{a}_{B/A})_{xyz}, (\mathbf{v}_{B/A})_{xyz}$ = relative acceleration and relative velocity of "B with respect to A" as measured by an observer attached to the rotating x, y, z frame of reference

$\dot{\mathbf{\Omega}}, \mathbf{\Omega}$ = angular acceleration and angular velocity of the x, y, z frame of reference

$\mathbf{r}_{B/A}$ = relative position of "B with respect to A"

Complicated spatial motion of the concrete bucket B occurs due to the rotation of the boom about the Z axis, motion of the carriage A along the boom, and extension and swinging of the cable AB. A translating-rotating x, y, z coordinate system can be established on the carriage, and a relative-motion analysis can then be applied to study this motion.

Equations 20–11 and 20–12 are identical to those used in Sec. 16.8 for analyzing relative plane motion.* In that case, however, application is simplified since $\mathbf{\Omega}$ and $\dot{\mathbf{\Omega}}$ have a *constant direction* which is always perpendicular to the plane of motion. For three-dimensional motion, $\dot{\mathbf{\Omega}}$ must be computed by using Eq. 20–6, since $\dot{\mathbf{\Omega}}$ depends on the change in *both* the magnitude and direction of $\mathbf{\Omega}$.

*Refer to Sec. 16.8 for an interpretation of the terms.

Procedure for Analysis

Three-dimensional motion of particles or rigid bodies can be analyzed with Eqs. 20–11 and 20–12 by using the following procedure.

Coordinate Axes

- Select the location and orientation of the X, Y, Z and x, y, z coordinate axes. Most often solutions are easily obtained if at the instant considered:

  (1) the origins are *coincident*
  (2) the axes are collinear
  (3) the axes are parallel

- If several components of angular velocity are involved in a problem, the calculations will be reduced if the x, y, z axes are selected such that only one component of angular velocity is observed in this frame ($\mathbf{\Omega}_{xyz}$) and the frame rotates with $\mathbf{\Omega}$ defined by the other components of angular velocity.

Kinematic Equations

- After the origin of the moving reference, A, is defined and the moving point B is specified, Eqs. 20–11 and 20–12 should be written in symbolic form as

$$\mathbf{v}_B = \mathbf{v}_A + \mathbf{\Omega} \times \mathbf{r}_{B/A} + (\mathbf{v}_{B/A})_{xyz}$$
$$\mathbf{a}_B = \mathbf{a}_A + \dot{\mathbf{\Omega}} \times \mathbf{r}_{B/A} + \mathbf{\Omega} \times (\mathbf{\Omega} \times \mathbf{r}_{B/A}) + 2\mathbf{\Omega} \times (\mathbf{v}_{B/A})_{xyz} + (\mathbf{a}_{B/A})_{xyz}$$

- If $\mathbf{r}_A$ and $\mathbf{\Omega}$ appear to *change direction* when observed from the fixed X, Y, Z reference use a set of primed reference axes, x', y', z' having $\mathbf{\Omega}' = \mathbf{\Omega}$ and Eq. 20–6 to determine $\dot{\mathbf{\Omega}}$ and the motion $\mathbf{v}_A$ and $\mathbf{a}_A$ of the origin of the moving x, y, z axes.

- If $(\mathbf{r}_{B/A})_{xyz}$ and $\mathbf{\Omega}_{xyz}$ appear to *change direction* as observed from x, y, z, then use a set of primed reference axes x', y', z' having $\mathbf{\Omega}' = \mathbf{\Omega}_{xyz}$ and Eq. 20–6 to determine $\dot{\mathbf{\Omega}}_{xyz}$ and the relative motion $(\mathbf{v}_{B/A})_{xyz}$ and $(\mathbf{a}_{B/A})_{xyz}$.

- After the final forms of $\dot{\mathbf{\Omega}}$, $\mathbf{v}_A$, $\mathbf{a}_A$, $\dot{\mathbf{\Omega}}_{xyz}$, $(\mathbf{v}_{B/A})_{xyz}$, and $(\mathbf{a}_{B/A})_{xyz}$ are obtained, numerical problem data may be substituted and the kinematic terms evaluated. The components of all these vectors may be selected either along the X, Y, Z axes or along x, y, z. The choice is arbitrary, provided a consistent set of unit vectors is used.

EXAMPLE 20–4

A motor and attached rod AB have the angular motions shown in Fig. 20–12. A collar C on the rod is located 0.25 m from A and is moving downward along the rod with a velocity of 3 m/s and an acceleration of 2 m/s². Determine the velocity and acceleration of C at this instant.

Solution

Coordinate Axes. The origin of the fixed X, Y, Z reference is chosen at the center of the platform, and the origin of the moving, x, y, z frame at point A, Fig. 20–12. Since the collar is subjected to two components of angular motion, $\boldsymbol{\omega}_p$ and $\boldsymbol{\omega}_M$, it will be viewed as having an angular velocity of $\boldsymbol{\Omega}_{xyz} = \boldsymbol{\omega}_M$ in x, y, z. The x, y, z axes will be attached to the platform so that $\boldsymbol{\Omega} = \boldsymbol{\omega}_p$.

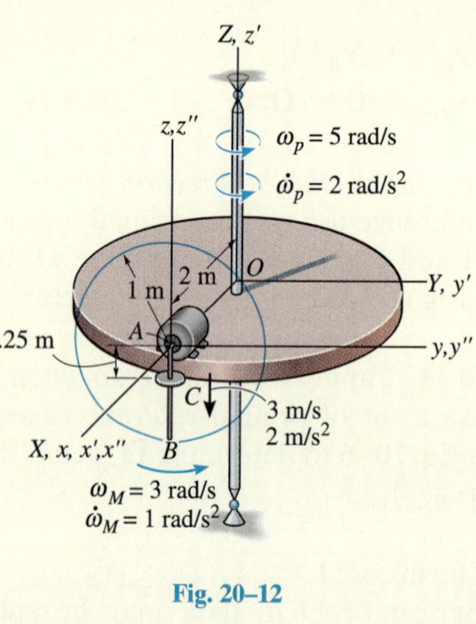

Fig. 20–12

Kinematic Equations. Equations 20–11 and 20–12, applied to points C and A, become

$$\mathbf{v}_C = \mathbf{v}_A + \mathbf{\Omega} \times \mathbf{r}_{C/A} + (\mathbf{v}_{C/A})_{xyz}$$
$$\mathbf{a}_C = \mathbf{a}_A + \dot{\mathbf{\Omega}} \times \mathbf{r}_{C/A} + \mathbf{\Omega} \times (\mathbf{\Omega} \times \mathbf{r}_{C/A}) + 2\mathbf{\Omega} \times (\mathbf{v}_{C/A})_{xyz} + (\mathbf{a}_{C/A})_{xyz}$$

Motion of A

Here $\mathbf{r}_A$ changes direction relative to $X\,Y\,Z$.
To find the time derivatives of $\mathbf{r}_A$ we will use a set of x', y', z' axes coincident with the X, Y, Z axes that rotate at $\mathbf{\Omega}' = \mathbf{\Omega} = \boldsymbol{\omega}_p$. Thus

$$\mathbf{\Omega} = \boldsymbol{\omega}_p = \{5\mathbf{k}\} \text{ rad/s } (\mathbf{\Omega} \text{ does not change direction relative to } X\,Y\,Z.)$$
$$\dot{\mathbf{\Omega}} = \dot{\boldsymbol{\omega}}_p = \{2\mathbf{k}\} \text{ rad/s}^2$$
$$\mathbf{r}_A = \{2\mathbf{i}\} \text{ m}$$
$$\mathbf{v}_A = \dot{\mathbf{r}}_A = (\dot{\mathbf{r}}_A)_{x'y'z'} + \boldsymbol{\omega}_p \times \mathbf{r}_A = \mathbf{0} + 5\mathbf{k} \times 2\mathbf{i} = \{10\mathbf{j}\} \text{ m/s}$$
$$\mathbf{a}_A = \ddot{\mathbf{r}}_A = [(\ddot{\mathbf{r}}_A)_{x'y'z'} + \boldsymbol{\omega}_p \times (\dot{\mathbf{r}}_A)_{x'y'z'}] + \dot{\boldsymbol{\omega}}_p \times \mathbf{r}_A + \boldsymbol{\omega}_p \times \dot{\mathbf{r}}_A$$
$$= [\mathbf{0} + \mathbf{0}] + 2\mathbf{k} \times 2\mathbf{i} + 5\mathbf{k} \times 10\mathbf{j} = \{-50\mathbf{i} + 4\mathbf{j}\} \text{ m/s}^2$$

Motion of C with Respect to A

Here $(\mathbf{r}_{C/A})_{xyz}$ changes direction relative to $x\,y\,z$. To find the time derivatives of $(\mathbf{r}_{C/A})_{xyz}$ use a set of x'', y'', z'' axes that rotate at $\mathbf{\Omega}'' = \mathbf{\Omega}_{xyz} = \boldsymbol{\omega}_M$. Thus

$$\mathbf{\Omega}_{xyz} = \boldsymbol{\omega}_M = \{3\mathbf{i}\} \text{ rad/s } (\mathbf{\Omega}_{xyz} \text{ does not change direction relative to } x\,y\,z.)$$
$$\dot{\mathbf{\Omega}}_{xyz} = \dot{\boldsymbol{\omega}}_M = \{1\mathbf{i}\} \text{ rad/s}^2$$
$$(\mathbf{r}_{C/A})_{xyz} = \{-0.25\mathbf{k}\} \text{ m}$$
$$(\mathbf{v}_{C/A})_{xyz} = (\dot{\mathbf{r}}_{C/A})_{xyz} = (\dot{\mathbf{r}}_{C/A})_{xyz} + \boldsymbol{\omega}_M \times (\mathbf{r}_{C/A})_{xyz}$$
$$= -3\mathbf{k} + [3\mathbf{i} \times (-0.25\mathbf{k})] = \{0.75\mathbf{j} - 3\mathbf{k}\} \text{ m/s}$$
$$(\mathbf{a}_{C/A})_{xyz} = (\ddot{\mathbf{r}}_{C/A})_{xyz} = [(\ddot{\mathbf{r}}_{C/A})_{xyz} + \boldsymbol{\omega}_M \times (\dot{\mathbf{r}}_{C/A})_{xyz}] + \dot{\boldsymbol{\omega}}_M \times (\mathbf{r}_{C/A})_{xyz} + \boldsymbol{\omega}_M \times (\mathbf{r}_{C/A})_{xyz}$$
$$= [-2\mathbf{k} + 3\mathbf{i} \times (-3\mathbf{k})] + (1\mathbf{i}) \times (-0.25\mathbf{k}) + (3\mathbf{i}) \times (0.75\mathbf{j} - 3\mathbf{k})$$
$$= \{18.25\mathbf{j} + 0.25\mathbf{k}\} \text{ m/s}^2$$

Motion of C

$$\mathbf{v}_C = \mathbf{v}_A + \mathbf{\Omega} \times \mathbf{r}_{C/A} + (\mathbf{v}_{C/A})_{xyz}$$
$$= 10\mathbf{j} + [5\mathbf{k} \times (-0.25\mathbf{k})] + (0.75\mathbf{j} - 3\mathbf{k})$$
$$= \{10.8\mathbf{j} - 3\mathbf{k}\} \text{ m/s} \qquad \textit{Ans.}$$
$$\mathbf{a}_C = \mathbf{a}_A + \dot{\mathbf{\Omega}} \times \mathbf{r}_{C/A} + \mathbf{\Omega} \times (\mathbf{\Omega} \times \mathbf{r}_{C/A}) + 2\mathbf{\Omega} \times (\mathbf{v}_{C/A})_{xyz} + (\mathbf{a}_{C/A})_{xyz}$$
$$= (-50\mathbf{i} + 4\mathbf{j}) + [2\mathbf{k} \times (-0.25\mathbf{k})] + 5\mathbf{k} \times [5\mathbf{k} \times (-0.25\mathbf{k})]$$
$$+ 2[5\mathbf{k} \times (0.75\mathbf{j} - 3\mathbf{k})] + (18.25\mathbf{j} + 0.25\mathbf{k})$$
$$= \{-57.5\mathbf{i} + 22.2\mathbf{j} + 0.25\mathbf{k}\} \text{ m/s}^2 \qquad \textit{Ans.}$$

EXAMPLE 20–5

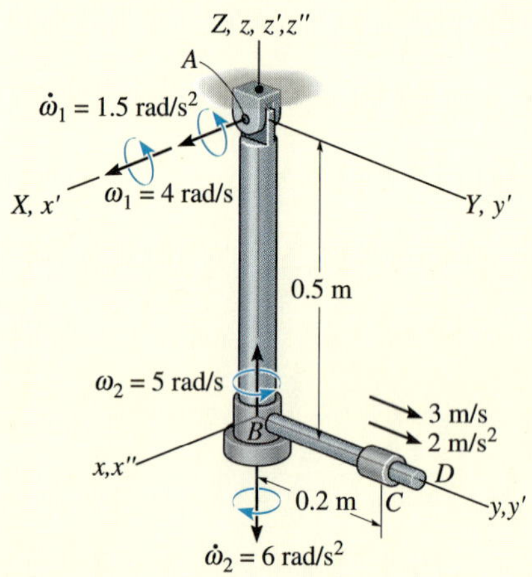

Fig. 20–13

The pendulum shown in Fig. 20–13 consists of two rods; AB is pin supported at A and swings only in the Y–Z plane, whereas a bearing at B allows the attached rod BD to spin about rod AB. At a given instant, the rods have the angular motions shown. Also, a collar C, located 0.2 m from B, has a velocity of 3 m/s and an acceleration of 2 m/s^2 along the rod. Determine the velocity and acceleration of the collar at this instant.

Solution I

Coordinate Axes. The origin of the fixed X, Y, Z frame will be chosen at A. Motion of the collar is conveniently observed from B, so the origin of the x, y, z frame is located at this point. We will choose $\mathbf{\Omega} = \boldsymbol{\omega}_1$ and $\mathbf{\Omega}_{xyz} = \boldsymbol{\omega}_2$.

Kinematic Equations

$$\mathbf{v}_C = \mathbf{v}_B + \mathbf{\Omega} \times \mathbf{r}_{C/B} + (\mathbf{v}_{C/B})_{xyz}$$
$$\mathbf{a}_C = \mathbf{a}_B + \dot{\mathbf{\Omega}} \times \mathbf{r}_{C/B} + \mathbf{\Omega} \times (\mathbf{\Omega} \times \mathbf{r}_{C/B}) + 2\mathbf{\Omega} \times (\mathbf{v}_{C/B})_{xyz} + (\mathbf{a}_{C/B})_{xyz}$$

Motion of B

To find the time derivatives of $\mathbf{r}_B$ let the x', y', z' axes rotate with $\mathbf{\Omega} = \boldsymbol{\omega}_1$. Then

$$\mathbf{\Omega} = \boldsymbol{\omega}_1 = \{4\mathbf{i}\} \text{ rad/s} \qquad \dot{\mathbf{\Omega}} = \dot{\boldsymbol{\omega}}_1 = \{1.5\mathbf{i}\} \text{ rad/s}^2$$
$$\mathbf{r}_B = \{-0.5\mathbf{k}\} \text{ m}$$
$$\mathbf{v}_B = \dot{\mathbf{r}}_B = (\dot{\mathbf{r}}_B)_{x'y'z'} + \boldsymbol{\omega}_1 \times \mathbf{r}_B = \mathbf{0} + 4\mathbf{i} \times (-0.5\mathbf{k}) = \{2\mathbf{j}\} \text{ m/s}$$
$$\mathbf{a}_B = \ddot{\mathbf{r}}_B = [(\ddot{\mathbf{r}}_B)_{x'y'z'} + \boldsymbol{\omega}_1 \times (\dot{\mathbf{r}}_B)_{x'y'z'}] + \dot{\boldsymbol{\omega}}_1 \times \mathbf{r}_B + \boldsymbol{\omega}_1 \times \dot{\mathbf{r}}_B$$
$$= [\mathbf{0} + \mathbf{0}] + 1.5\mathbf{i} \times (-0.5\mathbf{k}) + 4\mathbf{i} \times 2\mathbf{j} = \{0.75\mathbf{j} + 8\mathbf{k}\} \text{ m/s}^2$$

Motion of C with Respect to B

To find the time derivatives of $(\mathbf{r}_{C/B})_{xyz}$, let the x'', y'', z'' axes rotate with $\mathbf{\Omega}_{xyz} = \boldsymbol{\omega}_2$. Then

$$\mathbf{\Omega}_{xyz} = \boldsymbol{\omega}_2 = \{5\mathbf{k}\} \text{ rad/s} \qquad \dot{\mathbf{\Omega}}_{xyx} = \dot{\boldsymbol{\omega}}_2 = \{-6\mathbf{k}\} \text{ rad/s}^2$$
$$(\mathbf{r}_{C/B})_{xyz} = \{0.2\mathbf{j}\}\text{m}$$
$$(\mathbf{v}_{C/B})_{xyz} = (\dot{\mathbf{r}}_{C/B})_{xyz} = (\dot{\mathbf{r}}_{C/B})_{x''y''z''} + \boldsymbol{\omega}_2 \times (\mathbf{r}_{C/B})_{xyz} = 3\mathbf{j} + 5\mathbf{k} \times 0.2\mathbf{j} = \{-1\mathbf{i} + 3\mathbf{j}\} \text{ m/s}$$
$$(\mathbf{a}_{C/B})_{xyz} = (\ddot{\mathbf{r}}_{C/B})_{xyz} = [(\ddot{\mathbf{r}}_{C/B})_{x''y''z''} + \boldsymbol{\omega}_2 \times (\dot{\mathbf{r}}_{C/B})_{x''y''z''}] + \dot{\boldsymbol{\omega}}_2 \times (\mathbf{r}_{C/B})_{xyz} + \boldsymbol{\omega}_2 \times (\dot{\mathbf{r}}_{C/B})_{xyz}$$
$$= (2\mathbf{j} + 5\mathbf{k} \times 3\mathbf{j}) + (-6\mathbf{k} \times 0.2\mathbf{j}) + [5\mathbf{k} \times (-1\mathbf{i} + 3\mathbf{j})]$$
$$= \{-28.8\mathbf{i} - 3\mathbf{j}\} \text{ m/s}^2$$

Motion of C

$$\mathbf{v}_C = \mathbf{v}_B + \mathbf{\Omega} \times \mathbf{r}_{C/B} + (\mathbf{v}_{C/B})_{xyz} = 2\mathbf{j} + 4\mathbf{i} \times 0.2\mathbf{j} + (-1\mathbf{i} + 3\mathbf{j})$$
$$= \{-1\mathbf{i} + 5\mathbf{j} + 0.8\mathbf{k}\} \text{ m/s} \qquad \textit{Ans.}$$

$$\mathbf{a}_C = \mathbf{a}_B + \dot{\mathbf{\Omega}} \times \mathbf{r}_{C/B} + \mathbf{\Omega} \times (\mathbf{\Omega} \times \mathbf{r}_{C/B}) + 2\mathbf{\Omega} \times (\mathbf{v}_{C/B})_{xyz} + (\mathbf{a}_{C/B})_{xyz}$$
$$= (0.75\mathbf{j} + 8\mathbf{k}) + (1.5\mathbf{i} \times 0.2\mathbf{j}) + [4\mathbf{i} \times (4\mathbf{i} \times 0.2\mathbf{j})]$$
$$+ 2[4\mathbf{i} \times (-1\mathbf{i} + 3\mathbf{j})] + (-28.8\mathbf{i} - 3\mathbf{j})$$
$$= \{-28.8\mathbf{i} - 5.45\mathbf{j} + 32.3\mathbf{k}\} \text{ m/s}^2 \qquad \textit{Ans.}$$

Solution II

Coordinate Axes. Here we will let the x, y, z axes rotate at

$$\mathbf{\Omega} = \boldsymbol{\omega}_1 + \boldsymbol{\omega}_2 = \{4\mathbf{i} + 5\mathbf{k}\} \text{ rad/s}$$

Then $\mathbf{\Omega}_{xyz} = \mathbf{0}$.

Motion of B

From the constraints of the problem $\boldsymbol{\omega}_1$ does not change direction relative to X, Y, Z; however, the direction of $\boldsymbol{\omega}_2$ is changed by $\boldsymbol{\omega}_1$. Thus, to obtain $\dot{\mathbf{\Omega}}$ consider x', y', z' axes coincident with the X, Y, Z axes at A, such that $\dot{\mathbf{\Omega}} = \boldsymbol{\omega}_1$. Then taking the time derivative of its components,

$$\begin{aligned}\dot{\mathbf{\Omega}} = \dot{\boldsymbol{\omega}}_1 + \dot{\boldsymbol{\omega}}_2 &= [(\dot{\boldsymbol{\omega}}_1)_{x'y'z'} + \boldsymbol{\omega}_1 \times \boldsymbol{\omega}_1] + [(\dot{\boldsymbol{\omega}}_2)_{x'y'z'} + \boldsymbol{\omega}_1 \times \boldsymbol{\omega}_2] \\ &= [1.5\mathbf{i} + \mathbf{0}] + [-6\mathbf{k} + 4\mathbf{i} \times 5\mathbf{k}] = \{1.5\mathbf{i} - 20\mathbf{j} - 6\mathbf{k}\} \text{ rad/s}^2\end{aligned}$$

Also, $\boldsymbol{\omega}_1$ changes the direction of $\mathbf{r}_B$ so that the time derivatives of $\mathbf{r}_B$ can be computed using the primed axes defined above. Hence,

$$\begin{aligned}\mathbf{v}_B = \dot{\mathbf{r}}_B &= (\dot{\mathbf{r}}_B)_{x'y'z'} + \boldsymbol{\omega}_1 \times \mathbf{r}_B \\ &= \mathbf{0} + 4\mathbf{i} \times (-0.5\mathbf{k}) = \{2\mathbf{j}\} \text{ m/s} \\ \mathbf{a}_B = \ddot{\mathbf{r}}_B &= [(\ddot{\mathbf{r}}_B)_{x'y'z'} + \boldsymbol{\omega}_1 \times (\dot{\mathbf{r}}_B)_{x'y'z'}] + \dot{\boldsymbol{\omega}}_1 \times \mathbf{r}_B + \boldsymbol{\omega}_1 \times \dot{\mathbf{r}}_B \\ &= [\mathbf{0} + \mathbf{0}] + 1.5\mathbf{i} \times (-0.5\mathbf{k}) + 4\mathbf{i} \times 2\mathbf{j} = \{0.75\mathbf{j} + 8\mathbf{k}\} \text{ m/s}^2\end{aligned}$$

Motion of C with Respect to B

$$\begin{aligned}\mathbf{\Omega}_{xyz} &= \mathbf{0} \\ \dot{\mathbf{\Omega}}_{xyz} &= \mathbf{0} \\ (\mathbf{r}_{C/B})_{xyz} &= \{0.2\mathbf{j}\} \text{ m} \\ (\mathbf{v}_{C/B})_{xyz} &= \{3\mathbf{j}\} \text{ m/s} \\ (\mathbf{a}_{C/B})_{xyz} &= \{2\mathbf{j}\} \text{ m/s}^2\end{aligned}$$

Motion of C

$$\begin{aligned}\mathbf{v}_C &= \mathbf{v}_B + \mathbf{\Omega} \times \mathbf{r}_{C/B} + (\mathbf{v}_{C/B})_{xyz} \\ &= 2\mathbf{j} + [(4\mathbf{i} + 5\mathbf{k}) \times (0.2\mathbf{j})] + 3\mathbf{j} \\ &= \{-1\mathbf{i} + 5\mathbf{j} + 0.8\mathbf{k}\} \text{ m/s} \qquad \textit{Ans.}\end{aligned}$$

$$\begin{aligned}\mathbf{a}_C &= \mathbf{a}_B + \dot{\mathbf{\Omega}} \times \mathbf{r}_{C/B} + \mathbf{\Omega} \times (\mathbf{\Omega} \times \mathbf{r}_{C/B}) + 2\mathbf{\Omega} \times (\mathbf{v}_{C/B})_{xyz} + (\mathbf{a}_{C/B})_{xyz} \\ &= (0.75\mathbf{j} + 8\mathbf{k}) + [(1.5\mathbf{i} - 20\mathbf{j} - 6\mathbf{k}) \times (0.2\mathbf{j})] \\ &\quad + (4\mathbf{i} + 5\mathbf{k}) \times [(4\mathbf{i} + 5\mathbf{k}) \times 0.2\mathbf{j}] + 2[(4\mathbf{i} + 5\mathbf{k}) \times 3\mathbf{j}] + 2\mathbf{j} \\ &= \{-28.8\mathbf{i} - 5.45\mathbf{j} + 32.3\mathbf{k}\} \text{ m/s}^2 \qquad \textit{Ans.}\end{aligned}$$

Problems

20-31. Solve Example 20–5 such that the x, y, z axes move with curvilinear translation, $\mathbf{\Omega} = \mathbf{0}$, in which case the collar appears to have both an angular velocity $\mathbf{\Omega}_{xyz} = \boldsymbol{\omega}_1 + \boldsymbol{\omega}_2$ and radial motion.

***20-32.** Solve Example 20–5 by fixing x, y, z axes to rod BD so that $\mathbf{\Omega} = \boldsymbol{\omega}_1 + \boldsymbol{\omega}_2$. In this case the collar appears only to move radially outward along BD; hence $\mathbf{\Omega}_{xyz} = \mathbf{0}$.

20-33. At a given instant, the antenna has angular motion $\omega_1 = 3$ rad/s and $\dot{\omega}_1 = 2$ rad/s^2 about the z axis. At this same instant $\theta = 30°$, the angular motion about the x axis is $\omega_2 = 1.5$ rad/s and $\dot{\omega}_2 = 4$ rad/s^2. Determine the velocity and acceleration of the signal horn A at this instant. The distance from O to A is $d = 3$ ft.

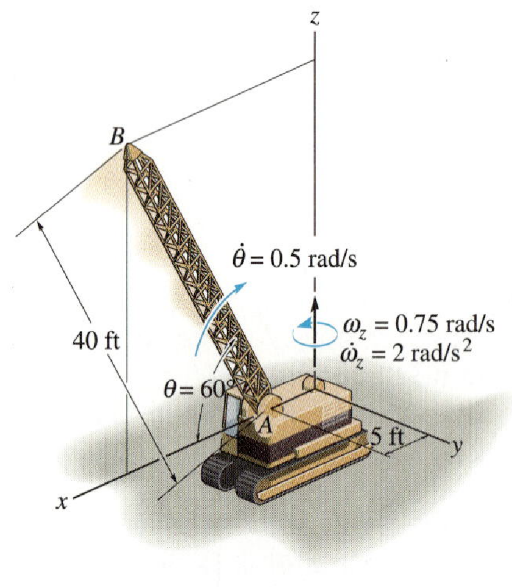

Prob. 20–34

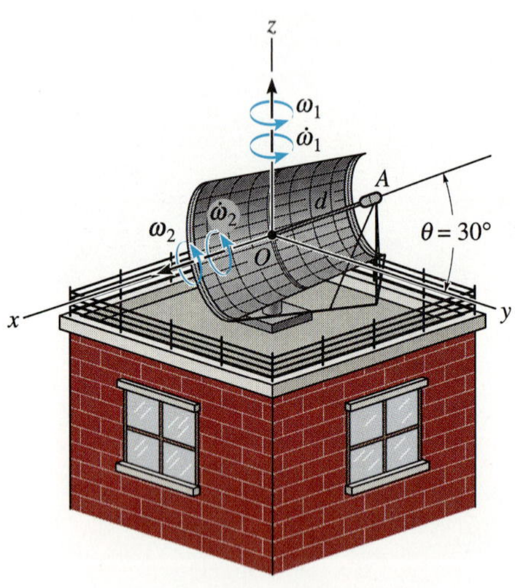

Prob. 20–33

20-34. The boom AB of the crane is rotating about the z axis with an angular velocity $\omega_z = 0.75$ rad/s and an angular acceleration of $\dot{\omega}_z = 2$ rad/s^2. At the same instant, $\theta = 60°$ and the boom is rotating upward at a constant rate $\dot{\theta} = 0.5$ rad/s. Determine the velocity and acceleration of the tip B of the boom at this instant.

20-35. At the instant shown, the cab of the excavator is rotating about the z axis with a constant angular velocity of $\omega_z = 0.3$ rad/s. At the same instant $\theta = 60°$, and the boom OBC has an angular velocity of $\dot{\theta} = 0.6$ rad/s, which is increasing at $\ddot{\theta} = 0.2$ rad/s^2, both measured relative to the cab. Determine the velocity and acceleration of point C on the grapple at this instant.

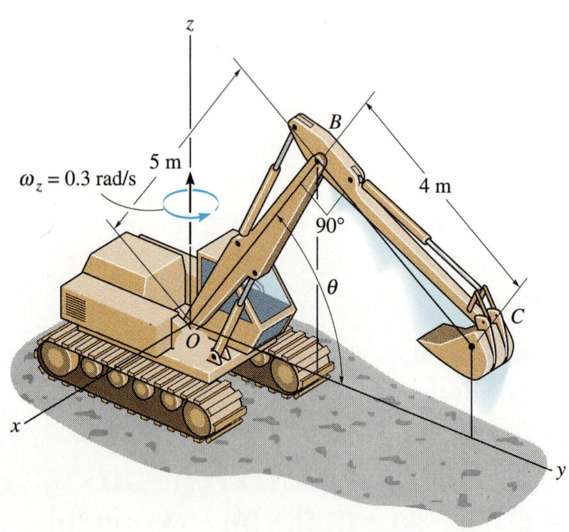

Prob. 20–35

***20-36.** At the instant shown, the frame of the excavator is traveling forward in the y direction with a velocity of 2 m/s and an acceleration of 1 m/s^2, while the cab is rotating about the z axis with an angular velocity of $\omega_z = 0.3$ rad/s, which is increasing at $\alpha_z = 0.4$ rad/s^2. At the same instant $\theta = 60°$, and the boom OBC has an angular velocity of $\dot{\theta} = 0.6$ rad/s, which is increasing at $\ddot{\theta} = 0.2$ rad/s^2, both measured relative to the cab. Determine the velocity and acceleration of point C on the grapple at this instant.

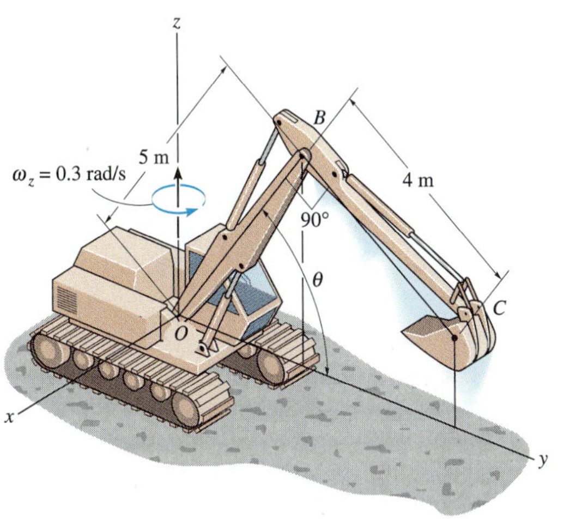

Prob. 20–36

20-37. At the instant shown, the boom is rotating about the z axis with an angular velocity $\omega_1 = 2$ rad/s and angular acceleration $\dot{\omega}_1 = 0.8$ rad/s^2. At this same instant the swivel is rotating at $\omega_2 = 3$ rad/s when $\dot{\omega}_2 = 2$ rad/s^2, both measured relative to the boom. Determine the velocity and acceleration of point P on the pipe at this instant.

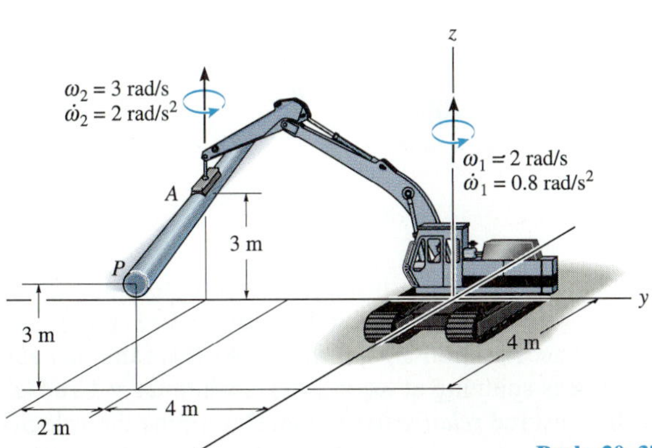

Prob. 20–37

20-38. At the instant shown, the frame of the brush cutter is traveling forward in the x direction with a constant velocity of 1 m/s, and the cab is rotating about the vertical axis with a constant angular velocity of $\omega_1 = 0.5$ rad/s. At the same instant the boom AB has an angular velocity of $\dot{\theta} = 0.8$ rad/s in the direction shown. Determine the velocity and acceleration of point B at the connection to the mower at this instant.

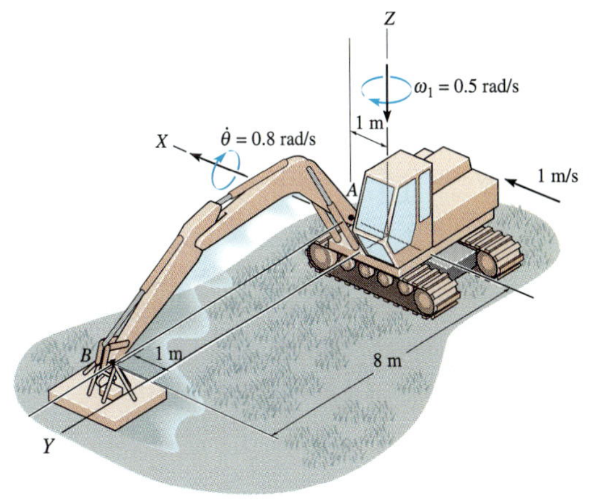

Prob. 20–38

20-39. At the instant shown, the frame of the brush cutter is traveling forward in the x direction with a constant velocity of 1 m/s, and the cab is rotating about the vertical axis with an angular velocity of $\omega_1 = 0.5$ rad/s, which is increasing at $\dot{\omega}_1 = 0.4$ rad/s^2. At the same instant the boom AB has an angular velocity of $\dot{\theta} = 0.8$ rad/s, which is increasing at $\ddot{\theta} = 0.9$ rad/s^2. Determine the velocity and acceleration of point B at the connection to the mower at this instant.

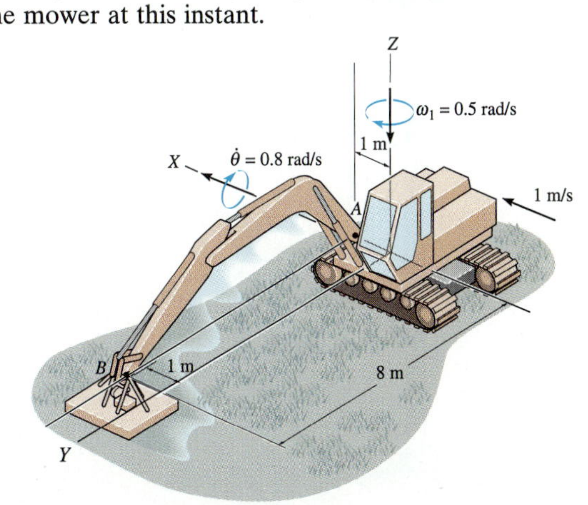

Prob. 20–39

***20–40.** At the instant shown, the helicopter is moving upwards with a velocity $v_H = 4$ ft/s and has an acceleration $a_H = 2$ ft/s^2. At the same instant the frame H, *not* the horizontal blade, is rotating about a vertical axis with a constant angular velocity $\omega_H = 0.9$ rad/s. If the tail blade B is rotating with a constant angular velocity $\omega_{B/H} = 180$ rad/s, measured relative to H, determine the velocity and acceleration of point P, located on the tip of the blade, at the instant the blade is in the vertical position.

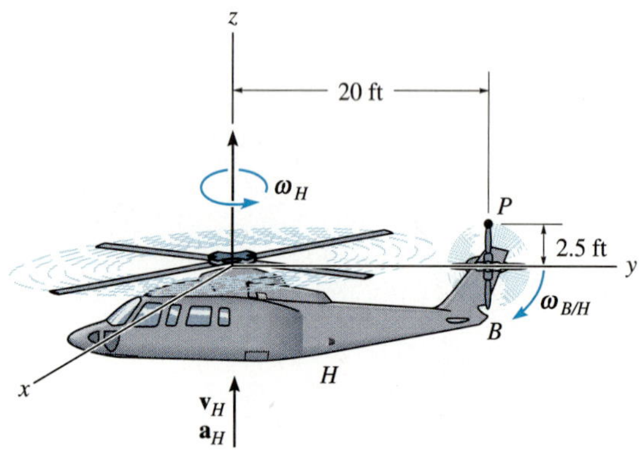

Prob. 20–40

20–41. At the instant shown, rod BD is rotating about the vertical axis with an angular velocity $\omega_{BD} = 7$ rad/s and an angular acceleration $\alpha_{BD} = 4$ rad/s^2. Also, $\theta = 60°$ and link AC is rotating downward such that $\dot{\theta} = 2$ rad/s and $\ddot{\theta} = 3$ rad/s^2. Determine the velocity and acceleration of point A on the link at this instant.

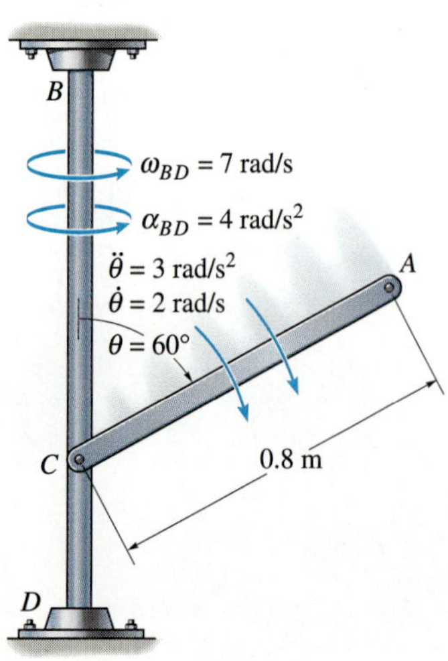

Prob. 20–41

20–42. At a given instant, the rod has the angular motions shown, while the collar C is moving down *relative* to the rod with a velocity of 6 ft/s and an acceleration of 2 ft/s^2. Determine the collar's velocity and acceleration at this instant.

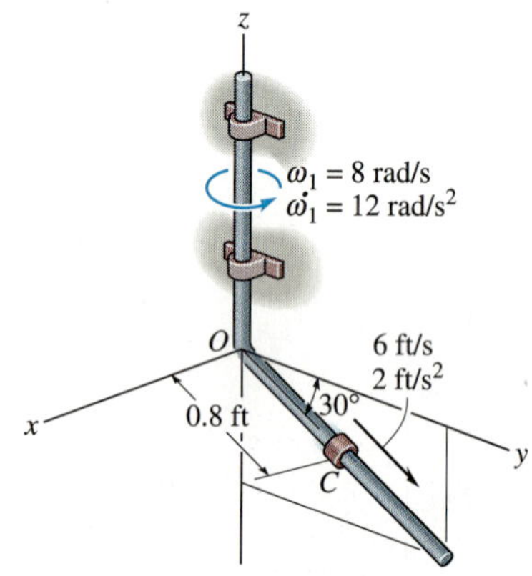

Prob. 20–42

20–43. The particle P slides around the circular hoop with a constant angular velocity of $\dot{\theta} = 6$ rad/s, while the hoop rotates about the x axis at a constant rate of $\omega = 4$ rad/s. If at the instant shown the hoop is in the x–y plane and the angle $\theta = 45°$, determine the velocity and acceleration of the particle at this instant.

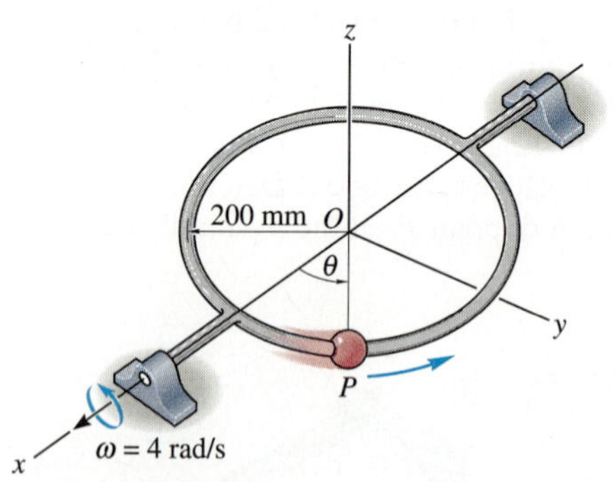

Prob. 20–43

***20–44.** At the given instant, the rod is spinning about the z axis with an angular velocity $\omega_1 = 3$ rad/s and angular acceleration $\dot{\omega}_1 = 4$ rad/s^2. At this same instant, the disk is spinning at $\omega_2 = 2$ rad/s when $\dot{\omega}_2 = 1$ rad/s^2, both measured *relative* to the rod. Determine the velocity and acceleration of point P on the disk at this instant.

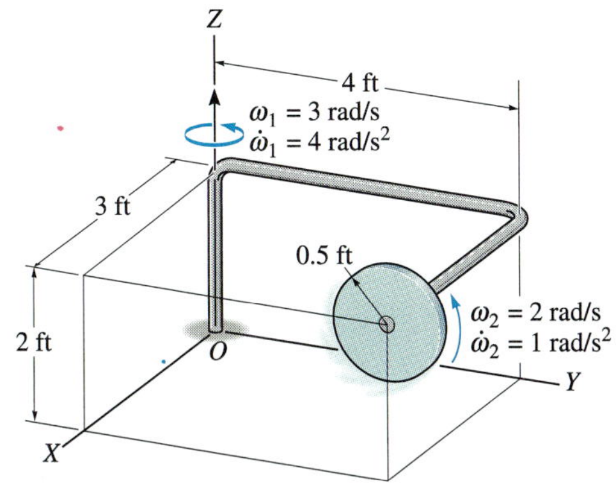

Prob. 20–44

20-45. At the instant shown, the base of the robotic arm is turning about the z axis with an angular velocity of $\omega_1 = 4$ rad/s, which is increasing at $\dot{\omega}_1 = 3$ rad/s^2. Also, the boom segment BC is rotating at a constant rate of $\omega_{BC} = 8$ rad/s. Determine the velocity and acceleration of the part C held in its grip at this instant.

20-46. At the instant shown, the base of the robotic arm is turning about the z axis with an angular velocity of $\omega_1 = 4$ rad/s, which is increasing at $\dot{\omega}_1 = 3$ rad/s^2. Also, the boom segment BC is rotating at $\omega_{BC} = 8$ rad/s, which is increasing at $\dot{\omega}_{BC} = 2$ rad/s^2. Determine the velocity and acceleration of the part C held in its grip at this instant.

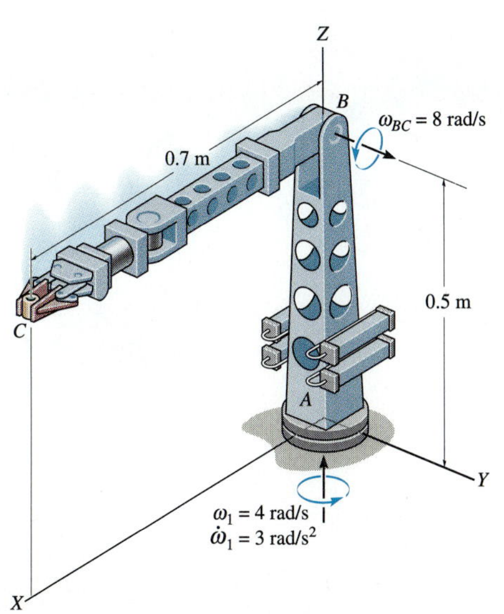

Probs. 20–45/46

20-47. At a given instant, the crane is moving along the track with a velocity $v_{CD} = 8$ m/s and acceleration of 9 m/s^2. Simultaneously, it has the angular motions shown. If the trolley T is moving outwards along the boom AB with a relative speed of 3 m/s and relative acceleration of 5 m/s^2, determine the velocity and acceleration of the trolley.

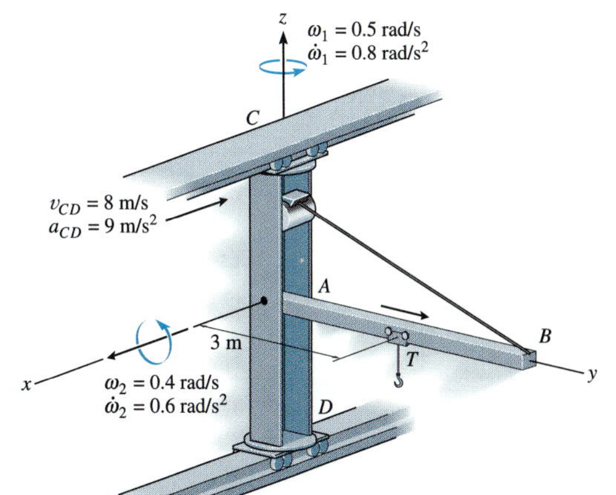

Prob. 20–47

***20-48.** The load is being lifted upward at a constant rate of 9 m/s relative to the crane boom AB. At the instant shown, the boom is rotating about the vertical axis at a constant rate of $\omega_1 = 4$ rad/s, and the trolley T is moving outward along the boom at a constant rate of $v_t = 2$ m/s. Furthermore, at this same instant the retractable arm supporting the load is vertical and is swinging in the y–z plane at an angular rate of 5 rad/s, with an increase in the rate of swing of 7 rad/s^2. Determine the velocity and acceleration of the center G of the load at this instant, i.e., when $s = 4$ m and $h = 3$ m.

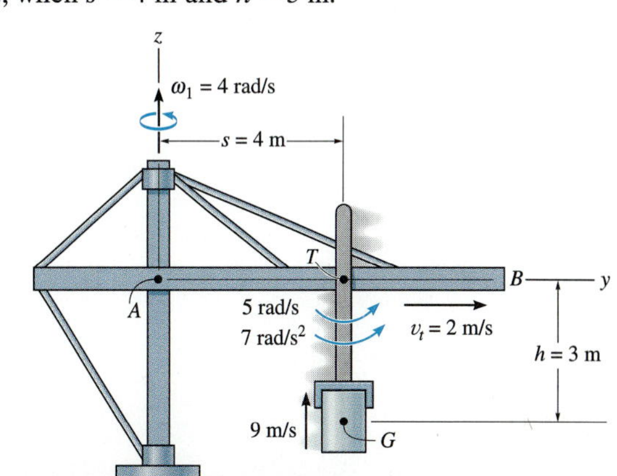

Prob. 20–48

The design of amusement-park rides requires a force analysis that depends on their three-dimensional motion.

CHAPTER 21

Three-Dimensional Kinetics of a Rigid Body

Chapter Objectives

- To introduce the methods for finding the moments of inertia and products of inertia of a body about various axes.
- To show how to apply the principles of work and energy and linear and angular momentum to a rigid body having three-dimensional motion.
- To develop and apply the equations of motion in three dimensions.
- To study the motion of a gyroscope and torque-free motion.

*21.1 Moments and Products of Inertia

When studying the planar kinetics of a body, it was necessary to introduce the moment of inertia I_G, which was computed about an axis perpendicular to the plane of motion and passing through the body's mass center G. For the kinetic analysis of three-dimensional motion it will sometimes be necessary to calculate six inertial quantities. These terms, called the moments and products of inertia, describe in a particular way the distribution of mass for a body relative to a given coordinate system that has a specified orientation and point of origin.

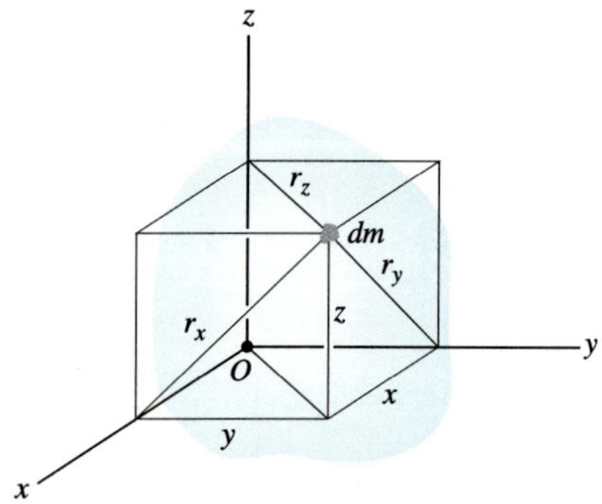

Fig. 21–1

Moment of Inertia. Consider the rigid body shown in Fig. 21–1. The *moment of inertia* for a differential element dm of the body about any one of the three coordinate axes is defined as the product of the mass of the element and the square of the shortest distance from the axis to the element. For example, as noted in the figure, $r_x = \sqrt{y^2 + z^2}$, so that the mass moment of inertia of dm about the x axis is

$$dI_{xx} = r_x^2\, dm = (y^2 + z^2)\, dm$$

The moment of inertia I_{xx} for the body is determined by integrating this expression over the entire mass of the body. Hence, for each of the axes, we may write

$$\begin{aligned} I_{xx} &= \int_m r_x^2\, dm = \int_m (y^2 + z^2)\, dm \\ I_{yy} &= \int_m r_y^2\, dm = \int_m (x^2 + z^2)\, dm \\ I_{zz} &= \int_m r_z^2\, dm = \int_m (x^2 + y^2)\, dm \end{aligned} \qquad (21\text{–}1)$$

Here it is seen that the moment of inertia is *always a positive quantity,* since it is the summation of the product of the mass dm, which is always positive, and the distances squared.

Product of Inertia. The *product of inertia* for a differential element dm is defined with respect to a set of *two orthogonal planes* as the product of the mass of the element and the perpendicular (or shortest) distances from the planes to the element. For example, with respect to the y–z and x–z planes, the product of inertia dI_{xy} for the element dm shown in Fig. 21–1 is

$$dI_{xy} = xy\, dm$$

Note also that $dI_{yx} = dI_{xy}$. By integrating over the entire mass, the product of inertia of the body for each combination of planes may be expressed as

$$\begin{aligned} I_{xy} &= I_{yx} = \int_m xy\, dm \\ I_{yz} &= I_{zy} = \int_m yz\, dm \\ I_{xz} &= I_{zx} = \int_m xz\, dm \end{aligned} \qquad (21\text{–}2)$$

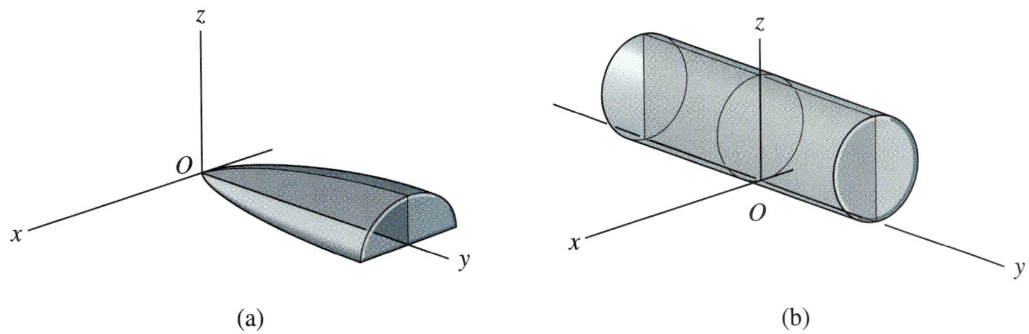

Fig. 21–2

Unlike the moment of inertia, which is always positive, the product of inertia may be positive, negative, or zero. The result depends on the signs of the two defining coordinates, which vary independently from one another. In particular, if either one or both of the orthogonal planes are *planes of symmetry* for the mass, the *product of inertia* with respect to these planes will be *zero.* In such cases, elements of mass will occur in *pairs* located on each side of the plane of symmetry. On one side of the plane the product of inertia for the element will be positive, while on the other side the product of inertia for the corresponding element will be negative, the sum therefore yielding zero. Examples of this are shown in Fig. 21–2. In the first case, Fig. 21–2*a*, the y–z plane is a plane of symmetry, and hence $I_{xy} = I_{xz} = 0$. Calculation of I_{yz} will yield a *positive* result, since all elements of mass are located using only positive y and z coordinates. For the cylinder, with the coordinate axes located as shown in Fig. 21–2*b*, the x–z and y–z planes are both planes of symmetry. Thus, $I_{xy} = I_{yz} = I_{zx} = 0$.

Parallel-Axis and Parallel-Plane Theorems. The techniques of integration which are used to determine the moment of inertia of a body were described in Sec. 17.1. Also discussed were methods to determine the moment of inertia of a composite body, i.e., a body that is composed of simpler segments, as tabulated on the inside back cover. In both of these cases the *parallel-axis theorem* is often used for the calculations. This theorem, which was developed in Sec. 17.1, is used to transfer the moment of inertia of a body from an axis passing through its mass center G to a parallel axis passing through some other point. In this regard, if G has coordinates x_G, y_G, z_G defined from the x, y, z axes, Fig. 21–3, then the parallel-axis equations used to calculate the moments of inertia about the x, y, z axes are

$$
\begin{aligned}
I_{xx} &= (I_{x'x'})_G + m(y_G^2 + z_G^2)\\
I_{yy} &= (I_{y'y'})_G + m(x_G^2 + z_G^2)\\
I_{zz} &= (I_{z'z'})_G + m(x_G^2 + y_G^2)
\end{aligned}
\tag{21–3}
$$

The products of inertia of a composite body are computed in the same manner as the body's moments of inertia. Here, however, the *parallel-plane theorem* is important. This theorem is used to transfer the products of inertia of the body from a set of three orthogonal planes passing through the body's mass center to a corresponding set of three parallel planes passing through some other point O. Defining the perpendicular distances between the planes as x_G, y_G, and z_G, Fig. 21–3, the parallel-plane equations can be written as

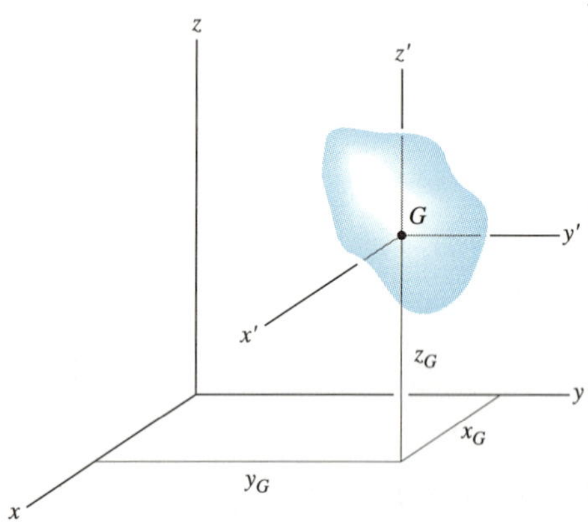

Fig. 21–3

$$
\begin{aligned}
I_{xy} &= (I_{x'y'})_G + m x_G y_G\\
I_{yz} &= (I_{y'z'})_G + m y_G z_G\\
I_{zx} &= (I_{z'x'})_G + m z_G x_G
\end{aligned}
\tag{21–4}
$$

The derivation of these formulas is similar to that given for the parallel-axis equation, Sec. 17.1.

Inertia Tensor. The inertial properties of a body are completely characterized by nine terms, six of which are independent of one another. This set of terms is defined using Eqs. 21–1 and 21–2 and can be written as

$$
\begin{pmatrix}
I_{xx} & -I_{xy} & -I_{xz}\\
-I_{yx} & I_{yy} & -I_{yz}\\
-I_{zx} & -I_{zy} & I_{zz}
\end{pmatrix}
$$

This array is called an *inertia tensor*. It has a unique set of values for a body when it is computed for each location of the origin O and orientation of the coordinate axes.

In general, for point O we can specify a unique axes inclination for which the products of inertia for the body are zero when computed with respect to these axes. When this is done, the inertia tensor is said to be "diagonalized" and may be written in the simplified form

The dynamics of the space shuttle as it orbits the earth can be predicted only if its moments and products of inertia are known.

$$
\begin{pmatrix}
I_x & 0 & 0\\
0 & I_y & 0\\
0 & 0 & I_z
\end{pmatrix}
$$

Here $I_x = I_{xx}$, $I_y = I_{yy}$, and $I_z = I_{zz}$ are termed the *principal moments of inertia* for the body, which are computed from the *principal axes of inertia.* Of these three principal moments of inertia, one will be a maximum and another a minimum of the body's moment of inertia.

Mathematical determination of the directions of principal axes of inertia will not be discussed here (see Prob. 21-21). There are many cases, however, in which the principal axes may be determined by inspection. From the previous discussion it was noted that if the coordinate axes are oriented such that *two* of the three orthogonal planes containing the axes are planes of *symmetry* for the body, then all the products of inertia for the body are zero with respect to the coordinate planes, and hence the coordinate axes are principal axes of inertia. For example, the x, y, z axes shown in Fig. 21–2*b* represent the principal axes of inertia for the cylinder at point O.

Moment of Inertia About an Arbitrary Axis. Consider the body shown in Fig. 21–4, where the nine elements of the inertia tensor have been computed for the x, y, z axes having an origin at O. Here we wish to determine the moment of inertia of the body about the Oa axis, for which the direction is defined by the unit vector $\mathbf{u}_a$. By definition $I_{Oa} = \int b^2\, dm$, where b is the *perpendicular distance* from dm to Oa. If the position of dm is located using $\mathbf{r}$, then $b = r \sin\theta$, which represents the *magnitude* of the cross product $\mathbf{u}_a \times \mathbf{r}$. Hence, the moment of inertia can be expressed as

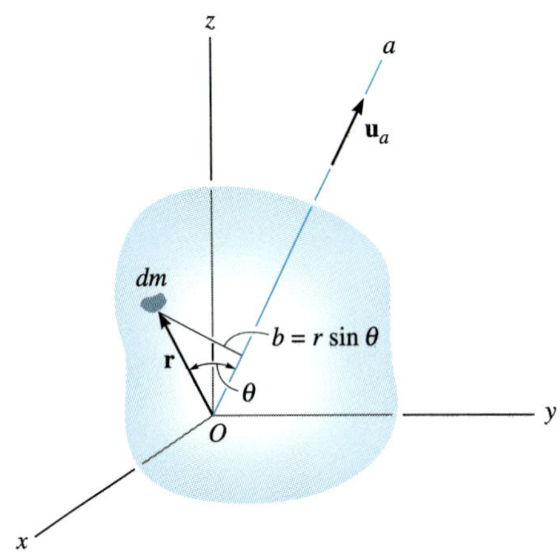

Fig. 21–4

$$I_{Oa} = \int_m |(\mathbf{u}_a \times \mathbf{r})|^2\, dm = \int_m (\mathbf{u}_a \times \mathbf{r})\cdot(\mathbf{u}_a \times \mathbf{r})\, dm$$

Provided $\mathbf{u}_a = u_x\mathbf{i} + u_y\mathbf{j} + u_z\mathbf{k}$ and $\mathbf{r} = x\mathbf{i} + y\mathbf{j} + z\mathbf{k}$, so that $\mathbf{u}_a \times \mathbf{r} = (u_y z - u_z y)\mathbf{i} + (u_z x - u_x z)\mathbf{j} + (u_x y - u_y x)\mathbf{k}$, then, after substituting and performing the dot-product operation, we can write the moment of inertia as

$$\begin{aligned} I_{Oa} &= \int_m [(u_y z - u_z y)^2 + (u_z x - u_x z)^2 + (u_x y - u_y x)^2]\, dm \\ &= u_x^2 \int_m (y^2 + z^2)\, dm + u_y^2 \int_m (z^2 + x^2)\, dm + u_z^2 \int_m (x^2 + y^2)\, dm \\ &\qquad - 2u_x u_y \int_m xy\, dm - 2u_y u_z \int_m yz\, dm - 2u_z u_x \int_m zx\, dm \end{aligned}$$

Recognizing the integrals to be the moments and products of inertia of the body, Eqs. 21–1 and 21–2, we have

$$I_{Oa} = I_{xx}u_x^2 + I_{yy}u_y^2 + I_{zz}u_z^2 - 2I_{xy}u_x u_y - 2I_{yz}u_y u_z - 2I_{zx}u_z u_x \qquad (21\text{–}5)$$

Thus, if the inertia tensor is specified for the x, y, z axes, the moment of inertia of the body about the inclined Oa axis can be found by using Eq. 21–5. For the calculation the direction cosines u_x, u_y, u_z of the axes must be determined. These terms specify the cosines of the coordinate direction angles α, β, γ made between the positive Oa axis and the positive x, y, z axes, respectively (see Appendix C).

EXAMPLE 21–1

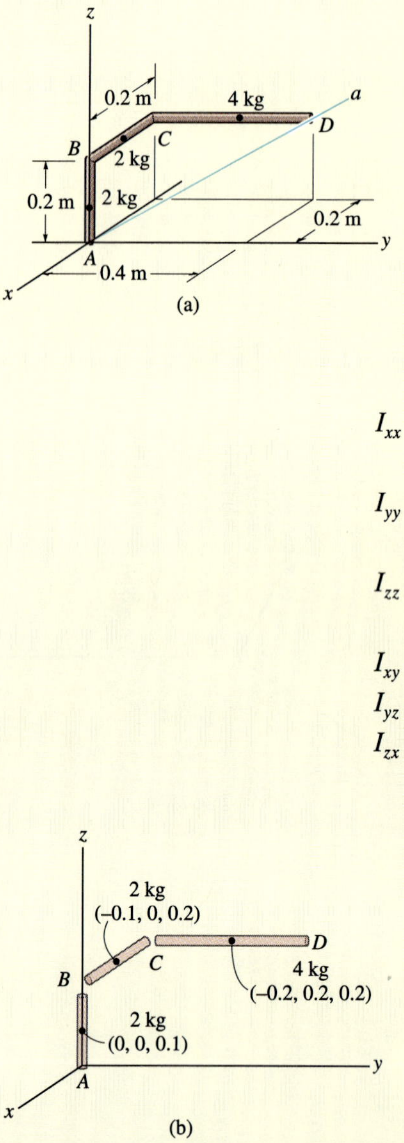

Fig. 21–5

Determine the moment of inertia of the bent rod shown in Fig. 21–5*a* about the *Aa* axis. The mass of each of the three segments is shown in the figure.

Solution

Before applying Eq. 21–5, it is first necessary to determine the moments and products of inertia of the rod about the x, y, z axes. This is done using the formula for the moment of inertia of a slender rod, $I = \frac{1}{12}ml^2$, and the parallel-axis and parallel-plane theorems, Eqs. 21–3 and 21–4. Dividing the rod into three parts and locating the mass center of each segment, Fig. 21–5*b*, we have

$$I_{xx} = [\tfrac{1}{12}(2)(0.2)^2 + 2(0.1)^2] + [0 + 2(0.2)^2] + [\tfrac{1}{12}(4)(0.4)^2 + 4((0.2)^2 + (0.2)^2)] = 0.480\ \text{kg}\cdot\text{m}^2$$

$$I_{yy} = [\tfrac{1}{12}(2)(0.2)^2 + 2(0.1)^2] + [\tfrac{1}{12}(2)(0.2)^2 + 2((-0.1)^2 + (0.2)^2)] + [0 + 4((-0.2)^2 + (0.2)^2)] = 0.453\ \text{kg}\cdot\text{m}^2$$

$$I_{zz} = [0 + 0] + [\tfrac{1}{12}(2)(0.2)^2 + 2(0.1)^2] + [\tfrac{1}{12}(4)(0.4)^2 + 4((-0.2)^2 + (0.2)^2)] = 0.400\ \text{kg}\cdot\text{m}^2$$

$$I_{xy} = [0 + 0] + [0 + 0] + [0 + 4(-0.2)(0.2)] = -0.160\ \text{kg}\cdot\text{m}^2$$

$$I_{yz} = [0 + 0] + [0 + 0] + [0 + 4(0.2)(0.2)] = 0.160\ \text{kg}\cdot\text{m}^2$$

$$I_{zx} = [0 + 0] + [0 + 2(0.2)(-0.1)] + [0 + 4(0.2)(-0.2)] = -0.200\ \text{kg}\cdot\text{m}^2$$

The *Aa* axis is defined by the unit vector

$$\mathbf{u}_{Aa} = \frac{\mathbf{r}_D}{r_D} = \frac{-0.2\mathbf{i} + 0.4\mathbf{j} + 0.2\mathbf{k}}{\sqrt{(-0.2)^2 + (0.4)^2 + (0.2)^2}} = -0.408\mathbf{i} + 0.816\mathbf{j} + 0.408\mathbf{k}$$

Thus,

$$u_x = -0.408 \qquad u_y = 0.816 \qquad u_z = 0.408$$

Substituting these results into Eq. 21–5 yields

$$\begin{aligned} I_{Aa} &= I_{xx}u_x^2 + I_{yy}u_y^2 + I_{zz}u_z^2 - 2I_{xy}u_xu_y - 2I_{yz}u_yu_z - 2I_{zx}u_zu_x \\ &= 0.480(-0.408)^2 + (0.453)(0.816)^2 + 0.400(0.408)^2 \\ &\quad -2(-0.160)(-0.408)(0.816) - 2(0.160)(0.816)(0.408) \\ &\quad -2(-0.200)(0.408)(-0.408) \\ &= 0.169\ \text{kg}\cdot\text{m}^2 \end{aligned}$$

Ans.

Problems

21-1. Show that the sum of the moments of inertia of a body, $I_{xx} + I_{yy} + I_{zz}$, is independent of the orientation of the x, y, z axes and thus depends only on the location of the origin.

21-2. Determine the moment of inertia of the cone with respect to a vertical $\bar{y}$ axis passing through the cone's center of mass. What is the moment of inertia about a parallel axis y' passing through the diameter of the base of the cone? The cone has a mass m.

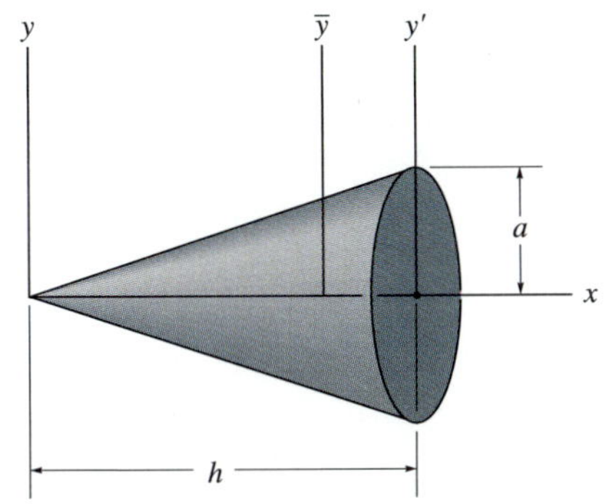

Prob. 21–2

21-3. Determine the moments of inertia I_x and I_y of the paraboloid of revolution. The mass of the paraboloid is m.

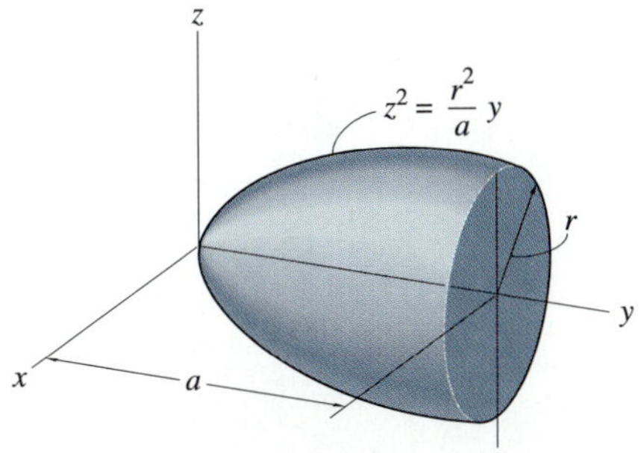

Prob. 21–3

***21-4.** Determine the product of inertia I_{xy} of the body formed by revolving the shaded area about the line $x = 5$ ft. Express the result in terms of the density of the material, ρ.

21-5. Determine the moment of inertia I_y of the body formed by revolving the shaded area about the line $x = 5$ ft. Express the result in terms of the density of the material, ρ.

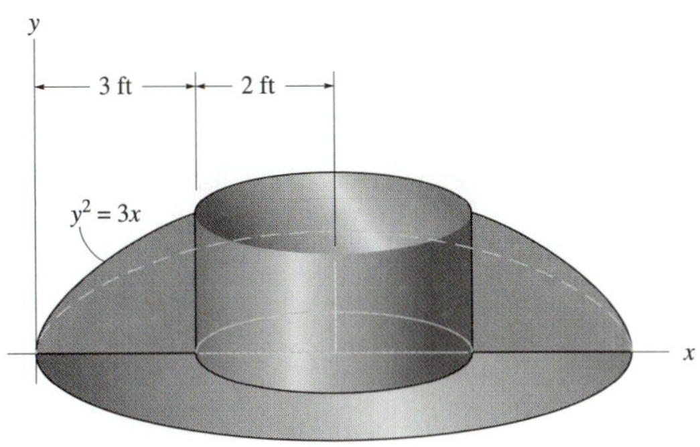

Probs. 21–4/5

21-6. Determine by direct integration the product of inertia I_{yz} for the homogeneous prism. The density of the material is ρ. Express the result in terms of the mass m of the prism.

21-7. Determine by direct integration the product of inertia I_{xy} for the homogeneous prism. The density of the material is ρ. Express the result in terms of the mass m of the prism.

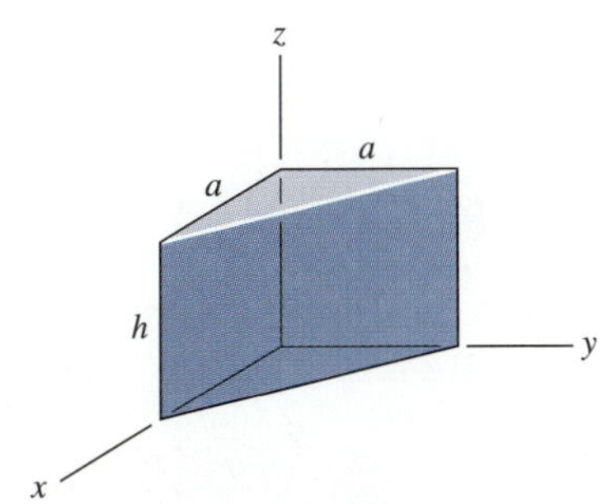

Probs. 21–6/7

***21-8.** Determine the product of inertia I_{xy} for the homogeneous tetrahedron. The density of the material is ρ. Express the result in terms of the total mass m of the solid. *Suggestion:* Use a triangular element of thickness dz and then express dI_{xy} in terms of the size and mass of the element using the result of Prob. 21-7.

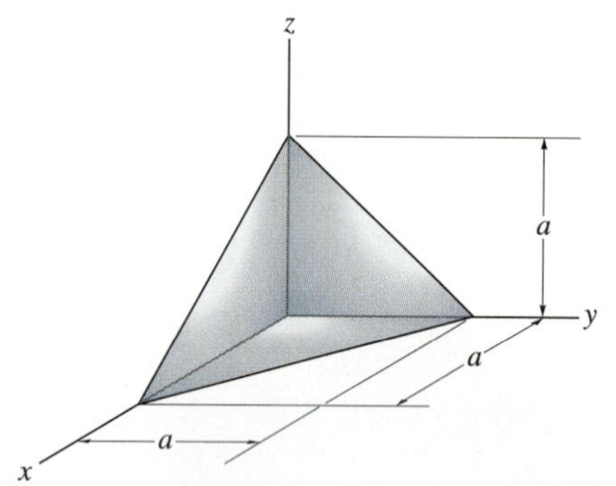

Prob. 21–8

21-9. Determine the mass moment of inertia of the homogeneous block with respect to its centroidal x' axis. The mass of the block is m.

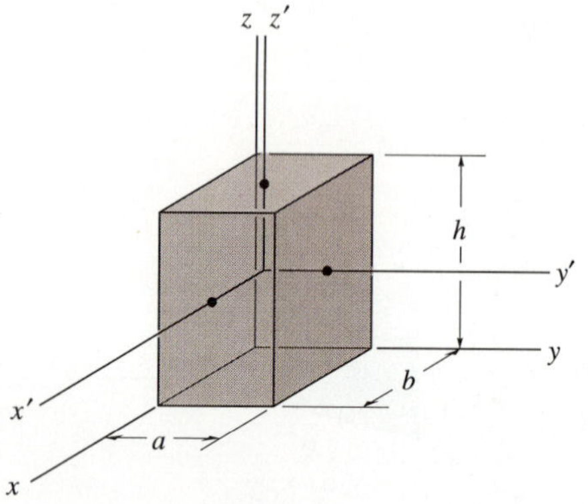

Prob. 21–9

21-10. Determine the moments of inertia for the homogeneous cylinder of mass m with respect to the x', y', z' axes.

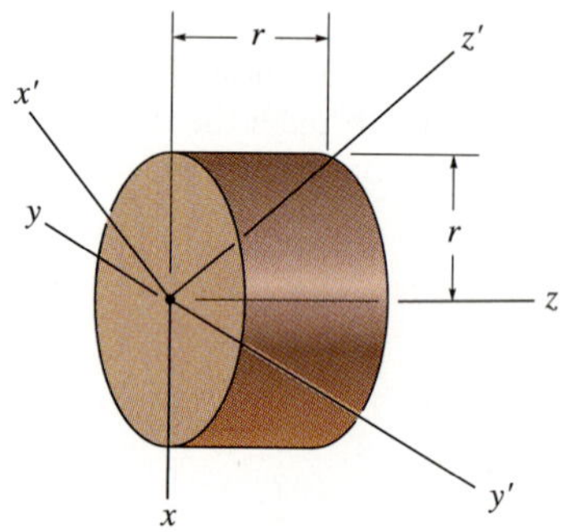

Prob. 21–10

21-11. Determine the moments of inertia about the x, y, z axes of the rod assembly. The rods have a mass of 0.75 kg/m.

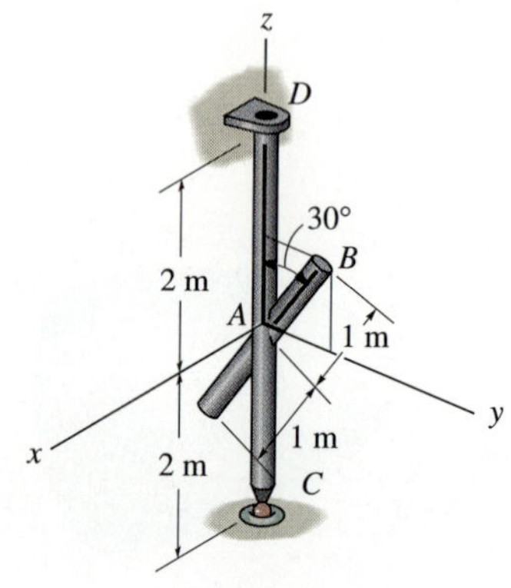

Prob. 21–11

***21-12.** Determine the moment of inertia of the disk about the axis of shaft AB. The disk has a mass of 15 kg.

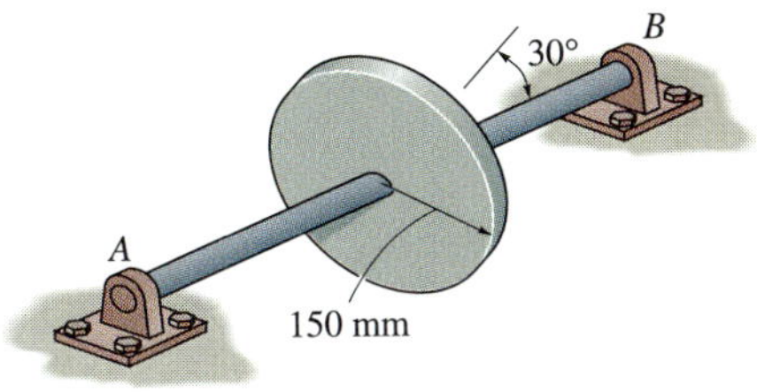

Prob. 21–12

21-13. The bent rod has a weight of 1.5 lb/ft. Locate the center of gravity $G(\bar{x}, \bar{y})$ and determine the principal moments of inertia $I_{x'}$, $I_{y'}$, and $I_{z'}$ of the rod with respect to the x', y', z' axes.

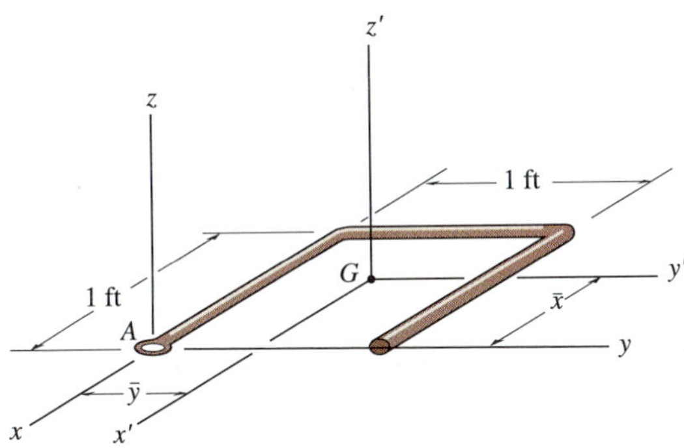

Prob. 21–13

21-14. Determine the moment of inertia of both the 1.5-kg rod and 4-kg disk about the z' axis.

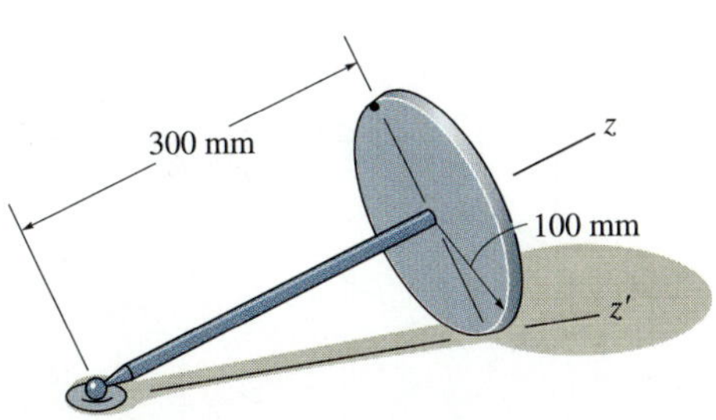

Prob. 21–14

21-15. The top consists of a cone having a mass of 0.7 kg and a hemisphere of mass 0.2 kg. Determine the moment of inertia I_z when the top is in the position shown.

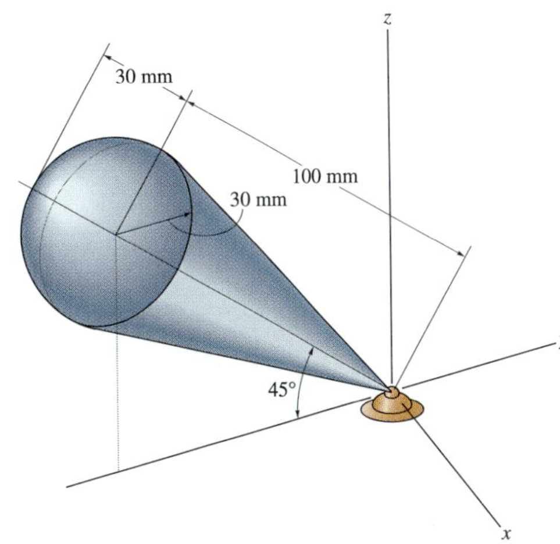

Prob. 21–15

***21-16.** Determine the moment of inertia about the z axis of the assembly which consists of the 1.5-kg rod CD and the 7-kg disk.

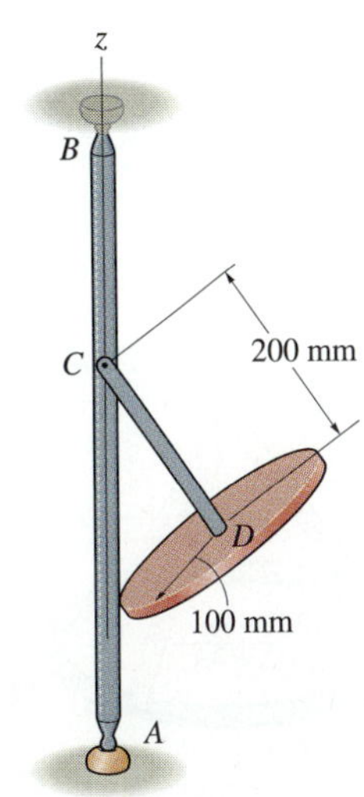

Prob. 21–16

21-17. The thin plate has a weight of 5 lb and each of the four rods weighs 3 lb. Determine the moment of inertia of the assembly about the z axis.

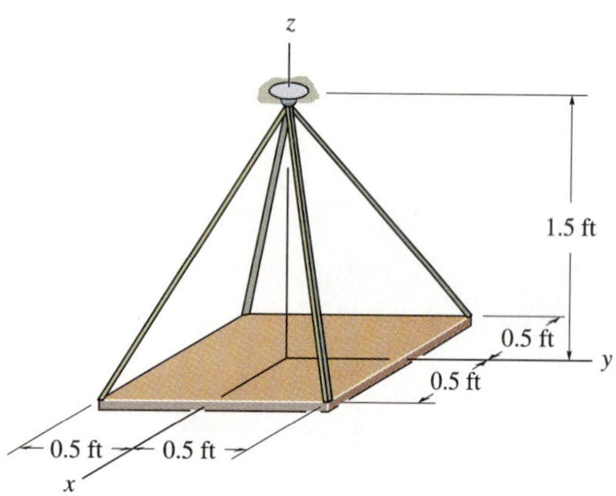

Prob. 21–17

21-18. Determine the moment of inertia of the rod-and-thin-ring assembly about the z axis. The rods and ring have a mass of 2 kg/m.

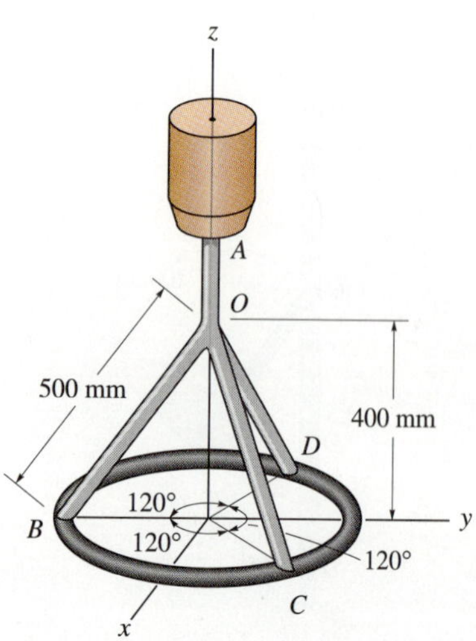

Prob. 21–18

21-19. Determine the moment of inertia of the rod-and-disk assembly about the x axis. The disks each have a weight of 12 lb. The two rods each have a weight of 4 lb, and their ends extend to the rims of the disks.

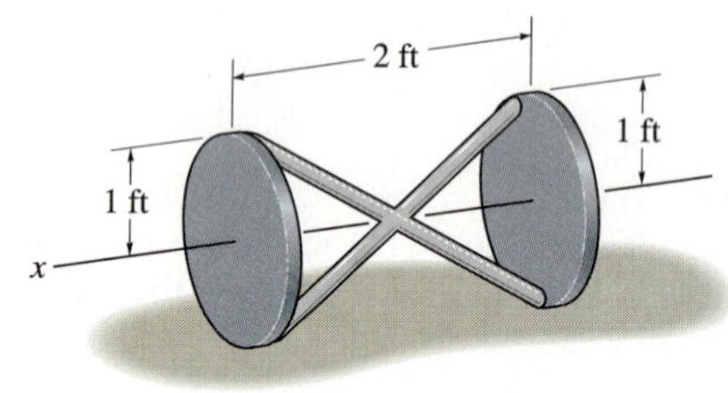

Prob. 21–19

***21-20.** The bent rod has a mass of 4 kg/m. Determine the moment of inertia of the rod about the Oa axis.

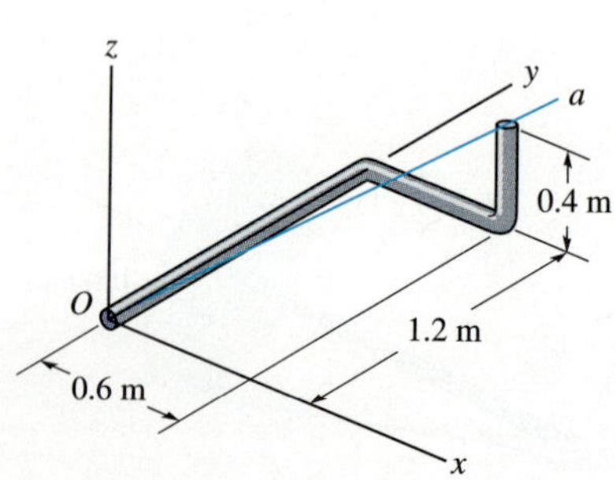

Prob. 21–20

★ 21.2 Angular Momentum

In this section we will develop the necessary equations used to determine the angular momentum of a rigid body about an arbitrary point. This formulation will provide a means for developing both the principle of impulse and momentum and the equations of rotational motion for a rigid body.

Consider the rigid body in Fig. 21–6, which has a mass m and center of mass at G. The X, Y, Z coordinate system represents an inertial frame of reference, and hence, its axes are fixed or translate with a constant velocity. The angular momentum as measured from this reference will be computed relative to the arbitrary point A. The position vectors $\mathbf{r}_A$ and $\boldsymbol{\rho}_A$ are drawn from the origin of coordinates to point A and from A to the ith particle of the body. If the particle's mass is m_i, the angular momentum about point A is

$$(\mathbf{H}_A)_i = \boldsymbol{\rho}_A \times m_i\mathbf{v}_i$$

where $\mathbf{v}_i$ represents the particle's velocity measured from the X, Y, Z coordinate system. If the body has an angular velocity $\boldsymbol{\omega}$ at the instant considered, $\mathbf{v}_i$ may be related to the velocity of A by applying Eq. 20–7, i.e.,

$$\mathbf{v}_i = \mathbf{v}_A + \boldsymbol{\omega} \times \boldsymbol{\rho}_A$$

Thus,

$$\begin{aligned}(\mathbf{H}_A)_i &= \boldsymbol{\rho}_A \times m_i(\mathbf{v}_A + \boldsymbol{\omega} \times \boldsymbol{\rho}_A)\\ &= (\boldsymbol{\rho}_A\, m_i) \times \mathbf{v}_A + \boldsymbol{\rho}_A \times (\boldsymbol{\omega} \times \boldsymbol{\rho}_A)\, m_i\end{aligned}$$

Summing all the particles of the body requires an integration, and since $m_i \rightarrow dm$, we have

$$\mathbf{H}_A = \left(\int_m \boldsymbol{\rho}_A\, dm\right) \times \mathbf{v}_A + \int_m \boldsymbol{\rho}_A \times (\boldsymbol{\omega} \times \boldsymbol{\rho}_A)\, dm \qquad (21\text{–}6)$$

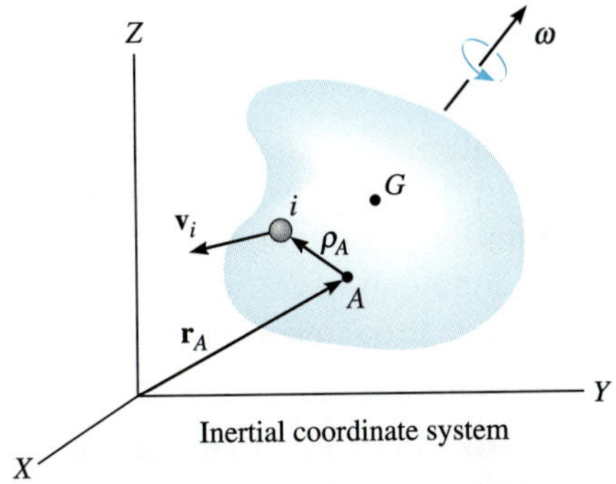

Fig. 21–6

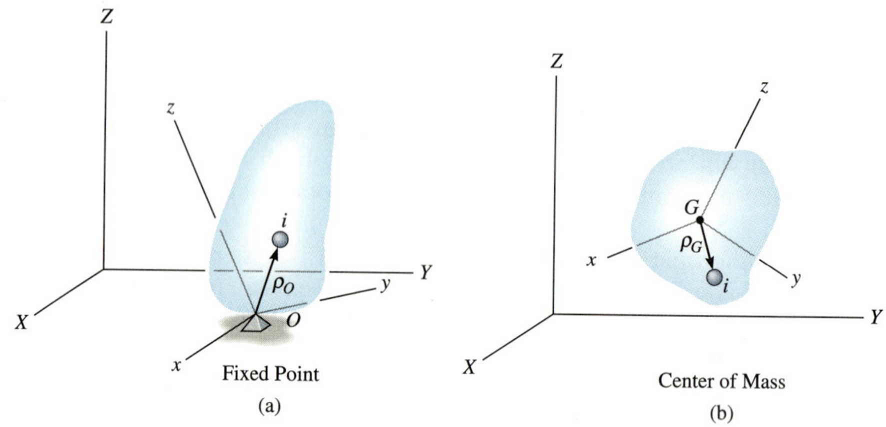

Fig. 21–7

Fixed Point *O*. If A becomes a *fixed point O* in the body, Fig. 21–7*a*, then $\mathbf{v}_A = \mathbf{0}$ and Eq. 21–6 reduces to

$$\mathbf{H}_O = \int_m \boldsymbol{\rho}_O \times (\boldsymbol{\omega} \times \boldsymbol{\rho}_O)\, dm \tag{21–7}$$

Center of Mass *G*. If A is located at the *center of mass G* of the body, Fig. 21–7*b*, then $\int_m \boldsymbol{\rho}_A\, dm = \mathbf{0}$ and

$$\mathbf{H}_G = \int_m \boldsymbol{\rho}_G \times (\boldsymbol{\omega} \times \boldsymbol{\rho}_G)\, dm \tag{21–8}$$

Arbitrary Point *A*. In general, A may be some point other than O or G, Fig. 21–7*c*, in which case Eq. 21–6 may nevertheless be simplified to the following form (see Prob. 21–22).

$$\mathbf{H}_A = \boldsymbol{\rho}_{G/A} \times m\mathbf{v}_G + \mathbf{H}_G \tag{21–9}$$

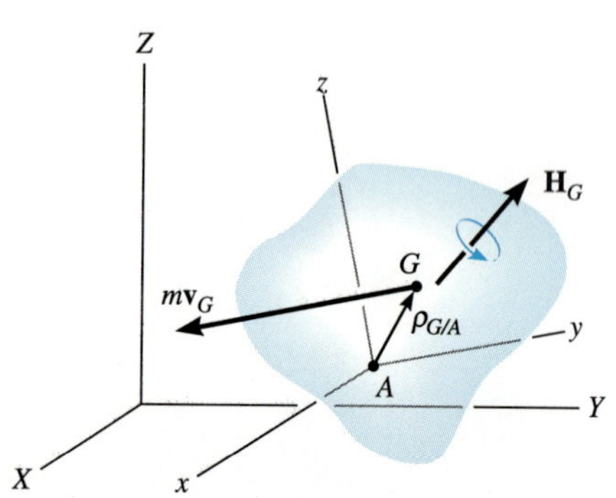

Arbitrary Point

(c)

Here the angular momentum consists of two parts–the moment of the linear momentum $m\mathbf{v}_G$ of the body about point A added (vectorially) to the angular momentum $\mathbf{H}_G$. Equation 21–9 may also be used for computing the angular momentum of the body about a fixed point O; the results, of course, will be the same as those computed using the more convenient Eq. 21–7.

Rectangular Components of H. To make practical use of Eqs. 21–7 through 21–9, the angular momentum must be expressed in terms of its scalar components. For this purpose, it is convenient to choose a second set of x, y, z axes having an arbitrary orientation relative to the X, Y, Z axes, Fig. 21–7, and for a general formulation, note that Eqs. 21–7 and 21–8 are both of the form

$$\mathbf{H} = \int_m \boldsymbol{\rho} \times (\boldsymbol{\omega} \times \boldsymbol{\rho})\ dm$$

Expressing $\mathbf{H}$, $\boldsymbol{\rho}$, and $\boldsymbol{\omega}$ in terms of x, y, and z components, we have

$$H_x\mathbf{i} + H_y\mathbf{j} + H_z\mathbf{k} = \int_m (x\mathbf{i} + y\mathbf{j} + z\mathbf{k}) \times [(\omega_x\mathbf{i} + \omega_y\mathbf{j} + \omega_z\mathbf{k}) \times (x\mathbf{i} + y\mathbf{j} + z\mathbf{k})]\ dm$$

Expanding the cross products and combining terms yields

$$\begin{aligned} H_x\mathbf{i} + H_y\mathbf{j} + H_z\mathbf{k} = {} & \left[\omega_x \int_m (y^2 + z^2)\ dm - \omega_y \int_m xy\ dm - \omega_z \int_m xz\ dm\right]\mathbf{i} \\ & + \left[-\omega_x \int_m xy\ dm + \omega_y \int_m (x^2 + z^2)\ dm - \omega_z \int_m yz\ dm\right]\mathbf{j} \\ & + \left[-\omega_x \int_m zx\ dm - \omega_y \int_m yz\ dm + \omega_z \int_m (x^2 + y^2)\ dm\right]\mathbf{k} \end{aligned}$$

Equating the respective $\mathbf{i}$, $\mathbf{j}$, $\mathbf{k}$ components and recognizing that the integrals represent the moments and products of inertia, we obtain

$$\begin{aligned} H_x &= I_{xx}\omega_x - I_{xy}\omega_y - I_{xz}\omega_z \\ H_y &= -I_{yx}\omega_x + I_{yy}\omega_y - I_{yz}\omega_z \\ H_z &= -I_{zx}\omega_x - I_{zy}\omega_y + I_{zz}\omega_z \end{aligned} \qquad (21\text{–}10)$$

These three equations represent the scalar form of the $\mathbf{i}$, $\mathbf{j}$, $\mathbf{k}$ components of $\mathbf{H}_O$ or $\mathbf{H}_G$ (given in vector form by Eqs. 21–7 and 21–8). The angular momentum of the body about the arbitrary point A, other than the fixed point O or the center of mass G, may also be expressed in scalar form. Here it is necessary to use Eq. 21–9 to represent $\boldsymbol{\rho}_{G/A}$ and $\mathbf{v}_G$ as Cartesian vectors, carry out the cross-product operation, and substitute the components, Eqs. 21–10, for $\mathbf{H}_G$.

Equations 21–10 may be simplified further if the x, y, z coordinate axes are oriented such that they become *principal axes of inertia* for the body at the point. When these axes are used, the products of inertia $I_{xy} = I_{yz} = I_{zx} = 0$, and if the principal moments of inertia about the x, y, z axes are represented as $I_x = I_{xx}$, $I_y = I_{yy}$, and $I_z = I_{zz}$, the three components of angular momentum become

$$H_x = I_x\omega_x \qquad H_y = I_y\omega_y \qquad H_z = I_z\omega_z \qquad (21\text{–}11)$$

The motion of the astronaut is controlled by use of small directional jets attached to his or her space suit. The impulses these jets provide must be carefully specified in order to prevent tumbling and loss of orientation.

Principle of Impulse and Momentum. Now that the formulation of the angular momentum for a body has been developed, the *principle of impulse and momentum,* as discussed in Sec. 19.2, may be used to solve kinetic problems which involve *force, velocity, and time.* For this case, the following two vector equations are available:

$$m(\mathbf{v}_G)_1 + \Sigma \int_{t_1}^{t_2} \mathbf{F}\, dt = m(\mathbf{v}_G)_2 \qquad (21\text{–}12)$$

$$(\mathbf{H}_O)_1 + \Sigma \int_{t_1}^{t_2} \mathbf{M}_O\, dt = (\mathbf{H}_O)_2 \qquad (21\text{–}13)$$

In three dimensions each vector term can be represented by three scalar components, and therefore a total of *six scalar equations* can be written. Three equations relate the linear impulse and momentum in the x, y, z directions, and the other three equations relate the body's angular impulse and momentum about the x, y, z axes. Before applying Eqs. 21–12 and 21–13 to the solution of problems, the material in Secs. 19.2 and 19.3 should be reviewed.

★21.3 Kinetic Energy

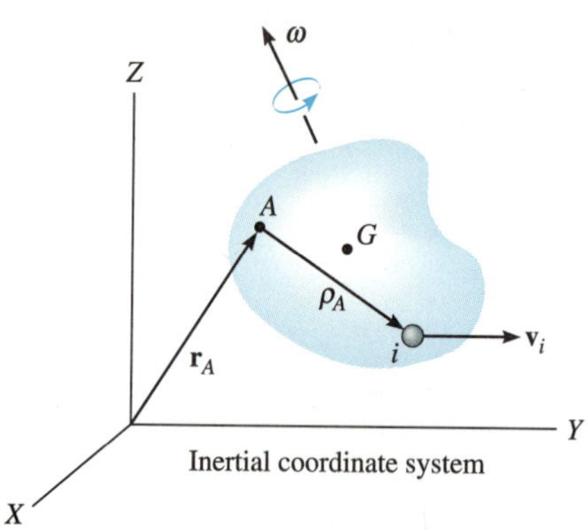

Fig. 21–8

In order to apply the principle of work and energy to the solution of problems involving general rigid body motion, it is first necessary to formulate expressions for the kinetic energy of the body. In this regard, consider the rigid body shown in Fig. 21–8, which has a mass m and center of mass at G. The kinetic energy of the ith particle of the body having a mass m_i and velocity $\mathbf{v}_i$, measured relative to the inertial X, Y, Z frame of reference, is

$$T_i = \tfrac{1}{2} m_i v_i^2 = \tfrac{1}{2} m_i (\mathbf{v}_i \cdot \mathbf{v}_i)$$

Provided the velocity of an arbitrary point A in the body is known, $\mathbf{v}_i$ may be related to $\mathbf{v}_A$ by the equation $\mathbf{v}_i = \mathbf{v}_A + \boldsymbol{\omega} \times \boldsymbol{\rho}_A$, where $\boldsymbol{\omega}$ is the angular velocity of the body, measured from the X, Y, Z coordinate system, and $\boldsymbol{\rho}_A$ is a position vector drawn from A to i. Using this expression for $\mathbf{v}_i$, the kinetic energy for the particle may be written as

$$\begin{aligned} T_i &= \tfrac{1}{2} m_i (\mathbf{v}_A + \boldsymbol{\omega} \times \boldsymbol{\rho}_A) \cdot (\mathbf{v}_A + \boldsymbol{\omega} \times \boldsymbol{\rho}_A) \\ &= \tfrac{1}{2} (\mathbf{v}_A \cdot \mathbf{v}_A)\, m_i + \mathbf{v}_A \cdot (\boldsymbol{\omega} \times \boldsymbol{\rho}_A)\, m_i + \tfrac{1}{2} (\boldsymbol{\omega} \times \boldsymbol{\rho}_A) \cdot (\boldsymbol{\omega} \times \boldsymbol{\rho}_A)\, m_i \end{aligned}$$

The kinetic energy for the entire body is obtained by summing the kinetic energies of all the particles of the body. This requires an integration, and since $m_i \to dm$, we get

$$T = \tfrac{1}{2} m (\mathbf{v}_A \cdot \mathbf{v}_A) + \mathbf{v}_A \cdot \left(\boldsymbol{\omega} \times \int_m \boldsymbol{\rho}_A\, dm \right) + \tfrac{1}{2} \int_m (\boldsymbol{\omega} \times \boldsymbol{\rho}_A) \cdot (\boldsymbol{\omega} \times \boldsymbol{\rho}_A)\, dm$$

The last term on the right may be rewritten using the vector identity $\mathbf{a} \times \mathbf{b} \cdot \mathbf{c} = \mathbf{a} \cdot \mathbf{b} \times \mathbf{c}$, where $\mathbf{a} = \boldsymbol{\omega}$, $\mathbf{b} = \boldsymbol{\rho}_A$, and $\mathbf{c} = \boldsymbol{\omega} \times \boldsymbol{\rho}_A$. The final result is

$$T = \tfrac{1}{2}m(\mathbf{v}_A \cdot \mathbf{v}_A) + \mathbf{v}_A \cdot \left(\boldsymbol{\omega} \times \int_m \boldsymbol{\rho}_A \, dm\right)$$

$$+ \tfrac{1}{2}\boldsymbol{\omega} \cdot \int_m \boldsymbol{\rho}_A \times (\boldsymbol{\omega} \times \boldsymbol{\rho}_A) \, dm \qquad (21\text{–}14)$$

This equation is rarely used because of the computations involving the integrals. Simplification occurs, however, if the reference point A is either a fixed point O or the center of mass G.

Fixed Point O. If A is a *fixed point O* in the body, Fig. 21–7*a*, then $\mathbf{v}_A = \mathbf{0}$, and using Eq. 21–7, we can express Eq. 21–14 as

$$T = \tfrac{1}{2}\boldsymbol{\omega} \cdot \mathbf{H}_O$$

If the x, y, z axes represent the principal axes of inertia for the body, then $\boldsymbol{\omega} = \omega_x\mathbf{i} + \omega_y\mathbf{j} + \omega_z\mathbf{k}$ and $\mathbf{H}_O = I_x\omega_x\mathbf{i} + I_y\omega_y\mathbf{j} + I_z\omega_z\mathbf{k}$. Substituting into the above equation and performing the dot product operations yields

$$T = \tfrac{1}{2}I_x\omega_x^2 + \tfrac{1}{2}I_y\omega_y^2 + \tfrac{1}{2}I_z\omega_z^2 \qquad (21\text{–}15)$$

Center of Mass G. If A is located at the *center of mass G* of the body, Fig. 21–7*b*, then $\int \boldsymbol{\rho}_A \, dm = \mathbf{0}$ and, using Eq. 21–8, we can write Eq. 21–14 as

$$T = \tfrac{1}{2}mv_G^2 + \tfrac{1}{2}\boldsymbol{\omega} \cdot \mathbf{H}_G$$

In a manner similar to that for a fixed point, the last term on the right side may be represented in scalar form, in which case

$$T = \tfrac{1}{2}mv_G^2 + \tfrac{1}{2}I_x\omega_x^2 + \tfrac{1}{2}I_y\omega_y^2 + \tfrac{1}{2}I_z\omega_z^2 \qquad (21\text{–}16)$$

Here it is seen that the kinetic energy consists of two parts; namely, the translational kinetic energy of the mass center, $\tfrac{1}{2}mv_G^2$, and the body's rotational kinetic energy.

Principle of Work and Energy. Having formulated the kinetic energy for a body, the *principle of work and energy* may be applied to solve kinetics problems which involve *force, velocity, and displacement.* For this case only one scalar equation can be written for each body, namely,

$$T_1 + \Sigma U_{1-2} = T_2 \qquad (21\text{–}17)$$

Before applying this equation, the material in Chapter 18 should be reviewed.

EXAMPLE 21–2

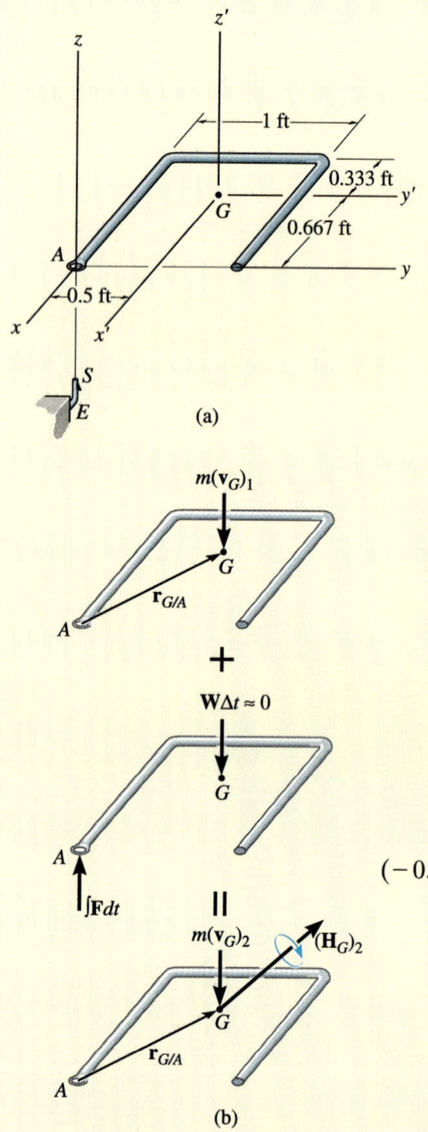

Fig. 21–9

The rod in Fig. 21–9*a* has a weight of 1.5 lb/ft. Determine its angular velocity just after the end *A* falls onto the hook at *E*. The hook provides a permanent connection for the rod due to the spring-lock mechanism *S*. Just before striking the hook the rod is falling downward with a speed $(v_G)_1 = 10$ ft/s.

Solution

The principle of impulse and momentum will be used since impact occurs.

Impulse and Momentum Diagrams. Fig. 21–9*b*. During the short time Δt, the impulsive force **F** acting at *A* changes the momentum of the rod. (The impulse created by the rod's weight **W** during this time is small compared to $\int \mathbf{F}\,dt$, so that it is neglected, i.e., the weight is a nonimpulsive force.) Hence, the angular momentum of the rod is *conserved* about point *A* since the moment of $\int \mathbf{F}\,dt$ about *A* is zero.

Conservation of Angular Momentum. Equation 21–9 must be used for computing the angular momentum of the rod, since *A* does not become a *fixed point* until *after* the impulsive interaction with the hook. Thus, with reference to Fig. 21–9*b*, $(\mathbf{H}_A)_1 = (\mathbf{H}_A)_2$, or

$$\mathbf{r}_{G/A} \times m(\mathbf{v}_G)_1 = \mathbf{r}_{G/A} \times m(\mathbf{v}_G)_2 + (\mathbf{H}_G)_2 \qquad (1)$$

From Fig. 21–9*a*, $\mathbf{r}_{G/A} = \{-0.667\mathbf{i} + 0.5\mathbf{j}\}$ ft. Furthermore, the primed axes are principal axes of inertia for the rod because $I_{x'y'} = I_{x'z'} = I_{z'y'} = 0$. Hence, from Eqs. 21–11, $(\mathbf{H}_G)_2 = I_{x'}\omega_x\mathbf{i} + I_{y'}\omega_y\mathbf{j} + I_{z'}\omega_z\mathbf{k}$. The principal moments of inertia are $I_{x'} = 0.0272$ slug ft^2, $I_{y'} = 0.0155$ slug·ft^2, $I_{z'} = 0.0427$ slug·ft^2 (see Prob. 21–13). Substituting into Eq. 1, we have

$$(-0.667\mathbf{i} + 0.5\mathbf{j}) \times \left[\left(\frac{4.5}{32.2}\right)(-10\mathbf{k})\right] = (-0.667\mathbf{i} + 0.5\mathbf{j}) \times \left[\left(\frac{4.5}{32.2}\right)(-v_G)_2\mathbf{k}\right]$$
$$+\ 0.0272\omega_x\mathbf{i} + 0.0155\omega_y\mathbf{j} + 0.0427\omega_z\mathbf{k}$$

Expanding and equating the respective **i**, **j**, and **k** components yields

$$-0.699 = -0.0699(v_G)_2 + 0.0272\omega_x \qquad (2)$$
$$-0.932 = -0.0932(v_G)_2 + 0.0155\omega_y \qquad (3)$$
$$0 = 0.0427\omega_z \qquad (4)$$

Kinematics. There are four unknowns in the above equations; however, another equation may be obtained by relating $\boldsymbol{\omega}$ to $(\mathbf{v}_G)_2$ using *kinematics*. Since $\omega_z = 0$ (Eq. 4) and after impact the rod rotates about the fixed point *A*, Eq. 20–3 may be applied, in which case $(\mathbf{v}_G)_2 = \boldsymbol{\omega} \times \mathbf{r}_{G/A}$, or

$$-(v_G)_2\mathbf{k} = (\omega_x\mathbf{i} + \omega_y\mathbf{j}) \times (-0.667\mathbf{i} + 0.5\mathbf{j})$$
$$-(v_G)_2 = 0.5\omega_x + 0.667\omega_y \qquad (5)$$

Solving Eqs. 2 through 5 simultaneously yields

$$(\mathbf{v}_G)_2 = \{-8.41\mathbf{k}\}\ \text{ft/s} \qquad \boldsymbol{\omega} = \{-4.09\mathbf{i} - 9.55\mathbf{j}\}\ \text{rad/s} \qquad \textit{Ans.}$$

EXAMPLE 21–3

A 5-N·m torque is applied to the vertical shaft CD shown in Fig. 21–10a, which allows the 10-kg gear A to turn freely about CE. Assuming that gear A starts from rest, determine the angular velocity of CD after it has turned two revolutions. Neglect the mass of shaft CD and axle CE and assume that gear A can be approximated by a thin disk. Gear B is fixed.

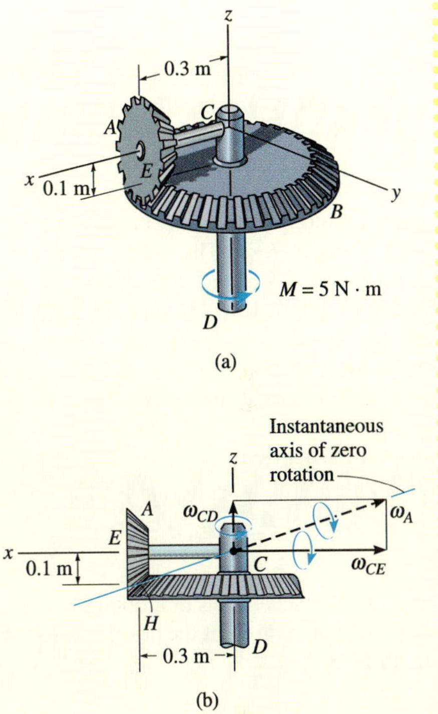

Fig. 21–10

Solution

The principle of work and energy may be used for the solution. Why?

Work. If shaft CD, the axle CE, and gear A are considered as a system of connected bodies, only the applied torque **M** does work. For two revolutions of CD, this work is $\Sigma U_{1-2} = (5 \text{ N·m})(4\pi \text{ rad}) = 62.83$ J.

Kinetic Energy. Since the gear is initially at rest, its initial kinetic energy is zero. A kinematic diagram for the gear is shown in Fig. 21–10b. If the angular velocity of CD is taken as $\boldsymbol{\omega}_{CD}$, then the angular velocity of gear A is $\boldsymbol{\omega}_A = \boldsymbol{\omega}_{CD} + \boldsymbol{\omega}_{CE}$. The gear may be imagined as a portion of a massless extended body which is rotating about the *fixed point C*. The instantaneous axis of rotation for this body is along line CH, because both points C and H on the body (gear) have zero velocity and must therefore lie on this axis. This requires that the components $\boldsymbol{\omega}_{CD}$ and $\boldsymbol{\omega}_{CE}$ be related by the equation $\omega_{CD}/0.1 \text{ m} = \omega_{CE}/0.3 \text{ m}$ or $\omega_{CE} = 3\omega_{CD}$. Thus,

$$\boldsymbol{\omega}_A = -\omega_{CE}\mathbf{i} + \omega_{CD}\mathbf{k} = -3\omega_{CD}\mathbf{i} + \omega_{CD}\mathbf{k} \qquad (1)$$

The x, y, z axes in Fig. 21–10a represent *principal axes of inertia* at C for the gear. Since point C is a fixed point of rotation, Eq. 21–15 may be applied to determine the kinetic energy, i.e.,

$$T = \tfrac{1}{2}I_x\omega_x^2 + \tfrac{1}{2}I_y\omega_y^2 + \tfrac{1}{2}I_z\omega_z^2 \qquad (2)$$

Using the parallel-axis theorem, the moments of inertia of the gear about point C are as follows:

$$I_x = \tfrac{1}{2}(10 \text{ kg})(0.1 \text{ m})^2 = 0.05 \text{ kg·m}^2$$
$$I_y = I_z = \tfrac{1}{4}(10 \text{ kg})(0.1 \text{ m})^2 + 10 \text{ kg}(0.3 \text{ m})^2 = 0.925 \text{ kg·m}^2$$

Since $\omega_x = -3\omega_{CD}$, $\omega_y = 0$, $\omega_z = \omega_{CD}$, Eq. 2 becomes

$$T_A = \tfrac{1}{2}(0.05)(-3\omega_{CD})^2 + 0 + \tfrac{1}{2}(0.925)(\omega_{CD})^2 = 0.6875\omega_{CD}^2$$

Principle of Work and Energy. Applying the principle of work and energy, we obtain

$$T_1 + \Sigma U_{1-2} = T_2$$
$$0 + 62.83 = 0.6875\omega_{CD}^2$$
$$\omega_{CD} = 9.56 \text{ rad/s} \qquad \textit{Ans.}$$

Problems

21-21. If a body contains *no planes of symmetry,* the principal moments of inertia can be determined mathematically. To show how this is done, consider the rigid body which is spinning with an angular velocity $\boldsymbol{\omega}$, directed along one of its principal axes of inertia. If the principal moment of inertia about this axis is I, the angular momentum can be expressed as $\mathbf{H} = I\boldsymbol{\omega} = I\omega_x\mathbf{i} + I\omega_y\mathbf{j} + I\omega_z\mathbf{k}$. The components of $\mathbf{H}$ may also be expressed by Eqs. 21–10, where the inertia tensor is assumed to be known. Equate the **i, j,** and **k** components of both expressions for $\mathbf{H}$ and consider ω_x, ω_y, and ω_z to be unknown. The solution of these three equations is obtained provided the determinant of the coefficients is zero. Show that this determinant, when expanded, yields the cubic equation

$$I^3 - (I_{xx} + I_{yy} + I_{zz})I^2 + (I_{xx}I_{yy} + I_{yy}I_{zz} + I_{zz}I_{xx} - I_{xy}^2 - I_{yz}^2 - I_{zx}^2)I - (I_{xx}I_{yy}I_{zz} - 2I_{xy}I_{yz}I_{zx} - I_{xx}I_{yz}^2 - I_{yy}I_{zx}^2 - I_{zz}I_{xy}^2) = 0$$

The three positive roots of I, obtained from the solution of this equation, represent the principal moments of inertia I_x, I_y, and I_z.

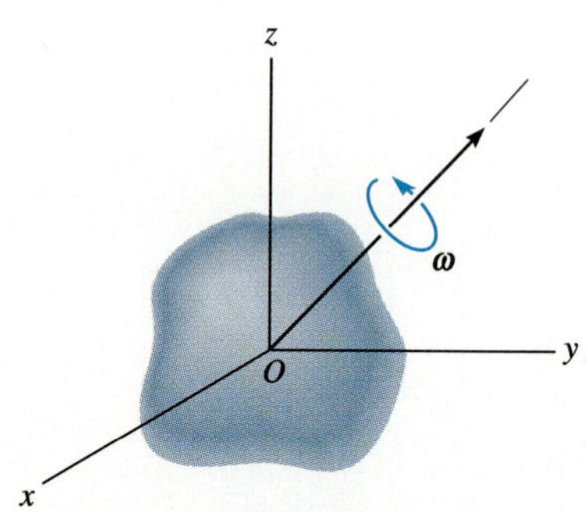

Prob. 21–21

21-22. Show that if the angular momentum of a body is determined with respect to an arbitrary point A, then $\mathbf{H}_A$ can be expressed by Eq. 21–9. This requires substituting $\boldsymbol{\rho}_A = \boldsymbol{\rho}_G + \boldsymbol{\rho}_{G/A}$ into Eq. 21–6 and expanding, noting that $\int \boldsymbol{\rho}_G\, dm = \mathbf{0}$ by definition of the mass center and $\mathbf{v}_G = \mathbf{v}_A + \boldsymbol{\omega} \times \boldsymbol{\rho}_{G/A}$.

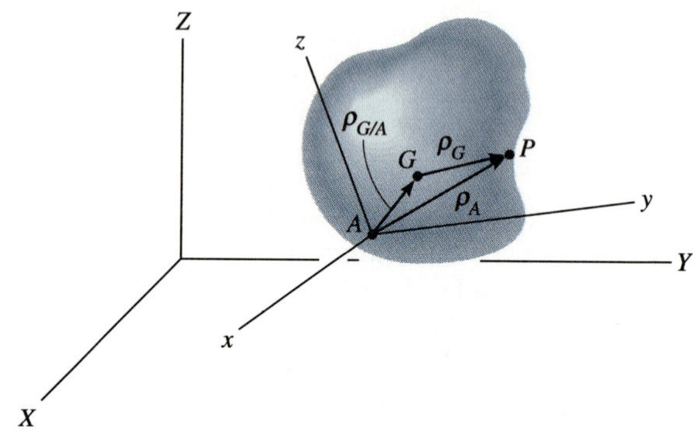

Prob. 21–22

21-23. Gear A has a mass of 5 kg and a radius of gyration of $k_z = 75$ mm. Gears B and C each have a mass of 200 g and a radius of gyration about the axis of their connecting shaft of 15 mm. If the gears are in mesh and C has an angular velocity of $\boldsymbol{\omega}_C = \{15\mathbf{j}\}$ rad/s, determine the total angular momentum for the system of three gears about point A.

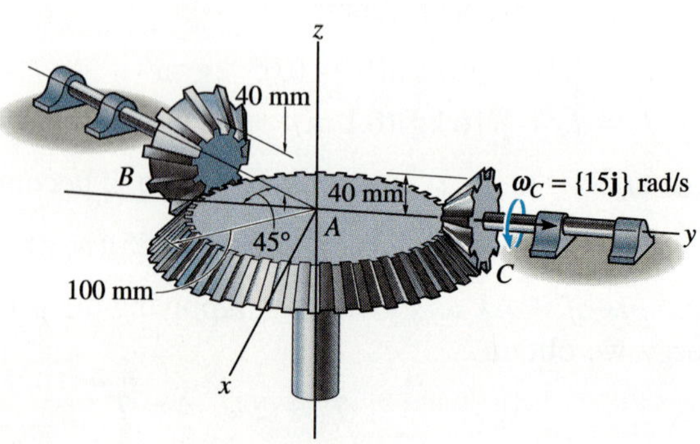

Prob. 21–23

***21-24.** The 4-lb rod AB is attached to the disk and collar using ball-and-socket joints. If the disk has a constant angular velocity of 2 rad/s, determine the kinetic energy of the rod when it is in the position shown. Assume the angular velocity of the rod is directed perpendicular to the axis of the rod. *Hint:* The motion has been specified in Prob. 20-22.

21-25. Determine the angular momentum of rod AB in Prob. 21-24 about its mass center at the instant shown. *Hint:* The motion has been specified in Prob. 20-22.

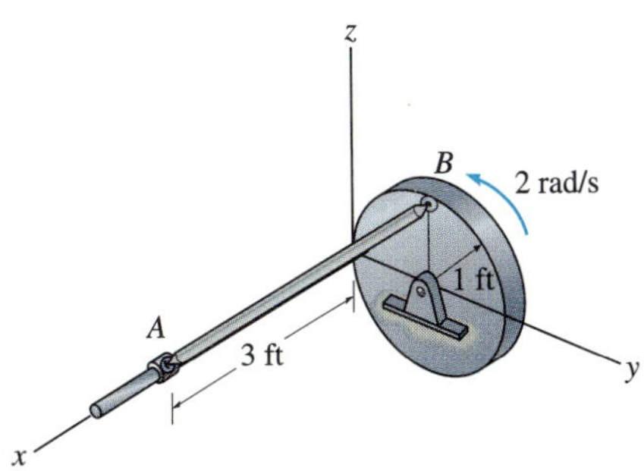

Probs. 21–24/25

21-26. The cone has a mass m and rolls without slipping on the conical surface so that it has an angular velocity about the vertical axis of $\boldsymbol{\omega}$. Determine the kinetic energy of the cone due to this motion.

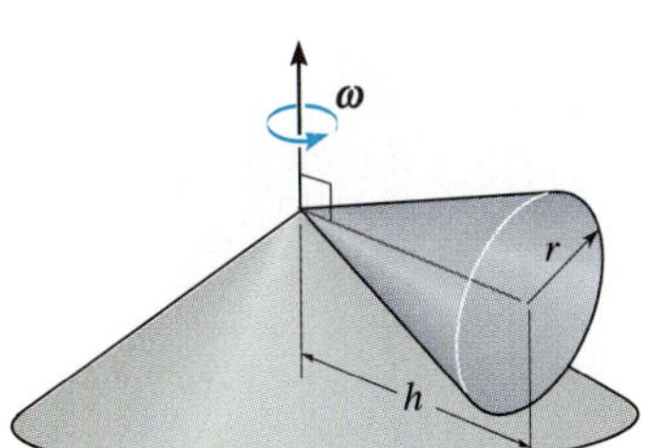

Prob. 21–26

21-27. A thin plate, having a mass of 4 kg, is suspended from one of its corners by a ball-and-socket joint O. If a stone strikes the plate perpendicular to its surface at an adjacent corner A with an impulse of $\mathbf{I}_s = \{-60\mathbf{i}\}$ N·s, determine the instantaneous axis of rotation for the plate and the impulse created at O.

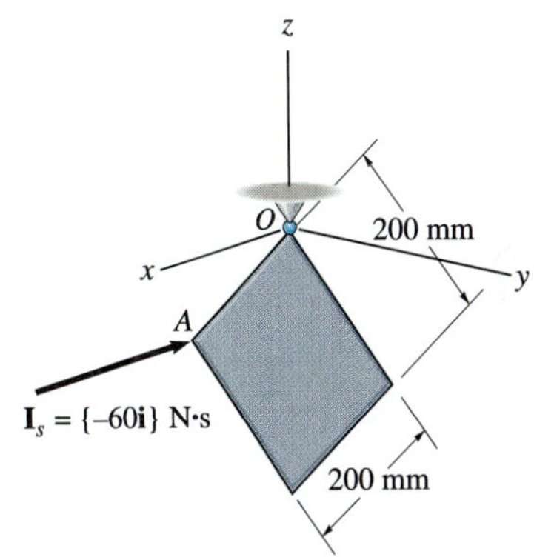

Prob. 21–27

***21-28.** The 5-kg disk is connected to the 3-kg slender rod. If the assembly is attached to a ball-and-socket joint at A and the 5-N·m couple moment is applied, determine the angular velocity of the rod about the z axis after the assembly has made two revolutions about the z axis starting from rest. The disk rolls without slipping.

21-29. The 5-kg disk is connected to the 3-kg slender rod. If the assembly is attached to a ball-and-socket joint at A and the 5-N·m couple moment gives it an angular velocity about the z axis of $\omega_z = 2$ rad/s, determine the magnitude of the angular momentum of the assembly about A.

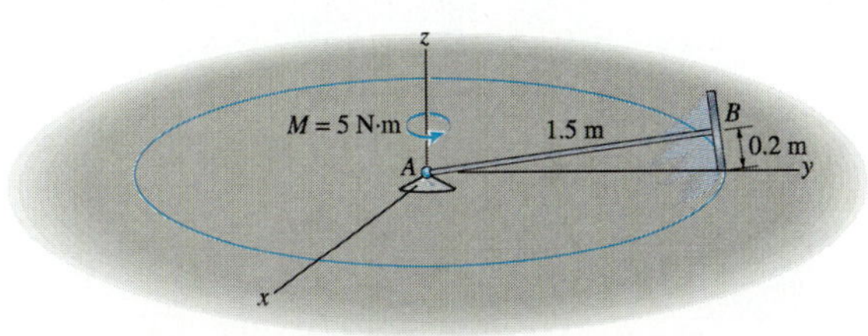

Probs. 21–28/29

21-30. Each of the two disks has a weight of 10 lb. The axle AB weighs 3 lb. If the assembly is rotating about the z axis at $\omega_z = 6$ rad/s, determine its angular momentum about the z axis and its kinetic energy. The disks roll without slipping.

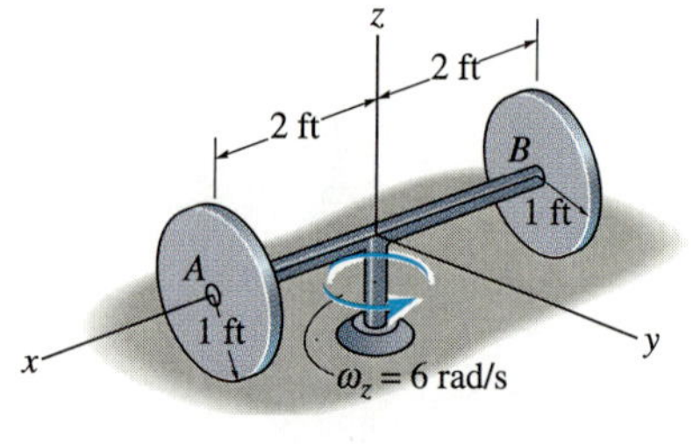

Prob. 21–30

21-31. The 2-kg thin disk is connected to the slender rod which is fixed to the ball-and-socket joint at A. If it is released from rest in the position shown, determine the spin of the disk about the rod when the disk reaches its lowest position. Neglect the mass of the rod. The disk rolls without slipping.

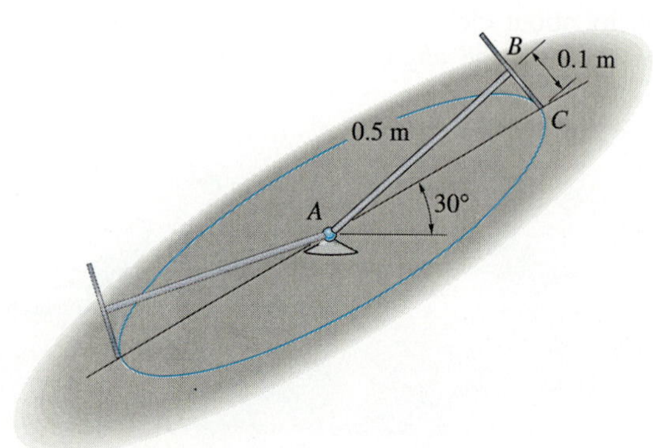

Prob. 21–31

***21-32.** The assembly consists of a 4-kg rod AB which is connected to link OA and to the collar at B by ball-and-socket joints. When $\theta = 0°$, $y = 600$ mm, the system is at rest, the spring is unstretched, and a 7-N·m couple moment is applied to the link at O. Determine the angular velocity of the link at the instant $\theta = 90°$. Neglect the mass of the link.

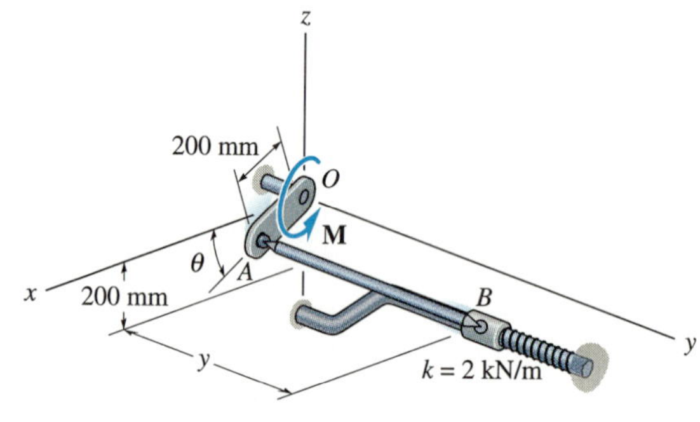

Prob. 21–32

21-33. The 25-lb thin plate is suspended from a ball-and-socket joint at O. A 0.2-lb projectile is fired with a velocity of $\mathbf{v} = \{-300\mathbf{i} - 250\mathbf{j} + 300\mathbf{k}\}$ ft/s into the plate and becomes embedded in the plate at point A. Determine the angular velocity of the plate just after impact and the axis about which it begins to rotate.

21-34. Solve Prob. 21-33 if the projectile emerges from the plate with a velocity of 275 ft/s in the same direction.

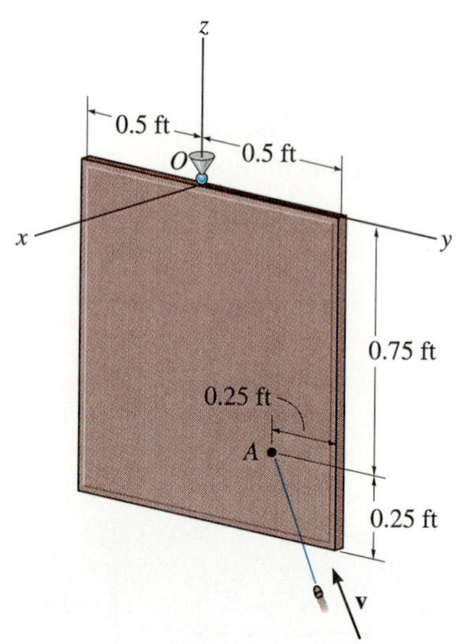

Probs. 21–33/34

21-35. The 4-lb rod AB is attached to the rod BC and collar A using ball-and-socket joints. If BC has a constant angular velocity of 2 rad/s, determine the kinetic energy of AB when it is in the position shown. Assume the angular velocity of AB is directed perpendicular to the axis of AB.

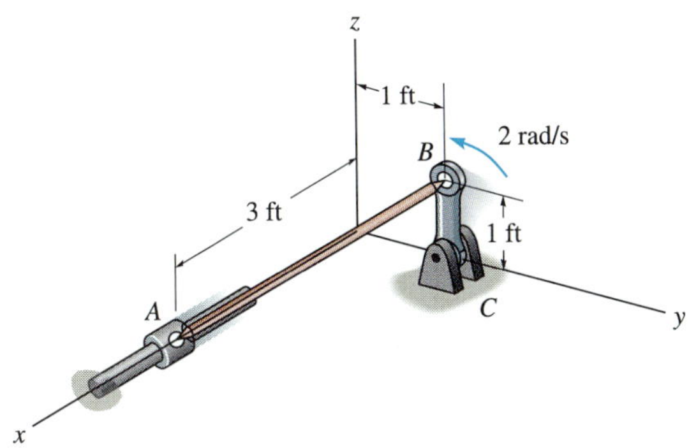

Prob. 21–35

***21-36.** The rod assembly is supported at G by a ball-and-socket joint. Each segment has a mass of 0.5 kg/m. If the assembly is originally at rest and an impulse of $\mathbf{I} = \{-8\mathbf{k}\}$ N·s is applied at D, determine the angular velocity of the assembly just after the impact.

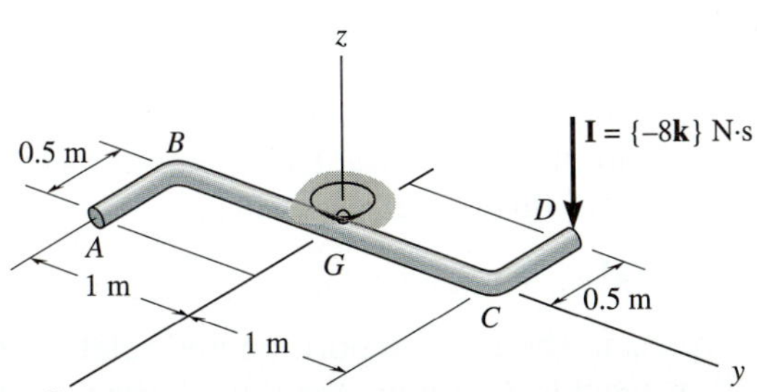

Prob. 21–36

21-37. The 15-lb plate is subjected to a force $F = 8$ lb which is always directed perpendicular to the face of the plate. If the plate is originally at rest, determine its angular velocity after it has rotated one revolution (360°). The plate is supported by ball-and-socket joints at A and B.

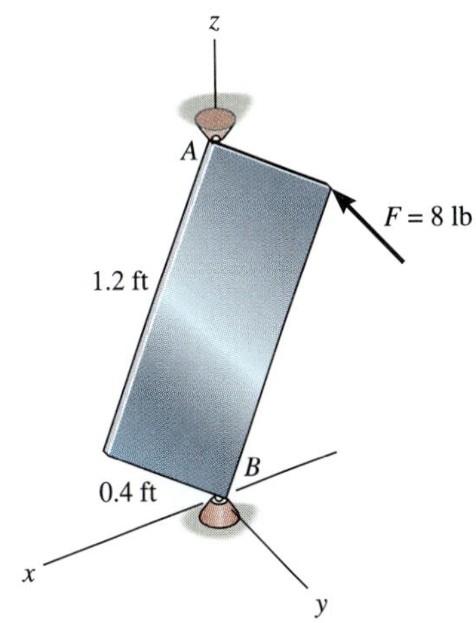

Prob. 21–37

21-38. The rod assembly has a mass of 2.5 kg/m and is rotating with a constant angular velocity of $\boldsymbol{\omega} = \{2\mathbf{k}\}$ rad/s when the looped end at C encounters a hook at S, which provides a permanent connection. Determine the angular velocity of the assembly immediately after impact.

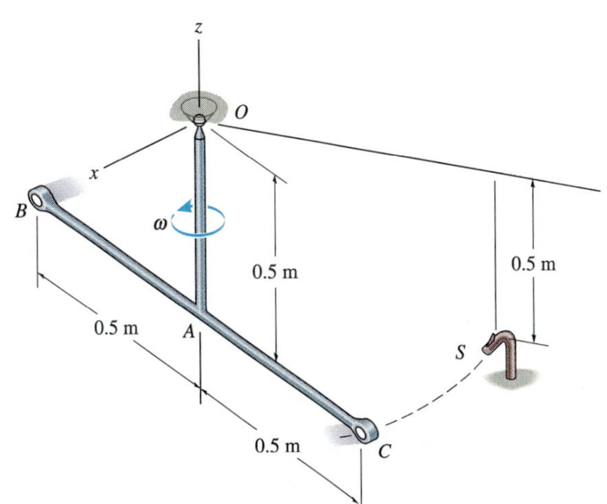

Prob. 21–38

★21.4 Equations of Motion

Having become familiar with the techniques used to describe both the inertial properties and the angular momentum of a body, we can now write the equations which describe the motion of the body in their most useful forms.

Equations of Translational Motion. The *translational motion* of a body is defined in terms of the acceleration of the body's mass center, which is measured from an inertial X, Y, Z reference. The equation of translational motion for the body can be written in vector form as

$$\Sigma \mathbf{F} = m\mathbf{a}_G \qquad (21\text{–}18)$$

or by the three scalar equations

$$\begin{aligned} \Sigma F_x &= m(a_G)_x \\ \Sigma F_y &= m(a_G)_y \\ \Sigma F_z &= m(a_G)_z \end{aligned} \qquad (21\text{–}19)$$

Here, $\Sigma \mathbf{F} = \Sigma F_x\mathbf{i} + \Sigma F_y\mathbf{j} + \Sigma F_z\mathbf{k}$ represents the sum of all the external forces acting on the body.

Equations of Rotational Motion. In Sec. 15.6, we developed Eq. 15–17, namely,

$$\Sigma \mathbf{M}_O = \dot{\mathbf{H}}_O \qquad (21\text{–}20)$$

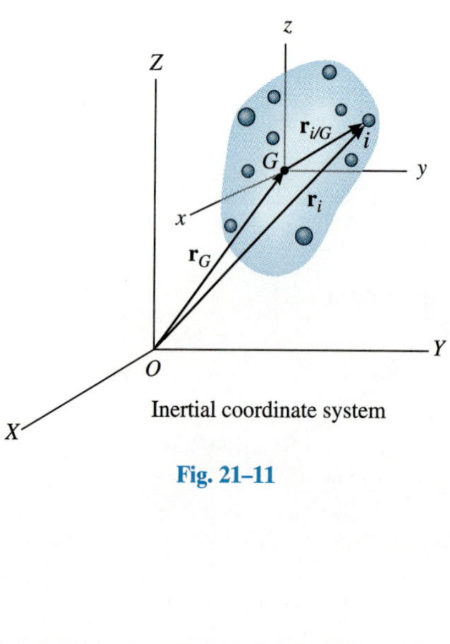

Fig. 21–11

which states that the sum of the moments about a fixed point O of all the external forces acting on a system of particles (contained in a rigid body) is equal to the time rate of change of the total angular momentum of the body about point O. When moments of the external forces acting on the particles are summed about the system's *mass center G*, one again obtains the same simple form of Eq. 21–20, relating the moment summation $\Sigma \mathbf{M}_G$ to the angular momentum $\mathbf{H}_G$. To show this, consider the system of particles in Fig. 21–11, where X, Y, Z represents an inertial frame of reference and the x, y, z axes, with origin at G, *translate* with respect to this frame. In general, G is *accelerating*, so by definition the translating frame is *not* an inertial reference. The angular momentum of the ith particle with respect to this frame is, however,

$$(\mathbf{H}_i)_G = \mathbf{r}_{i/G} \times m_i\mathbf{v}_{i/G}$$

where $\mathbf{r}_{i/G}$ and $\mathbf{v}_{i/G}$ represent the relative position and relative velocity of the ith particle with respect to G. Taking the time derivative we have

$$(\dot{\mathbf{H}}_i)_G = \dot{\mathbf{r}}_{i/G} \times m_i\mathbf{v}_{i/G} + \mathbf{r}_{i/G} \times m_i\dot{\mathbf{v}}_{i/G}$$

By definition, $\mathbf{v}_{i/G} = \dot{\mathbf{r}}_{i/G}$. Thus, the first term on the right side is zero since the cross product of equal vectors is zero. Also, $\mathbf{a}_{i/G} = \dot{\mathbf{v}}_{i/G}$, so that

$$(\dot{\mathbf{H}}_i)_G = (\mathbf{r}_{i/G} \times m_i \mathbf{a}_{i/G})$$

Similar expressions can be written for the other particles of the body. When the results are summed, we get

$$\dot{\mathbf{H}}_G = \Sigma(\mathbf{r}_{i/G} \times m_i \mathbf{a}_{i/G})$$

Here $\dot{\mathbf{H}}_G$ is the time rate of change of the total angular momentum of the body computed relative to point G.

The relative acceleration for the ith particle is defined by the equation $\mathbf{a}_{i/G} = \mathbf{a}_i - \mathbf{a}_G$, where $\mathbf{a}_i$ and $\mathbf{a}_G$ represent, respectively, the accelerations of the ith particle and point G measured with respect to the *inertial frame of reference.* Substituting and expanding, using the distributive property of the vector cross product, yields

$$\dot{\mathbf{H}}_G = \Sigma(\mathbf{r}_{i/G} \times m_i \mathbf{a}_i) - (\Sigma m_i \mathbf{r}_{i/G}) \times \mathbf{a}_G$$

By definition of the mass center, the sum $(\Sigma m_i \mathbf{r}_{i/G}) = (\Sigma m_i)\bar{\mathbf{r}}$ is equal to zero, since the position vector $\bar{\mathbf{r}}$ relative to G is zero. Hence, the last term in the above equation is zero. Using the equation of motion, the product $m_i \mathbf{a}_i$ may be replaced by the resultant *external force* $\mathbf{F}_i$ acting on the ith particle. Denoting $\Sigma \mathbf{M}_G = (\Sigma \mathbf{r}_{i/G} \times \mathbf{F}_i)$, the final result may be written as

$$\Sigma \mathbf{M}_G = \dot{\mathbf{H}}_G \tag{21–21}$$

The rotational equation of motion for the body will now be developed from either Eq. 21–20 or 21–21. In this regard, the scalar components of the angular momentum $\mathbf{H}_O$ or $\mathbf{H}_G$ are defined by Eqs. 21–10 or, if principal axes of inertia are used either at point O or G, by Eqs. 21–11. If these components are computed about x, y, z axes that are *rotating* with an angular velocity $\mathbf{\Omega}$, which may be *different* from the body's angular velocity $\boldsymbol{\omega}$, then the time derivative $\dot{\mathbf{H}} = d\mathbf{H}/dt$, as used in Eqs. 21–20 and 21–21, must account for the rotation of the x, y, z axes as measured from the inertial X, Y, Z axes. Hence, the time derivative of $\mathbf{H}$ must be determined from Eq. 20–6, in which case Eqs. 21–20 and 21–21 become

$$\begin{aligned} \Sigma \mathbf{M}_O &= (\dot{\mathbf{H}}_O)_{xyz} + \mathbf{\Omega} \times \mathbf{H}_O \\ \Sigma \mathbf{M}_G &= (\dot{\mathbf{H}}_G)_{xyz} + \mathbf{\Omega} \times \mathbf{H}_G \end{aligned} \tag{21–22}$$

Here $(\dot{\mathbf{H}})_{xyz}$ is the time rate of change of $\mathbf{H}$ measured from the x, y, z reference.

There are three ways in which one can define the motion of the x, y, z axes. Obviously, motion of this reference should be chosen to yield the simplest set of moment equations for the solution of a particular problem.

x, y, z Axes Having Motion $\boldsymbol{\Omega} = \mathbf{0}$. If the body has general motion, the x, y, z axes may be chosen with origin at G, such that the axes only *translate* relative to the inertial X, Y, Z frame of reference. Doing this would certainly simplify Eq. 21–22, since $\boldsymbol{\Omega} = \mathbf{0}$. However, the body may have a rotation $\boldsymbol{\omega}$ about these axes, and therefore the moments and products of inertia of the body would have to be expressed as *functions of time.* In most cases this would be a difficult task, so that such a choice of axes has restricted value.

x, y, z Axes Having Motion $\boldsymbol{\Omega} = \boldsymbol{\omega}$. The x, y, z axes may be chosen such that they are *fixed in and move with the body.* The moments and products of inertia of the body relative to these axes will be *constant* during the motion. Since $\boldsymbol{\Omega} = \boldsymbol{\omega}$, Eqs. 21–22 become

$$\begin{aligned} \Sigma \mathbf{M}_O &= (\dot{\mathbf{H}}_O)_{xyz} + \boldsymbol{\omega} \times \mathbf{H}_O \\ \Sigma \mathbf{M}_G &= (\dot{\mathbf{H}}_G)_{xyz} + \boldsymbol{\omega} \times \mathbf{H}_G \end{aligned} \tag{21–23}$$

We may express each of these vector equations as three scalar equations using Eqs. 21–10. Neglecting the subscripts O and G yields

$$\begin{aligned} \Sigma M_x &= I_{xx}\,\dot{\omega}_x - (I_{yy} - I_{zz})\omega_y\omega_z - I_{xy}(\dot{\omega}_y - \omega_z\omega_x) - I_{yz}(\omega_y^2 - \omega_z^2) \\ &\quad - I_{zx}\,(\dot{\omega}_z + \omega_x\omega_y) \\ \Sigma M_y &= I_{yy}\,\dot{\omega}_y - (I_{zz} - I_{xx})\omega_z\omega_x - I_{yz}\,(\dot{\omega}_z - \omega_x\omega_y) - I_{zx}(\omega_z^2 - \omega_x^2) \\ &\quad - I_{xy}\,(\dot{\omega}_x + \omega_y\omega_z) \\ \Sigma M_z &= I_{zz}\,\dot{\omega}_z - (I_{xx} - I_{yy})\omega_x\omega_y - I_{zx}\,(\dot{\omega}_x - \omega_y\omega_z) - I_{xy}(\omega_x^2 - \omega_y^2) \\ &\quad - I_{yz}(\dot{\omega}_y + \omega_z\omega_x) \end{aligned} \tag{21–24}$$

Notice that for a rigid body symmetric with respect to the x–y reference plane, and undergoing general plane motion in this plane, $I_{xz} = I_{yz} = 0$, and $\omega_x = \omega_y = d\omega_x/dt = d\omega_y/dt = 0$. Equations 21–24 reduce to the form $\Sigma M_x = \Sigma M_y = 0$, and $\Sigma M_z = I_{zz}\alpha_z$ (where $\alpha_z = \dot{\omega}_z$), which is essentially the third of Eqs. 17–16 or 17–17 depending on the choice of point O or G for summing moments.

If the x, y, and z axes are chosen as *principal axes of inertia,* the products of inertia are zero, $I_{xx} = I_x$, etc., and Eqs. 21–24 reduce to the form

$$\begin{aligned} \Sigma M_x &= I_x\dot{\omega}_x - (I_y - I_z)\omega_y\omega_z \\ \Sigma M_y &= I_y\dot{\omega}_y - (I_z - I_x)\omega_z\omega_x \\ \Sigma M_z &= I_z\dot{\omega}_z - (I_x - I_y)\omega_x\omega_y \end{aligned} \tag{21–25}$$

This set of equations is known historically as the *Euler equations of motion,* named after the Swiss mathematician Leonhard Euler, who first developed them. They apply *only* for moments summed about either point O or G.

When applying these equations it should be realized that $\dot{\omega}_x$, $\dot{\omega}_y$, $\dot{\omega}_z$ represent the time derivatives of the magnitudes of the x, y, z components of $\boldsymbol{\omega}$ as observed from x, y, z. Since the x, y, z axes are rotating at $\boldsymbol{\Omega} = \boldsymbol{\omega}$, then, from Eq. 20–6, it should be noted that $\dot{\boldsymbol{\omega}} = (\dot{\boldsymbol{\omega}})_{xyz} + \boldsymbol{\omega} \times \boldsymbol{\omega}$. Since $\boldsymbol{\omega} \times \boldsymbol{\omega} = \mathbf{0}$, $\dot{\boldsymbol{\omega}} = (\dot{\boldsymbol{\omega}})_{xyz}$. This important result indicates that the required time derivative of $\boldsymbol{\omega}$ can be obtained either by first finding the components of $\boldsymbol{\omega}$ along the x, y, z axes when these axes are oriented in a general position *and then* taking the time derivative of the magnitudes of these components, i.e., $(\dot{\boldsymbol{\omega}})_{xyz}$, or by finding the time derivative of $\boldsymbol{\omega}$ with respect to the X, Y, Z axes, i.e., $\dot{\boldsymbol{\omega}}$, and then determining the components $\dot{\omega}_x$, $\dot{\omega}_y$, $\dot{\omega}_z$. In practice, it is generally easier to compute $\dot{\omega}_x$, $\dot{\omega}_y$, $\dot{\omega}_z$ on the basis of finding $\dot{\boldsymbol{\omega}}$. See Example 21–5.

x, y, z Axes Having Motion $\boldsymbol{\Omega} \neq \boldsymbol{\omega}$. To simplify the calculations for the time derivative of $\boldsymbol{\omega}$, it is often convenient to choose the x, y, z axes having an angular velocity $\boldsymbol{\Omega}$ which is different from the angular velocity $\boldsymbol{\omega}$ of the body. This is particularly suitable for the analysis of spinning tops and gyroscopes which are *symmetrical* about their spinning axes.* When this is the case, the moments and products of inertia remain constant during the motion.

Equations 21–22 are applicable for such a set of chosen axes. Each of these two vector equations may be reduced to a set of three scalar equations which are derived in a manner similar to Eqs. 21–25,† i.e.,

$$\begin{aligned} \Sigma M_x &= I_x \dot{\omega}_x - I_y \Omega_z \omega_y + I_z \Omega_y \omega_z \\ \Sigma M_y &= I_y \dot{\omega}_y - I_z \Omega_x \omega_z + I_x \Omega_z \omega_x \\ \Sigma M_z &= I_z \dot{\omega}_z - I_x \Omega_y \omega_x + I_y \Omega_x \omega_y \end{aligned} \qquad (21\text{–}26)$$

Here Ω_x, Ω_y, Ω_z represent the x, y, z components of $\boldsymbol{\Omega}$, measured from the inertial frame of reference, and $\dot{\omega}_x$, $\dot{\omega}_y$, $\dot{\omega}_z$ must be determined relative to the x, y, z axes that have the rotation $\boldsymbol{\Omega}$. See Example 21–6.

Any one of these sets of moment equations, Eqs. 21–24, 21–25, or 21–26, represents a series of three first-order nonlinear differential equations. These equations are "coupled," since the angular-velocity components are present in all the terms. Success in determining the solution for a particular problem therefore depends upon what is unknown in these equations. Difficulty certainly arises when one attempts to solve for the unknown components of $\boldsymbol{\omega}$, given the external moments as functions of time. Further complications can arise if the moment equations are coupled to the three scalar equations of translational motion, Eqs. 21–19. This can happen because of the existence of kinematic constraints which

*A detailed discussion of such devices is given in Sec. 21.5.

†See Prob. 21–41.

relate the rotation of the body to the translation of its mass center, as in the case of a hoop which rolls without slipping. Problems necessitating the simultaneous solution of differential equations generally require application of numerical methods with the aid of a computer. In many engineering problems, however, one is required to determine the applied moments acting on the body, given information about the motion of the body. Fortunately, many of these types of problems have direct solutions, so that there is no need to resort to computer techniques.

Procedure for Analysis

Problems involving the three-dimensional motion of a rigid body can be solved using the following procedure.

Free-Body Diagram

- Draw a *free-body diagram* of the body at the instant considered and specify the x, y, z coordinate system. The origin of this reference must be located either at the body's mass center G, or at point O, considered fixed in an inertial reference frame and located either in the body or on a massless extension of the body.
- Unknown reactive forces can be shown having a positive sense of direction.
- Depending on the nature of the problem, decide what type of rotational motion $\boldsymbol{\Omega}$ the x, y, z coordinate system should have, i.e., $\boldsymbol{\Omega} = \mathbf{0}$, $\boldsymbol{\Omega} = \boldsymbol{\omega}$, or $\boldsymbol{\Omega} \neq \boldsymbol{\omega}$. When choosing, one should keep in mind that the moment equations are simplified when the axes move in such a manner that they represent principal axes of inertia for the body at all times.
- Compute the necessary moments and products of inertia for the body relative to the x, y, z axes.

Kinematics

- Determine the x, y, z components of the body's angular velocity and compute the time derivatives of $\boldsymbol{\omega}$.
- Note that if $\boldsymbol{\Omega} = \boldsymbol{\omega}$ then, $\dot{\boldsymbol{\omega}} = (\dot{\boldsymbol{\omega}})_{xyz}$, and we can either find the components of $\boldsymbol{\omega}$ along the x, y, z axes when the axes are oriented in a general position, and then take the time derivative of the magnitudes of these components, $(\dot{\boldsymbol{\omega}})_{xyz}$; or we can find the time derivative of $\boldsymbol{\omega}$ with respect to the X, Y, Z axes, $\dot{\boldsymbol{\omega}}$, and then determine the components $\dot{\omega}_x$, $\dot{\omega}_y$, $\dot{\omega}_z$.

Equations of Motion

- Apply either the two vector equations 21–18 and 21–22, or the six scalar component equations appropriate for the x, y, z coordinate axes chosen for the problem.

EXAMPLE 21–4

The gear shown in Fig. 21–12*a* has a mass of 10 kg and is mounted at an angle of 10° with a rotating shaft having negligible mass. If $I_z = 0.1\ \text{kg}\cdot\text{m}^2$, $I_x = I_y = 0.05\ \text{kg}\cdot\text{m}^2$, and the shaft is rotating with a constant angular velocity of $\omega = 30$ rad/s, determine the reactions that the bearing supports A and B exert on the shaft at the instant shown.

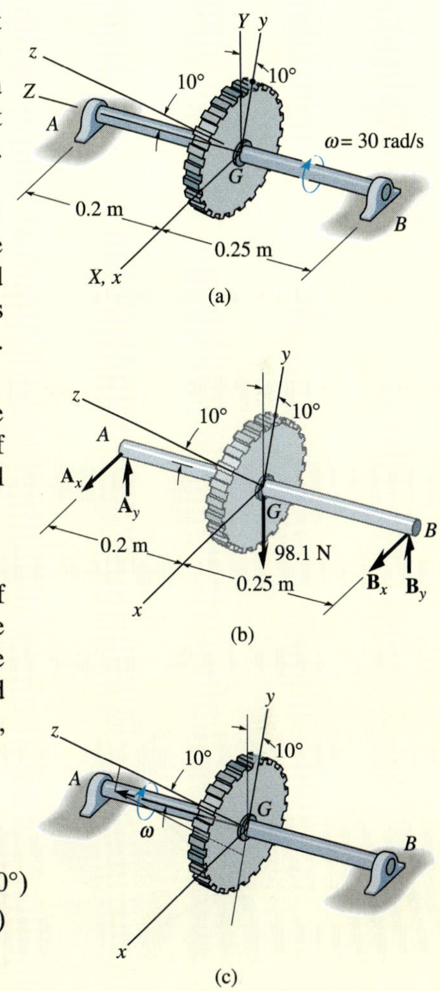

Fig. 21–12

Solution

Free-Body Diagram. Fig. 21–12*b*. The origin of the x, y, z coordinate system is located at the gear's center of mass G, which is also a fixed point. The axes are fixed in and rotate with the gear, since these axes will then always represent the principal axes of inertia for the gear. Hence $\boldsymbol{\Omega} = \boldsymbol{\omega}$.

Kinematics. As shown in Fig. 21–12*c*, the angular velocity $\boldsymbol{\omega}$ of the gear is constant in magnitude and is always directed along the axis of the shaft AB. Since this vector is measured from the X, Y, Z inertial frame of reference, for any position of the x, y, z axes,

$$\omega_x = 0 \qquad \omega_y = -30 \sin 10° \qquad \omega_z = 30 \cos 10°$$

These components remain constant for any general orientation of the x, y, z axes, and so $\dot{\omega}_x = \dot{\omega}_y = \dot{\omega}_z = 0$. Also note that since $\boldsymbol{\Omega} = \boldsymbol{\omega}$, then $\dot{\boldsymbol{\omega}} = (\dot{\boldsymbol{\omega}})_{xyz}$ and we can find these time derivatives relative to the X, Y, Z axes. In this regard $\boldsymbol{\omega}$ has a constant magnitude and direction $(+Z)$ and so $\dot{\boldsymbol{\omega}} = \mathbf{0}$, and so $\dot{\omega}_x = \dot{\omega}_y = \dot{\omega}_z = 0$. Furthermore, since G is a fixed point, $(a_G)_x = (a_G)_y = (a_G)_z = 0$.

Equations of Motion. Applying Eqs. 21–25 ($\boldsymbol{\Omega} = \boldsymbol{\omega}$) yields

$$\Sigma M_x = I_x\dot{\omega}_x - (I_y - I_z)\omega_y\omega_z$$

$$-(A_Y)(0.2) + (B_Y)(0.25) = 0 - (0.05 - 0.1)(-30 \sin 10°)(30 \cos 10°)$$

$$-0.2A_Y + 0.25B_Y = -7.70 \qquad (1)$$

$$\Sigma M_y = I_y\dot{\omega}_y - (I_z - I_x)\omega_z\omega_x$$

$$A_X(0.2) \cos 10° - B_X(0.25) \cos 10° = 0 + 0$$

$$A_X = 1.25B_X \qquad (2)$$

$$\Sigma M_z = I_z\dot{\omega}_z - (I_x - I_y)\omega_x\omega_y$$

$$A_X(0.2) \sin 10° - B_X(0.25) \sin 10° = 0 + 0$$

$$A_X = 1.25B_X$$

Applying Eqs. 21–19, we have

$$\Sigma F_X = m(a_G)_X; \qquad A_X + B_X = 0 \qquad (3)$$

$$\Sigma F_Y = m(a_G)_Y; \qquad A_Y + B_Y - 98.1 = 0 \qquad (4)$$

$$\Sigma F_Z = m(a_G)_Z; \qquad 0 = 0$$

Solving Eqs. 1 through 4 simultaneously gives

$$A_X = B_X = 0 \qquad A_Y = 71.6\text{ N} \qquad B_Y = 26.5\text{ N} \qquad \textit{Ans.}$$

EXAMPLE 21–5

The airplane shown in Fig. 21–13*a* is in the process of making a steady *horizontal* turn at the rate of ω_p. During this motion, the airplane's propeller is spinning at the rate of ω_s. If the propeller has two blades, determine the moments which the propeller shaft exerts on the propeller at the instant the blades are in the vertical position. For simplicity, assume the blades to be a uniform slender bar having a moment of inertia I about an axis perpendicular to the blades and passing through their center, and having zero moment of inertia about a longitudinal axis.

(a)

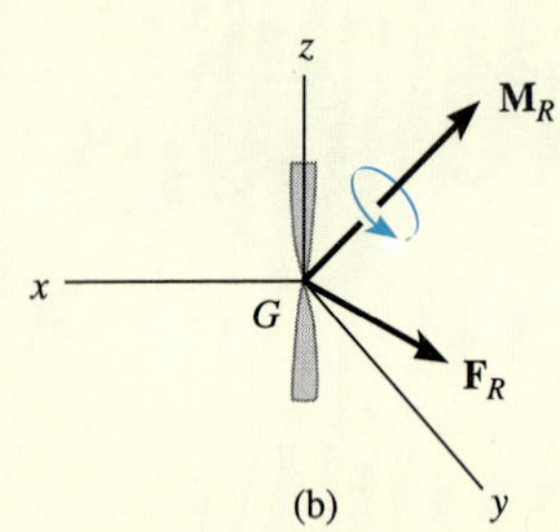

(b)

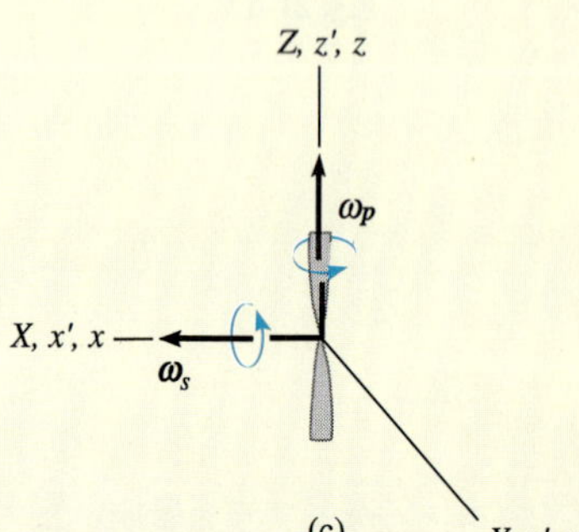

(c)

Fig. 21–13

Solution

Free-Body Diagram. Fig. 21–13*b*. The effect of the connecting shaft on the propeller is indicated by the resultants $\mathbf{F}_R$ and $\mathbf{M}_R$. (The propeller's weight is assumed to be negligible.) The x, y, z axes will be taken fixed to the propeller, since these axes always represent the principal axes of inertia for the propeller. Thus, $\boldsymbol{\Omega} = \boldsymbol{\omega}$. The moments of inertia I_x and I_y are equal ($I_x = I_y = I$) and $I_z = 0$.

Kinematics. The angular velocity of the propeller observed from the X, Y, Z axes, coincident with the x, y, z axes, Fig. 21–13*c*, is $\boldsymbol{\omega} = \boldsymbol{\omega}_s + \boldsymbol{\omega}_p = \omega_s\mathbf{i} + \omega_p\mathbf{k}$, so that the x, y, z components of $\boldsymbol{\omega}$ are

$$\omega_x = \omega_s \qquad \omega_y = 0 \qquad \omega_z = \omega_p$$

Since $\boldsymbol{\Omega} = \boldsymbol{\omega}$, then $\dot{\boldsymbol{\omega}} = (\dot{\boldsymbol{\omega}})_{xyz}$. The time derivative of $\dot{\boldsymbol{\omega}}$ can be determined with respect to the fixed X, Y, Z axes. To do this, Eq. 20–6 must be used since $\boldsymbol{\omega}$ is changing direction relative to X, Y, Z. Since $\boldsymbol{\omega} = \boldsymbol{\omega}_s + \boldsymbol{\omega}_p$, then $\dot{\boldsymbol{\omega}} = \dot{\boldsymbol{\omega}}_s + \dot{\boldsymbol{\omega}}_p$. The time rate of change of each of these components relative to the X, Y, Z axes can be obtained by using a third coordinate system x', y', z', which has an angular velocity $\boldsymbol{\Omega}' = \boldsymbol{\omega}_p$ and is coincident with the X, Y, Z axes at the instant shown. Thus

$$\begin{aligned}\dot{\boldsymbol{\omega}} &= (\dot{\boldsymbol{\omega}})_{x'y'z'} + \boldsymbol{\omega}_p \times \boldsymbol{\omega}\\ &= (\dot{\boldsymbol{\omega}}_s)_{x'y'z'} + (\dot{\boldsymbol{\omega}}_p)_{x'y'z'} + \boldsymbol{\omega}_p \times (\boldsymbol{\omega}_s + \boldsymbol{\omega}_p)\\ &= \mathbf{0} + \mathbf{0} + \boldsymbol{\omega}_p \times \boldsymbol{\omega}_s + \boldsymbol{\omega}_p \times \boldsymbol{\omega}_p\\ &= \mathbf{0} + \mathbf{0} + \omega_p\mathbf{k} \times \omega_s\mathbf{i} + \mathbf{0} = \omega_p\omega_s\mathbf{j}\end{aligned}$$

Since the X, Y, Z axes are coincident with the x, y, z axes at the instant shown, the components of $\dot{\boldsymbol{\omega}}$ along x y z are therefore

$$\dot{\omega}_x = 0 \qquad \dot{\omega}_y = \omega_p\omega_s \qquad \dot{\omega}_z = 0$$

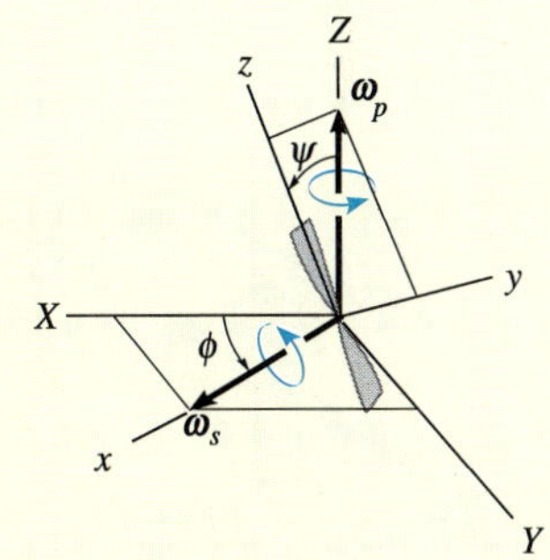

(d)

These same results can also be determined by direct calculation of $(\dot{\boldsymbol{\omega}})_{xyz}$; however, this will involve a bit more work. To do this, it will be necessary to view the propeller (or the x, y, z axes) in some *general position* such as shown in Fig. 21–13*d*. Here the plane has turned through an angle ϕ (phi) and the propeller has turned through an angle ψ (psi) relative to the plane. Notice that $\boldsymbol{\omega}_p$ is always directed along the fixed Z axis and $\boldsymbol{\omega}_s$ follows the x axis. Thus the general components of $\boldsymbol{\omega}$ are

$$\omega_x = \omega_s \qquad \omega_y = \omega_p \sin\psi \qquad \omega_z = \omega_p\cos\psi$$

Since ω_s and ω_p are constant, the time derivatives of these components become

$$\dot{\omega}_x = 0 \quad \dot{\omega}_y = \omega_p\cos\psi\,\dot{\psi} \qquad \omega_z = -\omega_p\sin\psi\,\dot{\psi}$$

But $\phi = \psi = 0°$ and $\dot{\psi} = \omega_s$ at the instant considered. Thus,

$$\omega_x = \omega_s \qquad \omega_y = 0 \qquad \omega_z = \omega_p$$

$$\dot{\omega}_x = 0 \qquad \dot{\omega}_y = \omega_p\omega_s \qquad \dot{\omega}_z = 0$$

which are the same results as those obtained previously.

Equations of Motion. Using Eqs. 21–25, we have

$$\Sigma M_x = I_x\dot{\omega}_x - (I_y - I_z)\omega_y\omega_z = I(0) - (I - 0)(0)\omega_p$$

$$M_x = 0 \qquad \textit{Ans.}$$

$$\Sigma M_y = I_y\dot{\omega}_y - (I_z - I_x)\omega_z\omega_x = I(\omega_p\omega_s) - (0 - I)\omega_p\omega_s$$

$$M_y = 2I\omega_p\omega_s \qquad \textit{Ans.}$$

$$\Sigma M_z = I_z\dot{\omega}_z - (I_x - I_y)\omega_x\omega_y = 0(0) - (I - I)\omega_s(0)$$

$$M_z = 0 \qquad \textit{Ans.}$$

EXAMPLE 21–6

The 10-kg flywheel (or thin disk) shown in Fig. 21–14*a* rotates (spins) about the shaft at a constant angular velocity of $\omega_s = 6$ rad/s. At the same time, the shaft is rotating (precessing) about the bearing at A with an angular velocity of $\omega_p = 3$ rad/s. If A is a thrust bearing and B is a journal bearing, determine the components of force reaction at each of these supports due to the motion.

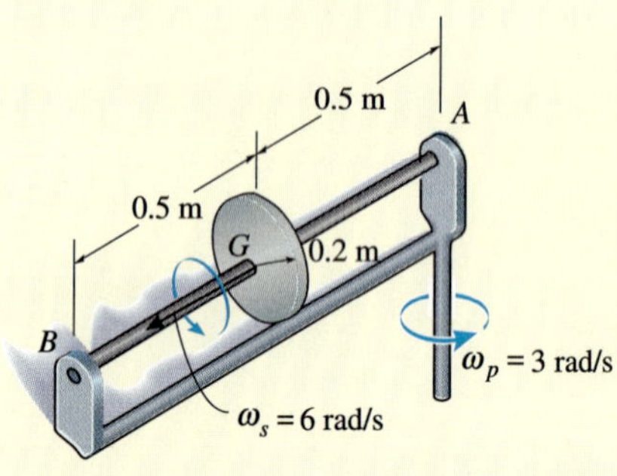

(a)

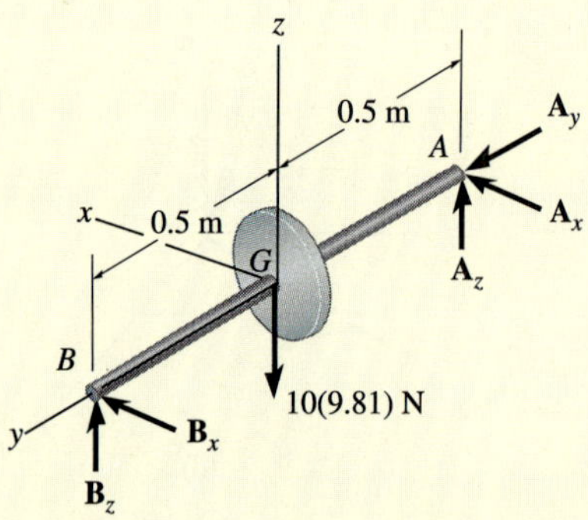

(b)

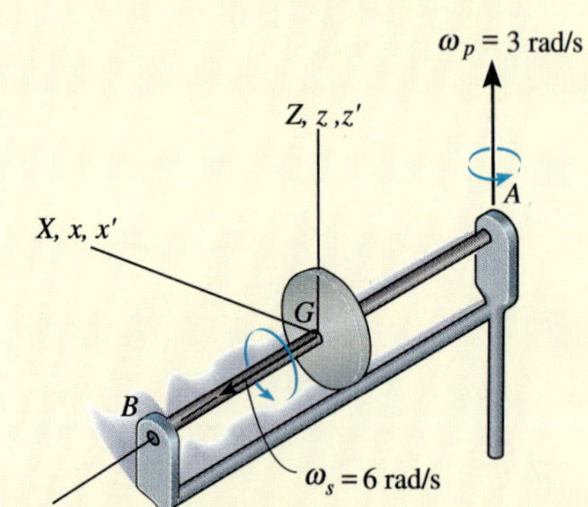

(c)

Fig. 21–14

Solution I

Free-Body Diagram. Fig. 21–14*b*. The origin of the x, y, z coordinate system is located at the center of mass G of the flywheel. Here we will let these coordinates have an angular velocity of $\mathbf{\Omega} = \boldsymbol{\omega}_p = \{3\mathbf{k}\}$ rad/s. Although the wheel spins relative to these axes, the moments of inertia *remain constant*,* i.e.,

$$I_x = I_z = \tfrac{1}{4}(10\text{ kg})(0.2\text{ m})^2 = 0.1\text{ kg}\cdot\text{m}^2$$
$$I_y = \tfrac{1}{2}(10\text{ kg})(0.2\text{ m})^2 = 0.2\text{ kg}\cdot\text{m}^2$$

Kinematics. From the coincident inertial X, Y, Z frame of reference, Fig. 21–14*c*, the flywheel has an angular velocity of $\boldsymbol{\omega} = \{6\mathbf{j} + 3\mathbf{k}\}$ rad/s, so that

$$\omega_x = 0, \quad \omega_y = 6\text{ rad/s}, \quad \omega_z = 3\text{ rad/s}$$

The time derivative of $\boldsymbol{\omega}$ must be determined relative to the x, y, z axes. In this case both $\boldsymbol{\omega}_p$ and $\boldsymbol{\omega}_s$ do not change and so

$$\dot{\omega}_x = 0, \quad \dot{\omega}_y = 0, \quad \dot{\omega}_z = 0$$

Equations of Motion. Applying Eqs. 21–26 ($\mathbf{\Omega} \neq \boldsymbol{\omega}$) yields

$$\Sigma M_x = I_x\dot{\omega}_x - I_y\Omega_z\omega_y + I_z\Omega_y\omega_z$$
$$-A_z(0.5) + B_z(0.5) = 0 - (0.2)(3)(6) + 0 = -3.6$$
$$\Sigma M_y = I_y\dot{\omega}_y - I_z\Omega_x\omega_z + I_x\Omega_z\omega_x$$
$$0 = 0 - 0 + 0$$
$$\Sigma M_z = I_z\dot{\omega}_z - I_x\Omega_y\omega_x + I_y\Omega_x\omega_y$$
$$A_x(0.5) - B_x(0.5) = 0 - 0 + 0$$

Applying Eqs. 21-19, we have

$$\Sigma F_X = m(a_G)_X; \qquad A_x + B_x = 0$$
$$\Sigma F_Y = m(a_G)_Y; \qquad A_y = -10(0.5)(3)^2$$
$$\Sigma F_Z = m(a_G)_Z; \qquad A_z + B_z - 10(9.81) = 0$$

Solving these equations, we obtain

$$A_x = 0 \quad A_y = -45.0\text{ N} \quad A_z = 52.6\text{ N} \qquad \textit{Ans.}$$
$$B_x = 0 \qquad B_z = 45.4\text{ N} \qquad \textit{Ans.}$$

Note that if the precession $\boldsymbol{\omega}_p$ had not occurred, the reactions at A and B would be equal to 49.05 N. In this case, however, the difference in the reactions is caused by the "gyroscopic moment" created whenever a spinning body precesses about another axis. We will study this effect in detail in the next section.

*This would not be true for the propeller in Example 21–5.

Solution II

This example can also be solved using Euler's equations of motion, Eqs. 21–25. In this case $\boldsymbol{\Omega} = \boldsymbol{\omega} = \{6\mathbf{j} + 3\mathbf{k}\}$ rad/s, and the time derivative $(\dot{\boldsymbol{\omega}})_{xyz}$ can be conveniently obtained with reference to the fixed X, Y, Z axes since $\dot{\boldsymbol{\omega}} = (\dot{\boldsymbol{\omega}})_{xyz}$. This calculation can be performed by choosing x', y', z' axes to have an angular velocity of $\boldsymbol{\Omega}' = \boldsymbol{\omega}_p$, Fig. 21–14c, so that

$$\dot{\boldsymbol{\omega}} = (\dot{\boldsymbol{\omega}})_{x'y'z'} + \boldsymbol{\omega}_p \times \boldsymbol{\omega} = \mathbf{0} + 3\mathbf{k} \times (6\mathbf{j} + 3\mathbf{k}) = \{-18\mathbf{i}\}\text{ rad/s}^2$$

$$\dot{\omega}_x = -18\text{ rad/s} \quad \dot{\omega}_y = 0 \quad \dot{\omega}_z = 0$$

The moment equations then become

$$\Sigma M_x = I_x\dot{\omega}_x - (I_y - I_z)\,\omega_y\omega_z$$

$$-A_z(0.5) + B_z(0.5) = 0.1(-18) - (0.2 - 0.1)(6)(3) = -3.6$$

$$\Sigma M_y = I_y\dot{\omega}_y - (I_z - I_x)\,\omega_z\omega_x$$

$$0 = 0 - 0$$

$$\Sigma M_z = I_z\dot{\omega}_z - (I_x - I_y)\,\omega_x\omega_y$$

$$A_x(0.5) - B_x(0.5) = 0 - 0$$

The solution then proceeds as before.

Problems

21-39. Derive the scalar form of the rotational equation of motion along the x axis when $\boldsymbol{\Omega} \neq \boldsymbol{\omega}$ and the moments and products of inertia of the body are *not constant* with respect to time.

***21-40.** Derive the scalar form of the rotational equation of motion along the x axis when $\boldsymbol{\Omega} \neq \boldsymbol{\omega}$ and the moments and products of inertia of the body are *constant* with respect to time.

21-41. Derive the Euler equations of motion for $\boldsymbol{\Omega} \neq \boldsymbol{\omega}$, i.e., Eqs. 21–26.

21-42. The 5-kg circular disk is mounted off center on a shaft which is supported by bearings at A and B. If the shaft is rotating at a constant rate of $\omega = 10$ rad/s, determine the vertical reactions at the bearings when the disk is in the position shown.

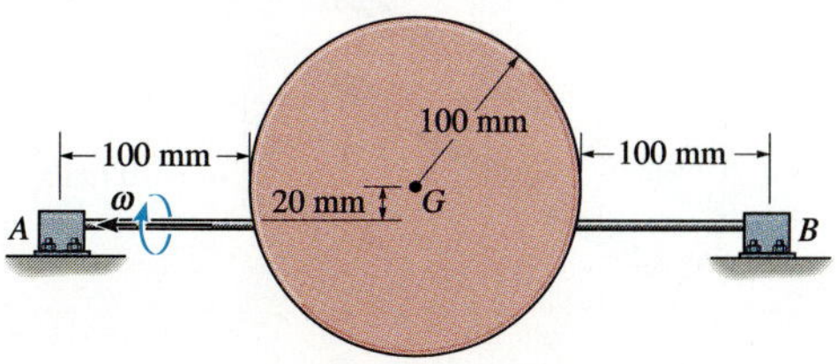

Prob. 21–42

21-43. The uniform plate has a mass of $m = 2$ kg and is given a rotation of $\omega = 4$ rad/s about its bearings at A and B. If $a = 0.2$ m and $c = 0.3$ m, determine the vertical reactions at the instant shown. Use the x, y, z axes shown and note that $I_{zx} = -\left(\dfrac{mac}{12}\right)\left(\dfrac{c^2 - a^2}{c^2 + a^2}\right)$.

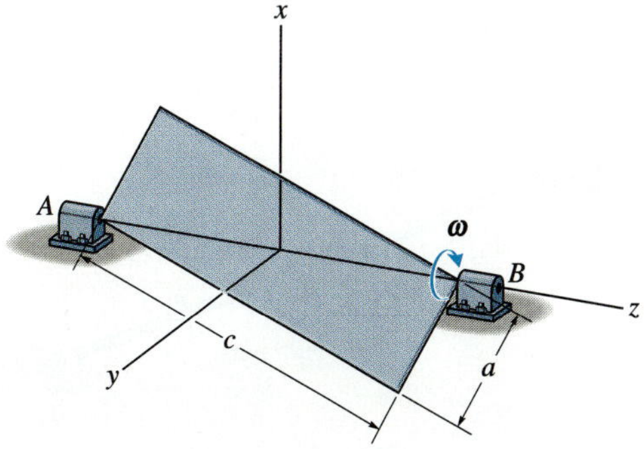

Prob. 21–43

***21–44.** The 20-lb disk is mounted on the horizontal shaft AB such that its plane forms an angle of 10° with the vertical. If the shaft rotates with an angular velocity of 3 rad/s, determine the vertical reactions developed at the bearings when the disk is in the position shown.

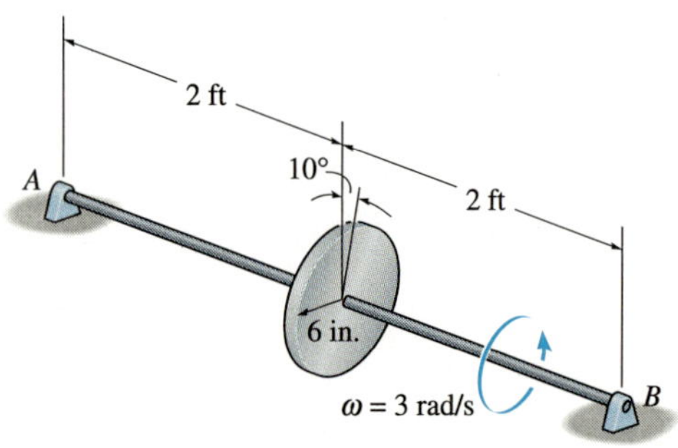

Prob. 21–44

21–45. The disk, having a mass of 3 kg, is mounted eccentrically on shaft AB. If the shaft is rotating at a constant rate of 9 rad/s, determine the reactions at the journal bearing supports when the disk is in the position shown.

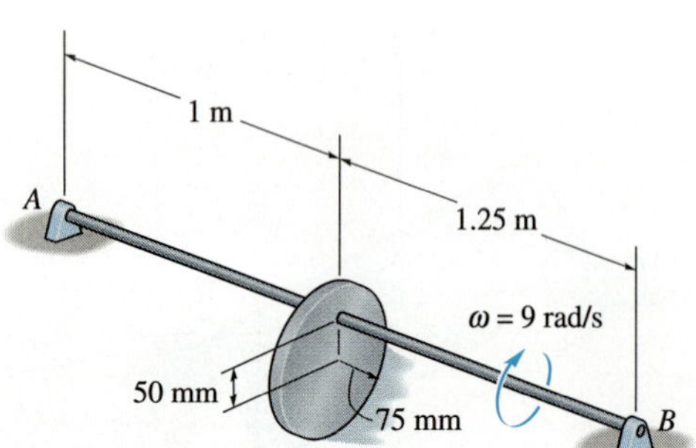

Prob. 21–45

21–46. The conical pendulum consists of a bar of mass m and length L that is supported by the pin at its end A. If the pin is subjected to a rotation $\boldsymbol{\omega}$, determine the angle θ that the bar makes with the vertical as it rotates. Also, determine the components of reaction at the pin.

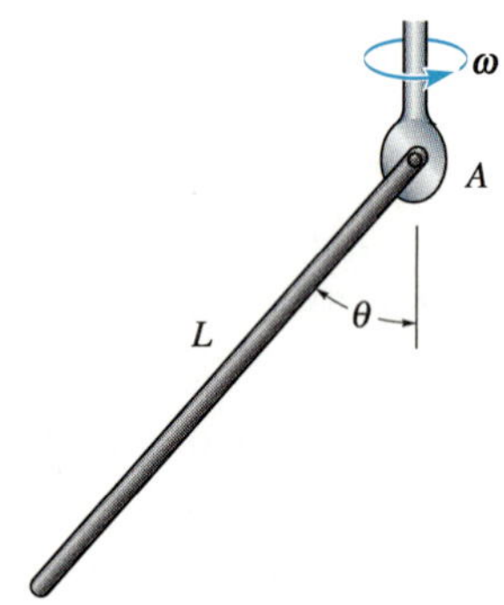

Prob. 21–46

21–47. The 5-kg rod AB is supported by a rotating arm. The support at A is a journal bearing, which develops reactions normal to the rod. The support at B is a thrust bearing, which develops reactions both normal to the rod and along the axis of the rod. Neglecting friction, determine the x, y, z components of reaction at these supports when the frame rotates with a constant angular velocity of $\omega = 10$ rad/s.

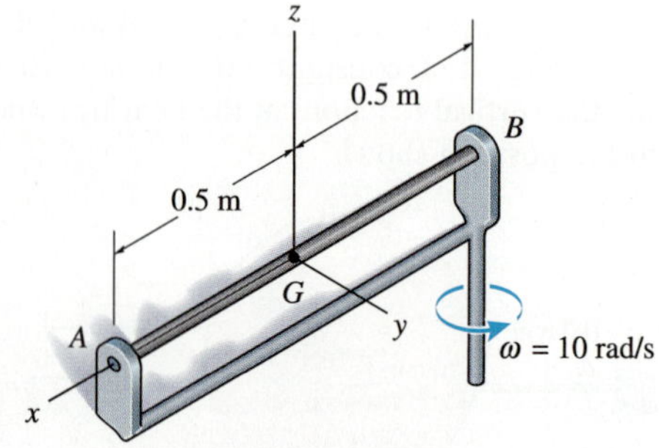

Prob. 21–47

***21-48.** The car is traveling around the curved road of radius ρ such that its mass center has a constant speed v_G. Write the equations of rotational motion with respect to the x, y, z axes. Assume that the car's six moments and products of inertia with respect to these axes are known.

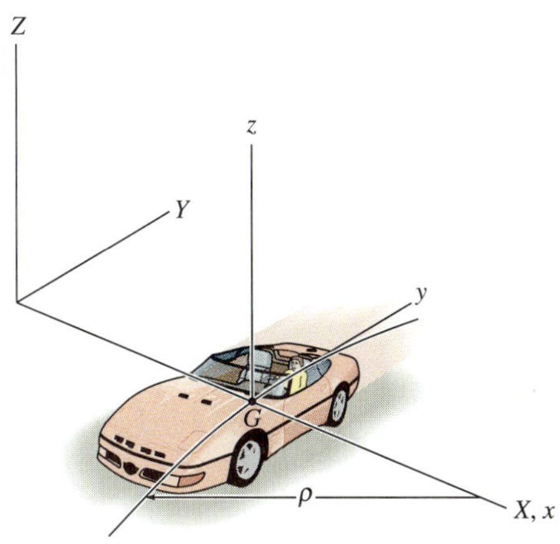

Prob. 21–48

21-49. The shaft is constructed from a rod which has a mass of 2 kg/m. Determine the components of reaction at the bearings A and B if at the instant shown the shaft is freely spinning and has an angular velocity of $\omega = 30$ rad/s. What is the angular acceleration of the shaft at this instant? Bearing A can support a component of force in the y direction, whereas bearing B cannot.

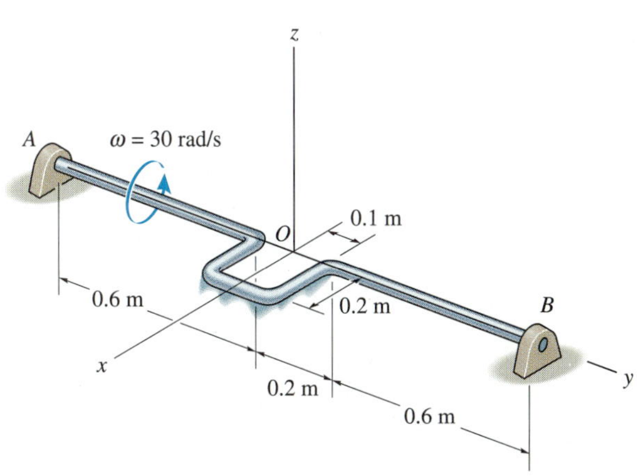

Prob. 21–49

21-50. A stone crusher consists of a large thin disk which is pin connected to a horizontal axle. If the axle is turning at a constant rate of 8 rad/s, determine the normal force which the disk exerts on the stones. Assume that the disk rolls without slipping and has a mass of 25 kg.

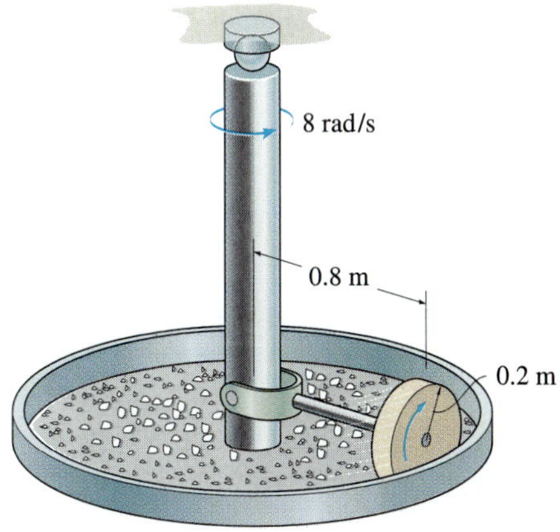

Prob. 21–50

21-51. The rod assembly has a weight of 5 lb/ft. It is supported at B by a smooth journal bearing, which develops x and y force reactions, and at A by a smooth thrust bearing, which develops x, y, and z force reactions. If a 50-lb·ft torque is applied along rod AB, determine the components of reaction at the bearings when the assembly has an angular velocity $\omega = 10$ rad/s at the instant shown.

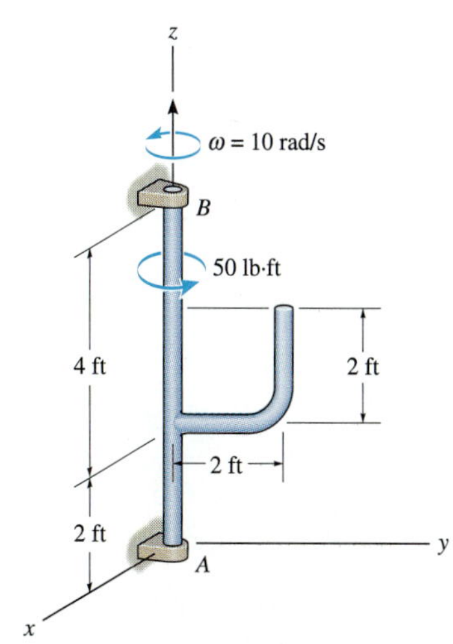

Prob. 21–51

***21-52.** The 25-lb disk is *fixed* to rod *BCD*, which has negligible mass. Determine the torque *T* which must be applied to the vertical shaft so that the shaft has an angular acceleration of $\alpha = 6\ \text{rad/s}^2$. The shaft is free to turn in its bearings.

21-53. Solve Prob. 21-52, assuming rod *BCD* has a weight of 2 lb/ft.

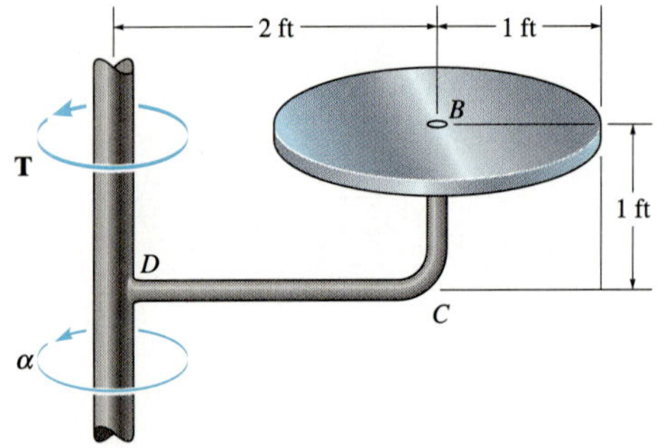

Probs. 21–52/53

21-54. The rod assembly has a mass of 1.5 kg/m. A resultant frictional moment **M**, developed at the journal bearings *A* and *B*, causes the shaft to decelerate at $1\ \text{rad/s}^2$ when it has an angular velocity $\omega = 6\ \text{rad/s}$. Determine the reactions at the bearings at the instant shown. What is the frictional moment *M* causing the deceleration? Establish the axes at *G* with $+y$ upward and $+z$ directed along the shaft axis toward *A*.

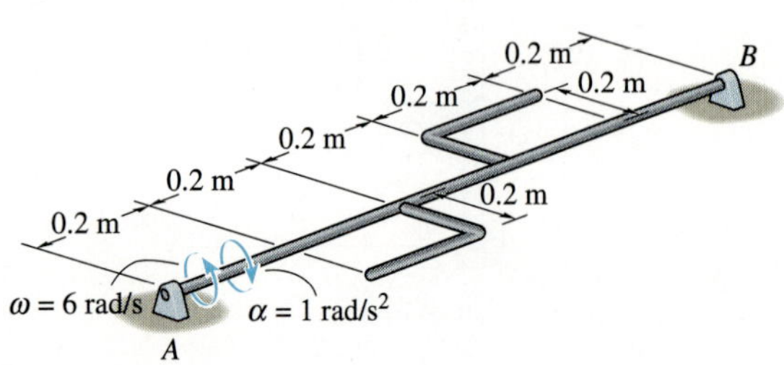

Prob. 21–54

21-55. A thin uniform plate having a mass of 0.4 kg is spinning with a constant angular velocity $\boldsymbol{\omega}$ about its diagonal *AB*. If the person holding the corner of the plate at *B* releases his finger, the plate will fall downward on its side *AC*. Determine the necessary couple moment *M* which if applied to the plate would prevent this from happening.

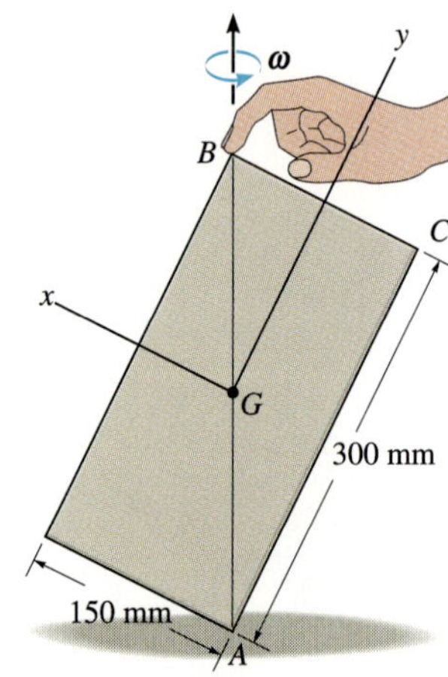

Prob. 21–55

***21-56.** The 15-lb cylinder is rotating about shaft *AB* with a constant angular speed $\omega = 4\ \text{rad/s}$. If the supporting shaft at *C*, initially at rest, is given an angular acceleration $\alpha_C = 12\ \text{rad/s}^2$, determine the components of reaction at the bearings *A* and *B*. The bearing at *A* cannot support a force component along the *x* axis, whereas the bearing at *B* does.

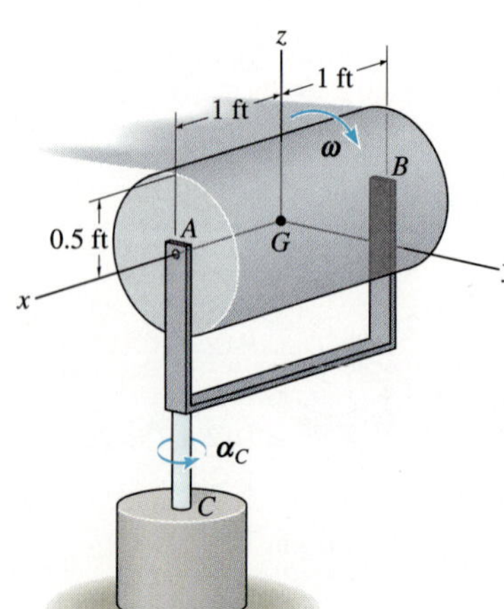

Prob. 21–56

21-57. The crankshaft is constructed from a rod which has a weight of 3 lb/ft. Determine its angular acceleration and the x and z components of reaction at bearings A and B at the instant the shaft is in the position shown.

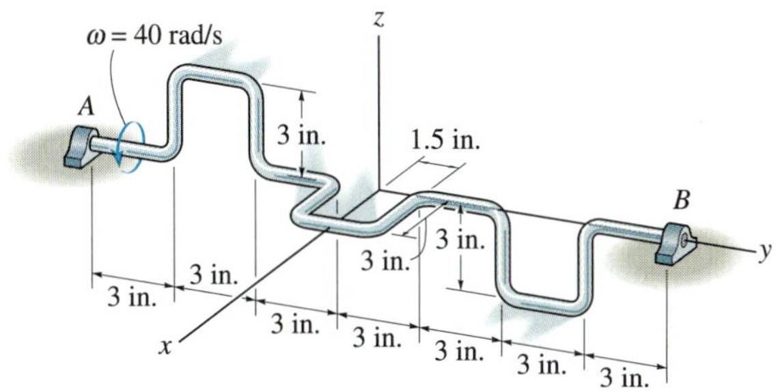

Prob. 21–57

21-58. The 4-kg slender rod AB is pinned at A and held at B by a cord. The axle CD is supported at its ends by ball-and-socket joints and is rotating with a constant angular velocity of 2 rad/s. Determine the tension developed in the cord and the magnitude of force developed at the pin A.

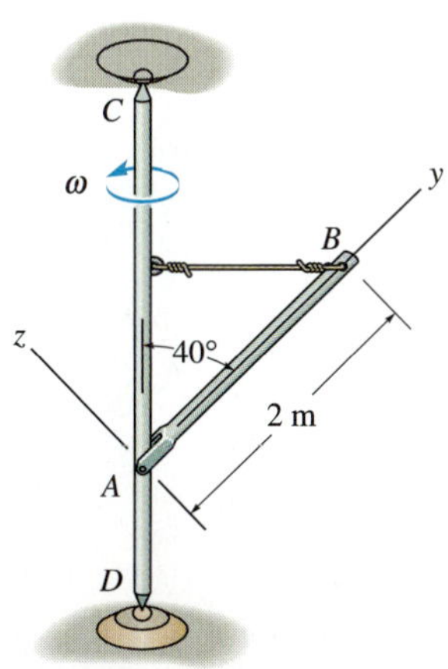

Prob. 21–58

21-59. Four spheres are connected to shaft AB. If $m_C = 1$ kg and $m_E = 2$ kg, determine the mass of D and F and the angles of the rods, θ_D and θ_F, so that the shaft is dynamically balanced, that is, so that the bearings at A and B exert only vertical reactions on the shaft as it rotates. Neglect the mass of the rods.

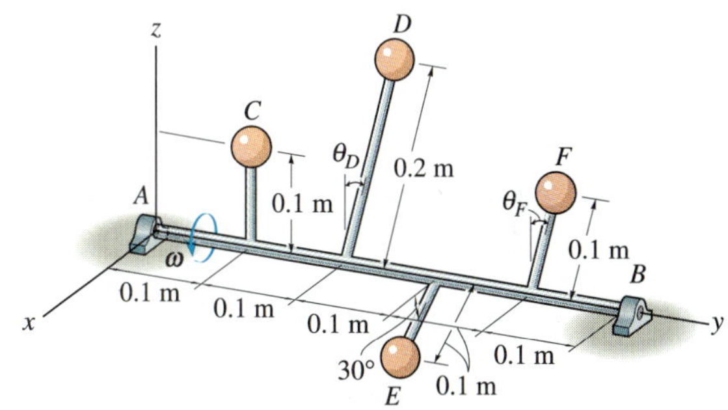

Prob. 21–59

***21-60.** Two uniform rods, each having a weight of 10 lb, are pin connected to the edge of a rotating disk. If the disk has a constant angular velocity $\omega_D = 4$ rad/s, determine the angle θ made by each rod during the motion, and the components of the force and moment developed at the pin A. *Suggestion:* Use the x, y, z axes oriented as shown.

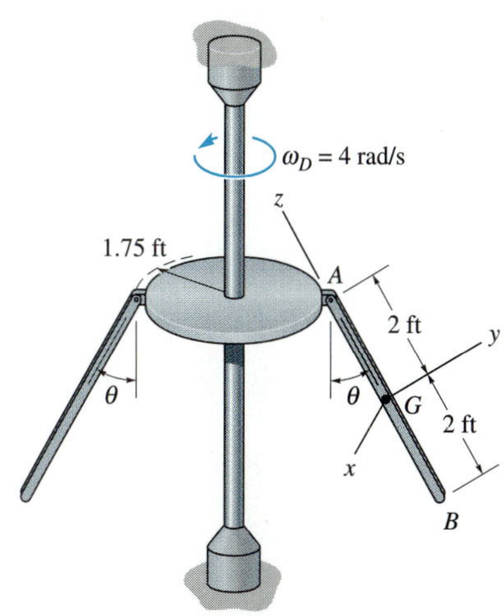

Prob. 21–60

★ 21.5 Gyroscopic Motion

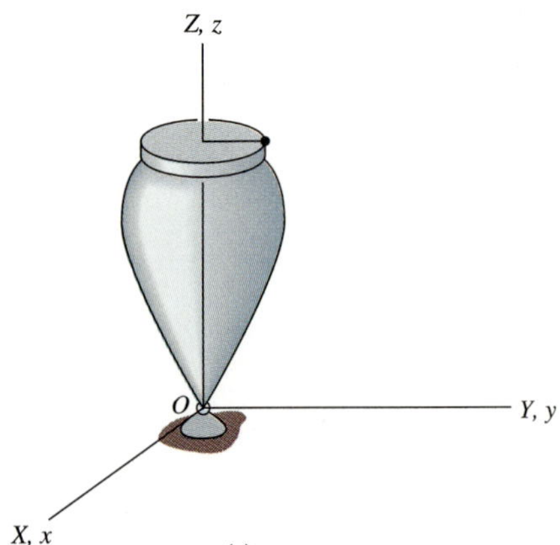

(a)

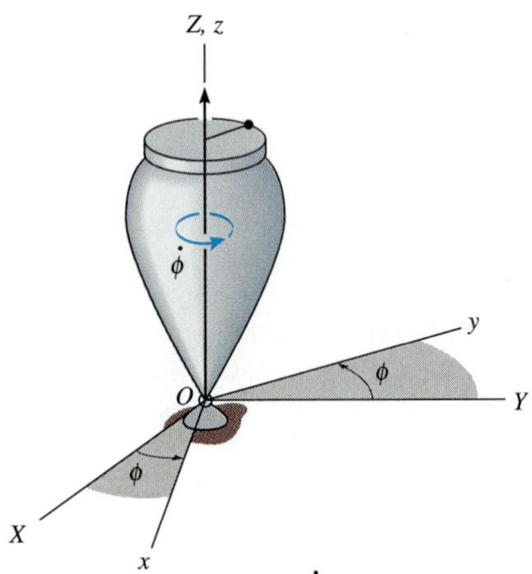

Precession $\dot{\phi}$

(b)

In this section the equations defining the motion of a body (top) which is symmetrical with respect to an axis and moving about a fixed point lying on the axis will be developed. These equations will then be applied to study the motion of a particularly interesting device, the gyroscope.

The body's motion will be analyzed using *Euler angles* ϕ, θ, ψ (phi, theta, psi). To illustrate how these angles define the position of a body, reference is made to the top shown in Fig. 21–15*a*. The top is attached to point O and has an orientation relative to the fixed X, Y, Z axes at some instant of time as shown in Fig. 21–15*d*. To define this final position, a second set of x, y, z axes will be needed. For purposes of discussion, assume that this reference is fixed in the top. Starting with the X, Y, Z and x, y, z axes in coincidence, Fig. 21–15*a*, the final position of the top is determined using the following three steps:

1. Rotate the top about the Z (or z) axis through an angle ϕ $(0 \leq \phi < 2\pi)$, Fig. 21–15*b*.
2. Rotate the top about the x axis through an angle θ $(0 \leq \theta \leq \pi)$, Fig. 21–15*c*.
3. Rotate the top about the z axis through an angle ψ $(0 \leq \psi < 2\pi)$ to obtain the final position, Fig. 20–15*d*.

The sequence of these three angles, ϕ, θ, then ψ, must be maintained, since finite rotations are *not vectors* (see Fig. 20–1). Although this is the case, the differential rotations $d\boldsymbol{\phi}$, $d\boldsymbol{\theta}$, and $d\boldsymbol{\psi}$ are vectors, and thus the angular velocity $\boldsymbol{\omega}$ of the top can be expressed in terms of the time derivatives of the Euler angles. The angular-velocity components $\dot{\phi}$, $\dot{\theta}$, and $\dot{\psi}$ are known as the *precession, nutation,* and *spin,* respectively. Their

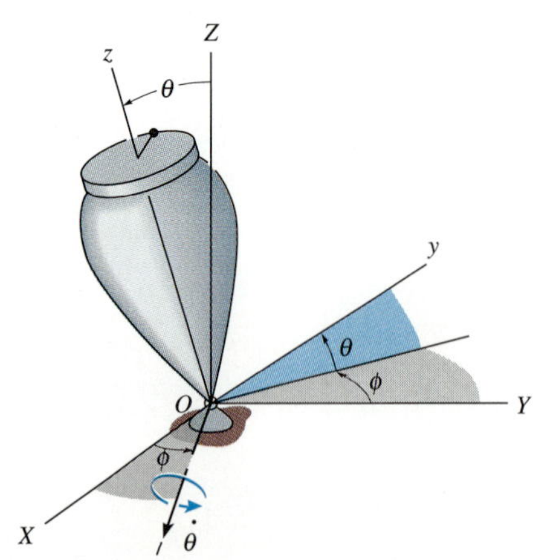

Nutation $\dot{\theta}$

(c)

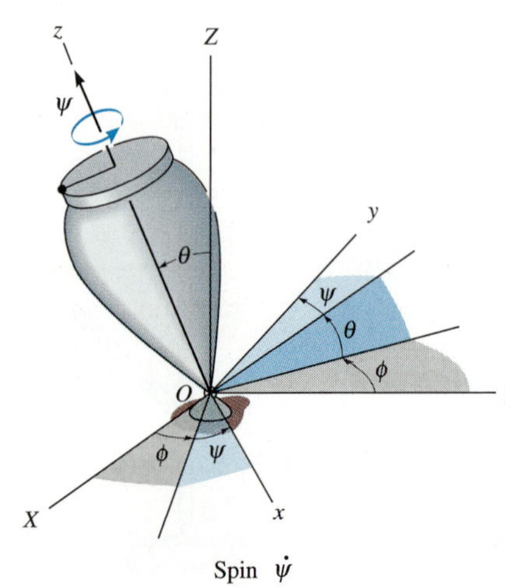

Spin $\dot{\psi}$

(d)

Fig. 21–15

positive directions are shown in Fig. 21–16. It is seen that these vectors are not all perpendicular to one another; however, $\boldsymbol{\omega}$ of the top can still be expressed in terms of these three components.

In our case the body (top) is symmetric with respect to the z or spin axis. If we consider the top to be oriented so that at the instant considered the spin angle $\psi = 0$ and *the x, y, z axes follow the motion of the body only in nutation and precession,* i.e., $\boldsymbol{\Omega} = \boldsymbol{\omega}_p + \boldsymbol{\omega}_n$, then the nutation and spin are always directed along the x and z axes, respectively, Fig. 21–16. Hence, the angular velocity of the body is specified only in terms of the Euler angle θ, i.e.,

$$\begin{aligned}\boldsymbol{\omega} &= \omega_x\mathbf{i} + \omega_y\mathbf{j} + \omega_z\mathbf{k}\\ &= \dot{\theta}\mathbf{i} + (\dot{\phi}\sin\theta)\mathbf{j} + (\dot{\phi}\cos\theta + \dot{\psi})\mathbf{k}\end{aligned} \qquad (21\text{–}27)$$

Since motion of the axes is not affected by the spin component,

$$\begin{aligned}\boldsymbol{\Omega} &= \Omega_x\mathbf{i} + \Omega_y\mathbf{j} + \Omega_z\mathbf{k}\\ &= \dot{\theta}\mathbf{i} + (\dot{\phi}\sin\theta)\mathbf{j} + (\dot{\phi}\cos\theta)\mathbf{k}\end{aligned} \qquad (21\text{–}28)$$

The x, y, z axes in Fig. 21–16 represent *principal axes of inertia* of the body for *any* spin of the body about these axes. Hence, the moments of inertia are constant and will be represented as $I_{xx} = I_{yy} = I$ and $I_{zz} = I_z$. Since $\boldsymbol{\Omega} \neq \boldsymbol{\omega}$, Eqs. 21–26 are used to establish the rotational equations of motion. Substituting into these equations the respective angular-velocity components defined by Eqs. 21–27 and 21–28, their corresponding time derivatives, and the moment of inertia components yields

$$\begin{aligned}\Sigma M_x &= I(\ddot{\theta} - \dot{\phi}^2\sin\theta\cos\theta) + I_z\dot{\phi}\sin\theta(\dot{\phi}\cos\theta + \dot{\psi})\\ \Sigma M_y &= I(\ddot{\phi}\sin\theta + 2\dot{\phi}\dot{\theta}\cos\theta) - I_z\dot{\theta}(\dot{\phi}\cos\theta + \dot{\psi})\\ \Sigma M_z &= I_z(\ddot{\psi} + \ddot{\phi}\cos\theta - \dot{\phi}\dot{\theta}\sin\theta)\end{aligned} \qquad (21\text{–}29)$$

Each moment summation applies only at the fixed point O or the center of mass G of the body. Since the equations represent a coupled set of nonlinear second-order differential equations, in general a closed-form solution may not be obtained. Instead, the Euler angles ϕ, θ, and ψ may be obtained graphically as functions of time using numerical analysis and computer techniques.

A special case, however, does exist for which simplification of Eqs. 21–29 is possible. Commonly referred to as *steady precession,* it occurs when the nutation angle θ, precession $\dot{\phi}$, and spin $\dot{\psi}$ all remain *constant.* Equations 21–29 then reduce to the form

$$\Sigma M_x = -I\dot{\phi}^2\sin\theta\cos\theta + I_z\dot{\phi}\sin\theta(\dot{\phi}\cos\theta + \dot{\psi}) \qquad (21\text{–}30)$$

$$\begin{aligned}\Sigma M_y &= 0\\ \Sigma M_z &= 0\end{aligned}$$

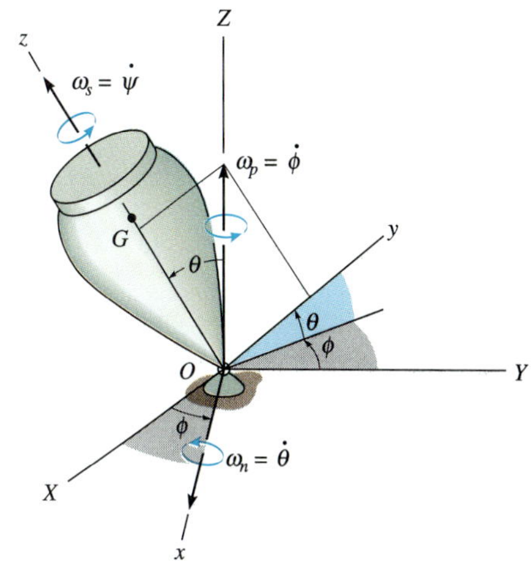

Fig. 21–16

Equation 21–30 may be further simplified by noting that, from Eq. 21–27, $\omega_z = \dot{\phi}\cos\theta + \dot{\psi}$, so that

$$\Sigma M_x = -I\dot{\phi}^2 \sin\theta\cos\theta + I_z\dot{\phi}\sin\theta\omega_z$$

or

$$\Sigma M_x = \dot{\phi}\sin\theta(I_z\omega_z - I\dot{\phi}\cos\theta) \qquad (21\text{–}31)$$

It is interesting to note what effects the spin $\dot{\psi}$ has on the moment about the x axis. In this regard consider the spinning rotor shown in Fig. 21–17. Here $\theta = 90°$, in which case Eq. 21–30 reduces to the form

$$\Sigma M_x = I_z\dot{\phi}\dot{\psi}$$

or

$$\Sigma M_x = I_z\Omega_y\omega_z \qquad (21\text{–}32)$$

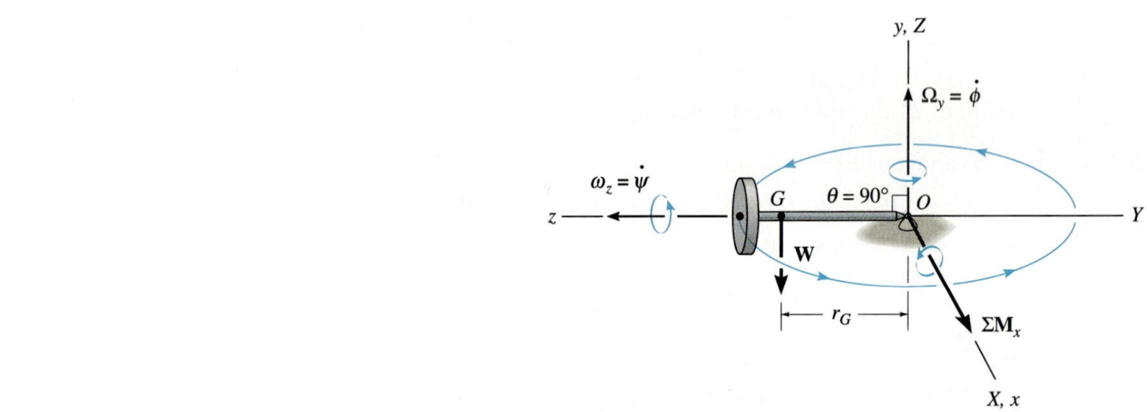

Fig. 21–17

From the figure it is seen that vectors $\Sigma\mathbf{M}_x$, $\boldsymbol{\Omega}_y$, and $\boldsymbol{\omega}_z$ all act along their respective *positive axes* and therefore are mutually perpendicular. Instinctively, one would expect the rotor to fall down under the influence of gravity! However, this is not the case at all, provided the product $I_z\Omega_y\omega_z$ is correctly chosen to counterbalance the moment $\Sigma M_x = Wr_G$ of the rotor's weight about O. This unusual phenomenon of rigid-body motion is often referred to as the *gyroscopic effect.*

Perhaps a more intriguing demonstration of the gyroscopic effect comes from studying the action of a *gyroscope,* frequently referred to as a *gyro.* A gyro is a rotor which spins at a very high rate about its axis of symmetry. This rate of spin is considerably greater than its precessional rate of rotation about the vertical axis. Hence, for all practical purposes, the angular momentum of the gyro can be assumed directed among its axis of spin. Thus, for the gyro rotor shown in Fig. 21–18, $\omega_z \gg \Omega_y$, and the magnitude of the angular momentum about point O, as determined from Eqs. 21–11, reduces to the form $H_O = I_z\omega_z$. Since both the magnitude and direction of $\mathbf{H}_O$ are constant as observed from x, y, z, direct application of Eq. 21–22 yields

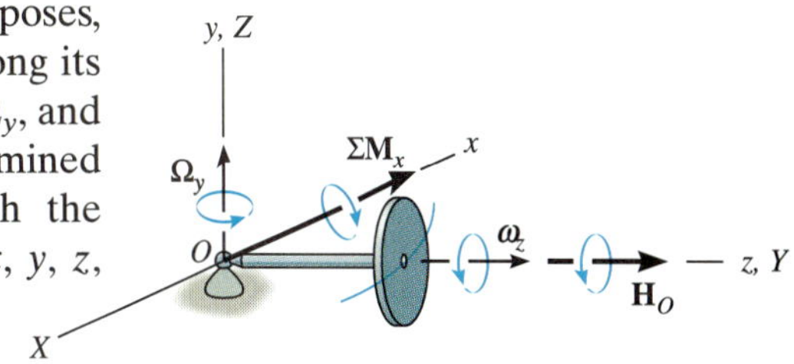

Fig. 21–18

$$\Sigma \mathbf{M}_x = \mathbf{\Omega}_y \times \mathbf{H}_O \tag{21–33}$$

Using the right-hand rule applied to the cross product, it is seen that $\mathbf{\Omega}_y$ always swings $\mathbf{H}_O$ (or $\boldsymbol{\omega}_z$) toward the sense of $\Sigma\mathbf{M}_x$. In effect, the *change in direction* of the gyro's angular momentum, $d\mathbf{H}_O$, is equivalent to the angular impulse caused by the gyro's weight about O, i.e., $d\mathbf{H}_O = \Sigma\mathbf{M}_x\,dt$, Eq. 21–20. Also, since $H_O = I_z\omega_z$ and ΣM_x, Ω_y, and H_O are mutually perpendicular, Eq. 21–33 reduces to Eq. 21–32.

When a gyro is mounted in gimbal rings, Fig. 21–19, it becomes *free* of external moments applied to its base. Thus, in theory, its angular momentum $\mathbf{H}$ will never precess but, instead, maintain its same fixed orientation along the axis of spin when the base is rotated. This type of gyroscope is called a *free gyro* and is useful as a gyrocompass when the spin axis of the gyro is directed north. In reality, the gimbal mechanism is never completely free of friction, so such a device is useful only for the local navigation of ships and aircraft. The gyroscopic effect is also useful as a means of stabilizing both the rolling motion of ships at sea and the trajectories of missiles and projectiles. Furthermore, this effect is of significant importance in the design of shafts and bearings for rotors which are subjected to forced precessions.

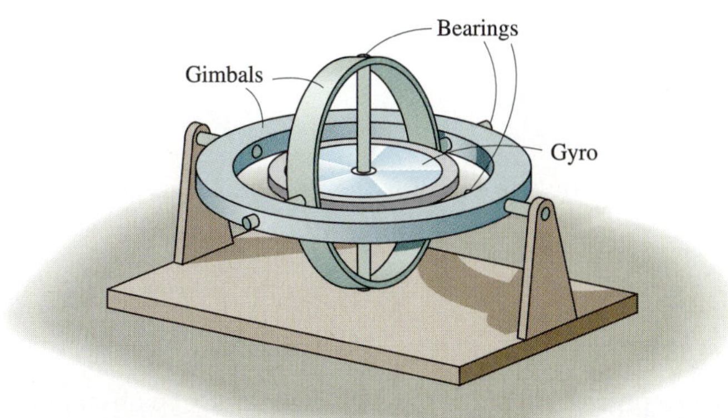

Fig. 21–19

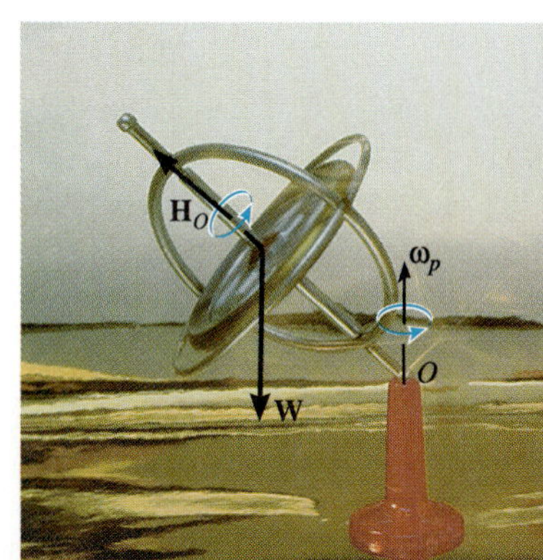

The spinning of the gyro within the frame of this toy gyroscope produces angular momentum $\mathbf{H}_O$ which is changing direction as the frame precesses $\boldsymbol{\omega}_p$ about the vertical axis. The gyroscope will not fall down since the moment of its weight $\mathbf{W}$ about the support is balanced by the change in the direction of $\mathbf{H}_O$.

EXAMPLE 21-7

The top shown in Fig. 21–20*a* has a mass of 0.5 kg and is precessing about the vertical axis at a constant angle of $\theta = 60°$. If it spins with an angular velocity $\omega_s = 100$ rad/s, determine the precessional velocity $\boldsymbol{\omega}_p$. Assume that the axial and transverse moments of inertia of the top are $0.45(10^{-3})$ kg·m² and $1.20(10^{-3})$ kg·m², respectively, measured with respect to the fixed point O.

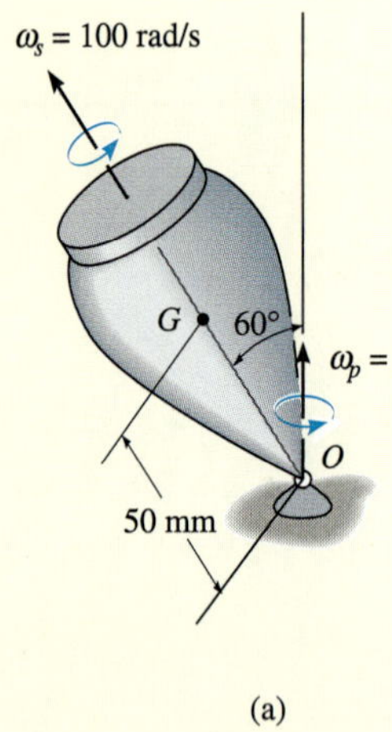

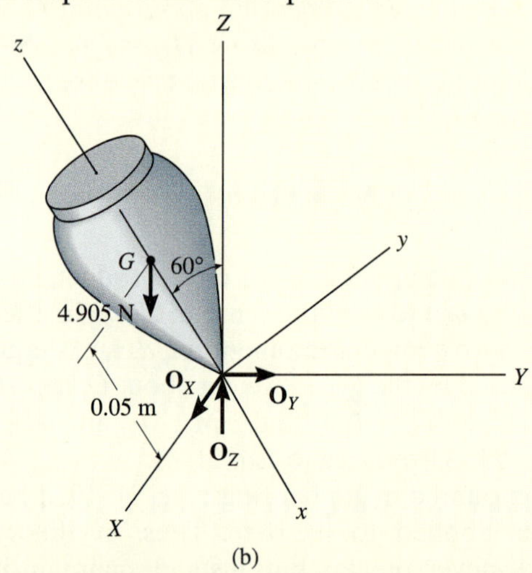

Fig. 21–20

Solution

Equation 21–30 will be used for the solution since the motion is *steady precession.* As shown on the free-body diagram, Fig. 21–20*b*, the coordinate axes are established in the usual manner, that is, with the positive z axis in the direction of spin, the positive Z axis in the direction of precession, and the positive x axis in the direction of the moment $\Sigma \mathbf{M}_x$ (refer to Fig. 21–16). Thus,

$$\Sigma M_x = -I\dot{\phi}^2 \sin\theta\cos\theta + I_z\dot{\phi}\sin\theta(\dot{\phi}\cos\theta + \dot{\psi})$$

$$4.905\text{ N}(0.05\text{ m})\sin 60° = -[1.20(10^{-3})\text{ kg}\cdot\text{m}^2\,\dot{\phi}^2]\sin 60°\cos 60° + [0.45(10^{-3})\text{ kg}\cdot\text{m}^2]\,\dot{\phi}\sin 60°(\dot{\phi}\cos 60° + 100\text{ rad/s})$$

or

$$\dot{\phi}^2 - 120.0\dot{\phi} + 654.0 = 0 \qquad (1)$$

Solving this quadratic equation for the precession gives

$$\dot{\phi} = 114\text{ rad/s} \qquad \text{(high precession)}$$ *Ans.*

and

$$\dot{\phi} = 5.72\text{ rad/s} \qquad \text{(low precession)}$$ *Ans.*

In reality, low precession of the top would generally be observed, since high precession would require a larger kinetic energy.

EXAMPLE 21-8

The 1-kg disk shown in Fig. 21–21*a* is spinning about its axis with a constant angular velocity $\omega_D = 70$ rad/s. The block at B has a mass of 2 kg, and by adjusting its position s one can change the precession of the disk about its supporting pivot at O. Determine the position s which will enable the disk to have a constant precessional velocity $\omega_p = 0.5$ rad/s about the pivot. Neglect the weight of the shaft.

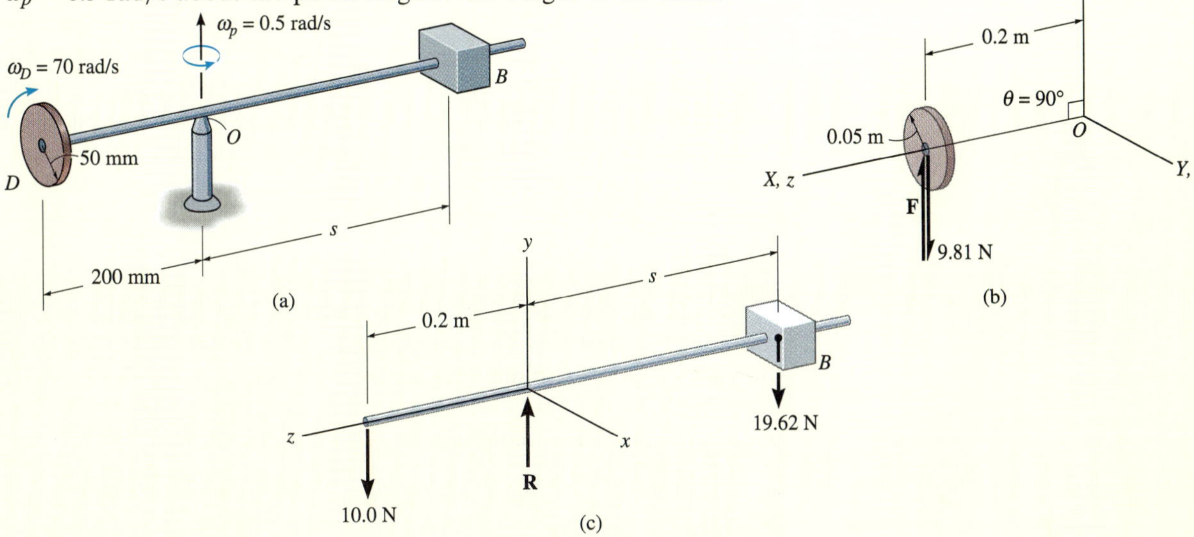

Fig. 21–21

Solution

The free-body diagram of the disk is shown in Fig. 21–21*b*, where **F** represents the force reaction of the shaft on the disk. The origin for both the x, y, z and X, Y, Z coordinate systems is located at point O, which represents a *fixed point* for the disk. (Although point O does not lie on the disk, imagine a massless extension of the disk to this point.) In the conventional sense, the Z axis is chosen along the axis of precession, and the z axis is along the axis of spin, so that $\theta = 90°$. Since the precession is *steady*, Eq. 21–31 may be used for the solution. This equation reduces to

$$\Sigma M_x = \dot{\phi} I_z \omega_z$$

which is the same as Eq. 21–32. Substituting the required data gives

$$9.81\text{ N}(0.2\text{ m}) - F(0.2\text{ m}) = 0.5\text{ rad/s}[\tfrac{1}{2}(1\text{ kg})(0.05\text{ m})^2](-70\text{ rad/s})$$
$$F = 10.0\text{ N}$$

As shown on the free-body diagram of the shaft and block B, Fig. 21–21*c*, summing moments about the x axis requires

$$(19.62\text{ N})s = (10.0\text{ N})(0.2\text{ m})$$
$$= 0.102\text{ m} = 102\text{ mm} \qquad \textit{Ans.}$$

★ 21.6 Torque-Free Motion

When the only external force acting on a body is caused by gravitation, the general motion of the body is referred to as *torque-free motion.* This type of motion is characteristic of planets, artificial satellites, and projectiles—provided the effects of air friction are neglected.

In order to describe the characteristics of this motion, the distribution of the body's mass will be assumed *axisymmetric.* The satellite shown in Fig. 21–22 is an example of such a body, where the z axis represents an axis of symmetry. The origin of the x, y, z coordinates is located at the mass center G, such that $I_{zz} = I_z$ and $I_{xx} = I_{yy} = I$ for the body. Since gravitation is the only external force present, the summation of moments about the mass center is zero. From Eq. 21–21, this requires the angular momentum of the body to be constant, i.e.,

$$\mathbf{H}_G = \text{constant}$$

At the instant considered, it will be assumed that the inertial frame of reference is oriented so that the positive Z axis is directed along $\mathbf{H}_G$ and the y axis lies in the plane formed by the z and Z axes, Fig. 21–22. The Euler angle formed between Z and z is θ, and therefore, with this choice of axes the angular momentum may be expressed as

$$\mathbf{H}_G = H_G \sin \theta \mathbf{j} + H_G \cos \theta \mathbf{k}$$

Furthermore, using Eqs. 21–11, we have

$$\mathbf{H}_G = I\omega_x \mathbf{i} + I\omega_y \mathbf{j} + I_z\omega_z \mathbf{k}$$

where ω_x, ω_y, ω_z represent the x, y, z components of the body's angular velocity. Equating the respective **i, j,** and **k** components of the above two equations yields

$$\omega_x = 0 \qquad \omega_y = \frac{H_G \sin \theta}{I} \qquad \omega_z = \frac{H_G \cos \theta}{I_z} \tag{21–34}$$

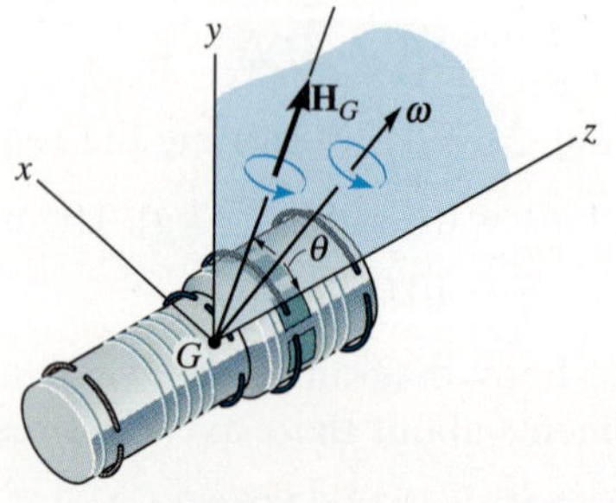

Fig. 21–22

or

$$\boldsymbol{\omega} = \frac{H_G \sin\theta}{I}\mathbf{j} + \frac{H_G \cos\theta}{I_z}\mathbf{k} \tag{21–35}$$

In a similar manner, equating the respective **i, j, k** components of Eq. 21–27 to those of Eq. 21–34, we obtain

$$\dot{\theta} = 0$$

$$\dot{\phi}\sin\theta = \frac{H_G \sin\theta}{I}$$

$$\dot{\phi}\cos\theta + \dot{\psi} = \frac{H_G \cos\theta}{I_z}$$

Solving, we get

$$\begin{aligned} \theta &= \text{const} \\ \dot{\phi} &= \frac{H_G}{I} \\ \dot{\psi} &= \frac{I - I_z}{I I_z} H_G \cos\theta \end{aligned} \tag{21–36}$$

Thus, for torque-free motion of an axisymmetrical body, the angle θ formed between the angular-momentum vector and the spin of the body remains constant. Furthermore, the angular momentum $\mathbf{H}_G$, precession $\dot{\phi}$, and spin $\dot{\psi}$ for the body remain constant at all times during the motion. Eliminating H_G from the second and third of Eqs. 21–36 yields the following relationship between the spin and precession:

$$\dot{\psi} = \frac{I - I_z}{I_z}\dot{\phi}\cos\theta \tag{21–37}$$

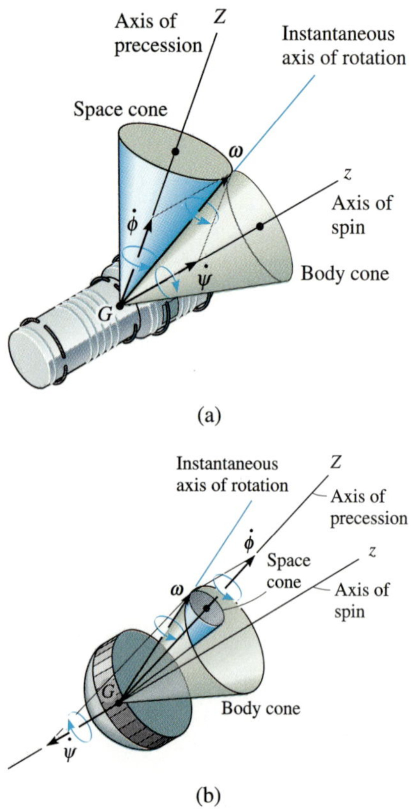

Fig. 21–23

As shown in Fig. 21–23*a*, the body precesses about the Z axis, which is fixed in direction, while it spins about the z axis. These two components of angular motion may be studied by using a simple cone model, introduced in Sec. 20.1. The *space cone* defining the precession is fixed from rotating, since the precession has a fixed direction, while the *body cone* rotates around the space cone's outer surface without slipping. On this basis, an attempt should be made to imagine the motion. The interior angle of each cone is chosen such that the resultant angular velocity of the body is directed along the line of contact of the two cones. This line of contact represents the instantaneous axis of rotation for the body cone, and hence the angular velocity of both the body cone and the body must be directed along this line. Since the spin is a function of the moments of inertia I and I_z of the body, Eq. 21–36, the cone model in Fig. 21–23*a* is satisfactory for describing the motion, provided $I > I_z$. Torque-free motion which meets these requirements is called *regular precession.* If $I < I_z$, the spin is negative and the precession positive. This motion is represented by the satellite motion shown in Fig. 21–23*b* ($I < I_z$). The cone model may again be used to represent the motion; however, to preserve the correct vector addition of spin and precession to obtain the angular velocity $\boldsymbol{\omega}$, the inside surface of the body cone must roll on the outside surface of the (fixed) space cone. This motion is referred to as *retrograde precession.*

Satellites are often given a spin before they are launched. If their angular momentum is not collinear with the axis of spin they will exhibit precession. In the photo on the left regular precession will occur since $I > I_z$, and in the photo on the right, retrograde precession will occur since $I < I_z$.

EXAMPLE 21–9

The motion of a football is observed using a slow-motion projector. From the film, the spin of the football is seen to be directed 30° from the horizontal, as shown in Fig. 21–24*a*. Also, the football is precessing about the vertical axis at a rate $\dot{\phi} = 3$ rad/s. If the ratio of the axial to transverse moments of inertia of the football is $\frac{1}{3}$, measured with respect to the center of mass, determine the magnitude of the football's spin and its angular velocity. Neglect the effect of air resistance.

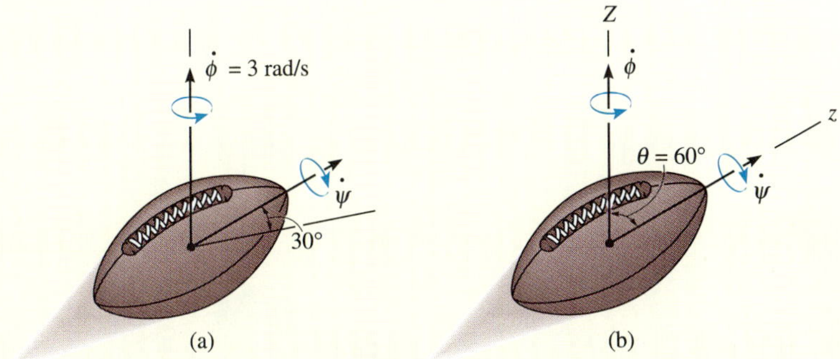

Fig. 21–24

Solution

Since the weight of the football is the only force acting, the motion is torque-free. In the conventional sense, if the z axis is established along the axis of spin and the Z axis along the precession axis, as shown in Fig. 21–24*b*, then the angle $\theta = 60°$. Applying Eq. 21–37, the spin is

$$\dot{\psi} = \frac{I - I_z}{I_z}\dot{\phi}\cos\theta = \frac{I - \frac{1}{3}I}{\frac{1}{3}I}(3)\cos 60°$$

$$= 3 \text{ rad/s} \qquad \textit{Ans.}$$

Using Eqs. 21–34, where $H_G = \dot{\phi}I$ (Eq. 21–36), we have

$$\omega_x = 0$$

$$\omega_y = \frac{H_G \sin\theta}{I} = \frac{3I\sin 60°}{I} = 2.60 \text{ rad/s}$$

$$\omega_z = \frac{H_G\cos\theta}{I_z} = \frac{3I\cos 60°}{\frac{1}{3}I} = 4.50 \text{ rad/s}$$

Thus,

$$\omega = \sqrt{(\omega_x)^2 + (\omega_y)^2 + (\omega_z)^2}$$
$$= \sqrt{(0)^2 + (2.60)^2 + (4.50)^2}$$
$$= 5.20 \text{ rad/s} \qquad \textit{Ans.}$$

Problems

21-61. Show that the angular velocity of a body, in terms of Euler angles ϕ, θ, and ψ, may be expressed as $\omega = (\dot{\phi} \sin\theta \sin\psi + \dot{\theta}\cos\psi)\mathbf{i} + (\dot{\phi}\sin\theta\cos\psi - \dot{\theta}\sin\psi)\mathbf{j} + (\dot{\phi}\cos\theta + \dot{\psi})\mathbf{k}$, where **i**, **j**, and **k** are directed along the x, y, z axes as shown in Fig. 21–15*d*.

21-62. A thin rod is initially coincident with the Z axis when it is given three rotations defined by the Euler angles $\phi = 30°$, $\theta = 45°$, and $\psi = 60°$. If these rotations are given in the order stated, determine the coordinate direction angles α, β, γ of the axis of the rod with respect to the X, Y, and Z axes. Are these directions the same for any order of the rotations? Why?

21-63. The rotor assembly on the engine of a jet airplane consists of the turbine, drive shaft, and compressor. The total mass is 700 kg, the radius of gyration about the shaft axis is $k_{AB} = 0.35$ m, and the mass center is at G. If the rotor has an angular velocity $\omega_{AB} = 1000$ rad/s, and the plane is pulling out of a vertical curve while traveling at 250 m/s, determine the components of reaction at the bearings A and B due to the gyroscopic effect.

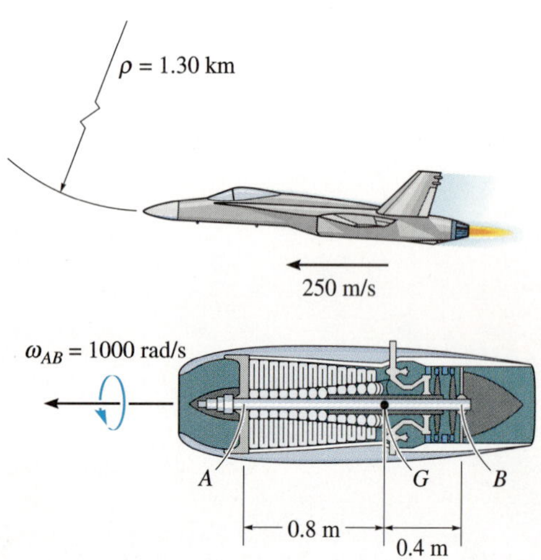

Prob. 21–63

***21-64.** The 30-lb wheel rolls without slipping. If it has a radius of gyration $k_{AB} = 1.2$ ft about its axle AB, and the vertical drive shaft is turning at 8 rad/s, determine the normal reaction the wheel exerts on the ground at C.

21-65. The 30-lb wheel rolls without slipping. If it has a radius of gyration $k_{AB} = 1.2$ ft about its axle AB, determine its angular velocity $\boldsymbol{\omega}$ so that the normal reaction at C becomes 60 lb.

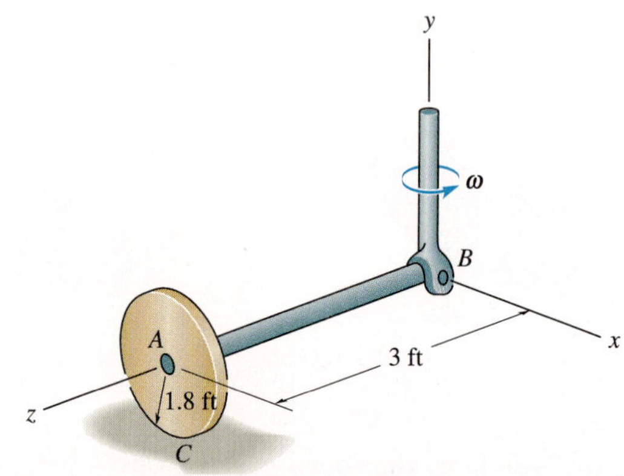

Probs. 21–64/65

21-66. The 20-kg disk is spinning about its center at $\omega_s = 20$ rad/s while the supporting axle is rotating at $\omega_y = 6$ rad/s. Determine the gyroscopic moment caused by the force reactions which the pin A exerts on the disk due to the motion.

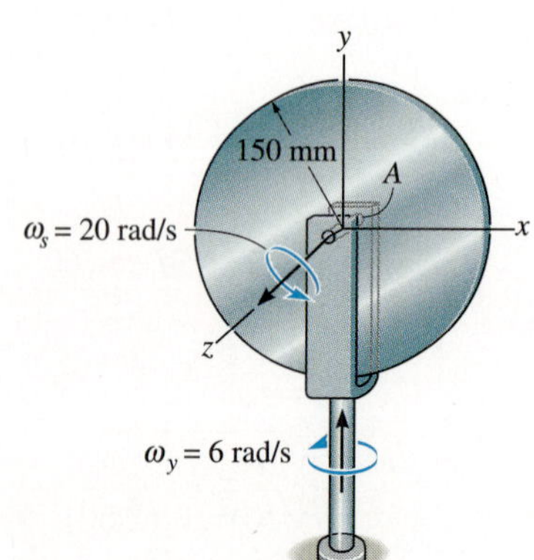

Prob. 21–66

21-67. The motor weighs 50 lb and has a radius of gyration of 0.2 ft about the z axis. The shaft of the motor is supported by bearings at A and B, and is turning at a constant rate of $\boldsymbol{\omega}_s = \{100\mathbf{k}\}$ rad/s, while the frame has an angular velocity of $\boldsymbol{\omega}_y = \{2\mathbf{j}\}$ rad/s. Determine the moment which the bearing forces at A and B exert on the shaft due to this motion.

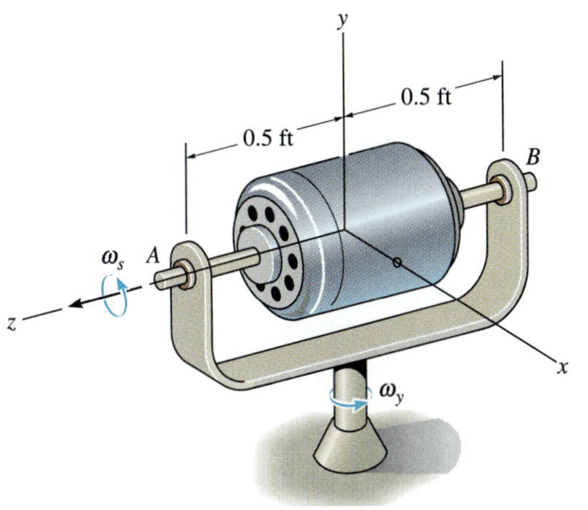

Prob. 21–67

***21-68.** The homogeneous cone has a mass of 7 kg and a vertex angle of 90°. If the cone rolls on the horizontal surface without slipping, determine the greatest precessional speed ω_p it can have before the tip A starts to rise from the surface.

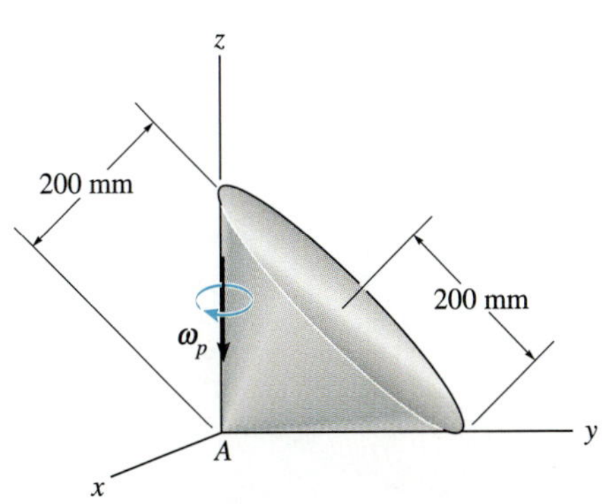

Prob. 21–68

21-69. The top has a mass of 90 g, a center of mass at G, and a radius of gyration $k = 18$ mm about its axis of symmetry. About any transverse axis acting through point O the radius of gyration is $k_t = 35$ mm. If the top is pinned at O and the precession is $\omega_p = 0.5$ rad/s, determine the spin ω_s.

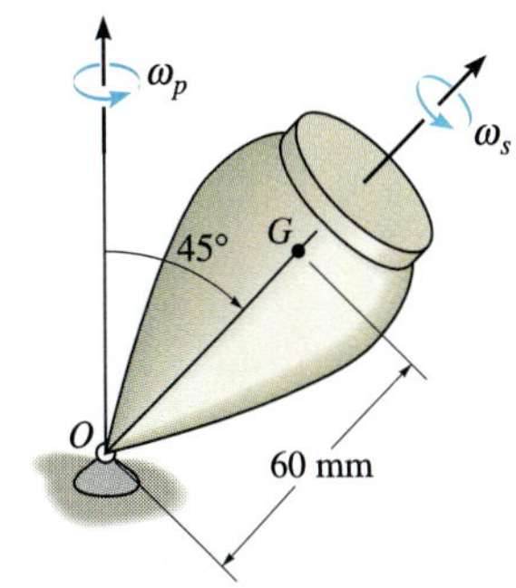

Prob. 21–69

21-70. The top consists of a thin disk that has a weight of 8 lb and a radius of 0.3 ft. The rod has a negligible mass and a length of 0.5 ft. If the top is spinning with an angular velocity $\omega_s = 300$ rad/s, determine the steady-state precessional angular velocity ω_p of the rod when $\theta = 40°$.

21-71. Solve Prob. 21-70 when $\theta = 90°$.

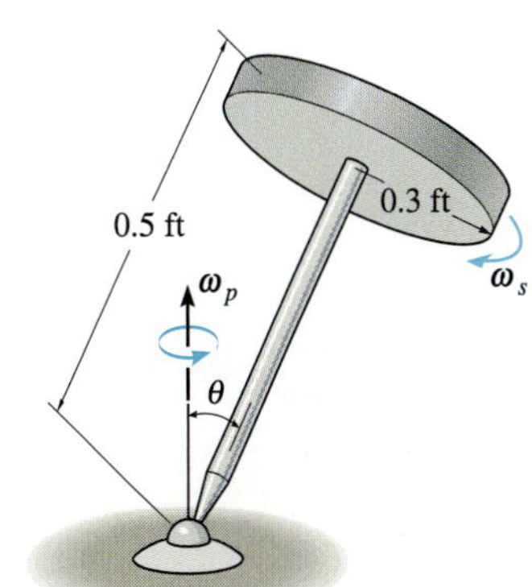

Probs. 21–70/71

*21-72. The top has a mass of 3 lb and can be considered as a solid cone. If it is observed to precess about the vertical axis at a constant rate of 5 rad/s, determine its spin.

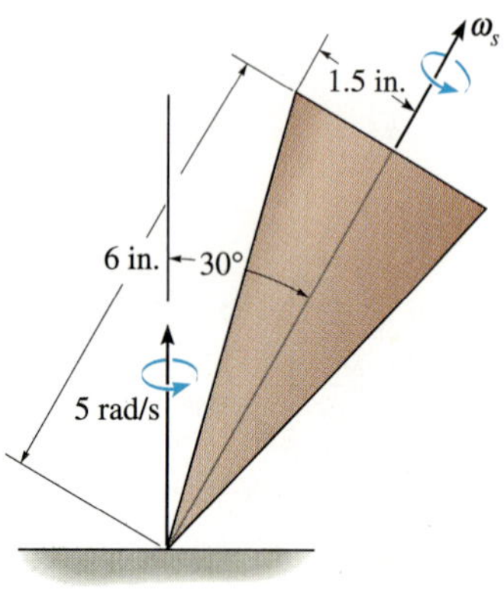

Prob. 21–72

21-73. The gyroscope consists of a uniform 450-g disk D which is attached to the axle AB of negligible mass. The supporting frame has a mass of 180 g and a center of mass at G. If the disk is rotating about the axle at ω_D = 90 rad/s, determine the constant angular velocity ω_p at which the frame precesses about the pivot point O. The frame moves in the horizontal plane.

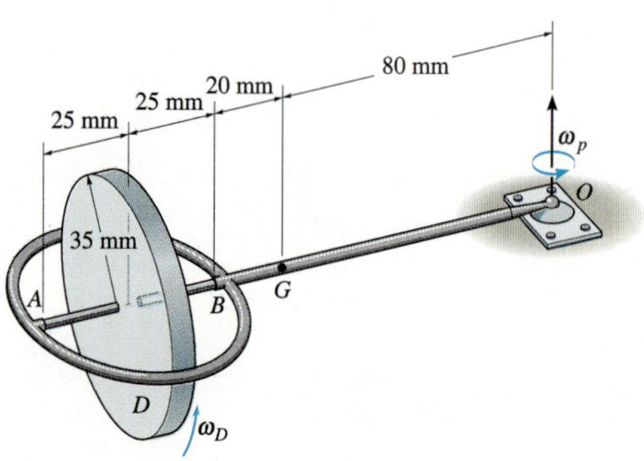

Prob. 21–73

21-74. The car is traveling at $v_C = 100$ km/h around the horizontal curve having a radius of 80 m. If each wheel has a mass of 16 kg, a radius of gyration $k_G = 300$ mm about its spinning axis, and a radius of 400 mm, determine the difference between the normal forces of the rear wheels, caused by the gyroscopic effect. The distance between the wheels is 1.30 m.

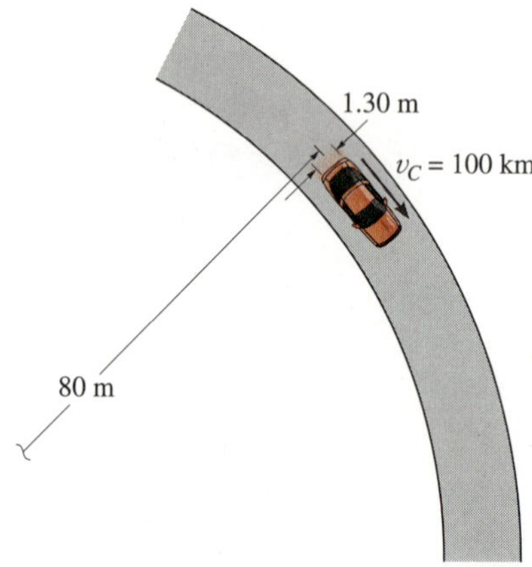

Prob. 21–74

21-75. The projectile shown is subjected to torque-free motion. The transverse and axial moments of inertia are I and I_z, respectively. If θ represents the angle between the precessional axis Z and the axis of symmetry z, and β is the angle between the angular velocity $\boldsymbol{\omega}$ and the z axis, show that β and θ are related by the equation $\tan\theta = (I/I_z)\tan\beta$.

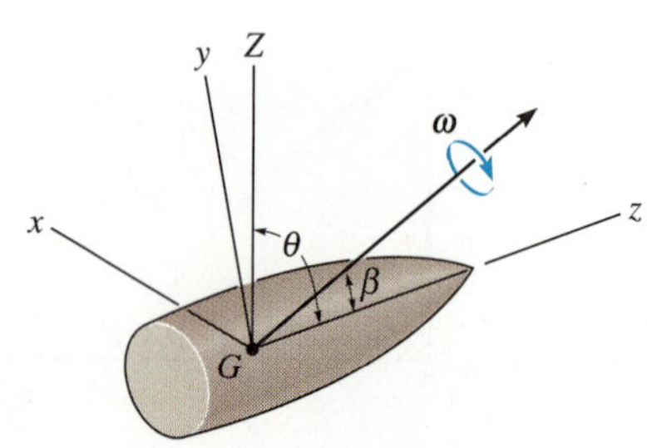

Prob. 21–75

*21-76. The radius of gyration about an axis passing through the axis of symmetry of the 2.5-Mg satellite is $k_z = 2.3$ m, and about any transverse axis passing through the center of mass G, $k_t = 3.4$ m. If the satellite has a steady-state precession of two revolutions per hour about the Z axis, determine the rate of spin about the z axis.

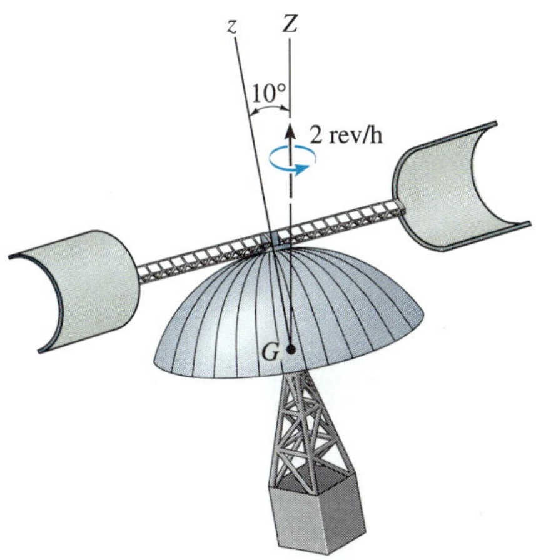

Prob. 21–76

21-77. The projectile has a mass of 0.9 kg and axial and transverse radii of gyration of $k_z = 20$ mm and $k_t = 25$ mm, respectively. If it is spinning at $\omega_s = 6$ rad/s when it leaves the barrel of a gun, determine its angular momentum. Precession occurs about the Z axis.

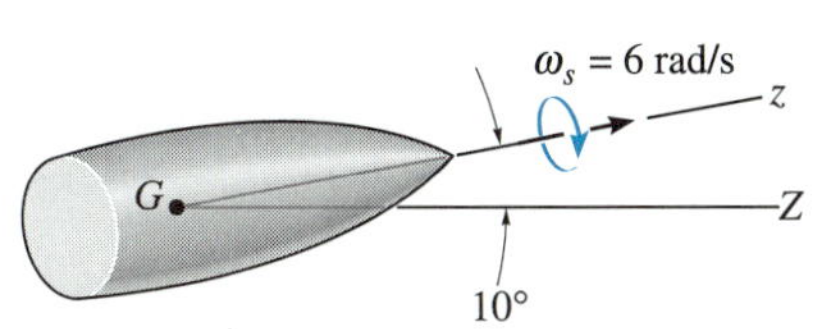

Prob. 21–77

21-78. The satellite has a mass of 1.8 Mg, and about axes passing through the mass center G the axial and transverse radii of gyration are $k_z = 0.8$ m and $k_t = 1.2$ m, respectively. If it is spinning at $\omega_s = 6$ rad/s when it is launched, determine its angular momentum. Precession occurs about the Z axis.

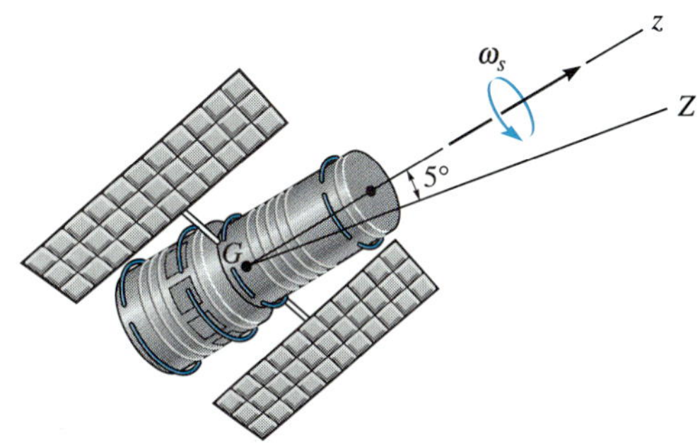

Prob. 21–78

21-79. The radius of gyration about an axis passing through the axis of symmetry of the 1.6-Mg space capsule is $k_z = 1.2$ m, and about any transverse axis passing through the center of mass G, $k_t = 1.8$ m. If the capsule has a known steady-state precession of two revolutions per hour about the Z axis, determine the rate of spin about the z axis.

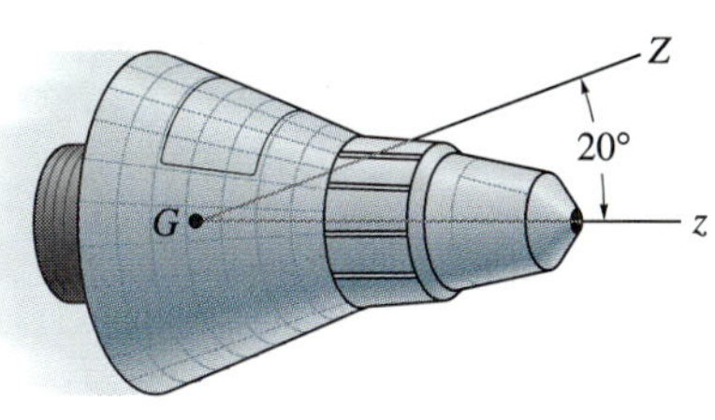

Prob. 21–79

*21-80. The rocket has a mass of 4 Mg and radii of gyration $k_z = 0.85$ m and $k_y = 2.3$ m. It is initially spinning about the z axis at $\omega_z = 0.05$ rad/s when a meteoroid M strikes it at A and creates an impulse $\mathbf{I} = \{300\mathbf{i}\}$ N·s. Determine the axis of precession after the impact.

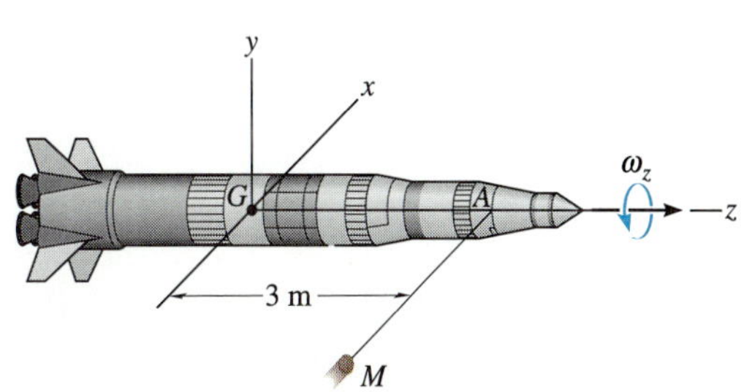

Prob. 21–80

The analysis of vibrations plays an important role in the study of the behavior of structures subjected to earthquakes.

CHAPTER 22

Vibrations

Chapter Objectives

- To discuss undamped one-degree-of-freedom vibration of a rigid body using the equation of motion and energy methods.
- To study the analysis of undamped forced vibration and viscous damped forced vibration.
- To introduce the concept of electrical circuit analogs to study vibrational motion.

* 22.1 Undamped Free Vibration

A *vibration* is the periodic motion of a body or system of connected bodies displaced from a position of equilibrium. In general, there are two types of vibration, free and forced. *Free vibration* occurs when the motion is maintained by gravitational or elastic restoring forces, such as the swinging motion of a pendulum or the vibration of an elastic rod. *Forced vibration* is caused by an external periodic or intermittent force applied to the system. Both of these types of vibration may be either damped or undamped. *Undamped* vibrations can continue indefinitely because frictional effects are neglected in the analysis. Since in reality both internal and external frictional forces are present, the motion of all vibrating bodies is actually *damped.*

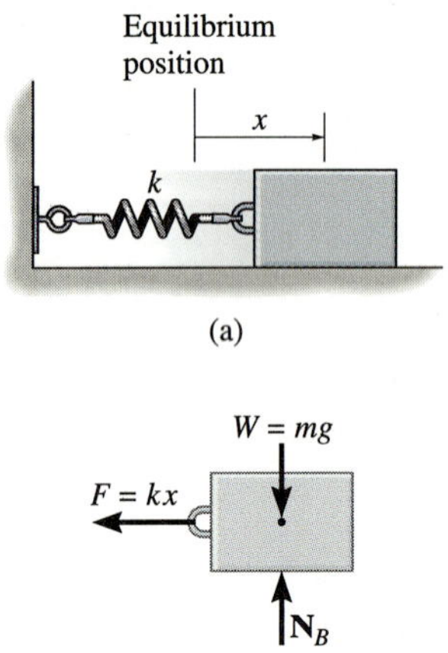

(a)
(b)
Fig. 22–1

The simplest type of vibrating motion is undamped free vibration, represented by the model shown in Fig. 22–1*a*. The block has a mass m and is attached to a spring having a stiffness k. Vibrating motion occurs when the block is released from a displaced position x so that the spring pulls on the block. The block will attain a velocity such that it will proceed to move out of equilibrium when $x = 0$, and provided the supporting surface is smooth, oscillation will continue indefinitely.

The time-dependent path of motion of the block may be determined by applying the equation of motion to the block when it is in the displaced position x. The free-body diagram is shown in Fig. 22–1*b*. The elastic restoring force $F = kx$ is always directed toward the equilibrium position, whereas the acceleration **a** is assumed to act in the direction of *positive displacement*. Noting that $a = d^2x/dt^2 = \ddot{x}$, we have

$$\overset{+}{\rightarrow}\Sigma F_x = ma_x; \qquad -kx = m\ddot{x}$$

Note that the acceleration is proportional to the block's displacement. Motion described in this manner is called *simple harmonic motion*. Rearranging the terms into a "standard form" gives

$$\ddot{x} + p^2x = 0 \tag{22–1}$$

The constant p is called the *circular frequency*, expressed in rad/s, and in this case

$$p = \sqrt{\frac{k}{m}} \tag{22–2}$$

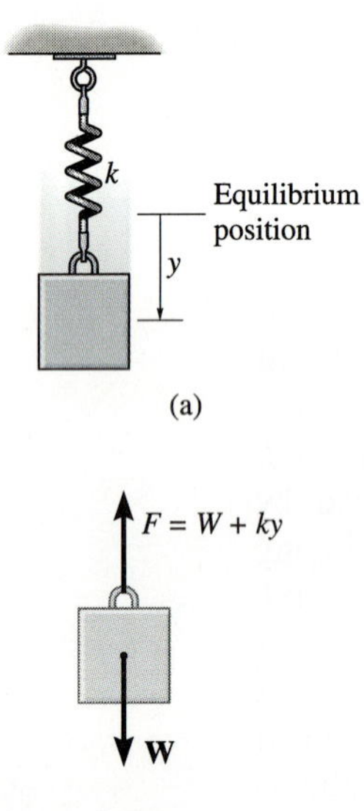

(a)
(b)
Fig. 22–2

Equation 22–1 may also be obtained by considering the block to be suspended and measuring the displacement y from the block's *equilibrium position*, Fig. 22–2*a*. When the block is in equilibrium, the spring exerts an upward force of $F = W = mg$ on the block. Hence, when the block is displaced a distance y downward from this position, the magnitude of the spring force is $F = W + ky$, Fig. 22–2*b*. Applying the equation of motion gives

$$+\downarrow\Sigma F_y = ma_y; \qquad -W - ky + W = m\ddot{y}$$

or

$$\ddot{y} + p^2y = 0$$

which is the same form as Eq. 22–1, where p is defined by Eq. 22–2.

Equation 22–1 is a homogeneous, second-order, linear, differential equation with constant coefficients. It can be shown, using the methods of differential equations, that the general solution is

$$x = A \sin pt + B \cos pt \qquad (22\text{–}3)$$

where A and B represent two constants of integration. The block's velocity and acceleration are determined by taking successive time derivatives, which yields

$$v = \dot{x} = Ap \cos pt - Bp \sin pt \qquad (22\text{–}4)$$

$$a = \ddot{x} = -Ap^2 \sin pt - Bp^2 \cos pt \qquad (22\text{–}5)$$

When Eqs. 22–3 and 22–5 are substituted into Eq. 22–1, the differential equation is indeed satisfied, showing that Eq. 22–3 is the solution to Eq. 22–1.

The integration constants A and B in Eq. 22–3 are generally determined from the initial conditions of the problem. For example, suppose that the block in Fig. 22–1*a* has been displaced a distance x_1 to the right from its equilibrium position and given an initial (positive) velocity $\mathbf{v}_1$ directed to the right. Substituting $x = x_1$ at $t = 0$ into Eq. 22–3 yields $B = x_1$. Since $v = v_1$ at $t = 0$, using Eq. 22–4 we obtain $A = v_1/p$. If these values are substituted into Eq. 22–3, the equation describing the motion becomes

$$x = \frac{v_1}{p} \sin pt + x_1 \cos pt \qquad (22\text{–}6)$$

Equation 22–3 may also be expressed in terms of simple sinusoidal motion. Let

$$A = C \cos \phi \qquad (22\text{–}7)$$

and

$$B = C \sin \phi \qquad (22\text{–}8)$$

where C and ϕ are new constants to be determined in place of A and B. Substituting into Eq. 22–3 yields

$$x = C \cos \phi \sin pt + C \sin \phi \cos pt$$

Since $\sin(\theta + \phi) = \sin \theta \cos \phi + \cos \theta \sin \phi$, then

$$x = C \sin(pt + \phi) \qquad (22\text{–}9)$$

If this equation is plotted on an x-versus-pt axis, the graph shown in Fig. 22–3 is obtained. The maximum displacement of the block from its

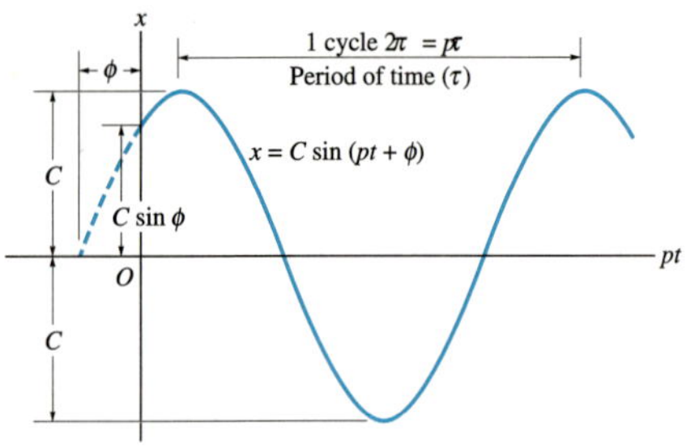

Fig. 22–3

equilibrium position is defined as the *amplitude* of vibration. From either the figure or Eq. 22–9 the amplitude is C. The angle ϕ is called the *phase angle* since it represents the amount by which the curve is displaced from the origin when $t = 0$. The constants C and ϕ are related to A and B by Eqs. 22–7 and 22–8. Squaring and adding these two equations, the amplitude becomes

$$C = \sqrt{A^2 + B^2} \tag{22–10}$$

If Eq. 22–8 is divided by Eq. 22–7, the phase angle is

$$\phi = \tan^{-1}\frac{B}{A} \tag{22–11}$$

Note that the sine curve, Eq. 22–9, completes one *cycle* in time $t = \tau$ (tau) when $p\tau = 2\pi$, or

$$\boxed{\tau = \frac{2\pi}{p}} \tag{22–12}$$

This length of time is called a *period,* Fig. 22–3. Using Eq. 22–2, the period may also be represented as

$$\tau = 2\pi\sqrt{\frac{m}{k}} \tag{22–13}$$

The *frequency* f is defined as the number of cycles completed per unit of time, which is the reciprocal of the period:

$$\boxed{f = \frac{1}{\tau} = \frac{p}{2\pi}} \tag{22–14}$$

or

$$f = \frac{1}{2\pi}\sqrt{\frac{k}{m}} \tag{22–15}$$

The frequency is expressed in cycles/s. This ratio of units is called a *hertz* (Hz), where 1 Hz = 1 cycle/s = 2π rad/s.

When a body or system of connected bodies is given an initial displacement from its equilibrium position and released, it will vibrate with a definite frequency known as the *natural frequency.* Provided the body has a single degree of freedom, that is, it requires only one coordinate to specify completely the position of the system at any time, then the vibrating motion of the body will have the same characteristics as the simple harmonic motion of the block and spring just presented. Consequently, the body's motion is described by a differential equation of the same "standard form" as Eq. 22–1, i.e.,

$$\boxed{\ddot{x} + p^2x = 0} \tag{22–16}$$

Hence, if the circular frequency p of the body is known, the period of vibration τ, natural frequency f, and other vibrating characteristics of the body can be established using Eqs. 22–3 through 22–15.

Important Points

- Free vibration occurs when the motion is maintained by gravitational or elastic restoring forces.
- The amplitude is the maximum displacement of the body.
- The period is the time required to complete one cycle.
- The frequency is the number of cycles completed per unit of time, where 1 Hz = 1 cycle/s.
- Only one position-coordinate system is needed to describe the location of a one-degree-of-freedom system.

Procedure for Analysis

As in the case of the block and spring, the circular frequency p of a rigid body or system of connected rigid bodies having a single degree of freedom can be determined using the following procedure:

Free-Body Diagram

- Draw the free-body diagram of the body when the body is displaced by a *small amount* from its equilibrium position.
- Locate the body with respect to its equilibrium position by using an appropriate *inertial coordinate* q. The acceleration of the body's mass center $\mathbf{a}_G$ or the body's angular acceleration $\boldsymbol{\alpha}$ should have a sense which is in the *positive direction* of the position coordinate.
- If the rotational equation of motion $\Sigma M_P = \Sigma(\mathcal{M}_k)_P$ is to be used, then it may be beneficial to also draw the kinetic diagram since it graphically accounts for the components $m(\mathbf{a}_G)_x$, $m(\mathbf{a}_G)_y$, and $I_G\boldsymbol{\alpha}$, and thereby makes it convenient for visualizing the terms needed in the moment sum $\Sigma(\mathcal{M}_k)_P$.

Equation of Motion

- Apply the equation of motion to relate the elastic or gravitational *restoring* forces and couple moments acting on the body to the body's accelerated motion.

Kinematics

- Using kinematics, express the body's accelerated motion in terms of the second time derivative of the position coordinate, $\ddot{q}$.
- Substitute the result into the equation of motion and determine p by rearranging the terms so that the resulting equation is of the "standard form," $\ddot{q} + p^2 q = 0$.

EXAMPLE 22–1

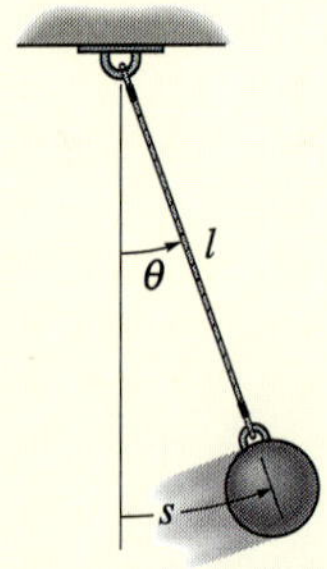

(a)

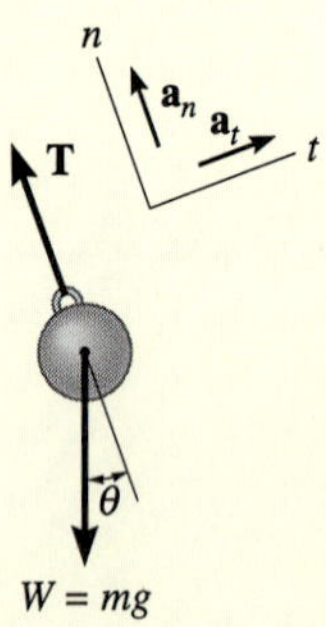

(b)

Fig. 22–4

Determine the period of vibration for the simple pendulum shown in Fig. 22–4*a*. The bob has a mass m and is attached to a cord of length l. Neglect the size of the bob.

Solution

Free-Body Diagram. Motion of the system will be related to the position coordinate $(q =)\ \theta$, Fig. 22–4*b*. When the bob is displaced by an angle θ, the *restoring force* acting on the bob is created by the weight component $mg \sin \theta$. Furthermore, $\mathbf{a}_t$ acts in the direction of *increasing* s (or θ).

Equation of Motion. Applying the equation of motion in the *tangential direction,* since it involves the restoring force, yields

$$+\nearrow \Sigma F_t = ma_t; \qquad -mg \sin \theta = ma_t \qquad (1)$$

Kinematics. $a_t = d^2s/dt^2 = \ddot{s}$. Furthermore, s may be related to θ by the equation $s = l\theta$, so that $a_t = l\ddot{\theta}$. Hence, Eq. 1 reduces to

$$\ddot{\theta} + \frac{g}{l} \sin \theta = 0 \qquad (2)$$

The solution of this equation involves the use of an elliptic integral. For *small displacements,* however, $\sin \theta \approx \theta$, in which case

$$\ddot{\theta} + \frac{g}{l}\theta = 0 \qquad (3)$$

Comparing this equation with Eq. 22–16 ($\ddot{x} + p^2x = 0$), it is seen that $p = \sqrt{g/l}$. From Eq. 22–12, the period of time required for the bob to make one complete swing is therefore

$$\tau = \frac{2\pi}{p} = 2\pi\sqrt{\frac{l}{g}} \qquad \textit{Ans.}$$

This interesting result, originally discovered by Galileo Galilei through experiment, indicates that the period depends only on the length of the cord and not on the mass of the pendulum bob or the angle θ.

The solution of Eq. 3 is given by Eq. 22–3, where $p = \sqrt{g/l}$ and θ is substituted for x. Like the block and spring, the constants A and B in this problem may be determined if, for example, one knows the displacement and velocity of the bob at a given instant.

EXAMPLE 22-2

The 10-kg rectangular plate shown in Fig. 22–5*a* is suspended at its center from a rod having a torsional stiffness $k = 1.5\ \text{N}\cdot\text{m/rad}$. Determine the natural period of vibration of the plate when it is given a small angular displacement θ in the plane of the plate.

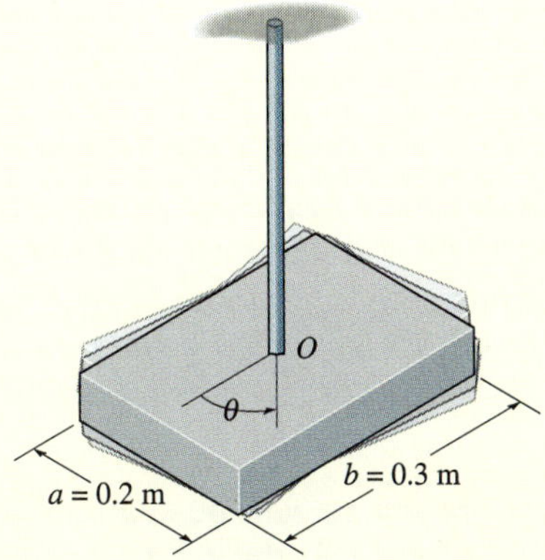

(a)

Solution

Free-Body Diagram. Fig. 22–5*b*. Since the plate is displaced in its own plane, the torsional *restoring* moment created by the rod is $M = k\theta$. This moment acts in the direction opposite to the angular displacement θ. The angular acceleration $\ddot{\theta}$ acts in the direction of positive θ.

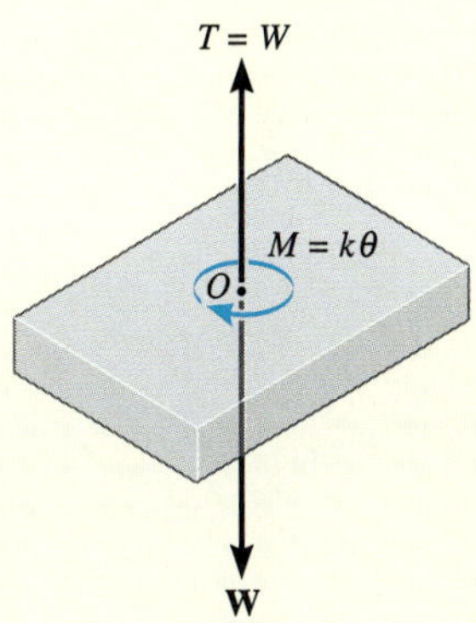

(b)

Fig. 22–5

Equation of Motion

$$\Sigma M_O = I_O\alpha; \qquad -k\theta = I_O\ddot{\theta}$$

or

$$\ddot{\theta} + \frac{k}{I_O}\theta = 0$$

Since this equation is in the "standard form," the circular frequency is $p = \sqrt{k/I_O}$.

From the table on the inside back cover, the moment of inertia of the plate about an axis coincident with the rod is $I_O = \frac{1}{12}m(a^2 + b^2)$. Hence,

$$I_O = \frac{1}{12}(10\ \text{kg})[(0.2\ \text{m})^2 + (0.3\ \text{m})^2] = 0.108\ \text{kg}\cdot\text{m}^2$$

The natural period of vibration is therefore,

$$\tau = \frac{2\pi}{p} = 2\pi\sqrt{\frac{I_O}{k}} = 2\pi\sqrt{\frac{0.108}{1.5}} = 1.69\ \text{s} \qquad \textit{Ans.}$$

EXAMPLE 22–3

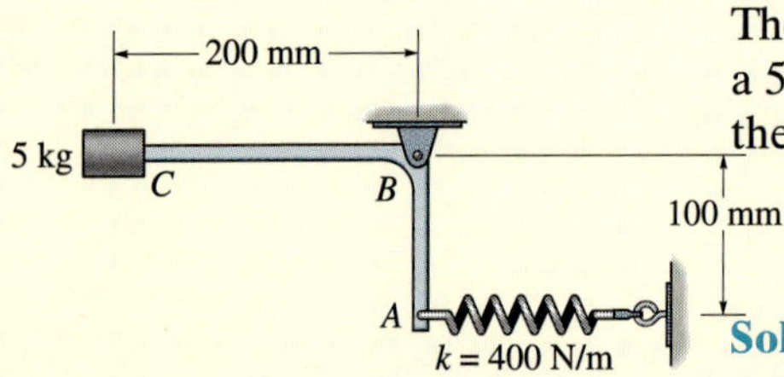

(a)

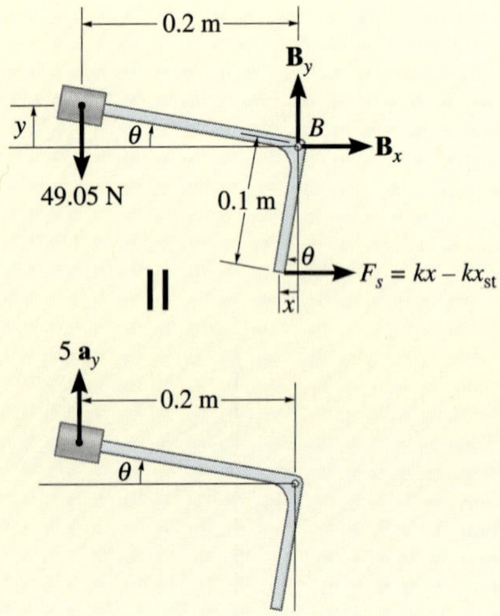

(b)

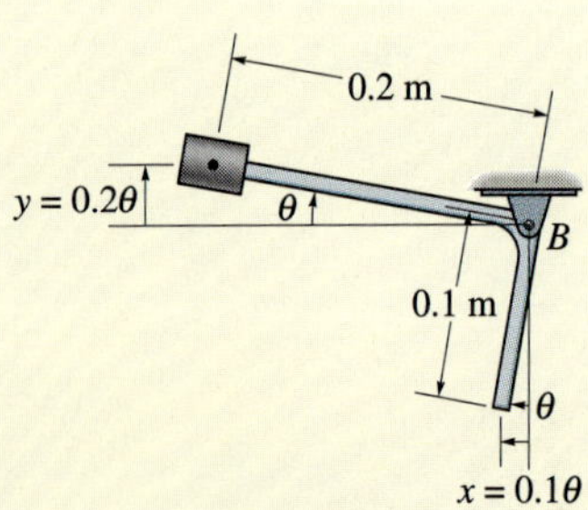

(c)

Fig. 22–6

The bent rod shown in Fig. 22–6*a* has a negligible mass and supports a 5-kg collar at its end. Determine the natural period of vibration for the system.

Solution

Free-Body and Kinetic Diagrams. Fig. 22–6*b*. Here the rod is displaced by a small amount θ from the equilibrium position. Since the spring is subjected to an initial compression of x_{st} for equilibrium, then when the displacement $x > x_{st}$ the spring exerts a force of $F_s = kx - kx_{st}$ on the rod. To obtain the "standard form," Eq. 22–16, $5\mathbf{a}_y$ acts *upward,* which is in accordance with positive θ displacement.

Equation of Motion. Moments will be summed about point B to eliminate the unknown reaction at this point. Since θ is small,

$$\circlearrowleft +\Sigma M_B = \Sigma(\mathcal{M}_k)_B;$$

$$kx(0.1\text{ m}) - kx_{st}(0.1\text{ m}) + 49.05\text{ N}(0.2\text{ m}) = -(5\text{ kg})a_y(0.2\text{ m})$$

The second term on the left side, $-kx_{st}(0.1\text{ m})$, represents the moment created by the spring force which is necessary to hold the collar in *equilibrium,* i.e., at $x = 0$. Since this moment is equal and opposite to the moment 49.05(0.2) created by the weight of the collar, these two terms cancel in the above equation, so that

$$kx(0.1) = -5a_y(0.2) \qquad (1)$$

Kinematics. The positions of the spring and the collar may be related to the angle θ, Fig. 22–6*c*. Since θ is small, $x = (0.1\text{ m})\theta$ and $y = (0.2\text{ m})\theta$. Therefore, $a_y = \ddot{y} = 0.2\ddot{\theta}$. Substituting into Eq. 1 yields

$$400(0.1\theta)0.1 = -5(0.2\ddot{\theta})0.2$$

Rewriting this equation in the "standard form" gives

$$\ddot{\theta} + 20\theta = 0$$

Compared with $\ddot{x} + p^2x = 0$ (Eq. 22–16), we have

$$p^2 = 20 \qquad p = 4.47\text{ rad/s}$$

The natural period of vibration is therefore

$$\tau = \frac{2\pi}{p} = \frac{2\pi}{4.47} = 1.40\text{ s} \qquad \textit{Ans.}$$

EXAMPLE 22–4

A 10-lb block is suspended from a cord that passes over a 15-lb disk, as shown in Fig. 22–7*a*. The spring has a stiffness $k = 200$ lb/ft. Determine the natural period of vibration for the system.

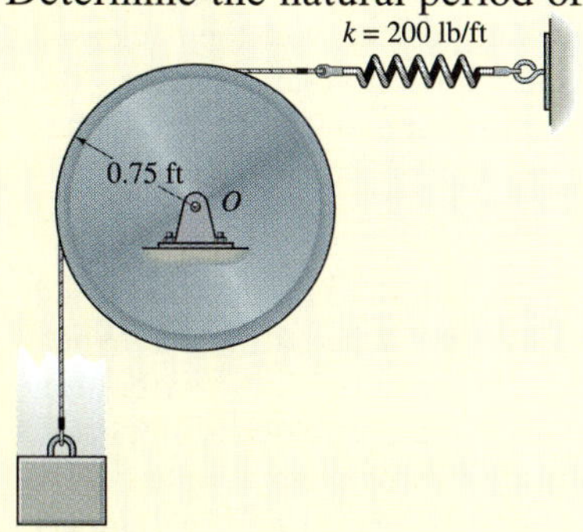

(a)

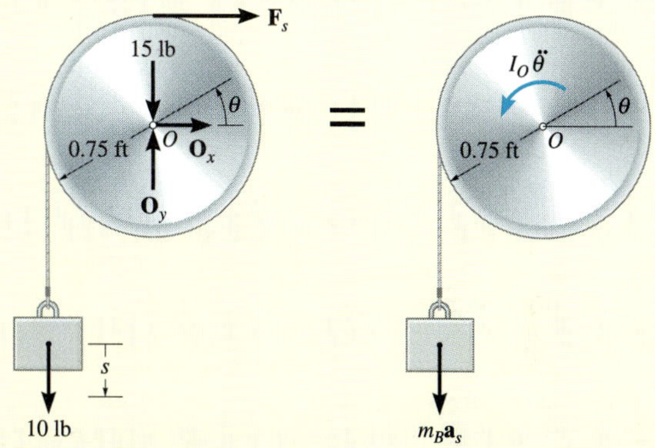

(b)

Solution

Free-Body and Kinetic Diagrams. Fig. 22–7*b*. The *system* consists of the disk, which undergoes a rotation defined by the angle θ, and the block, which translates by an amount s. The vector $I_O\ddot{\theta}$ acts in the direction of *positive* θ, and consequently $m_B\mathbf{a}_s$ acts downward in the direction of *positive s*.

Equation of Motion. Summing moments about point O to eliminate the reactions $\mathbf{O}_x$ and $\mathbf{O}_y$, realizing that $I_O = \frac{1}{2}mr^2$, yields

$$\circlearrowright+\Sigma M_O = \Sigma(\mathcal{M}_k)_O;$$

$$10\text{ lb}(0.75\text{ ft}) - F_s(0.75\text{ ft})$$
$$= \frac{1}{2}\left(\frac{15\text{ lb}}{32.2\text{ ft/s}^2}\right)(0.75\text{ ft})^2\ddot{\theta} + \left(\frac{10\text{ lb}}{32.2\text{ ft/s}^2}\right)a_s(0.75\text{ ft}) \quad (1)$$

Kinematics. As shown on the kinematic diagram in Fig. 22–7*c*, a small positive displacement θ of the disk causes the block to lower by an amount $s = 0.75\theta$; hence, $a_s = \ddot{s} = 0.75\ddot{\theta}$. When $\theta = 0°$, the spring force required for *equilibrium* of the disk is 10 lb, acting to the right. For position θ, the spring force is $F_s = (200\text{ lb/ft})(0.75\theta\text{ ft}) + 10\text{ lb}$. Substituting these results into Eq. 1 and simplifying yields

$$\ddot{\theta} + 368\theta = 0$$

Hence,

$$p^2 = 368 \qquad p = 19.2\text{ rad/s}$$

Therefore, the natural period of vibration is

$$\tau = \frac{2\pi}{p} = \frac{2\pi}{19.2} = 0.328\text{ s} \qquad \textit{Ans.}$$

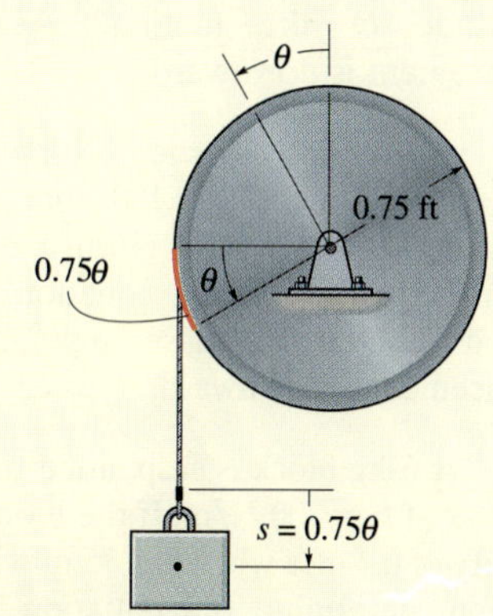

(c)

Fig. 22–7

Problems

22-1. A spring is stretched 175 mm by an 8-kg block. If the block is displaced 100 mm downward from its equilibrium position and given a downward velocity of 1.50 m/s, determine the differential equation which describes the motion. Assume that positive displacement is downward. Also, determine the position of the block when $t = 0.22$ s.

22-2. When a 2-kg block is suspended from a spring, the spring is stretched a distance of 40 mm. Determine the natural frequency and the period of vibration for a 0.5-kg block attached to the same spring.

22-3. A block having a weight of 8 lb is suspended from a spring having a stiffness $k = 40$ lb/ft. If the block is pushed $y = 0.2$ ft upward from its equilibrium position and then released from rest, determine the equation which describes the motion. What are the amplitude and the natural frequency of the vibration? Assume that positive displacement is downward.

***22-4.** A spring has a stiffness of 800 N/m. If a 2-kg block is attached to the spring, pushed 50 mm above its equilibrium position, and released from rest, determine the equation that describes the block's motion. Assume that positive displacement is downward.

22-5. A 2-kg block is suspended from a spring having a stiffness of 800 N/m. If the block is given an upward velocity of 2 m/s when it is displaced downward a distance of 150 mm from its equilibrium position, determine the equation which describes the motion. What is the amplitude of the motion? Assume that positive displacement is downward.

22-6. A spring is stretched 200 mm by a 15-kg block. If the block is displaced 100 mm downward from its equilibrium position and given a downward velocity of 0.75 m/s, determine the equation which describes the motion. What is the phase angle? Assume that positive displacement is downward.

22-7. A 6-kg block is suspended from a spring having a stiffness of $k = 200$ N/m. If the block is given an upward velocity of 0.4 m/s when it is 75 mm above its equilibrium position, determine the equation which describes the motion and the maximum upward displacement of the block measured from the equilibrium position. Assume that positive displacement is downward.

***22-8.** A 3-kg block is suspended from a spring having a stiffness of $k = 200$ N/m. If the block is pushed 50 mm upward from its equilibrium position and then released from rest, determine the equation that describes the motion. What are the amplitude and the natural frequency of the vibration? Assume that positive displacement is downward.

22-9. A platform, having an unknown mass, is supported by *four* springs, each having the same stiffness k. When nothing is on the platform, the period of vertical vibration is measured as 2.35 s; whereas if a 3-kg block is supported on the platform, the period of vertical vibration is 5.23 s. Determine the mass of a block placed on the (empty) platform which causes the platform to vibrate vertically with a period of 5.62 s. What is the stiffness k of each of the springs?

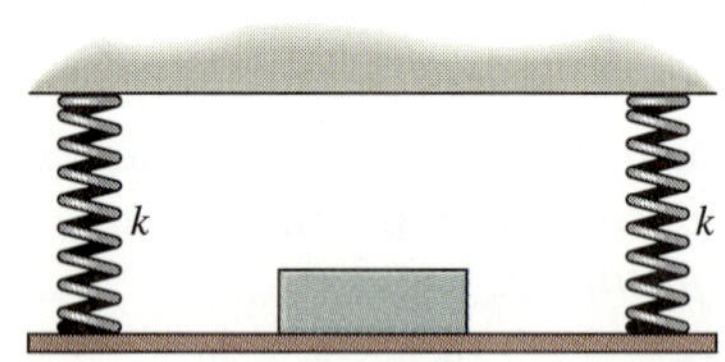

Prob. 22–9

22-10. A pendulum has a 0.4-m-long cord and is given a tangential velocity of 0.2 m/s toward the vertical from a position $\theta = 0.3$ rad. Determine the equation which describes the angular motion.

22-11. Determine to the nearest degree the maximum angular displacement of the bob in Prob. 22-10 if it is initially displaced $\theta = 0.2$ rad from the vertical and given a tangential velocity of 0.4 m/s away from the vertical.

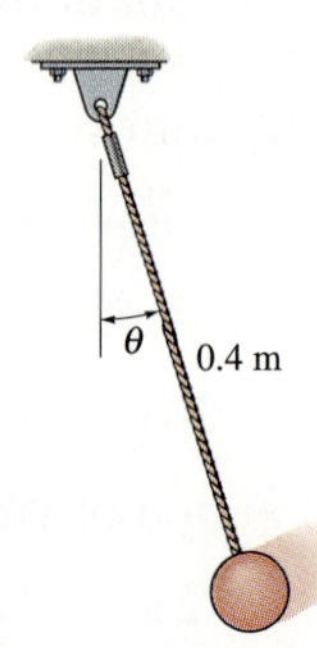

Probs. 22–10/11

*22-12. The pendulum of the Charpy impact machine is used to test the energy absorption of materials during impact. (See Example 19–5.) The pendulum weighs 46.7 lb, and its center of gravity is located at G. Through experiment, it is found that the pendulum will undergo 25 oscillations (back and forth) in 17 seconds. Determine the moment of inertia with respect to the pin at O.

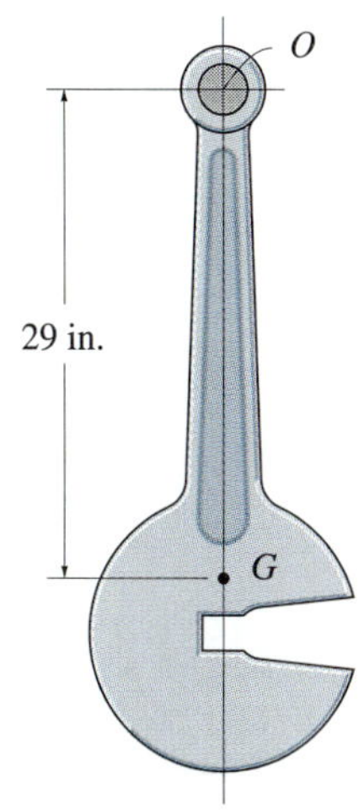

Prob. 22–12

22-13. The body of arbitrary shape has a mass m, mass center at G, and a radius of gyration about G of k_G. If it is displaced a slight amount θ from its equilibrium position and released, determine the natural period of vibration.

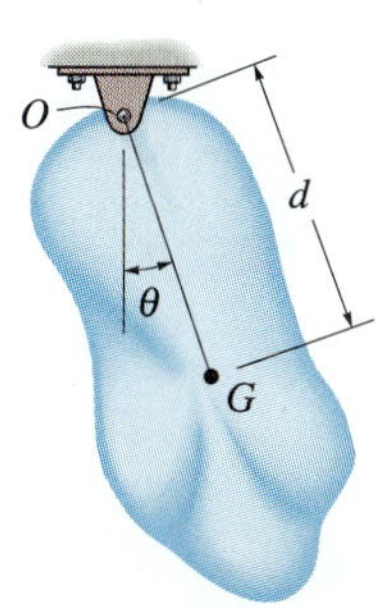

Prob. 22–13

22-14. The thin hoop of mass m is supported by a knife-edge. Determine the natural period of vibration for small amplitudes of swing.

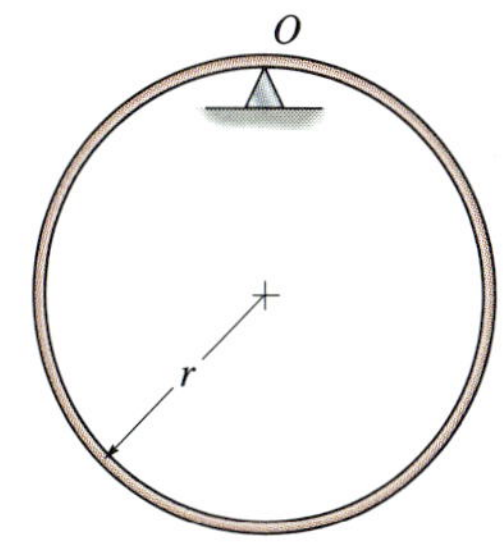

Prob. 22–14

22-15. The circular disk has a mass m and is pinned at O. Determine the natural period of vibration if it is displaced a small amount and released.

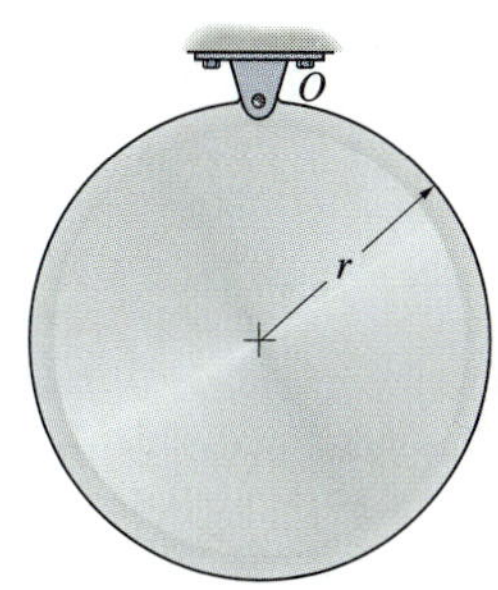

Prob. 22–15

*22-16. The square plate has a mass m and is suspended at its corner by the pin O. Determine the natural period of vibration if it is displaced a small amount and released.

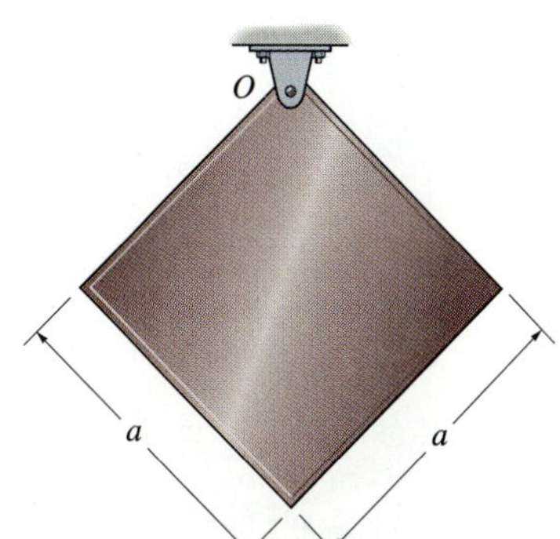

Prob. 22–16

22-17. The slender rod has a mass of 0.2 kg and is supported at O by a pin and at its end A by two springs, each having a stiffness $k = 4\ \text{N/m}$. The period of vibration of the rod can be set by fixing the 0.5-kg collar C to the rod at an appropriate location along its length. If the springs are originally unstretched when the rod is vertical, determine the position y of the collar so that the natural period of vibration becomes $\tau = 1$ s. Neglect the size of the collar.

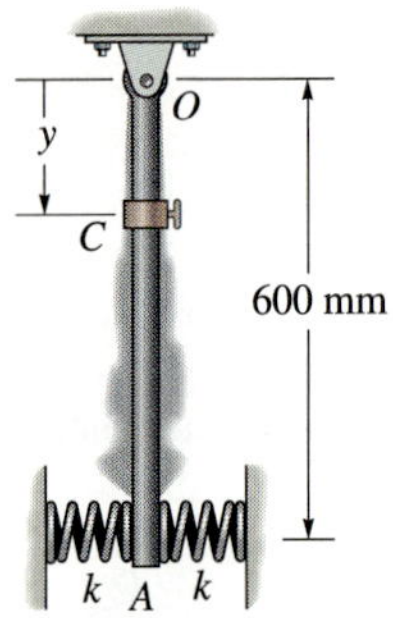

Prob. 22–17

22-18. Determine the *torsional stiffness* k of rod AB if the 4-kg thin rectangular plate has a natural period of vibration of $\tau = 0.3$ s as it oscillates around the axis of the rod. *Hint:* The torsional stiffness is defined from $M = k\theta$ and is measured in N·m/rad.

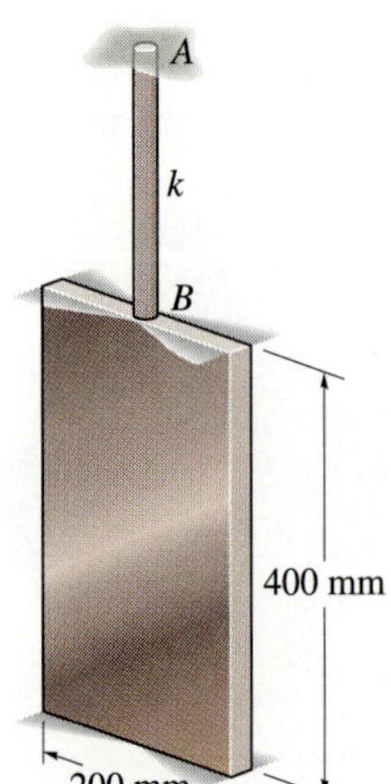

Prob. 22–18

22-19. If the lower end of the 30-kg slender rod is displaced a small amount and released from rest, determine the natural frequency of vibration. Each spring has a stiffness of $k = 500$ N/m and is unstretched when the rod is hanging vertically.

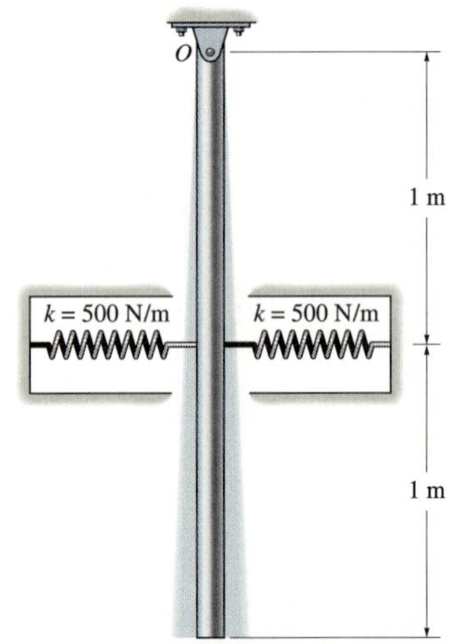

Prob. 22–19

***22-20.** The disk, having a weight of 15 lb, is pinned at its center O and supports the block A that has a weight of 3 lb. If the belt which passes over the disk is not allowed to slip at its contacting surface, determine the natural period of vibration of the system.

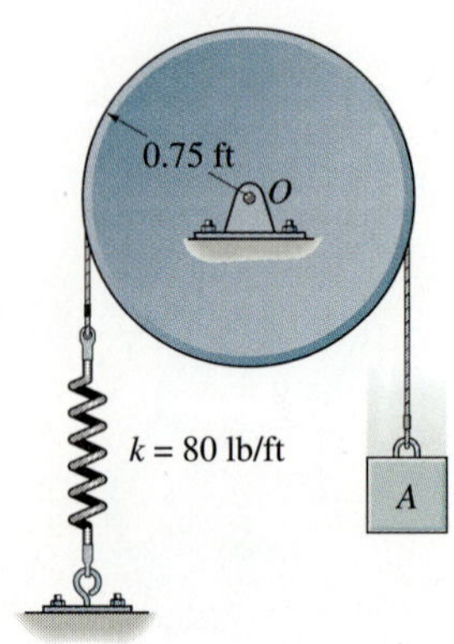

Prob. 22–20

22-21. The 6-lb weight is attached to the rods of negligible mass. Determine the natural frequency of vibration of the weight when it is displaced slightly from the equilibrium position and released.

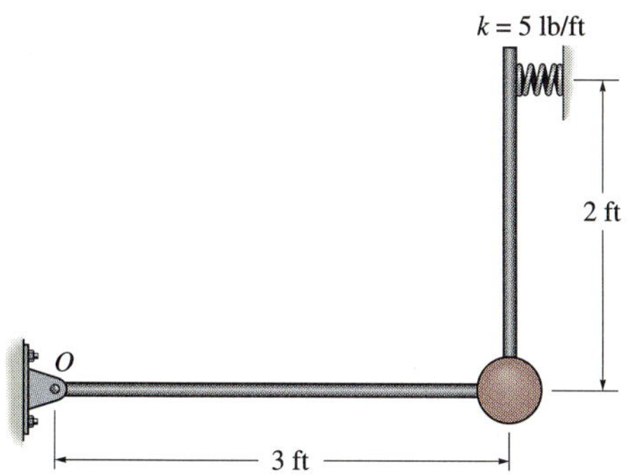

Prob. 22–21

22-22. The bell has a mass of 375 kg, a center of mass at G, and a radius of gyration about point D of $k_D = 0.4$ m. The tongue consists of a slender rod attached to the inside of the bell at C. If an 8-kg mass is attached to the end of the rod, determine the length l of the rod so that the bell will "ring silent," i.e., so that the natural period of vibration of the tongue is the same as that of the bell. For the calculation, neglect the small distance between C and D and neglect the mass of the rod.

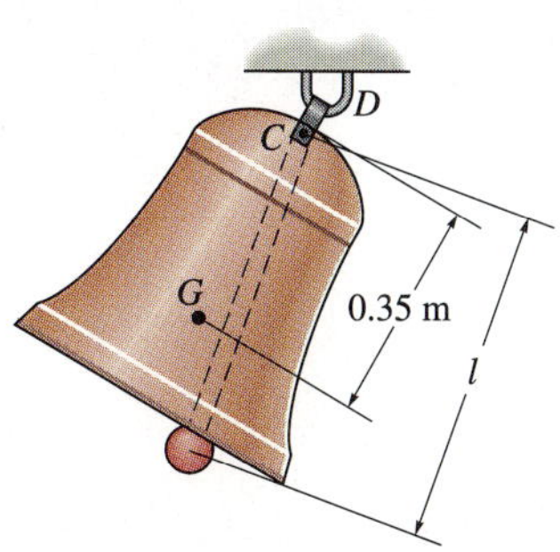

Prob. 22–22

22-23. The block has a mass m and is supported by a rigid bar of negligible mass. If the spring has a stiffness k, determine the natural period of vibration for the block.

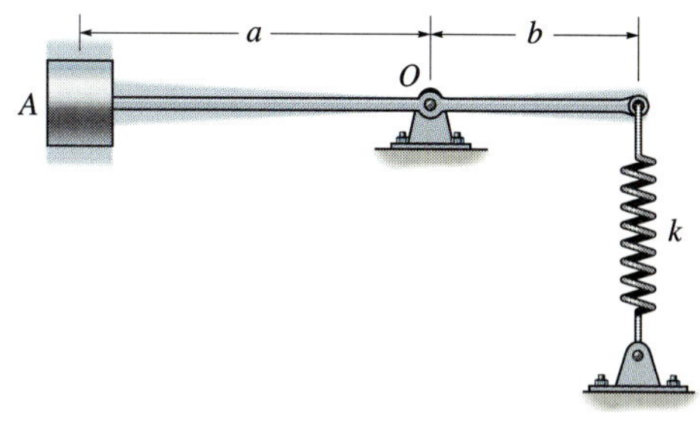

Prob. 22–23

***22-24.** The bar has a length l and mass m. It is supported at its ends by rollers of negligible mass. If it is given a small displacement and released, determine the natural frequency of vibration.

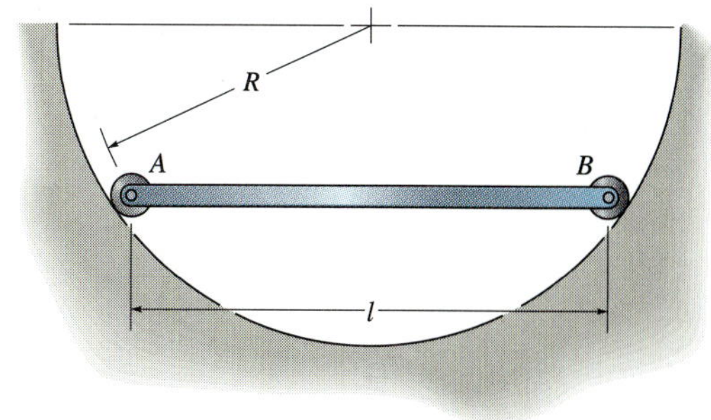

Prob. 22–24

22-25. Determine the natural frequency for small oscillations of the 10-lb sphere when the rod is displaced a slight distance and released. Neglect the size of the sphere and the mass of the rod. The spring has an unstretched length of 1 ft.

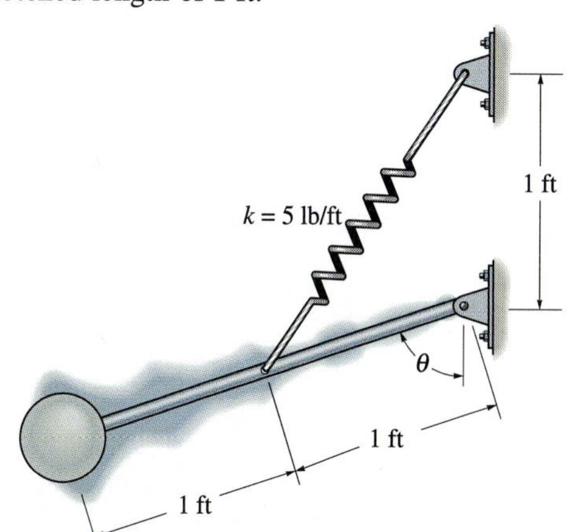

Prob. 22–25

★22.2 Energy Methods

The simple harmonic motion of a body, discussed in the previous section, is due only to gravitational and elastic restoring forces acting on the body. Since these types of forces are *conservative,* it is also possible to use the conservation of energy equation to obtain the body's natural frequency or period of vibration. To show how to do this, consider again the block and spring in Fig. 22–8. When the block is displaced an arbitrary amount x from the equilibrium position, the kinetic energy is $T = \frac{1}{2}mv^2 = \frac{1}{2}m\dot{x}^2$ and the potential energy is $V = \frac{1}{2}kx^2$. By the conservation of energy equation, Eq. 14–21, it is necessary that

$$T + V = \text{constant}$$

$$\frac{1}{2}m\dot{x}^2 + \frac{1}{2}kx^2 = \text{constant} \tag{22–17}$$

The differential equation describing the *accelerated motion* of the block can be obtained by *differentiating* this equation with respect to time; i.e.,

$$m\dot{x}\ddot{x} + kx\dot{x} = 0$$

$$\dot{x}(m\ddot{x} + kx) = 0$$

Since the velocity $\dot{x}$ is not *always* zero in a vibrating system,

$$\ddot{x} + p^2x = 0 \qquad p = \sqrt{k/m}$$

which is the same as Eq. 22–1.

If the energy equation is written for a *system of connected bodies,* the natural frequency or the equation of motion can also be determined by time differentiation. Here it is *not necessary* to dismember the system to account for reactive and connective forces which do no work.

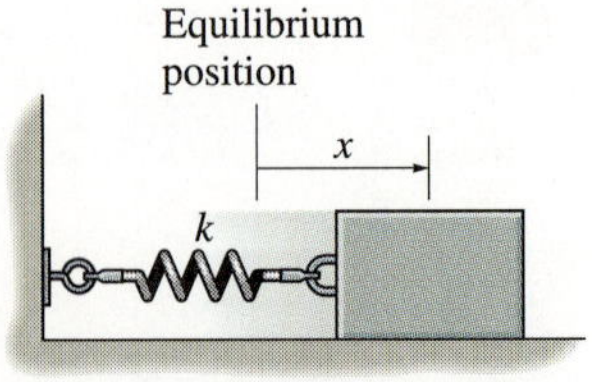

Fig. 22–8

The suspension of a railroad car consists of a set of springs which are mounted between the frame of the car and the wheel truck. This will give the car a natural frequency of vibration which can be determined.

Procedure for Analysis

The circular frequency p of a body or system of connected bodies can be determined by applying the conservation of energy equation using the following procedure.

Energy Equation

- Draw the body when it is displaced by a *small amount* from its equilibrium position and define the location of the body from its equilibrium position by an appropriate position coordinate q.
- Formulate the equation of energy for the body, $T + V =$ constant, in terms of the position coordinate.
- In general, the kinetic energy must account for both the body's translational and rotational motion, $T = \frac{1}{2}mv_G^2 + \frac{1}{2}I_G\omega^2$, Eq. 18–2.
- The potential energy is the sum of the gravitational and elastic potential energies of the body, $V = V_g + V_e$, Eq. 18–16. In particular, V_g should be measured from a datum for which $q = 0$ (equilibrium position).

Time Derivative

- Take the time derivative of the energy equation using the chain rule of calculus and factor out the common terms. The resultant differential equation represents the equation of motion for the system. The value of p is obtained after rearranging the terms in the "standard form," $\ddot{q} + p^2q = 0$.

EXAMPLE 22–5

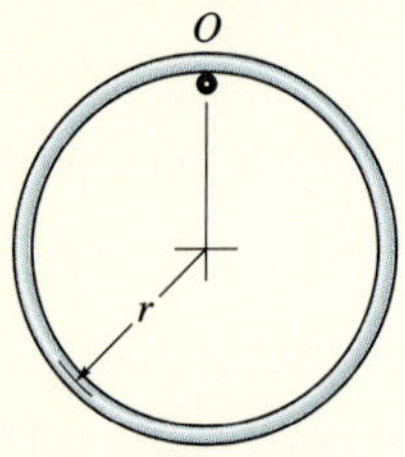

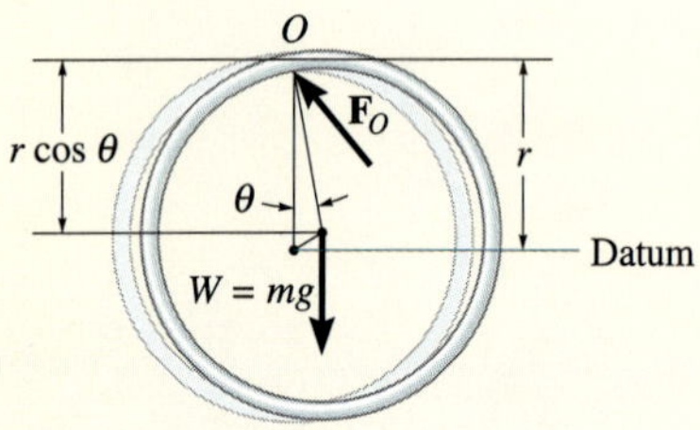

Fig. 22–9

The thin hoop shown in Fig. 22–9*a* is supported by a peg at *O*. Determine the period of oscillation for small amplitudes of swing. The hoop has a mass *m*.

Solution

Energy Equation. A diagram of the hoop when it is displaced a small amount $(q =)\theta$ from the equilibrium position is shown in Fig. 22–9*b*. Using the table on the inside back cover and the parallel-axis theorem to determine I_O, we can express the kinetic energy as

$$T = \tfrac{1}{2}I_O\omega^2 = \tfrac{1}{2}[mr^2 + mr^2]\dot{\theta}^2 = mr^2\dot{\theta}^2$$

If a horizontal datum is placed through the center of gravity of the hoop when $\theta = 0$, then the center of gravity moves upward $r(1 - \cos\theta)$ in the displaced position. For *small angles*, $\cos\theta$ may be replaced by the first two terms of its power series expansion, $\cos\theta = 1 - \theta^2/2 + \cdots$. Therefore, the potential energy is

$$V = mgr\left[1 - \left(1 - \frac{\theta^2}{2}\right)\right] = mgr\frac{\theta^2}{2}$$

The total energy in the system is

$$T + V = mr^2\dot{\theta}^2 + mgr\frac{\theta^2}{2}$$

Time Derivative

$$mr^2 2\dot{\theta}\ddot{\theta} + mgr\theta\dot{\theta} = 0$$
$$mr\dot{\theta}(2r\ddot{\theta} + g\theta) = 0$$

Since $\dot{\theta}$ is not always equal to zero, from the terms in parentheses,

$$\ddot{\theta} + \frac{g}{2r}\theta = 0$$

Hence,

$$p = \sqrt{\frac{g}{2r}}$$

so that

$$\tau = \frac{2\pi}{p} = 2\pi\sqrt{\frac{2r}{g}} \qquad \textit{Ans.}$$

EXAMPLE 22-6

A 10-kg block is suspended from a cord wrapped around a 5-kg disk, as shown in Fig. 22–10*a*. If the spring has a stiffness $k = 200$ N/m, determine the natural period of vibration for the system.

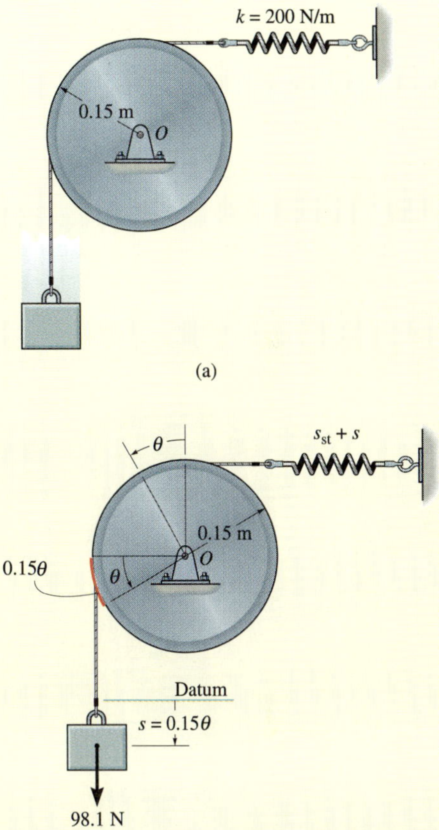

Fig. 22–10

Solution

Energy Equation. A diagram of the block and disk when they are displaced by respective amounts s and θ from the equilibrium position is shown in Fig. 22–10*b*. Since $s = (0.15\text{ m})\theta$, the kinetic energy of the system is

$$T = \tfrac{1}{2}m_b v_b^2 + \tfrac{1}{2}I_O\omega_d^2$$
$$= \tfrac{1}{2}(10\text{ kg})[(0.15\text{ m})\dot{\theta}]^2 + \tfrac{1}{2}[\tfrac{1}{2}(5\text{ kg})(0.15\text{ m})^2](\dot{\theta})^2$$
$$= 0.141(\dot{\theta})^2$$

Establishing the datum at the equilibrium position of the block and realizing that the spring stretches s_{st} for equilibrium, we can write the potential energy as

$$V = \tfrac{1}{2}k(s_{st} + s)^2 - Ws$$
$$= \tfrac{1}{2}(200\text{ N/m})[s_{st} + (0.15\text{ m})\theta]^2 - 98.1\text{ N}[(0.15\text{ m})\theta]$$

The total energy for the system is, therefore,

$$T + V = 0.141(\dot{\theta})^2 + 100(s_{st} + 0.15\theta)^2 - 14.72\theta$$

Time Derivative

$$0.281(\dot{\theta})\ddot{\theta} + 200(s_{st} + 0.15\theta)0.15\dot{\theta} - 14.72\dot{\theta} = 0$$

Since $s_{st} = 98.1/200 = 0.4905$ m, the above equation reduces to the "standard form"

$$\ddot{\theta} + 16\theta = 0$$

so that

$$p = \sqrt{16} = 4\text{ rad/s}$$

Thus,

$$\tau = \frac{2\pi}{p} = \frac{2\pi}{4} = 1.57\text{ s} \qquad \textit{Ans.}$$

Problems

22-26. Solve Prob. 22-13 using energy methods.

22-27. Solve Prob. 22-15 using energy methods.

***22-28.** Solve Prob. 22-16 using energy methods.

22-29. Solve Prob. 22-20 using energy methods.

22-30. Determine the differential equation of motion of the block of mass m when it is displaced slightly and released. Motion occurs in the vertical plane. The springs are attached to the block.

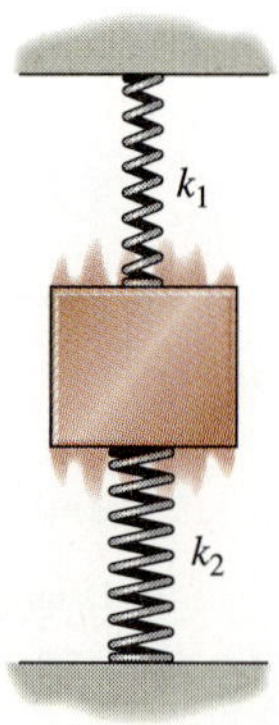

Prob. 22–30

22-31. Determine the differential equation of motion of the 3-kg block when it is displaced slightly and released. The surface is smooth and the springs are originally unstretched.

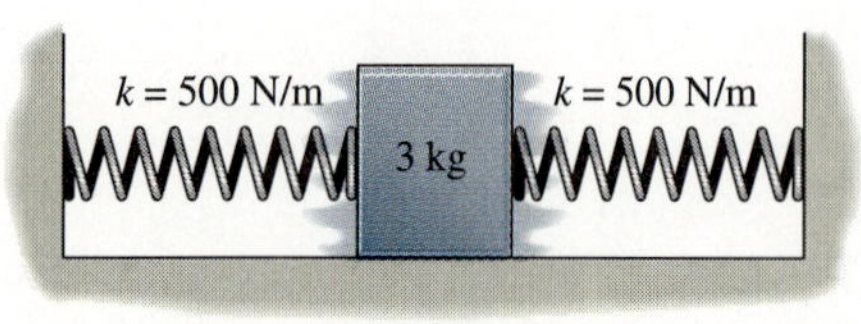

Prob. 22–31

***22-32.** The machine has a mass m and is uniformly supported by *four* springs, each having a stiffness k. Determine the natural period of vertical vibration.

Prob. 22–32

22-33. If the disk has a mass of 8 kg, determine the natural frequency of vibration. The springs are originally unstretched.

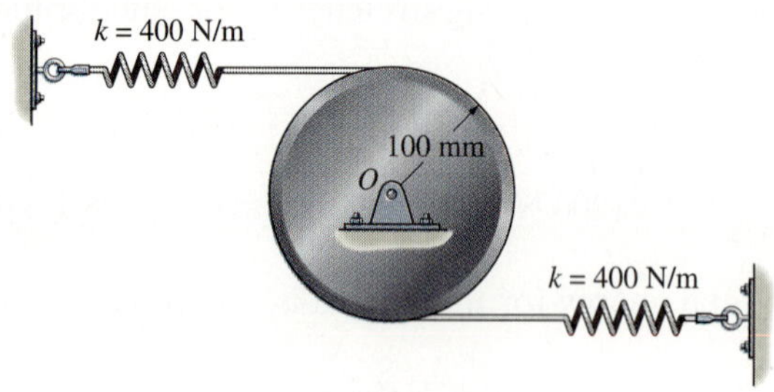

Prob. 22–33

22-34. Determine the natural period of vibration of the pendulum. Consider the two rods to be slender, each having a weight of 8 lb/ft.

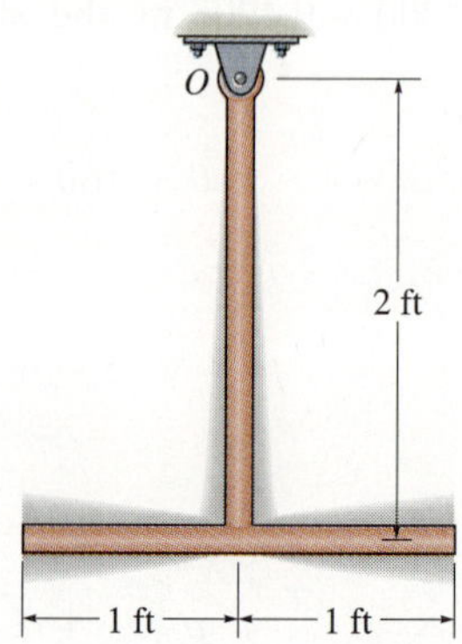

Prob. 22–34

22-35. Determine the natural period of vibration of the disk having a mass m and radius r. Assume the disk does not slip on the surface of contact as it oscillates.

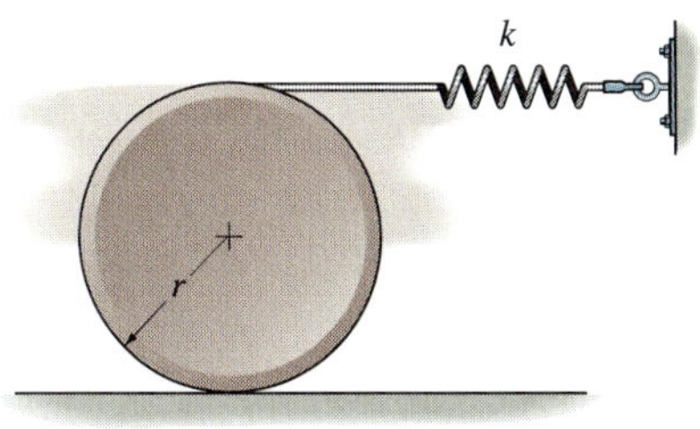

Prob. 22–35

***22-36.** The bar has a mass m and is held in the horizontal position using the two springs. If the end of the bar is given a small displacement, determine the natural period of vibration.

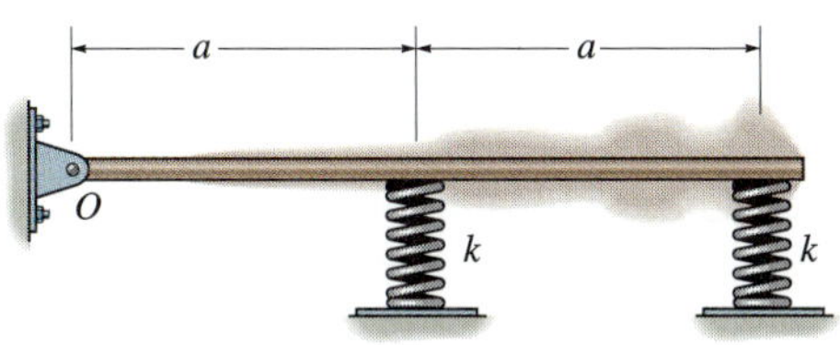

Prob. 22–36

22-37. The slender rod has a mass m and is pinned at its end O. When it is vertical, the springs are unstretched. Determine the natural period of vibration.

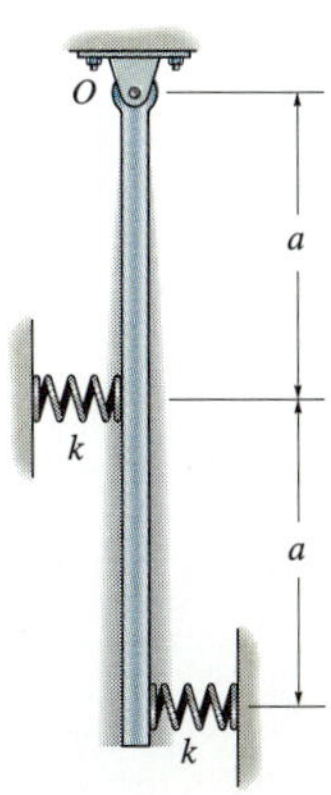

Prob. 22–37

22-38. The 5-lb sphere is attached to a rod of negligible mass and rests in the horizontal position. Determine the natural frequency of vibration. Neglect the size of the sphere.

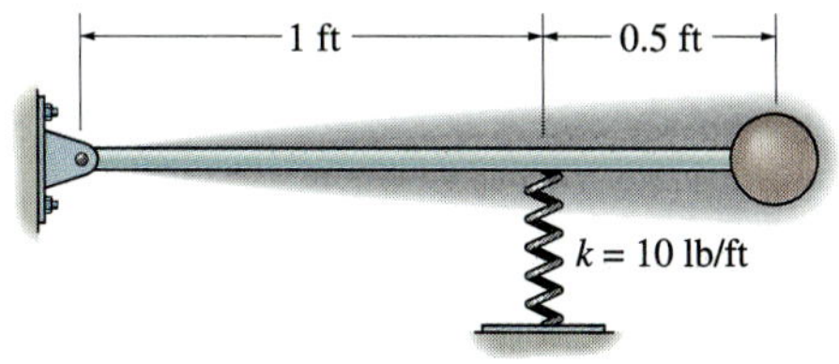

Prob. 22–38

22-39. Determine the natural period of vibration of the 10-lb semicircular disk. $I_O = \frac{1}{2}mr^2$.

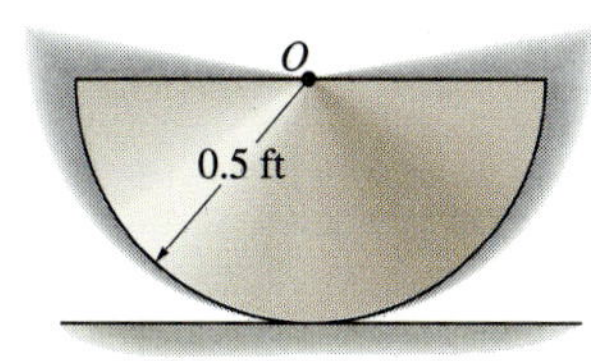

Prob. 22–39

***22-40.** The semicircular disk has a mass m and radius r, and it rolls without slipping in the semicircular trough. Determine the natural period of vibration of the disk if it is displaced slightly and released. *Hint:* $I_O = \frac{1}{2}mr^2$.

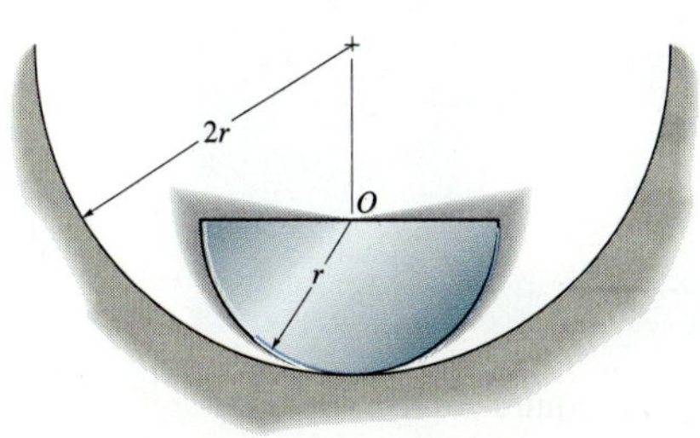

Prob. 22–40

★22.3 Undamped Forced Vibration

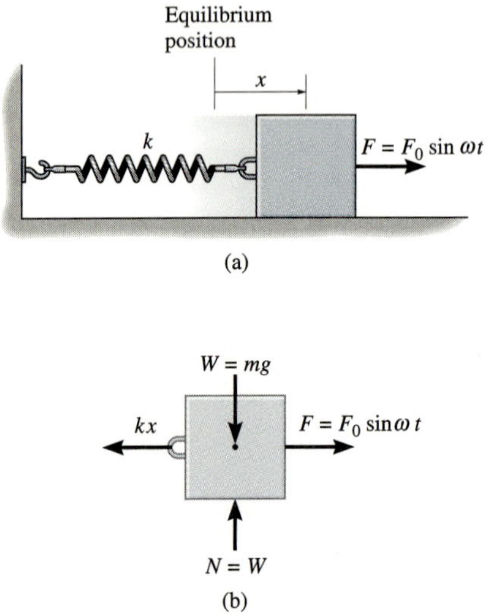

Fig. 22–11

Undamped forced vibration is considered to be one of the most important types of vibrating motion in engineering work. The principles which describe the nature of this motion may be used to analyze the forces which cause vibration in many types of machines and structures.

Periodic Force. The block and spring shown in Fig. 22–11*a* provide a convenient model which represents the vibrational characteristics of a system subjected to a periodic force $F = F_0 \sin \omega t$. This force has an amplitude of F_0 and a *forcing frequency* ω. The free-body diagram for the block when it is displaced a distance x is shown in Fig. 22–11*b*. Applying the equation of motion yields

$$\xrightarrow{+}\Sigma F_x = ma_x; \qquad F_0 \sin \omega t - kx = m\ddot{x}$$

or

$$\ddot{x} + \frac{k}{m}x = \frac{F_0}{m}\sin \omega t \tag{22–18}$$

This equation is referred to as a nonhomogeneous second-order differential equation. The general solution consists of a complementary solution, x_c, *plus* a particular solution, x_p.

The *complementary solution* is determined by setting the term on the right side of Eq. 22–18 equal to zero and solving the resulting homogeneous equation, which is equivalent to Eq. 22–1. The solution is defined by Eq. 22–3, i.e.,

$$x_c = A \sin pt + B \cos pt \tag{22–19}$$

where p is the circular frequency, $p = \sqrt{k/m}$, Eq. 22–2.

Since the motion is periodic, the *particular solution* of Eq. 22–18 may be determined by assuming a solution of the form

$$x_p = C \sin \omega t \tag{22–20}$$

where C is a constant. Taking the second time derivative and substituting into Eq. 22–18 yields

$$-C\omega^2 \sin \omega t + \frac{k}{m}(C \sin \omega t) = \frac{F_0}{m}\sin \omega t$$

Factoring out $\sin \omega t$ and solving for C gives

$$C = \frac{F_0/m}{(k/m) - \omega^2} = \frac{F_0/k}{1 - (\omega/p)^2} \tag{22–21}$$

Substituting into Eq. 22–20, we obtain the particular solution

$$x_p = \frac{F_0/k}{1 - (\omega/p)^2}\sin \omega t \tag{22–22}$$

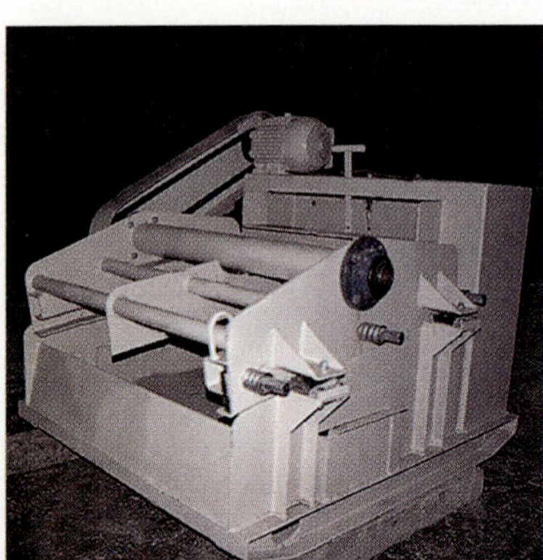

Shaker tables require forced vibration and are used to separate out materials.

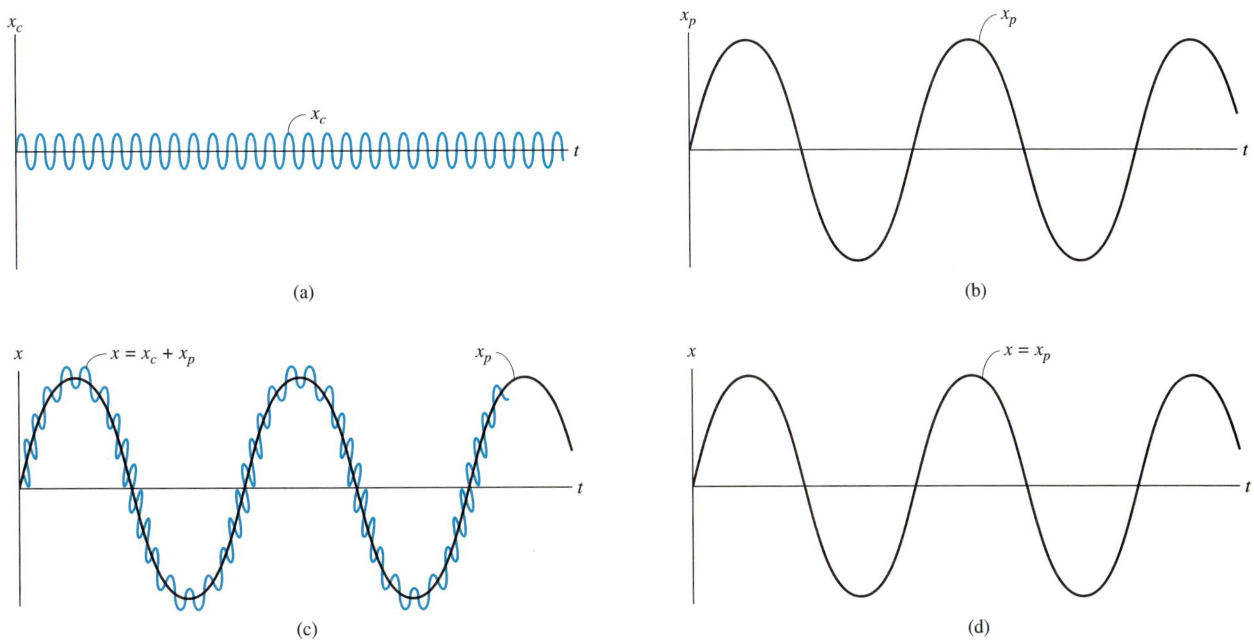

Fig. 22–12

The *general solution* is therefore

$$x = x_c + x_p = A \sin pt + B \cos pt + \frac{F_0/k}{1 - (\omega/p)^2} \sin \omega t \quad (22\text{–}23)$$

Here x describes two types of vibrating motion of the block. The *complementary solution* x_c defines the *free vibration,* which depends on the circular frequency $p = \sqrt{k/m}$ and the constants A and B, Fig. 22–12a. Specific values for A and B are obtained by evaluating Eq. 22–23 at a given instant when the displacement and velocity are known. The *particular solution* x_p describes the *forced vibration* of the block caused by the applied force $F = F_0 \sin \omega t$, Fig. 22–12b. The resultant vibration x is shown in Fig. 22–12c. Since all vibrating systems are subject to *friction,* the free vibration, x_c, will in time dampen out. For this reason the free vibration is referred to as *transient,* and the forced vibration is called *steady-state,* since it is the only vibration that remains, Fig. 22–12d.

From Eq. 22–21 it is seen that the *amplitude* of forced vibration depends on the *frequency ratio* ω/p. If the *magnification factor* MF is defined as the ratio of the amplitude of steady-state vibration, $(x_p)_{\max}$, to the static deflection F_0/k, which would be produced by the amplitude of the periodic force F_0, then, from Eq. 22–22,

The soil compactor operates by forced vibration developed by an internal motor. It is important that the forcing frequency not be close to the natural frequency of vibration, which is determined when the motor is turned off; otherwise resonance will occur and the machine will become uncontrollable.

$$\text{MF} = \frac{(x_p)_{max}}{F_0/k} = \frac{1}{1 - (\omega/p)^2} \tag{22–24}$$

This equation is graphed in Fig. 22–13. Note that for $\omega \approx 0$, the MF ≈ 1. In this case, because of the very low frequency $\omega \ll p$, the magnitude of the force **F** changes slowly and so the vibration of the block will be in phase with the applied force **F**. If the force or displacement is applied with a frequency close to the natural frequency of the system, i.e., $\omega/p \approx 1$, the amplitude of vibration of the block becomes extremely large. This occurs because the force **F** is applied to the block so that it always follows the motion of the block. This condition is called *resonance,* and in practice, resonating vibrations can cause tremendous stress and rapid failure of parts.* When the cyclic force $F_0 \sin \omega t$ is applied at high frequencies ($\omega > p$), the value of the MF becomes negative, indicating that the motion of the block is out of phase with the force. Under these conditions, as the block is displaced to the right, the force acts to the left, and vice versa. For extremely high frequencies ($\omega \gg p$) the block remains almost stationary, and hence the MF is approximately zero.

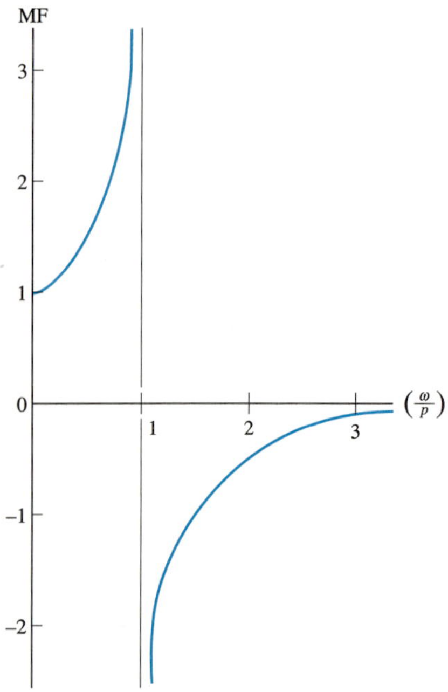

Fig. 22–13

Periodic Support Displacement. Forced vibrations can also arise from the periodic excitation of the support of a system. The model shown in Fig. 22–14*a* represents the periodic vibration of a block which is caused by harmonic movement $\delta = \delta_0 \sin \omega t$ of the support. The free-body diagram for the block in this case is shown in Fig. 22–14*b*. The coordinate x is measured from the point of zero displacement of the support, i.e., when the radial line OA coincides with OB, Fig. 22–14*a*. Therefore, general displacement of the spring is $(x - \delta_0 \sin \omega t)$. Applying the equation of motion yields

$$\xrightarrow{+} \Sigma F_x = ma_x; \qquad -k(x - \delta_0 \sin \omega t) = m\ddot{x}$$

or

$$\ddot{x} + \frac{k}{m}x = \frac{k\delta_0}{m}\sin \omega t \tag{22–25}$$

By comparison, this equation is identical to the form of Eq. 22–18, *provided* F_0 *is replaced* by $k\delta_0$. If this substitution is made into the solutions defined by Eqs. 22–21 to 22–23, the results are appropriate for describing the motion of the block when subjected to the support displacement $\delta = \delta_0 \sin \omega t$.

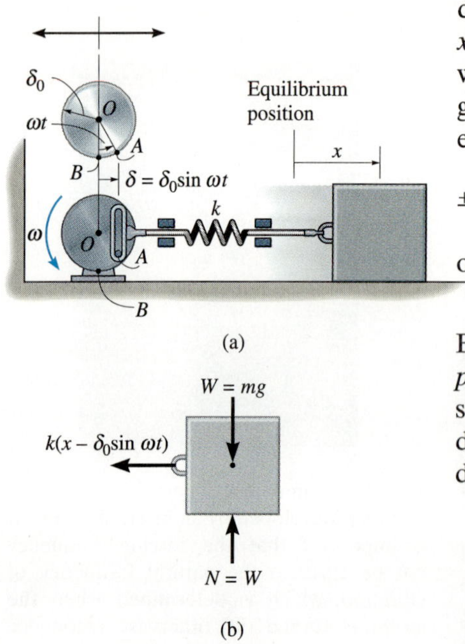

Fig. 22–14

*A swing has a natural period of vibration, as determined in Example 22–1. If someone pushes on the swing only when it reaches its highest point, neglecting drag or wind resistance, resonance will occur since the natural and forcing frequencies will be equu

EXAMPLE 22-7

The instrument shown in Fig. 22–15 is rigidly attached to a platform P, which in turn is supported by *four* springs, each having a stiffness $k = 800$ N/m. Initially the platform is at rest when the floor is subjected to a displacement $\delta = 10 \sin(8t)$ mm, where t is in seconds. If the instrument is constrained to move vertically and the total mass of the instrument and platform is 20 kg, determine the vertical displacement y of the platform as a function of time, measured from the equilibrium position. What floor vibration is required to cause resonance?

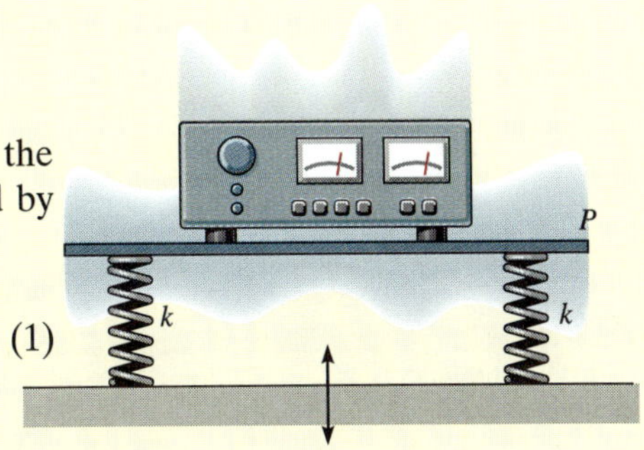

Fig. 22–15

Solution

Since the induced vibration is caused by the displacement of the supports, the motion is described by Eq. 22–23, with F_0 replaced by $k\delta_0$, i.e.,

$$y = A \sin pt + B \cos pt + \frac{\delta_0}{1 - (\omega/p)^2} \sin \omega t \qquad (1)$$

Here $\delta = \delta_0 \sin \omega t = 10 \sin(8t)$ mm, so that

$$\delta_0 = 10 \text{ mm} \qquad \omega = 8 \text{ rad/s}$$

$$p = \sqrt{\frac{k}{m}} = \sqrt{\frac{4(800 \text{ N/m})}{20 \text{ kg}}} = 12.6 \text{ rad/s}$$

From Eq. 22–22, with $k\delta_0$ replacing F_0, the amplitude of vibration caused by the floor displacement is

$$(y_p)_{\max} = \frac{\delta_0}{1 - (\omega/p)^2} = \frac{10}{1 - [(8 \text{ rad/s})/(12.6 \text{ rad/s})]^2} = 16.7 \text{ mm} \qquad (2)$$

Hence, Eq. 1 and its time derivative become

$$y = A \sin(12.6t) + B \cos(12.6t) + 16.7 \sin(8t)$$
$$\dot{y} = A(12.6) \cos(12.6t) - B(12.6) \sin(12.6t) + 133.3 \cos(8t)$$

The constants A and B are evaluated from these equations. Since $y = 0$ and $\dot{y} = 0$ at $t = 0$, then

$$0 = 0 + B + 0 \qquad B = 0$$
$$0 = A(12.6) - 0 + 133.3 \qquad A = -10.5$$

The vibrating motion is therefore described by the equation

$$y = -10.5 \sin(12.6t) + 16.7 \sin(8t) \qquad \textit{Ans.}$$

Resonance will occur when the amplitude of vibration caused by the floor displacement approaches infinity. From Eq. 2, this requires

$$\omega = p = 12.6 \text{ rad/s} \qquad \textit{Ans.}$$

★ 22.4 Viscous Damped Free Vibration

The vibration analysis considered thus far has not included the effects of friction or damping in the system, and as a result, the solutions obtained are only in close agreement with the actual motion. Since all vibrations die out in time, the presence of damping forces should be included in the analysis.

In many cases damping is attributed to the resistance created by the substance, such as water, oil, or air, in which the system vibrates. Provided the body moves slowly through this substance, the resistance to motion is directly proportional to the body's speed. The type of force developed under these conditions is called a *viscous damping force*. The magnitude of this force is expressed by an equation of the form

$$F = c\dot{x} \tag{22–26}$$

where the constant c is called the *coefficient of viscous damping* and has units of N·s/m or lb·s/ft.

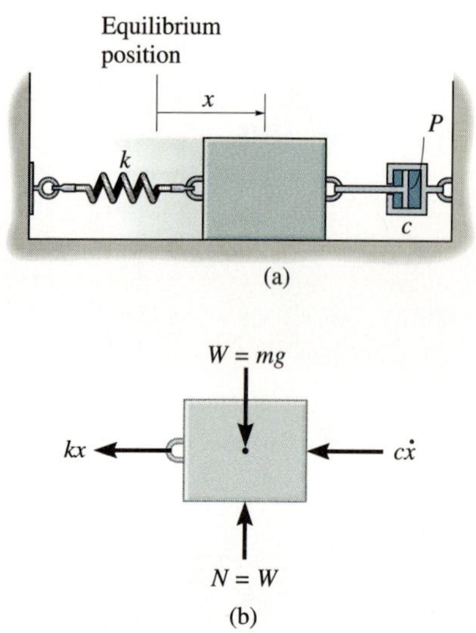

Fig. 22–16

The vibrating motion of a body or system having viscous damping may be characterized by the block and spring shown in Fig. 22–16*a*. The effect of damping is provided by the *dashpot* connected to the block on the right side. Damping occurs when the piston P moves to the right or left within the enclosed cylinder. The cylinder contains a fluid, and the motion of the piston is retarded since the fluid must flow around or through a small hole in the piston. The dashpot is assumed to have a coefficient of viscous damping c.

If the block is displaced a distance x from its equilibrium position, the resulting free-body diagram is shown in Fig. 22–16*b*. Both the spring force kx and the damping force $c\dot{x}$ oppose the forward motion of the block, so that applying the equation of motion yields

$\overset{+}{\rightarrow}\Sigma F_x = ma_x;$ $\qquad -kx - c\dot{x} = m\ddot{x}$

or

$$m\ddot{x} + c\dot{x} + kx = 0 \tag{22–27}$$

This linear, second-order, homogeneous, differential equation has solutions of the form

$$x = e^{\lambda t}$$

where e is the base of the natural logarithm and λ (lambda) is a constant. The value of λ may be obtained by substituting this solution into Eq. 22–27, which yields

$$m\lambda^2 e^{\lambda t} + c\lambda e^{\lambda t} + ke^{\lambda t} = 0$$

or

$$e^{\lambda t}(m\lambda^2 + c\lambda + k) = 0$$

Since $e^{\lambda t}$ is never zero, a solution is possible provided

$$m\lambda^2 + c\lambda + k = 0$$

Hence, by the quadratic formula, the two values of λ are

$$\begin{aligned}\lambda_1 &= -\frac{c}{2m} + \sqrt{\left(\frac{c}{2m}\right)^2 - \frac{k}{m}} \\ \lambda_2 &= -\frac{c}{2m} - \sqrt{\left(\frac{c}{2m}\right)^2 - \frac{k}{m}}\end{aligned} \qquad (22\text{–}28)$$

The general solution of Eq. 22–27 is therefore a linear combination of exponentials which involves both of these roots. There are three possible combinations of λ_1 and λ_2 which must be considered. Before discussing these combinations, however, we will first define the *critical damping coefficient* c_c as the value of c which makes the radical in Eqs. 22–28 equal to zero; i.e.,

$$\left(\frac{c_c}{2m}\right)^2 - \frac{k}{m} = 0$$

or

$$c_c = 2m\sqrt{\frac{k}{m}} = 2mp \qquad (22\text{–}29)$$

Here the value of p is the circular frequency $p = \sqrt{k/m}$, Eq. 22–2.

Overdamped System. When $c > c_c$, the roots λ_1 and λ_2 are both real. The general solution of Eq. 22–27 may then be written as

$$x = Ae^{\lambda_1 t} + Be^{\lambda_2 t} \qquad (22\text{–}30)$$

Motion corresponding to this solution is *nonvibrating*. The effect of damping is so strong that when the block is displaced and released, it simply creeps back to its original position without oscillating. The system is said to be *overdamped*.

Critically Damped System. If $c = c_c$, then $\lambda_1 = \lambda_2 = -c_c/2m = -p$. This situation is known as *critical damping*, since it represents a condition where c has the smallest value necessary to cause the system to be nonvibrating. Using the methods of differential equations, it may be shown that the solution to Eq. 22–27 for critical damping is

$$x = (A + Bt)e^{-pt} \qquad (22\text{–}31)$$

Underdamped System. Most often $c < c_c$, in which case the system is referred to as *underdamped*. In this case the roots λ_1 and λ_2 are complex numbers, and it may be shown that the general solution of Eq. 22–27 can be written as

$$x = D[e^{-(c/2m)t}\sin(p_d t + \phi)] \tag{22–32}$$

where D and ϕ are constants generally determined from the initial conditions of the problem. The constant p_d is called the *damped natural frequency* of the system. It has a value of

$$p_d = \sqrt{\frac{k}{m} - \left(\frac{c}{2m}\right)^2} = p\sqrt{1 - \left(\frac{c}{c_c}\right)^2} \tag{22–33}$$

where the ratio c/c_c is called the *damping factor.*

The graph of Eq. 22–32 is shown in Fig. 22–17. The initial limit of motion, D, diminishes with each cycle of vibration, since motion is confined within the bounds of the exponential curve. Using the damped natural frequency p_d, the period of damped vibration may be written as

$$\tau_d = \frac{2\pi}{p_d} \tag{22–34}$$

Since $p_d < p$, Eq. 22–33, the period of damped vibration, τ_d, will be greater than that of free vibration, $\tau = 2\pi/p$.

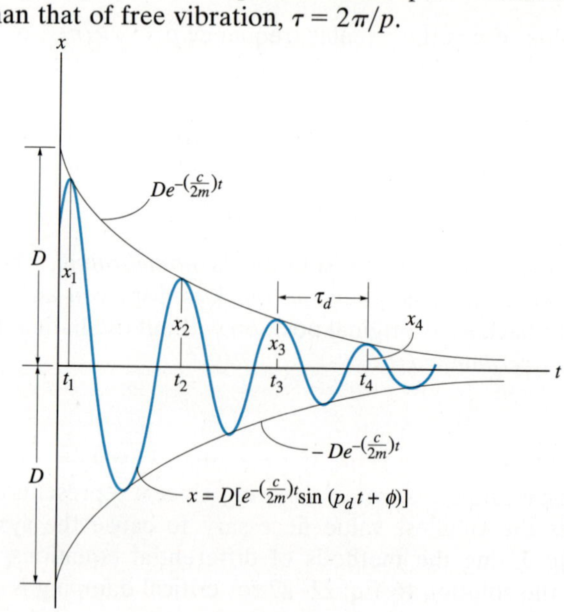

Fig. 22–17

★ 22.5 Viscous Damped Forced Vibration

The most general case of single-degree-of-freedom vibrating motion occurs when the system includes the effects of forced motion and induced damping. The analysis of this particular type of vibration is of practical value when applied to systems having significant damping characteristics.

If a dashpot is attached to the block and spring shown in Fig. 22–11*a*, the differential equation which describes the motion becomes

$$m\ddot{x} + c\dot{x} + kx = F_0 \sin \omega t \tag{22–35}$$

A similar equation may be written for a block and spring having a periodic support displacement, Fig. 22–14*a*, which includes the effects of damping. In that case, however, F_0 is replaced by $k\delta_0$. Since Eq. 22–35 is nonhomogeneous, the general solution is the sum of a complementary solution, x_c, and a particular solution, x_p. The complementary solution is determined by setting the right side of Eq. 22–35 equal to zero and solving the homogeneous equation, which is equivalent to Eq. 22–27. The solution is therefore given by Eq. 22–30, 22–31, or 22–32, depending on the values of λ_1 and λ_2. Because all systems contain friction, however, this solution will dampen out with time. Only the particular solution, which describes the *steady-state vibration* of the system, will remain. Since the applied forcing function is harmonic, the steady-state motion will also be harmonic. Consequently, the particular solution will be of the form

$$x_p = A' \sin \omega t + B' \cos \omega t \tag{22–36}$$

The constants A' and B' are determined by taking the necessary time derivatives and substituting them into Eq. 22–35, which after simplification yields

$$(-A'm\omega^2 - cB'\omega + kA') \sin \omega t + \\ (-B'm\omega^2 + cA'\omega + kB') \cos \omega t = F_0 \sin \omega t$$

Since this equation holds for all time, the constant coefficients of sin ωt and cos ωt may be equated; i.e.,

$$-A'm\omega^2 - cB'\omega + kA' = F_0$$
$$-B'm\omega^2 + cA'\omega + kB' = 0$$

Solving for A' and B', realizing that $p^2 = k/m$, yields

$$A' = \frac{(F_0/m)(p^2 - \omega^2)}{(p^2 - \omega^2)^2 + (c\omega/m)^2}$$
$$B' = \frac{-F_0(c\omega/m^2)}{(p^2 - \omega^2)^2 + (c\omega/m)^2} \qquad (22\text{–}37)$$

It is also possible to express Eq. 22–36 in a form similar to Eq. 22–9,

$$x_p = C' \sin(\omega t - \phi') \qquad (22\text{–}38)$$

in which case the constants C' and ϕ' are

$$C' = \frac{F_0/k}{\sqrt{[1 - (\omega/p)^2]^2 + [2(c/c_c)(\omega/p)]^2}}$$

(22–39)

$$\phi' = \tan^{-1}\left[\frac{2(c/c_c)(\omega/p)}{1 - (\omega/p)^2}\right]$$

The angle ϕ' represents the phase difference between the applied force and the resulting steady-state vibration of the damped system.

The *magnification factor* MF has been defined in Sec. 22.3 as the ratio of the amplitude of deflection caused by the forced vibration to the deflection caused by a static force $\mathbf{F}_0$. From Eq. 22–38, the forced vibration has an amplitude of C'; thus,

$$\text{MF} = \frac{C'}{F_0/k} = \frac{1}{\sqrt{[1 - (\omega/p)^2]^2 + [2(c/c_c)(\omega/p)]^2}} \qquad (22\text{–}40)$$

The MF is plotted in Fig. 22–18 versus the frequency ratio ω/p for various values of the damping factor c/c_c. It can be seen from this graph that the magnification of the amplitude increases as the damping factor decreases. Resonance obviously occurs only when the damping factor is zero and the frequency ratio equals 1.

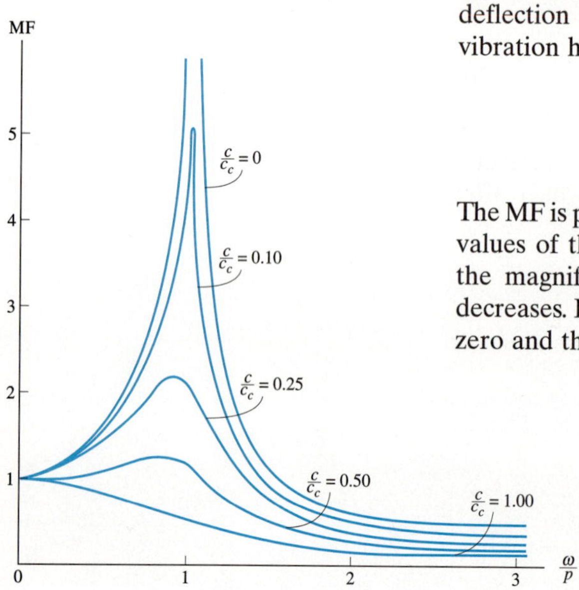

Fig. 22–18

EXAMPLE 22-8

The 30-kg electric motor shown in Fig. 22–19 is supported by *four* springs, each spring having a stiffness of 200 N/m. If the rotor R is unbalanced such that its effect is equivalent to a 4-kg mass located 60 mm from the axis of rotation, determine the amplitude of vibration when the rotor is turning at $\omega = 10$ rad/s. The damping factor is $c/c_c = 0.15$.

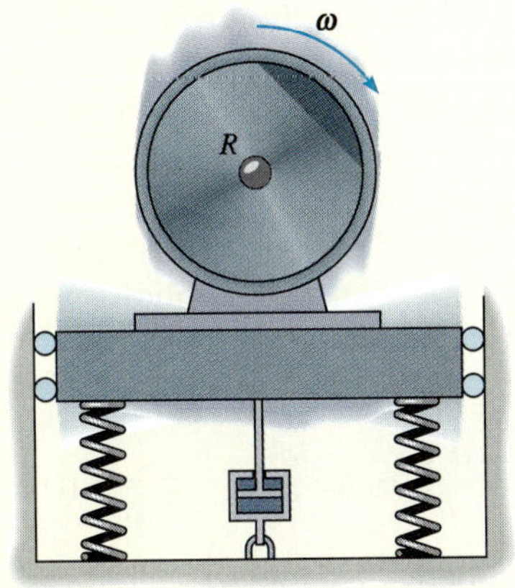

Fig. 22–19

Solution

The periodic force which causes the motor to vibrate is the centrifugal force due to the unbalanced rotor. This force has a constant magnitude of

$$F_0 = ma_n = mr\omega^2 = 4\text{ kg}(0.06\text{ m})(10\text{ rad/s})^2 = 24\text{ N}$$

Since $F = F_0 \sin \omega t$, where $\omega = 10$ rad/s, then

$$F = 24 \sin 10t$$

The stiffness of the entire system of four springs is $k = 4(200\text{ N/m}) = 800\text{ N/m}$. Therefore, the circular frequency of vibration is

$$p = \sqrt{\frac{k}{m}} = \sqrt{\frac{800\text{ N/m}}{30\text{ kg}}} = 5.16\text{ rad/s}$$

Since the damping factor is known, the steady-state amplitude may be determined from the first of Eqs. 22–39, i.e.,

$$C' = \frac{F_0/k}{\sqrt{[1-(\omega/p)^2]^2 + [2(c/c_c)(\omega/p)]^2}}$$

$$= \frac{24/800}{\sqrt{[1-(10/5.16)^2]^2 + [2(0.15)(10/5.16)]^2}}$$

$$= 0.0107\text{ m} = 10.7\text{ mm}$$ *Ans.*

★ 22.6 Electrical Circuit Analogs

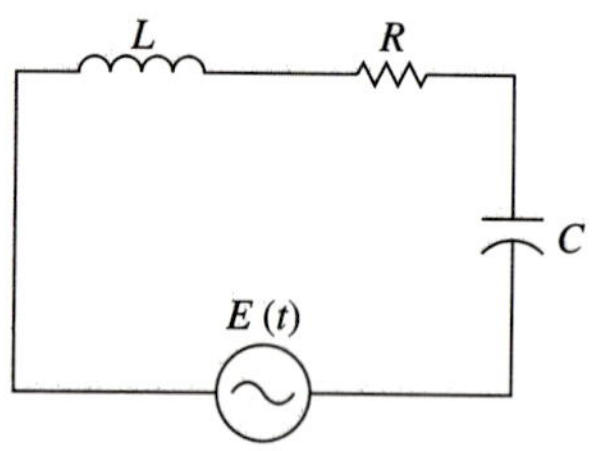

(a)

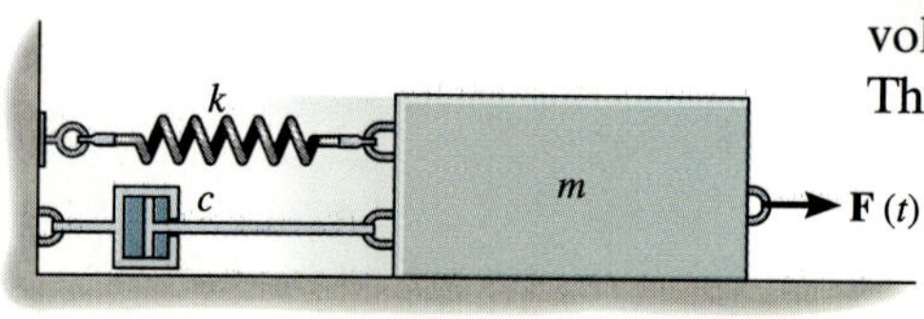

(b)

Fig. 22–20

The characteristics of a vibrating mechanical system may be represented by an electric circuit. Consider the circuit shown in Fig. 22–20*a*, which consists of an inductor L, a resistor R, and a capacitor C. When a voltage $E(t)$ is applied, it causes a current of magnitude i to flow through the circuit. As the current flows past the inductor the voltage drop is $L(di/dt)$, when it flows across the resistor the drop is Ri, and when it arrives at the capacitor the drop is $(1/C)\int i\,dt$. Since current cannot flow past a capacitor, it is only possible to measure the charge q acting on the capacitor. The charge may, however, be related to the current by the equation $i = dq/dt$. Thus, the voltage drops which occur across the inductor, resistor, and capacitor may be written as $L\,d^2q/dt^2$, $R\,dq/dt$, and q/C, respectively. According to Kirchhoff's voltage law, the applied voltage balances the sum of the voltage drops around the circuit. Therefore,

$$L\frac{d^2q}{dt^2} + R\frac{dq}{dt} + \frac{1}{C}q = E(t) \qquad (22\text{–}41)$$

Consider now the model of a single-degree-of-freedom mechanical system, Fig. 22–20*b*, which is subjected to both a general forcing function $F(t)$ and damping. The equation of motion for this system was established in the previous section and can be written as

$$m\frac{d^2x}{dt^2} + c\frac{dx}{dt} + kx = F(t) \qquad (22\text{–}42)$$

By comparison, it is seen that Eqs. 22–41 and 22–42 have the same form, and hence mathematically the problem of analyzing an electric circuit is the same as that of analyzing a vibrating mechanical system. The analogs between the two equations are given in Table 22–1.

This analogy has important application to experimental work, for it is much easier to simulate the vibration of a complex mechanical system using an electric circuit, which can be constructed on an analog computer, than to make an equivalent mechanical spring-and-dashpot model.

Table 22–1 Electrical–Mechanical Analogs

Electrical		Mechanical	
Electric charge	q	Displacement	x
Electric current	i	Velocity	dx/dt
Voltage	$E(t)$	Applied force	$F(t)$
Inductance	L	Mass	m
Resistance	R	Viscous damping coefficient	c
Reciprocal of capacitance	$1/C$	Spring stiffness	k

Problems

22-41. The block shown in Fig. 22–16 has a mass of 20 kg, and the spring has a stiffness $k = 600$ N/m. When the block is displaced and released, two successive amplitudes are measured as $x_1 = 150$ mm and $x_2 = 87$ mm. Determine the coefficient of viscous damping, c.

22-42. If the block-and-spring model is subjected to the impressed force $F = F_0 \cos \omega t$, show that the differential equation of motion is $\ddot{x} + (k/m)x = (F_0/m) \cos \omega t$, where x is measured from the equilibrium position of the block. What is the general solution of this equation?

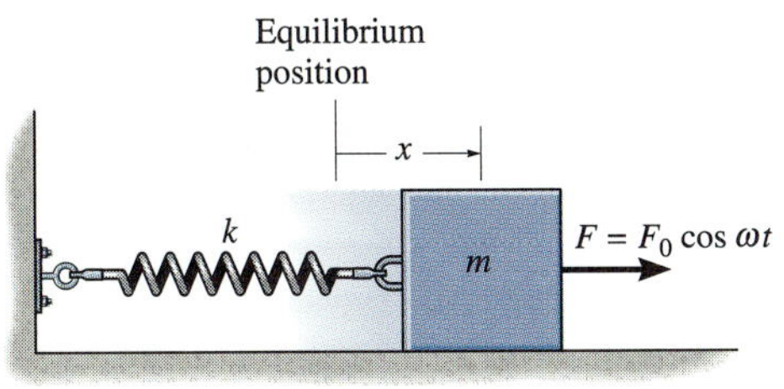

Prob. 22–42

22-43. A 4-kg block is suspended from a spring that has a stiffness of $k = 600$ N/m. The block is drawn downward 50 mm from the equilibrium position and released from rest when $t = 0$. If the support moves with an impressed displacement of $\delta = (10 \sin 4t)$ mm, where t is in seconds, determine the equation that describes the vertical motion of the block. Assume positive displacement is downward.

***22-44.** If the block is subjected to the impressed force $F = F_0 \cos \omega t$, show that the differential equation of motion is $\ddot{y} + (k/m)y = (F_0/m) \cos \omega t$, where y is measured from the equilibrium position of the block. What is the general solution of this equation?

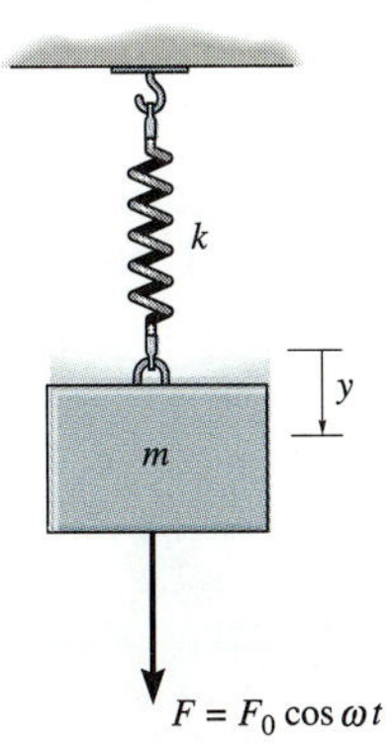

Prob. 22–44

22-45. The spring shown stretches 6 in. when it is loaded with a 50-lb weight. Determine the equation which describes the position of the weight as a function of time when the weight is pulled 4 in. below its equilibrium position and released from rest. The weight is subjected to the impressed force of $F = (-7 \sin 2t)$ lb, where t is in seconds.

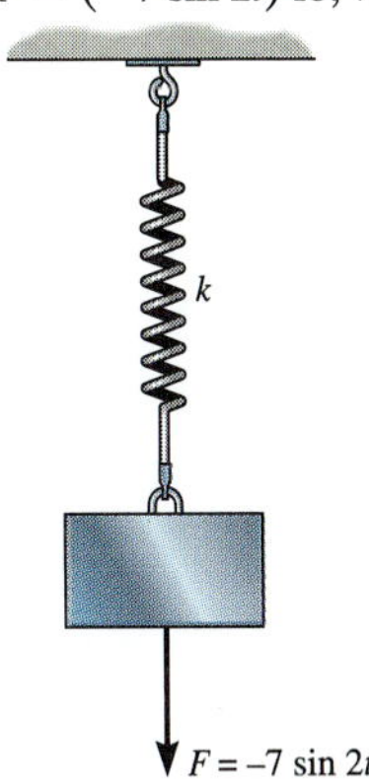

Prob. 22–45

22-46. A block having a mass of 0.8 kg is suspended from a spring having a stiffness of 120 N/m. If a dashpot provides a damping force of 2.5 N when the speed of the block is 0.2 m/s, determine the period of free vibration.

22-47. A 5-kg block is suspended from a spring having a stiffness of 300 N/m. If the block is acted upon by a vertical force $F = (7 \sin 8t)$ N, where t is in seconds, determine the equation which describes the motion of the block when it is pulled down 100 mm from the equilibrium position and released from rest at $t = 0$. Assume that positive displacement is downward.

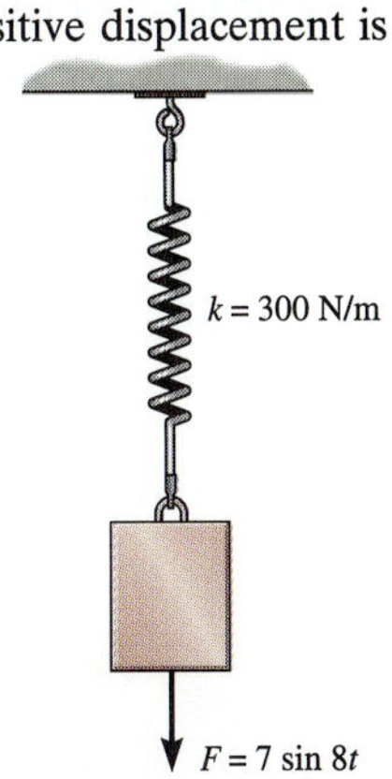

Prob. 22–47

***22-48.** The electric motor has a mass of 50 kg and is supported by *four springs,* each spring having a stiffness of 100 N/m. If the motor turns a disk D which is mounted eccentrically, 20 mm from the disk's center, determine the angular rotation ω at which resonance occurs. Assume that the motor only vibrates in the vertical direction.

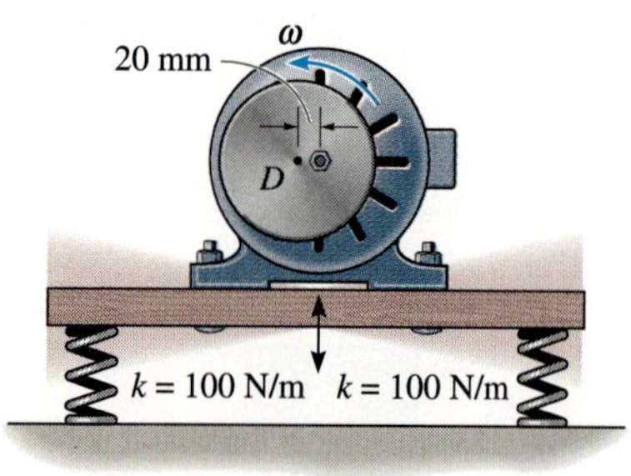

Prob. 22–48

22-49. The fan has a mass of 25 kg and is fixed to the end of a horizontal beam that has a negligible mass. The fan blade is mounted eccentrically on the shaft such that it is equivalent to an unbalanced 3.5-kg mass located 100 mm from the axis of rotation. If the static deflection of the beam is 50 mm as a result of the weight of the fan, determine the angular speed of the fan at which resonance will occur. *Hint:* See the first part of Example 22–8.

Prob. 22–49

22-50. The fan has a mass of 25 kg and is fixed to the end of a horizontal beam that has a negligible mass. The fan blade is mounted eccentrically on the shaft such that it is equivalent to an unbalanced 3.5-kg mass located 100 mm from the axis of rotation. If the static deflection of the beam is 50 mm as a result of the weight of the fan, determine the amplitude of steady-state vibration of the fan when the angular velocity of the fan is 10 rad/s. *Hint:* See the first part of Example 22–8.

22-51. What will be the amplitude of steady-state vibration of the fan in Prob. 22-50 if the angular velocity of the fan is 18 rad/s? *Hint:* See the first part of Example 22–8.

Probs. 22–50/51

***22-52.** The electric motor turns an eccentric flywheel which is equivalent to an unbalanced 0.25-lb weight located 10 in. from the axis of rotation. If the static deflection of the beam is 1 in. because of the weight of the motor, determine the angular velocity of the flywheel at which resonance will occur. The motor weighs 150 lb. Neglect the mass of the beam.

22-53. What will be the amplitude of steady-state vibration of the motor in Prob. 22–52 if the angular velocity of the flywheel is 20 rad/s?

22-54. Determine the angular velocity of the flywheel in Prob. 22-52 which will produce an amplitude of vibration of 0.25 in.

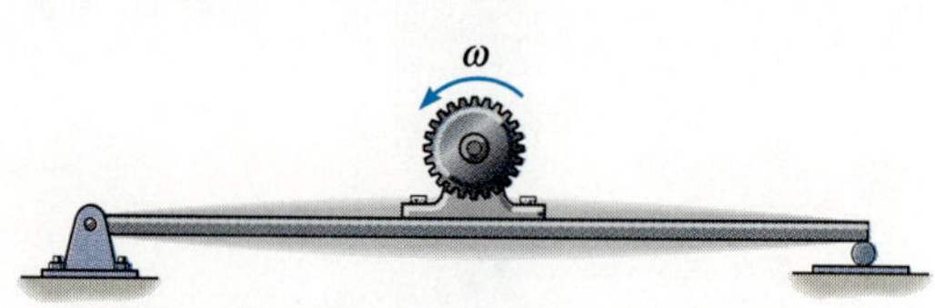

Probs. 22–52/53/54

22-55. The engine is mounted on a foundation block which is spring-supported. Describe the steady-state vibration of the system if the block and engine have a total weight of 1500 lb and the engine, when running, creates an impressed force $F = (50 \sin 2t)$ lb, where t is in seconds. Assume that the system vibrates only in the vertical direction, with the positive displacement measured downward, and that the total stiffness of the springs can be represented as $k = 2000$ lb/ft.

***22-56.** Determine the rotational speed ω of the engine in Prob. 22-55 which will cause resonance.

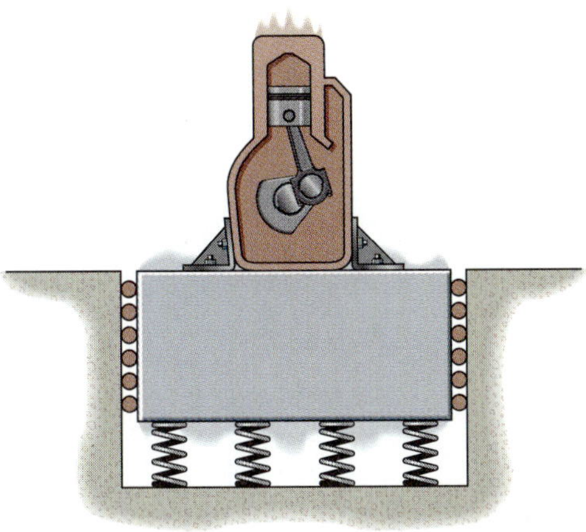

Probs. 22–55/56

22-57. The block, having a weight of 12 lb, is immersed in a liquid such that the damping force acting on the block has a magnitude of $F = (0.7|v|)$ lb, where v is in ft/s. If the block is pulled down 0.62 ft and released from rest, determine the position of the block as a function of time. The spring has a stiffness of $k = 53$ lb/ft. Assume that positive displacement is downward.

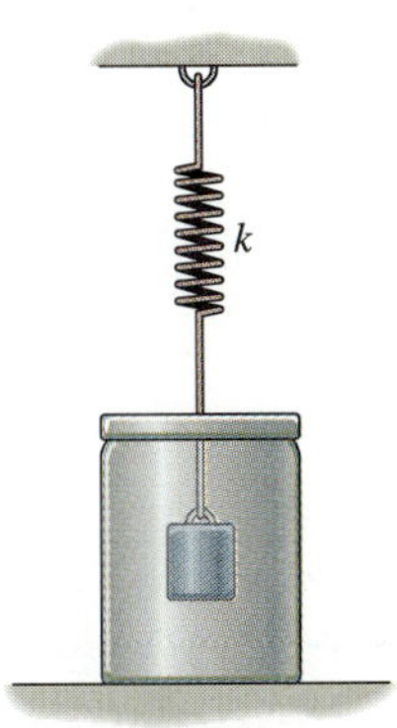

Prob. 22–57

22-58. A 7-lb block is suspended from a spring having a stiffness of $k = 75$ lb/ft. The support to which the spring is attached is given simple harmonic motion which may be expressed as $\delta = (0.15 \sin 2t)$ ft, where t is in seconds. If the damping factor is $c/c_c = 0.8$, determine the phase angle ϕ of forced vibration.

22-59. Determine the magnification factor of the block, spring, and dashpot combination in Prob. 22-58.

***22-60.** The 20-kg block is subjected to the action of the harmonic force $F = (90 \cos 6t)$ N, where t is in seconds. Write the equation which describes the steady-state motion.

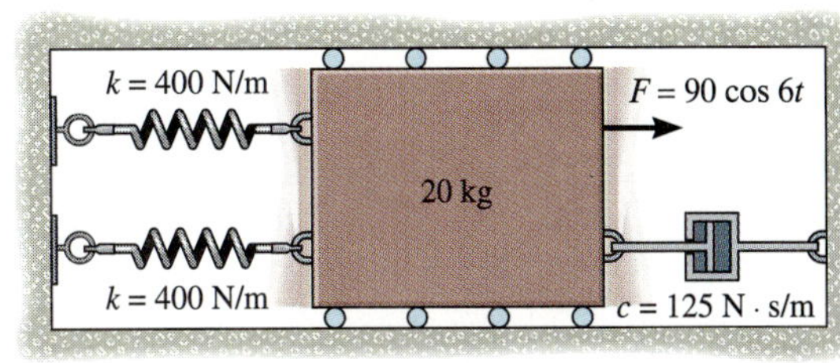

Prob. 22–60

22-61. The 200-lb electric motor is fastened to the midpoint of the simply supported beam. It is found that the beam deflects 2 in. when the motor is not running. The motor turns an eccentric flywheel which is equivalent to an unbalanced weight of 1 lb located 5 in. from the axis of rotation. If the motor is turning at 100 rpm, determine the amplitude of steady-state vibration. The damping factor is $c/c_c = 0.20$. Neglect the mass of the beam.

Prob. 22–61

22-62. A block having a mass of 7 kg is suspended from a spring that has a stiffness $k = 600$ N/m. If the block is given an upward velocity of 0.6 m/s from its equilibrium position at $t = 0$, determine its position as a function of time. Assume that positive displacement of the block is downward and that motion takes place in a medium which furnishes a damping force $F = (50|v|)$ N, where v is in m/s.

22-63. The damping factor, c/c_c, may be determined experimentally by measuring the successive amplitudes of vibrating motion of a system. If two of these maximum displacements can be approximated by x_1 and x_2, as shown in Fig. 22–17, show that the ratio $\ln x_1/x_2 = 2\pi(c/c_c)/\sqrt{1 - (c/c_c)^2}$. The quantity $\ln x_1/x_2$ is called the *logarithmic decrement.*

***22-64.** The small block at A has a mass of 4 kg and is mounted on the bent rod having negligible mass. If the rotor at B causes a harmonic movement $\delta_B = (0.1 \cos 15t)$ m, where t is in seconds, determine the amplitude of vibration of the block.

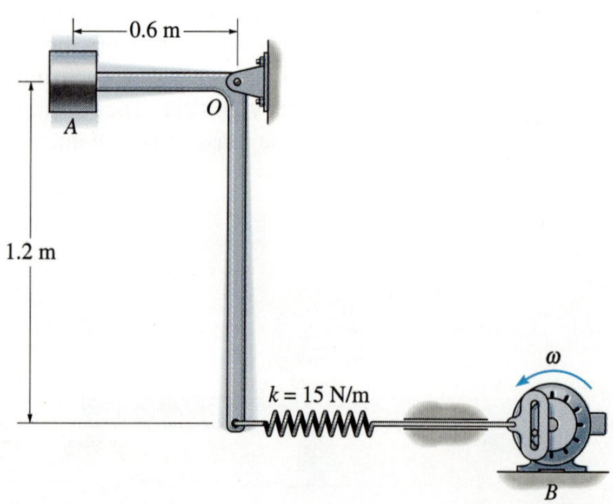

Prob. 22–64

22-65. Draw the electrical circuit that is equivalent to the mechanical system shown. Determine the differential equation which describes the charge q in the circuit.

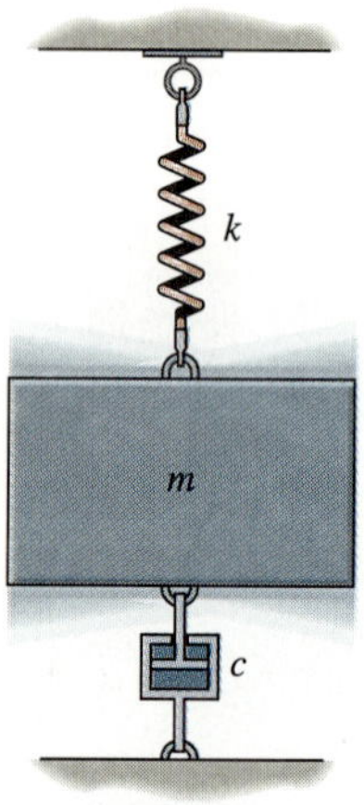

Prob. 22–65

22-66. Determine the mechanical analog for the electrical circuit. What differential equations describe the mechanical and electrical systems?

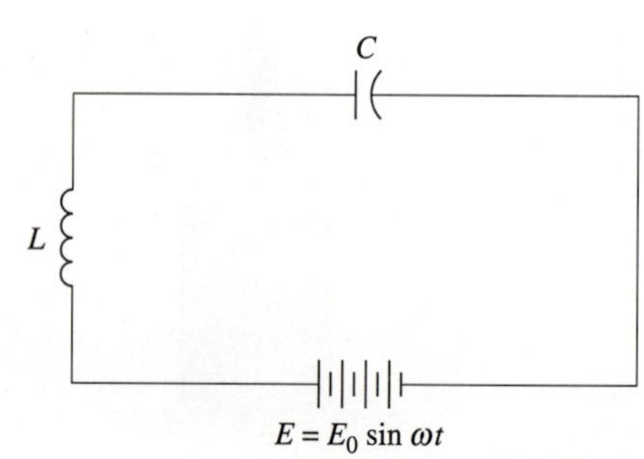

Prob. 22–66

22-67. Draw the electrical circuit that is equivalent to the mechanical system shown. Determine the differential equation which describes the charge q in the circuit.

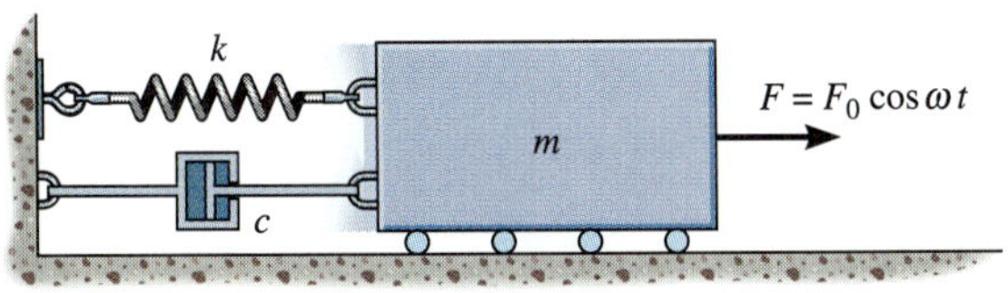

Prob. 22–67

***22-68.** Draw the electrical circuit that is equivalent to the mechanical system shown. What is the differential equation which describes the charge q in the circuit?

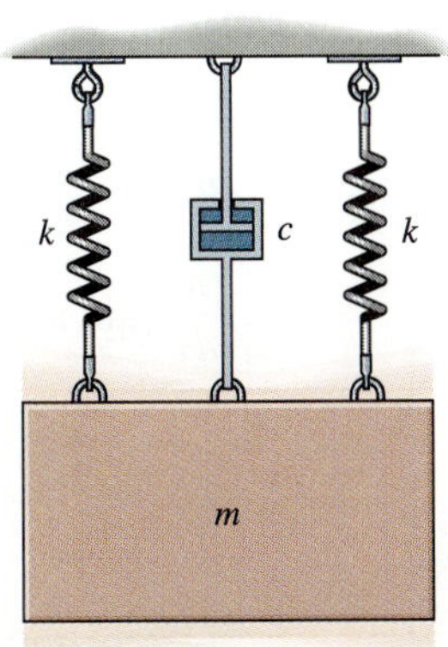

Prob. 22–68

22-69. Determine the differential equation of motion for the damped vibratory system shown. What type of motion occurs?

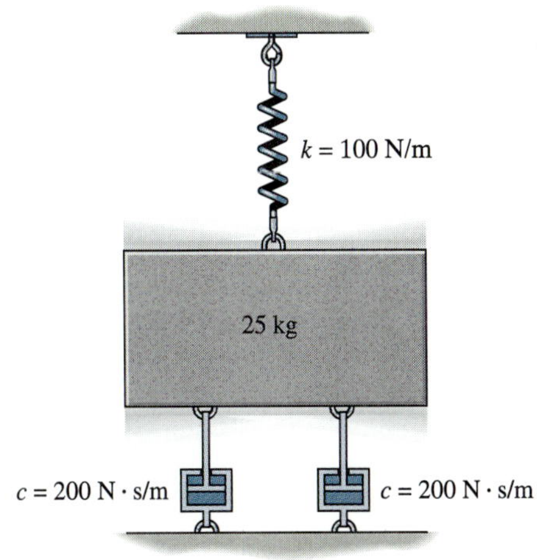

Prob. 22–69

APPENDIX A

Mathematical Expressions

Quadratic Formula

If $ax^2 + bx + c = 0$, then $x = \dfrac{-b \pm \sqrt{b^2 - 4ac}}{2a}$

Hyperbolic Functions

$$\sinh x = \frac{e^x - e^{-x}}{2}, \quad \cosh x = \frac{e^x + e^{-x}}{2}, \quad \tanh x = \frac{\sinh x}{\cosh x}$$

Trigonometric Identities

$$\sin \theta = \frac{A}{C}, \quad \csc \theta = \frac{C}{A}$$

$$\cos \theta = \frac{B}{C}, \quad \sec \theta = \frac{C}{B}$$

$$\tan \theta = \frac{A}{B}, \quad \cot \theta = \frac{B}{A}$$

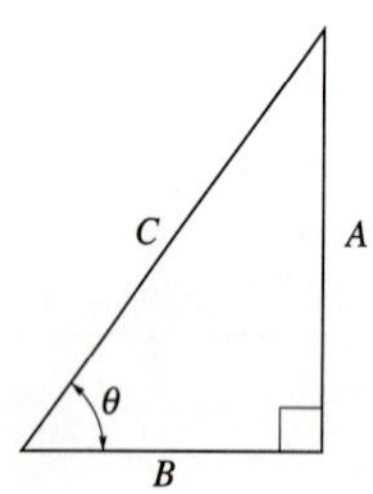

$$\sin^2 \theta + \cos^2 \theta = 1$$

$$\sin(\theta \pm \phi) = \sin\theta\cos\phi \pm \cos\theta\sin\phi$$

$$\sin 2\theta = 2\sin\theta\cos\theta$$

$$\cos(\theta \pm \phi) = \cos\theta\cos\phi \mp \sin\theta\sin\phi$$

$$\cos 2\theta = \cos^2\theta - \sin^2\theta$$

$$\cos\theta = \pm\sqrt{\frac{1 + \cos 2\theta}{2}}, \quad \sin\theta = \pm\sqrt{\frac{1 - \cos 2\theta}{2}}$$

$$\tan\theta = \frac{\sin\theta}{\cos\theta}$$

$$1 + \tan^2\theta = \sec^2\theta \qquad 1 + \cot^2\theta = \csc^2\theta$$

Power-Series Expansions

$$\sin x = x - \frac{x^3}{3!} + \cdots \qquad \sinh x = x + \frac{x^3}{3!} + \cdots$$

$$\cos x = 1 - \frac{x^2}{2!} + \cdots \qquad \cosh x = 1 + \frac{x^2}{2!} + \cdots$$

Derivatives

$$\frac{d}{dx}(u^n) = nu^{n-1}\frac{du}{dx}$$

$$\frac{d}{dx}(uv) = u\frac{dv}{dx} + v\frac{du}{dx}$$

$$\frac{d}{dx}\left(\frac{u}{v}\right) = \frac{v\dfrac{du}{dx} - u\dfrac{dv}{dx}}{v^2}$$

$$\frac{d}{dx}(\cot u) = -\csc^2 u\frac{du}{dx}$$

$$\frac{d}{dx}(\sec u) = \tan u \sec u\frac{du}{dx}$$

$$\frac{d}{dx}(\csc u) = -\csc u \cot u\frac{du}{dx}$$

$$\frac{d}{dx}(\sin u) = \cos u\frac{du}{dx}$$

$$\frac{d}{dx}(\cos u) = -\sin u\frac{du}{dx}$$

$$\frac{d}{dx}(\tan u) = \sec^2 u\frac{du}{dx}$$

$$\frac{d}{dx}(\sinh u) = \cosh u\frac{du}{dx}$$

$$\frac{d}{dx}(\cosh u) = \sinh u\frac{du}{dx}$$

Integrals

$$\int x^n\,dx = \frac{x^{n+1}}{n+1} + C, n \neq -1$$

$$\int \frac{dx}{a+bx} = \frac{1}{b}\ln(a+bx) + C$$

$$\int \frac{dx}{a+bx^2} = \frac{1}{2\sqrt{-ba}}\ln\left[\frac{a+x\sqrt{-ab}}{a-x\sqrt{-ab}}\right] + C, \quad ab < 0$$

$$\int \frac{x\,dx}{a+bx^2} = \frac{1}{2b}\ln(bx^2+a) + C,$$

$$\int \frac{x^2\,dx}{a+bx^2} = \frac{x}{b} - \frac{a}{b\sqrt{ab}}\tan^{-1}\frac{x\sqrt{ab}}{a} + C, \quad ab > 0$$

$$\int \frac{dx}{a^2-x^2} = \frac{1}{2a}\ln\left[\frac{a+x}{a-x}\right] + C, \quad a^2 > x^2$$

$$\int \sqrt{a+bx}\,dx = \frac{2}{3b}\sqrt{(a+bx)^3} + C$$

$$\int x\sqrt{a+bx}\,dx = \frac{-2(2a-3bx)\sqrt{(a+bx)^3}}{15b^2} + C$$

$$\int x^2\sqrt{a+bx}\,dx = \frac{2(8a^2-12abx+15b^2x^2)\sqrt{(a+bx)^3}}{105b^3} + C$$

$$\int \sqrt{a^2-x^2}\,dx = \frac{1}{2}\left[x\sqrt{a^2-x^2} + a^2\sin^{-1}\frac{x}{a}\right] + C, \quad a > 0$$

$$\int x\sqrt{a^2-x^2}\,dx = -\frac{1}{3}\sqrt{(a^2-x^2)^3} + C$$

$$\int x^2\sqrt{a^2-x^2}\,dx = -\frac{x}{4}\sqrt{(a^2-x^2)^3} + \frac{a^2}{8}\left(x\sqrt{a^2-x^2} + a^2\sin^{-1}\frac{x}{a}\right) + C, \quad a > 0$$

$$\int \sqrt{x^2 \pm a^2}\,dx = \frac{1}{2}\left[x\sqrt{x^2\pm a^2} \pm a^2\ln\left(x+\sqrt{x^2\pm a^2}\right)\right] + C$$

$$\int x\sqrt{x^2\pm a^2}\,dx = \frac{1}{3}\sqrt{(x^2\pm a^2)^3} + C$$

$$\int x^2\sqrt{x^2\pm a^2}\,dx = \frac{x}{4}\sqrt{(x^2\pm a^2)^3} \mp \frac{a^2}{8}x\sqrt{x^2\pm a^2} - \frac{a^4}{8}\ln\left(x+\sqrt{x^2\pm a^2}\right) + C$$

$$\int \frac{dx}{\sqrt{a+bx}} = \frac{2\sqrt{a+bx}}{b} + C$$

$$\int \frac{x\,dx}{\sqrt{x^2\pm a^2}} = \sqrt{x^2\pm a^2} + C$$

$$\int \frac{dx}{\sqrt{a+bx+cx^2}} = \frac{1}{\sqrt{c}}\ln\left[\sqrt{a+bx+cx^2} + x\sqrt{c} + \frac{b}{2\sqrt{c}}\right] + C, \quad c > 0$$

$$= \frac{1}{\sqrt{-c}}\sin^{-1}\left(\frac{-2cx-b}{\sqrt{b^2-4ac}}\right) + C, \quad c > 0$$

$$\int \sin x\,dx = -\cos x + C$$

$$\int \cos x\,dx = \sin x + C$$

$$\int x\cos(ax)\,dx = \frac{1}{a^2}\cos(ax) + \frac{x}{a}\sin(ax) + C$$

$$\int x^2\cos(ax)\,dx = \frac{2x}{a^2}\cos(ax) + \frac{a^2x^2-2}{a^3}\sin(ax) + C$$

$$\int e^{ax}\,dx = \frac{1}{a}e^{ax} + C$$

$$\int x\,e^{ax}\,dx = \frac{e^{ax}}{a^2}(ax-1) + C$$

$$\int \sinh x\,dx = \cosh x + C$$

$$\int \cosh x\,dx = \sinh x + C$$

APPENDIX B

Numerical and Computer Analysis

Occasionally the application of the laws of mechanics will lead to a system of equations for which a closed-form solution is difficult or impossible to obtain. When confronted with this situation, engineers will often use a numerical method which in most cases can be programmed on a microcomputer or "programmable" pocket calculator. Here we will briefly present a computer program for solving a set of linear algebraic equations, and three numerical methods which can be used to solve an algebraic or transcendental equation, evaluate a definite integral, and solve an ordinary differential equation. Application of each method will be explained by example, and an associated computer program written in Microsoft BASIC, which is designed to run on most personal computers, is provided.* A text on numerical analysis should be consulted for further discussion regarding a check of the accuracy of each method and the inherent errors that can develop from the methods.

B.1 Linear Algebraic Equations

Application of the equations of static equilibrium or the equations of motion sometimes requires solving a set of linear algebraic equations. The computer program listed in Fig. B–1 can be used for this purpose. It is based on the method of a Gaussian elimination and can solve at

*Similar types of programs can be written or purchased for programmable pocket calculators.

```
1 PRINT"Linear system of equations":PRINT
2 DIM A(10,11)
3 INPUT"Input number of equations : ",N
4 PRINT
5 PRINT"A  coefficients"
6 FOR I = 1 TO N
7 FOR J = 1 TO N
8 PRINT "A(";I;",";J;
9 INPUT")=",A(I,J)
10 NEXT J
11 NEXT I
12 PRINT
13 PRINT"B  coefficients"
14 FOR I = 1 TO N
15 PRINT "B(";I;
16 INPUT")=",A(I,N+1)
17 NEXT I
18 GOSUB 25
19 PRINT
20 PRINT"Unknowns"
21 FOR I = 1 TO N
22 PRINT "X(";I;")=";A(I,N+1)
23 NEXT I
24 END
25 REM Subroutine Guassian
26 FOR M=1 TO N
27 NP=M
28 BG=ABS(A(M,M))
29 FOR I = M TO N
30 IF ABS(A(I,M))<=BG THEN 33
31 BG=ABS(A(I,M))
32 NP=I
33 NEXT I
34 IF NP=M THEN 40
35 FOR I = M TO N+1
36 TE=A(M,I)
37 A(M,I)=A(NP,I)
38 A(NP,I)=TE
39 NEXT I
40 FOR I = M+1 TO N
41 FC=A(I,M)/A(M,M)
42 FOR J = M+1 TO N+1
43 A(I,J)=A(I,J)-FC*A(M,J)
44 NEXT J
45 NEXT I
46 NEXT M
47 A(N,N+1)=A(N,N+1)/A(N,N)
48 FOR I = N-1 TO 1 STEP -1
49 SM=0
50 FOR J=I+1 TO N
51 SM=SM+A(I,J)*A(J,N+1)
52 NEXT J
53 A(I,N+1)=(A(I,N+1)-SM)/A(I,I)
54 NEXT I
55 RETURN
```

Fig. B–1

most 10 equations with 10 unknowns. To do so, the equations should first be written in the following general format:

$$A_{11}x_1 + A_{12}x_2 + \cdots + A_{1n}x_n = B_1$$
$$A_{21}x_1 + A_{22}x_2 + \cdots + A_{2n}x_n = B_2$$
$$\vdots$$
$$A_{n1}x_1 + A_{n2}x_2 + \cdots + A_{nn}x_n = B_n$$

The "A" and "B" coefficients are "called" for when running the program. The output presents the unknowns $x_1, \ldots, x_n$.

EXAMPLE B–1

Solve the two equations

$$3x_1 + x_2 = 4$$
$$2x_1 - x_2 = 10$$

Solution

When the program begins to run, it first calls for the number of equations (2); then the A coefficients in the sequence $A_{11} = 3$, $A_{12} = 1$, $A_{21} = 2$, $A_{22} = -1$; and finally the B coefficients $B_1 = 4$, $B_2 = 10$. The output appears as

Unknowns

$X(1) = 2.8$ *Ans.*

$X(2) = -4.4$ *Ans.*

B.2 Simpson's Rule

Simpson's rule is a numerical method that can be used to determine the area under a curve given as a graph or as an explicit function $y = f(x)$. Likewise, it can be used to find the value of a definite integral which involves the function $y = f(x)$. To do so, the area must be subdivided into an *even number* of strips or intervals having a width h. The curve between three consecutive ordinates is approximated by a parabola, and the entire area or definite integral is then determined from the formula

$$\int_{x_0}^{x_n} f(x)\,dx \approx \frac{h}{3}\left[y_0 + 4(y_1 + y_3 + \cdots + y_{n-1}) + 2(y_2 + y_4 + \cdots + y_{n-2}) + y_n\right] \quad \text{(B–1)}$$

The computer program for this equation is given in Fig. B–2. For its use, we must first specify the function (on line 6 of the program). The upper and lower limits of the integral and the number of intervals are called for when the program is executed. The value of the integral is then given as the output.

```
1 PRINT"Simpson's rule":PRINT
2 PRINT" To execute this program :":PRINT
3 PRINT"    1- Modify right-hand side of the equation given below,
4 PRINT"       then press RETURN key"
5 PRINT"    2- Type  RUN 6":PRINT:EDIT 6
6 DEF FNF(X)=LOG(X)
7 PRINT:INPUT" Enter Lower Limit = ",A
8 INPUT" Enter Upper Limit = ",B
9 INPUT" Enter Number (even) of Intervals = ",N%
10 H=(B-A)/N%:AR=FNF(A):X=A+H
11 FOR J%=2 TO N%
12 K=2*(2-J%+2*INT(J%/2))
13 AR=AR+K*FNF(X)
14 X=X+H:NEXT J%
15 AR=H*(AR+FNF(B))/3
16 PRINT" Integral = ",AR
17 END
```

Fig. B–2

EXAMPLE B–2

Evaluate the definite integral

$$\int_2^5 \ln x \, dx$$

Solution

The interval $x_0 = 2$ to $x_6 = 5$ will be divided into six equal parts ($n = 6$), each having a width $h = (5 - 2)/6 = 0.5$. We then compute $y = f(x) = \ln x$ at each point of subdivision.

n	x_n	y_n
0	2	0.693
1	2.5	0.916
2	3	1.099
3	3.5	1.253
4	4	1.386
5	4.5	1.504
6	5	1.609

Thus, Eq. B–1 becomes

$$\begin{aligned}\int_2^5 \ln x \, dx &\approx \frac{0.5}{3}[0.693 + 4(0.916 + 1.253 + 1.504) \\ &\quad + 2(1.099 + 1.386) + 1.609] \\ &\approx 3.66\end{aligned}$$

Ans.

This answer is equivalent to the exact answer to three significant figures. Obviously, accuracy to a greater number of significant figures can be improved by selecting a smaller interval h (or larger n).

Using the computer program, we first specify the function ln x, line 6 in Fig. B–2. During execution, the program input requires the upper and lower limits 2 and 5 and the number of intervals $n = 6$. The output appears as

Integral = 3.66082 *Ans.*

B.3 The Secant Method

The secant method is used to find the real roots of an algebraic or transcendental equation $f(x) = 0$. The method derives its name from the fact that the formula used is established from the slope of the secant line to the graph $y = f(x)$. This slope is $[f(x_n) - f(x_{n-1})]/(x_n - x_{n-1})$, and the secant formula is

$$x_{n+1} = x_n - f(x_n)\left[\frac{x_n - x_{n-1}}{f(x_n) - f(x_{n-1})}\right] \tag{B–2}$$

For application it is necessary to provide two initial guesses, x_0 and x_1, and thereby evaluate x_2 from Eq. B–2 ($n = 1$). One then proceeds to reapply Eq. B–2 with x_1 and the calculated value of x_2 and obtain x_3 ($n = 2$), etc., until the value $x_{n+1} \approx x_n$. One can see this will occur if x_n is approaching the root of the function $f(x) = 0$, since the correction term on the right of Eq. B–2 will tend toward zero. In particular, the larger the slope, the smaller the correction to x_n, and the faster the root will be found. On the other hand, if the slope is very small in the neighborhood of the root, the method leads to large corrections for x_n, and convergence to the root is slow and may even lead to a failure to find it. In such cases other numerical techniques must be used for solution.

A computer program based on Eq. B–2 is listed in Fig. B–3. We must first specify the function on line 7 of the program. When the program is executed, two initial guesses, x_0 and x_1, must be entered in order to approximate the solution. The output specifies the value of the root. If it cannot be determined, this is so stated.

```
1 PRINT"Secant method":PRINT
2 PRINT" To execute this program :":PRINT
3 PRINT"     1) Modify right hand side of the equation given below,"
4 PRINT"        then press RETURN key."
5 PRINT"     2) Type  RUN 7"
6 PRINT:EDIT 7
7 DEF FNF(X)=.5*SIN(X)-2*COS(X)+1.3
8 INPUT"Enter point #1 =",X
9 INPUT"Enter point #2 =",X1
10 IF X=X1 THEN 14
11 EP=.00001:TL=2E-20
12 FP=(FNF(X1)-FNF(X))/(X1-X)
13 IF ABS(FP)>TL THEN 15
14 PRINT"Root can not be found.":END
15 DX=FNF(X1)/FP
16 IF ABS(DX)>EP THEN 19
17 PRINT "Root = ";X1;"          Function evaluated at this root = ";FNF(X1)
18 END
19 X=X1:X1=X1-DX
20 GOTO 12
```

Fig. B–3

EXAMPLE B–3

Determine the root of the equation

$$f(x) = 0.5 \sin x - 2 \cos x + 1.30 = 0$$

Solution

Guesses of the initial roots will be $x_0 = 45°$ and $x_1 = 30°$. Applying Eq. B–2,

$$x_2 = 30° - (-0.1821)\frac{(30° - 45°)}{(-0.1821 - 0.2393)} = 36.48°$$

Using this value in Eq. B–2, along with $x_1 = 30°$, we have

$$x_3 = 36.48° - (-0.0108)\frac{36.48° - 30°}{(-0.0108 + 0.1821)} = 36.89°$$

Repeating the process with this value and $x_2 = 36.48°$ yields

$$x_4 = 36.89° - (0.0005)\left[\frac{36.89° - 36.48°}{(0.0005 + 0.0108)}\right] = 36.87°$$

Thus $x = 36.9°$ is appropriate to three significant figures.

If the problem is solved using the computer program, first we specify the function, line 7 in Fig. B–3. During execution, the first and second guesses must be entered in radians. Choosing these to be 0.8 rad and 0.5 rad, the result appears as

Root = 0.6435022

Function evaluated at this root = 1.66893E-06

This result converted from radians to degrees is therefore

$$x = 36.9° \qquad \textit{Ans.}$$

B.4 Runge-Kutta Method

The Runge-Kutta method is used to solve an ordinary differential equation. It consists of applying a set of formulas which are used to find specific values of y for corresponding incremental values h in x. The formulas given in general form are as follows:

First-Order Equation. To integrate $\dot{x} = f(t, x)$ step by step, use

$$x_{i+1} = x_i + \frac{1}{6}(k_1 + 2k_2 + 2k_3 + k_4) \tag{B-3}$$

where

$$\begin{aligned} k_1 &= hf(t_i, x_i) \\ k_2 &= hf\left(t_i + \frac{h}{2}, x_i + \frac{k_1}{2}\right) \\ k_3 &= hf\left(t_i + \frac{h}{2}, x_i + \frac{k_2}{2}\right) \\ k_4 &= hf(t_i + h, x_i + k_3) \end{aligned} \tag{B-4}$$

Second-Order Equation. To integrate $\ddot{x} = f(t, x, \dot{x})$, use

$$\begin{aligned} x_{i+1} &= x_i + h\left[\dot{x}_i + \frac{1}{6}(k_1 + k_2 + k_3)\right] \\ \dot{x}_{i+1} &= \dot{x}_i + \frac{1}{6}(k_1 + 2k_2 + 2k_3 + k_4) \end{aligned} \tag{B-5}$$

where

$$\begin{aligned} k_1 &= hf(t_i, x_i, \dot{x}_i) \\ k_2 &= hf\left(t_i + \frac{h}{2}, x_i + \frac{h}{2}\dot{x}_i, \dot{x}_i + \frac{k_1}{2}\right) \\ k_3 &= hf\left(t_i + \frac{h}{2}, x_i + \frac{h}{2}\dot{x}_i + \frac{h}{4}k_1, \dot{x}_i + \frac{k_2}{2}\right) \\ k_4 &= hf\left(t_i + h, x_i + h\dot{x}_i + \frac{h}{2}k_2, \dot{x}_i + k_3\right) \end{aligned} \tag{B-6}$$

To apply these equations, one starts with initial values $t_i = t_0$, $x_i = x_0$ and $\dot{x}_i = \dot{x}_0$ (for the second-order equation). Choosing an increment h for t_0, the four constants k are computed, and these results are substituted into Eq. B–3 or B–5 in order to compute $x_{i+1} = x_1$, $\dot{x}_{i+1} = x_1$, corresponding to $t_{i+1} = t_1 = t_0 + h$. Repeating this process using $t_1, x_1, \dot{x}_1$ and h, values for $x_2, \dot{x}_2$ and $t_2 = t_1 + h$ are then computed, etc.

Computer programs which solve first- and second-order differential equations by this method are listed in Figs. B–4 and B–5, respectively. In order to use these programs, the operator specifies the function $\dot{x} = f(t, x)$ or $\ddot{x} = f(t, x, \dot{x})$ (line 7), the initial values t_0, x_0, $\dot{x}_0$ (for second-order equation), the final time t_n, and the step size h. The output gives the values of t, x, and $\dot{x}$ for each time increment until t_n is reached.

```
1 PRINT"Runge-Kutta Method for 1-st order Differential Equation":PRINT
2 PRINT" To execute this program :":PRINT
3 PRINT"   1) Modify right hand side of the equation given below,"
4 PRINT"        then Press RETURN key"
5 PRINT"   2) Type  RUN 7"
6 PRINT:EDIT 7
7 DEF FNF(T,X)=5*T+X
8 CLS:PRINT" Initial Conditions":PRINT
9 INPUT"Input  t    = ",T
10 INPUT"        x    = ",X
11 INPUT"Final  t    = ",T1
12 INPUT"step size  = ",H:PRINT
13 PRINT"         t             x"
14 IF T>=T1+H THEN 23
15 PRINT USING"######.#####";T;X
16 K1=H*FNF(T,X)
17 K2=H*FNF(T+.5*H,X+.5*K1)
18 K3=H*FNF(T+.5*H,X+.5*K2)
19 K4=H*FNF(T+H,X+K3)
20 T=T+H
21 X=X+(K1+K2+K2+K3+K3+K4)/6
22 GOTO 14
23 END
```

Fig. B–4

```
1 PRINT"Runge-Kutta Method for 2-nd order Differential Equation":PRINT
2 PRINT" To execute this program :":PRINT
3 PRINT"   1) Modify right hand side of the equation given below,"
4 PRINT"        then Press RETURN key"
5 PRINT"   2) Type  RUN 7"
6 PRINT:EDIT 7
7 DEF FNF(T,X,XD)=
8 INPUT"Input  t    = ",T
9 INPUT"        x    = ",X
10 INPUT"     dx/dt = ",XD
11 INPUT"Final  t    = ",T1
12 INPUT"step size  = ",H:PRINT
13 PRINT"         t             x            dx/dt"
14 IF T>=T1+H THEN 24
15 PRINT USING"######.#####";T;X;XD
16 K1=H*FNF(T,X,XD)
17 K2=H*FNF(T+.5*H,X+.5*H*XD,XD+.5*K1)
18 K3=H*FNF(T+.5*H,X+(.5*H)*(XD+.5*K1),XD+.5*K2)
19 K4=H*FNF(T+H,X+H*XD+.5*H*K2,XD+K3)
20 T=T+H
21 X=X+H*XD+H*(K1+K2+K3)/6
22 XD=XD+(K1+K2+K2+K3+K3+K4)/6
23 GOTO 14
24 END
```

Fig. B–5

EXAMPLE B–4

Solve the differential equations $\dot{x} = 5t + x$. Obtain the results for two steps using time increments of $h = 0.02$ s. At $t_0 = 0, x_0 = 0$.

Solution

This is a first-order equation, so Eqs. B–3 and B–4 apply. Thus, for $t_0 = 0, x_0 = 0, h = 0.02$, we have

$$k_1 = 0.02(0 + 0) = 0$$
$$k_2 = 0.02[5(0.01) + 0] = 0.001$$
$$k_3 = 0.02[5(0.01) + 0.0005] = 0.00101$$
$$k_4 = 0.02[5(0.02) + 0.00101] = 0.00202$$
$$x_1 = 0 + \tfrac{1}{6}[0 + 2(0.001) + 2(0.00101) + 0.00202] = 0.00101$$

Using the values $t_1 = 0 + 0.02 = 0.02$ and $x_1 = 0.00101$ with $h = 0.02$, the value for x_2 is now computed from Eqs. B–3 and B–4.

$$k_1 = 0.02[5(0.02) + 0.00101] = 0.00202$$
$$k_2 = 0.02[5(0.03) + 0.00202] = 0.00304$$
$$k_3 = 0.02[5(0.03) + 0.00253] = 0.00305$$
$$k_4 = 0.02[5(0.04) + 0.00406] = 0.00408$$
$$x_2 = 0.00101 + \tfrac{1}{6}[0.00202 + 2(0.00304) + 2(0.00305) + 0.00408]$$
$$= 0.00405$$ *Ans.*

To solve this problem using the computer program in Fig. B–4, the function is first entered on line 7, then the data $t_0 = 0, x_0 = 0, t_n = 0.04$, and $h = 0.02$ is specified. The results appear as

t	x
0.00000	0.00000
0.02000	0.00101
0.04000	0.00405

Ans.

APPENDIX C

Vector Analysis

The following discussion provides a brief review of vector analysis. A more detailed treatment of these topics is given in *Engineering Mechanics: Statics.*

Vector. A vector, **A,** is a quantity which has magnitude and direction, and adds according to the parallelogram law. As shown in Fig. C–1, **A** = **B** + **C,** where **A** is the *resultant vector* and **B** and **C** are *component vectors.*

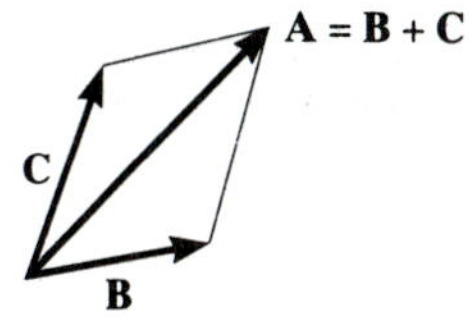

Fig. C–1

Unit Vector. A unit vector, $\mathbf{u}_A$, has a magnitude of one "dimensionless" unit and acts in the same direction as **A.** It is determined by dividing **A** by its magnitude A, i.e,

$$\mathbf{u}_A = \frac{\mathbf{A}}{A} \tag{C–1}$$

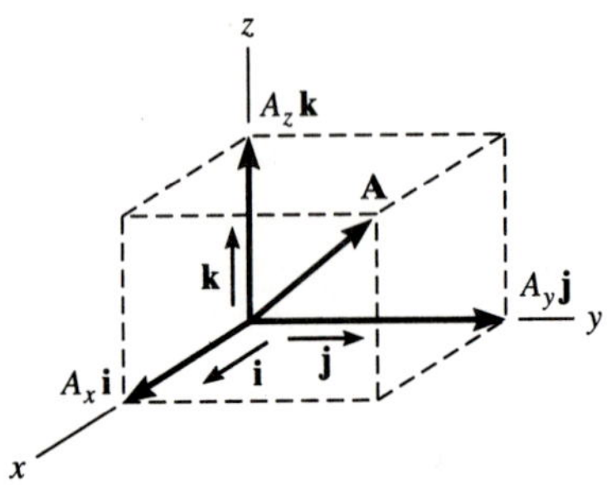

Fig. C–2

Cartesian Vector Notation. The directions of the positive x, y, z axes are defined by the Cartesian unit vectors **i, j, k,** respectively.

As shown in Fig. C–2, vector **A** is formulated by the addition of its x, y, z components as

$$\mathbf{A} = A_x\mathbf{i} + A_y\mathbf{j} + A_z\mathbf{k} \qquad \text{(C–2)}$$

The *magnitude* of **A** is determined from

$$A = \sqrt{A_x^2 + A_y^2 + A_z^2} \qquad \text{(C–3)}$$

The *direction* of **A** is defined in terms of its *coordinate direction angles,* α, β, γ, measured from the *tail* of **A** to the *positive* x, y, z axes, Fig. C–3. These angles are determined from the *direction cosines* which represent the **i, j, k** components of the unit vector $\mathbf{u}_A$; i.e., from Eqs. C–1 and C–2.

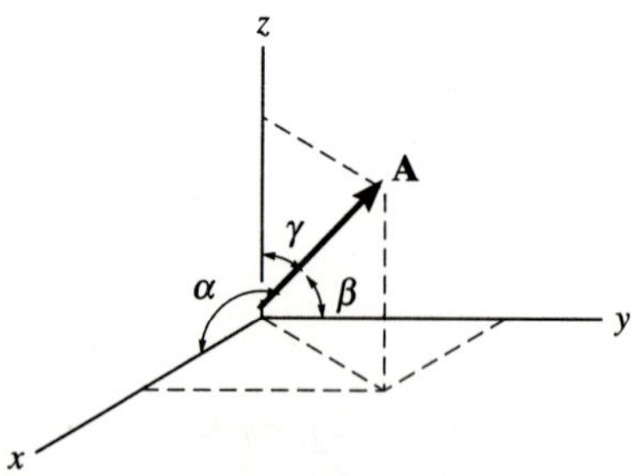

Fig. C–3

$$\mathbf{u}_A = \frac{A_x}{A}\mathbf{i} + \frac{A_y}{A}\mathbf{j} + \frac{A_z}{A}\mathbf{k} \qquad \text{(C–4)}$$

so that the direction cosines are

$$\cos\alpha = \frac{A_x}{A} \qquad \cos\beta = \frac{A_y}{A} \qquad \cos\gamma = \frac{A_z}{A} \qquad \text{(C–5)}$$

Hence, $\mathbf{u}_A = \cos\alpha\mathbf{i} + \cos\beta\mathbf{j} + \cos\gamma\mathbf{k}$, and using Eq. C–3, it is seen that

$$\cos^2\alpha + \cos^2\beta + \cos^2\gamma = 1 \qquad \text{(C–6)}$$

The Cross Product. The cross product of two vectors **A** and **B,** which yields the resultant vector **C,** is written as

$$\mathbf{C} = \mathbf{A} \times \mathbf{B} \qquad \text{(C–7)}$$

and reads **C** equals **A** "cross" **B.** The *magnitude* of **C** is

$$C = AB\sin\theta \qquad \text{(C–8)}$$

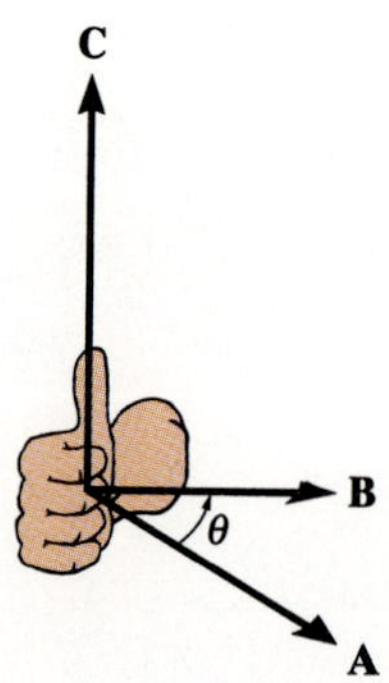

Fig. C–4

where θ is the angle made between the *tails* of **A** and **B** ($0° \leq \theta \leq 180°$). The *direction* of **C** is determined by the right-hand rule, whereby the fingers of the right hand are curled *from* **A** *to* **B** and the thumb points in the direction of **C,** Fig. C–4. This vector is perpendicular to the plane containing vectors **A** and **B.**

The vector cross product is *not* commutative, i.e., $\mathbf{A} \times \mathbf{B} \neq \mathbf{B} \times \mathbf{A}$. Rather,

$$\mathbf{A} \times \mathbf{B} = -\mathbf{B} \times \mathbf{A} \qquad \text{(C–9)}$$

The distributive law is valid; i.e.,

$$\mathbf{A} \times (\mathbf{B} + \mathbf{D}) = \mathbf{A} \times \mathbf{B} + \mathbf{A} \times \mathbf{D} \qquad \text{(C–10)}$$

And the cross product may be multiplied by a scalar m in any manner; i.e.,

$$m(\mathbf{A} \times \mathbf{B}) = (m\mathbf{A}) \times \mathbf{B} = \mathbf{A} \times (m\mathbf{B}) = (\mathbf{A} \times \mathbf{B})m \qquad \text{(C–11)}$$

Equation C–7 can be used to find the cross product of any pair of Cartesian unit vectors. For example, to find $\mathbf{i} \times \mathbf{j}$, the magnitude is $(i)(j) \sin 90° = (1)(1)(1) = 1$, and its direction $+\mathbf{k}$ is determined from the right-hand rule, applied to $\mathbf{i} \times \mathbf{j}$, Fig. C–2. A simple scheme shown in Fig. C–5 may be helpful in obtaining this and other results when the need arises. If the circle is constructed as shown, then "crossing" two of the unit vectors in a *counterclockwise* fashion around the circle yields a *positive* third unit vector, e.g., $\mathbf{k} \times \mathbf{i} = \mathbf{j}$. Moving *clockwise*, a *negative* unit vector is obtained, e.g., $\mathbf{i} \times \mathbf{k} = -\mathbf{j}$.

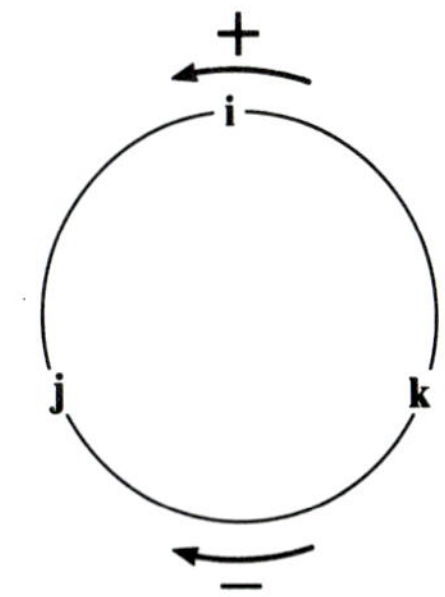

Fig. C–5

If $\mathbf{A}$ and $\mathbf{B}$ are expressed in Cartesian component form, then the cross product, Eq. C–7, may be evaluated by expanding the determinant

$$\mathbf{C} = \mathbf{A} \times \mathbf{B} = \begin{vmatrix} \mathbf{i} & \mathbf{j} & \mathbf{k} \\ A_x & A_y & A_z \\ B_x & B_y & B_z \end{vmatrix} \qquad \text{(C–12)}$$

which yields

$$\mathbf{C} = (A_yB_z - A_zB_y)\mathbf{i} - (A_xB_z - A_zB_x)\mathbf{j} + (A_xB_y - A_yB_x)\mathbf{k}$$

Recall that the cross product is used in statics to define the moment of a force $\mathbf{F}$ about point O, in which case

$$\mathbf{M}_O = \mathbf{r} \times \mathbf{F} \qquad \text{(C–13)}$$

where $\mathbf{r}$ is a position vector directed from point O to *any point* on the line of action of $\mathbf{F}$.

The Dot Product. The dot product of two vectors $\mathbf{A}$ and $\mathbf{B}$, which yields a scalar, is defined as

$$\mathbf{A} \cdot \mathbf{B} = AB \cos \theta \qquad \text{(C–14)}$$

and reads $\mathbf{A}$ "dot" $\mathbf{B}$. The angle θ is formed between the *tails* of $\mathbf{A}$ and $\mathbf{B}$ $(0° \le \theta \le 180°)$.

The dot product is commutative; i.e.,

$$\mathbf{A} \cdot \mathbf{B} = \mathbf{B} \cdot \mathbf{A} \qquad \text{(C–15)}$$

The distributive law is valid; i.e.,

$$\mathbf{A} \cdot (\mathbf{B} + \mathbf{D}) = \mathbf{A} \cdot \mathbf{B} + \mathbf{A} \cdot \mathbf{D} \qquad \text{(C–16)}$$

And scalar multiplication can be performed in any manner; i.e.,

$$m(\mathbf{A}\cdot\mathbf{B}) = (m\mathbf{A})\cdot\mathbf{B} = \mathbf{A}\cdot(m\mathbf{B}) = (\mathbf{A}\cdot\mathbf{B})m \qquad \text{(C–17)}$$

Using Eq. C–14, the dot product between any two Cartesian vectors can be determined. For example, $\mathbf{i}\cdot\mathbf{i} = (1)(1)\cos 0° = 1$ and $\mathbf{i}\cdot\mathbf{j} = (1)(1)\cos 90° = 0$.

If **A** and **B** are expressed in Cartesian component form, then the dot product, Eq. C–14, can be determined from

$$\mathbf{A}\cdot\mathbf{B} = A_xB_x + A_yB_y + A_zB_z \qquad \text{(C–18)}$$

The dot product may be used to determine the *angle θ formed between two vectors.* From Eq. C–14,

$$\theta = \cos^{-1}\left(\frac{\mathbf{A}\cdot\mathbf{B}}{AB}\right) \qquad \text{(C–19)}$$

It is also possible to find the *component of a vector in a given direction* using the dot product. For example, the magnitude of the component (or projection) of vector **A** in the direction of **B**, Fig. C–6, is defined by $A\cos\theta$. From Eq. C–14, this magnitude is

$$A\cos\theta = \mathbf{A}\cdot\frac{\mathbf{B}}{B} = \mathbf{A}\cdot\mathbf{u}_B \qquad \text{(C–20)}$$

where $\mathbf{u}_B$ represents a unit vector acting in the direction of **B**, Fig. C–6.

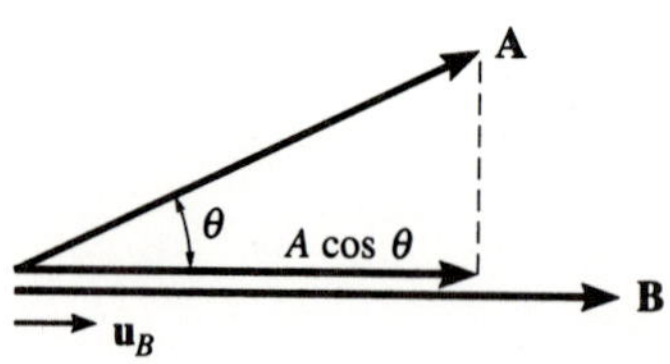

Fig. C–6

Differentiation and Integration of Vector Functions. The rules for differentiation and integration of the sums and products of scalar functions also apply to vector functions. Consider, for example, the two vector functions $\mathbf{A}(s)$ and $\mathbf{B}(s)$. Provided these functions are smooth and continuous for all s, then

$$\frac{d}{ds}(\mathbf{A}+\mathbf{B}) = \frac{d\mathbf{A}}{ds} + \frac{d\mathbf{B}}{ds} \qquad \text{(C–21)}$$

$$\int(\mathbf{A}+\mathbf{B})\,ds = \int\mathbf{A}\,ds + \int\mathbf{B}\,ds \qquad \text{(C–22)}$$

For the cross product,

$$\frac{d}{ds}(\mathbf{A}\times\mathbf{B}) = \left(\frac{d\mathbf{A}}{ds}\times\mathbf{B}\right) + \left(\mathbf{A}\times\frac{d\mathbf{B}}{ds}\right) \qquad \text{(C–23)}$$

Similarly, for the dot product,

$$\frac{d}{ds}(\mathbf{A}\cdot\mathbf{B}) = \frac{d\mathbf{A}}{ds}\cdot\mathbf{B} + \mathbf{A}\cdot\frac{d\mathbf{B}}{ds} \qquad \text{(C–24)}$$

APPENDIX D

Review for the Fundamentals of Engineering Examination

The Fundamentals of Engineering (FE) exam is given semiannually by the National Council of Engineering Examiners (NCEE) and is one of the requirements for obtaining a Professional Engineering License. A portion of this exam contains problems in dynamics, and this appendix provides a review of the subject matter most often asked on this exam. Before solving any of the problems, you should review the sections indicated in each chapter in order to become familiar with the boldfaced definitions and the procedures used to solve the various types of problems. Also, review the example problems in these sections.

The following problems are arranged in the same sequence as the topics in each chapter. Besides helping as preparation for the FE exam, these problems also provide additional examples for general practice of the subject matter. Partial solutions and answers to *all the problems* are given at the back of this appendix.

Chapter 12—Review Sections 12.1, 12.4–12.6, 12.8–12.9

D-1. The position of a particle is $s = (0.5t^3 + 4t)$ ft, where t is in seconds. Determine the velocity and the acceleration of the particle when $t = 3$ s.

D-2. After traveling a distance of 100 m, a particle reaches a velocity of 30 m/s, starting from rest. Determine its constant acceleration.

D-3. A particle moves in a straight line such that $s = (12t^3 + 2t^2 + 3t)$ m, where t is in seconds. Determine the velocity and acceleration of the particle when $t = 2$ s.

D-4. A particle moves along a straight line such that $a = (4t^2 - 2)$ m/s², where t is in seconds. When $t = 0$, the particle is located 2 m to the left of the origin, and when $t = 2$ s, it is 20 m to the left of the origin. Determine the position of the particle when $t = 4$ s.

D-5. Determine the speed at which the basketball at A must be thrown at the angle of 30° so that it makes it to the basket at B. At what speed does it pass through the hoop?

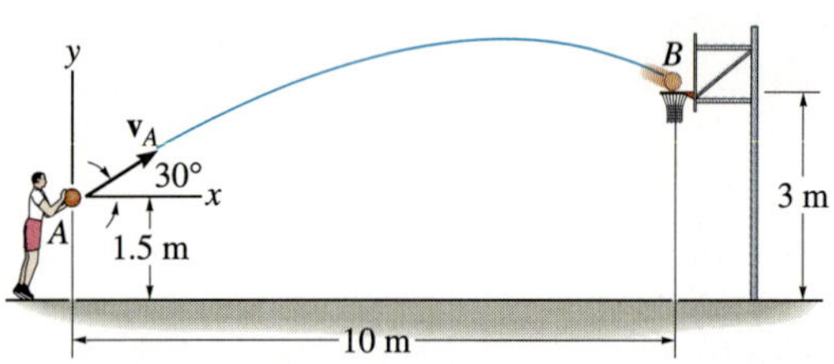

Prob. D–5

D-6. A particle moves with curvilinear motion in the x-y plane such that the y component of motion is described by the equation $y = (7t^3)$ m, where t is in seconds. If the particle starts from rest at the origin when $t = 0$, and maintains a *constant* acceleration in the x direction of 12 m/s², determine the particle's speed when $t = 2$ s.

D-7. Water is sprayed at an angle of 90° from the slope at 20 m/s. Determine the range R.

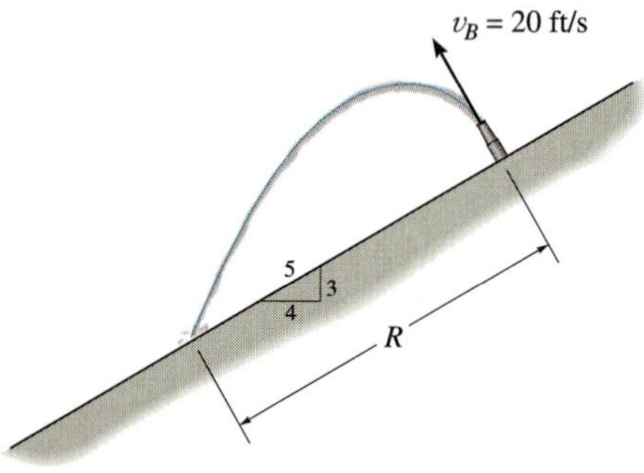

Prob. D–7

D-8. An automobile is traveling with a *constant speed* along a horizontal circular curve that has a radius of $\rho = 250$ m. If the magnitude of acceleration is $a = 1.5$ m/s², determine the speed at which the automobile is traveling.

D-9. A boat is traveling along a circular path having a radius of 30 m. Determine the magnitude of the boat's acceleration if at a given instant the boat's speed is $v = 6$ m/s and the rate of increase in speed is $\dot{v} = 2$ m/s².

D-10. A train travels along a horizontal circular curve that has a radius of 600 m. If the speed of the train is uniformly increased from 40 km/h to 60 km/h in 5 s, determine the magnitude of the acceleration at the instant the speed of the train is 50 km/h.

D-11. At a given instant, the automobile has a speed of 25 m/s and an acceleration of 3 m/s² acting in the direction shown. Determine the radius of curvature of the path and the rate of increase of the automobile's speed.

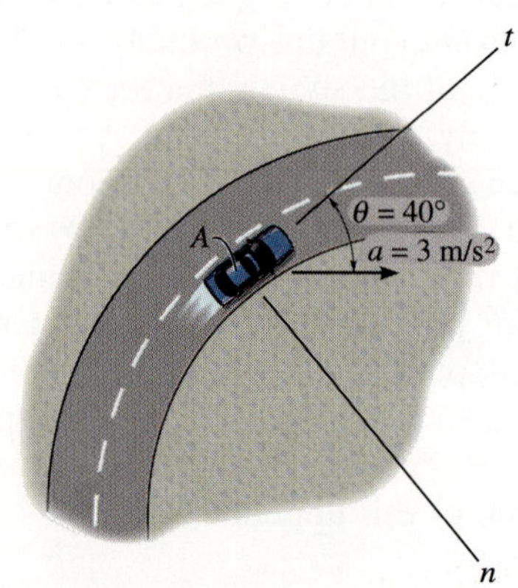

Prob. D–11

D-12. At the instant shown, cars A and B are traveling at the speeds shown. If B is accelerating at 1200 km/h^2 while A maintains a constant speed, determine the velocity and acceleration of A with respect to B.

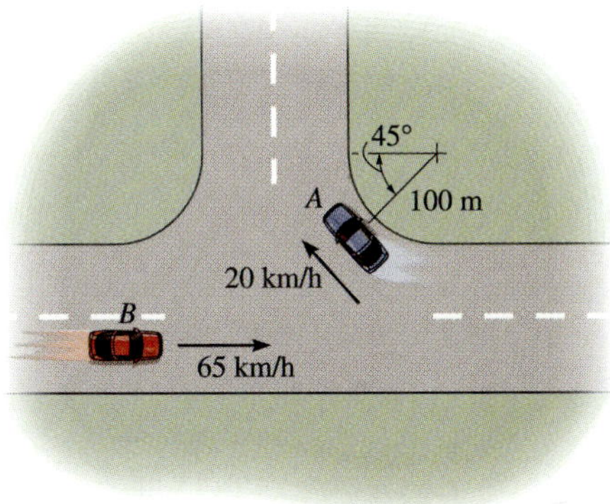

Prob. D–12

D-13. Determine the speed of point P on the cable in order to lift the platform at 2 m/s.

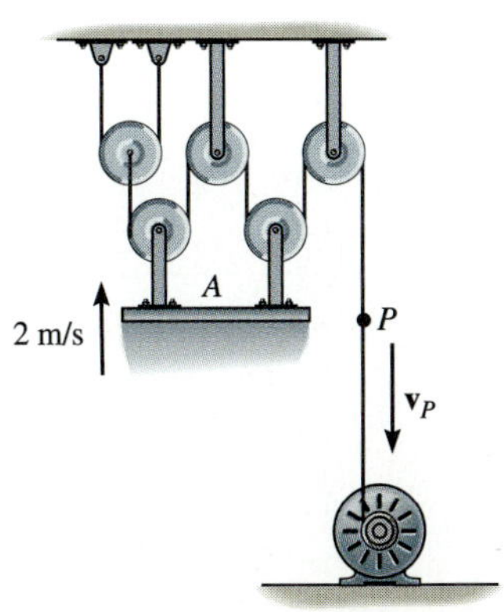

Prob. D–13

Chapter 13—Review Sections 13.1–13.5

D-14. The effective weight of a man in an elevator varies between 130 lb and 170 lb while he is riding in the elevator. When the elevator is at rest the man weighs 153 lb. Determine how fast the elevator car can accelerate, going up and going down.

D-15. Neglecting friction and the mass of the pulley and cord, determine the acceleration at which the 4-kg block B will descend. What is the tension in the cord? Block A has a mass of 2 kg.

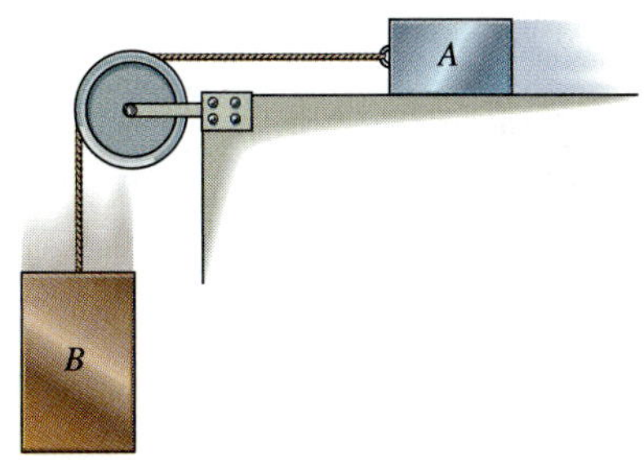

Prob. D–15

D-16. The blocks are suspended over a pulley by a rope. Neglecting the mass of the rope and the pulley, determine the acceleration of both blocks and the tension in the rope.

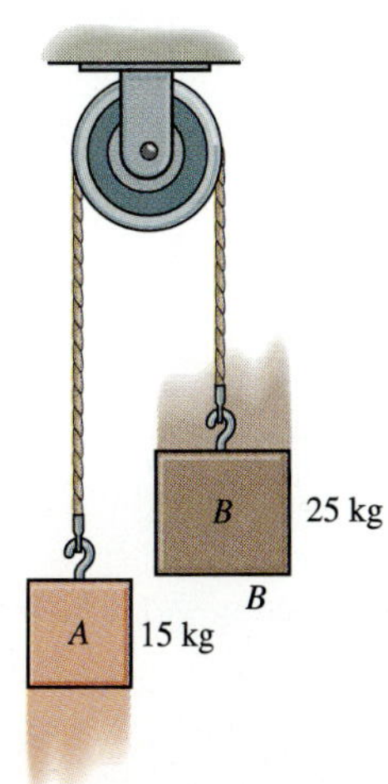

Prob. D–16

D-17. Block B rests upon a smooth surface. If the coefficients of static and kinetic friction between A and B are $\mu_s = 0.4$ and $\mu_k = 0.3$, respectively, determine the acceleration of each block if $P = 6$ lb.

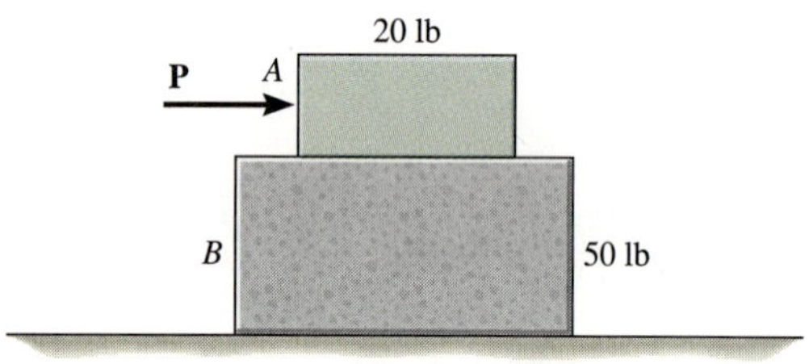

Prob. D–17

D-18. The block rests at a distance of 2 m from the center of the platform. If the coefficient of static friction between the block and the platform is $\mu_s = 0.3$, determine the maximum speed which the block can attain before it begins to slip. Assume the angular motion of the disk is slowly increasing.

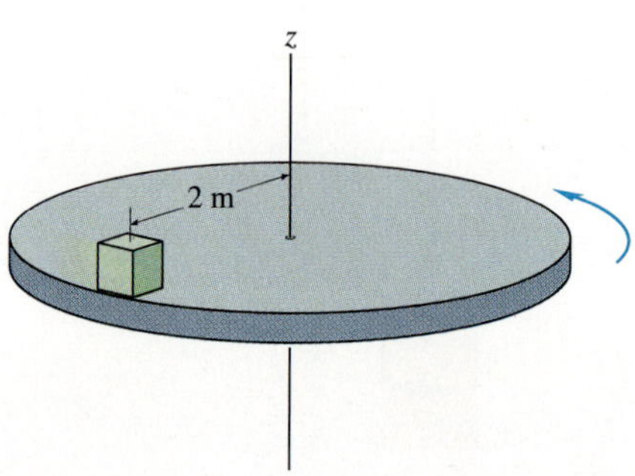

Prob. D–18

D-19. Determine the maximum speed that the jeep can travel over the crest of the hill and not lose contact with the road.

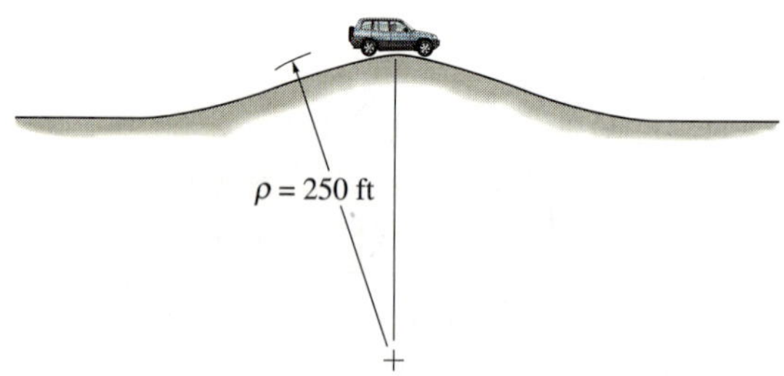

Prob. D–19

D-20. A pilot weighs 150 lb and is traveling at a constant speed of 120 ft/s. Determine the normal force he exerts on the seat of the plane when he is upside down at A. The loop has a radius of curvature of 400 ft.

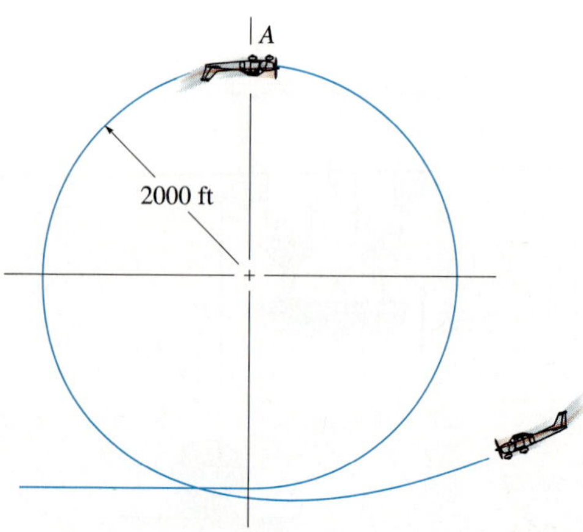

Prob. D–20

D-21. The sports car is traveling along a 20° banked road having a radius of curvature of $\rho = 500$ ft. If the coefficient of static friction between the tires and the road is $\mu_s = 0.2$, determine the maximum safe speed for travel so no slipping occurs. Neglect the size of the car.

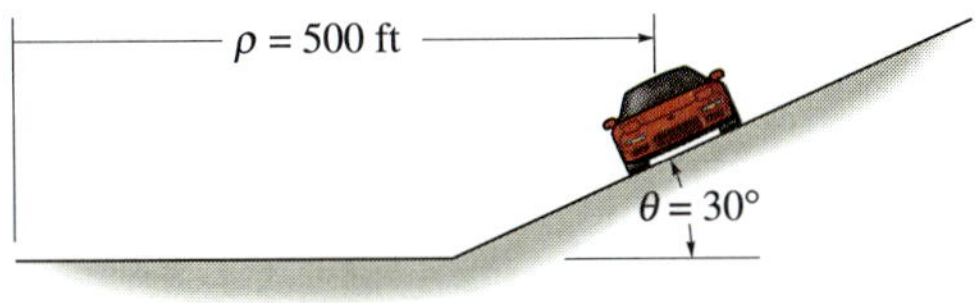

Prob. D–21

D-22. The 5-lb pendulum bob B is released from rest when $\theta = 0°$. Determine the tension in string BC immediately after it is released and when the pendulum reaches point D, where $\theta = 90°$.

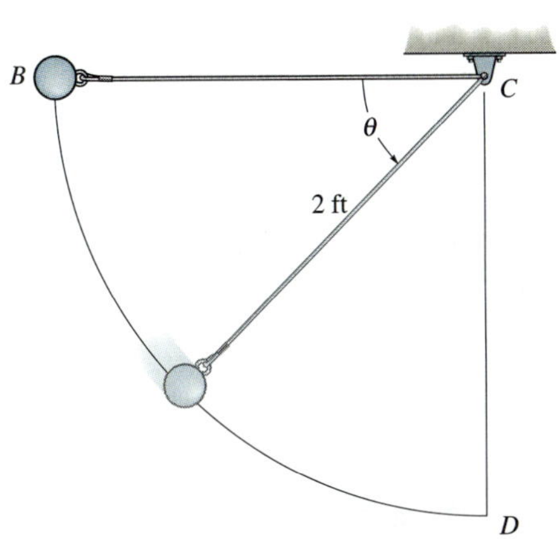

Prob. D–22

Chapter 14—Review All Sections

D-23. A 15 000-lb freight car is pulled along a horizontal track. If the car starts from rest and attains a velocity of 40 ft/s after traveling a distance of 300 ft, determine the total work done on the car in this distance if the rolling frictional force between the car and track is 80 lb.

D-24. The 20-lb block resting on the 30° inclined plane is acted upon by a 40-lb force. If the block's initial velocity is 5 ft/s down the plane, determine its velocity after it has traveled 10 ft down the plane. The coefficient of kinetic friction between the block and the plane is $\mu_k = 0.2$.

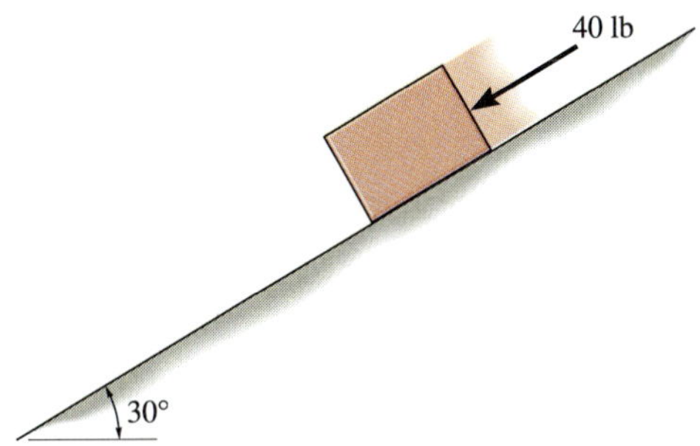

Prob. D–24

D-25. The 3-kg block is subjected to the action of the two forces shown. If the block starts from rest, determine the distance it has moved when it attains a velocity of 10 m/s. The coefficient of kinetic friction between the block and the surface is $\mu_k = 0.2$.

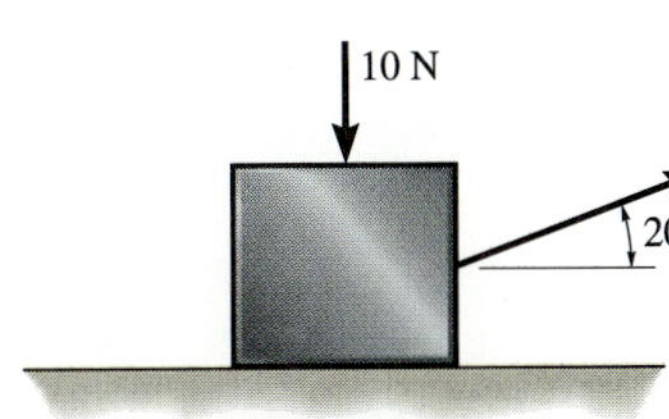

Prob. D–25

D-26. The 6-lb ball is to be fired from rest using a spring having a stiffness of $k = 40$ lb/ft. Determine how far the spring must be compressed so that when the ball reaches a height of 8 ft it has a velocity of 6 ft/s.

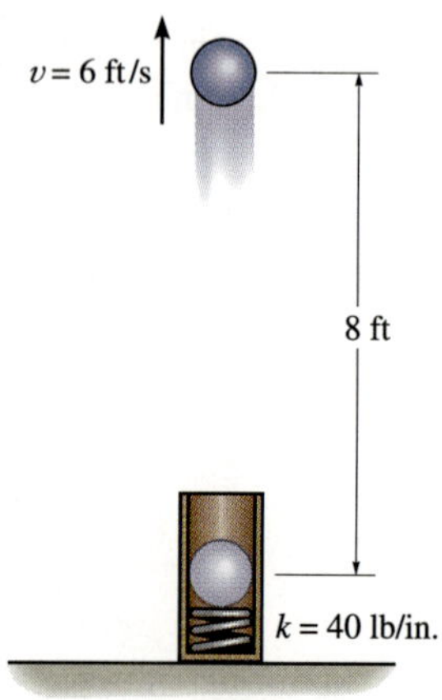

Prob. D–26

D-27. The 2-kg pendulum bob is released from rest when it is at A. Determine the speed of the bob and the tension in the cord when the bob passes through its lowest position, B.

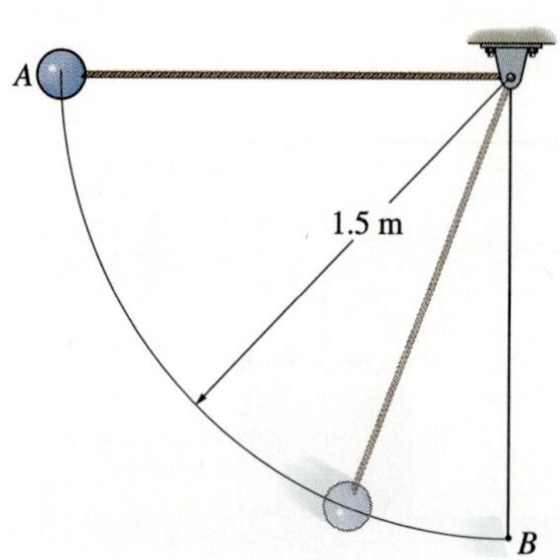

Prob. D–27

D-28. The 5-lb collar is released from rest at A and travels along the frictionless guide. Determine the speed of the collar when it strikes the stop B. The spring has an unstretched length of 0.5 ft.

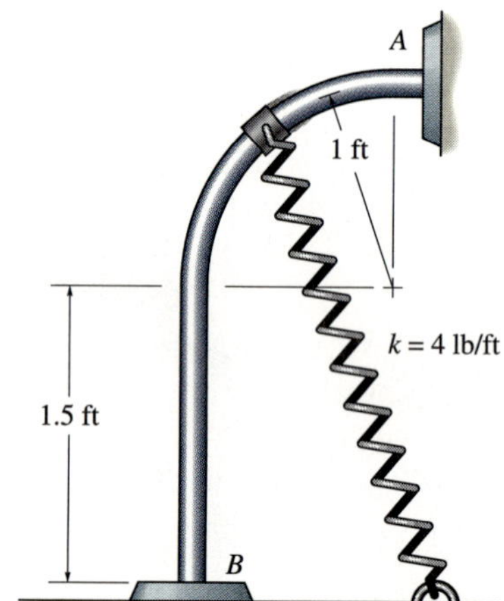

Prob. D–28

D-29. The 2-kg collar is given a downward velocity of 4 m/s when it is at A. If the spring has an unstretched length of 1 m and a stiffness of $k = 30$ N/m, determine the velocity of the collar at $s = 1$ m.

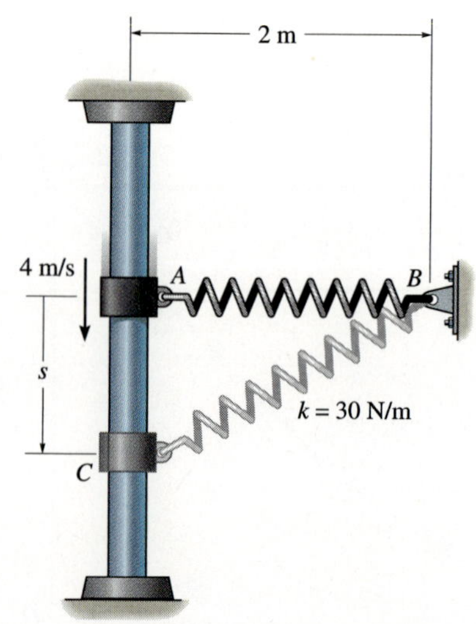

Prob. D–29

Chapter 15—Review Sections 15.1–15.4

D-30. A 30-ton engine exerts a constant horizontal force of $40(10^3)$ lb on a train having three cars that have a total weight of 250 tons. If the rolling resistance is 10 lb per ton for both the engine and cars, determine how long it takes to increase the speed of the train from 20 ft/s to 30 ft/s. What is the driving force exerted by the engine wheels on the tracks?

D-31. A 5-kg block is moving up a 30° inclined plane with an initial velocity of 3 m/s. If the coefficient of kinetic friction between the block and the plane is $\mu_k = 0.3$, determine how long a 100-N horizontal force must act on the block in order to increase the velocity of the block to 10 m/s up the plane.

D-32. The 10-lb block A attains a velocity of 1 ft/s in 5 seconds, starting from rest. Determine the tension in the cord and the coefficient of kinetic friction between block A and the horizontal plane. Neglect the weight of the pulley. Block B has a weight of 8 lb.

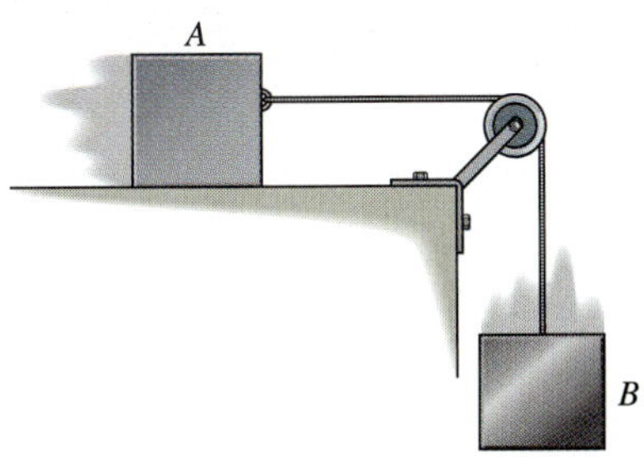

Prob. D–32

D-33. Determine the velocity of each block 10 seconds after the blocks are released from rest. Neglect the mass of the pulleys.

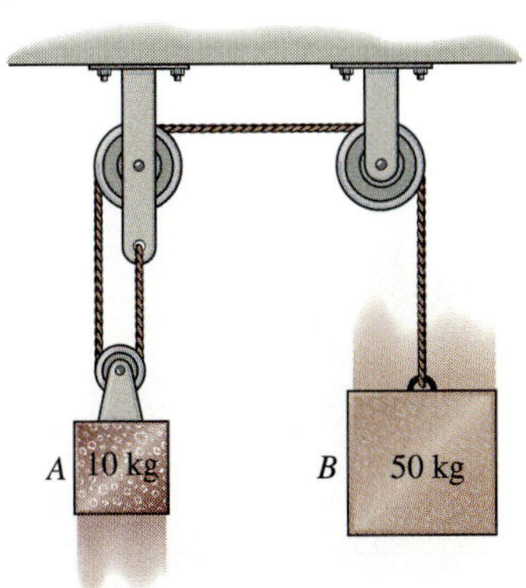

Prob. D–33

D-34. The two blocks have a coefficient of restitution of $e = 0.5$. If the surface is smooth, determine the velocity of each block after impact.

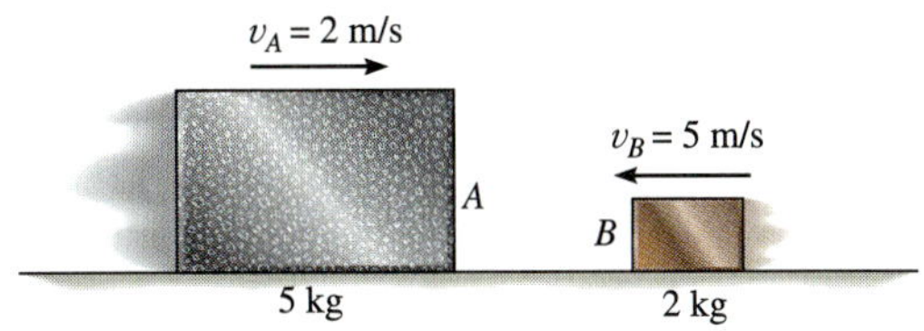

Prob. D–34

D-35. A 6-kg disk A has an initial velocity of $(v_A)_1 = 20$ m/s and strikes head-on disk B that has a mass of 24 kg and is originally at rest. If the collision is perfectly elastic, determine the speed of each disk after the collision and the impulse which disk A imparts to disk B.

D-36. Blocks A and B weigh 5 lb and 10 lb, respectively. After striking block B, A slides 2 in. to the right, and B slides 3 in. to the right. If the coefficient of kinetic friction between the blocks and the surface is $\mu_k = 0.2$, determine the coefficient of restitution between the blocks. Block B is originally at rest.

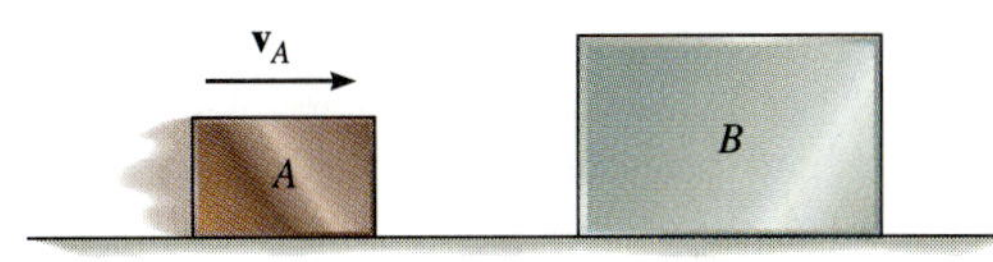

Prob. D–36

D-37. Disk A weighs 2 lb and is sliding on the smooth horizontal plane with a velocity of 3 ft/s. Disk B weighs 11 lb and is initially at rest. If after the impact A has a velocity of 1 ft/s, directed along the positive x axis, determine the speed of disk B after impact.

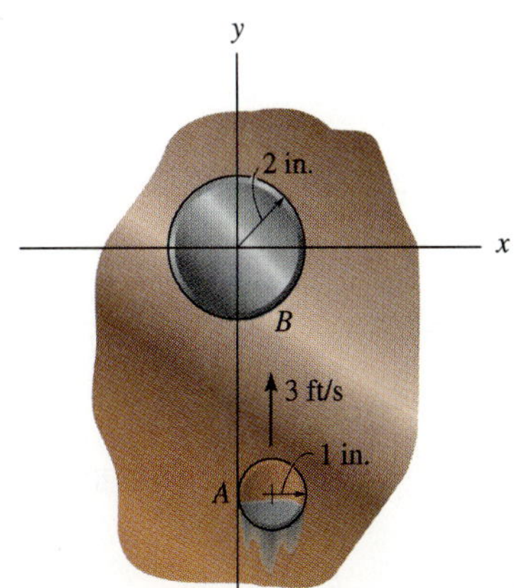

Prob. D–37

Chapter 16—Review Sections 16.3, 16.5–16.7

D-38. If gear A is rotating clockwise with an angular velocity of $\omega_A = 3$ rad/s, determine the angular velocities of gears B and C. Gear B is one unit, having radii of 2 in. and 5 in.

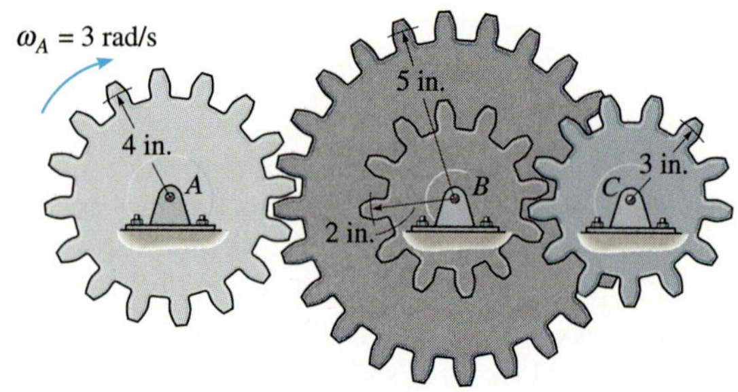

Prob. D–38

D-39. The spin drier of a washing machine has a constant angular acceleration of 2 rev/s^2, starting from rest. Determine how many turns it makes in 10 seconds and its angular velocity when $t = 5$ s.

D-40. Starting from rest, point P on the cord has a constant acceleration of 20 ft/s^2. Determine the angular acceleration and angular velocity of the disk after it has completed 10 revolutions. How many revolutions will the disk turn after it has completed 10 revolutions and P continues to move downward for 4 seconds longer?

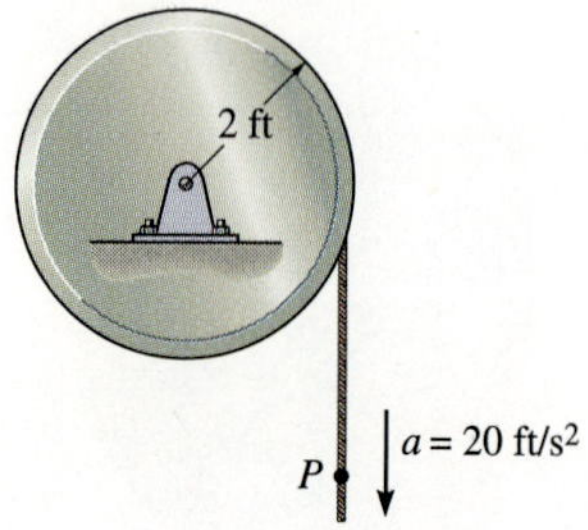

Prob. D–40

D-41. The center of the wheel has a velocity of 3 m/s. At the same time, it is slipping and has a clockwise angular velocity of $\omega = 2$ rad/s. Determine the velocity of point A at the instant shown.

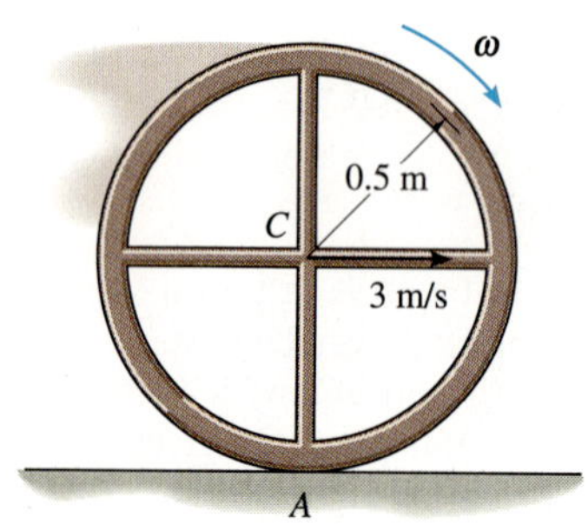

Prob. D–41

D-42. A cord is wrapped around the inner core of the gear and it is pulled with a constant velocity of 2 ft/s. Determine the velocity of the center of the gear, C.

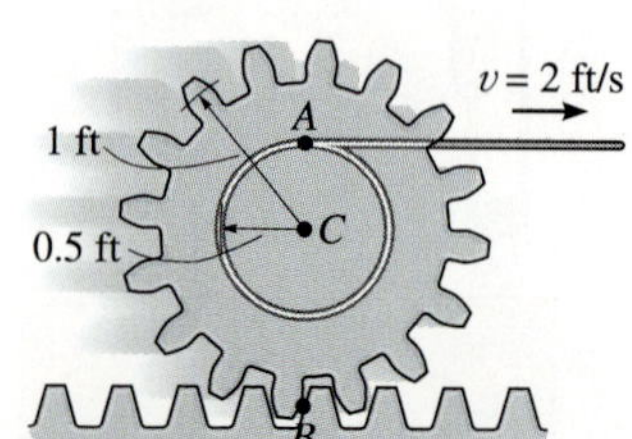

Prob. D–42

D-43. The center of the wheel is moving to the right with a speed of 2 m/s. If no slipping occurs at the ground, A, determine the velocity of point B at the instant shown.

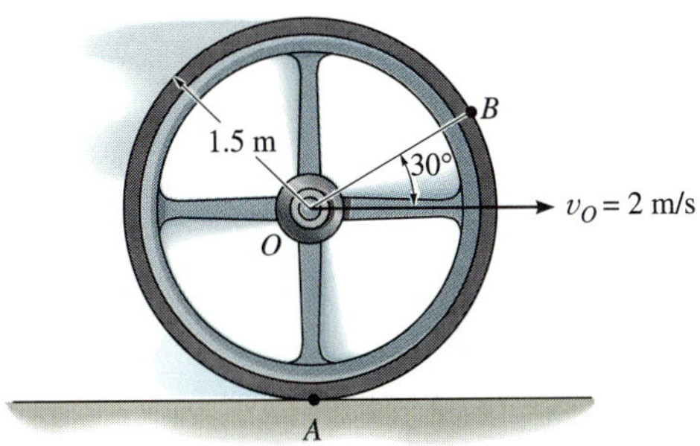

Prob. D–43

D-44. If the velocity of the slider block at B is 2 ft/s to the left, compute the velocity of the block at A and the angular velocity of the rod at the instant shown.

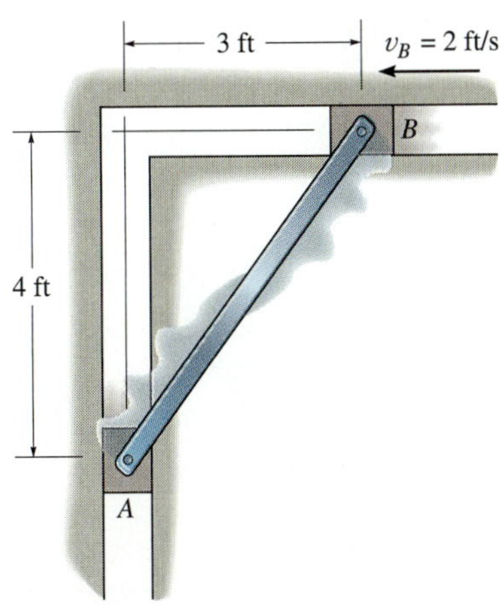

Prob. D–44

D-45. Determine the angular velocity of link AB at the instant shown.

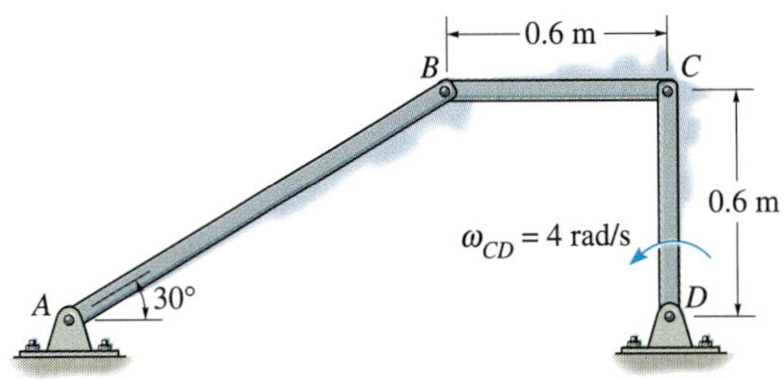

Prob. D–45

D-46. When the slider block C is in the position shown, the link AB has a clockwise angular velocity of 2 rad/s. Determine the velocity of block C at this instant.

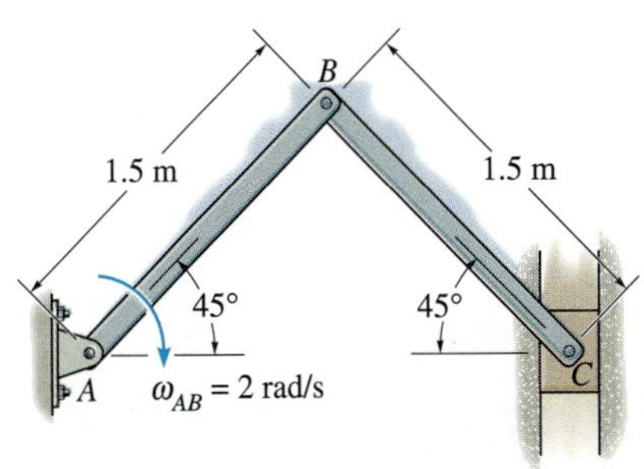

Prob. D–46

D-47. The center of the pulley is being lifted vertically with an acceleration of 3 m/s^2, and at the instant shown its velocity is 2 m/s. Determine the accelerations of points A and B. Assume that the rope does not slip on the pulley's surface.

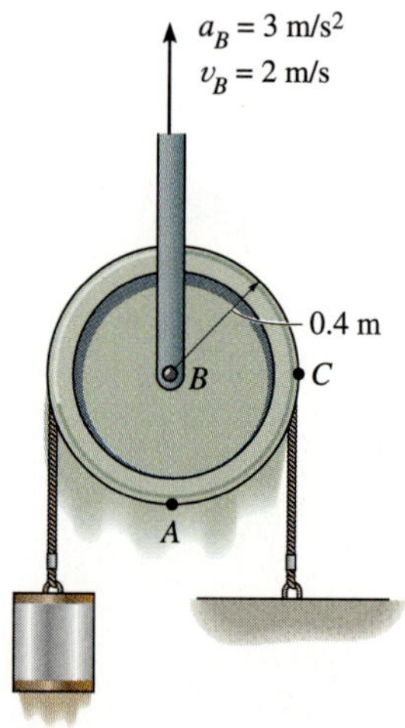

Prob. D–47

D-48. At a given instant, the slider block A has the velocity and deceleration shown. Determine the acceleration of block B and the angular acceleration of the link at this instant.

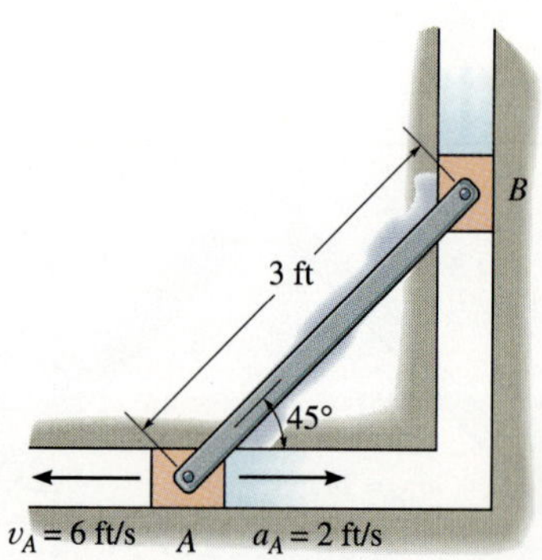

Prob. D–48

Chapter 17—Review All Sections

D-49. The 3500-lb car has a center of mass located at G. Determine the normal reactions of both front and both rear wheels on the road and the acceleration of the car if it is rolling freely down the incline. Neglect the weight of the wheels.

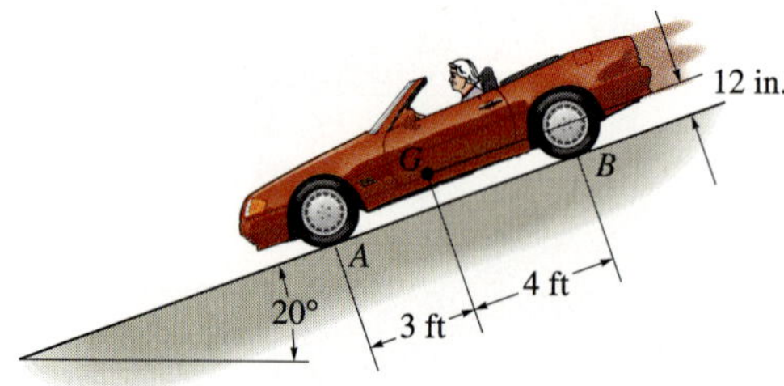

Prob. D–49

D-50. The 20-lb link AB is pinned to a moving frame at A and held in a vertical position by means of a string BC which can support a maximum tension of 10 lb. Determine the maximum acceleration of the link without breaking the string. What are the corresponding components of reaction at the pin A?

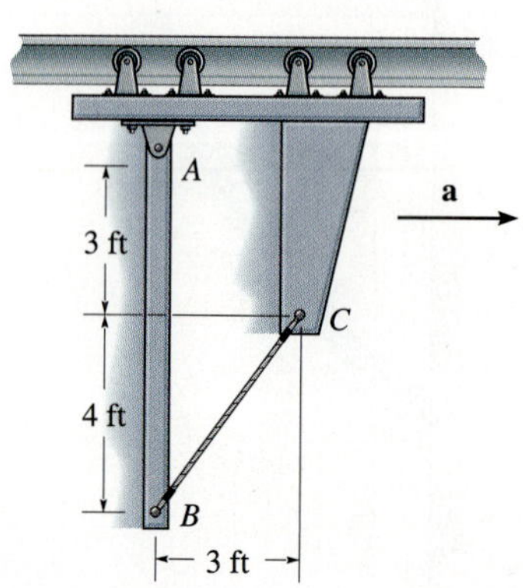

Prob. D–50

D-51. The 50-lb triangular plate is released from rest. Determine the initial angular acceleration of the plate and the horizontal and vertical components of reaction at B. The moment of inertia of the plate about the pinned axis B is $I_B = 2.30 \text{ slug} \cdot \text{ft}^2$.

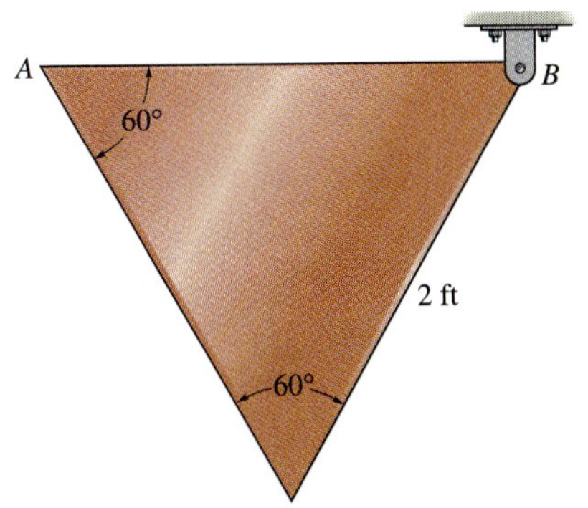

Prob. D–51

D-52. The 20-kg slender rod is pinned at O. Determine the reaction at O just after the cable is cut.

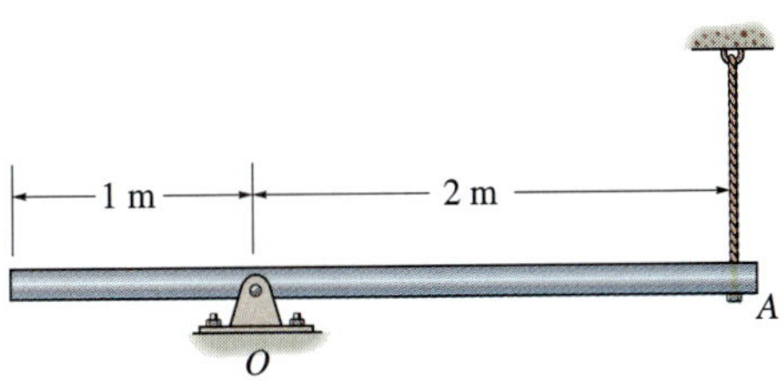

Prob. D–52

D-53. The 20-kg wheel has a radius of gyration of $k_G = 0.8$ m. Determine the angular acceleration of the wheel if no slipping occurs.

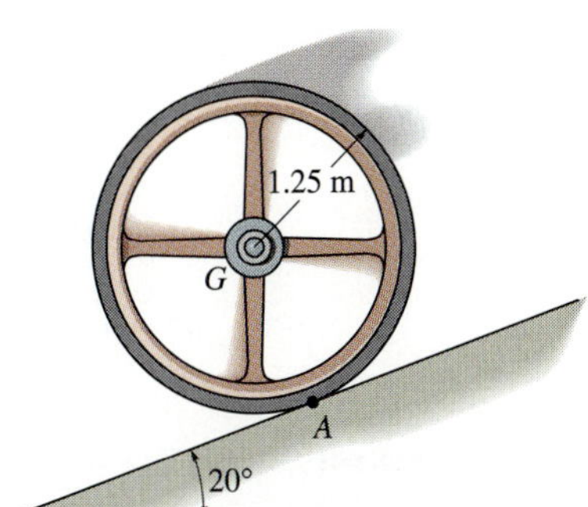

Prob. D–53

D-54. The 15-kg wheel has a wire wrapped around its inner hub and is released from rest on the inclined plane, for which the coefficient of kinetic friction is $\mu_k = 0.1$. If the centroidal radius of gyration of the wheel is $k_G = 0.8$ m, determine the angular acceleration of the wheel.

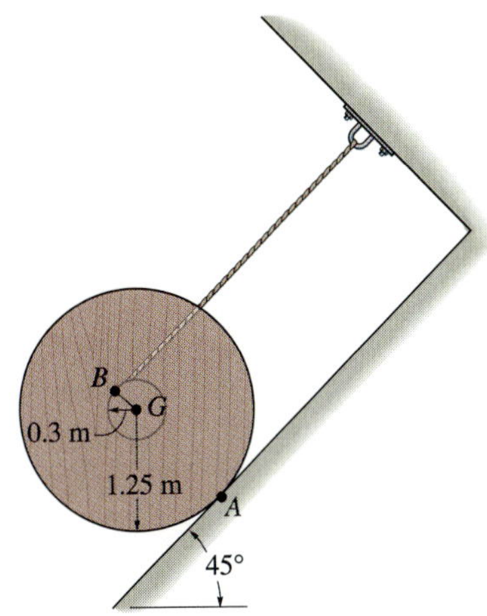

Prob. D–54

D-55. The 2-kg gear is at rest on the surface of a gear rack. If the rack is suddenly given an acceleration of 5 ft/s^2, determine the initial angular acceleration of the gear. The radius of gyration of the gear is $k_G = 0.3$ m.

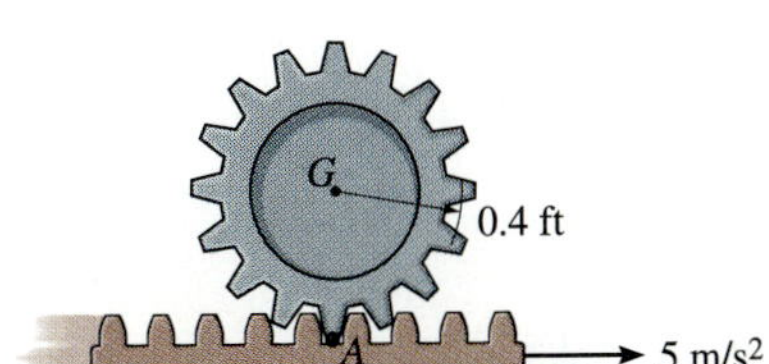

Prob. D–55

Solutions and Answers

D-1. $v = \dfrac{ds}{dt} = 1.5t^2 + 4\big|_{t=3} = 17.5 \text{ ft/s}$ ***Ans.***

$a = \dfrac{dv}{dt} = 3t\big|_{t=3} = 9 \text{ ft/s}^2$ ***Ans.***

D-2. $(30)^2 = (0)^2 + 2a(100 - 0)$

$a = 4.5 \text{ m/s}^2$ ***Ans.***

D-3. $v = \dfrac{ds}{dt} = 36t^2 + 4t + 3\big|_{t=2} = 155 \text{ m/s}$ ***Ans.***

$a = \dfrac{dv}{dt} = 72t + 4\big|_{t=2} = 148 \text{ m/s}^2$ ***Ans.***

D-4. $v = \int (4t^2 - 2)\, dt$

$v = \dfrac{4}{3}t^2 - 2t + C_1$

$s = \int\left(\dfrac{4}{3}t^3 - 2t + C_1\right) dt$

$s = \dfrac{1}{3}t^4 - t^2 + C_1 t + C_2$

$t = 0, s = -2, C_2 = -2$

$t = 2, s = -20, C_1 = -9.67$

$t = 4, s = 28.7 \text{ m}$ ***Ans.***

D-5. $\overset{+}{\rightarrow} s = s_0 + v_0 t$

$10 = 0 + v_A \cos 30^\circ\, t$

$+\uparrow\ s = s_0 + v_0 t + \dfrac{1}{2}a_c t^2$

$3 = 1.5 + v_A \sin 30^\circ\, t + \dfrac{1}{2}(-9.81)\, t^2$

$t = 0.933,\ v_A = 12.4 \text{ m/s}$ ***Ans.***

D-6. $v = v_0 + a_c t$

$v_x = 0 + 12(2) = 24 \text{ m/s}$

$v_y = \dfrac{dy}{dt} = 21t^2\big|_{t=2} = 84 \text{ m/s}$

$v = \sqrt{(24)^2 + (84)^2} = 87.4 \text{ m/s}$ ***Ans.***

D-7. $(\overset{+}{\leftarrow})\ s = s_0 + v_0 t$

$R\left(\dfrac{4}{5}\right) = 0 + 20\left(\dfrac{3}{5}\right)t$

$(+\uparrow)\ s = s_0 + v_0 t + \dfrac{1}{2}a_c t^2$

$-R\left(\dfrac{3}{5}\right) = 0 + 20\left(\dfrac{4}{5}\right)t + \dfrac{1}{2}(-9.81)t^2$

$t = 5.10 \text{ s}$

$R = 76.5 \text{ m}$ ***Ans.***

D-8. $a_t = 0$

$a_n = a = 1.5 = \dfrac{v^2}{250};\ v = 19.4 \text{ m/s}$ ***Ans.***

D-9. $a_t = 2 \text{ m/s}^2$

$a_n = \dfrac{v^2}{\rho} = \dfrac{(6)^2}{30} = 1.20 \text{ m/s}^2$

$a = \sqrt{(2)^2 + (1.20)^2} = 2.33 \text{ m/s}^2$ ***Ans.***

D-10. $a_t = \dfrac{\Delta v}{\Delta t} = \dfrac{60 - 40}{[(5 - 0)/3600]} = 14\,400 \text{ km/h}^2$

$a_n = \dfrac{v^2}{\rho} = \dfrac{(50)^2}{0.6} = 4167 \text{ km/h}^2$

$a = \sqrt{(14.4)^2 + (4.167)^2}\, 10^3 = 15.0(10^3) \text{ km/h}^2$

Ans.

D-11. $a_t = 3 \cos 40^\circ = 2.30 \text{ m/s}^2$ ***Ans.***

$a_n = \dfrac{v^2}{\rho};\ 3 \sin 40^\circ = \dfrac{(25)^2}{\rho},\ \rho = 324 \text{ m}$ ***Ans.***

D-12. $\mathbf{v}_A = \mathbf{v}_B + \mathbf{v}_{A/B}$

$-20 \cos 45^\circ\, \mathbf{i} + 20 \sin 45^\circ\, \mathbf{j} = 65\mathbf{i} + \mathbf{v}_{A/B}$

$\mathbf{v}_{A/B} = -79.14\mathbf{i} + 14.14\mathbf{j}$

$\mathbf{v}_{A/B} = \sqrt{(-79.14)^2 + (14.14)^2} = 80.4 \text{ km/h}$ ***Ans.***

$\mathbf{a}_A = \mathbf{a}_B + \mathbf{a}_{A/B}$

$\dfrac{(20)^2}{0.1} \cos 45^\circ\, \mathbf{i} + \dfrac{(20)^2}{0.1} \sin 45^\circ \mathbf{j} = 1200\mathbf{i} + \mathbf{a}_{A/B}$

$\mathbf{a}_{A/B} = 1628\mathbf{i} + 2828\mathbf{j}$

$a_{A/B} = \sqrt{(1628)^2 + (2828)^2} = 3.26(10^3) \text{ km/h}^2$ ***Ans.***

D-13. $4\, s_A + s_P = l$

$v_P = -4\, v_A = -4(-2) = 8 \text{ m/s}$ ***Ans.***

D-14. $+\uparrow \Sigma F_y = ma_y;\ 170 - 153 = \dfrac{153}{32.2}a$

$a = 3.58 \text{ ft/s}^2 \uparrow$ ***Ans.***

$+\downarrow \Sigma F_y = ma_y;\ 153 - 130 = \dfrac{153}{32.2}a'$

$a' = 4.84 \text{ ft/s}^2 \downarrow$ ***Ans.***

D-15. Block B:

$+\downarrow \Sigma F_y = ma_y;\ 4(9.81) - T = 4a$

Block A:

$\overset{+}{\leftarrow} \Sigma F_x = ma_x;\ T = 2a$

$T = 13.1 \text{ N} \quad a = 6.54 \text{ m/s}^2$ ***Ans.***

D-16. Block A:
$+\downarrow \Sigma F_y = ma_y;\ 15(9.81) - T = -15a$
Block B:
$+\downarrow \Sigma F_y = ma_y;\ 25(9.81) - T = 25a$
$a = 2.45\ \text{m/s}^2,\ T = 184\ \text{N}$ ***Ans.***

D-17. Blocks A and B:
$\overset{+}{\rightarrow} \Sigma F_x = ma_x;\ 6 = \dfrac{70}{32.2}a;\ a = 2.76\ \text{ft/s}^2$
Check if slipping occurs between A and B.
$\overset{+}{\rightarrow} \Sigma F_x = ma_x;\ 6 - F = \dfrac{20}{32.2}(2.76);$
$F = 4.29\ \text{lb} < 0.4(20) = 8\ \text{lb}$
$a_A = a_B = 2.76\ \text{m/s}^2$ ***Ans.***

D-18. $\Sigma F_n = m\dfrac{v^2}{\rho};\ (0.3)m(9.81) = m\dfrac{v^2}{2}$
$v = 2.43\ \text{m/s}$ ***Ans.***

D-19. $+\downarrow \Sigma F_n = ma_n;\ m\,(32.2) = m\left(\dfrac{v^2}{250}\right)$
$v = 89.7\ \text{ft/s}$ ***Ans.***

D-20. $+\downarrow \Sigma F_n = ma_n;\ 150 + N_p = \dfrac{150}{32.2}\left(\dfrac{(120)^2}{400}\right)$
$N_p = 17.7\ \text{lb}$ ***Ans.***

D-21. $\overset{+}{\leftarrow} \Sigma F_n = ma_n;\ N_c \sin 30^\circ + 0.2\,N_c \cos 30^\circ = m\dfrac{v^2}{500}$
$+\uparrow \Sigma F_b = 0;$
$N_c \cos 30^\circ - 0.2\,N_c \sin 30^\circ - m(32.2) = 0$
$v = 119\ \text{ft/s}$ ***Ans.***

D-22. At B:
$\overset{+}{\rightarrow} \Sigma F_n = ma_n,\quad T = \left(\dfrac{5}{32.2}\right)\left(\dfrac{(0)^2}{2}\right) = 0$ ***Ans.***
In the general position,
$+\nearrow \Sigma F_n = ma_n;\ T - 5\sin\theta = \left(\dfrac{5}{32.2}\right)\left(\dfrac{v^2}{2}\right)$
$\searrow + \Sigma F_t = ma_t;\ 5\cos\theta = \dfrac{5}{32.2}\left(\dfrac{v\,dv}{2\,d\theta}\right)$
$\displaystyle\int_0^{90^\circ} 64.4\cos\theta\, d\theta = \int_0^v v\, dv$
$v = 11.3\ \text{ft/s},$
When $\theta = 90^\circ$;
$T = 15\ \text{lb}$ ***Ans.***

D-23. $T_1 + \Sigma U_{1-2} = T_2$
$0 + U_{1-2} - 80(300) = \dfrac{1}{2}\left(\dfrac{15000}{32.2}\right)(40)^2$
$U_{1-2} = 397(10^3)\ \text{ft}\cdot\text{lb}$ ***Ans.***

D-24. $T_1 + \Sigma U_{1-2} = T_2$
$\dfrac{1}{2}\left(\dfrac{20}{32.2}\right)(5)^2 + 40(10) -$
$(0.2)(20\cos 30^\circ)(10) + 20(10\sin 30^\circ) = \dfrac{1}{2}\left(\dfrac{20}{32.2}\right)v^2$
$v = 39.0\ \text{ft/s}$ ***Ans.***

D-25. $+\uparrow \Sigma F_y = ma_y;\ N_b + 100\sin 20^\circ - 10 - 3(9.81) = 0$
$N_b = 5.23\ \text{N}$
$T_1 + \Sigma U_{1-2} = T_2$
$0 + (100\cos 20^\circ)d - 0.2(5.23)d = \dfrac{1}{2}(3)(10)^2$
$d = 1.61\ \text{m}$ ***Ans.***

D-26. $T_1 + V_1 = T_2 + V_2$
$0 + \dfrac{1}{2}(40)(x)^2 = \dfrac{1}{2}\left(\dfrac{6}{32.2}\right)(6)^2 + 6(8)$
$x = 1.60\ \text{ft}$ ***Ans.***

D-27. $T_A + V_A = T_B + V_B$
$0 + 2(9.81)(1.5) = \dfrac{1}{2}(2)(v_B)^2 + 0$
$v_B = 5.42\ \text{m/s}$ ***Ans.***
$+\uparrow \Sigma F_n = ma_n;\ T - 2(9.81) = 2\left(\dfrac{(5.42)^2}{1.5}\right)$
$T = 58.9\ \text{N}$ ***Ans.***

D-28. $T_A + V_A = T_B + V_B$
$0 + \dfrac{1}{2}(4)(2.5 - 0.5)^2 + 5(2.5)$
$= \dfrac{1}{2}\left(\dfrac{5}{32.2}\right)v_B^2 + \dfrac{1}{2}(4)(1 - 0.5)^2$
$v_B = 16.0\ \text{ft/s}$ ***Ans.***

D-29. $T_1 + V_1 = T_2 + V_2$
$\dfrac{1}{2}(2)(4)^2 + \dfrac{1}{2}(30)(2 - 1)^2$
$= \dfrac{1}{2}(2)(v)^2 - 2(9.81)(1) + \dfrac{1}{2}(30)(\sqrt{5} - 1)^2$
$v = 5.26\ \text{m/s}$ ***Ans.***

D-30. Three cars:

$\xrightarrow{+} mv_1 + \Sigma\int F dt = mv_2$

$$\frac{250(2000)}{32.2}(20) + 40(10^3)(t) - 10(250)t = \frac{250(2000)}{32.2}(30)$$

$t = 4.14$ s **Ans.**

Engine:

$\xrightarrow{+} mv_1 + \Sigma\int F dt = mv_2$

$$\frac{30(2000)}{32.2}(20) + F(4.14) - 10(30)(4.14) = \frac{30(2000)}{32.2}(30)$$

$F = 4800$ lb **Ans.**

D-31. $+\nwarrow \Sigma F_y = 0;\ N_b - 5(9.81)\cos 30° - 100 \sin 30° = 0$

$N_b = 92.48$ N

$+\nearrow mv_1 + \Sigma\int F dt = mv_2$

$5(3) + (100 \cos 30°)t - 5(9.81)\sin 30°(t) - 0.3(92.48)t = 5(10)$

$t = 1.02$ s. **Ans.**

D-32. Block B:

$(+\downarrow)\ mv_1 + \int F dt = mv_2$

$$0 + 8(5) - T(5) = \frac{8}{32.2}(1)$$

$T = 7.95$ lb **Ans.**

Block A:

$(\xrightarrow{+})\ mv_1 + \int F dt = mv_2$

$$0 + 7.95(5) - \mu_k(10)(5) = \frac{10}{32.2}(1)$$

$\mu_k = 0.789$ **Ans.**

D-33. $2\,s_A + s_B = l$

$2v_A = -v_B$

$+\downarrow m(v_A)_1 + \int F dt = m(v_A)_2$

$0 + 10(9.81)(10) - 2T(10) = 10(v_A)_2$

$+\downarrow m(v_B)_1 + \int F dt = m(v_B)_2$

$0 + 50(9.81)(10) - T(10) = 50(v_B)_2$

$T = 70.1$ N

$(v_A)_2 = -42.0$ m/s $= 42.0$ m/s $\uparrow$ **Ans.**

$(v_B)_2 = 84.1$ m/s $\downarrow$ **Ans.**

D-34. $\xrightarrow{+} \Sigma mv_1 = \Sigma mv_2;\ 5(2) - 2(5) = 5(v_A)_2 + 2(v_B)_2$

$$\xrightarrow{+} e = \frac{(v_B)_2 - (v_A)_2}{(v_A)_1 - (v_B)_1};\ 0.5 = \frac{(v_B)_2 - (v_A)_2}{2 - (-5)}$$

$(v_A)_2 = -1$ m/s $= 1$m/s $\leftarrow$ **Ans.**

$(v_B)_2 = 2.5$ m/s $\rightarrow$ **Ans.**

D-35. $\xrightarrow{+} \Sigma mv_1 = \Sigma mv_2;\ 6(20) + 0 = 6(v_A)_2 + 24(v_B)_2$

$$\xrightarrow{+} e = \frac{(v_B)_2 - (v_A)_2}{(v_A)_1 - (v_B)_1};\ 1 = \frac{(v_B)_2 - (v_A)_2}{20 - 0}$$

$(v_A)_2 = -12$ m/s $= 12$ m/s $\leftarrow$ **Ans.**

$(v_B)_2 = 8$ m/s $\rightarrow$ **Ans.**

Disk A:

$mv_1 + \int F dt = mv_2;\ 6(20) - \int F dt = -6(12)$

$\int F dt = 192$ N·s **Ans.**

D-36. After collision: $T_1 + \Sigma U_{1-2} = T_2$

$$\frac{1}{2}\left(\frac{5}{32.2}\right)(v_A)_2^2 - 0.2(5)\left(\frac{2}{12}\right) = 0$$

$(v_A)_2 = 1.465$ ft/s

$$\frac{1}{2}\left(\frac{10}{32.2}\right)(v_B)_2^2 - 0.2(10)\left(\frac{3}{12}\right) = 0$$

$(v_B)_2^2 = 1.794$ ft/s

$\Sigma mv_1 = \Sigma mv_2$

$$\frac{5}{32.2}(v_A)_1 + 0 = \frac{5}{32.2}(1.465) + \frac{10}{32.2}(1.794)$$

$(v_A)_1 = 5.054$

$$e = \frac{(v_B)_2 - (v_A)_2}{(v_A)_1 - (v_B)_1} = \frac{1.794 - 1.465}{5.054 - 0} = 0.0652$$ **Ans.**

D-37. $\Sigma m(v_x)_1 = \Sigma m(v_x)_2$

$$0 + 0 = \frac{2}{32.2}(1) + \frac{11}{32.2}(v_{Bx})_2$$

$(v_{Bx})_2 = -0.1818$ ft/s

$\Sigma m(v_y)_1 = \Sigma m(v_y)_2$

$$\frac{2}{32.2}(3) + 0 = 0 + \frac{11}{32.2}(v_{By})_2$$

$(v_{By})_2 = 0.545$ ft/s

$(v_B)_2 = \sqrt{(-0.1818)^2 + (0.545)^2} = 0.575$ ft/s **Ans.**

D-38. $\omega_B(5) = 3(4)$

$\omega_B = 2.40$ rad/s $\circlearrowleft$ **Ans.**

$\omega_C(3) = 2.40(2)$

$\omega_C = 1.60$ rad/s $\circlearrowright$ **Ans.**

D-39. $\theta = \theta_0 + \omega_0 t + \frac{1}{2}\alpha_c t^2$

$\theta = 0 + 0 + \frac{1}{2}(2)(10)^2 = 100$ rev **Ans.**

$\omega = \omega_0 + \alpha_c t$

$\omega = 0 + 2(5) = 10$ rev/s **Ans.**

D-40. $\alpha = \frac{a_t}{r} = \frac{20}{2} = 10\text{ rad/s}^2$ ***Ans.***
10 rev $= 20\pi$ rad
$\omega^2 = \omega_0^2 + 2\,\alpha_c(\theta - \theta_0)$
$\omega^2 = 0 + 2(10)(20\pi - 0)$
$\omega = 35.4$ rad/s ***Ans.***
$\theta = \theta_0 + \omega_0 t + \frac{1}{2}\alpha_c t^2$
$\theta = 0 + 35.4(4) + \frac{1}{2}(10)(4)^2 = 222$ rad
$\theta = \frac{222}{2\pi} = 35.3$ rev. ***Ans.***

D-41. $\mathbf{v}_A = \mathbf{v}_C + \boldsymbol{\omega} \times \mathbf{r}_{A/C}$
$\mathbf{v}_A\mathbf{i} = 3\mathbf{i} + (-2\mathbf{k}) \times (-0.5\mathbf{j})$
$v_A = 2$ m/s $\rightarrow$ ***Ans.***

D-42. $\mathbf{v}_A = \mathbf{v}_B + \boldsymbol{\omega} \times \mathbf{r}_{A/B}$
$2\mathbf{i} = \mathbf{0} + (-\omega\mathbf{k}) \times (1.5\mathbf{j})$
$\omega = 1.33$ rad/s $\circlearrowright$
$\mathbf{v}_C = \mathbf{v}_B + \boldsymbol{\omega} \times \mathbf{r}_{C/B}$
$\mathbf{v}_C\mathbf{i} = \mathbf{0} + (-1.33\mathbf{k}) \times (1\mathbf{j})$
$v_C = 1.33$ ft/s $\rightarrow$ ***Ans.***

D-43. $\mathbf{v}_O = \mathbf{v}_A + \boldsymbol{\omega} \times \mathbf{r}_{O/A}$
$2\mathbf{i} = \mathbf{0} + (-\omega\mathbf{k}) \times 1.5\mathbf{j}$
$\omega = 1.33$ rad/s $\circlearrowright$
$\mathbf{v}_B = \mathbf{v}_O + \boldsymbol{\omega} \times \mathbf{r}_{B/O}$
$\mathbf{v}_B = 2\mathbf{i} + (-1.33\mathbf{k}) \times (1.5 \cos 30°\,\mathbf{i} + 1.5 \sin 30°\,\mathbf{j})$
$\mathbf{v}_B = \{3\mathbf{i} - 1.73\mathbf{j}\}$ m/s ***Ans.***

D-44. $\mathbf{v}_A = \mathbf{v}_B + \boldsymbol{\omega} \times \mathbf{r}_{A/B}$
$-v_A\mathbf{j} = -2\mathbf{i} + \omega\mathbf{k} \times (-3\mathbf{i} - 4\mathbf{j})$
$v_A = 1.5$ ft/s $\downarrow$ ***Ans.***
$\omega = 0.5$ rad/s $\circlearrowleft$ ***Ans.***

D-45. $v_C = 4(0.6) = 2.4$ m/s $\leftarrow$
$\mathbf{v}_B = \mathbf{v}_C + \boldsymbol{\omega} \times \mathbf{r}_{B/C}$
$-v_B \sin 30°\,\mathbf{i} + v_B \cos 30°\,\mathbf{j}$
$= -2.4\mathbf{i} + (\omega_{BC}\,\mathbf{k}) \times (-0.6\mathbf{i})$
$v_B = 4.80$ m/s $\nwarrow$ $\omega_{BC} = 6.93$ rad/s $\circlearrowright$
$\omega_{AB} = \frac{4.80}{0.6/\sin 30°} = 4$ rad/s $\circlearrowleft$ ***Ans.***

D-46. $v_B = 2(1.5) = 3$ m/s $\searrow$
$\mathbf{v}_C = \mathbf{v}_B + \boldsymbol{\omega} \times \mathbf{r}_{C/B}$
$-v_C\,\mathbf{j} = 3 \cos 45°\,\mathbf{i} - 3 \sin 45°\,\mathbf{j}$
$+ (\omega\mathbf{k}) \times (1.5 \cos 45°\mathbf{i} - 1.5 \sin 45°\,\mathbf{j})$
$v_C = 4.24$ m/s $\downarrow$ ***Ans.***
$\omega = 2$ rad/s $\circlearrowright$

D-47. $\omega = \frac{v_B}{r_{B/C}} = \frac{2}{0.4} = 5$ rad/s $\circlearrowright$
$\alpha = \frac{a_B}{r_{B/C}} = \frac{3}{0.4} = 7.5$ rad/s^2 $\circlearrowright$
$\mathbf{a}_A = \mathbf{a}_B + \boldsymbol{\alpha} \times \mathbf{r}_{A/B} - \omega^2\,\mathbf{r}_{A/B}$
$\mathbf{a}_A = 3\mathbf{j} + (-7.5\mathbf{k}) \times (-0.4\mathbf{j}) - (5)^2\,(-0.4\mathbf{j})$
$\mathbf{a}_A = \{-3\mathbf{i} + 13\mathbf{j}\}$ m/s^2 ***Ans.***
$a_B = 3$ m/s ***Ans.***

D-48. $\mathbf{v}_B = \mathbf{v}_A + \boldsymbol{\omega} \times \mathbf{r}_{B/A}$
$-\mathbf{v}_B\mathbf{j} = -6\mathbf{i} + \omega\mathbf{k} \times (3 \cos 45°\,\mathbf{i} + 3 \sin 45°\,\mathbf{j})$
$\omega = 2.828$ rad/s, $\circlearrowright$ $v_B = 6$ ft/s $\downarrow$
$\mathbf{a}_B = \mathbf{a}_A + \boldsymbol{\alpha} \times \mathbf{r}_{B/A} - \omega^2\,\mathbf{r}_{B/A}$
$-a_B\mathbf{j} = 2\mathbf{i} + (\alpha\mathbf{k}) \times (3 \cos 45°\,\mathbf{i} + 3 \sin 45°\,\mathbf{j})$
$- (2.828)^2\,(3 \cos 45°\,\mathbf{i} + 3 \sin 45°\,\mathbf{j})$
$a_B = 31.9$ ft/s^2 $\downarrow$ ***Ans.***
$\alpha = 7.06$ rad/s^2 $\circlearrowright$ ***Ans.***

D-49. $+\swarrow\ \Sigma F_x = m(a_G)_x;\ 3500 \sin 20° = \frac{3500}{32.2}\,a_G$
$a_G = 11.0$ ft/s^2 ***Ans.***
$+\nwarrow\ \Sigma F_y = m(a_G)_y;\ N_A + N_B - 3500 \cos 20° = 0$
$\circlearrowright + \Sigma M_G = 0;\ N_B(4) - N_A(3) = 0$
$N_A = 1879$ lb ***Ans.***
$N_B = 1410$ lb ***Ans.***

D-50. $\circlearrowright + \Sigma M_A = \Sigma(M_k)_A;\ 10\left(\frac{3}{5}\right)(7) = \frac{20}{32.2}\,a(3.5)$
$a = 19.32$ m/s^2 ***Ans.***
$\overset{+}{\rightarrow}\ \Sigma F_x = m(a_G)_x;\ A_x + 10\left(\frac{3}{5}\right) = \frac{20}{32.2}(19.32)$
$A_x = 6$ lb ***Ans.***
$+\uparrow\ \Sigma F_y = m(a_G)_y;\ A_y - 20 + 10\left(\frac{4}{5}\right) = 0$
$A_y = 12$ lb ***Ans.***

D-51. $\circlearrowright + \Sigma M_B = I_B \alpha;\ 50(1) = 2.30\alpha$

$\alpha = 21.7\ \text{rad/s}^2$ ***Ans.***

$$a_G = \left(\frac{1}{\cos 30^\circ}\right)(21.7) = 25.1\ \text{m/s}^2$$

$\xrightarrow{+} \Sigma F_x = m(a_G)_x;$

$$B_x = \left(\frac{50}{32.2}\right) 25.1 \sin 30^\circ = 19.5\ \text{lb}$$ ***Ans.***

$$+\uparrow \Sigma F_y = m(a_G)_y;\ B_y - 50 = -\left(\frac{50}{32.2}\right) 25.1 \cos 30^\circ$$

$B_y = 16.2\ \text{lb}$ ***Ans.***

D-52. $\circlearrowleft + \Sigma M_O = I_O \alpha;$

$$20(9.81)(0.5) = \left[\frac{1}{12}(20)(3)^2 + 20(0.5)^2\right]\alpha$$

$\alpha = 4.90\ \text{rad/s}^2$

$\xrightarrow{+} \Sigma F_x = m(a_G)_x;\ O_x = 0$ ***Ans.***

$+\uparrow \Sigma F_y = m(a_G)_y;\ O_y - 20(9.81) = -20(4.90)(0.5)$

$O_y = 147\text{N}$ ***Ans.***

D-53. No slipping, so that

$a_G = 1.25\alpha$

$\circlearrowright + \Sigma M_A = \Sigma (M_k)_A;\ 20(9.81) \sin 20^\circ\ (1.25)$

$= 20(0.8)^2\alpha + (20\ a_G)(1.25)$

$\alpha = 1.90\ \text{rad/s}^2$ ***Ans.***

$a_G = 2.38\ \text{m/s}^2$

D-54. $+\nwarrow \Sigma F_y = 0;\ N_A - 15(9.81) \cos 45^\circ = 0;$

$N_A = 104.1\ \text{N}$

$a_G = \alpha(0.3)$

$\circlearrowright + \Sigma M_B = \Sigma (M_k)_B;\ (0.1)(104.1)(1.55) -$

$15(9.81) \sin 45^\circ(0.3) = -15(0.8)^2\alpha - 15(a_G)(0.3)$

$a_G = 0.413\ \text{m/s}^2$

$\alpha = 1.38\ \text{rad/s}^2$ ***Ans.***

D-55. $\xrightarrow{+} \Sigma F_x = m(a_G)_x;\ F_A = 2\ a_G$

$\circlearrowright + \Sigma M_G = I_G \alpha;\ F_A(0.4) = 2(0.3)^2\alpha$

$\mathbf{a}_A = \mathbf{a}_G + \boldsymbol{\alpha} \times \mathbf{r}_{A/G} - \omega^2 \mathbf{r}_{A/G}$

$5\mathbf{i} = a_G\mathbf{i} + (\alpha\mathbf{k}) \times (-0.4\mathbf{j}) - \mathbf{0}$

$5 = a_G + 0.4\alpha$

$F_A = 3.60\ \text{N}$

$a_G = 1.80\ \text{m/s}^2$

$\alpha = 8\ \text{rad/s}^2$ ***Ans.***

Answers to Selected Problems

Chapter 12

12–1. 59.5 ft/s, 1.29 s **12–2.** 13 m/s, 76 m, 8.33 s
12–3. 20 ft **12–5.** 0, 80.7 m
12–6. $v_{avg} = 4$ m/s, $(v_{sp})_{avg} = 6$ m/s, 18 m/s^2
12–7. $v|_{t=3} = 60$ m/s, $v_{avg} = 15$ m/s, $(v_{sp})_{avg} = 19.7$ m/s
12–9. $v_{avg} = 0$, $(v_{sp})_{avg} = 3$ m/s, 2 m/s^2
12–10. $v_{avg} = 0.333$ m/s, $(v_{sp})_{avg} = 1$ m/s
12–11. **a)** 240 mm, **b)** 60 mm/s, **c)** 20 mm/s^2 (for all t)
12–13. 21.9 s
12–14. $s = -18$ ft, $s_T = 46$ ft
12–15. 54 m **12–17.** 41.6 m, 10.4 m/s
12–18. 1.06 m/s^2 **12–19.** 16.9 ft
12–21. $v = -(2.667s^3 + 7333.3)^{\frac{1}{2}}$ m/s
12–22. $v = (20e^{-2t})$ m/s, $a = (-40e^{-2t})$ m/s^2, $s = 10(1 - e^{-2t})$ m
12–23. $v = (20 - 2s)$ m/s, 10 m
12–25. 1.29 m/s **12–26.** 314 m, 72.5 m/s
12–27. 3.56 m
12–29. $v = \left(2kt + \left(\dfrac{1}{v_0^2}\right)\right)^{-\frac{1}{2}}$,

$$s = \frac{1}{k}\left[\left(2kt + \left(\frac{1}{v_0^2}\right)\right)^{\frac{1}{2}} - \frac{1}{v_0}\right]$$

12–30. 32 s **12–31.** $2s$, 20.4 m
12–33. **a)** 45.5 m/s, **b)** 100 m/s
12–34. 11.2 km/s
12–35. $v = -R\sqrt{\dfrac{2g_0(y_0 - y)}{(R + y)(R + y_0)}}$, 3.02 km/s ↓
12–37. 980 m **12–42.** 33.3 s
12–43. 11.25 s **12–46.** 144 m
12–47. 3600 ft, 30 ft/s **12–49.** 8 ft/s^2
12–50. 37.8 ft/s, 1.13(10^3) ft
12–51. 917 m **12–53.** 87.5 m
12–55. 13.1 s, 523 ft **12–58.** 9.88 s
12–59. 120 m/s, 41.9 s **12–62.** 11.1 m/s^2, −25 m/s^2
12–63. 4.06 s **12–65.** 4.48 m/s^2, 7.04 m/s^2
12–66. 5.31 m/s^2, $\alpha = 53.0°$, $\beta = 37.0°$, $\gamma = 90.0°$
12–67. $y^2 + x^2 - 4y = 0$, 2 ft/s, 2 ft/s^2, 3.37 ft
12–69. (4 ft, 3 ft), $v_A = 4.47$ ft/s, $v_B = 8.06$ ft/s
12–70. $\Delta r = 3.61$ km, $d = 5$ km
12–71. $\mathbf{v}_{avg} = \{10\mathbf{i} - 10\mathbf{j}\}$ m/s
12–73. $v = \sqrt{c^2k^2 + b^2}$, $a = ck^2$
12–74. $a_x = -k \sin kt$, $a_y = k \sin kt$
12–75. $a_y = c$, $a_x = \dfrac{c}{2k}(y + ct^2)$
12–77. $v_x = v_0\left[1 + \left(\dfrac{\pi}{L}c\right)^2 \cos^2\left(\dfrac{\pi}{L}x\right)\right]^{-\frac{1}{2}}$,

$$v_y = \frac{v_0 \pi c}{L}\left(\cos\frac{\pi}{L}x\right)\left[1 + \left(\frac{\pi}{L}c\right)^2 \cos^2\left(\frac{\pi}{L}x\right)\right]^{-\frac{1}{2}}$$

12–78. 204 m, 41.8 m/s, 4.66 m/s^2
12–79. $v_0 = 15.3$ ft/s, $v_B = 81.7$ ft/s
12–81. $h = 2.87$ m, $s_x = 19.9$ m
12–82. $R = \dfrac{v_0^2}{g}\sin 2\theta$, $t = \dfrac{2v_0}{g}\sin\theta$, $h = \dfrac{v_0^2}{2g}\sin^2\theta$
12–83. 45° **12–85.** 67.7 ft/s, 58.9°
12–86. 26.0° (below horizontal), 89.4° (above horizontal)
12–87. 89.7 ft/s
12–89. $d = \dfrac{v_0^2}{g\cos\theta}(\sin 2\phi - 2\tan\theta\cos^2\phi)$,

$$\phi = \frac{1}{2}\tan^{-1}(-\operatorname{ctn}\theta)$$

12–90. 66.3 ft **12–91.** 19.4 m/s, 4.54 s
12–93. 25.4 ft **12–94.** 81.9 ft/s
12–95. 4.52 ft
12–97. $x = 5.15$ ft, $y = 1.33$ ft
12–98. $x = 32.3$ ft, $y = 6.17$ ft, 71.8 ft/s
12–99. $R = \left(\dfrac{v_0^2\sin 2\theta}{2g}\right) + \dfrac{v_0}{2g}\sqrt{v_0^2\sin^2 2\theta + 8gh\cos^2\theta}$
12–101. 0.488 m/s^2 **12–102.** 4.22 m/s^2
12–103. 7.42 ft/s^2 **12–105.** 1.96 m/s, 0.930 m/s^2
12–106. 4.58 m/s, 0.653 m/s^2 **12–107.** 5.99 Mm
12–109. 15.1 ft/s^2, 14 ft **12–110.** 19.9 m/s, 24.2 m/s^2
12–111. 19.9 ft/s, 16.0 ft/s^2
12–113. $v_t = 2$ m/s, $v_n = 0$, $a_t = 0$, $a_n = \dfrac{4\sin x}{(1 + \cos^2 x)^{3/2}}$
12–114. 322 mm/s^2 ∠26.6° **12–115.** 51.3 m/s^2, 10.6°
12–117. 5.02 m/s^2 **12–118.** 51.1 ft/s, 10.3 ft/s^2
12–119. 15.8 s, 35.7 m/s, 12.9 m/s^2
12–121. 8.61 m/s^2
12–122. $y = -0.0766x^2$ (parabola), 8.37 m/s, $a_n = 9.38$ m/s^2, $a_t = 2.88$ m/s^2
12–123. 10.4 m/s^2
12–125. 10.1 s, 47.6 m/s, 11.8 m/s^2
12–126. $a_n = 1.77$ m/s^2, $a_t = 4$ m/s^2
12–127. 1.21 s
12–129. $v_x = \dfrac{v_0}{\sqrt{1 + b^2}}$, $v_y = \dfrac{v_0 b}{\sqrt{1 + b^2}}$, $a_n = \dfrac{2c\,v_0^2}{(1 + b^2)^{3/2}}$
12–130. 1.5 m/s, 0.117 m/s^2
12–131. 43.0 m/s, 6.52 m/s^2 **12–133.** 104 s
12–134. $\mathbf{u} = 0.581\mathbf{i} + 0.161\mathbf{j} + 0.798\mathbf{k}$
12–135. $\alpha = 52.5°$, $\beta = 142°$, $\gamma = 85.1°$, or $\alpha = 128°$, $\beta = 37.9°$, $\gamma = 94.9°$
12–137. 464 ft/s, 43.2(10^3) ft/s^2
12–138. 14.3 in./s^2 **12–139.** 12 ft/s, 0.646 ft/s^2

12–141. $v_r = 4t \cos t^2, v_\theta = 4t \sin t^2,$
$a_r = 4 \cos t^2 - 16t^2 \sin t^2,$
$a_\theta = 4 \sin t^2 + 16t^2 \cos t^2$
12–142. 2.32 m/s^2
12–143. $v_r = ae^{at}, v_\theta = e^{at}, a_r = e^{at}(a^2 - 1), a_\theta = 2ae^{at}$
12–145. 0.075 rad/s, 2.25 ft/s^2
12–146. 48.3 in./s^2
12–147. $v_r = -\frac{1}{2}a\theta^{-\frac{3}{2}}\dot{\theta}, v_\theta = a\theta^{-\frac{1}{2}}\dot{\theta},$
$a_r = a\left[\left(\frac{3}{4}\theta^{-2} - 1\right)\theta^{-\frac{1}{2}}\dot{\theta}^2 - \frac{1}{2}\theta^{-\frac{3}{2}}\ddot{\theta}\right],$
$a_\theta = a\left[\ddot{\theta} - \frac{\dot{\theta}^2}{\theta}\right]\theta^{-\frac{1}{2}}$
12–149. $v_r = -16.9$ m/s, $v_\theta = 4.87$ m/s, $a_r = -89.4$ m/s^2, $a_\theta = -53.7$ m/s^2
12–150. 12.6 m/s, 83.2 m/s^2
12–151. $\mathbf{v} = \{-116\mathbf{u}_r - 163\,\mathbf{u}_z\}$ mm/s, $\mathbf{a} = \{-5.81\mathbf{u}_r - 8.14\,\mathbf{u}_z\}$ mm/s^2
12–153. 0.217 m/s^2
12–154. $v_r = -0.212$ m/s, $v_\theta = 0.812$ m/s, $a_r = -0.703$ m/s^2, $a_\theta = 0.0838$ m/s^2
12–155. 1.30 rad/s, 2.55 m/s^2
12–157. $v_r = 0.955$ in./s, $v_\theta = 21$ in./s, $a_r = -63$ in./s^2, $a_\theta = 5.73$ in./s^2
12–158. $v_r = 0.955$ in./s, $v_\theta = 21$ in./s, $a_r = -62.4$ in./s^2, $a_\theta = 19.7$ in./s^2
12–159. 164 mm/s, 659 mm/s^2
12–161. 1.32 m/s **12–162.** 8.66 m/s^2
12–163. $v_r = -\frac{20}{\sqrt{1 + \theta^2}}, v_\theta = \frac{20\theta}{\sqrt{1 + \theta^2}},$
$v_r = -14.1$ ft/s, $v_\theta = 14.1$ ft/s
12–165. $v_r = 0.242$ m/s, $v_\theta = 0.943$ m/s, $a_r = -2.32$ m/s^2, $a_\theta = 1.74$ m/s^2
12–166. 86.6 ft/s, 266 ft/s^2
12–167. $v_r = -24.2$ ft/s, $v_\theta = 25.3$ ft/s
12–169. $v = 0.242$ m/s, $a = 0.169$ m/s^2, $(v_A)_x = -0.162$ m/s, $(v_A)_y = 0.180$ m/s
12–170. 20.0 ft/s, 74.5 ft/s^2
12–171. $v_r = 0, v_\theta = 12$ ft/s, $a_r = -216$ ft/s^2, $a_\theta = 0$
12–173. 0.5 m/s ↑ **12–174.** 0.5 m/s ↑
12–175. 6 m/s ↗ **12–177.** 16 in. →
12–178. 4 ft/s → **12–179.** 1 ft/s ↑, 0.5 ft/s^2 ↓
12–181. 1 ft/s ↑ **12–182.** 30 ft/s ↑
12–183. **a)** 1.22 s, **b)** 2.90 m/s ↑
12–185. 2 m/s ↑
12–186. $v_C = (6 \sec \theta)$ ft/s →
12–187. 1.20 ft/s ↓, 1.11 ft/s^2 ↑
12–189. 3.81 ft/s ↑ **12–190.** 8 ft/s ↑, 6.80 ft/s^2 ↑
12–191. 2.5 ft/s ↑, 2.44 ft/s^2 ↑
12–193. 13 ft/s, 22.6°, 66.7 ft
12–194. 1.07 s, 5.93 m/s → **12–195.** 2.40 m/s, 50.3° θ⦪
12–197. 28.5 mi/h, 44.5° ⦣θ, 3418 mi/h^2, 80.6° ⦣θ
12–198. 3351 mi/h^2, 19.1° ⦣θ
12–199. 13.4 m/s, 31.7° θ⦪, 4.32 m/s^2, 79.0° ⦩θ
12–201. 26.5 mi/h, 40.9° ⦣, 1.79(10^3) mi/h^2, 72.0° ⦣
12–202. 120 km/h ↓, 4000 km/h^2, 0.716°θ⦪
12–203. 26.0 m/s, $\theta = 30°$ ⦨, 1.98 m/s^2, $\phi = 35.2°$ ⦪
12–205. 96.1 mi/h, 5.39°
12–206. $v_C = 13.0$ ft/s, $v_{B/A} = 37.0$ ft/s
12–207. 11.1 ft, 36.4 ft/s

Chapter 13

13–1. 41.7 nN **13–2.** 46.2 ft/s, 66.2 ft
13–3. 1.20 kN, 16.7 m **13–5.** 0.559
13–6. 5.60 kN
13–7. $v_1 = v_0\left(\frac{1 - \mu_k \operatorname{ctn} \theta}{1 + \mu_k \operatorname{ctn} \theta}\right)^{\frac{1}{2}}$
13–9. 7.23 m/s ↑ **13–10.** 4.25 kN
13–11. 5.98 kip **13–13.** 745 ft
13–14. 27.9 kN **13–15.** 12.9 m
13–17. $v = \left(\frac{F_0 t_0}{\pi m}\right)\left(1 - \cos\left(\frac{\pi t}{t_0}\right)\right),$
$v_{max} = \frac{2F_0 t_0}{\pi m}, s = \left(\frac{F_0 t_0}{\pi m}\right)\left(t - \frac{t_0}{\pi}\sin\left(\frac{\pi t}{t_0}\right)\right)$
13–18. 14.8 ft/s^2 **13–19.** 64.4 ft
13–21. 206 lb **13–22.** 11.9 ft/s
13–23. 158 N **13–25.** 3.62 m/s
13–26. 0.75 m/s^2 ↑, 1.32 kN
13–27. 1.29 ft/s^2, 9.70 ft
13–29. $T = mg \cos \theta (\sin \theta - \mu_k \cos \theta)$
13–30. 43.1 s, 186 ft **13–31.** 30 m/s
13–33. $a_A = 6.44$ ft/s^2, $a_B = 19.3$ ft/s^2
13–34. 0.519 s
13–35. $A_x = 35.6$ lb, $A_y = 236$ lb, 678 lb · ft
13–37. 4.28 ft/s^2 **13–38.** 14.6 ft/s
13–39. $v = \frac{1}{m}\sqrt{1.09F_0^2 t^2 + 2F_0\, tmv_0 + m^2 v_0^2},$
$x = \frac{y}{0.3} + v_0\left(\sqrt{\frac{2m}{0.3F_0}}\right)y^{\frac{1}{2}}$
13–41. 7.59 ft/s^2 **13–42.** $P = 2mg \tan \theta$
13–43. $P = 2mg\left(\frac{\sin \theta + \mu_s \cos \theta}{\cos \theta - \mu_s \sin \theta}\right)$
13–45. 57.0 N **13–46.** 1.63 kN
13–47. 1.80 kN
13–49. $N = 0$ when $x = d, v = \sqrt{\frac{kd^2}{(m_A + m_B)}}$
13–50. $N = 0, x = d$
13–51. $v = \sqrt{v_0^2 - 2gr_0\left(1 - \frac{r_0}{r}\right)}, r_{max} = \frac{2gr_0^2}{2gr_0 - v_0^2},$

$v_{esc} = \sqrt{2gr_0}$, $t = \dfrac{2}{3r_0\sqrt{2g}}(r_{max}^{\frac{3}{2}} - r_0^{\frac{3}{2}})$

13–53. 24.4 m/s **13–54.** 12.2 m/s
13–55. 1.59 kN, 282 m **13–57.** 16.1 Mm/h
13–59. 1.57 kN **13–61.** $F_n = 323$ lb, $F_t = 0$
13–62. $F_n = 287$ lb, $F_t = 96.2$ lb
13–63. $N = 1.73$ kN, $F = 291$ N
13–65. $F = 1.63$ kN, $N_s = 720$ N
13–66. $F = 1.23$ kN, $N_s = 924$ N
13–67. 182 lb **13–69.** 3.36 m/s² ↙, 361 N
13–70. 0, 104 N **13–73.** 2.83 m/s
13–74. 1.63 m/s, 7.36 N **13–75.** 5.88 lb, 23.0 ft/s²
13–77. $a_t = g\left(\dfrac{x}{\sqrt{1+x^2}}\right)$, $v = \sqrt{v_0^2 + gx^2}$,
$N = \dfrac{m}{\sqrt{1+x^2}}\left[g - \dfrac{(v_0^2 + gx^2)}{(1+x^2)}\right]$
13–78. 180 N, 5.44 m/s² **13–79.** 0.581 m
13–81. $\theta = 15.0°$, $\phi = 12.6°$ **13–82.** 38.6 N, 96.6 N
13–83. $F_r = -240$ N, $F_\theta = 66$ N, $F_z = -3.28$ N
13–85. 210 N
13–86. $F_{frict} = 68$ N, $N_C = 153$ N
13–87. $(F_z)_{max} = 547$ N, $(F_z)_{min} = 434$ N
13–89. $N_C = 1.96$ N, $F = 1.62$ N
13–90. $N_C = 3.03$ N, $F = 1.81$ N
13–91. $N_P = 3.72$ N, $F = 7.44$ N
13–93. $N = 59.3$ N, $P = 17.8$ N
13–94. −0.0155 lb **13–95.** −0.0108 lb
13–97. $P = 1.68$ lb, $N_s = 4.77$ lb
13–98. $P = 2.75$ lb, $N_s = 3.08$ lb
13–99. $F = 5.07$ kN, $N = 2.74$ kN
13–101. 4.00 rad/s, 8 N **13–102.** 0.300 lb
13–103. $\theta = \tan^{-1}\left(\dfrac{4r_c\,\dot{\theta}_0^2}{g}\right)$
13–105. 509 lb
13–106. $N = 0.328$ N, $F = 10.5$ N
13–107. $N = 5.76$ N, $F = 6.48$ N
13–109. 85.3 lb **13–110.** 113 lb
13–111. $v_r = 2.50$ m/s, $v_\theta = 2$ m/s
13–114. $v_B = v_0 = 7.71$ km/s, $v_A = 4.63$ km/s
13–115. 7.82 km/s, 1.97 h
13–117. **a)** 194(10³) mi, **b)** 392(10³) mi, **c)** 194(10³) mi $< r <$ 392(10³) mi, **d)** $r >$ 392(10³) mi
13–118. 11.5 Mm/h, 27.3 Mm
13–119. 30.8 km/s, $\dfrac{1}{r} = 0.502\,(10^{-12})\cos\theta + 6.11\,(10^{-12})$
13–121. 42.2 Mm, 3.07 km/s **13–122.** 23.9(10³) ft/s
13–123. 7.30(10³) ft/s, 1.69 h
13–125. $v_A = 6.11$ km/s, $\Delta v_B = -2.37$ km/s
13–126. 9.60 km/s, 6.50 Mm
13–127. **a)** 2.43 km/s, **b)** 6.38 km/s, **c)** $T_c = 2.19$ h, $T_e = 6.78$ h

Chapter 14

14–1. $U_W = 4.12$ kJ, $U_{N_P} = -2.44$ kJ, difference accounts for change in kinetic energy
14–2. 744 lb, 18.0(10³) ft · lb
14–3. 17.7 ft/s **14–5.** 0.365 ft/s
14–6. 2.12 km/s (approx.) **14–7.** 0.933 m
14–9. 12 m **14–10.** 15.4 m/s
14–11. 7.59 in. **14–13.** $v_A = 0.771$ ft/s
14–14. 2.64 m/s, 115 N **14–15.** 0.313 ft
14–17. 1.37 m/s **14–18.** 1.77 m/s
14–19. 1.67 m/s
14–21. $v_B = 24.0$ ft/s, $N_B = 7.18$ lb, $v_C = 16.0$ ft/s, $N_C = 1.18$ lb
14–22. 4.52 m/s
14–23. $v_B = 7.22$ ft/s, $N_B = 27.1$ lb, $v_C = 17.0$ ft/s, $N_C = 133$ lb, $v_D = 18.2$ ft/s
14–25. 42.2 ft/s, 50.6 lb, 26.2 ft/s²
14–26. 30.0 m/s, 130 m **14–27.** 3.81 ft
14–29. 36.2 ft **14–30.** 642 lb/ft, 18.0 ft/s
14–31. 2.83 m, 7.67 m/s
14–33. $U_{1-2} = GM_em\left(\dfrac{1}{r_1} - \dfrac{1}{r_2}\right)$
14–34. 5.80 ft/s **14–35.** 43.9 N
14–37. 6.89 m/s **14–38.** 179 mm
14–39. 3.84 m/s, 203 mm **14–41.** 200 kW
14–42. 4.44 s **14–43.** 181 ft
14–45. 8.86° **14–46.** 119 hp
14–47. 47.7 hp **14–49.** 0.285 m/s
14–50. 29.2 ft/s **14–51.** 29.0 ft/s
14–53. 2.05 hp **14–54.** 0.754 hp
14–55. 22.2 kW **14–57.** 2.84 hp
14–58. 1.63 hp **14–59.** 4.86 m/s
14–61. 0.193 hp **14–62.** 58.1 kW
14–63. 0.231 hp **14–65.** 7.59 in.
14–66. 1.37 m/s
14–67. $v_B = 24.0$ ft/s, $N_B = 7.18$ lb, $v_C = 16.0$ ft/s, $N_C = 1.18$ lb
14–69. $v_B = 7.22$ ft/s, $N_B = 27.1$ lb, $v_C = 17.0$ ft/s, $N_C = 133$ lb, $v_D = 18.2$ ft/s
14–70. 3.84 m/s, 5.89 N **14–71.** 25.4 ft/s
14–73. 106 ft/s **14–74.** 0.988 ft
14–75. 3.30 m/s **14–77.** 3.04 lb
14–78. 11.1 kN/m
14–79. $v_B = 10.6$ m/s, $T_B = 142$ N, $v_C = 9.47$ m/s, $T_C = 48.7$ N
14–81. 7.58 m/s, 1.56 kN, 2.90 kN
14–82. 17.7 ft/s **14–83.** 114 ft/s, 29.1 lb

14–85. 49.0 mm **14–86.** 22.3°, 0.587 m
14–87. (17.7 ft, 8.83 ft), 48.5 ft/s
14–89. 8.53 m, 10 m/s **14–90.** 80.2 ft/s, 952 lb
14–91. $v_0 = \sqrt{g(5\rho - 2h)}$
14–93. 34.1 ft/s, 7.84 lb, -20.4 ft/s^2
14–94. 150 ft/s, 62.4 lb
14–95. $x_A = 0.516$ ft, $x_B = 0.387$ ft
14–98. 34.8 Mm/h

Chapter 15

15–1. 29.4 ft/s **15–2.** 24.7 N · s
15–3. 0.439 s
15–5. $I_C = 6.55$ N · ms, $I_U = 6.05$ N · ms
15–6. 0.564 m/s **15–7.** 0.311 lb · s
15–9. 121 m/s **15–10.** 150 ft/s
15–11. 4.64 s **15–13.** 70.2 lb · s
15–14. 0.810 lb · s **15–15.** 1.93 s
15–17. $F = 24$ kN, $T = 24$ kN
15–18. 5.59 kip, 4.78 kip **15–19.** 162 N · s
15–21. 91.4 ft/s ← **15–22.** 42.4 ft/s
15–23. $v_B = 21.5$ ft/s ↓, $v_A = 21.5$ ft/s ↑
15–25. $(v_A)_2 = 20.7$ m/s ↓, $(v_B)_2 = 5.18$ m/s ↑
15–26. 16.1 m/s **15–27.** 1.92 m/s
15–29. 4.90 s **15–30.** 10.1 ft/s
15–31. 0.596 s **15–33.** 733 m/s
15–34. 5.04 m/s ← **15–35.** 9.52°
15–37. 0.178 m/s, 771 N **15–38.** 0.115 ft/s, 3.06 s
15–39. 20.7 ft/s, 38.6 lb · s
15–41. 3.48 ft/s, 504 lb, 0.216 s
15–42. 3.18 m **15–43.** 1.65 m
15–45. $v_2 = \sqrt{v_1^2 + 2gh},\ \theta_2 = \sin^{-1}\left(\dfrac{v_1 \sin \theta_1}{\sqrt{v_1^2 + 2gh}}\right)$
15–46. **a)** 1.17 ft/s ←, **b)** 1.17 ft/s ←
15–47. $v_B = 1.33$ m/s, $v_A = 0.667$ m/s, 2.5 s
15–49. 3.33 ft/s, 0.518 s **15–50.** 0.518 s, 0.863 ft
15–51. 0.364 ft ← **15–53.** 4.8 ft
15–54. 0.221 m **15–55.** 0.901
15–57. $(v_B)_2 = 1.73$ m/s ←, $(v_A)_2 = 4.53$ m/s ←
15–58. $(v_A)_2 = 1.04$ ft/s, $(v_B)_3 = 0.964$ ft/s, $(v_C)_3 = 11.9$ ft/s
15–61. 0.226 s **15–62.** 0.329 ft
15–63. 45.2 lb
15–65. $t_{\text{tot}} = \sqrt{\dfrac{2h}{g}}\left(\dfrac{1+e}{1-e}\right)$
15–66. 1.90 ft, 16.9 ft/s **15–67.** 21.8 mm
15–69. 1.53 m **15–70.** 1.68 kN
15–71. $(v_P)_2 = 14.3$ ft/s, $(v_H)_2 = 5.96$ ft/s, 3.52 ft
15–73. 0.940 m/s **15–74.** 34.8 mm
15–75. 42.8 ft/s ←, 2.49 kip
15–77. **a)** 8.81 m/s, 10.5°∠, **b)** 4.62 m/s, 20.3°∠, **c)** 3.96 m
15–78. 1.31 ft **15–79.** 1.50 m/s
15–81. $(v_A)_2 = 6.90$ m/s, $(v_B)_2 = 75.7$ m/s
15–82. $(v_A)_2 = 3.80$ m/s←, $(v_B)_2 = 6.51$ m/s, 68.6°∠
15–83. 29.3 ft/s
15–85. $(v_A)_2 = 4.60$ m/s, $(v_B)_2 = 3.16$ m/s, 0.708 m
15–86. 4.16 kN
15–87. $(v_A)_2 = 2.46$ ft/s, $(v_B)_2 = 3.09$ ft/s
15–89. 90°
15–90. $(H_A)_O = -1.50$ slug · ft^2/s, $(H_B)_O = 1.74$ slug · ft^2/s, $(H_C)_O = 0.396$ slug · ft^2/s
15–91. $\mathbf{H}_O = \{42\mathbf{i} + 21\mathbf{k}\}$ kg · m^2/s
15–93. $(H_A)_P = -57.6$ kg · m^2/s, $(H_B)_P = 94.4$ kg · m^2/s, $(H_C)_P = -41.2$ kg · m^2/s
15–94. $\mathbf{H}_O = \{-16.8\mathbf{i} + 14.9\mathbf{j} - 23.6\mathbf{k}\}$ slug · ft^2/s
15–95. $\mathbf{H}_P = \{-14.3\mathbf{i} - 9.32\mathbf{j} - 34.8\mathbf{k}\}$ slug · ft^2/s
15–97. $(H_A)_O = 72.0$ kg · m^2/s ↻, $(H_B)_O = 59.5$ kg · m^2/s ↻,
15–98. $(H_A)_P = 66.0$ kg · m^2/s ↻, $(H_B)_P = 73.9$ kg · m^2/s ↻
15–99. $6.76(10)^6$ kg · m^2/s **15–101.** 9.50 m/s
15–102. 19.3 ft/s **15–103.** 10.2 km/s, 13.8 Mm
15–105. 20.2 ft/s, 6.36 ft **15–106.** 11.9 s
15–107. 13.8 ft/s **15–110.** 21.9 m/s, 20.9°
15–111. $F_f = 19.6$ lb, $N_f = 174$ lb
15–113. $C_x = 95.8$ N, $D_y = 38.9$ N, $C_y = 118$ N
15–114. $F = 6.24$ N, $P = 3.12$ N
15–115. 858 N
15–117. $F_x = 19.5$ lb, $F_y = 1.96$ lb
15–118. 402 N **15–119.** 9.41 ft, 217 lb
15–121. 30.0 hp **15–122.** 22.4 lb
15–123. 9.72 N **15–125.** 580 ft/s
15–126. 25.0 m/s **15–127.** 0.0476 m/s^2
15–129. $2.07(10^3)$ ft/s **15–130.** 11.5 kN
15–131. $F = mv\,(gt + v)$
15–133. $v = \left(\dfrac{\rho A v_0^2 - F}{\rho A v_0}\right) \ln\left(\dfrac{M + m_0}{M + m_0 - \rho A v_0 t}\right),$
$v_{\max} = \left(\dfrac{\rho A v_0^2 - F}{\rho A v_0}\right) \ln\left(\dfrac{M + m_0}{M}\right)$
15–134. $v = \sqrt{\dfrac{2}{3} g \left(\dfrac{y^3 - h^3}{y^2}\right)}$

Review 1

R1–1. 5.47 s, 15.0 m **R1–2.** 9.32 m
R1–3. **a)** 834 mm, **b)** 1.12 m/s, 0.450 m/s^2
R1–5. $F_x = 237$ N←, $F_y = 772$ N ↑
R1–6. $N = 277$ lb, $F = 13.4$ lb
R1–7. 47.5° **R1–9.** 0.343 m/s^2, 0.436 m/s^2
R1–10. 8 s, 320 ft

R1–11. $v_B = 3.33$ ft/s ↑, $v_{B/C} = 13.3$ ft/s ↑
R1–13. 9.22 ft/s, 3.04 ft · lb **R1–14.** 1.52 ft, 0.739 s
R1–15. 0.379 m/s → **R1–17.** 4.32 m/s, 5.85 m/s
R1–18. 28.5 mi/h, 44.5°∠, $3.42(10^3)$ mi/h^2, 80.6°∠
R1–19. $3.35(10^3)$ mi/h^2, 19.1°∠
R1–21. 20.4 ft/s **R1–22.** 0.969 m/s
R1–23. 1.48 m/s **R1–25.** 18 ft/s ↓
R1–26. 0.6 ft/s ↑ **R1–27.** 21.8 m/s
R1–29. 10.1 ft/s **R1–30.** 2.13 ft/s
R1–31. $N_C = 24.8$ N, $F = 24.8$ N
R1–33. **a)** −30.5 m, **b)** 56.0 m, **c)** 10 m/s
R1–34. 14.1 m/s **R1–35.** 5.43 m
R1–37. 2.36 m/s **R1–38.** 2.34 m/s
R1–39. $v_A = 1.54$ m/s, $v_B = 4.62$ m/s
R1–41. $(v_B)_2 = \frac{1}{3}\sqrt{2gh}(1 + e)$
R1–42. 0.125 m/s ↓ **R1–43.** 4.90 lb
R1–45. 3 m **R1–46.** 13.9 ft/s
R1–47. 11.7 ft/s **R1–49.** 9.82 m/s, 2.30 m/s^2
R1–50. 21 m/s ↑

Chapter 16

16–1. 2 ft/s, 8.02 ft/s^2
16–2. 22 ft/s, $(a_A)_t = 12.0$ ft/s^2, $(a_A)_n = 242$ ft/s^2
16–3. 22.0 ft/s, $(a_B)_t = 9.00$ ft/s^2, $(a_B)_n = 322$ ft/s^2
16–5. 22.3 rad/s
16–6. 2.09 rad/s^2, 0.667 rev
16–7. 62.5 s
16–9. $v_A = v_E = 40$ mm/s, $v_w = 34.6$ mm/s
16–10. 158 ft, 375 ft/s^2 **16–11.** 3.80 rad/s, 217 ft/s^2
16–13. **a)** 16.5 ft/s, **b)** 40.5 ft, **c)** 8 rad/s^2
16–14. 89.6 rad/s **16–15.** 1.44 in.
16–17. 29.0 in./s, 412 in./s^2
16–18. 3 m/s, $(a_E)_t = 2.70$ m/s^2, $(a_E)_n = 600$ m/s^2
16–19. 0.185 m/s, $(a_E)_t = 5.81$ m/s^2, $(a_E)_n = 2.28$ m/s^2
16–21. 484 rev/min
16–22. 36 m, 12 m/s, 720 m/s^2
16–23. 25.8 m, 13.8 m/s, $(a_P)_t = 0.9$ m/s^2, $(a_P)_n = 952$ m/s^2
16–25. $(a_P)_n = 4267$ m/s^2, $(a_P)_t = 20.7$ m/s^2
16–26. 0.160 rad/s **16–27.** 0.150 rad/s
16–29. 8.49 rad/s, 0.6 m/s
16–30. 28.6 rad/s, 24.1 rad, 7.16 m/s, 205 m/s^2
16–31. $r_A = 31.8$ mm, $r_B = 31.8$ mm, 1.91 per minute
16–33. −19.2 rad/s, −183 rad/s^2
16–34. $v_y = 1.5 \cot\theta$
16–35. 4.62 rad/s ↺, 4.62 m/s ↓
16–37. $\omega = \dfrac{-v\sin\phi}{l\cos(\phi - \theta)}$, $\alpha = \dfrac{-v^2\sin^2\phi\sin(\phi - \theta)}{l^2\cos^3(\phi - \theta)}$
16–38. 18.5 ft/s ← **16–39.** $v = -r\omega\sin\theta$
16–41. $\omega' = -\dfrac{(R - r)\omega}{r}$, $\alpha' = -\dfrac{(R - r)\alpha}{r}$
16–42. $\omega' = \dfrac{(R + r)\omega}{r}$, $\alpha' = \dfrac{(R + r)\alpha}{r}$
16–43. $\omega = \dfrac{rv_A}{y\sqrt{y^2 - r^2}}$, $\alpha = \dfrac{rv_A^2(2y^2 - r^2)}{y^2(y^2 - r^2)^{\frac{3}{2}}}$
16–45. $\omega = \dfrac{h}{h^2 + x^2}v_B$, $\alpha = \dfrac{-2xh}{(h^2 + x^2)^2}v_B^2$
16–46. $\omega = \dfrac{v}{2r}$
16–47. $v_B = \dfrac{15\,\omega\sin\theta}{(34 - 30\cos\theta)^{\frac{1}{2}}}$,

$$a_B = \frac{15(\omega^2\cos\theta + \alpha\sin\theta)}{(34 - 30\cos\theta)^{\frac{1}{2}}} - \frac{225\,\omega^2\sin^2\theta}{(34 - 30\cos\theta)^{\frac{3}{2}}}$$

16–49. $v_C = b\omega\left[\dfrac{\left(\frac{b}{l}\right)\sin\theta\cos\theta}{\sqrt{1 - \left(\frac{b}{l}\right)^2\sin^2\theta}}\right] + b\omega\sin\theta$,

$$a_C = b\omega^2\left[\frac{\left(\frac{b}{l}\right)\left(\cos 2\theta + \left(\frac{b}{l}\right)^2\sin^4\theta\right)}{\left(1 - \left(\frac{b}{l}\right)^2\sin^2\theta\right)^{\frac{3}{2}}} + \cos\theta\right]$$

16–50. $\phi = \theta$, $\omega = \dfrac{2v}{h}\cos\theta$
16–51. 2.40 m/s → **16–53.** 20 rad/s, 2 ft/s →
16–54. 1.33 ft/s → **16–55.** 4 ft/s →
16–57. 5.18 ft/s →
16–58. 2.45 m/s ↙, 7.81 rad/s ↺
16–59. 3.54 ft/s ↓ **16–61.** 330 rad/s ↺
16–62. 330 in./s, 55.6° θ∡
16–63. $\omega_P = 5$ rad/s ↻, $\omega_A = 1.67$ rad/s ↺
16–65. 1.34 m/s ← **16–66.** 2.5 ft/s ←
16–67. 2.45 rad/s ↺, 2.20 ft/s ←
16–69. 9 m/s ← **16–70.** 5.33 rad/s
16–71. 2.40 ft/s ↑ **16–73.** 6.67 rad/s ↺
16–74. 16 ft/s →
16–75. $\omega_B = 90$ rad/s ↻, $\omega_A = 180$ rad/s ↺
16–77. 1.36 m/s → **16–78.** 2.40 m/s →
16–79. 1.33 ft/s →
16–81. 7.81 rad/s ↺, 2.45 m/s ↙
16–82. 8 m/s ←
16–83. $\omega_P = 5$ rad/s ↻, $\omega_A = 1.67$ rad/s ↺
16–85. 2.5 ft/s ← **16–86.** 63.4° ∠, 4.47 m/s
16–87. $\omega_{BC} = 2.83$ rad/s, 2.83 ft/s, $\omega_{AB} = 2.83$ rad/s

16–89. $v_B = 7.33$ ft/s $\rightarrow$, $v_A = 2.83$ ft/s, 45° ∠

16–90. 6 rad/s ↺, 4.76 m/s, 40.9° ∠

16–91. $\omega = 1.5\dfrac{v}{r}$

16–93. 11.4 rad/s ↻

16–94. $\omega_S = 57.5$ rad/s ↺, $\omega_{OA} = 10.6$ rad/s ↺

16–95. $\omega_S = 15.0$ rad/s, $\omega_R = 3.00$ rad/s

16–97. 4 rad/s ↺ **16–98.** 0.897 m/s ↗

16–99. 0.518 m/s ↘

16–101. 5.72 m/s, 36.2° θ∠

16–102. $v_B = 8$ ft/s ↑, $v_C = 2.93$ ft/s ↓

16–103. 57.7 rad/s ↺

16–105. 5 ft/s^2 ↗, 3.33 rad/s^2 ↻

16–106. $a_C = 1.80$ ft/s^2, 33.7° ∠θ

16–107. 1.64 ft/s ↘, 1.18 ft/s^2 ∠

16–109. 36.2 rad/s^2 ↻

16–110. 3 rad/s^2 ↻, 13.9 ft/s^2, 59.7° θ∠

16–111. 8 ft/s ↑, 21.4 ft/s^2, 2.39° ∠θ

16–113. 2.25 m/s^2, 32.6° ∠θ

16–114. 2.02° θ∠, 10.0 m/s^2 ∠

16–115. 4.83 m/s^2, 84.1° ∠ **16–117.** 2.41 m/s^2 ↓

16–118. $\alpha_{AB} = 0.75$ rad/s^2 ↺, $\alpha_{BC} = 3.94$ rad/s^2 ↺

16–119. 2 rad/s ↻, 7.68 rad/s^2 ↻

16–121. 9.38 in./s ∠(3-4-5), 54.7 in./s^2 ∠(3-4-5)

16–122. $v_B = 4v \rightarrow$, $v_A = 2\sqrt{2}v$ 45° ∠, $a_B = \dfrac{2v^2}{r} \downarrow$,

$a_A = \dfrac{2v^2}{r} \rightarrow$

16–123. 2.53 m/s^2

16–125. $a_A = 1.34$ m/s^2, 26.6° ∠, 1.5 rad/s^2 ↺,
$a_B = 1.65$ m/s^2 $\rightarrow$

16–126. $a_A = 73.0$ in./s^2, 80.5° θ∠, 18 rad/s^2↻,
$a_B = 113$ in./s^2 $\rightarrow$ **16–127.** 36 rad/s^2 ↻

16–129. 1 rad/s ↻, 10.9 rad/s^2 ↻

16–130. 7.17 rad/s ↻, rad/s ↻, 23.1 rad/s^2 ↺

16–131. $\mathbf{v}_B = \{0.6\mathbf{i} + 2.4\mathbf{j}\}$ m/s,
$\mathbf{a}_B = \{-14.2\mathbf{i} + 8.40\mathbf{j}\}$ m/s^2

16–133. $\mathbf{v}_A = \{-2.50\mathbf{i} + 2.00\mathbf{j}\}$ ft/s,
$\mathbf{a}_A = \{-3.00\mathbf{i} + 1.75\mathbf{j}\}$ ft/s^2

16–134. 0.833 rad/s ↺

16–135. 8 ft/s ←, 32 ft/s^2 ↓

16–137. 0, 37.5 rad/s^2 ↺

16–138. 1.89 rad/s ↺, 15.6 rad/s^2 ↺

16–139. 10 rad/s ↻, 24 rad/s^2 ↻

16–141. 0, 14.4 rad/s^2 ↻

16–142. 0.667 rad/s ↺, 3.08 rad/s^2 ↻

16–143. 3.22 rad/s ↻, 7.26 rad/s^2 ↻

16–145. 0.866 rad/s ↺, 3.23 rad/s^2 ↺

16–146. $(\mathbf{v}_{B/A})_{xyz} = \{31.0\mathbf{i}\}$ m/s,
$(\mathbf{a}_{B/A})_{xyz} = \{-14.0\mathbf{i} - 206\mathbf{j}\}$ m/s^2

16–147. 0, 9.24 rad/s^2 ↺

Chapter 17

17–1. $I_y = \dfrac{1}{3}ml^2$ **17–2.** $I_z = mR^2$

17–3. $I_x = \dfrac{3}{10}mr^2$ **17–5.** 1.56 in.

17–6. $I_x = \dfrac{2}{5}mr^2$ **17–7.** $I_x = \dfrac{1}{3}ma^2$

17–9. 2.25 slug · ft^2 **17–10.** $I_x = \dfrac{93}{70}mb^2$

17–11. $I_z = \dfrac{m}{10}a^2$

17–13. 118 slug · ft^2 **17–14.** 282 slug · ft^2

17–15. 7.67 kg · m^2 **17–17.** $I_O = \dfrac{1}{2}ma^2$

17–18. 1.58 slug · ft^2 **17–19.** 3.15 ft

17–21. 7.23 g · m^2 **17–22.** 5.64 slug · ft^2

17–23. 0.402 slug · ft^2 **17–25.** $a_{max} = 4.73$ m/s^2

17–26. $A_x = 7.36$ kN, $A_y = 14.7$ kN, $a_G = 4.90$ m/s^2

17–27. $F_{CB} = 41.9$ kN, $A_x = 41.9$ kN, $A_y = 28.3$ kN

17–29. $N_A = 568$ N, $N_B = 544$ N

17–30. 3.96 m/s^2 **17–31.** 6.78 m/s^2

17–33. 1.41 m/s^2, 1.38 m/s^2 **17–34.** 18.6°, 1.52 kN

17–35. 1.33 m/s^2, 2.38 kN **17–37.** 33.3 lb

17–38. 17.5 s, 11.3 s **17–39.** 6.74 kN · m, 1.64

17–41. $P_{max} = \dfrac{mgb}{2(d - \frac{h}{2})}$

17–43. 2.26 m/s^2, $N_B = 226$ N, $N_A = 559$ N, $N_B = 454$ N

17–45. $P = 24$ lb, 3.22 ft/s^2, $N_B = 14.7$ lb, $N_A = 65.3$ lb

17–46. 53.3 lb

17–47. $N'_A = 383$ N, $N'_B = 620$ N

17–49. 0.587 rad/s^2 **17–50.** 187 N

17–51. $N_E = 9.87$ lb, $V_E = 4.86$ lb, $M_E = 7.29$ lb · ft

17–53. $A_x = 14.9$ lb, 8.05 rad/s^2, $A_y = 2.50$ lb

17–54. $O_x = 0$, $O_y = 1.94$ lb, 14.9 rad/s^2

17–55. 20.8 rad/s

17–57. $P = 39.6$ N, $N_A = N_B = 325$ N

17–58. 4.78 rad/s **17–59.** 9.90 s

17–61. $O_x = 54.8$ N, $O_y = 41.2$ N

17–62. $\alpha = -\dfrac{2\mu_k}{mr}\left(\dfrac{mg - \mu_k P}{\cos\theta\,(1 + \mu_k^2)}\right) + \dfrac{2P}{mr}$

17–63. **a)** 28.0 N, **b)** 91.1 N **17–66.** 2.67 ft, 0

17–67. 0.0600 kg · m^2 **17–69.** 96.6 rad/s

17–70. 3.60 rad/s^2, $A_x = 163$ N, $A_y = 214$ N

17–71. 380 rad/s **17–73.** 193 N, 3.11 s

17–74. 891 rad/s^2

17–75. $\omega = \sqrt{\dfrac{3g}{L}(1 - \cos\theta)}$,
$O_y = mg\,(1 - 1.5\cos\theta + 1.5\cos^2\theta - 0.75\sin^2\theta)$
$O_x = mg\sin\theta\,(2.25\cos\theta - 1.5)$

17–77. $N = wx\left[\cos\theta + \dfrac{\omega^2}{2g}(2L - x)\right]$, $V = wx\sin\theta$, $M = \dfrac{wx^2}{2}\sin\theta$

17–78. 0.233 lb · ft

17–79. $T_B = 100$ N, $T_A = 102$ N

17–81. $T_A = 1.73$ lb, $T_C = 2.31$ lb, 6.68 ft/s

17–82. $a = \dfrac{g(m_B - m_A)}{\left(\frac{1}{2}M + m_B + m_A\right)}$

17–83. $a = \dfrac{8}{25}(2 - \mu_k)g$

17–85. 800 rad/s **17–86.** 17.6 rad/s

17–89. 46.9°

17–90. 0.0982 rad/s^2, $a_A = 0.217$ m/s^2, $a_B = 0.691$ m/s^2

17–91. 0.0769 m/s^2, 0.741 rad/s^2

17–93. 5.01 rad/s^2 **17–94.** 5.01 rad/s^2

17–95. 1.30 rad/s^2 **17–97.** 15.6 rad/s^2

17–98. 4.65 rad/s^2, 359 N **17–99.** 4.10 rad/s^2

17–101. 1.15 rad/s^2 **17–102.** $T_A = \dfrac{4}{7}W$

17–103. 4.18 rad/s^2, 43.3 lb **17–105.** 5.62 rad/s^2, 196 N

17–106. 90 lb, 0.229 rad/s^2 **17–107.** 4.06 ft

17–109. $F_T = 40.2$ lb, $N_T = 329$ lb

17–110. 6.21 rad/s^2 **17–111.** 13.4 rad/s^2

17–113. 7.55 rad/s^2 ↓, 0.755 m/s^2 ↓, 45.3 N

17–114. 0.0769

17–115. $(a_G)_x = 1.82$ m/s^2 ←, $(a_G)_y = 1.69$ m/s^2 ↓, 0.283 rad/s^2 ↓

Chapter 18

18–2. 6.99 ft · lb **18–3.** 10.4 ft · lb

18–5. 3.82 ft · lb **18–6.** 4.78 rad/s

18–7. 6.68 ft/s, $T_C = 2.31$ lb, $T_A = 1.73$ lb

18–9. 2.02 rad/s **18–10.** 1.78 rad/s

18–11. 6.92 rad/s **18–13.** 2.83 rad/s

18–14. 6.41 ft/s **18–15.** 9.42 m/s

18–17. 9.75 m **18–18.** 4.97 m/s

18–19. 5.34 m/s **18–21.** 0.836 rad/s

18–22. 11.0 rad/s **18–23.** 2.00 m

18–25. 0.859 m **18–26.** 1.32 rad/s

18–27. 7.08 rad/s

18–29. 0.891 rev, same result

18–30. 4.28 rad/s **18–31.** 5.05 ft/s

18–33. 11.9 ft/s **18–34.** 6.95 ft/s

18–35. 2.83 rad/s **18–37.** 4.28 rad/s

18–38. 11.9 ft/s **18–39.** 6.95 ft/s

18–41. 9.48 rad/s, $A_x = 20.9$ lb, $A_y = 5.00$ lb

18–42. 6.00 rad/s **18–43.** 2.03 rad/s

18–45. 39.3° **18–46.** 19.8 rad/s

18–47. 14.9 rad/s **18–49.** 3.50 m/s, 45° ↗

18–50. 4.79 rad/s **18–51.** 11.9 ft/s

18–53. 3.07 ft/s **18–54.** 41.8 rad/s

18–55. 39.3 rad/s **18–57.** 2.30 rad/s

18–58. 25.4° **18–59.** 85.1 rad/s

18–61. 299 mm **18–62.** 2.44 ft

18–63. 1.82 rad/s

Chapter 19

19–5. 20.8 rad/s **19–6.** 9.90 s

19–7. 96.6 rad/s **19–9.** 193 N, 3.11 s

19–10. 18.7 rad/s

19–11. 2.25 rad/s, 1.25(10^3) ft/s

19–13. 245 rad/s **19–14.** 2 m/s, 3.90 rad/s

19–15. 0.833 kg · m^2/s **19–17.** 11.4 rad/s

19–18. 47.3 rad/s **19–19.** 127 rad/s

19–21. $\omega_A = 1.70$ rad/s, $\omega_B = 5.10$ rad/s

19–22. **a)** 68.7 rad/s, **b)** 66.8 rad/s, **c)** 68.7 rad/s

19–23. $\omega_0 = \dfrac{v_G}{r}$ **19–25.** $y = \dfrac{2}{3}l$

19–26. 0.533 m **19–27.** $h = \dfrac{7}{5}r$

19–29. 0.557 m/s **19–30.** 15.2 kN · s

19–31. 1.59 m/s

19–33. $\omega = \dfrac{m_A k_A^2 \omega_A + m_B k_B^2 \omega_B}{m_A k_A^2 + m_B k_B^2}$

19–34. 0.175 rad/s

19–35. **a)** 0.0210 rad/s, **b)** 0

19–37. 0.190 rad/s

19–38. **a)** 0, **b)** $\dfrac{I}{I_z}\omega$, **c)** $\dfrac{2I}{I_z}\omega$

19–39. **a)** $\dfrac{I}{I_z}\omega$, **b)** 0, **c)** $\dfrac{I}{I_z}\omega$

19–41. 6.45 rad/s

19–42. **a)** $\omega = \dfrac{1}{4}\omega_0$, **b)** 1 rad/s

19–43. $\omega = \dfrac{1}{4}\omega_0$ **19–45.** 7.37 rad/s

19–46. $\theta = \tan^{-1}\left[\sqrt{\dfrac{7}{5}e}\right]$ **19–47.** 66.9°

19–49. 3.81 rad/s **19–50.** 39.8°

19–51. 22.4° **19–53.** 0.500 ft

19–54. $\omega_1 = 1.02\sqrt{\dfrac{g}{r}}$ **19–55.** 0.980 ft

Review 2

R2–1. $\omega_P = 20$ rad/s, $\omega_D = 6.67$ rad/s

R2–2. $\omega_P = 24$ rad/s, $\omega_D = 5.33$ rad/s

R2–3. 3.82 rad/s **R2–5.** 2 ft, 6.82 rad/s

R2–6. 12.5 m/s^2 ← **R2–7.** 45 ft/s →

R2–9. 3.46 m/s **R2–10.** 13.3 rad/s

R2–11. 12 rad/s ↓ **R2–13.** 0

R2–14. 0
R2–15. 132 rad/s
R2–17. 1.32 s
R2–18. 2.19 rad/s ↰
R2–19. 4.17 rad/s
R2–21. 4.00 m/s
R2–22. 53.5 ft/s^2, 1.66
R2–23. 4.94 ft/s^2
R2–25. 6.67 rad/s
R2–26. $\alpha = \dfrac{2\,mg}{R(M+2m)},\ h = \dfrac{mg}{M+2m}t^2$
R2–27. 3.89 rad/s
R2–29. 12.7 rad/s
R2–30. $F = 2.5\,mg\sin\theta,\ N = 0.25\,mg\cos\theta,$ $\theta = \tan^{-1}(0.1\mu_s)$
R2–31. $(\omega_S)_2 = \dfrac{5g\sin\theta}{7r}t,\ (\omega_C)_2 = \dfrac{2g\sin\theta}{3r}t$
R2–33. 1.47 rad/s ↰, 4.93 rad/s^2 ↻
R2–34. 2.66 rad/s^2 ↻
R2–35. 1.08 rad/s, 4.39 ft/s
R2–37. 73.3 rad/s^2, 0.296 s
R2–38. 30.7 rad/s
R2–39. 0.0708 rad/s
R2–41. 3.72 rad/s
R2–42. 4.45°
R2–43. $a_A = 56.2$ ft/s^2 ↓, $a_B = 40.2$ ft/s^2, 53.3° θ◸
R2–45. 2.09 rad/s^2, 0.667 rev
R2–46. 282 ft/s^2
R2–47. 10.4 s
R2–49. $A_x = 89.2$ N ←, $A_y = 66.9$ N ↑, 1.26 s
R2–50. 0.3 m/s^2, $N_B = 7.09$ kN, $N_A = 5.43$ kN

Chapter 20

20–1. $\boldsymbol{\omega} = \omega_x\,\mathbf{i} + \omega_y\,\mathbf{j} + \omega_s\,\mathbf{k},\ \alpha = \omega_y\,\omega_s\,\mathbf{i} - \omega_x\,\omega_s\,\mathbf{j}$
20–2. $\boldsymbol{\omega} = \{5.66\,\mathbf{j} + 6.26\,\mathbf{k}\}$ rad/s, $\boldsymbol{\alpha} = \{-3.39\,\mathbf{i}\}$ rad/s^2
20–3. $\boldsymbol{\omega} = \{2\,\mathbf{i} + 3\,\mathbf{k}\}$ rad/s, $\boldsymbol{\alpha} = \{6\,\mathbf{i} + 6\,\mathbf{j} + 4\,\mathbf{k}\}$ rad/s^2
20–5. $\boldsymbol{\omega} = \{13.9\,\mathbf{i} + 8.80\,\mathbf{k}\}$ rad/s, $\boldsymbol{\alpha} = \{-1.73\,\mathbf{i} + 11.1\,\mathbf{j} + 11.0\,\mathbf{k}\}$ rad/s^2
20–6. 41.2 rad/s, 4.00 m/s, 400 rad/s^2, 100 m/s^2
20–7. $\boldsymbol{\omega} = \{-8.0\,\mathbf{j} + 4.0\,\mathbf{k}\}$ rad/s, $\boldsymbol{\alpha} = \{32\,\mathbf{i}\}$ rad/s^2
20–9. $\boldsymbol{\omega} = \{-8.24\,\mathbf{j}\}$ rad/s, $\boldsymbol{\alpha} = \{24.7\,\mathbf{i} - 5.49\,\mathbf{j}\}$ rad/s^2
20–10. $\boldsymbol{\omega}_A = \left(\dfrac{r_C}{h_1}\right)\left(\dfrac{r_B h_1 \omega}{r_C h_2 + r_B h_1}\right)\mathbf{j} + \left(\dfrac{r_B h_1 \omega}{r_C h_2 + r_B h_1}\right)\mathbf{k}$
20–11. $\mathbf{v}_C = \{2.60\,\mathbf{i} + 4.00\,\mathbf{j}\}$ m/s, $\mathbf{a}_C = \{0.532\,\mathbf{i} + 5.39\,\mathbf{j} - 2.28\,\mathbf{k}\}$ m/s^2
20–13. $v_B = 0,\ v_C = 0.283$ m/s, $a_B = 1.13$ m/s^2, $a_C = 1.60$ m/s^2
20–14. $\mathbf{v}_A = \{-8.66\,\mathbf{i} + 8.00\,\mathbf{j} - 13.9\,\mathbf{k}\}$ ft/s, $\mathbf{a}_A = \{-4.00\,\mathbf{i} - 7.71\,\mathbf{j} - 3.20\,\mathbf{k}\}$ ft/s^2
20–15. $\mathbf{v}_A = \{-8.66\,\mathbf{i} + 8.00\,\mathbf{j} - 13.9\,\mathbf{k}\}$ ft/s, $\mathbf{a}_A = \{-24.8\,\mathbf{i} + 8.29\,\mathbf{j} - 30.9\,\mathbf{k}\}$ ft/s^2
20–17. $\omega_A = 47.8$ rad/s, $\omega_B = 7.78$ rad/s
20–18. $\mathbf{v}_B = \{10\,\mathbf{k}\}$ ft/s
20–19. $\mathbf{a}_B = \{110\,\mathbf{k}\}$ ft/s^2
20–21. $\mathbf{a}_B = \{-96.5\,\mathbf{i}\}$ ft/s^2
20–22. $\mathbf{v}_A = \{0.667\,\mathbf{i}\}$ ft/s, $\mathbf{a}_A = \{-0.148\,\mathbf{i}\}$ ft/s^2
20–23. $\mathbf{a}_A = \{1.19\,\mathbf{i}\}$ ft/s^2
20–25. $\boldsymbol{\omega}_{BC} = \{0.769\,\mathbf{i} - 2.31\,\mathbf{j} + 0.513\,\mathbf{k}\}$ rad/s, $\mathbf{v}_B = \{-0.333\,\mathbf{j}\}$ m/s
20–26. $\boldsymbol{\omega} = \{-2\,\mathbf{i}\}$ rad/s
20–27. 4.71 ft/s, $\boldsymbol{\omega}_{AB} = \{1.17\,\mathbf{i} + 1.27\,\mathbf{j} - 0.779\,\mathbf{k}\}$ rad/s
20–29. $\boldsymbol{\omega} = \{1.50\,\mathbf{i} + 2.60\,\mathbf{j} + 2.00\,\mathbf{k}\}$ rad/s, $\mathbf{v}_C = \{10.4\,\mathbf{i} - 7.79\,\mathbf{k}\}$ ft/s
20–30. $\mathbf{a}_C = \{99.6\,\mathbf{i} - 117\,\mathbf{k}\}$ ft/s^2, $\boldsymbol{\alpha} = \{10.4\,\mathbf{i} + 30.0\,\mathbf{j} + 3\mathbf{k}\}$ rad/s^2
20–31. $\mathbf{v}_C = \{-1.00\,\mathbf{i} + 5.00\,\mathbf{j} + 0.800\,\mathbf{k}\}$ m/s, $\mathbf{a}_C = \{-28.8\,\mathbf{i} - 5.45\,\mathbf{j} + 32.3\,\mathbf{k}\}$ m/s^2
20–33. $\mathbf{v}_A = \{-7.79\,\mathbf{i} - 2.25\,\mathbf{j} + 3.90\,\mathbf{k}\}$ ft/s, $\mathbf{a}_A = \{8.30\,\mathbf{i} - 35.2\,\mathbf{j} + 7.02\,\mathbf{k}\}$ ft/s^2
20–34. $\mathbf{v}_B = \{-17.3\,\mathbf{i} + 18.8\,\mathbf{j} + 10.0\,\mathbf{k}\}$ ft/s, $\mathbf{a}_B = \{-19.1\,\mathbf{i} + 24.0\,\mathbf{j} - 8.66\,\mathbf{k}\}$ ft/s^2
20–35. $\mathbf{v}_C = \{-1.79\,\mathbf{i} - 1.40\,\mathbf{j} + 3.58\,\mathbf{k}\}$ m/s, $\mathbf{a}_C = \{0.839\,\mathbf{i} - 3.15\,\mathbf{j} + 0.354\,\mathbf{k}\}$ m/s^2
20–37. $\mathbf{v}_P = \{2\,\mathbf{i} + 20\,\mathbf{j}\}$ m/s, $\mathbf{a}_P = \{-101\,\mathbf{i} - 14.8\,\mathbf{j}\}$ m/s^2
20–38. $\mathbf{v}_B = \{5\,\mathbf{i} - 0.5\,\mathbf{j} + 6.4\,\mathbf{k}\}$ m/s, $\mathbf{a}_B = \{-0.25\,\mathbf{i} - 7.12\,\mathbf{j}\}$ m/s^2
20–39. $\mathbf{v}_B = \{5\,\mathbf{i} - 0.5\,\mathbf{j} + 6.4\,\mathbf{k}\}$ m/s, $\mathbf{a}_B = \{2.95\,\mathbf{i} - 7.52\,\mathbf{j} + 7.20\,\mathbf{k}\}$ m/s^2
20–41. $\mathbf{v}_A = \{-4.85\,\mathbf{i} + 0.800\,\mathbf{j} - 1.39\,\mathbf{k}\}$ m/s, $\mathbf{a}_A = \{-14.0\,\mathbf{i} - 35.5\,\mathbf{j} - 3.68\,\mathbf{k}\}$ m/s^2
20–42. $\mathbf{v}_C = \{-5.54\,\mathbf{i} + 5.20\,\mathbf{j} - 3.00\,\mathbf{k}\}$ ft/s, $\mathbf{a}_C = \{-91.5\,\mathbf{i} - 42.6\,\mathbf{j} - 1.00\,\mathbf{k}\}$ ft/s^2
20–43. $\mathbf{v}_P = \{-0.849\,\mathbf{i} + 0.849\,\mathbf{j} + 0.566\,\mathbf{k}\}$ m/s, $\mathbf{a}_P = \{-5.09\,\mathbf{i} - 7.35\,\mathbf{j} + 6.79\,\mathbf{k}\}$ m/s^2
20–45. $\mathbf{v}_C = \{2.80\,\mathbf{j} - 5.60\,\mathbf{k}\}$ m/s, $\mathbf{a}_C = \{-56\,\mathbf{i} + 2.1\,\mathbf{j}\}$ m/s^2
20–46. $\mathbf{v}_C = \{2.80\,\mathbf{j} - 5.60\,\mathbf{k}\}$ m/s, $\mathbf{a}_C = \{-56\,\mathbf{i} + 2.1\,\mathbf{j} - 1.40\,\mathbf{k}\}$ m/s^2
20–47. $\mathbf{v}_T = \{-9.50\,\mathbf{i} + 3\,\mathbf{j} + 1.20\,\mathbf{k}\}$ m/s, $\mathbf{a}_T = \{-14.4\,\mathbf{i} + 3.77\,\mathbf{j} + 4.20\,\mathbf{k}\}$ m/s^2

Chapter 21

21–2. $I_{\bar{y}} = \dfrac{3m}{80}(h^2 + 4a^2),\ I_{y'} = \dfrac{m}{20}(2h^2 + 3a^2)$
21–3. $I_y = \dfrac{1}{3}mr^2,\ I_x = \dfrac{m}{6}(r^2 + 3a^2)$
21–5. $I_y = 4.48(10^3)\rho$
21–6. $I_{yz} = \dfrac{m}{6}ah$
21–7. $I_{xy} = \dfrac{m}{12}a^2$
21–9. $I_{x'} = \dfrac{m}{12}(a^2 + h^2)$
21–10. $I_{z'} = \dfrac{13}{24}mr^2,\ I_{x'} = \dfrac{13}{24}mr^2,\ I_{y'} = \dfrac{7mr^2}{12}$
21–11. $I_x = 4.50$ kg · m^2, $I_y = 4.38$ kg · m^2, $I_z = 0.125$ kg · m^2
21–13. $\bar{y} = 0.5$ ft, $\bar{x} = -0.667$ ft, $I_{x'} = 0.0272$ slug · ft^2, $I_{y'} = 0.0155$ slug · ft^2, $I_{z'} = 0.0427$ slug · ft^2
21–14. $I_{z'} = 0.0595$ kg · m^2
21–15. $3.54(10^{-3})$ kg · m^2
21–17. 0.0880 slug · ft^2
21–18. 0.429 kg · m^2
21–19. 0.455 slug · ft^2
21–23. $\mathbf{H} = \{-477(10^{-6})\,\mathbf{i} + 198(10^{-6})\,\mathbf{j} + 0.169\,\mathbf{k}\}$ kg · m^2/s
21–25. $\mathbf{H}_G = \{0.0207\,\mathbf{i} - 0.00690\,\mathbf{j} + 0.0690\,\mathbf{k}\}$ slug · ft^2/s
21–26. $T = \dfrac{9\,mh^2}{20}\left[1 + \dfrac{r^2}{6h^2}\right]\omega^2$

21–27. $\mathbf{u}_{IA} = 0.141\,\mathbf{j} - 0.990\,\mathbf{k}$, $\int \mathbf{F}_0 dt = \{8.57\,\mathbf{i}\}$ N · s
21–29. 26.9 kg · m²/s
21–30. $\mathbf{H}_z = \{16.6\,\mathbf{k}\}$ slug · ft²/s, 72.1 lb · ft
21–31. 26.2 rad/s
21–33. $\boldsymbol{\omega} = \{-2.16\,\mathbf{i} + 5.40\,\mathbf{j} + 7.20\,\mathbf{k}\}$ rad/s, $\mathbf{u}_A = -0.233\,\mathbf{i} + 0.583\,\mathbf{j} + 0.778\,\mathbf{k}$
21–34. $\boldsymbol{\omega} = \{-0.954\,\mathbf{i} + 2.38\,\mathbf{j} - 3.18\,\mathbf{k}\}$ rad/s, $\mathbf{u}_A = -0.233\,\mathbf{i} + 0.583\,\mathbf{j} + 0.778\,\mathbf{k}$
21–35. 0.0920 ft · lb
21–37. 58.4 rad/s
21–38. $\boldsymbol{\omega} = \{-0.444\,\mathbf{j} + 0.444\,\mathbf{k}\}$ rad/s
21–39. $\Sigma M_x = \frac{d}{dt}(I_x\omega_x - I_{xy}\omega_y - I_{xz}\omega_z) - \Omega_z(I_y\omega_y - I_{yz}\omega_z - I_{yx}\omega_x) + \Omega_y(I_z\omega_z - I_{zx}\omega_x - I_{zy}\omega_y)$
21–41. $\Sigma M_x = I_x\dot{\omega}_x - I_y\Omega_z\omega_y + I_z\Omega_y\omega_z$
21–42. $F_A = F_B = 19.5$ N
21–43. $A_x = 9.64$ N, $B_x = 9.98$ N
21–45. $A_x = B_x = 0$, $A_z = 23.1$ N, $B_z = 18.5$ N
21–46. $\theta = \cos^{-1}\left(\frac{3g}{2L\omega^2}\right)$
21–47. $B_x = -250$ N, $A_y = B_y = 0$, $A_z = B_z = 24.5$ N
21–49. $\dot{\omega}_y = 58.9$ rad/s², $A_y = 0$, $A_x = B_x = 72.0$ N, $A_z = B_z = 12.9$ N
21–50. 405 N
21–51. $B_y = -41.4$ lb, $B_x = -12.5$ lb, $A_x = -15.6$ lb, $A_y = -51.8$ lb, $A_z = 50$ lb
21–53. 23.4 lb · ft
21–54. 0.0320 N · m, $A_y = 13.2$ N, $B_y = 13.3$ N, $A_x = -1.08$ N, $B_x = 1.08$ N
21–55. $-0.9\omega^2$ mN · m
21–57. $\dot{\omega}_y = 51.5$ rad/s², $A_x = -8.97$ lb, $B_x = -9.66$ lb, $A_z = -6.07$ lb, $B_z = 15.2$ lb
21–58. $T = 23.3$ N, $F_A = 41.3$ N
21–59. $\theta_D = 139°$, $m_D = 0.661$ kg, $\theta_F = 40.9°$, $m_F = 1.32$ kg
21–62. $\alpha = 90°$, $\beta = 135°$, $\gamma = 45°$, No
21–63. $A_y = 13.7$ kN, $B_y = -13.7$ kN
21–65. 6.34 rad/s
21–66. 27.0 N · m
21–67. 12.4 lb · ft, $M_y = 0$, $M_z = 0$
21–69. $3.63(10^3)$ rad/s
21–70. 1.21 rad/s (low precession), 76.3 rad/s (high precession)
21–71. 1.19 rad/s
21–73. 27.9 rad/s
21–74. 53.4 N
21–77. $6.09(10^{-3})$ kg · m²/s
21–78. 12.5 Mg · m²/s
21–79. 2.35 rev/h

Chapter 22

22–1. $\ddot{y} + 56.1y = 0$, 0.192 m
22–2. 4.985 Hz, 0.201 s
22–3. 2.02 Hz, $y = -0.2\cos 12.7t$, 0.2 ft
22–5. $x = 0.1\sin(20t) + 0.150\cos(20t)$, 0.180 m
22–6. $x = 0.107\sin(7.00t) + 0.100\cos(7.00t)$, 43.0°
22–7. $x = -0.0693\sin(5.77t) - 0.075\cos(5.77t)$, 0.102 m
22–9. 1.36 N/m, 3.58 kg
22–10. $\theta = -0.101\sin(4.95t) + 0.3\cos(4.95t)$
22–11. 16°
22–13. $\tau = 2\pi\sqrt{\frac{k_G^2 + d^2}{gd}}$
22–14. $\tau = 2\pi\sqrt{\frac{2r}{g}}$
22–15. $\tau = 2\pi\sqrt{\frac{3r}{2g}}$
22–17. 503 mm
22–18. 5.85 N · m/rad
22–19. 0.905 Hz
22–21. 0.550 Hz
22–22. 0.457 m
22–23. $\tau = 2\pi\left(\frac{a}{b}\right)\sqrt{\frac{m}{k}}$
22–25. 0.597 Hz
22–26. $\tau = 2\pi\sqrt{\frac{k_G^2 + d^2}{gd}}$
22–27. $\tau = 2\pi\sqrt{\frac{3r}{2g}}$
22–29. 0.401 s
22–30. $\ddot{y} + \frac{(k_1 + k_2)}{m}y = 0$
22–31. $\ddot{x} + 333x = 0$
22–33. 2.25 Hz
22–34. 1.52 s
22–35. $\tau = 3.85\sqrt{\frac{m}{k}}$
22–37. $\tau = \frac{4\pi}{\sqrt{3}}\left(\frac{ma}{5ka + mg}\right)^{\frac{1}{2}}$
22–38. 0.851 Hz
22–39. 0.970 s
22–41. 18.9 N · s/m
22–42. $x = A\sin pt + B\cos pt + \frac{F_0/k}{1 - \left(\frac{\omega}{p}\right)^2}\cos\omega t$
22–43. $y = (-3.66\sin 12.25t + 50\cos 12.25t + 11.2\sin 4t)$ mm
22–45. $y = (0.0186\sin 8.02t + 0.333\cos 8.02t - 0.0746\sin 2t)$ ft
22–46. 0.666 s
22–47. $y = (361\sin 7.75t + 100\cos 7.75t - 350\sin 8t)$ mm
22–49. 14.0 rad/s
22–50. 14.6 mm
22–51. 35.5 mm
22–53. 0.490 in.
22–54. 19.0 rad/s
22–55. $x_p = (0.0276\sin 2t)$ ft
22–57. $y = 0.622\,[e^{-0.939t}\sin(11.9t + 1.49)]$
22–58. 9.89°
22–59. 0.997
22–61. 0.0269 in.
22–62. $y = \{-0.0702\,[e^{-3.57t}\sin(8.54t)]\}$ m
22–65. $L\ddot{q} + R\dot{q} + \frac{1}{C}q = 0$
22–66. $m\ddot{x} + kx = F_0\sin\omega t$, $L\ddot{q} + \frac{1}{C}q = E_0\sin\omega t$
22–67. $L\ddot{q} + R\dot{q} + \left(\frac{1}{C}\right)q = E_0\cos\omega t$
22–69. $\ddot{y} + 16\dot{y} + 4y = 0$, overdamped

Index

Geometric Properties of Line and Area Elements

Centroid Location	Centroid Location	Area Moment of Inertia

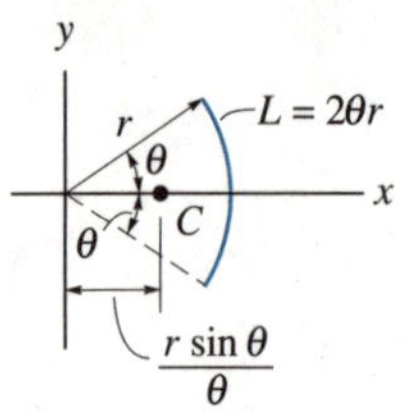

Circular arc segment

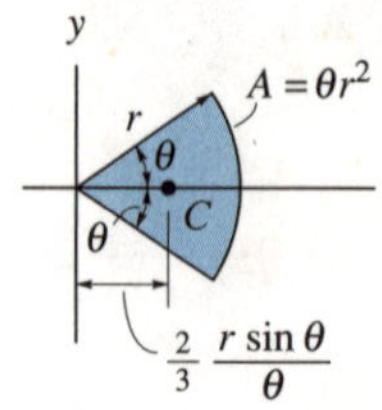

Circular sector area

$$I_x = \frac{1}{4} r^4(\theta - \frac{1}{2} \sin 2\theta)$$

$$I_y = \frac{1}{4} r^4(\theta + \frac{1}{2} \sin 2\theta)$$

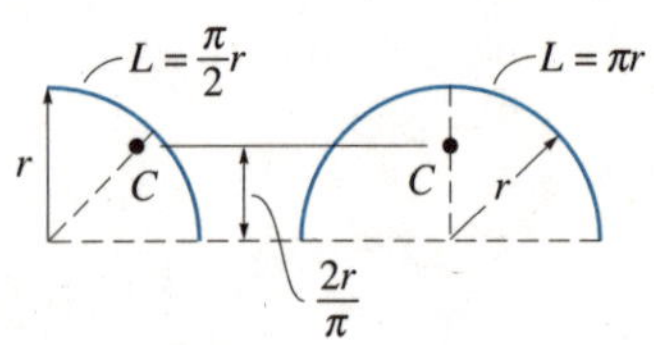

Quarter and semicircular arcs

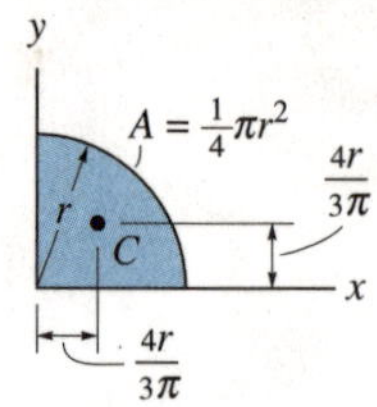

Quarter circular area

$$I_x = \frac{1}{16}\pi r^4$$

$$I_y = \frac{1}{16}\pi r^4$$

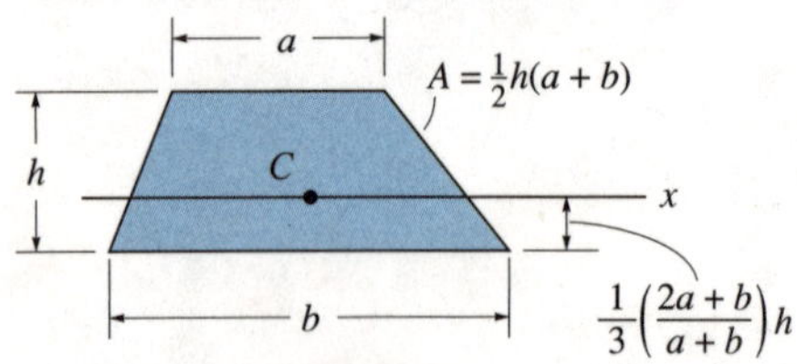

Trapezoidal area

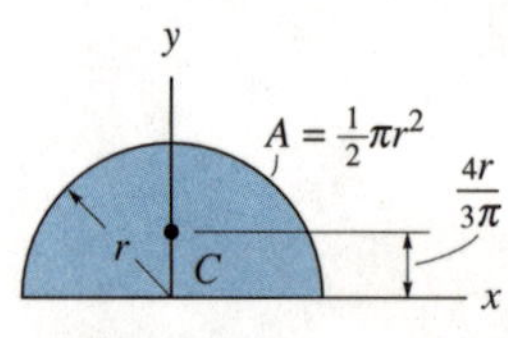

Semicircular area

$$I_x = \frac{1}{8}\pi r^4$$

$$I_y = \frac{1}{8}\pi r^4$$

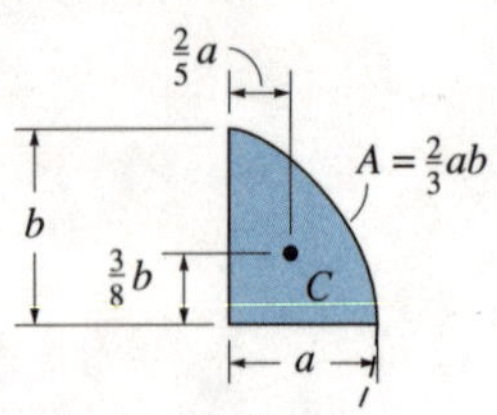

Semiparabolic area

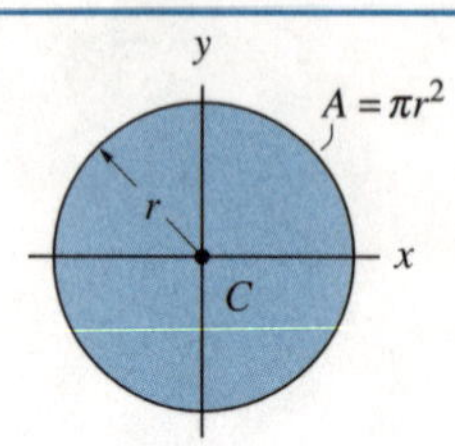

Circular area

$$I_x = \frac{1}{4}\pi r^4$$

$$I_y = \frac{1}{4}\pi r^4$$

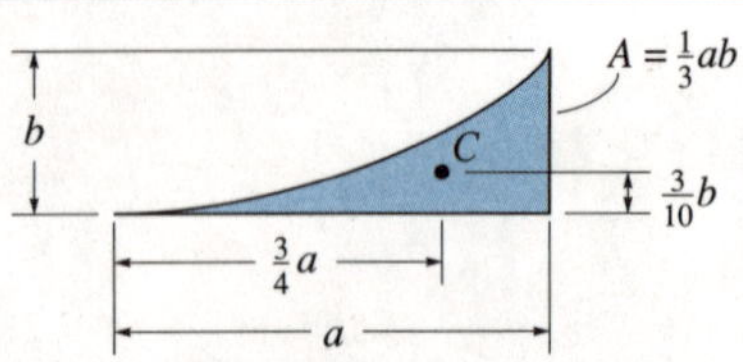

Exparabolic area

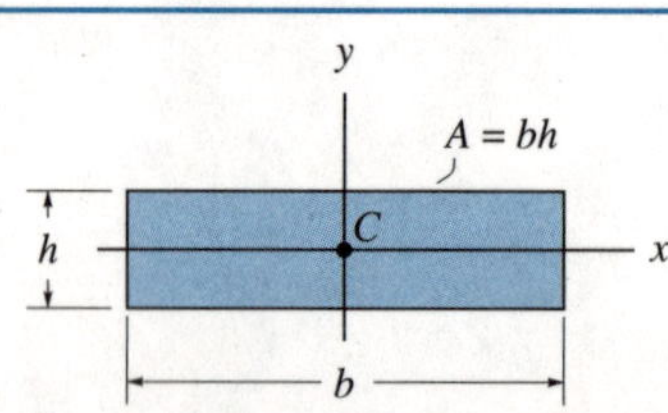

Rectangular area

$$I_x = \frac{1}{12} bh^3$$

$$I_y = \frac{1}{12} hb^3$$

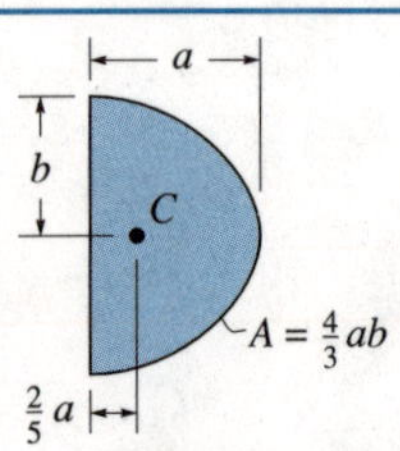

Parabolic area

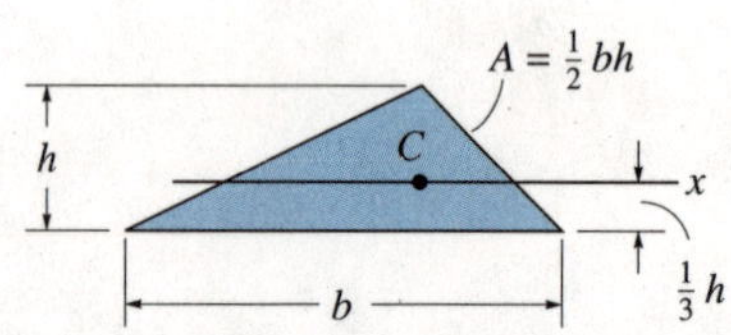

Triangular area

$$I_x = \frac{1}{36} bh^3$$